AF601752

Algorithms and Computation in Mathematics · Volume 29

More information about this series at http://www.springer.com/series/3339

Dirk Hachenberger · Dieter Jungnickel

Topics in Galois Fields

Dirk Hachenberger
Department of Mathematics
University of Augsburg
Augsburg, Germany

Dieter Jungnickel
Department of Mathematics
University of Augsburg
Augsburg, Germany

ISSN 1431-1550
Algorithms and Computation in Mathematics
ISBN 978-3-030-60804-0 ISBN 978-3-030-60806-4 (eBook)
https://doi.org/10.1007/978-3-030-60806-4

Mathematics Subject Classification: 11TXX, 12E20, 15B33, 94A55

This Springer imprint is published by the registered company Springer Nature Switzerland AG
The registered company address is: Gewerbestrasse 11, 6330 Cham, Switzerland

To my parents, Carola and Rudi

DIRK HACHENBERGER

To the memory of my mother

DIETER JUNGNICKEL

Preface

The idea of writing a book *Topics in Galois Fields* originated a long time ago, maybe in 1999, when the *Fifth International Conference on Finite Fields and Applications* was hosted by our home university in Augsburg, Germany. Although we could draw on our two previous monographs, *Finite Fields – Structure and Arithmetics* [211] by Dieter Jungnickel and *Finite Fields – Normal Bases and Completely Free Elements* [161] by Dirk Hachenberger, this proved to be a decidedly long-term project, which of course had to do with numerous other duties that had to be fulfilled. We are therefore very grateful to our publisher for their patience and for their trust that this work would be finished – eventually.

Dating back to the early work of Gauss and Galois, finite fields nowadays have become a very flourishing and appealing topic, not just in their own right, but also because of their great importance in modern applications. Finite fields play an essential role both in more theoretical areas like finite geometries, combinatorics and the analysis of algorithms as well as in all areas of information and communication theory, such as signal processing, coding theory, cryptography and computer science. Concrete applications include such diverse topics as data bases, linear feedback shift registers, pseudo-random sequences, compact discs as well as DVDs and Blue-Ray discs, concert hall acoustics, radar and sonar, spread spectrum communication, antenna design, x-ray imaging, spectroscopy, both symmetric and public-key cryptosystems, digital signatures, access control, and the design of telephone or computer networks.

On the theoretical side, a wealth of interesting and deep results have been established in recent years, partially motivated by the need to efficiently perform arithmetics in large finite fields. As the reader will see, there is a quite striking interplay between arithmetical and structural results: questions of efficiency in computing with finite fields turn out to pose new theoretical problems which in turn influence even the hardware implementation for finite field arithmetics. So the area is also well suited to serve as an example for the practical applicability of abstract algebra and to give the student an impression of what research is (or may be) about.

When extending the contents of our above mentioned monographs, one of our initial intentions has been the presentation of some important highlights which have not yet found their way into textbooks, but nevertheless have influenced a lot of current research. These comprise

- the primitive normal basis theorem of Lenstra and Schoof,
- the theorem on the existence of primitive elements in affine hyperplanes, by Cohen,
- as well as the Niederreiter method for factoring polynomials over finite fields.

Of course, when selecting these advanced topics from the huge area of Galois fields, we followed our own taste.

In addition, we have also tried to provide streamlined and/or clearer proofs for many results. A particular highlight is our treatment of Berlekamp's factorization algorithm for univariate polynomials over Galois fields: we introduce Berlekamp algebras in a novel way which provides a deeper understanding of the theoretical background and allows us to investigate how far this approach works for general fields.

In contrast to the approach of [211], we decided to present the whole topic in as self-contained a manner as possible, so that the present book can also be used for a first introduction to finite fields. The only necessary prerequisites are standard topics from first year studies, such as basic set theory, relations and mappings, linear algebra and, occasionally, some calculus.

We have also included quite a few exercises which complement and extend some of the material covered in the text. However, as this book is not intended as an introductory textbook for self-study (at least not primarily), but as an advanced monograph, we have neither aimed to give extensive exercises nor included solutions.

The resulting book consists of 14 chapters, which may be divided into four parts:

- The first part comprises the first four chapters and provides the foundations of finite fields. In Chapter 1, we introduce the fundamental algebraic structures as well as the results from elementary number theory which we require later. This is followed by an exposition of fundamental properties of polynomials and formal power series in Chapter 2, with emphasis on Möbius inversion and minimal polynomials of endomorphisms of vector spaces. After that, Chapter 3 covers the basic theory of finite fields, including existence and uniqueness theorems as well as the main structural results. This basic theory is extended by a more advanced topic in Chapter 4, where we study the algebraic closure of Galois fields. Already these four basic chapters contain a considerable amount of fundamental material which turns out to be very useful and is not covered in current established textbooks.
- The second part is formed by Chapters 5 and 6 and addresses the important basic problems of determining irreducible polynomials of any given degree over any given finite ground field, and of performing the factorization of univariate polynomials over finite fields. Concerning the latter, in addition to the classical method of Berlekamp, the more recent method of Niederreiter is presented in detail. Moreover, we investigate the connections between these two approaches.

- The third part comprises Chapters 7 to 9 and is motivated by some of the applications mentioned above. In the first of these three chapters, we consider several aspects of matrices over finite fields, leading, among other topics, to the Discrete Fourier Transform. A detailed study of the interaction between theoretical results on various types of bases used to represent finite extension fields and their application to efficiently performing the arithmetic operations is provided in Chapter 8, with emphasis on hardware implementations. This part is concluded with a detailed study of linear feedback shift registers and the sequences produced by such devices in Chapter 9 and includes the construction of pseudo-random sequences and of some cyclic difference sets, as well as a study of the linear complexity profile of arbitrary (finite) sequences over Galois fields.
- Chapter 10 is an interlude supplementing the second part. It deals with a further basic – but somewhat more advanced – topic, namely the theory of both additive and multiplicative finite field characters and their interplay via character sums, in particular, Gauss sums. These are important tools when considering, for instance, combinatorial questions like the existence of primitive elements with specified additional properties, which will be addressed in the final two chapters of the book. Moreover, we extend the Discrete Fourier Transform to abelian groups in general, which allows us to prove some results on sequences and difference sets complementing those obtained in Chapter 9.
- The fourth and final part consists of Chapters 11 through 14 and considers some advanced topics from the theory of Galois fields. Our first aim is the explicit determination of a normal element for every finite field extension and the construction of trace-compatible sequences of such elements for the algebraic closure of an arbitrary Galois field. After that, these results are strengthened in Chapter 12 by constructing completely normal elements for every finite field extension and trace-compatible sequences of such elements for the algebraic closure of any given finite ground field. The penultimate Chapter 13 provides a complete proof of the celebrated primitive normal basis theorem which does not rely on the use of computers: a standard pocket calculator turns out to be sufficient to verify the necessary computational steps. The final chapter mainly deals with another important result on primitive elements: essentially, every affine hyperplane in any given finite field extension contains a primitive element; here some computer support is still unavoidable.

We are indebted to Dr. Martin Peters from our publisher, Springer, for accompanying and supporting this long-standing project. We also thank the large number of students who have attended our lecture courses on finite fields and their applications over the years for their steady attention and interest. Special thanks are due to Lukas Graf who has supported us with computations needed for the final chapter of this book.

Augsburg,
August 2020

Dirk Hachenberger
Dieter Jungnickel

Contents

Chapter 1
Basic Algebraic Structures and Elementary Number Theory

Abstract In this preliminary chapter, we lay the algebraic foundations which we believe are necessary to understand the theory of finite fields and their applications. Although most of this chapter is certainly covered by many text books on Algebra, we decided to include this background material for the convenience of the reader. This not only serves to present the material in a manner as self-contained as possible but also to fix notations used throughout this book.

1.1 Basics on Monoids

We start with a **binary operation** on a non-empty set M. This is a mapping from $M \times M$ to M, which is usually denoted either by $\cdot$ (multiplicative notation) or by $+$ (additive notation). For $a, b \in M$, we write $a \cdot b$ or ab instead of $\cdot(a,b)$ and $a+b$ instead of $+(a,b)$. In the first three sections, where we consider abstract monoids or groups, we shall mainly adopt the multiplicative notation, although the additive monoid of the natural numbers, the additive group of the integers, or the additive group of a finite field are important examples.

A binary operation is called **commutative** if the **commutative law** is satisfied:

$$ab = ba \quad \text{for all } a, b \in M. \tag{1.1}$$

For $a, b, c \in M$, let $a(bc) := \cdot(a, \cdot(b,c))$. This means that terms in brackets have to be evaluated first. A binary operation is said to be **associative** if the **associative law** is satisfied:

$$(ab)c = a(bc) \quad \text{for all } a, b, c \in M. \tag{1.2}$$

In fact, associativity implies that brackets in terms like $(a((bc)d)(ef)g)h$ are redundant, as it equals $((((((ab)c)d)e)f)g)h$, that is, the evaluation from left to right.

Definition 1.1.1. A non-empty set M together with an associative binary (multiplicatively written) operation is called a **monoid**, provided there exists an element

D. Hachenberger and D. Jungnickel, *Topics in Galois Fields*,
Algorithms and Computation in Mathematics 29,
https://doi.org/10.1007/978-3-030-60806-4_1

$e \in M$ such that

$$eg = g = ge \quad \text{for all } g \in M. \tag{1.3}$$

The element e is called an **identity element** (or **neutral element**) of M.

A monoid M is said to be **commutative** if the underlying binary operation is commutative. It is called **finite** if the cardinality $|M|$ of M is finite. □

Obviously, the identity element e of a monoid is uniquely determined, and we therefore speak of the monoid $(M, \cdot, e)$ with underlying set M, binary operation $\cdot$ and identity element e.

In an additively written monoid M, the identity element is also called the **zero element** or just the **zero** of M; it is then usually denoted by 0. Similarly, in a multiplicatively written monoid, the identity element is also called the **unit element**; it is usually denoted by 1.

Example 1.1.2. The most fundamental examples for monoids are $(\mathbb{N}, +, 0)$ and $(\mathbb{N}, \cdot, 1)$, where $\mathbb{N}$ denotes the set of natural numbers (that is, the set of nonnegative integers)[1] and where $+$ and $\cdot$ are the usual addition and multiplication on $\mathbb{N}$. We write $\mathbb{N}^* := \mathbb{N} \setminus \{0\}$ for the set of positive integers. Since $n \cdot m \neq 0$ for all non-zero $n, m \in \mathbb{N}$, we see that $(\mathbb{N}^*, \cdot, 1)$ is likewise a commutative monoid. □

Remark 1.1.3. The fundamental **principle of induction** states that the set $\mathbb{N}$ of natural numbers is equipped with an injective mapping succ, called the **successor function**, which satisfies the following property:

If $S \subseteq \mathbb{N}$ such that $0 \in S$ and such that $x \in S$ implies $\mathrm{succ}(x) \in S$, then $S = \mathbb{N}$.

The unit element of $\mathbb{N}$ is the successor of the zero element of $\mathbb{N}$, that is, $\mathrm{succ}(0) = 1$. More general, $\mathrm{succ}(n) = n + 1$ for every $n \in \mathbb{N}$. □

Definition 1.1.4. Let $(M, \cdot, 1)$ be a (multiplicatively written) monoid. Then $u \in M$ is called a **unit** or an **invertible element** provided there exists a $v \in M$ such that

$$uv = 1 = vu. \tag{1.4}$$

The set of all units of M will be denoted by $U(M)$. (Note that the unit element 1 is a very special unit, as $1 \cdot 1 = 1$.)

In the case where $U(M) = M$, so that every element of M is invertible, M is said to be a **group**. A commutative monoid which is a group is also called an **abelian group**. □

Proposition 1.1.5. *Let $(M, \cdot, 1)$ be a monoid and $u \in M$.*

- *Assume that there are $x, y \in M$ such that $xu = 1 = uy$. Then $x = y$, and u is a unit in M.*
- *If $u \in U(M)$, then there is a unique $v \in M$ such that $uv = 1 = vu$.*

---

[1] Like Jacobson [204] or MacLane and Birkhoff [251] and other textbooks in Algebra, we consider 0 to be a natural number. While this is not absolutely standard, we find it convenient.

Proof. The first assertion holds because of $x = x \cdot 1 = x \cdot (uy) = (xu) \cdot y = 1 \cdot y = y$, and the second assertion is an immediate consequence of the first one. □

For a unit u of a monoid $(M, \cdot, 1)$, the unique element v satisfying $uv = 1 = vu$ is called the **inverse** of u. In multiplicative notation, the inverse of $u \in U(M)$ is usually denoted by u^{-1} or by $\frac{1}{u}$. In contrast, for additively written monoids the inverse of a unit u is written as $-u$. Note that the identity element of a monoid is self-inverse: $1^{-1} = 1$, respectively $-0 = 0$.

From $uu^{-1} = 1 = u^{-1}u$, we see that the inverse of a unit $u \in U(M)$ is also a unit, and moreover $(u^{-1})^{-1} = u$. Using additive notation, the latter reads $-(-u) = u$.

Example 1.1.6. The group of units of $(\mathbb{N}, +, 0)$ is $\{0\}$, and the group of units of $(\mathbb{N}, \cdot, 1)$ is $\{1\}$.

The most fundamental example of a (commutative) group is $(\mathbb{Z}, +, 0)$, where $\mathbb{Z}$ denotes the set of integers and where $+$ is the usual addition. With respect to the usual multiplication, $(\mathbb{Z}, \cdot, 1)$ is a commutative monoid, and its group of units is just $\{-1, 1\}$. Let $\mathbb{Z}^* := \mathbb{Z} \setminus \{0\}$. Then $(\mathbb{Z}^*, \cdot, 1)$ is also a monoid, with group of units $U(\mathbb{Z}^*) = \{-1, 1\}$. □

Definition 1.1.7. A non-empty subset S of a monoid $(M, \cdot, 1)$ is said to be **closed** provided that $ab \in S$ for all $a, b \in S$, that is, if the restriction of $\cdot$ to $S \times S$ is a binary operation on S. If additionally $1 \in S$, then S is called a **submonoid** of M. (Note that $(S, \cdot, 1)$ is a monoid in this case, too.)

Assume that $(M, \cdot, 1)$ is a group, and let S be a submonoid of M satisfying $u^{-1} \in S$ for all $u \in S$. Then S is called a **subgroup** of M; this is usually denoted as $S \leq M$. □

Example 1.1.8. The set of units of a monoid $(M, \cdot, 1)$ is closed: given $a, b \in U(M)$, we have

$$(ab)(b^{-1}a^{-1}) = 1 = (b^{-1}a^{-1})(ab),$$

hence $ab \in U(M)$. As 1 is a unit, $U(M)$ is a submonoid of M. By definition, $U(U(M)) = U(M)$, so that $(U(M), \cdot, 1)$ actually is a group. In view of this fact, $U(M)$ is usually called the **group of units** of M.

Obviously, the singleton set $\{1\}$ is a submonoid of any given monoid $(M, \cdot, 1)$, the **trivial submonoid** of M. If M is a group, then $\{1\}$ is a subgroup, the **trivial subgroup** of M. □

Next, we consider the important concept of a homomorphism between monoids:

Definition 1.1.9. Consider two (multiplicatively written) monoids $(M, \cdot, 1_M)$ and $(N, \cdot, 1_N)$. A mapping $\psi \colon M \to N$ is called a **monoid homomorphism** if it has the following two properties:

- $\psi(xy) = \psi(x)\psi(y)$ for all $x, y \in M$,
- $\psi(1_M) = 1_N$.

If M and N are actually groups, one speaks of a **group homomorphism** instead. □

The concept of a homomorphism is fundamental for all kinds of algebraic structures, such as monoids, groups, rings, modules, vector spaces, or algebras, and will play an important role throughout this book in various contexts.[2] In all of these cases, an injective homomorphism is called a **monomorphism**, a surjective homomorphism is called an **epimorphism**, and a bijective homomorphism is called an **isomorphism**. If the domain and the range of a homomorphism coincide, it is said to be an **endomorphism**. Finally, a bijective endomorphism is an **automorphism**.

Lemma 1.1.10. *Let $\psi\colon M \to N$ be any mapping from a monoid $(M,\cdot,1_M)$ to some monoid $(N,\cdot,1_N)$ satisfying $\psi(xy) = \psi(x)\psi(y)$ for all $x,y \in M$, and assume that $\psi(1_M)$ is a unit of N.*

Then ψ is a monoid homomorphism, and the image $\operatorname{im}\psi := \{\psi(a)\colon a \in M\}$ of M under ψ is a submonoid of N. Moreover, units in M are mapped to units in N:

$$\psi(u)^{-1} = \psi(u^{-1}) \quad \textit{for all } u \in U(M).$$

Proof. We have

$$1_N\psi(1_M) = \psi(1_M) = \psi(1_M 1_M) = \psi(1_M)\psi(1_M),$$

and multiplying from the right by $\psi(1_M)^{-1}$ yields $1_N = \psi(1_M)$. Thus ψ is indeed a monoid homomorphism. Now let $x,y \in \operatorname{im}\psi$, and choose $u,v \in M$ with $\psi(u) = x$ and $\psi(v) = y$. Then $xy = \psi(u)\psi(v) = \psi(uv)$, and therefore $\operatorname{im}\psi$ is closed and hence a submonoid, as $1_N = \psi(1_M)$.

Finally, let u be a unit in M. Then

$$1_N = \psi(1_M) = \psi(uu^{-1}) = \psi(u)\psi(u^{-1}),$$

and similarly $1_N = \psi(u^{-1})\psi(u)$. Thus $\psi(u)$ is a unit in N with inverse $\psi(u^{-1})$, and the second assertion follows from Proposition 1.1.5. □

In the special case where $(M,\cdot,1_M)$ and $(N,\cdot,1_N)$ are groups (so that $\psi\colon M \to N$ is a group homomorphism), Lemma 1.1.10 shows that $\operatorname{im}\psi$ is a subgroup of N.

Remark 1.1.11. Given a monoid $(M,\cdot,1_M)$ and an element $x \in M$, let

$$x^0 := 1_M, \quad x^1 := x \quad \text{and, recursively,} \quad x^{n+1} := x^n \cdot x \text{ (for } n \in \mathbb{N}^*).^3$$

By the induction principle, x^n is (uniquely) defined for all $n \in \mathbb{N}$. Thus we obtain a mapping

$$\psi_x\colon (\mathbb{N},+,0) \to (M,\cdot,1_M), \quad n \mapsto x^n,$$

and one can show by induction that ψ_x is a monoid homomorphism, that is,

$$x^{n+m} = x^n \cdot x^m \quad \text{for all } n,m \in \mathbb{N}.$$

[2] For example, in Section 1.3 we are going to study group homomorphisms in more detail.

[3] In additive notation, this becomes $0\cdot x := 0_M$, $1\cdot x := x$ and, recursively, $(n+1)\cdot x := n\cdot x + x$.

By definition, $\psi_x(0) = 1_M$, hence $\operatorname{im}\psi_x = \{x^n : n \in \mathbb{N}\}$ is a submonoid of M, by Lemma 1.1.10. On the other hand, if S is a submonoid of M containing x, then $\operatorname{im}\psi_x \subseteq S$. Consequently, with respect to set inclusion, $\operatorname{im}\psi_x$ is the smallest submonoid of M containing x; it is therefore called the **submonoid generated by** x. We shall follow standard notation and write $\langle x \rangle$ for $\operatorname{im}\psi_x$. If actually $\langle x \rangle = M$, we say that M is a **cyclic monoid** generated by x.

In view of the induction principle, the submonoid of $(\mathbb{N},+,0)$ generated by 1 is all of $\mathbb{N}$, so that $(\mathbb{N},+,0)$ is a cyclic monoid. □

Remark 1.1.12. Let us again consider the situation in Remark 1.1.11 and suppose additionally that x is a unit in M. Then $\langle x \rangle \subseteq U(M)$, that is, x^n is invertible for every $n \in \mathbb{N}$, and we write $x^{-n} := (x^n)^{-1}$ for $n \in \mathbb{N}$.[4] Then x^z is defined for all integers z, and ψ_x extends to a mapping

$$\gamma_x : (\mathbb{Z},+,0) \to (U(M),\cdot,1_M), \quad z \mapsto x^z.$$

One can check that $x^{z+w} = x^z \cdot x^w$ holds for all $z,w \in \mathbb{Z}$, so that γ_x actually is a group homomorphism. With respect to set inclusion, the image $\operatorname{im}\gamma_x$ of γ_x is the smallest subgroup of $U(M)$ containing x, called the **subgroup generated by** x. By an abuse of notation, one again writes $\langle x \rangle$ instead of $\operatorname{im}\gamma_x$: for units, $\langle x \rangle$ denotes the subgroup generated by x, and not just the (possibly smaller) monoid generated by x. We are going to study the mappings γ_x in more detail in Section 1.6 for the special case of groups. □

Let us return to the first part of the proof of Lemma 1.1.10. In essence, the argument given there boils down to the fact that multiplication by a unit can result in a **cancellation**: $x \in U(M)$ and $xy = xz$ imply $y = z$, by multiplying both sides from the left by x^{-1}. This concept turns out to be very important even in the more general case where x is not necessarily a unit and will be studied both in the remainder of this section and also in Section 1.2.

Definition 1.1.13. A monoid $(M,\cdot,1)$ is said to be a **monoid with cancellation** provided that the following **cancellation law** holds for all $x,y,z \in M$: if $xy = xz$ or $yx = zx$, then $y = z$. □

The basic examples for monoids with cancellation are provided by $(\mathbb{N},+,0)$ and $(\mathbb{N}^*,\cdot,1)$ as well as $(\mathbb{Z}^*,\cdot,1)$.

Now let $\psi : M \to N$ be any mapping from a monoid $(M,\cdot,1_M)$ to some monoid $(N,\cdot,1_N)$ with cancellation, and assume only that $\psi(xy) = \psi(x)\psi(y)$ holds for all $x,y \in M$. Then

$$1_N\psi(1_M) = \psi(1_M) = \psi(1_M 1_M) = \psi(1_M)\psi(1_M)$$

implies $1_N = \psi(1_M)$, and therefore the conclusions of Lemma 1.1.10 also hold under this weaker assumption on ψ (but stronger assumption on N).

[4] In additive notation, this reads $(-n)\cdot x := -(nx)$.

Trivially, every group is a monoid with cancellation. More generally, if $(G,\cdot,1)$ is a group and $S \subseteq G$ a submonoid of G, then $(S,\cdot,1)$ is a monoid with cancellation. Conversely, we now show that any commutative monoid with cancellation can be viewed as a submonoid of a suitable abelian group; for a generalization, we refer to the exercises.

Theorem 1.1.14. *Let $(M,\cdot,1)$ be a commutative monoid with cancellation and S a submonoid of M. Then there exist a commutative monoid M_S with cancellation and a monoid monomorphism $\psi\colon M \to M_S$ such that the image of S under ψ is contained in the group of units of M_S.*

Proof. Let us define a relation $\sim$ on the cartesian product $M \times S$ by

$$(a,b) \sim (a',b') \;:\Leftrightarrow\; ab' = ba'.$$

The commutativity of M implies that $\sim$ is reflexive and symmetric. We claim that $\sim$ is also transitive and therefore an equivalence relation on $M \times S$. In order to see this, let $(a,b) \sim (c,d)$ and $(c,d) \sim (e,f)$. Then $ad = bc$, and multiplying with f from the right gives $adf = bcf$. As $cf = de$, we obtain $adf = bde$. Since $\cdot$ is assumed to be commutative, this gives $afd = bed$. cancelling d shows $af = be$, which means $(a,b) \sim (e,f)$.

For $a \in M$ and $b \in S$, let $[a,b]$ be the equivalence class of (a,b), and denote the set of all equivalence classes by M_S. Now assume $(a,b) \sim (a',b')$ and $(c,d) \sim (c',d')$. Then $ab' = ba'$ and $cd' = dc'$, and therefore

$$(ac)(b'd') = (ab')(cd') = (ba')(dc') = (bd)(a'c'),$$

so that $(ac,bd) \sim (a'c',b'd')$. (Note that $bd, b'd' \in S$, as S is assumed to be closed.) Thus we may define a binary operation on the set M_S of all equivalence classes of $\sim$ by putting

$$[a,b]\cdot[c,d] \;:=\; [ac,bd].$$

Clearly, this operation is associative and commutative and has $[1,1]$ as an identity element, so that $(M_S,\cdot,[1,1])$ is a commutative monoid.

We next check that M_S is a monoid with cancellation. Thus let $[a,b]\cdot[c,d] = [a,b]\cdot[e,f]$. Then $[ac,bd] = [ae,bf]$ and therefore $acbf = bdae$. Since M is commutative, this gives $abcf = abde$, and cancelling ab implies $cf = de$, that is, $[c,d] = [e,f]$.

Finally, observe that two pairs $(a,1)$ and $(c,1)$ (with $a,c \in M$) are equivalent only if $a = c$. Thus the mapping

$$\psi\colon M \to M_S, \quad a \mapsto [a,1]$$

is injective and, in view of $[a,1]\cdot[c,1] = [ac,1]$, even a monoid monomorphism. Note that $[s,1]$ is invertible in M_S for all $s \in S$, as $(s,s) \sim (1,1)$ and therefore $[s,1][1,s] = [s,s] = [1,1]$. This establishes also the final assertion. □

It is usual to refer to the monomorphism ψ in Theorem 1.1.14 as an **embedding** of M into M_S and to identify M with its embedding $\psi(M)$ in M_S. The special case $S = M$ gives the following result:

Corollary 1.1.15. *Any commutative monoid with cancellation can be embedded into a commutative group.* □

In the situation of Corollary 1.1.15, we write $Q(M)$ instead of M_M and call $Q(M)$ the **quotient group** of M.

Example 1.1.16. (1) The quotient group of $(\mathbb{N}, +, 0)$ is the additive group $(\mathbb{Z}, +, 0)$ of the integers.

(2) The quotient group of $(\mathbb{N}^*, \cdot, 1)$ is the multiplicative group $(\mathbb{Q}^+, \cdot, 1)$ of the positive rational numbers.

(3) The quotient group of $(\mathbb{Z}^*, \cdot, 1)$ is the multiplicative group $(\mathbb{Q}^*, \cdot, 1)$ of the non-zero rational numbers.

(4) Let $S = p\mathbb{Z}^* = \{pz : z \in \mathbb{Z}^*\}$, where $p \in \mathbb{Z}$ is a prime number. Then S is a closed subset of $M = \mathbb{Z}^*$ and the monoid M_S constructed in the proof of Theorem 1.1.14 consists of all rational numbers $\frac{a}{b} \in \mathbb{Q}^*$ with denominator a power of p. □

The following result shows that the quotient group $Q(M)$ of a commutative monoid M with cancellation is the (unique) smallest group into which M can be embedded:

Theorem 1.1.17. *Let $(M, \cdot, 1)$ be a commutative monoid with cancellation and $Q(M)$ its quotient group. If $\eta : M \to G$ is a monoid monomorphism from M into a (multiplicatively written) group G, then η extends to a uniquely determined group monomorphism $\bar{\eta} : Q(M) \to G$.*

Proof. Suppose that $\gamma \colon Q(M) \to G$ is a group monomorphism extending $\eta : M \to G$. As $[a,b] = [a,1] \cdot [1,b] = [a,1] \cdot [b,1]^{-1}$ for all $a,b \in M$, we have

$$\gamma([a,b]) = \gamma([a,1])\gamma([b,1])^{-1} = \eta(a)\eta(b)^{-1},$$

so that γ is uniquely determined by η.

In order to prove the existence of such an extension, we first define a mapping $\eta' : M \times M \to G$ by $\eta'((a,b)) := \eta(a)\eta(b)^{-1}$. Consider the equivalence relation $\backsim$ on $M \times M$ given by

$$(a,b) \backsim (c,d) :\Leftrightarrow \eta'((a,b)) = \eta'((c,d)).$$

Then $(a,b) \backsim (c,d)$ if and only if $\eta(a)\eta(b)^{-1} = \eta(c)\eta(d)^{-1}$, which means $\eta(ad) = \eta(cb)$. As η is injective, this gives $ad = cb$ and hence $ad = bc$, as M is commutative. This shows $(a,b) \sim (c,d)$ with $\sim$ as in the proof of Theorem 1.1.14, leading to the definition of $Q(M)$. We conclude that $\sim$ and $\backsim$ coincide and that the mapping

$$\bar{\eta} : Q(M) \to G,\ [a,b] \mapsto \eta(a)\eta(b)^{-1}$$

is well-defined and extends η. It turns out that $\bar{\eta}$ is a group homomorphism. Moreover, $\bar{\eta}$ is injective, since $\bar{\eta}([a,b]) = 1_G$ implies $\eta(a) = \eta(b)$ and hence $a = b$. □

Example 1.1.18. Consider the monoid $(\mathbb{N},+,0)$ and a group $(G,\cdot,1)$, and let $g \in G$. By Remark 1.1.11 the mapping $\psi_g : \mathbb{N} \to G$, $n \mapsto g^n$ is a monoid homomorphism. By Theorem 1.1.17, ψ_g can be extended (uniquely) to a group homomorphism of $(\mathbb{Z},+,0)$ to G. Note that this unique extension is in fact the mapping γ_g considered in Remark 1.1.12, that is, the mapping $z \mapsto g^z$. □

Exercises

Exercise 1.1.19. The **center** of a monoid $(M,\cdot,1)$ is the set

$$C(M) := \{c \in M : cx = xc \text{ for all } x \in M\}.$$

Show that $C(M)$ is a submonoid of M and that $C(M) \cap U(M)$ is a subgroup of the group $U(M)$ of units of M. □

Exercise 1.1.20. Prove that the assertion of Theorem 1.1.14 also holds under the weaker assumption that $(M,\cdot,1)$ is a (not necessarily commutative) monoid with cancellation and that $S \subseteq C(M)$ is a submonoid of the center of M. □

1.2 Unique Factorization in Monoids

In this section, we study the structure of commutative monoids with cancellation. This is fundamental for the theory of rings (in particular, their multiplicative structure) to be considered in Sections 1.4 and 1.5. We use multiplicative notation throughout and start by introducing the concept of divisibility in an arbitrary commutative monoid.

Definition 1.2.1. Let $(M,\cdot,1)$ be a commutative monoid and let $a,b \in M$.

- If there is an $r \in M$ such that $ra = b$, then a is called a **divisor** of b, and b is said to be a **multiple** of a. We also say that a **divides** b and write $a \mid b$.
- If $a \mid b$ and $b \mid a$, then a and b are called **associates** in M. We write $a \approx b$ in this case.
- If $d \in M$ such that $d \mid a$ and $d \mid b$, then d is called a **common divisor** of a and b. Let $\mathrm{CD}(a,b)$ denote the set of all common divisors of a and b. If $d \in \mathrm{CD}(a,b)$ and if $c \mid d$ for every $c \in \mathrm{CD}(a,b)$, then d is called a **greatest common divisor** of a and b. We write $\mathrm{GCD}(a,b)$ for the set of all greatest common divisors of a and b.
- If $e \in M$ such that $a \mid e$ and $b \mid e$, then e is called a **common multiple** of a and b. Let $\mathrm{CM}(a,b)$ denote the set of all common multiples of a and b. If $e \in \mathrm{CM}(a,b)$

and if $e \mid f$ for every $f \in \mathrm{CM}(a,b)$, then e is called a **least common multiple** of a and b. The set of all least common multiples of a and b is denoted by $\mathrm{LCM}(a,b)$. □

Recall that $U(M)$ is the group of units of the monoid M. If $u \in U(M)$ and $a \in M$, then $a \mid u$ only if a is also a unit in M. On the other hand, $u \mid x$ for all $u \in U(M)$ and all $x \in M$ and therefore $U(M) \subseteq \mathrm{CD}(a,b)$ for all $a,b \in M$. Furthermore, ab is always a common multiple of a and b. We note that the sets $\mathrm{GCD}(a,b)$ and $\mathrm{LCM}(a,b)$ may be empty. Later in this section, we will consider additional assumptions on M which guarantee that these sets are non-empty.

Because of the existence of the unit element, the divisibility relation $\mid$ on a commutative monoid M is reflexive. The associative law implies that $\mid$ is transitive. Consequently, reflexivity and transitivity also hold for the association relation $\approx$. Since $\approx$ is symmetric by definition, it is even an equivalence relation on M. For any $a \in M$, we denote its equivalence class by $[a] := \{x \in M : a \approx x\}$ and put $U(M) \cdot a := \{ua : u \in U(M)\}$. In the case of a commutative monoid with cancellation, one has the following nice description of these equivalence classes:

Proposition 1.2.2. *Let $(M,\cdot,1)$ be a commutative monoid. Then $U(M) = [1]$ und $U(M) \cdot a \subseteq [a]$ for all $a \in M$. Moreover, if M is a monoid with cancellation, then even $U(M) \cdot a = [a]$ for all $a \in M$.*

Proof. Let $a \in M$ and $u \in U(M)$. Obviously, $a \mid ua$. On the other hand, $ua \mid a$ as $u^{-1} \cdot (ua) = a$. Hence $U(M) \cdot a \subseteq [a]$. Next let $a = 1$ and $x \in [1]$, so that there is an $r \in M$ with $rx = 1$. Since M is commutative, x is a unit, and we conclude $[1] = U(M)$.

Now let M be a commutative monoid with cancellation, and let $x \in [a]$ for some $a \in M$. Then there are elements $r,s \in M$ such that $x = ra$ and $a = sx$ and therefore $a = (sr)a$. Cancellation gives $1 = rs$, hence r and s are units in M and thus $x = ra \in U(M) \cdot a$. □

Note that $[u] = U(M)$ holds for every unit u in a commutative monoid M. Now assume that M is a commutative monoid with cancellation, and let $a,b \in M$. Then $\mathrm{GCD}(a,b)$ is either empty or equal to $U(M) \cdot d$, where d is any greatest common divisor of a and b. Similarly, $\mathrm{LCM}(a,b)$ is either empty or equals $U(M) \cdot e$, where e is any least common multiple of a and b.

Remark 1.2.3. Consider two elements a and b in a commutative monoid M with cancellation, and assume that $a \mid b$. Then there is a unique $r \in M$ such that $ra = b$, for $sa = b = ra$ implies $r = s$. This unique element r is usually denoted by $\frac{b}{a}$.[5] Of course, this notation does not imply that a has to be invertible in M. However, when a^{-1} is the inverse of a in the quotient group $Q(M)$ of M, then indeed $r = b \cdot a^{-1}$. This follows from $(a,b) \sim (r,1)$ (see the proof of Theorem 1.1.14), as the divisibility $a \mid b$ just means $[a,b] = [r,1]$ for some (unique) r.

If $d \mid a$ and $f \in M$, then $f \mid \frac{a}{d}$ if and only if $df \mid a$. Then we have

[5] If M is written additively, this becomes $r = b - a$.

$$\frac{\left(\frac{a}{d}\right)}{f} = \frac{a}{(df)},$$

for short $\frac{a}{df}$.[6] □

Proposition 1.2.4. *Let $\overline{M} = M/U(M) := \{[a] : a \in M\}$ be the set of all equivalence classes of associated elements in a commutative monoid $(M, \cdot, 1)$ with cancellation. Then the definition*

$$[a] \cdot [b] := [ab] \quad \textit{for all } a, b \in M$$

*turns $\overline{M}$ into a commutative monoid $(\overline{M}, \cdot, [1])$ with cancellation, called the **factor monoid** of $(M, \cdot, 1)$, for which $\{[1]\}$ is the only unit.*

Proof. Let $a, a', b, b' \in M$ such that $a \approx a'$ and $b \approx b'$. By Proposition 1.2.2, there are units $u, v \in M$ such that $ua = a'$ and $vb = b'$. As $U(M)$ is closed, we obtain

$$a'b' = (ua)(vb) = (uv)(ab) \in U(M) \cdot ab,$$

that is, $[ab] = [a'b']$. Hence the multiplication on $\overline{M}$ is well-defined. The associativity and the commutativity of M carry over to $\overline{M}$, and $[1]$ plays the role of the identity element. This shows that $(\overline{M}, \cdot, [1])$ is a commutative monoid.

Now let $[x]$ be any unit in $\overline{M}$, and let $[y] \in \overline{M}$ with $[1] = [x] \cdot [y] = [xy]$. Then $u = xy \in U(M)$, and $1 = x \cdot (yu^{-1})$ shows that x already is a unit in M, so that $[x] = U(M) = [1]$. Thus the group of units of $\overline{M}$ is indeed trivial.

It remains to show that $\overline{M}$ allows cancellation. Let $a, b, c \in M$ with $[a] \cdot [b] = [a] \cdot [c]$. Then $[ab] = [ac]$, and therefore $u(ab) = ac$ for some unit u in M. Using commutativity and cancelling a gives $ub = c$, that is, $[b] = [c]$. □

For example, the factor monoid of $(\mathbb{Z}^*, \cdot, 1)$ is $(\mathbb{N}^*, \cdot, 1)$.

Remark 1.2.5. Recall that a relation $\preceq$ on a non-empty set N for which $a \preceq b$ and $b \preceq a$ always imply $a = b$ is said to be **anti-symmetric**. If $\preceq$ is also reflexive and transitive, then $\preceq$ is called a **partial order** on N, and $(N, \preceq)$ is a **partially ordered set** or, for short, a **poset**.

The most fundamental examples are given by $\mathbb{N}$ and $\mathbb{Z}$ together with the usual order $\leq$, and by the power set $\mathscr{P}(S)$ of a set S together with set inclusion $\subseteq$. □

Remark 1.2.6. Let $(M, \cdot, 1)$ be a commutative monoid with cancellation. Then the divisibility relation $\mid$ on the corresponding factor monoid $(\overline{M}, \cdot, [1])$ is anti-symmetric by Propositions 1.2.2 and 1.2.4, as $U(\overline{M})$ is trivial. Therefore $\overline{M}$ also carries the structure of a partially ordered set. As $a \mid b$ implies $ax \mid bx$ for all $x \in M$, divisibility is compatible with the multiplication on $\overline{M}$, resulting in a canonical partially ordered monoid $(\overline{M}, \cdot, [1], \mid)$.

We shall return to this situation at the beginning of Chapter 2 when we study rings of formal power series. There the property of local finiteness will be important; see Definition 1.2.8. More generally, a partially ordered set $(N, \preceq)$ is said to be **locally**

[6] In additive notation, this reads $(a-d)-f = a-(d+f)$, or $a-d-f$ for short.

finite provided that the set $\{x \in N : x \preceq a\}$ is finite for every $a \in N$. For instance, $(\mathbb{N}, \leq)$ is locally finite, whereas $(\mathbb{Z}, \leq)$ is not. □

Now let $(M, \cdot, 1)$ be any commutative monoid with cancellation, and let $\mathscr{R}$ be a system of representatives of equivalence classes with respect to $\approx$. Then every $a \in M$ has a unique representation in the form $a = ur$ with $u \in U(M)$ and $r \in \mathscr{R}$. We now consider the question whether (and how) elements from $\mathscr{R}$ can be further decomposed in M. Of course, when M is a group, then $U(M) = M$ is the only $\approx$-equivalence class (represented by 1, for instance) and there is nothing left to do. We may therefore restrict attention to the case where $M \setminus U(M)$ is not empty.

For every $r \in M$, let $D(r)$ denote the set of divisors of r. For $r \in M \setminus U(M)$, the disjoint union $U(M) \mathbin{\dot\cup} U(M)r$ is always a subset of $D(r)$. The case of equality is dealt with in the following definition:

Definition 1.2.7. Let $(M, \cdot, 1)$ be a commutative monoid with cancellation, and let $r \in M \setminus U(M)$. Then r is called **irreducible** if each divisor of r is either a unit or an associate of r. We will denote the set of irreducible elements of M by $I(M)$. □

Definition 1.2.8. A commutative monoid $(M, \cdot, 1)$ with cancellation is called **locally finite** provided that for every $r \in M$ the set $D(r)$ of divisors is a union of a finite number of association classes, that is, $D(r) \cap \mathscr{R}$ is a finite set (where $\mathscr{R}$ is a system of representatives of equivalence classes with respect to $\approx$). □

Definition 1.2.9. Let M be a commutative monoid $(M, \cdot, 1)$ with cancellation.

- A sequence $(a_n)_{n \in \mathbb{N}}$ of elements of M is called a **divisor sequence** provided that $a_{n+1} \mid a_n$ for all $n \in \mathbb{N}$.
- We say that M satisfies the **chain condition** if, for any divisor sequence $(a_n)_{n \in \mathbb{N}}$ in M, there is an $m \in \mathbb{N}$ such that $a_m \approx a_j$ for all $j > m$. For short, M is also called a **c-monoid**. □

Of course, a locally finite monoid is also a c-monoid. However, under the weaker assumption of the chain condition, one can already show that any $a \in M$ can be written as a product of finitely many irreducible elements. Also, if M is not a group, every non-unit a has an irreducible divisor, so that $I(M) \neq \emptyset$.

Theorem 1.2.10. *Let a be an arbitrary element of a c-monoid M.*[7] *Then there exist $k \in \mathbb{N}$ and irreducible elements $r_1, \ldots, r_k \in M$ such that $a = r_1 \cdots r_k = \prod_{i=1}^{k} r_i$.*

Proof. We first show that every non-unit a in R has an irreducible divisor r. Assume otherwise. Then no divisor of a can have an irreducible divisor. Since $a_0 := a$ is neither a unit nor irreducible, there are **proper** divisors a_1 and b_1 of a_0 (that is, neither a_1 nor b_1 is an associate of a_0) such that $a_1 b_1 = a_0$. In particular, a_1 is not a unit. Since a_1 is not irreducible, there exist proper divisors a_2 and b_2 of a_1

[7] Strictly speaking, we should require a to be a non-unit. By a convenient (and common) abuse of language, we will view units as empty products of irreducible elements, so that they correspond to the case $k = 0$ of the assertion.

with $a_1 = a_2 b_2$. Continuing in this manner, we obtain a divisor sequence $(a_n)_{n\in\mathbb{N}}$ in M such that a_n is not an associate of a_{n+1} (for all n), which contradicts the chain condition.

Hence there exists an irreducible divisor r_1 of a_0. Assume that a is not a product of finitely many irreducible elements of M. Then this also holds for $a_1 := \frac{a_0}{r_1}$. In particular, $a_1 \in M \setminus U(M)$. Choose any irreducible divisor r_2 of a_1. Then $a_2 := \frac{a_1}{r_2}$ cannot be the product of finitely many irreducible elements, either. Continuing like this, we again obtain a divisor sequence for which a_n is never an associate of a_{n+1}, contradicting the chain condition. □

In order to prove a uniqueness result for the decomposition into irreducible elements, we need the following stronger concept:

Definition 1.2.11. Let $(M,\cdot,1)$ be a commutative monoid with cancellation. An element $r \in M \setminus U(M)$ is called a **prime element** or simply a **prime** provided that

$$r \mid ab \;\Rightarrow\; r \mid a \vee r \mid b \quad \text{for all } a,b \in M.$$

The set of prime elements of M will be denoted by $P(M)$. □

Proposition 1.2.12. *Every prime element in a commutative monoid with cancellation is irreducible.*

Proof. Let p be a prime element and assume that $d \mid p$, say $rd = p$. Then $p \mid rd$, and therefore $p \mid d$ or $p \mid r$. In the first of these cases, p and d are associates. In the second case, d is a unit, as then $r = ps$ for some s, so that $p = rd = (ps)d = p(sd)$, and hence $sd = 1$. □

Proposition 1.2.12 raises the question under which conditions the sets of irreducible and of prime elements coincide, and when this set is non-empty. We will obtain a satisfactory answer to this problem in Theorem 1.2.14. For this, we need one further definition:

Definition 1.2.13. A commutative monoid M with cancellation is called **factorial** provided that every element $a \in M$ has a factorization $a = p_1 \cdots p_\ell$ into prime elements $p_1,\dots,p_\ell \in M$.[8] □

If the factorial monoid M is not a group, then every non-unit has a prime divisor, and thus $P(M)$ is non-empty.

We can now prove the following major result: in a factorial monoid M, every element has a factorization as a product of finitely many irreducible elements, and this factorization is unique up to association.

Theorem 1.2.14. *Let $(M,\cdot,1)$ be a factorial monoid. Then the following hold:*

---

[8] As for irreducible elements, we use the convention to view units as empty products of primes, so that they correspond to the case $\ell = 0$.

(1) *Let $a \in M$ with $a = p_1 \cdots p_\ell$, where $\ell \in \mathbb{N}$ and $p_1, \ldots, p_\ell$ are prime elements. Assume also $a = r_1 \cdots r_k$ for some $k \in \mathbb{N}$ and irreducible elements $r_1, \ldots, r_k$. Then $k = \ell$, and there is a permutation β of $\{1, \ldots, \ell\}$ such that p_i and $r_{\beta(i)}$ are associates for $i = 1, \ldots, \ell$.*

(2) *Every irreducible element of M is a prime element, and thus $\mathbf{P}(M) = \mathbf{I}(M)$.*

(3) *M is locally finite.*

Proof. For every $n \in \mathbb{N}$, let P_n denote the set of all $a \in M$ which can be written as the product of at most n prime elements of M. By convention, $P_0 = U(M)$, so that (1) holds for $n = 0$. Assume inductively that (1) holds for all elements of P_n (for some $n \in \mathbb{N}$) and consider any element $a \in P_{n+1}$, say $a = p_1 \cdots p_\ell$ for some $\ell \leq n+1$ and primes $p_1, \ldots, p_\ell$. Assume also $a = r_1 \cdots r_k$ for some $k \in \mathbb{N}$ and irreducible elements $r_1, \ldots, r_k$. Then p_ℓ divides $r_1 \cdots r_k$. As p_ℓ is prime, there is some i such that p_ℓ divides r_i, without loss of generality, $i = k$. As r_k is irreducible, p_ℓ and r_k have to be associates. We may therefore cancel p_ℓ to obtain an identity of the form $a' := p_1 \cdots p_{\ell-1} = u \cdot r_1 \cdots r_{k-1}$, where $a' \in P_n$ and where u is a unit. Then $r_1' := ur_1$ is an irreducible element, and we obtain a representation $a' = r_1' \cdot r_2 \cdots r_{k-1}$ of a as a product of irreducible elements. Using induction, we conclude $\ell - 1 = k - 1$, where, after an appropriate reordering, p_i and r_i are associates for every $i = 1, \ldots, k-1$. As r_k and p_ℓ are likewise associates, this establishes the validity of (1).

Since M is factorial, an application of (1) to any irreducible element a shows in particular that every irreducible element is a prime, and hence (2) holds.

We finally prove (3). For this, let $\mathscr{R}$ again be a set of representatives for the equivalence classes of $\approx$ on M, and consider any element $a \in M$. In view of (1), there exist a unique $u \in U(M)$, a unique $\ell \in \mathbb{N}$, and unique primes in $\mathscr{R}$ such that $a = u \cdot p_1 \cdots p_\ell$. Now let d be any divisor of a. For $\ell = 0$ (that is, if a is a unit), $d \in U(M)$ as well, and therefore a has only one divisor, up to association.

In the general case, let $a = db$ with $b \in M$, and write d as $d = v \cdot r_1 \cdots r_n$ and b as $b = w \cdot s_1 \cdots s_m$ with units v, w and primes $r_1, \ldots, r_n, s_1, \ldots, s_m \in \mathscr{R}$. Then

$$u \cdot p_1 \cdots p_\ell = a = (vw) \cdot r_1 \cdots r_n \cdot s_1 \cdots s_m.$$

Now (1) gives $\ell = n + m$, hence $n \leq \ell$, and after a suitable reordering we may assume $p_i = r_i$ for $i = 1, \ldots, n$. Thus there exist a unique subset L of $\{1, \ldots, \ell\}$ and a unique $v \in U(M)$ such that $d = v \cdot \prod_{i \in L} p_i$. This shows that the number of pairwise non-associated divisors d of a is at most 2^ℓ and therefore finite. □

Remark 1.2.15. Let a be an arbitrary element of a factorial monoid $(M, \cdot, 1)$, and consider a factorization $a = u \cdot p_1 \cdots p_\ell$ with $u \in U(M)$ and prime elements $p_1, \ldots, p_\ell$. Grouping these primes into classes of associates shows the existence of a unique $m \in \mathbb{N}$, a unique (up to association) finite list of prime elements $q_1, \ldots, q_m \in M$ which are pairwise not associated, and unique positive integers $\alpha_1, \ldots, \alpha_m$ such that

$$a = \prod_{i=1}^{m} q_i^{\alpha_i}.$$

Here each factor $q_i^{\alpha_i}$ is called a **prime power**, and the α_i are called the **multiplicities** of the primes q_i in a. This representation of a is called the **prime power factorization** of a. □

Example 1.2.16. Let us consider a few examples.

(1) For our first (rather trivial) example, we switch to additive notation. The additive monoid $(\mathbb{N},+,0)$ of natural numbers is factorial. It contains a unique irreducible (hence prime) element, namely 1, and the prime power factorization of $n \in \mathbb{N}$ is just $n = n \cdot 1 = 1 + \cdots + 1$.

(2) The multiplicatively written monoid $(\mathbb{N}^*, \cdot, 1)$ is likewise factorial. Here the prime elements are just the ordinary prime numbers in $\mathbb{N}$; it is well known that there are infinitely many prime numbers. Similarly, the multiplicative monoid $(\mathbb{Z}^*, \cdot, 1)$ is also factorial. We will prove these facts in Section 1.6, when we study the ring structure of $\mathbb{Z}$.

(3) Let $p \in \mathbb{Z}$ be a prime and $\mathbb{Q}_p^*$ the set of all non-zero rational numbers with a denominator not divisible by p. Then $(\mathbb{Q}_p^*, \cdot, 1)$ is a factorial monoid which has, up to association, a unique irreducible element, namely p. Any $\lambda \in \mathbb{Q}_p^*$ can be written as $\frac{a}{b} \cdot p^k$, where $a,b \in \mathbb{Z}^*$ and where $k \in \mathbb{N}$, which gives the prime power factorization of λ in $\mathbb{Q}_p^*$. □

We conclude this section by investigating the existence of greatest common divisors and of least common multiples in commutative monoids with cancellation. The following definition is particularly important:

Definition 1.2.17. Let M be a commutative monoid with cancellation, and let a and b be two elements of M for which $\mathrm{GCD}(a,b)$ is non-empty.[9] Then a and b are said to be **relatively prime** provided that $\mathrm{GCD}(a,b) = U(M)$. □

The proof of the following simple result is left to the reader as Exercise 1.2.23:

Lemma 1.2.18. *Let a and b be two elements of a commutative monoid M with cancellation, and assume that d is a greatest common divisor of a and b. Then the following hold:*

(1) $\mathrm{GCD}(\frac{a}{d}, \frac{b}{d}) = U(M)$.

(2) $\frac{ab}{d}$ *is a least common multiple of a and b.* □

Proposition 1.2.19. *Let M be a commutative monoid with cancellation, and assume* $\mathrm{GCD}(a,b) \neq \emptyset$ *for all* $a,b \in M$. *Then every irreducible element of M is a prime element:* $I(M) = P(M)$.

Proof. Assume otherwise, and let $r \in M$ be an irreducible element dividing some product ab, but neither a nor b. The irreducibility of r gives

$$\mathrm{GCD}(r,a) = U(M) = \mathrm{GCD}(r,b).$$

[9] For instance, this always holds when M is factorial, as we will see in Proposition 1.2.20.

Choose some greatest common divisor d of rb and ab. Then $b \mid d$ and thus $\frac{d}{b}$ is a common divisor of r and a. Since a and r are relatively prime, $\frac{d}{b}$ is a unit. This implies $b \approx d$ and therefore $\mathrm{GCD}(rb, ab) = U(M) \cdot b$.

Given $x, y, z \in M$, let $v \in \mathrm{GCD}(x, y)$ and $w \in \mathrm{GCD}(y, z)$. Then it is not difficult to check $\mathrm{GCD}(v, z) = \mathrm{GCD}(x, w)$. Applying this with $x = r$, $y = rb$ and $z = ab$, we may take $v = r$ and $w = b$ to obtain $\mathrm{GCD}(r, ab) = \mathrm{GCD}(r, b) = U(M)$. Thus r and ab are relatively prime, a contradiction. □

In particular, $I(M) = P(M)$ whenever M is a c-monoid such that any two elements of M have a greatest common divisor. If additionally M is not a group, then $P(M)$ is not empty.

Proposition 1.2.20. *Any two elements of a factorial monoid have both a greatest common divisor and a least common multiple.*

Proof. Let M be the underlying monoid and consider the prime power factorizations of two given elements $a, b \in M$, say

$$a = \prod_{i=1}^{m} q_i^{\alpha_i} \cdot \prod_{i=1}^{k} p_i^{\beta_i} \quad \text{and} \quad b = \prod_{i=1}^{m} q_i^{\gamma_i} \cdot \prod_{i=1}^{l} r_i^{\delta_i},$$

where $q_1, \ldots, q_m$, $p_1, \ldots, p_k$, and $r_1, \ldots, r_l$ are primes which are pairwise non-associated. Let μ_i be the minimum of α_i and γ_i for all $i = 1, \ldots, m$. Then $\prod_{i=1}^{m} q_i^{\mu_i}$ is a greatest common divisor of a and b, and hence $\prod_{i=1}^{m} q_i^{\alpha_i + \gamma_i - \mu_i} \cdot \prod_{i=1}^{k} p_i^{\beta_i} \cdot \prod_{i=1}^{l} r_i^{\delta_i}$ is a least common multiple of a and b, by Lemma 1.2.18. □

Exercises

Exercise 1.2.21. Let $(M, \cdot, 1)$ be a commutative monoid with cancellation. Show that the following assertions are equivalent:

(1) M is a c-monoid and $\mathrm{GCD}(a, b)$ is non-empty for every $a, b \in M$.
(2) M is a c-monoid and $I(M) = P(M)$.
(3) M is factorial.
(4) M is locally finite and $\mathrm{GCD}(a, b)$ is non-empty for every $a, b \in M$. □

Exercise 1.2.22. Assume that M is a factorial monoid and let $a, b, c \in M$ such that a and b are relatively prime and $a \mid bc$. Show that a divides c. □

Exercise 1.2.23. Prove Lemma 1.2.18. □

1.3 Basics on Groups

In this section, we present some basic results from group theory. We will use multiplicative notation throughout.

Recall from Definition 1.1.7 that a non-empty subset U of a group $(G,\cdot,1)$ is called a subgroup if $1 \in U$ and if $a,b \in U$ always imply $ab \in U$ and $a^{-1} \in U$. The following lemma shows how to combine these three requirements into just one condition:

Lemma 1.3.1. *Let U be a non-empty subset of a group $(G,\cdot,1)$. Then U is a subgroup of G if and only if $ab^{-1} \in U$ for all $a,b \in U$.*

Proof. Obviously, the condition in question holds if U is any subgroup of G. Conversely, let U be a non-empty subset of G and assume $ab^{-1} \in U$ for all $a,b \in U$. Choose any element $a \in U$; then $1 = aa^{-1} \in U$. Using this shows $1 \cdot a^{-1} = a^{-1} \in U$ for all $a \in U$. Finally, given arbitrary elements $a,b \in U$, we have $b^{-1} \in U$ and therefore also $a(b^{-1})^{-1} = ab \in U$. □

Remark 1.3.2. Given any subgroup U of a group G, one defines a binary relation $\sim_U$ on G by

$$a \sim_U b \;:\Leftrightarrow\; a^{-1}b \in U. \tag{1.5}$$

Then $\sim_U$ is reflexive, as $a^{-1}a = 1 \in U$ for all $a \in G$. Moreover, $\sim_U$ is symmetric, since $(a^{-1}b)^{-1} = b^{-1}a$ for all $a,b \in G$. Finally, if $a^{-1}b$ and $b^{-1}c \in U$, then $(a^{-1}b)(b^{-1}c) = a^{-1}c \in U$, and therefore $\sim_U$ is also transitive. Thus $\sim_U$ is an equivalence relation on G.

It is easy to check that the equivalence class of $a \in G$ is the set $aU := \{au : u \in U\}$; we leave this as an exercise. The equivalence classes are called **left cosets** of U in G, and the set of all left cosets of U is denoted by G/U. □

Let U be a subgroup of $(G,\cdot,1)$, and let $a,b \in G$. By a basic property of equivalence relations, $aU = bU$ if and only if $aU \cap bU \neq \emptyset$. Moreover, the mapping $aU \to bU$, $au \mapsto bu$ is a bijection from aU to bU, and therefore any two left cosets of U in G have the same cardinality. In the case where G is a finite group, this gives the following well known result:

Theorem 1.3.3 (Theorem of Lagrange). *Let G be a finite group, and let U be any subgroup of G. Then $|U|$ divides $|G|$ and $|G| = |G/U| \cdot |U|$.* □

The number $|G/U|$ of left cosets of U in G is usually written as $[G:U]$ and called the **index** of U in G. The cardinality of a finite group is also called its **order**.

In general, there may be divisors d of the order $|G|$ of a finite group G for which no subgroup U with $|U| = d$ exists. However, finite abelian groups always contain subgroups for all divisors of their order. In Section 1.6, we will study the particular subclass of finite cyclic groups G; for this case, there is exactly one subgroup U of order d for each divisor of $|G|$.

Remark 1.3.4. A **right coset** of a subgroup U in a group G is a set of the form $Ub = \{ub : u \in U\}$, where $b \in G$. It is easy to see that the right cosets are the equivalence classes of the relation $\sim^U$ defined by

$$a \sim^U b \;:\Leftrightarrow\; ab^{-1} \in U. \qquad \square \tag{1.6}$$

Definition 1.3.5. A subgroup U of a group $(G,\cdot,1)$ is called **normal** provided that $gU = Ug$ (equivalently, $g^{-1}Ug = U$) holds for all $g \in G$. □

Of course, if G is abelian, then every subgroup of G is normal. In general, the normal subgroups of a group G can be described via the group homomorphisms with domain G. For this, we need a definition:

Definition 1.3.6. Let $(G,\cdot,1_G)$ and $(H,\cdot,1_H)$ be two groups, and let $\psi\colon G \to H$ be a group homomorphism. Then the **kernel** of ψ is defined as

$$\ker\psi := \{g \in G\colon \psi(g) = 1_H\}. \qquad \square$$

Proposition 1.3.7. *Let G and H be groups and $\psi\colon G \to H$ a group homomorphism. Then the kernel $\ker\psi$ is a normal subgroup of G.*

The pre-image $\psi^{-1}(y) = \{g \in G\colon \psi(g) = y\}$ of an element $y \in \operatorname{im}\psi$ under ψ is given by $v \cdot \ker\psi$, where $v \in G$ is any element with $\psi(v) = y$. In particular, ψ is injective if and only if the kernel of ψ is trivial, that is, $\ker\psi = \{1_G\}$.

Proof. In view of the remark following Definition 1.1.13, we have $\psi(1_G) = 1_H$, hence $\ker\psi$ is not empty. If $x,y \in \ker\psi$, then $\psi(xy^{-1}) = \psi(x)\psi(y)^{-1} = 1_H$, and therefore $\ker\psi$ is a subgroup of G, by Lemma 1.3.1. Moreover, one has

$$\psi(g^{-1}xg) = \psi(g)^{-1}\psi(x)\psi(g) = \psi(g)^{-1}\psi(g) = 1_H \quad \text{for all } g \in G,$$

and thus $\ker\psi$ is indeed a normal subgroup. The remaining assertions are left to the reader as a simple exercise. □

If N is a normal subgroup of G, we can define a binary operation $\cdot$ on the set G/N of left cosets of N in G by setting

$$aN \cdot bN := abN. \tag{1.7}$$

Note that the operation $\cdot$ is indeed well-defined: if au, bv are arbitrary elements of aN and bN, respectively, then

$$aubv = a(bb^{-1})ubv = ab(b^{-1}ub)v \in abN,$$

as $b^{-1}ub$ is an element of N. Moreover, G/N is a group with respect to $\cdot$, where $N = 1N$ is the identity element of G/N.

Definition 1.3.8. Let N be a normal subgroup of a group G, and let $\cdot$ be the multiplication of left cosets defined in (1.7). Then $(G/N,\cdot)$ is called the **factor group** of G modulo N, and the group epimorphism with kernel N defined by

$$\nu_N\colon G \to G/N, \;\; g \mapsto gN \tag{1.8}$$

is said to be the **natural epimorphism** from G to G/N. □

An alternative characterisation of normal subgroups is given in Exercise 1.3.17. Our next result relates the subgroups of a factor group G/N to the subgroups of G.

Proposition 1.3.9. *Let N be a normal subgroup of a group $(G,\cdot,1)$. Then:*

(1) *For every subgroup U of G, the set $UN := \{ux\colon u \in U, x \in N\}$ is a subgroup of G containing N.*

(2) *If V is a subgroup of G containing N, then N is a normal subgroup of V and V/N is a subgroup of G/N. Moreover, V/N is a normal subgroup of G/N if and only if V is a normal subgroup of G.*

(3) *Let $\mathscr{W}_N$ be the set of all subgroups V of G such that $N \subseteq V$, and let $\mathscr{U}_N$ be the set of all subgroups of G/N. For $V \in \mathscr{W}_N$, let $\pi_N(V) := V/N$. Then π_N is a bijection from $\mathscr{W}_N$ to $\mathscr{U}_N$. Moreover, the restriction of π_N to the set of normal subgroups of G containing N is a bijection with the set of normal subgroups of G/N.*

Proof. Parts (1) and (2) are easy to prove and will be left as an exercise.

Given any subgroup U of G, we obtain a subgroup $i_N(U) := UN/N$ of G/N, by (1) and (2). Conversely, the set $j_N(H) := \{x \in G\colon xN \in H\}$ is a subgroup of G containing N, for all subgroups H of G/N.

Note that always $U \leq j_N(i_N(U))$ and that equality holds if and only if N is contained in U. On the other hand, $i_N(j_N(H)) = H$ holds for all subgroups H of G/N. This proves that the restriction π_N of i_N to the set of subgroups of G containing N is a bijection to the set of subgroups of G/N (with inverse j_N), and it is easy to check that these mappings preserve normality. □

Theorem 1.3.10 (Homomorphism theorem). *Let $(G,\cdot,1_G)$ and $(H,\cdot,1_H)$ be two groups, and let $\psi : G \to H$ be a group homomorphism. Then there exists a unique group homomorphism $\sigma\colon G/\ker\psi \to H$ such that $\psi = \sigma \circ \nu_{\ker\psi}$, where $\circ$ denotes the composition of mappings. Moreover, σ is injective and, if ψ is an epimorphism, even an isomorphism.*

Proof. For simplicity, write $N := \ker\psi$. Now assume that $\sigma\colon G/N \to H$ is a group homomorphism satisfying $\psi = \sigma \circ \nu_N$. Then $\sigma(gN) = (\sigma \circ \nu_N)(g) = \psi(g)$ for all $g \in G$, and thus σ is uniquely determined by ψ.

We therefore define a mapping $\sigma\colon G/N \to H$ accordingly, by $\sigma(gN) := \psi(g)$. Note that this makes sense, since $\sigma(gxN) = \psi(gx) = \psi(g)\psi(x) = \psi(g)$ for all $x \in N$. It is easy to see that σ is a group monomorphism, as its kernel consists of the unit element N of G/N, and that σ is surjective if and only if ψ is. □

Corollary 1.3.11 (First isomorphism theorem). *Let N and V be normal subgroups of a group G, where $N \subseteq V$. Then the factor groups G/V and $(G/N)/(V/N)$ are isomorphic and there exists a unique isomorphism $\sigma\colon G/V \to (G/N)/(V/N)$ satisfying $\sigma(gV) = (gN)(V/N)$ for all $g \in G$.*

Proof. By Proposition 1.3.9, V/N is a normal subgroup of G/N. Let $\nu\colon G \to G/N$ and $\mu\colon G/N \to (G/N)/(V/N)$ be the associated natural epimorphisms. Then the composition

$$\phi := \mu \circ \nu\colon G \to (G/N)/(V/N)$$

is an epimorphism satisfying $\phi(g) = (gN)(V/N)$. Moreover, $g \in \ker\phi$ if and only if $gN \in V/N$, which holds if and only if $g \in V$. Finally, there is a unique isomorphism

$\sigma\colon G/V \to (G/N)/(V/N)$ satisfying $\sigma(gVU) = \phi(g)$, by the homomorphism theorem. □

We conclude this section with some basic facts concerning permutation groups. Given any non-empty set X, one denotes by $\mathrm{Sym}(X)$ (or, more commonly, just S_X) the set of all **permutations** (that is, bijective mappings) on X. Clearly, $\mathrm{Sym}(X)$ is a group with respect to the composition of mappings, with the identity map $\mathrm{id}_X : x \mapsto x$ as identity element. This group is called the **symmetric group** on X; it is non-abelian whenever X has at least three distinct elements (exercise).

Definition 1.3.12. Let $(G,\cdot,1)$ be a group and X a non-empty set. Then any group homomorphism

$$\pi\colon (G,\cdot,1) \to (\mathrm{Sym}(X),\circ,\mathrm{id}_X)$$

is called a **permutation representation** of G on X. One also says that G **acts** on X as a group of permutations. Such an action is said to be **faithful** when π is a monomorphim. □

If π is a faithful action of G on X, then the image of G under π is a subgroup of $\mathrm{Sym}(X)$ isomorphic to G, by the homomorphism theorem 1.3.10. For $x \in X$ and $g \in G$, it is convenient to write simply gx for the image $\pi(g)(x)$ of x under the permutation $\pi(g)$. As π is a group homomorphism, this gives $(gh)x = g(hx)$ for all $g,h \in G$ and all $x \in X$.[10]

Any action of a group G on a set X induces an equivalence relation $\sim$ on X by putting

$$x \sim y \iff \exists g \in G\colon gx = y \qquad \text{for } x,y \in X.$$

The equivalence class $Gx := \{gx\colon g \in G\}$ of x is called the **orbit** of x under G. Given any $x \in X$, one also defines a subgroup $G_x := \{g \in G\colon gx = x\}$ of G, the **stabilizer** of x in G. There is a fundamental relation between the stabilizer of an element and its orbit:

Proposition 1.3.13. *Let G be a group acting on a set X, and let $x \in X$. Then there exists a bijection from the orbit Gx of x under G to the set G/G_x of left cosets of the stabilizer of x in G.*

Proof. Define a mapping $\omega : G \to Gx$ by putting $\omega(g) = gx$. For $a,b \in G$, one has $\omega(a) = \omega(b)$ if and only if $ax = bx$, that is, if and only if $(a^{-1}b)x = x$. This means that $a^{-1}b$ stabilizes x, that is, $a^{-1}b \in G_x$, which in turn is equivalent to $aG_x = bG_x$. Thus $\omega(a) = \omega(b)$ if and only if $a \sim_U b$, where $U = G_x$ and $\sim_U$ is as in (1.5).

Obviously, ω is surjective. Thus ω induces a bijection between $\mathscr{L}$ and Gx, where $\mathscr{L}$ is a set of representatives of the left cosets of G_x in G, and therefore also a bijection between G/G_x and Gx. □

---

[10] An alternative common notation, especially in algebra texts, is x^g. Note that then x^{gh} means the image of x resulting from first applying $\pi(g)$ and then $\pi(h)$, reversing the order compared to the notation used in this book. In this notation, one should then also compose mappings in the reverse order, compared to the composition $\circ$ as used here.

Corollary 1.3.14. *Let G be a finite group acting on a set X. Then*

$$|G| = |Gx| \cdot |G_x|$$

for every $x \in X$.

Proof. Since G is finite, Gx and G_x are likewise finite, and $|Gx| = |G/G_x| = [G:G_x]$, by Proposition 1.3.13. Now the assertion follows from Theorem 1.3.3. □

Exercises

Exercise 1.3.15. Let A and B be two finite subgroups of a group G, and consider the set $AB = \{ab\colon a \in A, b \in B\}$.

(1) Show that

$$|AB| = \frac{|A| \cdot |B|}{|A \cap B|}.$$

(2) Prove that AB is a subgroup of G provided that at least one of the two subgroups involved is normal (in particular, in the abelian case). Then the subgroup AB is called the **product** of A and B. (If additive notation is used, one speaks of the **sum** $A+B$ of A and B.)

Exercise 1.3.16. An equivalence relation $\sim$ on a group $(G,\cdot,1)$ is called a **congruence relation** provided that $\sim$ is compatible with the binary operation of G, that is, if $a \sim b$ and $c \sim d$ imply $ac \sim bd$. In this context, the equivalence classes are also called **congruence classes**. Prove the following two properties of congruence relations:

(1) $a \sim b \Rightarrow a^{-1} \sim b^{-1}$ for all $a, b \in G$.

(2) The congruence class N of the identity element 1 of G is a normal subgroup of G, and $\sim$ coincides with $\sim_N$. □

Exercise 1.3.17. Let U be a subgroup of a group $(G,\cdot,1)$. Show that $\sim_U$ is a congruence relation on G if and only if U is a normal subgroup of G. Moreover, in this case $\sim_U$ and $\sim^U$ coincide. □

Exercise 1.3.18 (Second isomorphism theorem). Let N be a normal subgroup of a group G, and let U be a subgroup of G. Prove that $U \cap N$ is a normal subgroup of U and show the existence of a unique isomorphism $\sigma\colon U/(U \cap N) \to UN/N$ satisfying $\sigma(u(U \cap N)) = uN$ for all $u \in U$. □

Exercise 1.3.19. Let $(G,\cdot,1)$ be a group. Prove the following facts:

(1) Let $\mathscr{U}$ be any non-empty collection of subgroups of G. Then the intersection $\cap_{U \in \mathscr{U}} U$ of all $U \in \mathscr{U}$ is likewise a subgroup of G.

(2) Let $Y \subseteq G$ and put

$$\langle Y \rangle := \bigcap_{Y \subseteq U \leq G} U. \tag{1.9}$$

Show that $\langle Y \rangle$ is a subgroup of G, and that $\langle Y \rangle \subseteq H$ for all subgroups H of G containing Y.

(3) Assume that G is abelian and that $Y = \{g_1, ..., g_k\}$ is a finite set. Show that then $\langle g_1, \ldots, g_k \rangle := \langle Y \rangle$ is given by

$$\{g_1^{z_1} \cdots g_k^{z_k} : z_1, \ldots, z_k \in \mathbb{Z}\};$$

see Remark 1.1.12 for the special case $k = 1$. □

Exercise 1.3.20. Let G and H be two (multiplicatively written) groups.

(1) Show that the set $G \times H$ is a group with respect to the componentwise multiplication given by

$$(g,h) \cdot (g',h') := (gg',hh').$$

One calls the resulting group the (external) **direct product** of G and H.

(2) In the situation of part (2) of Exercise 1.3.15, assume that both A and B are normal subgroups of G and that $A \cap B = \{1\}$. Prove that AB is a normal subgroup of G which is isomorphic to the direct product $A \times B$. One therefore says that AB is the (internal) **direct product** of the subgroups A and B.

It is quite common to write $A \otimes B$ for direct products, instead of $A \times B$ or AB. In additive notation, one speaks of **direct sums** instead of direct products and uses the notation $A \oplus B$ instead of $A + B$. □

Exercise 1.3.21. Generalize the construction in Exercise 1.3.20 to an arbitrary collection of groups G_i (indexed by some index set I). □

1.4 Basics on Rings

In this section, we collect some basic results on rings.

Definition 1.4.1. A **ring** is a non-empty set R together with two binary operations, $+$ (addition) and $\cdot$ (multiplication), which satisfy the following axioms:

(1) $(R,+,0)$ is an abelian group (with zero element 0).

(2) $(R,\cdot,1)$ is a monoid (with unit element 1).[11]

(3) The **distributive laws** hold:[12]

---

[11] Some authors only require that the multiplication is associative; in this case, what we call a ring is referred to as a **ring with 1** or a **unitary ring**.

[12] As usual, the multiplication is considered to have a higher priority than the addition, so that the term $ab + ac$ is to be evaluated as $(ab) + (ac)$.

$$a(b+c) = ab+ac \text{ and } (a+b)c = ac+bc \quad \text{for all } a,b,c \in R.$$

If the multiplication is commutative, R is said to be a **commutative ring**. As usual, R is called **finite** provided that $|R|$ is finite. □

In the following result, we list the most basic properties of rings; the simple proofs may be left as exercises.

Lemma 1.4.2. *Let $(R,+,\cdot,0,1)$ be a ring. Then:*

(1) $a\cdot 0 = 0 = 0\cdot a$ *for all $a \in R$.*

(2) $a(-b) = (-a)b = -(ab)$ *for all $a,b \in R$, and one simply writes $-ab$ for this product.*

(3) $(-a)(-b) = ab$ *for all $a,b \in R$.*

(4) $(-1)a = -a = a(-1)$ *for all $a \in R$.* □

Given any singleton set $R = \{x\}$, one can define a (trivial) ring structure on R by putting $x+x := x =: x\cdot x$. Then, of course, the zero and the unit elements coincide. Such an object is therefore called a **trivial ring**. However, as soon as R is a ring containing some element $a \neq 0$, we also have $1 \neq 0$, as $0\cdot a = 0 \neq a = 1\cdot a$. Then $1 \in R^*$, where we write $R^* := R\setminus\{0\}$ for the set of all non-zero elements of R.

Definition 1.4.3. Let $(R,+,\cdot,0,1)$ be a ring. Then $u \in R$ is (multiplicatively) **invertible** or a **unit** provided that there exists an element $v \in R$ with $uv = 1 = vu$, that is, if u is a unit in the multiplicative monoid $(R,\cdot,1)$. The set $U(R)$ of all invertible elements of R is called the **group of units** of R. □

If R is a non-trivial ring, then $U(R) \subseteq R^* := R\setminus\{0\}$, as $0\cdot v = 0$ for all $v \in R$. The case where $U(R) = R^*$ – that is, where every non-zero element of R is a unit – gives rise to an important special class of rings:

Definition 1.4.4. Let R be a non-trivial ring and $U(R)$ its group of units, and assume $U(R) = R^*$. Then R is called a **skew field** or a **division ring**. If, in addition, R is commutative, then R is called a **field**. Finally, a finite field is usually called a **Galois field**. □

The term *Galois field* is chosen in honour of Evariste Galois, since his fundamental paper [123] from 1830 is generally viewed as the starting point of the systematic study of finite fields; see [292, Section 1.1] for a good account of the history of finite fields.

In Sections 1.1 and 1.2, we have considered monoids with cancellation. For rings, this leads to the following definition:

Definition 1.4.5. Let $(R,+,\cdot,0,1)$ be a non-trivial ring. An element $x \in R^*$ is said to be a **zero divisor** provided that there exists an element $y \in R^*$ such that $xy = 0$ or $yx = 0$. A non-trivial ring without zero divisors is called a **domain**. □

Remark 1.4.6. Note that a ring R is a domain provided that $xy = 0$ implies $x = 0$ or $y = 0$, that is, if R^* is a submonoid of the multiplicative monoid $(R, \cdot, 1)$. In particular, any skew field is a domain: if $xy = 0$ and $x \neq 0$, we may multiply from the left by x^{-1} to obtain $y = 0$; similarly, if $y \neq 0$, multiplication from the right with y^{-1} yields $x = 0$. In the case of a skew field $(F, +, \cdot, 0, 1)$, one refers to $(F^*, \cdot, 1)$ as the **multiplicative group** of F. □

Lemma 1.4.7. *Let $(R, +, \cdot, 0, 1)$ be a domain. Then $(R^*, \cdot, 1)$ is a monoid with cancellation.*

Proof. We have already noted that R^* is a submonoid of $(R, \cdot, 1)$. Now let $x, y, z \in R^*$ and assume $xz = yz$. Then $(x - y)z = 0$, hence $x - y = 0$ or $z = 0$. As $z \in R^*$, we conclude $x = y$. Similarly, $zx = zy$ also implies $x = y$. □

At this point, we can already prove the following interesting result on finite domains:

Theorem 1.4.8. *Any finite domain is a skew field.*

Proof. Given any non-zero element u of a finite domain $(R, +, \cdot, 0, 1)$, we define a mapping $\varphi_u : R \to R$ by putting $\varphi_u(x) = xu$. Because of the distributive laws, φ_u is an epimorphism of the additive group $(R, +, 0)$. Since R has no zero divisors, the kernel of φ_u is the trivial subgroup $\{0\}$, so that φ_u is injective, by Proposition 1.3.7. As R is finite, this already implies that φ_u is also surjective. In particular, there is an element $x \in R$ with $xu = 1$, and thus $x \in R^*$. Similarly, one also obtains the existence of an element $y \in R^*$ with $uy = 1$. Now Proposition 1.1.5 implies that u is a unit in $(R^*, \cdot, 1)$. This shows $U(R) = R^*$, as claimed. □

A famous theorem established by Wedderburn [400] in 1905 states that a finite skew field is actually a field; we will prove this major result in Section 3.7. As a corollary of Theorem 1.4.8 and Wedderburn's theorem, the Galois fields are the only finite domains.

In Section 1.1 we have seen that every commutative monoid with cancellation can be embedded into a (commutative) group, for instance, the quotient group. This can be extended to embedding any commutative domain into a suitable field. To do so, we require the concept of a ring homomorphism:

Definition 1.4.9. Let R and T be two rings. A mapping $\psi : R \to T$ is called a **ring homomorphism** provided that the following two conditions hold:

- ψ is a group homomorphism of the additive groups $(R, +, 0_R)$ and $(T, +, 0_T)$.
- $\psi(xy) = \psi(x)\psi(y)$ for all $x, y \in R$.

The **kernel** of ψ is the set $\ker \psi := \{r \in R : \psi(r) = 0_T\}$, that is, the kernel of ψ when considered as a homomorphism of the additive groups.

The mapping $\omega : R \to T$ with $\omega(r) = 0_T$ for all $r \in R$ is called the **trivial ring homomorphism** from R to T. □

By Lemma 1.1.10, a ring homomorphism $\psi\colon R \to T$ satisfies $\psi(1_R) = 1_T$ provided that $\psi(1_R)$ is a unit in T, and this condition holds whenever T is a domain.

We can now prove the promised extension of Theorem 1.1.14 to domains:

Theorem 1.4.10. *Let $(R,+,\cdot,0,1)$ be a commutative domain and $S \subseteq R^*$ a submonoid of $(R^*,\cdot,1)$. Then there exist a commutative domain R_S and a ring monomorphism $\psi\colon R \to R_S$ such that $\psi(S) \subseteq U(R_S)$.*

Proof. By Lemma 1.4.7, $(R^*,\cdot,1)$ is a commutative monoid with cancellation. Thus we can define an equivalence relation $\sim$ on the set $R^* \times S$ as in the proof of Theorem 1.1.14:

$$(a,b) \sim (c,d) \;\;:\Leftrightarrow\;\; ad = bc.$$

Note that $\sim$ even extends to an equivalence relation on all of $R \times S$ via this definition. To see this, assume $(a,b) \sim (c,d)$, where $a = 0$. Then $ad = 0 \cdot d = 0$ gives $bc = 0$ and thus $c = 0$, as $b \in S \subseteq R^*$ and as R is a domain. Conversely, $(0,d) \sim (0,b)$ for all $b,d \in S$ by definition.

As before, we denote the equivalence class of (a,b) by $[a,b]$ and write R_S for the set of all equivalence classes. Since $1 \in S$ by assumption, $[0,1] = \{(0,x)\colon x \in S\}$. As in the proof of Theorem 1.1.14, we have a well-defined multiplication on R_S given by $[a,b] \cdot [c,d] := [ab,cd]$. In the present situation, we can also define an addition on R_S as follows:

$$[a,b] + [c,d] := [ad + bc, cd] \quad \text{for } [a,b],[c,d] \in R_S.$$

We need to check that this operation is indeed well-defined. Thus let $(a,b) \sim (a',b')$ and $(c,d) \sim (c',d')$. Then $ab' = a'b$, hence $dd'ab' = dd'a'b$; similarly, $cd' = c'd$ gives $bb'cd' = bb'c'd$. Consequently,

$$a'd'bd + b'c'bd = b'd'ad + b'd'bc,$$

and thus $(a'd' + b'c', b'd') \sim (ad + bc, bd)$.

It is routine to check that these two binary operations turn R_S into a commutative domain with zero element $[0,1]$ and unit element $[1,1]$ and that the mapping

$$\psi\colon R \to R_S, \;\; a \mapsto [a,1]$$

is a ring monomorphism. As in the proof of Theorem 1.1.14, $\psi(S)$ is contained in the unit group of the monoid R_S^*. □

In the special case $S = R^*$, the construction in the proof of Theorem 1.4.10 yields a field $Q(R)$, the **quotient field** of R; compare with Corollary 1.1.15.

Theorem 1.4.11. *Let R be a commutative domain with quotient field $Q(R)$, and let $\eta\colon R \to F$ be a ring monomorphism of R into some field F. Then η uniquely extends to a ring monomorphism $\bar{\eta}\colon Q(R) \to F$.*

Proof. Clearly, η restricts to a monomorphism from the monoid $(R^*,\cdot,1)$ into the multiplicative group $(F^*,\cdot,1)$ of F. By Theorem 1.1.17 and its proof, this restriction extends uniquely to a group monomorphism $\bar{\eta}$ from the quotient group $Q(R^*)$ of R^* into $(F^*,\cdot,1)$, where $\bar{\eta}([a,b]) = \eta(a)\eta(b)^{-1}$ for $a \neq 0$. In order to extend $\bar{\eta}$ to all of R, we have to put $\bar{\eta}([0,1]) := \eta(0) = 0$.

In view of $Q(R)^* = Q(R^*)$, it only remains to check that $\bar{\eta}$ respects the addition:

$$\begin{aligned}\bar{\eta}([a,b]+[c,d]) &= \bar{\eta}([ad+bc,bd]) \\ &= \frac{\eta(a)\eta(d)+\eta(b)\eta(c)}{\eta(b)\eta(d)} \\ &= \frac{\eta(a)}{\eta(b)}+\frac{\eta(c)}{\eta(d)} \\ &= \bar{\eta}([a,b])+\bar{\eta}([c,d]).\end{aligned}$$ □

We now turn our attention to the most important substructures of rings, namely ideals; these play a similar role for rings as normal subgroups do for groups.

Definition 1.4.12. A subgroup I of the additive group $(R,+,0)$ of a ring R is called an **ideal** of R provided that $rx \in I$ and $xr \in I$ whenever $r \in R$ and $x \in I$. □

The following partial analogue of Proposition 1.3.7 is rather obvious; the proof is left to the reader, see Exercise 1.4.24.

Proposition 1.4.13. *Let R and T be two rings, and let $\psi\colon R \to T$ be a ring homomorphism. Then the kernel of ψ is an ideal of R.* □

We next introduce analogues of the congruence relations on groups considered in Exercises 1.3.16 and 1.3.17. This will then lead to isomorphism theorems for rings.

Proposition 1.4.14. *Let I be an ideal in a ring $(R,+,\cdot,0,1)$ and define a relation $\equiv_I$ on R by*

$$a \equiv_I b \quad :\Leftrightarrow \quad a-b \in I. \tag{1.10}$$

*Then $\equiv_I$ is a **congruence relation** on R, that is, an equivalence relation compatible with the ring operations: if $a \equiv_I b$ and $c \equiv_I d$, then also $a+c \equiv_I b+d$ and $ac \equiv_I bd$ (for all $a,b,c,d \in R$).*

Proof. As I is a subgroup of the additive group $(R,+,0)$ of R, we have $x-y \in I$ if and only if $-x+y \in I$. Thus $\equiv_I$ coincides with the relation $\sim_I$ considered in Remark 1.3.2 (but now in additive notation), and hence $\equiv_I$ is indeed an equivalence relation on R.

Now let $a \equiv_I b$ and $c \equiv_I d$, that is, $a-b,\, c-d \in I$. As I is a subgroup, we obtain $(a+c)-(b+d) = (a-b)+(c-d) \in I$, which shows $a+c \equiv_I b+d$. As I is even an ideal, also $(a-b)c \in I$ and $b(c-d) \in I$, which implies $ac-bd = (a-b)c+b(c-d) \in I$ so that $ac \equiv_I bd$. □

Instead of $x \equiv_I y$ one usually writes $x \equiv y \bmod I$, which is read as *x is congruent to y modulo I* or as *x and y are congruent modulo I*.

Proposition 1.4.15. *Let I be an ideal in a ring $(R,+,\cdot,0,1)$ and define a multiplication on the factor group $(R/I,+,I)$ by*

$$(a+I)(b+I) := ab+I. \tag{1.11}$$

This turns R/I into a ring with unit element $1+I$ which is commutative if R is.

Proof. Note that the multiplication given in Equation (1.11) is well defined: given $i,j \in I$ and $a,b \in R$, we have

$$(a+i)(b+j) = ab+aj+ib+ij \in ab+I,$$

as $aj+bi+ij \in I$. Now the assertions are easily checked. □

The ring $(R/I,+,\cdot,I,1+I)$ is called the **factor ring** or the **quotient ring** of R modulo I. In view of Proposition 1.4.15, the natural group epimorphism ν_I from $(R,+,0)$ to $(R/I,+,I)$ is in fact a ring homomorphism, which gives analogues of Proposition 1.3.9, Theorem 1.3.10 and Corollary 1.3.11 for factor rings:

Theorem 1.4.16 (Homomorphism theorem). *Let $\psi\colon R \to T$ be a homomorphism between two rings $(R,+,\cdot,0_R,1_R)$ and $(T,+,\cdot,0_T,1_T)$. Then there exists a unique ring monomorphism $\sigma\colon R/\ker\psi \to T$ with $\psi = \sigma \circ \nu_{\ker\psi}$. If ψ is surjective, then σ is an isomorphism.*

Proof. Define σ as in the proof of Theorem 1.3.10. Then σ is a ring isomorphism, and the assertion follows from Theorem 1.3.10. □

Theorem 1.4.17 (First isomorphism theorem). *Let $\psi : R \to T$ be a homomorphism between two rings $(R,+,\cdot,0_R,1_R)$ and $(T,+,\cdot,0_T,1_T)$, and let J be an ideal of R containing I.*

Then $\pi_I(J) := J/I$ is an ideal of R/I, the factor ring $(R/I)/(J/I)$ is isomorphic to R/J, and there exists a unique isomorphism $\sigma : R/J \to (R/I)/(J/I)$ satisfying $\sigma(r+J) = (r+I)+(J/I)$ for all $r \in R$. Moreover, the mapping π_I induces a bijection of the set of ideals of R containing I onto the set of ideals of R/I.

Proof. Let i_I and j_I be defined as in the proof of Proposition 1.3.9 (for the normal subgroup I of $(R,+,0)$). Then the restriction of i_I to the set of ideals of R containing I and the restriction of j_I to the set of ideals in R/I are inverse to each other. Thus the assertion follows from Corollary 1.3.11. □

Remark 1.4.18. Let $\mathscr{I}$ be a non-empty set of ideals in a ring R. Then the intersection $\cap_{I\in\mathscr{I}} I$ of all members of $\mathscr{I}$ is again an ideal in R. Applying this construction to the set of all ideals containing a specified non-empty subset X of R gives the **ideal generated by** X. With respect to set inclusion, this is the smallest ideal of R containing X. We will use the notation $I(X)$ for this ideal.

An ideal I is said to be **finitely generated** if it is of the form $I = I(X)$ for some finite set $X = \{x_1,\dots,x_k\}$. In this case, one usually writes $(x_1,\dots,x_k)$ instead of

$I(X)$.[13] Ideals which are generated by a single element $x \in R$ will be of particular importance throughout this book; in this case, we will use either of the notations (x) and xR. Rings where *all* ideals are of this type will be studied in detail in the next section. □

For the remainder of the present section, we concentrate on commutative rings and investigate those ideals whose corresponding factor ring is a domain or even a field.

Definition 1.4.19. Let $I \neq R$ be an ideal of a commutative ring R. Then I is called

- a **maximal ideal** provided that the only ideals J of R with $I \subseteq J \subseteq R$ are $J = I$ and $J = R$, and
- a **prime ideal** provided that $xy \in I$ implies $x \in I$ or $y \in I$. □

Theorem 1.4.20. *Let I be an ideal in a commutative ring R. Then:*

(1) *I is a prime ideal if only if R/I is a domain.*

(2) *Let $u \in R$. Then the coset $u + I$ is a unit in the factor ring R/I if and only if R is the ideal generated by $I \cup \{u\}$.*

(3) *I is a maximal ideal if and only if R/I is a field.*

(4) *Every maximal ideal is also a prime ideal.*

Proof. (1) Let I be a prime ideal, and let $a, b \in R$ with $(a+I)(b+I) = I$. Then $ab \in I$, and hence $a \in I$ or $b \in I$. If $a \in I$, then $a + I = I$; and if $b \in I$, then $b + I = I$. Thus R/I is a domain.

Conversely, assume that R/I is a domain. Then $ab \in I$ gives $(a+I)(b+I) = ab + I = I$ and therefore $a + I = I$ or $b + I = I$, that is, $a \in I$ or $b \in I$.

(2) The coset $u + I$ is a unit in R/I if and only if $su \equiv 1 \bmod I$ for some $s \in R$. But then 1 and then also $r = r \cdot 1$ are in the ideal generated by $I \cup \{u\}$ (for all $r \in R$).

Conversely, assume that R is the ideal generated by $I \cup \{u\}$. Then there exist elements $r \in R$ and $j \in I$ such that $ru + j = 1$; cf. Exercise 1.4.26. This shows $(r+I)(u+I) = 1 + I$, so that u is a unit in R/I.

(3) Let I be a maximal ideal of R, and let $a + I$ be a non-zero element of R/I, that is, $a \in R \setminus I$. Then the ideal J generated by $I \cup \{a\}$ contains I as a proper subset, and the maximality of I gives $R = J$. Now (2) shows that $a + I$ is a unit in R/I. Since this holds for all $a \in R \setminus I$, the factor ring R/I is a field.

Conversely, let R/I be a field and J an ideal of R which properly contains I. Any element $a \in J \setminus I$ gives rise to a unit $a + I$ in R/I, again by (2). Given a, choose $s \in R$ with $(a+I)(s+I) = 1 + I$. Then $sa - 1 \in I \subseteq J$, and $sa \in J$ shows $1 \in J$ and therefore $r = r \cdot 1 \in J$ for all $r \in R$.

(4) This is an immediate consequence of (1) and (3), as every field is a domain. □

---

[13] This is a little unfortunate, as one has to see from the context if the notation is intended to mean an ideal or just a k-tuple of elements.

Finally, we introduce subrings and subfields. A subset S of a ring $(R,+,\cdot,0,1)$ is called a **subring** of R provided that $(S,+,0)$ is a subgroup of $(R,+,0)$ and that S is closed under multiplication – that is, $ab \in S$ whenever $a,b \in S$. Note that it is *not* required that the unit element 1 of R belongs to S. If this is actually the case (that is, if $(S,\cdot,1)$ is also a submonoid of $(R,\cdot,1)$), then S is called a **unitary subring** of R. The concept of a subring is not all that important in our text. However, the special case of a subfield of a field will play a fundamental role.

Definition 1.4.21. Let $(E,+,\cdot,0,1)$ be a field, and assume that $F \subseteq E$ is a subring of E such that $u^{-1} \in F$ for all non-zero $u \in F$. Then F is called a **subfield** of E.

Of course, restricting the operations of E to a subfield F turns F itself into a field. Therefore, E is also said to be an **extension field** of the field F. □

We will develop the basic theory of field extensions in Chapter 3. We leave the following useful characterization of subfields as Exercise 1.4.28 to the reader:

Theorem 1.4.22. *Let E be a field, and let F be a subset of E containing at least two distinct elements. Then F is a subfield of E if and only if $(u-v)w^{-1} \in F$ whenever $u,v,w \in F$ and $w \neq 0$.* □

The proof of Theorem 1.4.20 shows that the trivial ideal $\{0\}$ in a field F is already maximal. Consequently, the only ideals of a field F are the trivial ones, namely F and $\{0\}$, and we obtain the following result.

Proposition 1.4.23. *Let F and K be two fields, and let $\sigma : F \to K$ be a non-trivial ring homomorphism. Then:*

(1) *σ is a monomorphism.*

(2) *σ induces a group monomorphism from F^* to K^*.*

(3) *The image of σ is a subfield of K.*

Proof. Since σ is non-trivial, there exists an element $u \in F$ with $\sigma(u) \neq 0$, so that the kernel of σ is an ideal in F which is distinct from F. By Theorem 1.4.20, the kernel of σ is then equal to $\{0\}$, which proves (1).

The second assertion follows from Lemma 1.1.10 and the subsequent remark.

Finally, we apply Theorem 1.4.22 to prove (3). Let L denote the image of σ. Then $0,1 \in L$, and hence $|L| \geq 2$. Given $u,v,w \in L$ and $w \neq 0$, we may choose $a,b,c \in F$ with $\sigma(a) = u$, $\sigma(b) = v$ and $\sigma(c) = w$. Then $c \neq 0$ and

$$\frac{u-v}{w} = \frac{\sigma(a)-\sigma(b)}{\sigma(c)} = \sigma\Big(\frac{a-b}{c}\Big) \in L,$$

as required. □

Exercises

Exercise 1.4.24. Extend Proposition 1.4.13 to a full analogue of Proposition 1.3.7 and give a formal proof. □

Exercise 1.4.25. Give a ring theoretic analogue of the second isomorphism theorem for groups considered in Exercise 1.3.18. □

Exercise 1.4.26. Let X be a non-empty subset of a commutative ring R, and let $I(X)$ be the ideal generated by X. Show that $y \in I(X)$ if and only if there exist a positive integer ℓ, elements $x_1,\dots,x_\ell \in X$ and ring elements $r_1,\dots,r_\ell \in R$ such that $y = r_1x_1 + \cdots + r_\ell x_\ell$. □

Exercise 1.4.27. Consider two integers a and b. Give a description of $\mathrm{GCD}(a,b)$ and $\mathrm{LCM}(a,b)$ in terms of ideals in the ring $\mathbb{Z}$ of integers. □

Exercise 1.4.28. Prove Theorem 1.4.22. □

1.5 Principal Ideal Domains and Euclidean Domains

In this section, we use the results of Section 1.2 to study the divisibility relation in commutative rings, in particular in commutative domains.

Let $(R,+,\cdot,0,1)$ be an arbitrary commutative ring. Then we may apply the basic notions concerning divisibility introduced in Section 1.2 to the multiplicative monoid $(R,\cdot,1)$ of R. We use the same notation as in that section; in particular, we again denote the association class of $a \in R$ by $[a]$. Since $0 \cdot r = 0$ for all $r \in R$, we have $[0] = \{0\}$, and therefore $R^* = R \setminus \{0\}$ is a union of association classes.

Our first aim is a description of the basic properties of divisibility in terms of ideals, generalizing and extending Exercise 1.4.27. According to Exercise 1.4.26, the ideal $(x_1,\dots,x_k)$ generated by a finite subset $X = \{x_1,\dots,x_k\}$ of R is given explicitly by

$$(x_1,\dots,x_k) = \Big\{\sum_{i=1}^{k} r_i x_i : r_1,\dots,r_k \in R\Big\}.$$

To avoid any possible confusion with k-tuples over R, we will generally use the notation $Rx_1 + \cdots + Rx_k$ instead of $(x_1,\dots,x_k)$, except sometimes for the case $k=1$.

Definition 1.5.1. Let I be an ideal in a commutative ring R which is generated by a single element $x \in R$, that is, $I = (x) = Rx$. Then I is called a **principal ideal**. □

Proposition 1.5.2. *Let R be a commutative ring and $a,b \in R$. Then:*

(1) *$a \mid b$ if and only if $Rb \subseteq Ra$. In particular, $a \approx b$ if and only if $Ra = Rb$.*

(2) *If $Ra + Rb$ is a principal ideal, say Rd, then d is a greatest common divisor of a and b.*

(3) *Let $I := Ra \cap Rb$. If I is a principal ideal, say $I = Re$, then e is a least common multiple of a and b.*

Proof. (1) First assume $a \mid b$. Then there is an $r \in R$ with $ra = b$, so that $sb = sr \cdot a$ for all $s \in R$, showing $Rb \subseteq Ra$. Conversely, if $Rb \subseteq Ra$, then $b = 1 \cdot b \in Ra$; hence there is an $r \in R$ with $b = ra$, that is, $a \mid b$.

(2) Assume $Ra + Rb = Rd$. Then $d \mid a$ and $d \mid b$, as $Ra \subseteq Rd$ and $Rb \subseteq Rd$. Thus d is a common divisor of a and b. On the other hand, if c is any common divisor of a and b, then $a, b \in Rc$ and therefore $Rd = Ra + Rb \subseteq Rc$, so that $c \mid d$.

(3) The common multiples of a and b are exactly the elements of the ideal $I := Ra \cap Rb$. Now observe that e is a least common multiple of a and b if and only if $e \in I$ and $e \mid f$ for every $f \in I$, that is, for $I = Re$. □

From now on, we restrict attention to commutative domains R. In this case, $(R^*, \cdot, 1)$ is a commutative monoid with cancellation, by Lemma 1.4.7, and we may apply the corresponding results obtained in Section 1.2.

Definition 1.5.3. A commutative domain R is called **Noetherian** provided that every ideal of R is finitely generated. In the special case where every ideal of R is a principal ideal, R is called a **principal ideal domain**. □

Corollary 1.5.4. *Let R be a principal ideal domain. Then any two elements of R have a greatest common divisor and a least common multiple.*

More precisely, given $a, b \in R$, let $Ra + Rb = Rd$ and $Ra \cap Rb = Re$. Then one has $\mathrm{GCD}(a,b) = U(R) \cdot d$ and $\mathrm{LCM}(a,b) = U(R) \cdot e$.

Proof. In the case $a = b = 0$, the assertion holds with $d = e = 0$, and for $a \neq 0$ and $b = 0$, it holds with $d = a$ and $e = 0$. (Note that these trivial cases remain valid for an arbitrary commutative ring.)

Now let $a, b \in R^*$. Since R is a principal ideal domain, the existence of the required elements $d, e \in R$ is guaranteed, and the assertion follows from Proposition 1.2.2. □

Theorem 1.5.5. *Let R be a Noetherian domain. Then $(R^*, \cdot, 1)$ is a c-monoid.*

Proof. Consider a divisor sequence $(a_n)_{n \in \mathbb{N}}$ in R^*, that is, $a_{n+1} \mid a_n$ for all n. By Proposition 1.5.2, $Ra_n \subseteq Ra_{n+1}$ for all n, so that $(Ra_n)_{n \in \mathbb{N}}$ is an increasing sequence of principal ideals in R. Let $J := \bigcup_{n \in \mathbb{N}} Ra_n$. Then J is an ideal of R. As R is Noetherian, J is generated by finitely many elements of R, say by $x_1, \ldots, x_\ell$.

For $i = 1, \ldots, \ell$, choose an index $n(i)$ with $x_i \in Ra_{n(i)}$, and let m be the maximum of these $n(i)$. Then $x_1, \ldots, x_\ell \in Ra_m$ and therefore $J \subseteq Ra_m$. As $(Ra_n)_{n \in \mathbb{N}}$ is increasing, we have $J \subseteq Ra_m \subseteq Ra_j$ for all $j \geq m$. On the other hand, also $Ra_j \subseteq J$ (for all j), by the definition of J, so that $Ra_m = Ra_j$ for all $j \geq m$. By Proposition 1.5.2, a_m and a_j are associates for all $j \geq m$, and hence $(R^*, \cdot, 1)$ is indeed a c-monoid. □

We now consider the decomposition/factorization of elements of R^* into irreducible or prime elements, where R is a commutative domain. Thus let $U(R)$ be the group of units of R, and recall from Definitions 1.2.7 and 1.2.11 that $p \in R^* \setminus U(R)$ is irreducible if it admits only trivial divisors, namely the elements of $U(R) \cup U(R) \cdot p$, and a prime if $p \mid ab$ implies $p \mid a$ or $p \mid b$ (for $a, b \in R$). By a slight abuse of notation, we let $P(R)$ denote the set of all prime elements and $I(R)$ the set of all irreducible elements of R^*. Then $P(R) \subseteq I(R)$, by Proposition 1.2.12.

Proposition 1.5.6. *Let R be a commutative domain, and let $p, r \in R^* \setminus U(R)$. Then:*

(1) *p is a prime element if and only if Rp is a prime ideal.*

(2) *r is irreducible if and only if Rr is a maximal principal ideal, that is, if and only if $Rr \subseteq Rx \subseteq R$ always implies $Rx = Rr$ or $Rx = R$ (for $x \in R$).*

Proof. (1) $ab \in Rp$ means $p \mid ab$. Thus Rp is a prime ideal if and only if $p \mid ab$ implies $p \mid a$ or $p \mid b$.

(2) $Rr \subseteq Rx$ means $x \mid r$. Thus Rr is a maximal principal ideal if and only if $x \mid r$ implies $x \in U(R)$ or $x \approx r$. □

Combining Theorems 1.2.10 and 1.5.5 gives the following decomposition theorem:

Corollary 1.5.7. *Let R be a Noetherian domain, and let $a \in R^*$. Then there exist a natural number k and irreducible elements $r_1, \ldots, r_k \in R$ such that $a = r_1 \cdots r_k$.* □

As in Definition 1.2.13, we introduce the following terminology:

Definition 1.5.8. A commutative domain R is called **factorial** provided that, for every $a \in R$, there exist a natural number ℓ and prime elements $p_1, \ldots, p_\ell \in R$ such that $a = p_1 \cdots p_\ell$. □

Theorem 1.5.9. *Principal ideal domains are factorial.*

Proof. Let R be a principal ideal domain. We first show that $P(R) = I(R)$. As noted above, $P(R) \subseteq I(R)$. To check the reverse inclusion, let r be an irreducible element of R. By Proposition 1.5.6, Rr is a maximal ideal, and hence the factor ring R/Rr is a field; see Theorem 1.4.20. In particular, R/Rr is a domain, and therefore Rr is a prime ideal, again by Theorem 1.4.20, so that r is a prime by Proposition 1.5.6.

Since any principal ideal domain is Noetherian, the assertion follows from Corollary 1.5.7 and $P(R) = I(R)$. □

Remark 1.2.15 shows that any non-unit a of a principal ideal domain R can be uniquely decomposed into a product of prime powers (up to association). Moreover, the proof of Theorem 1.5.9 shows that R/Ra is a field if and only if a is a prime.

We now consider a special class of commutative domains, which turn out to be principal ideal domains.

Definition 1.5.10. A commutative domain R is called a **Euclidean domain** provided that there exists a mapping $\delta : R^* \to \mathbb{N}$ with the following properties:

(1) $\delta(x) \leq \delta(xy)$ for all $x, y \in R^*$.

(2) If $a, b \in R$ and $b \neq 0$, then there are elements $q, r \in R$ such that

$$a = qb + r \quad \text{and} \quad r = 0 \text{ or } \delta(r) < \delta(b).$$

Such a mapping δ is called a **Euclidean function** for R. □

Remark 1.5.11. The process of determining from (a,b) a pair (q,r) as in Definition 1.5.10 is referred to as a **division with remainder**, and the elements q and r are called the **quotient** and the **remainder** (when dividing a by b), respectively. We stress that, in general, the elements q and r do not have to be uniquely determined. Even in this case, it is convenient to assume the existence of two mappings `div` and `mod` from $R \times R^*$ to R, such that $a \mathtt{\ div\ } b$ is a quotient and $a \mathtt{\ mod\ } b$ is a remainder when dividing a by b.

Trivially, b divides a if $r = 0$. On the other hand, if $b \mid a$, say $sb = a$, then $sb = a = qb + r$ and therefore $r = (s-q)b$. We claim that necessarily $r = 0$ in this situation. Assume otherwise. Then $\delta(r) < \delta(b)$, and $\delta(b) \leq \delta((s-q)b) = \delta(r)$, by the first axiom for a Euclidean function, which is a contradiction. Hence $a \mathtt{\ mod\ } b = 0$ if and only if b divides a, in which case $a = (a \mathtt{\ div\ } b) \cdot b$.

More generally, let a and a' be two elements of R which leave the same remainder under division by b, that is, $a \mathtt{\ mod\ } b = a' \mathtt{\ mod\ } b$. Then one also writes $a \equiv a' \bmod b$. Note that this happens if and only if $a - a'$ is a multiple of b, and thus $a \equiv a' \bmod b$ is an alternative (and perhaps more intuitive) way of writing $b \mid a - a'$. □

Example 1.5.12. The two most important examples of Euclidean domains are as follows:

- The ring $\mathbb{Z}$ is a Euclidean domain with respect to the absolute value $|\cdot|$; this will be discussed in detail in the next section.
- If F is a field and x an indeterminate, then the polynomial ring $F[x]$ is a Euclidean domain with respect to the degree function; see Section 2.2. □

We note that the Euclidean domains which admit a constant mapping as a Euclidean function are precisely the fields. This is an immediate consequence of the following more general result:

Proposition 1.5.13. *Let (R, δ) be a Euclidean domain. Then δ attains its minimum exactly on the set of units of R.*

Proof. Let $x \in R^*$. Then $\delta(1) \leq \delta(1 \cdot x) = \delta(x)$. If x is a unit, then $\delta(x) \leq \delta(xx^{-1}) = \delta(1)$. This shows that $\delta(x) = \delta(1)$ is minimal for all $x \in U(R)$.

Conversely, let $z \in R^*$ with $\delta(z) = \delta(1)$. Performing a division with remainder for the pair $(1,z)$ gives a pair (q,r) such that $1 = qz + r$ with $r = 0$ or $\delta(r) < \delta(z)$. Since $\delta(z) = \delta(1)$ is minimal, $\delta(r) < \delta(z)$ is impossible, and therefore $r = 0$. Thus $1 = qz$, and z is a unit in R. □

Theorem 1.5.14. *Euclidean domains are principal ideal domains.*

Proof. Let (R, δ) be a Euclidean domain and I an ideal of R. Since $\{0\}$ is a principal ideal, we may assume that I contains an element $b \neq 0$. Choose such an element with a minimal δ-value, and let $a \in I$ with $a \neq 0$. Then $\delta(b) \leq \delta(a)$, and dividing a by b gives a pair (q,r) with $a = qb + r$ and $r = 0$ or $\delta(r) < \delta(b)$. Since $a, b \in I$, also $r = a - qb \in I$, and the minimality of $\delta(b)$ gives $r = 0$. This shows that I is the principal ideal Rb. □

We remark that the converse of Theorem 1.5.14 does not hold: there exist principal ideal domains which are not Euclidean. The standard example for this is due to Motzkin [290], who established that the ring of integers in the algebraic number field $\mathbb{Q}(\sqrt{-19})$, that is, the subring of $\mathbb{R}$ formed by the numbers

$$\frac{a+b\sqrt{-19}}{2} \quad \text{with } a,b \in \mathbb{Z},\ a \equiv b \bmod 2,$$

is a principal ideal domain which is not Euclidean. This is a non-trivial result; an elementary exposition can be found in [60]. In this context, we also mention an interesting note by Greene [155] showing that principal ideal domains are nevertheless "almost Euclidean".

By Theorem 1.5.9, Euclidean domains are factorial. In particular, any two elements a and b have both a greatest common divisor d and a least common multiple e, by Corollary 1.5.4. In this situation, the Euclidean function together with the corresponding division with remainder can be used to determine both d and e effectively via the so-called *Euclidean algorithm.* This basically relies on the following observation:

Proposition 1.5.15. *Let a and b be two elements in a Euclidean domain (R,δ), and let (q,r) be a pair of elements of R satisfying $a = qb+r$ and $r = 0$ or $\delta(r) < \delta(b)$. Then* $\mathrm{GCD}(a,b) = \mathrm{GCD}(b,r)$.

Proof. If t is a common divisor of a and b, then t divides $r = a - qb$, and thus t is a common divisor of b and r. Conversely, if t is a common divisor of b and r, then t divides $a = qb + r$, so that t is a common divisor of a and b. This shows $\mathrm{CD}(a,b) = \mathrm{CD}(b,r)$ and therefore also $\mathrm{GCD}(a,b) = \mathrm{GCD}(b,r)$. □

Algorithm 1.5.16 (Euclidean algorithm).

- *Input:* A Euclidean domain (R,δ) with operators `div` and `mod` and two elements $a,b \in R$.
- *Output:* A greatest common divisor d of a and b.

(1) $s \leftarrow a,\ t \leftarrow b$;
(2) **while** $t \neq 0$ **do**
(3) $\quad r \leftarrow s$ `mod` t;
(4) $\quad s \leftarrow t,\ t \leftarrow r$
(5) **od**
(6) $d \leftarrow s$.

We note that this algorithm for computing greatest common divisors is correct in view of Proposition 1.5.15. Moreover, the termination is guaranteed as in each iteration of the **while** loop the δ-value of t decreases as long as $t \neq 0$. □

By Corollary 1.5.4, we have $Ra+Rb = Rd$, where d is a greatest common divisor of a and b (in any principal ideal domain, hence in particular in the present context of a Euclidean domain). Thus there are elements $x,y \in R$ such that $xa+yb=d$. The following extended version of the Euclidean algorithm not only determines a greatest common divisor of a and b but also two such elements $x,y \in R$.

Algorithm 1.5.17 (Extended Euclidean algorithm).

- *Input:* A Euclidean domain (R,δ) with operators `div` and `mod` and two elements $a,b \in R$.
- *Output:* A greatest common divisor d of a and b and two elements $x,y \in R$ such that $xa+yb=d$.

(1) $(\rho,\sigma,\tau) \leftarrow (a,1,0)$, $(r,s,t) \leftarrow (b,0,1)$;
(2) **while** $r \neq 0$ **do**
(3) $\quad (\bar{r},\bar{s},\bar{t}) \leftarrow (r,s,t)$;
(4) $\quad q \leftarrow \sigma \texttt{ div } r$, $(r,s,t) \leftarrow (\rho,\sigma,\tau) - (q\bar{r},q\bar{s},q\bar{t})$;
(5) $\quad (\rho,\sigma,\tau) \leftarrow (\bar{r},\bar{s},\bar{t})$;
(6) **od**
(7) $d \leftarrow \rho$, $x \leftarrow \sigma$, $y \leftarrow \tau$. □

Theorem 1.5.18. *Let a and b be two elements in a Euclidean domain (R,δ). Then Algorithm 1.5.17 terminates with a greatest common divisor d of a and b and two elements $x,y \in R$ such that $xa+yb=d$.*

Proof. As in the case of Algorithm 1.5.16, the algorithm will terminate, since $\delta(r)$ decreases in each iteration of the loop (see below). The correctness of the algorithm rests on showing that the condition

$$(*) \qquad sa+tb=r \text{ and } \sigma a+\tau b=\rho$$

is a **loop invariant**, that is, it holds throughout the entire course of the algorithm. This is obvious after the initialization in Step (1). We now use induction on the number of iterations performed to show that (*) is indeed preserved by each execution of the **while** loop. To see this, assume that (*) holds at the beginning of such an iteration and note that the then current values of (r,s,t) are used to define $(\bar{r},\bar{s},\bar{t})$ in Step (3). After executing Step (4), we indeed obtain

$$\begin{aligned} sa+tb &= (\sigma - q\bar{s})a + (\tau - q\bar{t})b \\ &= (\sigma a + \tau b) - q(\bar{s}a + \bar{t}b) \\ &= \rho - q\bar{r} = r \end{aligned}$$

for the updated values of (r,s,t). Note that also $r=0$ or $\delta(r) < \delta(\bar{r})$ as $q = \rho \texttt{ div } \bar{r}$ and $r = \rho - q\bar{r} = \rho \texttt{ mod } \bar{r}$, which justifies our initial remark. Moreover, (ρ,σ,τ) is then also updated in Step (5) by overwriting it with the old values of (r,s,t), so that also the second requirement in (*) holds at the end of the iteration.

Finally, at the termination of the algorithm, we have $r=0$ and put $d=\rho$, $x=\sigma$ and $y=\tau$ in Step (7). Because of (*), we then have $d=xa+yb$, as desired. The fact that d is a greatest common divisor of a and b follows from Proposition 1.5.15, as for Algorithm 1.5.16. □

Remark 1.5.19. Let R be any domain, and let I be an ideal of R. By Theorem 1.4.20, an element $u \in R$ gives rise to a unit $u+I$ in the factor ring R/I (we also say that *u is a unit modulo I*) if and only if the ideal generated by $I \cup \{u\}$ is all of R. In the special case of a principal ideal $I = Rb$, we see that u is a unit modulo Rb if and only if $R = Ru + Rb$. In this case, u and b are relatively prime, that is, $\mathrm{GCD}(u,b) = U(R)$, and there exist $x, y \in R$ with $xu + yb = 1$.

We may rewrite this condition as $xu = 1 - yb \in 1 + Rb$, that is, $xu \equiv 1 \bmod Rb$ (or, for short, $xu \equiv 1 \bmod b$). This is equivalent to $(x+Rb)\cdot(u+Rb) = 1+Rb$ and therefore $x+Rb$ is the inverse of $u+Rb$ in R/Rb (or, for short, x is inverse to u modulo b).

Now let b be an element of a Euclidean domain R, and suppose that we can recognize the units of R and even determine their inverses. Then, given any element $u \in R$, an application of the extended Euclidean algorithm with input (u,b) yields a representation $x'u + y'b = d$, where $d \in \mathrm{GCD}(u,b)$. As noted above, u is a unit modulo b if and only if d is a unit in R. In this case, a division by d gives $xu + yb = 1$, where $x = \frac{x'}{d}$ and $y = \frac{y'}{d}$. Altogether, this shows how units in R/Rb can be recognized and inverted (provided that we can do so for R).

This is of particular interest when b is a prime element, that is, when the factor ring R/Rb is a field F: it shows how one may invert elements in F^*, at least in principle. □

Exercises

Exercise 1.5.20. Let (R, δ) be a Euclidean domain, and let $a, b \in R^*$. Show the following:

(1) $a \approx b$ implies $\delta(a) = \delta(b)$.

(2) If $\delta(a) = \delta(b)$ and $a \mid b$, then $a \approx b$. □

Exercise 1.5.21. Show directly that any Euclidean domain is factorial (without using Theorem 1.5.9). □

1.6 The Ring of Integers, Cyclic Groups, and Orders

In the present section, we consider the ring $(\mathbb{Z}, +, \cdot, 0, 1)$ of integers in more detail and prove some basic facts from Number Theory, which are then applied to cyclic groups.

We start by recalling that $\mathbb{Z}$ is a commutative domain with group of units $U(\mathbb{Z}) = \{-1, 1\}$. The factor monoid $\overline{\mathbb{Z}}$ of $(\mathbb{Z}, \cdot, 1)$ (see Proposition 1.2.4) is isomorphic to the multiplicative monoid $(\mathbb{N}, \cdot, 1)$ of natural numbers. The **absolute value** of $x \in \mathbb{Z}$ is defined by

$$|x| := \begin{cases} x & \text{if } x \geq 0, \\ -x & \text{if } x < 0, \end{cases} \tag{1.12}$$

and the **sign function** on $\mathbb{Z}$ is given by

$$\mathrm{sgn}\colon \mathbb{Z} \mapsto \mathbb{Z},\ x \mapsto \begin{cases} 1 & \text{if } x > 0, \\ 0 & \text{if } x = 0, \\ -1 & \text{if } x < 0. \end{cases} \tag{1.13}$$

Note that both $|\cdot|$ and sgn are endomorphisms of the monoid $(\mathbb{Z}, \cdot, 1)$, and that $|x| = \mathrm{sgn}(x) \cdot x$ holds for all $x \in \mathbb{Z}$.

Theorem 1.6.1. *The ring of integers is a Euclidean domain with the absolute value $|\cdot|$ (restricted to $\mathbb{Z}^*$) as Euclidean function. Moreover, for every pair $(a,b) \in \mathbb{Z} \times \mathbb{Z}^*$, there exists a unique pair (q,r) of integers such that $a = qb + r$ and $0 \le r < |b|$.*

Proof. Obviously, $0 < |x| \le |x| \cdot |y| = |xy|$ for all $x, y \in \mathbb{Z}^*$. Given a pair (a,b), let $n \in \mathbb{N}$ be maximal such that $n|b| \le |a|$ and let $m := |a| - n|b|$. Then $0 \le m < |b|$ and $|a| = n|b| + m$, that is, $\mathrm{sgn}(a) \cdot a = n \cdot \mathrm{sgn}(b) \cdot b + m$. Multiplication by $\mathrm{sgn}(a)$ gives $a = q'b + r'$ with $q' = n \cdot \mathrm{sgn}(a) \cdot \mathrm{sgn}(b)$ and $r' = \mathrm{sgn}(a) \cdot m$. Moreover, $r' = 0$ or $|r'| = m < |b|$. If $r' < 0$, let $r := r' + |b|$ and $q := q' - 1$. Then $a = qb + r$ and $0 < r < |b|$, establishing that $(\mathbb{Z}, |\cdot|)$ is a Euclidean domain.

If (s,t) is a further pair of integers such that $a = sb + t$ and $0 \le t < |b|$, then $qb + r = sb + t$ yields $(q - s)b = t - r$. Without loss of generality let $r \le t$. Then

$$0 \le |q - s| \cdot |b| = |t - r| = t - r \le t < |b|.$$

This implies $|q - s| = 0$, that is, $q = s$ and then also $r = t$, as claimed. □

For all $x, y \in \mathbb{Z}$, there is a unique non-negative greatest common divisor of x and y, which will henceforth be denoted by $\gcd(x,y)$. Similarly, there also is a unique non-negative least common multiple of x and y, denoted by $\mathrm{lcm}(x,y)$. Note that $\mathrm{lcm}(x,y) = |xy|/\gcd(x,y)$ for $x, y \in \mathbb{Z}^*$.

Next, we characterize the subgroups of the additive group of integers:

Theorem 1.6.2. *Let U be a subgroup of $(\mathbb{Z}, +, 0)$. Then U is an ideal of the ring $(\mathbb{Z}, +, \cdot, 0, 1)$, and there is a unique $n \in \mathbb{N}$ such that $U = \mathbb{Z}n = \{zn\colon z \in \mathbb{Z}\}$.*

Proof. As the trivial subgroup $U = \{0\}$ is the principal ideal $\mathbb{Z} \cdot 0$, we may assume that U contains an element $u \ne 0$. Then also $\mathrm{sgn}(u) \cdot u = |u| \in U$, and therefore $U \cap \mathbb{N}^* \ne \emptyset$. Let n be the minimal element in $U \cap \mathbb{N}^*$. Then U contains the subgroup generated by n, which is just $\mathbb{Z}n$. Given any element $x \in U$, perform division with remainder for the pair (x,n) to obtain a unique pair (q,r) such that $x = qn + r$ and $0 \le r < n$. Then $r = x - qn \in U \cap \mathbb{N}$ and hence $r = 0$ (as n was minimal in $U \cap \mathbb{N}^*$), so that $x = qn \in \mathbb{Z}n$, which shows $U = \mathbb{Z}n$. Finally, if $\mathbb{Z}m = \mathbb{Z}n$, then $n \approx m$ and therefore $m \in \{n, -n\}$. □

As discussed in Remark 1.5.11, we write $x \equiv y \bmod n$ when $n \mid x - y$, that is, when $x - y$ is in the ideal $\mathbb{Z}n$.

Definition 1.6.3. For $n \in \mathbb{N}^*$, the factor ring $\mathbb{Z}/\mathbb{Z}n$ is called the ring of **residues modulo** n; it is commonly denoted by $\mathbb{Z}_n$. □

Proposition 1.6.4. *Let $n \in \mathbb{N}^*$. Then the ring $\mathbb{Z}_n$ of residues modulo n is finite: it has exactly n elements.*

Proof. By Theorem 1.6.1, given any $x \in \mathbb{Z}$, there is a unique $r \in \{0, 1, ..., n-1\}$ with $x \bmod n = r$ (where mod is the remainder operator, see Remark 1.5.11). Thus n divides $x - r$, and therefore $x \in r + \mathbb{Z}n$ for this unique r. □

Definition 1.6.5. For $n \in \mathbb{N}^*$, let $\phi(n)$ denote the number of units in the residue ring $\mathbb{Z}_n$. The function ϕ is generally called the **Euler totient function** or the **Euler phi function**. □

Remark 1.6.6. Recall from Remark 1.5.19 that $u \in \mathbb{Z}$ is a unit modulo n (which means that the residue class $u + \mathbb{Z}n$ is a unit in $\mathbb{Z}_n$) if and only if $\mathbb{Z}u + \mathbb{Z}n = \mathbb{Z}$, that is, if and only if $\gcd(u, n) = 1$. Since $\gcd(u, n) = \gcd(n, u \bmod n)$, we see that $\phi(n)$ is the number of all r in the residue system $\{0, 1, \ldots, n-1\}$ which are relatively prime to n.

Recall that we have also seen in Remark 1.5.19 how one may compute the inverse of such an element in $\mathbb{Z}_n$, using the extended Euclidean algorithm. This is the only non-obvious computational task in the factor ring $\mathbb{Z}_n$, if we take the usual arithmetic operations in $\mathbb{Z}$ for granted. □

Let us denote the set of positive primes in $\mathbb{Z}$ by $P^+(\mathbb{Z})$, so that

$$P^+(\mathbb{Z}) = \{2, 3, 5, 7, 11, 13, 17, 19, 23, 29, 31, 37, 41, 43, \ldots\}.$$

Note that $p \in P^+(\mathbb{Z})$ if and only if $\mathbb{Z}_p$ is a field, which holds if and only if $\phi(p) = p - 1$, as then each of the $p - 1$ non-zero elements of $\mathbb{Z}_p$ is invertible.

The following result is one of the most celebrated theorems proved by the ancient Greeks:

Theorem 1.6.7 (Euclid). *The ring of integers has infinitely many primes.*

Proof. Let $p_1, \ldots, p_k$ be the list of the first k positive primes in $\mathbb{Z}$ (for instance, $k = 14$ for the partial list displayed above). Put $m := \prod_{i=1}^{k} p_i$, and let r be any prime divisor of $m + 1$. Since $m + 1$ and m are relatively prime, r cannot be a member of the list $p_1, \ldots, p_k$. □

As a consequence, the list of the corresponding fields $\mathbb{Z}_p$ (for $p \in P^+(\mathbb{Z})$) shows the existence of infinitely many non-isomorphic Galois fields. For the convenience of the reader, we list all 168 primes $p < 1000$ in Table 1.1 below.[14]

We now apply the preceding results to study cyclic groups, that is, groups G that can be generated by a single element g; see Remark 1.1.12.

Definition 1.6.8. A (multiplicatively written) group G is called **cyclic** if there exists an element $g \in G$ such that

[14] We have taken this table from *The Prime Pages* [370], a very useful website which is mainly concerned with up-to-date prime number records.

Table 1.1 List of the 168 primes less than 1000.

2	3	5	7	11	13	17	19	23	29
31	37	41	43	47	53	59	61	67	71
73	79	83	89	97	101	103	107	109	113
127	131	137	139	149	151	157	163	167	173
179	181	191	193	197	199	211	223	227	229
233	239	241	251	257	263	269	271	277	281
283	293	307	311	313	317	331	337	347	349
353	359	367	373	379	383	389	397	401	409
419	421	431	433	439	443	449	457	461	463
467	479	487	491	499	503	509	521	523	541
547	557	563	569	571	577	587	593	599	601
607	613	617	619	631	641	643	647	653	659
661	673	677	683	691	701	709	719	727	733
739	743	751	757	761	769	773	787	797	809
811	821	823	827	829	839	853	857	859	863
877	881	883	887	907	911	919	929	937	941
947	953	967	971	977	983	991	997		

$$G = \langle g \rangle = \{g^z : z \in \mathbb{Z}\}.$$

Any such element g is called a **generator** of G. □

As noted in Remark 1.1.12, the mapping

$$\gamma_g : (\mathbb{Z}, +, 0) \to G, \quad z \mapsto g^z \tag{1.14}$$

is a group homomorphism, and hence the cyclic groups are exactly the homomorphic images of the additive group of the integers. Taking for G (now in additive notation) $(\mathbb{Z}, +, 0)$ itself, we obtain the following immediate consequence of Theorem 1.6.2:

Corollary 1.6.9. *All subgroups of $(\mathbb{Z}, +, 0)$ are cyclic. In particular, $(\mathbb{Z}, +, 0)$ itself is cyclic. The generators of the subgroup $\mathbb{Z}n$ are n and $-n$.* □

We now study the cyclic subgroups of a group G in more detail by considering the kernels of the homomorphisms γ_g.

Definition 1.6.10. Let g be an element of a (multiplicatively written) group G, let γ_g be the group homomorphism defined in (1.14), and let $\ker \gamma_g = \mathbb{Z}n$ (where $n \in \mathbb{N}$). Then g is said to have **(finite) order** n if $n \neq 0$, and we use the notation $\mathrm{ord}(g) = n$ in this case. If $n = 0$, we say that g has **infinite order** and write $\mathrm{ord}(g) = \infty$. □

It is clear that g has infinite order if and only if γ_g is injective, in which case $\langle g \rangle$ is isomorphic to the additive group of integers. In all other cases, $\langle g \rangle$ is a finite group.

Proposition 1.6.11. *Let $g \in G$ be an element of finite order n. Then $\langle g \rangle$ is isomorphic to $(\mathbb{Z}_n, +, 0)$, and hence the cardinality of $\langle g \rangle$ equals $n = \mathrm{ord}(g)$. Moreover, $\langle g \rangle = \{1, g, g^2, \ldots, g^{n-1}\}$.*

Proof. By hypothesis, the kernel of γ_g is equal to $\mathbb{Z}n = \mathbb{Z}/\mathbb{Z}n$. Now the first assertion is immediate from the homomorphism theorem (see 1.3.10).

Given any element $z \in \mathbb{Z}$, there exist integers q and r such that $z = qn + r$ and $0 \leq r < n$. Thus $g^z = g^{qn} \cdot g^r = (g^n)^q \cdot g^r = g^r$, as $g^n = 1$. Hence $g^z = g^{z \bmod n}$ for all $z \in \mathbb{Z}$, which gives the second assertion. $\square$

Thus we see that the additive groups of residues $\mathbb{Z}_n = \{k + n\mathbb{Z} : k \in \mathbb{N},\, 0 \leq k < n\}$ (with $n \geq 1$) together with $(\mathbb{Z}, +, 0)$ provide all isomorphism types of cyclic groups.

Remark 1.6.12. Let g be an element of a (multiplicatively written) finite group G. By Theorem 1.3.3, $\operatorname{ord}(g) = |\langle g \rangle|$ divides $|G|$. As $g^z = 1$ if and only if $\operatorname{ord}(g)$ divides $z \in \mathbb{Z}$ (by Definition 1.6.10), we obtain

$$g^{|G|} = 1 \quad \text{for all } g \in G. \quad \square \tag{1.15}$$

Let us apply (1.15) to the group of units of a residue ring $\mathbb{Z}_m$:

Theorem 1.6.13 (Euler). *Let $m \in \mathbb{N}^*$ and $a \in \mathbb{Z}$, and assume that a and m are relatively prime. Then $a^{\phi(m)} \equiv 1 \bmod m$.*

Proof. By Remark 1.6.6, a is a unit modulo m, that is, $a + \mathbb{Z}m \in U(\mathbb{Z}_m)$, as $\gcd(a, m) = 1$. Also, $|U(\mathbb{Z}_m)| = \phi(m)$ by definition, and therefore (1.15) implies $(a + \mathbb{Z}m)^{\phi(m)} = 1 + \mathbb{Z}m$, which is equivalent to the assertion. $\square$

In particular, the prime case of Theorem 1.6.13 gives a famous result of Fermat:

Theorem 1.6.14 (Fermat). *Let p be a prime and $a \in \mathbb{Z}$. Then $a^p \equiv a \bmod p$.*

Proof. The assertion is trivial for $p \mid a$. Thus assume that p does not divide a. Then $\gcd(a, p) = 1$ and a is a unit modulo p. As noted above, $\phi(p) = p - 1$, as p is a prime. Hence Theorem 1.6.13 gives $a^{p-1} \equiv 1 \bmod p$. Multiplying both sides of this congruence by a yields the assertion. $\square$

Next we wish to determine the subgroup structure of an arbitrary finite cyclic group. For this, we require the following useful lemma on the orders of powers of a given group element:

Lemma 1.6.15. *Let $(G, \cdot, 1)$ be a group, and let $g \in G$ be an element with finite order. Then*

$$\operatorname{ord}(g^k) = \frac{\operatorname{ord}(g)}{\gcd(\operatorname{ord}(g), k)} \quad \textit{for all } k \in \mathbb{Z}.$$

Proof. Let us write $n = \operatorname{ord}(g)$ and $d = \gcd(n, k)$. Since $(g^k)^{n/d} = (g^n)^{k/d} = 1$, the order ℓ of g^k has to divide $\frac{n}{d}$. On the other hand, $g^{k\ell} = (g^k)^\ell = 1$ shows that n divides $k\ell$, so that $\frac{n}{d}$ divides $\frac{k}{d} \cdot \ell$. By Lemma 1.2.18, $\frac{n}{d}$ and $\frac{k}{d}$ are relatively prime, and hence $\frac{n}{d}$ divides ℓ, proving $\ell = \frac{n}{d}$. $\square$

Remark 1.6.16. In particular, Lemma 1.6.15 shows that $\mathrm{ord}(g^k) = \mathrm{ord}(g)$ holds if and only if k and $n = \mathrm{ord}(g)$ are relatively prime. Hence the generators of the subgroup $\langle g \rangle$ are exactly the powers g^k with $k \in \{1,\ldots,n\}$ and $\gcd(n,k) = 1$. Thus a cyclic group with n elements has exactly $\phi(n)$ generators; note that this also follows from Proposition 1.6.11 and Definition 1.6.5. In fact, the group isomorphism

$$(\mathbb{Z}_n,+,0) \to (\langle g \rangle,\cdot,1), \quad k+\mathbb{Z}n \mapsto g^k$$

restricts to a bijection between the set of units of the residue ring $\mathbb{Z}_n$ and the set of generators of $\langle g \rangle$. □

Theorem 1.6.17. *Let $G = \langle g \rangle$ be a finite cyclic group with n elements, and let d be a divisor of n. Then there exists a unique subgroup U_d of G with order d. In fact, U_d is the cyclic subgroup generated by $g^{n/d}$, and the generators of U_d are precisely the elements of G with order d.*

Proof. The assertion is trivial if $U = \{1\}$. Thus we may assume the existence of an element $u \neq 1$ in U. Let k be the least integer $\neq 1$ such that $g^k \in U$, and let g^ℓ be an arbitrary element of U. Division with remainder gives integers q and r such that $\ell = qk + r$ and $0 \leq r < k$. Then $g^r = g^{\ell - qk} = g^\ell \cdot (g^k)^{-q} \in U$, and therefore $r = 0$ by the choice of k. Hence k divides ℓ, and U is the cyclic subgroup of G generated by g^k. Now $g^n = 1 \in U$ shows that k has to divide n. By Proposition 1.6.11 and Lemma 1.6.15, U has $\frac{n}{k} = \mathrm{ord}(g^k)$ elements. Thus U is indeed a cyclic subgroup of G of the form described in the assertion, for a suitable divisor of n.

Conversely, let d be any divisor of n, and let U_d be the set of all elements of G with order dividing d. Then U_d is a subgroup of G, since G is abelian. By the first part of the proof, U_d is cyclic. Let h be an arbitrary generator of U_d. Since $h^d = 1$ by the definition of U_d, the order of h (that is, the cardinality of U_d) has to divide d. On the other hand, $g^{n/d}$ has order d by Lemma 1.6.15 and therefore $\langle g^{n/d} \rangle$ is a subgroup of U_d (with cardinality d). Hence U_d has cardinality precisely d and is generated by $g^{n/d}$, completing the proof. □

The arguments used in the proof of Theorem 1.6.17 also lead to the following basic result for the Euler phi function:

Proposition 1.6.18. *For every positive integer n, one has*

$$\sum_{d|n} \phi(d) = n, \tag{1.16}$$

where the sum runs over all divisors $d \in \mathbb{N}$ of n.

Proof. Let G be a cyclic group with n elements, and recall that the order of every element of G divides n. Therefore the sets $\Omega_d := \{h \in G \colon \mathrm{ord}(h) = d\}$ (where d runs over all divisors of n) partition G, which implies $\sum_{d|n} |\Omega_d| = n$. By Theorem 1.6.17, an element h of G generates the unique (cyclic) subgroup U_d of G with d elements if and only if $\mathrm{ord}(h) = d$. Thus Ω_d consists of the $\phi(d)$ generators of U_d, and the assertion follows. □

Next, we prove an important lemma on the order of products in abelian groups:

Lemma 1.6.19. *Let G be a multiplicatively written group, and let g and h be two commuting elements (that is, $gh = hg$) of G with finite orders m and n, respectively. Then gh has finite order dividing mn, and equality holds provided that m and n are relatively prime.*

Proof. As g and h commute, we have $(gh)^{mn} = g^{mn}h^{mn} = 1$, and therefore the order of gh is finite and a divisor of mn.

Now assume that m and n are relatively prime, and let ℓ denote the order of gh. Then $(gh)^\ell = g^\ell h^\ell = 1$, and therefore $a = g^\ell = h^{-\ell}$ belongs to the intersection of the groups $\langle g \rangle$ and $\langle h \rangle$. Thus the order of a has to be a common divisor of m and n, so that $\mathrm{ord}(a) = 1$, that is, $a = g^\ell$ is the identity element of G. As g has order m, we conclude that m divides ℓ. Using also $(h^{-1})^\ell = 1$, a similar argument shows that n likewise is a divisor of ℓ. Using again $\gcd(m,n) = 1$, we see that mn divides ℓ, so that $\ell = \mathrm{ord}(gh) = mn$. □

Given commuting group elements g and h with finite orders m and n, respectively, one might suspect that the order of gh is the least common multiple of m and n. However, in general this is not correct: for a trivial example, take $h = g^{-1} \neq 1$, so that $\mathrm{ord}(h) = \mathrm{ord}(g) \neq 1$ and $\mathrm{ord}(gh) = 1 \neq \mathrm{ord}(g)$. Nevertheless, the group generated by g and h always contains *some* element with order $\mathrm{lcm}(m,n)$, as we will show in Theorem 1.6.21 below.

See Exercise 1.6.31 for a simple restriction on the possible orders of the product of two commuting group elements. The subsequent, considerably more difficult, Exercise 1.6.32 contains necessary and sufficient conditions for all orders that can occur.

In the remainder of this section, we study orders of elements in arbitrary finite groups. We begin with the following useful concept:

Definition 1.6.20. The **exponent** $\exp G$ of a finite group G is the least common multiple of the orders of all its elements. □

Theorem 1.6.21. *Let G be an abelian group. Then:*

- *If G contains elements with (finite) orders m and n, respectively, G also contains an element with order* $\mathrm{lcm}(m,n)$.
- *If G is finite, there exists an element of G which has order* $\exp G$.

Proof. We will write G multiplicatively. Note that it suffices to prove the first assertion, as the second assertion then follows using induction on $|G|$. Thus consider two (arbitrary) elements g and h of G which have orders m and n, respectively. We show that the subgroup $\langle g,h \rangle$ generated by g and h contains an element x with order $\ell := \mathrm{lcm}(m,n)$.

Let p be any prime divisor of ℓ, and let p^a be the largest power of p dividing ℓ. Without loss of generality, we may assume $p^a \mid m$. Then $x_p := g^{m/p^a}$ is an element of order p^a, by Lemma 1.6.15. Proceeding in this manner for all prime divisors p of ℓ, we obtain a collection of elements x_p whose product x lies in $\langle g,h \rangle$ and has order ℓ, by Lemma 1.6.19 and induction on the number of distinct prime divisors of ℓ. □

In many applications, one needs to be able to compute orders of elements in a finite group G. To do so for a specified $g \in G$, we require two pieces of information:

- a positive integer n such that $\operatorname{ord}(g)$ divides n;
- the list of all prime divisors of n.

In view of Remark 1.6.12, we may always take n to be the order of G, but sometimes we might have a better bound available (for instance, when G is abelian and we happen to know its exponent). The second requirement is equivalent to saying that we need the canonical prime power factorization of n.

We will present a reasonably efficient algorithm based on the following simple auxiliary result:

Lemma 1.6.22. *Let g be an element of some finite group G, and let n be any positive integer such that $g^n = 1$. Moreover, let p be a prime divisor of n and q the highest power of p dividing n. Put $m := n/q$, and let $g_p := g^m$ and $h := g^q$. Then:*

$$\operatorname{ord}(h) \mid m, \ \operatorname{ord}(g_p) \mid q \ \textit{ and } \ \operatorname{ord}(g) = \operatorname{ord}(h) \cdot \operatorname{ord}(g_p).$$

Proof. By hypothesis, $n = mq$, where m and q are relatively prime and where $g^n = 1$. This yields the divisibility assertions as well as the following decomposition of $\operatorname{ord}(g)$:

$$\operatorname{ord}(g) = \gcd(\operatorname{ord}(g), q) \cdot \gcd(\operatorname{ord}(g), m).$$

By Lemma 1.6.15,

$$\operatorname{ord}(g_p) = \frac{\operatorname{ord}(g)}{\gcd(m, \operatorname{ord}(g))} = \gcd(\operatorname{ord}(g), q)$$

and

$$\operatorname{ord}(h) = \frac{\operatorname{ord}(g)}{\gcd(\operatorname{ord}(g), q)} = \gcd(\operatorname{ord}(g), m),$$

which gives the desired formula for $\operatorname{ord}(g)$. □

Algorithm 1.6.23.

- *Input:* An element g in a finite group G, a positive integer n such that $g^n = 1$, and the set L of distinct prime divisors of n.
- *Output:* $\operatorname{ord}(g)$.

(1) $\operatorname{ord} \leftarrow 1$; $N \leftarrow n$; $\gamma \leftarrow g$;
(2) **while** $\gamma \neq 1$ **do**
(3) choose $p \in L$; $L \leftarrow L \setminus \{p\}$;
(4) let q be the highest power of p dividing N; $m \leftarrow N/q$; $h \leftarrow \gamma^m$;
(5) **while** $h \neq 1$ **do** $h \leftarrow h^p$; $\operatorname{ord} \leftarrow \operatorname{ord} \cdot p$;
(6) $\gamma \leftarrow \gamma^q$; $N \leftarrow m$;
(7) $\operatorname{ord}(g) \leftarrow \operatorname{ord}$. □

In Exercise 1.6.27, we ask the reader to verify that Algorithm 1.6.23 is correct. We now consider an example:

Example 1.6.24. Let $G = U(\mathbb{Z}_M)$ be the group of units modulo M, where $M = 343000$, we will apply Algorithm 1.6.23 to compute the order $\mathrm{ord}_M(3)$ of 3 modulo M. (Note that 3 and M are indeed relatively prime.)

We use the obvious choice $n := |G| = \phi(M)$, and hence we need to evaluate $\phi(M)$. This can be easily done using well-known properties of the Euler function ϕ, as soon as the prime power factorization of M is known. (The reader not familiar with this classical result will find the procedure explained in Remark 1.9.13 below.) As $M = 2^3 \cdot 5^3 \cdot 7^3$, we obtain

$$n = 7^2 \cdot (7-1) \cdot 5^2 \cdot (5-1) \cdot 2^2 = 117600,$$

so that $L = \{2,3,5,7\}$. The computations performed by Algorithm 1.6.23 with these data (of course, modulo M) are summarized in the following table:

p	m	q	h	ord	γ	N
				1	3	117600
2	3675	32	259307			
			115249	2		
			1	4	48841	3675
5	147	25	192081			
			27401	20		
			1	100	265001	147
3	49	3	4801			
			1	300	154001	49
7	1	49	154001			
			49001	2100		
			1	14700	1	1

Thus $\mathrm{ord}_{343000}(3) = 14700$. □

The preceding example should make it obvious that we are still missing a vital ingredient for an efficient implementation of Algorithm 1.6.23: we need to be able to perform (modular) exponentiations with rather large exponents. For this task, one uses the following standard method for exponentiation in an arbitrary monoid:

Algorithm 1.6.25 (Square and multiply).

- *Input:* an element b in some (multiplicatively written) monoid G, and an exponent $e > 1$.
- *Output:* $x := b^e$.

(1) $x \leftarrow 1$, $y \leftarrow b$, $c \leftarrow e$;
(2) **while** $c \neq 0$ **do**
(3)   **if** c is odd **then** $x \leftarrow xy$, $c \leftarrow c-1$ **fi**;

(4) $\quad y \leftarrow y^2,\ c \leftarrow c/2.$
(5) **od** □

Proposition 1.6.26. *Let b be an element of some monoid G, and let $e \geq 2$ be any integer. Algorithm 1.6.25 allows to compute b^e with at most $2\lceil \log_2 e \rceil$ multiplications in G.*

Proof. The correctness of Algorithm 1.6.25 follows from the observation that each step preserves the loop invariant $xy^c = b^e$. In order to determine how many multiplications are needed to perform this algorithm, we use the binary representation of the current value of c, that is,

$$c = (c_{k-1}, \ldots, c_0), \quad \text{where } c = 2^{k-1}c_{k-1} + \cdots + 2c_1 + c_0. \tag{1.17}$$

The validity of the condition in Step (3) amounts to checking whether $c_0 = 1$. In this case, we perform one multiplication and replace c with $c' := c - 1$. Note that the binary representation of c' is then obtained from (1.17) by changing the last bit c_0 to 0. In Step (4), we perform a squaring (which is, of course, also a multiplication) and replace c with $c' := c/2$; this time, the binary representation of c' is obtained from (1.17) by discarding the last bit c_0. These remarks show that one needs altogether either s (when e is a power of 2) or $s - 1$ squarings in Step (4), where $s := \lceil \log_2 e \rceil$, and that the number of multiplications in Step (3) equals the number of entries 1 in the binary representation of the exponent e. Altogether, this gives at most $2s$ multiplications. □

Exercises

Exercise 1.6.27. Prove that Algorithm 1.6.23 is correct, by using Lemma 1.6.22 and the loop invariant $\gamma^N = 1$ and $\mathrm{ord}(\gamma) \cdot \mathrm{ord} = \mathrm{ord}(g)$. □

Exercise 1.6.28. Let G be the group of units in the residue ring $\mathbb{Z}_n$, where $n = 496,125$. Use Algorithm 1.6.23 to determine the order of 2 in G. □

Exercise 1.6.29. Work out the computations performed by Algorithm 1.6.25 to compute the powers b^{31}, b^{32} and b^{33}. □

Exercise 1.6.30. Let G be a (multiplicatively written) group of order n. Devise a method for computing the inverse of an element $g \in G$. □

Exercise 1.6.31. Prove the following restriction on the order of the product of two commuting group elements g and h with finite orders m and n, respectively:

$$\frac{mn}{d^2} \mid \mathrm{ord}(gh) \mid \frac{mn}{d} = \mathrm{lcm}(m,n),$$

where $d = \gcd(m,n)$. □

Exercise 1.6.32. Prove the following (considerably more demanding) strengthening of Exercise 1.6.31: Let m and n be positive integers, put $d := \gcd(m,n)$, and let f be the largest divisor of d for which one has $\gcd(f, \frac{m}{d}) = \gcd(f, \frac{n}{d}) = 1$. Then there exist a finite abelian group G and elements $g, h \in G$ of orders m and n, respectively, such that $\operatorname{ord}(gh) = k$ if and only if k satisfies

$$\frac{mn}{df} \mid k \mid \frac{mn}{d} = \operatorname{lcm}(m,n).$$

(An elementary proof of this result is given in [213].) □

Exercise 1.6.33. Let $p \in \mathbb{N}$ be a prime, and let r be any positive integer. Show that $\phi(p^r) = (p-1)p^{r-1}$, where ϕ is the Euler phi function. □

1.7 Orders of Residues modulo n

In the present section, we develop further fundamental results from Number Theory which will be needed later in this book, including the structure of the group of units modulo a prime power. We begin with a few important concepts:

Definition 1.7.1. Let n be a positive integer.

- n is said to be **square-free** if it is not divisible by the square of any prime, that is, if 1 is the only square dividing n.
- The **square part** of n is the largest square dividing n.
- The **square-free part** $\operatorname{sf}(n)$ of n (sometimes also called the **core** of n)[15] is the quotient of n and the square part of n. In other words, $\operatorname{sf}(n)$ is the smallest positive number m for which n/m is a square.
- The **radical** $\operatorname{rad}(n)$ of n (sometimes also called the **square-free kernel** of n)[16] is the product of the distinct prime divisors of n. By convention, $\operatorname{rad}(1) = 1$. □

For instance, the radical of $360 = 2^3 \cdot 3^2 \cdot 5$ is $2 \cdot 3 \cdot 5 = 30$, whereas the square-free part is $2 \cdot 5 = 10$, and the square part is $2^2 \cdot 3^2 = 36$. We remark that no polynomial time algorithm is known for computing either the radical or the square-free part of a positive integer. As Exercise 1.7.15 shows, these two notions are closely related.

In what follows, we will also require a further, more technical concept:

Definition 1.7.2. Let $n \geq 2$ be a positive integer, and let $N \neq 0$ be any integer. The n-**part** of N is the largest divisor d of N for which $\operatorname{rad}(d)$ divides $\operatorname{rad}(n)$, that is, the largest divisor d of N which is a product of prime divisors of n. We will denote this divisor d by $\operatorname{pt}_n(N)$.[17] □

---

[15] Some authors also use the notation n' instead of $\operatorname{sf}(n)$.

[16] Unfortunately, occasionally one finds the term "square-free part" used to signify the radical of an integer. We note that the term "radical" is standard, though.

[17] Of course, we may assume N to be a positive integer ≥ 2, too: trivially, $\operatorname{pt}_n(1) = 1$ and $\operatorname{pt}_n(-N) = \operatorname{pt}_n(N)$ for all n.

We collect three obvious properties of n-parts as Exercise 1.7.16 and proceed by considering the r-part of q in the case where r is a prime. In later applications to finite fields, q will always be a prime power, but for now we just assume that q is some integer ≥ 2.

Lemma 1.7.3. *Let $q \in \mathbb{N}$ with $q \geq 2$ and assume that r is a prime divisor of $q-1$. Then the following hold:*

(1) *If r is odd, then $\mathrm{pt}_r(q^{r^\ell} - 1) = r^\ell \cdot \mathrm{pt}_r(q-1)$ for all $\ell \in \mathbb{N}$.*

(2) *If $r = 2$ and $q \equiv 1 \bmod 4$, then $\mathrm{pt}_2(q^{2^\ell} - 1) = 2^\ell \cdot \mathrm{pt}_2(q-1)$ for all $\ell \in \mathbb{N}$.*

(3) *If $r = 2$ and $q \equiv 3 \bmod 4$, then $\mathrm{pt}_2(q^{2^\ell} - 1) = 2^{\ell-1} \cdot \mathrm{pt}_2(q^2-1)$ for all $\ell \in \mathbb{N}^*$.*

Proof. The first two assertions are proved using induction on ℓ. The induction basis $\ell = 0$ is trivially correct. For the induction step from ℓ to $\ell+1$, we put $Q = q^{r^\ell}$ and consider the r-part of $Q^r - 1$.

We deal with the second assertion first. Thus let $r = 2$ and assume $q \equiv 1 \bmod 4$. By induction, $\mathrm{pt}_2(Q-1) = 2^\ell \cdot \mathrm{pt}_2(q-1)$. By hypothesis, 4 divides $Q-1$, so that $Q+1 \equiv 2 \bmod 4$. This gives

$$\mathrm{pt}_2(Q^2-1) = \mathrm{pt}_2(Q+1) \cdot \mathrm{pt}_2(Q-1) = 2 \cdot \mathrm{pt}_2(Q-1) = 2^{\ell+1} \cdot \mathrm{pt}_2(q-1),$$

proving (2).

Now let r be odd and note that

$$q^{r^{\ell+1}} - 1 = Q^r - 1 = (Q-1) \cdot (1 + Q + \cdots + Q^{r-1}).$$

By hypothesis, r divides $Q-1$. Thus $Q \equiv 1 \bmod r$, so that $1 + Q + \cdots + Q^{r-1} \equiv 0 \bmod r$. As $\mathrm{pt}_r(Q-1)$ divides $Q-1$, we see that $r \cdot \mathrm{pt}_r(Q-1)$ divides $\mathrm{pt}_r(Q^r-1)$. For simplicity, let us write $R = \mathrm{pt}_r(Q-1)$ and $Q = kR+1$, where k is not divisible by r. Using the binomial theorem, we obtain

$$Q^r - 1 = (kR+1)^r - 1 = \binom{r}{1} kR + S \ \text{ with } S = \sum_{j=2}^{r} \binom{r}{j} k^j R^j.$$

Note that $\binom{r}{j} k^j R^j$ is divisible by rR^2 for $2 \leq j \leq r-1$ and that $k^r R^r$ is divisible by R^3, which is a multiple of rR^2. Hence S is divisible by rR^2 and therefore also by $r^2 R$, so that

$$Q^r - 1 \equiv k \cdot rR \not\equiv 0 \mod r^2 R,$$

since r does not divide k. Altogether, we have shown that $Q^r - 1$ is a multiple of rR, but not of $r^2 R$. Thus $\mathrm{pt}_r(Q^r-1) = rR = r \cdot \mathrm{pt}_r(Q-1)$, and (1) follows with induction.

It remains to prove (3). Thus let $q \equiv 3 \bmod 4$ and $\ell \geq 1$. Then $q^{2^\ell} - 1 = Q^{2^{\ell-1}} - 1$, where now $Q := q^2$. As $q^2 \equiv 1 \bmod 4$, an application of (2) gives $\mathrm{pt}_2(Q^{2^{\ell-1}} - 1) = 2^{\ell-1} \mathrm{pt}_2(Q-1)$, as claimed. □

We also note the following simple but useful result:

Lemma 1.7.4. *Let q and s be positive integers, where $q \geq 2$, and let r be a prime dividing $q-1$ but not s. Then* $\mathrm{pt}_r(q-1) = \mathrm{pt}_r(q^s-1)$.

Proof. We may assume $s \geq 2$. By hypothesis, $1+q+\cdots+q^{s-1} \equiv s \not\equiv 0 \bmod r$ and $q \equiv 1 \bmod r$. Now $q^s - 1 = (q-1)\cdot(1+q+\cdots+q^{s-1})$ gives the assertion. □

We now apply Lemma 1.7.3 to obtain information about the multiplicative order of units modulo a prime power. Thus let $n \in \mathbb{N}$ with $n \geq 2$ and recall from Remark 1.5.19 that an integer q is a unit modulo n (that is, $q+n\mathbb{Z}$ belongs to the group $U(\mathbb{Z}_n)$ of units in the residue ring $\mathbb{Z}_n$) if and only if q and n are relatively prime. Then the (multiplicative) order of q modulo n is the order of $q+n\mathbb{Z}$ in $U(\mathbb{Z}_n)$; we will use the notation $\mathrm{ord}_n(q)$ instead of $\mathrm{ord}(q+n\mathbb{Z})$. Note that $\mathrm{ord}_n(q)$ is the least integer $d \geq 1$ such that q^d-1 is divisible by n. Moreover, $\mathrm{ord}_n(q)$ divides $\phi(n)$, by Theorem 1.3.3 and Definition 1.6.5.

Corollary 1.7.5. *Let $q \in \mathbb{N}$ with $q \geq 2$, and let r be a prime dividing $q-1$. If either r is odd or $r=2$ and $q \equiv 1 \bmod 4$, then*

$$\mathrm{ord}_{r^\ell}(q) = r^{\ell-\min(k,\ell)} \quad \textit{for all } \ell \in \mathbb{N}^*,$$

where $\mathrm{pt}_r(q-1) = r^k$.

Proof. Note that the assertion holds when $\ell \leq k$: then r^ℓ divides $q-1$, and hence $\mathrm{ord}_{r^\ell}(q) = 1 = r^{\ell-\min(k,\ell)}$.

Now let $\ell > k$, so that $r^k = \mathrm{pt}_r(q-1)$ is a proper divisor of r^ℓ. We have to check that $\mathrm{ord}_{r^\ell}(q) = r^{\ell-\min(k,\ell)} = r^{\ell-k}$. By Lemma 1.7.3,

$$\mathrm{pt}_r(q^{r^{\ell-k}}-1) = r^{\ell-k}\cdot \mathrm{pt}_r(q-1) = r^{\ell-k}\cdot r^k = r^\ell,$$

so that $\mathrm{ord}_{r^\ell}(q)$ divides $r^{\ell-k}$. On the other hand,

$$\mathrm{pt}_r(q^{r^{\ell-1-k}}-1) = r^{\ell-1-k}\cdot \mathrm{pt}_r(q-1) = r^{\ell-1-k}\cdot r^k = r^{\ell-1},$$

and hence $\mathrm{ord}_{r^\ell}(q)$ cannot divide $r^{\ell-1-k}$. □

Corollary 1.7.6. *Let $q \in \mathbb{N}$ with $q \equiv 3 \bmod 4$, and write $\mathrm{pt}_2(q^2-1) = 2^k$. Then $k \geq 3$ and*

$$\mathrm{ord}_{2^\ell}(q) = 2^{\ell+1-\min(k,\ell)} \quad \textit{for all } \ell \in \mathbb{N}\setminus\{0,1\}.$$

Proof. By hypothesis, 4 divides $q+1$ and therefore $q^2-1 = (q-1)(q+1)$ is a multiple of 8, so that $k \geq 3$. Also, $\mathrm{ord}_4(q) = 2$, and hence 2 divides the order of q modulo 2^ℓ for all $\ell \geq 2$. Together with Corollary 1.7.5, this shows $\mathrm{ord}_{2^\ell}(q) = 2\cdot \mathrm{ord}_{2^\ell}(q^2) = 2\cdot 2^{\ell-\min(k,\ell)}$. □

We are now ready to determine the structure of the group of units modulo a prime power,[18] provided we are willing to use one detail still missing at this point:

[18] Determining the structure of the group of units modulo an arbitrary integer $n > 2$ reduces to this special case; see Remark 1.10.9.

the multiplicative group of a Galois field $\mathbb{Z}/p\mathbb{Z}$, where p is a prime, is cyclic. We will just take this fact for granted now, as we will establish a more general result in Section 3.2 (of course, without using the results of the present section).

Theorem 1.7.7. *Let $r \in \mathbb{N}$ be a prime, and let ℓ be a positive integer. Then the following hold:*

(1) *If r is odd or $\ell \leq 2$, then $U(\mathbb{Z}_{r^\ell})$ is cyclic of order $(r-1)r^{\ell-1}$.*

(2) *If $r = 2$ and $\ell \geq 3$, then $U(\mathbb{Z}_{2^\ell})$ is isomorphic to the direct product of a cyclic group of order 2 with a cyclic group of order $2^{\ell-2}$.*

Proof. For $\ell = 1$, $\mathbb{Z}_r$ is a finite field with r elements, and the multiplicative group $U(\mathbb{Z}_r) = \mathbb{Z}_r^*$ is cyclic of order $r-1$ by Theorem 3.2.5. If $r = 2$ and $\ell = 2$, then -1 is a unit modulo 4 and $\mathrm{ord}_4(-1) = 2$. Since $U(\mathbb{Z}_4)$ has $\phi(4) = 2$ elements, it is generated by -1 and therefore a cyclic group of order 2.

Next, we consider the case where r is odd and $\ell \geq 2$. We will use an integer x of order $r-1$ modulo r (which exists by the case $\ell = 1$) and an integer y of the form $y = 1 + ar$, where a is not divisible by r, to construct an integer of order $(r-1)r^{\ell-1}$ modulo r^ℓ. By Exercise 1.6.33, this is the cardinality $\phi(r^\ell)$ of $U(\mathbb{Z}_{r^\ell})$, and hence the group of units modulo r^ℓ is indeed cyclic.

As x has order $r-1$ modulo r, it is clear that $\mathrm{ord}_{r^\ell}(x)$ has to be a multiple of $r-1$. Since $x^{r-1} - 1$ is divisible by r, an application of Corollary 1.7.5 with $q = x^{r-1}$ shows that $\mathrm{ord}_{r^\ell}(x)$ has the form $(r-1)r^j$ for some $j \leq \ell - 1$, so that x^{r^j} has order $r-1$ modulo r^ℓ. On the other hand, $\mathrm{pt}_r(y-1) = r$ by our choice of y, and therefore $\mathrm{ord}_{r^\ell}(y) = r^{\ell-1}$, again using Corollary 1.7.5. Since $r-1$ and $r^{\ell-1}$ are relatively prime, Lemma 1.6.19 shows that the product $x^{r^j}y$ has order $(r-1)r^{\ell-1}$ modulo r^ℓ, as desired. This concludes the proof of (1).

Finally, let $r = 2$ and $\ell \geq 3$, and consider 5 modulo 2^ℓ. Using $\mathrm{pt}_2(5-1) = 4$ in Lemma 1.7.3, we obtain $\mathrm{pt}_2(5^{2^m} - 1) = 2^{m+2}$ for all $m \geq 0$. Therefore $\mathrm{ord}_{2^\ell}(5) = 2^{\ell-2}$, that is, $5 + 2^\ell\mathbb{Z}$ generates a subgroup of order $2^{\ell-2}$ of the residues modulo 2^ℓ. Trivially, $-1 + 2^\ell\mathbb{Z}$ generates a subgroup of order 2, and we claim that these two subgroups have trivial intersection.

Assume otherwise, that is, $5^k \equiv -1 \bmod 2^\ell$ for some k. This gives $5^k \equiv -1 \bmod 4$, a contradiction. Thus the two subgroups of $U(\mathbb{Z}_{2^\ell})$ in question indeed intersect trivially, and hence their product has order $\mathrm{ord}_{2^\ell}(5) \cdot \mathrm{ord}_{2^\ell}(-1) = 2^{\ell-1}$ (by Exercise 1.3.15), which is the cardinality $\phi(2^\ell)$ of $U(\mathbb{Z}_{2^\ell})$ (by Exercise 1.6.33). This establishes also assertion (2). □

In the case where $U(\mathbb{Z}_{r^\ell})$ is cyclic, any generator $y + r^\ell\mathbb{Z}$ of $U(\mathbb{Z}_{r^\ell})$ is called a **primitive element** of $\mathbb{Z}_{r^\ell}$; we also say that y is a **primitive root** modulo r^ℓ. In this context, we mention an interesting strengthening of part (1) of Theorem 1.7.7:

Proposition 1.7.8. *Let r be an odd prime, and assume that z is a primitive root modulo r^2, that is, $r \nmid z$ and $\mathrm{ord}_{r^2}(z) = r(r-1)$. Then z is in fact a primitive root modulo r^ℓ for all $\ell \in \mathbb{N}^*$.*

Proof. Let us denote $\operatorname{ord}_r(z)$ by s, so that s divides $r-1$ and r divides z^s-1. Since $\operatorname{ord}_r(z)$ divides $\operatorname{ord}_{r^\ell}(z)$ for every $\ell \in \mathbb{N}^*$, we see that $\operatorname{ord}_{r^\ell}(z) = s \cdot \operatorname{ord}_{r^\ell}(z^s)$, where $\operatorname{ord}_{r^\ell}(z^s)$ is a divisor of $r^{\ell-1}$ by Corollary 1.7.5. In particular, the case $\ell = 2$ shows $s = r-1$ and $\operatorname{ord}_{r^2}(z^s) = r$, as z is a primitive root modulo r^2 by hypothesis. But then $\operatorname{pt}_r(z^{r-1}-1) = r$ and therefore $\operatorname{ord}_{r^\ell}(z^{r-1}) = r^{\ell-1}$ for all $\ell \geq 2$, by another application of Corollary 1.7.5. □

Up to now, we have only considered the n-part of an integer q in the special case where n is a prime dividing $q-1$. We now study the more general situation where n can be an arbitrary integer ≥ 2 for which the radical $\operatorname{rad}(n)$ divides $q-1$:

Proposition 1.7.9. *Let n, q and m be positive integers with $n, q \geq 2$ and assume that $\operatorname{rad}(n)$ divides $q-1$. Then the following hold:*

(1) *If n or m is odd or if $q \equiv 1 \bmod 4$, then $\operatorname{pt}_n(q^m-1) = \operatorname{pt}_n(m) \cdot \operatorname{pt}_n(q-1)$.*
(2) *If n and m are even and $q \equiv 3 \bmod 4$, then $\operatorname{pt}_n(q^m-1) = \frac{1}{2} \cdot \operatorname{pt}_n(m) \cdot \operatorname{pt}_n(q^2-1)$.*

Proof. Let $r_1, \ldots, r_t$ be the distinct prime divisors of n. By Exercise 1.7.16,

$$\operatorname{pt}_n(q^m-1) = \operatorname{pt}_{\operatorname{rad}(n)}(q^m-1) = \prod_{i=1}^{t} \operatorname{pt}_{r_i}(q^m-1).$$

Assume first that n or m is odd or that $q \equiv 1 \bmod 4$, and consider any prime divisor r of n. Write $m = m' \cdot R$, where R is a power of r (possibly $R = 1$) and where m' and r are relatively prime. Then, with $Q = q^R$,

$$\operatorname{pt}_r(q^m-1) = \operatorname{pt}_r(Q^{m'}-1) = \operatorname{pt}_r(Q-1) = \operatorname{pt}_r(q^R-1) = R \cdot \operatorname{pt}_r(q-1),$$

where we have used Lemmas 1.7.4 and 1.7.3. Thus $\operatorname{pt}_r(q^m-1) = \operatorname{pt}_r(m) \cdot \operatorname{pt}_r(q-1)$, as R is the largest power of r dividing m. Applying this observation for all prime divisors of n yields

$$\begin{aligned}
\operatorname{pt}_n(q^m-1) &= \prod_{i=1}^{t} \left(\operatorname{pt}_{r_i}(m) \cdot \operatorname{pt}_{r_i}(q-1)\right) \\
&= \prod_{i=1}^{t} \operatorname{pt}_{r_i}(m) \cdot \prod_{i=1}^{t} \operatorname{pt}_{r_i}(q-1) \\
&= \operatorname{pt}_n(m) \cdot \operatorname{pt}_n(q-1),
\end{aligned}$$

establishing (1).

The proof of (2) proceeds in a similar manner. If r is an odd prime divisor of n, we again have $\operatorname{pt}_r(q^m-1) = \operatorname{pt}_r(m) \cdot \operatorname{pt}_r(q-1)$. By hypothesis, r divides $q-1$, so that $q+1$ is not divisible by r and hence also $\operatorname{pt}_r(q^m-1) = \operatorname{pt}_r(m) \cdot \operatorname{pt}_r(q^2-1)$.

It remains to consider the prime divisor 2 of n. In analogy to the case of odd prime divisors, we now write $m = m' \cdot S$, where S is a power of 2 and m' is odd. Then Lemmas 1.7.4 and 1.7.3 give

$$\mathrm{pt}_2(q^m - 1) = \mathrm{pt}_2(q^S - 1) = \frac{1}{2}S \cdot \mathrm{pt}_2(q^2 - 1).$$

Again, we apply these observations for all prime divisors of n and obtain

$$\mathrm{pt}_n(q^m - 1) = \frac{1}{2} \cdot \mathrm{pt}_n(m) \cdot \mathrm{pt}_n(q^2 - 1),$$

as required. □

As in the case where n is a prime power, we may restate the preceding results in terms of orders modulo n:

Proposition 1.7.10. *Let Q and n be integers with $Q, n \geq 2$, and assume that* $\mathrm{rad}(n)$ *divides $Q-1$. Then the following hold:*

(1) *If n is odd or $Q \equiv 1 \bmod 4$, then*

$$\mathrm{ord}_n(Q) = \frac{n}{\gcd(\mathrm{pt}_n(Q-1), n)}.$$

(2) *If $n \equiv 0 \bmod 4$ and $Q \equiv 3 \bmod 4$, then*

$$\mathrm{ord}_n(Q) = \frac{2n}{\gcd(\mathrm{pt}_n(Q^2-1), n)}.$$

(3) *If $n \equiv 2 \bmod 4$ and $Q \equiv 3 \bmod 4$, then*

$$\mathrm{ord}_n(Q) = \mathrm{ord}_{n/2}(Q) = \frac{n}{\gcd(\mathrm{pt}_n(Q-1), n)}.$$

Proof. We use the following special case of Lemma 1.6.19 without further comment: if $a, b \in \mathbb{N}^*$ with $\gcd(Q, ab) = 1$ and $\gcd(a, b) = 1$, then $\mathrm{ord}_{ab}(Q)$ is the least common multiple of $\mathrm{ord}_a(Q)$ and $\mathrm{ord}_b(Q)$.

Write $n = 2^{\ell_0} \cdot n'$ where n' is odd, and let $n' = \prod_{i=1}^{t} r_i^{\ell_i}$ be the prime power factorization of n'. Then, with $r_0 := 2$,

$$\mathrm{ord}_n(Q) = \mathrm{lcm}\Big(\mathrm{ord}_{r_i^{\ell_i}}(Q) : i = 0, 1, \ldots, t\Big) = \prod_{i=0}^{t} \mathrm{ord}_{r_i^{\ell_i}}(Q),$$

and

$$\mathrm{ord}_n(Q) = \mathrm{lcm}\left(\mathrm{ord}_{n'}(Q), \mathrm{ord}_{2^{\ell_0}}(Q)\right) = \mathrm{ord}_{n'}(Q) \cdot \mathrm{ord}_{2^{\ell_0}}(Q).$$

Now let r be any prime divisor of n and put $R := \mathrm{pt}_r(n)$. By hypothesis, r divides $Q-1$, and thus $\mathrm{ord}_R(Q)$ is a power of r, by Corollaries 1.7.5 and 1.7.6.

Let $r_i^{k_i} = \mathrm{pt}_{r_i}(Q-1)$ for $i = 0, \ldots, t$. Then $k_i \geq 1$ for $1 \leq i \leq t$, by definition (whereas we may have $k_0 = 0$), and thus

$$(*) \qquad \mathrm{ord}_{r_i^{\ell_i}}(Q) = r_i^{\ell_i - \min(k_i, \ell_i)} = \frac{r_i^{\ell_i}}{\gcd(\mathrm{pt}_{r_i}(Q-1), r_i^{\ell_i})},$$

by Corollary 1.7.5. Note that the same formula also holds for $i=0$ provided that $Q\equiv 1 \bmod 4$. Multiplying all these identities immediately leads to the desired formula in case (1).

In order to show (2), we apply case (1) to the pair (n,Q^2). Note that $\mathrm{ord}_{n'}(Q^2)=\mathrm{ord}_{n'}(Q)$; this follows from $(*)$ and Lemma 1.7.4, since n' is odd. In view of

$$\mathrm{ord}_{2^{\ell_0}}(Q) = 2\cdot\mathrm{ord}_{2^{\ell_0}}(Q^2),$$

we obtain the formula claimed in case (2).

It remains to deal with case (3). Here $\ell_0=2$, $n'=\frac{n}{2}$ and $\mathrm{ord}_2(Q)=1$, so that

$$\mathrm{ord}_n(Q) = \mathrm{ord}_{n'}(Q) = \frac{n'}{\gcd(\mathrm{pt}_{n'}(Q-1),n')},$$

by case (1). Since $\mathrm{pt}_n(Q-1)=2\cdot\mathrm{pt}_{n'}(Q-1)$ and $n=2n'$, we indeed obtain the formula in (3). □

Finally, let us consider the case where $q-1$ is not necessarily divisible by the radical of n:

Remark 1.7.11. Let q and n be relatively prime integers ≥ 2, and let $s=\mathrm{ord}_{\mathrm{rad}(n)}(q)$ be the multiplicative order of q modulo the radical of n. Then $\mathrm{ord}_n(q)=s\cdot\mathrm{ord}_n(q^s)$; this follows from Lemma 1.6.15, as $\mathrm{ord}_n(q)$ is a multiple of $s=\mathrm{ord}_{\mathrm{rad}(n)}(q)$. Trivially, q^s-1 is divisible by $\mathrm{rad}(n)$. Hence, after computing s, one may take $Q:=q^s$ and apply Proposition 1.7.10 in order to determine $\mathrm{ord}_n(Q)$, and then multiplying this value by s gives $\mathrm{ord}_n(q)$. □

We conclude this section with a basic algorithm which determines the m-part of an integer N, where $m,N\geq 2$, without making use of the prime power factorization of the numbers m and N. This method, which is called **Algorithm r**, is due to Lüneburg [248].

Algorithm 1.7.12 (Algorithm r).

- *Input:* Two integers $m,N\geq 2$.
- *Output:* The m-part $\mathrm{pt}_m(N)$ of N.

(1) $r\leftarrow N,\ \delta\leftarrow\gcd(r,m)$;
(2) **while** $\delta\neq 1$ **do**
(3) $\quad r\leftarrow r/\delta$;
(4) $\quad \delta\leftarrow\gcd(r,m)$
(5) **od**
(6) $\mathrm{pt}_m(N)\leftarrow N/r$.

Proposition 1.7.13. *Algorithm 1.7.12 correctly determines the m-part of N.*

Proof. We show that, after termination, the value of the variable r satisfies the following three properties:

(a) r divides N,
(b) r and m are relatively prime,
(c) every prime divisor of N/r divides m.

This means that r is the largest divisor of N which is relatively prime to m, so that N/r is the m-part of N, as desired. In view of this, r will be called the **complementary m-part** of N.

Trivially, conditions (a) and (c) are satisfied after the initialization in Step (1). Now assume that (a) and (c) hold when entering the **while** loop at a certain point during the execution of the algorithm. Let $r' := r/\delta$, where δ is the greatest common divisor of m and the current value of r. Then r' divides N, since r divides N by assumption; hence r' satisfies condition (a). We claim that r' also satisfies condition (c); thus we need to check that every prime s dividing $N/r' = \delta N/r$ also divides m. This is obvious if s divides δ, as δ divides m; and if s divides N/r, it follows from the assumption that r satisfies condition (c).

Since the value of r decreases with every execution of the **while** loop, the algorithm has to terminate. In view of (2), the greatest common divisor of m and the final value of r is $\delta = 1$; then r also satisfies condition (b), as claimed. □

Remark 1.7.14. It should be clear that the concepts considered in this section – in particular, radicals and m-parts – can be generalized to arbitrary principal ideal domains; see the remarks following Definition 13.4.2. In particular, we will repeatedly require these more general notions for polynomial rings over a finite field.

Similarly, Algorithm r can be formulated whenever we have an efficient way of computing greatest common divisors, that is, for arbitrary Euclidean domains. For instance, we shall use this algorithm in Section 11.9 in the version for polynomial rings over a finite field. For an application in the context of finite abelian groups, we refer to Exercises 1.7.19 and 1.7.20 below. □

Exercises

Exercise 1.7.15. Let n be a positive integer. Prove $\mathrm{rad}(n) = \mathrm{sf}(n)\mathrm{sf}\big(n/\mathrm{rad}(n)\big)$. □

Exercise 1.7.16. Let $n \geq 2$ be a positive integer, and let M and N be arbitrary integers $\neq 0$. Check that

- $\mathrm{pt}_n(MN) = \mathrm{pt}_n(M)\mathrm{pt}_n(N)$.
- $\mathrm{pt}_n(N) = \mathrm{pt}_{\mathrm{rad}(n)}(N)$.
- $\mathrm{pt}_n(N) = \prod_{i=1}^{\ell} \mathrm{pt}_{r_i}(N)$, where $r_1, \ldots, r_\ell$ are the distinct prime divisors of n. □

Exercise 1.7.17. Show that Corollary 1.7.6 does not hold for $\ell = 1$. □

Exercise 1.7.18. Use the approach outlined in Remark 1.7.11 to prove the following generalization of Corollary 1.7.5:

Let $q \in \mathbb{N}$ with $q \geq 2$, let r be any prime not dividing q, and put $s := \mathrm{ord}_r(q)$. If either r is odd or $r = 2$ and $q \equiv 1 \bmod 4$, then

$$\mathrm{ord}_{r^\ell}(q) = s \cdot r^{\ell - \min(k,\ell)} \quad \textit{for all } \ell \in \mathbb{N}^*,$$

where $\mathrm{pt}_r(q^s - 1) = r^k$. □

Exercise 1.7.19. Starting with two integers $m > 2$ and $n \geq 2$, determine two further integers m_1 and n_2 as follows:

- let $m_0 := m/\gcd(m,n)$ and $n_0 := n/\gcd(m,n)$;
- let m_1 be the complementary n_0-part of m, and let n_1 be the complementary m_0-part of n;
- finally, let $n_2 := n_1/\gcd(m_1,n_1)$.

Prove that m_1 and n_2 are relatively prime and that the product $m_1 n_2$ is the least common multiple of m and n. □

Exercise 1.7.20. Let G be a finite abelian group and $g,h \in G$. Apply Exercise 1.7.19 to determine an element $a \in G$ satisfying $\mathrm{ord}(a) = \mathrm{lcm}(\mathrm{ord}(g), \mathrm{ord}(h))$. □

1.8 Basics on Modules

The basic theory of modules is a very important tool in the study of finite fields. For this reason, we consider modules over a commutative ring in this section. As we will see, modules generalize both abelian groups and commutative rings.

We start with a commutative ring $(R,+,\cdot,0,1)$ and an abelian group $(V,+,0_V)$. For the time being, we will call any mapping of the form $R \times V \to V$ a **pairing**; such a mapping is usually denoted by the symbol $\cdot$. Given $v \in V$ and $r \in R$, one writes $r \cdot v$ or simply rv for the image of the pair (r,v) under the pairing $\cdot$. Although the same symbol $\cdot$ is used to denote both pairings and the ring multiplication, it will always be clear from the context which operation is performed. A similar remark also applies to the additions in R and in V, respectively.

Definition 1.8.1. Let $(R,+,\cdot,0,1)$ be a commutative ring and $(V,+,0_V)$ an abelian group. A pairing $R \times V \to V$ is called a **scalar multiplication** provided it satisfies the following conditions:

(1) $r(v+w) = rv + rw$ for all $r \in R$ and all $v,w \in V$.
(2) $(r+s)v = rv + sv$ for all $r,s \in R$ and all $v \in V$.
(3) $(rs)v = r(sv)$ for all $r,s \in R$ and all $v \in V$.
(4) $1v = v$ for all $v \in V$.

A **module** over R or, more concisely, an ***R*-module** is an abelian group V together with a scalar multiplication.[19] □

[19] In this book, module elements are multiplied by ring elements from the left. If one wants to stress this notational convention, one calls V a **left** R-module. Of course, one can also define **right** R-modules in a completely analogous manner.

Example 1.8.2. Let us mention the most basic examples of modules:

- If F is a field and V is an F-module, then V is called a **vector space** over F. Throughout this text, we will assume familiarity with the (basic) theory of vector spaces as studied in Linear Algebra. Some more advanced facts on vector spaces and algebras will be summarized where needed.
- Every abelian group is a $\mathbb{Z}$-module. This is implicit in Section 1.6.
- If S is any subring of a commutative ring R, then R carries the structure of an S-module: the restriction of the multiplication in R to $S \times R$ is a scalar multiplication. In particular, R is itself an R-module. □

A few basic properties of R-modules are listed in Exercise 1.8.14. In view of the last of these, one simply writes $-rv$ instead of $r(-v)$.

Next, let us consider substructures and then structure preserving mappings in the context of modules.

Definition 1.8.3. Let $(V, +, 0_V)$ be a module over a commutative ring $(R, +, \cdot, 0, 1)$. A subset U of V is called an ***R*-submodule** (or simply a **submodule**) of V if U is a subgroup of $(V, +, 0_V)$ satisfying $ru \in U$ for all $r \in R$ and all $u \in U$. □

An alternative characterization of submodules is given in Exercise 1.8.15.

Example 1.8.4. Consider the commutative ring R itself as an R-module. Then the R-submodules of R are just the ideals of R. □

Example 1.8.5. Let R be a commutative ring and $(V, +, 0_V)$ an R-module. Let us list some important classes of submodules:

(1) For every $v \in V$, the set
$$Rv := \{rv : r \in R\}$$
is a submodule of V, namely the submodule **generated** by v; see also Remark 1.8.6 below.

(2) For every ideal I of R, one defines a subset M_I of V as follows:
$$M_I := \{v \in V : av = 0_V \text{ for all } a \in I\}.$$
Now let $u, w \in M_I$; $r, s \in R$; and $a \in I$. Then
$$a \cdot (ru + sw) = (ar)u + (as)w = r(au) + s(aw) = r \cdot 0_V + s \cdot 0_V = 0_V,$$
and hence $ru + sw \in M_I$. By Exercise 1.8.15, M_I is a submodule of V, namely the submodule **annihilated** by I. In the special case of a principal ideal $I = Ra$, we also write M_a instead of M_I; thus $M_a = \{v \in V : av = 0_V\}$.

(3) Assume that R is actually a domain, and let
$$T(V) := \{v \in V : av = 0_V \text{ for some } a \in R^*\}.$$

We now use Exercise 1.8.15 to show that $T(V)$ is a submodule of V. Thus let $r,s \in R$ and $u,w \in T(V)$, say $au = bw = 0_V$ with $a,b \in R^*$. Then

$$(ab)\cdot(ru+sw) = (br)(au)+(as)(bw) = 0_V.$$

As R is a domain, $ab \neq 0$ and therefore $ru+sw \in T(V)$, as required. One calls $T(V)$ the **torsion submodule** of V, and V is said to be a **torsion module** if $T(V) = V$. □

The observations on ideals in Remark 1.4.18 extend to general submodules as follows:

Remark 1.8.6. Consider a module V over a commutative ring R, and let $\mathscr{U}$ be a collection of R-submodules of V. Then $\bigcap_{U \in \mathscr{U}} U$ is likewise an R-submodule of V. Given any subset X of V, let

$$\mathrm{mdl}(X) := \bigcap_{X \subseteq U \subseteq V} U, \tag{1.18}$$

where the intersection is performed over all submodules U of V with $X \subseteq U$. Then $\mathrm{mdl}(X)$ is said to be the submodule of V **generated** by X. Note that $\mathrm{mdl}(X)$ is the smallest submodule of V containing X with respect to set inclusion.[20]

As for ideals, it is routine to show that $v \in \mathrm{mdl}(X)$ holds if and only if there exist an $\ell \in \mathbb{N}$, elements $x_1,\ldots,x_\ell \in X$, and ring elements $r_1,\ldots,r_\ell \in R$ such that $v = r_1x_1+\cdots+r_\ell x_\ell$. In this case, one also says that v is a **linear combination** of (a finite number of) elements in X. In particular, if $X = \{x_1,\ldots,x_k\}$ is itself finite,

$$\mathrm{mdl}(x_1,\ldots,x_k) := \mathrm{mdl}(X) = Rx_1+\cdots+Rx_k.$$

The special case $k = 1$ gives the R-submodules of V which are generated by a single element; see Example 1.8.5. □

Definition 1.8.7. A submodule U of a module V over a commutative ring R is said to be **finitely generated** provided that $U = \mathrm{mdl}(X)$ for some finite subset X of V. In particular, in the case where $U = \mathrm{mdl}(u)$ is generated by a single element, one speaks of a **cyclic submodule** and calls u a **generator** of U.

In the special case where V is itself generated by a single element, one calls V a **cyclic** R-module. Finally, V is called a **Noetherian** R-module if every submodule of V is finitely generated. □

Definition 1.8.8. Assume that $(V,+,0_V)$ and $(W,+,0_W)$ are modules over a commutative ring $(R,+,\cdot,0,1)$.

---

[20] In contrast to the special case of ideals, there seems to be no standard notation for the submodule generated by a set $X \subseteq V$. It would be tempting to extend the bracket notation used for ideals to the more general situation, but we prefer to introduce the more elaborate notation $\mathrm{mdl}(X)$ for the sake of clarity.

- A group homomorphism $\psi: V \to W$ of the underlying additive groups of V and W is said to be an R-**module homomorphism** provided that $\psi(rv) = r\psi(v)$ holds for all $r \in R$ and all $v \in V$.
- The kernel of an R-module homomorphism $\psi: V \to W$ is the kernel of ψ as a group homomorphism, that is, $\ker \psi = \{v \in V : \psi(v) = 0_W\}$. □

The results summarized in the following theorem are in complete analogy to the corresponding results for monoids, groups, and rings. Therefore we leave a formal proof as Exercise 1.8.16 to the reader.

Theorem 1.8.9. *Let $(V,+,0_V)$ and $(W,+,0_W)$ be modules over the commutative ring $(R,+,\cdot,0,1)$. Furthermore, let $\psi: V \to W$ be an R-module homomorphism and U an R-submodule of V. Then:*

(1) *The kernel of ψ is an R-submodule of V.*

(2) *The image of ψ is an R-submodule of W.*

(3) *The factor group $(V/U,+,U)$ is an R-module V/U with respect to the scalar multiplication $R \times V/U \to V/U$ defined by*

$$(r, v+U) \mapsto rv+U.$$

This module is called the ***factor module*** *of V modulo U.*

(4) *The natural epimorphism ν_U of $(V,+,0_V)$ onto $(V/U,+,U)$ is an epimorphism of R-modules.*

(5) *There exists a unique R-module monomorphism $\sigma: V/\ker\psi \to W$ such that $\psi = \sigma \circ \nu_{\ker\psi}$. If ψ is surjective, then σ is an isomorphism.* ***(Homomorphism theorem)***

(6) *There is a bijection between the submodules of V containing U and the submodules of V/U which associates any submodule Y of V containing U with the submodule Y/U of V/U. Moreover, $(V/U)/(Y/U) \cong V/Y$, and there exists a unique isomorphism $\sigma: V/Y \to (V/U)/(Y/U)$ satisfying $\sigma(v+Y) = (v+U)+(Y/U)$ for all $v \in V$.* ***(First isomorphism theorem)***

(7) *If L is a submodule of V, then $U+L$ is a submodule of V containing L, and $L \cap U$ is a submodule of U. Moreover, there exists a unique R-module isomorphism $\sigma: U/(U \cap L) \to (U+L)/L$ satisfying $\sigma(u+(U \cap L))) = v+L$ for all $u \in U$.* ***(Second isomorphism theorem)*** □

Remark 1.8.10. Let us consider a fundamental class of examples for R-module homomorphisms, in analogy to corresponding facts for groups noted in Remark 1.1.12 and Section 1.6. For this, let R be a commutative ring and consider R as an R-module, as in Example 1.8.2. Let $(V,+,0_V)$ be any further R-module. Given any element $v \in V$, define a mapping

$$\Psi_v : R \to V, \quad r \mapsto rv. \tag{1.19}$$

Then Ψ_v is an R-module homomorphism, by the definition of the scalar multiplication, and the image of Ψ_v is the (cyclic) submodule of V generated by v. Moreover, the kernel $\{r \in R : rv = 0\}$ of Ψ_v is an R-submodule of R and hence an ideal of R; see Example 1.8.4. □

Definition 1.8.11. In the situation of Remark 1.8.10, the kernel of Ψ_v is called the **order ideal** of v; we will denote this ideal by $\mathscr{O}(v)$. Moreover, v is called a **torsion element** of V if its order ideal is nontrivial, that is, if $\mathscr{O}(v) \neq \{0\}$. □

Remark 1.8.12. Given any subset X of V, the **annihilator ideal** of X is defined as $\mathscr{A}(X) := \cap_{v \in X} \mathscr{O}(v)$. In particular, the annihilator ideal of V is the set $\mathscr{A}(V)$ of all $r \in R$ such that $rv = 0$ for all $v \in V$; note that V is a torsion module when this ideal is non-trivial. The converse also holds, provided that R is a domain and that V is finitely generated (but not in general); see Exercises 1.8.19 and 1.8.20. □

We will consider order ideals and annihilators of modules over principal ideal domains in detail in the next section. Let us finish the present section with a few remarks on (direct) sums of R-modules; this extends the corresponding results for groups given in Exercises 1.3.20 and 1.3.21 to modules.

Remark 1.8.13. Let $(V, +, 0_V)$ be a module over a commutative ring $(R, +, \cdot, 0, 1)$, and let $\left(U_j\right)_{j \in J}$ be a family of R-submodules of V, where J is an arbitrary non-empty index set. Then the **sum** $\sum_{j \in J} U_j$ is the submodule of V generated by all these submodules:

$$\sum_{j \in J} U_j := \operatorname{mdl}\Big(\bigcup_{j \in J} U_j\Big).$$

Thus $y \in \sum_{j \in J} U_j$ if and only if there exist a finite subset K of J and elements $u_k \in U_k$ for all $k \in K$ such that $y = \sum_{k \in K} u_k$.

In particular, consider the case of just two submodules U and W of V. Then $U + W = \{u + w : u \in U, w \in W\}$ is the submodule defined on the subgroup $U + W$ of $(V, +)$ by restricting the scalar multiplication on $R \times V$ to $R \times (U + W)$. As in the case of groups, we speak of a **direct sum** and write $U \oplus W$ instead of $U + W$ if $U \cap W = \{0_V\}$ holds.

Direct sums are generalized to arbitrary families of submodules as follows. We call $S = \sum_{j \in J} U_j$ the **direct sum** of the submodules U_j provided that $S_i \cap U_i = \{0_V\}$ for every $i \in J$, where $S_i := \sum_{j \in J \setminus \{i\}} U_j$. In this case, one uses the notation $\oplus_{j \in J} U_j$ instead of $\sum_{j \in J} U_j$. □

Exercises

Exercise 1.8.14. Let $(V +, 0_V)$ be a module over the commutative ring $(R, +, \cdot, 0, 1)$. Prove the following properties:

(1) $r \cdot 0_V = 0_V$ for all $r \in R$.

(2) $0 \cdot v = 0_V$ for all $v \in V$.

(3) $r(-v) = (-r)v = -(rv)$ for all $r \in R$ and all $v \in V$. □

Exercise 1.8.15. Let V be a module over a ring R. Show that a non-empty subset U of V is a submodule if and only if $ru + sw \in U$ whenever $u, w \in U$ and $r, s \in R$. □

Exercise 1.8.16. Prove Theorem 1.8.9. □

Exercise 1.8.17. Let V be a module over a commutative ring R, and let $(U_j)_{j \in J}$ be a family of R-submodules of V. Let $S = \sum_{j \in J} U_j$. Prove that the following assertions are equivalent:

(1) $S = \bigoplus_{j \in J} U_j$.

(2) For every $y \in S$, there exist a unique finite subset K of J and unique elements $u_k \in U_k$ (for $k \in K$) such that $y = \sum_{k \in K} u_k$. □

Exercise 1.8.18. Extend the definition of external direct sums of groups in Exercises 1.3.20 and 1.3.21 to modules. □

Exercise 1.8.19. Let V be a finitely generated torsion module over a commutative domain R. Prove that the annihilator ideal $\mathscr{A}(V)$ of V is non-trivial.

Hint: Let V be generated by $x_1, \ldots, x_\ell \in V$, and choose elements $r_1, \ldots, r_\ell \in R^*$ such that $r_i x_i = 0$ for $i = 1, \ldots, \ell$. Show that the product $r_1 \cdots r_\ell$ is in $\mathscr{A}(V)$. □

Exercise 1.8.20. Let p be a prime, and consider the cyclic groups $\mathbb{Z}_{p^k}$ as $\mathbb{Z}$-modules (for all $k = 1, 2, \ldots$). Show that the direct sum of these (infinitely many) $\mathbb{Z}$-modules is a torsion module which has a trivial annihilator ideal. □

1.9 Torsion and Cyclic Modules over Principal Ideal Domains

In this section, we will generalize the results on cyclic groups obtained in Section 1.6 to cyclic modules over a principal ideal domain.

Let us start with a module V over an arbitrary commutative ring $(R, +, \cdot, 0, 1)$. To simplify the notation, we will henceforth denote the zero elements of both the underlying ring R and the module V by the same symbol 0; it will always be clear from the context which element is intended.

Given elements $a \in R$ and $v \in V$, we have defined sets Ra and Rv, namely the principal ideal in R generated by a and the submodule of V generated by v. For the sake of clarity, we shall write aR instead of Ra from now on (which is permissible, as R is assumed to be commutative). In part (2) of Example 1.8.5, we have introduced the submodule $M_a = \{v \in V : av = 0\}$ annihilated by the principal ideal aR. We now consider the connections between this type of submodules and the divisor relation in R, beginning with a simple general observation:

Lemma 1.9.1. *Let V be a module over a commutative ring R, let $a, b \in R$, and assume $a \mid b$. Then the corresponding submodules M_a and M_b of V satisfy $M_a \subseteq M_b$.*

Proof. By hypothesis, $a = rb$ for some $r \in R$. Given any element $x \in M_a$, we have $bx = r(ax) = r \cdot 0 = 0$, hence $x \in M_b$. □

Under the assumption that R is a principal ideal domain, much more can be said.

Proposition 1.9.2. *Let V be a module over a principal ideal domain R, and let $a, b \in R$. Then the corresponding submodules M_a and M_b of V satisfy the following identities:*

(1) $M_a \cap M_b = M_d$, *where d is any greatest common divisor of a and b.*

(2) $M_a + M_b = M_e$, *where e is any least common multiple of a and b.*

(3) $M_{ab} = M_a \oplus M_b$ *provided that a and b are relatively prime.*

Proof. (1) By Lemma 1.9.1, $M_d \subseteq M_a \cap M_b$ even holds for arbitrary common divisors d of a and b. By hypothesis, actually $d \in \mathrm{GCD}(a,b)$, so that we have $aR + bR = dR$ by Corollary 1.5.4. Thus there are elements $\alpha, \beta \in R$ such that $\alpha a + \beta b = d$. Now let $v \in M_a \cap M_b$. Then $dv = \alpha(av) + \beta(bv) = 0$, and hence also $M_a \cap M_b \subseteq M_d$.

(2) By Lemma 1.9.1, $M_a + M_b \subseteq M_e$ even holds for arbitrary common multiples e of a and b. By hypothesis, actually $e \in \mathrm{LCM}(a,b)$, so that $aR \cap bR = eR$ by Corollary 1.5.4. More precisely, Lemma 1.2.18 gives $e = ab/d$, where d is a greatest common divisor of a and b. Put $a' := a/d$ and $b' := b/d$, so that $e = ab' = a'b$. Here a' and b' are relatively prime, and hence there exist $\alpha, \beta \in R$ such that $\alpha a' + \beta b' = 1$. Now let $v \in M_e$. Then

$$v = 1 \cdot v = \alpha(a'v) + \beta(b'v) \in M_a + M_b,$$

as $a'v \in M_b$ and $b'v \in M_a$ because of $ba'v = ab'v = ev = 0$. Hence we also have $M_a + M_b \subseteq M_e$.

(3) Let a and b be relatively prime. Then ab is a least common multiple of a and b, and there exist elements $\alpha, \beta \in R$ with $\alpha a + \beta b = 1$. Hence part (2) gives $M_a + M_b = M_{ab}$, and this is indeed a direct sum: for $u \in M_a \cap M_b$, we obtain $u = 1 \cdot u = \alpha(au) + \beta(bu) = 0$, so that $M_a \cap M_b = \{0\}$. □

The situation in part (3) of Proposition 1.9.2 admits a simple but important generalization. We first need a definition:

Definition 1.9.3. Let R be a principal ideal domain, and let $a \in R$. A finite subset $\Delta = \{d_1, \ldots, d_\ell\}$ of R is called a **decomposition** of a provided that $a = \prod_{i=1}^{\ell} d_i$ and that d_i and d_j are relatively prime for all $i \neq j$. □

Corollary 1.9.4. *Let V be a module over a principal ideal domain R. Let $a \in R$, and let $\Delta \subseteq R$ be a decomposition of a. Then $M_a = \bigoplus_{d \in \Delta} M_d$.* □

As an important special case of Corollary 1.9.4, we obtain the **primary decomposition** of a module with non-trivial annihilator ideal:

Theorem 1.9.5. *Let V be a torsion module over a principal ideal domain R, and assume that the annihilator ideal of V is non-trivial, so that $\mathscr{A}(V) = \varepsilon R$ for some element $\varepsilon \in R^*$. Then*

$$V = \bigoplus_{i=1}^{\ell} M_{p_i^{n_i}},$$

where $\varepsilon = \prod_{i=1}^{\ell} p_i^{n_i}$ is the prime power factorization of ε. □

Remark 1.9.6. Theorem 1.9.5 generalizes to modules V over an arbitrary factorial domain R as follows. Given any prime element p of R, the union

$$M_{p^\infty} := \bigcup_{n\in\mathbb{N}} M_{p^n} \tag{1.20}$$

is a submodule of V, which is called the **primary module** of V corresponding to p. Let q be any further prime which is not an associate of p. Then $M_{p^\infty} \cap M_{q^\infty} = \{0\}$, so that the sum $M_{p^\infty} + M_{q^\infty}$ is a direct sum.

Now choose a system of representatives $P^+(R)$ for the set of primes of R (that is, a maximal set of pairwise non-associated primes in R). Then $\bigoplus_{p\in P^+(R)} M_{p^\infty}$ is the torsion submodule of V. In particular, if V is a torsion module, we obtain the **primary decomposition** $V = \bigoplus_{p\in P^+(R)} M_{p^\infty}$ of V.

In view of Exercises 1.8.19 and 1.8.20, this situation indeed generalizes the one considered in Theorem 1.9.5. Let us clarify this connection by assuming once again that $\mathscr{A}(V) = \varepsilon R$ for some non-zero ε. In this case, $M_{p^\infty} = \{0\}$ whenever $p \in P^+(R)$ does not divide ε; and $M_{p^\infty} = M_{p^n}$ if p divides ε and p^n is the highest power of p dividing ε, that is, if n is the multiplicity of p in the prime power factorization of ε. Thus we indeed recover the decomposition in Theorem 1.9.5. □

Our next definition concerns the order ideals introduced in Definition 1.8.11 for the special case of modules over a principal ideal domain.

Definition 1.9.7. Let V be a module over a principal ideal domain R, and let $v \in V$. Then any generator λ of the order ideal $\mathscr{O}(v)$ of v is called an ***R*-order** of v. We will use the notation $\mathrm{Ord}_R(v)$ or simply $\mathrm{Ord}(v)$ for λ.[21] □

The following basic facts on the orders of module elements generalize corresponding results for the special case of abelian groups.

Lemma 1.9.8. *Let V be a module over a principal ideal domain R, let $v \in V$ be a torsion element, and let r be any element of R. Then $\mathrm{Ord}_R(rv) = \frac{\lambda}{d}$, where $\mathrm{Ord}_R(v) = \lambda \neq 0$ and where d is a greatest common divisor of r and λ.*

Proof. Let $r = \alpha d$ and $\delta = \frac{\lambda}{d}$. Then $\delta r v = \alpha d \delta v = \alpha\lambda v = 0$, and therefore $\delta \in \mathscr{O}(rv)$, hence $\delta R \subseteq \mathscr{O}(rv)$.

Conversely, let $s \in \mathscr{O}(rv)$. Then $(sr)v = 0$, and therefore $\lambda = \delta d$ divides $sr = s\alpha d$. Hence $\delta \mid s\alpha$, and since δ and α are relatively prime, we obtain $\delta \mid s$. Thus $s \in \delta R$, proving $\mathscr{O}(rv) \subseteq \delta R$. □

[21] This makes sense as λ is uniquely determined up to association.

Lemma 1.9.9. *Let x and y be two torsion elements in a module V over a principal ideal domain R, and assume that $\mathrm{Ord}_R(x)$ and $\mathrm{Ord}_R(y)$ are relatively prime. Then*

$$\mathrm{Ord}_R(x+y) = \mathrm{Ord}_R(x)\cdot\mathrm{Ord}_R(y).$$

Proof. Let $\mathrm{Ord}_R(x) = a$ and $\mathrm{Ord}_R(y) = b$, that is, $\mathscr{O}(x) = aR$ and $\mathscr{O}(y) = bR$ (where $a,b \neq 0$), and write $\mathscr{O}(x+y) = tR$. Since $ab(x+y) = abx + aby = 0$, we have $ab \in \mathscr{O}(x+y)$, so that $t \mid ab$.

Conversely, $t(x+y) = 0$ shows that $tx = -ty =: z$ belongs to the submodule $Rx \cap Ry$ of $M_a \cap M_b$. As a and b are assumed to be relatively prime, $M_a \cap M_b = \{0\}$ by Proposition 1.9.2. Thus $z = 0$ and we conclude $tx = 0 = ty$, which yields $a \mid t$ and $b \mid t$. Hence t is divisible by the least common multiple ab of a and b.

Altogether, we have proved that t and ab are associates, so that $\mathscr{O}(x+y) = abR$, as claimed. □

Theorem 1.9.10. *Let V be a torsion module over a principal ideal domain R, and let $\mathscr{A}(V) = \varepsilon R$ where $\varepsilon \neq 0$. Then:*

(1) *There exists an element $x \in V$ with R-order ε, that is, $\mathscr{O}(x) = \mathscr{A}(V)$.*

(2) *For each divisor λ of ε, there exists an element $x_\lambda \in V$ with R-order λ.*

(3) *For each divisor λ of ε, one has $\mathscr{A}(M_\lambda) = \lambda R$.*

(4) *If λ and ρ are divisors of ε, then $M_\lambda = M_\rho$ holds if and only if λ and ρ are associates.*

Proof. We may assume that ε is not a unit in R, for otherwise we have the trivial case $V = \{0\}$.

Let p^n be the largest power of p dividing ε, where p is any prime divisor of ε, and consider the primary component M_{p^n} of V; see Remark 1.9.6. As M_{p^n} is annihilated by p^n, the order ideal of each $x \in M_{p^n}$ has the form $p^{i(x)}R$ with $0 \leq i(x) \leq n$. Suppose that $i(x) < n$ for all $x \in M_{p^n}$. Then $M_{p^n} = M_{p^{n-1}}$ and, with $\lambda = \varepsilon/p^n$,

$$V = M_{p^n} \oplus M_\lambda = M_{p^{n-1}} \oplus M_\lambda,$$

as λ is not divisible by p. Note that each $z \in V$ has a unique representation as $z = x + y$ with $x \in M_{p^n}$ and $y \in M_\lambda$. Moreover,

$$p^{n-1}\lambda z = p^{n-1}\lambda x + p^{n-1}\lambda y = 0.$$

Therefore V is annihilated by $p^{n-1}\lambda$, and hence $\varepsilon = p^n\lambda$ divides $p^{n-1}\lambda$. As p is not a unit, this is a contradiction, which proves the existence of an element $x \in M_{p^n}$ with R-order p^n.

Now let $\prod_{i=1}^{\ell} p_i^{n_i}$ be the prime power factorization of ε. As we have just seen, we may choose an element $x_i \in M_{p_i^{n_i}}$ with $\mathrm{Ord}_R(x_i) = p_i^{n_i}$ for $i = 1, \ldots, \ell$. Then Lemma 1.9.9 and induction show that $x := \sum_{i=1}^{\ell} x_i$ has R-order ε. This proves (1).

Next let λ be any divisor of ε, and put $\delta = \varepsilon/\lambda$. By part (1), we may choose an element $x \in V$ with R-order ε. Then Lemma 1.9.8 gives $\mathrm{Ord}_R(\delta x) = \lambda R$. This proves (2), taking $x_\lambda = \delta x$.

Now assume that the annihilator ideal of M_λ is αR. Then $\alpha \mid \lambda$, by the definition of M_λ. As the element x_λ constructed in (2) satisfies $\alpha x_\lambda = 0$, we conclude that $\lambda \approx \alpha$, and therefore $\mathscr{A}(M_\lambda) = \lambda R$. This proves (3), and then (4) is an immediate consequence. □

We now use the preceding results to characterize all elements of V having a given R-order, which then leads to a generalization of Euler's totient function if V is finite.

Remark 1.9.11. Let V be a torsion module over a principal ideal domain R with non-trivial annihilator ideal $\mathscr{A}(V) = \varepsilon R$, as in Theorems 1.9.5 and 1.9.10. With the same notation as before, any $v \in V$ has a unique decomposition as a sum $v_1 + \cdots + v_\ell$, where v_i is an element of the primary component $M_{p_i^{n_i}}$ for $i = 1, \ldots, \ell$.

Let λ be any divisor of ε, say $\lambda = \prod_{i=1}^{\ell} p_i^{m_i}$, with $0 \le m_i \le n_i$ for all i. By Lemma 1.9.9, v has R-order λ if and only if $\mathrm{Ord}_R(v_i) = p_i^{m_i}$ for all $i = 1, \ldots, \ell$, which holds if and only if $v_i \in M_{p_i^{m_i}} \setminus M_{p_i^{m_i-1}}$ when $m_i > 0$ and $v_i = 0$ when $m_i = 0$. □

Remark 1.9.12. Assume that V is finite in Remark 1.9.11 and define a function ϕ_V on the set of all divisors λ of ε such that $\phi_V(\lambda)$ is the number of elements in V with R-order λ. The decomposition of V in Theorem 1.9.5 shows that ϕ_V is **multiplicative**: writing λ as in Remark 1.9.11, we have

$$\phi_V(\lambda) = \prod_{i=1}^{\ell} \phi_V(p_i^{m_i}),$$

where $\phi_V(p_i^{m_i}) = \phi_V(1) = 1$ when $m_i = 0$ and $\phi_V(p_i^{m_i}) = \left|M_{p_i^{m_i}}\right| - \left|M_{p_i^{m_i-1}}\right|$ when $m_i > 0$. □

Remark 1.9.13. As noted in Example 1.8.2, every abelian group G is a $\mathbb{Z}$-module. We now assume that G is a finite and discuss the preceding results for this special case. Note first that $\mathscr{A}(G) = \varepsilon\mathbb{Z}$, where ε is the exponent of G introduced in Definition 1.6.20, so that Theorem 1.6.21 is contained in Theorem 1.9.10. Now let k be any divisor of ε. Then $\phi_G(k)$ is just the number of elements in G which have order k. By Theorem 1.9.5, every finite abelian group is (in additive notation) a direct sum of primary abelian groups, that is, of abelian groups of prime power order. In the language of Group Theory, these primary groups are called the **Sylow subgroups** of G.

Assume now that G is actually a finite cyclic group, without loss of generality the additive group of a residue ring $\mathbb{Z}_n$. Then $\varepsilon = n$ and ϕ_G is just the Euler totient function ϕ, restricted to the set of divisors of n. In particular, we conclude from Remark 1.9.12 that ϕ is multiplicative: if $n = \prod_{i=1}^{\ell} p_i^{a_i}$ is the prime power factorization of n, one has

$$\phi(n) = \prod_{i=1}^{\ell} \phi(p_i^{a_i}) = \prod_{i=1}^{\ell} (p-1)p^{a_i-1}, \tag{1.21}$$

where we have used Exercise 1.6.33 for the second equality. This allows an explicit evaluation of the Euler phi function, provided that we can determine the prime power factorization of the number n in question. □

For the rest of this section, we concentrate on the important case of cyclic (torsion) modules, as these are of particular interest for the theory of finite fields. We begin with a basic observation:

Proposition 1.9.14. *Let V be a cyclic module over a principal ideal domain R. Then all submodules and all factor modules of V are likewise cyclic.*

Proof. By hypothesis, $V = Rv$ for some $v \in V$. Given any submodule U of V, consider the ideal $I_U(v) := \{r \in R \colon rv \in U\}$ of R and let η be a generator of this ideal. As $\eta v \in U$, the cyclic module $R(\eta v)$ generated by ηv is contained in U. We now show that equality holds, which will establish the assertion for submodules. Thus let $y \in U$, say $y = rv$. Then $r \in I_U(v)$, and hence $r = \eta s$ for some $s \in R$, which gives $y = (\eta s)v = s(\eta v)$, as desired.

Since v is a generator of V, the coset $v + U$ is a generator of V/U, and therefore the factor module V/U is also cyclic. □

The following result generalizes Theorem 1.6.17:

Theorem 1.9.15. *Let V be a cyclic torsion module over a principal ideal domain R, say $\mathscr{A}(V) = \varepsilon R$ where $\varepsilon \neq 0$. Then:*

(1) *An element $x \in V$ is a generator of V if and only if $\mathscr{O}(x) = \mathscr{A}(V)$, that is, if and only if $\mathrm{Ord}_R(x) = \varepsilon$.*

(2) *The submodules of V are precisely the submodules of the form M_λ, where λ is a divisor of ε.*

(3) *For every divisor λ of ε, the submodule M_λ is cyclic. The generators of M_λ are precisely the elements of V with R-order λ.*

Proof. Let x be a generator of V, and let y be any element of V. Then there is an $r \in R$ such that $y = rx$, and the order ideal of x is contained in the order ideal of y. Therefore $\mathscr{O}(x) \subseteq \bigcap_{y \in V} \mathscr{O}(y) = \mathscr{A}(V) \subseteq \mathscr{O}(x)$, so that $\mathscr{O}(x) = \mathscr{A}(V) = \varepsilon R$.

Assume conversely that $\mathscr{O}(z) = \mathscr{A}(V)$ holds for some $z \in V$. Let x be a generator of V, and write $z = sx$ (with $s \in R$). By Lemma 1.9.8, $\varepsilon = \mathrm{Ord}_R(x)$ and s are relatively prime. Thus there are σ, u in R such that $\sigma s + \varepsilon u = 1$, which gives

$$x = 1 \cdot x = \sigma s x + u \varepsilon x = \sigma s x = \sigma z \in Rz.$$

Hence z is indeed a generator of V. This proves part (1).

Now it is easy to prove (2) and the generalization (3) of (1). By Proposition 1.9.14, every submodule U of V is cyclic. Let rx be a generator of U, where $r \in R$ and x is a generator of V, and let d be a greatest common divisor of r and ε. By part (1) and Lemma 1.9.8, $\mathscr{A}(U) = \mathscr{O}(rx)$ is generated by $\lambda := \frac{\varepsilon}{d}$, and thus U is a submodule of M_λ. On the other hand, M_λ is itself cyclic and $\mathscr{A}(M_\lambda) = \lambda R$. By part (1) (applied to the cyclic module M_λ), the generators of M_λ are precisely the elements in M_λ whose order ideal is λR. In particular, $U = R(rx) = M_\lambda$, as claimed. Finally, note that every element of V with R-order λ actually is in M_λ. □

As a consequence of Theorem 1.9.15, we can generalize the observations in Remark 1.6.6 as follows:

Proposition 1.9.16. *Let V be a cyclic torsion module over a principal ideal domain R, say $\mathscr{A}(V) = \varepsilon R$ where $\varepsilon \neq 0$. Then the generators of V correspond bijectively to the units of the factor ring $R/\varepsilon R$.*

Proof. Choose a fixed generator v of the cyclic R-module V. Applying part (5) of Theorem 1.8.9 to the mapping Ψ_v defined in (1.19) shows $V \cong R/R\varepsilon$. By Lemma 1.9.8 and part (3) of Theorem 1.9.15, a multiple rv of v generates V if and only if ε and r are relatively prime, which means that $r + \varepsilon R$ is a unit in $R/\varepsilon R$. Moreover, $sv = tv$ if and only if $(s-t)v = 0$, that is, if and only if $s \equiv t \bmod \varepsilon$. □

Remark 1.9.17. Let us discuss an alternative approach to generalizing the Euler totient function to modules, namely via counting the generators of finite cyclic R-modules. Let R be a principal ideal domain, denote the set of all $a \in R$ for which the factor ring R/aR is finite by F_R, and define ϕ_R by evaluating the number of units in the corresponding residue rings:

$$\phi_R : F_R \to \mathbb{N}, \ \ a \mapsto |U(R/aR)|.$$

Now let V be any cyclic R-module such that $\mathscr{A}(V) = aR$, with $a \in F_R$. Then V is finite and has precisely $\phi_R(a)$ generators, by Proposition 1.9.16. (Note that we may use the factor module R/aR itself as an explicit example for such a cyclic module V.) Moreover, ϕ_R and the function ϕ_V introduced in Remark 1.9.12 coincide on the set of divisors of a. In particular, this establishes the multiplicativity of ϕ_R as a function on F_R. □

The multiplicativity of functions such as ϕ_R will play an important role in the next section. We conclude the present section with a simple characterization of the cyclic modules among the finite modules.

Proposition 1.9.18. *Let V be a finite module over a principal ideal domain R. Then the residue ring $R/\mathscr{A}(V)$ is likewise finite, and V is cyclic if and only if $|V| = |R/\mathscr{A}(V)|$.*

Proof. Let $\mathscr{A}(V) = \varepsilon R$. By Theorem 1.9.10, we may choose an element $x \in V$ with $\mathscr{O}(x) = \mathscr{A}(V) = \varepsilon R$. Then $Rx \cong R/\mathscr{A}(V) = R/\varepsilon R$, as in the proof of Proposition 1.9.16. Since V is finite, so is its submodule Rx, and hence $R/\mathscr{A}(V) = |Rx|$ is finite and bounded by $|V|$. Obviously, equality is attained if and only if V is a cyclic module with generator x. □

Exercises

Exercise 1.9.19. Give a detailed proof for the results sketched in Remark 1.9.6. □

Exercise 1.9.20. Convince yourself that Lemmas 1.6.15 and 1.6.19 are special cases of Lemmas 1.9.8 and 1.9.9, respectively. □

Exercise 1.9.21. Determine all integers $n \geq 2$ for which $\phi(n) \in \{6, 12, 24, 48\}$. Which integers satisfy $\phi(n) = n - 1$?

We remark that the second question is really very simple, whereas the first one is more difficult; in particular, the case $\phi(n) = 48$ constitutes a challenge. □

1.10 The Chinese Remainder Theorem

In this final section of Chapter 1, we discuss the following problem for modules V over a commutative ring R: given submodules $U_1, \dots, U_\ell$ of V and elements $v_1, \dots, v_\ell \in V$ (where $\ell \geq 2$), can we find an element $x \in V$ satisfying

$$x \equiv v_i \bmod U_i \quad \text{for } i = 1, \dots, \ell,$$

(that is, $x - v_i \in U_i$) for all i? An answer to this question is provided by the Chinese remainder theorem.

We begin by introducing a fundamental mapping underlying this problem in the following more elaborate version of a special case of Exercise 1.8.18; we leave the simple proof as Exercise 1.10.13.

Proposition 1.10.1. *Let V be a module over a commutative ring R, let $U_1, \dots, U_\ell$ be submodules of V (where $\ell \geq 2$), and consider the cartesian product $W = \times_{i=1}^{\ell} V/U_i$ of the factor modules V/U_i (as sets). Then W becomes an R-module with respect to componentwise addition and componentwise scalar multiplication, namely the direct sum $W = \oplus_{i=1}^{\ell} V/U_i$ of the factor modules V/U_i. Moreover, the mapping*

$$\Gamma \colon V \to V/U_1 \oplus \dots \oplus V/U_\ell, \; x \mapsto (x + U_1, \dots, x + U_\ell) \tag{1.22}$$

is a module homomorphism with kernel $\bigcap_{i=1}^{\ell} U_i$. □

Thus the problem stated at the beginning of this section amounts to deciding whether or not the ℓ-tuple $(v_1 + U_1, \dots, v_\ell + U_\ell)$ belongs to the image of Γ (for a specified $(v_1, \dots, v_\ell) \in V^\ell$). The Chinese remainder theorem gives a sufficient criterion guaranteeing a positive answer for *all* choices of $(v_1, \dots, v_\ell)$, that is, it gives a condition which forces Γ to be surjective.

Let us first state a general version for modules which is not particularly illuminating; following this, we will present a much nicer criterion for the special case of rings, that is, for $V = R$.

Theorem 1.10.2. *Let V be a module over a commutative ring R, let $U_1,\dots,U_\ell$ be submodules of V (where $\ell \geq 2$), and assume*

$$U_{k+1} + \bigcap_{i=1}^{k} U_i = V \quad \textit{for } k = 1,\dots,\ell-1. \tag{1.23}$$

Then the mapping Γ defined in (1.22) *is surjective.*

Proof. We will use induction on ℓ. For the induction basis $\ell = 2$, let $U := U_1$ and $W := U_2$. The hypothesis $U + W = V$ allows us to write any two specified elements $v_1, v_2 \in V$ in the form $v_1 = u_1 + w_1$ and $v_2 = u_2 + w_2$, with $u_1, u_2 \in U$ and $w_1, w_2 \in W$. Put $x := u_2 + w_1$. Then $x - v_1 = u_2 - u_1 \in U$ and $x - v_2 = w_1 - w_2 \in W$, so that Γ maps x to $(v_1 + U, v_2 + W)$.

Now let $\ell > 2$ and assume that the assertion holds for $\ell - 1$. Let $v_1,\dots,v_\ell \in V$ be given. Then there is an element $y \in V$ such that $y + U_i = v_i + U_i$ for $i = 1,\dots,\ell-1$. Put $U := \bigcap_{i=1}^{\ell-1} U_i$ and $W := U_\ell$. By hypothesis, $V = U + W$, and thus the induction basis guarantees the existence of some $x \in V$ such that $x + U = y + U$ and $x + W = v_\ell + W$. Then U_i contains both $x - y$ and $y - v_i$ and hence also their sum $x - v_i$, for all $i = 1,\dots,\ell-1$. Finally, we also have $x - v_\ell \in U_\ell$, as $W = U_\ell$. Thus x is the desired preimage of $(v_1,\dots,v_\ell)$ under Γ. □

We now turn our attention to the special case where V is the ring R itself, so that the submodules are just the ideals of R; see Example 1.8.4. We require a few ring theoretic preliminaries before we can give the promised nicer version of Theorem 1.10.2 in this situation.

Definition 1.10.3. Let R be a commutative ring.

- Two ideals I and J of R are said to be **relatively prime** provided that $I + J = R$ (that is, if $1 \in I + J$).
- Let $I_1,\dots,I_\ell$ be ideals of R, where $\ell \geq 2$. The **product** $\prod_{i=1}^{\ell} I_i$ of the I_j is the ideal of R generated by all products $a_1 \cdots a_\ell$ with $a_j \in I_j$ for $j = 1,\dots,\ell$. □

Note that the first part of Definition 1.10.3 generalizes the concept of relatively prime elements of R (which was introduced in Definition 1.2.17 in the more general setting of monoids) to ideals: two elements a and b of R are relatively prime if and only if the principal ideals aR and bR are relatively prime.

Lemma 1.10.4. *Let $I_1,\dots,I_\ell$ be ideals of a commutative ring R. Then the product $\prod_{i=1}^{\ell} I_i$ is a subideal of the intersection $\bigcap_{i=1}^{\ell} I_i$.*

Proof. Let $a = \prod_{i=1}^{\ell} a_i$, where $a_j \in I_j$ for all j. Then $a \in Ra_j \subseteq I_j$ for all j and therefore $a \in \bigcap_{i=1}^{\ell} I_i$. By definition, these elements a generate $\prod_{i=1}^{\ell} I_i$, and thus $\prod_{i=1}^{\ell} I_i \subseteq \bigcap_{i=1}^{\ell} I_i$. □

Proposition 1.10.5. *Let J and $I_1,\dots,I_\ell$ be ideals of a commutative ring R, and assume that J and I_j are relatively prime for all $j = 1,\dots,\ell$. Then:*

(1) *The ideals J and $\prod_{i=1}^{\ell} I_i$ are relatively prime.*

(2) *The ideals J and $\bigcap_{i=1}^{\ell} I_i$ are relatively prime.*

Proof. In view of Lemma 1.10.4, it suffices to prove the first assertion. For this, we check that 1 is in $J + \prod_{i=1}^{\ell} I_i$, using induction on ℓ. The induction basis $\ell = 1$ is trivial, as J and I_1 are relatively prime by hypothesis.

Now let $\ell \geq 2$ and assume that the assertion holds for the ideals J and $I_1, \ldots, I_{\ell-1}$. Thus there are elements $u \in J$ and $x \in \prod_{i=1}^{\ell-1} I_i$ such that $u + x = 1$. As J and I_ℓ are relatively prime by hypothesis, there are also elements $v \in J$ and $y \in I_\ell$ such that $y + v = 1$. This gives

$$1 = (u+x)(v+y) = (uy+xv+uv)+xy \in J + \prod_{i=1}^{\ell} I_i. \quad \square$$

Theorem 1.10.6 (Chinese remainder theorem). *Let $I_1, \ldots, I_\ell$ be ideals of a commutative ring R (where $\ell \geq 2$), and assume that these ideals are pairwise relatively prime, that is, $I_i + I_j = R$ for all $i, j = 1, \ldots, \ell$ with $i \neq j$. Then the mapping*

$$\Gamma \colon R \to \oplus_{i=1}^{\ell} R/I_i, \quad x \mapsto (x + I_1, \ldots, x + I_\ell)$$

is a ring epimorphism with kernel $\bigcap_{i=1}^{\ell} I_i$.

Proof. This is the special case $V = R$ and $U_i = I_i$ (for $i = 1, \ldots, \ell$) of Theorem 1.10.2. Note that the hypothesis on the ideals I_j indeed translates into the required condition (1.23), since

$$I_{k+1} + \bigcap_{i=1}^{k} I_i \supseteq I_{k+1} + \prod_{i=1}^{k} I_i = R$$

for $k = 1, \ldots, \ell - 1$, because of Lemma 1.10.4 and part (1) of Proposition 1.10.5. $\square$

Theorem 1.10.6 allows an even nicer formulation when R is a principal ideal domain, using ring elements instead of ideals:

Corollary 1.10.7. *Let $b_1, \ldots, b_\ell$ be elements of a principal ideal domain R which are pairwise relatively prime, and let $(a_1, \ldots, a_\ell) \in R^\ell$ be arbitrary. Then there exists an element $x \in R$ satisfying*

$$x \equiv a_i \bmod b_i \text{ (that is, } b_i \mid x - a_i) \quad \text{for } i = 1, \ldots, \ell.$$

Moreover, x is uniquely determined modulo the product $b_1 \cdots b_\ell$.

Proof. We apply Theorem 1.10.6 to the ideals $I_1 = Rb_1, \ldots, I_\ell = Rb_\ell$ of R. As noted before, the hypothesis that the b_i are pairwise relatively prime translates into the condition $I_i + I_j = R$ (for all $i \neq j$). This gives the existence of the desired element x and it only remains to prove the uniqueness assertion. For this, we will use that the associated map $\Gamma \colon R \to \oplus_{i=1}^{\ell} R/Rb_i$ obviously has kernel $\bigcap_{i=1}^{k} Rb_i$.

Thus we need to show that the ideal $\bigcap_{i=1}^{k} Rb_i$ is generated by $b_1 \cdots b_\ell$. By Corollary 1.5.4, the generators of any ideal $Rb_i \cap Rb_j$ (with $i \neq j$) are the least common multiples of b_i and b_j, that is, the associates of b_ib_j, since b_i and b_j are assumed to be relatively prime. This establishes our claim for $\ell = 2$ and allows us to use induction on ℓ. Thus let $\ell \geq 3$. By part (2) of Proposition 1.10.5, the ideals $\bigcap_{i=1}^{\ell-1} Rb_i$ and Rb_ℓ are likewise relatively prime. By induction, we may assume that $\bigcap_{i=1}^{\ell-1} Rb_i$ is generated by $\prod_{i=1}^{\ell-1} b_i$, and then the induction basis $\ell = 2$ guarantees that $\bigcap_{i=1}^{\ell} Rb_i$ is indeed generated by $\prod_{i=1}^{\ell} b_i$. □

Let us also state the classical version of the Chinese remainder theorem, namely for the ring of integers, explicitly:

Corollary 1.10.8. *Let $n = n_1 n_2 \cdots n_\ell$, where the n_i are integers ≥ 2, and assume* $\gcd(n_i, n_j) = 1$ *whenever $i \neq j$. Then the residue ring $\mathbb{Z}_n$ is isomorphic to the direct sum of the rings $\mathbb{Z}_{n_i}$:*

$$\mathbb{Z}_n \cong \bigoplus_{i=1}^{\ell} \mathbb{Z}_{n_i}.$$

Proof. This follows by combining the homomorphism theorem for rings (see Theorem 1.4.16) with Corollary 1.10.7 and its proof. More precisely, these results show that the ring epimorphism

$$\Gamma : \mathbb{Z} \to \bigoplus_{i=1}^{\ell} \mathbb{Z}_{n_i}, \quad x \mapsto (x \bmod n_1, \ldots, x \bmod n_\ell) \tag{1.24}$$

induces the desired isomorphism. □

Remark 1.10.9. Let us have a closer look at the situation in Corollary 1.10.8. Clearly, the ring isomorphism Γ' induced by the epimorphism Γ restricts to an isomorphism

$$U(\mathbb{Z}_n) \to U\Big(\bigoplus_{i=1}^{\ell} n\mathbb{Z}_{n_i}\Big) = \bigotimes_{i=1}^{\ell} U(\mathbb{Z}_{n_i})$$

of the (multiplicative) groups of units. As we have determined the structure of the group $U(\mathbb{Z}_{p^m})$ for an arbitrary prime power p^m in Theorem 1.7.7, the preceding observation gives us the structure of $U(\mathbb{Z}_n)$ in general. Moreover, this yields a further proof for the multiplicativity of the Euler phi function:

$$\phi(n) = |U(\mathbb{Z}_n)| = \prod_{i=1}^{\ell} |U(\mathbb{Z}_{n_i})| = \prod_{i=1}^{\ell} \phi(n_i),$$

since the n_i are pairwise relatively prime by hypothesis. □

At this point, we digress to establish a (perhaps a little surprising) result on generators of finite cyclic groups which will only be needed in Chapter 13: given any such group G and any subgroup U of G, the number of generators contained in the coset of U determined by a specified generator g does not depend on the choice of g. For instance, consider $G = \mathbb{Z}_{30}$ and the subgroup U of order 6, which is generated by $5+30\mathbb{Z}$. Then the $\phi(30)=8$ generators of G determine all four non-trivial cosets of U, namely the cosets of

$$1+30\mathbb{Z},\ 7+30\mathbb{Z},\ 13+30\mathbb{Z} \text{ and } 19+30\mathbb{Z},$$

and each of these cosets contains exactly two of the generators. In general, one has the following formula which only depends on the orders of G and U:

Lemma 1.10.10. *Let G be a cyclic group of order n, let m be a divisor of n, and let $U=U_m$ be the subgroup of order m of G. Write the prime power factorizations of n and m in the form*

$$n = \prod_{i=1}^{k} r_i^{a_i} \cdot \prod_{j=1}^{\ell} s_j^{b_j} \quad \textit{and} \quad m = \prod_{i=1}^{k} r_i^{a_i} \cdot \prod_{j=1}^{\ell} s_j^{c_j},$$

where $0 \le c_j < b_j$ for all $j=1,\dots,\ell$. (Thus the r_i are those prime divisors of n which have the same multiplicity in m as in n.) Then the number of generators of G which are contained in the coset of U determined by some specified generator g of G is given by

$$\prod_{i=1}^{k} \phi(r_i^{a_i}) \cdot \prod_{j=1}^{\ell} s_j^{c_j} = \phi\Big(\prod_{i=1}^{k} r_i^{a_i}\Big) \cdot \prod_{j=1}^{\ell} s_j^{c_j}.$$

Proof. Up to isomorphism, we may assume that G is the additive group of the residue ring $\mathbb{Z}_n$; see Proposition 1.6.11. In view of Corollary 1.10.8, the generator g has a unique representation in the form

$$g = \sum_{i=1}^{k} g_i + \sum_{j=1}^{\ell} h_j \quad \text{with } \operatorname{ord}(g_i) = r_i^{a_i} \text{ and } \operatorname{ord}(h_j) = s_j^{b_j}$$

for all i,j. Similarly, every element $u \in U$ has a unique representation

$$u = \sum_{i=1}^{k} u_i + \sum_{j=1}^{\ell} v_j \quad \text{with } \operatorname{ord}(u_i) \mid r_i^{a_i} \text{ and } \operatorname{ord}(v_j) \mid s_j^{c_j}$$

for all i,j. Then

$$u+g = \sum_{i=1}^{k} (u_i+g_i) + \sum_{j=1}^{\ell} (v_j+h_j),$$

and hence $u+g$ is a generator of G if and only if $\operatorname{ord}(u_i+g_i)=r_i^{a_i}$ and $\operatorname{ord}(v_j+h_j)=s_j^{b_j}$ for all i and all j. Note that $\operatorname{ord}(v_j+h_j)=s_j^{b_j}$ holds for every choice of v_j, as $c_j<b_j$ (for all j). On the other hand, since u_i and g_i may both have order $r_i^{a_i}$, the sum u_i+g_i has order $r_i^{a_i}$ if and only if u_i+g_i is not contained in the subgroup of order $r_i^{a_i-1}$ of G, which leaves (for all i) exactly

$$r_i^{a_i} - r_i^{a_i-1} = \phi(r_i^{a_i})$$

choices for u_i for which the sum $u_i + g_i$ has order $r_i^{a_i}$. □

Remark 1.10.11. We note that the isomorphism assertion in Corollary 1.10.8 and the observations in Remark 1.10.9 extend from the ring of integers to arbitrary principal ideal domains R as follows. Let $a \in R$, and let $a = \prod_{i=1}^{\ell} d_i$ be any decomposition of a; see Definition 1.9.3. Then R/aR and $\oplus_{i=1}^{\ell} R/d_iR$ are isomorphic rings, and the Chinese remainder homomorphism Γ induces a group isomorphism

$$U(R/aR) \to U(\otimes_{i=1}^{\ell} R/d_iR) = \otimes_{i=1}^{\ell} U(R/d_iR)$$

between their respective groups of units. In particular, if a belongs to the set F_R defined in Remark 1.9.17, then all the d_i are likewise in F_R. Thus

$$\phi_R(a) = |U(R/aR)| = \prod_{i=1}^{\ell} |U(R/d_iR)| = \prod_{i=1}^{\ell} \phi_R(d_i),$$

which yields a further proof for the multiplicativity of the corresponding Euler function for R. □

Remark 1.10.12. Let R and $a = \prod_{i=1}^{\ell} d_i$ be as in Remark 1.10.11. By Theorem 1.9.5, any cyclic R-module V with annihilator ideal $\mathscr{A}(V) = aR$ has a decomposition

$$V = M_{d_1} \oplus \cdots \oplus M_{d_\ell},$$

and V and R/aR are isomorphic as R-modules. Moreover, M_{d_i} is a cyclic module annihilated by d_iR, and hence $M_{d_i} \cong R/d_iR$ for $i = 1, \ldots, \ell$. Thus the decompositon of V corresponds to the Chinese remainder theorem for R/aR. In particular, the corresponding decomposition of generators of V is mirrored by the corresponding decomposition of the unit group of R/aR; see Remark 1.9.11. □

Exercises

Exercise 1.10.13. Give a detailed proof for Proposition 1.10.1. □

Exercise 1.10.14. Let G and U be as in Lemma 1.10.10.

(1) Investigate the distribution of generators of G over cosets of U for the subgroup U of order 6 of $G = \mathbb{Z}_{60}$.

(2) Give a general formula for the number of cosets of U which contain a generator of G.

(3) Prove that *all* non-trivial cosets of U contain generators of G if and only if the index $[G : U]$ is a prime. □

Exercise 1.10.15. Provide formal proofs for the results sketched in Remarks 1.10.11 and 1.10.12 . □

Exercise 1.10.16. We return to the situation already considered in Exercise 1.6.28; thus let G be the group of units in $\mathbb{Z}_{496125}$. Use Corollary 1.10.8 together with Theorems 1.6.21 and 1.7.7 to show that 2 is an element with maximal order in G. □

Chapter 2
Basics on Polynomials

Abstract This chapter presents the fundamental properties of polynomials and, more generally, formal power series. Besides standard topics such as the evaluation of polynomials, roots, formal derivatives, interpolation and the like, we also consider two more advanced topics which will be important for later chapters, namely Möbius inversion and endomorphisms of vector spaces and their minimal polynomials.

2.1 Formal Power Series and Möbius Inversion

If one wants to define polynomials in a precise manner, one needs to introduce them as a special type of formal power series. Therefore, we first consider these more general objects in the present section; the basic ingredients are a commutative ring $(R,+,\cdot,0,1)$ and a suitable partially ordered set. We begin by recalling a few well-known facts concerning the set R^X of all mappings from X to R, where X is an arbitrary nonempty set (independent of any ordering). With respect to the **pointwise addition** defined by

$$(f+g)(x) := f(x)+g(x) \quad \text{for all } x \in X \text{ and all } f,g \in R^X, \tag{2.1}$$

R^X becomes an abelian group. Here the zero element is the **zero mapping** with $x \mapsto 0$ for all $x \in R$, and the additive inverse of $f \in R^X$ is the mapping $-f$ defined by $(-f)(x) := -f(x)$.

One also defines a **scalar multiplication** by

$$rf : X \to R,\ x \mapsto rf(x) \quad \text{for all } r \in R \text{ and } f \in R^X, \tag{2.2}$$

which turns R^X into an R-module. Of course, R^X can also be made into a commutative ring in a natural way, with respect to the **pointwise multiplication** given by

D. Hachenberger and D. Jungnickel, *Topics in Galois Fields*,
Algorithms and Computation in Mathematics 29,
https://doi.org/10.1007/978-3-030-60806-4_2

$$(f \cdot g)(x) := f(x) \cdot g(x) \quad \text{for all } x \in X \text{ and all } f, g \in R^X. \tag{2.3}$$

Moreover, the pointwise multiplication and the scalar multiplication are compatible in the sense that

$$r(f \cdot g) = (rf) \cdot g = f \cdot (rg) \quad \text{for all } r \in R \text{ and all } f, g \in R^X. \tag{2.4}$$

When the underlying ring is actually a field F (so that F^X is a vector space), one speaks of an F-**algebra**.

In order to define formal power series, one does not use the pointwise multiplication, but another type of multiplication, the so-called **convolution**. For this, the set X has to be equipped with a suitable partial order. We change our notation accordingly and also recall some facts from Section 1.2.

Definition 2.1.1. Let $(N, \cdot, 1_N)$ be a (commutative) monoid with cancellation. We say that N is **simple**, provided that 1_N is the only unit in N. □

Recall from Section 1.2 that the divisibility relation $|$ gives a partial order on any simple monoid N with cancellation. For example, N might arise as the factor monoid $\overline{M} = M/U(M)$ of some arbitrary commutative monoid $(M, \cdot, 1_M)$ with cancellation, where $U(M)$ is the group of units of M. Moreover, if $(M, |)$ is locally finite (see Definition 1.2.8), then $(N, |)$ is again locally finite, and, if $(M, \cdot, 1_M)$ is factorial, then $(N, \cdot, 1_N)$ is likewise factorial. By Theorem 1.2.14, the property of being factorial implies the local finiteness.

Definition 2.1.2. Let $(N, \cdot, 1_N)$ be a locally finite simple commutative monoid with cancellation, and let $(R, +, \cdot, 0, 1)$ be a commutative ring. The **convolution** $f \star g$ of $f, g \in R^N$ is defined by

$$(f \star g)(m) := \sum_{d|m} f(d) g\left(\frac{m}{d}\right), \tag{2.5}$$

where the sum is taken over all divisors d of m.[1] □

Theorem 2.1.3. *Let R be a commutative ring and $(N, \cdot, 1_N)$ a locally finite simple commutative monoid with cancellation. Then the set R^N of all mappings from N to R carries the structure of a commutative ring with respect to pointwise addition and convolution. This ring is denoted by $R[[N]]$.*

Proof. As noted before, R^N is an abelian group with respect to pointwise addition. We next show that the convolution $\star$ gives rise to a commutative monoid on R^N. The commutativity and associativity are trivial to check:

$$(f \star g)(m) = \sum_{\substack{x,y \in N \\ xy = m}} f(x) g(y) = (g \star f)(m)$$

[1] Note that $\{d \in N : d|m\}$ is finite for every $m \in N$ by hypothesis. If one uses additive notation for the monoid N, the convolution takes the form $(f \star g)(m) := \sum_{d|m} f(d) g(m-d)$.

and

$$(f\star(g\star h))(m)=\sum_{\substack{x,y,z\in N\\ xyz=m}} f(x)g(y)h(z)=((f\star g)\star h)(m)$$

for all $f,g,h\in R^N$ and all $m\in N$. Finally, the element $\varepsilon\in R^N$ defined by

$$\varepsilon(m)=\begin{cases}1 & \text{if } m=1_N,\\ 0 & \text{otherwise}\end{cases}$$

is the identity element with respect to $\star$, and thus $(R^N,\star,\varepsilon)$ is indeed a commutative monoid. It only remains to check the distributive law:

$$\begin{aligned}(f\star(g+h))(m) &= \sum_{\substack{x,y\in N\\ xy=m}} f(x)(g(y)+h(y))\\ &= \sum_{\substack{x,y\in N\\ xy=m}} f(x)g(y) + \sum_{\substack{x,y\in N\\ xy=m}} f(x)h(y)\\ &= ((f\star g)+(f\star h))(m)\end{aligned}$$

for all $f,g,h\in R^N$ and all $m\in N$. □

Definition 2.1.4. With R and N as in Theorem 2.1.3, the ring $R[[N]]$ is called the **ring of formal power series** over R and N, and any element of $R[[N]]$ is called a **formal power series**.

The most important case occurs when the underlying simple monoid is $(\mathbb{N},+,0)$. Then it is usual to use the notation $R[[x]]$ and to call this the **ring of formal power series** over R in the **indeterminate** x and any $f\in R[[x]]$ a **formal power series** in x over R. □

We postpone a discussion of the meaning of the symbol x in $R[[x]]$ to the next section and continue by describing the unit group of $R[[N]]$ for the case where N is factorial; in particular, N is a locally finite commutative monoid with cancellation (see Section 1.2).

Theorem 2.1.5. *Let $(R,+,\cdot,0,1)$ be a commutative ring and $(N,\cdot,1_N)$ a simple factorial monoid. Then $f\in R^N$ is a unit in $R[[N]]$ if and only if $f(1_N)$ is a unit in R.*

Proof. Assume first that $f\in U(R[[N]])$ and let g be the inverse of f, that is, $f\star g=\varepsilon$. As 1_N is divisible only by itself, this gives $1=\varepsilon(1_N)=(f\star g)(1_N)=f(1_N)g(1_N)$. Therefore $f(1_N)$ is a unit in R, with inverse $g(1_N)$.

Conversely, assume that $u:=f(1_N)$ is a unit in R. As in the proof of Theorem 1.2.14, let $P_0:=U(N)=\{1_N\}$, and for $n\in\mathbb{N}^*$ let P_n be the set of all $a\in N$ such that $a=p_1\cdot\ldots\cdot p_\ell$ for some $\ell\in\mathbb{N}^*$ with $1\le\ell\le n$, where $p_1,\ldots,p_\ell\in N$ are primes.

We now define a mapping $g\in R^N$ recursively (via induction on n) as follows. First, let $g(1_N)=u^{-1}$, so that g is defined on P_0. By definition, the set P_1 consists of all prime elements of N. For $p\in P_1$, we require

$$0 = \varepsilon(p) = (f \star g)(p) = f(1_N)g(p) + f(p)g(1_N),$$

which holds for

$$g(p) := -\frac{f(p)}{u^2}.$$

For the general case, let $n \in \mathbb{N}^*$ and $m \in P_n \setminus P_{n-1}$, and assume inductively that g is already defined on P_{n-1} and that $(f \star g)(d) = \varepsilon(d)$ holds for all $d \in P_{n-1}$. As N is simple, we have $d \in P_{n-1}$ for every proper divisor d of m (that is, for every $d|m$ with $d \neq m$). Now let

$$g(m) := \left(-\frac{1}{u}\right) \cdot \sum_{\substack{d|m \\ d \neq 1_N}} f(d) \cdot g\left(\frac{m}{d}\right).$$

Then

$$(f \star g)(m) = \sum_{\substack{d|m \\ d \neq 1_N}} f(d)g\left(\frac{m}{d}\right) + f(1_N)g(m) = 0.$$

Since N is factorial, $\cup_{n \in \mathbb{N}} P_n = N$. Thus g is defined on all of N and satisfies $f \star g = \varepsilon$, so that f is a unit in $R[[N]]$. □

Example 2.1.6. As a very special case, Theorem 2.1.5 shows that the constant mapping

$$\zeta : N \to R, \;\; m \mapsto 1$$

is a unit in $R[[N]]$; it is called the **zeta function** of $R[[N]]$. The inverse of the zeta function $\zeta \in R[[N]]$ is denoted by μ and called the **Möbius function** of $R[[N]]$. □

The concepts of the classical Möbius function and Möbius inversion (a topic we will treat soon in a very general setting) were introduced in 1832 and 1857 in fundamental papers of August Ferdinand Möbius [278] and Julius Wilhelm Richard Dedekind [104]. Nowadays, general Möbius and zeta functions play an important role in Algebraic Combinatorics, in particular in the study of partially ordered sets; see, for instance, Aigner [5]. We cannot venture into a deeper study of these functions, but will at least establish an important basic property of the Möbius function. For this, we require the following definition:

Definition 2.1.7. Let $(R, +, \cdot, 0, 1)$ be a commutative ring and $(N, \cdot, 1_N)$ a simple factorial monoid. A formal power series $f \in R[[N]]$ is called **multiplicative** provided that $f(k\ell) = f(k) \cdot f(\ell)$ whenever k and ℓ are relatively prime. □

Example 2.1.8. We have already seen one important example of a multiplicative function, namely the Euler totient function ϕ. For instance, this property of ϕ may be obtained as a consequence of the Chinese remainder theorem; see Remark 1.10.9. In the context of formal power series, the zeta functions introduced in Example 2.1.6 give trivial examples. A considerably more interesting general example is provided by the associated Möbius functions; this is an immediate consequence of a general result on multiplicative functions which we will prove next. □

Proposition 2.1.9. *Let $(R,+,\cdot,0,1)$ be a commutative ring and $(N,\cdot,1_N)$ a simple factorial monoid. Then the following hold for $R[[N]]$:*

(1) *$f(1_N)=1$ for every invertible multiplicative formal power series $f\in R[[N]]$.*

(2) *If $f,g\subset R[[N]]$ are multiplicative, then $f\star g$ is likewise multiplicative.*

(3) *If $f\in R[[N]]$ is multiplicative and invertible, then the inverse of f is again multiplicative.*

(4) *The set of all multiplicative invertible formal power series in $R[[N]]$ forms an abelian group with respect to the convolution $\star$.*

Proof. Let f be multiplicative. Then $f(1_N)=f(1_N\cdot 1_N)=f(1_N)\cdot f(1_N)$. If f is invertible, $f(1_N)$ has to be a unit, by Theorem 2.1.5. This shows the validity of (1).

In what follows, m and m' will denote any two elements of N which are relatively prime. Note that the set of divisors of mm' then is just the set of all products dd', where d and d' are divisors of m and m', respectively. Using this observation, (2) follows from an easy computation:

$$\begin{aligned}(f\star g)(mm') &= \sum_{d|m,\,d'|m'} f(dd')g\left(\frac{mm'}{dd'}\right)\\ &= \sum_{d|m,\,d'|m'} f(d)f(d')g\left(\frac{m}{d}\right)g\left(\frac{m'}{d'}\right)\\ &= \left(\sum_{d|m} f(d)g\left(\frac{m}{d}\right)\right)\cdot\left(\sum_{d'|m'} f(d')g\left(\frac{m'}{d'}\right)\right)\\ &= (f\star g)(m)\cdot(f\star g)(m').\end{aligned}$$

Next, let f be multiplicative and invertible, and denote the inverse of f by g. We now use induction on n to show

$$g(m)g(m') = g(mm') \quad \text{whenever } mm'\in P_n,$$

where the P_n are as in the proof of Theorem 2.1.5. This is clear for $n=0$, as then $m=m'=1_N$ and $g(1_N)=f(1_N)^{-1}=1$ in view of (1). For the induction step, assume $mm'\in P_n$, where $n\geq 1$. Since $\varepsilon=f\star g$ is multiplicative, the equation

$$(f\star g)(mm') = (f\star g)(m)\cdot(f\star g)(m')$$

gives

$$(*)\qquad \sum_{d|m,\,d'|m'} f(dd')g\left(\frac{mm'}{dd'}\right) = \sum_{d|m,\,d'|m'} f(d)g\left(\frac{m}{d}\right)f(d')g\left(\frac{m'}{d'}\right).$$

As f is multiplicative by hypothesis, $f(dd') = f(d)f(d')$ for all choices of d and d'; by the induction hypothesis, we also get the analogous identity for g whenever $dd' \neq 1_N$, since $\frac{mm'}{dd'}$ is a proper divisor of mm' in these cases and thus belongs to P_{n-1}. Hence all terms in $(*)$ cancel, with the possible exception of the case $d = d' = 1_N$, as 1_N is the only unit in N. This implies the desired identity $g(m)g(m') = g(mm')$ and establishes (3) and, together with (2), also (4). □

The following fundamental result has many applications.

Theorem 2.1.10 (Möbius inversion formula). *Consider the ring $R[[N]]$ of formal power series, where R is a commutative ring and $(N, \cdot, 1_N)$ a simple factorial monoid. With every $f \in R^N$, we associate another mapping $S_f \in R^N$ defined by*

$$S_f(m) := \sum_{d|m} f(d) \quad \textit{for all } m \in N.$$

Then the following identity holds for all $m \in N$:

$$f(m) = \sum_{d|m} \mu(d) \cdot S_f\left(\frac{m}{d}\right),$$

where μ is the Möbius function of $R[[N]]$.

Proof. Using the definition of the zeta function of $R[[N]]$, we get

$$S_f(m) = \sum_{d|m} f(d) \cdot 1 = \sum_{d|m} f(d) \cdot \zeta\left(\frac{m}{d}\right) = (f \star \zeta)(m) \quad \text{for all } m \in N.$$

Since $R[[N]]$ is commutative, this shows $S_f = f \star \zeta = \zeta \star f$, and multiplying from the left with the inverse μ of ζ gives

$$\mu \star S_f = (\mu \star \zeta) \star f = \varepsilon \star f = f.$$

But $f = \mu \star S_f$ means $f(m) = \sum_{d|m} \mu(d) \cdot S_f(\frac{m}{d})$ for all $m \in N$. □

The classical (and arguably most important) example occurs if we choose $\mathbb{Z}$ as the underlying ring and $(\mathbb{N}^*, \cdot, 1)$ as the underlying simple factorial monoid. In this case, the Möbius function of $\mathbb{Z}[[\mathbb{N}^*]]$ and the Möbius inversion formula reduce to the corresponding notions from classical Number Theory. This leads to some well-known facts which can be derived from the construction given in the proof of Theorem 2.1.5. For this, we first introduce some notation:

Notation 2.1.11. Let $n = \prod_{i=1}^{\ell} p_i^{a_i}$ be the prime power factorization of $n \in \mathbb{N}^*$. We will denote the number ℓ of distinct prime divisors of n by $\omega(n)$. Note that n is square-free if and only if $a_i = 1$ holds for $i = 1, \ldots, \ell$, by Definition 1.7.1. □

Proposition 2.1.12. *Let $\mu : \mathbb{N}^* \to \mathbb{Z}$ be the classical Möbius function. Then the following hold:*

(1) $\mu(1) = 1$.

(2) $\sum_{d|n} \mu(d) = 0$ *for* $n > 1$.

(3) *If* n *is square-free, then* $\mu(n) = (-1)^{\omega(n)}$.

(4) *If* n *is divisible by the square of a prime, then* $\mu(n) = 0$.

Proof. Recall that μ is the inverse of the corresponding classical zeta function ζ. As $\zeta(n) = 1$ for all $n \in \mathbb{N}^*$, assertions (1) and (2) are immediate from $\mu \star \zeta = \varepsilon$.

Next, let p be any prime. Then

$$0 = \sum_{d|p^a} \mu(d) = \sum_{i=0}^{a} \mu(p^i) = 1 + \sum_{i=1}^{a} \mu(p^i) \quad \text{for all } a \in \mathbb{N}^*.$$

Now $a = 1$ gives $\mu(p) = 1$, and induction shows $\mu(p^a) = 0$ for $a \geq 2$. This establishes (3) and (4) for the special case where n is a prime power, and then the general case follows from the multiplicativity of the Möbius function established in Proposition 2.1.9. □

The following application of the classical Möbius inversion formula gives an alternative description of the Euler totient function introduced in Definition 1.6.5 and a corresponding solution to Exercise 1.6.33.

Corollary 2.1.13. *Let* ϕ *be the Euler totient function. Then*

$$\phi(n) = \sum_{d|n} \mu(d) \cdot \frac{n}{d} \quad \text{for all } n \in \mathbb{N}^*.$$

In particular, when $n = p^a$ *is a power of a prime* p *(with* $a \geq 1$*), then*

$$\phi(p^a) = p^{a-1} \cdot (p-1).$$

Proof. We view ϕ as a formal power series in $\mathbb{Z}[[\mathbb{N}^*]]$ and apply Theorem 2.1.10 with $f := \phi$. In view of Proposition 1.6.18,

$$S_\phi(n) = \sum_{d|n} \phi(d) = n \quad \text{for all } n \in \mathbb{N}^*,$$

and therefore

$$\phi(n) = \sum_{d|n} \mu(d) \cdot S_\phi\left(\frac{n}{d}\right) = \sum_{d|n} \mu(d) \cdot \frac{n}{d} \quad \text{for all } n \in \mathbb{N}^*.$$

Finally, let $n = p^a > 1$ be a power of a prime p. Then Proposition 2.1.12 implies $\phi(p^a) = \sum_{i=0}^{a} \mu(p^i) \cdot p^{a-i} = p^a - p^{a-1}$, as claimed. □

We will see various further applications of Möbius inversion throughout this book. For applications in the theory of finite fields, we need to generalize the notion of the ring of formal power series a bit further. To this end, we consider a commutative ring $(R, +, \cdot, 0, 1)$ and an R-module V. For an arbitrary non-empty set X, let

V^X be the set of all mappings from X to V. As in the case of mappings over the ring R discussed at the beginning of this section, V^X is an abelian group with respect to pointwise addition and even an R-module with respect to the scalar multiplication defined by

$$(rf)(x) := rf(x) \quad \text{for all } f \in V^X, \text{ all } r \in R \text{ and all } x \in X. \tag{2.6}$$

Again, we specialize the set X to a locally finite simple commutative monoid $(N, \cdot, 1_N)$ with cancellation, for instance a factorial monoid. Consider the ring of formal power series over R and N and recall that the mapping ε defined in the proof of Theorem 2.1.3 is the identity element of $R[[N]]$. The mapping

$$R \to R[[N]], \quad r \mapsto r\varepsilon \tag{2.7}$$

is a monomorphism of rings; we therefore simply identify the ring elements with their images under this mapping. Similarly, every module element $v \in V$ corresponds to the mapping $\chi_v \in V^N$, where $\chi_v(1_N) := v$ and $\chi_v(m) := 0$ for $m \neq 1_N$. The following theorem shows that the R-module structure of V can be extended to an $R[[N]]$-module structure on V^N.

Theorem 2.1.14. *Let R be a commutative ring and $R[[N]]$ the ring of formal power series, where $(N, \cdot, 1_N)$ is a simple factorial monoid.*[2] *Furthermore, let V be any R-module. Then the set V^N of all mappings from N to V becomes an $R[[N]]$-module by defining the scalar multiplication as*

$$(f \diamond \gamma)(m) := \sum_{d \mid m} f(d)\gamma\left(\frac{m}{d}\right) \quad \textit{for all } f \in R[[N]], \textit{ all } \gamma \in V^N \textit{ and all } m \in N. \tag{2.8}$$

Proof. The proof is completely analogous to that of Theorem 2.1.3 (which actually is the case where $V = R[[N]]$ is considered as an $R[[N]]$-module). The details are left as an exercise. □

Example 2.1.15. Let us consider the special case where the underlying module is a multiplicatively written abelian group $(G, \cdot, 1)$ (that is, a multiplicatively written $\mathbb{Z}$-module) and where N is $(\mathbb{N}^*, \cdot, 1)$. Then the corresponding scalar multiplication reads as follows:

$$(f \diamond \gamma)(n) := \prod_{d \mid n} \gamma\left(\frac{n}{d}\right)^{f(d)} \quad \text{for all } f \in \mathbb{Z}[[\mathbb{N}^*]], \text{ all } \gamma \in G^{\mathbb{N}^*} \text{ and all } n \in \mathbb{N}^*. \tag{2.9}$$

The reader should check this in detail. □

We now consider Möbius inversion in this more general context and obtain the following result.

[2] In fact it would suffice to require a locally finite commutative monoid with cancellation.

Theorem 2.1.16. *Let R, N and V be as in Theorem 2.1.14, and let μ be the Möbius function of $R[[N]]$. With any $\gamma \in V^N$, we associate another mapping $S_\gamma \in V^N$ by*

$$S_\gamma(m) := \sum_{d|m} \gamma(d) \quad \text{for all } m \in N.$$

Then

$$\gamma(m) = \sum_{d|m} \mu(d) \cdot S_\gamma\left(\frac{m}{d}\right) \quad \text{for all } m \in N.$$

Proof. Since $d \mapsto \frac{m}{d}$ is a bijection on the set of divisors of m, we have

$$S_\gamma(m) = \sum_{d|m} \gamma(d) = \sum_{d|m} \gamma\left(\frac{m}{d}\right) = \sum_{d|m} \zeta(d) \cdot \gamma\left(\frac{m}{d}\right)$$

for all $m \in N$, where ζ is the zeta function of $R[[N]]$. This just means $S_\gamma = \zeta \diamond \gamma$, and therefore

$$\gamma = \varepsilon \diamond \gamma = (\mu \star \zeta) \diamond \gamma = \mu \diamond (\zeta \diamond \gamma) = \mu \diamond S_\gamma,$$

showing $\gamma(m) = \sum_{d|m} \mu(d) \cdot S_\gamma\left(\frac{m}{d}\right)$ for all $m \in N$. □

Exercises

Exercise 2.1.17. What are the non-invertible multiplicative functions over a domain?

2.2 Polynomial Rings

We again consider the ring $R[[N]]$ of formal power series over a commutative ring $(R, +, \cdot, 0, 1)$ with respect to a simple monoid $(N, \cdot, 1_N)$ which is factorial (or at least commutative with cancellation and locally finite). In the present section, we will study the subring of polynomials in $R[[N]]$.

Definition 2.2.1. The **support** $\operatorname{supp}(f)$ of a formal power series $f \in R[[N]]$ is the set of all $m \in N$ such that $f(m) \neq 0$. We denote the subset of all $f \in R[[N]]$ whose support is a finite set by $R[N]$. □

Theorem 2.2.2. *Let N and R be as above. Then $R[N]$ is both a subring and an R-submodule of $R[[N]]$.*

Proof. First of all, $\operatorname{supp}(f+g) \subseteq \operatorname{supp}(f) \cup \operatorname{supp}(g)$ for all $f, g \in R^N$, so that $R[N]$ is a subgroup of the additive group of $R[[N]]$. As for the convolution of two elements f and g from $R[N]$, let

$$F := \bigcup_{m \in \operatorname{supp}(f)} \{d \in N : d|m\} \text{ and } G := \bigcup_{m \in \operatorname{supp}(g)} \{d \in N : d|m\}.$$

Then $\operatorname{supp}(f \star g) \subseteq \{d+e \colon d \in F, e \in G\}$ is finite, as F and G are finite. Trivially, $\varepsilon \in R[N]$, so that $R[N]$ is a subring of $R[[N]]$. As $R[N]$ is closed with respect to the multiplication with scalars defined in Equation (2.2), it is also an R-submodule of $R[[N]]$. □

Definition 2.2.3. With R and N as above, $R[N]$ is called the **ring of polynomials** over R and N. The elements of $R[N]$ are called **polynomials**. □

In what follows, we shall be concerned with the most important special case, where the underlying simple monoid is $(\mathbb{N}, +, 0)$, the additive monoid of the natural numbers. We therefore have to switch from multiplicative to additive notation. Recall that one writes $R[[x]]$ instead of $R[[\mathbb{N}]]$ in this case. Similarly, the notation $R[x]$ is used instead of $R[\mathbb{N}]$. The symbol x is said to be an **indeterminate** or a **variable**, and $R[x]$ is called the **polynomial ring over R in the indeterminate** x.

Remark 2.2.4. Let us collect the basic vocabulary concerning polynomials over a commutative ring R as above.

- Let $f \in R[x]$. Then each $f(n)$ (where $n \in \mathbb{N}$) is called a **coefficient** of f. One usually writes f_n instead of $f(n)$.
- The zero mapping is simply written as 0; it is called the **zero polynomial** in this context.
- For every $r \in R$, the polynomial $r\varepsilon$ is called the **constant polynomial** corresponding to r; of course, every non-zero constant polynomial has support $\{0\}$. The set of constant polynomials forms a subring of $R[x]$ which is isomorphic to R, and hence one usually writes r instead of $r\varepsilon$. In particular, the identity polynomial ε is then simply written as 1.
- Let f be a non-zero polynomial in $R[x]$, so that the support of f is non-empty and finite. The **degree** of f, written as $\deg(f)$ or simply $\deg f$, is the largest $n \in \mathbb{N}$ such that $f_n \neq 0$ (with respect to the natural order $\leq$ on $\mathbb{N}$); then f_n is the **leading coefficient** of f. If f has leading coefficient 1, it is called a **monic polynomial**.
- So far, the zero polynomial has no degree. One might leave its degree undefined, but it is convenient to put $\deg(0) := -\infty$. Here $-\infty$ is a symbol which is assumed to satisfy $-\infty + n = -\infty$ and $-\infty < n$ for all $n \in \mathbb{N}$.

From now on, we shall write the convolution of polynomials f, g in $R[x]$ simply as an ordinary multiplication: $f \cdot g$ or fg. □

Remark 2.2.5. We now give an interpretation of the indeterminate x which leads to the standard notation for polynomials and, more generally, power series. Consider the special polynomial $e \in R[x]$ defined by

$$e_n := \begin{cases} 1 & \text{if } n = 1, \\ 0 & \text{if } n \neq 1; \end{cases}$$

in other words, e is the unique monic polynomial with support $\{1\}$. It is easy to show by induction that e^ℓ, the ℓ-fold convolution of e with itself, is the unique monic polynomial with support $\{\ell\}$ for all $\ell \geq 1$; in view of the usual convention $e^0 = 1$, this also holds for $\ell = 0$. Note that every polynomial f can be written in the form

$$f = \sum_{\ell \in \mathrm{supp}(f)} f_\ell e^\ell, \tag{2.10}$$

and this representation of f in powers of e is unique. (Of course, for the zero polynomial the empty sum has to be interpreted as 0.) Thus the R-module $R[x]$ is generated by the powers of e, that is, by the submonoid $\{e^\ell : \ell \in \mathbb{N}\}$ of $(R[x], \cdot, 1)$.

One now interprets the indeterminate x as the polynomial e, leading to the usual notation

$$f(x) = \sum_{\ell=0}^{\deg f} f_\ell x^\ell \tag{2.11}$$

for polynomials in $R[x]$. Sometimes it is also convenient to write f in the form

$$f(x) = \sum_{n \in \mathbb{N}} f_n x^n \quad \text{or} \quad f(x) = \sum_{n=0}^{\infty} f_n x^n,$$

where ∞ is a symbol such that $n < \infty$ for all $n \in \mathbb{N}$. The latter notation is usually extended to power series $f \in R[[x]]$ which are not necessarily polynomials, which makes sense as all information on the mapping f is encoded in its coefficients.[3] In this notation, the convolution of two power series f and g reads as follows:

$$f(x)g(x) = \sum_{n=0}^{\infty} \Big(\sum_{d=0}^{n} f_d g_{n-d} \Big) x^n. \quad \square \tag{2.12}$$

We now consider the special case where R is a commutative domain.

Proposition 2.2.6. *Let f and g be polynomials in $R[x]$, where R is a commutative domain. Then:*

(1) $\deg(f+g) \leq \max(\deg f, \deg g)$.

(2) $\deg(fg) = \deg f + \deg g$ **(degree formula)**, *and the leading coefficient of fg equals $f_{\deg f} \cdot g_{\deg g}$.*

Proof. Since $f \cdot 0 = 0$ and $f + 0 = f$, the formulas are correct when one of the polynomials is the zero polynomial.

Now let $f, g \neq 0$, and write $m = \deg f$ and $n = \deg g$. Without loss of generality, we may assume $m \leq n$. Then $\mathrm{supp}(f+g) \subseteq \{0, 1, \ldots, n\}$. If $m < n$, then $\deg(f+g) = n$ and g_n is the leading coefficient of $f+g$. If $m = n$ and $g_n \neq -f_n$, then again $\deg(f+g) = n$ and $f_n + g_n$ is the leading coefficient of $f+g$. Finally, in the case

[3] Of course, this is merely a *formal* sum, which explains the terminology *formal power series*.

$g_n = -f_n$ the leading coefficients cancel to 0 and therefore $\deg(f+g) < n$. This establishes (1).[4]

Multiplying f and g gives

$$f(x)g(x) = \sum_{k=0}^{\infty} \left(\sum_{j=0}^{k} f_j g_{k-j} \right) \cdot x^k.$$

Assume $f_j g_{k-j} \neq 0$. Then $f_j \neq 0$ and $g_{k-j} \neq 0$, since R is a domain. Because of $\deg f = m$ and $\deg g = n$, we conclude

$$k = j + (k-j) \leq m+n,$$

with equality only for $j = m$ and $k - j = n$. Using this observation, it is easy to check the validity of (2). □

Proposition 2.2.7. *Assume that R is a commutative domain. Then:*

(1) *$R[x]$ is likewise a commutative domain.*

(2) *The units of $R[x]$ are the units of R, when R is considered as the subring of constant polynomials.*

Proof. Both assertions follow easily from the degree formula for the multiplication of polynomials given in Proposition 2.2.6. First, let f and g be any two non-zero polynomials with degrees m and n, respectively. Then fg has degree $m+n \geq 0$ and thus $fg \neq 0$.

Now assume that f is a unit in $R[x]$, and let g be its inverse. Then $fg = 1$ has degree 0 and therefore $0 = \deg(fg) = m+n$ shows $m = n = 0$. Hence f and g are constant polynomials, that is, $f = f_0$ and $g = g_0$. This gives $1 = fg = f_0 g_0$, and therefore $f = f_0$ is a unit in R. □

Remark 2.2.8. By Theorem 2.1.5, any polynomial $f \in R[x]$ with $\deg f \geq 1$ and $f_0 \in U(R)$ is invertible as a formal power series in $R[[x]]$, even though it is not invertible as a polynomial in $R[x]$.

The famous **Hilbert basis theorem** (proved by Hilbert [183] in 1890) states that the polynomial ring $R[x]$ over a Noetherian domain is again Noetherian. Similarly, by a result going back to Gauss, $R[x]$ is factorial whenever R is factorial. The reader may find proofs for both these results in many books on (Commutative) Algebra, for instance, in the textbook by Goodman [151], which is freely available online. □

Finally, we consider the special case of a polynomial ring $F[x]$ over a field F. This case yields an important class of Euclidean domains and is therefore particularly interesting.

Theorem 2.2.9. *Let F be a field. Then the degree function is a Euclidean function for the polynomial ring $F[x]$, and hence $F[x]$ is a Euclidean domain.*

[4] For this part, it is not even necessary to assume that R is a domain.

Proof. By definition, the degree function maps the set $F[x]^*$ of non-zero polynomials onto the set $\mathbb{N}$ of natural numbers. Since F is a domain, Proposition 2.2.6 gives $\deg(fg) = \deg f + \deg g$ for non-zero polynomials f and g, and therefore $\deg f \leq \deg(fg)$ for all $f, g \in F[x]^*$.

Given polynomials $f(x), g(x) \in F[x]$ with $g(x) \neq 0$ we need to determine polynomials $q(x)$ and $r(x)$ satisfying $f(x) = q(x)g(x) + r(x)$ and $\deg r < \deg g$.[5] If $f(x) = 0$, we may take $q(x) = r(x) = 0$. If $\deg f < \deg g$, we can take $q(x) = 0$ and $r(x) = f(x)$. Thus assume $f(x) \neq 0$ and $\deg f \geq \deg g$, and write $m := \deg g$ and $n := \deg f$. We proceed using induction on $n - m$.

If $n - m = 0$, that is, $n = m$, we consider the polynomial $h(x) := \frac{f_n}{g_n} \cdot g(x)$. Then $\deg h = n$ and $u(x) := f(x) - h(x)$ has a strictly smaller degree than $f(x)$, since the leading coefficients of h and f agree. Now $r(x) := u(x)$ and $q(x) := \frac{f_n}{g_n}$ give

$$q(x) \cdot g(x) + r(x) = h(x) + u(x) = f(x)$$

with $\deg r < \deg f = \deg g$, as claimed.

Now let $\ell > 0$ and assume by induction that a division with remainder of $f(x)$ by $g(x)$ is possible for all polynomials f with degree n such that $n - m < \ell$. Take

$$h(x) := \frac{f_n}{g_m} \cdot x^{n-m} \cdot g(x).$$

Then the degree of h is $(n - m) + \deg g = n = \deg f$, and its leading coefficient equals $(f_n/g_m) \cdot g_m = f_n$, the leading coefficient of f. Hence the polynomial $u(x) := f(x) - h(x)$ has degree strictly smaller than n. By induction, there exist polynomials $s(x)$ and $r(x)$ with $\deg r < \deg g$ and $f(x) - h(x) = u(x) = s(x)g(x) + r(x)$. This gives $f(x) = q(x)g(x) + r(x)$ with $q(x) = \frac{f_n}{g_m} \cdot x^{n-m} + s(x)$, as desired. □

Note that the above proof of Theorem 2.2.9 is constructive and shows how to actually perform division with remainder:

Algorithm 2.2.10 (Polynomial division with remainder).

- *Input:* A field F and two polynomials $f(x), g(x) \in F[x]$ with $g \neq 0$.
- *Output:* Polynomials $q(x)$ and $r(x)$ with $f(x) = q(x)g(x) + r(x)$ and $\deg r < \deg g$.

(1) $q(x) \leftarrow 0,\ r(x) \leftarrow f(x)$;
(2) **while** $r(x) \neq 0$ and $\deg g < \deg r$ **do**
(3) $\quad t(x) \leftarrow r_{\deg r} g_{\deg g}^{-1} x^{\deg r - \deg g}$;
(4) $\quad q(x) \leftarrow q(x) + t(x)$;
(5) $\quad r(x) \leftarrow r(x) - t(x)g(x)$;
(6) **od**.

[5] Note that this covers the case when $r = 0$ is the zero polynomial, since $\deg(0) = -\infty$ by definition – one of the reasons for introducing that convention.

Remark 2.2.11. It is important to note that the quotient and the remainder occurring in the polynomial division of $f(x)$ by $g(x)$ are uniquely determined. Let $f,g \in F[x]$ with $g \neq 0$ and assume that $q_1(x)$, $q_2(x)$, $r_1(x)$ and $r_2(x)$ are polynomials satisfying

$$f(x) = q_1(x)g(x) + r_1(x) = q_2(x)g(x) + r_2(x),$$

where $\deg r_1 < \deg g$ and $\deg r_2 < \deg g$. Then $(q_1(x) - q_2(x)) \cdot g(x) = r_2(x) - r_1(x)$. Now let $h(x) := q_1(x) - q_2(x)$. If $h \neq 0$, then

$$\deg g \leq \deg(hg) = \deg(r_2 - r_1) \leq \max(\deg r_2, \deg r_1) < \deg g,$$

a contradiction. Hence $h = 0$, that is, $q_1 = q_2$. But then also $r_2 - r_1 = 0$, establishing the uniqueness.

As in the case of the division with remainder in $\mathbb{Z}$, one writes $r(x) = f(x) \bmod g(x)$ for the unique remainder, and $q(x) = f(x) \texttt{ div } g(x)$ for the unique quotient. In particular, $g(x)$ divides $f(x)$ if and only if $f(x) \bmod g(x) = 0$. □

Exercises

Exercise 2.2.12. Show that $R[x,y]$ is isomorphic to $R[x][y]$ and conclude that $R[x,y]$ is a commutative domain whenever R is such a domain. □

Exercise 2.2.13. Give a formal proof for the correctness of Algorithm 2.2.10, using the loop invariant $q(x)g(x) + r(x) = f(x)$. □

2.3 The Algebra of Univariate Polynomials over a Field

The present section summarizes the basic properties of the polynomial ring $F[x]$ in the indeterminate x, where F is a field.

By Proposition 2.2.7, the group of units of $F[x]$ is equal to F^*, the multiplicative group of the field F. Consequently, if $g(x)$ and $h(x)$ are two non-zero polynomials which are associated (that is, $g(x) \mid h(x)$ and $h(x) \mid g(x)$), then there is an element $\lambda \in F^*$ such that $\lambda g(x) = h(x)$. Thus, each association class of non-zero polynomials contains a unique monic polynomial. We shall denote the set of all monic polynomials in $F[x]$ as $F[x]_{\text{mon}}$.

Observation 2.3.1. We now list a few simple facts about $F[x]_{\text{mon}}$:

(1) $F[x]_{\text{mon}}$ is a submonoid of $(F[x]^*, \cdot, 1)$.

(2) $(F[x]_{\text{mon}}, \cdot, 1)$ is a simple factorial monoid.

(3) $F[x]^*$ decomposes into the product of the submonoids F^* and $F[x]_{\text{mon}}$, whose intersection is $\{1\}$. Therefore, $F[x]_{\text{mon}}$ is a complete set of representatives for the classes of the factor monoid $F[x]^*/F^*$.

(4) By Theorems 2.2.9 and 1.5.14, $F[x]$ is a Euclidean domain and hence also a principal ideal domain. Because of (1), every non-zero ideal I of $F[x]$ contains a unique monic polynomial $g(x)$ generating the ideal I; this polynomial is called the **monic generator** of I. □

Observation 2.3.2 (canonical factorization of polynomials). By Theorem 1.5.9, every principal ideal domain is a factorial domain. Thus the set $F[x]^*$ of all non-zero polynomials is a factorial commutative monoid, and hence its prime elements coincide with its irreducible elements, by Theorem 1.2.14; naturally, these are called the **irreducible polynomials**. In view of Observation 2.3.1 (4), given any polynomial $f(x) \in F[x]$, there exist a unique $\alpha \in F^*$, a unique $\ell \in \mathbb{N}^*$, unique distinct irreducible monic polynomials $h_1(x),\ldots,h_\ell(x)$ and unique integers $a_1,\ldots,a_\ell \geq 1$ (called the **multiplicities**) such that

$$f(x) = \alpha \cdot h_1(x)^{a_1} \cdots h_\ell(x)^{a_\ell};$$

this factorization is usually called the **canonical factorization** of f. □

Remark 2.3.3. In view of the degree formula in Proposition 2.2.6, every polynomial of degree 1 is necessarily irreducible; such a polynomial is called a **linear polynomial**. In particular, the monic linear polynomials are the polynomials of the form $x-\lambda$ for some $\lambda \in F$. Whether or not $F[x]$ admits further irreducible polynomials depends on the structure of F, as the following examples show:

- Let $\mathbb{C}$ be the field of complex numbers. By a well-known result from classical Algebra, the only irreducible polynomials in $\mathbb{C}[x]$ are the linear ones.
- For the field $\mathbb{R}$ of real numbers, all non-linear irreducible polynomials have degree 2.
- If F is a finite field, there exist irreducible polynomials with degree n in $F[x]$ for every $n \in \mathbb{N}^*$, as we shall see in Theorem 3.5.4. □

Observation 2.3.4. By Theorem 2.2.2, $F[x]$ is an F-algebra. Note that $F[x]$ is not finitely generated as an F-vector space; indeed, the infinite set consisting of the **monomials** x^n ($n \in \mathbb{N}$) forms a basis, the so-called **canonical basis** for $F[x]$. Now consider any ideal $I = (f)$ in $F[x]$ generated by a non-constant polynomial $f(x)$. Then the factor algebra $V := F[x]/I$ is a finitely generated F-vector space with dimension equal to the degree of f, say n. More precisely, one has the following:

(1) The cosets

$$1+I,\ x+I,\ldots,x^{n-1}+I$$

form a basis for V, which is again called the **canonical basis** of V over F.

(2) The set

$$F[x]_{<n} := \{g(x) \in F[x] \colon \deg g < n\}$$

of polynomials with degree strictly less that n provides a complete set of coset representatives of $I = (f)$ in $F[x]$. Note that $F[x]_{<n}$ is itself a subspace of $F[x]$ with dimension n.

(3) As an F-vector space,

$$F[x] = F[x]_{<\deg f} \oplus (f).$$

(4) A coset $g(x) + (f)$ is a unit in $F[x]/(f)$ if and only if g and f are relatively prime. In fact, the inverse of $g(x) + (f)$ can then be computed efficiently by applying the extended Euclidean algorithm 1.5.17 to the pair $(f(x), g(x))$: as f and g are relatively prime, the algorithm returns polynomials $a(x)$ and $b(x)$ such that

$$a(x)f(x) + b(x)g(x) = \gcd(f(x), g(x)) = 1,$$

so that $b(x) + (f)$ is the desired inverse of $g(x) + (f)$.

(5) In view of (4) and Theorem 1.4.20, the factor ring $F[x]/(f)$ is a field if and only if (f) is a maximal ideal in $F[x]$, which holds if and only if f is irreducible. □

Observation 2.3.5. We now specialize the results on the Chinese remainder theorem obtained in Section 1.9 to polynomial algebras. For this, suppose

$$f(x) = \prod_{i=1}^{k} g_i(x),$$

where $g_1(x), \ldots, g_k(x)$ are pairwise relatively prime. Then the mapping

$$\Psi : \begin{cases} F[x] \to \bigoplus_{i=1}^{k} F[x]/(g_i) \\ h(x) \mapsto \big(h(x) \bmod g_1(x), \ldots, h(x) \bmod g_k(x)\big) \end{cases}$$

is an epimorphism of rings (in fact of F-algebras) with kernel (f). Consequently, Ψ induces an isomorphism

$$\overline{\Psi} : F[x]/(f) \to \bigoplus_{i=1}^{k} F[x]/(g_i)$$

of F-algebras. In particular, this isomorphism gives an isomorphism between the unit groups of $F[x]/(f)$ and $\bigoplus_{i=1}^{k} F[x]/(g_i)$. Since the latter is equipped with the pointwise multiplication, its group of units is the direct product of the unit groups of $F[x]/(g_i)$ for $i = 1, \ldots, k$. □

In the remainder of this section, we assume that F is a finite field and consider the Euler and Möbius functions for this case. As the reader will notice, the situation is quite parallel to the classical case studied in Number Theory and already dealt with in Section 2.1.

Definition 2.3.6. Let $f(x)$ be any non-constant monic polynomial over a finite field F and denote the number of units in the factor ring $F[x]/(f)$ by $\phi_F(f)$; we also use the convention $\phi_F(1) := 1$. One calls ϕ_F the **Euler function for the ring** $F[x]$. □

Observation 2.3.7. By Observation 2.3.5, ϕ_F is multiplicative; that is, with the notation in 2.3.5,

$$\phi_F(f) = \prod_{i=1}^{k} \phi_F(g_i).$$

(Note that this is a special case of Remark 1.9.17.) Next, consider a polynomial of the form $g(x) = h(x)^m \in F[x]$, where $h(x)$ is monic and irreducible with degree $\ell \geq 1$ and where $m \in \mathbb{N}^*$. Let $V := F[x]_{<\ell}$ be the ℓ-dimensional F-subspace of $F[x]$ formed by the polynomials with degree strictly less than ℓ, and put

$$W := \{h(x) \cdot a(x) \colon a(x) \in F[x],\ \deg(a) \leq (m-1)\ell\}.$$

Then one has

$$F[x]_{<m\ell} = V \oplus W,$$

as a consequence of the division with remainder of polynomials $b(x)$ with degree $< m\ell$ by $h(x)$. Since $F[x]_{<m\ell}$ is a complete set of representatives of polynomials modulo $g(x) = h(x)^m$ and since $h(x)$ is irreducible, we see that $b(x) \in F[x]_{<m\ell}$ is a unit modulo $g(x)$ if and only if $v \neq 0$ in the unique decomposition $b(x) = v(x) + w(x)$ with $v(x) \in V$ and $w(x) \in W$, which implies

$$\phi_F(h^m) = |V \setminus \{0\}| \cdot |W| = \big(|F|^{\ell} - 1\big) \cdot |F|^{(m-1)\ell},$$

that is,

$$|F|^{m \cdot \deg h} - |F|^{(m-1) \cdot \deg h} = |F|^{(m-1) \cdot \deg h} \cdot \big(|F|^{\deg h} - 1\big).$$

Altogether, these observations lead to an explicit formula for $\phi_F(f)$ as soon as we can determine the canonical factorization of the polynomial f over F. □

The Euler function ϕ_F of the simple factorial monoid $F[x]_{\mathrm{mon}}$ of all monic polynomials with coefficients from the finite field F defines a mapping from $F[x]_{\mathrm{mon}}$ to $\mathbb{N}^*$ and may therefore be viewed as a power series in $\mathbb{Z}[[F[x]_{\mathrm{mon}}]]$. Let μ_F be the Möbius function of this domain of power series. As noted in Example 2.1.8, μ_F is a multiplicative function. Thus

$$\mu_F(f) = \prod_{i=1}^{\ell} \mu_F(g_i),$$

when $f(x) = \prod_{i=1}^{\ell} g_i(x)$ with relatively prime polynomials $g_i(x)$. Moreover, $\mu_F(1) = 1$ and $\sum_{h|f} \mu_F(h) = 0$ for any $f \neq 1$, since μ_F is the inverse of the zeta function ζ_F. In particular, if $g(x)$ is irreducible and $a \in \mathbb{N}^*$, then

$$0 = \sum_{h|g^a} \mu_F(h) = \sum_{j=0}^{a} \mu_F(g^j)$$

shows $\mu_F(g) = -1$ and $\mu_F(g^j) = 0$ for $j \geq 2$. Altogether, this gives the following:

Proposition 2.3.8. *Let F be a finite field and $f(x) \in F[x]_{\mathrm{mon}}$. Denote the number of distinct monic irreducible factors of $f(x)$ by $\omega_F(f)$, and let μ_F be the Möbius function of $\mathbb{Z}[[F[x]_{\mathrm{mon}}]]$. Then*

$$\mu_F(f) = (-1)^{\omega_F(f)} \quad \textit{for all } f(x) \in F[x]_{\text{mon}}. \quad \square$$

In what follows, we let a subscript of the form $a \mid f$ indicate that a union, a sum or a product, respectively, is formed over all monic divisors $a(x)$ of $f(x)$ in $F[x]$.

Proposition 2.3.9. *Let F be a finite field and $f(x) \in F[x]_{\text{mon}}$. Then*

$$\sum_{a|f} \phi_F(a) = |F|^{\deg f}$$

and

$$\phi_F(f) = \sum_{h|f} \mu_F\left(\frac{f}{h}\right)|F|^{\deg h}. \tag{2.13}$$

Proof. Assume first that $f(x) = h(x)^m$ is a power of a monic irreducible polynomial $h(x) \in F[x]$. Then Observation 2.3.7 gives the desired result:

$$\begin{aligned}\sum_{a|f} \phi_F(a) &= 1 + \sum_{j=1}^{m} \phi_F(h^j) = 1 + \sum_{j=1}^{m} \left(|F|^{j\cdot\deg h} - |F|^{(j-1)\cdot\deg h}\right)\\ &= 1 + |F|^{m\cdot\deg h} - |F|^0 = |F|^{\deg f}.\end{aligned}$$

Now assume $f(x) = g(x) \cdot h(x)$ with relatively prime monic polynomials $g(x), h(x) \in F[x]$. Using induction on the degree of the polynomials in question, we may assume $\sum_{a|g} \phi_F(a) = |F|^{\deg g}$ and $\sum_{b|h} \phi_F(b) = |F|^{\deg h}$. Then

$$|F|^{\deg f} = |F|^{\deg g} \cdot |F|^{\deg h} = \sum_{a|g}\sum_{b|h} \phi_F(a)\phi_F(b).$$

Using the multiplicativity of ϕ_F, we obtain $\phi_F(a)\phi_F(b) = \phi_F(ab)$. Since $g(x)$ and $h(x)$ are relatively prime, the monic divisors of $f(x)$ are in one to one correspondence with the pairs $(a(x), b(x))$, where $a(x)$ is a monic divisor of $g(x)$ and $b(x)$ is a monic divisor of $h(x)$. Altogether, this gives

$$|F|^{\deg f} = \sum_{a|g}\sum_{b|h} \phi_F(a)\phi_F(b) = \sum_{c|f} \phi_F(c).$$

Now the first assertion follows by induction on the number of distinct monic irreducible factors of $f(x)$. Finally, Equation (2.13) is obtained by applying Möbius inversion; see Theorem 2.1.10. $\square$

We will return to the study of polynomials in the context of field extensions in later sections.

Exercises

Exercise 2.3.10. Check the details of the assertions collected in Observations 2.3.1 through 2.3.7. $\square$

2.4 Evaluation of Polynomials, Roots, and Formal Derivatives

In the present section, we continue the study of basic properties of univariate polynomials over a field F. We begin with the concept of evaluating polynomials in $F[x]$ at elements coming from some F-algebra.

Definition 2.4.1. Let F be a field and a an element of some F-algebra A. Given any polynomial $f(x) = \sum_{i=0}^{n} f_i x^i \in F[x]$, the **evaluation** of $f(x)$ at a is defined by

$$f(a) := \sum_{i=0}^{n} f_i a^i \in A. \tag{2.14}$$

Thus a gives rise to a mapping Γ_a, the **evaluation homomorphism** at a:

$$\Gamma_a \colon F[x] \to A, \quad f(x) \mapsto f(a); \tag{2.15}$$

in case $f(a) = 0$, one calls a a **root** (or a **zero**) of f in A. □

It is easily checked that Γ_a is indeed an F-algebra homomorphism, that the image $\operatorname{im}\Gamma_a$ is the F-subalgebra of A generated by a, and that the kernel of Γ_a is an ideal in $F[x]$. Moreover, Γ_a is injective if and only if its kernel consists of the zero polynomial only. We shall use the notation $\operatorname{alg}_F(a)$ instead of $\operatorname{im}\Gamma_a$.

Definition 2.4.2. Assume that the homomorphism Γ_a defined in (2.15) is not injective. Then the monic generator of $\ker\Gamma_a$ is called the **minimal polynomial** of a over F. We will use the notation $\operatorname{mpol}_a(x)$ – or, when we wish to emphasize the underlying field – $\operatorname{mpol}_{a,F}(x)$.[6] One says that a has **finite degree** over F; more precisely, the degree of the minimal polynomial $\operatorname{mpol}_a(x)$ is also called the **degree** of a over F and denoted by $\deg_F(a)$. □

Again, we mention a few simple facts which the reader should check. By definition, the minimal polynomial of an element a of A of finite degree is the monic polynomial of least degree in $F[x]$ which admits a as a root. Moreover, any polynomial having a as a root is a multiple of the minimal polynomial of a. For $f(x)$ and $g(x)$ in $F[x]$, one has $f(a) = g(a)$ if and only if $\operatorname{mpol}_a(x)$ divides $f(x) - g(x)$, that is, $f(x) \equiv g(x) \bmod \operatorname{mpol}_a(x)$.

Theorem 2.4.3. *Let F be a field, a an element of some F-algebra A, and Γ_a the evaluation homomorphism at a. Then $\operatorname{alg}_F(a)$ is a commutative F-algebra.*

Moreover, $\operatorname{alg}_F(a)$ is isomorphic to $F[x]$ if and only if Γ_a is injective; in this case, $\{a^m : m \in \mathbb{N}\}$ is a basis for $\operatorname{alg}_F(a)$ over F.

Otherwise (if Γ_a is not injective), $\operatorname{alg}_F(a)$ is finitely generated and has dimension $n := \deg(\operatorname{mpol}_a)$ over F; in this case, $\operatorname{alg}_F(a)$ admits $\{e, a, \ldots, a^{n-1}\}$ as a basis, where e is the unit element of A.

---

[6] It should be noted that the simpler notations $\langle a \rangle$ instead of $\operatorname{alg}_F(a)$ and m_a instead of mpol_a are quite usual in the literature, but we prefer clarity to brevity here.

Proof. By the homomorphism theorem for F-algebras, $\mathrm{alg}_F(a)$ is isomorphic to the factor algebra $F[x]/\ker(\Gamma_a)$. In particular, if Γ_a is injective, then $\mathrm{alg}_F(a)$ is isomorphic to the polynomial ring $F[x]$, and therefore not finitely generated as an F-vector space. In this case, $\{a^m : m \in \mathbb{N}\}$ is a basis of $\mathrm{alg}_F(a)$.

Now assume that Γ_a is not injective. Then $F[x]/\ker(\Gamma_a) = F[x]/(\mathrm{mpol}_a)$ has dimension n over F, where n is the degree of the minimal polynomial of a over F. This is also the F-dimension of $\mathrm{alg}_F(a)$, and therefore

$$e = \Gamma_a(1),\ a = \Gamma_a(x), \ldots, a^{n-1} = \Gamma_a(x^{n-1})$$

form a basis of $\mathrm{alg}_F(a)$. □

Proposition 2.4.4. *Let $a \in A$ have finite degree over F and let $f(x) \in F[x]$. Then $f(a)$ is a unit in $\mathrm{alg}_F(a)$ if and only if $f(x)$ and $\mathrm{mpol}_a(x)$ are relatively prime.*

Proof. Since Γ_a induces an isomorphism between $F[x]/(\mathrm{mpol}_a)$ and $\mathrm{alg}_F(a)$, the image $f(a)$ is a unit in $\mathrm{alg}_F(a)$ if and only if $f(x)$ is a unit modulo $\mathrm{mpol}_a(x)$. Now Observation 2.3.4 (4) gives the desired criterion. □

Remark 2.4.5. Assume that the criterion in Proposition 2.4.4 is satisfied. Since $F[x]$ is a Euclidean domain, the inverse of $f(x)$ modulo $\mathrm{mpol}_a(x)$ can be determined efficiently using the extended Euclidean algorithm (see part (4) of Observation 2.3.4), and the isomorphism Γ_a allows us to transform this into a computation of the inverse of $f(a)$ in $\mathrm{alg}_F(a)$ as follows. The algorithm provides polynomials $g(x)$ and $h(x)$ in $F[x]$ such that $1 = g(x)f(x) + h(x) \cdot \mathrm{mpol}_a(x)$, and evaluation at a gives

$$e \;=\; g(a)f(a) + h(a)\mathrm{mpol}_a(a) \;=\; g(a)f(a).$$

Thus $g(a)$ is the desired inverse of $f(a)$ in $\mathrm{alg}_F(a)$.

Conversely, assume that $f(a)$ is a unit in $\mathrm{alg}_F(a)$, say $b \cdot f(a) = e$ for a suitable $b \in \mathrm{alg}_F(a)$. Let $b = g(a)$ for some polynomial $g(x)$. Then $\Gamma_a(f(x)g(x)) = e = \Gamma_a(1)$ and therefore $f(x)g(x) \equiv 1$ modulo $\mathrm{mpol}_a(x)$. This shows that $f(x)$ is a unit modulo $\mathrm{mpol}_a(x)$ and therefore relatively prime to $\mathrm{mpol}_a(x)$. □

In what follows, we will discuss three particularly important special instances for the F-algebra A:

- First (and most naturally), we view F itself as an F-algebra.
- Then, generalizing the first instance, we consider any extension field of F as an F-algebra.
- Finally, another important example is given by the F-algebra $\mathrm{End}_F(V)$ of F-linear mappings on an F-vector space V. This will be the topic of Section 2.6.

Example 2.4.6. Consider F as an F-algebra and let $\lambda \in F$. Obviously, the minimal polynomial of λ is just $x - \lambda$. Moreover, Γ_λ is surjective, as $\Gamma_\lambda(x - \lambda + \beta) = \beta$ for $\beta \in F$. □

Observation 2.4.7. Evaluating a polynomial f at $\lambda \in F$ just gives the remainder when dividing f by $x - \lambda$:

$$\Gamma_\lambda(f(x)) = f(\lambda) \equiv f \mod (x - \lambda)$$

for every $f \in F[x]$. In order to see this, we divide f by $x - \lambda$ with remainder, say $f(x) = q(x) \cdot (x - \lambda) + r(x)$ with $\deg r < \deg(x - \lambda) = 1$. Then $r(x)$ is a constant polynomial, say $r(x) = \alpha \in F$, and

$$\Gamma_\lambda(f(x)) = f(\lambda) = q(\lambda) \cdot (\lambda - \lambda) + \alpha = \alpha,$$

which indeed gives $f(\lambda) = \alpha = r(x) = f(x) \bmod (x - \lambda)$.

In particular, $\lambda \in F$ is a root of f if and only if $x - \lambda$ divides $f(x)$, that is, if and only if there exists a polynomial $h(x)$ such that $f(x) = (x - \lambda) \cdot h(x)$. Note that then $\deg h = \deg f - 1$. □

Remark 2.4.8. Let us briefly return to the general case. Assume that $a \in A$ has finite degree over F. Because of $\Gamma_a(x) = a$, we see that a is a unit in $\mathrm{alg}_F(a)$ if and only if x and $\mathrm{mpol}_a(x)$ are relatively prime. Since x is an irreducible polynomial of $F[x]$, this holds if and only if x does not divide $\mathrm{mpol}_a(x)$, which is equivalent to the fact that 0 is not a root of $\mathrm{mpol}_a(x)$ in F.

Now let $\mathrm{mpol}_a(x) = \sum_{i=0}^{n} \alpha_i x^i$ with $\alpha_n = 1$. Evaluating $\mathrm{mpol}_a(x)$ at 0 gives $\mathrm{mpol}_a(0) = \alpha_0$. If this is non-zero, then $\mathrm{mpol}_a(a) = 0$ gives

$$e = \frac{1}{\alpha_0} \cdot \alpha_0 \cdot e = -\frac{1}{\alpha_0} \cdot \left(\sum_{i=1}^{n} \alpha_i a^i \right),$$

where e is the unit element of A. Therefore

$$\sum_{i=1}^{n} \left(-\frac{\alpha_i}{\alpha_0} \right) \cdot a^{i-1}$$

is the inverse of a in $\mathrm{alg}_F(a)$. □

We now turn to the evaluation of polynomials at elements of extension fields. For this, we first recall the relevant notions:

Definition 2.4.9. Let F and E be fields such that $F \subseteq E$. Then F is called a **subfield** of E, while E is called an **extension field** of F; we use the notation E/F to indicate such a **field extension**. Note that E is an F-algebra in this situation, in particular an F-vector space. The **degree** of E over F is the dimension of E as an F-vector space and is usually denoted by $[E : F]$. The field extension E/F is called **finite** if $[E : F]$ is finite. Any field K with $F \subseteq K \subseteq E$ is called an **intermediate field** of E/F. □

Next, we introduce some vocabulary concerning the evaluation of polynomials over F at field elements from E:

Definition 2.4.10. Consider a field extension E/F, and let $v \in E$. Then v is called **transcendental** over F if Γ_v, the evaluation homomorphism at v, is injective. Otherwise, v is said to be **algebraic** (of degree $\deg_F(v)$) over F. If every $v \in E$ is algebraic over F, then E is called an **algebraic extension** of F. □

Trivially, E/F is algebraic whenever E/F has finite degree. We will study algebraic extensions of Galois fields in detail in Chapter 4. We now proceed by strengthening Theorem 2.4.3 in the special case of field extensions.

Theorem 2.4.11. *Let E/F be a field extension, and let $v \in E$ be algebraic over F. Then the minimal polynomial of v is an irreducible polynomial in $F[x]$, and* $\mathrm{alg}_F(v)$ *is an intermediate field of E/F.*

Proof. Write $g(x) := \mathrm{mpol}_v(x)$, and let $f, h \in F[x]$ be polynomials satisfying $g(x) = f(x)h(x)$. Since E is a domain, $0 = g(v) = f(v)h(v)$ gives $f(v) = 0$ or $h(v) = 0$, say $f(v) = 0$. Then $f(x) \in \ker(\Gamma_v)$, hence $g(x)$ divides $f(x)$. Thus $f(x)$ and $g(x)$ are associates, and therefore $h(x)$ is a unit in $F[x]$, that is, a non-zero constant polynomial. This shows that $\mathrm{mpol}_v(x)$ is irreducible in $F[x]$.

Since $\mathrm{mpol}_v(x)$ is irreducible, $\ker(\Gamma_v) = (\mathrm{mpol}_v)$ is a maximal ideal in $F[x]$; see Observation 2.3.4. Now $\mathrm{alg}_F(v)$ is isomorphic to $F[x]/\ker(\Gamma_v)$, and hence $\mathrm{alg}_F(v)$ is a field, by Theorem 1.4.20. As $\mathrm{alg}_F(v)$ contains F (as the image of the constant polynomials under Γ_v), we see that $\mathrm{alg}_F(v)$ is an intermediate field of E/F. □

We note that $\lambda \in E$ is an element of F if and only if it has degree 1 over F, that is, if the minimal polynomial of λ is $x - \lambda$.

Our next result concerns the maximal number of roots that a polynomial over F can have in some extension field:

Proposition 2.4.12. *Consider a field extension E/F, and let $f(x) \in F[x]$ be a non-zero polynomial. Then $f(x)$ has at most n distinct roots in E, where $n = \deg f$.*

Proof. As $f(x)$ is also a polynomial in $E[x]$, it suffices to show that any polynomial $g(x)$ of degree n in $E[x]$ can have at most n distinct roots in E. First assume $n = 0$; then g is a non-zero constant polynomial and has no roots in E. Now let $n \geq 1$. If $v \in E$ is a root of g, then $x - v$ divides g in $E[x]$ (see Observation 2.4.7), say $g(x) = (x - v)h(x)$, where h has degree $n - 1$. Using induction on n, we conclude that h has at most $n - 1$ distinct roots in E. If w is any root of g distinct from v, then $0 = g(w) = (w - v)h(w)$, and therefore w is a root of h, since E is a domain. Thus g can have at most $n - 1$ roots in E which are distinct from v, and the assertion follows. □

Remark 2.4.13. As in the proof of Proposition 2.4.12, let $v \in E$ be a root of some non-zero polynomial $g(x) \in E[x]$. Then $x - v$ is an irreducible divisor of g. By the uniqueness of the decomposition of g in $E[x]$ into a product of powers of irreducible polynomials (see Observation 2.3.2), there is a largest integer $k \geq 1$ such that $(x - v)^k$ divides g, that is, $g(x) = (x - v)^k h(x)$ and $h(v) \neq 0$. This integer k is called the **multiplicity** of the root v, and v is said to be a **multiple root** of g if $k \geq 2$.

Assume that g has m distinct roots $v_1, \ldots, v_m \in E$ with multiplicities $k_1, \ldots, k_m$, respectively. Since $(x-v_i)^{k_i}$ and $(x-v_j)^{k_j}$ are relatively prime for $i \neq j$, we see that $\prod_{i=1}^m (x-v_i)^{k_i}$ divides g in $E[x]$, and therefore $\sum_{i=1}^m k_i \leq \deg g$. Thus g can have at most $\deg g$ roots in E even when the roots are counted with their multiplicity. □

Definition 2.4.14. Let E/F is a field extension, and $f(x)$ a polynomial of degree at least 1 over F. One says that f **splits over** E **into linear factors** if there are elements $v_1, \ldots, v_n \in E$ such that $f(x) = \prod_{i=1}^n (x-v_i)$ in $E[x]$. Note that the v_i do not have to be distinct. □

In Section 3.1, we will consider the "splitting field" of a polynomial. We conclude this section by introducing the useful concept of the formal derivative of a polynomial, which allows us to obtain results concerning the multiplicity of roots.

Definition 2.4.15. Let $f(x) = \sum_{i=0}^n f_i x^i$ be a non-constant polynomial in $F[x]$, where F is a field. The **formal derivative** – or, for short, just the **derivative** – of f is the polynomial

$$f'(x) := \sum_{i=1}^n i f_i x^{i-1} \in F[x].$$

If $f(x)$ is a constant polynomial, then its derivative is defined to be the zero polynomial. □

Observation 2.4.16. Let us collect some useful properties of formal derivatives:

(1) $(f(x)+g(x))' = f'(x)+g'(x)$ for all $f, g \in F[x]$;
(2) $(\lambda f(x))' = \lambda f'(x)$ for all $f \in F[x]$ and all $\lambda \in F$;
(3) $(f(x)g(x))' = f'(x)g(x)+f(x)g'(x)$ for all $f, g \in F[x]$.

The first two assertions are obvious, while the third may be proved for the cases where $g(x) = x^k$ using induction on k; then the general case follows in combination with the first two parts. □

Proposition 2.4.17. *Let $f(x) \in F[x]$ be a polynomial, E an extension field of F, and $v \in E$ a root of f. Then v is a multiple root of f in E if and only if $f'(v) = 0$.*

Proof. Let k be the multiplicity of the root v and assume $k \geq 2$. Thus there is a polynomial $g(x) \in E[x]$ such that $f(x) = (x-v)^2 g(x)$. Then the derivative of $f(x)$ satisfies

$$f'(x) = 2(x-v)g(x) + (x-v)^2 g'(x) = (x-v) \cdot \big(2g(x) + (x-v)g'(x)\big),$$

so that v is also a root of f'.

Conversely, assume $f'(v) = f(v) = 0$ and let $h(x) \in E[x]$ be such that $f(x) = h(x)(x-v)$. Then $f'(x) = h'(x)(x-v) + h(x)$. This gives $0 = f'(v) = h(v)$, so that v is a multiple root of f. □

Corollary 2.4.18. *Let $f(x) \in F[x]$ be a non-constant polynomial, E an extension field of F, and assume that $f(x)$ splits over E into linear factors. Then f has a multiple root in E if and only if f and its derivative f' are not relatively prime in $F[x]$.*

Proof. First let $v \in E$ be a multiple root of $f(x)$, and let $g(x)$ be the minimal polynomial of v over F. Then g is a non-constant polynomial dividing both f and f', since $f(v) = f'(v) = 0$. Therefore f and f' are not relatively prime.

Conversely, assume that a non-constant polynomial $h(x) \in F[x]$ is a monic common divisor of f and f'. As a divisor of f, the polynomial h splits over E into linear factors, and hence there is some $u \in E$ with $h(u) = 0$. But then also $f(u) = 0 = f'(u)$, and u is a multiple root of f in E by Proposition 2.4.17. □

Exercises

Exercise 2.4.19. Check the observations following Definitions 2.4.1 and 2.4.2. □

Exercise 2.4.20. In Remark 2.4.5, we dealt with elements $b = f(a)$ which are units in $\mathrm{alg}_F(a)$. In this context, it is natural to ask what happens if we consider arbitrary units in an F-algebra A.

Prove that the inverse of an arbitrary invertible element $a \in A$ belongs to the algebra $\mathrm{alg}_F(a)$ generated by a. Consequently, one has $b^{-1} \in \mathrm{alg}_F(a)$ whenever $b \in \mathrm{alg}_F(a)$ is an element which is invertible in A. □

Exercise 2.4.21. Give a detailed proof for part (3) of Observation 2.4.16. Moreover, prove the **chain rule** for formal derivatives: $(f \circ g)'(x) = f'(g(x)) \cdot g'(x)$. □

Exercise 2.4.22. Let F be an arbitrary field. Show that there is a unique way of extending the formal derivative from the polynomial ring $F[x]$ to the quotient field $F(x)$ of $F[x]$ (that is, the **field of rational functions** over F in the indeterminate x) if one requires that the rules in Observation 2.4.16 carry over to this quotient field. The resulting formula for the derivative of a rational function, namely

$$\left(\frac{a(x)}{b(x)}\right)' = \frac{a'(x)b(x) - a(x)b'(x)}{b(x)^2},$$

is usually called the **quotient rule**. □

2.5 Interpolation

In this section we consider interpolation polynomials, a topic combining the Chinese remainder theorem with the evaluation of polynomials. In what follows, let F be any field. We start with the following generalization of part (3) of Observation 2.4.16:

Lemma 2.5.1. *Let $h(x) \in F[x]$ be a polynomial and suppose that $h(x) = \prod_{i=1}^{\ell} g_i(x)$, where the polynomials $g_i(x) \in F[x]$ are not necessarily relatively prime. Then*

$$h'(x) = \sum_{i=1}^{\ell} \left[g_i'(x) \prod_{\substack{j=1 \\ j \neq i}}^{\ell} g_j(x) \right] \tag{2.16}$$

Proof. This follows from the product formula in part (3) of Observation 2.4.16, by using induction on ℓ. □

We now consider the special case of Lemma 2.5.1 where $g_i(x) = x - u_i$ for distinct elements $u_1, \ldots, u_\ell$ of F and $h(x) = \prod_{i=1}^{\ell}(x - u_i)$. Then

$$h'(x) = \sum_{i=1}^{\ell} \left[(x-u_i)' \cdot \prod_{\substack{j=1 \\ j \neq i}}^{\ell} (x - u_j) \right] = \sum_{k=1}^{\ell} H_k(x), \tag{2.17}$$

where

$$H_k(x) := h(x) \,\texttt{div}\, (x - u_k) = \prod_{\substack{i=1 \\ i \neq k}}^{\ell} (x - u_i). \tag{2.18}$$

By Proposition 2.4.17, $h'(u_j) \neq 0$ for all j, since h has no multiple roots. Applying the Chinese remainder mapping

$$\Psi : \begin{cases} F[x] & \to \bigoplus_{i=1}^{\ell} F[x]/(x - u_i) \\ f(x) & \mapsto (f(x) \,\texttt{mod}\, (x - u_1), \ldots, f(x) \,\texttt{mod}\, (x - u_\ell)) \end{cases} \tag{2.19}$$

to this situation gives $\Psi(f(x)) = (f(u_1), \ldots, f(u_\ell)) \in F^\ell$ for every $f(x) \in F[x]$. Moreover, the kernel of Ψ is the ideal (h) generated by $h(x)$; also, Ψ is surjective since $F[x]/(h)$ has F-dimension $\deg h = \ell$, which is also the dimension of $\bigoplus_{i=1}^{\ell} F[x]/(x - u_i)$. Altogether, this establishes the following result:

Theorem 2.5.2. *Let $u_1, \ldots, u_\ell$ be distinct elements of a field F, and let $(\beta_1, \ldots, \beta_\ell)$ be an arbitrary ℓ-tuple in F^ℓ. Then there is a unique polynomial $g(x) \in F[x]$ with $\deg g < \ell$ and $g(u_i) = \beta_i$ for $i = 1, \ldots, \ell$.* □

Definition 2.5.3. The polynomial $g(x)$ in Theorem 2.5.2 is called the **interpolation polynomial** through the points $(u_1, \beta_1), \ldots, (u_\ell, \beta_\ell)$ of $F \times F$. □

Observation 2.5.4 (Lagrange interpolation formula). Let $\overline{\Psi}$ denote the F-algebra isomorphism from $F[x]/(h)$ to F^ℓ induced by the mapping Ψ in (2.19), where F^ℓ is equipped with the pointwise multiplication and where $F[x]/(h)$ is represented as $F[x]_{<\ell}$. We denote the canonical basis of F^ℓ by $\{\mathbf{e}^1, \ldots, \mathbf{e}^\ell\}$, that is,

$$e_j^k = \begin{cases} 1 & \text{if } k = j, \\ 0 & \text{if } k \neq j, \end{cases}$$

and put

$$e_k(x) := \overline{\Psi}^{-1}(\mathbf{e}^k) \in F[x]_{<\ell} \quad \text{for } i = 1, \ldots, \ell.$$

Then

$$g(x) := \sum_{k=1}^{\ell} \beta_k e_k(x)$$

is the interpolation polynomial through the points $(u_1, \beta_1), \ldots, (u_\ell, \beta_\ell)$. The basis $\{e_1(x), \ldots, e_\ell(x)\}$ of $F[x]_{<\ell}$ is often called the **spectral basis** corresponding to the elements $u_1, \ldots, u_\ell$ of F. □

We now provide an explicit formula for the polynomials in the spectral basis:

Theorem 2.5.5. *Let $g_i(x) = x - u_i$ for distinct elements $u_1, \ldots, u_\ell$ of F and $h(x) = \prod_{i=1}^{\ell}(x - u_i)$. Then, using the same notation as in (2.17) and (2.18),*

$$e_k(x) = \frac{1}{h'(u_k)} \cdot H_k(x) = \frac{1}{h'(u_k)} \cdot \prod_{\substack{i=1 \\ i \neq k}}^{\ell} (x - u_j) \tag{2.20}$$

for all $k = 1, \ldots, \ell$.

Proof. Fix k and consider the polynomial $f(x) := \frac{1}{h'(u_k)} \cdot H_k(x) \in F[x]_{<\ell}$. Then $f(u_j) = 0$ for $j \neq k$, since $x - u_j$ divides $H_k(x)$. Also,

$$h'(u_k) = \sum_{j=1}^{\ell} H_j(u_k) = H_k(u_k)$$

gives $f(u_k) = \frac{1}{h'(u_k)} \cdot H_k(u_k) = 1$. These observations show $\overline{\Psi}(f) = \mathbf{e}^k$. As k was arbitrary, we obtain the desired identity (2.20) for all k. □

Remark 2.5.6. Note that the polynomials in the spectral basis give a decomposition of the unit element of $F[x]/(h)$ as a sum of **pairwise orthogonal idempotents**:

- $e_j(x) \cdot e_k(x) \equiv 0 \bmod h(x)$ for all $j \neq k$,
- $e_j(x)^2 \equiv e_j(x) \bmod h(x)$ for all j,
- $\sum_{i=1} e_i(x) \equiv 1 \bmod h(x)$.

This will be used in the proof of the normal basis theorem in Section 3.9. □

Exercises

Exercise 2.5.7. Provide the details of the inductive proof of Lemma 2.5.1. □

2.6 Vector Spaces with Endomorphisms

In this final section, we consider the evaluation of polynomials at endomorphisms of vector spaces, as announced in Section 2.4. Thus let F be a field and V an F-vector space. We denote the set of all vector space homomorphisms from V into itself by $\mathrm{End}_F(V)$. It is well-known that $\mathrm{End}_F(V)$ carries the structure of an F-algebra, with the composition of F-linear mappings as multiplication. As at the beginning of Section 2.4, consider the mapping

$$\Gamma_\tau\colon F[x] \to \mathrm{End}_F(V),\ \ f(x) \mapsto f(\tau), \tag{2.21}$$

where now $\tau \in \mathrm{End}_F(V)$. Let $f(x) = \sum_{j=0}^{n} f_j x^j$. Then $f(\tau)$ is the F-linear mapping given explicitly by

$$f(\tau)(\mathbf{v}) = \sum_{j=0}^{n} f_j \tau^j(\mathbf{v}) \quad \text{for all } \mathbf{v} \in V,$$

where $f_0\tau^0(\mathbf{v}) = f_0\mathbf{v}$. If $\tau \in \mathrm{End}_F(V)$ has finite degree over F, then the minimal polynomial $\mathrm{mpol}_\tau(x)$ of τ is the monic polynomial f of least degree such that $f(\tau)$ is the zero mapping on V.

We are mainly interested in the case where V is finitely generated, say with dimension m. Then the F-algebra $\mathrm{End}_F(V)$ is likewise finitely generated and has dimension m^2 over F, and hence $\mathrm{alg}_F(\tau)$ is also finitely generated for every endomorphism τ. By a fundamental result from Linear Algebra,[7] the minimal polynomial of any $\tau \in \mathrm{End}_F(V)$ has degree at most m (instead of the trivial upper bound m^2). This is a consequence of the Cayley-Hamilton theorem stating that τ always is a root of its characteristic polynomial, which has degree $m = \dim_F V$ by its definition. A different argument for the bound in question will be given later in this section.

Remark 2.6.1. Representing the elements of $\mathrm{End}_F(V)$ as square matrices of size (m,m) with respect to some basis of V gives an isomorphism from $\mathrm{End}_F(V)$ to the F-algebra $F^{(m,m)}$ of all (m,m)-matrices over F, equipped with the matrix multiplication. Then the minimal polynomial of an endomorphism τ of V corresponds to the minimal polynomial of the associated matrix $A \in F^{(m,m)}$, that is, the monic polynomial g of least degree such that $g(A)$ is the zero matrix in $F^{(m,m)}$. We recall that any root λ of g with $\lambda \in F$ is an **eigenvalue** of A: there exists a non-zero vector $\mathbf{v} \in V$ such that $A\mathbf{v} = \lambda\mathbf{v}$, that is, an **eigenvector** for the eigenvalue λ.

It is well-known that conjugate matrices $A, B \in F^{(m,m)}$ always have the same minimal polynomial. This shows that a polynomial may, in general, have infinitely many roots in an algebra, in contrast to the case of field extensions considered in Section 2.4. □

We now turn to a fundamental result in Linear Algebra: $(V,+,0)$ admits the structure of an $F[x]$-module with respect to any given $\tau \in \mathrm{End}_F(V)$.

[7] For this and the subsequent results from Linear Algebra, we refer the reader to the textbook by Hoffmann and Kunze [190].

Observation 2.6.2. Given an endomorphism τ of an F-vector space V, one defines a pairing of the vectors in V with the polynomials in $F[x]$ as follows:

$$f(x)\cdot\mathbf{v} := f(\tau)(\mathbf{v}) \text{ for all } \mathbf{v}\in V \text{ and all } f(x)\in F[x]. \tag{2.22}$$

This turns V into an $F[x]$-module which will be denoted by (V,τ); it will always be clear from the context which endomorphism τ is used. Instead of $f(x)\cdot\mathbf{v}$, we often use the simpler notation $\mathbf{v}^f$.

Now fix a vector $\mathbf{v}\in V$. Then the pair $(\tau,\mathbf{v})$ gives rise to the mapping

$$\Gamma_{\tau,\mathbf{v}}\colon F[x]\to V,\ f(x)\mapsto f(\tau)(\mathbf{v}) = \mathbf{v}^f,$$

which is a homomorphism between the two $F[x]$-modules $F[x]$ and (V,τ). □

We assume from now on that V is finitely generated, say $\dim_F V = m$, so that (V,τ) is a torsion module. In view of the notation used in Sections 1.8 and 2.4, we introduce the following terminology:

Definition 2.6.3. Let τ be an F-endomorphism of the F-vector space V, where V is finitely generated. The monic generator of the kernel of $\Gamma_{\tau,\mathbf{v}}$ (which is an ideal in $F[x]$) is called the **minimal polynomial** of $\mathbf{v}$ with respect to τ; it will be denoted by $\mathrm{mpol}_{\tau,\mathbf{v}}(x)$. Alternatively, this polynomial is also called the **τ-order** of $\mathbf{v}$ and written as $\mathrm{Ord}_\tau(\mathbf{v})$. □

Remark 2.6.4. The image of $\Gamma_{\tau,\mathbf{v}}$ is the $F[x]$-submodule of V generated by $\mathbf{v}$; it will be denoted by $\mathrm{mdl}_\tau(\mathbf{v})$. This submodule of (V,τ) is isomorphic to the factor module (that is, factor ring) $F[x]/(\mathrm{Ord}_\tau(\mathbf{v}))$ and therefore has F-dimension equal to $\deg(\mathrm{Ord}_\tau(\mathbf{v}))$, which is at most $m = \dim_F V$. □

Theorem 2.6.5. *Let V be an m-dimensional vector space over a field F and let τ be any F-endomorphism of V. Then the minimal polynomial of τ has degree at most m.*

Proof. Consider V as an (F,τ)-module. Then V is a torsion module, and the annihilator ideal of V is generated by $\mathrm{mpol}_\tau(x)$. Thus $\mathrm{Ord}_\tau(\mathbf{v})$ divides $\mathrm{mpol}_\tau(x)$ for every $\mathbf{v}\in V$. Moreover, an application of Theorem 1.9.10 gives the existence of some vector $\mathbf{w}\in V$ such that

$$\mathrm{Ord}_\tau(\mathbf{w}) = \mathrm{mpol}_{\tau,\mathbf{w}}(x) = \mathrm{mpol}_\tau(x).$$

Let n be the degree of $\mathrm{Ord}_\tau(\mathbf{w})$. Then $\mathbf{w},\tau(\mathbf{w}),\dots,\tau^{n-1}(\mathbf{w})$ are linearly independent, and hence $n\le \dim_F V = m$. □

Remark 2.6.6. Of course, all results from Section 1.8 may now be applied to the present situation. For the later application to normal basis generators of finite fields, we summarize the main facts as follows:

- For any monic divisor g of $\mathrm{mpol}_\tau(x)$, let

$$(V,\tau)_g := \{\mathbf{v}\in V\colon g(\tau)(\mathbf{v}) = \mathbf{0}\}.$$

Then $(V,\tau)_g$ is an (F,τ)-submodule of (V,τ), namely the (F,τ)-submodule annihilated by $g(x)$.

- By Theorem 1.9.5, the canonical factorization

$$\mathrm{mpol}_\tau(x) = \prod_{i=1}^{\ell} h_i(x)^{a_i}$$

of $\mathrm{mpol}_\tau(x)$ into irreducible polynomials yields the primary decomposition

$$(V,\tau) = \bigoplus_{i=1}^{\ell} (V,\tau)_{h_i^{a_i}}.$$

Therefore every vector $\mathbf{v} \in V$ has a unique representation of the form $\mathbf{v} = \sum_{i=1}^{\ell} \mathbf{v}_i$ with $\mathbf{v}_i \in (V,\tau)_{h_i^{a_i}}$ for all i. □

Remark 2.6.7. When calculating with τ-orders, the following two formulas are of fundamental importance:

- Let $\mathbf{v} \in V$ and $g(x) \in F[x]$. Then

$$\mathrm{Ord}_\tau(\mathbf{v}^g) = \frac{\mathrm{Ord}_\tau(\mathbf{v})}{\gcd(\mathrm{Ord}_\tau(\mathbf{v}), g(x))};$$

see Lemma 1.9.8.

- Let $\mathbf{u}, \mathbf{v} \in V$ and assume that $\mathrm{Ord}_\tau(\mathbf{u})$ and $\mathrm{Ord}_\tau(\mathbf{v})$ are relatively prime. Then

$$\mathrm{Ord}_\tau(\mathbf{u}+\mathbf{v}) = \mathrm{Ord}_\tau(\mathbf{u}) \cdot \mathrm{Ord}_\tau(\mathbf{v});$$

see Lemma 1.9.9. □

We now assume that (V,τ) is a **cyclic module**, that is $\deg(\mathrm{mpol}_\tau) = m = \dim_F V$; one also says that τ is a **cyclic endomorphism**.

Observation 2.6.8. Let (V,τ) be a cyclic module. Then a vector $\mathbf{v} \in V$ generates the $F[x]$-module (V,τ) if and only if its τ-order satisfies $\mathrm{Ord}_\tau(\mathbf{v}) = \mathrm{mpol}_\tau(x)$, in which case $(V,\tau) = \mathrm{mdl}_\tau(\mathbf{v})$ is isomorphic to $F[x]/(\mathrm{mpol}_\tau)$.

Now fix a generator $\mathbf{w}$ of (V,τ). Given any vector $\mathbf{v} \in V$, there is a unique polynomial $g(x) \in F[x]_{<m}$ with $\mathbf{v} = \mathbf{w}^g$. In particular, $\mathbf{v} = \mathbf{w}^g$ likewise generates (V,τ) if and only if g and mpol_τ are relatively prime, and thus the generators of (V,τ) correspond to the units of the factor ring $F[x]/(\mathrm{mpol}_\tau)$.

More generally, every (F,τ)-submodule of (V,τ) is of the form $(V,\tau)_g$ for some monic divisor g of mpol_τ in $F[x]$. The primary decomposition given in Remark 2.6.6 above reflects the Chinese remainder isomorphism between $F[x]/(\mathrm{mpol}_\tau)$ and $\bigoplus_{i=1}^{\ell} F[x]/(h_i^{a_i})$. As noted there, any $\mathbf{v} \in V$ decomposes uniquely as $\mathbf{v} = \sum_{i=1}^{\ell} \mathbf{v}_i$ with $\mathbf{v}_i \in (V,\tau)_{h_i^{a_i}}$ for all i. Here $\mathbf{v}$ is a generator of (V,τ) if and only if $\mathrm{Ord}_\tau(\mathbf{v}_i) = h_i(x)^{a_i}$ for all i. □

Observation 2.6.9. Let us consider the special case where F is a finite field. Then we may count the number of generators of a cyclic module (V,τ) and, more generally, the number of vectors $\mathbf{v}$ such that $\mathrm{Ord}_\tau(\mathbf{v}) = g$ for any given monic divisor g of the minimal polynomial mpol_τ. Note that the latter is also the number of generators of the submodule $(V,\tau)_g$ and therefore equal to $\phi_F(g)$, the number of units in the residue ring $F[x]/(g)$; see Section 2.3. Using the complete decomposition of mpol_τ as in Remark 2.6.6 then shows that the total number of generators of (V,τ) is given by

$$\phi_F(\mathrm{mpol}_\tau) = \prod_{i=1}^{\ell} \phi_F(h_i^{a_i}) = \prod_{i=1}^{\ell} \left(|F|^{a_i \cdot \deg h_i} - |F|^{(a_i-1)\cdot \deg h_i}\right). \quad \square$$

Chapter 3
Field Extensions and the Basic Theory of Galois Fields

Abstract The present chapter is devoted to the basic theory of finite fields, including existence and uniqueness theorems as well as the main structural results. For this purpose, we also extend the fundamental material covered in Chapters 1 and 2 by proving several results on field extensions in general (in particular, in the first two sections).

3.1 The Splitting Field of a Polynomial

Let F be an arbitrary field and $f(x)$ any non-constant polynomial over F. The main result of the present section guarantees the existence of a *smallest* extension field E over F such that $f(x)$ splits over E into linear factors. Moreover, this field E has finite degree and is unique up to isomorphism, which motivates calling E *the* splitting field of f over F.

We start with the following basic result on extensions with finite degree:

Lemma 3.1.1. *Let E/F be a field extension, and let K be an intermediate field of E/F. Then E/F has finite degree if and only if both E/K and K/F have finite degree. In this case, $[E:F]=[E:K]\cdot[K:F]$.*

Proof. If E/F has finite degree, then E/K has finite degree, since $F\subseteq K$ and any F-basis of E generates E as a K-vector space; and K/F has finite degree since K is an F-subspace of E.

Now assume that both E/K and K/F have finite degree, and let B be an F-basis for K and C a K-basis for E. We claim that the product set

$$BC := \{bc\colon\, b\in B,\, c\in C\}$$

is a basis for E over F. This will establish the assertion in view of $|BC|=|B\times C|$, which holds as the mapping $B\times C\to BC$, $(b,c)\mapsto bc$ is easily seen to be injective. (Assume $bc=b'c'$, that is, $bc-b'c'=0$. If $c\neq c'$, then $b=b'=0$, since $b,b'\in K$ and

D. Hachenberger and D. Jungnickel, *Topics in Galois Fields*,
Algorithms and Computation in Mathematics 29,
https://doi.org/10.1007/978-3-030-60806-4_3

C is a K-basis of E, contradicting the linear independence of K over F. Therefore $c = c'$, and $b = b'$.)

We now check that BC is linearly independent over F. Assume $\sum_{bc \in BC} \lambda_{bc} bc = 0$, where all $\lambda_{bc} \in F$, that is,

$$\sum_{c \in C} \Big(\sum_{b \in B} \lambda_{bc} b \Big) c = 0.$$

As the inner sums are elements from K, the linear independence of C over K shows $\sum_{b \in B} \lambda_{bc} b = 0$ for all $c \in C$. Then the linear independence of B over F gives $\lambda_{bc} = 0$ for all $b \in B$ and all $c \in C$, as claimed.

Finally, consider an arbitrary element $v \in E$. Then there are scalars α_c in K with $v = \sum_{c \in C} \alpha_c c$. For each α_c, there exist scalars $\beta_{b,c} \in F$ such that $\alpha_c = \sum_{b \in B} \beta_{b,c} b$. This gives $v = \sum_{b \in B, c \in C} \beta_{b,c} bc$, and therefore BC generates E as F-vector space.

Thus BC is indeed a basis for E over F, and E/F has degree $[E:F] = |BC| = [E:K] \cdot [K:F]$. □

It is worthwhile to state the following fact established in the proof of Lemma 3.1.1 explicitly:

Corollary 3.1.2. *Let K be an intermediate field of a field extension E/F with finite degree, and assume that B is an F-basis for K and C a K-basis for E. Then BC is an F-basis for E.* □

We next show that any non-constant polynomial $f(x) \in F[x]$ has a root in a suitable extension field with finite degree:

Lemma 3.1.3. *Let $f(x) \in F[x]$ be a non-constant polynomial, and let $h(x) \in F[x]$ be any irreducible factor of $f(x)$. Then $K := F[x]/(h)$ is a field, and the canonical embedding*

$$F \to K, \quad \lambda \mapsto \lambda + (h)$$

of F into K gives a field extension K/F with finite degree $[K:F] = \deg h$. Moreover, the coset $x + (h)$ is a root of f in K.

Proof. By Observation 2.3.4, $F[x]/(h)$ is a field extension with F-basis

$$1 + (h),\ x + (h), \ldots,\ x^{m-1} + (h),$$

where $m = \deg h$. Evaluating f at the coset $x + (h)$ gives

$$f(x + (h)) = f(x) + (h) = (h) = 0$$

in K, since h divides f. □

As a consequence of Lemma 3.1.3, any non-constant polynomial $f(x)$ splits into linear factors in some extension field with finite degree:

Proposition 3.1.4. *Let F be a field and $f(x) \in F[x]$ any polynomial with degree $n \geq 1$. Then there exists a field extension E/F with degree $[E : F] \leq n!$ such that f splits over E into linear factors.*[1]

Proof. The assertion is trivial for $n = 1$: then f is linear and we may take $E = F$. For the general case $n \geq 2$, we choose any irreducible divisor $h(x)$ of f. By Lemma 3.1.3, there is a field extension K of F with $[K : F] = \deg h \leq \deg f = n$ such that h has a root v in K. Then also $f(v) = 0$, and we may write $f(x) = (x - v)g(x)$ for some $g(x) \in K[x]$ with degree $n - 1$. By induction on n, there exists an extension field E over K of degree $[E : K] \leq (n-1)!$ such that g splits into linear factors over E. Thus f also splits into linear factors over E, and the degree of E/F equals $[E : K] \cdot [K : F] \leq n \cdot (n-1)! = n!$, by Lemma 3.1.1. □

As mentioned before, we want to find a *smallest* extension of F for which a given polynomial splits into linear factors. Formally, we need the following

Definition 3.1.5. Let $f(x) \in F[x]$ be any non-constant polynomial. A field extension E of F is called a **splitting field** for f over F if the following three properties hold:

- The degree of E/F is finite.
- f splits in $E[x]$ into linear factors.
- If K is any intermediate field of E/F such that f splits in $K[x]$ into linear factors, then $K = E$. □

The proof of the existence of splitting fields requires some further preparation:

Observation 3.1.6. Let E be a field and $\mathcal{K}$ a collection of subfields of E. Then $\bigcap_{K \in \mathcal{K}} K$ is again a subfield of E. In particular, the intersection of *all* subfields of E is a subfield, which is called the **prime field** of E.[2]

Now let E/F be a field extension and S a subset of E. The intersection of all intermediate fields K of E/F with $S \subseteq K$ is denoted by $F(S)$. Thus $F(S)$ is itself an intermediate field of E/F, in fact the smallest subfield of E containing F and S as subsets; one calls $F(S)$ the subfield of E **generated by** S **over** F. If $S = \{v_1, \ldots, v_m\}$ is finite, we also write $F(v_1, \ldots, v_m)$ instead of $F(S)$.

In particular, let $S = \{v\}$ be a singleton set, so that $F(v)$ is the intermediate field of E/F generated by v over F. If v is transcendental over F, then $F(v)$ is the quotient field of $\mathrm{alg}_F(v)$ and therefore isomorphic to the rational function field $F(x)$ in the indeterminate x over F, since $\mathrm{alg}_F(v)$ is isomorphic to the F-algebra $F[x]$ in this case. Otherwise, v is algebraic over F and $F(v) = \mathrm{alg}_F(v)$, since the F-subalgebra $\mathrm{alg}_F(v)$ generated by v then is a field; see Theorem 2.4.11. □

Theorem 3.1.7. *Let F be a field and $f(x) \in F[x]$ a non-constant polynomial. Then there exists a splitting field for f over F.*

---

[1] As usual, $n!$ denotes the **factorial** of n, that is, $0! := 1$ and $n! := n \cdot (n-1)! = n(n-1)\cdots 1$ for $n \geq 1$.

[2] We shall return to this subfield in the next section.

Proof. By Proposition 3.1.4, there is an extension L/F of finite degree such that f splits over L into linear factors. Let $R \subseteq L$ be the set of roots of f in L. Then $E := F(R)$ is an intermediate field of L/F, and therefore E/F also has finite degree. Obviously, $f(x)$ splits over E into linear factors. Now assume that K is any intermediate field of E/F such that f splits over K into linear factors. Then $F \cup R \subseteq K$ and therefore $E = F(R) \subseteq K$, hence $K = E$. This shows that E is a splitting field of $f(x)$ over F. □

We are now going to prove that splitting fields are unique up to isomorphism, which will be immediate from Theorem 3.1.9 below. We require an auxiliary result and begin with the following simple observation. Let F and F' be fields, and let $\sigma : F \to F'$ be an isomorphism. Then σ has a natural extension to an isomorphism between the polynomial rings $F[x]$ and $F'[x]$ (which is again denoted by σ):

$$\sigma\Big(\sum_{i=0}^{n} f_i x^i\Big) := \sum_{i=0}^{n} \sigma(f_i)x^i.$$

Lemma 3.1.8. *Let F and F' be fields and $\sigma : F \to F'$ an isomorphism. Moreover, let E/F be a field extension and $v \in E$ an algebraic element with minimal polynomial* $\mathrm{mpol}_v(x)$ *over F, and put $K := F(v)$ and $K' := F'[x]/(\sigma(\mathrm{mpol}_v))$. Then K' is an extension field of F', and there exists an isomorphism $\rho : K \to K'$ extending σ.*

Proof. Note that $\sigma(\mathrm{mpol}_v(x))$ is irreducible in $F'[x]$, and thus K' is an extension field of F'. Trivially, $[K' : F'] = [K : F] = n$, where $n := \deg(\mathrm{mpol}_v)$. Then $\{1, v, \ldots, v^{n-1}\}$ is a basis of K/F, and any element $w = \sum_{i=0}^{n-1} \lambda_i v^i \in E$ can be identified with the polynomial $\sum_{i=0}^{n-1} \lambda_i x^i \in F[x]$ with degree at most $n-1$. Now

$$\rho(w) := \sum_{i=0}^{n-1} \sigma(\lambda_i)x^i + (\sigma(\mathrm{mpol}_v))$$

gives the desired extension of σ. □

Theorem 3.1.9. *Let F and F' be fields and $\sigma : F \to F'$ an isomorphism. Moreover, let $f(x) \in F[x]$ be a non-constant polynomial, and assume that E/F and E'/F' are splitting fields for $f(x)$ and $\sigma(f(x))$, respectively. Then there exists an isomorphism $\tau : E \to E'$ which extends σ.*

Proof. We use induction on the number m of roots of f which are not contained in the ground field F. If $m = 0$, then F and F' are themselves splitting fields and there is nothing to show.

Now assume $m \geq 1$ and let $v \in E \setminus F$ be a root of f in E with minimal polynomial $\mathrm{mpol}_v(x)$ over F. Then $\mathrm{mpol}_v(x)$ divides $f(x)$, say $f(x) = \mathrm{mpol}_v(x)g(x)$, and the degree of $\mathrm{mpol}_v(x)$ is at least 2. Let $v' \in E'$ be a root of $\sigma(\mathrm{mpol}_v(x))$ in E'. In view of Lemma 3.1.8, we may extend σ to an isomorphism ρ of $F(v)$ to $F'(v') = F'[x]/(\sigma(\mathrm{mpol}_v))$. By induction, ρ extends to an isomorphism $\tau : E \to E'$, since f has fewer roots in $E \setminus F(v)$ than in $E \setminus F$. This gives the desired extension of σ. □

We conclude this section with two related results on field extensions.

Definition 3.1.10. A field extension E/F is called **simple** if there exists an element $v \in E$ such that $F(v) = E$; such an element v is called a **generator** for E/F. □

The following fundamental result due to Ernst Steinitz [361] characterizes the simple field extensions with finite degree.

Theorem 3.1.11 (Steinitz's theorem). *A field extension E/F with finite degree is simple if and only if the number of intermediate fields of E/F is finite.*

Proof. The assertion is trivial if $E = F$, and thus we may assume $E \neq F$.

First let E/F be a simple extension, say $E = F(u)$, and consider the minimal polynomial $\mathrm{mpol}_{u,F}(x)$ of u over F. Given any monic polynomial divisor $a(x)$ of $\mathrm{mpol}_{u,F}$ in $E[x]$, we obtain an intermediate field $L_a := F(C_a)$ of E/F, where C_a is the set of coefficients of a. Note that the number of intermediate fields of this type is finite, as $\mathrm{mpol}_{u,F}$ has at most 2^n monic divisors in $E[x]$, where $n = \deg(\mathrm{mpol}_{u,F})$. Hence it suffices to show that there are no other intermediate fields of E/F.

Thus let K be any intermediate field of E/F, and let $h(x) := \mathrm{mpol}_{u,K}(x)$ be the minimal polynomial of u over K. Then h is a monic divisor of $\mathrm{mpol}_{u,F}(x)$ in $E[x]$. Consider the intermediate field L_h. Since $C_h \subseteq K$, we have $L_h \subseteq K$. Let $g(x) = \mathrm{mpol}_{u,L_h}(x)$ be the minimal polynomial of u over L_h. Then h divides g, as $L_h \subseteq K$. On the other hand, $h(x) \in L_h[x]$ by the definition of L_h, and therefore actually $h = g$. In view of $K(u) = E = L_h(u)$, this implies $[E : L_h] = \deg g = \deg h = [E : K]$. Now $L_h \subseteq K$ gives $K = L_h$, as desired.

Conversely, let the number of intermediate fields of E/F be finite. We first assume that E and F are infinite fields and choose an arbitrary element $u \in E \setminus F$. If $E = F(u)$, there is nothing to prove. Thus let $v \in E \setminus F(u)$ and let $L := F(u,v)$ be the intermediate field generated by u and v. We denote the set of all intermediate fields of E/F by $\mathscr{K}$ and consider the mapping

$$F \to \mathscr{K}, \quad \lambda \mapsto F(u + \lambda v).$$

Note that this mapping cannot be injective, since the number of intermediate fields of E/F is finite, whereas F is infinite. Thus there are distinct elements $\lambda, \mu \in F$ with $K := F(u + \lambda v) = F(u + \mu v)$. Then

$$v = \frac{(u + \lambda v) - (u + \mu v)}{\lambda - \mu} \in K \quad \text{and} \quad u = (u + \lambda v) - \lambda v \in K,$$

so that $F(u,v) = K = F(u + \lambda v)$ is a simple extension of F. As $[E : F(u,v)]$ is strictly smaller than $[E : F(u)]$, the assertion follows by induction on the degree of the field extensions considered.

Finally, the case where F and E are finite fields will be an immediate consequence of Theorem 3.2.12. □

Definition 3.1.12. Let E/F be a field extension, and let K and L be intermediate fields of E/F. Then the smallest intermediate field of E/F containing both K and L (that is, the intersection of all intermediate fields containing both K and L) is denoted by KL and called the **compositum** of K and L. Moreover, K and L are said to be **linearly disjoint** provided that $K \cap L = F$ and $KL = E$. □

Lemma 3.1.13. *Let E/F be a field extension with finite degree, and let K and L be linearly disjoint intermediate fields of E/F such that $[K:F] = [E:L]$. Then every F-basis of K is an L-basis of E, and every F-basis of L is a K-basis of E.*

Proof. It suffices to check the first assertion. Thus let B be a basis of K/F and consider $\langle B \rangle_L$, the L-span of B. Since $1 \in K = \langle B \rangle_F \subseteq \langle B \rangle_L$ and $L \subseteq \langle B \rangle_L$, we obtain $E = KL \subseteq \langle B \rangle_L$ and hence $E = \langle B \rangle_L$. Now the hypothesis $|B| = [K:F] = [E:L]$ shows that B is an L-basis for E. □

Combining Lemma 3.1.13 with Corollary 3.1.2 yields the following useful result:

Corollary 3.1.14. *Let E/F be a field extension with finite degree, and let K and L be linearly disjoint intermediate fields of E/F such that $[K:F] = [E:L]$. Assume that B and C are F-bases for K and L, respectively. Then BC is an F-basis for E.* □

3.2 Basics on Fields and Field Extensions

In this section, we discuss some further basic facts concerning general fields and field extensions. In particular, we will consider roots of unity, the characteristic and the prime field of a field, Galois automorphisms, and perfect fields.

We start by investigating the multiplicative group $(K^*, \cdot, 1)$ of an arbitrary field K. For simplicity, we will just write K^* throughout.

Observation 3.2.1. Let K be a field and u any element of K^*. Then the mapping

$$\alpha_u : \mathbb{Z} \to K^*, \quad z \to u^z \tag{3.1}$$

is a homomorphism from the additive group of the ring $\mathbb{Z}$ of integers into the multiplicative group K^* of K, and the image of α_u is the subgroup of K^* generated by u; see Section 1.6. By Theorem 1.6.2, there is a unique $n \in \mathbb{N}$ such that the kernel of α_u equals $n\mathbb{Z}$.

- If $n \neq 0$, then n is called the **multiplicative order** of u (written as $\mathrm{ord}(u) = n$), as the image of $\mathbb{Z}$ under α_u is a (cyclic) subgroup of K^* with n elements in this case.
- If $n = 0$, then u is said to have **infinite order**, and we write $\mathrm{ord}(u) = \infty$; then the image of $\mathbb{Z}$ under α_u is isomorphic to $(\mathbb{Z}, +, 0)$. Of course, this can only occur when K is an infinite field. □

The elements $u \in K^*$ for which the associated mapping α_u is not injective are of special interest:

Definition 3.2.2. Let K be a field and u an element of K^* with finite order. Then u is called a **root of unity** in K. More specifically, if n is any positive multiple of $\mathrm{ord}(u)$ (so that $u^n = 1$), then u is said to be an **n-th root of unity**. In the special case $n = \mathrm{ord}(u)$, one calls u a **primitive n-th root of unity**. □

We can now prove an important structural restriction for the multiplicative group of a field:

Theorem 3.2.3. *Let K be an arbitrary field. Then any finite subgroup of K^* is necessarily cyclic.*

Proof. Let U be a finite subgroup of K^*, and denote the exponent of U by e. By definition, $\mathrm{ord}(u)$ divides e for every $u \in U$, and therefore $u^e = 1$ or, equivalently, $u^e - 1 = 0$ for all $u \in U$. Thus the polynomial $x^e - 1 \in K[x]$ has at least $|U|$ distinct roots in K, which implies $|U| \leq e = \deg(x^e - 1)$, by Proposition 2.4.12. On the other hand, there is an element $v \in U$ with $\mathrm{ord}(v) = e$, by Theorem 1.6.21. Hence $|U| = e$, and U is the cyclic group generated by v. □

Corollary 3.2.4. *Let K be a field, and let U and V be finite subgroups of K^*. Then $UV := \{uv \colon u \in U, v \in V\}$ is likewise a finite subgroup of K^*. Moreover, $|UV| = \mathrm{lcm}(|U|, |V|)$ and $|U \cap V| = \gcd(|U|, |V|)$. In particular, K^* contains at most one subgroup of any prescribed finite order.*

Proof. Clearly, UV is a finite subgroup of K^*, since K^* is commutative. Consider the mapping $\gamma \colon U \times V \to UV$ with $(u, v) \mapsto uv$. Since $U \times V$ is a group with respect to componentwise multiplication, γ is an epimorphism, with kernel $U \cap V$. Hence

$$|UV| = \frac{|U \times V|}{|U \cap V|} = \frac{|U||V|}{|U \cap V|} = \frac{mn}{|U \cap V|},$$

where $m := |U|$ and $n := |V|$. Since K is a field, UV is cyclic by Theorem 3.2.3. Hence U and V are the unique subgroups of UV of orders m and n, respectively. Consequently, $|U \cap V| = \gcd(m, n)$ and thus $|UV| = mn/\gcd(m, n) = \mathrm{lcm}(m, n)$. In particular, this gives $U = V$ for the special case $m = n$. □

Theorem 3.2.3 contains the following important result on the structure of finite fields:

Theorem 3.2.5. *The multiplicative group of a finite field is cyclic.* □

Definition 3.2.6. Let K be a finite field. Then every generator of its multiplicative group is called a **primitive element** of K. □

By Proposition 1.6.18, the number of primitive elements of a finite field K equals $\phi(q-1)$, where $q = |K|$ is the cardinality of K and where ϕ is Euler's totient function. Hence we get an explicit formula for the number of primitive elements, provided that we can compute the prime power factorization of $q - 1$; see Remark 1.9.13.

Next, we consider the analogue of the mapping given in (3.1) for the additive group of a field K: given any element $u \neq 0$ in K, we define a mapping

$$\beta_u : \mathbb{Z} \to K, \quad z \mapsto zu. \tag{3.2}$$

This is a group homomorphism from the additive group of the ring of integers into the additive group of K, and the image of β_u is the subgroup of $(K, +, 0)$ generated by u. The kernel of β_u is a subgroup of $(\mathbb{Z}, +, 0)$, say $p_u\mathbb{Z}$, where $p_u \in \mathbb{N}$. Now let v be a further non-zero element of K. Then

$$p_u v = (p_u u)(u^{-1}v) = 0 \cdot (u^{-1}v) = 0,$$

so that p_v is divisible by p_u. By symmetry, p_u also divides p_v. Thus there exists a constant $p \in \mathbb{N}$ such that $\ker \beta_u = p\mathbb{Z}$ for every $u \in K^*$. This leads to the following definition:

Definition 3.2.7. Let K be a field. The unique number $p \in \mathbb{N}$ such that $p\mathbb{Z} = \ker \beta_v$ for every non-zero $v \in K$ is called the **characteristic** of K and denoted by $\operatorname{char} K$. We say that K has **positive characteristic** if $\operatorname{char} K > 0$, and **characteristic zero** otherwise. In addition, the characteristic of an integral domain is defined to be the characteristic of its quotient field. □

If the field K has characteristic zero, then β_u is injective for all non-zero $u \in K$ and its image is isomorphic to the additive group of $\mathbb{Z}$. In particular, K has to have infinite order in this case. In other words, any finite field necessarily has positive characteristic.

Proposition 3.2.8. *Assume that K is a field with positive characteristic p. Then p is a prime number.*

Proof. Putting $u = 1$ in (3.2) gives

$$\beta_1(xy) = (xy) \cdot 1 = (x \cdot 1) \cdot (y \cdot 1) = \beta_1(x) \cdot \beta_1(y),$$

and hence β_1 is even a ring homomorphism from $\mathbb{Z}$ to K. By hypothesis, the kernel of β_1 is the ideal $p\mathbb{Z}$, where $p \in \mathbb{N}^*$. Of course, $p \neq 1$, for otherwise $1 = 1 \cdot 1 = 0$ in K, a contradiction. Now assume $p = mn$ with $m, n \in \mathbb{N}$, $m, n \neq 1$, so that $0 = \beta_1(p) = \beta_1(mn) = \beta_1(m)\beta_1(n)$. Since K is a domain, $\beta_1(m) = 0$ or $\beta_1(n) = 0$, and therefore $p \mid n$ or $p \mid m$, a contradiction. □

The proof of Proposition 3.2.8 shows that the image of β_1 is a subfield P of K which is isomorphic to $\mathbb{Z}_p := \mathbb{Z}/p\mathbb{Z}$, the field of residues modulo the prime p (when K has positive characteristic p). Note that P is the subfield of K generated by 1, and therefore the intersection of all subfields of K.

On the other hand, when K has characteristic zero, the image of β_1 is isomorphic to the ring $\mathbb{Z}$. In this case, β_1 extends to a unique monomorphism $\overline{\beta}_1$ from the quotient field $\mathbb{Q}$ of $\mathbb{Z}$ to K, by Theorem 1.4.11, and the image of $\overline{\beta}_1$ is a subfield P of K isomorphic to $\mathbb{Q}$; again, P is the intersection of all subfields of K.

Definition 3.2.9. Let K be a field. Then the intersection of all subfields of K is called the **prime field** of K. □

Thus the prime field of a field K is either isomorphic to the field $\mathbb{Q}$ of rational numbers (with characteristic zero) or to one of the fields $\mathbb{Z}_p$ of residues modulo p, where $p \in \mathbb{N}$ is a prime (with characteristic p). Let us note some consequences of these observations for Galois fields:

Corollary 3.2.10. *Let $p \in \mathbb{N}$ be a prime. If K is a Galois field with p elements, then K is isomorphic to the residue field $\mathbb{Z}_p$. In particular, any two fields with p elements are isomorphic.* □

We postpone the generalization of Corollary 3.2.10 for arbitrary Galois fields to the next section and just give an immediate necessary condition for the existence of Galois fields now:

Theorem 3.2.11. *Let K be a finite field. Then there exist a prime p and a positive integer m such that K has cardinality p^m.*

Proof. Let p be the characteristic of K, and let P be its prime field. By hypothesis, K/P is a field extension of finite degree $m := [K : P]$. Thus K is an m-dimensional vector space over P, and hence $|K| = p^m$. □

We will show in Section 3.3 that there exists a unique (up to isomorphism) finite field with q elements for every prime power $q \geq 2$.

Theorem 3.2.12. *Let K be a finite field. Then K is a simple extension over its prime field P.*

Proof. By Theorem 3.2.5, K^* is cyclic. This gives $K = P(u)$, where u is any primitive element for K. □

Next, we provide a further simple but useful consequence of Theorem 3.2.5 for finite fields with odd characteristic:

Proposition 3.2.13. *Let K be a finite field with q elements, and assume that q is odd. Then the set*

$$K^{\square} := \{v^2 : v \in K, v \neq 0\}$$

of non-zero squares of K is the unique subgroup of index 2 of the multiplicative group K^*. Moreover, $-1 \in K^{\square}$ if and only if $q \equiv 1 \bmod 4$.

Proof. Consider the map γ on K^* which sends every element to its square. Clearly, γ is a group homomorphism, and the kernel of γ is the set of roots of the polynomial $x^2 - 1$, which splits as $(x-1)(x+1)$ over K. Thus the kernel of γ is $\{1, -1\}$ and has cardinality 2, since $1 \neq -1$, as K has odd characteristic. On the other hand, the image of γ is $K^{\square}$, and the homomorphism theorem for groups (Theorem 1.3.10) yields

$$|K^{\square}| = |K^*/\ker\gamma| = \frac{q-1}{2}.$$

Now assume $\alpha^2 = -1$ for some $\alpha \in K$. Then α has order 4 in K^* (since $\alpha^4 = 1$ but $\alpha^2 = -1 \neq 1$), and thus 4 has to divide $q-1$. Conversely, let this condition be satisfied. Then we may select a primitive element ω of K and put $\alpha := \omega^{(q-1)/4}$, so that $\alpha^2 = \omega^{(q-1)/2}$ has order 2. Since -1 is the only element of K^* with order 2, we conclude $\alpha^2 = -1$, and thus $-1 \in K^{\square}$. □

We continue with an interesting application of the binomial theorem for fields with positive characteristic.

Theorem 3.2.14. *Let K be a field with positive characteristic p, and let $t \in \mathbb{N}$. Then the mapping*

$$\sigma_t : K \to K, \quad w \mapsto w^{p^t}$$

is an injective ring homomorphism fixing the prime field of K elementwise. Moreover, ψ_t is an automorphism of K provided that K is finite.

Proof. Let $a, b \in K$. Trivially,

$$\sigma_t(ab) = (ab)^{p^t} = a^{p^t} \cdot b^{p^t} = \sigma_t(a)\sigma_t(b).$$

In order to prove $\sigma_t(a+b) = \sigma_t(a) + \sigma_t(b)$, we use the well-known binomial formula

$$(a+b)^p = \sum_{i=0}^{p} \binom{p}{i} a^i b^{p-i} = \sum_{i=0}^{p} \frac{p!}{i!(p-i)!} \cdot a^i b^{p-i}.$$

For $i \in \{1, \ldots, p-1\}$, the binomial coefficient $\binom{p}{i}$ is divisible by p, as p is a prime. Since p is the characteristic of K, this implies $(a+b)^p = a^p + b^p$. By induction, $(a+b)^{p^t} = a^{p^t} + b^{p^t}$ for any power p^t of p. (Note that the case $t = 0$ also holds, since σ_0 is the identity on K.)

Thus σ_t is indeed a ring homomorphism, and σ_t is injective as $a^{p^t} = 0$ implies $a = 0$. Since 0 and 1 are fixed by σ_t, the prime field of K (which is generated by 1) is fixed elementwise. The final assertion holds, since any injective map on a finite set is a bijection. □

Definition 3.2.15. Let E/F be a field extension. Then an automorphism of E fixing every element of F is said to be a **Galois automorphism** of E/F. Clearly, the set of all Galois automorphisms of E/F forms a group with respect to the composition of mappings, the **Galois group** $\mathrm{Gal}(E/F)$ of E/F. □

Example 3.2.16. Let P be the prime field of some field E. Then every automorphism σ of E is a Galois automorphism of E/P, since 1 generates P and is fixed by σ.

Now let K be a finite field with characteristic p. Then Theorem 3.2.14 shows that $\sigma_t : K \to K$, $w \mapsto w^{p^t}$ is a Galois automorphism of K/P for every $t \in \mathbb{N}$. □

We shall study the Galois group of an extension of finite fields in Section 3.4 in more detail. Also, some basic results on the general theory of so-called Galois extensions will be given in Section 3.8.

We conclude this section with an important concept in the theory of splitting fields which depends on the characteristic of the underlying field; our treatment follows that of Jacobson [204].

Definition 3.2.17. Let F be a field. An irreducible polynomial $g(x) \in F[x]$ is called **separable** if it has no multiple roots in its splitting field. If every irreducible polynomial in $F[x]$ is separable, one says that F is **perfect**.

Given a field extension E/F, an element $v \in E$ is called **separable** over F provided that v is algebraic over F and $\text{mpol}_{v,F}(x)$ is separable. The extension E/F is called **separable** if every $v \in E$ is separable over F. □

Note that a field F is perfect if and only if all its algebraic extensions are separable. As we shall see, this is in some sense the typical case. Let us assume the existence of some irreducible polynomial $f(x) \in F[x]$ which is *not* separable, so that f has multiple roots in its splitting field. By Corollary 2.4.18, f and its formal derivative f' have a non-trivial common factor in $F[x]$, and hence f divides f' (since f is irreducible). As the degree of f is strictly smaller than that of f', we conclude that f' is the zero polynomial.

Now let $f(x) = \sum_{j=0}^{m} f_j x^j$, where $m \geq 2$. Then $f'(x) = \sum_{j=1}^{m} j f_j x^{j-1}$, and $f' = 0$ implies $jf_j = 0$ for $j = 1, \ldots, m$. If $f_j \neq 0$ for some j, we must have $j = 0$ in F, and hence F has a positive characteristic dividing j, say p. We conclude that f necessarily has the form

$$f(x) = f_0 + \sum_{k=1}^{\ell} \alpha_k x^{kp}.$$

We claim that F cannot be a finite field. Otherwise Theorem 3.2.14 shows the existence of elements $\beta_k \in F$ with $\beta_k^p = \alpha_k$ (for $k = 1, \ldots, \ell$) and of an element g_0 with $g_0^p = f_0$, which gives

$$f(x) = \left(g_0 + \sum_{k=1}^{\ell} \beta_k x^k\right)^p,$$

contradicting the irreducibility of f. Thus we have proved the following preliminary result:

Proposition 3.2.18. *A field which is not perfect is necessarily an infinite field with positive characteristic. In particular, all finite fields are perfect.* □

For the sake of completeness, we include an explicit example for a non-separable irreducible polynomial. For this, we need the following auxiliary result, which is of independent interest:

Proposition 3.2.19. *Let F be a field with positive characteristic p, and let $b \in F$. Then either $x^p - b$ is irreducible, or b is a p-th power in F and $x^p - b$ is a p-th power in $F[x]$.*

Proof. Assume that $f(x) = x^p - b$ is reducible in $F[x]$, say $f(x) = g(x)h(x)$, where g is a monic polynomial of degree m with $1 \leq m \leq p-1$. Let β be a root of g in its splitting field. Then

$$\beta^p - b = f(\beta) = g(\beta)h(\beta) = 0$$

gives $b = \beta^p$. Thus

$$x^p - b = x^p - \beta^p = (x - \beta)^p = g(x)h(x),$$

and hence $g(x) = (x-\beta)^m$, so that $\beta^m \in F$. As $\gcd(m,p) = 1$, we have $1 = sm + tp$ for suitable integers s and t, which gives

$$\beta = \beta^{sm+tp} = (\beta^m)^s \cdot (\beta^p)^t = (\beta^m)^s \cdot b^t \in F.$$

Thus $b = \beta^p$ is indeed a p-th power in F. □

Example 3.2.20. Let F be the rational function field $\mathbb{Z}_p(t)$ in the indeterminate t over $\mathbb{Z}_p$, and consider the polynomial $f(x) := x^p - t$ over F. We claim that f is irreducible in $F[x]$. By Proposition 3.2.19, it suffices to show that t is not a p-th power in F. Assume otherwise, say

$$t = \left(\frac{a_0 + a_1 t + \cdots + a_m t^m}{b_0 + b_1 t + \cdots + b_n t^n}\right)^p$$

in F, which gives the relation

$$\left(b_0^p + b_1^p t^p + \cdots + b_n^p t^{np}\right) \cdot t = a_0^p + a_1^p t^p + \cdots + a_m^p t^{mp}$$

in $\mathbb{Z}_p[t]$. Now the linear independence of the monomials in $\mathbb{Z}_p[t]$ gives the contradiction $b_0 = b_1 = \cdots = b_n = 0$. Thus f is indeed irreducible in $F[x]$. Also, f has multiple roots, since $f' = 0$; in fact, f is the p-th power $f(x) = (x - \tau)^p$ over its splitting field E, where τ is a root of f in $E[x]$. □

Now it is also easy to complete the partial characterization of the perfect fields given in Proposition 3.2.18, using a connection to Theorem 3.2.14:

Theorem 3.2.21. *Let F be an infinite field of positive characteristic p, and let $F^{(p)}$ denote the set of all p-th powers in F, that is, the image of the monomorphism σ_1 defined in Theorem 3.2.14. Then F is perfect if and only if $F^{(p)} = F$ (that is, if σ_1 is actually an automorphism of F).*

Proof. Assume $F^{(p)} \neq F$, and choose an element $b \in F \setminus F^{(p)}$. By Proposition 3.2.19, $f(x) := x^p - b$ is irreducible. Then f is not separable (since $f' = 0$), and hence F is not perfect.

Conversely, assume that F is not perfect and let $f(x)$ be a non-separable irreducible polynomial over F. As we have seen before, f has the form

$$f(x) = f_0 + \sum_{k=1}^{\ell} \alpha_k x^{kp}.$$

Since f is irreducible, it cannot be a p-th power, which means that at least one of its coefficients is not a p-th power in F, so that $F^{(p)} \neq F$. □

3.3 Finite Fields: Existence and Uniqueness

This section contains the fundamental structural results on finite fields which have already been mentioned. Let us first summarize what we have established up to now. Thus let K be some finite field. Then:

- The multiplicative group K^* of K is cyclic (Theorem 3.2.5).
- K is a finite extension over its prime field P, and therefore $|K| = p^m$, where $p = |P|$ is the characteristic of K and $m = [K : P]$ is the degree of K over P (Theorem 3.2.11). Moreover, P is isomorphic to $\mathbb{Z}_p$, the field of residues modulo p (Corollary 3.2.10).
- K/P is a simple field extension, that is, $K = P(u)$ for a suitable element $u \in K$. Here u can be chosen as a primitive element of K, that is, as a generator of the multiplicative group of K (Theorem 3.2.12).
- Let u be an arbitrary generator of K over P. Then the evaluation homomorphism $\Gamma_u\colon P[x] \to K$, $f(x) \mapsto f(u)$ is surjective. Therefore K is isomorphic to $P[x]/(g)$, where $g(x) := \operatorname{mpol}_u(x)$ is the minimal polynomial of u over P; thus g is a monic irreducible polynomial with degree m (see the proof of Theorem 2.4.11).

Since the multiplicative group of K is cyclic of order $p^m - 1$, we have $u^{p^m-1} = 1$ for every $u \in K^*$. Therefore the monic polynomial $x^{p^m-1} - 1 \in P[x]$ splits over K into linear factors:

$$x^{p^m-1} - 1 = \prod_{v \in K^*} (x - v), \tag{3.3}$$

which implies the fundamental identity

$$x^{|K|} - x = x^{p^m} - x = \prod_{v \in K} (x - v). \tag{3.4}$$

This shows that K is a splitting field for $x^{p^m} - x$ over P and suggests constructing finite fields as splitting fields.

Theorem 3.3.1. *Let p be any prime and m any positive integer. Then there exists a finite field with cardinality p^m.*

Proof. We write $q := p^m$. Let P be the residue field $\mathbb{Z}_p$, and let K be the set of roots of the polynomial $x^q - x$ over P in its splitting field L. The derivative of $x^q - x$ is $qx^{q-1} - 1 = -1$ and therefore relatively prime to $x^q - x$. By Corollary 2.4.18, $x^q - x$ has no multiple roots in L, and therefore $|K| = q$. Note that P is a subset of K, since $\lambda^p = \lambda$ for every $\lambda \in P$. Now let $u, v, w \in K$ with $w \neq 0$. Then

$$\left(\frac{u-v}{w}\right)^q = \frac{u^q - v^q}{w^q} = \frac{u-v}{w},$$

and hence also $\frac{u-v}{w} \in K$. Now Theorem 1.4.22 shows that K is a subfield of L and thus coincides with L. Therefore, the splitting field of $x^q - x$ is a field with q elements. □

In view of the uniqueness of splitting fields, the following result is not surprising:

Theorem 3.3.2. *Let K and K' be two finite fields with p^m elements each, where $p \in \mathbb{N}$ is a prime and m a positive integer. Then K and K' are isomorphic.*

Proof. Let P and P' be the prime fields of K and K' respectively. By Corollary 3.2.10, both P and P' are isomorphic to the residue field $\mathbb{Z}_p$; let σ be an isomorphism from P to P'. As we have seen, both K and K' are splitting fields for $x^{p^m} - x$, considered as a polynomial in $P[x]$ and $P'[x]$, respectively. Hence σ extends to an isomorphism between K and K', by Theorem 3.1.9. □

Remark 3.3.3. In view of the importance of Theorem 3.3.2, we will make the argument used to prove this result a little more explicit. Let K, K', P, P' and σ be as in the above proof. Then σ maps the unit element of P to the unit element of P', that is, $\sigma(1) = 1$, where we identify these two elements for convenience.

Now let $u \in K$ be a primitive element with minimal polynomial $g(x) := \mathrm{mpol}_u(x)$ over P, and put $h(x) := \sigma(g(x)) \in P'[x]$. Then h is monic and irreducible, since any decomposition $h(x) = a(x)b(x)$ in $P'[x]$ would give rise to a corresponding decomposition $g(x) = \sigma^{-1}(a(x))\sigma^{-1}(b(x))$ in $P[x]$. Since $g(u) = 0$, the polynomial $g(x)$ divides $x^{p^m} - x$ in $P[x]$, and therefore $h(x)$ divides $\sigma(x^{p^m} - x) = x^{p^m} - x$ in $P'[x]$. But $x^{p^m} - x$ splits over K' as $\prod_{w \in K'}(x - w)$, and therefore $h(x)$ has a root $v \in K'$. The kernel of the P'-algebra homomorphism

$$\Gamma_v\colon P'[x] \to K', \quad f(x) \mapsto f(v)$$

is the ideal generated by $h(x)$, and its image $P'(v)$ is an intermediate field of K'/P' with degree $\deg h = m = [K' : P']$; thus $K' = P'(v)$. As $\sigma\big(f(x) \bmod g(x)\big) = \sigma(f(x)) \bmod h(x)$ for every $f(x) \in P[x]$, the mapping

$$\tau\colon K \to K', \quad \sum_{j=0}^{m-1} \lambda_j u^j \mapsto \sum_{j=0}^{m-1} \sigma(\lambda_j) v^j$$

is an isomorphism between K and K'. □

Notation 3.3.4. Let $q \neq 1$ be any prime power. Then the unique (up to isomorphism) finite field with q elements is called the **Galois field** with q elements; it is denoted by $\mathrm{GF}(q)$ or by $\mathbb{F}_q$.[3] □

We conclude this section with a summary of the main results on finite fields, which also serves to fix the notation which will be used throughout.

Theorem 3.3.5. *Let p be an arbitrary prime, let m and n be arbitrary positive integers, and write $q := p^m$. Then:*

(1) *There exists a (unique) finite field F with q elements, namely* $\mathrm{GF}(q)$.

[3] Even though the notation $\mathbb{F}_q$ is more common, we prefer to use $\mathrm{GF}(q)$ in this book for better readability.

(2) *For every finite field F, there exists a (simple) extension E/F with degree n.*

(3) *For every finite field F, there exists a monic irreducible polynomial of degree n in $F[x]$.*

(4) *If E/F is an extension of finite fields with $|F| = q$ and $|E| = q^n$, then E is isomorphic to $F[x]/(f)$, where f is an arbitrary monic irreducible polynomial with degree n over F.*

Proof. Part (1) holds in view of Theorems 3.3.1 and 3.3.2. For (2), let $F = q$ and apply the argument in the proof of Theorem 3.3.1 to the polynomial $x^{q^n} - x$ over F. Then the splitting field of this polynomial is a field with q^n elements, that is, an n-dimensional extension over F.

For part (3), let E/F be an extension as in (2) and v a primitive element of E. Then $\Gamma_v : F[x] \to E$, $f(x) \mapsto f(v)$ is surjective, and its kernel is an ideal which is generated by a monic irreducible polynomial with degree n over F.

Finally, (4) follows from the fact that any two finite fields with the same cardinality are isomorphic. □

Alternatively, one could prove the existence of finite fields for all prime powers by directly establishing the existence of irreducible polynomials for any given degree. We shall do so in Section 3.5 by giving a formula for the number of monic irreducible polynomials, after studying automorphisms and intermediate fields of extensions of Galois fields in the next section.

3.4 Finite Fields: Extensions and Galois Automorphisms

In this section, we establish further fundamental properties of an extension E/F of Galois fields. We start by considering the intermediate fields of E/F. This requires a simple auxiliary result:

Lemma 3.4.1. *Let q be an integer different from $0, 1$ and -1, and let d and n be two positive integers. Then $q^d - 1$ divides $q^n - 1$ if and only if d divides n.*

Proof. Assume first that $d \mid n$. With $Q = q^d$ and $k = \frac{n}{d}$, we have

$$q^n - 1 = Q^k - 1 = (Q - 1) \cdot \left(\sum_{j=0}^{k-1} Q^j \right),$$

and therefore $Q - 1 = q^d - 1$ divides $q^n - 1$.

Conversely, assume that $q^d - 1$ divides $q^n - 1$ and write $n = ad + r$ with $0 \leq r < d$. Since

$$q^n - 1 = (q^d - 1)q^{n-d} + (q^{n-d} - 1),$$

the integer $q^d - 1$ also divides $q^{n-d} - 1$. By induction, $q^d - 1$ divides $q^r - 1$. As $r < d$, we conclude $q^r - 1 = 0$. Therefore $r = 0$, so that d divides n. □

Theorem 3.4.2. *Consider an extension E/F of Galois fields, where $|F| = q$, and let n be the degree of E/F. Then any intermediate field of E/F has cardinality q^d for some divisor d of n. Moreover, for every divisor d of n there exists a unique intermediate field K_d of E/F with $|K_d| = q^d$.*

Proof. First let K be any intermediate field of E/F. Then the degree formula in Lemma 3.1.1 shows that $[K:F]$ divides $n = [E:F]$, and therefore $|K| = q^d$ where $d = [K:F]$ is a divisor of n. Moreover,

$$K \subseteq \{u \in E \colon u^{q^d} - u = 0\},$$

since $v^{q^d-1} = 1$ for every $v \in K^*$. Thus all q^d elements of K have to be roots of the polynomial $x^{q^d} - x$ of degree q^d over F in E, and therefore K is the only intermediate field of E/F with q^d elements.

It remains to show that such an intermediate field with q^d elements exists for every divisor d of n. By Lemma 3.4.1, $q^d - 1$ divides $q^n - 1$, which is the cardinality of the multiplicative group E^* of E. Since E^* is cyclic, there is a unique subgroup U of E^* with order $q^d - 1$. Then $u^{|U|} = u^{q^d-1} = 1$ for every $u \in U$, so that

$$x^{q^d} - x = \prod_{w \in L} (x - w),$$

where $L := U \cup \{0\} \subseteq E$. Note $F \subseteq L$, as $w^q = w$ for all $w \in F$. As in the proof of Theorem 3.3.1, one sees that L is actually a field. This establishes the existence of the desired intermediate field with cardinality q^d. □

Remark 3.4.3. Let E/F be an extension of Galois fields with degree n, and let k and ℓ be divisors of n. Consider the intermediate fields K_ℓ and K_m of E/F with degrees ℓ and m over F, respectively. Then

$$K_\ell \cap K_m = K_{\gcd(\ell,m)}$$

is the intermediate field of degree $\gcd(\ell,m)$ over F, and the compositum of K_ℓ and K_m is the intermediate field $K_{\operatorname{lcm}(\ell,m)}$ of degree $\operatorname{lcm}(\ell,m)$ over F.

In the particular case $\gcd(\ell,m) = 1$ and $n = \ell m$, the intermediate fields K_ℓ and K_m are linearly disjoint, and the additional assumption

$$[K_\ell : F] = \ell = \frac{n}{m} = [E : K_m]$$

of Lemma 3.1.13 is satisfied. Consequently, any F-basis for K_ℓ is a K_m-basis for E, and vice versa. □

Proposition 3.4.4. *Let E/F be an extension of Galois fields, w an element of E, and $K = F(w)$ the intermediate field of E/F generated by w. Then K is the splitting field of the minimal polynomial* $\operatorname{mpol}_w(x)$ *of w over F.*

Proof. Denote the degree of $\operatorname{mpol}_w(x)$ by d. Then $|K| = |F(w)| = q^d$, and therefore d divides n. Moreover, $\operatorname{mpol}_w(x)$ divides $x^{q^d} - x$, since $w^{q^d} - w = 0$. Finally, $x^{q^d} - x$

splits over K as $\prod_{u\in K}(x-u)$, so that $\mathrm{mpol}_w(x)$ also splits in $K[x]$ into linear factors. Hence $K=F(w)$ is indeed the splitting field of $\mathrm{mpol}_w(x)$. □

We now give an explicit description for the set of roots of the minimal polynomials considered in Proposition 3.4.4. Recall that such a polynomial cannot have multiple roots, by Proposition 3.2.18.

Proposition 3.4.5. *Let E/F be an extension of Galois fields, where $F=\mathrm{GF}(q)$, and let w be an element of E with degree d over F. Then the roots of the minimal polynomial* $\mathrm{mpol}_w(x)$ *of w in E are the distinct elements*

$$w,\, w^q,\, w^{q^2},\ldots,\, w^{q^{d-1}},$$

and hence mpol_w *splits in $E[x]$ as*

$$\mathrm{mpol}_w(x) \;=\; \prod_{j=0}^{d-1}\left(x-w^{q^j}\right).$$

Proof. Because of $\lambda^q=\lambda$ for all $\lambda\in F$, the mapping $\sigma_j\colon E\to E,\ u\mapsto u^{q^j}$ is a Galois automorphism of E/F for all $j\in\mathbb{N}$; see Theorem 3.2.14. Now let $f(x)=\sum_{i=0}^m f_ix^i$ be a polynomial over F, and assume that w is root of f. Applying σ_j to the condition $f(w)=0$ gives

$$f(w^{q^j}) = \sum_{i=0}^m f_iw^{iq^j} = \left(\sum_{i=0}^m f_iw^i\right)^{q^j} = f(w)^{q^j} = 0$$

for all $j\in\mathbb{N}$. This shows $\{w^{q^j}: j\in\mathbb{N}\}\subseteq R$, where R is the set of roots of mpol_w in E. Since $w^{q^d}=w$, we obtain $w^{q^j}=w^{q^{j \bmod d}}$ for all $j\in\mathbb{N}$ and thus

$$\{w^{q^j}:\ j\in\mathbb{N}\} \;=\; \{w^{q^r}:\ r\in\mathbb{N},\ 0\le r\le d-1\}.$$

As mpol_w has degree d, it only remains to check that $w, w^q,\ldots,w^{q^{d-1}}$ are distinct. Assume otherwise, say $w^{q^i}=w^{q^j}$, where $0\le i<j\le d-1$. Then

$$(w^{q^j})^{q^{d-i}} = w^{q^j\cdot q^{d-i}} = w^{q^{j+d-i}} = w^{q^d\cdot q^{j-i}} = (w^{q^d})^{q^{j-i}} = w^{q^{j-i}},$$

and, by an analogous calculation, $(w^{q^i})^{q^{d-i}}=w^{q^{i-i}}=w$. Consequently, $w^{q^{j-i}}=w$. But then

$$f(w)^{q^{j-i}} = f(w)\ \text{ for every } f(x)\in F[x],$$

and hence $u^{q^{j-i}}=u$ for all $u\in F(w)$. Thus the polynomial $x^{q^{j-i}}-x$ has at least $|F(w)|=q^d$ roots in E, which contradicts our assumption $0<j-i<d$. Hence $w, w^q,\ldots,w^{q^{d-1}}$ are indeed distinct. □

Remark 3.4.6. Let E, F and w be as in Proposition 3.4.5. The distinct roots of mpol_w are usually called the **conjugates** of w over F. More generally, one says that two elements of E are **conjugate** if they have the same minimal polynomial. Clearly, being conjugate is an equivalence relation on E, and Proposition 3.4.5 shows that we might alternatively define two elements u and v of E to be conjugate if $v = u^{q^j}$ for some $j \in \mathbb{N}$.

Now assume $E = \mathrm{GF}(q^n)$, and let $\mathscr{R}$ be a system of representatives for the conjugacy classes in E. Then

$$x^{q^n} - x = \prod_{v \in \mathscr{R}} \mathrm{mpol}_v(x)$$

is the decomposition of $x^{q^n} - x$ into monic irreducible polynomials over F. Note that every factor in this decomposition has a degree dividing n.

In the next section, we will show that every monic polynomial $g(x) \in F[x]$ whose degree divides n in fact occurs as $\mathrm{mpol}_v(x)$ for some $v \in \mathscr{R}$. Consequently, for any positive integer n, the polynomial $x^{q^n} - x$ splits over F into the product of *all* monic irreducible polynomials in $F[x]$ with degree dividing n. □

Remark 3.4.7. Consider the extension E/F with $F = \mathrm{GF}(q)$ and $E = \mathrm{GF}(q^n)$. For every divisor d of n, let $\delta_q(d)$ denote the number of elements of E with degree d over F. (Note that this number only depends on the cardinality q of F, justifying the notation.) By definition,

$$\sum_{d|n} \delta_q(d) = q^n,$$

and an application of the Möbius inversion formula in Theorem 2.1.10 shows

$$\delta_q(n) = \sum_{d|n} \mu\left(\frac{n}{d}\right) \cdot q^d,$$

which is the number of elements $u \in E$ generating E over F (that is, $F(u) = E$). Since $\mathrm{mpol}_v(x) = \mathrm{mpol}_u(x)$ if and only if u and v are conjugate, the number of distinct monic irreducible polynomials in $F[x]$ which occur as the minimal polynomial of some generator of E is given by

$$\frac{1}{n} \cdot \delta_q(n) = \frac{1}{n} \cdot \sum_{d|n} \mu\left(\frac{n}{d}\right) \cdot q^d.$$

As mentioned in Remark 3.4.6, we will see soon that $\frac{1}{n} \cdot \delta_q(n)$ is in fact the number of *all* monic irreducible polynomials in $F[x]$ with degree n. □

In the proof of Proposition 3.4.5, we have used certain Galois automorphisms σ_j of the extension E/F. The next result shows that these automorphisms already give the entire Galois group of E/F:

Theorem 3.4.8. *Let E/F be an extension of finite fields, where $F = \mathrm{GF}(q)$ and $E = \mathrm{GF}(q^n)$. Then the Galois group $\mathrm{Gal}(E/F)$ is cyclic of order n and is generated by the* **Frobenius automorphism**

$$\sigma : E \to E, \quad w \mapsto w^q. \tag{3.5}$$

Proof. As noted at the beginning of the proof of Proposition 3.4.5, $\sigma = \sigma_1$ is a Galois automorphism of E/F, and hence $\mathrm{Gal}(E/F)$ contains all powers of σ. Now $\sigma^n(w) = w^{q^n} = w$ for all $w \in E$, that is, σ^n is the identity on E, and therefore the order of σ divides n. It remains to show that $\sigma, \sigma^2, \ldots, \sigma^{n-1}, \sigma^n = \sigma^0 = \mathrm{id}$ are distinct, and that every Galois automorphism β of E/F has to be one of these powers of σ.

For this purpose, we choose a primitive element w of E and apply Proposition 3.4.5 in this special case. We conclude that the roots of the minimal polynomial mpol_w of w over F are the conjugates

$$w = \sigma^0(w),\, w^q = \sigma(w), \ldots, w^{q^{n-1}} = \sigma^{n-1}(w)$$

of w, and hence $\sigma^0, \sigma, \sigma^2, \ldots, \sigma^{n-1}$ are indeed distinct. Now let $\beta \in \mathrm{Gal}(E/F)$ and write $\mathrm{mpol}_w(x) = \sum_{i=0}^n \lambda_i x^i$. Then

$$\mathrm{mpol}_w(\beta(w)) = \sum_{i=0}^n \lambda_i \beta(w)^i = \beta\Big(\sum_{i=0}^n \lambda_i w^i\Big) = \beta(0) = 0,$$

and therefore $\beta(w) = w^{q^k} = \sigma^k(w)$ for some $k \in \{0, \ldots, n-1\}$. As w is a primitive element, this implies $\beta(u) = \sigma^k(u)$ for all $u \in E$ and thus $\beta = \sigma^k$. □

Combining Theorems 3.4.2 and 3.4.8, we obtain the following result:

Theorem 3.4.9 (Galois correspondence for extensions of finite fields). *Let E/F be an extension of finite fields. Then the intermediate fields of E/F correspond bijectively to the subgroups of $\mathrm{Gal}(E/F)$.*

More precisely, let $n := [E : F]$, and let K be the intermediate field of E/F with cardinality q^d, where $d \mid n$. Then the Galois group of E/K is generated by σ^d, where σ is the Frobenius automorphism of E/F; thus σ^d is the Frobenius automorphism of E/K and has order n/d. Moreover, the restriction of σ to K gives rise to the Frobenius automorphism of K/F, which has order d. □

Remark 3.4.10. We may now rephrase the assertion of Proposition 3.4.5 as follows. For any $w \in E$, the roots of the minimal polynomial mpol_w are given by the orbit of w under the Galois group $\mathrm{Gal}(E/F)$. More precisely, mpol_w splits in $F(w)[x]$ as $\prod_\gamma (x - \gamma(w))$, where γ runs over the Galois group of $F(w)/F$. □

We conclude this section with the following important application of the preceding results, which will be generalized in Theorem 3.5.9.

Corollary 3.4.11. *Let E/F be an extension of Galois fields, let K be an intermediate field, and let $u \in E$ generate E over F. Then the minimal polynomial $\mathrm{mpol}_{u,F}(x)$ of u over F splits in $K[x]$ into $[K : F]$ irreducible polynomials of degree $[E : K]$ each.*

Proof. Write $n = [E:F]$ and $d = [K:F]$. The conjugates of u over F split into d conjugacy classes $C_0, \ldots, C_{d-1}$ over K corresponding to the cosets of $\mathrm{Gal}(E/K)$ as a subgroup of $\mathrm{Gal}(E/F)$, namely

$$C_i = \left\{u^{q^i},\, u^{q^{i+d}},\, u^{q^{i+2d}}, \ldots,\, u^{q^{i+(n-d)}}\right\} \quad \text{for } i = 0, \ldots, d-1.$$

Hence

$$\mathrm{mpol}_{u,F}(x) = \prod_{i=0}^{d-1} \mathrm{mpol}_{u^{q^i},K}(x) \quad \text{in } K[x],$$

where

$$\mathrm{mpol}_{u^{q^i},K}(x) = \prod_{v \in C_i} (x - v) \quad \text{in } E[x]. \quad \square$$

3.5 Finite Fields: Basics on Irreducible Polynomials

In this section, we will prove some fundamental results on irreducible polynomials over finite fields. We begin with the explicit statement of a result contained in Section 3.4, as it will be used in this form in many situations.

Proposition 3.5.1. *Let f be an irreducible polynomial of degree $n \geq 1$ over $F = \mathrm{GF}(q)$, and let α be a root of f in some extension field of F. Then the field $E = F(\alpha)$ is the splitting field of f and has degree n over F, so that $E \cong \mathrm{GF}(q^n)$. Moreover, the roots of f are distinct, and f splits over E as follows:*

$$f = (x - \alpha)(x - \alpha^q) \cdots (x - \alpha^{q^{n-1}}). \tag{3.6}$$

Proof. Since f is irreducible, it has to be the minimal polynomial of α. Now all assertions follow from Proposition 3.4.5. $\square$

Let E/F be an extension of Galois fields, where $F = \mathrm{GF}(q)$ and $E = \mathrm{GF}(q^n)$. We already noted that the decomposition of $x^{q^n} - x$ over F into the product of irreducible polynomials is given by

$$x^{q^n} - x = \prod_{v \in \mathscr{R}} \mathrm{mpol}_v(x), \tag{3.7}$$

where $\mathscr{R}$ is a system of representatives of conjugacy classes of elements of E; see Remark 3.4.6. As promised there, we now prove that *every* monic irreducible polynomial $g(x) \in F[x]$ with degree dividing n indeed occurs in this product:

Proposition 3.5.2. *Let $f(x)$ be an irreducible polynomial with degree m over $F = \mathrm{GF}(q)$. Then f divides the polynomial $x^{q^n} - x \in F[x]$ if and only if m divides n.*

Proof. First assume that f divides $x^{q^n} - x$ and consider the splitting field $E = \mathrm{GF}(q^n)$ of $x^{q^n} - x$ over F. Then f has a root ω in E, and $F(\omega)$ is the splitting field of f, by Proposition 3.5.1. As $[F(\omega):F] = m$, Lemma 3.1.1 shows that m divides n.

Conversely, let m be a divisor of n, say $m = n/d$. As before, $F(\omega)$ is an extension field with degree m over F and thus $F(\omega) \cong \mathrm{GF}(q^m)$. By Theorem 3.3.5, $F(\omega)$ admits an extension $E/F(\omega)$ with degree d. Then $E \cong \mathrm{GF}(q^n)$, so that ω is a common root of f and $x^{q^n} - x$. Hence f divides $x^{q^n} - x$, since f is irreducible. □

As a consequence of Proposition 3.5.2, we obtain the following irreducibility criterion:

Corollary 3.5.3. *Let $f(x)$ be a monic polynomial of degree n over* GF(q). *Then f is irreducible if and only if* $\gcd\left(x^{q^d} - x, f(x)\right) = 1$ *for all d with $1 \le d \le \frac{n}{2}$.*

Proof. Let f be reducible, and let g be a non-trivial factor of f with degree at most $n/2$. By Proposition 3.5.2, g divides one of the polynomials $x^{q^d} - x$ with $d \le \frac{n}{2}$ and thus $\gcd\left(x^{q^d} - x, f(x)\right) \neq 1$. The converse also follows from Proposition 3.5.2. □

The following result provides an alternative proof for the existence of irreducible polynomials of any given degree over any finite field, and thereby an alternative proof for the existence of finite fields of any prime power order $q \ge 2$.

Theorem 3.5.4. *Let F be a finite field with q elements, and let n be any positive integer. Then the number $i_q(n)$ of all monic irreducible polynomials in $F[x]$ with degree n is given by*

$$i_q(n) = \frac{1}{n} \cdot \sum_{d|n} \mu\left(\frac{n}{d}\right) q^d. \tag{3.8}$$

In particular, there exists some monic irreducible polynomial $f(x) \in F[x]$ with degree n and hence also a field extension E/F with degree n.

Proof. For any $d \in \mathbb{N}^*$, we let $I_{d,q}(x) \in F[x]$ denote the product of all monic irreducible polynomials in $F[x]$ of degree d.[4] Since $x^{q^n} - x$ has no multiple roots, Proposition 3.5.2 implies

$$\prod_{d|n} I_{d,q}(x) = x^{q^n} - x.$$

Comparing the degrees of these polynomials, we obtain

$$\sum_{d|n} d \cdot i_q(d) = q^n,$$

and Möbius inversion (Theorem 2.1.10) gives the desired formula

$$n \cdot i_q(n) = \sum_{d|n} \mu(d) q^{n/d} = \sum_{d|n} \mu\left(\frac{n}{d}\right) q^d.$$

Finally, the following crude estimate suffices to prove $i_q(n) \neq 0$:

---

[4] As we do not assume the existence of irreducible polynomials of any given degree n at this point, we will use the standard interpretation of the product in question as 1 if it should be empty. Of course, we will show that this cannot occur.

$$i_q(n) = \frac{1}{n} \cdot \sum_{d|n} \mu\left(\frac{n}{d}\right) q^d \geq \frac{1}{n} \cdot \left(q^n - \sum_{d|n, d \neq n} q^d\right)$$

$$\geq \frac{1}{n} \cdot (q^n - q^{n-1} - \cdots - q) = \frac{1}{n} \cdot \left(q^n - \frac{q(q^{n-1}-1)}{q-1}\right),$$

which is indeed positive. □

Remark 3.5.5. Theorem 3.5.4 in fact shows that the number $i_q(n)$ of all monic irreducible polynomials with degree n over $\mathrm{GF}(q)$ is about q^n/n, as the largest error term is at most $\frac{1}{2}q^{n/2}$. Since there are exactly q^n monic polynomials of degree n over $\mathrm{GF}(q)$, the probability that a randomly selected monic polynomial of degree n over a finite field F turns out to be irreducible is roughly $1/n$. Thus the probability of finding irreducible polynomials by repeatedly selecting a monic polynomial of the desired degree at random is quite good. Of course, to turn this idea into a probabilistic algorithm requires a sufficiently efficient way of testing the selected polynomials for irreducibility. □

Remark 3.5.6. We also want to point out an inductive argument providing the existence of extensions of degree n for any given finite field F, where n is any given positive integer. Assume that F has q elements.

Choose a prime r dividing n, and consider the polynomial $x^{q^r} - x$ over F which has derivative -1 and therefore no multiple roots, by Corollary 2.4.18. Now let $f(x) \in F[x]$ be a monic irreducible divisor of $x^{q^r} - x$. Then $f(x)$ has degree 1 or r. Note that the product of the monic linear factors of $x^{q^r} - x$ over F is $\prod_{v \in F}(x - v) = x^q - x$. As $q^r > q$, the degree of the polynomial $I_{r,q}(x) := (x^{q^r} - x)/(x^q - x)$ is greater than 1, hence $i_q(r) \geq 1$. Now let $g(x)$ be any divisor of $I_{r,q}(x)$ of degree r. Then $K = F[x]/(g)$ is a field with q^r elements, which can be considered as an extension of F. We next consider the pair $(K, \frac{n}{r})$. By induction on the number of prime divisors of n, we may assume that there is an extension E of K with degree $\frac{n}{r}$. Then E is a finite field with q^n elements which can be considered as an extension of F. □

We next note that Möbius inversion can also be used to obtain a formula for the polynomials $I_{n,q}(x)$ introduced in the proof of Theorem 3.5.4.

Proposition 3.5.7. *Let F be a finite field with q elements, and let n be any positive integer. Then the product $I_{n,q}(x)$ of all monic irreducible polynomials in $F[x]$ of degree n is given by*

$$I_{n,q}(x) = \prod_{d|n} \left(x^{q^d} - x\right)^{\mu(n/d)}.$$

Proof. View $I_{n,q}(x)$ as an element of the multiplicative monoid $F[x]_{\mathrm{mon}}$ of all monic polynomials over F, and let $F(x)_{\mathrm{mon}}$ denote the quotient group of this monoid. Then $F(x)_{\mathrm{mon}}$ is a (multiplicatively written) abelian group, that is, a $\mathbb{Z}$-module. Now consider the mapping $I_F : n \mapsto I_{n,q}(x)$ from $\mathbb{N}^*$ to $F(x)_{\mathrm{mon}}$, and recall that the set of all mappings from $\mathbb{N}^*$ to $F(x)_{\mathrm{mon}}$ carries the structure of an $\mathbb{Z}[[\mathbb{N}^*]]$-module; see Section 2.1. The desired formula follows from Theorem 2.1.16 (using multiplicative notation), as obviously $x^{q^n} - x = S_{I_F}(n)$. □

Example 3.5.8. As an example, let us apply the preceding results to determine the three irreducible polynomials of degree 4 over $F = \mathrm{GF}(2)$. Proposition 3.5.7 yields

$$I_{4,q}(x) = (x^{16} - x)(x^4 - x)^{-1} = x^{12} + x^9 + x^6 + x^3 + 1.$$

We first guess one factor of $I_{4,q}(x)$. Note that such a polynomial clearly has to be irreducible when considered as a polynomial over $\mathbb{Z}$. Assuming some knowledge of classical Algebra, an obvious guess is the classical cyclotomic polynomial[5]

$$\Phi_5(x) = x^4 + x^3 + x^2 + x + 1.$$

As $\Phi_5(x)$ is relatively prime to both $x^4 - x = x(x+1)(x^2+x+1)$ and $x^2 - x = x(x+1)$, Corollary 3.5.3 shows that $\Phi_5(x)$ indeed remains irreducible over $\mathrm{GF}(2)$. Next, we obtain a second irreducible polynomial $f(x)$ of degree 4 as follows:

$$f(x) := \Phi_5(x+1) = x^4 + x^3 + 1.$$

Finally, dividing $I_{4,q}(x)$ by $f(x)\Phi_5(x)$ gives the third irreducible polynomial $g(x) = x^4 + x + 1$. (We note that this final computational step could be avoided by using the notion of reciprocal polynomials, which we will introduce in Section 5.1.) □

We conclude this section with an important generalization of Corollary 3.4.11 which describes the behavior of an irreducible polynomial over $\mathrm{GF}(q)$ when considered over an *arbitrary* finite extension field of $\mathrm{GF}(q)$.

Theorem 3.5.9. *Consider the Galois field $F = \mathrm{GF}(q)$, let $f(x)$ be a monic irreducible polynomial of degree n over F, and let k be any positive integer. Considered as a polynomial in $K := \mathrm{GF}(q^k)$, f splits into d distinct irreducible factors of degree $\frac{n}{d}$ each, where $d = \gcd(n,k)$. In particular, f remains irreducible over K if and only if $\gcd(n,k) = 1$.*

Proof. Let $\ell := \mathrm{lcm}(n,k)$ be the least common multiple of n and k, and consider an ℓ-dimensional extension $L \cong \mathrm{GF}(q^\ell)$ over F. Then L/F contains (copies of) $E = \mathrm{GF}(q^n)$ and $K = \mathrm{GF}(q^k)$ as intermediate fields. As noted in Remark 3.4.3, the intersection of K and E is the intermediate field M with degree $\gcd(n,k)$ over F.[6] By Corollary 3.4.11, f splits over M into the product of $\gcd(n,k)$ polynomials of degree $\frac{n}{\gcd(n,k)}$. In particular, this establishes the assertion for the special case $k = n$. (Of course, in this case we even obtain the explicit factorization of f into linear factors from Proposition 3.5.1.)

Now let $k \neq n$ and pick any of the $\gcd(n,k)$ irreducible factors over M, say $g(x)$. It remains to show that g remains irreducible when considered as a polynomial in $K[x]$. Choose a root α of g in E, so that g is the minimal polynomial of α over M. Since $F(\alpha) = E$, we obtain $E \subseteq K(\alpha)$ and therefore $K(\alpha) = L$, as there is no proper subfield of L containing both K and E. Thus the degree of α over K is

---

[5] See Exercise 3.5.11 for an alternative approach avoiding this assumption. We shall treat cyclotomic polynomials over finite fields in detail in the next section.

[6] Note that E and K are linearly disjoint over M, again by Remark 3.4.3.

$$[L:K] = \frac{\operatorname{lcm}(n,k)}{k} = \frac{n}{\gcd(n,k)} = \deg g.$$

Now $g(\alpha) = 0$ shows that g is also the minimal polynomial of α over K, so that g is indeed irreducible over K. □

Exercises

Exercise 3.5.10. Give a strengthening of Corollary 3.5.3 in terms of the smallest prime divisor r of n. □

Exercise 3.5.11. Show that GF(16) contains a primitive fifth root of unity and use this fact to devise an alternative approach (not depending on knowing the classical cyclotomic polynomial $\Phi_5(x)$ as in Example 3.5.8) to finding the irreducible polynomial $x^4+x^3+x^2+x+1$ over $\mathbb{Z}_2$. □

Exercise 3.5.12. Let E be an extension field with degree n of a finite field F with q elements. Prove the following alternative formula for $I_{d,q}(x)$, where d is any divisor of n:

$$I_{d,q}(x) = \prod_{\substack{w \in E \\ F(w)=K_d}} (x-w),$$

where K_d denotes the unique intermediate field of E/F with degree d over F. □

3.6 Cyclotomic Polynomials and Cyclotomic Field Extensions

The present section is devoted to a particularly important class of polynomials, namely the cyclotomic polynomials. Even though these polynomials are well-known objects in classical Algebra, we will give a self-contained treatment here. We start with an arbitrary field F and let n be any positive integer.

Proposition 3.6.1. *Consider the polynomial $x^n - 1$ over the field F. If F has positive characteristic p, let $n = p^a \cdot n'$, where n' is not divisible by p and $a \in \mathbb{N}$, so that*[7]

$$x^n - 1 = (x^{n'} - 1)^{p^a}.$$

Otherwise, let $n' = n$. In either case, $x^{n'} - 1$ has no multiple roots in its splitting field.

Proof. As the characteristic of F does not divide n', the derivative $n'x^{n'-1}$ of $x^{n'} - 1$ is non-zero. Since $n'x^{n'-1}$ has no roots in common with $x^{n'} - 1$, these two polynomials are relatively prime. Hence $x^{n'} - 1$ has no multiple roots in its splitting field, by Corollary 2.4.18. □

[7] Recall that exponentiation with p is a monomorphism in this case, see Theorem 3.2.14.

From now on, we always assume that the characteristic of the underlying field F does not divide the degree n of $x^n - 1$. Of course, this holds for all n when F has characteristic zero.

Definition 3.6.2. The splitting field of $x^n - 1$ over F is called the n-th **cyclotomic extension** of F and will be denoted by $F^{(n)}$. □

Note that any root of $x^n - 1$ in $F^{(n)}$ is an n-th root of unity. More precisely, $F^{(n)}$ is the smallest extension of F containing a primitive n-th root of unity (and therefore all n-th roots of unity).

Proposition 3.6.3. *Let $F = \mathrm{GF}(q)$, and assume that q and n are relatively prime. Then the degree of $F^{(n)}$ over F equals $\mathrm{ord}_n(q)$, the order of q modulo n (that is, the smallest positive integer k such that $q^k \equiv 1 \bmod n$).*

Proof. Note that $\mathrm{GF}(q^m)$ contains a primitive n-th root of unity if and only if n divides $q^m - 1$, which holds in view of Theorems 3.2.5 and 1.6.17. Therefore, the degree k of the splitting field $F^{(n)}$ of $x^n - 1$ over F has to be the smallest integer m with this property. □

In particular, the multiplicative group of $F^{(n)}$ contains the n-th roots of unity as its unique subgroup of order n. We will denote this cyclic subgroup of $F^{(n)}$ by U_n, and the set of all primitive n-th roots of unity in U_n by C_n. Note that $|C_n| = \phi(n)$, where ϕ is the Euler function.

Definition 3.6.4. Let F be an arbitrary field and n a positive integer which is not divisible by the characteristic of F. Then

$$\Phi_n(x) := \prod_{\omega \in C_n} (x - \omega)$$

is called the n-th **cyclotomic polynomial** over F. □

According to the preceding definition, Φ_n is a polynomial with degree $\phi(n)$ over the cyclotomic extension $F^{(n)}$ of F. However, Φ_n is actually always a polynomial in $F[x]$, which justifies speaking of the cyclotomic polynomials over F. In Theorem 3.6.10, we will show an even stronger result: the cyclotomic polynomials have coefficients in the prime field P of F. For this purpose, we will establish some useful identities for cyclotomic polynomials, which also lead to a reasonably efficient algorithm for computing Φ_n. We start with the following general formula:

Proposition 3.6.5. *Let F be an arbitrary field, and let n be a positive integer which is not divisible by the characteristic of F. Then*

$$\Phi_n(x) = \prod_{d \mid n} (x^d - 1)^{\mu(n/d)}.$$

Proof. Note that the sets C_d with $d \mid n$ form a partition of the group U_n of all n-th roots of unity over F, so that

$$x^n - 1 = \prod_{d|n} \Phi_d(x) \tag{3.9}$$

in $F^{(n)}[x]$. Applying Möbius inversion (in the multiplicative version already used in the proof of Proposition 3.5.7) yields the desired expression for the cyclotomic polynomials in terms of the polynomials of the form $x^d - 1$. □

Example 3.6.6. Trivially, $\Phi_1(x) = x - 1$. Now consider an integer n of the form $n = r^\ell$, where r is a prime different from the characteristic of F and where $\ell \geq 1$. Then

$$\Phi_{r^\ell}(x) = \prod_{j=0}^{\ell} (x^{r^j} - 1)^{\mu(r^{\ell-j})} = \frac{x^{r^\ell} - 1}{x^{r^{\ell-1}} - 1} = \sum_{i=0}^{r-1} x^{i \cdot r^{\ell-1}}. \tag{3.10}$$

In particular,

$$\Phi_r(x) = x^{r-1} + x^{r-2} + \cdots + x + 1 \tag{3.11}$$

for every prime $r \neq \operatorname{char} F$. Note that $\Phi_{r^\ell}(x)$ is in indeed a polynomial over the prime field P of F and that

$$\Phi_{r^\ell}(x) = \Phi_r(x^{r^{\ell-1}}) \tag{3.12}$$

for every $\ell \in \mathbb{N}^*$, which is a very special case of Proposition 3.6.8 below. □

We now turn to the promised identities for cyclotomic polynomials. Throughout, we work over a given field F and assume that the characteristic of F does not divide d for any of the Φ_d under consideration. We begin with the following useful auxiliary result:

Lemma 3.6.7. *$\Phi_{mn}(x)$ divides $\Phi_n(x^m)$ for all positive integers m and n.*

Proof. Let ζ be any root of Φ_{mn}, that is, any primitive mn-th root of unity. Then ζ^m is a primitive n-th root of unity, and thus $\Phi_n(\zeta^m) = 0$. Hence all roots of Φ_{mn} are also roots of $\Phi_n(x^m)$. Since Φ_{mn} has only simple roots, we obtain the assertion. □

Proposition 3.6.8. *Let m and n be positive integers, and assume* $\operatorname{rad}(m) \mid n$*; that is, every prime divisor of m also divides n. Then*

$$\Phi_{mn}(x) = \Phi_n(x^m). \tag{3.13}$$

Proof. By Lemma 3.6.7, Φ_{mn} divides $\Phi_n(x^m)$, and hence it suffices to show that these two monic polynomials have the same degree. Recall that Φ_d has degree $\phi(d)$. In view of Remark 1.9.13,

$$\phi(d) = d \cdot \prod_r \left(1 - \frac{1}{r}\right),$$

where the product runs over all primes r dividing d. Since mn and n have the same prime divisors (by hypothesis), this identity yields

$$\deg \Phi_{mn}(x) = \phi(mn) = m\phi(n) = m \cdot \deg \Phi_n(x) = \deg \Phi_n(x^m),$$

as claimed. □

Proposition 3.6.9. *Let n be a positive integer and r a prime not dividing n. Then*

$$\Phi_{nr}(x) = \frac{\Phi_n(x^r)}{\Phi_n(x)}. \tag{3.14}$$

Proof. By Lemma 3.6.7, $\Phi_{nr}(x)$ divides $\Phi_n(x^r)$. Now let ζ be any primitive n-th root of unity. Then ζ^r is again a primitive n-th root of unity, since r does not divide n, and therefore $\Phi_n(x)$ divides $\Phi_n(x^r)$. By definition, Φ_r and Φ_{nr} are relatively prime, so that the polynomial $\Phi_n(x)\Phi_{nr}(x)$ divides $\Phi_n(x^r)$, and hence it suffices to show that these two monic polynomials have the same degree. Since r does not divide n, the multiplicativity of the Euler function yields

$$\begin{aligned} \deg\left(\Phi_n(x)\Phi_{nr}(x)\right) &= \phi(n) + \phi(nr) = \phi(n)(1+\phi(r)) \\ &= r \cdot \phi(n) = \deg \Phi_n(x^r), \end{aligned}$$

as claimed. □

The preceding two results already suffice to prove that the cyclotomic polynomials are indeed polynomials over the prime field P of F:

Theorem 3.6.10. *Let F be a field and n a positive integer which is not divisible by the characteristic of F. Then the n-th cyclotomic polynomial $\Phi_n(x)$ has coefficients in the prime field P of F.*

Proof. We use induction on n. Trivially, $\Phi_1(x) = x - 1 \in P[x]$. Now let $n > 1$ and assume inductively $\Phi_m(x) \in P[x]$ for every proper divisor m of n. We choose a prime r dividing n and put $m := n/r$. Assume first that r divides m. Then

$$\Phi_n(x) = \Phi_{mr}(x) = \Phi_m(x^r),$$

by Proposition 3.6.8. By induction, $\Phi_m(x)$ has coefficients in P, so that $\Phi_n(x)$ also has coefficients in P. Finally, the case where r does not divide m follows in the same way, as then $\Phi_n(x) \cdot \Phi_m(x) = \Phi_m(x^r)$, by Proposition 3.6.9. □

Remark 3.6.11. Assume that F has characteristic zero. Then the prime field of F is the field $\mathbb{Q}$ of rational numbers and therefore the ring of integers is a subset of F. In this case, one can even show that all cyclotomic polynomials $\Phi_n(x)$ over F have *integral* coefficients. This follows by a minor variation of the proof of Theorem 3.6.10, where we now assume inductively that $\Phi_m(x)$ has integral coefficients for every proper divisor m of n. The case where r divides m follows exactly as before.

Finally, if r does not divide m, we again use $\Phi_n(x) \cdot \Phi_m(x) = \Phi_m(x^r)$. By induction, $\Phi_m(x)$ and hence also $\Phi_m(x^r)$ have coefficients in $\mathbb{Z}$. Since $\Phi_m(x)$ is a monic polynomial, the polynomial division of $\Phi_m(x^r)$ by $\Phi_m(x)$ is entirely performed in $\mathbb{Z}[x]$, and hence the resulting quotient $\Phi_n(x)$ also has integral coefficients. □

We continue with some further identities for cyclotomic polynomials. The following result is a more general version of Proposition 3.6.9.

Proposition 3.6.12. *Let m and t be positive integers, and assume that m and t are relatively prime. Then*

$$\Phi_m(x^t) = \prod_{d|t} \Phi_{md}(x). \tag{3.15}$$

Proof. Assume first that $t = r^k$ is a prime power. Then the formula is correct for $k = 1$, by Proposition 3.6.9. We now use induction on k. Thus let $k \geq 2$ and put $s := r^{k-1}$ and $y := x^s$. Then

$$\Phi_m\left(x^{r^k}\right) = \Phi_m(y^r) = \Phi_{mr}(y) \cdot \Phi_m(y) = \Phi_{mr}(x^s) \cdot \Phi_m(x^s).$$

Since $s = r^{k-1}$, every prime divisor of s divides mr and hence Proposition 3.6.8 gives $\Phi_{mr}(x^s) = \Phi_{mrs}(x) = \Phi_{mr^k}(x)$. On the other hand, by induction,

$$\Phi_m(x^s) = \Phi_m(x^{r^{k-1}}) = \prod_{j=0}^{k-1} \Phi_{mr^j}(x).$$

Combining these observations yields

$$\Phi_m(x^{r^k}) = \Phi_{mr^k}(x) \cdot \prod_{j=0}^{k-1} \Phi_{mr^j}(x) = \prod_{j=0}^{k} \Phi_{mr^j}(x) = \prod_{d|r^k} \Phi_{md}(x),$$

which establishes the assertion for the case $t = r^k$.

For the general case, we use induction on the number of distinct prime divisors of t. Choose any prime r dividing t and write $t = r^k \cdot s$, where s is not divisible by r. Let $y := x^s$. Then the previous case gives

$$\Phi_m(x^t) = \Phi_m\left(y^{r^k}\right) = \prod_{j=0}^{k} \Phi_{mr^j}(y) = \prod_{j=0}^{k} \Phi_{mr^j}(x^s).$$

As $\gcd(s, mr^j) = 1$ for every j and as the number of distinct prime divisors of s is strictly smaller than that of t, we obtain by induction

$$\Phi_{mr^j}(x^s) = \prod_{e|s} \Phi_{mr^j e}(x) \quad \text{for } j = 0, \ldots, k.$$

Altogether, $\Phi_m(x^t) = \prod_{j=0}^{k} \prod_{e|s} \Phi_{mr^j e}(x) = \prod_{d|t} \Phi_{md}(x)$, as claimed. □

We conclude with the following general identity:

Theorem 3.6.13. *Let m and t be positive integers which are not divisible by the characteristic of the underlying field F. Write $t = \ell \cdot s$, where every prime divisor of ℓ divides m and where m and s are relatively prime. Then*

$$\Phi_m(x^t) = \prod_{d|s} \Phi_{md\ell}(x).$$

Proof. Let $y := x^s$. Then

$$\Phi_m(x^t) = \Phi_m(y^\ell) = \prod_{d|s} \Phi_{md}(y) = \prod_{d|s} \Phi_{md}(x^\ell) = \prod_{d|s} \Phi_{md\ell}(x),$$

by Propositions 3.6.8 and 3.6.12. □

We now show how one may apply the preceding identities to compute the cyclotomic polynomial $\Phi_n(x)$ in a reasonably efficient way, even though no polynomial algorithm is known for this task; our presentation follows Lüneburg [247]. In principle, $\Phi_n(x)$ could be computed by using the formula given in Proposition 3.6.5; however, this method is rather cumbersome and definitely not advisable in practice. A better strategy is first finding the prime factorization of n (which one needs anyway) and then proceeding recursively.

Algorithm 3.6.14. Let n be any positive integer which is not divisible by the characteristic of the underlying field F. Then (the coefficients of) the cyclotomic polynomial $\Phi_n(x)$ can be computed over the prime field P of F as follows.

Step 1. Compute the canonical prime power factorization of n, say $n = p_1^{e_1} \cdots p_s^{e_s}$.
Step 2. $i \leftarrow 1,\ r \leftarrow p_i,\ f(x) \leftarrow x^{r-1} + \cdots + 1$.
Step 3. If $i < s$, put $i \leftarrow i+1$; otherwise, go to Step 5.
Step 4. $r \leftarrow p_i$ and $f(x) \leftarrow f(x^r)/f(x)$. Go to Step 3.
Step 5. $m \leftarrow p_1 \cdots p_s,\ \Phi_n(x) \leftarrow f(x^{n/m})$. □

Note that Algorithm 3.6.14 is correct. In Step 2, we initialize $f(x)$ as $\Phi_{p_1}(x)$, which is correct by Example 3.6.6. After this, the loop in Steps 3 and 4 computes $\Phi_{\mathrm{rad}(n)}(x)$, where $\mathrm{rad}(n) = p_1 \cdots p_s$ is the radical of n; this follows by recursively applying Proposition 3.6.12, with t taking the prime values $p_2, \dots, p_s$. Finally, Step 5 indeed gives $\Phi_n(x)$, by Proposition 3.6.8.

Example 3.6.15. As an example, let us use Algorithm 3.6.14 to compute the cyclotomic polynomials $\Phi_{80}(x)$ and $\Phi_{400}(x)$, considered over the field of rational numbers. Here $f(x) = x+1$ in Step 2, and then $f(x)$ is replaced by

$$\Phi_{\mathrm{rad}(n)}(x) = \Phi_{10}(x) = \frac{f(x^5)}{f(x)} = \frac{x^5+1}{x+1} = x^4 - x^3 + x^2 - x + 1$$

in Steps 3 and 4. Then Step 5 yields $\Phi_{80}(x) = x^{32} - x^{24} + x^{16} - x^8 + 1$ and $\Phi_{400}(x) = x^{160} - x^{120} + x^{80} - x^{40} + 1$. □

When n is even, Algorithm 3.6.14 (and the computation in the preceding Example) can be simplified by using $\Phi_{2n}(x) = \Phi_n(-x)$; see Exercise 3.6.20. There are further interesting (or curious) facts regarding the cyclotomic polynomials over the rationals. For instance, all coefficients of $\Phi_n(x)$ belong to $\{0, 1, -1\}$ provided that $n \leq 104$; this follows from the stronger result that the property in question holds whenever n has at most two odd prime divisors. On the other hand, $\Phi_{105}(x)$ involves the coefficient -2 (see https://en.wikipedia.org/wiki/Cyclotomic_polynomial):

$$\begin{aligned}\Phi_{105}(x) &= x^{48}+x^{47}+x^{46}-x^{43}-x^{42}-2x^{41}-x^{40}-x^{39}\\ &\quad +x^{36}+x^{35}+x^{34}+x^{33}+x^{32}+x^{31}-x^{28}-x^{26}\\ &\quad -x^{24}-x^{22}-x^{20}+x^{17}+x^{16}+x^{15}+x^{14}+x^{13}\\ &\quad +x^{12}-x^{9}-x^{8}-2x^{7}-x^{6}-x^{5}+x^{2}+x+1.\end{aligned}$$

More generally, it can be shown that, for every positive integer t, there is some n for which $\Phi_n(x)$ involves the coefficient $-t$. For proofs of these facts, the interested reader may consult Lüneburg [247].

By a well-known result from classical Algebra, the cyclotomic polynomials are irreducible over $\mathbb{Q}$. As we will not need this result, we do not include a proof here and just refer the reader to the literature; see, for instance, Goodman [151], Ireland and Rosen [203] or Jacobson [204]. We now determine how these polynomials split when considered over finite fields:

Proposition 3.6.16. *Let $F = \mathrm{GF}(q)$, and let d be a positive integer which is not divisible by the characteristic of F. Then the cyclotomic polynomial $\Phi_d(x)$ splits in $F[x]$ into $\phi(d)/\mathrm{ord}_d(q)$ monic irreducible polynomials of degree $\mathrm{ord}_d(q)$ each.*

Proof. By Proposition 3.6.3, the d-th cyclotomic field $F^{(d)}$ over F is the Galois field E with q^ℓ elements, where $\ell = \mathrm{ord}_d(q)$. Let ζ be any primitive d-th root of unity. Then $\mathrm{mpol}_\zeta(x)$ has degree ℓ, and the roots of ζ are the ℓ primitive d-th roots of unity $\zeta, \zeta^q, \ldots, \zeta^{q^{\ell-1}}$, by Proposition 3.5.1. Thus the set C_d of all primitive d-th roots of unity splits into conjugacy classes with cardinality ℓ under the action of the Galois group of E/F. Any such class gives rise to exactly one monic factor of $\Phi_d(x)$, which is irreducible over F and has degree ℓ. Because of $|C_d| = \phi(d)$, the total number of irreducible factors of $\Phi_d(x)$ over F is indeed $\phi(d)/\mathrm{ord}_d(q)$. □

Remark 3.6.17. For later applications, we need a simple result relating primitive m-th and n-th roots of unity, where m is not divisible by the characteristic of F and where n divides m. Obviously, U_n then is a subgroup of U_m, and the mapping

$$\pi_n^m \colon U_m \to U_n, \quad u \mapsto u^{m/n} \tag{3.16}$$

is a group epimorphism. Of course, π_n^m maps every primitive m-th root of unity to a primitive n-th root of unity, that is, $\pi_n^m(C_m) \subseteq C_n$. However, it is not immediately obvious whether or not equality holds, as an arbitrary pre-image of a primitive n-th root of unity is not necessarily a primitive m-th root of unity. Nevertheless, the next result guarantees that there always exists at least one such pre-image. □

Proposition 3.6.18. *Let m be a positive integer which is not divisible by the characteristic of $F = \mathrm{GF}(q)$, let n be a divisor of m, and let ζ be any primitive n-th root* *of unity in $F^{(n)}$. Then there exists a primitive m-th root of unity $\eta \in F^{(m)}$ such that $\eta^{m/n} = \zeta$.*

Proof. We put $t := m/n$ and distinguish three cases. Assume first that every prime divisor of t divides n. Then

$$\Phi_m(x) = \Phi_{nt}(x) = \Phi_n(x^t),$$

by Proposition 3.6.8. Hence the pre-image of C_n under π_n^m is entirely contained in C_m in this case.

Next, let t and n be relatively prime. Then the mapping $u \mapsto u^t$ is an automorphism of U_n, and we may choose a primitive n-th root of unity α such that $\alpha^t = \zeta$. Note that U_t is the kernel of the homomorphism π_n^m, so that

$$\left(\pi_n^m\right)^{-1}(\zeta) = \alpha U_t = \{\alpha\beta : \beta \in U_t\}.$$

Now choose β as a primitive t-th root of unity. Then $\eta := \alpha\beta$ has order $nt = m$, since n and t are relatively prime, and we have found the desired pre-image η of ζ.

Finally, in the general case, let $t = sn'$, where each prime divisor of n' divides n and where s and n are relatively prime. By the first case, there is a primitive (nn')-th root of unity γ such that $\gamma^{n'} = \zeta$, and then the second case (applied to γ) gives a primitive m-th root of unity η such that $\eta^s = \gamma$. Then $\eta^t = \eta^{sn'} = \gamma^{n'} = \zeta$. □

We conclude this section with the following result which will be used later to obtain an explicit formula for the number of normal elements in any finite extension E/F of Galois fields; see Theorem 3.10.5. In what follows, we write ϕ_q for the Euler function of the ring $\mathrm{GF}(q)[x]$; see Observation 2.3.7.

Proposition 3.6.19. *Let n be a positive integer and $F = \mathrm{GF}(q)$, where q is a power of the prime p. Write $n = p^a \cdot m$, where $a \in \mathbb{N}$ and where m is not divisible by p. Then the number of units $\phi_q(x^n - 1)$ in the residue ring $F[x]/(x^n - 1)$ is given by*

$$q^{m\cdot(p^a-1)} \cdot \prod_{d|m} \left(q^{\mathrm{ord}_d(q)} - 1\right)^{\frac{\phi(d)}{\mathrm{ord}_d(q)}}.$$

Proof. We have

$$x^n - 1 = (x^m - 1)^{p^a} = \prod_{d|m} \Phi_d(x)^{p^a},$$

where each $\Phi_d(x)$ is square-free and splits into $\phi(d)/\mathrm{ord}_d(q)$ irreducible polynomials of degree $\mathrm{ord}_d(q)$ each, by Proposition 3.6.16. Hence the multiplicativity of the function ϕ_q gives

$$\phi_q(x^n-1) = \prod_{d|m}\left(q^{p^a\cdot\mathrm{ord}_d(q)} - q^{(p^a-1)\cdot\mathrm{ord}_d(q)}\right)^{\frac{\phi(d)}{\mathrm{ord}_d(q)}}$$

$$= \prod_{d|m} q^{(p^a-1)\cdot\phi(d)}\cdot\left(q^{\mathrm{ord}_d(q)}-1\right)^{\frac{\phi(d)}{\mathrm{ord}_d(q)}}$$

$$= \left(\prod_{d|m} q^{\phi(d)}\right)^{p^a-1}\cdot\prod_{d|m}\left(q^{\mathrm{ord}_d(q)}-1\right)^{\frac{\phi(d)}{\mathrm{ord}_d(q)}}.$$

Now the assertion follows from $\prod_{d|m} q^{\phi(d)} = q^{\sum_{d|m}\phi(d)} = q^m$. □

Exercises

Exercise 3.6.20. Let $n\geq 3$ be an odd positive integer. Show $\Phi_{2n}(x) = \Phi_n(-x)$.

Exercise 3.6.21. Let $n\geq 2$. Show $\Phi_n(x) = x^{\phi(n)}\Phi_n(x^{-1})$.

Exercise 3.6.22. Let E be a field and U a finite subgroup of the multiplicative group E^* of E. Show that

$$\prod_{u\in U}(x-\beta u) = x^{|U|}-\beta^{|U|}$$

holds for every $\beta\in E^*$. □

3.7 Wedderburn's Theorem

We recall from Definition 1.4.4 that a ring R is called a skew field (or a division ring) provided that every non-zero element is a unit in R. In this section, we will prove the celebrated result of Wedderburn [400] that there are no proper finite skew fields. We present the (by now standard) proof given by Witt [406].

Theorem 3.7.1 (Wedderburn's theorem). *Every finite skew field is commutative and thus isomorphic to a Galois field* GF(q) *for some prime power* q.

Proof. Let E be a finite skew field. We split the proof into several steps.

Step 1. Consider the set

$$F := \{\lambda\in E : \lambda x = x\lambda \text{ for every } x\in E\}.$$

Obviously, 0 and 1 are contained in F, and $\lambda-\mu$ and $\lambda\mu$ are in F whenever $\lambda,\mu\in F$. If $\lambda\in F$ is non-zero, then

$$(y^{-1}\lambda^{-1}y)^{-1} = y^{-1}\lambda y = \lambda \quad \text{for all } y\in E^*,$$

which shows $\lambda^{-1} = y^{-1}\lambda^{-1}y$ for all $y \in E^*$, and thus also $\lambda^{-1} \in F$. Hence F is a (commutative) field, which is called the **center** of E. As E is finite, $F = \mathrm{GF}(q)$ for some prime power q. Then E is a vector space over F, say with dimension n, and therefore $|E| = q^n$. Clearly, the assertion of the theorem follows if we can show $n = 1$. By way of contradiction, assume $n > 1$.

Step 2. We define an equivalence relation on the multiplicative group E^* of E (the *conjugacy* relation in the group theoretic sense) by

$$x \sim y \;:\Longleftrightarrow\; \exists z \in E^* : z^{-1}xz = y$$

and denote the equivalence class of $y \in E^*$ by C_y. Note that $|C_y| = 1$ if and only if y belongs to the center F of E.

Given any $y \in E^*$, we also define the **normalizer** of y in E by

$$N_y := \{x \in E : xy = yx\};$$

trivially, $F \subseteq N_y$. Using similar arguments as in Step 1, one shows that N_y is a skew subfield of E. Hence any N_y is an F-vector space and thus $|N_y| = q^{n(y)}$, where $n(y)$ denotes the dimension of that space. Since the multiplicative group N_y^* is a subgroup of E^*, we see that $q^{n(y)} - 1 = |N_y^*|$ divides $q^n - 1 = |E^*|$, and therefore $n(y)$ divides n, by Lemma 3.4.1. Moreover, the conjugacy class C_y is a complete set of (left) coset representatives of N_y^* in E^*, so that $|C_y|$ equals the index $[E^* : N_y^*]$ of N_y^* in E^*.

Now enumerate the elements of F^* as $\lambda_1, \ldots, \lambda_{q-1}$, and let $y_1, \ldots, y_s$ be a complete system of representatives for the equivalence classes with at least 2 elements. Considering the partition of E^* into equivalence classes yields the following **class equation**:

$$\begin{aligned} q^n - 1 = |E^*| &= \sum_{j=1}^{q-1} |C_{\lambda_j}| + \sum_{i=1}^{s} |C_{y_i}| \\ &= q - 1 + \sum_{i=1}^{s} [E^* : N_{y_i}^*] \\ &= q - 1 + \sum_{i=1}^{s} \frac{q^n - 1}{q^{n(y_i)} - 1}. \end{aligned}$$

Step 3. We now use the polynomial $x^n - 1$ and its cyclotomic divisors, considered as polynomials over the field $\mathbb{Q}$ of rational numbers. Obviously,

$$\frac{x^n - 1}{x^{n(y)} - 1} = \prod_{d|n,\, d\nmid n(y)} \Phi_d(x)$$

for all $y \in E^* \setminus F^*$. Here $n(y) \neq n$, so that the cyclotomic polynomial $\Phi_n(x)$ occurs as a factor, and therefore $\Phi_n(x)$ divides both $x^n - 1$ and $(x^n - 1)/(x^{n(y_i)} - 1)$ for $i = 1, \ldots, s$. By Remark 3.6.11, $\Phi_n(x)$ has integral coefficients, and evaluating the preceding polynomial identity at q shows (together with the class equation from Step 2) that $\Phi_n(q)$ has to divide

$$q^n - 1 - \sum_{i=1}^{s} \frac{q^n - 1}{q^{n(y_i)} - 1} = q - 1.$$

Step 4. For the final step, we use an argument involving the absolute value of complex roots of unity. Note that $\Phi_n(x)$ splits over $\mathbb{C}$ as

$$\Phi_n(x) = \prod_{\zeta \in C_n} (x - \zeta),$$

where C_n is the set of primitive n-th roots of unity in $\mathbb{C}$. The real and the imaginary parts of any $\zeta \in C_n$ have the form $a_k := \cos \frac{2\pi k}{n}$ and $b_k := \sin \frac{2\pi k}{n}$, respectively, where $k \in \{0, 1, \ldots, n-1\}$ is relatively prime to n. Taking absolute values gives

$$|q - \zeta|^2 = (q - a_k)^2 + b_k^2 = q^2 - 2a_k q + a_k^2 + b_k^2 = q^2 - 2a_k q + 1.$$

Because of our assumption $n \geq 2$, we have $-1 \leq a_k = \cos \frac{2\pi k}{n} < 1$ for every k and therefore

$$|q - \zeta| > \sqrt{q^2 - 2q + 1} = q - 1.$$ [8]

Evaluating the cyclotomic polynomial $\Phi_n(x)$ over $\mathbb{C}$ at q and taking absolute values then yields

$$|\Phi_n(q)| = \prod_{\zeta \in C_n} |q - \zeta| > (q-1)^{|C_n|} = (q-1)^{\phi(n)} \geq q - 1.$$

Thus $|\Phi_n(q)|$ cannot divide $q - 1$, which contradicts the result of Step 3. □

Remark 3.7.2. Wedderburn's theorem can be strengthened by weakening its hypothesis as follows. The associative law and the existence of multiplicative inverses (which form part of the requirement that E^* is a group) may be replaced by the following condition: given any element $y \neq 0$, there exists an element $i(y)$ such that $(xy)i(y) = x$ holds for all $x \in E$. This result (which combines two results known as the Skornyakov-San Soucie and the Artin-Zorn theorems, respectively) is – like Wedderburn's theorem itself – of considerable geometric interest, since it is used in the theory of projective planes. We refer the interested reader to the monograph of Hughes and Piper [198], which contains a detailed study of the interaction between algebraic structures generalizing fields and symmetry properties of the associated projective planes. □

[8] Alternatively, one may argue more geometrically: all these roots ζ lie on the unit circle in the complex plane $\mathbb{C}$, but not on the real line in $\mathbb{C}$, whereas the integer $q - 1$ is in the positive part of the real line.

3.8 Dedekind's Independence Theorem and Galois Extensions

The main aim of this section is to provide some basic results on field extensions with finite degree which admit the maximum possible number of Galois automorphisms. As a preparation, we first consider multiplicative characters of fields and prove Dedekind's famous independence theorem for such mappings.

Let E/F be a field extension with finite degree, view E as an F-vector space, and let V be any further finite dimensional F-vector space. By a well-known fact from Linear Algebra, the linear mappings from V into E again form an F-vector space,[9] which is denoted by $\mathrm{Hom}_F(V,E)$ and has dimension $\dim_F V \cdot \dim_F E$; of course, this also holds for an arbitrary F-vector space E. The additional assumption that E is in fact an extension field of F allows us to endow $\mathrm{Hom}_F(V,E)$ with the structure of an E-vector space by defining a scalar multiplication as follows:

$$(a\tau)(v) := a \cdot \tau(v) \quad \text{for all } v \in V, \tag{3.17}$$

where $a \in E$ and $\tau \in \mathrm{Hom}_F(V,E)$. The following simple result determines the dimension of $\mathrm{Hom}_F(V,E)$ as an E-vector space by providing a canonical basis:

Lemma 3.8.1. *Let B be a basis of the vector space V over F, and define linear mappings $\varepsilon_b : V \to E$ (for $b \in B$) by the requirement*

$$\varepsilon_b(c) := \begin{cases} 1 & \text{for } c = b, \\ 0 & \text{otherwise.} \end{cases}$$

Then $\{\varepsilon_b : b \in B\}$ is a basis for $\mathrm{Hom}_F(V,E)$ over E, and hence the dimension of $\mathrm{Hom}_F(V,E)$ over E equals the dimension of V over F.

Proof. Assume $\sum_{b\in B} \lambda_b \varepsilon_b = 0$, where $\lambda_b \in E$ for all b. Then

$$0 = \Big(\sum_{b\in B} \lambda_b \varepsilon_b\Big)(c) = \lambda_c \quad \text{for all } c \in B,$$

so that the ε_b are linearly independent over E. Similarly, $\tau = \sum_{b\in B} \tau(b)\varepsilon_b$ for every $\tau \in \mathrm{Hom}_F(V,E)$. □

Definition 3.8.2. Let E and K be fields. A group homomorphism χ from the multiplicative group $(E^*,\cdot,1)$ into the multiplicative group $(K^*,\cdot,1)$ is called a **multiplicative character** from E into K.

Now let $\chi_1,\ldots,\chi_n$ be multiplicative characters and $a_1,\ldots,a_n \in K$. One calls $\chi_1,\ldots,\chi_n$ **linearly independent** provided that

$$\sum_{i=1}^{n} a_i \chi_i(u) = 0 \quad \text{for all } u \in E^* \tag{3.18}$$

[9] Of course, this is with respect to pointwise addition and pointwise scalar multiplication, as in Section 2.1.

only holds for $a_1 = a_2 = \cdots = a_n = 0$. □

Theorem 3.8.3 (Dedekind independence theorem). *Let E and K be fields, and let $\chi_1, \ldots, \chi_n$ be distinct multiplicative characters from E into K. Then $\chi_1, \ldots, \chi_n$ are linearly independent over K.*

Proof. We proceed by induction on n. If $n = 1$, then $a\chi(u) = 0$ for all $u \in E^*$ implies $0 = a\chi(1) = a \cdot 1 = a$. Now let $n \geq 2$ and assume the validity of the theorem for $n-1$. Suppose that $\chi_1, \ldots, \chi_n$ are distinct multiplicative characters from E into K such that $\sum_{i=1}^{n} a_i \chi_i(u) = 0$ for all $u \in E^*$, where $a_1, \ldots, a_n \in K$ and at least one $a_i \neq 0$. Then the induction hypothesis shows that in fact all $a_i \neq 0$.

Since χ_1 and χ_2 are distinct, there exists an element $w \in E^*$ such that $\chi_1(w) \neq \chi_2(w)$. Replacing u by uw in (3.18) and multiplying (3.18) by $\chi_1(w)$, respectively, yields the following two equalities (for all $u \in E^*$):

$$\sum_{i=1}^{n} a_i \chi_i(u) \chi_i(w) = 0,$$

$$\sum_{i=1}^{n} a_i \chi_i(u) \chi_1(w) = 0.$$

Subtracting these two identities gives

$$\sum_{i=2}^{n} \big(a_i(\chi_i(w) - \chi_1(w)\big)\chi_i(u) = 0 \quad \text{for all } u \in E^*,$$

which means that $\chi_2, \ldots, \chi_n$ are linearly dependent, as $a_2(\chi_2(w) - \chi_1(w)) \neq 0$. This contradicts the induction hypothesis and finishes the proof. □

We now apply the Dedekind independence theorem in the special case where the χ_i are Galois automorphisms of some field extension E/F.

Proposition 3.8.4. *Let E/F be a field extension with finite degree n. Then the Galois group of E over F has at most n elements.*

Proof. Every Galois automorphism of E/F is an F-endomorphism of E which restricts to a group homomorphism of the multiplicative group E^*, that is, a multiplicative character from E into E. By Dedekind's independence theorem, distinct Galois automorphisms are linearly independent over E and therefore also linearly independent in $\mathrm{End}_F(E)$, where $\mathrm{End}_F(E)$ is considered as an E-vector space. Hence

$$n = [E : F] = \dim_E(\mathrm{End}_F(E)) \geq |\mathrm{Gal}(E/F)|,$$

by Lemma 3.8.1. □

The preceding inequality suggests considering those field extensions which admit the maximal possible number of Galois automorphisms:

Definition 3.8.5. A field extension E/F with finite degree n is called a **Galois extension** if the Galois group of E/F has cardinality n. A Galois extension E/F is called **cyclic** or **abelian** if its Galois group has the respective property. □

Remark 3.8.6. Let E/F be a Galois extension of degree n. Then $\dim_F(\mathrm{End}_F(E)) = |\mathrm{Gal}(E/F)|$, by the proof of Proposition 3.8.4. Therefore the elements of the Galois group of E/F form an E-basis of $\mathrm{End}_F(E)$ in this case. □

Example 3.8.7. Consider an extension E/F of finite fields, say $F = \mathrm{GF}(q)$ and $E = \mathrm{GF}(q^n)$. By Theorem 3.4.8, E/F is a cyclic Galois extension, and its Galois group is generated by the Frobenius automorphism σ. In view of Remark 3.8.6, $\mathrm{id}_E, \sigma, \ldots, \sigma^{n-1}$ form a basis for $\mathrm{End}_F(E)$ over E. □

In the remainder of this section, we will summarize the fundamental properties of arbitrary Galois extensions, without including proofs; see, for instance, Jacobson [205] or Goodman [151]. For the special case of finite fields, these results have been established in the preceding sections.

We begin with a simple result on arbitrary extensions with finite degree (not necessarily Galois). Let E/F be such an extension, and let $G := \mathrm{Gal}(E/F)$. For every subgroup H of G, the set

$$\mathscr{F}(H) := \{w \in E \colon h(w) = w \text{ for all } h \in H\}$$

is an intermediate field of E/F. Conversely, for every intermediate field K of E/F, the set

$$\mathscr{G}(K) := \{g \in G \colon g(w) = w \text{ for all } w \in K\} = \mathrm{Gal}(E/K)$$

is a subgroup of G. Such a pair $(\mathscr{F}, \mathscr{G})$ s called a **Galois pairing** and has the following basic properties:

- $H \subseteq \mathscr{G} \circ \mathscr{F}(H)$ and $\mathscr{F} \circ \mathscr{G} \circ \mathscr{F}(H) = \mathscr{F}(H)$ for every subgroup H of G.
- $K \subseteq \mathscr{F} \circ \mathscr{G}(K)$ and $\mathscr{G} \circ \mathscr{F} \circ \mathscr{G}(K) = \mathscr{G}(K)$ for every intermediate field K of E/F.

In the special case where E/F is a Galois extension, $(\mathscr{F}, \mathscr{G})$ it is called the **Galois correspondence**, in view of the following major result.

Theorem 3.8.8 (Fundamental theorem on Galois extensions). *Let E/F be a Galois extension with Galois group G. Then:*

(1) *The mappings $\mathscr{F}$ and $\mathscr{G}$ in the associated Galois pairing are bijective and inverse to each other, that is,*

- $H = \mathscr{G} \circ \mathscr{F}(H)$ *and* $[E : \mathscr{F}(H)] = |H|$ *for every subgroup H of G;*
- $K = \mathscr{F} \circ \mathscr{G}(K)$ *and* $[K : F] = [G : \mathscr{G}(K)]$ *for every intermediate field K of E/F.*

(2) *E/K is a Galois extension for every intermediate field K.*

(3) *For an intermediate field K, the extension K/F is Galois if and only if $\mathscr{G}(K)$ is a normal subgroup of G. In this case,* $\mathrm{Gal}(K/F)$ *is isomorphic to the factor group $G/\mathrm{Gal}(E/K)$, and the restriction of the elements of G to K is an epimorphism $G \to \mathrm{Gal}(K/F)$ with kernel* $\mathrm{Gal}(E/K)$.

(4) *The extension E/F is simple.* □

We also mention the following characterization of Galois extensions:

Proposition 3.8.9. *E/F is a Galois extension if and only if $\mathscr{F}(G) = F$, where $G =$* $\mathrm{Gal}(E/F)$. □

Finally, we discuss an alternative characterization of Galois extensions. For this, we require a further concept complementing that of separability introduced in Definition 3.2.17.

Definition 3.8.10. A field extension E/F is called **normal** provided that every irreducible polynomial $f(x) \in F[x]$ with a root in E splits over E into linear factors, so that E contains a splitting field of f over F. □

It can be shown that E/F is normal if E is the splitting field of some polynomial over F. It is easy to see that Galois extensions have both properties in question:

Lemma 3.8.11. *Every Galois extension is both normal and separable.*

Proof. Let E/F be a Galois extension, and let $f(x)$ be an irreducible polynomial over F with a root $w \in E$; we may assume f to be monic, so that $f(x) = \mathrm{mpol}_w(x)$. Now let $K := F(w)$, let H be the Galois group of E/K, and let $\mathscr{R}$ be a system of coset representatives of H in $G = \mathrm{Gal}(E/F)$. Then $\gamma(w) \neq \delta(w)$ whenever γ and δ are distinct elements of $\mathscr{R}$. Moreover, $f(\gamma(w)) = \gamma(f(w)) = \gamma(0) = 0$ for all $\gamma \in \mathscr{R}$. Since $[K : F] = |\mathscr{R}| = \deg f$, we see that

$$f(x) = \prod_{\gamma \in \mathscr{R}} (x - \gamma(w))$$

splits over E into linear factors. Hence E/F is both normal and separable. □

In fact, the converse of Lemma 3.8.11 also holds:

Theorem 3.8.12. *A field extension E/F with finite degree is a Galois extension if and only if it is separable and normal.* □

In particular, let E/F be an abelian Galois extension. Then K/F is a Galois extension for every intermediate field K of E/F, and hence $F(v)$ is a splitting field for $\mathrm{mpol}_v(x)$ over F for every $v \in E$.

We end this section with a further remark which will be useful later, when considering normal bases.

Remark 3.8.13. Let E/F be a Galois extension with Galois group G, let K and L be two intermediate fields of E/F, and consider the compositum KL of K and L introduced in Definition 3.1.12. Then

$$\mathrm{Gal}(E/KL) = \mathrm{Gal}(E/K) \cap \mathrm{Gal}(E/L);$$

in particular, $\mathrm{Gal}(E/K)$ and $\mathrm{Gal}(E/L)$ intersect trivially provided that $KL = E$.

Now assume that the set

$$\mathrm{Gal}(E/K) \cdot \mathrm{Gal}(E/L) := \{\alpha\beta : \alpha \in \mathrm{Gal}(E/K) \text{ and } \beta \in \mathrm{Gal}(E/L)\}$$

is a subgroup of $\mathrm{Gal}(E/F)$.[10] Then

$$\mathrm{Gal}(E/K \cap L) = \mathrm{Gal}(E/K) \cdot \mathrm{Gal}(E/L);$$

in particular, $\mathrm{Gal}(E/F) = \mathrm{Gal}(E/K) \cdot \mathrm{Gal}(E/L)$ provided that $K \cap L = F$.

Finally, suppose both $KL = E$ and $K \cap L = F$. Then

$$[K : F] = [\mathrm{Gal}(E/F) : \mathrm{Gal}(E/K)] = |\mathrm{Gal}(E/L)| = [E : L],$$

and any basis of K/F is a basis of E/L, since the hypothesis in Lemma 3.1.13 is then satisfied. □

3.9 The Normal Basis Theorem

In the present section, we establish the existence of a particularly interesting type of bases for Galois extensions.

Definition 3.9.1. Let E/F be a Galois extension with Galois group G, and let w be an element of E for which the set $G(w) := \{\gamma(w) : \gamma \in G\}$ of **conjugates** of w under G is a basis for E/F. Then w is called a **normal element** and $G(w)$ a **normal basis** for E/F, and the minimal polynomial $\mathrm{mpol}_w(x)$ is said to be a **normal polynomial** over F. □

Note that $G(w)$ is the set of roots of the minimal polynomial of w over F, by Proposition 3.4.5, and that $E = F(w)$ when w is a normal element for E/F. Before we consider the existence problem for normal elements, we provide a useful general characterization of bases for E/F in terms of the Galois group $\mathrm{Gal}(E,F)$:

Proposition 3.9.2. *Let E/F be a Galois extension with degree n and Galois group $G = \{\gamma_1, \ldots, \gamma_n\}$. Then the elements $w_1, \ldots, w_n \in E$ form a basis for E/F if and only if*

---

[10] For instance, this holds when one of these groups is a normal subgroup of $\mathrm{Gal}(E/F)$, in particular when $\mathrm{Gal}(E/F)$ is abelian.

$$M := \left(\gamma_i(w_j)\right)_{i,j=1,\ldots,n}$$

is an invertible matrix in $E^{(n,n)}$.

Proof. First let $w_1,\ldots,w_n$ be a basis for E/F, and assume $\mathbf{a}M = \mathbf{0}$ for some vector $\mathbf{a} = (a_1,\ldots,a_n) \in E^n$, that is,

$$\sum_{i=1}^{n} a_i\gamma_i(w_j) = 0 \quad \text{for } j = 1,\ldots,n.$$

Trivially, this gives

$$0 = \sum_{j=1}^{n} c_j \cdot \Big(\sum_{i=1}^{n} a_i\gamma_i(w_j)\Big) = \sum_{i=1}^{n} a_i\gamma_i\Big(\sum_{j=1}^{n} c_j w_j\Big)$$

for all $c_1,\ldots,c_n \in F$. As $w_1,\ldots,w_n$ is a basis of E over F, this shows $\sum_{i=1}^{n} a_i\gamma_i(u) = 0$ for all $u \in E$, and hence Theorem 3.8.3 gives $\mathbf{a} = \mathbf{0}$. Thus M has rank n and is therefore invertible.

Conversely, assume that M is invertible, and let $\sum_{j=1}^{n} \lambda_j w_j = 0$ for some $\lambda_1,\ldots,\lambda_n$ in F. Applying any γ_i to this equation gives $\sum_{j=1}^{n} \lambda_j\gamma_i(w_j) = 0$. By definition of M, this means

$$M \cdot (\lambda_1,\ldots,\lambda_n)^T = \mathbf{0}$$

(where the upper case T denotes the transposition of a vector or a matrix, as usual). Since M is assumed to be invertible, we conclude $\lambda_j = 0$ for all j. Thus $w_1,\ldots,w_n$ are linearly independent and hence a basis for E/F. □

Corollary 3.9.3. *Let E/F be a Galois extension with degree n and Galois group $G = \{\gamma_1,\ldots,\gamma_n\}$. Then $w \in E$ is a normal element for E/F if and only if*

$$M(w) := \left((\gamma_i\gamma_j)(w)\right)_{i,j=1,\ldots,n}$$

is an invertible matrix in $E^{(n,n)}$. □

We now turn to the main goal of the present section, namely the following famous result which is due to Hensel [182] in the special case of finite fields and to Noether [310] and Deuring [105] in general.

Theorem 3.9.4 (Normal basis theorem). *Every Galois extension admits a normal basis.*

Proof. Let $n := [E : F]$. We split the proof into the following two cases:

- *Case 1.* The Galois group of E/F is cyclic.[11]

[11] In particular, Case 1 establishes the normal basis theorem for finite fields; we will investigate this situation in more detail in the next section. There is an alternative proof for this special case of the normal basis theorem which is completely elementary but quite lengthy; it is due to Ore [314] and rests on the theory of linearized polynomials (see Definition 3.11.3) developed in his earlier paper [313]. This proof can be found in several text books, for instance, in Berlekamp [30] or Lidl and Niederreiter [243].

- *Case 2.* The ground field F has sufficiently many elements. More precisely, we assume the condition $|F| > n(n-1)$.

Since the Galois groups of extensions of finite fields are always cyclic, these two (overlapping) cases cover all possible situations for a Galois extension.[12] The details require some effort.

Proof for Case 1

Let $G := \mathrm{Gal}(E/F)$ be the (cyclic) Galois group of E/F and choose a generator σ for G. Thus

$$G = \{\sigma^0 = \mathrm{id}_E, \sigma^1 = \sigma, \ldots, \sigma^{n-1}\}.$$

We view σ as an F-endomorphism on E and make use of the results developed in Section 2.6. According to Observation 2.6.2, we may turn E into an $F[x]$-module (E,σ) by putting

$$f(x)\cdot v := f(\sigma)(v) \quad \text{for all } v \in E \text{ and all } f(x) \in F[x].$$

Then (E,σ) is a torsion module, and the annihilator ideal of (E,σ) is generated by the minimal polynomial $\mathrm{mpol}_\sigma(x)$ of σ. In the present situation, $\mathrm{mpol}_\sigma(x) = x^n - 1$, since $\sigma^n = \mathrm{id}_E$ and since the elements of G are linearly independent over F (as shown in the proof of Proposition 3.8.4).

Now let v be any element of E. Again using Observation 2.6.2, the mapping

$$\Gamma_{\sigma,v}\colon F[x] \to E, \quad f(x) \mapsto f(\sigma)(v) \tag{3.19}$$

is a homomorphism between the two $F[x]$-modules $F[x]$ and (E,σ). The kernel of $\Gamma_{\sigma,v}$ is generated by the minimal polynomial $\mathrm{mpol}_{\sigma,v}(x)$ of v with respect to σ, see Definition 2.6.3. Moreover, there always exists an element $w \in E$ such that $\mathrm{mpol}_{\sigma,w}(x) = \mathrm{mpol}_\sigma(x)$, as shown in the proof of Theorem 2.6.5. Let $w \in E$ be such an element, that is, $\mathrm{mpol}_{\sigma,w}(x) = x^n - 1$. Then $G(w) = \{\sigma^0(w), \sigma(w), \ldots, \sigma^{n-1}(w)\}$ is a linearly independent set and thus the desired normal basis for E/F. This completes the proof for Case 1.

Proof for Case 2

We now come to the proof for the case where the cardinality of the ground field is infinite. In fact, the argument (which is due to Artin [10]) applies as soon as the ground field has cardinality exceeding $n(n-1)$.

Let $G = \{\gamma_1, \ldots, \gamma_n\}$ be the Galois group of E/F. We use that E/F is a simple field extension, by part (4) of Theorem 3.8.8. Thus let $E = F(v)$ with $v \in E$ and

---

[12] It is possible to prove the normal basis theorem without such a case distinction; see, for instance, Blessenohl [43]. We have decided to follow the more usual approach, as the special case of finite fields is covered by Case 1 and thus admits a comparatively simple proof. The proof of Case 2 has been included in this book just for the sake of completeness.

denote the minimal polynomial of v over F by $g(x)$. By Lemma 3.8.11, E/F is separable. In particular, g splits over E into linear factors, and its roots are exactly the conjugates of v under G:

$$g(x) = \prod_{i=1}^{n} \big(x - \gamma_i(v)\big). \tag{3.20}$$

We now use the results developed in Section 2.5 and consider the ring $R_g := E[x]/(g)$ of residues modulo g of polynomials with coefficients in E. Since the roots of $g(x)$ are pairwise different, the corresponding linear factors $x - \gamma_i(v)$ are relatively prime. Thus R_g is isomorphic to the E-algebra

$$\bigoplus_{i=1}^{n} E[x]/\big(x - \gamma_i(v)\big), \tag{3.21}$$

and an explicit isomorphism is induced by the mapping

$$\Psi\colon \begin{cases} E[x] \to E^n = \bigoplus_{i=1}^{n} E, \\ a(x) \mapsto \big(a(\gamma_1(v)), \ldots, a(\gamma_n(v))\big), \end{cases}$$

that is, the evaluation of polynomials $a(x) \in E[x]$ at the roots of $g(x)$; this is a homomorphism of E-algebras, where E^n is equipped with the componentwise multiplication. The kernel of Ψ is the ideal generated by the monic polynomial $h(x)$ of least degree in $E[x]$ such that $h(\gamma_i(v)) = 0$ for $i = 1, \ldots, n$, and thus $h(x) = \prod_{i=1}^{n}(x - \gamma_i(v)) = g(x)$. As $\dim_E(E[x]/(g)) = \deg g = n$, the homomorphism Ψ has to be surjective, and therefore Ψ induces the desired isomorphism

$$\overline{\Psi}\colon \begin{cases} R_g \to \bigoplus_{i=1}^{n} E[x]/\big(x - \gamma_i(v)\big), \\ a(x) + (g) \mapsto \Big(a(x) + \big(x - \gamma_1(v)\big), \ldots, \big(a(x) + (x - \gamma_n(v))\big)\Big). \end{cases}$$

For $k = 1, \ldots, n$, let $e_k(x)$ be the unique polynomial of degree at most $n-1$ in $E[x]$ satisfying $e_k(\gamma_k(v)) = 1$ and $e_k(\gamma_\ell(v)) = 0$ for all $\ell = 1, \ldots, n$ with $\ell \neq k$, according to Observation 2.5.4. By Theorem 2.5.5, the e_k are given explicitly by

$$e_k(x) = \frac{1}{g'(\gamma_k(v))} \cdot \frac{g(x)}{x - \gamma_k(v)},$$

where g' is the formal derivative of g. These polynomials have degree $n-1$ and form the spectral basis corresponding to the elements $\gamma_1(v), \ldots, \gamma_n(v)$. Thus they give rise to a decomposition of the unit element in $E[x]/(g)$ as a sum of pairwise orthogonal idempotents (see Remark 2.5.6), that is,

- $e_k(x) \cdot e_\ell(x) \equiv 0 \bmod g(x)$ for all $k \neq \ell$,
- $e_k(x)^2 \equiv e_k(x) \bmod g(x)$ for all k,
- $\sum_{i=1}^{n} e_i(x) \equiv 1 \bmod g(x)$.

After these preparations, we are ready to consider normal bases. For simplicity, we will write $[ij]$ for the unique index $\ell \in \{1,\dots,n\}$ such that $\gamma_i\gamma_j = \gamma_\ell$. Let $D(x)$ be the determinant of the matrix

$$M := \left(e_{[ij]}(x)\right)_{i,j=1,\dots,n},$$

so that $D(x)$ is a polynomial in $E[x]$ (which, of course, depends on the choice of v). As $\deg e_k = n-1$ for every k, the well-known formula[13]

$$\det M = \sum_{\pi \in S_n} (-1)^{\mathrm{sgn}(\pi)} \cdot \prod_{i=1}^{n} e_{[i\pi(i)]}(x)$$

shows $\deg D \le n(n-1)$, since each product in this sum has degree $n(n-1)$.

We claim that $D(x)$ cannot be the zero polynomial. In order to see this, we consider the matrix

$$MM^T = \left(\sum_{j=1}^{n} e_{[kj]}(x) \cdot e_{[j\ell]}(x)\right)_{k,\ell=1,\dots,n}.$$

First let $k = \ell$. Then $\gamma_k\gamma_j = \gamma_\ell\gamma_j$ for all j, and therefore the (k,k)-entry is

$$\sum_{j=1}^{n} e_{[kj]}(x)^2 \equiv \sum_{j=1}^{n} e_{[kj]}(x) \equiv 1 \mod g(x).$$

Now let $k \neq \ell$. Then $\gamma_k\gamma_j \neq \gamma_\ell\gamma_j$ for all j, and the (k,ℓ)-entry is congruent to 0 mod $g(x)$, since every summand satisfies $e_{[kj]}(x) \cdot e_{[j\ell]}(x) \equiv 0 \mod g(x)$. Consequently, the polynomial $D(x)^2 = \det(MM^T)$ is congruent to 1 mod $g(x)$, which establishes our claim.

Finally, the hypothesis on the cardinality of F shows $|F| > n(n-1) \ge \deg D$. Thus there exists an element $\lambda \in F$ with $D(\lambda) \neq 0$. Since λ is fixed by every $\gamma \in G$ and since g has coefficients in F, we obtain (with $\gamma_1 = \mathrm{id}_E$)

$$e_k(\lambda) = \frac{g(\lambda)}{g'(\gamma_k(v)) \cdot (\lambda - \gamma_k(v))} = \gamma_k\left(\frac{g(\lambda)}{g'(v) \cdot (\lambda - v)}\right) = \gamma_k(e_1(\lambda)). \tag{3.22}$$

Now choose $w := e_1(\lambda)$. Then Equation (3.22) shows that $D(\lambda)$ is the determinant of the matrix

$$\left(e_{[ij]}(\lambda)\right)_{i,j=1,\dots,n} = \left((\gamma_i\gamma_j)(w)\right)_{i,j=1,\dots,n}.$$

As $D(\lambda) \neq 0$, Corollary 3.9.3 guarantees that w is a normal element for E/F. This establishes Theorem 3.9.4 also for Case 2 and thus completes the proof. □

In the remainder of this section, we apply Lemmas 3.1.1 and 3.1.13 to obtain some further interesting results on normal bases.

---

[13] As usual, S_n denotes the symmetric group on $\{1,\dots,n\}$, and $\mathrm{sgn}(\pi)$ denotes the sign of a permutation π.

Proposition 3.9.5. *Let E/F be a Galois extension, and let K and L be intermediate fields satisfying $KL = E$ and $K \cap L = F$. Assume in addition that K/F is likewise a Galois extension, and let $u \in K$ be a normal element for K/F. Then u is also normal for E/L.*

Proof. As mentioned in Remark 3.8.13, the hypothesis that K and L are linearly disjoint implies $[K : F] = [E : L]$, and therefore any F-basis for K/F is also an L-basis for E/L, by Lemma 3.1.13. We apply this for the special case of a normal basis $H(u) = \{\gamma(u) : \gamma \in H\}$ for K/F, where $H = \mathrm{Gal}(K/F)$.

By Theorem 3.8.8, the restriction of $G := \mathrm{Gal}(E/F)$ to K yields an epimorphism from G onto H with kernel $\mathrm{Gal}(E/K)$. By Remark 3.8.13, this restriction even gives an isomorphism between $\mathrm{Gal}(E/L)$ and H, since K and L are linearly disjoint. Hence

$$H(u) = \{\delta(u) : \delta \in \mathrm{Gal}(E/L)\}$$

is also a normal basis for E/L. □

We are now able to prove the following result which plays an important role in many constructions of normal bases:

Theorem 3.9.6 (Reduction theorem for normal elements). *Let E/F be a Galois extension with Galois group G, and let K and L be intermediate fields satisfying $KL = E$ and $K \cap L = F$. Assume in addition that both K/F and L/F are likewise Galois extensions, and let v and w be normal elements for K/F and L/F, respectively. Then vw is a normal element for E/F.*

Proof. Let $\gamma \in H = \mathrm{Gal}(E/K)$ and $\eta \in U = \mathrm{Gal}(E/L)$. As $\eta(v) \in K$ is fixed by H and as $w \in L$ is fixed by U, we obtain

$$\eta(v) \cdot \gamma(w) = \gamma(\eta(v)) \cdot \gamma(\eta(w)) = (\gamma\eta)(vw).$$

Since K and L are linearly disjoint, we have $G = HU$ and $H \cap U = \{id_E\}$, by Remark 3.8.13. This gives

$$\begin{aligned} H(v) \cdot U(w) &= \{\eta(v)\gamma(w) : \gamma \in H,\, \eta \in U\} \\ &= \{(\gamma\eta)(vw) : \gamma \in H,\, \eta \in U\} \\ &= \{g(vw) : g \in G\} = G(vw). \end{aligned}$$

By Corollary 3.1.14, $G(vw)$ is a (normal) basis for E/F. □

In view of Remark 3.4.3, we obtain the following useful consequence of Theorem 3.9.6 for the finite case:

Theorem 3.9.7. *Let E/F be an extension of Galois fields where $F = \mathrm{GF}(q)$ and $E = \mathrm{GF}(q^n)$. Moreover, assume $n = k\ell$, where k and ℓ are relatively prime. Let K and L be the intermediate fields with degrees k and ℓ over F, respectively. If u is normal for K/F and v is normal for L/F, then uv is normal for E/F.* □

3.10 Basics on Normal Bases for Finite Fields

In this brief section, we will review (the proof of) the normal basis theorem given in 3.9.4 for the special case of finite fields, in combination with results from Sections 2.6 and 3.6. We pursue two aims in doing so: to fix some notations which will be used from now on, and to prepare later results on normal bases of various types.

Throughout this section, we consider the field extension E/F, where $F = \mathrm{GF}(q)$ and $E = \mathrm{GF}(q^n)$ have characteristic p. Since the Galois group of E/F is the cyclic group generated by the Frobenius automorphism $\sigma\colon w \mapsto w^q$, we may use the results on cyclic endomorphisms from Section 2.6.

We begin by summarizing some basic facts on normal elements for E/F. Recall from Definition 2.6.3 that $\mathrm{mpol}_{\sigma,v}(x)$ is called the σ-order of $v \in E$. For the present case of finite fields, we introduce the following alternative notation:

Notation 3.10.1. Let σ be the Frobenius automorphism of the extension E/F. Then $\mathrm{mpol}_{\sigma,v}(x)$ is also called the **q-order** of v, and we use the more suggestive notation $\mathrm{Ord}_q(v)$ instead of $\mathrm{mpol}_{\sigma,v}(x)$. □

We note the following consequence of Case 1 in the proof of Theorem 3.9.4:

Theorem 3.10.2. *Consider an extension E/F of Galois fields, where $F = \mathrm{GF}(q)$ and $E = \mathrm{GF}(q^n)$. Then there exists a normal element w for E/F, that is,*

$$w,\, w^q, \ldots, w^{q^{n-1}}$$

form a basis for E over F. Moreover, an element $v \in E$ is normal for E/F if and only if $\mathrm{Ord}_q(v) = x^n - 1$. □

Observation 3.10.3. We again write (E,σ) to denote the $F[x]$-module structure of E with respect to σ, as in Case 1 in the proof of Theorem 3.9.4, and list some consequences of the results in Section 2.6:

(1) The (F,σ)-submodules of (E,σ) correspond bijectively to the monic divisors $g(x)$ of $x^n - 1$ in $F[x]$. Given such a divisor g, the corresponding submodule is

$$(E,\sigma)_g := \{v \in E\colon g(\sigma)(v) = 0\}.$$

(2) All (F,σ)-submodules of (E,σ) are cyclic, and the generators of the submodule $(E,\sigma)_g$ are exactly the $v \in E$ with $\mathrm{Ord}_q(v) = g$.

(3) Since $(E,\sigma)_g$ is isomorphic to $F[x]/(g)$, the number of elements of E with q-order g equals the number $\phi_q(g)$ of units in the residue ring $F[x]/(g)$, where ϕ_q is the Euler function for the domain $F[x]$ introduced in Definition 2.3.6. □

Observation 3.10.4. As in Section 3.6, let $n = p^a \cdot n'$, where n' is not divisible by p and $a \in \mathbb{N}$. Then

$$x^n - 1 = \prod_{d|n'} \Phi_d(x)^{p^a},$$

which gives rise to a decomposition of (E,σ):

$$(E,\sigma) = \bigoplus_{d|n'} (E,\sigma)_{\Phi_d^{p^a}}.$$

Thus any $w \in E$ can be written uniquely as a sum

$$w = \sum_{d|n'} w_d \quad \text{with } w_d \in (E,\sigma)_{\Phi_d^{p^a}} \text{ for every } d \mid n'.$$

Moreover, w is a normal element if and only if each w_d generates $(E,\sigma)_{\Phi_d^{p^a}}$, that is, if and only if

$$\mathrm{Ord}_q(w_d) = \Phi_d(x)^{p^a} \quad \text{for every } d \mid n'.$$

By Proposition 3.6.16, the cyclotomic polynomial $\Phi_d(x)$ splits over F into distinct irreducible divisors with degree $\mathrm{ord}_d(q)$ each, say

$$\Phi_d(x) = f_1(x)\cdots f_\delta(x) \quad \text{with } \delta := \phi(d)/\mathrm{ord}_d(q).$$

Consequently, the submodule corresponding to $\Phi_d^{p^a}(x)$ can be decomposed as

$$(E,\sigma)_{\Phi_d^{p^a}} = \bigoplus_{i=1}^{\delta} (E,\sigma)_{f_i^{p^a}},$$

and any element w_d of this submodule has a unique decomposition

$$w_d = \sum_{i=1}^{\delta} w_{d,i} \quad \text{with } w_{d,i} \in (E,\sigma)_{f_i^{p^a}} \text{ for } i = 1,\ldots,\delta.$$

Moreover, w_d is a generator of $(E,\sigma)_{\Phi_d^{p^a}}$ if and only if

$$\mathrm{Ord}_q(w_{d,i}) = f_i(x)^{p^a} \text{ for } i = 1,\ldots,\delta. \quad \square$$

Finally, an application of Proposition 3.6.19 gives the following formula of Ore [314] for the number of normal elements:

Theorem 3.10.5. *Consider an extension of Galois fields E/F, where $F = \mathrm{GF}(q)$ and $E = \mathrm{GF}(q^n)$. As before, let $n = p^a \cdot n'$, where n' is not divisible by p and $a \in \mathbb{N}$. Then the number of normal elements for E/F is given by*

$$\phi_q(x^n - 1) = q^{n'\cdot(p^a-1)} \cdot \prod_{d|n'} \left(q^{\mathrm{ord}_d(q)} - 1\right)^{\frac{\phi(d)}{\mathrm{ord}_d(q)}}. \quad \square$$

Exercises

Exercise 3.10.6. Use Theorem 3.10.5 to give explicit formulas for $\phi_q(x^3-1)$ depending on q mod 3. □

3.11 Finite Fields: Endomorphisms and q-Polynomials

In Section 3.8, we have used the E-vector space of all F-endomorphisms of E to study the Galois group of a field extension E/F with finite degree. In the present section, we study this vector space in more detail for the special case of finite fields. As a major example, we will consider certain additive q-analogues of the cyclotomic polynomials and obtain results similar to those in Section 3.6.

All this will rest on the results on interpolation polynomials given in Section 2.5, and thus it makes sense to begin by considering the general problem of representing mappings from $F=\mathrm{GF}(q)$ into itself via (interpolation) polynomials. Instead of the standard notation F^F for the set of all mappings $F\to F$, we will use the more suggestive notation $\mathrm{map}(F)$. We start with the following simple

Observation 3.11.1. Let $F=\mathrm{GF}(q)$. Then $\mathrm{map}(F)$ is an F-vector space with dimension q (with respect to pointwise addition and scalar multiplication). The **canonical basis** of $\mathrm{map}(F)$ consists of all mappings of the form

$$e^u\colon F\to F,\quad v\mapsto\begin{cases}1 & \text{if } v=u,\\ 0 & \text{otherwise,}\end{cases}$$

where $u\in F$. □

As F is finite, each $\alpha\in\mathrm{map}(F)$ can be described by the interpolation polynomial $g_\alpha(x)\in F[x]$ through the points $(v,\alpha(v))$ with $v\in F$; see Section 2.5. Since any such interpolation polynomial has degree at most $q-1$, the vector space $\mathrm{map}(F)$ corresponds to the vector space $F[x]_{<q}$ of all polynomials in $F[x]$ with degree strictly less than q. Note that two arbitrary polynomials g and h in $F[x]$ induce the same mapping on F if and only if $\prod_{u\in F}(x-u)=x^q-x$ divides $g-h$.

Theorem 3.11.2. *Let $\alpha\in\mathrm{map}(F)$, where $F=\mathrm{GF}(q)$. Then the interpolation polynomial $g_\alpha(x)\in F[x]$ corresponding to α satisfies*

$$g_\alpha(x)=\sum_{u\in F}\alpha(u)\cdot\left(1-(x-u)^{q-1}\right).$$

Proof. We use the spectral basis of the space $F[x]_{<q}$ corresponding to the canonical basis of $\mathrm{map}(F)$. For every $u\in F$, define a polynomial e_u by

$$e_u(x):=1-(x-u)^{q-1}.$$

Let $v\in F$. Then

$$e_u(v) = \begin{cases} 1 & \text{if } v = u, \\ 0 & \text{otherwise,} \end{cases}$$

as the reader may easily check. Thus $e_u(x)$ induces the mapping e^u on F, and the e_u form the desired spectral basis.

Now the interpolation formula of Lagrange (see Remark 2.5.4) shows that the polynomial $g_\alpha(x) \in F[x]_{<q}$ corresponding to $\alpha \in \text{map}(F)$ is $\sum_{u \in F} \alpha(u) e_u(x)$, as claimed. □

We now use interpolation polynomials to study endomorphisms of the extension field $E = \text{GF}(q^n)$ over $F = \text{GF}(q)$. According to Remark 3.8.6, the elements of the Galois group of E/F provide a canonical basis for the E-vector space $\text{End}_F(E)$ of all F-linear mappings on E. As $\text{Gal}(E/F)$ is generated by the Frobenius automorphism σ, any $\tau \in \text{End}_F(E)$ has a unique representation in the form

$$\tau = \sum_{i=0}^{n-1} t_i \sigma^i \quad \text{with } t_i \in E \text{ for } i = 1, \ldots, n-1.$$

Then

$$\tau(v) = \sum_{i=0}^{n-1} t_i \sigma^i(v) = \sum_{i=0}^{n-1} t_i v^{q^i} \quad \text{for all } v \in E,$$

and hence the mapping $\tau \in \text{End}_F(E)$ is induced by the (interpolation) polynomial $\sum_{i=0}^{n-1} t_i x^{q^i}$. This fact motivates introducing the following concept:

Definition 3.11.3. Let $E = \text{GF}(q^n)$, and let $g(x) = \sum_{i=0}^{m} g_i x^i$ be a polynomial over E. Then the polynomial

$$A_g(x) := \sum_{i=0}^{m} g_i x^{q^i}$$

is called the q**-polynomial** (or the **linearized polynomial**) associated with g. □

With this terminology, we may summarize the preceding observations as follows:

Proposition 3.11.4. *Let* $F = \text{GF}(q)$ *and* $E = \text{GF}(q^n)$. *Then the mapping*

$$\eta : E[x]_{<n} \to \text{End}_F(E), \quad \sum_{i=0}^{n-1} t_i x^i \mapsto \sum_{i=0}^{n-1} t_i \sigma^i \tag{3.23}$$

is an isomorphism between the E*-vector spaces* $E[x]_{<n}$ *and* $\text{End}_F(E)$. *Moreover, the* q*-polynomial* $A_g(x)$ *associated with a polynomial* $g(x) \in E[x]_{<n}$ *is the interpolation polynomial for the endomorphism* $\eta(g)$ *on* E. □

We now consider the situation at hand from a slightly different point of view. Recall from Section 2.6 that

$$\Gamma_\sigma : F[x] \to \text{End}_F(V), \quad f(x) \mapsto f(\sigma)$$

is a homomorphism between F-algebras with kernel $(\text{mpol}_\sigma) = (x^n - 1)$. Therefore Γ_σ induces an isomorphism between $F[x]_{<n}$ and the F-subalgebra $\text{alg}_F(\sigma)$ of $\text{End}_F(E)$ generated by σ (considered as F-vector spaces). In view of $\Gamma_\sigma(f(x)) = \sum_{i=0}^{n-1} f_i\sigma^i$ for every $f(x) \in F[x]_{<n}$, we see that the restriction of Γ_σ to $F[x]_{<n}$ equals the restriction of η to $F[x]_{<n}$. This gives the following result:

Proposition 3.11.5. *The q-polynomial $A_f(x)$ associated with $f(x) \in F[x]_{<n}$ is the interpolation polynomial for the endomorphism $\eta(f) \in \text{alg}_F(\sigma)$.* □

Remark 3.11.6. Consider $E = \text{GF}(q^n)$ as an $F[x]$-module with respect to σ, as in Section 3.10. By Observation 3.10.3, any monic divisor $g(x)$ of $x^n - 1$ in $F[x]$ corresponds to the (F,σ)-submodule $(E,\sigma)_g = \{v \in E : g(\sigma)(v) = 0\}$. Thus $(E,\sigma)_g$ is precisely the set of roots of the q-polynomial $A_g(x)$ associated with g. □

In order to distinguish more clearly between the two kinds of polynomials we deal with in this context – namely, q-orders and interpolation polynomials – we introduce a second indeterminate z and write the q-orders in this variable. For instance, we have $A_{z^n-1}(x) = x^{q^n} - x$ in this notation.

We now come to the main topic of this section, namely additive q-analogues of the cyclotomic polynomials. Recall from Definition 3.6.4 that the d-th cyclotomic polynomial $\Phi_d(x)$ over F is the product of all linear factors $x - \omega$, where ω runs over the primitive d-th roots of unity. By definition, all these ω are in the splitting field $F^{(d)}$ of $x^d - 1$, and we form a product running over all elements of order d in the multiplicative group of $F^{(d)}$ (or in the multiplicative group of any extension field of that splitting field). For the additive analogue, we consider a corresponding product running over all elements with a fixed q-order $g(z)$, where g is a monic polynomial in $F[z]$ which is not a multiple of z; these polynomials were introduced by Hachenberger [159].

Definition 3.11.7. Let $F = \text{GF}(q)$, and let $g(z) \in F[z]$ be a monic polynomial which is not a multiple of z. The **order of z modulo** $g(z)$, denoted as $\text{ord}_g(z)$, is the order of the unit z in the residue ring $F[z]/(g(z))$:

$$\text{ord}_g(z) := \min\{\ell \in \mathbb{N}^*\colon\ g(z) \mid z^\ell - 1\}.$$

Then all elements with q-order g lie in $K := \text{GF}(q^k)$, where $k = \text{ord}_g(z)$, and we define a subset of K collecting these elements:

$$\Omega_g := \{v \in K\colon\ \text{Ord}_q(v) = g(z)\}. \tag{3.24}$$

Finally, we call the polynomial

$$\Psi_g(x) := \prod_{v \in \Omega_g} (x - v) \tag{3.25}$$

the **Psi polynomial** of $g(z)$ over F. □

Observation 3.11.8. We now collect some simple but useful facts concerning the Psi polynomials, emphasizing relevant analogies to results on cyclotomic polynomials:

(1) According to Definition 3.11.7, $\Psi_g(x)$ is a polynomial in $K[x]$. As in the case of the cyclotomic polynomials, Ψ_g is actually always a (monic) polynomial in $F[x]$. This follows by observing that each set Ω_g is invariant under the Frobenius automorphism of K/F.

(2) By definition, Ω_g has cardinality $\phi_q(g)$. Therefore $\Psi_g(x)$ splits over F into the product of

$$\frac{|\Omega_g|}{\mathrm{ord}_g(z)} = \frac{\phi_q(g)}{\mathrm{ord}_g(z)}$$

irreducible polynomials of degree $\mathrm{ord}_g(z)$ each, in analogy with Proposition 3.6.16.

(3) Note that it is immaterial whether we consider the set Ω_g as a subset of K (as in Definition 3.11.7) or of any extension field of K. In particular, we may replace K with any field $\mathrm{GF}(q^m)$, where m is a multiple of k.

(4) Similarly, it does not matter whether we consider the submodule

$$(K,\rho)_g = \{v \in K\colon g(\rho)(v) = 0\}$$

of the $F[x]$-module (K,ρ) or the submodule

$$(L,\tau)_g = \{v \in L\colon g(\tau)(v) = 0\}$$

of the $F[x]$-module (L,τ) for some extension field L of K, where ρ and τ denote the Frobenius automorphisms of K/F and L/F, respectively. This follows as ρ is the restriction of τ to K and as the underlying sets of these two modules agree, since they both coincide with the set of roots of the q-polynomial $A_g(x)$ associated with $g(z)$; see Remark 3.11.6. Thus we may view this module associated with g in the context of any extension field of K and denote it simply by V_g. In particular, we may again replace K with any field $\mathrm{GF}(q^m)$, where m is a multiple of k.

(5) By definition, the cyclotomic polynomial $\Phi_{q^n-1}(x)$ is the product of all minimal polynomials of primitive elements for E/F. Similarly, $\Psi_{z^n-1}(x)$ is the product of all normal polynomials for E/F; see Theorem 3.10.2. Note that each normal basis for E/F gives rise to the set of roots of exactly one monic irreducible factor of $\Psi_{z^n-1}(x)$ in $F[x]$.

(6) As noted in Equation (3.9), $x^n - 1$ is the product of the cyclotomic polynomials $\Phi_d(x)$, where d runs over the divisors of n. As an additive q-analogue, we have[14]

$$E = \dot{\bigcup_{g|z^n-1}} \Omega_g \quad \text{and hence} \quad x^{q^n} - x = \prod_{g|z^n-1} \Psi_g(x). \tag{3.26}$$

(7) For a fixed polynomial $g(z)$, we obtain the following analogue of (3.26):

[14] Here $\dot{\bigcup}$ denotes a disjoint union, as usual. In what follows, a subscript of the form $h \mid f$ will always indicate that the respective union, product or sum runs over all *monic* divisors $h(z)$ of $f(z)$, where $f(z), h(z) \in F[z]$.

$$V_g = \bigcup_{f|g} \Omega_f \quad \text{and hence} \quad A_g(x) = \prod_{f|g} \Psi_f(x). \tag{3.27}$$

(8) Applying Möbius inversion to Equation (3.9) gave us the product formula for the cyclotomic polynomial $\Phi_n(x)$ in Proposition 3.6.5. It is possible to proceed in an analogous manner using the formula in Equation (3.26) to obtain a q-analogue for the Psi polynomials. We leave this to the reader (see Exercise 3.11.20) and proceed by applying this approach to the formula in Equation (3.27) instead. □

Proposition 3.11.9. *Let $g(z) \in F[z]$ be a monic polynomial which is not a multiple of z. Then*

$$\Psi_g(x) = \prod_{f|g} A_f(x)^{\mu_q(g/f)}, \tag{3.28}$$

where μ_q denotes the Möbius function of the ring $\mathbb{Z}[[F[z]_{\text{mon}}]]$ of formal power series.[15]

Proof. The desired product formula for $\Psi_g(x)$ follows by applying Möbius inversion to the formula in Equation (3.27); we will explain this in some detail. As in the proofs of Propositions 3.5.7 and 3.6.5, we use the multiplicative version of Theorem 2.1.16. Here the underlying simple monoid is the multiplicative monoid $F[z]_{\text{mon}}$ of all monic polynomials in the variable z with coefficients from $F = \text{GF}(q)$, and we consider the ring $\mathbb{Z}[[F[z]_{\text{mon}}]]$ of formal power series. Note that the quotient group $F(z)_{\text{mon}}$ of $F[z]_{\text{mon}}$ is a module with respect to $\mathbb{Z}[[F[z]_{\text{mon}}]]$.

Now we view the operators Ψ and A as mappings from $F[z]_{\text{mon}}$ to $F(z)_{\text{mon}}$ and apply Theorem 2.1.16. In this situation, $S_\gamma(m)$ corresponds to $A_g(x)$, and we obtain

$$\Psi_g(x) = \prod_{f|g} A_{g/f}(x)^{\mu_q(f)} = \prod_{f|g} A_f(x)^{\mu_q(g/f)},$$

as claimed. □

In the remainder of this section, we establish identities for the Psi polynomials which can facilitate the explicit computation of such polynomials; this is similar to the approach for determining the cyclotomic polynomials used in Section 3.6.

It will be convenient to agree on some notational conventions. From now on, we let $\mathbb{M}_z^*$ denote the set of all monic polynomials in $F[z]$ which are not divisible by z. Given any such polynomial $g(z)$, we denote the module defined on the set of roots of the q-polynomial $A_g(x)$ associated with g by V_g, as explained in part (4) of Observation 3.11.8. As we shall only deal with a finite number of polynomials in $\mathbb{M}_z^*$ simultaneously (namely, divisors of some given polynomial $g(z)$), we may as well assume that all q-polynomials $f(z)$ under consideration lead to orders $\text{ord}_f(z)$ dividing a common positive integer n, so that all sets Ω_f and all modules V_f considered are contained in $E = \text{GF}(q^n)$. As before, the Frobenius automorphism of E/F

[15] We choose the notation μ_q in analogy to the notation ϕ_q used for the Euler function of this domain.

will be denoted by σ. Finally, we will denote the product of the distinct irreducible factors of a polynomial $f(z)$ by $\mathrm{rad}(f)$.[16]

We now introduce a special type of module epimorphisms, in analogy to the group epimorphisms defined in Equation (3.16) to study primitive roots of unity.

Observation 3.11.10. Let $f(z)$ and $g(z)$ be polynomials in $\mathbb{M}_z^*$, where f divides g, and define a mapping Π_f^g as follows:

$$\Pi_f^g : V_g \to V_f, \quad v \mapsto v^{g/f} := \left(\frac{g(z)}{f(z)}\right)(\sigma)(v). \tag{3.29}$$

Note that $V_f \subseteq V_g$ and that $\Pi_f^g(v)$ indeed is in V_f for all $v \in V_g$, since $\mathrm{Ord}_q(v)$ divides $g(z)$ so that the q-order of $\Pi_f^g(v)$ divides $g/(g/f) = f$. Obviously, Π_f^g is a homomorphism of (F,σ)-modules, and its kernel is the (F,σ)-submodule $V_{g/f}$. Consequently, if $\Pi_f^g(w) = v$, then the pre-image of v under Π_f^g is equal to $w + V_h$, where $h := g/f$.

As Π_f^g is an F-linear mapping, the dimension formula shows that the image of Π_f^g has F-dimension

$$\dim_F V_g - \dim_F V_{g/f} = \deg g - \deg(g/f) = \deg f = \dim_F V_f,$$

and therefore Π_f^g is indeed an epimorphism. Finally, one easily checks

$$\Pi_f^g = \Pi_f^k \circ \Pi_k^g \quad \text{for all } f,g,k \in \mathbb{M}_z^* \text{ with } f \mid k \mid g; \tag{3.30}$$

this transitivity property will turn out to be quite useful. □

We now prove three results concerning the pre-image of the set Ω_f of generators of V_f under Π_f^g, which will be the basis for proving the desired identities for the Psi polynomials. We begin with a very simple fact, namely the following q-analogue of the observation $\pi_n^m(C_m) \subseteq C_n$ in Remark 3.6.17:

Lemma 3.11.11. *Let $f(z)$ and $g(z)$ be polynomials in $\mathbb{M}_z^*$ with $f \mid g$. Then Π_f^g maps every generator of V_g to a generator of V_f, that is, $\Pi_f^g(\Omega_g) \subseteq \Omega_f$.*

Proof. Let w be any element in Ω_g and $u = \Pi_f^g(w) = w^h$, where $h(z) := g(z)/f(z)$. Then the order formula in Remark 2.6.7 gives

$$\mathrm{Ord}_q(u) = \frac{\mathrm{Ord}_q(w)}{\gcd\left(h(z), \mathrm{Ord}_q(w)\right)} = \frac{f(z)h(z)}{\gcd\left(h(z), f(z)h(z)\right)} = f(z),$$

as claimed. □

Proposition 3.11.12. *Let $f(z)$ and $g(z)$ be polynomials in $\mathbb{M}_z^*$ with $f \mid g$, and assume $\mathrm{rad}(g) = \mathrm{rad}(f)$. Then $\left(\Pi_f^g\right)^{-1}(\Omega_f) = \Omega_g$.*

[16] Note that this is the analogue of the radical of a positive integer introduced in Definition 1.7.1. Thus, we will again use the term "radical" for this notion, as there seems to be no standard term in the polynomial situation.

Proof. Let $h(z) := g(z)/f(z)$, and assume $h \neq 1$. (Otherwise, the assertion is trivial.) Let u be an arbitrary generator of V_f and pick any pre-image w of u, that is, $\Pi_f^g(w) = w^h = u$. (Recall that Π_f^g is surjective.) Then the order formula in Remark 2.6.7 yields

$$\mathrm{Ord}_q(w) = \mathrm{Ord}_q(w^h) \cdot \gcd\big(h(z), \mathrm{Ord}_q(w)\big) = f(z) \cdot \gcd\big(h(z), \mathrm{Ord}_q(w)\big).$$

Assume first that $h(z)$ is irreducible. Then the hypothesis $\mathrm{rad}(g) = \mathrm{rad}(f)$ shows that h divides f and hence also $\mathrm{Ord}_q(w)$. This gives $\mathrm{Ord}_q(w) = f(z)h(z) = g(z)$, so that w is a generator of V_g, which proves the assertion in this special case.

For the general case, we use induction on the number of irreducible divisors of h in $F[z]$, counted with multiplicity. Choose any such irreducible divisor $r(z)$, and put $s(z) := h(z)/r(z)$. By the transitivity formula (3.30),

$$\Pi_f^g = \Pi_f^{fs} \circ \Pi_{fs}^g.$$

The special case above shows $\big(\Pi_{fs}^g\big)^{-1}(\Omega_{fs}) = \Omega_g$, as $g/fs = r$ is irreducible, and induction gives $\big(\Pi_f^{fs}\big)^{-1}(\Omega_f) = \Omega_{fs}$, as $\mathrm{rad}(fs) = \mathrm{rad}(f)$. Combining these observations yields the assertion. □

Proposition 3.11.13. *Let $f(z)$ and $g(z)$ be polynomials in $\mathbb{M}_z^*$ with $f \mid g$, and assume that $h(z) := g(z)/f(z)$ and $f(z)$ are relatively prime. Then*

$$[\Pi_f^g]^{-1}(\Omega_f) = \dot{\bigcup_{a|h}} \Omega_{fa}.$$

Proof. We begin with the following auxiliary observation:

$$\mathrm{Ord}_q(u+y) = \mathrm{Ord}_q(u) \cdot \mathrm{Ord}_q(y) \quad \text{for all } u \in V_f \text{ and all } y \in V_h, \tag{3.31}$$

which holds by the second part of Remark 2.6.7, since h and f are relatively prime.

Now let u be any element in $\Omega_f \subseteq V_g$, and let $v = \Pi_f^g(u) = u^h$. Then the order formula in the first part of Remark 2.6.7 gives

$$\mathrm{Ord}_q(v) = \frac{\mathrm{Ord}_q(u)}{\gcd\big(h(z), \mathrm{Ord}_q(u)\big)} = \frac{f(z)}{\gcd\big(h(z), f(z)\big)} = f(z),$$

as h and f are relatively prime. Thus the restriction of Π_f^g to V_f leaves the set Ω_f invariant, and we claim that it actually permutes the elements of Ω_f. It suffices to check that $\Pi_f^g(u) = \Pi_f^g(u')$ for $u, u' \in \Omega_f$ implies $u = u'$. Note that $\Pi_f^g(u) = \Pi_f^g(u')$ means $(u - u')^h = 0$, that is, $u - u' \in V_h$. Then $u' = u + y$ for some $y \in V_h$, and Equation (3.31) shows

$$\mathrm{Ord}_q(u) = \mathrm{Ord}_q(u') = \mathrm{Ord}_q(u+y) = \mathrm{Ord}_q(u) \cdot \mathrm{Ord}_q(y).$$

This forces $\mathrm{Ord}_q(y) = 1$, so that $y = 0$ and hence $u = u'$, as claimed.

We have now shown that each $v \in \Omega_f$ can be written as $v = u^h$ for some $u \in \Omega_f$. Then the pre-image of v under Π_f^g is $u + V_h$, as noted in Observation 3.11.10. Using Equation (3.31) again, we obtain $u + y \in \Omega_{fa}$ whenever $y \in V_h$ has q-order $a(z)$. This shows that each monic divisor $a(z)$ of $h(z)$ leads to exactly $\phi_q(a)$ elements of Ω_{fa} which are mapped to any specified $v \in \Omega_f$ under Π_f^g. We conclude that Ω_{fa} has to be a subset of the pre-image of Ω_f under Π_f^g, as $\phi_q(fa) = \phi_q(a)\phi_q(f)$ (by the multiplicativity of the Euler function ϕ_q). Since this holds for all monic divisors a of h, we obtain

$$\dot{\bigcup_{a|h}} \Omega_{fa} \subseteq [\Pi_f^g]^{-1}(\Omega_f).$$

On the other hand, $\sum_{a|h} \phi_q(a) = |V_h| = q^{\deg h}$, and therefore

$$\sum_{a|h} |\Omega_{fa}| = \phi_q(f) \cdot q^{\deg h} = |\Omega_f| \cdot |V_h|,$$

showing that the two sets are equal. □

As a consequence of the preceding two propositions, we can now prove the following q-analogues of Propositions 3.6.8 and 3.6.12 for the Psi polynomials.

Corollary 3.11.14. *Let $f(z)$ and $g(z)$ be polynomials in $\mathbb{M}_z^*$ with $f \mid g$, put $h(z) := g(z)/f(z)$, and assume* $\mathrm{rad}(g) = \mathrm{rad}(f)$. *Then $\Psi_g(x)$ is the composition of $\Psi_f(x)$ with the q-polynomial associated with $h(z)$, that is,*

$$\Psi_g(x) = \Psi_f(A_h(x)).$$

Proof. We first show that all elements of Ω_g are roots of $\Psi_f(A_h(x))$, so that $\Psi_g(x)$ divides $\Psi_f(A_h(x))$. Thus let $v \in \Omega_g$. By Lemma 3.11.11, $u := \Pi_f^g(v) \in \Omega_f$, and therefore $\Psi_f(u) = 0$. In view of Proposition 3.11.5, we obtain

$$u = \Pi_f^g(v) = v^h = h(\sigma)(v) = A_h(v),$$

and thus v is a root of $\Psi_f(A_h(x))$, as claimed.

Now it suffices to check that the monic polynomials $\Psi_g(x)$ and $\Psi_f(A_h(x))$ have the same degree. By definition, $\Psi_g(x)$ has degree $|\Omega_g| = \phi_q(g)$. On the other hand, the degree of $\Psi_f(A_h(x))$ is $\deg \Psi_f \cdot \deg A_h = \phi_q(f) \cdot |V_h|$. By Observation 3.11.8 and Proposition 3.11.12, $\Omega_g = \Omega_f + V_h$, and the assertion follows. □

Corollary 3.11.15. *Let $f(z)$ and $g(z)$ be polynomials in $\mathbb{M}_z^*$ with $f \mid g$, and assume that $h(z) := g(z)/f(z)$ and $f(z)$ are relatively prime. Then*

$$\Psi_f(A_h(x)) = \prod_{a|h} \Psi_{fa}(x).$$

Proof. We use the same strategy as in the proof of Corollary 3.11.14 and check that $\prod_{a|h} \Psi_{fa}(x)$ divides $\Psi_f(A_h(x))$ and that both polynomials have the same degree.

Let $a(z)$ be any monic divisor of $h(z)$, and let w be any element with q-order fa. As f and a are relatively prime, we may write w as $u+v$, where $\mathrm{Ord}_q(u) = f(z)$ and $\mathrm{Ord}_q(v) = a(z)$; see Proposition 1.9.2. Then

$$A_h(w) = h(\sigma)(w) = w^h = u^h + v^h = u^h.$$

Using Equation (3.31) from the proof of Proposition 3.11.13 with $y = 0$ gives $\mathrm{Ord}_q(u^h) = \mathrm{Ord}_q(u) = f(z)$, and hence $A_h(w)$ is a root of $\Psi_f(x)$. As this holds for all choices of a and for all $w \in \Omega_{fa}$, we obtain the desired divisibility result.

As in the proof of Corollary 3.11.14, the degree of the polynomial $\Psi_f(A_h(x))$ equals $\phi_q(f) \cdot |V_h| = |\Omega_f + V_h|$. By (the proof of) Proposition 3.11.13, $\Omega_f + V_h$ is the disjoint union of the sets Ω_{fa}, where $a(z)$ runs over all monic divisors of $h(z)$. This proves that $\prod_{a|h} \Psi_{fa}(x)$ also has degree $|\Omega_f + V_h|$. □

Finally, we settle the general case by proving the following q-analogue of Theorem 3.6.13:

Theorem 3.11.16. *Let $f(z)$ and $g(z)$ be polynomials in $\mathbb{M}_z^*$ with $f \mid g$. Put $h(z) := g(z)/f(z)$ and write $h(z) = c(z)d(z)$, where* $\mathrm{rad}(d)$ *divides* $\mathrm{rad}(f)$ *and where c and f are relatively prime. Then*

$$\Psi_f(A_h(x)) = \prod_{a|c} \Psi_{fa}(A_d(x)).$$

Proof. Since $c(z)$ and $d(z)$ are relatively prime, the submodule V_h decomposes as $V_c \oplus V_d$; see Proposition 1.9.2. This decomposition is reflected in the associated q-polynomials as

$$A_h(x) = A_c(A_d(x)) = A_d(A_c(x)),$$

since Π_c^h maps V_h onto V_d and Π_d^h maps V_h onto V_c. Now let $y := A_d(x)$. Then

$$\Psi_f(A_h(x)) = \Psi_f(A_c(A_d(x))) = \Psi_f(A_c(y)) = \prod_{a|c} \Psi_{fa}(y),$$

by Corollary 3.11.15. Since $\mathrm{rad}(d)$ divides $\mathrm{rad}(fa)$ for every divisor a of c, we obtain $\Psi_{fa}(y) = \Psi_{fa}(A_d(x))$ for all a. This results in the desired formula. □

The arguments used in the preceding proofs may also be applied to establish the following q-analogue of Proposition 3.6.18.

Theorem 3.11.17. *Let $f(z)$ and $g(z)$ be polynomials in $\mathbb{M}_z^*$ with $f \mid g$, and let $u \in \Omega_f$. Then there exists an element $w \in \Omega_g$ such that $\Pi_f^g(w) = u$.*

Proof. Write $h(z) = g(z)/f(z) = c(z)d(z)$ as in Theorem 3.11.16. Then Proposition 3.11.12 gives $[\Pi_f^{fd}]^{-1}(u) = \Omega_{fd}$. Now let $v \in [\Pi_f^{fd}]^{-1}(u)$. By (the proof of) Proposition 3.11.13, there exists an element $v' \in \Omega_{fd}$ with $\Pi_{fd}^g(v') = v$, since $c(z) = g(z)/[f(z)d(z)]$ and $f(z)d(z)$ are relatively prime. Thus $[\Pi_{fd}^g]^{-1}(v) = v' + V_c$, and we see that there are $\phi_q(c)$ elements in $[\Pi_{fd}^g]^{-1}(v)$ with q-order $f(z)d(z)c(z) = g(z)$.

We can now apply the transitivity formula (3.30) to obtain a stronger result than asserted, namely that

$$[\Pi_f^g]^{-1}(u) = [\Pi_{fd}^g]^{-1}([\Pi_f^{fd}]^{-1}(u))$$

contains exactly $|V_d| \cdot \phi_q(c) = \phi_q(c) \cdot q^{\deg d}$ elements with q-order $g(z)$. □

Exercises

Exercise 3.11.18. Consider two elements a and b of a finite field E and the mapping

$$\gamma: E \to E, \quad v \mapsto \begin{cases} av^{-1} & \text{if } v \neq 0, \\ b & \text{otherwise.} \end{cases}$$

Determine the interpolation polynomial associated with this mapping according to Section 2.5. □

Exercise 3.11.19. Prove the following alternative description for the polynomial $g_\alpha(x)$ appearing in Theorem 3.11.2:

$$g_\alpha(x) = \alpha(0) + \sum_{j=0}^{q-2} (-1)^{j+1} \binom{q-1}{j} \left(\sum_{u \in F} \alpha(u) u^j \right) x^{q-1-j}.$$

Exercise 3.11.20. Apply Möbius inversion to the formula in (3.26) to obtain an alternative product formula for the Psi polynomials. □

Exercise 3.11.21. Use Proposition 2.3.8 to make Equation (3.28) in Proposition 3.11.9 a little more explicit. □

Exercise 3.11.22. Consider the situation in Lemma 3.11.11 and show that $\Pi_f^g(w)$ cannot belong to Ω_f when $w \in V_g \setminus \Omega_g$. □

Exercise 3.11.23. Show that the restriction of Π_f^g to V_f is an automorphism of V_f in the situation of Proposition 3.11.13. □

3.12 The Trace and the Norm

This section deals with two particularly important mappings which may be associated with the Galois group of any given Galois extension:

Definition 3.12.1. Let E/F be a Galois extension with Galois group G. Then the mappings

$$\mathrm{Tr}_{E/F}:\begin{cases}E\to F\\ v\mapsto \sum_{g\in G} g(v)\end{cases} \quad\text{and}\quad \mathrm{Norm}_{E/F}:\begin{cases}E\to F\\ v\mapsto \prod_{g\in G} g(v)\end{cases} \tag{3.32}$$

are called the **trace function** and the **norm function** of E/F, respectively. If F is the prime field of E, one often uses the terms **absolute trace** and **absolute norm** instead.

In the special case $F = \mathrm{GF}(q)$ and $E = \mathrm{GF}(q^n)$, Theorem 3.4.8 shows that the trace and norm functions can also be defined more explicitly as follows:

$$\mathrm{Tr}_{E/F}(v) = \sum_{j=0}^{n-1} v^{q^j} \quad\text{and}\quad \mathrm{Norm}_{E/F}(v) = \prod_{j=0}^{n-1} v^{q^j} = v^{(q^n-1)/(q-1)} \tag{3.33}$$

for all $v \in E$. □

It is not immediately obvious that the trace and norm functions just introduced indeed map E to the ground field F. This fact is the first of some basic properties of these two functions which we collect in the following result.

Proposition 3.12.2. *Let E/F be a Galois extension with Galois group G. Then:*

(1) $\mathrm{Tr}_{E/F}$ *and* $\mathrm{Norm}_{E/F}$ *are mappings into* F.

(2) $\mathrm{Tr}_{E/F}$ *is a surjective* F*-linear mapping.*

(3) *The restriction of* $\mathrm{Norm}_{E/F}$ *to the multiplicative group* E^* *of* E *is a group homomorphism into* F^*. *Moreover, this homomorphism is surjective provided that* E *and* F *are finite.*

(4) $\mathrm{Tr}_{E/F}(v) = \mathrm{Tr}_{E/F}(g(v))$ *and* $\mathrm{Norm}_{E/F}(v) = \mathrm{Norm}_{E/F}(g(v))$ *for all* $v \in E$ *and all* $g \in G$.

(5) $\mathrm{Tr}_{E/F}(\lambda) = n\lambda$ *and* $\mathrm{Norm}_{E/F}(\lambda) = \lambda^n$ *for all* $\lambda \in F$, *where* $n = [E:F]$.

Proof. Let $h \in G$. Then

$$h(\mathrm{Tr}_{E/F}(v)) = \sum_{g\in G}(hg)(v) = \mathrm{Tr}_{E/F}(v)$$

and

$$h(\mathrm{Norm}_{E/F}(v)) = \prod_{g\in G}(hg)(v) = \mathrm{Norm}_{E/F}(v).$$

Thus both $\mathrm{Tr}_{E/F}(v)$ and $\mathrm{Norm}_{E/F}(v)$ are fixed by G and hence belong to F, by part (1) of Theorem 3.8.8. This establishes assertions (1) and (4).

Regarding (2) and (3), it is clear that $\mathrm{Tr}_{E/F}$ is a linear mapping and that $\mathrm{Norm}_{E/F}$ yields a group homomorphism from the multiplicative group E^* of E into F^*. It remains to check the surjectivity of these mappings.

As the trace function is F-linear, the image of E has to be either F or $\{0\}$. The latter case would mean $\sum_{g\in G} g = 0$, which contradicts the linear independence of the Galois automorphisms established in Proposition 3.8.4. Thus the trace is indeed surjective.

Now consider the norm function and assume that E and F are finite, say $F = \mathrm{GF}(q)$ and $E = \mathrm{GF}(q^n)$. Choose any primitive element v for E. Then

$$\mathrm{Norm}_{E/F}(v) = v^{1+q+\cdots+q^{n-1}} = v^{(q^n-1)/(q-1)}$$

has multiplicative order $q-1$, that is, $\mathrm{Norm}_{E/F}(v)$ is a primitive element for F. Hence $\mathrm{Norm}_{E/F}$ is likewise surjective (in the finite case).[17]

Finally, assertion (5) is an immediate consequence of $g(\lambda) = \lambda$ for every $g \in G$ and every $\lambda \in F$. □

Proposition 3.12.3. *Let E/F be a Galois extension with Galois group G. Assume $F(w) = E$ for some element $w \in E$, and let* $\mathrm{mpol}_w(x) = x^n + \sum_{j=0}^{n-1} \gamma_j x^j$ *be the minimal polynomial of w over F. Then*

$$\gamma_0 = (-1)^n \cdot \mathrm{Norm}_{E/F}(w) \quad \textit{and} \quad \gamma_{n-1} = -\mathrm{Tr}_{E/F}(w).$$

Proof. By Lemma 3.8.11, $\mathrm{mpol}_w(x)$ splits over E as $\prod_{g\in G}(x - g(w))$. Hence

$$\gamma_0 = \prod_{g\in G}(-g(w)) = (-1)^n \cdot \mathrm{Norm}_{E/F}(w),$$

$$\gamma_{n-1} = \sum_{g\in G}(-g(w)) = -\mathrm{Tr}_{E/F}(w). \quad \square$$

Next, we characterize the elements of the kernels of $\mathrm{Tr}_{E/F}$ and $\mathrm{Norm}_{E/F}$, respectively, for the special case of cyclic Galois extensions. We begin with the trace:

Proposition 3.12.4. *Let E/F be a cyclic Galois extension of degree n with Galois group G generated by σ. Then $w \in E$ satisfies* $\mathrm{Tr}_{E/F}(w) = 0$ *if and only if w is of the form $w = \sigma(v) - v$ for some $v \in E$.*

Proof. First let $w = \sigma(v) - v$ for some $v \in E$. Then

$$\mathrm{Tr}_{E/F}(w) = \mathrm{Tr}_{E/F}(\sigma(v)) - \mathrm{Tr}_{E/F}(v) = 0,$$

by Proposition 3.12.2.

For the converse, we use the **trace polynomial** for E/F, that is, the polynomial $t(x)$ defined by

$$t(x) := \frac{x^n - 1}{x - 1} = x^{n-1} + \cdots + x + 1.$$

Then

$$\mathrm{Tr}_{E/F}(w) = \sum_{i=0}^{n-1} \sigma^i(w) = t(\sigma)(w) \quad \text{for all } w \in E.$$

Now assume $\mathrm{Tr}_{E/F}(w) = 0$. By Theorem 3.9.4, we may choose a normal element u for E/F. Then there is a polynomial $f(x) \in F[x]$ such that $w = f(\sigma)(u)$, and $\mathrm{Tr}_{E/F}(w) = 0$ gives

[17] It should be noted that the norm is, in general, not surjective; see Exercise 3.12.12.

$$[t(x) \cdot f(x)](\sigma)(u) = 0.$$

Hence the (F,σ)-order $x^n - 1$ of u divides $t(x)f(x)$, so that f has to be a multiple of $x-1$, say $f(x) = (x-1) \cdot g(x)$. Now put $v := g(\sigma)(u)$. Then

$$\sigma(v) - v = (x-1)(\sigma)(v) = [(x-1)g(x)](\sigma)(u) = f(\sigma)(u) = w. \quad \square$$

The corresponding result for the norm is known as *Hilbert's Satz 90*, as it is indeed the 90-th theorem in Hilbert's classical report on Algebraic Number Theory [184]. The proof is a little more involved than for the case of the trace function and rests on the following auxiliary result for general Galois extensions which – according to Jacobson [205] – is due to Emmy Noether.

Lemma 3.12.5. *Let E/F be a Galois extension with Galois group G, and let*

$$\eta : G \to E^*, \quad \alpha \mapsto u_\alpha$$

be a mapping satisfying **Noether's equation**

$$u_{\beta\alpha} = \beta(u_\alpha)u_\beta \quad \text{for all } \alpha, \beta \in G. \tag{3.34}$$

Then there exists an element $v \in E^*$ such that $u_\beta = v \cdot \beta(v)^{-1}$ for all $\beta \in G$.

Proof. In view of Theorem 3.8.3, we may choose an element $w \in E$ satisfying

$$v := \sum_{\alpha \in G} u_\alpha \alpha(w) \neq 0.$$

Now we use Equation (3.34) to compute $\beta(v)$ for any given $\beta \in G$:

$$\begin{aligned}\beta(v) &= \sum_{\alpha\in G} \beta(u_\alpha)(\beta\alpha)(w) = \sum_{\alpha\in G} u_{\beta\alpha}u_\beta^{-1} \cdot (\beta\alpha)(w) \\ &= u_\beta^{-1} \cdot \sum_{\alpha\in G} u_{\beta\alpha}(\beta\alpha)(w) = u_\beta^{-1} \cdot \sum_{\gamma\in G} u_\gamma \gamma(w) = u_\beta^{-1} \cdot v,\end{aligned}$$

as desired. $\square$

Proposition 3.12.6 (Hilbert's Satz 90). *Let E/F be a Galois extension of degree n with cyclic Galois group G generated by σ. Then $w \in E$ satisfies $\mathrm{Norm}_{E/F}(w) = 1$ if and only if w is of the form $w = v \cdot \sigma(v)^{-1}$ for some $v \in E^*$.*

Proof. First assume $w = v \cdot \sigma(v)^{-1}$ for some $v \in E^*$. Then

$$\begin{aligned}\mathrm{Norm}_{E/F}(w) &= \mathrm{Norm}_{E/F}(v) \cdot \mathrm{Norm}_{E/F}\big(\sigma(v)^{-1}\big) \\ &= \mathrm{Norm}_{E/F}(v) \cdot \big(\mathrm{Norm}_{E/F}(v)\big)^{-1} = 1,\end{aligned}$$

by Proposition 3.12.2.

For the converse, we let $w \in E$ be any element with $\mathrm{Norm}_{E/F}(w) = 1$. We will use Lemma 3.12.5 with the following choice for the mapping η, where we simplify the notation by writing u_i instead of u_{σ^i}:

$$u_i := w\sigma(w)\sigma^2(w)\cdots\sigma^{i-1}(w) \quad \text{for } i = 1,\ldots,n.$$

We have to check that Noether's Equation (3.34) is satisfied, which now reads

$$u_{i+j} = \sigma^j(u_i)u_j \quad \text{for all } i,j = 1,\ldots,n,$$

where the indices are to be read modulo n. Suppose $i+j \le n$. Then indeed

$$\sigma^j(u_i)u_j = \left(\sigma^j(w)\sigma^{j+1}(w)\cdots\sigma^{j+i-1}(w)\right)\cdot\left(w\sigma(w)\cdots\sigma^{j-1}(w)\right) = u_{i+j}.$$

As $u_n = \mathrm{Norm}_{E/F}(w) = 1$, the same identity also holds in the case $i+j > n$. Thus Noether's equation is indeed satisfied, and we conclude $w = u_1 = v\cdot\sigma(v)^{-1}$ for some $v \in E^*$. □

In the case of finite fields, Proposition 3.12.6 allows a short direct proof not relying on Lemma 3.12.5, see Exercise 3.12.13. On the other hand, one may also give a computational proof for Proposition 3.12.4 by using an additive analogue of Lemma 3.12.5, which allows one to avoid the use of the normal basis theorem; see, for instance, Jacobson [205]. As a final remark, we state the following explicit version of the preceding results for the finite case for later reference:

Corollary 3.12.7. *Consider an extension E/F of Galois fields, where $F = \mathrm{GF}(q)$ and $E = \mathrm{GF}(q^n)$. Then*

- *$w \in E$ satisfies $\mathrm{Tr}_{E/F}(w) = 0$ if and only if $w = v^q - v$ for some $v \in E$;*
- *$w \in E$ satisfies $\mathrm{Norm}_{E/F}(w) = 1$ if and only if $w = v^{q-1}$ for some $v \in E^*$.* □

We conclude this section with two important results concerning transitivity properties of the trace and norm functions:

Theorem 3.12.8. *Let E/F be a Galois extension, and assume that K is an intermediate field of E/F such that K/F is likewise a Galois extension. Then the following two identities hold for all $w \in E$:*

$$\mathrm{Tr}_{E/F}(w) = \mathrm{Tr}_{K/F}(\mathrm{Tr}_{E/K}(w)) \quad \textit{and} \quad \mathrm{Norm}_{E/F}(w) = \mathrm{Norm}_{K/F}(\mathrm{Norm}_{E/K}(w)).$$

Proof. We use the fundamental properties of Galois extensions given in Theorem 3.8.8. First, E/K is a Galois extension, and $H := \mathrm{Gal}(E/K)$ is a subgroup of G. As K/F is also a Galois extension, H is in fact a normal subgroup of G and $\mathrm{Gal}(K/F)$ is isomorphic to the factor group G/H. Moreover, $g(K) = K$ for every $g \in G$ and the mapping $G \to \mathrm{Gal}(K/F)$ sending every $g \in G$ to its restriction to K is a group epimorphism with kernel H; thus every system $\mathscr{R}$ of representatives of cosets of H in G induces the Galois group of K/F on K. Using these facts, we obtain

$$\mathrm{Tr}_{E/F}(w) = \sum_{g\in G} g(w) = \sum_{r\in\mathscr{R}}\sum_{h\in H}(rh)(w)$$
$$= \sum_{r\in\mathscr{R}} r\Big(\sum_{h\in H} h(w)\Big) = \mathrm{Tr}_{K/F}(\mathrm{Tr}_{E/K}(w)),$$

as claimed. The identity for the norm is proved in a similar way. □

Proposition 3.12.9. *Let E/F and K be as in Theorem 3.12.8. Then:*

- *If $w \in E$ is a normal element for E/F, then $\mathrm{Tr}_{E/K}(w)$ is a normal element for K/F.*
- *Assume that E and F are Galois fields and that w is a primitive element of E. Then $\mathrm{Norm}_{E/K}(w)$ is a primitive element of K.*

Proof. Let $G = \mathrm{Gal}(E/F), H = \mathrm{Gal}(E/K)$ and $u := \mathrm{Tr}_{E/K}(w)$. As noted in the proof of Theorem 3.12.8, $g(K) = K$ for every $g \in G$ and the restriction of G to K gives a bijection of $\mathscr{R}$ to the Galois group of K/F for every system of representatives $\mathscr{R}$ for the cosets of H in G. Now assume $\sum_{r\in\mathscr{R}} a_r r(u) = 0$ for some elements $a_r \in F$. Then

$$0 = \sum_{r\in\mathscr{R}} a_r r\Big(\sum_{h\in H} h(w)\Big) = \sum_{r\in\mathscr{R}}\sum_{h\in H} a_r(rh)(w).$$

Since $\{rh : r \in \mathscr{R}, h \in H\} = G$ and since w is a normal element for E/F, we obtain $a_r = 0$ for all r. Thus u is a normal element for K/F, which establishes the first assertion.

For the second assertion, let $F = \mathrm{GF}(q)$ and $E = \mathrm{GF}(q^n)$, and assume that w is a primitive element of E. Then

$$\mathrm{Norm}_{E/K}(w) = w^{1+q^k+\cdots+q^{(\ell-1)k}} = w^{(q^n-1)/(q^k-1)},$$

where $k = [K : F]$ and $\ell = n/k$. Thus $\mathrm{Norm}_{E/K}(w)$ has order $q^k - 1$ and is indeed a primitive element of K. □

Finally, we prove the following converse of Proposition 3.12.9 for finite fields:

Theorem 3.12.10. *Let E/F be an extension of Galois fields, and let K be an intermediate field. Then:*

- *Let ζ be any primitive element of K. Then there exists a primitive element η of E such that $\mathrm{Norm}_{E/K}(\eta) = \zeta$.*
- *Let u be any normal element for K/F. Then there exists a normal element w for E/F such that $\mathrm{Tr}_{E/K}(w) = u$.*

Proof. Let $F = \mathrm{GF}(q)$, $E = \mathrm{GF}(q^n)$ and $K = \mathrm{GF}(q^k)$, where k divides n. According to Equation (3.33),

$$\mathrm{Norm}_{E/K}(v) = v^{(q^n-1)/(q^k-1)} \quad \text{for all } v \in E.$$

Now the first assertion follows by applying Proposition 3.6.18 in the situation where ζ is a primitive (q^k-1)-th root of unity and η a primitive (q^n-1)-th root of unity.

For the second assertion, we use the results on Psi polynomials established in Section 3.11. Note that z^k-1 divides z^n-1, as k divides n. By hypothesis, the element $u \in K$ has q-order z^k-1; see part (5) of Observation 3.11.8. By Theorem 3.11.17, there exists an element $w \in E$ with q-order z^n-1 such that

$$\Pi_{z^k-1}^{z^n-1}(w) = u,$$

so that w is a pre-image of u which is normal for E/F. Note that $(z^n-1)/(z^k-1) = \sum_{i=0}^{m-1} z^{ki}$, where $m = n/k$. This gives

$$u = \Pi_{z^k-1}^{z^n-1}(w) = \Big(\sum_{i=0}^{m-1} z^{ki}\Big)(\sigma)(w) = \sum_{i=0}^{m-1} w^{q^{ki}} = \mathrm{Tr}_{E/K}(w),$$

as desired. □

Exercises

Exercise 3.12.11. Give an explicit formula for the trace and norm functions of the Galois extension $\mathbb{C}/\mathbb{R}$. □

Exercise 3.12.12. Let $m > 1$ be a square-free integer. Determine the Galois group of $\mathbb{Q}(\sqrt{m})/\mathbb{Q}$ and give an explicit formula for the trace and norm functions of this extension. Apply this to show that -1 cannot occur as a norm in the special case where $m \equiv 3 \bmod 4$ is a prime. □

Exercise 3.12.13. Give a direct proof for Corollary 3.12.7 based on the explicit description of the trace and norm in Equation (3.33). □

3.13 Finite Fields: Duality

In this section, we consider a bilinear form associated with the trace function, which leads to the fundamental concept of duality. This is one of the central ideas in the theory of finite fields: it turns out to be a very useful technical tool which can be applied in surprisingly many situations, and there also is a direct connection to the efficient hardware implementation of multiplication.

We begin by recalling a few standard facts from Linear Algebra. Let F be a field and V a finitely generated F-vector space, say with dimension n.

The F-vector space $\mathrm{Hom}_F(V,F)$ of all F-linear mappings from V to F, that is, the space of all **linear forms** of V is usually denoted by V^* and called the **dual space** of V. The vector spaces V and V^* are isomorphic. (However, there is no canonical isomorphism in this case.)

A **bilinear form** on V is a mapping $\beta: V \times V \to F$ such that $\beta(\mathbf{u},\cdot)$ and $\beta(\cdot,\mathbf{w})$ are linear forms for all $\mathbf{u},\mathbf{w} \in V$. A bilinear form β is called **non-degenerate** if $\beta(\mathbf{u},\cdot)$ and $\beta(\cdot,\mathbf{w})$ are not the zero form, unless $\mathbf{u} = \mathbf{0}$ or $\mathbf{w} = \mathbf{0}$, respectively. Finally, β is said to be **symmetric** provided that $\beta(\mathbf{u},\mathbf{w}) = \beta(\mathbf{w},\mathbf{u})$ for all $\mathbf{u},\mathbf{w} \in V$.

Proposition 3.13.1. *Let β be a non-degenerate bilinear form on the F-vector space V, and let $B = \{\mathbf{b}_1,\dots,\mathbf{b}_n\}$ be a basis of V. Then there exists a unique basis $C = \{\mathbf{c}_1,\dots,\mathbf{c}_n\}$ of V such that*

$$\beta(\mathbf{b}_i,\mathbf{c}_j) = \delta_{i,j} := \begin{cases} 1 & \text{if } i = j, \\ 0 & \text{if } i \neq j. \end{cases}$$

Proof. For every $\mathbf{b}_k$ in B, we define a linear form $b_k^*: V \to F$ by

$$b_k^*: \sum_{i=1}^{n} \lambda_i \mathbf{b}_i \mapsto \lambda_k.$$

Then the set $B^* := \{b_1^*,\dots,b_n^*\}$ forms a basis of V^*. Since β is non-degenerate, the mapping

$$\psi: V \mapsto V^*, \quad \mathbf{v} \mapsto \beta(\cdot,\mathbf{v})$$

is an isomorphism between V and its dual space V^*. We now put $\mathbf{c}_j := \psi^{-1}(b_j^*)$ for $j = 1,\dots,n$. Then $C = \{\mathbf{c}_1,\dots,\mathbf{c}_n\}$ is a basis of V, since B^* is a basis of V^*, and it has the desired properties:

$$\beta(\mathbf{b}_i,\mathbf{c}_j) = \psi(\mathbf{c}_j)(\mathbf{b}_i) = \psi(\psi^{-1}(b_j^*))(\mathbf{b}_i) = b_j^*(\mathbf{b}_i) = \delta_{i,j}.$$

It remains to check the uniqueness of C. Thus let $\mathbf{d}_1,\dots,\mathbf{d}_n$ be any basis of V such that $\beta(\mathbf{b}_i,\mathbf{d}_j) = \delta_{i,j}$. Then $\psi(\mathbf{d}_j)(\mathbf{b}_i) = \delta_{i,j}$ for all i and j, and therefore $\psi(\mathbf{d}_j) = b_j^*$, that is, $\mathbf{d}_j = \mathbf{c}_j$. □

Definition 3.13.2. Let β be a non-degenerate symmetric bilinear form. Then the basis C in Proposition 3.13.1 is called the **dual basis** of B (in view of its construction from a basis of the dual space V^*).[18] By symmetry, B is also the dual basis of C. □

Remark 3.13.3. Let β be a non-degenerate symmetric bilinear form on the F-vector space V, and let U be a subspace of V. Then

$$U^{\perp} := \{\mathbf{w} \in V : \beta(\mathbf{u},\mathbf{w}) = 0 \text{ for all } \mathbf{u} \in U\}$$

is a subspace of V; it is called the subspace **orthogonal** to U (with respect to β). Since β is non-degenerate, one has

$$\dim_F U^{\perp} = \dim_F V - \dim_F U$$

and, by symmetry, $(U^{\perp})^{\perp} = U$ for all subspaces U. □

[18] In the context of duality, the ordering of the bases indicated by the indices is important. Nevertheless, we will generally continue to write bases as sets.

Proposition 3.13.4. *Let β be a non-degenerate symmetric bilinear form on the F-vector space V, where the characteristic of F is not equal to 2. Then there exists a basis $B = \{\mathbf{b}_1, \dots, \mathbf{b}_n\}$ of V such that $\beta(\mathbf{b}_i, \mathbf{b}_j) = 0$ whenever $i \neq j$.*[19]

Proof. We use induction on the dimension n of V. The case $n = 1$ is trivial. Thus let $n \geq 2$ and assume that the assertion holds for all instances with dimension $\leq n-1$. We first show that there is a vector $\mathbf{v}$ with $\beta(\mathbf{v}, \mathbf{v}) \neq 0$; here the assumption on the characteristic is used. Assume otherwise. Then

$$0 = \beta(\mathbf{u}+\mathbf{w}, \mathbf{u}+\mathbf{w}) = \beta(\mathbf{u}, \mathbf{u}) + 2\beta(\mathbf{u}, \mathbf{w}) + \beta(\mathbf{w}, \mathbf{w}) = 2\beta(\mathbf{u}, \mathbf{w})$$

for all $\mathbf{u}, \mathbf{w} \in V$. Since $2 \neq 0$, this gives $\beta(\mathbf{u}, \mathbf{w}) = 0$ for all $\mathbf{u}, \mathbf{w} \in V$, so that β would be degenerate.

Now choose any $\mathbf{v} \in V$ with $\beta(\mathbf{v}, \mathbf{v}) \neq 0$ and consider the space $W = \langle \mathbf{v} \rangle^{\perp}$ orthogonal to $\mathbf{v}$. Then W has dimension $n-1$. We now restrict β to $W \times W$, and use the induction hypothesis to choose a basis $\mathbf{b}_1, \dots, \mathbf{b}_{n-1}$ of W such that $\beta(\mathbf{b}_i, \mathbf{b}_j) = 0$ for all $i \neq j$. Hence we obtain the desired basis for the whole vector space V by taking $\mathbf{b}_n := \mathbf{v}$. □

Definition 3.13.5. Let β be a non-degenerate symmetric bilinear form on an F-vector space V, and let $B = \{\mathbf{b}_1, \dots, \mathbf{b}_n\}$ be a basis of V such that

$$\beta(\mathbf{b}_i, \mathbf{b}_j) = 0 \quad \text{whenever } i \neq j.$$

Then B is called an **orthogonal basis** of V (with respect to β). Now $\beta(\mathbf{b}_i, \mathbf{b}_i) \neq 0$ for all i, as β is non-degenerate, and B is called an **orthonormal basis** if actually $\beta(\mathbf{b}_i, \mathbf{b}_i) = 1$ for all i. □

Remark 3.13.6. Let β be a non-degenerate symmetric bilinear form on V, and let $B = \{\mathbf{b}_1, \dots, \mathbf{b}_n\}$ be an orthogonal basis. Consider a basis vector $\mathbf{b}_k$, write $\omega := \beta(\mathbf{b}_k, \mathbf{b}_k) \in F^*$ and assume that $\omega = \lambda^2$ is a square in F^*. Then we may replace $\mathbf{b}_k$ by $\mathbf{b}_k' := \frac{1}{\lambda}\mathbf{b}_k$ in order to obtain

$$\beta(\mathbf{b}_k', \mathbf{b}_k') = \frac{1}{\lambda^2}\beta(\mathbf{b}_k, \mathbf{b}_k) = \frac{1}{\lambda^2} \cdot \omega = 1.$$

Similarly, if ω is a non-square, we may replace $\mathbf{b}_k$ to change ω by an arbitrary quadratic factor. What may be achieved in this way clearly depends on the structure of the non-squares in F^*. For instance, all elements are squares if F is either $\mathbb{C}$ or a perfect (in particular, a finite) field of characteristic 2, so that we can then always change ω to 1. For the reals, it is clear that one may achieve entries ± 1, and a well-known result from classical Linear Algebra – namely, Sylvester's law of inertia – states that the number of positive entries does not depend on the choice of B. For the case of a finite field of odd characteristic, this problem will be settled in Theorem 3.13.14 below. □

[19] As we shall see soon, such bases may also exist in the case of characteristic 2. However, this depends on the specific bilinear form β, see Exercise 3.13.23.

We now introduce the main object of the present section: a non-degenerate symmetric bilinear form associated with the trace function of a Galois extension E/F, where we view E as an n-dimensional vector space over F.

Proposition 3.13.7. *Let E/F be a Galois extension. Then the mapping*

$$E \times E \to F, \quad (u,v) \mapsto \mathrm{Tr}_{E/F}(uv) \tag{3.35}$$

is a non-degenerate symmetric bilinear form.

Proof. Let $\tau(u,v) := \mathrm{Tr}_{E/F}(uv)$. Trivially, $\tau(u,v) = \tau(v,u)$ for all $u,v \in E$. Moreover,

$$\mathrm{Tr}_{E/F}((\lambda u + \rho w)v) = \lambda \mathrm{Tr}_{E/F}(uv) + \rho \mathrm{Tr}_{E/F}(wv)$$

for all $\lambda, \rho \in F$ and all $u,v,w \in E$. Thus τ is indeed a symmetric bilinear form. Now let $u \in E^*$. Then $\{uv : v \in E\} = E$ and hence $\mathrm{Tr}_{E/F}(uv) \neq 0$ for some $v \in E$, by Proposition 3.12.2. This shows that τ is non-degenerate. □

Definition 3.13.8. The symmetric bilinear form τ introduced in Proposition 3.13.7 is called the **trace bilinear form** associated with E/F. □

The trace bilinear form allows the following alternative characterization of bases, which is sometimes more useful than the one via the Galois group given in Proposition 3.9.2. To simplify the notation, we just write Tr instead of $\mathrm{Tr}_{E/F}$ from now on, provided that no intermediate fields are involved.

Proposition 3.13.9. *Let E/F be a Galois extension. Then the elements $b_1,\ldots,b_n$ form a basis for E/F if and only if $\Delta_{E/F}(b_1,\ldots,b_n) \neq 0$, where*

$$\Delta_{E/F}(b_1,\ldots,b_n) = \det \begin{pmatrix} \mathrm{Tr}(b_1b_1) & \mathrm{Tr}(b_1b_2) & \ldots & \mathrm{Tr}(b_1b_n) \\ \mathrm{Tr}(b_2b_1) & \mathrm{Tr}(b_2b_2) & \ldots & \mathrm{Tr}(b_2b_n) \\ \vdots & \vdots & \ddots & \vdots \\ \mathrm{Tr}(b_nb_1) & \mathrm{Tr}(b_nb_2) & \ldots & \mathrm{Tr}(b_nb_n) \end{pmatrix} \tag{3.36}$$

is the **discriminant** *of $b_1,\ldots,b_n$.*

Proof. First let $b_1,\ldots,b_n$ be a basis for E/F. We have to show $\Delta_{E/F}(b_1,\ldots,b_n) \neq 0$. It suffices to check that the columns of the matrix in (3.36) are linearly independent. Thus suppose

$$\lambda_1 \mathrm{Tr}(b_ib_1) + \cdots + \lambda_n \mathrm{Tr}(b_ib_n) = 0 \quad \text{for } i = 1,\ldots,n, \tag{3.37}$$

where $\lambda_1,\ldots,\lambda_n \in F$. Write $c := \lambda_1 b_1 + \cdots + \lambda_n b_n$. Then (3.37) shows $\mathrm{Tr}(b_ic) = 0$ for $i = 1,\ldots,n$. Therefore $\mathrm{Tr}(vc) = 0$ for all $v \in E$, as $b_1,\ldots,b_n$ span E. By Proposition 3.13.7, $c = 0$ and hence $\lambda_1 = \cdots = \lambda_n = 0$, since $b_1,\ldots,b_n$ are linearly independent.

Conversely, assume $\Delta_{E/F}(b_1,\ldots,b_n) \neq 0$ and suppose $\lambda_1 b_1 + \cdots + \lambda_n b_n = 0$ for some elements $\lambda_1,\ldots,\lambda_n$ of F. Then

$$0 = \mathrm{Tr}\Big(b_i \cdot \Big(\sum_{j=1}^{n} \lambda_j b_j\Big)\Big) = \sum_{j=1}^{n} \lambda_j \mathrm{Tr}(b_i b_j) \quad \text{for } i = 1,\ldots,n,$$

which implies the validity of (3.37). Since the columns of the matrix in (3.36) are linearly independent, we conclude $\lambda_1 = \cdots = \lambda_n = 0$, and hence $b_1,\ldots,b_n$ form a basis for E/F. □

We now consider the concept of duality introduced before in the special case of the trace bilinear form. Here the following terminology is standard:

Definition 3.13.10. Let $B = \{b_1,\ldots,b_n\}$ be a basis for a Galois extension E/F, and consider the trace bilinear form τ associated with E/F. Let $C = \{c_1,\ldots,c_n\}$ be the dual basis of B with respect to τ (see Proposition 3.13.1), that is,

$$\tau(b_i,c_j) = \mathrm{Tr}_{E/F}(b_i c_j) = \delta_{i,j} \quad \text{for } i,j = 1,\ldots,n.$$

Then C is called the **trace-dual basis** of B for E/F.

Now assume

$$\mathrm{Tr}_{E/F}(b_i b_j) = 0 \quad \text{for all } i \neq j. \tag{3.38}$$

Then B is said to be a **trace-orthogonal basis** for E/F. If additionally

$$\mathrm{Tr}_{E/F}(b_i^2) = 1 \quad \text{for all } i, \tag{3.39}$$

one says that B is a **self-dual basis** for E/F.

Finally, let B be trace-orthogonal and assume $\mathrm{Tr}_{E/F}(b_i^2) = 1$ with possibly one exception. Then B is called an **almost self-dual basis** for E/F. □

Of course, it is an interesting question whether or not one can always find a self-dual basis for E/F, perhaps even with additional interesting properties. While such questions have been investigated for general Galois extensions, we will from now on restrict ourselves to the special case of finite fields (with the exception of Theorem 3.13.19, which is easily established for arbitrary Galois extensions). For the general case, the interested reader may consult [22] and [21].

For the remainder of this section, we let $F = \mathrm{GF}(q)$ and $E = \mathrm{GF}(q^n)$. We always assume (tacitly) $n \geq 2$, as the case $n = 1$ is trivial. As we shall see, E/F does not always admit a self-dual basis, which motivated considering the weaker notions introduced in Definition 3.13.10. Our major goal is a proof of the following two results which settle the existence question for almost self-dual and self-dual bases, respectively, for all prime powers q and all integers n.

Theorem 3.13.11. *There always exists an almost self-dual basis of* $\mathrm{GF}(q^n)$ *over* $\mathrm{GF}(q)$. □

Theorem 3.13.12 (Self-dual basis theorem). *A self-dual basis of* $\mathrm{GF}(q^n)$ *over* $\mathrm{GF}(q)$ *exists if and only if either q is even or both n and q are odd.* □

The original proofs of these existence theorems are rather lengthy and involved. (This holds even in the case of the weaker existence problem for trace-orthogonal bases.) We shall present considerably simpler proofs due to Jungnickel, Menezes and Vanstone [216]. We need to distinguish the cases of even and of odd characteristic and begin with the even case, which was settled by Lempel [231] for $q=2$ and by Seroussi and Lempel [348] in general.

Proposition 3.13.13. *If q is an even prime power, then there exists a self-dual basis of $E = \mathrm{GF}(q^n)$ over $F = \mathrm{GF}(q)$ for all n.*

Proof. Since q is even, one has $\mathrm{Tr}(a^2) = \mathrm{Tr}(a)^2$ for every $a \in E$, and thus condition (3.39) is equivalent to

$$\mathrm{Tr}(b_i) = 1 \quad \text{for } i = 1, \dots, n.$$

We will give a recursive construction for the elements of the desired self-dual basis. We begin by choosing an arbitrary element $b_1 \neq 1$ in E with $\mathrm{Tr}(b_1) = 1$, which is possible by Proposition 3.12.2 (as we may assume $n \geq 2$). Now let $k \leq n-1$ and suppose that we have already constructed elements $b_1, \dots, b_k$ satisfying

$$\mathrm{Tr}(b_i b_j) = \delta_{i,j} \text{ for } i, j = 1, \dots, k \quad \text{and} \quad b_1 + \dots + b_k \neq 1; \tag{3.40}$$

the additional condition in (3.40) is of no interest in itself, but serves to carry out the inductive step. It is easily verified that $b_1, \dots, b_k$ are linearly independent over F. We now have to find an element $b_{k+1} \in E$ with $\mathrm{Tr}(b_{k+1}) = 1$ satisfying

$$\mathrm{Tr}(b_i b_{k+1}) = 0 \quad \text{for } i = 1, \dots, k \tag{3.41}$$

and

$$b_1 + \dots + b_{k+1} \neq 1 \quad \text{if } k \leq n-2. \tag{3.42}$$

In order to do so, we consider the linear subspace S of E generated by $\{b_1, \dots, b_k\}$ and its orthogonal space $S^\perp$ with respect to the trace bilinear form, that is,

$$S^\perp = \{v \in E \colon \mathrm{Tr}(vb_i) = 0 \text{ for } i = 1, \dots, k\}.$$

We first show that $S^\perp$ is not contained in the hyperplane $T = \{v \in E \colon \mathrm{Tr}(v) = 0\}$, that is, the kernel of the trace function. Assume otherwise, and note $1 \in T^\perp$ (by definition of $\perp$ and T). Then $1 \in T^\perp \subseteq (S^\perp)^\perp = S$, say

$$1 = \lambda_1 b_1 + \dots + \lambda_k b_k \quad \text{with } \lambda_1, \dots, \lambda_k \in F. \tag{3.43}$$

Multiplying Equation (3.43) by b_i and taking the trace shows $\lambda_i = 1$ for $i = 1, \dots, k$, which contradicts (3.40).

Hence $S^\perp$ is indeed not contained in the hyperplane T and therefore intersects T in a subspace of dimension $n-k-1$, as S has dimension k so that $S^\perp$ has dimension $n-k$. Thus each coset of T in E contains $q^{n-k-1} \neq 0$ elements of $S^\perp$; in particular, this holds for the coset $T_1 = \{w \in E \colon \mathrm{Tr}(w) = 1\}$, proving that T_1 contains an element b_{k+1} satisfying (3.41). Moreover, $q^{n-k-1} > 1$ provided that $k \leq n-2$, so that we may also satisfy condition (3.42) in this case. (We note in passing that $b_1 + \cdots + b_{n-1} = 1$ in the terminal case $k = n-1$.) This gives the desired recursive construction of a self-dual basis $B = \{b_1, \ldots, b_n\}$ for E/F. □

In order to deal with the case of odd characteristic, we need the following well-known result on symmetric bilinear forms over $\mathrm{GF}(q)$.

Theorem 3.13.14. *Let $F = \mathrm{GF}(q)$, where q is odd, and let $n \geq 2$ be an integer. Then there are exactly two equivalence classes of non-degenerate symmetric bilinear forms on F^n, which may be represented by the identity matrix I and the matrix $M = \mathrm{diag}(1, \ldots, 1, \eta)$, where η is an arbitrary non-square in F.*

Proof. Let $\beta\colon F^n \times F^n \to F$ be an arbitrary non-degenerate symmetric bilinear form. By Proposition 3.13.4, β can be represented by a diagonal matrix, that is, we may choose a basis $B = \{b_1, \ldots, b_n\}$ of F^n satisfying

$$\begin{aligned} &\beta(b_i, b_i) = \omega_i \neq 0 \quad \text{for } i = 1, \ldots, n \qquad \text{and} \\ &\beta(b_i, b_j) = 0 \quad \text{for } i, j = 1, \ldots, n \text{ and } i \neq j. \end{aligned} \tag{3.44}$$

As explained in Remark 3.13.6, we may change the value of ω_i in (3.44) by an arbitrary non-zero square λ_i^2. By Proposition 3.2.13, the non-zero squares form a subgroup $F^{\Box}$ of F^* with index 2. Hence we may assume $\omega_i \in \{1, \eta\}$ for all i, where η may be chosen as an arbitrary non-square. Note that the set

$$A = \{a^2 + b^2 \colon \ a, b \in F, \ b \neq 0\}$$

has to contain a non-square: otherwise, we would have $F^{\Box} \subseteq A \subseteq F^{\Box} \cup \{0\}$, and then $F^{\Box} \cup \{0\}$ would be a proper subgroup of $(F, +, 0)$ with $\frac{q-1}{2} + 1 > \frac{q}{2}$ elements, which is absurd. Therefore, we may select a non-square of the form $\eta' = 1 + a^2$; we then choose $\eta := 1/\eta'$.

Now assume $\omega_i = \omega_j = \eta$ for two distinct indices i and j. Then we may replace b_i and b_j by $c_i := b_i + ab_j$ and $c_j := ab_i - b_j$, so that

$$\beta(c_i, c_i) = \beta(c_j, c_j) = \eta(1 + a^2) = 1 \quad \text{and} \quad \beta(c_i, c_j) = 0.$$

This shows that values $\omega_i = \eta$ may be replaced in pairs by values $\omega_i = 1$, and hence β can be represented by either I or M. Finally, these two matrices represent non-equivalent forms, since their determinants differ by a non-square. □

We now easily obtain the following result due to Jungnickel, Menezes and Vanstone [216].

Proposition 3.13.15. *Let q be an odd prime power. Then there always exists an almost self-dual basis of* $\mathrm{GF}(q^n)$ *over* $\mathrm{GF}(q)$.

Proof. By Proposition 3.13.7, the trace bilinear form associated to E/F is non-degenerate. Since q is odd, we may apply Theorem 3.13.14 and choose a basis B for which this bilinear form is represented by either I or M. □

The validity of Theorem 3.13.11 now follows from Propositions 3.13.13 and 3.13.15. With a little more work, one can also settle the existence problem for self-dual bases in odd characteristic in a similar way. To do so, we need two simple auxiliary results. The first of these is merely an easy computation, which we leave to the reader.

Lemma 3.13.16. *Let $B = \{b_1, \dots, b_n\}$ and $C = \{c_1, \dots, c_n\}$ be two bases of* $\mathrm{GF}(q^n)$ *over* $\mathrm{GF}(q)$*, and consider the matrices*

$$M_B = \begin{pmatrix} b_1 & b_2 & \dots & b_n \\ b_1^q & b_2^q & \dots & b_n^q \\ \vdots & \vdots & \ddots & \vdots \\ b_1^{q^{n-1}} & b_2^{q^{n-1}} & \dots & b_n^{q^{n-1}} \end{pmatrix} \quad \textit{and} \quad M_C = \begin{pmatrix} c_1 & c_2 & \dots & c_n \\ c_1^q & c_2^q & \dots & c_n^q \\ \vdots & \vdots & \ddots & \vdots \\ c_1^{q^{n-1}} & c_2^{q^{n-1}} & \dots & c_n^{q^{n-1}} \end{pmatrix}.$$

Then $M_B^T M_C = \left(\mathrm{Tr}(b_i c_j)\right)_{i,j=1,\dots,n}$. □

Lemma 3.13.17. *Let $B = \{b_1, \dots, b_n\}$ be any basis for $E = \mathrm{GF}(q^n)$ over $F = \mathrm{GF}(q)$, where q is odd, and consider the matrix M_B defined in Lemma 3.13.16. Then* $\det M_B$ *is an element of F if and only if n is odd.*

Proof. Note that $\det M_B$ belongs to F if and only if

$$\det M_B = (\det M_B)^q = \det M_B^{(q)},$$

where $M_B^{(q)}$ denotes the matrix obtained from M_B by replacing each entry with its q-th power. Now $M_B^{(q)}$ also arises from M_B via a cyclic permutation of its n rows:

$$M_B^{(q)} = \begin{pmatrix} b_1^q & b_2^q & \dots & b_n^q \\ b_1^{q^2} & b_2^{q^2} & \dots & b_n^{q^2} \\ \vdots & \vdots & \ddots & \vdots \\ b_1^{q^{n-1}} & b_2^{q^{n-1}} & \dots & b_n^{q^{n-1}} \\ b_1 & b_2 & \dots & b_n \end{pmatrix}.$$

Hence $\det M_B^{(q)} = (-1)^{n-1} \cdot \det M_B$. As B is a basis of E, Proposition 3.9.2 shows $\det M_B \neq 0$. Therefore $\det M_B \in F$ if and only if $(-1)^{n-1} = 1$, which means that n is odd (as q is odd). □

We can now give the promised simple proof for the following result due to Seroussi and Lempel [348].

Proposition 3.13.18. *Let q be an odd prime power. Then there exists a self-dual basis of $E = \mathrm{GF}(q^n)$ over $F = \mathrm{GF}(q)$ if and only if n is likewise odd.*

Proof. By definition, E/F admits a self-dual basis if and only if the trace bilinear form τ can be represented by the identity matrix I. By Theorem 3.13.14, this is equivalent to requiring that $\det A$ is a square in F for any matrix A representing τ. Now choose an arbitrary basis $B = \{b_1, \ldots, b_n\}$, and consider the matrix M_B defined in Lemma 3.13.16. Applying this result with $C = B$, we see that the matrix $A := M_B^T M_B$ represents τ with respect to B. Thus $\det A = (\det M_B)^2$ is always a square in E, and it is actually a square in F if and only if n is odd, by Lemma 3.13.17. □

The validity of Theorem 3.13.12 now follows from Propositions 3.13.13 and 3.13.18.

In the remainder of this section, we consider the interaction of the concepts of duality and normality. We begin with a simple but important observation.

Theorem 3.13.19. *Let E/F be any Galois extension. Then the trace-dual basis of a normal basis is again a normal basis.*

Proof. We write the Galois group G of E/F as $G = \{g_1, \ldots, g_n\}$, where g_1 is the identity. Let b be any normal element for E/F and label the elements in the associated normal basis $B = \{b_1, \ldots, b_n\}$ correspondingly, that is, $b_i = g_i(b)$ for $i = 1, \ldots, n$.

Now let $C = \{c_1, \ldots, c_n\}$ be the trace-dual basis of B, so that $\mathrm{Tr}(b_i c_j) = \delta_{i,j}$. As G permutes the elements of B and as $g_i(b_1) = b_i$ by our labelling, the following holds for all i and j:

$$\begin{aligned}\mathrm{Tr}\big(b_i \cdot g_j(c_1)\big) &= g_j\Big(\mathrm{Tr}\big(g_j^{-1}(b_i) \cdot c_1\big)\Big) = \mathrm{Tr}\big(g_j^{-1}(b_i) \cdot c_1\big) \\ &= \mathrm{Tr}\big(g_j^{-1}(g_i(b_1)) \cdot c_1\big) = \delta_{i,j} = \mathrm{Tr}\big(b_i \cdot c_j\big).\end{aligned}$$

(For the penultimate equality, note that $(g_j^{-1} \circ g_i)(b_1) = b_1$ if and only if $i = j$.) Hence the linear forms $\mathrm{Tr}(\cdot, g_j(c_1))$ and $\mathrm{Tr}(\cdot, c_j)$ coincide, and thus $g_j(c_1) = c_j$. As this holds for all j, we see that C is the normal basis generated by c_1. □

It is natural to ask whether there are bases combining the two properties we have studied, that is, self-dual normal bases. This is a considerably more difficult problem, and the known proofs are rather involved and technical. Here one has the following necessary conditions which combine Proposition 3.13.18 with a result due to Morii and Imamura [286].

Proposition 3.13.20. *Assume the existence of a self-dual normal basis for* $\mathrm{GF}(q^n)$ *over* $\mathrm{GF}(q)$. *Then, necessarily, either q is even and n is not a multiple of* 4, *or both q and n are odd.*

Proof. First let q be odd. Then the assertion is an immediate consequence of Proposition 3.13.18, as no self-dual basis can exist if n is even.

We now investigate the situation where q is even and n is a multiple of 4, say $n = 4m$. We have to show that no normal basis for $E = \mathrm{GF}(q^n)$ over $F = \mathrm{GF}(q)$ can be self-dual in this case. Thus let α be any normal element for E/F. We put

$$\gamma := \sum_{i=0}^{m-1} \alpha^{q^{4i}} = \mathrm{Tr}_{E/K}(\alpha),$$

where $K = \mathrm{GF}(q^4)$ is the intermediate field of E/F with degree 4 over F.[20] Using $\alpha^{q^{4m}} = \alpha$, one easily checks

$$\gamma^{q^4} = \gamma \quad \text{and} \quad \gamma^{q^2} \neq \gamma. \tag{3.45}$$

We now consider the element γ^{q+1} and compute its trace $\mathrm{Tr}_{K/F}(\gamma^{q+1})$ over F in two different ways. Using (3.45) several times, we get

$$\begin{aligned}
\mathrm{Tr}_{K/F}(\gamma^{q+1}) &= \gamma^{q+1} + \gamma^{q(q+1)} + \gamma^{q^2(q+1)} + \gamma^{q^3(q+1)} \\
&= \gamma^{q+1} \cdot (1 + \gamma^{q^2-1} + \gamma^{q^3+q^2-q-1} + \gamma^{q^3-q}) \\
&= \gamma^{q+1} \cdot (1 + \gamma^{q^2-1}) \cdot (1 + \gamma^{q^3-q}) \neq 0.
\end{aligned}$$

Again using (3.45), we also obtain

$$\begin{aligned}
\mathrm{Tr}_{K/F}(\gamma^{q+1}) &= \mathrm{Tr}_{K/F}(\gamma \cdot \gamma^q) = \mathrm{Tr}_{K/F}\Big(\gamma \cdot \sum_{j=0}^{m-1} \alpha^{q^{4j+1}}\Big) \\
&= \mathrm{Tr}_{K/F}\Big(\sum_{j=0}^{m-1} \gamma^{q^{4j}} \alpha^{q^{4j+1}}\Big) = \mathrm{Tr}_{K/F}\Big(\sum_{i,j=0}^{m-1} \alpha^{q^{4i+4j}} \alpha^{q^{4j+1}}\Big) \\
&= \sum_{h=0}^{3} \sum_{i,j=0}^{m-1} \alpha^{q^{4i+4j+h}} \alpha^{q^{4j+h+1}} = \sum_{i=0}^{m-1} \sum_{k=0}^{4m-1} \alpha^{q^{4i+k}} \alpha^{q^{k+1}} \\
&= \sum_{i=0}^{m-1} \sum_{k=0}^{4m-1} (\alpha^{q^{4i}} \alpha^q)^{q^k} = \sum_{i=0}^{m-1} \mathrm{Tr}_{E/F}(\alpha^{q^{4i}} \alpha^q),
\end{aligned}$$

and not all terms $\mathrm{Tr}_{E/F}(\alpha^{q^{4i}} \alpha^q)$ in the final sum can be 0, as $\mathrm{Tr}_{K/F}(\gamma^{q+1}) \neq 0$. This shows that the normal basis generated by α is not self-dual. □

Some years later, Lempel and Weinberger [234] managed to show that the necessary conditions for the existence of a self-dual normal basis of $\mathrm{GF}(q^n)$ over $\mathrm{GF}(q)$ given in Proposition 3.13.20 are in fact sufficient. As the constructions needed to prove their result heavily rely on the use of circulant matrices, they will be postponed to Chapter 7.

We conclude this section with some simple observations which are useful for actual constructions.

Proposition 3.13.21. *Consider the extension E/F of Galois fields, where $F = \mathrm{GF}(q)$ and $E = \mathrm{GF}(q^n)$. Assume that $n = k\ell$, where k and ℓ are relatively prime, and let K and L be the intermediate fields of E/F with degree k and ℓ, respectively. Finally, let $B = \{b_1, \dots, b_k\}$ be a basis for K/F and $C = \{c_1, \dots, c_\ell\}$ a basis for L/F. Then:*

[20] It may be noted that γ is a normal element of K/F, since α is normal in E/F, by Proposition 3.12.9. However, we will not require this fact here.

(1) *$BC := \{b_i c_j : i = 1,\ldots,k;\ j = 1,\ldots,\ell\}$ is a basis for E/F.*

(2) $\mathrm{Tr}_{E/F}(uv) = \mathrm{Tr}_{K/F}(u)\mathrm{Tr}_{L/F}(v)$ *for all $u \in K$ and all $v \in L$.*

(3) *If both B and C are trace-orthogonal, then BC is likewise trace-orthogonal.*

(4) *If both B and C are self-dual, then BC is likewise self-dual.*

(5) *If both B and C are normal, then BC is likewise normal.*

Proof. Part (1) is a consequence of Corollary 3.1.14, since K and L are linear disjoint in the present situation (see also Remark 3.4.3).

Now let $u \in K$ and $v \in L$. The transitivity of the trace function (see Theorem 3.12.8) gives

$$\mathrm{Tr}_{E/F}(uv) = \mathrm{Tr}_{K/F}\big(\mathrm{Tr}_{E/K}(uv)\big).$$

Since $u \in K$, we have $\mathrm{Tr}_{E/K}(uv) = u\mathrm{Tr}_{E/K}(v)$. Moreover, the restriction of the Galois group of E/K to L induces the Galois group of L/F. (Let φ be the restriction of σ^k to L, where σ is the Frobenius automorphism of E/F. Since the restriction of σ to L has order ℓ, and since k and ℓ are relatively prime, φ has order ℓ as well and therefore generates the Galois group of L/F.) Hence $\mathrm{Tr}_{E/K}(v) = \mathrm{Tr}_{L/F}(v)$, since $v \in L$. Altogether, this gives

$$\mathrm{Tr}_{E/F}(uv) = \mathrm{Tr}_{K/F}\big(u\mathrm{Tr}_{L/F}(v)\big) = \mathrm{Tr}_{K/F}(u) \cdot \mathrm{Tr}_{L/F}(v),$$

since $\mathrm{Tr}_{L/F}(v) \in F$. This proves (2).

Assume that B and C are both trace-orthogonal, and let (i,j) and (s,t) be distinct pairs from $\{1,\ldots,k\} \times \{1,\ldots,\ell\}$. In view of (2),

$$\mathrm{Tr}_{E/F}([b_i c_j] \cdot [b_s c_t]) = \mathrm{Tr}_{E/F}([b_i b_s] \cdot [c_j c_t]) = \mathrm{Tr}_{K/F}(b_i b_s) \cdot \mathrm{Tr}_{L/F}(c_j c_t) = 0.$$

Thus the basis BC is likewise trace-orthogonal. If B and C are even self-dual, then

$$\mathrm{Tr}_{E/F}([b_i c_j] \cdot [b_i c_j]) = \mathrm{Tr}_{E/F}(b_i^2 \cdot c_j^2) = \mathrm{Tr}_{K/F}(b_i^2) \cdot \mathrm{Tr}_{L/F}(c_j^2) = 1$$

for all pairs (i,j), and thus BC is a self-dual basis for E/F. This establishes parts (3) and (4).

Finally, part (5) follows from Theorem 3.9.7, since BC is the set of conjugates of $b_1 c_1$ under the Galois group of E/F in this case. □

For later reference, we state the following immediate consequence of Proposition 3.13.21 explicitly:

Corollary 3.13.22. *Assume that u and v generate self-dual normal bases for* $\mathrm{GF}(q^m)$ *and* $\mathrm{GF}(q^n)$ *over $F = \mathrm{GF}(q)$, respectively, where m and n are relatively prime. Then $w = uv$ generates a self-dual normal basis for* $\mathrm{GF}(q^{mn})$ *over F.* □

Exercises

Exercise 3.13.23. Let F be any field with characteristic 2. Find a non degenerate symmetric bilinear form β on the 2-dimensional vector space F^2 for which there is no orthogonal basis. Generalize your example to higher dimensions. □

Exercise 3.13.24. Check the assertion of Lemma 3.13.16. □

Chapter 4
The Algebraic Closure of a Galois Field

Abstract In the present chapter, we study the algebraic closure of a Galois field. For this purpose, we also provide some basic results on algebraic extensions, extending the material covered in Chapters 2 and 3.

4.1 Preliminaries on Algebraic Extensions

In this section, we collect some fundamental results on algebraic extensions. Let us first recall some basic definitions from Section 2.4.

Given a field extension E/F and an element $v \in E$, the evaluation homomorphism at v is defined by

$$\Gamma_v \colon F[x] \to E, \quad f(x) \mapsto f(v).$$

Recall that v is said to be algebraic over F provided that Γ_v is not injective; in this case, the unique monic generator $\mathrm{mpol}_v(x)$ of $\ker \Gamma_v$ is called the minimal polynomial of v over F. In what follows, the minimal polynomial will also be denoted by $\mathrm{mpol}_{v,F}(x)$, if we wish to emphasize the underlying field. The extension E/F is **algebraic** if every $v \in E$ is algebraic over F. We also say that a field extension E/F is **finite** if the degree $[E:F]$ is finite.

Lemma 4.1.1. *Every field extension E/F with finite degree is algebraic. Conversely, if E/F is an algebraic extension and $S \subseteq E$ is finite, then $F(S)/F$ is finite.*

Proof. The first assertion is clear, as the intermediate field $F(v)$ of E/F generated by $v \in E$ is isomorphic to $F[x]/\ker \Gamma_v$. Since E/F is finite and $[F(v):F] \le [E:F]$, the kernel of Γ_v has to be non-trivial.

The partial converse in the second assertion obviously holds if $S = \{u\}$ is a singleton. We proceed using induction on the cardinality of S. Choose some $u \in S$, put $S' := S \setminus \{u\}$, and let $K = F(u)$ and $L = K(S')$. Since E/K is algebraic, L/K is a finite extension by induction. As K/F is a finite extension, Lemma 3.1.1 shows that

D. Hachenberger and D. Jungnickel, *Topics in Galois Fields*,
Algorithms and Computation in Mathematics 29,
https://doi.org/10.1007/978-3-030-60806-4_4

L/F is likewise finite. Since $F(S)$ is an intermediate field of L/F, we conclude that $F(S)$ is a finite extension over F. □

Proposition 4.1.2. *Consider a field extension E/F, and let K be an intermediate field. Then E/F is algebraic if and only if both E/K and K/F are algebraic.*

Proof. It is clear that E/F can only be algebraic if both E/K and K/F are algebraic. For the converse, let u be an arbitrary element of E; by hypothesis, u is algebraic over K. Let $S \subseteq K$ be the set of coefficients of $\text{mpol}_{u,K}(x)$, and consider the subfield $L := F(S)$ of K. By Lemma 4.1.1, L/F is a finite extension. Moreover, $[L(u) : L] = [K(u) : K]$, since $\text{mpol}_{u,K}(x) = \text{mpol}_{u,L}(x)$ (by the definition of L); now Lemma 3.1.1 shows that the extension $L(u)/F$ is finite. Since $F(u)$ is an intermediate field of $L(u)/F$, we obtain that $F(u)/F$ is also finite, so that u is algebraic over F. □

Corollary 4.1.3. *Consider a field extension E/F and assume that $u, v \in E$ such that u is algebraic over F and v is algebraic over $F(u)$. Then v is also algebraic over F.*

Proof. The extension $F(u,v)/F$ is finite, as $F(u,v)/F(u)$ and $F(u)/F$ are finite. □

Theorem 4.1.4. *Let E/F be a field extension, and put*

$$\text{acl}(E/F) := \{u \in E : \ u \text{ is algebraic over } F\}.$$

Then $\text{acl}(E/F)$ *is an intermediate field of E/F which is an algebraic extension of F. Moreover,* $\text{acl}(E/F)$ *contains all elements of E which are algebraic over* $\text{acl}(E/F)$.

Proof. By definition, $F \subseteq \text{acl}(E/F) \subseteq E$. Now let $u, v, w \in \text{acl}(E/F)$, where $w \neq 0$. Applying Corollary 4.1.3 twice shows that $F(u,v,w)$ is an algebraic extension over F and hence finite, by Lemma 4.1.1. As $\frac{u-v}{w} \in F(u,v,w)$, we conclude that $\frac{u-v}{w}$ is algebraic over F, and thus $\frac{u-v}{w} \in \text{acl}(E/F)$. This shows that $\text{acl}(E/F)$ is a subfield of E and hence an intermediate field of E/F. By the definition of $\text{acl}(E/F)$, the extension $\text{acl}(E/F)/F$ is algebraic.

Now assume that $w \in E$ is algebraic over $K := \text{acl}(E/F)$, so that $K(w)/K$ is an algebraic extension. We have already established that K/F is also an algebraic extension, and therefore the extension $K(w)/F$ is likewise algebraic, by Proposition 4.1.2. Thus w is even algebraic over F, and hence $w \in \text{acl}(E/F)$, as claimed. □

Definition 4.1.5. Let E/F be a field extension. Then the intermediate field $\text{acl}(E/F)$ defined in Theorem 4.1.4 is called the **(relative) algebraic closure** of F in E. In the case $\text{acl}(E/F) = E$, one says that F is **algebraically closed** in E. □

Given any field F, we wish to determine an algebraic extension which is as large as possible. This motivates the following definitions.

Definition 4.1.6. A field K is called **algebraically closed** provided that every non-constant polynomial $g(x) \in K[x]$ has a root in K; in other words, the only monic irreducible polynomials in $K[x]$ should be the linear ones.

Now let $\widehat{F}/F$ be a field extension. Then $\widehat{F}$ is called an **algebraic closure** of F provided that $\widehat{F}$ is algebraically closed, but no proper intermediate field of $\widehat{F}/F$ is algebraically closed. □

Proposition 4.1.7. *Assume that $\widehat{F}$ is an algebraic closure of a field F. Then $\widehat{F}/F$ is an algebraic extension.*

Proof. Let $L := \mathrm{acl}(\widehat{F}/F)$ be the algebraic closure of F in $\widehat{F}$, and let $g(x) \in L[x]$ be any non-constant polynomial. Since $\widehat{F}$ is algebraically closed, g has a root in $\widehat{F}$, say u; then u is algebraic over L, as $g(x) \in L[x]$. Since L is the algebraic closure of F in $\widehat{F}$, we obtain $u \in L$, so that g has a root in L. As g was arbitrary, L is algebraically closed, and hence $\widehat{F} = L$, by the definition of an algebraic closure. Now Theorem 4.1.4 shows that $\widehat{F}$ is algebraic over F. □

Theorem 4.1.8. *Let C be any algebraically closed extension of a field F. Then there is a unique intermediate field $\widehat{F}$ of C/F which is an algebraic closure of F.*

Proof. We will verify that $L := \mathrm{acl}(C/F)$ is the unique algebraic closure of F in C/F. By Theorem 4.1.4, L is an algebraic extension of F. Now let $g(x) \in L[x]$. As C is algebraically closed, g splits over C into linear factors, and hence C contains a splitting field L_g of g over L. By Proposition 3.1.4, L_g/L is finite, so that L_g is an algebraic extension of L. Again using Theorem 4.1.4, we conclude $L_g = L$, and hence L is algebraically closed.

Finally, let K be an intermediate field of C/F which is an algebraic closure of F. By Proposition 4.1.7, K is an algebraic extension of F, and thus K is an intermediate field of L/F. It remains to show $L \subseteq K$. Thus let w be an arbitrary element of L. Since w is algebraic over F, it is also algebraic over K. As K is algebraically closed, the only monic irreducible polynomials in $K[x]$ are the linear ones. In particular, the minimal polynomial $\mathrm{mpol}_{w,K}(x)$ has to be a linear polynomial; thus $\mathrm{mpol}_{w,K}(x) = x - w$, which shows $w \in K$. □

Our next aim is to prove that any field admits an algebraic closure. For this, we require a generalization of Definition 3.1.5 and an interesting strengthening of Theorem 3.1.7.

Definition 4.1.9. Let F be a field, and let $\mathscr{P} \subseteq F[x]$ be a nonempty set of nonzero polynomials with coefficients in F. Then an extension field E of F is called a **splitting field** of $\mathscr{P}$ over F provided that every $f(x) \in \mathscr{P}$ splits over E into linear factors, and there is no proper subfield of E satisfying this condition. □

Theorem 4.1.10. *Let F be a field and $\mathscr{P} \subseteq F[x]$ be any nonempty set of non-constant polynomials. Then there exists a splitting field of $\mathscr{P}$ over F.*

Proof. If $\mathscr{P}$ is a finite set, one may take $g(x) := \prod_{f \in \mathscr{P}} f(x)$. By Theorem 3.1.7, there exists a splitting field of g over F. Obviously, this is also a splitting field of $\mathscr{P}$ over F.

Now let $\mathscr{P}$ be infinite and consider the system $\mathscr{P}_{\mathrm{fin}}$ of all finite subsets of $\mathscr{P}$. For $I \in \mathscr{P}_{\mathrm{fin}}$, let E_I be a splitting field of I over F. Given two fixed sets $I, J \in \mathscr{P}_{\mathrm{fin}}$ with $I \subseteq J$, we may assume $E_I \subseteq E_J$, since $\prod_{f \in I} f(x)$ divides $\prod_{h \in J} h(x)$. However, it is not quite obvious that we may actually assume $E_I \subseteq E_J$ for *any two* finite sets $I, J \subseteq \mathscr{P}$ with $I \subseteq J$; indeed, this requires the existence of some set Ω containing all fields E_I

as subsets. We will take this fact for granted here, as it follows from a fundamental result from Set Theory, namely Zorn's lemma; see, for instance, Ciesielski [76].

Thus we obtain a collection $\{E_I : I \in \mathscr{P}_{\text{fin}}\}$ of splitting fields E_I over F such that $E_I \subseteq E_J$ whenever $I \subseteq J$. We now define $L \subseteq \Omega$ by

$$L := \bigcup_{I \in \mathscr{P}_{\text{fin}}} E_I;$$

then L is a field, since $E_I \cup E_J \subseteq E_{I \cup J}$ for all $I, J \in \mathscr{P}_{\text{fin}}$. Given any $f(x) \in \mathscr{P}$, we may use the singleton $I = \{f(x)\} \in \mathscr{P}_{\text{fin}}$ to see that $f(x)$ splits over L into linear factors.

Finally, let K be an intermediate field of L/F such that every $f(x) \in \mathscr{P}$ splits over K into linear factors. Then $E_I \subseteq K$ for all $I \in \mathscr{P}_{\text{fin}}$, and hence $L \subseteq K$, so that $K = L$. Altogether, this shows that L is a splitting field of $\mathscr{P}$ over F. □

We are now ready to prove that every field has an algebraic closure. In Section 4.2, we will provide an explicit alternative construction for the special case of Galois fields.

Theorem 4.1.11. *Let F be a field. Then there exists an algebraic closure of F.*

Proof. By Theorem 4.1.10, there exists a splitting field C of the set $\mathscr{P} := F[x]_{\text{mon}}$ of all monic polynomials with coefficients from F. According to the proof of that result, we may assume $C = \bigcup_I E_I$, where I runs over all finite subsets of monic polynomials from $F[x]$ and where E_I is a splitting field of I. Thus C is a union of finite extensions E_I/F and therefore algebraic, since any $u \in C$ is in some E_I, and E_I/F is algebraic. We claim that C is an algebraic closure of F. Note that it suffices to show that C is algebraically closed, since any algebraic closure $K \subseteq C$ of F has to contain the splitting fields of all singletons $\{f(x)\} \in F[x]_{\text{mon}}$ and then also all E_I, where I runs over all finite subsets of $F[x]_{\text{mon}}$, so that $K = C$.

It remains to show that every non-constant polynomial has a root in C, and it obviously suffices to consider irreducible polynomials. Thus let g be any irreducible polynomial over C, and denote the set of coefficients of g by S. Then $F(S)$ is an algebraic extension of F, by Lemma 4.1.1. We consider the product $h(x)$ of all minimal polynomials $\text{mpol}_{w,F}(x)$, where w runs over the elements in S. Then h is a monic polynomial over F, and $E_h := E_{\{h(x)\}}$ is its splitting field. By the definition of h, we have $S \subseteq E_h$ and therefore $g(x) \in E_h[x]$. By Lemma 3.1.3, g has a root in some finite extension field K of E_h, say u. Since both K/E_h and E_h/F are finite, the extension $F(u)/F$ is also finite. Thus there is an irreducible polynomial in $F[x]$, namely $f(x) := \text{mpol}_{u,F}(x)$, such that u is a root of f in K. As g is an irreducible polynomial over E_h with root u and as $F \subseteq E_h$, we conclude that g divides f. Therefore, the splitting field $E_f \subseteq C$ of f over F contains a root of g, as claimed. □

Finally, we want to show that the algebraic closure of a field is uniquely determined up to isomorphism. This will be a simple consequence of the following preliminary result:

Proposition 4.1.12. *Let F be a field, C an algebraic closure of F, and K/F any algebraic extension of F. Then there exists a monomorphism $\alpha: K \to C$ fixing F elementwise.*

Proof. Let Ω be the set of all pairs (M, ω), where M is an intermediate field of K/F and where $\omega: M \to C$ is a monomorphism fixing F elementwise. Note that $\Omega \neq \emptyset$, since $(F, \iota) \in \Omega$, where $\iota(\lambda) = \lambda$ for all $\lambda \in F$. Moreover, Ω is partially ordered by defining

$$(M, \omega) \preceq (N, \sigma) \quad :\Leftrightarrow \quad M \subseteq N \text{ and } \omega \subseteq \sigma,$$

where $\omega \subseteq \sigma$ means that the restriction of σ to M equals ω. As in the proof of Theorem 4.1.10, we will have to appeal to Zorn's lemma. Thus let $\mathscr{C}$ be a chain in Ω (that is, a subset of Ω which is totally ordered with respect to $\preceq$), and put

$$Q := \bigcup_{(N,\omega) \in \mathscr{C}} N \quad \text{and} \quad \alpha := \bigcup_{(N,\omega) \in \mathscr{C}} \omega.$$

Note that α can be considered as a subset of $Q \times C$, since ω is a subset of $N \times C$ for each $(N, \omega) \in \mathscr{C}$. Clearly, Q is an intermediate field of K/F and $\alpha: Q \to C$ is a monomorphism fixing F elementwise. Thus (Q, α) lies in Ω and is an upper bound for $\mathscr{C}$, which shows that each chain in Ω has an upper bound in Ω. By Zorn's lemma, Ω has a maximal element: there is a pair $(L, \lambda) \in \Omega$ such that $(L, \lambda) \preceq (B, \beta)$ and $(B, \beta) \in \Omega$ imply $(L, \lambda) = (B, \beta)$.

The assertion follows if we can show $L = K$. Given any element $u \in K$, consider the minimal polynomial g of u over L. Applying λ to the coefficients of g, we obtain the polynomial $\lambda(g(x))$ in $C[x]$. As C is algebraically closed, $\lambda(g(x))$ has a root $w \in C$. Then one can extend λ to a monomorphism of $L(u)$ into C fixing F elementwise by putting by $\lambda(u) := w$; see Proposition 3.1.8. Now the maximality of (L, λ) in Ω gives $L = L(u)$, and therefore $u \in L$. □

Theorem 4.1.13. *Any two algebraic closures of a field F are isomorphic.*

Proof. Let C_1 and C_2 be two algebraic closures of F. In view of Propositions 4.1.7 and 4.1.12, there exists a monomorphism $\alpha: C_1 \to C_2$ fixing F elementwise. Then $\alpha(C_1)$ is an algebraically closed intermediate field of C_2/F, and Theorem 4.1.8 gives $\alpha(C_1) = C_2$. Thus α is the desired isomorphism between C_1 and C_2. □

4.2 Algebraic Extensions of Galois Fields

In this section, we provide a description of all algebraic extensions of any given Galois field. In particular, this leads to an explicit construction for the algebraic closure of any finite field which does not rely on the use of Zorn's lemma. All this rests on the following important concept:

Definition 4.2.1. Let S be a non-empty subset of $\mathbb{N}^*$, and assume that the following two conditions are satisfied:

- $n \in S$ implies $d \in S$ for all divisors d of n;
- $m, n \in S$ implies $\operatorname{lcm}(m,n) \in S$.

Then S is called a **Steinitz number**.[1] □

Throughout this section, F will denote the Galois field $\mathrm{GF}(q)$ with q elements. We start by considering an arbitrary extension C/F which is not necessarily algebraic. By Theorem 3.2.3 and Corollary 3.2.4, all finite subgroups of the multiplicative group C^* of C are cyclic, and C^* contains at most one subgroup of any given finite order. Combining this fact with Theorem 3.4.2 shows that there is at most one intermediate field K of C/F with a prescribed degree $[K:F] = n$, in which case K is isomorphic to $\mathrm{GF}(q^n)$. Put

$$N(C/F) := \{n \in \mathbb{N}^* : C/F \text{ has an intermediate field with degree } n \text{ over } F\}. \quad (4.1)$$

For $n \in N(C/F)$, the unique intermediate field of C/F with degree n over F will be denoted by E_n. Note that $N(C/F)$ is non-empty, since $E_1 = F$ so that $1 \in N(C/F)$. We first show that $N(C/F)$ is always a Steinitz number.

Theorem 4.2.2. *Consider an arbitrary extension C/F, where F is a Galois field. Then $N(C/F)$ is a Steinitz number.*

Proof. Let $n \in N(C/F)$, and let d be a divisor of n. By definition, C/F contains the intermediate field E_n with degree n over F. By Theorem 3.4.2, there is a unique intermediate field K of E_n/F with degree d over F. Hence $K = E_d$ is an intermediate field of C/F, so that $d \in N(C/F)$.

Now let $m, n \in N(C/F)$. Choose a primitive element $v \in E_n$ and consider the intermediate field $E_m(v)$ of C/F. The minimal polynomial of v over F has degree n and is irreducible over F, but splits over E_m into d distinct irreducible polynomials of degree n/d each, where $d = \gcd(n,m)$; see Theorem 3.5.9. Therefore,

$$[E_m(v):F] = [E_m(v):E_m] \cdot [E_m:F] = \frac{n}{d} \cdot m = \operatorname{lcm}(n,m),$$

so that $E_m(v) = E_{\operatorname{lcm}(n,m)}$ and thus $\operatorname{lcm}(n,m) \in N(C/F)$. □

Remark 4.2.3. Given any extension C/F of a Galois field F, there is a bijection ψ from its Steinitz number $N(C/F)$ to the set of all intermediate fields of C/F with finite degree over F, given by $\psi(n) = E_n$. Moreover,

- $\psi(\operatorname{lcm}(n,m)) = E_n E_m$ is the compositum of the fields E_n and E_m;
- $\psi(\gcd(n,m)) = E_n \cap E_m$.

This follows from the proof of Theorem 4.2.2; see also Remark 3.4.3. □

---

[1] Steinitz numbers are also called **closed subsets** of $\mathbb{N}^*$ or **supernatural numbers**. They were introduced by Steinitz [361] in his fundamental work on field extensions, where also the notion of perfect fields was introduced. Moreover, Steinitz proved the existence and uniqueness of algebraic closures in this paper.

Remark 4.2.4. Let E be a finite extension of a Galois field F, say $[E:F]=n$. By Theorem 3.4.2, the Steinitz number of E/F equals the set D_n of all divisors of n. Actually, any finite Steinitz number is of this form for some $n \in \mathbb{N}^*$, as is easily checked. Thus the set system of all Steinitz numbers can be viewed as a superset of the set $\mathbb{N}^*$ of the positive integers, by identifying $n \in \mathbb{N}^*$ with D_n. □

Our next aim is a description of the set of all algebraic intermediate fields of an arbitrary extension C/F in terms of Steinitz numbers. We start with an easy special case, namely the algebraic closure of F in C:

Proposition 4.2.5. *Consider an arbitrary field extension C/F, where $F=\mathrm{GF}(q)$, and let $N(C/F)$ be its Steinitz number. Then*

$$A := \bigcup_{n \in N(C/F)} E_n$$

is the algebraic closure $\mathrm{acl}(C/F)$ *of F in C.*

Proof. Let $v \in C$ be algebraic over F. Then $F(v)=E_n$ for some $n \in N(C/F)$, and therefore $\mathrm{acl}(C/F) \subseteq A$. Conversely, let $v \in A$. Then there exists an $m \in N(C/F)$ such that $v \in E_m$. Thus v has finite degree over F, so that v is algebraic over F, and hence also $A \subseteq \mathrm{acl}(C/F)$. □

Proposition 4.2.6. *Let C/F be a field extension, where $F=\mathrm{GF}(q)$, and let K be any intermediate field of C/F. Then K is algebraic over F if and only if it has the form*

$$k(S) := \bigcup_{n \in S} E_n,$$

where S is some Steinitz number which is a subset of the Steinitz number $N(C/F)$.

Moreover, the mapping $S \mapsto k(S)$ gives a bijection between the set of all Steinitz numbers contained in $N(C/F)$ and the set of all intermediate fields of C/F which are algebraic over F.

Proof. Note that an intermediate field K of C/F is algebraic over F if and only if it is contained in the algebraic closure of F in C; in this case, the algebraic closure of F in K equals K. Now let K be such an intermediate field, and put

$$S_K := \{n \in N(C/F) \colon E_n \subseteq K\}.$$

An application of Theorem 4.2.2 to the extension K/F shows that S_K is the Steinitz number $N(K/F)$, and Proposition 4.2.5 gives

$$K = \mathrm{acl}(K/F) = \bigcup_{n \in S_K} E_n = k(S_K).$$

Conversely, let S be any subset of $N(C/F)$ which is a Steinitz number. We first show that $k(S)$ is an intermediate field of C/F. For this, let $u,v,w \in k(S)$, where

$w \neq 0$. Then there exist $\ell, m, n \in S$ such that $u \in E_\ell$, $v \in E_m$, and $w \in E_n$. As S is a Steinitz number, $t := \operatorname{lcm}(\operatorname{lcm}(\ell, m), n)$ is likewise in S. Then u, v, w all belong to the field E_t, and hence $\frac{u-v}{w} \in E_t$. This shows $\frac{u-v}{w} \in k(S)$, so that $k(S)$ is a subfield of C. Moreover, $k(S)$ contains $F = E_1$, as $1 \in S$. Finally, note that any element $v \in k(S)$ is algebraic over F since $v \in E_n$ for some $n \in S$, so that v has finite degree over F. Thus $k(S)$ is indeed an algebraic extension of F.

We have already established $k(S_K) = K$ for any algebraic intermediate field of C/F, and it is easily checked that also $S_{k(S)} = S$ holds for any Steinitz number S contained in $N(C/F)$. Therefore, the mapping $S \mapsto k(S)$ is a bijection. □

We can now give a nice characterization of the algebraic closures of Galois fields in terms of Steinitz numbers:

Theorem 4.2.7. *Let $F = \mathrm{GF}(q)$ be the Galois field with q elements, and let A be an algebraic extension field of F. Then A is an algebraic closure of F if and only if its Steinitz number $N(A/F)$ equals the set $\mathbb{N}^*$ of all positive integers.*

Proof. First assume that A is an algebraic closure of F. By Theorem 3.3.5, there exists a monic irreducible polynomial $f(x) \in F[x]$ for any given degree $n \in \mathbb{N}^*$. Note that f has a root v in A, as A is assumed to be algebraically closed. Then $F(v) \subseteq A$ has degree n over F, and therefore $n \in N(A/F)$. This shows $N(A/F) = \mathbb{N}^*$.

Conversely, let A/F be an algebraic extension of F satisfying $N(A/F) = \mathbb{N}^*$. We have to show that any non-constant polynomial $f(x) \in A[x]$ has a root in A. Let C_f be the set of coefficients of f and consider the field $L = F(C_f)$, which is a finite extension of F and a subfield of A. Let $[L : F] = \ell$, that is, $L = E_\ell$. By definition, $f(x) \in L[x]$. Now let K be any extension field of L which contains a root of f. Then K has finite degree over L, say $[K : L] = k$, so that K is a field with $q^{k\ell}$ elements. Since $N(A/F) = \mathbb{N}^*$, there exists a (unique) intermediate field K' of A/F with degree $k\ell$ over F. By Theorem 3.3.5, K and K' are isomorphic, so that K' also contains a root of f. This shows that A is indeed algebraically closed. □

Corollary 4.2.8. *Let $F = \mathrm{GF}(q)$ be the Galois field with q elements, and let $\widehat{F}$ be an algebraic closure of F. Then*

$$\widehat{F} = \operatorname{acl}(\widehat{F}/F) = \bigcup_{n \in \mathbb{N}^*} E_n.$$

Moreover, the intermediate fields of $\widehat{F}/F$ are in a one-to-one correspondence with the Steinitz numbers.

Proof. This is an immediate consequence of Propositions 4.2.5 and 4.2.6, applied to the situation in Theorem 4.2.7. □

We are now ready to give the promised constructive proof for the existence (and also for the uniqueness) of an algebraic closure of an arbitrary Galois field. More generally, given any Galois field F and any Steinitz number S, we will show the existence and uniqueness of an algebraic extension A/F with $N(A/F) = S$.

Theorem 4.2.9. *Let F be any Galois field and S any Steinitz number. Then there exists an algebraic extension A/F such that $N(A/F) = S$.*

Proof. If S is finite, the result holds in view of Theorem 3.3.5 and Remark 4.2.4. Thus let S be infinite. Note that the intersection of any two Steinitz numbers is again a Steinitz number. In particular, the intersection of S with some finite Steinitz number D_m has to be a finite Steinitz number, and hence $S \cap D_m = D_\ell$ for some divisor ℓ of m, by Remark 4.2.4. We will call ℓ the greatest common divisor of S and m and use the notation $\gcd(S,m)$. Let

$$d_n := \gcd(S,n!) \quad \text{for all } n \in \mathbb{N}^*,$$

and consider the sequence $(d_n)_{n\in\mathbb{N}^*}$. Then d_n divides d_{n+1} for all n. Now we recursively construct a tower $(K_n)_{n\in\mathbb{N}^*}$ of extension fields of F with degree sequence (d_n) as follows. We let $K_1 := F$, which is a (trivial) extension of F with degree $d_1 = 1$. Assume that the extension K_n of degree d_n over F has already been constructed. Choose a monic irreducible polynomial $f_n(x) \in K_n[x]$ with degree d_{n+1}/d_n, put

$$K_{n+1} := K_n[x]/(f_n),$$

and view K_n as a subfield of K_{n+1} by embedding K_n as the set of constant polynomials modulo f_n. This results in the desired tower of fields:

$$F = K_1 \subseteq K_2 \subseteq \ldots \subseteq K_n \subseteq \ldots$$

Then

$$A := \bigcup_{n\in\mathbb{N}^*} K_n$$

is an extension field of F which is algebraic over F, since it is the union of a tower of field extensions with finite degrees. We claim that the Steinitz number $N(A/F)$ of A equals S.

Given $m \in N(A/F)$, let E_m be the intermediate field of A/F with degree m over F. Then $E_m \subseteq K_n$ for some $n \in \mathbb{N}^*$, so that m divides $\gcd(S,n!) = d_n = [K_n : F]$. This shows that $N(A/F)$ is a subset of S.

Conversely, let $k \in S$. Then k divides $d_k = \gcd(S,k!)$, and therefore K_k contains a subfield with degree k over F, proving that S is a subset of $N(A/F)$. Thus $S = N(A/F)$, as claimed. □

Using a similar approach as in the proof of Theorem 4.2.9, we can also establish the desired uniqueness result:

Theorem 4.2.10. *Let A and A' be two algebraic extensions of a Galois field F, and assume that their Steinitz numbers $N(A/F)$ and $N(A'/F)$ agree. Then there exists an isomorphism from A to A' fixing F elementwise.*

Proof. Define the positive integers $d_n := \gcd(S,n!)$ as in the proof of Theorem 4.2.9, and let K_n and K_n' be the intermediate fields with degree d_n over F of A/F and A'/F, respectively. Then $F = K_1 = K_1'$ and

$$A = \bigcup_{n\in\mathbb{N}^*} K_n \quad \text{and} \quad A' = \bigcup_{n\in\mathbb{N}^*} K'_n.$$

We now construct a sequence of isomorphisms $(\alpha_n : K_n \to K'_n)$ fixing F elementwise. Trivially, the identity map is such an isomorphism for $n = 1$. Now let $n \geq 1$ and an isomorphism $\alpha_n : K_n \to K'_n$ fixing F elementwise be given. As in Lemma 3.1.8, α_n can be extended to an isomorphism $\alpha_{n+1} : K_{n+1} \to K'_{n+1}$ fixing F elementwise. By induction, we obtain the desired sequence of isomorphisms. Finally, put $\alpha := \bigcup_{n\in\mathbb{N}^*} \alpha_n$, considered as a subset of $A \times A'$. Then α is an isomorphism between A and A' fixing F elementwise. □

Combining Theorems 4.2.7, 4.2.9 and 4.2.10 provides the desired alternative (constructive) proof for the existence and uniqueness of an algebraic closure of an arbitrary Galois field:

Corollary 4.2.11. *Every Galois field F has a unique (up to isomorphism) algebraic closure $\widehat{F}$.* □

4.3 Projective Systems and Galois Groups

In this section, we describe the Galois groups of arbitrary algebraic extensions of a Galois field. For this, we first need to introduce the following abstract concept, which will also be essential for the remaining two sections of this chapter:

Definition 4.3.1. Let $(M, \preceq)$ be a partially ordered set, and let $(X_m)_{m\in M}$ be a system of sets. For all $n, m \in M$ with $n \preceq m$, let $\varphi_{m,n} : X_m \to X_n$ be a mapping. The pair (X, φ) is called a **projective system** over M provided that the following two conditions hold:[2]

- $\varphi_{m,m} = \mathrm{id}_{X_m}$ is the identity on X_m;
- $\varphi_{n,k} \circ \varphi_{m,n} = \varphi_{m,k}$ for all $m, n, k \in M$ with $k \preceq n \preceq m$.

Then the **projective limit** $\mathrm{prolim}_{M,\preceq}(X, \varphi)$ of (X, φ) is the set of all sequences $y \in \times_{m\in M} X_m$ such that[3]

$$\varphi_{m,n}(y_m) = y_n \quad \text{for all } n, m \in M \text{ with } n \preceq m. \quad \square \tag{4.2}$$

Remark 4.3.2. The concept of a projective limit is of particular interest if the sets X_m are equipped with some algebraic structure, for instance that of a group, a ring, or a module. In these cases, the mappings $\varphi_{m,n}$ are usually required to preserve the given structure: they should be homomorphisms. Then the projective limit will inherit a corresponding structure via componentwise operations. This will be our main tool

[2] If we wish to emphasize the underlying partially ordered set, we also use the notation (M, X, φ).

[3] Recall that the cartesian product $\times_{m\in M} X_m$ is the set of all sequences $(y_m)_{m\in M}$ such that $y_m \in X_m$ for all $m \in M$.

for studying the Galois groups as well as the multiplicative and additive groups of the algebraic closures of Galois fields. □

Let us illustrate Definition 4.3.1 with a first example, which shows how one may view formal power series in terms of projective limits.

Example 4.3.3. Consider the ordered set $(\mathbb{N}, \leq)$, and let R be a commutative ring and $R[x]$ the polynomial ring in the indeterminate x. For each $m \in \mathbb{N}$, we choose X_m as the factor ring $R[x]/(x^m)$ and define the mappings $\varphi_{m,n} : X_m \to X_n$ for $n, m \in \mathbb{N}$ with $n \leq m$ as follows:

$$\varphi_{m,n} \colon g(x) + x^m R[x] \mapsto g(x) + x^n R[x].$$

Then condition (4.2) is satisfied, and the projective limit $\operatorname{prolim}_{\mathbb{N},\leq}(X, \varphi)$ becomes a ring if we define $y + z$ and yz by

$$(y+z)_n := y_n + z_n \quad \text{and} \quad (yz)_n := y_n z_n \qquad \text{for all } n \in \mathbb{N}.$$

This ring turns out to be isomorphic to the ring of formal power series over R. Indeed, it is easily checked that the mapping $\Psi \colon R[[x]] \to \operatorname{prolim}_{\mathbb{N},\leq}(X, \varphi)$ given by

$$\sum_{j=0}^{\infty} f_j x^j \mapsto \Big(\sum_{j=0}^{m} f_j x^j \Big)_{m \in \mathbb{N}}$$

is an isomorphism. □

We will now investigate the Galois group of an algebraic extension A/F of a Galois field $F = \mathrm{GF}(q)$. As in Section 4.2, let $N(A/F)$ be the Steinitz number associated with A/F and denote the unique intermediate field with q^n elements by E_n, for all $n \in N(A/F)$. We abbreviate $N(A/F)$ as $\mathscr{N}$ and consider the partially ordered set $(\mathscr{N}, |)$ with respect to divisibility.

As in the case of a finite extension over F, the mapping

$$\sigma : A \to A, \ \nu \mapsto \nu^q$$

is called the **Frobenius automorphism** of A; obviously, σ fixes F elementwise and is therefore a Galois automorphism of A/F. By Theorem 3.4.8, the Galois group of E_n/F is isomorphic to the cyclic group $\mathbb{Z}_n := \mathbb{Z}/n\mathbb{Z}$ and generated by the restriction of σ to E_n (for each $n \in \mathscr{N}$). This suggests to define a projective system over $\mathscr{N}$ by taking $X_n := \mathbb{Z}_n$ and choosing the required mappings for $m, n \in \mathscr{N}$ with $n \mid m$ as natural epimorphisms (in view of Remark 4.3.2):

$$\mathtt{mod}\,{}^{m}_{n} \colon \mathbb{Z}_m \to \mathbb{Z}_n, \quad a + m\mathbb{Z} \mapsto a + n\mathbb{Z}. \tag{4.3}$$

It is easily checked that this indeed gives a projective system $(\mathscr{N}, (\mathbb{Z}_n)_{n \in \mathscr{N}}, \mathtt{mod})$; we will denote the projective limit of this system by $\mathbb{Z}_{[\mathscr{N}]}$. Note that $\mathbb{Z}_{[\mathscr{N}]}$ is a subgroup of the cartesian product $\times_{n \in \mathscr{N}} \mathbb{Z}_n$ with respect to componentwise addition.

The following lemma provides a useful (and more intuitive) description of this projective limit. For this, we let $\mathbb{Z}_{\mathscr{N}}$ be the set of all sequences $f\colon \mathscr{N} \to \mathbb{Z}$ satisfying $f_m \equiv f_n \bmod n$ whenever $n \mid m$. Clearly, $\mathbb{Z}_{\mathscr{N}}$ is an abelian group with respect to componentwise addition.

Lemma 4.3.4. *Let $\mathscr{N}$ be any Steinitz number. Then the mapping*

$$\eta\colon \mathbb{Z}_{\mathscr{N}} \to \mathbb{Z}_{[\mathscr{N}]}, \quad f \mapsto (f_n + n\mathbb{Z})_{n\in\mathscr{N}}$$

is a group epimorphism with kernel

$$I_{\mathscr{N}} := \{f \in \mathbb{Z}_{\mathscr{N}}\colon f_n \equiv 0 \bmod n \text{ for all } n \in \mathscr{N}\}.$$

Hence $\mathbb{Z}_{[\mathscr{N}]}$ is isomorphic to the factor group $\mathbb{Z}_{\mathscr{N}}/I_{\mathscr{N}}$.

Proof. Let f be a sequence in $\mathbb{Z}_{\mathscr{N}}$, so that $f_m \equiv f_n \bmod n$ for all $n,m \in \mathscr{N}$ with $n \mid m$. Thus $f_m + n\mathbb{Z} = f_n + n\mathbb{Z}$ for all such n,m, hence

$$\mathtt{mod}\,_n^m(f_m + m\mathbb{Z}) = f_m + n\mathbb{Z} = f_n + n\mathbb{Z},$$

and therefore $\eta(f) \in \mathbb{Z}_{[\mathscr{N}]}$. It is clear that η respects the group structure and has $I_{\mathscr{N}}$ as its kernel.

It remains to check that η is surjective. Let $y \in \mathbb{Z}_{[\mathscr{N}]}$ be arbitrary. For all $n \in \mathscr{N}$, choose an integer a_n with $y_n = a_n + n\mathbb{Z}$. Since $\mathtt{mod}\,_n^m(y_m) = y_n$ for all $n,m \in \mathscr{N}$ with $n \mid m$, we obtain

$$\mathtt{mod}\,_n^m(a_m + m\mathbb{Z}) = a_m + n\mathbb{Z} = a_n + n\mathbb{Z},$$

that is, $a_m \equiv a_n \bmod n$ for all such n,m. Thus the sequence $(a_n)_{n\in\mathscr{N}}$ is a pre-image of y under η. □

We can now prove the main result of the present section:

Theorem 4.3.5. *Consider an algebraic extension A/F of a Galois field F, and let $\mathscr{N} = N(A/F)$ be the Steinitz number of A/F. Then the Galois group of A/F is isomorphic to the projective limit $\mathbb{Z}_{[\mathscr{N}]}$.*

Proof. As above, we write $F = \mathrm{GF}(q)$ and let σ be the Frobenius automorphism of A. Then σ leaves every intermediate field of A/F invariant; as noted before, the restriction of σ to E_n, where $n \in \mathscr{N}$, is the Frobenius automorphism of E_n/F and generates the Galois group of that extension.

We now use the sequences in $\mathbb{Z}_{\mathscr{N}}$ to generate further Galois automorphisms of A/F from σ as follows. Given any sequence $\psi \in \mathbb{Z}_{\mathscr{N}}$ and any $u \in A$, put

$$u^{\psi} := \sigma^{\psi_n}(u), \tag{4.4}$$

where n is the degree of u over F, that is, $F(u) = E_n$. Note that $\sigma^{\psi_n}(u) = \sigma^{\psi_m}(u)$ for all $m \in \mathscr{N}$ with $n \mid m$, as then $\psi_m \equiv \psi_n \bmod n$ and as σ^n induces the identity on E_n. We use this to check that the mapping $u \mapsto u^\psi$ is a Galois automorphism of A/F. Given $u,v \in A$, let $\ell \in \mathscr{N}$ be the least common multiple of the degrees of u and v. Then, in E_ℓ,

$$(u+v)^\psi = \sigma^{\psi_\ell}(u+v) = \sigma^{\psi_\ell}(u) + \sigma^{\psi_\ell}(v) = u^\psi + v^\psi,$$

and similarly

$$(uv)^\psi = \sigma^{\psi_\ell}(uv) = \sigma^{\psi_\ell}(u)\sigma^{\psi_\ell}(v) = u^\psi v^\psi.$$

Moreover, $u^\psi = u$ for all $u \in F$, as then $u^\psi = \sigma^{\psi_1}(u) = \sigma^0(u) = u$. Finally, $v^\psi = 0$ implies $v = 0$, so that the mapping $u \mapsto u^\psi$ is injective; it is also surjective, as each element $u \in A$ (with degree n, say) has a pre-image under the Galois automorphism induced by σ^{ψ_n} on E_n.

We claim that the preceding construction yields *all* Galois automorphisms of A/F. Thus let $\omega \in \mathrm{Gal}(E/F)$ be arbitrary. Then ω fixes all finite intermediate fields of A/F, since there is a unique such field E_n (with cardinality q^n) for all $n \in \mathscr{N}$. Consequently, the restriction ω_n of ω to E_n is a Galois automorphism of E_n/F and therefore equal to the restriction of $\sigma^{\psi(n)}$ to E_n for some $\psi(n) \in \{0,1,...,n-1\}$. This defines a mapping $\psi\colon \mathscr{N} \to \mathbb{Z}$ with $\psi(n) \in \{0,\dots,n-1\}$ for all $n \in \mathscr{N}$, and we want to show $\psi \in \mathbb{Z}_\mathscr{N}$. Thus let $n,m \in \mathscr{N}$ with $n \mid m$. Then $E_n \subseteq E_m$ and therefore ω_n is obtained by restricting ω_m to E_n, so that $\sigma^{\psi(m)} = \sigma^{\psi(n)}$ on E_n, which means $\psi(m) \equiv \psi(n) \bmod n$. Thus $\psi \in \mathbb{Z}_\mathscr{N}$ and ω arises from ψ, as desired.

In view of Lemma 4.3.4, it only remains to check that the mapping Ω which sends a sequence $\psi \in \mathbb{Z}_\mathscr{N}$ to the Galois automorphism of A/F defined in (4.4) is a group homomorphism with kernel $I_\mathscr{N}$. Let ψ and η in $\mathbb{Z}_\mathscr{N}$ and consider an element $u \in E$, say with degree n over F. Then

$$\begin{aligned}\Omega(\psi+\eta)(u) &= u^{\psi+\eta} = \sigma^{(\psi+\eta)_n}(u) = \sigma^{\psi_n+\eta_n}(u)\\ &= \sigma^{\psi_n}(\sigma^{\eta_n}(u)) = (u^\eta)^\psi = \big(\Omega(\psi)\circ\Omega(\eta)\big)(u),\end{aligned}$$

proving that Ω is indeed a homomorphism. Finally, we note that $\psi \in \mathbb{Z}_\mathscr{N}$ belongs to the kernel of Ω if and only if $\sigma^{\psi_{\deg u}}(u) = u$ for all $u \in A$, which is equivalent to $\psi_{\deg u} \equiv 0 \bmod \deg u$ for all u. By definition, the degrees of the elements of A form the Steinitz number $N(A/F) = \mathscr{N}$, and therefore the kernel of Ω is $I_\mathscr{N}$. □

As an immediate consequence of Theorems 4.3.5 and 4.2.7, we obtain the following result determining the Galois groups of the algebraic closures of Galois fields:

Corollary 4.3.6. *Let F be any Galois field, and let $\widehat{F}$ be the algebraic closure of F. Then the Galois group of $\widehat{F}/F$ is isomorphic to the projective limit $\mathbb{Z}_{[\mathbb{N}^*]}$.* □

Exercises

Exercise 4.3.7. Consider the algebraic closure $\widehat{F}$ of a finite field $F = \mathrm{GF}(q)$. Exhibit some Galois automorphism of $\widehat{F}/F$ which does not belong to the group generated by the Frobenius automorphism $\sigma \in \mathrm{Gal}(\widehat{F}/F)$. □

4.4 The Multiplicative Group of the Algebraic Closure

In this section, we study the multiplicative group of the algebraic closure $\widehat{F}$ of a Galois field $F = \mathrm{GF}(q)$ with characteristic p. This relies on and extends results from Section 3.6 on cyclotomic polynomials and cyclotomic field extensions over F.

Throughout this section, we use the notation

$$\mathbb{N}_p^* := \{k \in \mathbb{N}^* \colon k \text{ is not divisible by } p\};$$

note that $\mathbb{N}_p^*$ is a Steinitz number. We begin by collecting some results on the finite subgroups of F^* which either were established before or are easy consequences of previous results.

Theorem 4.4.1. *Let $F = \mathrm{GF}(q)$ be the Galois field with q elements, p its characteristic, and $\widehat{F}$ the algebraic closure of F. Then:*

(1) *Every finite subgroup of $\widehat{F}^*$ is cyclic and has order n for some $n \in \mathbb{N}_p^*$.*

(2) *For every $n \in \mathbb{N}_p^*$, there is a unique subgroup U_n of $\widehat{F}^*$ with order n.*

(3) *One has $U_n \subseteq U_m$ if and only if n divides m.*

(4) $U_n \cap U_m = U_{\gcd(n,m)}$ *for all $n, m \in \mathbb{N}_p^*$.*

(5) $U_n U_m = U_{\mathrm{lcm}(n,m)}$ *for all $n, m \in \mathbb{N}_p^*$.*

Proof. By Theorem 3.2.3 and Corollary 3.2.4, all finite subgroups of the multiplicative group $\widehat{F}^*$ are cyclic, and $\widehat{F}^*$ contains at most one subgroup of any given finite order. Now let U be any finite subgroup of $\widehat{F}^*$. By Corollary 4.2.8, $\widehat{F}$ is the union of its subfields $E_m \cong \mathrm{GF}(q^m)$, and hence U is a subgroup of E_m^* for some m. As q is a power of p, the order of U cannot be a multiple of p. Conversely, let $n \in \mathbb{N}_p^*$ and let $k := \mathrm{ord}_n(q)$ be the multiplicative order of q modulo n. Then the multiplicative group E_k^* of the subfield E_k of $\widehat{F}$ has order $q^k - 1$ (and is cyclic). Since n divides $q^k - 1$, there is a unique subgroup U of E_k^* with order n. This establishes parts (1) and (2). Now parts (3) to (5) follow easily from general results on cyclic groups. □

Remark 4.4.2. By part (2) of Theorem 4.4.1, $\widehat{F}$ contains $\phi(n)$ primitive n-th roots of unity, for all $n \in \mathbb{N}_p^*$. In other words, the splitting field $F^{(n)}$ of the n-th cyclotomic polynomial $\Phi_n(x) \in F[x]$ is the subfield E_k of $\widehat{F}$, where $k = \mathrm{ord}_n(q)$; see Proposition 3.6.3. □

In particular, $\widehat{F}$ contains primitive $(q^n - 1)$-th roots of unity for all n, that is, primitive elements for all its subfields $E_n \cong \mathrm{GF}(q^n)$. It is natural to ask whether there exist sequences of such elements which are, in some sense, compatible with each other. To make this question more precise, we introduce the following terminology:

Definition 4.4.3. Let $\widehat{F}$ be the algebraic closure of the Galois field $F = \mathrm{GF}(q)$, and consider a sequence $(y_n)_{n \in \mathbb{N}^*}$, where $y_n \in E_n$ for all $n \in \mathbb{N}^*$. Then $(y_n)_{n \in \mathbb{N}^*}$ is called **norm-compatible** provided that $\mathrm{Norm}_n^m(y_m) = y_n$ for all $m, n \in \mathbb{N}^*$ with $n \mid m$, where Norm_n^m denotes the norm function from E_m to E_n.

Assume in addition that y_n is a primitive element of E_n for all n. Then $(y_n)_{n\in\mathbb{N}^*}$ is called a **primitive element** for $\widehat{F}$. □

Theorem 4.4.4. *Let $\widehat{F}$ be the algebraic closure of a Galois field F. Then there exists a primitive element for $\widehat{F}$.*

Proof. Let $n,m \in \mathbb{N}^*$ with $n \mid m$, and let ρ be a primitive element of E_n. Then there exists a primitive element ζ of E_m such that $\mathrm{Norm}_n^m(\zeta) = \rho$, by Theorem 3.12.10. Using induction, we obtain a sequence $(w_\ell)_{\ell\in\mathbb{N}^*}$ such that w_ℓ is a primitive element of $E_{\ell!}$ and

$$\mathrm{Norm}_{\ell!}^{(\ell+1)!}(w_{\ell+1}) = w_\ell \quad \text{for all } \ell.$$

Now let $k \in \mathbb{N}^*$ be arbitrary, let $\ell(k)$ be the least positive integer in $\mathbb{N}^*$ for which k divides $\ell(k)!$, and put

$$v_k := \mathrm{Norm}_k^{\ell(k)!}(w_{\ell(k)}).$$

By Proposition 3.12.9, v_k is a primitive element of E_k. Given $n,m \in \mathbb{N}^*$ with $n \mid m$, the transitivity of the norm (see Theorem 3.12.8) yields

$$\begin{aligned}\mathrm{Norm}_n^m(v_m) &= \mathrm{Norm}_n^m\big(\mathrm{Norm}_m^{\ell(m)!}(w_{\ell(m)})\big) = \mathrm{Norm}_n^{\ell(m)!}(w_{\ell(m)})\\ &= \mathrm{Norm}_n^{\ell(n)!}\big(\mathrm{Norm}_{\ell(n)!}^{\ell(m)!}(w_{\ell(m)})\big) = \mathrm{Norm}_n^{\ell(n)!}(w_{\ell(n)}) = v_n.\end{aligned}$$

Thus $(v_n)_{n\in\mathbb{N}^*}$ is the desired norm-compatible sequence of primitive elements. □

For a deeper study of the structure of $\widehat{F}^*$, we again need a suitable projective limit. We will use the following projective system over $\mathbb{N}_p^*$: we choose X_k as the unique subgroup U_k of $\widehat{F}^*$ with order k, and as mappings $\varphi_{m,n}$ (where $n,m \in \mathbb{N}_p^*$ with $n \mid m$) we take the group epimorphisms

$$\pi_n^m : U_m \to U_n, \quad u \mapsto u^{m/n} \tag{4.5}$$

which were already used in Section 3.6.[4] Thus we consider the projective system[5]

$$\big(\mathbb{N}_p^*, (U_n)_{n\in\mathbb{N}_p^*}, \{\pi_n^m : n,m \in \mathbb{N}_p^*, n \mid m\}\big);$$

we will denote the projective limit of this system by $\mathrm{prolim}(\widehat{F}^*)$. Using this projective limit, we can now define generators for $\widehat{F}^*$ in a natural manner:

Definition 4.4.5. An element $u \in \mathrm{prolim}(\widehat{F}^*)$ is called a **generator** of $\widehat{F}^*$ provided that u_n generates U_n for all $n \in \mathbb{N}_p^*$. □

By Theorem 4.4.4, we already know the existence of primitive elements for $\widehat{F}$. Therefore, the existence of generators of $\widehat{F}^*$ is guaranteed by the following result connecting these two notions:

---

[4] As noted in the proof of Proposition 3.12.9, $\pi_{q^k-1}^{q^\ell-1}$ is just the norm mapping $\mathrm{Norm}_{E_l/E_k} = \mathrm{Norm}_k^\ell$.

[5] See Exercise 4.4.10.

Proposition 4.4.6. *The existence of a primitive element of $\widehat{F}$ implies the existence of a generator of $\widehat{F}^*$.*

Proof. Let $(y_n)_{n\in\mathbb{N}^*}$ be a primitive element of $\widehat{F}$. Given any $k\in\mathbb{N}_p^*$, put $\ell := \operatorname{ord}_k(q)$ and define

$$u_k := \pi_k^{q^\ell-1}(y_\ell) = y_\ell^{(q^\ell-1)/k}.$$

Then $\operatorname{ord}(u_k)=k$, that is, u_k generates U_k. Now let $a,b\in\mathbb{N}_p^*$ with $a\mid b$, and let α and β denote $\operatorname{ord}_a(q)$ and $\operatorname{ord}_b(q)$, respectively. Then, as $(y_n)_{n\in\mathbb{N}^*}$ is norm-compatible,

$$\begin{aligned}\pi_a^b(u_b) &= u_b^{b/a} = \left(y_b^{(q^\beta-1)/b}\right)^{b/a}\\ &= y_b^{(q^\beta-1)/a} = \left(y_b^{(q^\beta-1)/(q^\alpha-1)}\right)^{(q^\alpha-1)/a}\\ &= \left(\operatorname{Norm}_\alpha^\beta(y_b)\right)^{(q^\alpha-1)/a} = y_a^{(q^\alpha-1)/a} = u_a.\end{aligned}$$

Thus $u\in\operatorname{prolim}(\widehat{F}^*)$ is a generator of $\widehat{F}^*$. □

Our main goal in this section is to show that $\operatorname{prolim}(\widehat{F}^*)$ is a cyclic module over $\mathbb{Z}_{\mathbb{N}_p^*}$ and in fact isomorphic to $\mathbb{Z}_{[\mathbb{N}_p^*]}$, where we use the projective systems $(\mathcal{N},(\mathbb{Z}_n)_{n\in\mathcal{N}},\mathrm{mod})$ with projective limit $\mathbb{Z}_{[\mathcal{N}]}$ introduced in Section 4.3 in the special case of the Steinitz number $\mathcal{N}=\mathbb{N}_p^*$. This still requires some effort.

We first need to define a scalar multiplication which turns $\operatorname{prolim}(\widehat{F}^*)$ into an $\mathbb{Z}_{\mathbb{N}_p^*}$-module:

Lemma 4.4.7. *Given $u\in\operatorname{prolim}(\widehat{F}^*)$ and $f\in\mathbb{Z}_{\mathbb{N}_p^*}$, define $u^f\in\operatorname{prolim}(\widehat{F}^*)$ by*

$$(u^f)_n := u_n^{f_n},\ \text{where } n\in\mathbb{N}_p^*. \tag{4.6}$$

Then $\operatorname{prolim}(\widehat{F}^*)$ is an $\mathbb{Z}_{\mathbb{N}_p^*}$-module with respect to this scalar multiplication.

Proof. By definition, $f=(f_n)_{n\in\mathbb{N}_p^*}$ is a sequence of integers satisfying $f_m\equiv f_n$ for all $n,m\in\mathbb{N}_p^*$ with $n\mid m$. The reader may check that $\mathbb{Z}_{\mathbb{N}_p^*}$ is a commutative ring with respect to componentwise addition and multiplication:

$$(f+g)_n := f_n+g_n \ \text{ and }\ (fg)_n := f_ng_n \quad \text{for all } f,g\in\mathbb{Z}_{\mathbb{N}_p^*} \text{ and all } n\in\mathbb{N}_p^*.$$

By definition of the projective limit, $u=(u_n)_{n\in\mathbb{N}_p^*}$ satisfies $u_n\in U_n$ for all n and $\pi_n^m(u_m)=u_m^{m/n}=u_n$ for all $n,m\in\mathbb{N}_p^*$ with $n\mid m$. It is easily seen that $\operatorname{prolim}(\widehat{F}^*)$ is a commutative group with respect to componentwise multiplication:

$$(uv)_n := u_nv_n \quad \text{for all } u,v\in\operatorname{prolim}(\widehat{F}^*) \text{ and all } n\in\mathbb{N}_p^*.$$

Trivially, $(u^f)_n=u_n^{f_n}\in U_n$ for all $n\in\mathbb{N}_p^*$. Also, whenever $n\mid m$,

$$\pi_n^m((u^f)_m) = (u_m^{f_m})^{m/n} = (u_m^{m/n})^{f_m} = u_n^{f_m} = u_n^{f_n} = (u^f)_n,$$

as then $f_m \equiv f_n \bmod n$. Thus $u^f \in \operatorname{prolim}(\widehat{F}^*)$ for all $u \in \operatorname{prolim}(\widehat{F}^*)$ and all $f \in \mathbb{Z}_{\mathbb{N}_p^*}$. It is routine to check

$$u^{f+g} = u^f u^g, \ (uv)^f = u^f v^f \ \text{ and } \ (u^f)^g = u^{fg}$$

for all $f,g \in \mathbb{Z}_{\mathbb{N}_p^*}$ and all $u,v \in \operatorname{prolim}(\widehat{F}^*)$, where all operations in the limits are performed componentwise. Finally, if $e_n = 1$ for all $n \in \mathbb{N}_p^*$, then $u^e = u$ for all $u \in \operatorname{prolim}(\widehat{F}^*)$. Altogether, this proves the assertion. □

Next, we characterize the generators of $\widehat{F}^*$ in terms of module homomorphisms from $\mathbb{Z}_{\mathbb{N}_p^*}$ to the projective limit $\operatorname{prolim}(\widehat{F}^*)$:

Lemma 4.4.8. *Let u be any sequence in* $\operatorname{prolim}(\widehat{F}^*)$. *Then the mapping*

$$\Gamma_u\colon \mathbb{Z}_{\mathbb{N}_p^*} \to \operatorname{prolim}(\widehat{F}^*), \quad f \mapsto u^f$$

is a homomorphism of $\mathbb{Z}_{\mathbb{N}_p^*}$-modules. Moreover, Γ_u is surjective if and only if u is a generator of $\widehat{F}^*$.

Proof. Obviously, Γ_u is a module homomorphism. Now assume that u is a generator of $\widehat{F}^*$, and let $v \in \operatorname{prolim}(\widehat{F}^*)$. Since u_n generates U_n, there is an integer f_n such that $u_n^{f_n} = v_n$ (for all n). Now let $n,m \in \mathbb{N}_p^*$ with $n \mid m$. Then $u_m^{f_m} = v_m$ and $u_n^{f_n} = v_n$. In view of $u_m^{m/n} = u_n$, we obtain

$$(u_m^{m/n})^{f_n} = v_n = (u_m^{f_m})^{m/n},$$

which means

$$\frac{m}{n} f_n \equiv \frac{m}{n} f_m \mod \operatorname{ord}(u_m).$$

As u_m generates U_m, the order of u_m equals m. Consequently, $m = n \cdot \frac{m}{n}$ divides $\frac{m}{n}(f_n - f_m)$ and therefore $f_m \equiv f_n \bmod n$. This shows $f \in \mathbb{Z}_{\mathbb{N}_p^*}$, and thus Γ_u is indeed surjective.

Conversely, assume that Γ_u is surjective. Choose any generator w of $\operatorname{prolim}(\widehat{F}^*)$. As Γ_u is surjective, there is some $f \in \mathbb{Z}_{\mathbb{N}_p^*}$ such that $w = u^f$, that is, $u_n^{f_n} = w_n$ for all n. By definition of a generator, w_n generates U_n for all $n \in \mathbb{N}_p^*$. Now $w_n = u_n^{f_n}$ shows that u_n also generates U_n for all $n \in \mathbb{N}_p^*$. Thus u is indeed a generator of $\widehat{F}^*$. □

After all these preparations, the proof of our main result is now easy:

Theorem 4.4.9. *The projective limit* $\operatorname{prolim}(\widehat{F}^*)$ *is isomorphic to* $\mathbb{Z}_{[\mathbb{N}_p^*]}$.

Proof. Let u be any generator of $\widehat{F}^*$. By Lemma 4.4.8, u yields an epimorphism $\Gamma_u\colon \mathbb{Z}_{\mathbb{N}_p^*} \to \operatorname{prolim}(\widehat{F}^*)$. Note that $u_n^{f_n} = 1$ for all n if and only if $f_n \equiv 0 \bmod n$ for all n. Hence the kernel of Γ_u is the set $I_{\mathscr{N}}$ introduced in Lemma 4.3.4 (where $\mathscr{N} = \mathbb{N}_p^*$ in the present situation), so that $\operatorname{prolim}(\widehat{F})$ is isomorphic to $\mathbb{Z}_{N_p^*}/I_{N_p^*} \cong \mathbb{Z}_{[\mathbb{N}_p^*]}$. □

Exercises

Exercise 4.4.10. Check that $\left(\mathbb{N}_p^*, (U_n)_{n\in\mathbb{N}_p^*}, \{\pi_n^m : n,m \in \mathbb{N}_p^*, n \mid m\}\right)$ is indeed a projective system. □

4.5 The Additive Group of the Algebraic Closure

In this section, we investigate the additive structure of the algebraic closure $\widehat{F}$ of a Galois field $F = \mathrm{GF}(q)$, by considering the additive group of $\widehat{F}$ as an $F[x]$-module with respect to the Frobenius automorphism σ of $\widehat{F}/F$. This builds on the results on q-orders and Psi polynomials proved in Section 3.11, and we will use the notations and conventions introduced there.

Given any polynomial $g(z) \in \mathbb{M}_z^*$ (that is, any monic polynomial over F which is not a multiple of z), we have the objects Ω_g, $\Psi_g(x)$ and V_g introduced in Definition 3.11.7 and Observation 3.11.8. As noted in parts (4) and (5) of that Observation, we may replace the field $K = \mathrm{GF}(q^k)$ associated with a specific $g(z)$, where $k = \mathrm{ord}_g(z)$, by an arbitrary extension field of L of K. As we now choose $L = \widehat{F}$, we may view all Ω_g as subsets of $\widehat{F}$ and all V_g as submodules

$$(E,\sigma)_g = \{v \in \widehat{F} : g(\sigma)(v) = 0\}$$

of the $F[x]$-module $(\widehat{F},\sigma)$ simultaneously (for all $g \in \mathbb{M}_z^*$). We also recall that the Psi polynomials are in fact always polynomials over the ground field F.

We now proceed in analogy to the multiplicative case studied in Section 4.4. Here the module epimorphisms $\Pi_f^g : V_g \to V_f$ introduced in Observation 3.11.10 will play a fundamental role.

By Theorem 3.10.2, $\widehat{F}$ contains normal elements for all its subfields $E_n \cong \mathrm{GF}(q^n)$. Again, it is natural to ask whether there exist sequences of such elements which are compatible with each other. To answer this question, we proceed in analogy to Definition 4.4.3 and Theorem 4.4.4. As shown in the proof of Theorem 3.12.10, $\Pi_{z^n-1}^{z^m-1}$ is the trace mapping from E_m onto E_n whenever $n,m \in \mathbb{N}^*$ with $n \mid m$; for simplicity, we will use the notation Tr_n^m instead.

Definition 4.5.1. Let $\widehat{F}$ be the algebraic closure of the Galois field $F = \mathrm{GF}(q)$, and consider a sequence $(y_n)_{n\in\mathbb{N}^*}$, where $y_n \in E_n$ for all $n \in \mathbb{N}^*$. Then $(y_n)_{n\in\mathbb{N}^*}$ is called **trace-compatible** provided that $\mathrm{Tr}_n^m(y_m) = y_n$ for all $m,n \in \mathbb{N}^*$ with $n \mid m$.

Assume in addition that y_n is a normal element for E_n/F for all n. Then $(y_n)_{n\in\mathbb{N}^*}$ is called a **normal element** for $\widehat{F}/F$. □

Theorem 4.5.2. *Let $\widehat{F}$ be the algebraic closure of a Galois field F. Then there exists a normal element for $\widehat{F}/F$.*

Proof. Let $n,m \in \mathbb{N}^*$ with $n \mid m$, and let u be a normal element for E_n/F. Then there exists a normal element v for E_m/F such that $\mathrm{Tr}_n^m(v) = u$, by Theorem 3.12.10.

Using induction, we obtain a sequence $(w_\ell)_{\ell\in\mathbb{N}^*}$ such that w_ℓ is a normal element for $E_{\ell!}/F$ and

$$\mathrm{Tr}_{\ell!}^{(\ell+1)!}(w_{\ell+1}) = w_\ell \quad \text{for all } \ell.$$

Now let $k \in \mathbb{N}^*$ be arbitrary, let $\ell(k)$ be the least positive integer in $\mathbb{N}^*$ for which k divides $\ell(k)!$, and put

$$v_k := \mathrm{Tr}_k^{\ell(k)!}(w_{\ell(k)}).$$

By Proposition 3.12.9, v_k is a normal element for E_k/F. Given $n,m \in \mathbb{N}^*$ with $n \mid m$, the transitivity of the trace (see Theorem 3.12.8) yields

$$\begin{aligned}\mathrm{Tr}_n^m(v_m) &= \mathrm{Tr}_n^m\big(\mathrm{Tr}_m^{\ell(m)!}(w_{\ell(m)})\big) = \mathrm{Tr}_n^{\ell(m)!}(w_{\ell(m)}) \\ &= \mathrm{Tr}_n^{\ell(n)!}\big(\mathrm{Tr}_{\ell(n)!}^{\ell(m)!}(w_{\ell(m)})\big) = \mathrm{Tr}_n^{\ell(n)!}(w_{\ell(n)}) = v_n.\end{aligned}$$

Thus $(v_n)_{n\in\mathbb{N}^*}$ is the desired trace-compatible sequence of normal elements. □

For a deeper study of the additive structure of $\widehat{F}^*$, we once more need a suitable projective limit:

Observation 4.5.3. The transitivity formula (3.26) for the collection of mappings of the form Π_f^g shows that

$$\big(\mathbb{M}_z^*, (V_g)_{g\in\mathbb{M}_z^*}, \{\Pi_f^g : f(z), g(z) \in \mathbb{M}_z^*, f \mid g\}\big)$$

is a projective system. According to Definition 4.3.1, the elements of the projective limit $\mathrm{prolim}(\widehat{F},\sigma)$ of this system are those sequences $(v_g)_{g\in\mathbb{M}_z^*}$ with $v_g \in V_g$ which satisfy the condition

$$\Pi_f^g(v_g) = v_f \quad \text{for all } f(z), g(z) \in \mathbb{M}_z^* \text{ with } f \mid g.$$

Obviously, $\mathrm{prolim}(\widehat{F},\sigma)$ becomes an $F[x]$-module by applying the (scalar) multiplication componentwise. □

Using this projective limit, we can now define the following additive analogue of the generators for $\widehat{F}^*$ introduced in Definition 4.4.5:

Definition 4.5.4. An element $u \in \mathrm{prolim}(\widehat{F},\sigma)$ is called a **generator** of $\widehat{F}$ provided that u_f generates V_f as (F,σ)-module for all $f(z) \in \mathbb{M}_z^*$, that is, provided that $\mathrm{Ord}_q(u_f) = f(z)$ for all f. □

As is to be expected, one then has the following additive analogue of Proposition 4.4.6, which guarantees the existence of generators for $\widehat{F}$, in view of Theorem 4.5.2.

Proposition 4.5.5. *The existence of a normal element for $\widehat{F}$ implies the existence of a generator for $\widehat{F}$.*

Proof. Let y be a normal element of $\widehat{F}$. Given any $h(z) \in \mathbb{M}_z^*$, put $\ell := \mathrm{ord}_h(z)$ and define

$$u_h := y_\ell^{(z^\ell-1)/h} = \frac{z^\ell-1}{h(z)}(\sigma)(y_\ell) = \Pi_h^{z^\ell-1}(y_\ell).$$

Then $\mathrm{Ord}_q(u_h) = h(z)$, that is, u_h generates the (F,σ)-submodule V_h. Now let $f(z),g(z) \in \mathbb{M}_z^*$ with $f \mid g$, and let α and β denote $\mathrm{ord}_f(z)$ and $\mathrm{ord}_g(z)$, respectively. Then, as $(y_n)_{n\in\mathbb{N}^*}$ is trace-compatible,

$$\begin{aligned}
\Pi_f^g(u_g) = u_g^{g/f} &= \left(y_\beta^{(z^\beta-1)/g}\right)^{g/f} \\
&= y_\beta^{(z^\beta-1)/f} = \left(y_\beta^{(z^\beta-1)/(z^\alpha-1)}\right)^{(z^\alpha-1)/f} \\
&= \left(\mathrm{Tr}_\alpha^\beta(y_\beta)\right)^{(z^\alpha-1)/f} = y_\alpha^{(z^\alpha-1)/f} = u_f.
\end{aligned}$$

Thus $u \in \mathrm{prolim}(\widehat{F},\sigma)$ is a generator for $\widehat{F}$. □

In the remainder of this section, we will provide an external description of $\mathrm{prolim}(\widehat{F},\sigma)$, in analogy to Theorem 4.4.9. For this, we require the following polynomial analogue of the Steinitz numbers:

Definition 4.5.6. Let $\mathscr{P}$ be a non-empty subset of $\mathbb{M}_z^*$, and assume that the following two conditions hold:

- $g(z) \in \mathscr{P}$ implies $f(z) \in \mathscr{P}$ for each divisor f of g in $\mathbb{M}_z^*$;
- $g(z),h(z) \in \mathscr{P}$ implies $\mathrm{lcm}(g,h) \in \mathscr{P}$.

Then $\mathscr{P}$ is called a **closed subset** of $\mathbb{M}_z^*$. □

For a closed subset $\mathscr{P}$ of $\mathbb{M}_z^*$, let $F[x]_{\mathscr{P}}$ be the set of all sequences $s = (s_f)_{f\in\mathscr{P}}$ over $F[x]$ satisfying

$$s_g \equiv s_f \mod f(x) \quad \text{for all } f(z),g(z) \in \mathscr{P} \text{ with } f \mid g.$$

Obviously, $F[x]_{\mathscr{P}}$ is a commutative ring with respect to componentwise operations, which becomes an $F[x]$-module if we define a scalar multiplication by

$$a(x)\cdot s := \left(a(x)s_f\right)_{f\in\mathscr{P}}.$$

Moreover, the set $I_{\mathscr{P}}$ of all sequences $s \in F[x]_{\mathscr{P}}$ satisfying $s_f \equiv 0 \mod f(x)$ for all $f \in \mathscr{P}$ is both an ideal and an $F[x]$-submodule of $F[x]_{\mathscr{P}}$.

We now use quotient rings of the form $R_f := \mathbb{F}[x]/\left(f(x)\right)$ to define a projective system over $\mathscr{P}$ as follows:

$$\left(\mathscr{P}, \{R_f : f \in \mathscr{P}\}, \{\mathtt{Mod}_f^g : f(z),g(z) \in \mathscr{P}, f \mid g\}\right),$$

where

$$\mathtt{Mod}_f^g : R_g \to R_f, \quad a(x) + g(x)F[x] \mapsto a(x) + f(x)F[x],$$

and denote the projective limit of this system by $F[x]_{[\mathscr{P}]}$. Then one has the following result, the proof of which will be left to the reader as Exercise 4.5.12:

Lemma 4.5.7. *Let $\mathscr{P}$ be a closed subset of $\mathbb{M}_z^*$. Then the mapping*

$$\omega\colon \mathbb{F}[x]_{\mathscr{P}} \to \mathbb{F}[x]_{[\mathscr{P}]}, \quad s \mapsto \left(s_f + f(x)F[x]\right)_{f\in\mathscr{P}}$$

is a ring epimorphism with kernel $I_{\mathscr{P}}$. Hence $\mathbb{F}[x]_{[\mathscr{P}]}$ is isomorphic to the factor ring $\mathbb{F}[x]_{\mathscr{P}}/I_{\mathscr{P}}$. □

From now on, we restrict ourselves to the special case $\mathscr{P} = \mathbb{M}_z^*$. In analogy to Lemma 4.4.7, we define a scalar multiplication which turns $\operatorname{prolim}(\widehat{F},\sigma)$ into an $F[x]_{\mathbb{M}_z^*}$-module:

Lemma 4.5.8. *Given $v \in \operatorname{prolim}(\widehat{F},\sigma)$ and $s \in F[x]_{\mathbb{M}_z^*}$, define $v^s \in \operatorname{prolim}(\widehat{F},\sigma)$ by*

$$\left(v^s\right)_f := v_f^{s_f}, \text{ where } f(z) \in \mathbb{M}_z^*.$$

Then $\operatorname{prolim}(\widehat{F},\sigma)$ is an $F[x]_{\mathbb{M}_z^*}$-module with respect to this scalar multiplication.

Proof. By definition of the projective limit, $v_f \in V_f$ for all f, so that also $v_f^{s_f} = s_f(\sigma)(v_f) \in V_f$ for all f. Moreover,

$$\Pi_f^g\left((v^s)_g\right) = \left(v_g^{s_g}\right)^{g/f} = \left(v_g^{g/f}\right)^{s_g} = v_f^{s_g} = v_f^{s_f} = (v^s)_f,$$

since $s_g \equiv s_f \bmod f(x)$ and since V_f is annihilated by $f(x)$. Thus $v^s \in \operatorname{prolim}(\widehat{F},\sigma)$.

It is routine to check

$$v^{s+t} = v^s + v^t, \quad (u+v)^s = u^s + v^s \text{ and } (v^s)^t = v^{st}$$

for all $s,t \in \mathbb{F}[x]_{\mathbb{M}_z^*}$ and all $u,v \in \operatorname{prolim}(\widehat{F},\sigma)$. Finally, if $e_f = 1$ for all $f(z) \in \mathbb{M}_z^*$, then $v^e = v$ for all $v \in \operatorname{prolim}(\widehat{F},\sigma)$. Altogether, this proves the assertion. □

Now we can give the following characterization of the generators for $\widehat{F}$ in terms of module homomorphisms from $F[x]_{\mathbb{M}_z^*}$ to the projective limit $\operatorname{prolim}(\widehat{F},\sigma)$, in analogy to Lemma 4.4.8:

Lemma 4.5.9. *Let u be any sequence in $\operatorname{prolim}(\widehat{F},\sigma)$. Then the mapping*

$$\Gamma_u\colon F[x]_{\mathbb{M}_z^*} \to \operatorname{prolim}(\widehat{F},\sigma), \quad s \mapsto u^s$$

is a homomorphism of $F[x]_{\mathbb{M}_z^*}$-modules. Moreover, Γ_u is surjective if and only if u is a generator for $\widehat{F}$.

Proof. Obviously, Γ_u is a module homomorphism. Assume first that u is a generator for $\widehat{F}$, and let $v \in \operatorname{prolim}(\widehat{F})$. Since u_f generates V_f for each f, there are monic polynomials $s_f(x) \in F[x]$ such that $u_f^{s_f} = s_f(\sigma)(u_f) = v_f$ for all $f(z) \in \mathbb{M}_z^*$. Now let $f(z), g(z) \in \mathbb{M}_z^*$ with $f \mid g$. Then

$$u_g^{s_g} = v_g, \quad u_f^{s_f} = v_f \quad \text{and} \quad \Pi_f^g(u_g) = u_f,$$

and hence

$$\left(u_g^{g/f}\right)^{s_f} = v_f = \left(u_g^{s_g}\right)^{g/f},$$

that is,

$$\frac{g(x)}{f(x)} \cdot s_f(x) \equiv \frac{g(x)}{f(x)} \cdot s_g(x) \bmod g(x),$$

so that $g = f \cdot \frac{g}{f}$ divides $\frac{g}{f}(s_f - s_g)$. Therefore $s_g \equiv s_f \bmod f$, and $s \in F[x]_{\mathbb{M}_z^*}$. Thus Γ_u is indeed surjective.

Conversely, assume that Γ_u is surjective. Choose any generator w for $\operatorname{prolim}(\widehat{F})$. As Γ_u is surjective, there exists some $f \in F[x]_{\mathbb{M}_z^*}$ such that $w = u^f$, that is, $u_f^{s_f} = w_f$ for all f. By definition of a generator, $\operatorname{Ord}_q(w_f) = f(z)$ for all f, and therefore also $\operatorname{Ord}_q(u_f) = f(z)$ for all f. Thus u is indeed a generator of $\widehat{F}^*$. □

After all these preparations, it is a simple matter to prove the desired analogue of Theorem 4.4.9:

Theorem 4.5.10. *The projective limit* $\operatorname{prolim}(\widehat{F}, \sigma)$ *is isomorphic to* $F[x]_{[\mathbb{M}_z^*]}$.

Proof. Let u be any generator for $\widehat{F}$. Then u yields an epimorphism

$$\Gamma_u \colon F[x]_{\mathbb{M}_z^*} \to \operatorname{prolim}(\widehat{F}, \sigma),$$

by Lemma 4.5.9. Clearly, $u_f^{s_f} = 0$ holds for all f if and only if $s_f(x) \equiv 0 \bmod f(x)$ for all f. Thus the kernel of Γ_u is the set $I_{\mathscr{P}}$, and Lemma 4.5.7 gives the assertion. □

Exercises

Exercise 4.5.11. Let $\widehat{F}$ be the algebraic closure of $\mathrm{GF}(q)$. Show that every finite submodule of $(\widehat{F}, \sigma)$ is a module of the type V_g for some $g(z) \in \mathbb{M}_z^*$. Also, prove the following two identities for these submodules:

- $V_g \cap V_h = V_{\gcd(g,h)}$ for all $g, h \in \mathbb{M}_z^*$;
- $V_g + V_h = V_{\operatorname{lcm}(g,h)}$ for all $g, h \in \mathbb{M}_z^*$. □

Exercise 4.5.12. Use similar arguments as in the proof of Lemma 4.3.4 to prove Lemma 4.5.7. □

Chapter 5
Irreducible Polynomials over Finite Fields

Abstract In contrast to the theoretical construction of finite fields as splitting fields presented in Chapter 3, we now consider the explicit description of the extension field $\mathrm{GF}(q^n)$ of $F = \mathrm{GF}(q)$ as the factor ring $F[x]/(f)$ of the polynomial ring $F[x]$ with respect to an irreducible polynomial f of degree n. Not surprisingly, this will require irreducibility criteria for certain types of polynomials, in particular, for binomials and trinomials.

The closely related problems of testing an arbitrary monic polynomial f over F for irreducibility or, more generally, of decomposing f into irreducible polynomials will be postponed to Chapter 6. Thus the main topic of the present chapter is the construction of *some* irreducible polynomial of any given degree over any given Galois field $\mathrm{GF}(q)$, but we will also consider this problem for two important special types of polynomials (namely primitive and self-reciprocal polynomials) in the final two sections.

5.1 Extensions with Degree a Power of the Characteristic

Throughout this chapter, we let $F = \mathrm{GF}(q)$ be a fixed ground field with characteristic p; the prime field of F will be denoted by P, and the absolute trace mapping $\mathrm{Tr}_{F/P}$ simply by Tr.

It might be most elegant to approach our topic using the algebraic closure $\widehat{F}$ of F, as we will provide explicit constructions for *all* n-dimensional extensions of F, that is, for all the finite subfields E_n of $\widehat{F}$; see Section 4.2. For example, in Theorems 5.1.5 and 5.1.11 we will construct infinite sequences of irreducible polynomials. However, as one does not require the algebraic closure for these tasks, we will appeal only to splitting fields instead of $\widehat{F}$, in order to keep the treatment as elementary as possible.

In this first section, we deal with the case where n is a power of the characteristic p, say $n = p^k$. In the easiest case, namely $n = p$, there is a simple irreducibility criterion for trinomials of the special form $x^p - x - \beta$; here a polynomial is called a

D. Hachenberger and D. Jungnickel, *Topics in Galois Fields*, Algorithms and Computation in Mathematics 29, https://doi.org/10.1007/978-3-030-60806-4_5

trinomial if it has just three non-zero coefficients. The reader might wish to compare this result with the corresponding result for **binomials** – that is, polynomials with just two non-zero coefficients – over arbitrary fields with characteristic p in Proposition 3.2.19; of course, in the finite case only the second alternative in that result occurs.

Theorem 5.1.1. *Consider a trinomial g of the form $g(x) = x^p - x - \beta$ over $F = \mathrm{GF}(q)$. Then one of the following two alternatives occurs:*

- $\mathrm{Tr}(\beta) = 0$, *and g splits in $F[x]$ into p distinct linear factors.*
- $\mathrm{Tr}(\beta) \neq 0$, *and g is irreducible over F and has $\mathrm{GF}(q^p)$ as its splitting field.*

Proof. Let v be any root of g in its splitting field E, so that $v^p = v + \beta$. Then the p roots of g in E are given by the elements $v + \alpha$ with $\alpha \in P$, as

$$(v+\alpha)^p = v^p + \alpha^p = (v+\beta) + \alpha = (v+\alpha) + \beta.$$

We now check that these p roots belong to F if and only if $\mathrm{Tr}(\beta) = 0$. Using $v^p = v + \beta$ and induction gives

$$v^{p^\ell} = v + \beta + \beta^p + \cdots + \beta^{p^{\ell-1}} \quad \text{for all } \ell \in \mathbb{N}^*.$$

Now let $\ell = m$, where $q = p^m$. Then

$$v^q = v + \sum_{j=0}^{m-1} \beta^{p^j} = v + \mathrm{Tr}(\beta),$$

and therefore $v^q = v$ (that is, $v \in F$ and hence also $v + \alpha \in F$ for all $\alpha \in P$) if and only if $\mathrm{Tr}(\beta) = 0$, as claimed.

It remains to consider the case where $\mathrm{Tr}(\beta) \neq 0$. We again use $v^q = v + \mathrm{Tr}(\beta)$. By induction, $v^{q^\ell} = v + \ell \cdot \mathrm{Tr}(\beta)$ for all $\ell \in \mathbb{N}^*$. In particular, $v^{q^p} = v + p \cdot \mathrm{Tr}(\beta) = v$, so that $F(v) = \mathrm{GF}(q^p)$. Moreover,

$$\{v, v^q, \ldots, v^{q^{p-1}}\} = \{v, v + \mathrm{Tr}(\beta), \ldots, v + (p-1)\mathrm{Tr}(\beta)\} = v + P,$$

as $\mathrm{Tr}(\beta) \neq 0$. Now Proposition 3.5.1 shows that

$$g(x) = \prod_{\alpha \in P} (x - v - \alpha) = \prod_{i=0}^{p-1} \left(x - v^{q^i}\right)$$

is the minimal polynomial of v over F. Thus g is irreducible and $E = F(v) = \mathrm{GF}(q^p)$ is its splitting field. □

Remark 5.1.2. By Theorem 5.1.1, constructing an extension with degree p over F amounts to finding an element $\beta \in F$ with nonzero absolute trace. Of course, this can be done using methods from Linear Algebra. Recall that the absolute trace Tr is a non-trivial linear form, and therefore the set of all $\beta \in F$ with $\mathrm{Tr}(\beta) \neq 0$ contains exactly $p^m - p^{m-1}$ elements, where $m = [F : P]$. This is also the number of monic irreducible polynomials in $F[x]$ of the form $x^p - x - \beta$. In the simplest case $F = P$, one can take for β any element of F^*. In particular, the polynomial $x^p - x - 1$ is irreducible over $\mathrm{GF}(p)$ (for every prime p). □

If we want to obtain extensions with degree an arbitrary power p^k of p, it is natural to try constructing a sequence of irreducible polynomials with degrees $p, p^2, \ldots, p^k$ by applying Theorem 5.1.1 iteratively. For the first step $k = 2$, this requires finding an element $w \in \mathrm{GF}(q^p)$ with non-zero absolute trace. One might be tempted to use a root of an irreducible polynomial $g(x) = x^p - x - \beta$ over F for w, but usually this does not work; see Exercise 5.1.13. The following result provides a suitable choice for w in all cases:

Theorem 5.1.3. *Let $g(x) = x^p - x - \beta$ be an irreducible trinomial over $F = \mathrm{GF}(q)$, and let v be a root of g in $E = \mathrm{GF}(q^p)$. Then $\gamma := \beta v^{p-1}$ has absolute trace $-\mathrm{Tr}(\beta)$, and $x^p - x - \gamma$ is an irreducible trinomial over E.*

Proof. We evaluate the absolute trace of γ by using the transitivity formula for the trace. As v is a root of g, we may rewrite γ as follows:

$$\gamma = \beta v^{p-1} = \frac{\beta v^p}{v} = \frac{\beta(v+\beta)}{v} = \beta + \beta^2 v^{-1}.$$

This gives

$$\mathrm{Tr}_{E/F}(\gamma) = p\beta + \beta^2 \mathrm{Tr}_{E/F}(v^{-1}) = \beta^2 \mathrm{Tr}_{E/F}(v^{-1}), \tag{5.1}$$

and we need to compute $\mathrm{Tr}_{E/F}(v^{-1})$. For this, we determine the minimal polynomial of v^{-1} over F, which has degree p (since $F(v^{-1}) = F(v) = E$). In fact, it is the monic polynomial $h(x) := x^p + \beta^{-1}x^{p-1} - \beta^{-1}$, as

$$-\beta v^p \cdot h(v^{-1}) = -\beta v^p \left(v^{-p} + \beta^{-1} v^{1-p} - \beta^{-1}\right) = -\beta - v + v^p = g(v) = 0,$$

which shows that v^{-1} is indeed a root of h.[1] By Proposition 3.12.3, $\mathrm{Tr}_{E/F}(v^{-1})$ is the negative coefficient of x^{p-1} in $h(x)$, that is, $\mathrm{Tr}_{E/F}(v^{-1}) = -\beta^{-1}$. Substituting in (5.1) gives $\mathrm{Tr}_{E/F}(\gamma) = -\beta$ and hence $\mathrm{Tr}_{E/P}(\gamma) = -\mathrm{Tr}(\beta)$, as claimed. In view of Theorem 5.1.1, $\mathrm{Tr}_{E/P}(\gamma)$ is nonzero and $x^p - x - \gamma$ is irreducible over E. □

Example 5.1.4. Let us show how one might apply Theorem 5.1.3 to compute an irreducible polynomial with degree p^2 over $P = \mathrm{GF}(p)$. As noted in Remark 5.1.2, $f(x) := x^p - x - 1$ is an irreducible polynomial with degree p over P. Let v be any root of f in $E = \mathrm{GF}(p^p)$, so that $g_0(x) := x^p - x - v^{p-1}$ is an irreducible polynomial with degree p over E. Since the roots of f are the p conjugates of v, we see – more generally – that all p polynomials

$$g_i(x) := x^p - x - v^{p^i(p-1)} \quad \text{with } i = 0, \ldots, p-1$$

[1] The reader might wonder how we arrived at the explicit form of h. In fact, this is an application of a result on so-called *reciprocal polynomials* which always allows one to determine the minimal polynomial of an element v^{-1} from that of v; see Remark 5.1.10 or Theorem 5.7.1.

are irreducible over $E = P(v)$. Now we consider the product $g := g_0 \cdots g_{p-1}$ of these p polynomials. Note that g is invariant under the Frobenius automorphism of E/P, and therefore a polynomial in $P[x]$. Let w be any root of g_0. It is easily checked that the set of roots of g consists of the p^2 conjugates of w under the Frobenius automorphism of $E(w) = \mathrm{GF}(p^{p^2})$ over P, which shows that g is an irreducible polynomial with degree p^2 over P, as desired. Moreover, the coefficients of g may be computed explicitly by using the equation $f(v) = 0$ (at least in principle).

As a concrete example, we use this approach to determine an irreducible polynomial with degree 9 for the case $p = 3$. Here we obtain

$$\begin{aligned} g(x) &= (x^3 - x - v^2)(x^3 - x - v^6)(x^3 - x - v^{18}) \\ &= x^9 - \mathrm{Tr}_{E(w)/E}(v^2)x^6 + 2\mathrm{Tr}_{E(w)/E}(v^2)x^4 + (v^2v^6 + v^2v^{18} + v^6v^{18} - 1)x^3 \\ &\qquad - \mathrm{Tr}_{E(w)/E}(v^2)x^2 - (v^2v^6 + v^2v^{18} + v^6v^{18})x - \mathrm{Norm}_{E(w)/E}(v^2). \end{aligned}$$

We know from our theoretical considerations that g has coefficients in $P = \mathrm{GF}(3)$, and we would like to evaluate these coefficients explicitly. Proposition 3.12.2 suggests to determine the minimal polynomial $h(x)$ of v^2 over P for this task. Using the equation $f(v) = v^3 - v - 1 = 0$, we compute

$$v^6 = (v+1)^2 = 2(v^2+v) - v^2 + 1 = 2v^4 - v^2 + 1,$$

and hence

$$h(x) = x^3 + x^2 + x - 1 = (x - v^2)(x - v^6)(x - v^{18}).$$

Comparing coefficients yields

$$\mathrm{Tr}_{E(w)/E}(v^2) = -1, \;\; \mathrm{Norm}_{E(w)/E}(v^2) = 1 \;\text{ and }\; v^2v^6 + v^2v^{18} + v^6v^{18} = 1,$$

so that $g(x) = x^9 + x^6 + x^4 + x^2 - x - 1$. □

Theorem 5.1.3 leads to the following iterative method for constructing extensions with degree p^k over F for all $k \in \mathbb{N}^*$:

Theorem 5.1.5. *Let $F = \mathrm{GF}(q)$ be the Galois field with q elements, and let p be the characteristic of F. Choose an element β_0 in F with non-zero absolute trace, put $g_0(x) := x^p - x - \beta_0$, and construct a sequence $(g_k)_{k\in\mathbb{N}}$ of trinomials as follows. If β_k and $g_k(x) = x^p - x - \beta_k$ have already been constructed (for some $k \in \mathbb{N}$), let $\beta_{k+1} := \beta_k v_k^{p-1}$, where v_k is some root of g_k, and put $g_{k+1}(x) := x^p - x - \beta_{k+1}$.*

Then each g_k is an irreducible polynomial over an extension field $E^{(k)}$ of F with degree $[E^{(k)} : F] = p^k$.

Proof. We use induction on k. For the induction basis $k = 0$, the assertion holds by Theorem 5.1.1. Now let $k \geq 0$ and assume that g_k is an irreducible trinomial over an extension field $E^{(k)}$ of F with $[E^{(k)} : F] = p^k$. Again by Theorem 5.1.1, the splitting field of g_k is an extension with degree p over $E^{(k)}$ and hence an extension field $E^{(k+1)}$ of F with degree p^{k+1}. Now Theorem 5.1.3 shows that g_{k+1} is indeed an irreducible trinomial over $E^{(k+1)}$. □

In order to obtain an irreducible polynomial with degree p^{k+1} over F (where $k \in \mathbb{N}^*$), it would therefore be sufficient to determine the minimal polynomial of v_k over F, where v_k is a root of $g_k(x)$. Of course, this could be done by computing a linear dependence for the vectors $1, v_k, \ldots, v_k^{p^{k+1}}$ explicitly. We shall provide an alternative (much simpler) method for determining an irreducible polynomial over F with degree an arbitrary power of p later; see Theorem 5.1.11.

The computation of an irreducible polynomial with degree p^2 over $\mathrm{GF}(p)$ as in Example 5.1.4 is rather cumbersome for larger values of p, since it involves the multiplication of p trinomials over the extension field $\mathrm{GF}(p^p)$. We now present a generalization of Theorem 5.1.1 which is due to Varshamov [384] and often allows one to find an irreducible polynomial with degree mp over F explicitly from an irreducible polynomial with degree m, without performing any computations over the extension field $\mathrm{GF}(q^m)$. Note that the special choice $f(x) = x$ recovers Theorem 5.1.1 from this result.

Theorem 5.1.6. *Let $f(x) = x^m + f_{m-1}x^{m-1} + \cdots + f_1x + f_0$ be an irreducible polynomial with degree m over $F = \mathrm{GF}(q)$, and let $\gamma \in F$. Then the polynomial*

$$g(x) := f(x^p - x - \gamma) \in F[x]$$

with degree mp is irreducible over F if and only if the element $m\gamma - f_{m-1} \in F$ has non-zero absolute trace.

Proof. Put $E := \mathrm{GF}(q^m)$, and let $u \in E$ be a root of f. By Theorem 5.1.1, the polynomial

$$h(x) := x^p - x - (u + \gamma) \in E[x]$$

is irreducible over E if and only if the absolute trace of $u + \gamma$ is non-zero. Using the transitivity of the trace and Proposition 3.12.2, we obtain the criterion

$$\mathrm{Tr}_{E/P}(u+\gamma) = \mathrm{Tr}_{F/P}\big(\mathrm{Tr}_{E/F}(u+\gamma)\big) = \mathrm{Tr}(m\gamma - f_{m-1}) \neq 0$$

for the irreducibility of h. Now let w be a root of h, that is, $w^p - w - \gamma = u$. Thus w is also a root of g, and $F(w) = F(u,w) = E(w)$. Note that h is irreducible over E if and only if w has degree p over E, which holds if and only if w has degree mp over F, that is, if and only if g is the minimal polynomial of w over F. Altogether, we conclude that g is irreducible over F if and only if $\mathrm{Tr}(m\gamma - f_{m-1}) \neq 0$. □

If we choose $\gamma = 0$ in Theorem 5.1.6, we obtain a particularly simple irreducibility criterion:

Corollary 5.1.7. *Let $f(x) = x^m + f_{m-1}x^{m-1} + \cdots + f_1x + f_0$ be an irreducible polynomial with degree m over F. Then the polynomial $g(x) := f(x^p - x)$ is irreducible over F if and only if f_{m-1} has non-zero absolute trace.* □

Example 5.1.8. We can now give a much simpler construction for an irreducible polynomial with degree p^2 over $P = \mathrm{GF}(p)$ than that presented in Example 5.1.4. According to the proof of Theorem 5.1.3 (with $\beta = 1$), the polynomial $x^p + x^{p-1} - 1$ is irreducible over P. Then the criterion in Corollary 5.1.7 is trivially satisfied, and therefore

$$g(x) = f(x^p - x) = (x^p - x)^p + (x^p - x)^{p-1} - 1$$

is an irreducible polynomial with degree p^2 over P. With $y := x^{p-1}$, we compute

$$\begin{aligned}(x^p - x)^{p-1} &= \big(x(y-1)\big)^{p-1} = y(y-1)^{p-1} \\ &= y \cdot \frac{(y-1)^p}{y-1} = y \cdot (y^{p-1} + \cdots + y + 1),\end{aligned}$$

which gives

$$g(x) = x^{p^2} - x^p + \sum_{j=1}^{p} x^{j(p-1)} - 1.$$

As a concrete example, $p = 5$ yields the irreducible polynomial

$$g(x) = x^{25} + x^{20} + x^{16} + x^{12} + x^8 - x^5 + x^4 - 1$$

with degree 25 over GF(5). □

Next, we present a further result of Varshamov [384] which provides a simple iterative construction of irreducible polynomials with increasing degree over a fixed ground field. For this, we have to define the reciprocal of a polynomial, which has already been used implicitly in the proof of Theorem 5.1.3 (see the footnote on page 199).

Definition 5.1.9. Let $f(x)$ be a non-zero polynomial with degree n over F. Then the **reciprocal** of f is the polynomial $f^*(x) := x^n f(x^{-1})$. If $f(x) = f^*(x)$, one calls $f(x)$ a **self-reciprocal** polynomial (or a **palindrome**). □

Remark 5.1.10. Consider a polynomial $f(x) = \sum_{j=0}^{n} f_j x^j$ with degree n over F. Then

$$f^*(x) = x^n \cdot \Big(\sum_{j=0}^{n} f_j x^{-j}\Big) = \sum_{j=0}^{n} f_j x^{n-j} = \sum_{j=0}^{n} f_{n-j} x^j.$$

Thus $\deg f^* = \deg f$ if and only if $f_0 = f(0) \neq 0$.

Now assume that f is irreducible, and let v be a root of f. Then f^* is likewise irreducible, and v^{-1} is a root of $f^*(x)$. In other words, if f is the minimal polynomial of v over F, then the minimal polynomial of v^{-1} over F is given by $\frac{1}{f(0)} \cdot f^*(x)$. This is easily checked; see also Theorem 5.7.1 for a formal proof of a slightly stronger assertion. □

Theorem 5.1.11. *Let $f(x) = x^m + f_{m-1}x^{m-1} + \cdots + f_1 x + f_0$ be a monic irreducible polynomial with degree m over $F = \mathrm{GF}(p)$, and suppose that there is a non-zero element γ of F such that*

$$(m\gamma - f_{m-1}) \cdot f'(-\gamma) \neq 0,$$

where, as usual, f' denotes the formal derivative of f. Put $h_1(x) := f(x^p - x - \gamma)$ and construct a sequence $(h_k)_{k\in\mathbb{N}^*}$ of polynomials with $h_k(0) \neq 0$ over F as follows. If h_k has already been constructed for some $k \in \mathbb{N}^*$, let

$$h_{k+1}(x) := \frac{1}{h_k(0)} \cdot h_k^*(x^p - x - \gamma).$$

Then h_k is a monic irreducible polynomial with degree mp^k over F for all $k \in \mathbb{N}^*$.

Proof. We want to use induction on k. To make this work, we actually need to prove the following stronger assertion: h_k is a monic irreducible polynomial with degree mp^k (so that, in particular, $h_k(0) \neq 0$), and $h_k'(\lambda) \neq 0$ for all $\lambda \in F$.

For the induction basis $k = 1$, an application of Theorem 5.1.6 shows that h_1 is indeed irreducible, as $\mathrm{Tr}(m\gamma - f_{m-1}) = m\gamma - f_{m-1} \neq 0$ by hypothesis. Moreover, $h_1'(x) = f'(x^p - x - \gamma) \cdot (-1)$, and evaluating $h_1'(x)$ at any $\lambda \in F$ gives $h_1'(\lambda) = -f'(-\gamma) \neq 0$, again by hypothesis and using $\lambda^p = \lambda$ for $\lambda \in F$.

For the induction step, assume that h_k is a monic irreducible polynomial with degree $n := mp^k$ satisfying $h_k'(\lambda) \neq 0$ for all $\lambda \in F$ (for some $k \in \mathbb{N}^*$). Let us write

$$h_k(x) = x^n + b_{n-1}x^{n-1} + \cdots + b_1 x + b_0.$$

Then $h_{k+1}(x) = (h_k(0))^{-1} \cdot h_k^*(x^p - x - \gamma)$ is a monic polynomial with degree $np = mp^{k+1}$ and constant coefficient $\neq 0$. Now let v be a root of h_k. In view of Remark 5.1.10, the minimal polynomial of v^{-1} over F is

$$g(x) := \frac{1}{h_k(0)} \cdot h_k^*(x) = x^n + \frac{b_1}{b_0}x^{n-1} + \cdots + \frac{b_{n-1}}{b_0}x + \frac{1}{b_0}.$$

By Theorem 5.1.6, h_{k+1} is irreducible over F if and only if the absolute trace of $n\gamma - \frac{b_1}{b_0}$ is non-zero. Since n is divisible by the characteristic p and since $F = P$, this amounts to showing $b_1 \neq 0$, which indeed holds as $b_1 = h_k'(0) \neq 0$.

It remains to check that $h_{k+1}'(\lambda) \neq 0$ for all $\lambda \in F$. With the above notation, $h_{k+1}'(x) = g'(x^p - x - \gamma) \cdot (-1)$, and therefore

$$h_{k+1}'(\lambda) = -g'(-\gamma) = h_{k+1}'(0) \quad \text{for all } \lambda \in F.$$

Using the explicit form of g above and taking $p \mid n$ into account, we obtain

$$\begin{aligned} h'_{k+1}(0) &= -\sum_{j=1}^{n-1}(n-j)\cdot\frac{b_j}{b_0}\cdot(-\gamma)^{n-j-1} \\ &= \frac{1}{b_0}\cdot(-\gamma)^{n-2}\cdot\sum_{j=1}^{n-1} jb_j\cdot(-\gamma^{-1})^{j-1} \\ &= \frac{1}{b_0}\cdot(-\gamma)^{n-2}\cdot h'_k(-\gamma^{-1}) \neq 0, \end{aligned}$$

by the induction hypothesis. □

Example 5.1.12. As noted in Example 5.1.8, $f(x) := x^p + x^{p-1} - 1$ is an irreducible trinomial over $F = \mathrm{GF}(p)$. We apply Theorem 5.1.11 with this choice of f and with $\gamma = 1$. Then $f'(x) = -x^{p-2}$, and therefore

$$(m\gamma - f_{m-1})\cdot f'(-\gamma) = (p\cdot 1 - 1)\cdot f'(-1) = (-1)^{p-2} \neq 0.$$

Hence Theorem 5.1.11 gives a sequence $(h_k(x))_{k\in\mathbb{N}}$ of monic irreducible polynomials h_k with degree p^k, namely $h_0(x) = x^p - x - 1$ and $h_{k+1}(x) := \frac{1}{h_k(0)}\cdot h_k^*(x^p - x - 1)$ for all $k \in \mathbb{N}^*$.

For example, let $p = 2$. Then we obtain the sequence of monic irreducible polynomials starting with $h_0(x) = x^2 + x + 1$,

$$h_1(x) = (x^2+x+1)^2 + (x^2+x+1) + 1 = x^4 + x + 1,$$

and

$$h_2(x) = (x^2+x+1)^4 + (x^2+x+1)^3 + 1 = x^8 + x^6 + x^5 + x^4 + x^3 + x + 1. \quad \square$$

Exercises

Exercise 5.1.13. Let v be a root of an irreducible polynomial $g(x) = x^p - x - \beta$ over F, and let $E = \mathrm{GF}(q^p)$. Show that $\mathrm{Tr}(v) = 0$, unless $p = 2$ and $q = 2^m$, where m is odd. □

Exercise 5.1.14. Use Theorem 5.1.5 to determine the extension with degree 27 over the ternary field $F = \mathrm{GF}(3)$. Try to find an irreducible polynomial with degree 27 over F explicitly via this approach (not a pleasant task). Alternatively, use the method in Example 5.1.12 for $p = 3$. □

Exercise 5.1.15. Use Theorem 5.1.6 to show that

$$g(x) = (x^p - x + 1)^2 + 1 = x^{2p} - 2x^{p+1} + 2x^p + x^2 - 2x + 2$$

is an irreducible polynomial with degree $2p$ over $\mathrm{GF}(p)$, whenever $p \equiv 3 \bmod 4$. Also, give a similar construction for primes $p \equiv 1 \bmod 4$. □

5.2 Irreducible Binomials

Recall that a **binomial** is a polynomial of the form $x^n - \alpha$. In this section, we determine necessary and sufficient conditions for the existence of an *irreducible* binomial in $F[x]$, where $F = \mathrm{GF}(q)$ again is some finite field; as usual, we let p denote the characteristic of F. More precisely, we will solve the following two problems:

- Determine the set B_q of all $n \in \mathbb{N}^*$ for which there exists an irreducible binomial with degree n in $F[x]$.
- Given an integer $n \in B_q$, determine the set $B_q(n)$ of all $\alpha \in F$ for which $x^n - \alpha$ is irreducible over F.

Note that $1 \in B_q$ and that $B_q(1) = F = \mathrm{GF}(q)$ for all prime powers q, since $x - \alpha$ is irreducible for every $\alpha \in F$. Therefore, we assume $n \geq 2$ from now on. We begin with the following simple fact:

Observation 5.2.1. Assume that n is a multiple of the characteristic p, say $n = pn'$, and let $\alpha \in F$ be arbitrary. Then there exists an element $\beta \in F$ with $\beta^p = \alpha$, and we obtain

$$x^n - \alpha = x^{pn'} - \beta^p = \left(x^{n'} - \beta\right)^p,$$

and therefore $x^n - \alpha$ cannot be irreducible. Hence B_q is a subset of the set $\mathbb{N}_p^*$ of all positive integers which are not divisible by p. □

Settling the two problems above in general requires some effort. Before we start with this task, the reader might find it helpful to solve the two simplest cases, namely $n = 2$ and $n = 3$, for himself; see Exercise 5.2.11. We begin our investigation with the following basic criterion for the irreducibility of a binomial:

Proposition 5.2.2. *Consider the field $F = \mathrm{GF}(q)$ with characteristic p, and let n be an arbitrary integer in $\mathbb{N}_p^*$. Let e be a divisor of $q-1$, and let α be any element of order e in F^*.*

Then the splitting field of $x^n - \alpha$ has degree $\mathrm{ord}_{ne}(q)$ over F, and $x^n - \alpha$ is irreducible if and only if $\mathrm{ord}_{ne}(q) = n$. □

Proof. We first note that the formal derivative nx^{n-1} of $x^n - \alpha$ is not the zero polynomial, as $p \nmid n$ by hypothesis. Therefore $x^n - \alpha$ has no multiple roots in its splitting field E, by Corollary 2.4.18. Let R be the set of roots of $x^n - \alpha$ in E, and let $\beta \in R$ be arbitrary. Then $\beta^{ne} = \alpha^e = 1$, so that R consists of ne-th roots of unity over F.

As α is a primitive e-th root of unity in E (by hypothesis), Proposition 3.6.18 gives us a *primitive* ne-th root of unity η over F such that $\eta^n = \alpha$. Thus $R \cap C_{ne} \neq \emptyset$, where C_{ne} denotes the set of all primitive (ne)-th roots of unity over F, as in Section 3.6. This shows that the splitting field of $x^n - \alpha$ is the (ne)-th cyclotomic extension $F^{(ne)} = F(\eta)$ of F, which has degree $\mathrm{ord}_{ne}(q)$ over F, by Proposition 3.6.3.

For the second assertion, note that $\mathrm{ord}_{ne}(q) = n$ holds if and only if $F(\eta) = F^{(ne)}$ has degree n over F. As η is a root of the monic polynomial $x^n - \alpha$, this occurs if and only if $x^n - \alpha$ is the minimal polynomial of η, that is, if and only if $x^n - \alpha$ is irreducible over F. □

In view of Proposition 5.2.2, we now need to investigate the condition $\mathrm{ord}_{ne}(q) = n$ and derive more tractable criteria. The following equivalent number theoretic conditions will be the basis for solving the two problems stated above. For this, the concepts from Section 1.7, and in particular Proposition 1.7.10, will play an important role. Recall that $\mathrm{rad}(n)$ denotes the radical of n, that is, the product of all prime divisors of n.

Theorem 5.2.3. *Consider a field $F = \mathrm{GF}(q)$ with characteristic p and any integer $n \in \mathbb{N}_p^*$, and let e be a divisor of $q-1$. Then $\mathrm{ord}_{ne}(q) = n$ holds if and only if the following three conditions are satisfied:*

(a) *$(q-1)/e$ and n are relatively prime.*

(b) *$\mathrm{rad}(n)$ divides e.*

(c) *$q \equiv 1 \bmod 4$ if n is a multiple of 4.*

Proof. Assume that $\mathrm{ord}_{ne}(q) = n$ holds. Then $x^n - \alpha$ is irreducible over F, where α is an arbitrarily chosen element in F^* with order e, by Proposition 5.2.2. We apply this fact to check that conditions (a), (b) and (c) have to be satisfied. Assume otherwise, so that at least one of these three conditions is violated.

We first consider the case where condition (a) is violated, and choose some prime r dividing both n and $(q-1)/e$, say $n = r\ell$ and $q-1 = ret$. Then the group U formed by the r-th powers in F^* has order et and therefore contains all elements of order e. In particular, $\alpha \in U$, and hence $\alpha = b^r$ for some b. But then

$$x^n - \alpha = x^{r\ell} - b^r = (x^\ell - b) \cdot \Big(\sum_{j=0}^{r-1} b^{r-1-j} x^{j\ell} \Big),$$

so that $x^n - \alpha$ would be reducible. Thus condition (a) must hold.

Now let condition (b) be violated, and choose any prime r dividing n but not e. As condition (a) is satisfied, r does not divide $q-1$. Hence r has an inverse modulo $q-1$, say $rs \equiv 1 \bmod q-1$. Then, with $n = r\ell$ and $b := \alpha^s$,

$$x^n - \alpha = x^{r\ell} - \alpha^{rs} = x^{r\ell} - b^r,$$

and $x^n - \alpha$ would factor as in the previous case. Thus condition (b) also holds.

Finally, assume that condition (c) is violated, so that n is a multiple of 4 and $q \equiv 3 \bmod 4$. Since 4 divides ne, an application of Proposition 1.7.10 gives

$$\mathrm{ord}_{ne}(q) = 2 \cdot \frac{ne}{\gcd(\mathrm{pt}_{ne}(q^2-1), ne)}.$$

By definition, e divides $D := \mathrm{pt}_{ne}(q^2-1)$, say $D = de$. Then

$$\mathrm{ord}_{ne}(q) = 2 \cdot \frac{n}{\gcd(d,n)}.$$

Since n is even and $\mathrm{rad}(n)$ divides e (as condition (b) is satisfied), e is an even divisor of $q-1 \equiv 2 \bmod 4$. Hence $4 \mid d$, as q^2-1 is a multiple of 8. Thus 4 divides

$\gcd(d,n)$, and the preceding equation shows that $\mathrm{ord}_{ne}(q)$ divides $n/2$, contradicting the hypothesis $\mathrm{ord}_{ne}(q) = n$. Hence (c) holds as well.

Conversely, assume that all three conditions (a), (b) and (c) are satisfied. Then $\mathrm{rad}(ne) = \mathrm{rad}(e) \mid e \mid q-1$. Let $D := \mathrm{pt}_{ne}(q-1)$. Then $e \mid D = \mathrm{pt}_e(q-1) \mid q-1$. Now put $d := D/e$. Because of (c), we are in Case (1) or in Case (3) of Proposition 1.7.10, and therefore

$$\mathrm{ord}_{ne}(q) = \frac{ne}{\gcd(\mathrm{pt}_{ne}(q-1), ne)} = \frac{n}{\gcd(d,n)}.$$

As $d = D/e$ divides $(q-1)/e$, we obtain $\gcd(d,n) = 1$ because of (a), and hence $\mathrm{ord}_{ne}(q) = n$. □

The following simple observation provides a useful alternative formulation of Theorem 5.2.3:

Corollary 5.2.4. *With the same hypotheses as in Theorem 5.2.3, condition* (a) *may be replaced by the condition*

(a′) $\mathrm{pt}_n(q-1)$ *divides e.*

Proof. We verify that conditions (a) and (a′) are equivalent. Assume first that (a) holds, but that (a′) is violated. Then there exists a prime divisor of n which also divides $(q-1)/e$, which contradicts (a). Thus (a′) holds after all.

Conversely, assume that (a′) holds. Then

$$\frac{q-1}{e} \mid \frac{q-1}{\mathrm{pt}_n(q-1)}.$$

By the definition of the function pt_n, the fraction $(q-1)/\mathrm{pt}_n(q-1)$ is relatively prime to n. Then $(q-1)/e$ is also relatively prime to n, and therefore (a) holds. □

For later use, we mention a further simple consequence of Theorem 5.2.3:[2]

Corollary 5.2.5. *Let q be a prime power, and let n and e be integers* ≥ 2, *where* $\gcd(q,e) = 1$. *Put* $s := \mathrm{ord}_e(q)$ *and assume that the following three conditions are satisfied:*

(1) $(q^s - 1)/e$ *and n are relatively prime.*

(2) $\mathrm{rad}(n)$ *divides e.*

(3) $q^s \equiv 1 \bmod 4$ *if n is a multiple of* 4.

Then one has $\mathrm{ord}_{ne}(q) = ns$.

Proof. Note that n and q are relatively prime, because of $\gcd(q,e) = 1$ and condition (2). Therefore, we may apply Theorem 5.2.3 with q^s instead of q, which gives $\mathrm{ord}_{ne}(q^s) = n$. Since $s = \mathrm{ord}_e(q)$ divides $\mathrm{ord}_{ne}(q)$, we obtain $\mathrm{ord}_{ne}(q) = s \cdot \mathrm{ord}_{ne}(q^s) = sn$. □

---

[2] Corollary 5.2.5 actually holds for integers $q \geq 2$, without the assumption that q is a prime power. See, for instance, Jungnickel [211, Lemma 2.3.1] for a (rather lengthy, but elementary) proof.

Combining Proposition 5.2.2 and Theorem 5.2.3 gives the desired characterization of the irreducible binomials over finite fields. This result is basically due to Serret: it is contained in his 1866 book [351] for primes q.

Theorem 5.2.6. *Consider the field $F = \mathrm{GF}(q)$ with characteristic p. Let $n \in \mathbb{N}^*$, let e be a divisor of $q-1$, and let $\alpha \in F^*$ be any element with order e. Then the binomial $x^n - \alpha$ is irreducible over F if and only if the following four conditions are satisfied:*

(1) *p does not divide n.*

(2) $\mathrm{rad}(n)$ *divides e.*

(3) *$(q-1)/e$ and n are relatively prime.*

(4) $q \equiv 1 \bmod 4$ *if n is a multiple of 4.* □

Using Theorem 5.2.6, we are now able to give explicit descriptions of the sets B_q and $B_q(n)$ introduced at the beginning of this section.

Corollary 5.2.7. *Let $F = \mathrm{GF}(q)$ be a finite field with characteristic p, where q is even or $q \equiv 1 \bmod 4$. Then the set of all $n \in \mathbb{N}^*$ for which there exists an irreducible binomial with degree n in $F[x]$ is given by*

$$B_q = \{n \in \mathbb{N}_p^* : \mathrm{rad}(n) \mid q-1\}.$$

Now let n be any element of B_q with $n \geq 2$. Then $\alpha \in B_q(n)$ if and only if $\mathrm{pt}_n(q-1)$ *divides* $\mathrm{ord}(\alpha)$. *Moreover,*

$$|B_q(n)| = \phi(\mathrm{pt}_n(q-1)) \cdot \frac{q-1}{\mathrm{pt}_n(q-1)}.$$

Proof. In order to prove the first statement, let $M_q := \{n \in \mathbb{N}_p^* : \mathrm{rad}(n) \mid q-1\}$. It is immediate from Theorem 5.2.6 that $B_q \subseteq M_q$. Now let n be any element of M_q. We choose α as a primitive element of F, so that $e = \mathrm{ord}(\alpha) = q-1$. Then all the conditions in Theorem 5.2.6 are satisfied (note that $q \not\equiv 3 \bmod 4$ by hypothesis), and therefore $x^n - \alpha$ is irreducible over F. This shows $n \in B_q$ and hence $B_q = M_q$.

For the second assertion, let α be any element of F^*, say with order e. Assume first that $x^n - \alpha$ is irreducible. Then $\mathrm{pt}_n(q-1)$ has to divide e, by Proposition 5.2.2, Theorem 5.2.3 and Corollary 5.2.4. Conversely, assume that $\mathrm{pt}_n(q-1)$ divides e. Then Theorem 5.2.3 and Corollary 5.2.4 give $\mathrm{ord}_{ne}(q) = n$, so that $x^n - \alpha$ is irreducible by Proposition 5.2.2, that is, $\alpha \in B_q(n)$.

It remains to determine the cardinality of $B_q(n)$. Let $D := \mathrm{pt}_n(q-1)$ and write $q-1 = Dm$. Then $\gcd(D,m) = 1$, by the definition of pt_n. Therefore $U_D U_m$ is a decomposition of the multiplicative group F^* of F, that is, any $u \in F^*$ has a unique representation of the form $u = u_1 u_2$ with $u_1 \in U_D$ and $u_2 \in U_m$. Moreover, $\mathrm{ord}(u) = \mathrm{ord}(u_1) \cdot \mathrm{ord}(u_2)$. Consequently, $u \in B_q(n)$ if and only if $\mathrm{ord}(u_1) = D$, that is, if and only if u_1 is a primitive D-th root of unity. Hence $|B_q(n)| = \phi(D)m$, as claimed. □

We finally consider the case where $q \equiv 3 \bmod 4$. The proof of the following result is left to the reader as Exercise 5.2.13, since it is similar to that of Corollary 5.2.7.

Corollary 5.2.8. *Let $F = \mathrm{GF}(q)$ be a finite field with characteristic p, where $q \equiv 3 \bmod 4$. Then the set of all $n \in \mathbb{N}^*$ for which there exists an irreducible binomial with degree n in $F[x]$ is given by*

$$\{n : n \in \mathbb{N}_p^*,\ \mathrm{rad}(n) \mid (q-1)/2\} \cup \{2n : n \in \mathbb{N}_p^*,\ \mathrm{rad}(n) \mid (q-1)/2\}.$$

Now let n be any element of B_q with $n \geq 2$. Then $\alpha \in B_q(n)$ if and only if $\mathrm{pt}_n(q-1)$ divides $\mathrm{ord}(\alpha)$. Moreover,

$$|B_q(n)| = \phi(\mathrm{pt}_n(q-1)) \cdot \frac{q-1}{\mathrm{pt}_n(q-1)}. \qquad \square$$

As a final application of the methods in this section, we prove a useful result on the splitting of certain cyclotomic polynomials.

Proposition 5.2.9. *Let $F = \mathrm{GF}(q)$ be a finite field with characteristic p, let e be a positive integer which is not divisible by p, and put $s := \mathrm{ord}_e(q)$. Assume that the cyclotomic polynomial $\Phi_e(x)$ splits over F into ℓ distinct irreducible factors, say $f_1(x), \ldots, f_\ell(x)$. Moreover, let n be any positive integer satisfying the following conditions:*

(1) *$(q^s - 1)/e$ and n are relatively prime.*
(2) *$\mathrm{rad}(n)$ divides e.*
(3) *$q^s \equiv 1 \bmod 4$ if n is a multiple of 4.*

Then the cyclotomic polynomial $\Phi_{ne}(x)$ splits over F into the distinct irreducible factors $f_1(x^n), \ldots, f_\ell(x^n)$.

Proof. By Corollary 5.2.5, $\mathrm{ord}_{ne}(q) = ns$. Note that the hypothesis $\mathrm{rad}(n) \mid e$ implies

$$\prod_{i=1}^{\ell} f_i(x^n) = \Phi_e(x^n) = \Phi_{ne}(x) \quad \text{and} \quad \phi(ne) = n\phi(e), \tag{5.2}$$

see Proposition 3.6.8. Recall $\ell = \phi(e)/\mathrm{ord}_e(q)$, by Proposition 3.6.16. Using this proposition a second time, we see that $\Phi_{ne}(x)$ splits over F into

$$\frac{\phi(ne)}{\mathrm{ord}_{ne}(q)} = \frac{n\phi(e)}{ns} = \frac{\phi(e)}{s} = \frac{\phi(e)}{\mathrm{ord}_e(q)} = \ell$$

irreducible factors. Thus (5.2) gives the complete factorization of $\Phi_{ne}(x)$ over F, as claimed. $\square$

Example 5.2.10. It is easily checked that the cyclotomic polynomial Φ_{15} splits over $F = \mathrm{GF}(2)$ into the two irreducible factors

$$f_1(x) = x^4 + x + 1 \quad \text{and} \quad f_2(x) = x^4 + x^3 + 1;$$

see also Example 3.5.8. Now Proposition 5.2.9 (with $s = \mathrm{ord}_{15}(2) = 4$) shows that the canonical factorization of the cyclotomic polynomial Φ_{15n} over F is

$$\Phi_{15n}(x) = \left(x^{4n} + x^n + 1\right)\left(x^{4n} + x^{3n} + 1\right),$$

whenever n has the form $n = 3^a 5^b$, in particular for $n = 3,5,9,15,25$. □

Exercises

Exercise 5.2.11. Settle the existence problem for irreducible quadratic and cubic polynomials over $\mathrm{GF}(q)$ by proving the following two elementary facts:

- There exists an irreducible binomial with degree 2 over F if and only if q is odd. In this case, $B_q(2)$ is the set of all non-squares in F^*. Moreover, $\alpha \in B_q(2)$ if and only if $\mathrm{ord}(\alpha)$ does not divide $(q-1)/2$.
- There exists an irreducible binomial with degree 3 over F if and only if $q \equiv 1 \bmod 3$. In this case, $B_q(3)$ is the set of non-cubes in F^*. Moreover, $\alpha \in B_q(3)$ if and only if $\mathrm{ord}(\alpha)$ does not divide $(q-1)/3$. □

Exercise 5.2.12. Consider the field $F = \mathrm{GF}(q)$ with characteristic p and any integer $n \in \mathbb{N}_p^*$. Also, let e be a divisor of $q-1$, and let α be any element in F^* with order e. Write $n = n_0\overline{n}$, where $\gcd(\overline{n}, e) = 1$ and where every prime divisor of n_0 divides e. Show that

$$x^n - \alpha = \prod_{d \mid \overline{n}} \gcd\left(x^n - \alpha, \Phi_{den_0}(x)\right). \tag{5.3}$$

Moreover, prove that the degree of $\gcd(x^n - \alpha, \Phi_{den_0}(x))$ equals $n_0 \cdot \phi(d)$ (for every divisor d of $\overline{n}$). □

Exercise 5.2.13. Provide the details of the proof for Corollary 5.2.8. □

Exercise 5.2.14. Let r be a prime, m a positive integer, and α an element of $\mathrm{GF}(q)^*$ which is not an r-th power. Moreover, assume $q \equiv 1 \bmod 4$ if $r = 2$ and $m \geq 2$. Show that the binomial $x^{r^m} - \alpha$ is irreducible over $\mathrm{GF}(q)$. □

5.3 Extensions with Degree 2^m

We now turn our attention to extensions over $F = \mathrm{GF}(q)$ with degree r^m, where r is a prime distinct from the characteristic p of F. In this section, we consider the case $r = 2$, and in the next section that of an odd prime r. In view of Section 5.2, we may assume $q \equiv 3 \bmod 4$:

Observation 5.3.1. Let $q \equiv 1 \bmod 4$, so that $\mathrm{pt}_2(q-1) = 2^k$ for some $k \geq 2$, and choose an element $\alpha \in F^*$ with order e divisible by 2^k. Note that this is equivalent to requiring that α is a non-square in F^*. Now let $n = 2^m$, where $m \in \mathbb{N}^*$ is arbitrary. Then all conditions in Theorem 5.2.6 are satisfied, and therefore $x^{2^m} - \alpha$ is an irreducible binomial with degree 2^m over F. □

Thus let $q \equiv 3 \bmod 4$. Then -1 is a non-square in F, and therefore the cyclotomic polynomial $\Phi_4(x) = x^2 + 1$ is irreducible over F. Put $K = F(\iota)$, where ι is a root of Φ_4, so that K is an extension of F with degree 2. Let $\mathrm{pt}_2(q^2 - 1) = 2^k$, where $k \geq 3$ (as $q^2 = 1 \bmod 8$), and choose an element $\beta \in K^*$ with order e divisible by 2^k. Then the binomial $x^{2^{m-1}} - \beta$ is irreducible over K for every $m \in \mathbb{N}^*$, by Observation 5.3.1.

Now let ν be a root of $x^{2^{m-1}} - \beta$. Applying the Frobenius automorphism of K/F to $x^{2^{m-1}} - \beta$ yields the irreducible polynomial $x^{2^{m-1}} - \beta^q$ with root ν^q. Let α and γ denote the trace and norm of β over F, respectively; that is, $\alpha = \beta + \beta^q$ and $\gamma = \beta^{q+1}$. Then

$$f(x) := \left(x^{2^{m-1}} - \beta\right)\left(x^{2^{m-1}} - \beta^q\right) = x^{2^m} - \alpha x^{2^{m-1}} + \gamma$$

is a polynomial in $F[x]$ with root ν. Note that $K(\nu) = F(\nu)$ is an extension of F with degree 2^m. Therefore, f is the minimal polynomial of ν and hence irreducible over F. This establishes the following result:

Theorem 5.3.2. *Let $F = \mathrm{GF}(q)$ and $K = \mathrm{GF}(q^2)$, where $q \equiv 3 \bmod 4$, and assume that β is a non-square in K^*. Then*

$$x^{2^m} - \mathrm{Tr}_{K/F}(\beta)x^{2^{m-1}} + \mathrm{Norm}_{K/F}(\beta)$$

is an irreducible polynomial over F for every $m \in \mathbb{N}^*$. □

Corollary 5.3.3. *Let $F = \mathrm{GF}(q)$ with $q \equiv 3 \bmod 4$, and let β be a primitive 2^k-th root of unity over F, where $\mathrm{pt}_2(q^2 - 1) = 2^k$. Then*

$$x^{2^m} - (\beta + \beta^q)x^{2^{m-1}} - 1$$

is an irreducible polynomial over F for every $m \in \mathbb{N}^*$.

Proof. Note that $q + 1 = 2^{k-1}t$, where t is odd. Thus the norm of β is

$$\mathrm{Norm}(\beta) = \beta^{q+1} = \left(\beta^{2^{k-1}}\right)^t = (-1)^t = -1,$$

since $\beta^{2^{k-1}}$ is a primitive second root of unity. Now the assertion follows from Theorem 5.3.2. □

Example 5.3.4. Let us use the approach in the proof of Corollary 5.3.3 to determine the complete factorization of the cyclotomic polynomials $\Phi_{2^\ell}(x)$ with $\ell \geq 3$ over $F = \mathrm{GF}(q)$, where $q \equiv 3 \bmod 4$. (Note that $\Phi_2(x) = x + 1$ and $\Phi_4(x) = x^2 + 1$ are irreducible.) As before, let $\mathrm{pt}_2(q^2 - 1) = 2^k$.

Assume first that $\ell < k$, and let C_{2^ℓ} be the set of all primitive 2^ℓ-th roots of unity in $K = \mathrm{GF}(q^2)$. Choose a system of representatives R_ℓ for the conjugacy classes of C_{2^ℓ} under the Frobenius automorphism of K/F. Of course, the conjugacy class of $\beta \in R_\ell$ is just the set $\{\beta, \beta^q\}$. A similar argument as in the proof of Corollary 5.3.3 gives $\mathrm{Norm}_{K/F}(\beta) = 1$ for all $\beta \in R_\ell$ (as $\ell < k$). These observations show that

$$\begin{aligned}\Phi_{2^\ell}(x) &= \prod_{\beta\in R_\ell} (x-\beta)(x-\beta^q)\\ &= \prod_{\beta\in R_\ell} (x^2 - \mathrm{Tr}_{K/F}(\beta)x+1)\end{aligned}$$

is the canonical factorization of $\Phi_{2^\ell}(x)$ over F.

For $\ell = k$, we use the same argument as in the previous case. Note that now $\mathrm{Norm}_{K/F}(\beta) = -1$, as in the proof of Corollary 5.2.4. Thus

$$\Phi_{2^k}(x) = \prod_{\beta\in R_k} (x^2 - \mathrm{Tr}_{K/F}(\beta)x-1)$$

is the canonical factorization of $\Phi_{2^k}(x)$ over F.

It remains to consider the case $\ell > k$. We use the previous case $\ell = k$ and apply Proposition 5.2.9 with $e = 2^k$, $n = 2^{\ell-k}$ and $s = \mathrm{ord}_e(q) = 2$, which holds since $2^k \mid q^2-1$ but $2^k \nmid q-1$. This shows that

$$\Phi_{2^\ell}(x) = \prod_{\beta\in R_k} (x^{2^{\ell+1-k}} - \mathrm{Tr}_{K/F}(\beta)x^{2^{\ell-k}}-1)$$

is the canonical factorization of $\Phi_{2^\ell}(x)$ over F. □

As the preceding results show, it is essential to determine a non-square $\beta \in K^*$ if we want to construct irreducible polynomials with degree a power of 2 over $\mathrm{GF}(q)$, where $q \equiv 3 \bmod 4$ and $K = \mathrm{GF}(q^2)$.

Therefore, we conclude this section with a discussion of this problem. Its solution rests on the following explicit method for determining a square root of any square $\gamma \neq \pm 1$ in K^*, which is a slight generalization of a result given in Shoup [355] (for primes q).[3]

Proposition 5.3.5. *Let $K = \mathrm{GF}(q^2)$, where $q \equiv 3 \bmod 4$, and let ι be a primitive 4-th root of unity in K. Furthermore, let $\gamma \neq \pm 1$ be a square in K^* and define ω and β in K^* by*

$$\omega := \gamma^{(q-1)/2} \quad \text{and} \quad \beta := \begin{cases} (\omega+1)^{(q-1)/2}\cdot\gamma^{(q+1)/4} & \text{if } \omega\neq -1,\\ \iota\cdot\gamma^{(q+1)/4} & \text{otherwise.}\end{cases}$$

Then $\beta^2 = \gamma$. Moreover, if $\mathrm{ord}(\gamma)$ is divisible by 2, then $\mathrm{ord}(\beta) = 2\cdot\mathrm{ord}(\gamma)$.

Proof. Assume first that $\omega \neq -1$. Then

[3] Of course, a **square root** of an element α is an element β satisfying $\alpha = \beta^2$ and $\mathrm{ord}(\beta) = 2\cdot\mathrm{ord}(\alpha)$. We note that the restriction on γ poses no problem, as we obtain a square root ι of -1 automatically when we construct K as the factor ring $F[x]/(x^2-1)$, namely the residue class of x. Nevertheless, the formula in Proposition 5.3.5 actually also holds for $\gamma = \pm 1$. However, for $\gamma = 1$ this is of no interest, as the square root of 1 is -1; and for $\gamma = -1$, it is of no help, since we then are in the case $\omega = -1$ and thus already require a primitive 4-th root of unity in K for applying the formula. Therefore, this approach cannot be used to construct a square root of -1 within K.

$$\beta^2 = (\omega+1)^{q-1} \cdot \gamma^{(q+1)/2} = \frac{(\omega+1)^q}{\omega+1} \cdot \omega\gamma = \frac{\omega^{q+1}+\omega}{\omega+1} \cdot \gamma.$$

Since γ is a square in K^*, it belongs to the subgroup of order $(q^2-1)/2$ of K^*. Therefore, $\omega^{q+1} = \gamma^{(q^2-1)/2} = 1$, so that $\beta^2 = \gamma$. Now assume $\omega = -1$. Then

$$\beta^2 = \iota^2 \cdot \gamma^{(q+1)/2} = -\omega\gamma = \gamma,$$

which proves the first assertion.

It remains to check the statement on the order of β. Note that $\mathrm{ord}(\gamma)$ divides $\mathrm{ord}(\beta)$, as $\gamma = \beta^2$. More precisely, by Lemma 1.6.15,

$$\mathrm{ord}(\gamma) = \mathrm{ord}(\beta^2) = \frac{\mathrm{ord}(\beta)}{\gcd(\mathrm{ord}(\beta),2)}.$$

If $\mathrm{ord}(\gamma)$ is divisible by 2, then so is $\mathrm{ord}(\beta)$, and therefore

$$\mathrm{ord}(\beta) = \gcd(\mathrm{ord}(\beta),2) \cdot \mathrm{ord}(\gamma) = 2 \cdot \mathrm{ord}(\gamma). \qquad \square$$

Observation 5.3.6. Let $q \equiv 3 \bmod 4$ and $K = \mathrm{GF}(q^2)$, and write $\mathrm{pt}_2(q^2-1) = 2^k$. Then we may determine a non-square β in K^* as follows. Put $\gamma_1 := \iota$; this a square in K^* with $\mathrm{ord}(\gamma_1) = 4$, as $k \geq 3$. An application of Proposition 5.3.5 yields an element β of order 8 with $\beta^2 = \gamma_1$. Then β is a square if and only if $k \geq 4$. In this case, we put $\gamma_2 := \beta$ and continue in the same way until we reach an element $\beta \in K^*$ which has order 2^k; then we have found the desired non-square. □

5.4 Extensions with Degree an Odd Prime Power

In this section, let r be an odd prime which is different from the characteristic p of the underlying field $F = \mathrm{GF}(q)$. We will show how one may determine an extension with degree r^m over F for an arbitrary $m \in \mathbb{N}^*$. As in the case $r = 2$ dealt with in the previous section, we may assume $q \not\equiv 1 \bmod r$:

Observation 5.4.1. Let $q \equiv 1 \bmod r$, so that $\mathrm{pt}_r(q-1) = r^k$ for some $k \in \mathbb{N}^*$, and choose an element $\alpha \in F^*$ with order e divisible by r^k. Note that this is equivalent to requiring that α is not an r-th power in F^*. Now let $n = r^m$, where $m \in \mathbb{N}^*$ is arbitrary. Then all conditions in Theorem 5.2.6 are satisfied, and therefore $x^{r^m} - \alpha$ is an irreducible binomial with degree r^m over F. □

From now on, let $q \not\equiv 1 \bmod r$. We proceed in a similar manner as in Section 5.3 and choose a primitive r-th root of unity ζ in $K = \mathrm{GF}(q^s)$, where $s = \mathrm{ord}_r(q)$ (so that $s \mid r-1$). Thus $K = F(\zeta)$ is the r-th cyclotomic extension $F^{(r)}$ over F, by Proposition 3.6.3. Choose any element $\alpha \in K^*$ which is not an r-th power, which means that the order of α is divisible by $\mathrm{pt}_r(q^s-1)$. Then $x^{r^m} - \alpha$ is an irreducible binomial over K for every $m \in \mathbb{N}^*$, by Observation 5.4.1. We now investigate the roots of these polynomials, with the aim of obtaining a result analogous to Theorem 5.3.2. This turns out to be rather more involved.

Thus fix $m \in \mathbb{N}^*$, and let v be a root of the irreducible polynomial $x^{r^m} - \alpha$. Then v has degree r^m over K. As the order of α is divisible by $\mathrm{pt}_r(q^s - 1)$, we have $\zeta \in F(\alpha)$, so that $K = F(\zeta) = F(\alpha)$. Then $v^{r^m} = \alpha$ gives $K \subseteq F(v)$, and therefore $F(v) = K(v)$. In view of $[K:F] = s$ and $[K(v):K] = r^m$, we conclude $[F(v):F] = sr^m$, so that the minimal polynomial $g(x)$ of v over F has degree sr^m. By Proposition 3.5.1, the s conjugates of v under the action of the Frobenius automorphism of K/F are roots of g. Note that v^{q^j} is a root of $x^{r^m} - \alpha^{q^j}$, as v is a root of $x^{r^m} - \alpha$, and therefore

$$g(x) = \prod_{j=0}^{s-1} \left(x^{r^m} - \alpha^{q^j}\right).$$

This establishes the first part of the following result, but is not yet the desired analogue of Theorem 5.3.2: we now encounter the additional problem that $K(v)$ has degree sr^m, whereas we seek an extension $F(w)$ of F with degree r^m.

Theorem 5.4.2. *Let r be an odd prime and $F = \mathrm{GF}(q)$, where $q \not\equiv 1 \bmod r$, and let $K = \mathrm{GF}(q^s)$, where $s = \mathrm{ord}_r(q)$. Assume that $\alpha \in K^*$ is not an r-th power, and let m be any positive integer.*

Let v be a root of the polynomial $x^{r^m} - \alpha$. Then $K(v) = F(v)$ has degree sr^m over F, and $\prod_{j=0}^{s-1} \left(x^{r^m} - \alpha^{q^j}\right)$ is the minimal polynomial of v over F.

Finally, let E be the intermediate field of $K(v)/F$ with degree r^m, and put $w := \mathrm{Tr}_{K(v)/E}(v)$. Then w has degree r^m over F, and hence the minimal polynomial of w over F is an irreducible polynomial with degree r^m.

Proof. As we have already established the first assertion, it remains to prove the claims on w. As $w \in E$ by definition, this amounts to checking that $F(w) = E$. Assume otherwise, which means that $F(w)$ has degree r^k over F for some $k \leq m-1$. Then $K(w)$ is contained in the intermediate field L of $K(v)/F$ with degree sr^{m-1}. Note that $L = K(v^r)$: this follows by observing that v^r is a root of $x^{r^{m-1}} - \alpha$ and applying the first assertion to this situation. Hence v has degree r over L; in fact, the minimal polynomial of v over L is $x^r - v^r$.

Now put $Q := q^{r^m}$. Then $\mathrm{ord}_r(Q) = \mathrm{ord}_r(q) = s$. Let us divide Q^i by r with remainder, for $i = 0, \ldots, s-1$. This gives integers $a_i \geq 0$ and $b_i \in \{1, 2, ..., r-1\}$ such that $Q^i = a_i r + b_i$. Note that the b_i have to be distinct: otherwise, we would have $b_i = b_j$ for some $i < j$, and then $Q^j - Q^i = Q^i(Q^{j-i} - 1)$ would be divisible by r, which is impossible as r neither divides Q nor $Q^c - 1$ for $c \in \{1, ..., s-1\}$. We can now rewrite w as follows:

$$w = \mathrm{Tr}_{K(v)/E}(v) = \sum_{j=0}^{s-1} v^{Q^j} = \sum_{j=0}^{s-1} (v^r)^{a_j} \cdot v^{b_j},$$

which shows that v is a root of the non-zero polynomial

$$h(x) := \sum_{j=0}^{s-1} (v^r)^{a_j} \cdot x^{b_j} - w \in K(v^r)[x]$$

with degree at most $r-1$. This contradicts our previous observation that v has degree r over $L = K(v^r)$. □

Remark 5.4.3. It is natural to apply Theorem 5.4.2 in the special case where α is a primitive r^k-th root of unity, with $r^k = \mathrm{pt}_r(q^s - 1)$. In order to find such an element α, we need to determine an irreducible monic divisor $h(x)$ of $\Phi_{r^k}(x)$ over F. Then we can take $K := F[x]/\big(h(x)\big)$ and α as the residue class of x, which reduces the problem to the factorization of cyclotomic polynomials over F.

These observations lead to the following iterative procedure for obtaining an element α with the desired order. Recall that

$$\Phi_{r^k}(x) = \Phi_r(x^{r^{k-1}}) = \sum_{j=0}^{r-1} x^{jr^{k-1}};$$

see Example 3.6.6 and Proposition 3.6.8. Now let $\ell \in \mathbb{N}$ with $1 \le \ell \le k$. By Proposition 3.6.16, $\Phi_{r^\ell}(x)$ splits over F into

$$\frac{\phi(r^\ell)}{\mathrm{ord}_{r^\ell}(q)} = r^{\ell-1} \cdot \frac{r-1}{s}$$

irreducible polynomials with degree s each. Now proceed as follows:

- Let $g_0(x)$ be one of the $(r-1)/s$ monic irreducible factors of $\Phi_r(x)$ over F.
- Assume $k \ge 2$. As $g_0(x^r)$ divides $\Phi_{r^2}(x)$ and has degree rs, it splits over F into r irreducible polynomials with degree s each. Let $g_1(x)$ be one of these factors.
- Assume $k \ge 3$. Then $g_1(x^r)$ divides $\Phi_{r^3}(x)$ and also splits over F into r irreducible factors. Let $g_2(x)$ be such a factor.

Continuing in this manner (for $k \ge 4$), we finally reach an irreducible divisor of $\Phi_{r^k}(x)$ over F. This gives the desired explicit description of $K = \mathrm{GF}(q^s)$ as a factor ring as well as an element $\alpha \in K^*$ which is not an r-th power. ⊓

We conclude this section with two explicit examples for the application of Theorem 5.4.2.

Example 5.4.4. Take $q = 5$ and $r = 3$ in Theorem 5.4.2. Then $s = \mathrm{ord}_3(5) = 2$ and $\mathrm{pt}_r(q^2 - 1) = \mathrm{pt}_3(24) = 3$. Thus we choose $\alpha \in K = \mathrm{GF}(25)$ as a primitive third root of unity, that is, as a root of $\Phi_3(x) = x^2 + x + 1$ over $F = \mathrm{GF}(5)$.

Then the polynomial $x^{3^m} - \alpha$ is irreducible over K for every $m \in \mathbb{N}^*$. The conjugates of α over F are α and $\alpha^5 = \alpha^2$, which gives $\mathrm{Tr}_{K/F}(\alpha) = \alpha + \alpha^2 = -1$ and $\mathrm{Norm}_{K/F}(\alpha) = \alpha \cdot \alpha^2 = 1$. Hence

$$\big(x^{3^m} - \alpha\big)\big(x^{3^m} - \alpha^2\big) = x^{2\cdot 3^m} + x^{3^m} + 1$$

is an irreducible polynomial over F which is a factor of the 3^{m+1}-th cyclotomic polynomial over F.

Now let $v \in \mathrm{GF}(5^{2\cdot 3^m})$ be a root of $x^{3^m} - \alpha$, that is, a primitive 3^{m+1}-th root of unity. Then $F(w) = \mathrm{GF}(5^{3^m})$, where

$$w := v + v^{5^{3^m}}.$$

As $\mathrm{ord}(v) = 3^{m+1}$ and as 5 is a primitive root modulo 3^ℓ for every ℓ (by Proposition 1.7.8), 5^{3^m} has order 2 modulo 3^{m+1}, that is, $5^{3^m} \equiv -1 \bmod 3^{m+1}$. Hence

$$\mathrm{GF}(5^{3^m}) = \mathrm{GF}(5)(v + v^{-1})$$

for any primitive 3^{m+1}-th root of unity v. □

Of course, Example 5.4.4 was chosen to work nicely. The computations were not too bad, but we only determined an element w of the desired degree. If we actually want an explicit irreducible polynomial, we should also compute the minimal polynomial of w over F. Our second example illustrates that this can become unpleasant (at least when computing by hand) even for small values of q and r.

Example 5.4.5. We now take $q = 2$ and $r = 5$ in Theorem 5.4.2, so that $s = \mathrm{ord}_5(2) = 4$ and $\mathrm{pt}_r(q^4 - 1) = \mathrm{pt}_5(15) = 5$. Hence we choose $\alpha \in K = \mathrm{GF}(16)$ as a primitive fifth root of unity, that is, as a root of $\Phi_5(x) = x^4 + x^3 + x^2 + x + 1$ over $F = \mathrm{GF}(2)$. Then the polynomial $x^{5^m} - \alpha$ is irreducible over K for every $m \in \mathbb{N}^*$. We shall only discuss the easiest case $m = 1$.

Thus we consider the irreducible polynomial $g(x) = x^5 - \alpha$ over K, let v be a root of g, and put

$$w := \mathrm{Tr}_{K(v)/E}(v) = \sum_{j=0}^{3} v^{32^j} = v + v^7 + v^{24} + v^{18},$$

where E is the intermediate field of $K(v)/F$ with degree 5. Then the minimal polynomial $f(x)$ of w is given by

$$f(x) = \prod_{j=0}^{4} \left(x - w^{2^j}\right) = \prod_{j=0}^{4} \left(x - \left(v + v^7 + v^{18} + v^{24}\right)^{2^j}\right).$$

Of course, this expression may be evaluated explicitly, using that v is a primitive 25-th root of unity and $\Phi_{25}(x) = \Phi_5(x^5)$. This can even be done by hand, if one is sufficiently patient; alternatively, one could use a computer algebra system. The result is the irreducible binary polynomial $f(x) = x^5 + x^2 + 1$. □

5.5 Composition of Field Extensions

In this section we consider the problem of determining an extension of $F = \mathrm{GF}(q)$ with degree n for an arbitrary n. The following simple observation shows that this problem can be reduced to the case of extensions with degree a prime power, which we have already settled in the preceding sections.

Proposition 5.5.1. *Consider the finite field $F = \mathrm{GF}(q)$, and let u and v be elements with degrees k and ℓ over F, respectively. Assume that k and ℓ are relatively prime, and let $E := F(u,v)$. Then E has degree $k\ell$ over F. Moreover, both $u+v$ and uv have degree $k\ell$ over F, that is, $E = F(u+v) = F(uv)$.*

Proof. By Theorem 3.5.9, the minimal polynomial of v over F remains irreducible over the extension field $F(u)$ with degree k over F. This proves that $E = (F(u))(v)$ has degree $n = k\ell$ over F.

We now show $E = F(u+v)$. Consider any proper maximal subfield K of E. Then K cannot contain both u and v, say $u \notin K$, so that $E = K(u)$. Since the minimal polynomial of u over K divides that of u over F, we see that $[E:K]$ divides k. If we also have $v \notin K$, the same reasoning shows that $[E:K]$ also divides ℓ, contradicting the hypothesis $\gcd(k,\ell) = 1$. Hence $v \in K$, and then $u \notin K$ gives $u+v \notin K$. Therefore, $u+v$ cannot belong to any proper maximal subfield of E, which proves $E = F(u+v)$. Finally, $E = F(uv)$ follows by a similar argument. □

Remark 5.5.2. Let $n = \prod_{i=1}^{t} r_i^{a_i}$ be the prime power factorization of n. Then we can determine elements v_i with degree $r_i^{a_i}$ over $F = \mathrm{GF}(q)$ for all i, according to the results proved in Sections 5.1, 5.3 and 5.4. Let E be the extension of F obtained by adjoining all these elements, that is, $E = F(v_1,\dots,v_t)$. Then E has degree n over F, which follows by a repeated application of Proposition 5.5.1.

Now assume that we even know the minimal polynomial $f_i(x)$ of v_i over F. By Theorem 3.5.9, f_i remains irreducible over the intermediate field of E/F with degree $n_i := n/r_i^{a_i}$. Thus we can use the irreducible polynomials $f_1,\dots,f_t$ over F to construct E via the following chain of intermediate fields (with $m_0 = 1$, $E_{m_0} = F$ and $f_0 = 1$):

$$F = E_{m_0} \subset E_{m_1} \subset E_{m_2} \subset \cdots \subset E_{m_t} = E,$$

where $E_{m_k} = E_{m_{k-1}} / \big(f_{k-1}(x)\big)$ is the intermediate field with degree $m_k = \prod_{i=1}^{k} r_i^{a_i}$ of E/F, for $k = 1,\dots,t$. □

Let us return to the situation of Proposition 5.5.1. We might actually wish to determine an irreducible polynomial with degree $n = k\ell$ over $F = \mathrm{GF}(q)$ explicitly. Of course, this may be done by computing the minimal polynomial $h(x)$ of $u+v$, namely

$$h(x) = \prod_{i=0}^{k\ell-1} \big(x - (u+v)^{q^i}\big).$$

(Alternatively, we could also use the minimal polynomial of uv.) Unfortunately, the explicit computation of an irreducible polynomial over F as a minimal polynomial is, in general, not an easy task, as we have seen in Example 5.4.5. In the present situation, one can facilitate this task considerably, provided that we know the minimal polynomials of u and v over F: it is possible to derive the coefficients of the minimal polynomials of both $u+v$ and uv by just using operations over F, entirely avoiding computations in the extension fields $F(u)$ and $F(v)$.

In the remainder of this section, we will consider this situation in more detail. For this purpose, we introduce a further theoretical concept, which involves the Frobenius automorphism:

Definition 5.5.3. Let E be any extension field of $F = \mathrm{GF}(q)$. (Note that E does not have to be finite; for instance, E could be the algebraic closure $\widehat{F}$ of F.) Let $G \subseteq E$ be a nonempty subset which is invariant under the Frobenius automorphism $\sigma : x \mapsto x^q$, and assume that $(G, *)$ is an abelian group with respect to a binary operation $*$ which is compatible with σ, that is, $(a*b)^q = a^q * b^q$ for all $a, b \in G$.[4]

Any non-constant monic polynomial $f(x)$ over F which splits into linear factors over E and has all its roots in G is called a ***G*-polynomial**. Now let $f(x)$ and $g(x)$ be two G-polynomials, say

$$f(x) = \prod_{i=1}^{s}(x-\alpha_i)^{a_i} \quad \text{and} \quad g(x) = \prod_{j=1}^{t}(x-\beta_j)^{b_j}.$$

Then we define a polynomial $(f \circledast g)(x)$ over E as follows:

$$(f \circledast g)(x) := \prod_{i=1}^{s}\prod_{j=1}^{t}\left(x-\alpha_i * \beta_j\right)^{a_i b_j}. \tag{5.4}$$

We will call $f \circledast g$ the **root composition** of f and g.[5] □

By definition, all roots of the root composition $f \circledast g$ of two G-polynomials f and g belong to G. As both f and g are assumed to be polynomials over F, they are invariant under σ. This implies that $f \circledast g$ is also invariant under σ, so that $f \circledast g$ is actually a polynomial over F, which means that $f \circledast g$ is again a G-polynomial. Moreover, $f \circledast g$ has degree

$$\deg(f \circledast g) = \sum_{i=1}^{s}\sum_{j=1}^{t} a_i b_j = \Big(\sum_{i=1}^{s} a_i\Big)\cdot\Big(\sum_{j=1}^{t} b_j\Big) = \deg f \cdot \deg g.$$

Note that any irreducible polynomial over F with roots in G is a G-polynomial, by Proposition 3.5.1. It is now natural to ask whether or not the root composition of

[4] Of course, the most natural instances (and perhaps even the only interesting ones) are provided by taking for $(G, *)$ either the additive or the multiplicative group of the field E. The abstract definition given above will allow us to treat minimal polynomials of elements of either of the forms $u+v$ and uv simultaneously.

[5] This operation was introduced by Brawley and Carlitz [47] who used the term **composed product** instead.

two irreducible G-polynomials is again irreducible. This question is answered in the following result:

Theorem 5.5.4. *With the same notation as in Definition 5.5.3, let f and g be two G-polynomials with degrees k and ℓ, respectively. Then the G-polynomial $f \circledast g$ is irreducible over F if and only if both f and g are irreducible and* $\gcd(k,\ell) = 1$.

Proof. Assume first that $f \circledast g$ is irreducible. Then the distributive law in Exercise 5.5.6 shows that both f and g have to be irreducible. It remains to check that k and ℓ are relatively prime. Let α and β be roots of f and g, respectively. Then $\gamma := \alpha * \beta$ is a root of the irreducible polynomial $f \circledast g$ with degree $k\ell$. Thus α, β and γ have degrees k, ℓ and $k\ell$ over F, respectively. Now let $n := \mathrm{lcm}(k,\ell)$. Then

$$\gamma^{q^n} = (\alpha * \beta)^{q^n} = \alpha^{q^n} * \beta^{q^n} = \alpha * \beta = \gamma = \gamma^{q^{k\ell}}.$$

As γ has degree $k\ell$, we conclude that $k\ell$ divides n, and hence $\gcd(k,\ell) = 1$.

Conversely, assume that both f and g are irreducible and that $\gcd(k,\ell) = 1$. Let α, β, γ and n be as in the first part of the proof; note that now $n = k\ell$, as $\gcd(k,\ell) = 1$ by hypothesis. Then γ belongs to the intermediate field of E/F with degree n, since $\gamma^{q^n} = \gamma$, as before. Now let m be the degree of γ over F, so that $m \mid n$. We want to show that actually $m = n$. Using $\gamma^{q^m} = \gamma$, we obtain

$$\alpha * \beta = \gamma = \gamma^{q^{mk}} = \alpha^{q^{mk}} * \beta^{q^{mk}} = \alpha * \beta^{q^{mk}},$$

and therefore $\beta = \beta^{q^{mk}}$ (since $(G,*)$ is a group). This shows that the degree ℓ of β over F divides mk. Since ℓ and k are relatively prime, we conclude that ℓ divides m. By an analogous argument, k also divides m, so that $n = k\ell \mid m$. Hence γ has degree n over F, as claimed. Since γ is a root of the monic polynomial $f \circledast g$ with degree n, this polynomial has to be irreducible. □

We now specialize this setup to the situation in Proposition 5.5.1, so that $E = F(u,v)$, and restrict our attention to the natural choices for $*$ and G, that is, the addition in E and the multiplication in E^*. Let us denote the respective root compositions by $\oplus$ and $\odot$. Suppose that we know the minimal polynomials of u and v over F and note

$$\mathrm{mpol}_{u+v}(x) = \big(\mathrm{mpol}_u \oplus \mathrm{mpol}_v\big)(x) \quad \text{and} \quad \mathrm{mpol}_{uv}(x) = \big(\mathrm{mpol}_u \odot \mathrm{mpol}_v\big)(x).$$

This motivates discussing how one may determine the coefficients of both $f \oplus g$ and $f \odot g$, where f and g are monic irreducible polynomials over F, by using methods from Linear Algebra. We will merely sketch the procedure; in particular, we assume that the reader is familiar with the necessary concepts from this area (or willing to look them up – for instance, by consulting Hoffmann and Kunze [190]).

Let C_h denote the companion matrix of h, where $h(x)$ is a non-constant monic polynomial over F. Then h is the characteristic polynomial of C_h, that is, $h(x) = \det(xI_m - C_h)$, where $m = \deg h$ and I_m is the (m,m)-identity matrix. In particular, the roots of h are the eigenvalues of C_h. We then have the following two results, see Zhang [408, Theorem 6.19]:

- Let A denote the Kronecker product $C_f \otimes C_g$ of the companion matrices of f and g. Then the eigenvalues of A are precisely the products $\alpha\beta$, where α is a root of f and β a root of g. Moreover,

$$(f \odot g)(x) = \det\left(xI_{k\ell} - C_f \otimes C_g\right),$$

  where $k = \deg f$ and $\ell = \deg g$.
- Similarly,

$$(f \oplus g)(x) = \det\left(xI_{k\ell} - (C_f \otimes I_\ell + I_k \otimes C_g)\right).$$

We close this section with alternative formulas for $f \oplus g$ and $f \odot g$ which only involve determinants of (k,k)- and (ℓ,ℓ)-matrices. The following result is Theorem 3.27 in Menezes et al. [270], who attribute it to Blake, Gao and Mullin.

Result 5.5.5 *Let $f(x) = x^k + f_{k-1}x^{k-1} + \cdots + f_0$ and $g(x) = x^\ell + g_{\ell-1}x^{\ell-1} + \cdots + g_0$ be two monic irreducible polynomials over F, where $\gcd(k,\ell) = 1$. Then*

$$(f \odot g)(x) = \det\left(\sum_{j=0}^{\ell} g_j x^j C_f^{\ell-j}\right) = \det\left(\sum_{i=0}^{k} f_i x^i C_g^{k-i}\right)$$

and

$$(f \oplus g)(x) = \det\left(\sum_{j=0}^{\ell} g_j \left(xI_k - C_f\right)^j\right) = \det\left(\sum_{i=0}^{k} f_i \left(xI_\ell - C_g\right)^i\right). \qquad \square$$

Exercises

Exercise 5.5.6. Let $u(x), v(x), w(x) \in F[x]$ be G-polynomials, as in Definition 5.5.3. Prove the distributive law

$$u \circledast (v \cdot w) = (u \circledast v) \cdot (u \circledast w). \qquad \square$$

5.6 Primitive Polynomials

Chapter 8 will be devoted to the important problem of efficiently performing the arithmetic operations in an extension field $E = \mathrm{GF}(q^n)$ of $F = \mathrm{GF}(q)$. A standard approach is to construct E via an irreducible polynomial $f(x)$ over F with degree n and to use a so-called polynomial basis. In this situation, one generally prefers to choose f from the following particular class of irreducible polynomials which is the topic of the present section:

Definition 5.6.1. A monic irreducible polynomial with degree n over GF(q) is said to be a **primitive polynomial** if its roots are primitive elements for GF(q^n). □

Of course, in general an irreducible polynomial f over GF(q) will not be primitive. However, if we actually choose the minimal polynomial of some primitive element of GF(q^n) for f, we obtain by definition a primitive polynomial. Thus Theorem 3.5.4 and Theorem 3.2.5 have the following interesting consequence.

Corollary 5.6.2. *Given any prime power q and any positive integer n, there exists a primitive polynomial with degree n over* GF(q). □

Since splitting fields can be constructed explicitly (at least in theory), Corollary 5.6.2 may be considered to be a constructive result. However, it is most definitely not practical. In fact, no efficient deterministic algorithm for finding a primitive polynomial of an arbitrary degree n over GF(q) or, equivalently, a primitive element of GF(q^n) is known, and one usually relies on probabilistic algorithms. This will be discussed in more detail later, after presenting some theoretical results. We begin with a simple necessary condition for primitivity.

Proposition 5.6.3. *Let f be a primitive polynomial with degree n over $F =$ GF(q), and denote the constant term of f by γ. Then $(-1)^n\gamma$ is a primitive element for F.*

Proof. By Proposition 3.12.9, the norm of a primitive element has to be primitive. Now the assertion follows from Proposition 3.12.3. □

Recall that Theorem 3.5.9 allows us to search for an irreducible polynomial with degree n over GF(q^k) by restricting attention to polynomials over the smaller subfield GF(q) provided that $\gcd(n,k) = 1$. Obviously, this approach is of no use if we actually require a primitive polynomial, since a primitive polynomial over GF(q) cannot remain primitive when considered over GF(q^k). However, as we shall see in Example 5.6.8 below, the following simple consequence of Theorem 3.5.9 can be helpful in constructing primitive polynomials of a large degree over a given finite field by first considering primitive polynomials of smaller degree over a suitable extension field.

Proposition 5.6.4. *Let f be a primitive polynomial with degree nk over* GF(q). *Considered as a polynomial over* GF(q^k), *f splits into k distinct primitive polynomials with degree n.* □

At this point, we introduce another concept which may be viewed as a generalization of the notion of primitive polynomials.

Definition 5.6.5. Let f be an irreducible polynomial with degree n over $F =$ GF(q), and let α be a root of f in $E =$ GF(q^n). Then the order of α in E^* is also called the **order** of f, and is denoted by ord(f) or simply ord f. □

In view of Proposition 3.5.1, ord f does not depend on the choice of α, and thus the preceding definition makes sense. Clearly, ord f is a divisor of $q^n - 1$; in particular, the primitive polynomials with degree n are the polynomials with order q^n-1. As already noted in Observation 3.11.8, the cyclotomic polynomial $\Phi_{q^n-1}(x)$ is the product of all these primitive polynomials. More generally, the cyclotomic polynomial $\Phi_e(x)$ is the product of all irreducible polynomials over F with order e, which leads to the following reformulation of Proposition 5.2.9:

Proposition 5.6.6. *Let $F=\mathrm{GF}(q)$ be a finite field with characteristic p, let e be a positive integer which is not divisible by p, and put $s:=\mathrm{ord}_e(q)$. Let $f_1(x),\dots,f_\ell(x)$ be the distinct monic irreducible polynomials with order e over F, where $\ell=\phi(e)/s$. Moreover, let n be any positive integer satisfying the following conditions:*

(1) *$(q^s-1)/e$ and n are relatively prime.*

(2) *$\mathrm{rad}(n)$ divides e.*

(3) *$q^s\equiv 1 \bmod 4$ if n is a multiple of 4.*

Then $f_1(x^n),\dots,f_\ell(x^n)$ are the distinct monic irreducible polynomials with order en over F. □

We also note the following partial factorization of the product $I_{n,q}$ of all irreducible polynomials with degree n over $\mathrm{GF}(q)$:

Proposition 5.6.7. *Let q be a prime power and n a positive integer. Then*

$$I_{n,q}(x) = \prod_{m|q^n-1,\ \mathrm{ord}_m(q)=n} \Phi_m(x).$$

Proof. Let f be any irreducible polynomial with degree n and order m over $F=\mathrm{GF}(q)$, and let α be a root of f in $\mathrm{GF}(q^n)$. Then α is a primitive m-th root of unity and f is the minimal polynomial of α, so that $n=\deg f=\mathrm{ord}_m(q)$. Moreover, Φ_m is the product of all irreducible polynomials over F with order m, and all these polynomials have degree n over F. These observations imply the assertion. □

Example 5.6.8. By Theorem 3.5.4, there are exactly nine irreducible polynomials with degree 6 over $F=\mathrm{GF}(2)$. According to Proposition 5.6.7, a partial factorization of the product $I_{6,2}$ of these nine polynomials is given by

$$I_{6,2} = \Phi_{63}\Phi_{21}\Phi_9.$$

By Proposition 5.6.4, Φ_{63} factors into six primitive polynomials, Φ_{21} splits into two polynomials of order 21, and $\Phi_9=x^6+x^3+1$ is irreducible of order 9. Even with this partial factorization, the determination of the eight irreducible polynomials with degree 6 and order 21 or 63 seems an unpleasant task. Nevertheless, using a clever approach, this can be done quite easily.

The key is given by Theorem 3.5.9 which tells us that each of these eight polynomials splits into two irreducible factors with degree 3 over $K=\mathrm{GF}(4)$. Since it is trivial to check a polynomial with degree 3 for irreducibility (one just needs to check that there are no roots), this suggests to solve our problem by first determining the twenty irreducible polynomials with degree 3 over K. Two of these actually have coefficients in F, namely the factors x^3+x+1 and x^3+x^2+1 of Φ_7; they are of no interest for our present problem. The following simple observation shows that the determination of the remaining eighteen irreducible polynomials with degree 3 over K is indeed helpful. Let

$$f(x) = x^3+ax^2+bx+c \quad \text{and} \quad f^{\#}(x) = x^3+a^2x^2+b^2x+c^2,$$

that is, $f^{\#}$ is obtained by applying the Frobenius automorphism of K/F to f. It is clear that $f^{\#}$ is irreducible over K if and only if f is irreducible. Provided that not all coefficients of f belong to F, one has $f \neq f^{\#}$ and the product $ff^{\#}$ is an irreducible polynomial with degree 6 over F. To see this, note that α^2 is a root of $f^{\#}$ whenever α is a root of f; hence

$$f(x) = (x-\alpha)(x-\alpha^4)(x-\alpha^{16}) \quad \text{and} \quad f^{\#}(x) = (x-\alpha^2)(x-\alpha^8)(x-\alpha^{32}),$$

and therefore $ff^{\#}$ is the minimal polynomial of α over F.

It is now quite easy to find the desired polynomials. We use $K = \{0,1,\omega,1+\omega\}$, where ω is a root of x^2+x+1, that is, a primitive third root of unity. Note first that $f(x) = x^3+\omega$ is irreducible (which follows by an application of Theorem 5.2.6); here $f^{\#}(x) = x^3+(1+\omega)$ and we obtain the irreducible polynomial $\Phi_9 = x^6+x^3+1$ of order 9 already listed above.

For the remaining cases, we may use the (E,K)-norm to distinguish between polynomials of orders 21 and 63, respectively. Note that every root w of Φ_{21} has norm 1 over K, as

$$w \cdot w^4 \cdot w^{16} = w^{21} = 1,$$

so that the minimal polynomial of w over K has to have constant term 1; in contrast, the minimal polynomial of a root of Φ_{63} has constant term $\neq 1$. Applying this to the irreducible polynomials $g(x) = x^3+\omega x+1$ and $h(x) = x^3+\omega x^2+1$ leads to the two irreducible factors of Φ_{21} over F:

$$g^{\#}(x) = x^3+(1+\omega)x+1 \quad \text{gives} \quad gg^{\#}(x) = x^6+x^4+x^2+x+1,$$

and

$$h^{\#}(x) = x^3+(1+\omega)x^2+1 \quad \text{gives} \quad hh^{\#}(x) = x^6+x^5+x^4+x^2+1.$$

Any further irreducible polynomial k with degree 3 over K has to be primitive and leads to a primitive polynomial $kk^{\#}$ with degree 6 over F. We give just two examples and leave it to the reader to determine the remaining four factors of Φ_{63} over F in Exercise 5.6.14. We first take $k(x) = x^3+x^2+x+\omega$, so that $k^{\#}(x) = x^3+x^2+x+(1+\omega)$ and $kk^{\#}(x) = x^6+x^4+x^3+x+1$. Similarly, $k(x) = x^3+x+\omega$ gives $k^{\#}(x) = x^3+x+(1+\omega)$ and $kk^{\#}(x) = x^6+x^3+x^2+x+1$. □

In the remainder of this section, we discuss the construction of primitive elements and primitive polynomials. To do so, we need to compute powers and to determine the order of certain elements in $GF(q)^*$; recall that we have already seen how to perform these tasks in Algorithms 1.6.25 and 1.6.23, respectively.

The following method for computing a primitive element ω of $GF(q)^*$ goes back to Gauss. We first enumerate the elements of $GF(q)^*$ in a suitable way, say as $x_1, \ldots, x_{q-1}$. It should be pointed out that we do not require an actual list of all these elements, but we have to be able to always produce an element distinct from $x_1, \ldots, x_k$ as long as $k < q-1$. So what we really need is only a "potential listing" of the elements in $GF(q)^*$. This can, of course, be done easily. In the case where $q = p$ is a prime, we may simply take $x_i = i$. In the case of a proper extension of $GF(p)$, we may first specify any irreducible polynomial with degree n over $GF(p)$ to give an explicit description of $GF(p^n)$ and thus identify $GF(p^n)^*$ with the non-zero polynomials with degree $\leq n-1$ over $GF(p)$, and then list these in any convenient way – for instance, by ordering the n-tuples of coefficients lexicographically.

Algorithm 5.6.9.

- *Input:* A (potential) listing $x_1, \ldots, x_{q-1}$ of the elements of $GF(q)^*$.
- *Output:* A primitive element ω of $GF(q)^*$.

(1) $i \leftarrow 1,\ x \leftarrow x_i,\ \ e \leftarrow \mathrm{ord}(x)$;
(2) **while** $e \neq q-1$ **do**
(3)   **repeat**
(4) $i \leftarrow i+1,\ y \leftarrow x_i$
(5)   **until** $y^e \neq 1$;
(6) $f \leftarrow \mathrm{ord}(y),\ \ \ell \leftarrow \mathrm{lcm}(e, f)$;
(7)   Find positive integers a and b such that $x^a y^b$ has order ℓ;
(8) $x \leftarrow x^a y^b,\ \ e \leftarrow \ell$
(9) **od**
(10) $\omega \leftarrow x$. □

Algorithm 5.6.9 requires some comments. First of all, the algorithm will terminate, because we already know the existence of primitive elements so that, for some value of i, there is an element x_i of order $q-1$. As long as the order e of the current element x is still smaller than $q-1$, the **repeat** loop finds the next element y in the given listing which has an order not dividing e, so that y can be used to enlarge the current value of e in Steps (6) to (8). It remains to discuss how one may compute suitable integers a and b in Step (7).

A solution follows easily from the standard argument used to prove the existence of primitive elements of $GF(q)^*$; see Theorem 1.6.21. One first determines the prime power factorization of the orders of x and y, say

$$\mathrm{ord}(x) = p_1^{a_1} \cdots p_s^{a_s} \quad \text{and} \quad \mathrm{ord}(y) = p_1^{b_1} \cdots p_s^{b_s}; \tag{5.5}$$

note that $\mathrm{ord}(x)$ and $\mathrm{ord}(y)$ may be easily obtained in this form while performing Algorithm 1.6.23. Because of (5.5), we have

$$\ell = \operatorname{lcm}(e, f) = p_1^{\max(a_1, b_1)} \cdots p_s^{\max(a_s, b_s)}.$$

We now choose

$$a := \prod_{a_i < b_i} p_i^{a_i} \quad \text{and} \quad b := \prod_{b_i \le a_i} p_i^{b_i},$$

so that

$$\operatorname{ord}(x^a) = \prod_{a_i \ge b_i} p_i^{a_i} \quad \text{and} \quad \operatorname{ord}(y^b) = \prod_{b_i > a_i} p_i^{b_i}.$$

Since $\operatorname{ord}(x^a)$ and $\operatorname{ord}(y^b)$ are relatively prime, we obtain the desired conclusion

$$\operatorname{ord}(z) = \operatorname{ord}(x^a y^b) = \operatorname{ord}(x^a)\operatorname{ord}(y^b) = \prod_{i=1}^{s} p_i^{\max(a_i, b_i)} = \ell.$$

Alternatively, one may also compute appropriate values for a and b without determining the prime power factorization of $\operatorname{ord}(x)$ and $\operatorname{ord}(y)$; see Exercise 1.7.19.

Even though Algorithm 5.6.9 is reasonably efficient in practice, it is clearly not polynomial in n (where $q = p^n$ for a fixed prime p). In fact, no deterministic polynomial algorithm is known for constructing primitive elements. By a result of Shparlinski [356], such an element can always be found in time $O(q^{\frac{1}{4}+\varepsilon})$ (for an arbitrary $\varepsilon > 0$). Moreover, no polynomial algorithm for factoring large integers is known.

When q is a prime p, one usually chooses the sequence of primes to list the elements of $\mathrm{GF}(p)$, that is, one takes $x_1 = 2, x_2 = 3, x_3 = 5, \ldots$ As Pohst and Zassenhaus [322] state, experience shows that one will then reach a primitive root modulo p after considering about $\log p$ of the x_i. However, there seems to be no hope of proving this with the known methods of Number Theory, unless one assumes the validity of the (still unproven) generalized Riemann hypothesis. In this context we mention that it has be shown that the smallest primitive root modulo p occurs in an interval $[1, x]$, where x is about $\sqrt{p}$; see Landau [228].

Alternatively, one might want to repeatedly compute the order of a random element b of $\mathrm{GF}(q)^*$ until a primitive element is found. Clearly, the probability that b is primitive is $\phi(q-1)/(q-1)$, see Remark 1.6.16, and

$$\frac{\phi(q-1)}{q-1} = \prod_{r \mid q-1} \left(1 - \tfrac{1}{r}\right), \tag{5.6}$$

where r runs over the distinct prime divisors of $q-1$; this follows easily from Remark 1.9.13. Thus the effectiveness of a random selection depends heavily on the factorization of $q-1$. In this context, we note the following curious result mentioned by Koblitz [220].

Proposition 5.6.10. *There exists a sequence of primes p such that the probability that a random element of* $\mathrm{GF}(p)$ *is a primitive root modulo p approaches* 0.

Proof. Consider the sequence $n_k = k!$, and choose a sequence of primes p_k satisfying $p_k \equiv 1 \bmod n_k$. (This is possible by Dirichlet's theorem which states that there are infinitely many primes of the form $na+b$, whenever a and b are any two integers which are relatively prime; a proof may be found in many text books, see e.g. Ireland and Rosen [203].) Then every prime divisor of n_k also divides $p_k - 1$, and thus Equation (5.6) shows

$$\frac{\phi(p_k-1)}{p_k-1} \leq \prod_{r|n_k} \left(1-\tfrac{1}{r}\right).$$

As $k \to \infty$, this product approaches the infinite product $\prod_r \left(1-\frac{1}{r}\right)$, which is 0. □

The idea of randomly finding a primitive element can, of course, be made more efficient by combining it with Algorithm 5.6.9, that is, by choosing the elements x_i at random. This probabilistic method seems to be reasonably efficient in practice. Lüneburg [249] reports on his experience with finding primitive roots modulo "IBM primes" (that is, primes $p < 2^{32}$); the maximum number of random elements that had to be chosen was 4, and a primitive root was always found in less than 1 second of computing, including the time needed for factoring $p-1$ (with the slow computing devices of the mid-eighties!).

It is usually considered computationally advantageous to use a primitive polynomial with degree n over $F = \mathrm{GF}(p)$ in defining $E = \mathrm{GF}(q)$, where $q = p^n$. What we have done up to now shows how one can (at least in principle) find such a polynomial: one first constructs some irreducible polynomial f with degree n by the methods presented in the preceding sections, then one determines a primitive element ω for $E \cong \mathrm{GF}(p)[x]/(f)$, and finally one computes the minimal polynomial g of ω, namely

$$g(x) = \mathrm{mpol}_\omega(x) = \prod_{k=0}^{n-1} \left(x - \omega^{q^k}\right). \tag{5.7}$$

For this, one either needs to evaluate the product in Equation (5.7) explicitly, or one computes ω^n as a linear combination of the elements $1, \omega, \ldots, \omega^{n-1}$ of the polynomial basis associated with ω. The following example illustrates both approaches.

Example 5.6.11. Let us find a primitive polynomial with degree 4 over GF(3). We first choose an irreducible polynomial of this degree. Note that $3^4 - 1 = 5 \cdot 16$ and $\mathrm{ord}_5(3) = 4$; therefore, the cyclotomic polynomial

$$f(x) := \Phi_5(x) = x^4 + x^3 + x^2 + x + 1$$

is irreducible over GF(3) by Proposition 3.6.16. Of course, f only has order 5. We now use Algorithm 5.6.9 for finding a primitive element of $E = \mathrm{GF}(81)$; here we represent the elements of E by polynomials with degree ≤ 3 in x or, more precisely, by their residue classes modulo f.

Note that the order of the element $h+(f)$ of $E = \mathrm{GF}(3)[x]/(f)$ is just the smallest exponent e for which $h^e \equiv 1 \bmod f$. We first use those elements of E corresponding to linear polynomials. Of course, $x^5 \equiv 1 \bmod f$. We now compute[6] the order of

[6] All necessary computations were done using MAPLE, which allows one to solve these tasks both easily and quickly.

$x+1+(f)$ via Algorithm 1.6.23 and obtain

$$(x+1)^{40} \equiv 1 \bmod f, \quad (x+1)^{16} \equiv x^3 \bmod f \quad \text{and} \quad (x+1)^{20} \equiv 2 \bmod f,$$

so that $x+1+(f)$ has order 40. Since 40 is a multiple of 5, there is no need to perform any computations for Step 4 of Algorithm 5.6.9, and we simply replace our first element x by $x+1$. We next compute the order of $x+2+(f)$. This time, we find

$$(x+2)^{40} \equiv 2 \bmod f \quad \text{and} \quad (x+2)^{16} \equiv x^3 \bmod f,$$

showing that $\omega = x+2+(f)$ is a primitive element for E.

It remains to determine the corresponding primitive polynomial $g = \mathrm{mpol}_{\omega}$. Using the method given in (5.7), we obtain $g(z)$ as the remainder

$$g(z) = \big(z-(x+2)\big)\big(z-(x^3+2)\big)\big(z-(x^9+2)\big)\big(z-(x^{27}+2)\big) \ \mathtt{mod}\ f,$$

giving $g = z^4+2z^3+z^2+z+2$. Alternatively, we may put

$$g(z) = z^4+dz^3+cz^2+bz+a$$

with unknown coefficients a,b,c,d. The condition $g(x+2) \equiv 0 \bmod f$ yields the equation

$$0 \equiv (d+1)x^3+(c+2)x^2+(c+b+1)x+2d+c+2b+a \bmod f,$$

which immediately gives $d=2$, $c=1$, $b=1$ and $a=2$, confirming the result obtained before. □

Once a primitive polynomial f of a given degree n has been found, one may of course (for small values of q and n) determine all irreducible polynomials with degree n over $\mathrm{GF}(q)$ by successively computing the minimal polynomials of the powers of x modulo f. If only the irreducible polynomials of a given order e are desired, one considers only those powers of x which have order e in $(\mathrm{GF}(q)[x]/(f))^*$; note that the order of $x^k \ \mathtt{mod}\ f$ is $(q^n-1)/\gcd(k,q^n-1)$, by Lemma 1.6.15.

Example 5.6.12. We determine the $\phi(20)/4 = 2$ irreducible polynomials with degree 4 and order 20 over $\mathrm{GF}(3)$. According to Example 5.6.11, $x^4+2x^3+x^2+x+2$ is a primitive polynomial over $\mathrm{GF}(3)$; let ω be a root of this polynomial. As ω has order 80, the power ω^k has order 20 if and only if $\gcd(k,80)=4$, that is, if k is not divisible by 5 and satisfies $k \equiv 4 \bmod 8$. We therefore need to compute the minimal polynomial $g(x)$ of ω^4 (and thus also of ω^{12}, ω^{36} and ω^{28}) and the minimal polynomial $h(x)$ of ω^{44} (and thus also of ω^{52}, ω^{76} and ω^{68}). Using either of the methods discussed above, we obtain

$$g(x) = x^4+x^3+2x+1 \quad \text{and} \quad h(x) = x^4+2x^3+x+1.$$

In view of Remark 5.1.10, one actually needs to compute only one of these two irreducible polynomials with degree 4 and order 20, since they form a pair of reciprocal polynomials: $(\omega^4)^{-1} = \omega^{76}$ is a root of the reciprocal polynomial g^* of g. □

A detailed algorithm for finding a primitive polynomial with degree n over $\mathrm{GF}(p)$ including a complete set of PASCAL procedures is given by Lüneburg [250], who also discusses the computational performance of his programs. In Appendix E VI., he gives many examples, for instance, primitive polynomials with degrees 100 and 101 over $\mathrm{GF}(2)$. As Lüneburg notes, the running time heavily depends on the number of primes dividing $p^n - 1$. His book still provides an interesting (but not particularly easy) reading in computational algebra.

Remark 5.6.13. As mentioned above, no deterministic polynomial algorithm for constructing primitive elements is known. For practical applications, it is often possible to use an element of sufficiently large order instead of a primitive element. Therefore, the explicit construction of such elements has found considerable interest; a seminal paper in this direction is due to Gao [126]. Unfortunately, one usually can only guarantee lower bounds for the resulting orders which are much smaller than the expected actual order. The interested reader should consult Section 4.4 of the *Handbook of finite fields* [292] and the references given there for more on this topic. □

Exercises

Exercise 5.6.14. Use Example 5.6.11 and Remark 5.1.10 to determine all primitive polynomials with degree 4 over $\mathrm{GF}(3)$. □

Exercise 5.6.15. Check that $x^4 + x + 1$ is a primitive polynomial over $F = \mathrm{GF}(2)$; see also Example 3.5.8. Use this polynomial to construct $E = \mathrm{GF}(16)$ and to determine the minimal polynomials of all elements of E over F. □

5.7 Self-Reciprocal Irreducible Polynomials

In this section, we consider the construction of self-reciprocal irreducible polynomials over $\mathrm{GF}(q)$. We begin with two basic results on (self-)reciprocal polynomials which simplify the factorization of the cyclotomic polynomials and thus the determination of all irreducible polynomials of a given degree n over $\mathrm{GF}(q)$.

Recall from Definition 5.1.9 that a non-zero polynomial $f(x)$ with degree n is said to be **self-reciprocal** provided that $f(x) = f^*(x)$, where $f^*(x) := x^n f(x^{-1})$. The following result is a minor strengthening of an observation in Remark 5.1.10.

Theorem 5.7.1. *Let* $f(x) = x^n + a_{n-1}x^{n-1} + \cdots + a_1 x + a_0$ *be a monic irreducible polynomial with degree* n *and order* e *over* $\mathrm{GF}(q)$. *Then*

$$f^{\wedge}(x) := \frac{1}{a_0} f^*(x)$$

is again a monic irreducible polynomial with degree n and order e. Moreover, if f is the minimal polynomial of α, then $f^{\wedge}$ is the minimal polynomial of α^{-1}.

Proof. Let α be a root of f. By Proposition 3.5.1, we have

$$f(x) = (x-\alpha)(x-\alpha^q)\cdots(x-\alpha^{q^{n-1}}), \tag{5.8}$$

and therefore

$$\begin{aligned} f^*(x) &= x^n f(1/x) = (1-\alpha x)(1-\alpha^q x)\cdots(1-\alpha^{q^{n-1}} x) \\ &= (-1)^n \alpha^{1+q+\cdots+q^{n-1}} (x-\alpha^{-1})(x-\alpha^{-q})\cdots(x-\alpha^{-q^{n-1}}). \end{aligned}$$

Evaluating the constant term in Equation (5.8), we see $a_0 = (-1)^n \alpha^{1+q+\cdots+q^{n-1}}$ and therefore

$$f^*(x) = a_0(x-\alpha^{-1})(x-\alpha^{-q})\cdots(x-\alpha^{-q^{n-1}}). \tag{5.9}$$

Since f is the minimal polynomial of α, (5.9) shows that $f^{\wedge} = f^*/a_0$ is indeed the minimal polynomial of α^{-1} and therefore irreducible. Since the orders of α and α^{-1} in $\mathrm{GF}(q)^*$ agree, $f^{\wedge}$ has the same order as f. □

Theorem 5.7.2. *Let $f(x)$ be an irreducible polynomial with degree n over $F = \mathrm{GF}(q)$ satisfying $f = f^{\wedge}$, where $f(x) \neq x-1$ if q is odd.*

Then f is actually self-reciprocal, that is, $f = f^*$. Except for the linear polynomials $x+1$ and $x-1$, the degree of f has to be even, say $n = 2m$, and one has $\alpha^{-1} = \alpha^{q^m}$ for every root α of f.

Proof. Let $e := \operatorname{ord} f$, so that the roots of f are primitive e-th roots of unity. By Theorem 5.7.1, those primitive e-th roots of unity which actually are roots of f arise in pairs $\{\alpha, \alpha^{-1}\}$. Clearly, $\alpha = \alpha^{-1}$ only holds for $e \leq 2$, and then $f(x) = x+1$ or $f(x) = x-1$. In all other cases, the number of roots of f (that is, the degree of f) must be even, say $n = 2m$. Now let α be any root of f. As any two roots of f are conjugates, α^{-1} has to be the image of α under some power τ of the Frobenius automorphism σ of $F(\alpha)/F$. Then $\tau^2(\alpha) = \alpha$, and therefore $\tau = \sigma^m$, since α has degree $2m$ over F. This shows $\alpha^{-1} = \tau(\alpha) = \alpha^{q^m}$, as claimed, and also that the product of all conjugates of α equals 1. Thus $a_0 = 1$, and hence $f = f^{\wedge} = f^*$. □

In particular, we note the following immediate consequence of Theorems 5.7.1 and 5.7.2, which strengthens Proposition 3.6.16:

Corollary 5.7.3. *Let q be a prime power, and let n be a positive integer with $\gcd(n,q) = 1$ and $\operatorname{ord}_n(q) \neq 1$. Moreover, let $f(x)$ be any irreducible factor of $\Phi_n(x)$. Then either f is self-reciprocal (in which case $\deg f = \operatorname{ord}_n(q)$ is even), or $f^{\wedge} \neq f$ is likewise an irreducible factor of Φ_n.* □

Example 5.7.4. We re-examine Example 5.6.8 in the light of the preceding results. Thus we consider the irreducible polynomials with degree 3 over $K = \mathrm{GF}(4) = \{0, 1, \omega, \omega + 1\}$, where ω is a root of $x^2 + x + 1$.

We first note that the two irreducible factors $x^3 + x + 1$ and $x^3 + x^2 + 1$ of $\Phi_7(x)$ constitute a pair of reciprocal polynomials. The same holds for the factors

$$f(x) = x^3 + \omega \quad \text{and} \quad f^{\wedge}(x) = \frac{1}{\omega} f^*(x) = x^3 + (1 + \omega)$$

of $\Phi_9(x) = x^6 + x^3 + 1$, and for the factors

$$g(x) = x^3 + \omega x + 1 \quad \text{and} \quad g^{\wedge}(x) = g^*(x) = x^3 + \omega x^2 + 1$$

of $\Phi_{21}(x)$; note that g even yields all four irreducible factors of Φ_{21} over GF(4), namely $g, g^*, g^\#$ and $(g^*)^\#$. Similarly, the factors $k(x) = x^3 + x^2 + x + \omega$ and $\ell(x) = x^3 + x + \omega$ of $\Phi_{63}(x)$ yield four further irreducible monic polynomials which we did not list in Example 5.6.8, namely

$$k^{\wedge}(x) = x^3 + (1 + \omega)(x^2 + x + 1) \quad \text{and} \quad \ell^{\wedge}(x) = x^3 + (1 + \omega)(x^2 + 1)$$

as well as

$$(k^{\wedge})^\#(x) = x^3 + \omega(x^2 + x + 1) \quad \text{and} \quad (\ell^{\wedge})^\#(x) = x^3 + \omega(x^2 + 1). \qquad \square$$

Before turning to the general problem of constructing irreducible self-reciprocal polynomials over GF(q), let us consider a small example.

Example 5.7.5. We determine all monic irreducible self-reciprocal polynomials with degree 4 over GF(4). As before, let $\mathrm{GF}(4) = \{0, 1, \omega, \omega + 1\}$, where ω is a root of $x^2 + x + 1$. We first show that there are exactly four such polynomials:[7] if α is a root of a monic irreducible self-reciprocal polynomial with degree $n = 4$ over GF(4), then $\alpha^{4^2} = \alpha^{-1}$, by Theorem 5.7.2. Thus $\alpha^{17} = 1$, and therefore the minimal polynomial of α divides Φ_{17}, which splits over GF(4) into four irreducible polynomials (with degree 4).

Given this information, it is possible to find the four factors of Φ_{17} without too much trouble. Thus let $f(x) = x^4 + ax^3 + bx^2 + ax + 1$ be an irreducible self-reciprocal polynomial. Then $b \neq 0$ and $a + b \neq 1$, since otherwise 1 or ω would be a root of f; moreover, $a \neq 0$, since f cannot be a square. This leaves only six candidates for f, which naturally split into pairs $\{f, f^\#\}$ such that $ff^\#$ is a self-reciprocal polynomial with degree 8 over GF(2), where we use the same approach as in Example 5.6.8. The self-reciprocal polynomial $f(x) = x^4 + \omega x^3 + x^2 + \omega x + 1$ yields

$$f^\# = x^4 + (1 + \omega)x^3 + x^2 + (1 + \omega)x + 1 \quad \text{and} \quad ff^\# = x^8 + x^7 + x^6 + x^4 + x^2 + x + 1.$$

Similarly, $g(x) = x^4 + x^3 + \omega x^2 + x + 1$ and $h(x) = x^4 + \omega x^3 + \omega x^2 + \omega x + 1$ give

[7] Actually, this follows from the formula in Theorem 5.7.9 below, but we prefer to give an ad hoc argument now.

$$g^{\#} = x^4 + x^3 + (1+\omega)x^2 + x + 1 \quad \text{with} \quad gg^{\#} = x^8 + x^5 + x^4 + x^3 + 1$$

and

$$h^{\#} = x^4 + (1+\omega)(x^3 + x^2 + x) + 1 \quad \text{with} \quad hh^{\#} = x^8 + x^7 + x^5 + x^4 + x^3 + x + 1.$$

It only remains to decide which of these three pairs of polynomials consists of two reducible polynomials; there are two easy ways of doing so. Either one might recognize $hh^{\#}$ as the cyclotomic polynomial Φ_{15}, which splits into two irreducible factors by Proposition 3.6.16, namely $1+x+x^4$ and $1+x^3+x^4$; see Example 3.5.8. Or one might observe that the product of the four polynomials f, $f^{\#}$, g and $g^{\#}$ is $\Phi_{17}(x) = x^{16} + \cdots + x + 1$, proving that these four self-reciprocal polynomials have to be irreducible. Of course, $ff^{\#}$ and $gg^{\#}$ are then two irreducible self-reciprocal polynomials with degree 8 over GF(2). □

As we have seen in Theorems 5.7.1 and 5.7.2, an irreducible polynomial $f(x)$ with degree at least 2 is self-reciprocal if and only if its set of roots is closed under inversion; in this case, f necessarily has even degree, say $2d$. Moreover, the root α of f satisfies $\alpha^{-1} = \alpha^{q^d}$ and is therefore also a root of the polynomial $x^{q^d+1} - 1$. This motivates studying the factorization of binomials of this particular form.

Note that the only other irreducible polynomials for which the set of roots is closed under inversion are the linear polynomials $x+1$ and $x-1$. Since we prefer to avoid the necessity of dealing explicitly with these two polynomials in what follows, it is convenient to introduce the following notation:

$$M_{n,q}(x) := \begin{cases} (x^{q^n+1} - 1)/(x-1) & \text{if } q \text{ is even,} \\ (x^{q^n+1} - 1)/(x^2-1) & \text{if } q \text{ is odd.} \end{cases} \tag{5.10}$$

Our first aim is the determination of formulas for the product $SI_{n,q}$ as well as the number $si_q(n)$ of all self-reciprocal monic irreducible polynomials with degree $2n$ over GF(q). This is similar to the treatment of the corresponding questions for irreducible polynomials in general given in Section 3.5. The following auxiliary result gives the factorization of the polynomial $M_{n,q}$ in terms of the $SI_{d,q}$.

Lemma 5.7.6. *The polynomial $M_{n,q} \in \mathrm{GF}(q)[x]$ is the product of all self-reciprocal monic irreducible polynomials with degree $2d$ over* GF(q)*, where d runs over all divisors of n for which n/d is odd:*

$$M_{n,q} = \prod_{d \mid n,\, n/d \text{ odd}} SI_{d,q}. \tag{5.11}$$

Hence

$$q^n - \delta = \sum_{d \mid n,\, n/d \text{ odd}} 2d \cdot si_q(d), \tag{5.12}$$

where $\delta = 1$ if q is odd and $\delta = 0$ otherwise.

Proof. We start with a preliminary remark. Let d be a divisor of n and write $n = ad$. A simple computation shows that $q^d + 1$ divides $q^n + 1$ if a is odd; and if a is even, one has $\gcd(q^d + 1, q^n + 1) \mid 2$.

Now let $f(x)$ be any self-reciprocal monic irreducible polynomial with degree $2d$, where d divides n and n/d is odd. Let α be a root of f. As noted above, α is a root of $x^{q^d+1} - 1$, and hence f divides $x^{q^d+1} - 1$ and therefore also $x^{q^n+1} - 1$, since $q^d + 1$ divides $q^n + 1$ in this case. Thus f is a divisor of $M_{n,q}$.

Conversely, let f be any monic irreducible factor of $M_{n,q}$, and let α be a root of f. Then α satisfies $\alpha^{q^n} = \alpha^{-1}$, and thus the set of roots of f is closed under inversion. Hence f is self-reciprocal and has even degree, say $2d$. Then α generates $\mathrm{GF}(q^{2d})$, which therefore is a subfield of the splitting field $\mathrm{GF}(q^{2n})$ of $x^{q^n+1} - 1$; this shows that d divides n, say $n = ad$. It only remains to check that a is odd. But f divides both $x^{q^d+1} - 1$ and $x^{q^{ad}+1} - 1$, and the greatest common divisor of these two polynomials is $x^e - 1$, where $e = \gcd\left(q^{ad} + 1,\ q^d + 1\right)$. As mentioned above, $e \mid 2$ if a is even, in which case f would have to divide $x^2 - 1$, a contradiction. □

As in Section 3.5, we can now obtain an explicit formula for $SI_{n,q}$ by applying Möbius inversion. Actually, we require a minor variation of the classical Möbius inversion formula treated in Section 2.1, since only those divisors d of n for which n/d is odd appear in Equation (5.11). For this case, one has the following result.

Theorem 5.7.7. *Let f be any function from $\mathbb{N}^*$ into an additively written group G. Define a further mapping $\tilde{S}_f \colon \mathbb{N}^* \to G$ by*

$$\tilde{S}_f(n) := \sum_{d|n,\ n/d \text{ odd}} f(d).$$

Then f can be written in terms of $\tilde{S}_f$ as follows:

$$f(n) = \sum_{d|n,\ d \text{ odd}} \mu(d)\tilde{S}_f\left(\frac{n}{d}\right).$$

Proof. Using Proposition 2.1.12, the assertion follows via a simple computation:

$$\begin{aligned}\sum_{d|n,\ d \text{ odd}} \mu(d)\tilde{S}_f\left(\frac{n}{d}\right) &= \sum_{d|n,\ d \text{ odd}} \mu(d) \sum_{b|(n/d),\ n/db \text{ odd}} f(b) \\ &= \sum_{b|n,\ n/b \text{ odd}} f(b) \sum_{d|(n/b)} \mu(d) = f(n). \quad \square\end{aligned}$$

Proposition 5.7.8. *One has*

$$SI_{n,q}(x) = \prod_{d|n,\ d \text{ odd}} M_{n/d,q}(x)^{\mu(d)}.$$

Proof. We apply the multiplicative version of Theorem 5.7.7 with $G = F(x)^*$ (the multiplicative group of rational functions over $F = \mathrm{GF}(q)$) and $f(n) = SI_{n,q}(x)$. This gives the desired result, since here

$$\tilde{S}_f(n) = \sum_{d|n,\, n/d \text{ odd}} SI_{d,q} = M_{n,q},$$

by Lemma 5.7.6. □

Either by applying Möbius inversion to Equation (5.12) or by using Proposition 5.7.8, we also obtain a formula for the numbers $si_q(n)$:

$$si_q(n) = \frac{1}{2n} \sum_{d|n,\, d \text{ odd}} \mu(d)(q^{n/d} - \delta), \tag{5.13}$$

where δ is defined as in Lemma 5.7.6. This may be simplified a little by using the identity in Exercise 5.7.24, which shows that one may ignore the terms $-\delta$ in Equation (5.13), unless n is a power of 2. In this way, we obtain the following result due to Carlitz [68].

Theorem 5.7.9. *The number of self-reciprocal monic irreducible polynomials with degree* $2n$ *over* $\mathrm{GF}(q)$ *is given by*

$$si_q(n) = \begin{cases} \frac{1}{2n}(q^n - 1), & \text{if } q \text{ is odd and } n = 2^a \\ \frac{1}{2n} \sum_{d|n,\, d \text{ odd}} \mu(d) q^{n/d} & \text{otherwise.} \end{cases}$$ □

We remark that the formula in Theorem 5.7.9 admits a combinatorial interpretation in terms of "primitive self-complementary necklaces" of length n in q colors; then it makes sense even if q is not prime power, see Miller [277].

We now turn our attention to the problem of actually constructing self-reciprocal irreducible polynomials. For this, we consider the following transformation, which associates with every polynomial $f(x)$ with degree n over $\mathrm{GF}(q)$ a self-reciprocal polynomial $f^Q(x)$ with degree $2n$:

$$f^Q(x) := x^n f(x+x^{-1}). \tag{5.14}$$

We call f^Q the ***Q*-transform** of f. This transformation goes back to Carlitz [68] and was also studied by Miller [277], Andrews [8], Cohen [80] and Meyn [275]. We first show that every self-reciprocal polynomial with degree $2n$ is the Q-transform of a polynomial with degree n.

Lemma 5.7.10. *Let* $g(x)$ *be any self-reciprocal monic polynomial with degree* $2n$ *over* $F = \mathrm{GF}(q)$*. Then* g *is the* Q*-transform of a unique polynomial* $f(x)$ *with degree* n *over* F*. If* g *is irreducible, then* f *is likewise irreducible.*

Proof. We use induction on n to show

$$x^{-n} g(x) = f\big(x+x^{-1}\big) \tag{5.15}$$

for some polynomial f with degree n over F; the reader is asked to prove the uniqueness of f as Exercise 5.7.25. The case $n = 0$ is trivial. Thus let $n \geq 1$ and assume the validity of the assertion for all $k < n$. Let $g(x) = \sum_{j=0}^{2n} a_j x^j$. As g is self-reciprocal, $g(x) = x^{2n} g(x^{-1})$, and therefore $a_j = a_{2n-j}$ for $j = 0, \ldots, n-1$. It follows that $x^{-n} g(x)$ can be written in the form

$$x^{-n} g(x) = A_0 + \sum_{i=1}^{n} A_i \left(x^i + x^{-i}\right).$$

Now define a polynomial $h(x)$ by

$$h(x) := x^{-n} g(x) - A_n \left(x + x^{-1}\right)^n.$$

Clearly, we can write h in the form

$$h(x) = B_0 + \sum_{i=1}^{n-1} B_i \left(x^i + x^{-i}\right).$$

In the case $h = 0$, Equation (5.15) holds for $f = A_n x^n$. Hence we may assume $h \neq 0$. Let k be the largest index with $B_k \neq 0$. Then $x^k h(x)$ is a self-reciprocal polynomial with degree $2k < 2n$, and thus the induction hypothesis gives $h(x) = f_0(x + x^{-1})$ for some polynomial f_0 with degree k; this yields the validity of Equation (5.15) for $f = f_0 + A_n x^n$. The irreducibility assertion is obvious from the fact that the Q-transform is multiplicative. □

One may also give an explicit formula for f in terms of the coefficients of g. This can be done along the lines of Theorem 2 of Andrews [8] who uses a slightly different set-up. As the next result shows, the Q-transform does not necessarily preserve irreducibility. We use the notation $g^{\wedge}$ introduced in Theorem 5.7.1.

Lemma 5.7.11. *Let f be a monic irreducible polynomial over $F = \mathrm{GF}(q)$, where $\deg f \geq 2$. Then either f^Q is likewise irreducible, or f^Q is the product of a pair $g, g^{\wedge}$ of irreducible polynomials with degree n which are not self-reciprocal.*

Proof. Let α be any root of f^Q; then $\beta := \alpha + \alpha^{-1}$ is a root of f, by Equation (5.14). Since f is irreducible, β generates the extension field $E = \mathrm{GF}(q^n)$ of F. Hence $\beta^{q^n} = \beta$, and n is the smallest positive integer k satisfying $\beta^{q^k} = \beta$. After substituting for β, a short computation gives the equivalent condition

$$\left(\alpha^{q^n+1} - 1\right)\left(\alpha^{q^n-1} - 1\right) = 0, \tag{5.16}$$

and again n is the smallest positive integer for which a factorization as in (5.16) holds. Now either $\alpha^{q^n+1} = 1$ or $\alpha^{q^n-1} = 1$. In the first case, f^Q has to be irreducible by Lemma 5.7.6; and in the second case, f^Q has two irreducible factors g and h with degree n each. Such a factor cannot be self-reciprocal, since this would imply $\alpha^{q^{n/2}+1} = 1$, contradicting the minimality of n. As α^{-1} also is a root of f by

Equation (5.14), we conclude $g = \mathrm{mpol}_\alpha$ and $h = \mathrm{mpol}_{\alpha^{-1}}$, and therefore $h = g^\wedge$ by Theorem 5.7.1. □

We now need a criterion on f which allows us to decide which of the two cases in Lemma 5.7.11 occurs; such a criterion was given by Meyn [275]. We require the following auxiliary result which allows us to reduce our problem to testing a certain quadratic polynomial for irreducibility; this is a special case of a result due to Cohen [78].

Lemma 5.7.12. *Let f be a monic irreducible polynomial with degree $n \neq 1$ over $F = \mathrm{GF}(q)$. Then f^Q is likewise irreducible if and only if the polynomial*

$$g(x) := x^2 - \beta x + 1 \tag{5.17}$$

is irreducible over $E = \mathrm{GF}(q^n)$, where β is any root of f.

Proof. As in the proof of Lemma 5.7.11, β has the form $\beta = \alpha + \alpha^{-1}$, where α is a root of f^Q. This shows $g(x) = (x-\alpha)(x-\alpha^{-1})$. Hence $\beta \in F(\alpha)$, and α generates an extension of F with degree $2n$ if and only if g is irreducible over E. □

In view of Lemma 5.7.12, the answer to our problem clearly depends on the characteristic of F. We begin with the case of even characteristic, where the resulting criterion is particularly nice.

Theorem 5.7.13. *Let $f(x) = x^n + a_{n-1}x^{n-1} + \cdots + a_1 x + a_0$ be a monic irreducible polynomial with degree $n \geq 2$ over $F = \mathrm{GF}(q)$, where q is even. Then f^Q is likewise irreducible if and only if*

$$\mathrm{Tr}_{F/GF(2)}(a_1/a_0) = 1. \tag{5.18}$$

Proof. We apply the criterion in Lemma 5.7.12. The substitution $y = \beta^{-1}x$ shows that the quadratic polynomial g in Equation (5.17) is irreducible over $E = \mathrm{GF}(q^n)$ if and only if the polynomial $h(y) := y^2 + y + \beta^{-2}$ is irreducible. By Theorem 5.1.1, this is equivalent to the condition $\mathrm{Tr}_{E/GF(2)}(\beta^{-2}) = 1$, which simplifies to $\mathrm{Tr}_{E/GF(2)}(\beta^{-1}) = 1$. Using the transitivity formula for the trace function, we may write this condition as

$$\mathrm{Tr}_{E/GF(2)}(\beta^{-1}) = \mathrm{Tr}_{F/GF(2)}\big(\mathrm{Tr}_{E/F}(\beta^{-1})\big) = 1. \tag{5.19}$$

By Theorem 5.7.1, β^{-1} is a root of the monic irreducible polynomial $f^\wedge = a_0^{-1} f^*$, and thus $\mathrm{Tr}_{E/F}(\beta^{-1})$ is the coefficient of x^{n-1} in $f^\wedge$ by Proposition 3.12.3, that is, $\mathrm{Tr}_{E/F}(\beta^{-1})) = a_1/a_0$. Hence the condition (5.19) indeed reduces to the desired criterion (5.18). □

For $q = 2$, Theorem 5.7.13 specializes to a result already obtained by Varshamov and Garakov [385]:

Corollary 5.7.14. *Let $f(x) = x^n + a_{n-1}x^{n-1} + \cdots + a_1 x + 1$ be a monic irreducible polynomial with degree $n \geq 2$ over $\mathrm{GF}(2)$. Then f^Q is likewise irreducible if and only if $a_1 = 1$.* □

For odd q, we obtain the following less satisfactory criterion, which requires checking whether a certain element of $\mathrm{GF}(q)$ is a non-square.

Theorem 5.7.15. *Let $f(x) = x^n + a_{n-1}x^{n-1} + \cdots + a_1x + a_0$ be a monic irreducible polynomial with degree $n \geq 2$ over $F = \mathrm{GF}(q)$, where q is odd. Then f^Q is likewise irreducible if and only if $f(2) \cdot f(-2)$ is a non-square in F^*.*

Proof. Let β be a root of f. Again, we have to consider the quadratic polynomial g in Equation (5.17). Since q is odd, the standard formula for the solutions of a quadratic equation shows that g is irreducible over F if and only if $\beta^2 - 4$ is a non-square in E^*, where $E = \mathrm{GF}(q^n)$, which means that $(\beta^2-4)^{(q^n-1)/2} = -1$. Using the abbreviation N for $\mathrm{Norm}_{E/F}$, we compute

$$(\beta^2-4)^{(q^n-1)/2} = \left((\beta^2-4)^{(q^n-1)/(q-1)}\right)^{(q-1)/2} = N(\beta^2-4)^{(q-1)/2},$$

which implies that $\beta^2 - 4$ is a non-square in E^* if and only if $N(\beta^2-4)$ is a non-square in F^*.

We now use the factorization $N(\beta^2-4) = N(2-\beta) \cdot N(-2-\beta)$. By Proposition 3.12.3, $(-1)^n f(0) = N(\beta)$, since β is a root of f. Hence $2-\beta$ is a root of the (irreducible) polynomial $h(x) := f(2-x)$, and thus $N(2-\beta) = (-1)^n h(0) = (-1)^n f(2)$. Similarly, also $N(-2-\beta) = (-1)^n f(-2)$. Combining these observations gives the desired criterion. □

We conclude this section by investigating the question whether the Q-transform may be applied iteratively to produce infinite families of self-reciprocal irreducible polynomials. As the following results of Meyn [275] show, this is indeed the case provided that q is even.

Theorem 5.7.16. *Let $f(x) = x^{2n} + a_{2n-1}x^{2n-1} + \cdots + a_1x + 1$ be a self-reciprocal irreducible polynomial over $F = \mathrm{GF}(q)$, where q is even and $\mathrm{Tr}_{F/GF(2)}(a_1) = 1$. Then f^Q is likewise a self-reciprocal irreducible polynomial, and its linear coefficient b_1 satisfies $\mathrm{Tr}_{F/GF(2)}(b_1) = 1$.*

Proof. By Theorem 5.7.13, the assumption on the linear coefficient a_1 of f guarantees that the self-reciprocal polynomial f^Q is indeed irreducible. It remains to compute the absolute trace of the linear coefficient b_1 of f^Q. For this, let α be a root of f^Q. Then $\beta := \alpha + \alpha^{-1}$ is a root of f and, as we have seen in the proof of Lemma 5.7.12, also of the irreducible quadratic $g(x) = x^2 + \beta x + 1$. With $E = F(\beta)$ and $L = E(\alpha)$, Proposition 3.12.3 gives the following identities, where we also use that the linear and the second highest coefficients of a self-reciprocal polynomial agree:

$$\mathrm{Tr}_{L/E}(\alpha) = \beta, \; \mathrm{Tr}_{L/F}(\alpha) = b_1 \text{ and } \mathrm{Tr}_{E/F}(\beta) = a_1.$$

Using the transitivity formula for the trace function several times, these identities combine to give the desired result:

$$\mathrm{Tr}_F(b_1) = \mathrm{Tr}_F\left(\mathrm{Tr}_{L/F}(\alpha)\right) = \mathrm{Tr}_F\left(\mathrm{Tr}_{E/F}(\beta)\right) = \mathrm{Tr}_F(a_1) = 1,$$

where we have used the abbreviation Tr_F for the absolute trace $\mathrm{Tr}_{F/GF(2)}$. □

Corollary 5.7.17. *Let $h(x) = x^n + c_{n-1}x^{n-1} + \cdots + c_1 x + 1$ be an irreducible polynomial with degree $n \geq 2$ over $F = \mathrm{GF}(2)$, and assume $c_1 = c_{n-1} = 1$. Then an iterated application of the Q transform to h gives a series of self-reciprocal irreducible polynomials with degrees $2^m n$ for all positive integers m.*

Proof. As $c_1 = 1$, the self-reciprocal polynomial $f = h^Q$ is irreducible by Corollary 5.7.14, and the condition $c_{n-1} = 1$ implies that f has linear coefficient $a_1 = 1$. Hence the assertion follows by repeated application of Theorem 5.7.16. □

Following Meyn, we call an irreducible polynomial $h(x)$ of the form used in Corollary 5.7.17 a **type A** irreducible polynomial. We now exhibit some examples.

Example 5.7.18. For $n = 2$, we may use for h the irreducible polynomial $x^2 + x + 1$, so that $f(x) = x^4 + x^3 + x^2 + x + 1$; this example is already due to Wiedemann [404]. There is no appropriate choice for $n = 3$, since the only self-reciprocal irreducible polynomial with degree 6 is the cyclotomic polynomial $x^6 + x^3 + 1$ (which belongs to $h(x) = x^3 + x + 1$). For $n = 5$, we may take $h(x) = x^5 + x^4 + x^3 + x + 1$, which gives $f(x) = x^{10} + x^9 + x^5 + x + 1$. □

The case $n = 3$ in Example 5.7.18 is exceptional, as the following result of Niederreiter [300] shows:

Result 5.7.19 *Let $n \geq 2$ be a positive integer. If $n \neq 3$, then there exists a type A irreducible polynomial over $F = \mathrm{GF}(2)$.* □

Example 5.7.20. Even though the general proof of Result 5.7.19 is rather involved, we can give a simple indirect argument due to Meyn [275] to settle the special case $n = 2m$, where m is odd. Thus we assume that every monic irreducible polynomial with degree n over $F = \mathrm{GF}(2)$ with linear coefficient 1 is of **type B**, that is, the second highest coefficient is 0. Since there are no type A polynomials by our assumption, we conclude from Lemma 5.7.10 and Corollary 5.7.14 that the number of monic irreducible type B polynomials cannot be smaller than the number $si_2(2n)$ of self-reciprocal monic irreducible polynomials with degree $2n$ over F. Trivially, the reciprocal of a type B polynomial is not of type B, and hence the number $i_2(n)$ of monic irreducible polynomials with degree n over F has to be at least $2 \cdot si_2(2n)$ (not even taking into account irreducible polynomials for which both the linear and the second highest coefficient are 0). By Theorem 5.7.9,

$$2 \cdot si_2(2n) = \frac{1}{2m} \sum_{d|m} \mu(d) 2^{2m/d},$$

and therefore, by Theorem 3.5.4,

$$i_2(n) = \frac{1}{n} \sum_{d|n} \mu(d) 2^{n/d} = \frac{1}{2m} \sum_{d|m} \mu(d) 2^{2m/d} - \frac{1}{2m} \sum_{d|m} \mu(d) 2^{m/d} < 2 \cdot si_2(2n),$$

which is the desired contradiction. □

Meyn could not obtain an analogue of Corollary 5.7.17 in the case of an odd characteristic, since the criterion provided by Theorem 5.7.15 is difficult to apply recursively. We shall conclude this section with results due to Cohen [83] who managed to obtain such an analogue by using a minor modification of the Q-transform. Given a polynomial f with degree n, we may define a self-reciprocal polynomial f^R by

$$f^R(x) := (2x)^n f\left(\tfrac{1}{2}(x+x^{-1})\right) = 2^n f^Q\left(\tfrac{1}{2}x\right), \tag{5.20}$$

since q is assumed to be odd. We then have the following result.

Theorem 5.7.21. *Let $f(x) = x^n + c_{n-1}x^{n-1} + \cdots + c_1x + 1$ be an irreducible polynomial with degree $n \geq 2$ over $F = \mathrm{GF}(q)$, where q is odd, and assume that n is even if $q \equiv 3 \bmod 4$. Moreover, suppose also that*

$$f(1) \cdot f(-1) \textit{ is a non-square in } F^*. \tag{5.21}$$

Then repeated application of the transformation (5.20) *to f results in a sequence (f_m) of self-reciprocal irreducible polynomials with degree $2^m n$ for all $m \in \mathbb{N}^*$.*

Proof. Formally, we define the sequence (f_m) by $f_0 := f$ and $f_k := (f_{k-1})^R$. We first use induction to show that, for every positive integer k,

$$f_k(1) \cdot f_k(-1) = c_k f(1) \cdot f(-1) \quad \text{for some square } c_k \in F.$$

This is trivial for $k = 0$, where we just take $c_0 = 1$. For $k \geq 1$, Equation (5.20) gives

$$f_k(1) \cdot f_k(-1) = 2^{2n}(-1)^n f_{k-1}(1) \cdot f_{k-1}(-1) = 2^{2n}(-1)^n c_{k-1} f(1) \cdot f(-1),$$

and thus we may take $c_k = 2^{2n}(-1)^n c_{k-1}$. Note that c_k is indeed a square, as c_{k-1} is a square (by induction) and as $(-1)^n$ is also a square: by Proposition 3.2.13, -1 is a square if $q \equiv 1 \bmod 4$; and if $q \equiv 3 \bmod 4$, then n is even by hypothesis.

It follows that $f_k(1) \cdot f_k(-1)$ is a non-square for every k. Using induction, it now suffices to prove that f_1 is irreducible. To this end, put $g(x) := 2^n f(x/2)$. Then, in view of (5.20),

$$f_1(x) = f^R(x) = 2^n f^Q(x/2) = g^Q(x).$$

Now Theorem 5.7.15 shows the irreducibility of f_1, as

$$g(2) \cdot g(-2) = 2^{2n} f(1) \cdot f(-1)$$

is a non-square in F^*. □

We finally settle the existence problem for polynomials with the properties assumed in Theorem 5.7.21; this turns out to be simpler than the corresponding problem for even values of q.

Proposition 5.7.22. *Let q be an odd prime power and let $n \in \mathbb{N}^*$, where n is even if $q \equiv 3 \bmod 4$. Then there exists a monic irreducible polynomial f with degree n over $F = \mathrm{GF}(q)$ satisfying condition* (5.21).

Proof. By Theorem 5.7.9, the total number of self-reciprocal irreducible polynomials with degree $2n$ over F is known; proceeding in analogy to the proof of Theorem 3.5.4, one easily checks that this number is positive. Hence there always exists some such polynomial h. By Lemma 5.7.10 and Theorem 5.7.15, $h = g^Q$ for some irreducible polynomial g with degree n for which $g(2)g(-2)$ is not a square in F. By (5.20), we may write h as $h = f^R$, where we put $f(x) := 2^{-n}g(2x)$. But then $f(1)f(-1) = 2^{-2n}g(2)g(-2)$ is a non-square. □

Example 5.7.23. We obtain a series of irreducible polynomials with degree 2^m over GF(3) by using $f(x) = x^2 + x - 1$. Similarly, we may use $f(x) = x^2 + x + 2$ over GF(5). In general, for $q \equiv 1 \bmod 4$ we can start the recursive construction with $f = x^2 + 2cx + 1$, where c is an arbitrary element of GF(q) for which $c^2 - 1$ is a non-square. □

Series of irreducible polynomials as constructed in Corollary 5.7.17 and Theorem 5.7.21 are useful for iterated presentations of infinite extensions of finite fields; the interested reader should consult Brawley and Schnibben [48].

Exercises

Exercise 5.7.24. Use Proposition 2.1.12 to prove the following identity:

$$\sum_{d|n,\, d \text{ odd}} \mu(d) = \begin{cases} 1 & \text{if } n \text{ is a power of 2,} \\ 0 & \text{otherwise.} \end{cases} \tag{5.22}$$

Exercise 5.7.25. Prove the uniqueness of the polynomial f satisfying Equation (5.15). □

Chapter 6
Factorization of Univariate Polynomials over Finite Fields

Abstract The topic of the present chapter is the factorization of univariate polynomials over finite fields, with emphasis on the methods of Berlekamp and of Niederreiter. Nevertheless, some of the ideas work for perfect fields in general, and therefore we will restrict attention to finite fields only where necessary.

The first two sections are devoted to two particular partial factorizations of a polynomial, the square-free factorization and the distinct degree factorization. After that we develop the method of Berlekamp in Sections 6.3-6.5. In Sections 6.6, 6.8 and 6.9 we discuss the method of Niederreiter, and in Section 6.7 connections between the Berlekamp algebra and the Niederreiter space associated to a polynomial are studied.

6.1 The Square-Free Factorization of a Polynomial

Throughout this chapter, we consider the polynomial algebra $F[x]$ over a field F. Let $f(x)$ be an arbitrary non-constant monic polynomial over F. By the results of Section 2.3, there exist a unique $s \in \mathbb{N}^*$, unique distinct monic polynomials $u_1(x), \ldots, u_s(x)$ which are irreducible over F, and unique positive integers $\ell_1, \ldots, \ell_s$ (called the **multiplicities**) such that

$$f(x) = \prod_{i=1}^{s} u_i(x)^{\ell_i}. \tag{6.1}$$

This decomposition of f is referred to as the **canonical** or the **unique factorization** of f over F. We are going to discuss algorithms which determine this decomposition for a given monic polynomial f. As usual, we assume that f is given in terms of its coefficients: $f(x) = \sum_{i=0}^{n} f_i x^i$, where we always denote the degree of f as n and where $f_n = 1$. Our emphasis will be on finite coefficient fields F; nevertheless, we will work over arbitrary fields as far as possible.

D. Hachenberger and D. Jungnickel, *Topics in Galois Fields*,
Algorithms and Computation in Mathematics 29,
https://doi.org/10.1007/978-3-030-60806-4_6

Let us start with some general remarks, which are valid for any field F. A non-constant polynomial f is called **square-free** if $\ell_i = 1$ for each $i = 1,\ldots,s$ in its canonical factorization (6.1).[1] In general, the polynomial

$$\operatorname{rad}(f(x)) := \prod_{i=1}^{s} u_i(x) \tag{6.2}$$

already considered in Section 3.11 will be called the **radical** of f, and thus f is square-free if and only if $\operatorname{rad}(f) = f$.

As in Section 2.4, we let f' denote the formal derivative of f. Applying the rules in Observation 2.4.16 and Exercise 2.4.21 to the canonical factorization (6.1) of f gives

$$\begin{aligned} f'(x) &= \sum_{i=1}^{s} \Big(\big(u_i(x)^{\ell_i}\big)' \cdot \prod_{\substack{k=1\\ k\neq i}}^{s} u_k(x)^{\ell_k} \Big) \\ &= \sum_{i=1}^{s} \Big(\ell_i u_i(x)^{\ell_i - 1} \cdot u_i'(x) \cdot \prod_{\substack{k=1\\ k\neq i}}^{s} u_k(x)^{\ell_k} \Big) \\ &= \Big(\prod_{i=1}^{s} u_i(x)^{\ell_i - 1} \Big) \cdot \Big(\sum_{i=1}^{s} \ell_i u_i'(x) \cdot \prod_{\substack{k=1\\ k\neq i}}^{s} u_k(x) \Big) \\ &= \frac{f(x)}{\operatorname{rad}(f(x))} \cdot h(x), \end{aligned}$$

where the polynomial h is given by

$$h(x) = \sum_{j=1}^{s} \ell_j u_j'(x) \cdot \frac{\operatorname{rad}(f(x))}{u_j(x)}.$$

Thus $f/\operatorname{rad}(f)$ divides f' and is therefore a common divisor of f and f'. In order to determine the greatest common divisor of f and f', we still have to decide when a given irreducible factor u_i of f divides h. Since u_i divides $\operatorname{rad}(f)/u_j$ for every $j \neq i$, but not for $j = i$, this happens if and only if u_i divides $\ell_i u_i'(x)$. Since the degree of $u_i'(x)$ is strictly less than that of u_i, we conclude that u_i divides h if and only if $\ell_i u_i'(x)$ is the zero polynomial. This establishes the following preliminary result:

Proposition 6.1.1. *Let $f(x) \in F[x]$ be a non-constant monic polynomial with canonical factorization* (6.1)*. Let I denote the set of all indices i such that $\ell_i u_i'(x) \neq 0$, and let $I^c := \{1,\ldots,s\} \setminus I$ be its complement. Then*

$$\gcd(f(x), f'(x)) = \prod_{i\in I} u_i(x)^{\ell_i - 1} \cdot \prod_{i \in I^c} u_i(x)^{\ell_i}. \quad \square$$

[1] For convenience, we shall also consider constant polynomials to be square-free.

In view of Proposition 6.1.1, we need to consider the condition $\ell g'(x) = 0$, where $\ell \in \mathbb{N}^*$ and where g is a non-constant polynomial over F. Essentially, this has already been done in the discussion following Definition 3.2.17, but we will recapitulate the argument here. Obviously, $\ell g'(x) = 0$ can happen for two reasons:

Case 1: $\ell = 0$ in F.
Then F has to have positive characteristic, say p, and ℓ is a multiple of p.

Case 2: g' is the zero polynomial.
Let $g(x) = \sum_{j=0}^{m} g_j x^j$, where $m \geq 1$ and $g_m \neq 0$. Then $g'(x) = \sum_{j=1}^{m} j g_j x^{j-1} = 0$ if and only if $j g_j = 0$ for $j = 1, \ldots, m$. Because of $m \geq 1$ and $g_m \neq 0$, this again implies that F has positive characteristic, say p. (In other words, for fields with characteristic zero, the only polynomials with derivative 0 are the constant ones.) Moreover, p has to divide j for each j with $g_j \neq 0$. Therefore, g has the form $\overline{g}(x^p)$ for a uniquely determined polynomial $\overline{g}(x) \in F[x]$, say $\overline{g}(x) = \sum_{i=0}^{k} \gamma_i x^i$. Thus $g' = 0$ if and only if g has the form $g(x) = \sum_{i=0}^{k} \gamma_i x^{pi}$.

Now assume that F is a perfect field with characteristic p; see Definition 3.2.17. Then Theorem 3.2.21 shows that there exist unique elements $\delta_i \in F$ satisfying $\delta_i^p = \gamma_i$ (for every i), and therefore

$$g(x) = \left(\sum_{i=0}^{k} \delta_i x^i \right)^p,$$

and hence g is reducible.

We now apply the preceding observations to the situation in Proposition 6.1.1 to determine the set I in the special case of perfect fields:

Corollary 6.1.2. *With the setup in Proposition 6.1.1, assume that F is a perfect field. Then one has the following:*

- *If F has characteristic 0, then $\ell_i u_i'(x) \neq 0$ for all i. Thus $I = \{1, \ldots, s\}$ in this case, and therefore*

$$\gcd(f(x), f'(x)) = \prod_{i=1}^{s} u_i(x)^{\ell_i - 1}.$$

- *If F has positive characteristic p (for instance, F may be a finite field, by Proposition 3.2.18), then $u' \neq 0$ for every irreducible polynomial $u(x) \in F[x]$. Therefore, $i \in I$ if and only if ℓ_i is not divisible by p in this case, that is,*

$$\gcd(f(x), f'(x)) = \prod_{\substack{i=1 \\ p \nmid \ell_i}}^{s} u_i(x)^{\ell_i - 1} \cdot \prod_{\substack{i=1 \\ p \mid \ell_i}}^{s} u_i(x)^{\ell_i}.$$

In particular, f is square-free if and only if $\gcd(f, f') = 1$. □

Thus the only fields for which we cannot determine the set I in Proposition 6.1.1 in general are the non-perfect fields. Therefore, we assume from now on that F is actually perfect. The main aim of this section is the determination of a particular *partial* factorization of a polynomial f with canonical factorization as in (6.1) in this situation. Obviously, f has a unique representation in the form

$$f(x) = \prod_{j=1}^{m} h_j(x)^j, \tag{6.3}$$

where $h_1(x), \ldots, h_m(x)$ are square-free monic polynomials over F which are pairwise relatively prime and where $h_m(x) \neq 1$. In other words, the polynomial h_j is the product of all those monic irreducible divisors u_i of f whose multiplicity is precisely j (for every j). Note that some of the h_j with $j < m$ may be trivial (that is, $h_j(x) = 1$) and that $m = \max\{\ell_1, \ldots, \ell_s\}$. The decomposition of f in Equation (6.3) is called the **square-free factorization** of f. In particular, determining the square-free factorization of f yields its radical: $\operatorname{rad}(f) = \prod_{j=1}^{m} h_j(x)$.

As F is now assumed to be perfect, we have $u'(x) \neq 0$ for every monic irreducible polynomial u; this fact is indispensable in what follows. As we will see, the square-free factorization of $f(x) \in F[x]$ can then be determined explicitly by using only elementary tools: computing the derivative, performing polynomial divisions, calculating greatest common divisors, and inverting the Frobenius automorphism $w \mapsto w^p$ (when F has positive characteristic p). Similar to the book of Geddes, Czapor and Labahn [135], we will present two algorithms:

- Algorithm 6.1.4 yields the square-free factorization of an arbitrary polynomial $f(x) \in F[x]$ when F has characteristic 0, and of a certain proper subclass of polynomials when F has positive characteristic; see Definition 6.1.3 below.
- Algorithm 6.1.7 yields the square-free factorization of an arbitrary polynomial $f(x) \in F[x]$ also when F has positive characteristic. In fact, it essentially consists of a repeated application of a minor modification of the first algorithm, taking into account that certain non-constant polynomials have derivative 0 in this case (which makes it a little more technical).

Definition 6.1.3. Let F be a perfect field. Then a monic polynomial $g(x) \in F[x]$ is said to have **type 0** provided that either $g(x) = 1$ or that $\deg g \geq 1$ and the characteristic of F does not divide any of the multiplicities of the irreducible divisors of g. (Of course, this always holds when F has characteristic zero.) □

Algorithm 6.1.4 (Square-free factorization for type 0 polynomials).

- *Input:* A perfect field F and a monic non-constant polynomial $f(x)$ of type 0 over F.
- *Output:* The square-free factorization of $f(x)$ in a list of the form

$$L = [(h_{j_1}(x), j_1), \ldots, (h_{j_t}(x), j_t)],$$

where $h_{j_1}, \ldots, h_{j_t}$ are the non-constant polynomials h_j appearing in the factorization in Equation (6.3) and where $j_1, \ldots, j_t$ denote their respective multiplicities.

- *Remark:* The algorithm uses (for theoretical purposes only) an additional polynomial $L(x)$, which may be viewed as a global variable measuring the progress made in constructing the list L. To be precise, $L(x)$ is the product of all polynomials in the current list L (with multiplicities). Thus $L(x) = 1$ at the beginning and $L(x) = f(x)$ at the end of the algorithm.

(1) $L \leftarrow [\ ]$, $L(x) \leftarrow 1$, $i \leftarrow 1$, $a(x) \leftarrow f(x)$, $b(x) \leftarrow f'(x)$;
(2) $d(x) \leftarrow \gcd(a(x), b(x))$, $w(x) \leftarrow a(x)$ `div` $d(x)$;
(3) **while** $w(x) \neq 1$ **do**
(4) $\quad y(x) \leftarrow \gcd(w(x), d(x))$, $z(x) \leftarrow w(x)$ `div` $y(x)$;
(5) $\quad$ **if** $z(x) \neq 1$ **then** append the pair $(z(x), i)$ to L; $L(x) \leftarrow L(x) \cdot z(x)^i$ **fi**;
(6) $\quad i \leftarrow i + 1$, $w(x) \leftarrow y(x)$, $d(x) \leftarrow d(x)$ `div` $y(x)$
(7) **od**. □

Theorem 6.1.5. *Let F be a perfect field, and let $f(x)$ be a non-constant monic polynomial of type 0 over F. Then Algorithm 6.1.4 correctly determines the square-free factorization of f.*

Proof. The argument rests on showing that the condition

$$(*) \quad f(x) = L(x) \cdot w(x)^i \cdot d(x)$$

is a loop invariant, that is, it holds throughout the entire course of the algorithm. This is obvious after the initialization in Steps (1) and (2), and we will use induction on i to show that (*) is indeed preserved by each execution of the **while** loop. For this, we let $f(x) = \prod_{j=1}^{m} h_j(x)^j$ denote the (as yet unknown) square-free factorization of f, as in Equation (6.3). Since f has type 0, Corollary 6.1.2 gives

$$\gcd(f(x), f'(x)) = \prod_{i=1}^{s} u_i(x)^{\ell_i - 1} = \frac{f(x)}{\operatorname{rad}(f(x))} = \prod_{j=2}^{m} h_j(x)^{j-1}.$$

Consequently, after the initialization, we have $i = 1$ and

$$a(x) = f(x) = \prod_{j=i}^{m} h_j(x)^{j-i+1}$$

as well as

$$d(x) = \gcd(f(x), f'(x)) = \prod_{j=i+1}^{m} h_j(x)^{j-i}.$$

Since $d(x)$ divides $a(x)$, we also have

$$w(x) = a(x) \texttt{ div } d(x) = \prod_{j=i}^{m} h_j(x).$$

For the induction step, we assume that the control variable has reached some value $i \geq 1$. Suppose that (*) holds before the **while** loop is entered with this value i and that

$$w(x) = \prod_{j=i}^{m} h_j(x) \quad \text{and} \quad d(x) = \prod_{j=i+1}^{m} h_j(x)^{j-i}$$

at this point. (As we have just seen, this is the case for $i = 1$.) Note that $w(x) \neq 1$ as long as $i \leq m$, since $w(x)$ is then divisible by $h_m(x) \neq 1$, so that the **while** loop is actually performed.

Assume first that $i < m$. We claim that then the corresponding statements hold after the operations in the **while** loop have been executed. Using the induction hypothesis, we compute

$$y(x) = \gcd(w(x), d(x)) = \prod_{j=i+1}^{m} h_j(x)$$

and

$$z(x) = w(x) \ \mathtt{div} \ y(x) = h_i(x).$$

This shows that the correct term $h_i(x)$ with multiplicity i is appended at the end of the list L, provided that $z(x) \neq 1$. Moreover,

$$\begin{aligned} f(x) &= L(x) \cdot w(x)^i \cdot d(x) \\ &= L(x) \cdot (z(x) \cdot y(x))^i \cdot \left(y(x) \cdot \frac{d(x)}{y(x)} \right) \\ &= \left(L(x) \cdot z(x)^i \right) \cdot y(x)^{i+1} \cdot \frac{d(x)}{y(x)}. \end{aligned}$$

Altogether, we see that the induction assertion (with $i+1$ instead of i) is indeed satisfied, because of the updates $i \leftarrow i+1$, $w(x) \leftarrow y(x)$, $d(x) \leftarrow d(x) \ \mathtt{div} \ y(x)$ and $L(x) \leftarrow L(x) \cdot z(x)^i$ in the **while** loop.

In the case where $i = m$, the **while** loop is entered with $w(x) = h_m(x)$ and with $d(x) = 1$. Consequently, $y(x)$ is set to 1 and $z(x)$ attains the value $w(x) = h_m(x)$, so that $(h_m(x), m))$ is appended to the list L. Then i is increased to $m+1$ and $w(x)$ is updated to $y(x) = 1$, while $d(x)$ retains its value 1. As before, the formula $f(x) = L(x) \cdot w(x)^i \cdot d(x)$ remains valid, and now indeed $L(x) = f(x)$. Because of $w(x) = 1$, the algorithm terminates (with the correct output). □

Example 6.1.6. Assume that F is a perfect field with characteristic distinct from 2 and 3, and let $f(x) \in F[x]$ be a monic polynomial with canonical factorization of the form

$$u_1(x)^3 \cdot u_2(x) \cdot u_3(x)^8.$$

Note that

$$\begin{aligned} f'(x) = 3 \cdot u_1(x)^2 u_1'(x) \cdot u_2(x) \cdot u_3(x)^8 \\ + u_1(x)^3 \cdot u_2'(x) \cdot u_3(x)^8 \\ + 8 \cdot u_1(x)^3 \cdot u_2(x) \cdot u_3(x)^7 u_3'(x). \end{aligned}$$

Thus the **while** loop in Algorithm 6.1.4 is entered with

$$d(x) = \gcd(f(x), f'(x)) = u_1(x)^2 \cdot u_3(x)^7,$$

which also follows from Corollary 6.1.2, and with

$$w(x) = f(x) \texttt{ div } d(x) = u_1(x) \cdot u_2(x) \cdot u_3(x).$$

In order to distinguish the current values of the variables during the various executions of the **while** loop, we mark them with the index i of the control variable. So, with d and w as above, we initially have $d_1 = d$ and $w_1 = w$ (for simplicity, we omit the indeterminate x from now on). The first execution of the **while** loop gives

$$y_1 = u_1 u_3, \quad z_1 = u_2, \quad w_2 = u_1 u_3, \quad d_2 = u_1 u_3^6,$$

and the initially empty list L becomes $L = [(u_2, 1)]$. The second execution of the **while** loop gives

$$y_2 = u_1 u_3, \quad z_2 = 1, \quad w_3 = u_1 u_3, \quad d_3 = u_3^5,$$

and L remains unchanged. Continuing in this manner, we obtain the following data, where we omit stating L if the list does not change:

- $y_3 = u_3$, $z_3 = u_1$, $w_4 = u_3$, $d_4 = u_3^4$ and $L = [(u_2, 1), (u_1, 3)]$;
- $y_4 = u_3$, $z_4 = 1$, $w_5 = u_3$, $d_5 = u_3^3$;
- $y_5 = u_3$, $z_5 = 1$, $w_6 = u_3$, $d_6 = u_3^2$;
- $y_6 = u_3$, $z_6 = 1$, $w_7 = u_3$, $d_7 = u_3$;
- $y_7 = u_3$, $z_7 = 1$, $w_8 = u_3$, $d_8 = 1$;
- $y_8 = 1$, $z_8 = u_3$, $w_9 = 1$, $d_9 = 1$ and $L = [(u_2, 1), (u_1, 3), (u_3, 8)]$.

Since the variable w becomes 1 for the first time when $i = 9$, the algorithm stops at this point. As expected, we obtain

$$f(x) = h_1(x) \cdot h_3(x)^3 \cdot h_8(x)^8 \quad \text{with} \quad h_1 = u_1, \; h_3 = u_3, \; h_8 = u_8.$$

In this particular case, we have even obtained the canonical factorization of f (without actually knowing this in reality, since f is then given in terms of its coefficients). Note that this phenomenon always occurs when f is a polynomial of type 0 with distinct multiplicities.

In order to get an impression how Algorithm 6.1.4 performs in practice when f is given via its coefficients, the reader is invited to work out an example over GF(5); see Exercise 6.1.12. □

Our next goal is the square-free factorization of an arbitrary polynomial f over a perfect field F with positive characteristic p (in particular, a finite field). Again, let $f(x) = \prod_{j=1}^{m} h_j(x)^j$ be the (as yet unknown) square-free factorization of f. In view of Corollary 6.1.2, we split the index set as follows:

$$J := \{j \in \mathbb{N} \colon 1 \leq j \leq m, \; p \nmid j\} \quad \text{and} \quad J' := \{1, \ldots, m\} \setminus J. \tag{6.4}$$

Now let

$$f_0(x) := \prod_{j\in J} h_j(x)^j \quad \text{and} \quad f_1(x) := \prod_{j\in J'} h_j(x)^{j/p}.$$

Then $f(x) = f_0(x)\cdot f_1(x)^p = f_0(x)\cdot g(x^p)$ for the unique polynomial $g \in F[x]$ with $f_1(x)^p = g(x^p)$. Thus f_0 comprises all those factors $u_i(x)^{\ell_i}$ in Equation (6.1) with $p \nmid \ell_i$ (so that f_0 is of type 0), while $g(x^p)$ collects the remaining terms. Algorithmically, we first need to split f into these two parts $f_0(x)$ and $g(x^p)$, and then to determine $f_1(x)$ (essentially by inverting the Frobenius automorphism on F). Clearly, the square-free factorization of $f(x)$ is the product of the square-free factorizations of $f_0(x)$ and $f_1(x)^p$, and the latter is obtained from that of f_1 by multiplying each multiplicity by p.

Altogether, these observations lead to the following algorithm, which basically consists of repeated calls of a certain procedure SFF. This auxiliary procedure arises from a slight modification of Algorithm 6.1.4; it performs the splitting of f discussed above and determines the square-free factorization of the part f_0 of type 0.

Algorithm 6.1.7 (Square-free factorization over perfect fields with positive characteristic).

- *Input:* A non-constant monic polynomial f over a perfect field F with positive characteristic p (for instance, a finite field), a list L, and a parameter r.
- *Output:* The square-free factorization of f as a list similar to the one produced by Algorithm 6.1.4, but not necessarily ordered with respect to increasing multiplicities.
- *Remark:* As in Algorithm 6.1.4, an additional polynomial $L(x)$ is used to monitor the overall progress made in constructing the list L. The subsequent procedure is called in the main body of the algorithm (in general, repeatedly) in the form SFF$(c(x), L, r)$, where initially $L = [\ ]$, $c(x) = f(x)$ and $r = 0$.

(1) $L \leftarrow [\ ],\ L(x) \leftarrow 1,\ r \leftarrow 0$;
(2) SFF$(f(x), L, r)$;
(3) **while** $d(x) \neq 1$ **do**
(4) determine the polynomial $\delta(x) \in F[x]$ such that $\delta(x)^p = d(x)$;
(5) $r \leftarrow r+1$; SFF$(\delta(x), L, r)$
(6) **od**.

Procedure SFF$(c(x), L, r)$

(1) $i \leftarrow 1,\ a(x) \leftarrow c(x),\ b(x) \leftarrow c'(x)$;
(2) $d(x) \leftarrow \gcd(a(x), b(x)),\ w(x) \leftarrow a(x)$ div $d(x)$;
(3) **while** $w(x) \neq 1$ **do**
(4) $y(x) \leftarrow \gcd(w(x), d(x)),\ z(x) \leftarrow w(x)$ div $y(x)$;
(5) **if** $z(x) \neq 1$ and i mod $p \neq 0$ **then**
append the pair $(z(x), ip^r)$ to L; $L(x) \leftarrow L(x)\cdot z(x)^{ip^r}$ **fi**;
(6) $i \leftarrow i+1,\ w(x) \leftarrow y(x),\ d(x) \leftarrow d(x)$ div $y(x)$

(7) **od**. □

Theorem 6.1.8. *Let F be a perfect field with positive characteristic p, and let $f(x)$ be any non-constant monic polynomial over F. Then Algorithm 6.1.7 correctly determines the square-free factorization of f.*

Proof. We begin by analyzing the initial call of the procedure SFF (with $L = [\ ]$, $c(x) = f(x)$ and $r = 0$) and show that it accomplishes the following three tasks:

- it splits f as $f(x) = f_0(x) \cdot g(x^p)$, where f_0 has type 0;
- it determines the square-free factorization of the part f_0 (storing it in the list L);
- it terminates with $d(x) = g(x^p)$.

Note that SFF arises from Algorithm 6.1.4 by just modifying Step (5) slightly, and therefore the proof of these three properties will be quite similar to that of Theorem 6.1.5. In view of the entire course of the algorithm with its repeated calls of SFF, we will now write the loop invariant in the form

$$(**) \quad f(x) = L(x) \cdot w(x)^{ip^r} \cdot d(x)^{p^r},$$

which reduces to the previous form (*) for the initial call of SFF, since then $r = 0$. Let us write (as yet theoretically) f in the form $f(x) = f_0(x) \cdot g(x^p)$, as in the discussion preceding Algorithm 6.1.7. This gives $f'(x) = f_0'(x) \cdot g(x^p)$, and hence

$$\gcd(f(x), f'(x)) = \gcd(f_0(x), f_0'(x)) \cdot g(x^p).$$

As in the proof of Theorem 6.1.5, $\gcd(f_0(x), f_0'(x)) = f_0(x)/\mathrm{rad}(f_0(x))$ (since f_0 has type 0), and therefore now

$$\gcd(f(x), f'(x)) = \frac{f_0(x)}{\mathrm{rad}(f_0(x))} \cdot g(x^p) = \prod_{j \in J} h_j(x)^{j-1} \cdot \prod_{j \in J'} h_j(x)^j,$$

where J and J' are as in Equation (6.4). Consequently,

$$\frac{f(x)}{\gcd(f(x), f'(x))} = \mathrm{rad}(f_0(x)) = \prod_{j \in J} h_j(x).$$

In what follows, we will use the notation $m' := \max(J)$, with the convention $m' = 0$ if J is empty.

Because of $i = 1$ and $r = 0$, the loop invariant (**) holds after the initialization in Steps (1) and (2) of SFF, and the current data can be written as

$$d(x) = \Big(\prod_{\substack{j \in J \\ j \geq i+1}} h_j(x)^{j-i}\Big) \cdot g(x^p), \quad w(x) = \prod_{\substack{j \in J \\ j \geq i}} h_j(x) \quad \text{and} \quad L(x) = \prod_{\substack{j \in J \\ j < i}} h_j(x).$$

First assume $w(x) = 1$. Then $d(x) = f(x) = g(x^p)$ and the **while** loop is not entered at all. In this case, the initial call of SFF$(f(x), L, r)$ terminates directly, and the desired properties hold (with L still being empty).

Now let $w(x) \neq 1$, so that $m' \geq 1$. Then the **while** loop is entered as long as $i \leq m'$, as in the proof of Theorem 6.1.5. As there, we assume inductively that the loop invariant (**) holds and that d, w and L have the form given above for the current value i of the control variable when the loop is entered. We first consider the case $i < m'$. The operations in the loop yield

$$y(x) = \gcd(w(x), d(x)) = \prod_{\substack{j \in J \\ j \geq i+1}} h_j(x)$$

and $z(x) = w(x) \texttt{ div } y(x)$. Note that the value of z depends on whether or not p divides i. If i is not divisible by p, we obtain $z(x) = h_i(x)$, and the pair $(z(x), ip^r)$ is appended to L, provided that $z(x) \neq 1$. On the other hand, if i is divisible by p, then $z(x) = 1$, since $i \in J'$ in this case. In either case, we can rewrite the loop invariant as follows:

$$\begin{aligned} f(x) &= L(x) \cdot w(x)^{ip^r} \cdot d(x)^{p^r} \\ &= L(x) \cdot \big(z(x) \cdot y(x)\big)^{ip^r} \cdot \left(y(x) \cdot \frac{d(x)}{y(x)}\right)^{p^r} \\ &= \left(L(x) \cdot z(x)^{ip^r}\right) \cdot y(x)^{(i+1)p^r} \cdot \left(\frac{d(x)}{y(x)}\right)^{p^r}. \end{aligned}$$

It is now clear that the update rules for i, w, d and L are correct, as in the proof of Theorem 6.1.5.

Finally, assume that we enter the **while** loop with $i = m'$. Then we obtain $d(x) = g(x^p)$, and $y(x) = 1$, since $w(x)$ divides $f_0(x)$ and is therefore relatively prime to $d(x)$. Moreover, $z(x) = h_{m'}(x)$ is appended to L with the correct multiplicity $m'p^r$. Now $w(x)$ is updated to 1, and the initial call of SFF indeed terminates with L storing the square-free factorization of f_0 and with $d(x) = g(x^p)$.

Now it is not difficult to see that the overall algorithm is correct, since the **while** loop there is performed as long as $d(x) \neq 1$. Let us consider the situation after the initial call of SFF, assuming that $d(x) = g(x^p) \neq 1$. Then we extract the p-th root $\delta(x) = f_1(x)$ of $d(x) = g(x^p)$ in Step (4), and the subsequent call of SFF, namely SFF$(\delta(x), L, r)$ in Step (5), processes $\delta(x)$ in the same manner as $f(x)$ before.

In particular, we now deal with the square-free factorization of $\delta(x) = f_1(x)$, which gives that of $d(x) = f_1(x)^p$ by multiplying each multiplicity by p. As r was increased from 0 to 1 before this call of SFF, the relevant square-free factors of f_1 (that is, those with a multiplicity not divisible by p) are appended to the list with their correct multiplicities as factors of f. After this call of SFF has been processed, $d(x)$ contains that part of f_1 which is comprised by the factors which still have multiplicity divisible by p (so as factors of f, these factors have multiplicities divisible by p^2). If still $d(x) \neq 1$, the algorithm proceeds in the same manner by repeatedly extracting p-th roots und splitting the current $d(x)$ via SFF. Note that r is increased by 1 for each further call of SFF, which correctly multiplies the multiplicities by a further factor p each time. Thus the entire algorithm will terminate with the correct square-free factorization of f when finally $d(x) = 1$ has been reached. More formally, one may use induction on r and the loop invariant (**) to establish the correctness of Algorithm 6.1.7. □

Example 6.1.9. Let us illustrate Algorithm 6.1.7 with an example similar to the one discussed in Example 6.1.6: we again consider a (monic) polynomial $f(x)$ with a canonical factorization of the type

$$u_1(x)^3 \cdot u_2(x) \cdot u_3(x)^8.$$

However, we now assume that F is a perfect field with characteristic 2, so that f is no longer of type 0. Throughout, we use the same notational conventions as in Example 6.1.6.

Algorithm 6.1.7 is started with the call SFF$(f(x),[\ ],0)$. After the initialization, we have

$$d_1 = u_1^2 u_3^8 \quad \text{and} \quad w_1 = u_1 u_2.$$

The first execution of the **while** loop gives

$$y_1 = u_1,\ \ z_1 = u_2,\ \ d_2 = u_1 u_3^8,\ \ w_2 = u_1,$$

and the list L becomes $[(u_2,1)]$. The second execution of the **while** loop gives

$$y_2 = u_1,\ \ z_2 = 1,\ \ d_3 = u_3^8,\ \ w_3 = u_1,$$

but the list L remains $[(u_2,1)]$, since $i = 2$ is the characteristic of F. The third execution of the **while** loop gives

$$y_3 = 1,\ \ z_3 = u_1,\ \ d_4 = u_3^8,\ \ w_4 = 1,$$

and the list L becomes $[(u_2,1),(u_1,3)]$. Now, w attains the value 1 for the first time, so that the **while** loop terminates.

As the current $d(x)$-value is distinct from 1, the polynomial $\delta(x)$ with $\delta(x)^2 = d(x)$ is determined. Here $\delta(x) = u_3(x)^4$, since $d(x) = u_3(x)^8$. The procedure SFF is now called with

$$\mathrm{SFF}(u_3(x)^4,[(u_2,1),(u_1,3)],1).$$

The initialization gives $a = c = u_3^4$ and $b = c' = 0$, hence $d = \gcd(a,b) = u_3^4$ and $w = a \ \mathtt{div}\ d = 1$. Since $w = 1$, the **while** loop is not entered and this second call of SFF terminates. Again, the polynomial $\delta(x)$ with $\delta(x)^2 = d(x)$ is determined, since $d(x) \neq 1$. Now $\delta(x) = u_3(x)^2$, so that there is a third call of SFF:

$$\mathrm{SFF}(u_3(x)^2,[(u_2,1),(u_1,3)],2).$$

This time the initialization gives $a = c = u_3^2$ and $b = c' = 0$, hence $d = \gcd(a,b) = u_3^2$ and $w = a \ \mathtt{div}\ d = 1$. Again, the **while** loop is not executed, and the third call of SFF terminates. Now the algorithm determines $\delta(x) = u_3(x)$, since the current value of $d(x)$ is $u_3(x)^2$. The procedure SFF is called a fourth time, now as follows:

$$\mathrm{SFF}(u_3(x),[(u_2,1),(u_1,3)],3).$$

We obtain $a=c=u_3$ and $b=c'=u_3'$, hence $d=\gcd(a,b)=1$ and $w=a\ \mathtt{div}\ d=u_3$. Therefore, the **while** loop of this SFF is entered, and we get

$$y(x)=1 \quad \text{and} \quad z(x)=u_3(x).$$

The current value of i is 1, while r is equal to 3, so that $ip^r=8$, and the pair $(u_3,8)$ is appended to the current list L. After that, w and d both become equal to 1, so that the entire algorithm terminates with the correct output

$$L=[(u_2,1),(u_1,3),(u_3,8)].$$

For a concrete example, see Exercise 6.1.12. □

Exercises

Exercise 6.1.10. Let F be a perfect field and suppose that the multiplicities of the irreducible factors of f (over F) are pairwise distinct. Work out a strategy which determines the canonical factorization of f from its square-free factorization. □

Exercise 6.1.11. Let $g(x)$ be a monic irreducible polynomial over a perfect field F with characteristic $p=3$, and let $f(x)=g(x)^\ell$. Apply Algorithm 6.1.7 for the two cases $\ell=5$ and $\ell=6$. □

Exercise 6.1.12. Determine the square-free factorization of the polynomial

$$f(x)=x^{12}-2x^{11}+2x^{10}+x^9-2x^8-2x^7+x^6-x^4+2x^3+x^2+x-2$$

over GF(5) (using the information that f is a type 0 polynomial). □

Exercise 6.1.13. Determine the square-free factorization of the polynomial

$$x^{12}+x^{11}+\zeta x^{10}+(1+\zeta)x^9+(1+\zeta)x^8+x^4+x^3+\zeta x^2+(1+\zeta)x+1+\zeta$$

over $F=GF(4)$, where ζ is a primitive third root of unity in F, that is, ζ satisfies the relation $\zeta^2+\zeta+1=0$. □

6.2 The Distinct Degree Factorization

In this short section, we consider another kind of partial factorization for the special case of finite fields, which applies to square-free polynomials only. This rests on basic results from Section 3.5, in particular Corollary 3.5.3.

Thus let $f(x)$ be a monic square-free polynomial with degree $n\geq 1$ over $F=$ GF(q), and denote the product of all distinct monic irreducible divisors of f with degree ℓ by $f_\ell(x)$, for $\ell = 1, \ldots, n$. (By convention, $f_\ell(x) := 1$ if f has no irreducible divisor of degree ℓ over F.) Since f is assumed to be square-free, we have

$$f(x) = \prod_{\ell=1}^{n} f_\ell(x). \tag{6.5}$$

The decomposition (6.5) is called the **distinct degree factorization** of f.

Example 6.2.1. Let us consider the (square-free) polynomial

$$f(x) = x^{10} + x^9 + x^8 + x^3 + x^2 + x$$

over $F = \mathrm{GF}(2)$. Then

$$\begin{aligned} f_1(x) &= x^2 + x, \\ f_2(x) &= x^2 + x + 1, \\ f_3(x) &= x^6 + x^5 + x^4 + x^3 + x^2 + x + 1, \\ f_j(x) &= 1 \text{ for } j = 4, \ldots, 10, \end{aligned}$$

since $f(x) = x(x+1)(x^2+x+1)(x^3+x+1)(x^3+x^2+1)$ is the canonical factorization of f. □

In the proof of Theorem 3.5.4, we have defined $I_{d,q}(x)$ as the product of *all* monic irreducible polynomials over F with degree d (for every $d \in \mathbb{N}^*$). Thus

$$f_\ell(x) = \gcd\big(f(x), I_{\ell,q}(x)\big) \text{ for every } \ell. \tag{6.6}$$

As noted in the proof of Theorem 3.5.4,

$$x^{q^m} - x = \prod_{d|m} I_{d,q}(x) \quad \text{for all } m \in \mathbb{N}^*.$$

The following algorithm for computing the distinct degree factorization rests on a very simple observation: suppose that $g(x) \in F[x]$ is a monic square-free polynomial, and that $g_d(x) = 1$ for every proper divisor d of some positive integer m. Then

$$g_m(x) = \gcd\big(x^{q^m} - x, g(x)\big).$$

Concerning our given polynomial f, we start with $\ell = 1$ and obtain $f_1(x)$ as $\gcd(f(x), x^q - x)$. If $f_d(x)$ has already been determined for $d = 1, \ldots, \ell - 1$ (where $2 \le \ell \le n$), we obtain

$$f_\ell(x) = \gcd\big(x^{q^\ell} - x,\ f(x) \texttt{ div } f_1(x) \cdots f_{\ell-1}(x)\big).$$

In view of Corollary 3.5.3, we can actually stop this process as soon as we reach a value $\ell > n/2$. Formally, we obtain the following procedure:

Algorithm 6.2.2 (Distinct degree factorization over finite fields).

- *Input:* A square-free monic polynomial $f(x)$ with degree $n \geq 1$ over $F = \mathrm{GF}(q)$.
- *Output:* The distinct degree factorization of f as a list of the form

$$L = [(f_{i_1}(x), i_1), \ldots, (f_{i_m}(x), i_m)].$$

More precisely, $f(x) = \prod_{j=1}^{m} f_{i_j}(x)$, where $i_1 < \ldots < i_m$ and $f_{i_j}(x) \neq 1$ for all j.
- *Remark:* As for the algorithms in Section 6.1, we use an additional polynomial $L(x)$ to monitor the overall progress made in constructing the list L.

(1) $L \leftarrow [\ ]$, $\ell \leftarrow 1$, $L(x) \leftarrow 1$;
(2) $a(x) \leftarrow f(x)$, $h(x) \leftarrow x$;
(3) **while** $\ell \leq n/2$ **and** $a(x) \neq 1$ **do**
(4) $\quad h(x) \leftarrow h(x)^q \texttt{ mod } a(x)$, $z(x) \leftarrow \gcd(a(x), h(x) - x)$;
(5) $\quad$ **if** $z(x) \neq 1$ **then** append $(z(x), \ell)$ to L; $L(x) \leftarrow L(x) \cdot z(x)$,
$\qquad a(x) \leftarrow a(x) \texttt{ div } z(x)$, $h(x) \leftarrow h(x) \texttt{ mod } a(x)$ **fi**;
(6) $\quad \ell \leftarrow \ell + 1$
(7) **od**;
(8) **if** $a(x) \neq 1$ **then** append $(a(x), \deg a)$ to L **fi**. □

Theorem 6.2.3. *Algorithm 6.2.2 correctly determines the distinct degree factorization of any non-constant monic square-free polynomial f over $F = \mathrm{GF}(q)$.*

Proof. Assume first that f has degree 1. Then the **while** loop is not performed at all, and the algorithm immediately ends with the correct list $L = [(f(x), 1)]$. We may therefore assume $n \geq 2$, so that the **while** loop is entered at least once. This time, we use a particularly simple loop invariant, namely

$$(*) \quad f(x) = L(x) \cdot a(x),$$

which obviously holds after the initialization. The **while** loop is entered for the first time with $\ell = 1$, and we may write the data at this point in the form

$$a(x) = \prod_{j=\ell}^{n} f_j(x), \ L(x) = \prod_{j=1}^{\ell-1} f_j(x) \ \text{ and } \ h(x) = x^{q^{\ell-1}} \texttt{ mod } a(x)$$

(with the usual convention that the empty product for $L(x)$ is interpreted as 1).

Now suppose that the while loop is entered for some value $\ell \geq 1$ of the control variable (so that still $a(x) \neq 1$ at this point), and assume with induction that (*) and the conditions above hold for the then current data. Then Step (4) updates $h(x)$ to

$$h(x) = \left(x^{q^{\ell-1}} \texttt{ mod } a(x)\right)^q \texttt{ mod } a(x) = x^{q^\ell} \texttt{ mod } a(x)$$

and computes

$$z(x) = \gcd\left(a(x), h(x) - x\right) = \gcd\left(\prod_{j=\ell}^{n} f_j(x), x^{q^\ell} - x\right).$$

Since all irreducible monic divisors of $x^{q^\ell} - x$ over F have a degree dividing ℓ, this greatest common divisor is simply $f_\ell(x)$ (by definition). If $f_\ell(x) = z(x) \neq 1$, Step (5) correctly appends $(z(x), \ell)$ to the list L, updates the corresponding polynomials to

$$L(x) \cdot z(x) = \Big(\prod_{j=1}^{\ell-1} f_j(x)\Big) \cdot f_\ell(x) = \prod_{j=1}^{\ell} f_j(x)$$

and to $a(x) = \prod_{j=\ell+1}^{n} f_j(x)$, and finally replaces $h(x)$ with its remainder modulo this new value of $a(x)$.[2] Step (6) then increases the current value ℓ of the control variable to $\ell+1$. Altogether, this shows that (*) and the formulas for the data remain valid when ℓ is replaced by $\ell+1$, as claimed.

It remains to analyze the possibilities for the termination of the algorithm. If $a(x)$ has already reached the value 1 when the **while** loop terminates, all distinct degree factors of f have been determined and stored in L, and the algorithm simply stops. The only other possibility is that ℓ has exceeded $n/2$, but that the current $a(x)$ is still a non-trivial factor of f. In this case, $\deg a \leq n$ (trivially) and

$$\gcd\big(x^{q^\ell} - x, a(x)\big) = 1 \quad \text{for all } \ell \text{ with } 1 \leq \ell \leq n/2;$$

then Corollary 3.5.3 shows that a is actually irreducible over F. Moreover, the degree of a has to be larger than $n/2$, as all irreducible factors of f with degree $\ell \leq n/2$ have already been taken care of in the **while** loop. Thus Step (8) correctly appends the pair $(a(x), \deg a)$ to L, completing the distinct degree factorization of f. □

Example 6.2.4. Let us confirm the results stated in Example 6.2.1 for the case $q = 2$ and $f(x) = x^{10} + x^9 + x^8 + x^3 + x^2 + x$. By Corollary 6.1.2, f is indeed square-free, as $f'(x) = x^8 + x^2 + 1$ and $\gcd(f, f') = 1$. Thus we may apply Algorithm 6.2.2. The first execution of the **while** loop yields

$$z(x) = f_1(x) = \gcd(f(x), x^2 + x) = x^2 + x,$$
$$a(x) = x^8 + x^6 + x^5 + x^4 + x^3 + x^2 + 1.$$

The second execution of the **while** loop gives

$$z(x) = f_2(x) = \gcd(a(x), x^4 + x) = x^2 + x + 1,$$
$$a(x) = x^6 + x^5 + x^4 + x^3 + x^2 + x + 1.$$

The third (and final) execution of the **while** loop results in

$$z(x) = f_3(x) = \gcd(a(x), x^8 + x) = x^6 + x^5 + x^4 + x^3 + x^2 + x + 1, \;\; a(x) = 1,$$

and the algorithm terminates with the list

---

[2] While the algorithm would also work without the reduction of $h(x)$ modulo $a(x)$, it serves to keep the degree of h always as small as possible.

$$L = \left[(x^2+x,1),\,(x^2+x+1,2),\,(x^6+x^5+x^4+x^3+x^2+x+1,3)\right]. \quad \square$$

Exercises

Exercise 6.2.5. Determine the distinct degree factorization of $f(x) = x^{15}+1$ over GF(2) and compare your result with Exercise 5.6.15. □

6.3 Berlekamp Algebras

In the present and the subsequent two sections we are going to discuss the method of Berlekamp [27, 28] for determining the canonical factorization of polynomials over finite fields. Actually, several of the basic ideas can be explained in a much more general framework. As a first stage, one may even consider an arbitrary field F and use the Chinese remainder theorem together with generalizations of some results in Section 2.5 to obtain information about non-trivial divisors of a given polynomial $f(x) \in F[x]$ – at least theoretically; this will be our point of view in the present as well as in the next section.

Only in Section 6.5 we will add the assumption that the underlying field F has positive characteristic. Then the information in question can be obtained effectively by making use of a finite subfield Q of F together with the Frobenius automorphism on Q. In the most important case, Q is chosen as the prime subfield of F, and we will then use the notation P instead of Q (as earlier). Ultimately, this approach will provide us with an efficient probabilistic method for determining the canonical factorization of polynomials over finite fields.

Thus let $f(x)$ be a monic polynomial with degree $n \geq 1$ over an arbitrary field F. As before, we let

$$f(x) = \prod_{i=1}^{s} u_i(x)^{\ell_i} \tag{6.7}$$

be the (unknown) canonical factorization of f over F. We now introduce the following notation. For $i = 1,\ldots,s$, we put

$$U_i(x) := u_i(x)^{\ell_i} \quad \text{and} \quad \overline{U}_i(x) := \frac{f(x)}{U_i(x)} = \prod_{\substack{k=1\\k\neq i}}^{s} U_k(x). \tag{6.8}$$

We call the polynomials $U_i(x)$ the **primary factors** of f over F and the decomposition of f as

$$f(x) = \prod_{i=1}^{s} U_i(x)$$

the **primary factorization** of f. Note that $U_i(x) = u_i(x)$ for all i in the special case where f is square-free.

As usual, the F-vector space $F[x]_{<n}$ formed by the polynomials $g(x) \in F[x]$ with $\deg g < n$ will be identified with the canonical system of representatives of the factor algebra $F[x]/(f)$. By the Chinese remainder theorem (see Section 1.9), the F-algebra homomorphism

$$\Gamma: \begin{cases} F[x] & \to \bigoplus_{i=1}^{s} F[x]/(U_i) \\ g(x) & \mapsto \big(g(x) \bmod U_1(x), \ldots, g(x) \bmod U_s(x)\big) \end{cases}$$

is surjective, and its kernel is the ideal (f) of $F[x]$ generated by f. Thus Γ induces an isomorphism

$$\overline{\Gamma}: F[x]/(f) \to \bigoplus_{i=1}^{s} F[x]/(U_i).$$

In particular, given any s-tuple $(g_1(x), \ldots, g_s(x)) \in F[x]^s$ with $\deg g_i < \deg U_i$ for all i, there exists a unique polynomial $h(x) \in F[x]_{<n}$ such that

$$\Gamma(h(x)) = (g_1(x), \ldots, g_s(x)).$$

Observation 6.3.1. Since $U_i(x)$ and $\overline{U}_i(x)$ are relatively prime, there exists a unique polynomial $b_i(x) \in F[x]$ with $\deg b_i < \deg U_i$ such that $b_i(x)\overline{U}_i(x) \equiv 1 \bmod U_i(x)$. Note that the polynomial

$$e_i(x) := b_i(x)\overline{U}_i(x) \tag{6.9}$$

has degree $< n$ for all $i = 1, \ldots, s$. These polynomials form a so-called *system of pairwise orthogonal idempotents decomposing the unit element* for the factor algebra $F[x]/(f)$ in the following sense (compare with Remark 2.5.6): since $U_k(x)$ divides $\overline{U}_i(x)$ whenever $k \neq i$, we see that $\Gamma(e_i(x))$ is the canonical basis vector $e^i \in F^s$; that is, $e^i_k = 1$ if $k = i$, and $e^i_k = 0$ otherwise. Moreover,

- $e_i(x)e_j(x) \equiv 0 \bmod f(x)$ for all $i \neq j$,
- $e_i(x)^2 \equiv e_i(x) \bmod f(x)$ for all i,
- $e_1(x) + \cdots + e_s(x) = 1$.

The elements $e_i(x)$ are also called the **primitive idempotents** of the factor algebra $F[x]/(f)$, since any non-zero idempotent element[3] of this algebra can be written as $\sum_{i \in T} e_i(x)$ for some non-empty subset T of $\{1, \ldots, s\}$. The field elements 0 and 1 are said to be the **trivial idempotents**. □

We are now ready to introduce the notion of a *Berlekamp algebra*. This is usually done in a very different manner, and only under the assumption that F has positive characteristic. We have decided to use an alternative approach, since we want to emphasize which ideas work in general and where the assumption of a positive characteristic becomes necessary. In this way, the usual definition of a Berlekamp algebra for finite fields turns into a major result; see Theorem 6.5.1.

[3] Recall that an element a of an F-algebra is called an **idempotent** if $a^2 = a$. For properties of idempotents in a general (commutative) F-algebra, see the exercises at the end of this section.

Definition 6.3.2. In the situation described above, the **Berlekamp algebra** $\mathscr{B}_f$ of f is the F-subspace of $F[x]_{<n}$ generated by the primitive idempotents $e_1(x),\dots,e_s(x)$ of the factor algebra $F[x]/(f)$. □

Since the primitive idempotents $e_1(x),\dots,e_s(x)$ are linearly independent over F, the F-dimension of the vector space $\mathscr{B}_f$ is equal to s, the (unknown) number of distinct irreducible divisors of f. Note that the terminology introduced in Definition 6.3.2 makes sense, since $\mathscr{B}_f$ indeed becomes an F-algebra when the multiplication is considered modulo f. Moreover, this algebra turns out to be isomorphic to the s-tuple space F^s equipped with pointwise operations, and a canonical isomorphism between F^s and $\mathscr{B}_f$ is given by

$$\Gamma_f\colon\ F^s \to \mathscr{B}_f,\ (\lambda_1,\dots,\lambda_s) \mapsto \sum_{i=1}^{s} \lambda_i e_i(x), \tag{6.10}$$

which is easily checked by direct computation:

$$\begin{aligned}\Big(\sum_{i=1}^{s}\alpha_i e_i(x)\Big)\cdot\Big(\sum_{j=1}^{s}\beta_j e_j(x)\Big) &= \sum_{i=1}^{s}\sum_{j=1}^{s}\alpha_i\beta_j e_i(x)e_j(x)\\ &\equiv \sum_{k=1}^{s}\alpha_k\beta_k e_k(x)^2\\ &\equiv \sum_{k=1}^{s}\alpha_k\beta_k e_k(x) \mod f(x).\end{aligned}$$

Definition 6.3.3. We call $e_1(x),\dots,e_s(x)$ the **spectral basis** of $\mathscr{B}_f$. By definition, for every $h\in\mathscr{B}_f$ there exist unique elements $c_1,\dots,c_s\in F$ such that

$$h(x) = c_1e_1(x)+\cdots+c_se_s(x);$$

the c_i are called the **spectral coefficients** of h. We will denote the set of all spectral coefficients of h by C_h. □

Remark 6.3.4. Trivially, the set of all spectral coefficients of h satisfies $C_h\subseteq F$ and $|C_h|\le s$. For instance, $C_h=\{0,1\}$ whenever h is a non-trivial idempotent of $F[x]/(f)$.

One should realize that the spectral basis of the Berlekamp algebra $\mathscr{B}_f$ is intimately related to the factorization of f: if we would know this basis, we could immediately split f into its primary divisors, as $\gcd(e_i(x),f(x))=\overline{U}_i(x)$ and $U_i(x)=f(x)/\overline{U}_i(x)$ for all i. □

More generally, any *non-trivial* polynomial $h\in\mathscr{B}_f$ (see Remark 6.3.6 below) yields a non-trivial (partial) factorization of f. This rests on the following crucial result:

Theorem 6.3.5. *Let h be a polynomial in $\mathscr{B}_f$ with spectral coefficients $c_1,\dots,c_s$, and let $\lambda\in F$. Then one has* $\gcd(h(x)-\lambda,f(x))\neq 1$ *if and only if $\lambda\in C_h$. In this* *case,*

$$\gcd\big(h(x)-\lambda, f(x)\big) = \prod_{\substack{j=1\\ c_j=\lambda}}^{s} U_j(x).$$

Moreover,

$$f(x) = \prod_{\gamma\in C_h} \gcd\big(h(x)-\gamma, f(x)\big). \tag{6.11}$$

Proof. Using $\lambda = \lambda\cdot 1 = \lambda\cdot\sum_{i=1}^{s} e_i(x)$, we may write $h(x)-\lambda$ as

$$h(x)-\lambda = \sum_{i=1}^{s}(c_i-\lambda)e_i(x).$$

As $U_j(x)$ divides $e_i(x)$ whenever $j\neq i$, we obtain $\gcd(h(x)-\lambda, U_j(x))\neq 1$ if and only if $u_j(x)$ divides $(c_j-\lambda)e_j(x)$. Since $e_j(x)$ and $u_j(x)$ are relatively prime, this means $c_j=\lambda$, in which case even $U_j(x)$ divides $h(x)-\lambda$. This establishes the first assertion.

Now the second assertion follows as $U_j(x)$ divides $h(x)-c_j$ and as $h(x)-\lambda$ and $h(x)-\mu$ are relatively prime for distinct $\lambda,\mu\in F$. Thus the spectral coefficients of h do not have to be considered with their multiplicities: the set C_h is sufficient. □

Remark 6.3.6. Consider some polynomial $h\in\mathscr{B}_f$. Since $\deg h<\deg f$, Equation (6.11) gives a non-trivial (partial) factorization of $f(x)$ if and only if $h(x)-\gamma\neq 0$ for all $\gamma\in C_h$. Let us discuss this in more detail. Of course, the Berlekamp algebra contains the underlying field F as a subalgebra. In fact, the pre-image of F under the mapping Γ_f in Equation (6.10) is the **diagonal subalgebra**

$$\{(\lambda,\ldots,\lambda)\in F^s : \lambda\in F\}$$

of F^s. If h is a constant polynomial, say $h(x)=\mu\in F$, the set C_h is just the singleton $\{\mu\}$ and $f(x)$ divides $h(x)-\mu=0$. In all other cases, $h\in\mathscr{B}_f\setminus F$ and $s>1$, and then Equation (6.11) gives a non-trivial factorization of f. In view of these observations, F is also called the **trivial subalgebra** of $\mathscr{B}_f$.

Note that $\mathscr{B}_f=F$ is possible. This case plays a special role, since it occurs if and only if $s=1$, which means that f has just one primary factor. In view of the results in Section 6.1, we may always assume that f is actually square-free, provided that F is a perfect field. This shows that every algorithm capable of determining the F-dimension of the Berlekamp algebra $\mathscr{B}_f$ can be used to decide whether or not f is irreducible. □

Example 6.3.7. Let us consider the square-free polynomial $f(x)=x^5-2x^3-x^2+2$ over $F=\mathrm{GF}(5)$. Here the polynomial $h(x)=x^3-2x^2-2x$ is a non-trivial element of the Berlekamp algebra of f. (We shall deal with the problem of finding such a polynomial later.) When $\lambda=0$ or $\lambda=-1$, then $\gcd(h(x)-\lambda, f(x))=1$. All other choices of λ lead to non-trivial factors of f:

$$\begin{aligned}\gcd(h(x)-2,f(x)) &= x-1,\\ \gcd(h(x)-1,f(x)) &= x^2-2,\\ \gcd(h(x)+2,f(x)) &= x^2+x+1;\end{aligned}$$

in particular, the spectral coefficients of h are $1, 2$ and -2. This gives the non-trivial factorization

$$f(x) = (x-1)\cdot(x^2-2)\cdot(x^2+x+1),$$

which is in fact the canonical factorization of f over F. □

So far, we have considered just a single polynomial h in the Berlekamp algebra $\mathscr{B}_f$. Our next result deals with bases of $\mathscr{B}_f$ containing the trivial idempotent 1. It turns out that the availability of such a basis provides sufficient information for computing the primary factorization of f, which leads to a generic algorithm for this problem.

Theorem 6.3.8. *Let $f(x) = \prod_{i=1}^{s} U_i(x)$ be the primary factorization of $f \in F[x]$. Assume $s \geq 2$, let $B = \{h_0(x), h_1(x), \ldots, h_{s-1}(x)\}$ with $h_0(x) = 1$ be a basis for the Berlekamp algebra of f, and consider any two distinct indices $j, k \in \{1, \ldots, s\}$. Then there exist a polynomial $h \in B$ and two distinct elements $\gamma, \delta \in C_h$ such that*

$$h(x) \bmod U_j(x) = \gamma \quad \textit{and} \quad h(x) \bmod U_k(x) = \delta.$$

Proof. Assume otherwise. Then there exist elements $\alpha_i \in C_{h_i}$ such that

$$h_i(x) \bmod U_j(x) = h_i(x) \bmod U_k(x) = \alpha_i \quad \text{for } i = 0, \ldots, s-1,$$

where, of course, $\alpha_0 = 1$. Thus $U_j(x)U_k(x)$ divides $h_i(x) - \alpha_i$, since $U_j(x)$ and $U_k(x)$ are relatively prime. We now expand the idempotent $e_k(x)$ with respect to the given basis B of $\mathscr{B}_f$, say

$$e_k(x) = \mu_0 + \mu_1 h_1(x) + \cdots + \mu_{s-1} h_{s-1}(x)$$

with $\mu_0, \ldots, \mu_{s-1} \in F$. Reducing this modulo $U_j(x)U_k(x)$ gives

$$e_k(x) \equiv \omega \bmod U_j(x)U_k(x) \quad \text{with } \omega := \mu_0 + \mu_1\alpha_1 + \cdots + \mu_{s-1}\alpha_{s-1},$$

and we obtain the contradiction

$$0 = e_k(x) \bmod U_j(x) = \omega = e_k(x) \bmod U_k(x) = 1. \quad \square$$

Theorem 6.3.5, Remark 6.3.6 and Theorem 6.3.8 lead to the following generic algorithm for computing the primary factorization of a polynomial f. We leave a formal proof for the correctness of this algorithm to the reader.

Algorithm 6.3.9 (Generic primary factorization of a polynomial).

- *Input:* A field F, a non-constant monic polynomial $f(x) \in F[x]$, a basis B for the Berlekamp algebra $\mathscr{B}_f$ as in Theorem 6.3.8, and, for every index $i = 1, \ldots, s-1$ (where $s \geq 2$), the set C_{h_i} of all spectral coefficients of h_i.
- *Output:* The set $\{U_1(x), \ldots, U_s(x)\}$ of all primary factors of f.

(1) $\mathscr{F} \leftarrow \{f(x)\}$,
(2) **for** $i = 1$ **to** $s-1$ **do**
(3)   **for** $\lambda \in C_{h_i}$ **do**
(4)   **for** $g(x) \in \mathscr{F}$ **do**
(5) $d(x) \leftarrow \gcd(h_i(x) - \lambda, g(x))$;
(6)   **if** $d(x) \neq 1$ **and** $d(x) \neq g(x)$
(7)   **then** $\mathscr{F} \leftarrow (\mathscr{F} \setminus \{g(x)\}) \cup \{d(x), g(x) \texttt{ div } d(x)\}$
(8)   **fi**
(9)   **od**
(10)   **od**
(11) **od** □

In practice, it is a good idea to determine the radical $\operatorname{rad}(f)$ of f first, by using Algorithm 6.1.7. Then an application of Algorithm 6.3.9 to $\operatorname{rad}(f)$ produces the distinct irreducible factors $u_i(x)$ of f. Finally, the multiplicities of the u_i – and thus the primary factors U_i of f – can be determined easily, by using polynomial divisions.

Example 6.3.10. Let us return to the situation investigated in Example 6.3.7, that is, $f(x) = x^5 - 2x^3 - x^2 + 2$ and $F = \mathrm{GF}(5)$. In Example 6.5.4, we will see that $h_0(x) = 1$, $h_1(x) = x^4 + x^2$ and $h_2(x) = x^3 - 2x^2 - 2x$ form a basis for $\mathscr{B}_f$. (Note that $h_2(x)$ is the polynomial h which we already used in Example 6.3.7.) In particular, f has $s = 3$ distinct irreducible factors over F. We now apply Algorithm 6.3.9 starting with h_1.

The spectral coefficients of h_1 are 1, -1 and 2. Beginning with $\lambda = 1$, we obtain $\gcd(h_1(x) - 1, f(x)) = x^2 - 2$, and $\mathscr{F}$ becomes $\{x^2 - 2, x^3 - 1\}$. Taking $\lambda = -1$ gives

$$\gcd(h_1(x) + 1, x^2 - 2) = 1 \quad \text{and} \quad \gcd(h_1(x) + 1, x^3 - 1) = x^2 + x + 1,$$

and $\mathscr{F}$ is updated to $\{x^2 - 2, x - 1, x^2 + x + 1\}$. The final spectral coefficient $\lambda = 2$ of $h_1(x)$ yields nothing new, as $\gcd(h_1(x) - 2, x^2 - 2) = 1$, $\gcd(h_1(x) - 2, x - 1) = x - 1$ and $\gcd(h_1(x) - 2, x^2 + x + 1) = 1$. Since $s = 3$ and f is square-free, the set $\mathscr{F}$ now already contains the canonical factorization of f, so that considering h_2 does not change $\mathscr{F}$. Thus the algorithm terminates with $\mathscr{F} = \{x^2 - 2, x - 1, x^2 + x + 1\}$. □

Of course, the requirements for the input to Algorithm 6.3.9 lead to two fundamental questions:

- How can one compute a basis of the Berlekamp algebra $\mathscr{B}_f$ from the coefficients of f? In particular, how can one at least determine the F-dimension of $\mathscr{B}_f$, that is, the number of distinct irreducible divisors of f?

- Given a non-trivial polynomial $h \in \mathscr{B}_f$, how can one compute its spectral coefficients?

For fields with a positive characteristic, a solution of the first question was first proposed by Berlekamp [27] in 1967; this will be the main topic of Section 6.5. As for the second question, we will obtain a partial answer in the next section, and reconsider it for finite fields in the second part of Section 6.5.

Before we address these fundamental problems, we wish to emphasize that the basic setup considered so far allows some flexibility provided that F has a non-trivial subfield.

Definition 6.3.11. Let $f(x)$ be a non-constant monic polynomial over the field F, let $e_1(x), \ldots, e_s(x)$ be the primitive idempotents of the factor algebra $F[x]/(f)$, and assume that Q is a subfield of F. Then the Q-subspace $\mathscr{B}_f^Q$ of $F[x]_{<n}$ generated by $e_1(x), \ldots, e_s(x)$ is called the **Berlekamp algebra of f over Q**. In terms of spectral coefficients, this means

$$h \in \mathscr{B}_f^Q \;\Leftrightarrow\; C_h \subseteq Q \qquad \text{for } h \in \mathscr{B}_f. \tag{6.12}$$

The special case where Q is the prime field P of F is of particular interest. Then $\mathscr{B}_f^P$ is also called the **prime Berlekamp algebra** or the **absolute Berlekamp algebra** of f. □

Observation 6.3.12. Note that $\mathscr{B}_f^Q$ is just the image of the restriction of the canonical isomorphism Γ_f in Equation (6.10) to Q^s, and therefore a Q-subalgebra of $\mathscr{B}_f$. In particular, the Q-dimension of $\mathscr{B}_f^Q$ equals the F-dimension s of $\mathscr{B}_f$. Reconsidering Theorem 6.3.5, Remark 6.3.6, Theorem 6.3.8, and Algorithm 6.3.9, we see that all our results continue to hold when F is replaced by Q. □

Remark 6.3.13. Assume that F is a finite field which has a sufficiently small subfield Q. Then one may determine the set C_h of spectral coefficients of any polynomial $h \in \mathscr{B}_f^Q$ by an exhaustive search, using Equation (6.11). This idea will be discussed in more detail in the second part of Section 6.5. □

Exercises

Exercise 6.3.14. Let F be a field and $\mathscr{A}$ a commutative algebra over F. Show that the set $\mathscr{I}$ of all idempotents of $\mathscr{A}$ is a submonoid of the multiplicative monoid $(\mathscr{A}, \cdot, e)$ of $\mathscr{A}$, where e denotes the multiplicative identity of $\mathscr{A}$. □

Exercise 6.3.15. Let $f(x)$ be a monic polynomial over a finite field F. Determine the number of idempotents of the factor algebra $\mathscr{A} = F[x]/(f)$. □

Exercise 6.3.16. Let F be a field, V a finite dimensional vector space over F, and $\mathscr{A} = \mathrm{End}_F(V)$ the algebra of all F-endomorphisms of V. Show that $\tau \in \mathscr{A}$ is an idempotent if and only if τ is a projection onto some F-subspace of V. □

Exercise 6.3.17. Let F be the Galois field $\mathrm{GF}(q)$ and $E = \mathrm{GF}(q^n)$ its n-dimensional extension field. Apply results from Section 3.11 to characterize the idempotents of $\mathrm{End}_F(E)$ in terms of q-polynomials. □

6.4 Fundamental Properties of Spectral Coefficients

The present section is devoted to the study of the spectral coefficients of some polynomial h in a Berlekamp algebra $\mathscr{B}_f^Q$, where Q is a subfield of F and $f(x) \in F[x]$ (as before). Again, we will not impose any restrictions on F. Observe that h is, in general, a polynomial in $F[x]$: it does not need to have coefficients in Q. Our first result gives a useful description of the set $C_h \subseteq Q$ of all spectral coefficients of h as the set of roots of a certain polynomial:

Theorem 6.4.1. *Let $f(x)$ be a non-constant monic polynomial over some field F, and let $h \in \mathscr{B}_f^Q$, where Q is any subfield of F. Then the polynomial*

$$\rho_h(x) := \prod_{\gamma \in C_h} (x - \gamma) \tag{6.13}$$

is the monic polynomial $a(x) \in Q[x]$ of least degree such that $f(x)$ divides the composition $a(h(x))$ of a with h.

Proof. We make use of the following homomorphism between Q-algebras:

$$\Omega\colon\ Q[x] \to F[x]/(f),\ \ g(x) \mapsto g(h(x)) + (f).$$

By Equation (6.11),

$$f(x) = \prod_{\gamma \in C_h} \gcd(f(x), h(x) - \gamma).$$

Hence f divides the polynomial $\prod_{\gamma \in C_h} (h(x) - \gamma) = \rho_h(h(x))$. Thus ρ_h is in the kernel of Ω, which is therefore non-trivial. We denote the monic generator of this kernel by $\mu(x)$ and refer to μ as the **minimal polynomial of h modulo** f, since $\mu(x)$ is the monic polynomial of least degree in $Q[x]$ such that f divides $\mu(h(x))$. Now the assertion amounts to showing that μ coincides with ρ_h.

Note that ρ_h splits over Q into distinct linear factors, as $C_h \subseteq Q$. Since μ divides ρ_h, there is a subset R of C_h such that $\mu(x) = \prod_{\delta \in R}(x - \delta)$. By the definition of μ as the monic generator of the kernel of Ω, we see that f already divides

$$\mu(h(x)) = \prod_{\delta \in R} (h(x) - \delta).$$

According to Theorem 6.3.5, the spectral coefficients γ of h are precisely the elements of F satisfying $\gcd(h(x) - \gamma, f(x)) \neq 1$. This shows $R = C_h$ and establishes the assertion $\rho_h = \mu$. □

Remark 6.4.2. Using basic results from Linear Algebra, the minimal polynomial of h modulo f may be obtained as follows:

- Determine successively the polynomials

$$a_0(x) := 1,\ a_1(x) := h(x),\ a_2(x) := h(x)^2 \bmod f, \ldots, a_m(x) := h(x)^m \bmod f$$

  until this list becomes linearly dependent over Q for the first time.
- This yields unique elements $\alpha_0, \ldots, \alpha_{m-1} \in Q$ such that $a_m(x) = \sum_{i=0}^{m-1} \alpha_i a_i(x)$. Then

$$\rho_h(x) = x^m - \sum_{i=0}^{m-1} \alpha_i x^i,$$

  since $h(x)^m - \sum_{i=0}^{m-1} \alpha_i h(x)^i \equiv a_m(x) - \sum_{i=0}^{m-1} \alpha_i a_i(x) \equiv 0 \bmod f$. □

Example 6.4.3. Let us again return to the situation studied in Examples 6.3.7 and 6.3.10, that is, $f(x) = x^5 - 2x^3 - x^2 + 2$ over $F = \mathrm{GF}(5)$. As noted in 6.3.10, the polynomial $h(x) = x^4 + x^2$ belongs to $\mathscr{B}_f$. Since

$$\begin{aligned} h(x)^2 \bmod f &= -x^4 - x^3 + x^2 - 2x - 2, \\ h(x)^3 \bmod f &= -x^4 - 2x^3 - 2x^2 - x - 1, \end{aligned}$$

we see that 1, h, $h^2 \bmod f$ and $h^3 \bmod f$ are linearly dependent over F, while the first three of these polynomials are linearly independent. Now the relation

$$h(x)^3 - 2h(x)^2 - h(x) + 2 \equiv 0 \bmod f$$

gives $\rho_h(x) = x^3 - 2x - x + 2$, which factors over F as $(x-1)(x+1)(x-2)$. Thus the spectral coefficients of h are 1, -1 and 2, confirming the result obtained in Example 6.3.10. □

Remark 6.4.4. Let h be a non-trivial polynomial in a generalized Berlekamp algebra $\mathscr{B}_f^Q$. So far, we have reduced the determination of the spectral coefficients of h to the problem of finding all roots of the polynomial ρ_h associated with h in Theorem 6.4.1. Of course, the **root finding problem** is a special instance of the polynomial factorization problem. This poses the natural question how the (generalized) Berlekamp algebra of a square-free polynomial f looks like when f splits into linear factors. Thus let f be such a polynomial, so that $n = \deg f = s$, say

$$f(x) = \prod_{i=1}^{n} (x - \lambda_i).$$

Then the Berlekamp algebra $\mathscr{B}_f$ coincides with $F[x]_{<n}$, and therefore Theorem 6.3.5 can be applied for *any* polynomial h over F with $\deg h < n$. We also note that the primitive idempotents of f are explicitly given by

$$e_j(x) = \frac{1}{f'(\lambda_j)} \cdot \frac{f(x)}{x - \lambda_j} \quad \text{for } j = 1, \ldots, n;$$

this follows from Theorem 2.5.5, adapted to the notation of the present chapter.

The preceding observations indicate that the root finding problem lies at the heart of the more general factorization problem. For the case where F is a finite field, we will see in Section 6.5 how one can attack this problem efficiently, at least with a probabilistic approach. $\sqcap$

Next, we consider the relationship between the Berlekamp algebras $\mathscr{B}_f^Q$ and $\mathscr{B}_g^Q$, where $g(x) \in F[x]$ is some monic divisor of f.

Proposition 6.4.5. *Let F be a field and Q a subfield of F, and let $f(x)$ be a monic polynomial over F with exactly s distinct monic irreducible divisors in $F[x]$. Moreover, let $g(x)$ be a monic divisor of f in $F[x]$ with exactly t distinct monic irreducible divisors. Then the mapping*

$$\phi\colon\ F[x] \to F[x],\ \ h(x) \mapsto h(x) \ \mathtt{mod}\ g(x)$$

induces an algebra epimorphism ϕ_0 from $\mathscr{B}_f^Q$ onto $\mathscr{B}_g^Q$, and the kernel of ϕ_0 has F-dimension $s-t$.

Proof. With the notation in Equations (6.1) and (6.7), we let T denote the set of indices j (where $1 \le j \le s$) such that $u_j(x)$ divides $g(x)$. Then the canonical factorization of g has the form $g(x) = \prod_{j\in T} u_j(x)^{m_j}$ with $1 \le m_j \le \ell_j$ for all $j \in T$. Let $\varepsilon_i(x) := e_i(x) \ \mathtt{mod}\ g(x)$ for $i = 1,\dots,s$. Then

$$\varepsilon_i(x)^2 \equiv e_i(x)^2 \equiv e_i(x) \equiv \varepsilon_i(x) \bmod g \quad \text{for } i = 1,\dots,s.$$

For distinct indices i,k from $\{1,\dots,s\}$, one has

$$\varepsilon_i(x)\varepsilon_k(x) \equiv e_i(x)e_k(x) \equiv 0 \bmod g,$$

and, when $\ell \notin T$, $\varepsilon_\ell(x) \equiv e_\ell(x) \equiv 0 \bmod g$. Hence

$$\sum_{j\in T} \varepsilon_j(x) \equiv \sum_{i=1}^{s} \varepsilon_i(x) \equiv \sum_{i=1}^{s} e_i(x) \equiv 1 \bmod g,$$

and considering the degrees involved gives that even

$$\sum_{j\in T} \varepsilon_j(x) = \sum_{i=1}^{s} \varepsilon_i(x) = 1.$$

Altogether, this shows that the $\varepsilon_j(x)$ with $j \in T$ are the primitive idempotents of the algebra $F[x]/(g)$.

Now let $h(x)$ be a polynomial in $\mathscr{B}_f^Q$, and let $c_1,\dots,c_s$ be the spectral coefficients of h. Then

$$h(x) = \sum_{i=1}^{s} c_i e_i(x) \equiv \sum_{j\in T} c_j \varepsilon_j(x) \bmod g.$$

and therefore $h(x) \bmod g(x)$ belongs to $\mathscr{B}_g^Q$. This shows that ϕ induces an algebra homomorphism ϕ_0 from $\mathscr{B}_f^Q$ to $\mathscr{B}_g^Q$. Since ϕ_0 is the Q-linear extension of the mapping

$$e_i(x) \mapsto \varepsilon_i(x) \quad \text{for } i = 1, \ldots, s,$$

ϕ_0 is in fact onto, as claimed. Finally, the kernel of ϕ_0 is the Q-vector space spanned by the primitive idempotents $e_j(x)$ with $j \in \{1, \ldots, s\} \setminus T$, and therefore has dimension $s - t$. □

Remark 6.4.6. Note that the epimorphism ϕ_0 in the proof of Proposition 6.4.5 can be interpreted as a projection. It is interesting to observe that this projection is actually a bijection whenever g is (a multiple of) the radical $\mathrm{rad}(f(x))$ of f. In particular, any basis of $\mathscr{B}_f^Q$ gives rise to a basis of $\mathscr{B}_{\mathrm{rad}(f)}^Q$ by taking remainders modulo $\mathrm{rad}(f)$. □

Recall that $\mathrm{rad}(f)$ can be determined efficiently via Algorithm 6.1.7 provided that F is a perfect field. The preceding remark indicates once again that it is definitely worthwhile to begin by computing the square-free factorization of f in this case, and to apply the method of Berlekamp to the radicals of that partial factorization afterwards. Considering an arbitrary f is therefore mainly of theoretical interest: it allows to keep the presentation as general as possible.

Proposition 6.4.7. *Assume that $g(x) \in F[x]$ is a monic divisor of $f(x) \in F[x]$ such that $\overline{g}(x) := f(x)/g(x)$ and $g(x)$ are relatively prime. Then the mapping*

$$\Psi : F[x] \to F[x] \times F[x], \; v(x) \mapsto \big(v(x) \bmod g(x), v(x) \bmod \overline{g}(x)\big)$$

induces an isomorphism between $\mathscr{B}_f^Q$ and $\mathscr{B}_g^Q \times \mathscr{B}_{\overline{g}}^Q$.

Proof. Proposition 6.4.5 shows that Ψ maps $\mathscr{B}_f^Q$ to the algebra $\mathscr{B}_g^Q \times \mathscr{B}_{\overline{g}}^Q$. Of course, the restriction of Ψ to $\mathscr{B}_f^Q$ has a trivial kernel. Let t_1 and t_2 denote the number of distinct monic irreducible divisors of g and $\overline{g}$, respectively. Then

$$\dim_Q \mathscr{B}_f^Q = s = t_1 + t_2 = \dim_Q \mathscr{B}_g^Q + \dim_Q \mathscr{B}_{\overline{g}}^Q,$$

which shows that Ψ induces an isomorphism between $\mathscr{B}_f^Q$ and $\mathscr{B}_g^Q \times \mathscr{B}_{\overline{g}}^Q$. □

Observation 6.4.8. As an application of Proposition 6.4.7, we reconsider the generic factorization algorithm in 6.3.9 for an arbitrary subfield Q of F. Let $s = \dim_Q \mathscr{B}_f^Q$, and let $B = \{h_0(x), h_1(x), \ldots, h_{s-1}(x)\}$ with $h_0(x) = 1$ be a Q-basis for $\mathscr{B}_f^Q$.

- First assume $s = 1$. Then $f(x)$ has the form $u(x)^\ell$, where $u(x)$ is irreducible, and therefore the square-free factorization algorithms applied to f yields $u(x)$ and its multiplicity ℓ (provided that F is a perfect field).
- Now let $s \geq 2$. Choose any non-trivial element h of $\mathscr{B}_f^Q$, for instance one of the basis vectors $h_i(x)$ with $i \neq 0$, and assume that the spectral coefficients of h are available. Take any $\gamma \in C_h$. Then $g(x) := \gcd(h(x) - \gamma, f(x))$ is a non-trivial

divisor of f. In fact, g is a product of certain primary factors of f, and therefore $\overline{g}(x) := f(x)/g(x)$ is relatively prime to g. Then $a(x) := h(x) \bmod g(x)$ is a trivial element of $\mathscr{B}_g^Q$, since the only spectral coefficient of $a(x)$ is γ, by the definition of g. Moreover, the spectral coefficients of the polynomial $\overline{a}(x) := h(x) \bmod \overline{g}(x)$ in $\mathscr{B}_{\overline{g}}^Q$ are the elements of the set $C_h \setminus \{\gamma\}$.

- By Proposition 6.4.5, the polynomials $h_i(x) \bmod g(x)$ generate the Berlekamp algebra $\mathscr{B}_g^Q$ as a Q-vector space. A Q-basis of $\mathscr{B}_g^Q$ can therefore be obtained by applying the Gaussian algorithm to this system. Afterwards, the entire procedure underlying Theorem 6.3.5 may be applied to g instead of f, and the analogous statement holds, of course, for $\overline{g}$.

Once again, it is advisable to determine the square-free factorization of f first, and to apply this procedure only to the resulting square-free factors of f. □

The final result of this section concerns polynomial actions on Berlekamp algebras:

Theorem 6.4.9. *Let $f(x) \in F[x]$ be a monic polynomial with degree $n \geq 1$, let Q be a subfield of F, and let $\phi(x)$ be any polynomial in $Q[x]$. Then the mapping*

$$\overline{\phi}\colon \mathscr{B}_f^Q \to F[x]_{<n}, \quad h(x) \mapsto \phi(h(x)) \bmod f(x)$$

defines an action $\overline{\phi}$ of ϕ on the Berlekamp algebra $\mathscr{B}_f^Q$. Moreover, ϕ maps the spectral coefficients of any polynomial $h \in \mathscr{B}_f^Q$ to the spectral coefficients of $\overline{\phi}(h)$.

Proof. Write $\phi(x) = \sum_{i=0}^m a_i x^i$ and let $h \in \mathscr{B}_f^Q$. We need to check that $\overline{\phi}(h)$ again belongs to $\mathscr{B}_f^Q$. Thus we have to consider $\phi(h(x)) = \sum_{k=0}^m a_k h(x)^k$. If $h(x) = \sum_{i=1}^s c_i e_i(x)$ is the spectral representation of h (that is, $c_1, \ldots, c_s \in C_h \subseteq Q$), we get

$$\phi(h(x)) = \sum_{k=0}^m a_k \cdot \Big(\sum_{i=1}^s c_i e_i(x) \Big)^k.$$

As in (6.10) (but for the Berlekamp algebra of f over Q), the mapping

$$\Gamma_f\colon Q^s \to \mathscr{B}_f^Q, \quad (\lambda_1, \ldots, \lambda_s) \mapsto \sum_{i=1}^s \lambda_i e_i(x) \tag{6.14}$$

is a canonical isomorphism between Q-algebras. For the action $\overline{\phi}$ of the polynomial ϕ, this gives

$$\begin{aligned}\phi(h(x)) &= \sum_{k=0}^{m} a_k \cdot \Big(\sum_{i=1}^{s} c_i e_i(x)\Big)^k \\ &\equiv \sum_{k=0}^{m} a_k \cdot \Big(\sum_{i=1}^{s} c_i^k e_i(x)\Big) \\ &\equiv \sum_{i=1}^{s} \Big(\sum_{k=0}^{m} a_k c_i^k\Big) \cdot e_i(x) \\ &\equiv \sum_{i=1}^{s} \phi(c_i) e_i(x) \mod f(x).\end{aligned}$$

Note that $\phi(c_i) \in Q$ for $i = 1, \ldots, s$, as ϕ is a polynomial in $Q[x]$ and as $C_h \subseteq Q$. Thus $\overline{\phi}(h) \in \mathscr{B}_f^Q$. Moreover, the preceding calculation shows that the spectral coefficients of $\overline{\phi}(h)$ are given by $\phi(c_1), \ldots, \phi(c_s) \in Q$, which establishes both assertions. □

As we shall see in the next section, certain specific choices for ϕ in Theorem 6.4.9 turn out to be very useful when the underlying field F is a Galois field.

6.5 Factorization over Finite Fields: Berlekamp's Method

If we want to apply the results of the preceding two sections in practice, we require an effective method for producing a basis of a Berlekamp algebra $\mathscr{B}_f^Q$ from the coefficients of the polynomial f. In the case where F is a finite field, this can indeed be done via a certain linearization, which is the method introduced in 1967 by Berlekamp in his seminal paper [27].

We begin by considering a monic polynomial $f(x)$ over an arbitrary perfect field F with a positive characteristic p.[4] We may assume $n = \deg f \geq 2$. Let Q be a finite subfield of F, for instance the prime subfield P of F. Throughout, we will denote the cardinality of Q by p^r and use the basic notation introduced in Section 6.3. In particular, the number of distinct irreducible monic divisors of f is again denoted by s, the primary factors of f by $U_i(x)$, and the primitive idempotents of the algebra $F[x]/(f)$ by $e_i(x)$.

The following fundamental result of Berlekamp gives a characterization of $\mathscr{B}_f^Q$ and indicates how one may find a basis for this Q-algebra.

Theorem 6.5.1. *With the assumptions and notations above, the Berlekamp algebra $\mathscr{B}_f^Q$ is given by the set of all polynomials $h(x) \in F[x]_{<n}$ for which $h(x)^{p^r} - h(x)$ is a multiple of f:*

$$\mathscr{B}_f^Q = \{h(x) \in F[x] \colon \deg h < \deg f,\ h(x)^{|Q|} \equiv h(x) \bmod f(x)\}. \tag{6.15}$$

[4] Instead of requiring that F is perfect, it would actually be sufficient to assume that every irreducible factor of f over F is a separable polynomial, see Definition 3.2.17.

Proof. Since Q is a finite subfield of F, the product over all linear factors $x-\lambda$ with $\lambda \in Q$ is a polynomial, namely

$$\prod_{\lambda \in Q} (x-\lambda) = x^{|Q|} - x = x^{p^r} - x;$$

see Equation (3.4) in Section 3.3. Substituting any polynomial $h(x) \in F[x]$ gives

$$h(x)^{p^r} - h(x) = \prod_{\lambda \in Q} (h(x)-\lambda),$$

where $h(x)-\lambda$ and $h(x)-\mu$ are relatively prime whenever λ and μ are distinct.

Now let $h(x) \in F[x]_{<n}$ and assume that $h(x)^{p^r} - h(x)$ is a multiple of f. Then, for $j = 1,\ldots,s$, there is a unique element $\omega_j \in Q$ such that the primary divisor U_j divides $h(x)-\omega_j$, that is, $h(x) \bmod U_j(x) = \omega_j \in Q$. On the other hand,

$$\omega_j = \omega_j \cdot 1 \equiv \omega_j \cdot \Big(\sum_{i=1}^{s} e_i(x) \Big) = \sum_{i=1}^{s} \omega_j e_i(x) \equiv \omega_j e_j(x) \mod U_j,$$

which yields

$$h(x) \equiv \omega_j \equiv \omega_j e_j(x) \equiv \sum_{k=1}^{s} \omega_k e_k(x) \mod U_j,$$

as U_j divides e_k for $k \neq j$. Since this holds for all j, we obtain

$$h(x) \equiv \sum_{k=1}^{s} \omega_k e_k(x) \mod f.$$

Since h and all idempotents e_k belong to $F[x]_{<n}$, the preceding congruence implies $h(x) = \sum_{k=1}^{s} \omega_k e_k(x)$, so that $h \in \mathscr{B}_f^Q$, as claimed.

Conversely, assume $h \in \mathscr{B}_f^Q$, and let C_h be the set of spectral coefficients of h. By Theorem 6.3.5, $f(x) = \prod_{\lambda \in C_h} \gcd(h(x)-\lambda, f(x))$, so that f divides the polynomial

$$\prod_{\lambda \in Q} (h(x)-\lambda) = h(x)^{p^r} - h(x).$$

Thus h has the desired properties: $\deg h < n$ and $h^{p^r} \equiv h \bmod f$. □

For the remainder of this section, we assume that F itself is a finite field, say $F = \mathrm{GF}(q)$. In this situation, Theorem 6.5.1 can be used to compute a basis for the original Berlekamp algebra $\mathscr{B}_f$ as follows. Taking $Q = F$ in Theorem 6.5.1 shows that $\mathscr{B}_f$ is the kernel of the F-linear mapping

$$F[x]_{<n} \to F[x]_{<n}, \quad g(x) \mapsto g(x)^q - g(x) \bmod f(x). \tag{6.16}$$

Now let $b_0 := 1$ and $b_1(x) := x^q \bmod f(x)$, and define polynomials $b_k(x)$ recursively by putting $b_{j+1}(x) := b_j(x) b_1(x) \bmod f(x)$ for $j = 0,\ldots,n-2$. Then

$$(x^j)^q = (x^q)^j \equiv b_1(x)^j \equiv b_j(x) \bmod f(x) \quad \text{for } j = 0, 1, \ldots, n-1.$$

Thus the image of

$$\beta_f : F[x]_{<n} \to F[x]_{<n}, \; g(x) \mapsto g(x)^q \ \mathtt{mod}\ f(x)$$

is the subspace of $F[x]_{<n}$ generated by the polynomials $b_j(x)$. Writing the coefficients of these polynomials as

$$b_j(x) = B_{0,j} + B_{1,j}x + \cdots + B_{n-1,j}x^{n-1}$$

defines a matrix

$$B_f := (B_{i,j})_{i,j=0,\ldots,n-1} \in F^{(n,n)} \tag{6.17}$$

which represents β_f with respect to the canonical basis $\{1, x, x^2, \ldots, x^{n-1}\}$ of $F[x]_{<n}$. Hence Theorem 6.5.1 specializes to the following result:

Proposition 6.5.2. *Let $f(x)$ be a non-constant monic polynomial over a finite field F. Then the matrix B_f in Equation* (6.17) *has* 1 *as an eigenvalue. The Berlekamp algebra $\mathscr{B}_f$ of f is the corresponding eigenspace, and its F-dimension is the number of distinct irreducible divisors of f. In particular, if f is square-free, then f is irreducible if and only if this eigenspace is* 1*-dimensional.* □

Remark 6.5.3. If I denotes the (n,n)-identity matrix over F, then $\mathscr{B}_f$ is the kernel of the matrix $B_f - I$. Thus a basis for $\mathscr{B}_f$ can be determined efficiently by first computing B_f and then applying the Gaussian algorithm to $B_f - I$. In particular, the number of distinct irreducible factors of f equals the co-rank of $B_f - I$. □

Example 6.5.4. Let us once again return to the situation of Example 6.3.7, where $q = 5$ and $f(x) = x^5 - 2x^3 - x^2 + 2$. Then

$$\begin{aligned} x^5 &\equiv 2x^3 + x^2 - 2 \mod f, \\ x^{10} &\equiv -x^4 - x^3 + 2x + 1 \mod f, \\ x^{15} &\equiv -2x^4 - 2x^3 + 2x - 2 \mod f, \\ x^{20} &\equiv 2x^4 + x^3 + x^2 - 2x - 1 \mod f, \end{aligned}$$

so that

$$B_f - I = \begin{pmatrix} 0 & -2 & 1 & -2 & -1 \\ 0 & -1 & 2 & 2 & -2 \\ 0 & 1 & -1 & 0 & 1 \\ 0 & 2 & -1 & 2 & 1 \\ 0 & 0 & -1 & -2 & 1 \end{pmatrix}.$$

The kernel of this matrix is spanned by the three vectors

$$\begin{pmatrix}1\\0\\0\\0\\0\end{pmatrix},\quad \begin{pmatrix}0\\-2\\-2\\1\\0\end{pmatrix}\quad\text{and}\quad \begin{pmatrix}0\\0\\1\\0\\1\end{pmatrix},$$

which gives rise to the basis $\{1, x^3-2x^2-2x, x^4+x^2\}$ of $\mathscr{B}_f$ already used in Example 6.3.10. □

We have now seen that one may determine (a basis for) the Berlekamp algebra $\mathscr{B}_f$ of f in the finite case efficiently, but this still leaves a major problem. By Theorem 6.3.5, any non-trivial polynomial $h \in \mathscr{B}_f$ gives rise to a (partial) factorization of $f(x)$ as $\prod_{\gamma\in C_h} \gcd(h(x)-\gamma, f(x))$. Obviously, computing this factorization requires the set C_h of spectral coefficients of h. In principle, these coefficients could be obtained by trial and error: just take any one of the q field elements $\lambda \in F$ and test whether or not $\gcd(h(x)-\lambda, f(x)) \neq 1$. Clearly, this is only practical when q is sufficiently small, even if Theorem 6.4.1 is applied in order to describe the distinct spectral coefficients of h as the roots of the minimal polynomial of h modulo f. For large q, trial and error is not feasible, and we need to obtain information about C_h by other methods.

In the remainder of this section, we shall see how one may overcome this problem by using appropriate probabilistic approaches. First, we consider the situation where F has a (small) non-trivial subfield Q (so that F is not a prime field). This will result in a practical method for extension fields with a small characteristic p, in particular, for the binary case $p=2$. At the end of this section, we will modify these methods to deal with fields with a (large) odd characteristic.

Remark 6.5.5. Let $m=[F:Q]$, and let $\{\beta_0,\ldots,\beta_{m-1}\}$ be any basis for F/Q. Then the elements

$$\beta_i x^j \quad \text{with } i=0,\ldots,m-1;\ j=0,\ldots,n-1$$

form a Q-Basis for $F[x]_{<n}$. Similar to the approach leading to Proposition 6.5.2, one can then determine a matrix representing the Q-linear transformation in Equation (6.16) with respect to this basis. Moreover, the kernel of this mapping provides a basis for the Berlekamp algebra $\mathscr{B}_f^Q$. □

We will provide an alternative way for determining a basis for the Berlekamp algebra $\mathscr{B}_f^Q$ in Theorem 6.5.8 below, which in fact is a special case of Theorem 6.4.9. This alternative approach will lead us immediately to the probabilistic method of Cantor and Zassenhaus [61] for finding a polynomial h in the Berlekamp algebra of $f(x)$ such that C_h is a very small subset of the prime field of F. Let us first introduce the following useful concept.

Definition 6.5.6. Let $f(x)$ be a monic polynomial over $F=\mathrm{GF}(q)$, let Q be a proper subfield of F, and let h be any polynomial in $\mathscr{B}_f$. Then the (F,Q)**-trace** $\tau_{F/Q}(h)$ **of** h **modulo** f is defined by

$$\tau_{F/Q}(h(x)) := \sum_{j=0}^{m-1} h(x)^{p^{rj}} \mod f(x),$$

where $m = [F : Q]$ and $|Q| = p^r$. (Note that, in general, $\tau_{F/Q}(h)$ does not have coefficients in Q.) □

Remark 6.5.7. In actual computations, it is always desirable to keep the degrees of the polynomials involved as small as possible. In the case of the (F,Q)-trace of h modulo f, the following approach is suitable. We put $h_0(x) := h(x)$ and recursively $h_i(x) := h_{i-1}(x)^{p^r} \mathtt{~mod~} f$ for $i = 1,\ldots,m-1$. Then

$$h_i(x) \equiv h_0(x)^{p^{ri}} \mod f(x) \text{ for every } i,$$

so that $H(x) := \sum_{i=0}^{m-1} h_i(x)$ is the (F,Q)-trace of h modulo f; note that

$$h_{m-1}(x)^{p^r} \equiv h(x)^{p^{rm}} = h(x)^q \equiv h(x) = h_0(x) \mod f,$$

by Equation (6.15). The determination of H can be summarized in algorithmic notation as follows:

(1) $H(x) \leftarrow h(x)$;
(2) **for** $i = 1$ **to** $m-1$ **do**
(3) $\quad H(x) \leftarrow \left(H(x)^{p^r} \bmod f(x)\right) + h(x)$
(4) **od**

Of course, the intermediate results $H(x)^{p^r} \mathtt{~mod~} f(x)$ should be computed using square-and-multiply modulo f, see Algorithm 1.6.25. We now use the representation $\tau_{F/Q}(h(x)) = H(x)$ of the (F,Q)-trace to check that $\tau_{F/Q}(h(x))$ is in fact in $\mathscr{B}_f^Q$:

$$H(x)^{p^r} = \sum_{i=0}^{m-1} h_i(x)^{p^r} \equiv \sum_{i=0}^{m-1} h_0(x)^{p^{ri+r}} \equiv \sum_{j=1}^{m} h(x)^{p^{rj}} \equiv H(x) \mod f,$$

since $h(x)^{p^{rm}} \equiv h(x) \bmod f$. Now Theorem 6.5.1 gives the desired result.

In this context, it is interesting to note that the (F,Q)-trace $\tau_{F/Q}(h(x))$ is in fact obtained from a polynomial action on $\mathscr{B}_f$ as in Theorem 6.4.9 (applied with $Q = F$), where the underlying polynomial τ is the (F,Q)**-trace polynomial** given by

$$\tau(x) := \sum_{j=0}^{m-1} x^{p^{rj}}.$$

However, Theorem 6.4.9 only guarantees that the associated mapping $\overline{\tau}$ acts on $\mathscr{B}_f$. As we just observed, $\overline{\tau}$ actually maps $\mathscr{B}_f$ into $\mathscr{B}_f^Q$; this is due to the special form of the (F,Q)-trace mapping τ. □

Let us collect the essential facts about $\overline{\tau}$:

Theorem 6.5.8. *Let $f(x)$ be a monic polynomial over $F = \mathrm{GF}(q)$, let Q be a proper subfield of F, and consider the (F,Q)-trace mapping modulo f:*

$$\tau_{F/Q} \colon \mathscr{B}_f \to F[x]_{<n}, \quad h(x) \mapsto \sum_{j=0}^{m-1} h(x)^{p^{rj}} \bmod f,$$

where $m = [F : Q]$ and $|Q| = p^r$. Then $\tau_{F/Q}$ is Q-linear and maps $\mathscr{B}_f$ onto $\mathscr{B}_f^Q$. Moreover, for every $h \in \mathscr{B}_f$, the set of spectral coefficients of $H(x) := \tau_{F/Q}(h(x))$ is given by

$$C_H = \{\mathrm{Tr}_{F/Q}(c) \colon c \in C_h\},$$

where $\mathrm{Tr}_{F/Q}$ denotes the trace mapping for the field extension F/Q.

Proof. Let $h = \sum_{i=1}^{s} c_i e_i(x) \in \mathscr{B}_f$, so that $c_1, \ldots, c_s$ are the spectral coefficients of h. Using the fundamental properties of the primitive idempotents $e_i(x)$, we obtain

$$\begin{aligned} H(x) = \tau_{F/Q}(h(x)) &\equiv \sum_{j=0}^{m-1} \sum_{i=1}^{s} \left(c_i e_i(x)\right)^{p^{rj}} \equiv \sum_{j=0}^{m-1} \sum_{i=1}^{s} c_i^{p^{rj}} e_i(x) \\ &\equiv \sum_{i=1}^{s} \left(\sum_{j=0}^{m-1} c_i^{p^{rj}} \right) e_i(x) \equiv \sum_{i=1}^{s} \mathrm{Tr}_{F/Q}(c_i) e_i(x) \bmod f. \end{aligned}$$

Thus the spectral coefficients of H are indeed the (F,Q)-traces of the spectral coefficients of h; in particular, $\tau_{F/Q}(h(x)) \in \mathscr{B}_f^Q$.

Clearly, $\tau_{F/Q}$ is linear. Thus it only remains to check that $\tau_{F/Q}$ maps $\mathscr{B}_f$ onto $\mathscr{B}_f^Q$, which means that the image of $\tau_{F/Q}$ should have dimension $\dim_Q \mathscr{B}_f^Q = s$. To see this, we note that $\tau_{F/Q}\left(\sum_{i=1}^{s} c_i e_i(x)\right) = 0$ if and only if $\mathrm{Tr}_{F/Q}(c_i) = 0$ for all i. Hence the kernel of $\tau_{F/Q}$ is isomorphic to the s-fold cartesian product of the kernel of $\mathrm{Tr}_{F/Q}$, an $(m-1)$-dimensional Q-subspace of F (by Proposition 3.12.2). Thus the kernel of $\tau_{F/Q}$ has Q-dimension $s(m-1)$, and hence the image of $\tau_{F/Q}$ has Q-dimension

$$\dim_Q \mathscr{B}_f - s(m-1) = sm - s(m-1) = s,$$

as claimed. □

According to Theorem 6.5.8, we may obtain any $H \in \mathscr{B}_f^Q$ as the (F,Q)-trace modulo f of some $h \in \mathscr{B}_f$. If we wish to use such an H for computing a non-trivial (partial) factorization of f, namely

$$f(x) = \prod_{\lambda \in C_H} \gcd(H(x) - \lambda, f(x)),$$

we need to ensure that C_H is not a singleton set; see Remark 6.3.6. Thus the (F,Q)-trace of the set C_h of spectral coefficients of h should be non-constant. Since for every s-tuple $(c_1, \ldots, c_s) \in F^s$ there is a unique $h \in \mathscr{B}_f$ with $h(x) \equiv c_i \bmod U_i(x)$ for all i, this cannot hold for all choices of h. However, Theorem 6.5.8 and its proof lead to an easy probabilistic analysis of this situation:

Theorem 6.5.9. *Under the assumptions of Theorem 6.5.8, let some polynomial h be chosen uniformly at random from $\mathscr{B}_f$, and let H be its (F,Q)-trace modulo f.*

Then the probability that the set C_H of spectral coefficients of H contains at least two distinct elements (so that H gives rise to a non-trivial factorization of f) equals $1-1/|Q|^{s-1}$, where s is the number of distinct irreducible divisors of f over F.

Proof. By definition, $\mathscr{B}_f$ consists of all polynomials of the form $\sum_{i=1}^{s} c_i e_i(x)$, where the $e_i(x)$ are the primitive idempotents modulo f and where all c_i belong to F. Thus choosing h uniformly at random from $\mathscr{B}_f$ amounts to drawing the coefficient vector $(c_1,\dots,c_s)$ uniformly at random from F^s.

Recall that $H(x)=\tau_{F/Q}(h(x))$ does *not* lead to a non-trivial factorization of f if and only if H belongs to the diagonal subspace Q of $\mathscr{B}_f^Q$, that is, if $(c_1,\dots,c_s)$ is in the pre-image of Q under $\tau_{F/Q}$. By Theorem 6.5.8 and its proof, this pre-image has cardinality

$$|Q|\cdot|\ker(\tau_{F/Q})| = |Q|\cdot|Q|^{\dim_Q \ker(\tau_{F/Q})} = |Q|\cdot|Q|^{s(m-1)} = |Q|^{s(m-1)+1}.$$

Hence the probability that $H=\tau_{F/Q}(h)$ leads to a non-trivial factorization of f is given by

$$1-\frac{|Q|^{s(m-1)+1}}{|\mathscr{B}_f|} = 1-\frac{|Q|^{s(m-1)+1}}{|Q|^{sm}} = 1-\frac{1}{|Q|^{sm-s(m-1)-1}} = 1-\frac{1}{|Q|^{s-1}},$$

as $\dim_Q \mathscr{B}_f = sm$. □

Remark 6.5.10. We may assume $s\geq 2$. Then the probability determined in Theorem 6.5.9 is the larger the bigger the subfield Q is; in the extreme case $Q=F$, we obtain $1-1/|F|^{s-1}$ as the probability for a successful choice out of $\mathscr{B}_f$. However, we need the spectral coefficients of H in order to factorize f via Theorem 6.3.5. For this, it is advisable to choose Q as *small* as possible, that is, one should use the prime subfield P of F. Note that this still gives a probability

$$1-\frac{1}{p^{s-1}} \geq 1-\frac{1}{p} \geq \frac{1}{2}$$

for a successful choice of h, where p is the characteristic of F. If we choose a sequence $h_1(x),\dots,h_k(x)$ of k polynomials independently and uniformly at random from $\mathscr{B}_f$, the probability that none of the polynomials $\tau_{F/P}(h_i(x))$ leads to a non-trivial factorization of $f(x)$ is therefore at most $1/2^k$, which can be made arbitrarily small by choosing k large enough.

In particular, this approach works very well in the important special case where F has characteristic 2, so that P is the binary field. Then

$$f(x) = \gcd(H(x),f(x))\cdot\gcd(H(x)+1,f(x)) \quad \text{for all } H\in\mathscr{B}_f^P,$$

and the Berlekamp algebra $\mathscr{B}_f^P$ just consists of the idempotents of $F[x]/(f)$. □

Unfortunately, the trial and error method for determining C_H for $H = \tau_{F/Q}(h)$ outlined in Remark 6.5.10 is not practical when F has a large characteristic p; in particular, this problem occurs when F is itself a large prime field. Here one uses an approach due to Cantor and Zassenhaus [61], which is quite similar to using the trace function modulo f but relies on the multiplicative subgroup of F instead. As we now consider a field $F = \mathrm{GF}(q)$ with odd characteristic p, the multiplicative group F^* of F splits into the set of squares S and the set of non-squares N, both of which have cardinality $(q-1)/2$:

$$S = \{z \in F^* : z^{(q-1)/2} = 1\} \quad \text{and} \quad N = \{z \in F^* : z^{(q-1)/2} = -1\}; \tag{6.18}$$

see Proposition 3.2.13. Accordingly,

$$\begin{aligned} x^q - x = \sum_{\lambda \in F} (x-\lambda) &= x \cdot \prod_{\lambda \in S} (x-\lambda) \cdot \prod_{\nu \in N} (x-\nu) \\ &= x \cdot \left(x^{(q-1)/2} - 1\right) \cdot \left(x^{(q-1)/2} + 1\right). \end{aligned}$$

We now apply the polynomial action on $\mathscr{B}_f$ induced by the **quadratic character polynomial** $\pi(x) := x^{(q-1)/2}$; that is, given any polynomial $h(x) = \sum_{i=1}^{s} c_i e_i(x)$ in $\mathscr{B}_f$, we consider the polynomial

$$H(x) = \overline{\pi}(h) := h(x)^{(q-1)/2} \ \mathtt{mod}\ f(x).$$

Using the fundamental properties of the primitive idempotents modulo f, we obtain

$$H(x) \equiv \left(\sum_{i=1}^{s} c_i e_i(x)\right)^{(q-1)/2} \equiv \sum_{i=1}^{s} c_i^{(q-1)/2} e_i(x) \mod f(x).$$

Note that $c_i^{(q-1)/2} \in \{0, 1, -1\}$ for all i. Thus $H(x)$ is in fact a polynomial in the prime Berlekamp algebra $\mathscr{B}_f^P$, and there are just three possible spectral coefficients for H, namely 0, 1 and -1, so that

$$f(x) = \gcd(H(x), f(x)) \cdot \gcd(H(x)-1, f(x)) \cdot \gcd(H(x)+1, f(x)).$$

We then have the following analogue of Theorem 6.5.9:

Theorem 6.5.11. *Let $f(x)$ be a monic polynomial over $F = \mathrm{GF}(q)$, and assume that F has odd characteristic. Let some polynomial h be chosen uniformly at random from $\mathscr{B}_f$, and let $H(x) := h(x)^{(q-1)/2}$ $\mathtt{mod}$ $f(x)$.*

Then the probability that the set C_H of spectral coefficients of H contains at least two distinct elements (so that H gives rise to a non-trivial factorization of f) equals

$$1 - \frac{1}{q^s} - \frac{1}{2^{s-1}} \cdot \left(1 - \frac{1}{q}\right)^s,$$

where s is the number of distinct irreducible divisors of f over F.

Proof. As in the proof of Theorem 6.5.9, a random choice of $h(x) = \sum_{i=1}^{s} c_i e_i(x)$ from $\mathscr{B}_f$ amounts to drawing the corresponding spectral vector $(c_1, \dots, c_s)$ uniformly at random from F^s. As noted above,

$$C_H = \{\gamma^{(q-1)/2} \colon \gamma \in C_h\} \subseteq \{0, 1, -1\},$$

and hence the image of the action $\overline{\pi}$ of the quadratic character polynomial π on $\mathscr{B}_f$ consists of all polynomials $a(x) \in \mathscr{B}_f$ with $C_a \subseteq \{0, 1, -1\}$. Clearly, $H(x)$ does *not* lead to a non-trivial factorization of $f(x)$ if and only if

$$H(x) = \sum_{i=1}^{s} \lambda e_i(x) = \lambda \quad \text{with } \lambda \in \{0, 1, -1\}.$$

Thus we need to distinguish three cases:

- First assume $H(x) = 0$. Then $C_h = \{0\}$, which means $h(x) = 0$. Thus H has a unique pre-image h under $\overline{\pi}$ in this case.
- Now let $H(x) = 1$, which means that C_h consists of squares in F^*, so that $(c_1, \dots, c_s) \in S^s$; see Equation (6.18). This event occurs with probability

$$\left(\frac{|S|}{|F|}\right)^s = \left(\frac{q-1}{2q}\right)^s.$$

- Similarly, $H(x) = -1$ means that C_h consists of non-squares in F^*, so that $(c_1, \dots, c_s) \in N^s$. As $|N| = |S|$, this event also occurs with probability

$$\left(\frac{|S|}{|F|}\right)^s = \left(\frac{q-1}{2q}\right)^s.$$

Hence the probability that $H = \overline{\pi}(h)$ leads to a non-trivial factorization of f is given by

$$1 - \frac{1}{q^s} - 2 \cdot \left(\frac{q-1}{2q}\right)^s = 1 - \frac{1}{q^s} - \frac{1}{2^{s-1}} \cdot \left(1 - \frac{1}{q}\right)^s,$$

as claimed. □

Remark 6.5.12. Again, we may assume $s \geq 2$. By Theorem 6.5.11, the probability of a successful choice of $h \in \mathscr{B}_f$ is at least

$$1 - \frac{1}{q^2} - \frac{1}{2} \cdot \left(1 - \frac{1}{q}\right)^2 = \frac{1}{2} + \frac{2q-3}{2q^2} > \frac{1}{2},$$

as $q \geq 3$. Thus we may proceed as in Remark 6.5.10 and choose a sequence $h_1(x), \dots, h_k(x)$ of k polynomials independently and uniformly at random from $\mathscr{B}_f$. Then the probability that none of the polynomials $H_i = \overline{\pi}(h_i)$ leads to a non-trivial factorization of f is $< 1/2^k$. □

Exercises

Exercise 6.5.13. Consider the polynomial $f(x) = x^5 - x^4 + x - 1$ over the prime field $P = \mathrm{GF}(5)$. Determine a basis of the Berlekamp algebra $\mathscr{B}_f$ of f and use this to compute the canonical factorization of f over P. □

Exercise 6.5.14. Consider the polynomial $f(x) = x^5 - x^4 + x - 1$ introduced in Exercise 6.5.13 as a polynomial over $F = \mathrm{GF}(25)$. Use Remark 6.5.5 to determine a basis of the Berlekamp algebra $\mathscr{B}_f^P$, where again $P = \mathrm{GF}(5)$. □

Exercise 6.5.15. Consider the polynomial $f(x) = x^n - \omega$ over $F = \mathrm{GF}(q)$, where $q \equiv 1 \bmod n$ and $\omega \neq 0$. Determine a basis of the Berlekamp algebra $\mathscr{B}_f$ of f, depending on the parameter $m := \mathrm{ord}(\omega)$. □

6.6 Niederreiter Spaces

In the remainder of this chapter, we present a method for factorizing polynomials over finite fields first developed by Harald Niederreiter in the 1990s; see the series of papers by Niederreiter [303, 302, 304, 305], by Niederreiter and Göttfert [306, 307], and by Göttfert [154] for various aspects of this approach. Our treatment follows the same philosophy as our presentation of Berlekamp's method: we will first present the basic ideas in a framework which is as general as possible, and then explain why this approach gives a practical method in the special case of finite fields.

Let F be a perfect field, and let $f(x)$ be a monic polynomial over F.[5] In Berlekamp's method, the factor algebra $F[x]/(f)$ was the main tool: the canonical factorization of f turned out to be intimately related to the structure of that algebra via the Chinese remainder theorem. The setting for Niederreiter's method is another algebraic object, namely the field $F(x)$ of rational functions in the indeterminate x over F. We now associate with f the following F-vector space:

$$W_f := \left\{ \frac{g(x)}{f(x)} : g(x) \in F[x], \deg g < \deg f \right\}, \tag{6.19}$$

that is, the space of all rational functions with denominator f and with numerator a polynomial with strictly smaller degree. With the usual notation $n := \deg f$, we have $\dim_F W_f = n$, and the canonical F-basis of W_f consists of the rational functions $x^i/f(x)$ with $i = 0, \ldots, n-1$.

We begin with a direct decomposition of W_f which is related to the unique factorization of f. As before, we let $f(x) = \prod_{i=1}^{s} u_i(x)^{\ell_i}$ be the (unknown) canonical factorization of f over F and denote the primary divisors of f by $U_i(x) := u_i(x)^{\ell_i}$ and their cofactors by $\overline{U}_i(x) := f(x)/U_i(x)$.

[5] We will need the assumption that all irreducible factors of the polynomial under consideration are separable. For simplicity, we assume throughout that F is perfect.

Proposition 6.6.1. *With the notation above,* $W_f = \bigoplus_{i=1}^{s} W_{U_i}$.

Proof. Let $V := \sum_{i=1}^{s} W_{U_i}$, and consider any element $\sum_{i=1}^{s} g_i(x)/U_i(x)$ of V. Note that

$$\sum_{i=1}^{s} \frac{g_i(x)}{U_i(x)} = \sum_{i=1}^{s} \frac{g_i(x)\overline{U}_i(x)}{f(x)} = \frac{\sum_{i=1}^{s} g_i(x)\overline{U}_i(x)}{f(x)}$$

and

$$\deg(g_i \cdot \overline{U}_i) \leq \deg g_i + \deg \overline{U}_i < \deg U_i + \deg \overline{U}_i = \deg f$$

for all i. Thus $\sum_{i=1}^{s} g_i(x)/U_i(x) \in W_f$, and hence $V \subseteq W_f$.

Assume $\sum_{i=1}^{s} g_i(x)/U_i(x) = 0$. Then $0 = f \cdot \sum_{i=1}^{s} g_i(x)/U_i(x) = \sum_{i=1}^{s} g_i(x)\overline{U}_i(x)$, and hence

$$-g_j(x)\overline{U}_j(x) = \sum_{\substack{i=1 \\ i \neq j}}^{s} g_i(x)\overline{U}_i(x) \quad \text{for } j = 1, \ldots, s.$$

Since the right hand side of this identity is divisible by $U_j(x)$ and since $U_j(x)$ and $\overline{U}_j(x)$ are relatively prime, we see that $U_j(x)$ has to divide $g_j(x)$, and therefore actually $g_j(x) = 0$ (as $\deg g_j < \deg U_j$). As this holds for all j, we conclude that V is in fact the *direct* sum of the vector spaces W_{U_i}. Combining these observations gives

$$\deg f = \dim_F W_f \geq \dim_F V = \sum_{i=1}^{s} \dim_F W_{U_i} = \sum_{i=1}^{s} \deg U_i = \deg f,$$

and hence $W_f = V = \bigoplus_{i=1}^{s} W_{U_i}$, as claimed. □

The representation of g/f (where $\deg g < \deg f$) as $\sum_{i=1}^{s} g_i(x)/U_i(x)$ according to Proposition 6.6.1 is called the **partial fraction decomposition** of g/f.

The following definition of the Niederreiter subspace of W_f involves the derivatives of the irreducible factors u_i of f; note that $u_i'(x) \neq 0$ for all i, since F is assumed to be perfect. In order to gain more flexibility, we introduce a subfield Q of F as a further parameter (as we also did in Berlekamp's method).

Definition 6.6.2. In the situation just described, the **fractional Niederreiter space** $\bar{\mathcal{N}}_f^Q$ of f over Q is defined to be the Q-subspace of W_f generated by the rational functions $u_i'(x)/u_i(x)$:

$$\bar{\mathcal{N}}_f^Q := \left\{ \sum_{i=1}^{s} \lambda_i \frac{u_i'(x)}{u_i(x)} : \lambda_i \in Q \text{ for } i = 1, \ldots, s \right\}. \qquad \square$$

Remark 6.6.3. Note first that $\bar{\mathcal{N}}_f^Q$ is indeed a Q-subspace of W_f, since

$$\frac{u_i'(x)}{u_i(x)} = \frac{u_i'(x) \cdot u_i(x)^{\ell_i - 1}}{U_i(x)} \in W_{U_i} \subseteq W_f$$

for all i. By Proposition 6.6.1, the fractions $u_i'(x)/u_i(x)$ are linearly independent over Q, and therefore the Q-dimension of $\bar{\mathcal{N}}_f^Q$ is just the number of distinct monic irreducible divisors of f over F. We refer to $\{u_i'(x)/u_i(x)\colon i=1,\dots,s\}$ as the **fractional basis** of $\bar{\mathcal{N}}_f^Q$.

By definition, $\bar{\mathcal{N}}_f^Q$ depends only on the radical $\mathrm{rad}(f(x)) = \prod_{i=1}^s u_i(x)$ of f, but not on the multiplicities of the u_i:

$$\bar{\mathcal{N}}_f^Q = \bar{\mathcal{N}}_{\mathrm{rad}(f)}^Q \subseteq W_{\mathrm{rad}(f)}.$$

Moreover, $W_{\mathrm{rad}(f)} \subseteq W_f$, with equality if and only if f is square-free. □

In what follows, it will be convenient to switch from fractions to polynomials via the canonical F-vector space isomorphisms

$$W_f \to F[x]_{<n}, \ \frac{g(x)}{f(x)} \mapsto g(x) \quad \text{and} \quad W_{\mathrm{rad}(f)} \to F[x]_{<t}, \ \frac{a(x)}{\mathrm{rad}(f(x))} \mapsto a(x),$$

where t denotes the degree of $\mathrm{rad}(f)$. This leads to the following definition:

Definition 6.6.4. The **polynomial Niederreiter space** $\mathcal{N}_f^Q$ of f over Q is the Q-subspace of $F[x]_{<n}$ generated by the polynomials $u_i'(x)f(x)/u_i(x)$:

$$\mathcal{N}_f^Q := \Big\{\sum_{i=1}^s \lambda_i u_i'(x)\frac{f(x)}{u_i(x)} : \lambda_i \in Q \text{ for } i=1,\dots,s\Big\}. \qquad \square$$

Remark 6.6.5. In what follows, we write $\bar{u}_i(x) := \mathrm{rad}(f(x))/u_i(x)$ for $i=1,\dots,s$ and denote the span of a subset C of some vector space V over Q by $\langle C\rangle_Q$. By definition, the polynomial Niederreiter space of the radical of f over Q is

$$\mathcal{N}_{\mathrm{rad}(f)}^Q = \Big\langle u_i'(x)\frac{\mathrm{rad}(f(x))}{u_i(x)} : i=1,\dots,s\Big\rangle_Q = \Big\langle u_i'(x)\bar{u}_i(x) : i=1,\dots,s\Big\rangle_Q. \tag{6.20}$$

Thus $\mathcal{N}_{\mathrm{rad}(f)}^Q$ and $\mathcal{N}_f^Q$ are related by the simple identity

$$\mathcal{N}_f^Q = \frac{f(x)}{\mathrm{rad}(f(x))} \cdot \mathcal{N}_{\mathrm{rad}(f)}^Q. \tag{6.21}$$

In more detail, Proposition 6.6.1 and the definition of the Niederreiter spaces give the following identities:

$$\begin{aligned}
\mathcal{N}_f^Q &= \bigoplus_{i=1}^s \frac{f(x)}{U_i(x)} \cdot \mathcal{N}_{U_i}^Q = \bigoplus_{i=1}^s \frac{f(x)}{U_i(x)} \cdot \frac{U_i(x)}{u_i(x)} \cdot \mathcal{N}_{u_i}^Q \\
&= \bigoplus_{i=1}^s \frac{f(x)}{u_i(x)} \cdot \mathcal{N}_{u_i}^Q = \frac{f(x)}{\mathrm{rad}(f(x))} \cdot \bigoplus_{i=1}^s \frac{\mathrm{rad}(f(x))}{u_i(x)} \cdot \mathcal{N}_{u_i}^Q \\
&= \frac{f(x)}{\mathrm{rad}(f(x))} \cdot \mathcal{N}_{\mathrm{rad}(f)}^Q. \qquad \square
\end{aligned}$$

Definition 6.6.6. The **derived bases** for the Niederreiter spaces $\mathscr{N}_f^Q$ and $\mathscr{N}_{\mathrm{rad}(f)}^Q$ are defined as the sets $\{d_i(x) : i = 1,\ldots,s\}$ and $\{\delta_i(x) : i = 1,\ldots,s\}$, respectively, where

$$d_i(x) := u_i'(x)\frac{f(x)}{u_i(x)} \quad \text{and} \quad \delta_i(x) := u_i'(x)\frac{\mathrm{rad}(f(x))}{u_i(x)} = u_i'(x)\bar{u}_i(x)$$

for $i = 1,\ldots,s$. □

Observation 6.6.7. Note that $d_i(x) = f(x)\delta_i(x)/\mathrm{rad}(f(x))$ for $i = 1,\ldots,s$. Moreover, the derived basis for $\mathscr{N}_{\mathrm{rad}(f)}^Q$ is *orthogonal* in the sense that

$$\gcd(\delta_j(x), u_j(x)) = 1 \quad \text{and} \quad \delta_i(x) \equiv 0 \bmod u_j(x) \;\text{ if } i \neq j$$

for $i, j = 1,\ldots,s$.

For every polynomial $H \in \mathscr{N}_f^Q$, there exists a unique s-tuple $(c_1,\ldots,c_s) \in Q^s$ with $H(x) = \sum_{i=1}^s c_i d_i(x)$; we put $D_H := \{c_1,\ldots,c_s\}$ and call the c_i the **derived coefficients** of H. An analogous terminology is used with respect to the derived basis $\delta_1(x),\ldots,\delta_s(x)$ of $\mathscr{N}_{\mathrm{rad}(f)}^Q$. Using the bijection

$$\mathscr{N}_f^Q \to \mathscr{N}_{\mathrm{rad}(f)}^Q, \quad H(x) \mapsto H(x)\cdot\frac{\mathrm{rad}(f(x))}{f(x)},$$

we obtain $D_H = D_h$, where $H \in \mathscr{N}_f^Q$ and where $h(x) = H(x)\mathrm{rad}(f(x))/f(x)$ is the corresponding polynomial in $\mathscr{N}_{\mathrm{rad}(f)}^Q$. □

If we could manage to determine the derived basis of $\mathscr{N}_f^Q$, we would immediately obtain the canonical factorization of f, since $\gcd(f(x), d_i(x)) = f(x)/u_i(x)$ and therefore

$$\frac{f(x)}{\gcd(f(x), d_i(x))} = u_i(x) \quad \text{for } i = 1,\ldots,s.$$

(Here we use that $u_i'(x)$ and $u_i(x)$ are relatively prime, by our assumption that F is a perfect field.) Of course, we could also use the derived basis of $\mathscr{N}_{\mathrm{rad}(f)}^Q$ instead. Similar to the case of Berlekamp's method, it turns out that the canonical factorization of f can be obtained (at least in principle) from *any* basis of some polynomial Niederreiter space $\mathscr{N}_f^Q$. This is our next aim; we will first deal with the radical of f and consider the general case later.

Theorem 6.6.8. *Let $f(x)$ be a monic polynomial over a perfect field F, and let Q be any subfield of F. In addition, let h be a polynomial in $\mathscr{N}_{\mathrm{rad}(f)}^Q$ with set of derived coefficients $D_h = \{c_1,\ldots,c_s\}$, and let $\lambda \in Q$.*

Then $\gcd(h(x) - \lambda\,\mathrm{rad}(f)'(x), \mathrm{rad}(f(x))) \neq 1$ *if and only if $\lambda \in D_h$, in which case*

$$\gcd\left(h(x) - \lambda\,\mathrm{rad}(f)'(x), \mathrm{rad}(f(x))\right) = \prod_{\substack{i=1 \\ c_i=\lambda}}^{s} u_i(x).$$

Moreover,

$$\operatorname{rad}(f(x)) = \prod_{\lambda \in D_h} \gcd\left(h(x) - \lambda \operatorname{rad}(f)'(x), \operatorname{rad}(f(x))\right).$$

Proof. Using the factorization $\operatorname{rad}(f(x)) = \prod_{i=1}^{s} u_i(x)$, the derivative of $\operatorname{rad}(f(x))$ may be written as follows:

$$\operatorname{rad}(f)'(x) = \sum_{i=1}^{s} \left(u_i'(x) \cdot \prod_{\substack{j=1 \\ j \neq i}}^{s} u_j(x)\right) = \sum_{i=1}^{s} u_i'(x) \cdot \frac{\operatorname{rad}(f(x))}{u_i(x)} = \sum_{i=1}^{s} \delta_i(x).$$

Thus $\operatorname{rad}(f)'(x)$ belongs to the Niederreiter space $\mathcal{N}_{\operatorname{rad}(f)}^{Q}$, and we obtain

$$h(x) - \lambda \operatorname{rad}(f)'(x) = \sum_{i=1}^{s} (c_i - \lambda)\delta_i(x) \quad \text{for } \lambda \in Q.$$

As $u_j(x)$ divides $\delta_i(x)$ for $j \neq i$, we conclude that $u_i(x)$ divides $h(x) - \lambda \operatorname{rad}(f)'(x)$ if and only if it divides $(c_i - \lambda)\delta_i(x)$. In view of $\gcd(u_i(x), \delta_i(x)) = 1$, this happens if and only if $c_i = \lambda$, proving the first assertion. Now the second assertion follows by observing

$$\gcd(h(x) - \lambda \operatorname{rad}(f)'(x), h(x) - \mu \operatorname{rad}(f)'(x)) = 1 \quad \text{for } \lambda \neq \mu:$$

if $u_i(x)$ divides both $h(x) - \lambda \operatorname{rad}(f)'(x)$ and $h(x) - \mu \operatorname{rad}(f)'(x)$, then $\lambda = c_i = \mu$. □

Remark 6.6.9. Note that Theorem 6.6.8 does *not* lead to a non-trivial factorization of $\operatorname{rad}(f)$ if and only if $h(x) - \lambda \operatorname{rad}(f)'(x) = 0$ for some $\lambda \subset Q$. Thus $Q \cdot \operatorname{rad}(f)'(x)$ is a (trivial) subspace of the Niederreiter space $\mathcal{N}_{\operatorname{rad}(f)}^{Q}$. In particular, one has $\mathcal{N}_{\operatorname{rad}(f)}^{Q} = Q \cdot \operatorname{rad}(f)'(x)$ if and only if $s = 1$, that is, if and only if $\operatorname{rad}(f)$ is irreducible. □

Finally, it is a simple matter to generalize Theorem 6.6.8 from $\operatorname{rad}(f)$ to f itself:

Theorem 6.6.10. *Let $f(x)$ be a monic polynomial over a perfect field F, and let Q be any subfield of F. In addition, let H be a polynomial in $\mathcal{N}_f^Q$ with set of derived coefficients $D_H = \{c_1, \ldots, c_s\}$, and let $\lambda \in Q$. Then*

$$\gcd\left(H(x) - \lambda \cdot \frac{f(x)}{\operatorname{rad}(f(x))} \cdot \operatorname{rad}(f)'(x), f(x)\right) = \frac{f(x)}{\operatorname{rad}(f(x))} \cdot \prod_{\substack{i=1 \\ c_i = \lambda}}^{s} u_i(x).$$

Moreover,

$$f(x) = \prod_{\lambda \in D_H} \gcd\left(H(x) - \lambda \cdot \frac{f(x)}{\operatorname{rad}(f(x))} \cdot \operatorname{rad}(f)'(x), f(x)\right). \tag{6.22}$$

Proof. Let h be the polynomial in $\mathcal{N}^Q_{\mathrm{rad}(f)}$ corresponding to H according to Observation 6.6.7, that is, $H(x) = h(x) \cdot f(x)/\mathrm{rad}(f(x))$. Then Theorem 6.6.8 gives

$$\begin{aligned}
&\gcd\left(H(x) - \lambda \cdot \frac{f(x)}{\mathrm{rad}(f(x))} \cdot \mathrm{rad}(f)'(x), f(x)\right) \\
&= \frac{f(x)}{\mathrm{rad}(f(x))} \cdot \gcd\left(h(x) - \lambda \mathrm{rad}(f)'(x), \mathrm{rad}(f(x))\right) \\
&= \frac{f(x)}{\mathrm{rad}(f(x))} \cdot \prod_{\substack{i=1 \\ c_i=\lambda}}^{s} u_i(x),
\end{aligned}$$

as claimed. Now the second assertion is also immediate. □

Note that Equation (6.22) can also be written in the form

$$f(x) = \frac{f(x)}{\mathrm{rad}(f(x))} \cdot \prod_{\lambda \in D_h} \gcd\left(h(x) - \lambda \mathrm{rad}(f)'(x), \mathrm{rad}(f(x))\right),$$

since $D_H = D_h$.

As before for Berlekamp's method, the general results presented for the Niederreiter method mainly serve theoretical purposes: one should work with square-free polynomials in practical applications, after using the algorithms from Section 6.1.

Exercises

Exercise 6.6.11. In this exercise, the reader is asked to prove some further facts regarding the generalization of Theorem 6.6.8 to a general polynomial f (instead of its radical):

- f' belongs to $\mathcal{N}^Q_f$.
- The derived coefficients of $f'(x)$ coincide with the multiplicities $\ell_1, \ldots, \ell_s$ of the irreducible divisors $u_i(x)$ of f.
- Let $H(x) = \sum_{i=1}^s c_i d_i(x) \in \mathcal{N}^Q_f$ and $\lambda \in Q$. Then

$$\gcd\left(H(x) - \lambda f'(x), f(x)\right) = \frac{f(x)}{\mathrm{rad}(f(x))} \cdot \prod_{\substack{i=1 \\ c_i=\lambda \ell_i}}^{s} u_i(x).$$

While this alternative to the formula given in Theorem 6.6.10 has a simpler structure, it is not really useful in practice: one wants to obtain a nontrivial factorization of f just from the derived coefficients of $H \in \mathcal{N}^Q_f$, and not from terms including additionally the unknown multiplicities ℓ_i of the irreducible factors $u_i(x)$ of f. □

6.7 Links between Berlekamp Algebras and Niederreiter Spaces

The discussion so far shows that Niederreiter spaces and Berlekamp algebras have quite a few properties in common. The next step would be an analogue of Theorem 6.3.8 for the derived basis of a Niederreiter space, which would then lead to a procedure for the canonical factorization of the polynomial f similar to Algorithm 6.3.9; the interested reader might work this out as an exercise. We prefer to provide a direct link between Berlekamp algebras and Niederreiter spaces instead, again focussing on square-free polynomials.

Theorem 6.7.1. *Let $f(x)$ be a monic polynomial over a perfect field F, and let Q be any subfield of F. Then the mapping*

$$\alpha: \mathscr{B}^{Q}_{\mathrm{rad}(f)} \to \mathscr{N}^{Q}_{\mathrm{rad}(f)}, \quad h(x) \mapsto h(x)\mathrm{rad}(f)'(x) \bmod \mathrm{rad}(f(x))$$

is a canonical isomorphism between Q-vector spaces.

Proof. As before, we let $\varepsilon_1(x),\dots,\varepsilon_s(x)$ denote the spectral basis for the Berlekamp algebra $\mathscr{B}^{Q}_{\mathrm{rad}(f)}$ and $\delta_1(x),\dots,\delta_s(x)$ the derived basis for the polynomial Niederreiter space $\mathscr{N}^{Q}_{\mathrm{rad}(f)}$. As f is square-free, $\varepsilon_i(x) = \beta_i(x)\overline{u}_i(x)$, where $\deg\beta_i < \deg u_i$ and $\beta_i(x)\overline{u}_i(x) \equiv 1 \bmod u_i(x)$; see Observation 6.3.1.

Now consider some polynomial $h(x) = \sum_{i=1}^{s} c_i\varepsilon_i(x)$ in $\mathscr{B}^{Q}_{\mathrm{rad}(f)}$ and put $g(x) := \sum_{i=1}^{s} c_i\delta_i(x)$ and $\overline{g}(x) := h(x)\mathrm{rad}(f)'(x) \bmod \mathrm{rad}(f(x))$; we will establish the assertion by showing $\overline{g} = g$. To this end, we first compute

$$\begin{aligned} h(x)\mathrm{rad}(f)'(x) &= \Big(\sum_{i=1}^{s} c_i\beta_i(x)\overline{u}_i(x)\Big)\cdot\Big(\sum_{i=1}^{s} u_i'(x)\overline{u}_i(x)\Big) \\ &= \sum_{i,j=1}^{s} c_i\beta_i(x)\overline{u}_i(x)u_j'(x)\overline{u}_j(x). \end{aligned}$$

Now let $i,j,k \in \{1,\dots,s\}$ and note that $u_k(x)$ divides $\overline{u}_i(x)\overline{u}_j(x)$, unless $i=j=k$. Therefore, reducing the preceding identity modulo $u_k(x)$ gives

$$h(x)\mathrm{rad}(f)'(x) \equiv c_k\beta_k(x)\overline{u}_k(x)u_k'(x)\overline{u}_k(x) \equiv c_k u_k'(x)\overline{u}_k(x) \mod u_k(x)$$

for $k = 1,\dots,s$, where we have used $\beta_k(x)\overline{u}_k(x) \equiv 1 \bmod u_k(x)$. By Definition 6.6.6, $c_k u_k'(x)\overline{u}_k(x) = c_k\delta_k(x)$. As $u_k(x)$ divides $c_i\delta_i(x)$ for $i \neq k$, we conclude

$$h(x)\mathrm{rad}(f)'(x) \equiv g(x) \bmod u_k(x) \quad \text{for } k = 1,\dots,s$$

and hence, by the definition of $\overline{g}(x)$,

$$\overline{g}(x) \equiv h(x)\mathrm{rad}(f)'(x) \equiv g(x) \bmod \mathrm{rad}(f(x)).$$

As both $\overline{g}$ and g have degree $< \deg \mathrm{rad}(f)$, we see that these two polynomials actually agree, as claimed. □

Remark 6.7.2. The inverse of the canonical isomorphism α in Theorem 6.7.1 is given by

$$\alpha^{-1}\colon \mathscr{N}^{Q}_{\mathrm{rad}(f)} \to \mathscr{B}^{Q}_{\mathrm{rad}(f)}, \quad h(x) \mapsto h(x)\eta(x) \bmod \mathrm{rad}(f(x)), \tag{6.23}$$

where the polynomial $\eta(x)$ is the inverse of $\mathrm{rad}(f)'(x)$ modulo $\mathrm{rad}(f(x))$, which can be easily computed from the given data using the extended Euclidean algorithm. Therefore, properties of $\mathscr{B}^{Q}_{\mathrm{rad}(f)}$ and $\mathscr{N}^{Q}_{\mathrm{rad}(f)}$ can be transferred in either direction by using these two canonical isomorphisms. For instance, given some $h \in \mathscr{N}^{Q}_{\mathrm{rad}(f)}$, the determination of the derived coefficients of h is equivalent to that of the spectral coefficients of $\eta(x)h(x) \bmod \mathrm{rad}(f(x))$. □

As an example, we now transform Theorem 6.4.1 to the equivalent situation for Niederreiter spaces, which is also of independent interest for the symbolic integration of rational functions; see, for instance, Geddes et al. [135, Chapter 11].

Theorem 6.7.3. *Let $f(x)$ be a monic polynomial over a perfect field F, let Q be any subfield of F, and let $H(x) \in \mathscr{N}^{Q}_{f}$. Then*

$$\rho^{*}_{H}(x) := \prod_{\lambda \in D_H} (x-\lambda)$$

is the monic polynomial $\gamma(x) \in Q[x]$ of least degree such that

$$\gamma\Big(H(x)\cdot\frac{\mathrm{rad}(f(x))}{f(x)}\cdot\eta(x) \bmod \mathrm{rad}(f(x))\Big) = 0,$$

where $\eta(x)$ is the inverse of $\mathrm{rad}(f)'(x)$ *modulo* $\mathrm{rad}(f(x))$.

Proof. According to Observation 6.6.7, $f(x)/\mathrm{rad}(f(x))$ divides $H(x)$ and $h(x) := H(x)\mathrm{rad}(f(x))/f(x)$ is the polynomial in $\mathscr{N}^{Q}_{\mathrm{rad}(f)}$ corresponding to H; moreover, $D_H = D_h$. Then $b(x) := h(x)\eta(x) \bmod \mathrm{rad}(f(x))$ belongs to $\mathscr{B}^{Q}_{\mathrm{rad}(f)}$ and $C_b = D_h$, by the proof of Theorem 6.7.1. Applying Theorem 6.4.1 to this polynomial b shows that ρ_b is the monic polynomial of least degree in $Q[x]$ such that f divides $\rho_b(b(x))$. In view of

$$\rho_b(x) = \prod_{\lambda \in C_b}(x-\lambda) = \prod_{\lambda \in D_H}(x-\lambda) = \rho^{*}_{H}(x),$$

this establishes the assertion. □

Finally, we generalize Theorem 6.7.1 to the polynomial f itself:

Theorem 6.7.4. *Let $f(x)$ be a monic polynomial over a perfect field F, and let Q be any subfield of F. Then the mapping*

$$\beta\colon \mathscr{B}_f^Q \to \mathscr{N}_f^Q,\ H(x) \mapsto H(x)\cdot \mathrm{rad}(f)'(x)\cdot \frac{f(x)}{\mathrm{rad}(f(x))} \bmod f(x)$$

is a canonical isomorphism between Q-vector spaces.

Proof. We use Proposition 6.4.5 in combination with Remark 6.4.6 and Theorem 6.7.1 as follows:

$$\alpha_1\colon \mathscr{B}_f^Q \to \mathscr{B}_{\mathrm{rad}(f)}^Q,\ H(x) \mapsto H(x) \bmod \mathrm{rad}(f(x))$$

is an isomorphism between Q-algebras and both

$$\alpha\colon \mathscr{B}_{\mathrm{rad}(f)}^Q \to \mathscr{N}_{\mathrm{rad}(f)}^Q,\ h(x) \mapsto h(x)\cdot \mathrm{rad}(f)'(x) \bmod \mathrm{rad}(f(x))$$

and

$$\alpha_2\colon \mathscr{N}_{\mathrm{rad}(f)}^Q \to \mathscr{N}_f^Q,\ h(x) \mapsto h(x)\cdot \frac{f(x)}{\mathrm{rad}(f(x))}$$

are isomorphisms between Q-vector spaces. Moreover, the composition $\alpha_2 \circ \alpha \circ \alpha_1$ is a canonical isomorphism:

$$\sum_{i=1}^{s} c_i e_i(x) \xrightarrow{\alpha_1} \sum_{i=1}^{s} c_i \varepsilon_i(x) \xrightarrow{\alpha} \sum_{i=1}^{s} c_i \delta_i(x) \xrightarrow{\alpha_2} \sum_{i=1}^{s} c_i d_i(x).$$

Now it suffices to check that this composition is just the mapping β defined in the assertion. For this, let $H \in \mathscr{B}_f^Q$ and perform polynomial division with remainder twice to get first

$$H(x) = a(x)\mathrm{rad}(f(x)) + r(x) \quad \text{with } r(x) = H(x) \bmod \mathrm{rad}(f(x))$$

and then

$$r(x)\mathrm{rad}(f)'(x) = b(x)\mathrm{rad}(f(x)) + s(x) \quad \text{with } s(x) = r(x)\mathrm{rad}(f)'(x) \bmod \mathrm{rad}(f(x)),$$

so that

$$(\alpha_2 \circ \alpha \circ \alpha_1)(H(x)) = (\alpha_2 \circ \alpha)(r(x)) = \alpha_2(s(x)) = s(x)\frac{f(x)}{\mathrm{rad}(f(x))}.$$

On the other hand, also

$$\begin{aligned} H(x)\mathrm{rad}(f)'(x)\frac{f(x)}{\mathrm{rad}(f(x))} &= \big(a(x)\mathrm{rad}(f(x))+r(x)\big)\mathrm{rad}(f)'(x)\frac{f(x)}{\mathrm{rad}(f(x))} \\ &= a(x)\mathrm{rad}(f)'(x)f(x) + r(x)\mathrm{rad}(f)'(x)\frac{f(x)}{\mathrm{rad}(f(x))} \\ &= a(x)\mathrm{rad}(f)'(x)f(x) + \big(b(x)\mathrm{rad}(f(x))+s(x)\big)\frac{f(x)}{\mathrm{rad}(f(x))} \\ &= \big(a(x)\mathrm{rad}(f)'(x)+b(x)\big)f(x) + s(x)\frac{f(x)}{\mathrm{rad}(f(x))}, \end{aligned}$$

and hence

$$H(x)\mathrm{rad}(f)'(x)\frac{f(x)}{\mathrm{rad}(f(x))} \bmod f(x) = s(x)\frac{f(x)}{\mathrm{rad}(f(x))},$$

as claimed. □

Remark 6.7.5. It is interesting to note that the inverse of the mapping α_1 used in the proof of Theorem 6.7.4 provides a *lifting* from $\mathscr{B}^Q_{\mathrm{rad}(f)}$ to $\mathscr{B}^Q_f$; let us describe this mapping explicitly. As $\mathrm{rad}(f)$ and $\mathrm{rad}(f)'$ are relatively prime, the polynomials f and $\mathrm{rad}(f)'$ are likewise relatively prime. Let $\theta(x)$ be the inverse of $\mathrm{rad}(f)'(x)$ modulo $f(x)$. Then the inverse of the mapping β in Theorem 6.7.4 is

$$\beta^{-1}\colon \mathscr{N}^Q_f \to \mathscr{B}^Q_f,\ \ H(x) \mapsto H(x)\cdot\frac{\mathrm{rad}(f(x))}{f(x)}\cdot\theta(x) \bmod f(x).$$

The lifting from $\mathscr{B}^Q_{\mathrm{rad}(f)}$ to $\mathscr{B}^Q_f$ is therefore the composition $\beta^{-1}\circ\alpha_2\circ\alpha$, which maps $h\in\mathscr{B}^Q_{\mathrm{rad}(f)}$ to $h(x)\mathrm{rad}(f)'(x)\theta(x) \bmod f(x)$. □

If one wants to deal with the factorization problem for polynomials over finite fields in practice, it is useful to determine a basis for a Niederreiter space *without* relying on the connection to the corresponding Berlekamp algebras. This will be the topic of the next section.

6.8 Prime Niederreiter Spaces in Positive Characteristic

In this section, we present Niederreiter's method for computing a basis for the **prime Niederreiter space** $\mathscr{N}^P_f$ of a monic polynomial f over a perfect field F with positive characteristic p, where $P\cong\mathbb{Z}_p$ denotes the prime subfield of F. His approach relies on a linearization technique which is completely different from that of Berlekamp; its core is a P-linear mapping which will be defined next. This then leads to a certain differential equation over the field $\widehat{F}(x)$ of rational functions over the algebraic closure $\widehat{F}$ of F.[6]

Definition 6.8.1. With the notations above, the mapping

$$\Psi\colon \widehat{F}(x) \to \widehat{F}(x), \ \ y(x) \mapsto y^{(p-1)}(x) + y(x)^p$$

is called the **Niederreiter mapping** over F. Here $y^{(p-1)}(x)$ denotes the $(p-1)$-th (formal) derivative of the rational function $y(x)$, while $y(x)^p$ is the p-th power of $y(x)$, as usual. □

Now the main result of Niederreiter can be stated as follows, where the W-spaces are defined as in Equation (6.19):

Theorem 6.8.2. *Let $f(x)$ be a monic polynomial over a perfect field F, and let P be the prime subfield of F. Then*

$$\bar{\mathcal{N}}^{P}_{\mathrm{rad}(f)} = \ker \Psi \cap W_{\mathrm{rad}(f)}.$$

Proof. The proof is somewhat involved and will be split into several steps. It uses the notion of the **degree** of a rational function $y(x) = a(x)/b(x)$, which is defined as $\deg y := \deg a - \deg b$.[7]

Step 1. Assume $y(x) \in \ker \Psi$. Then $\deg y \leq -1$.

Let $y(x) = a(x)/b(x)$ be an arbitrary non-zero rational function in $\widehat{F}(x)$. By Exercise 2.4.22, the formal derivative of $y(x)$ is

$$y'(x) = \frac{a'(x)b(x) - a(x)b'(x)}{b(x)^2}.$$

Note that both $\deg a'b$ and $\deg ab'$ are bounded by $\deg a + \deg b - 1$, which gives

$$\deg y' = \deg(a'b - ab') - 2\deg b \leq \deg a - \deg b - 1 = \deg y - 1$$

and, with induction, $\deg(y^{(p-1)}) \leq \deg y - (p-1)$.

By assumption, $y^{(p-1)}(x) = -y(x)^p$, and comparing degrees shows

$$p \cdot \deg y = \deg(-y^p) = \deg y^{(p-1)} \leq \deg y - (p-1),$$

so that indeed $\deg y \leq -1$.

In view of Step 1, the simplest candidates for elements in the kernel of Ψ are rational functions of the form $\beta/(x-\alpha)$:

Step 2. Let $\alpha, \beta \in \widehat{F}$ and $y(x) := \beta/(x-\alpha)$. Then $y \in \ker \Psi$ if and only if $\beta \in P$.

---

[6] As our arguments will need splitting fields for a variety of polynomials over F, it is convenient – though not absolutely necessary – to make use of the algebraic closure of F.

[7] Note that this is well-defined: it does not depend on the representation of $y(x)$ as a fraction.

Using Wilson's theorem, $(p-1)! \equiv -1 \bmod p$, an easy computation gives

$$y^{(p-1)}(x) = \left(\frac{\beta}{x-\alpha}\right)^{(p-1)} = \frac{\beta \cdot \prod_{i=1}^{p-1}(-i)}{(x-\alpha)^p}$$
$$= \frac{\beta \cdot (-1)^{p-1} \cdot (p-1)!}{(x-\alpha)^p} = \frac{-\beta}{(x-\alpha)^p},$$

which implies

$$\Psi(y(x)) = \frac{-\beta}{(x-\alpha)^p} + \frac{\beta^p}{(x-\alpha)^p} = \frac{-\beta+\beta^p}{(x-\alpha)^p}.$$

Hence $\Psi(y(x)) = 0$ if and only if $\beta^p = \beta$, which means $\beta \in P$.

Next, we consider rational functions whose denominator is a product of distinct linear factors:

Step 3. Let $\alpha_1, \ldots, \alpha_m$ be distinct elements of $\widehat{F}$, put $u(x) := \prod_{i=1}^{m}(x-\alpha_i)$, and let $b(x) \in \widehat{F}[x]$ with $\deg b < m$. Then $b(x)/u(x) \in \ker \Psi$ if and only if $b(\alpha_i)/u'(\alpha_i) \in P$ for all i.

An application of the partial fraction decomposition in Proposition 6.6.1 to $b(x)/u(x)$ gives unique field elements $\beta_1, \ldots, \beta_m \in \widehat{F}$ such that

$$\frac{b(x)}{u(x)} = \sum_{i=1}^{m} \frac{\beta_i}{x-\alpha_i},$$

and hence, using the computation from Step 2,

$$\Psi\left(\frac{b(x)}{u(x)}\right) = \sum_{i=1}^{m} \Psi\left(\frac{\beta_i}{x-\alpha_i}\right) = \sum_{i=1}^{m} \frac{-\beta_i+\beta_i^p}{(x-\alpha_i)^p}.$$

As the denominators $(x-\alpha_i)^p$ are relatively prime, we see that $\Psi(b(x)/u(x)) = 0$ if and only if $\beta_i = \beta_i^p$ for every i, which means $\beta_i \in P$ for all i. Thus it suffices to show that $\beta_i = b(\alpha_i)/u'(\alpha_i)$ for all i. For this, we multiply the partial fraction decomposition of $b(x)/u(x)$ with $u(x)$ to obtain

$$b(x) = \sum_{i=1}^{m} \beta_i \cdot \frac{u(x)}{x-\alpha_i}.$$

Since α_j is a root of the polynomial $u(x)/(x-\alpha_i)$ whenever $i \neq j$, evaluating b at α_j gives

$$b(\alpha_j) = \beta_j \cdot \prod_{\substack{i=1 \\ i \neq j}}^{m} (\alpha_j - \alpha_i) \quad \text{for } j = 1, \ldots, m.$$

Evaluating the derivative $u'(x) = \sum_{i=1}^{m} u(x)/(x-\alpha_i)$ of u at α_j shows

$$u'(\alpha_j) = \prod_{\substack{i=1 \\ i \neq j}}^{m} (\alpha_j - \alpha_i) \neq 0,$$

since the α_k are distinct. Hence $\beta_j = b(\alpha_j)/u'(\alpha_j)$ for $j = 1, \ldots, m$, as claimed.

In the next step, we investigate fractions with an irreducible denominator:

Step 4. Let $u(x)$ be a monic irreducible polynomial with degree m over F, and let $b(x) \in F[x]$ have degree $< m$. Then $b(x)/u(x)$ belongs to $\ker \Psi$ *if and only if $b(x) = \beta u'(x)$ for some $\beta \in P$.*

Let $E \subseteq \widehat{F}$ be the splitting field of u, say $u(x) = \prod_{i=1}^{m} (x - \alpha_i)$, where the α_i are distinct (since F is perfect). By Step 3, $b(x)/u(x) \in \ker \Psi$ if and only if $\beta_i := b(\alpha_i)/u'(\alpha_i) \in P$ for all i. Clearly, this holds with $\beta_i = \beta$ for all i, provided that $b(x) = \beta u'(x)$.

Conversely, assume $\beta_i \in P$ for all i. Note that E/F is a Galois extension, since E is the splitting field of the separable polynomial u. Its Galois group G permutes the roots of u transitively: for $i = 1, \ldots, m$, there is a unique $\gamma_i \in G$ such that $\gamma_i(\alpha_1) = \alpha_i$. Since both $b(x)$ and $u'(x)$ have coefficients in the fixed field F of G and since $P \subseteq F$, we get

$$\beta_i = \frac{b(\alpha_i)}{u'(\alpha_i)} = \frac{b(\gamma_i(\alpha_1))}{u'(\gamma_i(\alpha_1))} = \frac{\gamma_i(b(\alpha_1))}{\gamma_i(u'(\alpha_1))} = \gamma_i\left(\frac{b(\alpha_1)}{u'(\alpha_1)}\right) = \gamma_i(\beta_1) = \beta_1$$

for $i = 1, \ldots, m$. We now put $\beta := \beta_1$ and use the partial fraction decomposition from Step 3 to obtain

$$b(x) = \sum_{i=1}^{m} \beta_i \frac{u(x)}{x - \alpha_i} = \beta \cdot \sum_{i=1}^{m} \frac{u(x)}{x - \alpha_i} = \beta u'(x),$$

as claimed.

Now the following result is an easy consequence:

Step 5. Let $h(x) \in F[x]$ with $\deg h < \deg \mathrm{rad}(f)$, *and assume $h(x)/\mathrm{rad}(f(x)) \in \bar{\mathcal{N}}^P_{\mathrm{rad}(f)}$. Then $h(x)/\mathrm{rad}(f(x)) \in$* $\ker \Psi$.

By the definition of $\bar{\mathcal{N}}^P_{\mathrm{rad}(f)}$, there are unique $c_1, \ldots, c_s \in P$ such that

$$\frac{h(x)}{\mathrm{rad}(f(x))} = \sum_{i=1}^{s} c_i \frac{u_i'(x)}{u_i(x)},$$

where $\prod_{i=1}^{s} u_i(x)$ is the canonical factorization of $\mathrm{rad}(f(x))$ over F. By Step 4, $c_i u_i'(x)/u_i(x) \in \ker \Psi$ for every i, and therefore also $h(x)/\mathrm{rad}(f(x)) \in \ker \Psi$.

We have now established $\bar{\mathcal{N}}^P_{\mathrm{rad}(f)} \subseteq \ker \Psi \cap W_{\mathrm{rad}(f)}$, and it only remains to check the reverse inclusion:

Step 6. Let $h(x) \in F[x]$ *with* $\deg h < \deg \mathrm{rad}(f)$, *and assume* $\Psi\big(h(x)/\mathrm{rad}(f(x))\big) = 0$. *Then* $h(x)/\mathrm{rad}(f(x)) \in \bar{\mathcal{N}}^P_{\mathrm{rad}(f)}$.

Using Proposition 6.6.1 again, we obtain unique polynomials $h_i(x) \in F[x]$ with $\deg h_i < \deg u_i$ for all i such that

$$\frac{h(x)}{\mathrm{rad}(f(x))} = \sum_{i=1}^{s} \frac{h_i(x)}{u_i(x)}.$$

For $i = 1, \ldots, m$, put $m_i := \deg u_i$ and write $u_i(x) = \prod_{j=1}^{m_i}(x - \alpha_{i,j})$ over its splitting field $E_i \subseteq \widehat{F}$. Another application of Proposition 6.6.1 gives unique field elements $\beta_{i,j} \in \widehat{F}$ such that

$$\frac{h(x)}{\mathrm{rad}(f(x))} = \sum_{i=1}^{s}\sum_{j=1}^{m_i} \frac{\beta_{i,j}}{x - \alpha_{i,j}}.$$

Using the calculation in Step 2 then gives

$$\Psi\Big(\frac{h(x)}{\mathrm{rad}(f(x))}\Big) = \sum_{i=1}^{s}\sum_{j=1}^{m_i} \frac{-\beta_{i,j} + \beta_{i,j}^p}{(x - \alpha_{i,j})^p},$$

where the $\alpha_{i,j}$ are distinct. Hence $\Psi(h(x)/\mathrm{rad}(f(x))) = 0$ implies $\beta_{i,j} \in P$ for all i, j, and therefore every summand $\beta_{i,j}/(x - \alpha_{i,j})$ is in $\ker \Psi$, by Step 2. Recombining the respective summands, we conclude

$$\frac{h_i(x)}{u_i(x)} = \sum_{j=1}^{m_i} \frac{\beta_{i,j}}{(x - \alpha_{i,j})} \in \ker \Psi \quad \text{for } i = 1, \ldots, m.$$

Now Step 4 shows $h_i(x) = \beta_i u_i'(x)$ for some $\beta_i \in P$, as $h_i \in F[x]$ and as u_i is irreducible over F; that is, $\beta_i = \beta_{i,j}$ for all $j = 1, \ldots, m_i$. As this holds for all i,

$$\frac{h(x)}{\mathrm{rad}(f(x))} = \sum_{i=1}^{s} \beta_i \frac{u_i'(x)}{u(x)} \in \bar{\mathcal{N}}^P_{\mathrm{rad}(f)}.$$

This concludes the proof of Step 6 and establishes the validity of the theorem. □

Remark 6.8.3. In the situation of Theorem 6.8.2, let f be a square-free polynomial. Switching from its fractional Niederreiter space $\bar{\mathcal{N}}_f^P$ to the polynomial version $\mathcal{N}_f^P$ as in Definition 6.6.4, we see that a polynomial $g(x) \in F[x]$ with $\deg g < n = \deg f$ belongs to $\mathcal{N}_f^P$ if and only if it satisfies the equation $\Psi(g(x)/f(x)) = 0$. Thus we need to investigate the kernel of the P-linear mapping defined by

$$\psi(g(x)) := f(x)^p \cdot \Psi\Big(\frac{g(x)}{f(x)}\Big) = f(x)^p \cdot \Big(\frac{g(x)}{f(x)}\Big)^{(p-1)} + g(x)^p,$$

where $g(x) \in F[x]_{<n}$. For this, we reconsider (Steps 2 and 3 in) the proof of Theorem 6.8.2: over the splitting field of f, the partial fraction decomposition of $g(x)/f(x)$ gives an expression of the form

$$\frac{g(x)}{f(x)} = \sum_{i=1}^{n} \frac{\beta_i}{x - \alpha_i}.$$

Then the $(p-1)$-th derivative of $g(x)/f(x)$ is

$$\left(\frac{g(x)}{f(x)}\right)^{(p-1)} = -\sum_{i=1}^{n} \frac{\beta_i}{(x-\alpha_i)^p},$$

and recombining these fractions results in

$$-\frac{\sum_{i=1}^{n} \beta_i \cdot \prod_{j=1,\, j\neq i}^{n} (x^p - \alpha_j^p)}{f(x)^p}.$$

Since F is perfect, the preceding observations show that

$$\psi(g(x)) = \sum_{i=1}^{n} \beta_i \cdot \prod_{\substack{j=1\\ j\neq i}}^{n} (x^p - \alpha_j^p) \;+\; g(x)^p$$

is a polynomial of the form $\gamma(x^p)$ for some $\gamma \in F[x]$.

Let us further assume that $F = P \cong \mathbb{Z}_p$, and let us represent ψ with respect to the canonical bases $\{1, x, \ldots, x^{n-1}\}$ of its domain and $\{1, x^p, \ldots, x^{(n-1)p}\}$ of its image, say by the matrix N_f. With this setup, an application of the Gaussian algorithm yields a basis for the Niederreiter space $\mathscr{N}_f^P$, which is just the kernel of N_f. □

Example 6.8.4. Let us illustrate the approach outlined in Remark 6.8.3 with an example for $p = 2$, so that $F = P$ is the binary field. Then the linear mapping ψ has the form

$$\psi(g(x)) = f(x)^2 \left(\frac{g(x)}{f(x)}\right)' + g(x)^2 = f(x)g'(x) + g(x)f'(x) + g(x)^2.$$

In particular, we obtain the following formulae for monomials:

$$\psi(x^{2m}) = x^{2m} f'(x) + x^{4m},$$

and

$$\psi(x^{2m+1}) = x^{2m} f(x) + x^{2m+1} f'(x) + x^{4m+2},$$

and expanding these terms for a specified polynomial f leads immediately to the matrix N_f with kernel $\mathscr{N}_f^P$. For a specific example, we take $f(x) = x^6 + x^4 + x + 1$, which is indeed square-free as $f'(x) = 1$. Applying ψ to x^k for $k = 0, \ldots, 5$ then yields

k	0	1	2	3	4	5
$\psi(x^k)$	0	$x^6 + x^4 + x^2 + 1$	$x^4 + x^2$	$x^8 + x^2$	$x^8 + x^4$	$x^8 + x^4$

Hence the desired representation matrix N_f for ψ with respect to the canonical bases $\{1,x,x^2,x^3,x^4,x^5\}$ and $\{1,x^2,x^4,x^6,x^8,x^{10}\}$ is given by

$$N_f = \begin{pmatrix} 0&1&0&0&0&0\\ 0&1&1&1&0&0\\ 0&1&1&0&1&1\\ 0&1&0&0&0&0\\ 0&0&0&1&1&1\\ 0&0&0&0&0&0 \end{pmatrix}.$$

Now the kernel of N_f leads to the following basis for the Niederreiter space $\mathcal{N}_f^P$:

$$h_0(x) = 1,\ \ h_1(x) = x^4+x^3+x^2,\ \ h_2(x) = x^5+x^3+x^2;$$

in particular, f has three distinct irreducible factors over F. Of course, $h_0 = f'$ is a trivial element of $\mathcal{N}_f^P$. An application of Theorem 6.6.8 to the two non-trivial polynomials h_1 and h_2 gives

$$\begin{aligned} f(x) &= \gcd\big(h_1(x), f(x)\big)\cdot\gcd\big(h_1(x)-f'(x), f(x)\big)\\ &= (x^2+x+1)\cdot(x^4+x^3+x^2+1), \end{aligned}$$

and

$$\begin{aligned} f(x) &= \gcd\big(h_2(x), f(x)\big)\cdot\gcd\big(h_2(x)-f'(x), f(x)\big)\\ &= (x^3+x+1)\cdot(x^3+1). \end{aligned}$$

Finally, we may compute the canonical factorization of f using a similar strategy as in Algorithm 6.3.9:

$$f(x) = (x+1)\cdot(x^2+x+1)\cdot(x^3+x+1). \qquad \square$$

Note that Theorem 6.8.2 only deals with the radical of the given polynomial f. By Remarks 6.6.3 and 6.6.5, the two fractional Niederreiter spaces $\bar{\mathcal{N}}_f^P$ and $\bar{\mathcal{N}}_{\mathrm{rad}(f)}^P$ coincide, whereas the two polynomial Niederreiter spaces $\mathcal{N}_f^P$ and $\mathcal{N}_{\mathrm{rad}(f)}^P$ differ by the factor $f(x)/\mathrm{rad}(f(x))$. We conclude this section with a corresponding generalization of Theorem 6.8.2.

Consider any polynomial H in $\mathcal{N}_f^P$. Then $h(x) := H(x)\cdot\mathrm{rad}(f(x))/f(x)$ is a polynomial in $\mathcal{N}_{\mathrm{rad}(f)}^P$, and hence Theorem 6.8.2 yields

$$\Psi\left(\frac{H(x)}{f(x)}\right) = \Psi\left(\frac{h(x)}{\mathrm{rad}(f(x))}\right) = 0.$$

This establishes $\bar{\mathcal{N}}_f^P \subseteq \ker\Psi\cap W_f$, as $\deg H < \deg f$. The following theorem (which we state in terms of the polynomial Niederreiter space) shows that the reverse inclusion also holds. This is more involved; our proof essentially follows Gao and von zur Gathen [128] and will merely be sketched.

Theorem 6.8.5. *Let $f(x)$ be a monic polynomial with degree n over a perfect field F with positive characteristic p, let $H(x)$ be any polynomial in $F[x]_{<n}$, and let P be the prime subfield of F. Then H belongs to the polynomial Niederreiter space $\mathscr{N}_f^P$ if and only if $\Psi(H(x)/f(x)) = 0$, where Ψ is the operator introduced in Definition 6.8.1.*

Proof. In view of the preceding remarks, it remains to prove that $H \in \mathscr{N}_f^P$ whenever $H(x) \in F[x]_{<n}$ and $H(x)/f(x) \in \ker \Psi$. For this, it suffices to show that H is divisible by $f(x)/\mathrm{rad}(f(x))$: then $\Psi(h(x)/\mathrm{rad}(f(x))) = 0$, where $h(x) := H(x) \cdot \mathrm{rad}(f(x))/f(x)$, and therefore $h(x) \in \mathscr{N}_{\mathrm{rad}(f)}^P$ and $H(x) = h(x) \cdot f(x)/\mathrm{rad}(f(x)) \in \mathscr{N}_f^P$.

The divisibility in question can be established using the following idea from [128]: put $d(x) := \gcd(f(x), H(x))$, and let

$$\mu_k(x) := \mathrm{rad}(f(x))^k \cdot \frac{f(x)}{d(x)} \cdot \left(\frac{H(x)}{f(x)}\right)^{(k)} \quad \text{for all } k \in \mathbb{N}.$$

By definition, $\mu_0(x) = H(x)/d(x)$ is a polynomial over F. Assume inductively that $\mu_k(x) \in F[x]$ for some k. Then an analysis of $\big(\mu_k(x) \cdot \mathrm{rad}(f(x))\big)'$ shows that also $\mu_{k+1}(x) \in F[x]$. Now the hypothesis $\Psi(H(x)/f(x)) = 0$ implies

$$-\left(\frac{H(x)}{d(x)}\right)^p = \left(\frac{f(x)}{\mathrm{rad}(f(x)) \cdot d(x)}\right)^{p-1} \cdot \mu_{p-1}(x).$$

It turns out that the right hand side is a polynomial over F which is divisible by every irreducible divisor u_i of f. Therefore, u_i has to divide the polynomial $H(x)/d(x)$ on the left hand side. Thus $\mathrm{rad}(f)$ divides $H(x)/d(x)$, and it follows that H is indeed a multiple of $f(x)/\mathrm{rad}(f(x))$. □

6.9 Concluding Remarks

We conclude this chapter with a brief outlook on further aspects of the factorization of polynomials. There is a considerable literature on this topic; we will restrict ourselves to just a few remarks. For a very brief survey listing many references, the reader might wish to have a look at Section 11.4 of the *Handbook of finite fields* [292].

We first note that the connections between Berlekamp algebras and Niederreiter spaces discussed in Section 6.7 were only discovered some time after Niederreiter's method was established. The recognition of the similarities for these two classes of objects initiated a general study of *factorization spaces* and the search for operators related to the factorization of polynomials. For more information on these topics, the reader may consult Fleischmann [118], Lee and Vanstone [230] and Wan [393].

Next, we present a rough outline of a generalization of Theorems 6.8.2 and 6.8.5, which provides a characterization of the generalized Niederreiter space $\mathcal{N}_f^{\bar{Q}}$ over some finite subfield $Q \neq P$ of the underlying field F. For details, we refer the reader to [302, 304].

Thus let F be a perfect field with positive characteristic p, and let Q be a subfield of F with cardinality p^r, where $r \geq 2$. Then the (p^r-1)-th derivative of *every* rational function $y(x)$ over F equals zero, and therefore the operator

$$\widehat{F}(x) \to \widehat{F}(x), \quad y(x) \mapsto y^{(p^r-1)}(x) + y(x)^{p^r}$$

is neither interesting nor helpful. Fortunately, one can overcome this problem by replacing formal derivatives with another type of derivative. The basic setup is as follows.

- The quotient field $F((x))$ of the ring $F[[x]]$ of formal power series in the indeterminate x over F is called the **field of formal Laurent series** over F.
- Now consider $F((x^{-1}))$, the field of formal Laurent series in the indeterminate x^{-1} over F. The elements of $F((x^{-1}))$ have the form

  $$\sum_{n=w}^{\infty} s_n x^{-n}, \quad \text{where } w \in \mathbb{Z}.$$

  Such a formal Laurent series involves finitely many terms x^m with $m \in \mathbb{N}$ and possibly infinitely many terms x^m with a negative exponent m. In particular, any rational function in a W-space as in Equation (6.19) can be viewed as an element of $F((x^{-1}))$.
- Finally, one defines the **Hasse-Teichmüller derivative** of order $k \in \mathbb{N}$ on $F((x^{-1}))$ as follows:

  $$\mathscr{H}^{(k)}\left(\sum_{n=w}^{\infty} s_n x^{-n}\right) := \sum_{n=w}^{\infty} \binom{-n}{k} \cdot s_n x^{-n-k},$$

  where

  $$\binom{-n}{k} := \frac{(-n)(-n-1)\cdots(-n-k-1)}{k!}.$$

With these concepts, we can state the following fundamental characterization theorem:

Result 6.9.1 *Let $f(x)$ be a monic polynomial with degree n over a perfect field F with positive characteristic p, and consider the space W_f of all rational functions over F of the form $g(x)/f(x)$ with* $\deg g < n$. *Moreover, let Q be a subfield of F with cardinality p^r, where $r \geq 2$.*

Then $y(x) \in W_f$ belongs to the fractional Niederreiter space $\mathcal{N}_f^{\bar{Q}}$ if and only if it satisfies the following differential equation:

$$\mathscr{H}^{(p^r-1)}(y(x)) - y(x)^{p^r} = 0. \qquad \square \tag{6.24}$$

Quite often, one also needs to be able to factor multivariate polynomials over finite fields. The reader interested in this topic should consult Section 11.5 of the *Handbook of finite fields* [292] and the references given there.

Finally, there are also many applications which require factoring univariate polynomials over the field $\mathbb{Q}$ of rational numbers. For this problem, the factorization of univariate polynomials over finite fields serves as an important auxiliary ingredient, via the process of *Hensel lifting*. A second important tool which is needed for this task is the computation of short vectors in lattices; this has applications also to numerous other fundamental problems. Both the so-called *Lenstra-Lenstra-Lovász lattice basis reduction algorithm* (or, for short, *LLL algorithm*) for finding short vectors and its application to factoring polynomials with rational coefficients come from the seminal 1982 paper by Lenstra, Lenstra and Lovász [235]. An excellent account of this topic may also be found in Part III of the monograph on computer algebra by von zur Gathen and Gerhard [388].

Chapter 7
Matrices over Finite Fields

Abstract In this chapter, we study several aspects of matrices over finite fields. We begin with some basic results on the rank and order of matrices and then apply some of these results to discuss an interesting theoretical topic, namely matrix representations of an extension field $\mathrm{GF}(q^n)$ over $\mathrm{GF}(q)$. Following this, we provide results on the numbers of matrices over $\mathrm{GF}(q)$ of various important special types – such as symmetric, orthogonal and circulant matrices – that will be needed later to study basis representations of extension fields. Last but not least, we also consider the Discrete Fourier Transform, which is not only a very useful tool for some results given in the present chapter, but has many interesting further applications.

7.1 Rank and Order of Matrices over Finite Fields

In this section, we collect several fundamental results concerning the rank and order of matrices over finite fields. For general background from Linear Algebra, we refer the reader to the book by Hoffman and Kunze [190] (as before).

Obviously, the number of (m,n)-matrices over $F = \mathrm{GF}(q)$ is q^{mn}, and these matrices form a vector space $F^{(m,n)}$ of dimension mn. In what follows, we shall mainly consider square matrices. The (n,n)-matrices over F also constitute an F-algebra $R = F^{(n,n)}$. The invertible matrices in R form a group, the **general linear group**; this group is usually denoted by either $\mathrm{GL}(n,q)$ or $\mathrm{GL}_n(q)$.

Note that the elements of $G = \mathrm{GL}(n,q)$ are exactly those matrices in R which transform the elements of a fixed ordered basis $B = (\alpha_1,\ldots,\alpha_n)$ of the vector space F^n over $F = \mathrm{GF}(q)$ into an ordered basis again. Hence the order of G agrees with the total number of distinct ordered bases of F^n over F, which is given in the following simple (and well-known) result.

Proposition 7.1.1. *The number $b(n,q)$ of ordered bases of the vector space $V = F^n$ over $F = \mathrm{GF}(q)$ is*

$$b(n,q) := |\mathrm{GL}(n,q)| = q^{n(n-1)/2}(q^n-1)(q^{n-1}-1)\cdots(q-1). \tag{7.1}$$

D. Hachenberger and D. Jungnickel, *Topics in Galois Fields*,
Algorithms and Computation in Mathematics 29,
https://doi.org/10.1007/978-3-030-60806-4_7

Proof. As k linearly independent vectors $\alpha_1, \ldots, \alpha_k$ span a k-dimensional subspace of V, they rule out exactly q^k possibilities for choosing a further vector not dependent on the given k vectors. Thus there are $q^n - 1$ ways of selecting the first basis vector α_1; then $q^n - q$ ways of selecting the second basis vector α_2; and, in general, $q^n - q^{k-1}$ ways of selecting the k-th basis vector α_k. Taking out powers of q gives the desired formula. □

It is clear that the matrices in $\mathrm{GL}(n,q)$ with determinant 1 form a subgroup, the **special linear group**, which is usually denoted by either $\mathrm{SL}(n,q)$ or $\mathrm{SL}_n(q)$. The order of this group is as follows:

Corollary 7.1.2. *The number of (n,n)-matrices over $F = \mathrm{GF}(q)$ with determinant* 1 *is*

$$|\mathrm{SL}(n,q)| = q^{n(n-1)/2}(q^n - 1)(q^{n-1} - 1)\cdots(q^2 - 1). \tag{7.2}$$

Proof. The determinant on $F^{(n,n)}$ induces a group epimorphism $\mathrm{GL}(n,q) \to F^*$ with kernel $\mathrm{SL}(n,q)$, and hence

$$|\mathrm{GL}(n,q)|/|\mathrm{SL}(n,q)| = q - 1,$$

by Theorem 1.3.10. In other words, the determinants of the matrices in $GL(n,q)$ are uniformly distributed over F^*. □

Our next aim is to generalize Proposition 7.1.1 by giving a formula for the number of all (m,n)-matrices with prescribed rank k over $\mathrm{GF}(q)$. This turns out to be closely related to the number of subspaces of a prescribed dimension in a vector space over $\mathrm{GF}(q)$. We need the following notation.

Definition 7.1.3. The number of k-dimensional subspaces of an n-dimensional vector space over $\mathrm{GF}(q)$, where $1 \le k \le n$, is denoted by $\left[{n \atop k}\right]_q$. These numbers are called the **Gaussian coefficients**. It is also usual to put $\left[{n \atop 0}\right]_q := 1$. □

Proposition 7.1.4. *The Gaussian coefficient $\left[{n \atop k}\right]_q$, where $1 \le k \le n$, is given explicitly as follows:*

$$\left[{n \atop k}\right]_q = \frac{(q^n - 1)(q^{n-1} - 1)\cdots(q^{n-k+1} - 1)}{(q^k - 1)(q^{k-1} - 1)\cdots(q - 1)}. \tag{7.3}$$

Proof. As in the proof of Proposition 7.1.1, there are exactly

$$(q^n - 1)(q^n - q)\cdots(q^n - q^{k-1}) = q^{k(k-1)/2}(q^n - 1)(q^{n-1} - 1)\cdots(q^{n-k+1} - 1) \tag{7.4}$$

ordered k-tuples of linearly independent vectors in $V = \mathrm{GF}(q)^n$ over $\mathrm{GF}(q)$. Each such k-tuple forms an ordered basis for some k-dimensional subspace of V, and each k-dimensional subspace has exactly $b(k,q)$ ordered bases. Thus the number of k-subspaces of V is indeed the quotient of these two quantities. □

The Gaussian coefficients may be viewed as an analogue of the usual binomial coefficients, where we deal with vector spaces over GF(q) and their subspaces instead of the Boolean algebras of sets and their subsets, with the Boolean case corresponding in some sense to $q = 1$; see Exercise 7.1.12 for an illustration of this phenomenon.

Theorem 7.1.5. *Let $V = \mathrm{GF}(q)^{(m,n)}$ be the vector space of all (m,n)-matrices over the field $F = \mathrm{GF}(q)$. Then the number $f(m,n,k)$ of matrices in V with rank k, where $k \le \min\{m,n\}$, is given by*

$$\begin{aligned} f(m,n,k) &= \begin{bmatrix} m \\ k \end{bmatrix}_q q^{k(k-1)/2}(q^n-1)(q^{n-1}-1)\cdots(q^{n-k+1}-1) \\ &= \begin{bmatrix} n \\ k \end{bmatrix}_q q^{k(k-1)/2}(q^m-1)(q^{m-1}-1)\cdots(q^{m-k+1}-1) \\ &= \frac{q^{k(k-1)/2}\prod_{i=0}^{k-1}(q^{m-i}-1)\prod_{i=0}^{k-1}(q^{n-i}-1)}{\prod_{i=0}^{k-1}(q^{k-i}-1)}. \end{aligned} \tag{7.5}$$

Proof. We partition the (m,n)-matrices A of rank k according to their column space. As A corresponds to a linear mapping from F^n to F^m, the number of such matrices with a prescribed k-dimensional column space U equals the number of linear transformations from F^n onto $U \cong F^k$, which in turn equals the number of (k,n)-matrices of rank k. Hence this number does not depend on the choice of U: it is simply the number of ordered k-tuples of linearly independent vectors in $V = \mathrm{GF}(q)^n$ over GF(q) given by (7.4). Therefore $f(m,n,k)$ equals this number multiplied by the number of k-dimensional subspaces of F^m, which gives the first equality in (7.5).

Now the second equality in (7.5) follows, as the (m,n)-matrices of rank k correspond bijectively to the (n,m)-matrices of rank k, via transposition. Finally, the third equality is obtained by substituting the value of the Gaussian coefficient according to Proposition 7.1.4 in either of the first two expressions. □

We now turn our attention to the possible orders of invertible matrices over finite fields. By Theorem 1.3.3, the order of an invertible (n,n)-matrix over GF(q) has to divide the order of GL(n,q) given in Equation (7.1). The following well-known result provides a much stronger upper bound.

Theorem 7.1.6. *Let A be an invertible (n,n)-matrix over $F = \mathrm{GF}(q)$. Then the order of A is at most $q^n - 1$. Moreover, equality holds if and only if the minimal and characteristic polynomials of A over F coincide and are primitive polynomials of degree n over F.*

Proof. Denote the order of A by k and consider the F-algebra $R = \{f(A) : f \in F[x]\}$. Then R is isomorphic to the truncated polynomial algebra $F[x]/(m_A)$, where (m_A) denotes the ideal generated by the minimal polynomial m_A of A. Write $r = \deg m_A$. By the Cayley-Hamilton theorem, m_A divides the characteristic polynomial χ_A of A (which has degree n), and thus we have $r \le n$. But R is an r-dimensional vector space over F and hence has exactly $q^r \le q^n$ elements. On the other hand, R contains at least

$k+1$ elements, namely the k distinct powers of A and the zero matrix. Thus we have $k \leq q^n - 1$. Moreover, we obtain equality in this bound if and only if $m_A = \chi_A$ is a primitive polynomial of degree n. □

We now state without proof an interesting result due to Niven [309], who also gave an algorithmic procedure for determining all possible orders of matrices over $\mathrm{GF}(q)$. His results are more technical and not required for our purposes.

Result 7.1.7 *The least common multiple of all orders of matrices in* $\mathrm{GL}(n,q)$ *– that is, the exponent of* $\mathrm{GL}(n,q)$ *– is* $p^e M$*, where* q *is a power of the prime* p*, where* e *is the least integer such that* $n \leq p^e$*, and where* M *is the least common multiple of* $q-1, q^2-1, \ldots, q^n-1$*. In particular, the order of any invertible* (n,n)*-matrix over* $F = \mathrm{GF}(q)$ *divides* $p^e M$*.* □

As we shall see in the next section, matrices achieving equality in Theorem 7.1.6 are of particular interest. Thus they have been given a name:

Definition 7.1.8. An invertible (n,n)-matrix A of order $q^n - 1$ over $\mathrm{GF}(q)$ is called a **Singer cycle**, and the group generated by A is said to be a **Singer subgroup** of $\mathrm{GL}(n,q)$. □

Next, we state two classical facts about Singer subgroups, again without proof; see, for instance, Huppert [200, §II.7].

Result 7.1.9 *Any two Singer subgroups of* $G = \mathrm{GL}(n,q)$ *are conjugate in* G*. The normalizer* $N = N_G(S)$ *of a Singer subgroup* S *of* G *is a semidirect product* $N = S \cdot \langle \sigma \rangle$*, where* σ *is the Frobenius automorphism of order* n *of the extension field* $\mathrm{GF}(q^n)$ *over* $\mathrm{GF}(q)$*.* □

We note the following simple application of Result 7.1.9.

Corollary 7.1.10. *The number* $s(n,q)$ *of Singer subgroups of* $G = \mathrm{GL}(n,q)$ *equals*

$$s(n,q) = \frac{q^{n(n-1)/2}}{n} \cdot \prod_{i=1}^{n-1} (q^i - 1), \tag{7.6}$$

and the number of Singer cycles in G *is given by* $s(n,q)\phi(q^n-1)$*, where* ϕ *is the Euler totient function.*

Proof. Since any two Singer subgroups of G are conjugate in G, the number of Singer subgroups equals the order of G divided by the order of the stabilizer of such a subgroup S. Since the stabilizer of S under the action of G by conjugation is just the normalizer $N_G(S)$ of S in G, we obtain (7.6) by substituting the respective values according to Proposition 7.1.1 and Result 7.1.9. As any Singer subgroup contains exactly $\phi(q^n-1)$ generators, we also obtain the number of Singer cycles stated in the assertion. □

More generally, the reader might wonder what the number of matrices in $\mathrm{GL}(n,q)$ of a given order k is. Determining these numbers is a considerably more difficult problem, and no simple expressions are known in general. We refer the interested reader to the survey by Morrison [288, 1.11] and the references given there.

We conclude this section with a method for constructing a matrix A in $\mathrm{GL}(n,q)$ of any desired order k dividing $q^n - 1$, provided that we have the extension field $\mathrm{GF}(q^n)$ at our disposal.

Example 7.1.11. Consider $E = \mathrm{GF}(q^n)$ as a vector space over $F = \mathrm{GF}(q)$, and let γ be any element of E. Then γ defines an F-linear mapping $M_\gamma \colon E \to E$ via

$$M_\gamma \colon \xi \mapsto \gamma\xi \quad \text{for } \xi \in E.$$

Of course, the case $\gamma = 0$ is not very interesting, as it just gives the trivial mapping sending all elements of E to 0. But for $\gamma \neq 0$, we obtain an automorphism of the F-vector space E. Note that the order of M_γ then equals the order of γ in E^*.

We now represent M_γ with respect to some basis B of E over F and consider the associated matrix $A_B(\gamma)$. Then $\gamma = 0$ just leads to the zero matrix, independent of the choice of B. For $\gamma \neq 0$, the matrix $A_B(\gamma)$ has order $\mathrm{ord}(\gamma)$ in $\mathrm{GL}(n,q)$. In particular, choosing γ as a primitive element ω for E^* gives a matrix $A_B(\omega)$ of the maximum possible order $q^n - 1$, that is, a Singer cycle. □

Exercises

Exercise 7.1.12. Prove the following identities for the Gaussian coefficients:

$$\begin{bmatrix} n \\ k \end{bmatrix}_q = \begin{bmatrix} n \\ n-k \end{bmatrix}_q \tag{7.7}$$

and

$$\begin{bmatrix} n \\ k \end{bmatrix}_q = q^k \begin{bmatrix} n-1 \\ k \end{bmatrix}_q + \begin{bmatrix} n-1 \\ k-1 \end{bmatrix}_q = \begin{bmatrix} n-1 \\ k \end{bmatrix}_q + q^{n-k} \begin{bmatrix} n-1 \\ k-1 \end{bmatrix}_q. \tag{7.8}$$

Exercise 7.1.13. Discuss how the Frobenius automorphism σ operates on the given Singer subgroup S when representing the normalizer $N_G(S)$ as a semidirect product in Result 7.1.9. Note that this depends on the choice of S; the simplest case arises by taking S as an embedding of E^* as in Example 7.1.11. □

7.2 Matrix Representations of Finite Fields

In this section, we discuss the possibility of representing finite fields of characteristic p as sets of matrices over $\mathrm{GF}(q)$, where q is some power of p. While this approach does not seem to be of any particular practical interest, it is perhaps more intuitive than the standard representation as factor rings of the polynomial ring $\mathrm{GF}(p)[x]$. Here is a formal definition.

Definition 7.2.1. Any subring of the full matrix ring $F^{(n,n)}$ over a field F which is itself a field is called a **matrix field** of **degree** n over F. □

We are here only interested in the special case $F = \mathrm{GF}(q)$. Using some basic Linear Algebra, it is rather easy to obtain a representation of the extension field $\mathrm{GF}(q^n)$ as a matrix field of degree n over F.

Theorem 7.2.2. $\mathrm{GF}(q^n)$ *can be represented as a matrix field of degree n over* $\mathrm{GF}(q)$.

Proof. We use the construction described in Example 7.1.11 and consider the set $R(B)$ of all matrices $A_B(\gamma)$ with $\gamma \in E$. Then $R(B)$ is a subfield of $F^{(n,n)}$ isomorphic to E: indeed, it is easily checked that the mapping $\gamma \mapsto A_B(\gamma)$ gives an isomorphism between E and $R(B)$. □

Combining Theorems 7.2.2 and 7.1.6 gives the following result.

Corollary 7.2.3. *The smallest degree of a representation of* $\mathrm{GF}(q^n)$ *as a matrix field over* $\mathrm{GF}(q)$ *is n.* □

Of course, there are many different matrix representations one may obtain via the method described in the proof of Theorem 7.2.2, depending on the choice of the basis B. We now discuss one particularly interesting possibility.

Example 7.2.4. As we have seen in Example 7.1.11 and in the proof of Theorem 7.2.2, the q^n matrices $A_B(\gamma)$ with $\gamma \in E$ form a matrix representation $R(B)$ of degree n for $E = \mathrm{GF}(q^n)$ over $F = \mathrm{GF}(q)$. Note that we may write $R(B)$ as the $q^n - 1$ powers of the Singer cycle $A_B(\omega)$ together with the zero matrix, where ω is a primitive element for E. The minimal polynomial m_ω of ω is a primitive polynomial of degree n over F. Let $m_\omega = x^n + a_{n-1}x^{n-1} + \cdots + a_1x + a_0$. We now choose the **polynomial basis** with respect to ω for B, that is, we take $B = (1, \omega, \omega^2, \ldots, \omega^{n-1})$. Then $A_B(\omega)$ is just the **companion matrix** of m_ω:

$$A_B(\omega) = \begin{pmatrix} 0 & 0 & \cdots & \cdots & 0 & -a_0 \\ 1 & 0 & & & 0 & -a_1 \\ 0 & 1 & \ddots & & \vdots & \vdots \\ 0 & 0 & \ddots & \ddots & \vdots & \vdots \\ \vdots & \vdots & & \ddots & 0 & -a_{n-2} \\ 0 & 0 & \cdots & 0 & 1 & -a_{n-1} \end{pmatrix}. \quad \square \tag{7.9}$$

Hence any primitive polynomial of degree n leads to a matrix representation for E over F of the smallest possible degree via the Singer cycle given in (7.9).

Essentially, all minimal degree representations may be obtained in this way. This is an immediate consequence of the following result, since any matrix representation of E has to contain some matrix of order $q^n - 1$, by Theorem 3.2.5.

Theorem 7.2.5. *Let $A \in \mathrm{GL}(n,q)$, and assume that A has order $q^n - 1$. Then*

$$R = \{A^i : i = 0, \ldots, q^n - 2\} \cup \{0\}$$

is a representation of $E = \mathrm{GF}(q^n)$ as a matrix field of degree n over $F = \mathrm{GF}(q)$. Moreover, A is similar to the companion matrix of some primitive polynomial of degree n over F, and $\det A$ *is a primitive element for F^*.*

Proof. By the proof of Theorem 7.1.6, the algebra $R \cong F[x]/(m_A)$ indeed consists of the powers of A together with the zero matrix, so that R is actually a field and therefore isomorphic to E. Moreover, $m_A = \chi_A$ is a primitive polynomial of degree n over F. Because of this fact, A has to be similar to the companion matrix M of m_A. If we write $m_A = x^n + a_{n-1}x^{n-1} + \cdots + a_1x + a_0$, the matrix M is just the matrix $A_B(\omega)$ in (7.9), where ω is some root of m_A in E. Therefore $\det A = \det A_B(\omega) = (-1)^n a_0$. On the other hand, by Proposition 3.12.3, also

$$(-1)^n a_0 = \mathrm{Norm}_{E/F}(\omega) = \omega^{(q^n-1)/(q-1)}.$$

But ω is a primitive element for E^*, and hence $\det A = (-1)^n a_0$ has order $q - 1$. Therefore $\det A$ is a primitive element for F^*. □

There are considerably more general results due to Willett [405], who classified all matrix fields of degree n over $\mathrm{GF}(q)$ using primitive polynomials over the underlying prime field $\mathrm{GF}(p)$. We shall present his results in the remainder of this section. Willett's classification rests on the following basic result.

Theorem 7.2.6. *Let F be a finite field with characteristic p, and let $S \subseteq F^{(n,n)}$ be a matrix field. Then there exist positive integers k and r, a matrix $A \in F^{(n,n)}$ and an idempotent matrix $J \in F^{(n,n)}$, such that A and J both have rank r and*

$$S = \{0, J\} \cup \{A^i : 1 \le i \le p^k - 2\}.$$

Moreover, $r \le n$ and $k \le mr$, where $|F| = p^m$.

Proof. Obviously, S has characteristic p. Let k be the dimension of S as a P-vector space, where P is the prime subfield of F, and let J be the unit element of the field S. Then $J^2 = J$ shows that $J \neq 0$ is an idempotent element, that is, a projection matrix. Hence there exists an invertible matrix $M \in F^{(n,n)}$ such that

$$M^{-1}JM = \mathrm{diag}(0, \ldots, 0, 1, \ldots, 1),$$

where the number $r \ge 1$ of entries 1 is the rank of J. Note that

$$S' := \{M^{-1}XM : X \in S\}$$

is a subring of $F^{(n,n)}$ which is isomorphic to S and hence likewise a matrix field. Since

$$(M^{-1}JM)(M^{-1}XM) = M^{-1}JXM = M^{-1}XM \quad \text{for all } X \in S,$$

every matrix in S' is a block diagonal matrix of the form $\operatorname{diag}(0,Y)$, where $Y \in F^{(r,r)}$; moreover, Y has rank r provided it is not the zero matrix.

As S' is a finite field, we may choose a matrix $D \in F^{(r,r)}$ such that $\operatorname{diag}(0,D)$ is a primitive element for S'. Now consider the mapping

$$\varphi_D \colon P[x] \to F^{(r,r)}, \; f(x) \mapsto f(D).$$

Then φ_D is a homomorphism between P-algebras, and the minimal polynomial $\nu_D(x)$ of D over P is a primitive polynomial of degree k. Moreover, the image of φ_D equals

$$\{D^i \colon 0 \le i \le p^k - 2\} \cup \{0\}.$$

Altogether, this gives

$$\begin{aligned} S \setminus \{0\} &= \{M \cdot \operatorname{diag}(0,D^i) \cdot M^{-1} \colon 0 \le i \le p^k - 2\} \\ &= \{J\} \cup \{M \cdot \operatorname{diag}(0,D)^i \cdot M^{-1} \colon 1 \le i \le p^k - 2\} \\ &= \{J\} \cup \{(M \cdot \operatorname{diag}(0,D) \cdot M^{-1})^i \colon 1 \le i \le p^k - 2\} \\ &= \{J\} \cup \{A^i \colon 1 \le i \le p^k - 2\}, \end{aligned}$$

where $A := M \cdot \operatorname{diag}(0,D) \cdot M^{-1}$. This completes the proof, as the restriction on r is immediate and that on k a consequence of Theorem 7.1.6. □

Remark 7.2.7. We stress that the matrix field S in Theorem 7.2.6 is not necessarily an F-subalgebra of $F^{(n,n)}$ (actually, not even a P-subalgebra), since it does not need to contain the identity matrix $I \in F^{(n,n)}$.

Now let J, D, ν_D and A be as in the proof of Theorem 7.2.6, and consider the mapping

$$\varphi_A \colon P[x] \to F^{(n,n)}, \; g(x) \mapsto g(A).$$

Let ν_A be the monic generator of its kernel, that is, let ν_A be the minimal polynomial of A when $F^{(n,n)}$ is considered as a P-algebra. We claim that ν_A divides

$$(\nu_D(x) - \nu_D(0)) \cdot \nu_D(x).$$

This is seen as follows, where we write $\nu_D(x) = \sum_{i=0}^{k} \lambda_i x^i$ with $\lambda_k = 1$; note that $\lambda_0 = \nu_D(0) \neq 0$, as the minimal polynomial ν_D is primitive. Put $h(x) := \nu_D(x) - \lambda_0$. Then

$$\begin{aligned} h(A) &= h(M^{-1} \cdot \operatorname{diag}(0,D) \cdot M) = M^{-1} \cdot h(\operatorname{diag}(0,D)) \cdot M \\ &= M^{-1} \cdot \operatorname{diag}(0,h(D)) \cdot M = M^{-1} \cdot \operatorname{diag}(0,-\lambda_0 I_r) \cdot M = -\lambda_0 J. \end{aligned}$$

Since $J^2 - J = 0$, we obtain

$$\frac{1}{\lambda_0^2} \cdot h(A)^2 + \frac{1}{\lambda_0} \cdot h(A) = 0,$$

which shows (after multiplication with λ_0^2) that A is indeed annihilated by

$$(\nu_D(x) - \lambda_0)^2 + \lambda_0 \cdot (\nu_D(x) - \lambda_0) = (\nu_D(x) - \lambda_0) \cdot \nu_D(x).$$

Note here that x divides $\nu_D(x) - \lambda_0$. If $J \neq I$, this takes into account that 0 is an eigenvalue of A. If $J = I$, already $\nu_D(A) = 0$ and therefore $\nu_A = \nu_D$.

The converse of Theorem 7.2.6 also holds: let us start with a matrix $D \in F^{(r,r)}$, where $1 \leq r \leq n$, which has a primitive minimal polynomial ν_D over P. Then $\mathrm{alg}_P(D) \subseteq F^{(r,r)}$ is a field with p^k elements, where $k = \deg \nu_D$. Embedding $F^{(r,r)}$ into $F^{(n,n)}$ as a set of block diagonal matrices then gives a subring S' of $F^{(n,n)}$ which is actually a field.

More generally, $S' = \{M^{-1}XM \colon X \in S\}$ will be a matrix field in $F^{(n,n)}$, where M can be an arbitrary element of $GL(n,F)$. In other words, up to similarity, the matrix fields in $F^{(n,n)}$ are precisely the subrings of the form

$$\{0\} \cup \left\{\mathrm{diag}(0, D^i) \colon 0 \leq i \leq p^k - 2\right\}, \tag{7.10}$$

where D is any matrix with the properties given above. □

For the explicit determination of all matrix fields in $\mathrm{GF}(q)^{(n,n)}$, one still has to describe all matrices $D \in F^{(r,r)}$ which are annihilated by a primitive polynomial over the prime field P. Using some results from Linear Algebra, this is quite easy in the case where $F = P$.

Lemma 7.2.8. *Let $D \in \mathrm{GF}(p)^{(r,r)}$ be a root of a primitive polynomial f of degree k over $\mathrm{GF}(p)$. Then r is a multiple of k and D is similar to a matrix of the form $\mathrm{diag}(A, \ldots, A)$, where A is the companion matrix of f.*

Proof. Note that $f = m_D$ is the minimal polynomial of D over $\mathrm{GF}(p)$. Up to similarity, we may assume that D is in rational normal form. Thus D is a block diagonal matrix formed by matrices which are companion matrices of polynomials dividing the minimal polynomial f. This implies the assertion, since f is irreducible. □

Combining Theorem 7.2.6 and Lemma 7.2.8, we obtain the desired description of all representations of $\mathrm{GF}(q)$ as a matrix field over $\mathrm{GF}(p)$.

Theorem 7.2.9. *Let $q = p^m$ be a power of a prime p. Up to similarity, every representation of $\mathrm{GF}(q)$ as a matrix field M over $\mathrm{GF}(p)$ has the form*

$$M = \{\mathrm{diag}(0, \ldots, 0, A^i, \ldots, A^i) \colon i = 0, \ldots, p^m - 2\} \cup \{0\},$$

where A is the companion matrix of some primitive polynomial of degree m over $\mathrm{GF}(p)$. □

The general case of matrix fields in $\mathrm{GF}(q)^{(n,n)}$ is more involved. Here we first have to use Theorem 3.5.9 to obtain the factorization of the primitive polynomial $f \in \mathrm{GF}(p)[x]$ annihilating D into distinct irreducible polynomials over $\mathrm{GF}(q)$. We summarize the resulting analogue of Lemma 7.2.8 in the following

Observation 7.2.10. Let $D \in \mathrm{GF}(q)^{(r,r)}$ be a root of a primitive polynomial f of degree k over $\mathrm{GF}(p)$, where $q = p^m$, and write $d = \gcd(m,k)$. By Theorem 3.5.9, f splits over $F = \mathrm{GF}(q)$ into d distinct irreducible polynomials of degree k/d, say

$$f = f_1 \cdots f_d.$$

Then the minimal polynomial of D over F has the form

$$m_D(x) = \prod_{i \in \Delta} f_i(x),$$

where Δ is a nonempty subset of $\{1,...,d\}$. Let $\Delta_s \subseteq \cdots \subseteq \Delta_1 = \Delta$ be such that the polynomials

$$\mu_j(x) := \prod_{i \in \Delta_j} f_i(x)$$

give the elementary divisors of D. Then, by the rational normal form, D is similar to the matrix $\mathrm{diag}(A_1,...,A_s)$, where each A_j is the companion matrix of μ_j. In particular,

$$r = \left(\sum_{j=1}^{s} |\Delta_j|\right) \cdot \frac{k}{d}$$

is a multiple of k/d. □

Let us conclude this section with an example illustrating Observation 7.2.10:

Example 7.2.11. By Example 5.6.8, the polynomial $f_0 = x^3 + x^2 + x + \omega$ is a primitive polynomial of degree 3 over GF(4), where we write $\mathrm{GF}(4) = \{0, 1, \omega, \omega^2\}$ and where ω is a root of $g = x^2 + x + 1$. According to Example 7.2.4, the companion matrix

$$T = \begin{pmatrix} 0 & 0 & \omega \\ 1 & 0 & 1 \\ 0 & 1 & 1 \end{pmatrix}$$

of f_0 generates a matrix field M isomorphic to $\mathrm{GF}(4^3) = \mathrm{GF}(64)$ in $\mathrm{GF}(4)^{(3,3)}$. Since T^2 also generates M^*, the minimal polynomial f_1 of T^2 is likewise primitive over GF(4). One easily checks $f_1 = x^3 + x^2 + x + (1 + \omega)$, using that f_0 is the minimal polynomial of T. Thus both T and T^2 are roots of the polynomial

$$f = f_0 f_1 = x^6 + x^4 + x^3 + x + 1,$$

which is therefore a primitive polynomial of degree 6 over GF(2). Then the matrix $D = \mathrm{diag}(T, T^2)$ is likewise a root of f and thus generates a matrix field L isomorphic to GF(64) in $R = \mathrm{GF}(4)^{(6,6)}$. Note that the minimal polynomial of D is f, which is reducible as a polynomial over GF(4).

On the other hand, L contains a matrix subfield K isomorphic to $\mathrm{GF}(4)$, and the minimal polynomial m_D of D over K is, of course, an irreducible polynomial of degree 3 in $K[x]$. Note that K is *not* the natural copy of $\mathrm{GF}(4)$ in the matrix ring R, that is, $K \neq \{\lambda I_6 : \lambda \in \mathrm{GF}(4)\}$; in fact,

$$K = \langle D^{21} \rangle \cup \{0\} = \{0, I_6, \Omega, \Omega^2\} \text{ with } \Omega = \mathrm{diag}(\omega, \omega, \omega, 1+\omega, 1+\omega, 1+\omega),$$

and the minimal polynomial m_D of D over K turns out to be the primitive polynomial $m_D = X^3 + X^2 + X + \Omega$ of degree 3 over K. Thus the minimal polynomials of D over K and over the given ground field $\mathrm{GF}(4)$, respectively, do not coincide. □

The matrix fields in $\mathrm{GF}(q)^{(n,n)}$ have also been characterized by Beard [23, 24]; however, his approach is more involved. The same author has also determined the number of matrix fields in $\mathrm{GF}(q)^{(n,n)}$ and, more precisely, the number of matrix fields of order p^m consisting of matrices of rank r in $\mathrm{GF}(q)^{(n,n)}$; see [25].

In view of Theorem 7.2.9, we will henceforth mainly be interested in representations of $\mathrm{GF}(p^n)$ as a matrix field of degree n over $\mathrm{GF}(p)$. Even then the rather explicit result of Theorem 7.2.6 still leaves some interesting questions. In particular, one might wonder whether one may find examples among the matrix fields conjugate to one of type (7.10) which consist of particularly "nice" matrices. For instance, by a result of Seroussi and Lempel [349], there always is a representation of $\mathrm{GF}(q^n)$ as a matrix field of degree n over $\mathrm{GF}(q)$ consisting of symmetric matrices only. As this topic is more difficult, we postpone its discussion to Section 7.7.

7.3 Circulant Matrices and Normal Bases

In this section, we present basic results on circulant matrices, which are also useful for the study of normal elements. We first show that all circulant (n,n)-matrices over an arbitrary field F form a ring isomorphic to the truncated polynomial ring $F[x]/(x^n - 1)$, which is in turn isomorphic to the group algebra[1] of the cyclic group of order n over F. After this, we will specialize to the case $F = \mathrm{GF}(q)$ and determine the order of the multiplicative subgroup $C(n,q)$ of all circulant matrices in $\mathrm{GL}(n,q)$, which turns out to be isomorphic to the group of units of the ring $R_{n,q} := \mathrm{GF}(q)[x]/(x^n - 1)$.

We remark that the ring $R_{n,q}$ is the most important algebraic structure used in Coding Theory, since its ideals can serve to describe all cyclic codes of length n over $\mathrm{GF}(q)$. The cyclic codes constitute the class of codes most widely used in practice because of their good behavior with respect to encoding and decoding; see, for instance, Berlekamp [30], Bierbrauer [37] or MacWilliams and Sloane [255].

---

[1] For the time being, we shall not make use of the latter isomorphism, so that the reader not familiar with group algebras can safely ignore this connection now. These matters will only be taken up later in the book; see Chapter 10.

Definition 7.3.1. An (n,n)-matrix $C = (c_{ij})_{i,j=0,\dots,n-1}$ with entries from a field F is said to be **circulant** if its rows are generated by successive cyclic shifts (to the right) of its first row. More formally, we require

$$c_{i+1,j+1} = c_{ij} \text{ for all } i,j = 0,\dots,n-1, \tag{7.11}$$

where all indices are taken modulo n. Thus C is determined by the entries $c_j := c_{0j}$ in its first row:

$$C = \begin{pmatrix} c_0 & c_1 & c_2 & \dots & c_{n-1} \\ c_{n-1} & c_0 & c_1 & \dots & c_{n-2} \\ c_{n-2} & c_{n-1} & c_0 & \dots & c_{n-3} \\ \vdots & \vdots & \vdots & \ddots & \vdots \\ c_1 & c_2 & c_3 & \dots & c_0 \end{pmatrix}. \tag{7.12}$$

We often denote such a matrix as $\mathrm{circ}(c_0,\dots,c_{n-1})$. □

We shall see that the set of all circulant (n,n)-matrices over F constitutes an n-dimensional F-algebra. To this purpose, let P be the permutation matrix $\mathrm{circ}(0,1,0,\dots,0)$, and consider the mapping

$$\Gamma_P \colon F[x] \to F^{(n,n)}, \quad c(x) \mapsto c(P).$$

The image of Γ_P is the F-subalgebra $\mathrm{alg}_F(P)$ of $F^{(n,n)}$ generated by P, and every element of $\mathrm{alg}_F(P)$ is a circulant matrix. The kernel of Γ_p is generated by $x^n - 1$, which is the minimal polynomial of P. Hence $\mathrm{alg}_F(P)$ is an n-dimensional subspace of $F^{(n,n)}$. Since the set of all circulant (n,n)-matrices over F also is an n-dimensional F-subspace, we see that $\mathrm{alg}_F(P)$ in fact comprises all circulant matrices. As $\mathrm{alg}_F(P)$ is isomorphic to $F[x]/(x^n-1)$, we have established the following fundamental result:

Theorem 7.3.2. *Let F be any field. Then the mapping τ which sends the matrix C in* (7.12) *to the coset of the polynomial $c(x) = c_0 + c_1x + \dots + c_{n-1}x^{n-1}$ is an isomorphism between the algebra of circulant (n,n)-matrices over F and the algebra $F[x]/(x^n-1)$.* □

To facilitate notation, we will from now on use the same symbol $f(x)$ to denote both the polynomial f over F and its coset in $F[x]/(x^n-1)$; it should always be clear from the context which meaning is intended. Next, we investigate the set of units of this ring.[2]

Proposition 7.3.3. *A polynomial $c(x) \in F[x]/(x^n-1)$ is a unit if and only if it is relatively prime to x^n-1.*

Proof. Let us write $R := F[x]/(x^n-1)$. Assume first that c and x^n-1 have a common factor g of degree at least 1. Then we have $ch = 0$, where $h = (x^n-1)/g$ is an element $\neq 0$ of R. Thus c is a zero divisor and therefore not a unit.

[2] The following result and the subsequent remark are special cases of material on algebras covered in Section 2.4. Nevertheless, it is useful to deal with them explicitly here.

Conversely, assume that c and $x^n - 1$ are relatively prime. Using the extended Euclidean algorithm for computing the greatest common divisor of two polynomials, we can determine polynomials $u, v \in F[x]$ satisfying $uc + v(x^n - 1) = 1$. Then the reduction d of u modulo $x^n - 1$ is the desired inverse of c in R. □

In view of Theorem 7.3.2, the proof of Proposition 7.3.3 shows that the inverse of a non-singular circulant matrix C is again circulant and gives a simple method for computing this inverse: if c is the polynomial corresponding to C, then the circulant matrix D corresponding to the polynomial d is the desired inverse of C.

We now specialize to the case $F = \mathrm{GF}(q)$ and determine the order of the multiplicative group $C(n,q)$ of all non-singular circulant (n,n)-matrices over $\mathrm{GF}(q)$. By Theorem 7.3.2, $C(n,q)$ is isomorphic to the group of units of the ring $R_{n,q} = \mathrm{GF}(q)[x]/(x^n - 1)$. Using the corresponding Euler function introduced in Definition 2.3.6, we obtain the following result, where we again write $\phi_q(f)$ instead of $\phi_{\mathrm{GF}(q)}(f)$.

Corollary 7.3.4. *The group $C(n,q)$ of all circulant invertible (n,n)-matrices over $\mathrm{GF}(q)$ has order $\phi_q(x^n - 1)$.* □

Comparing the preceding result with Theorem 3.10.5, it is not surprising that there is a close connection between normal elements of $\mathrm{GF}(q^n)$ over $\mathrm{GF}(q)$ and circulant matrices.

Proposition 7.3.5. *Let α be a normal element of $E = \mathrm{GF}(q^n)$ over $F = \mathrm{GF}(q)$, and let $C = (c_{ij})_{i,j=0,\dots,n-1}$ be an invertible matrix over F. Then the basis $\{\beta_0, \dots, \beta_{n-1}\}$ defined by*

$$\beta_i := \sum_{j=0}^{n-1} c_{ij}\alpha^{q^j} \quad \text{for } i = 0, \dots, n-1 \tag{7.13}$$

is likewise a normal basis for E/F if and only if C is a circulant matrix.

Proof. Because of (7.13), we have

$$\beta_i^q = \sum_{j=0}^{n-1} c_{ij}\alpha^{q^{j+1}} \quad \text{for } i = 0, \dots, n-1.$$

Hence we always obtain $\beta_i^q = \beta_{i+1}$ – so that β is a normal element for E/F – if and only if

$$\beta_i^q = \sum_{j=0}^{n-1} c_{ij}\alpha^{q^{j+1}} = \beta_{i+1} = \sum_{j=0}^{n-1} c_{i+1,j}\alpha^{q^j} = \sum_{j=0}^{n-1} c_{i+1,j+1}\alpha^{q^{j+1}}.$$

Clearly this holds if and only if $c_{ij} = c_{i+1,j+1}$ for all i and j, which means that C should be circulant. □

Note that Proposition 7.3.5 gives a very simple way of computing all normal bases of E/F provided that one such basis is already known, since then every normal basis of E/F is obtained from a suitable circulant matrix C. In particular, together with Corollary 7.3.4, we will obtain an alternative, more elementary and constructive proof of Theorem 3.10.5.

Corollary 7.3.6. *The number of normal elements of* $\mathrm{GF}(q^n)$ *over* $\mathrm{GF}(q)$ *equals the order* $\phi_q(x^n-1)$ *of the group* $C(n,q)$ *of invertible circulant* (n,n)*-matrices over* $\mathrm{GF}(q)$. □

We conclude this section with a concrete method for evaluating $\phi_q(x^n-1)$ which does not depend on the structure theory of general polynomial algebras dealt with in Section 2.3. For this purpose, we first state some useful general properties of ϕ_q, which are analogous to those of the Euler totient function.

Theorem 7.3.7. *Let q be a prime power. Then the function ϕ_q has the following two properties:*

$$\phi_q(fg) \;=\; \phi_q(f)\phi_q(g) \quad \textit{whenever } \gcd(f,g)=1; \tag{7.14}$$

$$\phi_q(f^k) = q^{nk}-q^{n(k-1)} = q^{nk}(1-q^{-n}) \quad \textit{if } f \textit{ is irreducible of degree } n. \tag{7.15}$$

Proof. If f and g are relatively prime, the Chinese remainder theorem (see 1.10.6) implies the following isomorphism of rings:

$$\mathrm{GF}(q)[x]/(fg) \;\cong\; \mathrm{GF}(q)[x]/(f)\oplus \mathrm{GF}(q)[x]/(g),$$

which immediately gives the validity of (7.14). For (7.15), note that there are q^{nk} polynomials of degree $\le nk-1$ in $\mathrm{GF}(q)[x]$. As f is irreducible, all these polynomials are relatively prime to f, except for the multiples of f. But there are exactly $q^{n(k-1)}$ multiples g of f which have degree $< nk$, since g/f can be any polynomial of degree $< n(k-1)$. □

Theorem 7.3.7 shows that it suffices to know the degrees of the irreducible factors of f to compute $\phi_q(f)$; the explicit factorization of f is not required.

Corollary 7.3.8. *Let f be a polynomial of degree n over* $\mathrm{GF}(q)$*, and assume that the distinct irreducible factors $f_1,\dots,f_r$ of f have degrees $n_1,\dots,n_r$ respectively. Then:*

$$\phi_q(f) = q^n(1-q^{-n_1})\cdots(1-q^{-n_r}); \tag{7.16}$$

$$\phi_q(f) \;=\; (q^{n_1}-1)\cdots(q^{n_r}-1) \quad \textit{if } \gcd(f,f')=1. \tag{7.17}$$

Proof. For (7.16), write $f=f_1^{k_1}f_2^{k_2}\cdots f_r^{k_r}$, so that $n=k_1n_1+\dots+k_rn_r$. Then Theorem 7.3.7 gives

$$\begin{aligned}\phi_q(f) &= \phi_q(f_1^{k_1})\cdots\phi_q(f_r^{k_r})\\ &= \left(q^{n_1k_1}(1-q^{-n_1})\right)\cdots\left(q^{n_rk_r}(1-q^{-n_r})\right)\\ &= q^n(1-q^{-n_1})\cdots(1-q^{-n_r}).\end{aligned}$$

Now assume $\gcd(f,f')=1$. By Corollary 6.1.2, f has no multiple roots in this case. Thus $k_1=\dots=k_r=1$, and we obtain the simplified formula (7.17). □

We now use Corollary 7.3.8 to compute $\phi_q(x^n-1)$ without explicitly factoring x^n-1:

Theorem 7.3.9. *Let q be a power of the prime p, let n be a positive integer and write $n=p^a m$, where p does not divide m. Then:*

$$\phi_q(x^n-1) = q^n \cdot \prod_{d|m}\left(1-q^{-\mathrm{ord}_d(q)}\right)^{\phi(d)/\mathrm{ord}_d(q)}, \tag{7.18}$$

where ϕ is Euler's totient function.

Proof. In order to apply Corollary 7.3.8, we need to determine the degrees of the distinct irreducible factors of x^n-1. This was already done in Section 3.6 and can be summarized as follows. One has $x^n-1=(x^m-1)^{p^a}$, and x^m-1 has no multiple roots. Moreover, x^m-1 is the product of the cyclotomic polynomials Φ_d, where d runs over all divisors of m. Finally, each Φ_d splits over $\mathrm{GF}(q)$ into $\phi(d)/\mathrm{ord}_d(q)$ irreducible factors of degree $\mathrm{ord}_d(q)$ each. □

Since the n powers $\alpha, \alpha^q, \ldots, \alpha^{q^{n-1}}$ of a normal element α of $\mathrm{GF}(q^n)$ over $\mathrm{GF}(q)$ lead to the same normal basis, combining Corollary 7.3.6 and Theorem 7.3.9 immediately gives the following result of Ore [314].

Theorem 7.3.10. *Let q be a power of the prime p, let n be a positive integer and write $n=p^a m$, where p does not divide m. Then the number of normal bases of $\mathrm{GF}(q^n)$ over $\mathrm{GF}(q)$ equals*

$$\frac{q^n}{n} \cdot \prod_{d|m}\left(1-q^{-\mathrm{ord}_d(q)}\right)^{\phi(d)/\mathrm{ord}_d(q)}. \tag{7.19}$$

Remark 7.3.11. In Remark 3.5.5, we noted that the number $i_q(n)$ of irreducible polynomials of degree n over $\mathrm{GF}(q)$ is approximately q^n/n (as the largest error term is at most $\frac{1}{2}q^{n/2}$), so that the probability that a randomly selected monic polynomial of degree n over a finite field turns out to be irreducible is roughly $1/n$. Theorem 7.3.10 implies that the number $n_q(n)$ of irreducible polynomials of degree n over $\mathrm{GF}(q)$ which are the minimal polynomials of a normal basis generator is also quite large. By Equation (7.19), we have

$$n_q(n) \geq \frac{1}{n}\cdot q^n \cdot \prod_{d|m}\left(1-\frac{1}{q}\right)^{\phi(d)} = \frac{1}{n}\cdot q^n \cdot \left(1-\frac{1}{q}\right)^m,$$

and equality occurs if and only if x^n-1 splits over F into linear factors. Hence the leading term of $n_q(n)$ is again q^n/n, and the largest error term is at most mq^{n-1}/n.

This indicates that $n_q(n)$ is indeed quite large; however, precise estimates are not easy to obtain. Von zur Gathen and Giesbrecht [389] proved that the probability that an arbitrarily chosen element of $\mathrm{GF}(q^n)$ is a normal basis generator is larger than $\frac{1}{16\log_q n}$; their proof is too involved to be included here. This result explains the fact that it is indeed not much harder to find an irreducible polynomial generating a normal basis than an arbitrary irreducible polynomial, both probabilistically and deterministically. For instance, it is also shown in [389] that a normal polynomial of degree n over GF(q) can be constructed deterministically with complexity $O^\wedge(n^4 q^{1/2})$, where the notation $O^\wedge$ means that logarithmic terms in n are neglected. The same authors also gave even faster probabilistic algorithms.

These results have been improved subsequently. Von zur Gathen and Shoup [390] gave a probabilistic algorithm for constructing normal bases with complexity $O^\wedge(n^2 + n\log q)$, and Poli [323] provided a deterministic algorithm with complexity $O^\wedge(n^3 + n\log q)$. Frandsen [119] established an estimate on the frequency $\nu_q(n) = \phi_q(x^n - 1)/q^n$ of normal elements in GF(q^n) which is asymptotically optimal up to a constant factor. He also gave the explicit lower bound

$$\nu_q(n) \geq \frac{1}{e\lceil \log_q n \rceil} \quad \text{for all } n, q \geq 2,$$

which is, in general, considerably stronger than the estimate of [389].

We will study the problem of explicitly constructing normal elements and normal polynomials in considerable detail in Chapter 11, using a theory of "cyclotomic modules". □

7.4 The Discrete Fourier Transform

In this section, we present an introductory discussion of a particularly important transformation method, namely the Discrete Fourier Transform. Later, we will put this into a more general theoretical framework in terms of characters of abelian groups; for the time being, we merely use roots of unity.

Definition 7.4.1. Let n be a positive integer, and let E be any field of characteristic not dividing n which contains n distinct n-th roots of unity. We choose a primitive n-th root of unity, say ζ, in E and define an (n,n)-matrix Z as follows: $Z = (\zeta^{ij})_{i,j=0,\dots,n-1}$. Explicitly, the **Fourier matrix** Z is given by

$$Z = \begin{pmatrix} 1 & 1 & 1 & \dots & 1 \\ 1 & \zeta & \zeta^2 & \dots & \zeta^{n-1} \\ 1 & \zeta^2 & \zeta^4 & \dots & \zeta^{2(n-1)} \\ \vdots & \vdots & \vdots & \ddots & \vdots \\ 1 & \zeta^{n-1} & \zeta^{2(n-1)} & \dots & \zeta^{(n-1)^2} \end{pmatrix}. \tag{7.20}$$

Given any polynomial $a(x) = a_{n-1}x^{n-1} + \cdots + a_1x + a_0$ in $R := E[x]/(x^n - 1)$,[3] one defines the **Fourier vector A** of $a(x)$ by

$$\mathbf{A} := \mathbf{a}Z, \tag{7.21}$$

where $\mathbf{a} := (a_0, a_1, \ldots, a_{n-1})$. Thus $\mathbf{A} = (A_0, A_1, \ldots, A_{n-1})$ with

$$A_j = \sum_{i=0}^{n-1} a_i \zeta^{ij} = a(\zeta^j) \quad \text{for } j = 0, \ldots, n-1. \tag{7.22}$$

The transformation (7.21) of R into E^n is called the **Discrete Fourier Transform** or, for short, the **DFT**. □

The fundamental importance of the DFT stems from the fact that it gives an algebra isomorphism between $R = E[x]/(x^n - 1)$ and the algebra E^n with componentwise operations. This can be derived from the Chinese remainder theorem, which we will leave as an exercise. We prefer to give a more direct proof here, as this will also provide a simple explicit form for the inverse of the DFT.

Theorem 7.4.2. *Let n be a positive integer, and let E be any field of characteristic not dividing n which contains a primitive n-th root of unity ζ. Then the DFT is an isomorphism from the algebra $R = E[x]/(x^n - 1)$ to the algebra E^n (with componentwise operations).*

Proof. Clearly, the DFT respects the addition of polynomials and the multiplication of a polynomial by a scalar. Since ζ^j is an n-th root of unity, we have

$$a(\zeta^j)b(\zeta^j) = (ab)(\zeta^j) = (ab \bmod x^n - 1)(\zeta^j) \quad \text{for } j = 0, \ldots, n-1;$$

hence (7.22) shows that the DFT also respects the multiplication of two polynomials in R. Finally, the DFT is a bijection, since the Fourier matrix Z is non-singular:

$$nZ^{-1} = \left(\zeta^{-ij}\right)_{i,j=0,\ldots,n-1}. \tag{7.23}$$

This is immediate from

$$\sum_{j=0}^{n-1} \zeta^{ij}\zeta^{-jk} = \sum_{j=0}^{n-1} \zeta^{j(i-k)} = n \cdot \delta_{ik},$$

which is easily checked using $x^n - 1 = (x-1) \cdot \sum_{\ell=0}^{n-1} x^\ell$. □

It is worthwhile to state the following direct consequence of the proof of Theorem 7.4.2 explicitly:

Corollary 7.4.3. *The polynomial $a(x) = a_{n-1}x^{n-1} + \cdots + a_1x + a_0 \in E[x]/(x^n - 1)$ can be recovered from its Fourier vector $\mathbf{A} = \mathbf{a}Z$ as follows:*

[3] As before, we shall denote both the polynomial $f \in E[x]$ and its image in R by the same symbol and refer to the elements of R simply as polynomials.

$$a_j = \frac{1}{n} \cdot \sum_{i=0}^{n-1} A_i \zeta^{-ij} \qquad \textit{for } j = 0, \ldots, n-1. \tag{7.24}$$

In other words, the **inverse DFT** *is the mapping from E^n to $E[x]/(x^n-1)$ induced by the matrix Z^{-1} defined via Equation* (7.23). □

The DFT has many interesting applications. The classical example is given by the Fast Fourier Transform (FFT) algorithms for the case $E = \mathbb{C}$, which are used for fast arithmetics (in particular, fast multiplication of polynomials and of large integers); see, for instance, Aho, Hopcroft and Ullman [4] or Nussbaumer [311]. The case of a finite field E is an important tool in Coding Theory, where it is generally known under the name **Mattson-Solomon polynomial**; see, for instance, Berlekamp [30], MacWilliams and Sloane [255] and, in particular, Blahut [39] who gives an extensive treatment of this topic.

We now specialize the DFT to finite fields and consider a somewhat more general setup. Let $F = \mathrm{GF}(q)$, let ζ be a primitive n-th root of unity over F, where n and q are relatively prime, and put $E = F(\zeta)$; we recall that E has been determined explicitly in Proposition 3.6.16. We apply the DFT to the ring $R = E[x]/(x^n-1)$. The following useful result characterizes the Fourier vectors of those polynomials in R which actually have coefficients in the ground field F.

Proposition 7.4.4. *Let $F = \mathrm{GF}(q)$, let ζ be a primitive n-th root of unity over F, where n and q are relatively prime, and put $E = F(\zeta)$. Then the polynomial*

$$a(x) = a_{n-1}x^{n-1} + \cdots + a_1 x + a_0 \in E[x]/(x^n-1)$$

actually belongs to $R_{n,q} = F[x]/(x^n-1)$ if and only if its Fourier vector $\mathbf{A} = \mathbf{a}Z$ satisfies the following condition:

$$A_j^q = A_{jq} \quad \textit{for } j = 0, \ldots, n-1\,. \tag{7.25}$$

Proof. Note that $a(x)$ has coefficients in F if and only if

$$a_i^q = a_i \quad \text{for } i = 0, \ldots, n-1. \tag{7.26}$$

First assume the validity of (7.26). Then (7.22) gives

$$A_j^q = \Big(\sum_{i=0}^{n-1} a_i \zeta^{ij}\Big)^q = \sum_{i=0}^{n-1} a_i \zeta^{i(jq)} = A_{jq} \quad \text{for } j = 0, \ldots, n-1,$$

that is, the validity of condition (7.25). Conversely, assume that this condition is satisfied. Then we obtain from (7.24)

$$a_i^q = \Big(\frac{1}{n} \cdot \sum_{j=0}^{n-1} A_j \zeta^{-ij}\Big)^q = \frac{1}{n} \cdot \sum_{j=0}^{n-1} A_{jq} \zeta^{-i(jq)} = a_i \quad \text{for } i = 0, \ldots, n-1,$$

that is, the validity of (7.26). □

Remark 7.4.5. Of course, one may also ask the complementary question to the one dealt with in Proposition 7.4.4, namely, which Fourier vectors in E^n even belong to F^n. Thus consider the Fourier vector A of some polynomial $a(x) \in E[x]/(x^n - 1)$. In view of Equation (7.22), A will actually be in F^n if and only if

$$a(x)^q \equiv a(x) \bmod x^n - 1,$$

that is, if and only if $a(x)$ belongs to the Berlekamp algebra of the polynomial $x^n - 1$ over F; see Theorem 6.5.1. □

Proposition 7.4.4 and Equation (7.22) show that the Fourier vector of a polynomial $a(x)$ over F is already determined by the values $a(\alpha)$, where α runs over a system of representatives of the conjugacy classes of n-th roots of unity. In other words, we have to select just one root α of each irreducible factor of $x^n - 1$. This observation motivates the following generalization of Theorem 7.4.2 in the case of finite fields, which can also be derived from the Chinese remainder theorem; again, we leave this as an exercise.

Theorem 7.4.6. *Let $F = \mathrm{GF}(q)$, and denote the irreducible factors of $x^n - 1$ by $f_1, \dots, f_r$ and their respective degrees by $n_1, \dots, n_r$. Moreover, assume that n and q are relatively prime. Then*

$$R_{n,q} \cong \mathrm{GF}(q^{n_1}) \oplus \mathrm{GF}(q^{n_2}) \oplus \cdots \oplus \mathrm{GF}(q^{n_r}). \tag{7.27}$$

Proof. Let ζ be a primitive n-th root of unity over F, put $E = F(\zeta)$, and select a root α_i of f_i for $i = 1, \dots, r$. Because of Theorem 7.4.2, the mapping

$$\tau\colon a(x) \mapsto \big(a(\alpha_1), a(\alpha_2), \dots, a(\alpha_r)\big)$$

is a homomorphism from $R_{n,q}$ into E^r. Since α_i is a root of the irreducible polynomial f_i of degree n_i for $i = 1, \dots, r$, we have $\mathrm{GF}(q^{n_i}) \cong F[x]/(f_i)$, and hence the image of τ is the algebra $A = \mathrm{GF}(q^{n_1}) \oplus \mathrm{GF}(q^{n_2}) \oplus \cdots \oplus \mathrm{GF}(q^{n_r})$. In view of Proposition 7.4.4, an arbitrary vector in A can have at most one extension to the Fourier vector $\mathbf{A}$ of some polynomial $a(x) \in R_{n,q}$, and therefore τ must be injective, since the DFT is an isomorphism. □

The isomorphism τ just described in the proof of Theorem 7.4.6 is called the **Discrete Fourier Transform** of $R_{n,q}$.

Corollary 7.4.7. *Let n and q be relatively prime. Then the number of units of $R_{n,q}$ is*

$$\phi_q(x^n - 1) = (q^{n_1} - 1) \cdots (q^{n_r} - 1),$$

where $n_1, \dots, n_r$ are the degrees of the irreducible factors of $x^n - 1$. □

Thus we have obtained $\phi_q(x^n - 1)$ and hence the order of the group $C(n,q)$ of invertible circulant (n,n)-matrices over $\mathrm{GF}(q)$ in a third way, provided that n and q are relatively prime. Moreover, the general case can be reduced to Corollary 7.4.7, as we shall see later. Further (less obvious) applications of the DFT will be given in Section 7.6, where we will use Theorem 7.4.2 to compute the order of an interesting subgroup of $C(n,q)$, and in Sections 9.8 and 9.9, where we will use the DFT to determine the linear complexities of certain periodic sequences.

Remark 7.4.8. We note that Theorem 7.4.6 can also be obtained in a more systematic, theoretical way. Let us sketch this connection, as it leads to important generalizations. The necessary background on the representation theory of algebras is quite standard; see, for instance, Curtis and Reiner [100] or Huppert [200]. Now one uses that the algebra $R_{n,q}$ – which is isomorphic to the group algebra of the cyclic group of order n over $\mathrm{GF}(q)$ – is semisimple, which follows from the hypothesis $\gcd(n,q) = 1$. Then Theorem 7.4.6 can be recognized as a rather special case of the Wedderburn theorem on the structure of semisimple algebras. It should not be too surprising that a considerably more elementary proof, such as the one presented above, is possible in the special case of finite fields. Moreover, this proof has the advantage of providing us with an explicit isomorphism between the algebras $R_{n,q}$ and $\mathrm{GF}(q^{n_1}) \oplus \mathrm{GF}(q^{n_2}) \oplus \cdots \oplus \mathrm{GF}(q^{n_r})$, namely the DFT.

It should also be mentioned that the Wedderburn theory of semisimple algebras can be used to consider a much more general version of the DFT for arbitrary (even non-commutative) semisimple group algebras EG. The DFT for arbitrary abelian groups G and some of its consequences will be considered in Chapter 10; both the cases $E = \mathbb{C}$ and $E = \mathrm{GF}(q)$ have important applications in Coding Theory and in Design Theory. For an interesting monograph on the DFT for general groups, the reader is referred to Beth [32]. Again, there also is considerable interest in FFT algorithms in this more general setting, in particular for the classical case $E = \mathbb{C}$; see [32] and, as examples of subsequent research, Baum and Clausen [19] and the references quoted there.

Finally, in the cyclic case, there is an interesting generalization of the DFT to the situation where n is divisible by the characteristic p of $\mathrm{GF}(q)$, the so-called Hasse transform; we refer the reader to Weinberger and Lempel [402] and, for more details, to Weinberger [401]. □

Exercises

Exercise 7.4.9. Use the Chinese remainder theorem (see 1.10.6) to give alternative proofs for Theorems 7.4.2 and 7.4.6. □

Exercise 7.4.10. Let Z be the Fourier matrix given by Equation (7.20). Determine the matrix Z^2. □

7.5 Symmetric and Skew-symmetric Matrices

In this section, we consider symmetric and skew-symmetric matrices; in particular, we will determine the numbers of invertible matrices of these types over $\mathrm{GF}(q)$.

These results are basic ingredients for enumerating orthogonal matrices and self-dual bases in the next section.

Definition 7.5.1. A matrix A over a field F is called **symmetric** if it satisfies the condition $A = A^T$, and **skew-symmetric** if it satisfies $A = -A^T$ and has diagonal entries 0 only. □

We remark that the second condition required for a skew-symmetric matrix has to be added in view of the special case where F has characteristic 2; in all other cases, it is an immediate consequence of the condition $A = -A^T$.

We now determine the number of invertible symmetric matrices over GF(q). Since we want to use an inductive argument, we will in fact solve the more general problem of computing the number $N_q(n,r)$ of symmetric matrices with rank r over GF(q). The following result is due to Carlitz [65] in the case of odd q; we present the proof of MacWilliams [252] which requires no restriction on the characteristic of F.

Theorem 7.5.2. *The number of symmetric (n,n)-matrices with rank r over* GF(q) *is given by*

$$N_q(n,r) = \prod_{i=1}^{s} \frac{q^{2i}}{q^{2i}-1} \cdot \prod_{j=0}^{r-1} (q^{n-j}-1), \tag{7.28}$$

where $s = \left\lfloor \frac{r}{2} \right\rfloor$.[4]

Proof. We use induction on n and note that Equation (7.28) holds for $n=1$, as $N_q(1,0)=1$ and $N_q(1,1)=q-1$. Thus assume the validity of (7.28) for a particular value of n and consider a symmetric $(n+1,n+1)$-matrix. Any such matrix has the form

$$A = \begin{pmatrix} z_0 & \mathbf{z} \\ \mathbf{z}^T & B \end{pmatrix},$$

where B is a symmetric (n,n)-matrix and where $\mathbf{z} = (z_1,\ldots,z_n)$. Denote the rank of B by t; we shall analyze the possibilities for the rank of A. First assume that $\mathbf{z}$ is a linear combination of the rows of B, say $\mathbf{z} = \mathbf{y}B$ with $\mathbf{y} = (y_1,\ldots,y_n)$. If also $z_0 = \mathbf{y}\mathbf{z}^T$, then A has rank t; obviously, there are q^t such cases corresponding to the q^t vectors $\mathbf{z}$ in the row space of B. On the other hand, if $z_0 \neq \mathbf{y}\mathbf{z}^T$, then A has rank $t+1$, and there are $(q-1)q^t$ such cases.

It remains to consider the case where $\mathbf{z}$ does not depend on the rows of B. Then A has rank $t+2$, as the submatrix

$$C = \begin{pmatrix} \mathbf{z} \\ B \end{pmatrix}$$

of A has rank $t+1$ and as the column $\mathbf{z}^T$ does not depend on the columns of $B = B^T$; there are $q(q^n - q^t)$ such cases. Using these observations, we obtain the following recurrence relations for $N_q(n+1,r)$:

[4] Here we use the standard convention that an empty product is to be interpreted as 1. Thus $N_q(n,0) = 1$ and $N_q(n,1) = q^n - 1$ for all $n \geq 1$. Of course, these values are obvious anyway.

$$N_q(n+1,1) = qN_q(n,1) + (q-1)N_q(n,0) \tag{7.29}$$

and, for $r \geq 2$,

$$\begin{aligned} N_q(n+1,r) = q^r N_q(n,r) + (q-1)q^{r-1}N_q(n,r-1) \\ + (q^{n+1} - q^{r-1})N_q(n,r-2). \end{aligned} \tag{7.30}$$

These formulae allow to compute the values $N_q(n+1,r)$ from the corresponding values for n, where we put $N_q(n,n+1) = 0$. The correctness of the closed form given in the assertion follows by substituting the values given by Equation (7.28) in the right hand sides of Equations (7.29) and (7.30): for $r = 1$, one easily obtains the desired result $N_q(n+1,1) = q^{n+1} - 1$; and for $r \geq 2$ elementary but tedious calculations (note that a case distinction is needed, depending on the parity of r) show that Equation (7.28) also holds for $n+1$ instead of n. □

Corollary 7.5.3. *The number of invertible symmetric (n,n)-matrices over* GF(q) *is given by*

$$N_q(n,n) = \prod_{i=1}^{n}(q^i - \varepsilon_i), \quad \textit{where } \varepsilon_i = \begin{cases} 0 & \textit{if } i \textit{ is even,} \\ 1 & \textit{otherwise.} \end{cases}$$

Proof. This follows easily by simplifying Equation (7.28) for $r = n$. □

We now turn our attention to skew-symmetric matrices over $F = \mathrm{GF}(q)$. Again, we will follow MacWilliams [252] and prove a considerably more general result, giving the number $N_q^0(n,r)$ of skew-symmetric matrices with rank r over F. For the case of odd characteristic, the result in question is already due to Carlitz [66]. For background, we first summarize some basic facts from Linear Algebra as follows:

Remark 7.5.4. Recall that symmetric matrices belong to symmetric bilinear forms, as discussed in some detail in Section 3.13. Similarly, skew-symmetric matrices A belong to alternating bilinear forms, that is, bilinear forms f satisfying

$$f(\mathbf{x},\mathbf{x}) = \mathbf{x}^T A\mathbf{x} = 0 \quad \text{for all } \mathbf{x}.$$

Such a form can always be represented by a diagonal matrix of the type

$$\mathrm{diag}(1,-1;1,-1;\ldots;1,-1;0,\ldots,0).$$

In particular, the rank of any skew-symmetric matrix has to be even. □

Theorem 7.5.5. *If r is odd, then there are no skew-symmetric (n,n)-matrices with rank r over* GF(q)*; and if $r = 2s$ is even, the number of such matrices is given by*

$$N_q^0(n,2s) = \prod_{i=1}^{s} \frac{q^{2i-2}}{q^{2i}-1} \cdot \prod_{j=0}^{2s-1}(q^{n-j}-1). \tag{7.31}$$

Proof. The proof is similar to that of Theorem 7.5.2. In view of Remark 7.5.4, it suffices to consider the case where $r = 2s$ is even. Again, we will use induction on n. Note that $N_q^0(n,0) = 1$ for all n; in particular, Equation (7.31) holds for $n = 1$. Thus assume the validity of Equation (7.31) for a given value of n and consider a skew-symmetric $(n+1,n+1)$-matrix A. Any such matrix has the form

$$A = \begin{pmatrix} 0 & \mathbf{z} \\ -\mathbf{z}^T & B \end{pmatrix},$$

where B is a skew-symmetric (n,n)-matrix and where $\mathbf{z} = (z_1,\ldots,z_n)$. Denote the rank of B by t; we shall analyze the possibilities for the rank of A. First assume that $\mathbf{z}$ is a linear combination of the rows of B, say $\mathbf{z} = \mathbf{y}B$ with $\mathbf{y} = (y_1,\ldots,y_n)$. Since B is skew-symmetric, we obtain $\mathbf{y}(-\mathbf{z}^T) = \mathbf{y}B\mathbf{y}^T = 0$, and therefore A likewise has rank t; clearly, there are q^t choices for $\mathbf{z}$.

Now assume that $\mathbf{z}$ does not depend on the rows of B. Then A will have rank $t+2$; there are $q^n - q^t$ such cases. Using these observations, we obtain the following recurrence relation for $N_q^0(n+1,2s)$ for $s \geq 1$:

$$N_q^0(n+1,2s) = q^{2s}N_q^0(n,2s) + (q^n - q^{2s-2})N_q^0(n,2s-2).$$

Using this equation and the induction hypothesis, one can check that Equation (7.31) also holds for $n+1$ instead of n. □

Corollary 7.5.6. *If n is odd, then there are no invertible skew-symmetric (n,n)-matrices over* GF(q)*; and if n is even, the number of such matrices is given by*

$$N_q^0(n,n) = \prod_{i=1}^{n-1}(q^i - \varepsilon_i), \quad \textit{where } \varepsilon_i = \begin{cases} 0 & \textit{if } i \textit{ is even,} \\ 1 & \textit{otherwise.} \end{cases}$$

Proof. This follows easily from Theorem 7.5.5 by simplifying Equation (7.31) for $n = 2s$. □

Exercises

Exercise 7.5.7. As noted in Remark 7.5.4, there are no skew-symmetric matrices of odd rank. Give an inductive proof for the special case $N_q^0(n,2s+1) = 0$ along the lines of the proof of Theorem 7.5.5. □

7.6 Orthogonal Matrices and Self-dual Bases

In this section, we determine the number of orthogonal (n,n)-matrices over $F =$ GF(q). As an application, we also obtain the number of self-dual bases of GF(q^n) over F. We begin by recalling the definition of an orthogonal matrix.

Definition 7.6.1. An (n,n)-matrix A is said to be **orthogonal** if it satisfies the condition $AA^T = I$. The multiplicative group of all orthogonal (n,n)-matrices over $\mathrm{GF}(q)$ will be denoted by $O(n,q)$. □

Remark 7.6.2. We remark that the preceding terminology might cause some confusion. The reader should realize that the term *orthogonal group* – which we were careful to avoid – usually refers to the group of isometries of an orthogonal geometry, that is, of a vector space equipped with a non-degenerate quadratic form. In the case of finite fields, no ambiguity arises if both q and n are odd, since any two non-degenerate quadratic forms are equivalent in this case, by Theorem 3.13.14. However, if q is odd and n is even, then there are two distinct orthogonal groups (of different orders), since there are exactly two equivalence classes of non-degenerate quadratic forms then, again by Theorem 3.13.14.

Finally, if q is even, an orthogonal geometry is by definition a symplectic geometry refined by an additional quadratic form. In particular, such a geometry must have even dimension and contains isotropic vectors only. Note that the orthogonal matrices introduced in Definition 7.6.1 correspond to the isometries leaving the symmetric bilinear form represented by the identity matrix I invariant. While this makes sense regardless whether or not n is even, we have two rather different notions for the case of even characteristic. In particular, the standard text books do not contain the order of $O(n,q)$ for this case. We refer the reader to Artin [11] or to Dieudonné [106] for more details on the classical groups and their geometries. □

Before we determine the number of orthogonal matrices, we establish the following connection between such matrices and self-dual bases which was observed in [216].

Proposition 7.6.3. *Let $B = \{\beta_1, \ldots, \beta_n\}$ be a self-dual basis for $E = \mathrm{GF}(q^n)$ over $F = \mathrm{GF}(q)$, and let $C = (c_{ij})_{i,j=1,\ldots,n}$ be any invertible matrix over F. Then the basis $A = \{\alpha_1, \ldots, \alpha_n\}$ defined by*

$$\alpha_i = \sum_{j=1}^{n} c_{ij}\beta_j \quad \text{for } i = 1, \ldots, n \tag{7.32}$$

is again a self-dual basis for E/F if and only if C is an orthogonal matrix. Consequently, the number $sd(n,q)$ of self-dual bases for E over F is given by

$$sd(n,q) = \frac{1}{n!}|O(n,q)|.$$

Proof. Using the self-duality of B, we see that A is likewise self-dual if and only if

$$\delta_{ij} = \mathrm{Tr}(\alpha_i\alpha_j) = \mathrm{Tr}\Big(\Big(\sum_{h=1}^{n} c_{ih}\beta_h\Big) \cdot \Big(\sum_{k=1}^{n} c_{jk}\beta_k\Big)\Big)$$

$$= \sum_{h,k=1}^{n} c_{ih}c_{jk}\mathrm{Tr}(\beta_h\beta_k) = \sum_{k=1}^{n} c_{ik}c_{jk}$$

for all $i, j = 1, \dots, n$, which holds if and only if $CC^T = I$. □

We now turn our attention to determining the order of the group $O(n,q)$; we will give elementary proofs following MacWilliams [252]. In her approach, the following definition is needed.

Definition 7.6.4. Let A be a symmetric invertible (n,n)-matrix over $\mathrm{GF}(q)$. Then any (invertible) (n,n)-matrix M satisfying $A = MM^T$ is called a **factor** of A.[5] □

The following simple result forms the basis for MacWilliams' approach to the groups $O(n,q)$.

Proposition 7.6.5. *Let $sf(n,q)$ denote the number of symmetric invertible (n,n)-matrices over* $\mathrm{GF}(q)$ *which admit a factor. Then*

$$|O(n,q)| = \frac{|\mathrm{GL}(n,q)|}{sf(n,q)}.$$

Proof. Consider any coset of $O = O(n,q)$ in the general linear group $G = \mathrm{GL}(n,q)$, say MO. Given the specific representative M, we obtain an invertible symmetric matrix A_M admitting a factor by putting $A_M := MM^T$. It now suffices to check that two matrices $M, N \in G$ yield identical symmetric matrices $A_M = A_N$ if and only if their cosets agree. But this is easy: $MM^T = NN^T$ is equivalent to $N^{-1}MM^T(N^T)^{-1} = I$, and therefore holds if and only if $N^{-1}M$ is an orthogonal matrix. □

This leaves the problem of determining the numbers $sf(n,q)$, which will require some effort. First, we characterize those invertible symmetric matrices which admit a factor. This turns out to depend on the characteristic of $\mathrm{GF}(q)$, and one of the two cases is quite simple:

Proposition 7.6.6. *Let A be an invertible symmetric matrix over $F = \mathrm{GF}(q)$, where q is odd. Then A admits a factor if and only if its determinant is a square in F^*.*

Proof. First assume $A = MM^T$. Then $\det A = (\det M)^2$ is a square. Conversely, let A be an invertible symmetric matrix whose determinant is a square. By Theorem 3.13.14, A is congruent to the identity matrix I. Thus there exists an invertible matrix L satisfying $LAL^T = I$, and hence L^{-1} is a factor of A. □

The case of even characteristic is rather more involved. Here we have the following result due to Albert [6].

Proposition 7.6.7. *Let A be an invertible symmetric (n,n)-matrix over $F = \mathrm{GF}(q)$, where q is even. Then A admits a factor if and only if it has at least one non-zero diagonal entry.*

---

[5] Seroussi and Lempel [348] consider a more general situation and show that *every* symmetric matrix A can be written in the form $A = MM^T$ for some (not necessarily square) matrix M; they also determine the minimum number of columns of such a matrix M. In this connection, we also mention another interesting paper: Weinberger and Lempel [402] characterize the symmetric circulant matrices which have a circulant factor.

Proof. Let $\mathbf{x} = (x_1, \dots, x_n)$ be any vector in $V = F^n$. As q is even and A is symmetric,

$$\mathbf{x}A\mathbf{x}^T = \sum_{i,j=1}^{n} a_{ij}x_ix_j = \sum_{i=1}^{n} a_{ii}x_i^2 = \Big(\sum_{i=1}^{n} b_ix_i\Big)^2, \tag{7.33}$$

where the b_i are the uniquely determined elements satisfying $b_i^2 = a_{ii}$, for $i = 1, \dots, n$. Hence the set of all isotropic vectors forms a subspace

$$V(A) = \{\mathbf{x} \in V \colon \mathbf{x}A\mathbf{x}^T = 0\}$$

of V, and $V(A)$ has dimension $n-1$ if and only if A contains at least one non-zero diagonal entry (otherwise all vectors of V are isotropic).

It is now easy to prove the necessity of the condition in question. Assume that A has a factor, say $A = MM^T$, so that we may write the condition $\mathbf{x}A\mathbf{x}^T = 0$ as $(\mathbf{x}M)(\mathbf{x}M)^T = 0$. Since M is invertible, the set of all vectors $\mathbf{x}M$ equals V. Hence $V(A)$ is the set of all vectors $\mathbf{y}$ satisfying $\mathbf{y}\mathbf{y}^T = 0$, and therefore is a subspace of dimension $n-1$. Thus A must contain a non-zero diagonal entry.

Conversely, assume that A contains a non-zero diagonal entry. We first show that we may reduce the assertion to the case where

$$a_{ii} = 1 \quad \text{for } i = 1, \dots, n. \tag{7.34}$$

To see this, it suffices to prove that A is congruent to a matrix with diagonal entries 1. Let S be the matrix of some non-singular linear transformation of V mapping $V(I)$ onto $V(A)$, which is possible, as $V(A)$ has dimension $n-1$. Thus we have

$$(\mathbf{x}S)A(\mathbf{x}S)^T = 0 \iff \mathbf{x}\mathbf{x}^T = 0$$

and therefore

$$V(I) = V(SAS^T). \tag{7.35}$$

In view of (7.33), the set of isotropic vectors with respect to an invertible symmetric matrix A is the kernel of some linear mapping from V into F and therefore determines the diagonal entries of A up to a scalar factor. Hence (7.35) shows that each diagonal entry of SAS^T equals some constant $c \neq 0$. Now let d be the unique element of F satisfying $d^2 = 1/c$ and put $R = dS$. Then each entry on the main diagonal of RAR^T equals 1; hence we may indeed assume the validity of (7.34).

We next claim that some principal minor A_{ii} of A has to be distinct from 0:[6]

$$A_{ii} \neq 0 \quad \text{for some } i \in \{1, \dots, n\}. \tag{7.36}$$

[6] We use the standard notation A_{ij} for the **principal minor** of A arising as the determinant of the matrix obtained by discarding the i-th row and the j-th column of A. Recall that the associated **cofactor** is defined as $C_{ij} = (-1)^{i+j}A_{ij}$, and that the **classical adjoint** adj A of A is the transpose of the matrix $C = (C_{ij})$ formed by the cofactors. Of course, in our situation (where F has characteristic 2) cofactors and principal minors coincide.

Assume otherwise. By a well-known result from Linear Algebra, the classical adjoint can be used to compute A^{-1}:

$$A^{-1} = \frac{1}{\det A} \cdot \operatorname{adj} A. \tag{7.37}$$

If we assume $A_{ii} = 0$ for all i, then (7.33) (applied to A^{-1}) and (7.37) show $V(A^{-1}) = V$. We then choose a vector $\mathbf{x} \notin V(A)$ and obtain the contradiction

$$0 \neq \mathbf{x}A\mathbf{x}^T = \mathbf{x}AA^{-1}A\mathbf{x}^T = (\mathbf{x}A)A^{-1}(\mathbf{x}A)^T = 0.$$

This proves the validity of (7.36). Without loss of generality, we may assume $i = 1$, so that

$$A_{11} \neq 0. \tag{7.38}$$

After these preparations, we can use induction on n to prove that A has a factor. The case $n = 1$ is trivial, since then $A = I$ by (7.34). Thus assume that the result is already known to hold for dimension $n-1$. Because of (7.38), A has the form

$$A = \begin{pmatrix} 1 & \mathbf{z} \\ \mathbf{z}^T & B \end{pmatrix},$$

where B is a non-singular symmetric $(n-1,n-1)$-matrix with diagonal entries 1 and where $\mathbf{z} = (z_2,\dots,z_n)$. Since B has rank $n-1$, $\mathbf{z}$ is a linear combination of the rows of B, say $\mathbf{z} = \mathbf{y}B$ with $\mathbf{y} = (y_2,\dots,y_n)$. Since A has rank n, the first row of A cannot depend on the remaining rows which implies $\mathbf{y}\mathbf{z}^T \neq 1$. Thus $w \neq 1$, where w is the unique element of F satisfying $w^2 = \mathbf{y}\mathbf{z}^T = \mathbf{y}B\mathbf{y}^T$. Because of the induction hypothesis, B has a factor, say $B = LL^T$. We now put

$$M = \begin{pmatrix} 1+w & \mathbf{y}L \\ \mathbf{0}^T & L \end{pmatrix}$$

and obtain the desired factorization of A:

$$MM^T = \begin{pmatrix} 1+w & \mathbf{y}L \\ \mathbf{0}^T & L \end{pmatrix} \cdot \begin{pmatrix} 1+w & \mathbf{0} \\ L^T\mathbf{y}^T & L^T \end{pmatrix} = \begin{pmatrix} 1+w^2+\mathbf{y}LL^T\mathbf{y}^T & \mathbf{y}LL^T \\ LL^T\mathbf{y}^T & LL^T \end{pmatrix} = A. \quad \square$$

We now combine the preceding results with those obtained in Section 7.5 to determine the numbers $sf(n,q)$ and the orders of the groups $O(n,q)$. We use the same notation as in that section and begin with the case of characteristic 2.

Theorem 7.6.8. *Let q be even. Then:*

$$sf(n,q) = \begin{cases} N_q(n,n) & \text{if } n \text{ is odd,} \\ (q^n-1)N_q^0(n,n) & \text{if } n \text{ is even,} \end{cases}$$

and

$$|O(n,q)| = \prod_{i=1}^{n-1}(q^i - \varepsilon_i), \quad \textit{where } \varepsilon_i = \begin{cases} 1 & \textit{if } i \textit{ is even,} \\ 0 & \textit{if } i \textit{ is odd.} \end{cases}$$

Proof. Proposition 7.6.7 gives $sf(n,q) = N_q(n,n) - N_q^0(n,n)$. In view of Corollaries 7.5.3 and 7.5.6, we readily obtain the first assertion. Then the second assertion follows from Proposition 7.6.5 by substituting the values for $N_q(n,n)$ and $N_q^0(n,n)$ given in 7.5.3 and 7.5.6 and the value of $|\mathrm{GL}(n,q)|$ given in Equation (7.1). □

Next, we settle the case where both q and n are odd.

Theorem 7.6.9. *Let q and n be odd. Then:*

$$sf(n,q) = N_q(n,n)/2,$$

and

$$|O(n,q)| = 2 \cdot \prod_{i=1}^{n-1}(q^i - \varepsilon_i), \quad \textit{where } \varepsilon_i = \begin{cases} 1 & \textit{if } i \textit{ is even,} \\ 0 & \textit{if } i \textit{ is odd.} \end{cases}$$

Proof. Consider any invertible symmetric matrix A, and choose an arbitrary non-square $m \in \mathrm{GF}(q)$. Then $\det mA$ is a non-square if and only if $\det A$ is a square, since n is odd. Thus the criterion given in Proposition 7.6.6 holds for exactly half of the invertible symmetric matrices, which proves the first assertion. Now the second assertion follows from Proposition 7.6.5 by substituting the value for $N_q(n,n)$ given in 7.5.3 and the value of $|\mathrm{GL}(n,q)|$ given in Equation (7.1). □

The preceding two results suffice to determine the number of self-dual bases of $\mathrm{GF}(q^n)$ over $\mathrm{GF}(q)$, a result due to Jungnickel, Menezes and Vanstone [216].[7]

Theorem 7.6.10. *The number $sd(n,q)$ of self-dual bases for $\mathrm{GF}(q^n)$ over $\mathrm{GF}(q)$ is given by*

$$sd(n,q) = \frac{\gamma}{n!} \cdot \prod_{i=1}^{n-1}(q^i - \varepsilon_i),$$

where

$$\varepsilon_i = \begin{cases} 1 & \textit{if } i \textit{ is even,} \\ 0 & \textit{if } i \textit{ is odd,} \end{cases} \qquad \textit{and} \qquad \gamma = \begin{cases} 2 & \textit{if } q \textit{ and } n \textit{ are odd,} \\ 1 & \textit{if } q \textit{ is even,} \\ 0 & \textit{otherwise.} \end{cases}$$

Proof. By Theorem 3.13.12, $\mathrm{GF}(q^n)$ always admits a self-dual basis over $\mathrm{GF}(q)$, unless q is odd and n is even. Thus the assertion is an immediate consequence of Proposition 7.6.3 and Theorems 7.6.8 and 7.6.9. □

---

[7] For even values of q, this problem was solved earlier by Imamura [202] via a direct enumeration (based on the method used in the proof of Proposition 3.13.13.)

Finally, for the sake of completeness, we include a formula for the order of $O(n,q)$ in the case where q is odd and n is even. This is the most involved case, but the methods used to establish it are the same as before. Therefore, we simply state the result in question and refer the reader to MacWilliams [252] for a proof.

Result 7.6.11 *Let q be odd and n even, say $n = 2s$. Then:*

$$|O(2s,q)| = 2q^{s(s-1)}(q^s+\varepsilon)\prod_{i=1}^{s-1}(q^{2i}-1),$$

where

$$\varepsilon = \begin{cases} 1 & \text{if } s \text{ is odd and } q \equiv 3 \bmod 4, \\ -1 & \text{otherwise.} \end{cases}$$

□

Exercises

Exercise 7.6.12. By Theorem 3.13.11, $\mathrm{GF}(q^n)$ always admits an almost self-dual basis over $\mathrm{GF}(q)$. Consider the case where no self-dual basis can exist, that is, assume that q is odd and n is even. Determine the number of almost self-dual bases in this case (see [216, p. 27]). □

7.7 Symmetric Matrix Representations of Finite Fields

This section – which is mainly based on the work of Seroussi and Lempel [349] – is devoted to an interesting theoretical application of the concept of duality. We return to the topic of Section 7.2 and consider representations of $E = \mathrm{GF}(q^n)$ as a matrix field M over $F = \mathrm{GF}(q)$, where we now require that M consists of symmetric matrices only. As we have seen there, the case of real interest is that of representations by (n,n)-matrices; hence we now restrict attention to this case. Such a representation of E as a field of symmetric (n,n)-matrices over F is called a **symmetric representation** for E.

First we need to introduce some auxiliary notation and two preliminary results relating the two coordinate representations of elements of E with respect to a dual pair of bases. Here and throughout this entire section, we simplify our notation for the trace function by writing just Tr instead of $\mathrm{Tr}_{E/F}$, where E and F always denote $\mathrm{GF}(q^n)$ and $\mathrm{GF}(q)$, respectively.

Definition 7.7.1. Let $B = (\beta_1,\ldots,\beta_n)$ and $C = (\gamma_1,\ldots,\gamma_n)$ be a dual pair of ordered bases for $E = \mathrm{GF}(q^n)$ over $F = \mathrm{GF}(q)$, that is, $\mathrm{Tr}(\beta_i\gamma_j) = \delta_{ij}$ for all i, j. Given any element $\xi = x_1\beta_1 + \cdots + x_n\beta_n$ of E, we write

$$r_B(\xi) := (x_1,\ldots,x_n)$$

and call the entries x_i of $r_B(\xi)$ the **primal coordinates** of ξ (with respect to B). Similarly, we define the **dual coordinates** of ξ as the coordinates of ξ with respect to the dual basis C:

$$r_C(\xi) := \big((\xi)_1, \ldots, (\xi)_n\big),$$

that is, $\xi = (\xi)_1\gamma_1 + \cdots + (\xi)_n\gamma_n$. □

As the following two basic results show, the connection between dual and primal coordinates is quite simple. These results will be very useful also in Chapter 8, when we study the arithmetics of finite field extensions; one reason for the practical importance of dual bases is the fact that they facilitate the determination of the coordinate representation of arbitrary elements.

Lemma 7.7.2. *Let $B = (\beta_1, \ldots, \beta_n)$ and $C = (\gamma_1, \ldots, \gamma_n)$ be a dual pair of ordered bases for E over F, and let ξ be any element of E. Then the primal coordinate x_i in $r_B(\xi)$ equals* $\mathrm{Tr}(\xi\gamma_i)$, *and the dual coordinate $(\xi)_i$ in $r_C(\xi)$ equals* $\mathrm{Tr}(\xi\beta_i)$.

Proof. Let $\xi = x_1\beta_1 + \cdots + x_n\beta_n$. By multiplying this equation with γ_i and taking the trace, we obtain the first assertion:

$$\mathrm{Tr}(\xi\gamma_i) = \mathrm{Tr}\big((x_1\beta_1 + \cdots + x_n\beta_n)\gamma_i\big) = x_1\mathrm{Tr}(\beta_1\gamma_i) + \cdots + x_n\mathrm{Tr}(\beta_n\gamma_i) = x_i,$$

by duality. The second assertion follows interchanging the roles of B and C. □

Conversely, one may use dual pairs of coordinate vectors to compute the trace of products:

Lemma 7.7.3. *Let $B = (\beta_1, \ldots, \beta_n)$ and $C = (\gamma_1, \ldots, \gamma_n)$ be a dual pair of ordered bases for E over F. Then*

$$\mathrm{Tr}(\xi\eta) = r_B(\xi)r_C(\eta)^T = r_C(\eta)r_B(\xi)^T \quad \textit{for all } \xi, \eta \in E.$$

Proof. The assertion is an immediate consequence of duality and the linearity of the trace function:

$$\begin{aligned}\mathrm{Tr}(\xi\eta) &= \mathrm{Tr}\Big(\big(\sum_{i=1}^{n} x_i\beta_i\big)\cdot\big(\sum_{j=1}^{n}(\eta)_j\gamma_j\big)\Big) \\ &= \sum_{i,j=1}^{n} x_i(\eta)_j\mathrm{Tr}(\beta_i\gamma_j) = \sum_{i=1}^{n} x_i(\eta)_i = r_B(\xi)r_C(\eta)^T. \quad \square\end{aligned}$$

In the proof of Theorem 7.2.2, we have seen that any basis B for E/F gives rise to a representation of E as a matrix field $R(B)$ of degree n by associating with each $\alpha \in E$ first the linear operator

$$M_\alpha\colon\ \xi \mapsto \alpha\xi \quad \text{for } \xi \in E$$

and then its matrix $A_B(\alpha)$ with respect to the basis B. Before we can characterize those bases B which yield a symmetric representation $R(B)$ and show that such bases always exist, we need some further preliminary facts.

We will use the following notation. Any ordered basis $B = (\beta_1, \dots, \beta_n)$ for E/F will also be considered as a row vector $\mathbf{b} = (\beta_1, \dots, \beta_n) \in E^n$. Given any invertible (n,n)-matrix S over F, the row vector $\mathbf{b}S$ again belongs to a basis for E/F; this basis will also be denoted as BS. Similarly, the basis with row vector $\lambda \mathbf{b}$, where $\lambda \in E^*$, will be written as λB. Finally, the dual basis of B will henceforth usually be denoted by B^*.

Lemma 7.7.4. *Let B and B^* be a pair of dual bases for E over F. Then:*

$$(\lambda B)^* = \lambda^{-1} B^* \quad \textit{for all } \lambda \in E^*,$$

and

$$(BS)^* = B^*(S^{-1})^T \quad \textit{for all } S \in \mathrm{GL}(n,q).$$

Proof. The first assertion is an immediate consequence of the definition of dual bases, whereas the second assertion is a little more involved. Given any basis X for E/F with associated vector $\mathbf{x} \in E^n$, the matrix M_X considered in Lemma 3.13.16 may now be written as follows:

$$M_X = \begin{pmatrix} \mathbf{x} \\ \mathbf{x}^{(q)} \\ \vdots \\ \mathbf{x}^{(q^{n-1})} \end{pmatrix},$$

where the notation $\mathbf{x}^{(k)}$ means that each entry of $\mathbf{x}$ is raised to its k-th power. By Lemma 3.13.16, two bases X and Y of E are dual to each other if and only if $M_Y^T M_X = I$, which is, of course, equivalent to $M_X M_Y^T = I$. We now apply this criterion with $X = BS$ and $Y = B^*(S^{-1})^T$. Since the entries of S belong to F, we obtain

$$M_X = \begin{pmatrix} \mathbf{b}S \\ \mathbf{b}^{(q)}S \\ \vdots \\ \mathbf{b}^{(q^{n-1})}S \end{pmatrix} = M_B S \quad \text{and} \quad M_Y = \begin{pmatrix} \mathbf{b}^*(S^{-1})^T \\ (\mathbf{b}^*)^{(q)}(S^{-1})^T \\ \vdots \\ (\mathbf{b}^*)^{(q^{n-1})}(S^{-1})^T \end{pmatrix} = M_{B^*}(S^{-1})^T,$$

and hence indeed $M_X M_Y^T = \left(M_B S\right)\left(M_{B^*}(S^{-1})^T\right)^T = M_B M_{B^*}^T = I$, since B and B^* are dual to each other. □

After these preparations, we may now give the desired characterization of symmetric representations.

Theorem 7.7.5. *Let B be an ordered basis for E over F. Then the corresponding matrix representation $R(B)$ of E is symmetric if and only if the dual basis of B satisfies $B^* = \lambda B$ for some $\lambda \in E^*$.*

Proof. Let $B = (\beta_1, \dots, \beta_n)$ and $B^* = (\gamma_1, \dots, \gamma_n)$. We claim that the entries of the matrices $A_B(\alpha) \in R(B)$ may be computed as follows:

$$\big(A_B(\alpha)\big)_{ij} = \mathrm{Tr}(\alpha\gamma_i\beta_j) \quad \text{for } i,j = 1,\dots,n. \tag{7.39}$$

To see this, note that the j-th column of $A_B(\alpha)$ is the coordinate vector of $\alpha\beta_j$ with respect to B. Hence Equation (7.39) is an immediate consequence of Lemma 7.7.2 applied with $\xi = \alpha\beta_j$.

Now assume $B^* = \lambda B$ for some $\lambda \in E$. Then, using (7.39) twice,

$$\big(A_B(\alpha)\big)_{ij} = \mathrm{Tr}\big(\alpha(\lambda\beta_i)\beta_j\big) = \mathrm{Tr}\big(\alpha(\lambda\beta_j)\beta_i\big) = (A_B(\alpha))_{ji}$$

for all $\alpha \in E$ and for $i,j = 1,\dots,n$. Hence $R(B)$ is indeed a symmetric representation of E.

Conversely, assume that $R(B)$ is symmetric. Using (7.39) again, we obtain

$$\mathrm{Tr}(\alpha\gamma_i\beta_j) = \big(A_B(\alpha)\big)_{ij} = \big(A_B(\alpha)\big)_{ji} = \mathrm{Tr}(\alpha\gamma_j\beta_i)$$

for all $\alpha \in E$ and for $i,j = 1,\dots,n$. Thus $\mathrm{Tr}\big(\alpha(\gamma_i\beta_j - \gamma_j\beta_i)\big) = 0$ for all $\alpha \in E$ and therefore

$$\gamma_i\beta_j = \gamma_j\beta_i \quad \text{for } i,j = 1,\dots,n,$$

as the trace bilinear form is non-degenerate by Proposition 3.13.7. Thus

$$\gamma_i\beta_i^{-1} = \gamma_j\beta_j^{-1} \quad \text{for } i,j = 1,\dots,n,$$

so that $\gamma_i\beta_i^{-1}$ is a constant $\lambda \in E^*$, which gives the desired result $B^* = \lambda B$. □

Next, we use the criterion given in Theorem 7.7.5 together with previous results on duality to settle the existence problem for symmetric representations.

Theorem 7.7.6. *Every finite extension of* GF(q) *admits a symmetric representation as a matrix field over* GF(q).

Proof. According to Theorem 7.7.6, we have to find a basis B for $E = \mathrm{GF}(q^n)$ over $F = \mathrm{GF}(q)$ satisfying $B^* = \lambda B$ for some $\lambda \in E^*$. If E admits a self-dual basis B over F, we may simply take $\lambda = 1$; by Theorem 3.13.12, this settles the cases where either q is even or both q and n are odd.

Thus let q be odd and n even. We choose a primitive element ω of E and consider the associated polynomial basis $W = (1,\omega,\dots,\omega^{n-1})$ for E over F as well as its dual basis $W^* = (\gamma_0,\dots,\gamma_{n-1})$. (Contrary to our usual convention, it is more convenient to use indices $0,\dots,n-1$ here, so that γ_i corresponds to the element ω^i of the primal basis W.) Now define a symmetric matrix $M = (m_{ij})$ with entries

$$m_{ij} := \mathrm{Tr}(\omega^{i+j+1}) \quad \text{for } i,j = 0,\dots,n-1.$$

Then M transforms W^* into the basis ωW, which is seen as follows. Let $\mathbf{w}^* \in E^n$ be the row vector associated with W^* and consider the j-th entry of the row vector $\mathbf{w}^*M$. Using Lemma 7.7.2, one computes

$$(\mathbf{w}^*M)_j = \sum_{i=0}^{n-1} m_{ij}\gamma_i = \sum_{i=0}^{n-1} \mathrm{Tr}(\omega^{j+1}\omega^i)\gamma_i = \sum_{i=0}^{n-1} (\omega^{j+1})_i\gamma_i = \omega^{j+1},$$

so that indeed $W^*M = \omega W$. In particular, M has to be invertible, as it describes a change of basis.

Now let us assume that M has a factor, say $M = SS^T$ for some $S \in \mathrm{GL}(n,q)$. Then the proof is easily finished: we put $B := W^*S$ and use Lemma 7.7.4 to obtain

$$\begin{aligned} B^* &= (W^*S)^* = W(S^{-1})^T = \left(\omega^{-1}W^*M\right)(S^{-1})^T \\ &= \omega^{-1}W^*SS^T(S^{-1})^T = \omega^{-1}W^*S = \omega^{-1}B, \end{aligned}$$

so that B satisfies the criterion of Theorem 7.7.5 with $\lambda = \omega^{-1}$.

It remains to check that M indeed has a factor. In view of Proposition 7.6.6, we need to show that $\det M$ is a square in F^*. To this purpose, we apply Lemma 3.13.16 to the matrices M_W and $M_{\omega W}$ associated with the bases $W = (1,\omega,\ldots,\omega^{n-1})$ and $\omega W = (\omega,\omega^2,\ldots,\omega^n)$, respectively, and obtain

$$M = \left(\mathrm{Tr}(\omega^{i+j+1})\right) = M_W^T M_{\omega W} = M_W^T(M_W D),$$

where $D = \mathrm{diag}(\omega,\omega^q,\ldots,\omega^{q^{n-1}})$. Hence

$$\det M = \det(M_W^T M_W)\cdot \det D.$$

By Lemma 3.13.17, $\det(M_W^T M_W) = (\det M_W)^2$ cannot be a square in the ground field F, since n is even. Moreover, $\det D$ is likewise a non-square in F:

$$\det D = \omega^{1+q+\cdots+q^{n-1}} = \omega^{(q^n-1)/(q-1)},$$

and this is a primitive element of F^*, as ω was chosen as a primitive element of E^*. Thus $\det M$ is the product of two non-squares in F and therefore indeed a square, by Proposition 3.2.13. □

The arguments used in the last part of the preceding proof can also be applied to obtain a complete description of all bases B giving a symmetric representation of E in the case where self-dual bases exist, that is, if either q is even or both q and n are odd.

Theorem 7.7.7. *Assume that either q is even, or both q and n are odd. Then a basis B for $F = \mathrm{GF}(q^n)$ over $E = \mathrm{GF}(q)$ satisfies $B^* = \lambda B$ for some $\lambda \in E^*$ (and thus yields a symmetric representation of E) if and only if λ is a square, say $\lambda = \mu^2$, and the basis μB is self-dual.*

Proof. Let $\lambda = \mu^2$, and assume that the basis μB is self-dual. Using Lemma 7.7.4, we obtain

$$B^* = \mu(\mu^{-1}B^*) = \mu(\mu B)^* = \mu(\mu B) = \lambda B.$$

Conversely, assume $B^* = \lambda B$. We begin by showing that λ is indeed a square, which is clear if q is even. Thus let q and n both be odd. By Lemma 3.13.16,

$$I = M_B^T M_{B^*} = M_B^T M_{\lambda B} = M_B^T M_B D,$$

where $D = \mathrm{diag}(\lambda, \lambda^q, \ldots, \lambda^{q^{n-1}})$, and therefore $(\det M_B)^2 \cdot \det D = 1$. This shows that $\det D = \lambda^{(q^n-1)/(q-1)}$ is a square (in F). As q and n are odd, $(q^n - 1)/(q-1)$ is also odd, and thus λ has to be a square in E, say $\lambda = \mu^2$. Finally, μB is indeed self-dual:

$$(\mu B)^* = \mu^{-1} B^* = \mu^{-1}(\lambda B) = \mu B,$$

by another application of Lemma 7.7.4. □

Seroussi and Lempel also considered matrix representations $R(B)$ of $\mathrm{GF}(q^n)$ over $\mathrm{GF}(q)$ which are **closed under transposition**, that is, $M \in R(B)$ always implies $M^T \in R(B)$. Since this result rests on arguments which are quite similar to those used above, we omit its proof.

Result 7.7.8 *Let B be a basis for $E = \mathrm{GF}(q^n)$ over $\mathrm{GF}(q)$ and assume that the associated matrix representation $R(B)$ is closed under transposition. Then:*

- *If either q is even or both q and n are odd, $R(B)$ is necessarily symmetric.*
- *If q is odd and n is even, then $R(B)$ is not symmetric if and only if $B^* = (\lambda B)^{q^{n/2}}$ for some $\lambda \in E^*$.* □

Finally, we mention another interesting result of Seroussi and Lempel [349] which gives a criterion when two bases induce the same representation; again, we omit the proof.

Result 7.7.9 *Let B and C be two bases for $\mathrm{GF}(q^n)$ over $\mathrm{GF}(q)$. Then $R(B) = R(C)$ if and only if there exist $\lambda \in E^*$ and $k \in \{0, \ldots, n-1\}$ such that $C = (\lambda B)^{q^k}$.* □

7.8 The Existence of Self-dual Normal Bases

In this section, we present a major result due to Lempel and Weinberger [234]: the necessary conditions for the existence of a self-dual normal basis given in Proposition 3.13.20 are in fact sufficient. The proof of this result is constructive and uses circulant matrices in an essential way.

Theorem 7.8.1 (Self-dual normal basis theorem). *A self-dual normal basis for $\mathrm{GF}(q^n)$ over $\mathrm{GF}(q)$ exists if and only if either q is even and n is not a multiple of 4 or both q and n are odd.* □

We remark that Theorem 7.8.1 can be generalized to arbitrary Galois extensions of finite degree; this is due to Bayer-Fluckiger and Lenstra, see [22, 21]. The proof of Theorem 7.8.1 follows by combining Proposition 3.13.20 with the subsequent two results.

Proposition 7.8.2. *If n is odd, then there exists a self-dual normal basis for $E = \mathrm{GF}(q^n)$ over $F = \mathrm{GF}(q)$.*

Proof. We will show that one may use any given normal basis to construct a self-dual normal basis. Let α be an arbitrary normal element for E/F, denote the associated normal basis by A, and consider the circulant matrix C over E with first row $\mathbf{a} := (\alpha, \alpha^q, \ldots, \alpha^{q^{n-1}})$:

$$C = \begin{pmatrix} \alpha & \alpha^q & \alpha^{q^2} & \ldots & \alpha^{q^{n-1}} \\ \alpha^{q^{n-1}} & \alpha & \alpha^q & \ldots & \alpha^{q^{n-2}} \\ \alpha^{q^{n-2}} & \alpha^{q^{n-1}} & \alpha & \ldots & \alpha^{q^{n-3}} \\ \vdots & \vdots & \vdots & \vdots & \vdots \\ \alpha^q & \alpha^{q^2} & \alpha^{q^3} & \ldots & \alpha \end{pmatrix}.$$

Note that C differs from the matrix M_A previously associated with A by a permutation of the last $n-1$ rows – actually, by reversing their order; see Lemma 3.13.16. As in that result, the matrix $D := C^T C$ satisfies

$$D = \mathrm{Tr}(\mathbf{a}^T \mathbf{a}), \tag{7.40}$$

where $\mathrm{Tr}(X)$ denotes the matrix obtained by applying the trace operator $\mathrm{Tr}_{E/F}$ to every entry of the matrix X. Hence $\det D \neq 0$, by Theorem 3.13.9.

In what follows, all indices are to be computed modulo n. We first evaluate the matrix $R := C^2$. This matrix is, of course, likewise circulant, and its first row $\mathbf{r} = (\rho_0, \ldots, \rho_{n-1})$ is given by

$$\rho_i = \sum_{j=0}^{n-1} \alpha^{q^j} \alpha^{q^{i-j}} = \sum_{j=0}^{n-1} (\alpha^{q^{j-1}} \alpha^{q^{(i-2)-(j-1)}})^q = \Big(\sum_{j=0}^{n-1} \alpha^{q^j} \alpha^{q^{(i-2)-j}}\Big)^q = \rho_{i-2}^q.$$

Hence

$$\rho_i = \omega^{q^{im}} \quad \text{for } i = 0, \ldots, n-1, \tag{7.41}$$

where $\omega := \rho_0$ and $m := (n+1)/2$. We also require the matrix $T := RR^T$, which is again circulant. Because of (7.41), its first row is $\mathbf{t} = (\tau_0, \ldots, \tau_{n-1})$, where

$$\tau_i = \sum_{j=0}^{n-1} \rho_j \rho_{n-i+j} = \sum_{j=0}^{n-1} \omega^{q^{jm}(1+q^{-im})} = \sum_{j=0}^{n-1} \omega^{q^j(1+q^{-im})} \tag{7.42}$$

for $i = 0, \ldots, n-1$; here the final equality holds as n and m are relatively prime. We now check

$$T = \mathrm{Tr}(\mathbf{w}^T \mathbf{w}), \text{ where } \mathbf{w} = \big(\omega, \omega^{q^{-m}}, \omega^{q^{-2m}}, \ldots, \omega^{q^{-(n-1)m}}\big). \tag{7.43}$$

Indeed, using Equation (7.42), the (i,j)-entry of T is

$$\begin{aligned} t_{i,j} &= \tau_{j-i} = \sum_{k=0}^{n-1} \omega^{q^k(1+q^{-(j-i)m})} = \sum_{k=0}^{n-1} \omega^{q^{k-im}(1+q^{-(j-i)m})} \\ &= \sum_{k=0}^{n-1} \left(\omega^{q^{-im}} \omega^{q^{-jm}}\right)^{q^k} = \mathrm{Tr}_{E/F}\left(\omega^{q^{-im}} \omega^{q^{-jm}}\right). \end{aligned}$$

Since circulant matrices commute by Theorem 7.3.2, we also have a second identity for T:

$$T = RR^T = C^2(C^2)^T = (C^T C)^2 = D^2. \tag{7.44}$$

As D is non-singular, T is likewise non-singular. In view of (7.43), we conclude from Theorem 3.13.9 that the vector $\mathbf{w}$ describes a basis W for E/F. We now claim that D^{-1} transforms W into a self-dual normal basis B for E/F; more precisely, we show that the first entry β_0 of the vector

$$\mathbf{b} = (\beta_0, \ldots, \beta_{n-1}) := \mathbf{w}D^{-1}$$

is a normal element generating a self-dual basis. Since D has entries in F by (7.40), $\mathbf{b}$ clearly belongs to a basis B. As D is symmetric, we obtain from (7.43) and (7.44)

$$\mathrm{Tr}(\mathbf{b}^T\mathbf{b}) = \mathrm{Tr}(D^{-1}\mathbf{w}^T\mathbf{w}D^{-1}) = D^{-1}\mathrm{Tr}(\mathbf{w}^T\mathbf{w})D^{-1} = D^{-1}TD^{-1} = I, \tag{7.45}$$

where the second equality follows from the linearity of the trace, since D has entries in F. Equation (7.45) shows that B is indeed a self-dual basis. Now write $D^{-1} = (e_{ij})$. As noted after Proposition 7.3.3, D^{-1} is again circulant. Using this fact, one computes

$$\begin{aligned} \beta_j &= \sum_{i=0}^{m-1} e_{ij}\omega^{q^{-im}} = \sum_{i=0}^{m-1} e_{i+1,j+1}\omega^{q^{-(i+1)m+m}} \\ &= \sum_{i=0}^{m-1} e_{i,j+1}\omega^{q^{-im+m}} = \sum_{i=0}^{m-1} \left(e_{i,j+1}\omega^{q^{-im}}\right)^{q^m} = (\beta_{j+1})^{q^m}. \end{aligned}$$

Repeated application of the preceding formula gives

$$\beta_j = (\beta_0)^{q^{m(n-j)}} \quad \text{for } j = 0, \ldots, n-1.$$

Hence the elements $\beta_0, \ldots, \beta_{n-1}$ are a permutation of the conjugates of β_0, as m and n are relatively prime. Thus β_0 is indeed a normal element, and B is the desired self-dual normal basis for E/F. □

Proposition 7.8.3. *If q is even and n is not a multiple of 4, then there exists a self-dual normal basis for $E = \mathrm{GF}(q^n)$ over $F = \mathrm{GF}(q)$.*

Proof. For odd n, the assertion is contained in Proposition 7.8.2. Thus assume that n is even. We first settle the case $n = 2$. As every normal element for E/F has non-zero trace, we may clearly choose such an element α with $\mathrm{Tr}_{E/F}(\alpha) = 1$. Then

$$\mathrm{Tr}_{E/F}(\alpha^{2q}) = \mathrm{Tr}_{E/F}(\alpha^2) = \mathrm{Tr}_{E/F}(\alpha)^2 = 1 \quad \text{and} \quad \mathrm{Tr}_{E/F}(\alpha\alpha^q) = 0,$$

since $\alpha\alpha^q$ satisfies $(\alpha\alpha^q)^{q-1} = \alpha^{q^2-1} = 1$ and therefore belongs to F. Hence $\{\alpha, \alpha^q\}$ is a self-dual normal basis for E/F.

It remains to consider the general case $n = 2m$, where m is odd. As before, we let α generate a self-dual normal basis for $\mathrm{GF}(q^2)$ over $\mathrm{GF}(q)$. Since m is odd, we may also choose an element β which generates a self-dual normal basis for $\mathrm{GF}(q^m)$ over $\mathrm{GF}(q)$. Then $\gamma = \alpha\beta$ generates the desired self-dual normal basis for E/F, by Corollary 3.13.22. □

Remark 7.8.4. If one wants to prove Theorem 7.8.1 in the case of even characteristic only, it is possible to avoid using Proposition 7.8.2 altogether. Instead of appealing to this result when q is even and n is odd, one may proceed as follows. By Theorem 7.3.10 and Corollary 7.3.8, the number of normal bases for E/F is

$$\frac{1}{n}\phi_q(x^n - 1) = \frac{1}{n}(q^{n_1} - 1)\cdots(q^{n_r} - 1),$$

where the irreducible factors of $x^n - 1$ have degrees $n_1, \ldots, n_r$, respectively. (Note that $x^n - 1$ has only simple roots in this case, since q is even and n is odd.) Hence the number of normal bases for E/F is odd, which immediately implies the existence of a self-dual normal basis.

Thus we have an alternative – albeit non-constructive – proof for Proposition 7.8.3. In the important special case $q = 2$, Theorem 7.8.9 below provides a reasonably easy method for actually computing a self-dual normal basis. □

Sometimes it is possible to give a simple explicit formula for a self-dual normal basis. The following result of Lempel and Seroussi [233] shows that this is the case whenever the degree of the extension equals the characteristic of the ground field. This is not all that surprising if one recalls from Section 5.1 that finding some irreducible polynomial of degree n over a Galois field of characteristic p is particularly easy for $n = p$. Indeed, the proof of the following result may be viewed as a variation of the methods used in that section.

Theorem 7.8.5. *Let q be a power of the prime p, and let β be a root of the polynomial $x^q - x - m$ over $P = \mathrm{GF}(p)$, where $m \neq 0$. Put*

$$\gamma := (\beta - \alpha)^{p-1} - 1,$$

where α is an arbitrary element of $F = \mathrm{GF}(q)$. Then γ generates a self-dual normal basis for $E = \mathrm{GF}(q^p)$ over F.

Proof. Note that $\beta \notin F$, and thus $\beta \neq \alpha$. We first prove that γ indeed generates an extension of degree p of F, by checking that the smallest positive integer k satisfying $\gamma^{q^k} = \gamma$ is $k = p$. Now $\beta^q = \beta + m$, and hence $\beta^{q^k} = \beta + km$. Therefore

$$\begin{aligned}\gamma^{q^k} &= \left((\beta-\alpha)^{p-1}-1\right)^{q^k} = (\beta+km-\alpha)^{p-1}-1 \\ &= \frac{(\beta+km-\alpha)^p-(\beta+km-\alpha)}{\beta+km-\alpha} = \frac{(\beta-\alpha)^p-(\beta-\alpha)}{\beta+km-\alpha},\end{aligned} \tag{7.46}$$

which equals γ only for $km=0$. Hence $\gamma^{q^k}=\gamma$ if and only if k is a multiple of p, proving our claim about γ. In view of Theorem 3.13.9, it now suffices to show

$$\mathrm{Tr}_{E/F}(\gamma\gamma^{q^k}) = 0 \quad \text{for } k=1,\ldots,p-1 \tag{7.47}$$

and

$$\mathrm{Tr}_{E/F}(\gamma^2) = 1. \tag{7.48}$$

In what follows, Tr always stands for $\mathrm{Tr}_{E/F}$. Using Equation (7.46), a simple computation establishes condition (7.47):

$$\begin{aligned}\mathrm{Tr}(\gamma\gamma^{q^k}) &= \sum_{i=0}^{p-1}\gamma^{q^i}\gamma^{q^{k+i}} = \sum_{i=0}^{p-1}\frac{\left((\beta-\alpha)^p-(\beta-\alpha)\right)^2}{(\beta-\alpha+im)(\beta-\alpha+im+km)} \\ &= \frac{\left((\beta-\alpha)^p-(\beta-\alpha)\right)^2}{km}\left(\sum_{i=0}^{p-1}\frac{1}{\beta-\alpha+im}-\sum_{i=0}^{p-1}\frac{1}{\beta-\alpha+(i+k)m}\right) \\ &= 0.\end{aligned}$$

Proving (7.48) needs more effort. First assume $p=2$. Then $m=1$ and indeed

$$\mathrm{Tr}(\gamma^2) = \mathrm{Tr}(\gamma)^2 = (\mathrm{Tr}(\beta+\alpha+1))^2 = \mathrm{Tr}(\beta)^2 = (\beta^q+\beta)^2 = 1.$$

Now let p be odd. Then $\gamma^2=(\gamma+1)^2-2(\gamma+1)+1$ and $\mathrm{Tr}(1)=0$, which implies $\mathrm{Tr}(\gamma^2)=\mathrm{Tr}\left((\gamma+1)^2\right)-2\mathrm{Tr}(\gamma+1)$. Hence it suffices to prove

$$\mathrm{Tr}(\gamma+1) = \mathrm{Tr}\left((\gamma+1)^2\right) = -1. \tag{7.49}$$

Checking (7.49) requires a couple of simple congruences for odd primes p. The first of these is

$$1^k+2^k+\cdots+(p-1)^k \equiv \begin{cases} 0 \bmod p & \text{if } k \text{ is not divisible by } p-1, \\ -1 \bmod p & \text{otherwise,}\end{cases}$$

which can be seen as follows. If k is not divisible by $p-1$, we may choose some a such that $a^k \not\equiv 1 \bmod p$. Then multiplication with a^k permutes the set of k-th powers, so that

$$a^k\left(1^k+2^k+\cdots+(p-1)^k\right) \equiv 1^k+2^k+\cdots+(p-1)^k \bmod p,$$

which establishes the first case of (7.8). The second case is even simpler, as then $a^k \equiv 1 \bmod p$ for $a=1,\ldots,p-1$.

The other congruences required are as follows:

$$\binom{p-1}{j} \equiv (-1)^j \mod p$$

and

$$\binom{2p-2}{j} \equiv \begin{cases} (-1)^j(j+1) \mod p & \text{for } j=0,\ldots,p-2, \\ 0 \mod p & \text{for } j=p-1. \end{cases}$$

We leave it to the reader to check these by simply evaluating the binomial coefficients in question (written as fractions of factorials in the standard way) modulo p. Using the preceding congruences together with (7.8), we first compute

$$\begin{aligned}
\mathrm{Tr}(\gamma+1) &= \sum_{i=0}^{p-1}(\gamma+1)^{q^i} = \sum_{i=0}^{p-1}(\beta+im-\alpha)^{p-1} \\
&= (\beta-\alpha)^{p-1} + \sum_{i=1}^{p-1}\sum_{j=0}^{p-1}\binom{p-1}{j}(\beta-\alpha)^j(im)^{p-1-j} \\
&= (\beta-\alpha)^{p-1} + \sum_{i=1}^{p-1}\sum_{j=0}^{p-1}(-1)^j(\beta-\alpha)^j(im)^{p-1-j} \\
&= (\beta-\alpha)^{p-1} + \sum_{j=0}^{p-1}\left((-1)^j(\beta-\alpha)^j m^{p-1-j}\sum_{i=1}^{p-1} i^{p-1-j}\right) \\
&= (\beta-\alpha)^{p-1} - (\beta-\alpha)^0 - (\beta-\alpha)^{p-1} = -1,
\end{aligned}$$

which establishes the first part of (7.49). In a similar manner, we also obtain the second part:

$$\begin{aligned}
\mathrm{Tr}\big((\gamma+1)^2\big) &= \sum_{i=0}^{p-1}\left((\gamma+1)^{q^i}\right)^2 = \sum_{i=0}^{p-1}(\beta+im-\alpha)^{2p-2} \\
&= \sum_{i=1}^{p-1}\left(\sum_{j=0}^{2p-2}\binom{2p-2}{j}(\beta-\alpha)^j(im)^{2p-2-j}\right) + (\beta-\alpha)^{2p-2} \\
&= \sum_{i=1}^{p-1}\sum_{j=0}^{p-2}\binom{2p-2}{j}\left((\beta-\alpha)^j(im)^{2p-2-j} + (\beta-\alpha)^{2p-2-j}(im)^j\right) \\
&\quad + \sum_{i=1}^{p-1}\binom{2p-2}{p-1}(\beta-\alpha)^{p-1}(im)^{p-1} + (\beta-\alpha)^{2p-2}
\end{aligned}$$

$$= \sum_{i=1}^{p-1} \sum_{j=0}^{p-2} (-1)^j (j+1) \left((\beta-\alpha)^j (im)^{2p-2-j} + (\beta-\alpha)^{2p-2-j} (im)^j \right)$$
$$+ (\beta-\alpha)^{2p-2}$$
$$= (\beta-\alpha)^{2p-2} + \sum_{j=0}^{p-2} \left((-1)^j (j+1)(\beta-\alpha)^j m^{2p-2-j} \right) \sum_{i=1}^{p-1} i^{2p-2-j}$$
$$+ \sum_{j=0}^{p-2} \left((-1)^j (j+1)(\beta-\alpha)^{2p-2-j} m^j \right) \sum_{i=1}^{p-1} i^j$$
$$= (\beta-\alpha)^{2p-2} - (\beta-\alpha)^0 - (\beta-\alpha)^{2p-2} = -1. \quad \square$$

Remark 7.8.6. It should be stressed that the element β needed in Theorem 7.8.5 can be determined efficiently. In fact, it suffices to know any element $\eta \in F$ with non-zero absolute trace, say $\mathrm{Tr}_{F/P}(\eta) = m \neq 0$, and to choose β as a root of $f = x^p - x - \eta$. Indeed, $\beta^p = \beta + \eta$ shows

$$\beta^q = \beta + \eta + \eta^p + \cdots + \eta^{p^{a-1}} = \beta + m,$$

where $q = p^a$. Note that f is irreducible by Theorem 5.1.1 and hence allows us to construct E from F. By Theorem 7.8.5, trinomials of the special form $x^p - x - \eta$ can in fact be used to determine generators for self-dual normal bases explicitly: in this special case, finding a self-dual normal basis for E/F is only marginally more involved than constructing E at all. $\square$

Blake, Gao and Mullin [41] also have a result similar to Theorem 7.8.5; these authors even obtain a self-dual normal basis of "low complexity", a concept which will be studied in Chapter 8.

For practical applications, the case $q = 2$ is of particular importance. Here Wang [396] gave a method which allows to determine a self-dual normal basis for $E = \mathrm{GF}(2^n)$ over $F = \mathrm{GF}(2)$ from an *arbitrary* normal basis A, whenever n is odd. As his method is somewhat simpler than that provided in the proof of Proposition 7.8.3, it will be included here: we will show how one explicitly constructs an invertible matrix which transforms A into the desired self-dual normal basis.

Thus let A be the normal basis generated by $\alpha \in E$, with corresponding vector $\mathbf{a} = (\alpha, \alpha^q, \ldots, \alpha^{q^{n-1}}) \in E^n$. By Proposition 7.3.5, the vector $\mathbf{b} = \mathbf{a}C$ also describes a normal basis B whenever C is an invertible circulant (n,n)-matrix over F. The key lemma for Wang's method is the following characterization of those circulant matrices C for which B is actually self-dual; this result even holds for arbitrary values of q and n.

Proposition 7.8.7. *Let α be a normal element for $E = \mathrm{GF}(q^n)$ over $F = \mathrm{GF}(q)$, put $\mathbf{a} := (\alpha, \alpha^q, \ldots, \alpha^{q^{n-1}})$, and let C be an invertible circulant (n,n)-matrix over F.*

Then the vector $\mathbf{b} = \mathbf{a}C$ *belongs to a self-dual normal basis* B *for* E/F *if and only if*

$$\mathrm{Tr}(\mathbf{a}^T\mathbf{a}) = D^T D, \tag{7.50}$$

where we write $D := C^{-1}$ *and where* $\mathrm{Tr}(X)$ *again denotes the matrix obtained by applying the trace operator* $\mathrm{Tr}_{E/F}$ *to all entries of the matrix* X.

Proof. By Proposition 7.3.5, B is a normal basis. Now B is self-dual if and only if

$$\mathrm{Tr}(\mathbf{b}^T\mathbf{b}) = I. \tag{7.51}$$

Since D has entries from F,

$$\mathrm{Tr}(\mathbf{a}^T\mathbf{a}) = \mathrm{Tr}\big((\mathbf{b}D)^T\mathbf{b}D\big) = \mathrm{Tr}\big(D^T(\mathbf{b}^T\mathbf{b})D\big) = D^T\,\mathrm{Tr}(\mathbf{b}^T\mathbf{b})D,$$

which shows that conditions (7.50) and (7.51) are indeed equivalent. □

Following Wang, we now proceed to produce an explicit solution for Equation (7.50) in the case where $q = 2$ and n is odd. Denote the first column of the circulant matrix D by $(d_0, \dots, d_{n-1})$. Then (7.50) is equivalent to the following system of equations, since the matrices in question are circulant (so that it suffices to consider the entries in the first row):

$$\sum_{k=0}^{n-1} d_k d_{k-j} = \mathrm{Tr}(\alpha^{2^j+1}) \quad \text{for } j = 0, \dots, n-1,$$

where all indices are computed modulo n. This system of equations may be simplified as follows. Note first that the substitution $\ell := k - j$ transforms the sum

$$\sum_{k=0}^{n-1} d_k d_{k-j} \quad \text{into} \quad \sum_{\ell=0}^{n-1} d_\ell d_{\ell+j}.$$

Also, applying the automorphism $x \mapsto x^{2^{n-j}}$ to α^{2^j+1} gives $\mathrm{Tr}(\alpha^{2^j+1}) = \mathrm{Tr}(\alpha^{2^{n-j}+1})$ for all j. Therefore, the equations for j and $n-j$ are identical. Moreover, the equation for $j = 0$ just reduces to $d_0 + \cdots + d_{n-1} = 1$, since squaring is the identity for $q = 2$ and $\mathrm{Tr}(\alpha^2) = \mathrm{Tr}(\alpha) = 1$ (because of the linear independence of the conjugates of α). These observations establish the following auxiliary result.

Lemma 7.8.8. *For* $q = 2$ *and odd* n, *condition (7.50) is equivalent to the following system of equations:*

$$d_0 + \cdots + d_{n-1} = 1; \tag{7.52}$$

$$\sum_{k=0}^{n-1} d_k d_{k+j} = \mathrm{Tr}(\alpha^{2^j+1}) \quad \textit{for } j = 1, \dots, m, \tag{7.53}$$

where $m = (n-1)/2$ *and where* $(d_0, \dots, d_{n-1})$ *is the first column of* D. □

Theorem 7.8.9. *Let α be a normal element for $E = \mathrm{GF}(2^n)$ over $F = \mathrm{GF}(2)$, where $n = 2m+1$, and denote the indices $j \in \{1,\dots,m\}$ satisfying $\mathrm{Tr}(\alpha^{2^j+1}) = 1$ by $n_1,\dots,n_t$, where $n_1 < n_2 < \dots < n_t$. Put $N := n_t$, and let $\varepsilon = 0$ if N is even and $\varepsilon = 1$ otherwise. Now define elements $d_0,\dots,d_{2m} \in F$ as follows:*

$$\begin{aligned} d_k &= 1 && \text{for } k \in \{0, N, (N+\varepsilon n)/2\}; \\ d_{(N+\varepsilon n \pm n_i)/2} &= 1 && \text{for all even } n_i \text{ with } i \in \{1,\dots,t-1\}; \\ d_{(N+(1-\varepsilon)n \pm n_i)/2} &= 1 && \text{for all odd } n_i \text{ with } i \in \{1,\dots,t-1\}; \\ d_k &= 0 && \text{for all other indices } k. \end{aligned}$$

Then $d_0,\dots,d_{2m}$ satisfy the system of equations in Lemma 7.8.8.

Proof. We remark first that the effect of using the terms involving ε is just to make sure that all indices appearing in the defining equations above are indeed integers. Since the case where N is odd is completely analogous to that of even N, we assume from now on that N is even, and hence $\varepsilon = 0$, which simplifies the notation. As an odd number of the d_k have been chosen as 1, Equation (7.52) is satisfied. We now consider a term on the left hand side of the j-th equation in (7.53) which is equal to 1, say

$$d_k d_{k+j} = 1. \tag{7.54}$$

Clearly, this implies

$$k = (N \pm n_h)/2 \quad \text{or} \quad k = (N+n\pm n_h)/2, \tag{7.55}$$

$$k+j = (N \pm n_i)/2 \quad \text{or} \quad k+j = (N+n\pm n_i)/2 \tag{7.56}$$

for suitable values of h and i; which of the respective two cases in (7.55) and (7.56) actually occurs only depends on the parities of n_h and n_i. We now show that terms satisfying (7.54) occur in pairs, and thus the contributions of the two pairs to the j-th equation in (7.53) cancel, unless $h = i$ in (7.55) and (7.56); remember that all indices are computed modulo n. There are four cases to consider:

Case 1. $k = (N \pm n_h)/2$ and $k+j = (N\pm n_i)/2$, hence $j = (\pm n_i \mp n_h)/2$. Then also $d_l d_{l+j} = 1$, where $l = (N \mp n_i)/2$ and $l+j = (N \mp n_h)/2$.

Case 2. $k = (N+n\pm n_h)/2$ and $k+j = (N+n\pm n_i)/2$, hence $j = (\pm n_i \mp n_h)/2$. Then also $d_l d_{l+j} = 1$, where $l = (N+n\mp n_i)/2$ and $l+j = (N+n\mp n_h)/2$.

Case 3. $k = (N\pm n_h)/2$ and $k+j = (N+n\pm n_i)/2$, hence $j = (n\pm n_i \mp n_h)/2$. Then also $d_l d_{l+j} = 1$, where $l = (N+n\mp n_i)/2$ and $l+j = (N\mp n_h)/2$.

Case 4. $k = (N+n\pm n_h)/2$ and $k+j = (N\pm n_i)/2$, hence $j = (n\pm n_i\mp n_h)/2$. Then also $d_l d_{l+j} = 1$, where $l = (N\mp n_i)/2$ and $l+j = (N+n\mp n_h)/2$.

Hence the only terms (7.54) which do not cancel are indeed those with $h = i$, in which case we necessarily have $j = n_i$, as the reader may easily check. In other words, the left hand side of the j-th equation in (7.53) equals 1 if and only if j

is one of the indices n_i, that is, if and only if $\mathrm{Tr}(\alpha^{2^j+1}) = 1$. This proves that all equations in (7.53) are satisfied for our choice of the d_i. □

Remark 7.8.10. As pointed out by Wang, his method may even be applied without the prior knowledge of a normal basis: just select an arbitrary element $\alpha \in \mathrm{GF}(2^n)$ until a normal element is found. Clearly one only needs to consider elements α with trace 1.

One then computes the matrix $\mathrm{Tr}(\mathbf{a}^T\mathbf{a})$ and solves equation (7.50) by the method given above. Note that Lemma 7.8.8 and Theorem 7.8.9 remain valid also if α is not a normal element; the only restriction on α which was actually used in solving (7.50) was the assumption that $\mathrm{Tr}(\alpha) = 1$. When a matrix D satisfying (7.50) has been determined, one checks whether α is actually a normal element by testing whether or not D is non-singular. This works as D is non-singular if and only if $\mathrm{Tr}(\mathbf{a}^T\mathbf{a}) = D^T D$ is non-singular, which in turn holds if and only if the conjugates of α form a basis, by Theorem 3.13.9.

In this case, we compute the inverse C of D and put $\mathbf{b} := \mathbf{a}C$, which gives us the desired self-dual normal basis. Otherwise, we select another element α with $\mathrm{Tr}(\alpha) = 1$ and repeat the preceding procedure until we find an α for which our matrix D is indeed invertible. We note that the necessary tests for invertibility and the actual computation of the inverse of D should, of course, be performed by exploiting the equivalence between circulant (n,n)-matrices and polynomials modulo $x^n - 1$. We leave it to the reader to give an explicit algorithm for the computation of a self-dual normal basis using Wang's method. □

We conclude this section with a connection to matrix representations. As in Sections 7.2 and 7.7, we let B be any basis for E/F and denote the corresponding representation of E as a matrix field of degree n by $R(B)$. Now let α generate a self-dual normal basis for E/F, and consider the corresponding matrix $A := A_B(\alpha)$ in $R(B)$. Then $\{A, A^q, \dots, A^{q^{n-1}}\}$ is a self-dual normal basis for the matrix field $R(B)$ over the ground field $\{\lambda I \colon \lambda \in F\}$ of scalar matrices.

It is possible to characterize the matrices $A \in \mathrm{GL}(n,q)$ which generate a self-dual normal basis for some matrix representation of E. The following result is due to Lempel [232] and will be stated without proof. As in Section 7.3, P denotes the circulant permutation matrix with first row $(0,1,0,\dots,0)$.

Result 7.8.11 *Let $A = (a_{ij})_{i,j=0,\dots,n-1}$ be an invertible (n,n)-matrix over $F = \mathrm{GF}(q)$. Then A generates a self-dual normal basis for a matrix representation $R(B)$ of $\mathrm{GF}(q^n)$ over F if and only if the following four conditions hold:*

- *A is symmetric;*
- *$a_{ij} = a_{i-j,-j}$ for $i,j = 0,\dots,n-1$;*
- $\sum_{j=0}^{n-1} a_{ij} = \begin{cases} \pm 1 & \text{if } i = 0, \\ 0 & \text{otherwise;} \end{cases}$
- *$A^q = P^T A P$.* □

7.9 Circulant Orthogonal Matrices and Self-dual Normal Bases

In this section, we determine the number $sdn(n,q)$ of self-dual normal bases for $\mathrm{GF}(q^n)$ over $\mathrm{GF}(q)$. As in the cases of normal bases and of self-dual bases, respectively, this problem can be reduced to finding the order of a suitable group, which is immediate by combining Propositions 7.3.5 and 7.6.3.

Proposition 7.9.1. *Let α generate a self-dual normal basis for $E = \mathrm{GF}(q^n)$ over $F = \mathrm{GF}(q)$, and let $C = (c_{ij})_{i,j=0,\dots,n-1}$ be an invertible matrix over F. Then the basis $B = (\beta_0,\dots,\beta_{n-1})$ defined by*

$$\beta_i := \sum_{j=0}^{n-1} c_{ij}\alpha^{q^j} \quad \text{for } i = 0,\dots,n-1$$

is likewise a self-dual normal basis for E/F if and only if C is an orthogonal circulant matrix. □

Corollary 7.9.2. *Assume $sdn(n,q) \neq 0$. Then*

$$sdn(n,q) = \frac{1}{n}|OC(n,q)|,$$

where $OC(n,q)$ denotes the group of orthogonal circulant (n,n)-matrices over $\mathrm{GF}(q)$. □

The order of $OC(n,q)$ was determined by MacWilliams [253] in the case where q is a prime; the general case was established by Byrd and Vaughan [56]. We shall present a more elementary and also more constructive proof due to Beth, Geiselmann and Jungnickel [33, 215]. We begin with the case where q and n are relatively prime; here the proof is similar to that given by MacWilliams, though the presentation can be made more elegant by using the Discrete Fourier Transform. After this, in the case where n is divisible by the characteristic p of $F = \mathrm{GF}(q)$, we employ the same method as MacWilliams, which requires non-trivial modifications when q is even.

Altogether, the determination of $|OC(n,q)|$ needs considerably more effort than that of $|O(n,q)|$, even though it is in some sense conceptually simpler, since we can exploit the isomorphism between the algebra of circulant (n,n)-matrices over F and the ring $R_{n,q} = F[x]/(x^n-1)$ described in Theorem 7.4.2. To do so, we need a simple observation characterizing those polynomials in $R_{n,q}$ which correspond to orthogonal matrices in $C(n,q)$.

Lemma 7.9.3. *Let C be the circulant (n,n)-matrix with first row $(c_0,c_1,\dots,c_{n-1})$ over $\mathrm{GF}(q)$, and let $c(x) = c_0 + c_1x + \dots + c_{n-1}x^{n-1}$ be the polynomial in $R_{n,q}$ corresponding to C. Then C is orthogonal if and only if*

$$c(x)c^T(x) = 1, \quad \text{where } c^T(x) = c_0 + c_{n-1}x + \dots + c_2x^{n-2} + c_1x^{n-1}.$$

Proof. It suffices to note that $c^T(x)$ is the polynomial associated with the transpose C^T of C. □

Remark 7.9.4. Because of $x^n - 1$ in $R_{n,q}$, we have $c^T(x) = c(x^{n-1}) = c(x^{-1})$; therefore the condition in Lemma 7.9.3 may also be written as

$$c(x)c(x^{-1}) = 1. \tag{7.57}$$

The polynomial $c^T(x)$ is said to be the **transpose** of $c(x)$. Note that c^T is just a cyclic shift of the reciprocal polynomial c^* introduced in Definition 5.1.9, as $c^T(x) = xc^*(x)$ in $R_{n,q}$.

A polynomial in $R_{n,q}$ with $c(x)c^T(x) = 1$ is called **orthogonal**, whereas a polynomial satisfying $c(x) = c^T(x)$ is said to be **symmetric**. □

By using the DFT, the determination of $|OC(n,q)|$ is rather easy if n and q are relatively prime: basically, one applies Theorem 7.4.6 in a straightforward manner. We now set up the notation needed to do so. Thus let $x-1$, possibly $x+1$ (this factor arises only if q is odd and n is even) and $f_1, \ldots, f_r$ be the irreducible factors of $x^n - 1$; thus the f_i are those irreducible factors of $x^n - 1$ which have roots of order ≥ 3. By Theorem 5.7.1, any irreducible factor f of $x^n - 1$ gives rise to a further (not necessarily distinct) irreducible factor $f^\wedge$ obtained by dividing the reciprocal polynomial f^* of f by the constant term of f; moreover, $f(\alpha) = 0$ implies $f^\wedge(\alpha^{-1}) = 0$.

Hence we may split the irreducible factors of $x^n - 1$ with roots of order ≥ 3 into those satisfying $f = f^\wedge$ (which are actually self-reciprocal and have even degree, see Theorem 5.7.2) and into a certain number of pairs $(f, f^\wedge)$ with $f \neq f^\wedge$. Without loss of generality, we put $r = s + 2t$ and assume

$$f_i^* = f_i \text{ for } i = 1, \ldots, s \quad \text{and} \quad (f_{s+j})^\wedge = f_{s+t+j} \neq f_{s+j} \text{ for } j = 1, \ldots, t. \tag{7.58}$$

We can now prove the following result.

Theorem 7.9.5. *Let n and q be relatively prime, let the irreducible factors of $x^n - 1$ with roots of order ≥ 3 over $F = \mathrm{GF}(q)$ be labelled as in (7.58), and denote their respective degrees by $\deg f_i = 2d_i$ for $i = 1, \ldots, s$ and $\deg f_{s+j} = \deg f_{s+t+j} = e_j$ for $j = 1, \ldots, t$. Then:*

$$|OC(n,q)| = \varepsilon \cdot \prod_{i=1}^{s} (q^{d_i} + 1) \cdot \prod_{j=1}^{t} (q^{e_j} - 1), \tag{7.59}$$

where

$$\varepsilon = \begin{cases} 1 & \text{if } q \text{ is even,} \\ 2 & \text{if } q \text{ and } n \text{ are odd,} \\ 4 & \text{if } q \text{ is odd and } n \text{ is even.} \end{cases}$$

Proof. In view of Lemma 7.9.3, we need to determine the number of polynomials $c(x)$ in $R_{n,q}$ satisfying (7.57). In order to apply the discrete Fourier transform to that equation, we select roots α_i of f_i for $i = 1, \ldots, s+2t$ in such a way that

$$\alpha_{s+t+j} = (\alpha_{s+j})^{-1} \quad \text{for } j = 1,\ldots,t. \tag{7.60}$$

We first consider the cases where either q is even or both q and n are odd. Then $x^n - 1$ is divisible by just $x-1$ (but, for odd q, not by $x+1$) and Theorem 7.4.6 yields the isomorphism

$$\tau\colon\ c(x) \mapsto \big(c(1), c(\alpha_1), c(\alpha_2), \ldots, c(\alpha_{s+2t})\big)$$

between $R_{n,q}$ and

$$A := \mathrm{GF}(q) \oplus \mathrm{GF}(q^{2d_1}) \oplus \cdots \oplus \mathrm{GF}(q^{2d_s}) \oplus \mathrm{GF}(q^{e_{s+1}}) \oplus \cdots \oplus \mathrm{GF}(q^{e_{s+2t}}).$$

We now apply τ to Equation (7.57) and obtain the equivalent condition

$$\tau(c(x))\tau(c(x^{-1})) = (1,\ldots,1)$$

in A, that is,

$$c(1)^2 = c(\alpha_i)c(\alpha_i^{-1}) = 1 \quad \text{for } i = 1,\ldots,s+2t. \tag{7.61}$$

In particular, (7.61) shows that every component of the Fourier vector $\mathbf{C}$ of $c(x)$ has to be invertible. For $i \in \{1,\ldots,s\}$, both α_i and α_i^{-1} are roots of the self-reciprocal polynomial f_i of degree $2d_i$. Then $\alpha_i^{-1} = \alpha_i^{q^{d_i}}$ by Theorem 5.7.2, and hence (7.61) yields the condition

$$c(\alpha_i)c(\alpha_i^{-1}) = c(\alpha_i)c(\alpha_i^{q^{d_i}}) = c(\alpha_i)^{q^{d_i}+1} = 1 \quad \text{for } i = 1,\ldots,s.$$

For $i = s+j$ with $j \in \{1,\ldots,t\}$, we obtain from (7.60) and (7.61) the condition

$$c(\alpha_{s+j})c(\alpha_{s+t+j}) = 1 \quad \text{for } j = 1,\ldots,t,$$

which shows that the value of $c(\alpha_{s+j})$ determines that of $c(\alpha_{s+t+j})$ in this case. Altogether, we see that the solutions of (7.61) are in a one-to-one correspondence with the vectors $(\beta_0, \beta_1, \ldots, \beta_{s+t})$ in

$$\mathrm{GF}(q) \oplus \mathrm{GF}(q^{2d_1}) \oplus \cdots \oplus \mathrm{GF}(q^{2d_s}) \oplus \mathrm{GF}(q^{e_{s+1}}) \oplus \cdots \oplus \mathrm{GF}(q^{e_{s+t}})$$

satisfying

$$\beta_0^2 = 1, \quad \beta_i^{q^{d_i}+1} = 1 \text{ for } i = 1,\ldots,s \quad \text{and} \quad \beta_{s+j} \neq 0 \text{ for } j = 1,\ldots,t.$$

Clearly, there are one or two choices for β_0 (depending on whether q is even or odd), $q^{d_i}+1$ choices for the β_i with $i \in \{1,\ldots,s\}$, and $q^{e_j}-1$ choices for the β_{s+j} with $j \in \{1,\ldots,t\}$. This yields formula (7.59) in the first two cases.

Finally, the case where q is odd and n is even follows by the same reasoning; one just has to take into account that $x^n - 1$ now has the additional irreducible factor $x+1$. Thus the Fourier vector $\mathbf{C}$ contains one further component, namely $c(-1)$, which has to satisfy the condition $c(-1)^2 = 1$. Clearly, there are two choices for $c(-1)$, and we obtain (7.59) also in this final case. □

It should be noted that there is no need to factor $x^n - 1$ explicitly in order to apply Theorem 7.9.5, since it suffices to determine the conjugacy classes of n-th roots of unity. For this, one chooses an arbitrary primitive n-th root of unity ζ over GF(q) and applies the Frobenius automorphism $\sigma\colon \xi \mapsto \xi^q$ to the powers of ζ. Of course, it suffices to work with the exponents, that is, to compute the orbits of the cyclic group $\mathbb{Z}_n$ of residues modulo n under the bijection $x \mapsto xq$ instead of actually writing down the orbits of powers of ζ under σ. An example should make this clear.

Example 7.9.6. Let us consider the case $q = 2$, $n = 15$. The orbits of $\mathbb{Z}_{15}$ under the bijection $x \mapsto 2x$ are as follows:

$$B_1 = \{0\},\ B_2 = \{1,2,4,8\},\ B_3 = \{3,6,12,9\},\ B_4 = \{5,10\},\ B_5 = \{7,14,13,11\}.$$

Of course, the orbit B_1 corresponds to the factor $x+1$ of $x^{15}+1$. Two of the remaining five orbits, namely B_3 and B_4, belong to self-reciprocal factors of degrees 4 and 2, respectively, of $x^{15}+1$, since they contain with each element x also the element $-x$, so that the corresponding conjugacy class is closed under inversion. The remaining two orbits B_2 and B_5 satisfy $B_5 = -B_2$ and thus belong to the conjugacy classes of two roots of unity which are inverse to each other; thus they give rise to a pair $(f, f^\wedge)$ of irreducible factors of degree 4 with $f \neq f^\wedge$. Hence Theorem 7.9.5 gives

$$|OC(15,2)| = (2^2+1)(2+1)(2^4-1) = 225.$$

We conclude this example with determining the irreducible factors of $x^{15}+1$ in GF$(2)[x]$, even though this is, as we have seen, not necessary in the present context. Obviously B_3 and B_4 belong to Φ_5 and Φ_3, respectively, whereas B_2 and B_5 have to belong to the two irreducible factors of Φ_{15} over GF(2). Using Proposition 3.6.12, we obtain

$$\Phi_{15} = \frac{\Phi_3(x^5)}{\Phi_3(x)} = \frac{x^{10}+x^5+1}{x^2+x+1} = x^8+x^7+x^5+x^4+x^3+x+1.$$

Since the two irreducible factors of degree 4 of Φ_{15} are reciprocals of each other, one easily finds $f = x^4+x+1$ and $f^\wedge = f^* = x^4+x^3+1$. □

It remains to consider $|OC(n,q)|$ for values of n which are multiples of the characteristic p of GF(q). Obviously, it suffices to consider the case $n = sp$ and to reduce the determination of $|OC(n,q)|$ to that of $|OC(s,q)|$. We first show that the ring $R_{sp,q}$ admits an epimorphism onto a subring which may be identified with $R_{s,q}$ and study some of its properties. We recall that an ideal I is said to be **nilpotent** of **index** p if it satisfies $I^p = \{0\}$.

Proposition 7.9.7. *Let q be a power of the prime p and s any positive integer. Then the ring $R_{sp,q}$ admits an epimorphism π onto a subring S isomorphic to $R_{s,q}$. Moreover, the kernel of π is a nilpotent principal ideal of index p with order $q^{s(p-1)}$,* *namely $I = (x^s - 1)$. Finally, the units of $R_{sp,q}$ are precisely the pre-images of the units in S under π.*

Proof. Consider the mapping $\pi\colon R_{sp,q} \to R_{sp,q}$ defined as follows:

$$\pi(a(x)) := \sum_{i=0}^{s-1} (a_i + a_{i+s} + \cdots + a_{i+(p-1)s}) x^{pi}, \tag{7.62}$$

where $a(x) = a_0 + a_1 x + \cdots + a_{sp-1} x^{sp-1}$. Clearly, the mapping $\rho\colon a(x) \mapsto a(x)^p$ is an endomorphism of $R := R_{sp,q}$, and the mapping ω defined by

$$\omega\colon \sum_{i=0}^{sp-1} b_i x^i \mapsto \sum_{i=0}^{sp-1} b_i^p x^i$$

is an automorphism of R. Using $x^{sp} = 1$ in R, we observe

$$a(x)^p = \sum_{i=0}^{s-1} (a_i^p + a_{i+s}^p + \cdots + a_{i+(p-1)s}^p) x^{pi}.$$

Hence $\pi = \omega^{-1} \circ \rho$, and thus π is an endomorphism of R. It is easily seen from Equation (7.62) that the image S of π is indeed isomorphic to $\mathrm{GF}[y]/(y^s - 1) \cong R_{s,q}$, by identifying x^p with y. Using (7.62) again, the kernel I of π is, by definition,

$$I = \{a \in R\colon a_i + a_{i+s} + \cdots + a_{i+(p-1)s} = 0 \text{ for } i = 0, \ldots, s-1\}. \tag{7.63}$$

Because of $\pi = \omega^{-1} \circ \rho$, we also have the following alternative description of I:

$$I = \{a \in R\colon \rho(a) = 0\} = \{a \in R\colon a^p = 0\},$$

as ω is an isomorphism of R. Hence $a \in I$ if and only if a^p is a multiple of $x^{sp} - 1 = (x^s - 1)^p$, that is, if and only if a is a multiple of $x^s - 1$. Therefore the description of I in the assertion is correct. Moreover, the product of any p elements of I is a multiple of $x^{sp} - 1$ and hence equals 0, so that I is a nilpotent ideal of index p. The assertion on the cardinality of I is immediate from (7.63).

Finally, the units of R are indeed the pre-images of the units in $S \cong R/I \cong R_{s,q}$ under π, since $I = (x^s - 1)$ is nilpotent; see Lemma 7.9.8 below. □

We note that already the weaker property that I is a **nil ideal** – that is, for every $a \in I$ there exists some positive integer $m = m(a)$ such that $a^m = 0$ – suffices to prove the assertion on the units of $R_{sp,q}$, as the following general result shows.

Lemma 7.9.8. *Let R be a commutative ring, and let I be a nilideal of R. Then $a \in R$ is a unit if and only if $a + I$ is a unit in R/I.*

Proof. Trivially, the image of a unit in R is a unit in $S := R/I$. Conversely, assume that $a + I$ is a unit in S. Then there exists an element $b \in R$ such that $ab \in 1 + I$, say $ab = 1 - i$. Since I is a nilideal, $i^m = 0$ for some positive integer m, which implies

$$(1-i)(1+i+\cdots+i^{m-1}) = 1-i^m = 1.$$

Thus $1-i$ is a unit in R, and then $ab = 1-i$ shows that a is also a unit. □

It is worthwhile pointing out the following immediate consequence of Proposition 7.9.7 and Theorem 7.3.2:

Corollary 7.9.9. *Let q be a power of the prime p, and s any positive integer. Then* $|C(sp,q)| = q^{s(p-1)}|C(s,q)|$. □

Of course, Corollary 7.9.9 is just a special case of the general results given in Section 7.3. However, by combining it with Corollary 7.4.7, one may in fact obtain the order of the group $C(n,q)$ of invertible circulant (n,n)-matrices over $\mathrm{GF}(q)$ in a different way, which provides a further proof for Theorem 7.3.9.

Observation 7.9.10. We note another interesting property of the epimorphism π defined in Equation (7.62). Let $a(x) = a_0 + a_1x + \cdots + a_{sp-1}x^{sp-1} \in R_{sp,q}$. Then $a^T(x) = a_0 + a_{sp-1}x + a_{sp-2}x^2 + \cdots + a_1x^{sp-1}$, and hence

$$\begin{aligned}\pi((a^T(x)) &= \sum_{i=0}^{s-1}(a_{sp-i} + a_{sp-i-s} + \cdots + a_{s-i})x^{pi} \\ &= \sum_{i=0}^{s-1}(a_{s-i} + a_{(s-i)+s} + \cdots + a_{(s-i)+(p-1)s})x^{pi} = (\pi(a(x))^T.\end{aligned}$$

Thus π is compatible with taking transposes:

$$\pi((a^T(x)) = (\pi(a(x))^T \quad \text{for all } a \in R_{sp,q}. \tag{7.64}$$

In particular, (7.64) implies that π maps orthogonal polynomials in $R_{sp,q}$ to orthogonal polynomials in $R_{s,q}$. However, an arbitrary pre-image of an orthogonal polynomial in $R_{s,q}$ will, in general, not be orthogonal. □

We shall now compute $|OC(sp,q)|$ by determining the number of orthogonal pre-images of a given orthogonal polynomial in $R_{s,q}$. We need to consider two cases, depending on the characteristic of $\mathrm{GF}(q)$. We begin with the odd case, which is a little simpler as it requires fewer case distinctions; here the proof by MacWilliams [253] for the special case where q is a prime carries over without problems.

Theorem 7.9.11. *Let q be a power of an odd prime p, and let s be any positive integer. Then*

$$|OC(sp,q)| = q^{s(p-1)/2}|OC(s,q)|.$$

Proof. We use the representation of $R_{s,q}$ as a subring $S \cong R/I$ of $R = R_{sp,q}$ as in Proposition 7.9.7. Consider an arbitrary orthogonal polynomial in S, say $\pi(c(x))$. Thus

$$(c+I)(c^T+I) = 1+I, \tag{7.65}$$

and we want to determine the number of polynomials $k \in I$ satisfying

$$(c-k)(c^T-k^T) = 1. \tag{7.66}$$

To this end, we introduce the following four subsets of I:

$$C := \{a \in I\colon a = a^T\},\ \ D := \{a \in I\colon a = -a^T\},$$
$$E := \{a - a^T : a \in I\},\ \ F := \{a + a^T : a \in I\}.$$

We claim that

$$C, D, E \text{ and } F \text{ all have cardinality } q^{s(p-1)/2}. \tag{7.67}$$

Let us first prove the validity of (7.67) for C. Thus we need to show that the number of symmetric polynomials $a \in R$ which are actually contained in I is $q^{s(p-1)/2}$. Using the definition of the transpose polynomial given in Lemma 7.9.3, one easily checks that the total number of symmetric polynomials in R is q^{α}, where

$$\alpha := \begin{cases} 1 + \frac{sp-1}{2} & \text{if } s \text{ is odd,} \\ 2 + \frac{sp-2}{2} & \text{if } s \text{ is even.} \end{cases}$$

Because of Equation (7.63), the conditions for a to belong to I are

$$a_i + a_{i+s} + \cdots + a_{i+(p-1)s} = 0 \quad \text{for } i = 0, \ldots, s-1. \tag{7.68}$$

Since $a = a^T$, we have $a_{sp-j} = a_j$ for $1 \le j \le sp/2$. Therefore, with the abbreviations $p' := (p-1)/2$ and $s' := s/2$ (for even s), the conditions in (7.68) become[8]

$$a_i + a_{i+s} + \cdots + a_{i+p's} + a_{p's-i} + \cdots + a_{2s-i} + a_{s-i} = 0 \ \text{ for } i \ne 0, s';$$
$$a_0 + 2a_s + 2a_{2s} + \cdots + 2a_{p's} = 0 \ \text{ for } i = 0;$$
$$2a_{s'} + 2a_{3s'} + \cdots + 2a_{(p-2)s'} + a_{ps'} = 0 \ \text{ for } i = s'.$$

Observe that the conditions of the first type for i and $i' = s - i$ always agree. This shows that the number of independent linear conditions imposed on the coefficients of a by (7.68) is

$$\beta := \begin{cases} 1 + \frac{s-1}{2} & \text{if } s \text{ is odd,} \\ 2 + \frac{s-2}{2} & \text{if } s \text{ is even.} \end{cases}$$

Hence the number of polynomials in C is $q^{\alpha-\beta} = q^{s(p-1)/2}$, as claimed.

Now note that two polynomials $a, b \in I$ satisfy $a - a^T = b - b^T$ if and only if $a - b \in C$. This implies $|I| = |E||C|$, since C is a subgroup of $(I, +)$ and E a system of coset representatives for C. We conclude that (7.67) also holds for E, as $|I| = q^{s(p-1)}$ by Proposition 7.9.7.

The assertion for D and F in (7.67) follows by a similar argument, with obvious modifications. Next observe $E \subseteq D$ and $F \subseteq C$; hence (7.67) implies

---

[8] Of course, the subsequent condition for s' is only present when s is even.

$$C = F \quad \text{and} \quad D = E. \tag{7.69}$$

After these preparations, we can attack the problem of counting the number of solutions of Equation (7.66). Reordering that equation as $cc^T - 1 = kc^T + k^T c - kk^T$, we see that the solutions for (7.66) correspond to the simultaneous solutions of the following two equations:

$$r = kc^T - \frac{kk^T}{2} \quad \text{with } k \in I; \tag{7.70}$$

$$cc^T - 1 = r + r^T \quad \text{with } r \in I. \tag{7.71}$$

By (7.65), $cc^T - 1 \in I$ and then clearly even $cc^T - 1 \in C$. Hence Equation (7.71) always has a solution, as $C = F$ by (7.69). Next note that any two solutions r and r' of Equation (7.71) differ by a polynomial in D, as then $r - r' = (r')^T - r^T$. Conversely, $r + a$ is also a solution of (7.71) whenever r is such a solution and $a \in D$. Now (7.67) shows that Equation (7.71) has exactly $q^{s(p-1)/2}$ solutions.

Therefore it suffices to show that, given any solution r of (7.71), Equation (7.70) has a unique solution for k. Since c is invertible by Proposition 7.9.7, Equation (7.70) is equivalent to

$$k = \frac{r}{c^T} + \frac{kk^T}{2c^T}. \tag{7.72}$$

We now substitute (7.72) into itself and obtain

$$k = \frac{r}{c^T} + \frac{1}{2c^T}\left(\frac{r}{c^T} + \frac{kk^T}{2c^T}\right)\left(\frac{r^T}{c} + \frac{kk^T}{2c}\right). \tag{7.73}$$

Iterating this procedure sufficiently often, we eventually obtain an expression for k which consists entirely of terms involving only c, r and their transposed polynomials, but not k and k^T. To see this, recall that the ideal I is nilpotent of index p by Proposition 7.9.7. Hence we may discard products of p or more elements of I which shows that, after at most $p - 2$ iterations, the resulting expression for k will indeed no longer contain any term involving k or k^T on the right hand side. This establishes our claim that Equation (7.70) has a unique solution for any given solution r of Equation (7.71). □

Example 7.9.12. It may be helpful to look at the last step of the preceding proof for the smallest case $p = 3$. Here we indeed require no further iteration beyond (7.73), since products containing at least three elements from I are 0. Explicitly, we obtain the following solution of Equation (7.70) in this case:

$$k = \frac{r}{c^T} + \frac{1}{2c^T}\left(\frac{r}{c^T} + \frac{kk^T}{2c^T}\right)\left(\frac{r^T}{c} + \frac{kk^T}{2c}\right) = \frac{r}{c^T} + \frac{rr^T}{2c(c^T)^2},$$

as $r, k, k^T \in I$. □

It remains to consider the analogue of Theorem 7.9.11 for even values of q. The first part of the proof uses the same ideas and is in fact somewhat simpler, which will allow us to be comparatively brief. In particular, the case needed for the application in Corollary 7.9.2 – that is, the case of odd values of s – presents no difficulties at all. However, the case of even s is rather more involved and actually gives a somewhat surprising result when $s \equiv 2 \bmod 4$; this case requires a non-trivial modification of MacWilliams' proof.

Theorem 7.9.13. *Let q be a power of 2, and let s be any positive integer. Then*

$$|OC(2s,q)| = \begin{cases} q^{(s+1)/2}|OC(s,q)| & \text{if } s \text{ is odd,} \\ 2q^{s/2}|OC(s,q)| & \text{if } s \equiv 2 \bmod 4, \\ q^{s/2}|OC(s,q)| & \text{if } s \equiv 0 \bmod 4. \end{cases}$$

Proof. We use the same general setup as in the proof of Theorem 7.9.11 (of course, with $p = 2$). This time, we define three subsets of I as follows:

$$C = \{a \in I\colon a = a^T\}, \ D = \{a \in C\colon a_0 = a_s = 0\}, \ E = \{a + a^T : a \in I\}.$$

We now claim

$$|C| = q|D| = \begin{cases} q^{(s+1)/2} & \text{if } s \text{ is odd,} \\ q^{(s+2)/2} & \text{if } s \text{ is even.} \end{cases} \tag{7.74}$$

First note that the number of polynomials $a \in R_{2s,q}$ with $a = a^T$ is q^{s+1}, since the symmetry of a imposes the $s-1$ conditions

$$a_i = a_{2s-i} \quad \text{for } i = 1,\ldots,s-1. \tag{7.75}$$

The conditions for such an a to belong to I are $a_i + a_{s+i} = 0$ for $i = 0,\ldots,s-1$, which may be written as

$$a_i + a_{s-i} = 0 \quad \text{for } i = 0,\ldots,s-1, \tag{7.76}$$

by (7.75). Clearly, (7.76) gives a total of $1 + \frac{s-1}{2}$ linearly independent equations if s is odd, but only $1 + \frac{s-2}{2}$ linearly independent equations if s is even, since the equation for $i = s/2$ is trivial in this case. This proves the assertion on $|C|$. Finally, $|C| = q|D|$, since D is the subset of those $a \in C$ satisfying $a_0 = a_s = 0$, whereas we only have the restriction $a_0 + a_s = 0$ for polynomials in C. Using $|I| = |E||C|$ and Proposition 7.9.7, Equation (7.74) also gives

$$|E| = \begin{cases} q^{(s-1)/2} & \text{if } s \text{ is odd,} \\ q^{(s-2)/2} & \text{if } s \text{ is even.} \end{cases} \tag{7.77}$$

In view of $E \subseteq D$, Equations (7.74) and (7.77) yield[9]

$$E = \begin{cases} D & \text{if } s \text{ is odd,} \\ \{a \in D\colon a_{s/2} = 0\} & \text{if } s \text{ is even} \end{cases} \tag{7.78}$$

Now let $u = \pi(c)$ be an arbitrary orthogonal polynomial in $R_{s,q}$ (so that $cc^T \in 1+I$), say $u(y) = u_0 + u_1 y + \cdots + u_{s-1} y^{s-1}$. Consider any polynomial $c+k$ ($k \in I$) in the coset of c. Then $c+k$ is orthogonal if and only if $(c+k)(c^T+k^T) = 1$. Because of $I^2 = 0$, this is equivalent to

$$cc^T + 1 = r + r^T, \quad \text{where } r = kc^T \in I. \tag{7.79}$$

Since c is invertible, (7.79) has a solution if and only if $cc^T + 1 \in I$ belongs to the proper subset E of I; in this case, we obtain exactly $|C|$ solutions. (Note that $r+a$ is also a solution if r is a solution and $a \in C$, and that any two solutions differ by some polynomial in C.) Hence we need to determine the cardinality of the set

$$X = X(s,q) := \{\pi(c) \in OC(s,q)\colon cc^T + 1 \in E\},$$

and the assertion is equivalent to showing

$$|X(s,q)| = \begin{cases} |OC(s,q)| & \text{if } s \text{ is odd,} \\ \frac{2}{q}|OC(s,q)| & \text{if } s \equiv 2 \bmod 4, \\ \frac{1}{q}|OC(s,q)| & \text{if } s \equiv 0 \bmod 4. \end{cases} \tag{7.80}$$

Without loss of generality, we may choose for c the canonical pre-image

$$c(x) = u_0 + u_1 x + \cdots + u_{s-1} x^{s-1} \tag{7.81}$$

of u. Then

$$c^T(x) = u_0 + u_{s-1} x^{s+1} + \cdots + u_2 x^{2s-2} + u_1 x^{2s-1},$$

and the product $b = cc^T$ has constant term $u_0^2 + u_1^2 + \cdots + u_{s-1}^2$, which is also the constant term of $uu^T \in R_{s,q}$. Since u is orthogonal, we obtain

$$b_0 = u_0^2 + u_1^2 + \cdots + u_{s-1}^2 = 1. \tag{7.82}$$

Moreover, $b_s = 0$ and therefore $cc^T + 1 = b + 1 \in D$. We can now apply (7.78). If s is odd, we conclude $cc^T + 1 \in E$ for all $u = \pi(c) \in OC(s,q)$, proving the validity of (7.80) for this case.

From now on, let s be even, say $s = 2t$. Then (7.78) yields

$$cc^T + 1 \in E \iff b_t = u_0 u_t + u_1 u_{t+1} + \cdots + u_{t-1} u_{2t-1} = 0. \tag{7.83}$$

[9] For the s even case, note that any polynomial of the form $a + a^T$ necessarily has coefficient 0 for $x^{s/2}$, whereas this is not required for polynomials in D.

Assume first that t is likewise even, say $t = 2m$, $s = 4m$. One easily checks that the q polynomials $v_\lambda := 1 + \lambda y^m + \lambda y^{3m} \in R_{s,q}$, where $\lambda \in F = \mathrm{GF}(q)$, are orthogonal and form a subgroup $G \cong (F,+)$ of $OC(s,q)$. Therefore $u_\lambda := v_\lambda u$ is again orthogonal for each $\lambda \in F$. The coefficients of u_λ are given by

$$u_{i,\lambda} = u_i + \lambda u_{i-m} + \lambda u_{i+m} \quad \text{for } i = 0, \ldots, 4m-1, \tag{7.84}$$

where all indices are computed modulo $4m$. We now require the coefficient $b_{2m,\lambda}$ of x^{2m} in the product $b_\lambda := c_\lambda c_\lambda^T$, where $c_\lambda(x) := u_{0,\lambda} + u_{1,\lambda}x + \cdots + u_{4m-1,\lambda}x^{4m-1}$ is the canonical pre-image of u_λ. Using (7.82), (7.83) and (7.84), a short computation gives

$$b_{2m,\lambda} = \sum_{i=0}^{2m-1} u_{i,\lambda} u_{i+2m,\lambda} = \sum_{i=0}^{2m-1} u_i u_{i+2m} + \lambda^2 \cdot \sum_{i=0}^{4m-1} u_i^2 = b_{2m} + \lambda^2.$$

Hence $b_{2m,\lambda} = 0$ for the unique $\lambda \in F$ satisfying $\lambda^2 = b_{2m}$, which shows that exactly one of the q polynomials u_λ in the coset Gu has a canonical pre-image c_λ satisfying the condition $c_\lambda c_\lambda^T + 1 \in E$. This establishes the validity of (7.80) when s is a multiple of 4.

Finally, it remains to consider the case where t is odd, that is, $s \equiv 2 \bmod 4$. Here the coefficient b_t of x^t in cc^T satisfies

$$b_t = \alpha(\alpha+1) \quad \text{with } \alpha := u_0 + u_2 + \cdots + u_{2t-2}, \tag{7.85}$$

which is seen as follows. As u is orthogonal, the coefficient of x^{2t-k} in uu^T has to be 0 for $k \neq 0$, so that

$$\sum_{i=0}^{2t-1} u_i u_{i+k} = 0 \quad \text{for } k = 1, \ldots, 2t-1,$$

where all indices are computed modulo $2t$. We now add the preceding equations for $k = 1, 3, \ldots, t-2$ and rearrange terms (where we take into account that t is odd and put $t' := (t-1)/2$):

$$\begin{aligned} 0 &= \sum_{i=0}^{2t-1} \sum_{j=1}^{t'} u_i u_{i+2j-1} = \sum_{i=0}^{t-1} \sum_{j=1}^{t'} u_{2i} u_{2i+2j-1} + \sum_{i=0}^{t-1} \sum_{j=1}^{t'} u_{2i+1} u_{2i+2j} \\ &= \sum_{i=0}^{t-1} \sum_{j=1}^{t'} u_{2i}\big(u_{2i+(2j-1)} + u_{2i-(2j-1)}\big) = \sum_{i=0}^{t-1} u_{2i} u_{2i+t} + \sum_{i=0}^{t-1} \sum_{j=1}^{t} u_{2i} u_{2j-1} \\ &= \sum_{i=0}^{t-1} u_{2i} u_{2i+t} + \Big(\sum_{i=0}^{t-1} u_{2i}\Big)\Big(\sum_{j=1}^{t} u_{2j-1}\Big) = b_t + \alpha(\alpha+1); \end{aligned}$$

here we have used the formula for b_t in (7.83) and $1 = u(1) = u_0 + u_1 + \cdots + u_{2t-1}$ in the final step. (Note that the orthogonality of u gives $u(1)^2 = 1$.) This establishes the desired formula (7.85). We now consider the polynomials

$$u_\lambda := u + \lambda j \quad \text{with } j := 1 + y + \cdots + y^{2t-1} \quad (\text{for } \lambda \in F).$$

It is easily checked that both $j(u+u^T)$ and $jj^T = j^2$ are equal to 0, and therefore the orthogonality of u shows that each u_λ is likewise orthogonal. As in the previous case, we require the coefficient $b_{t,\lambda}$ of x^t in the product $b_\lambda := c_\lambda c_\lambda^T$, where

$$c_\lambda(x) := (u_0 + \lambda) + (u_1 + \lambda)x + \cdots + (u_{2t-1} + \lambda)x^{2t-1}$$

is the canonical pre-image of u_λ. Using (7.83) for c_λ instead of c, we get

$$b_{t,\lambda} = (\alpha + t\lambda)(\alpha + t\lambda + 1) = (\alpha + \lambda)(\alpha + \lambda + 1) \quad \text{for all } \lambda \in F,$$

since t is odd. Hence $b_{t,\lambda} = 0$ if and only if $\lambda \in \{\alpha, \alpha+1\}$, which shows that exactly two of the q polynomials u_λ lead to canonical pre-images satisfying the criterion $c_\lambda c_\lambda^T + 1 \in E$. This establishes the validity of (7.80) also in the final case $s \equiv 2 \bmod 4$. □

Example 7.9.14. We stress that the proof of Theorem 7.9.13 is constructive. As an illustration, we determine the 6144 elements of $OC(12,4)$. By Theorem 7.9.5, the group $OC(3,4)$ has order 3 and thus obviously corresponds to the three orthogonal polynomials

$$f_1 := 1, \ f_2 := x \text{ and } f_3 := x^2$$

in $R_{3,4}$. By Theorem 7.9.13 and its proof, each of these polynomials lifts to 16 orthogonal polynomials in $R_{6,4}$ (since $s = 3$ is odd), which altogether correspond to the 48 matrices in $OC(6,4)$. By (7.76), the coefficients a_i of the 16 symmetric polynomials needed to actually construct the elements of $OC(6,4)$ have to satisfy the conditions $a_0 = a_3$ and $a_1 = a_2$. All such polynomials form the 2-dimensional vector space $C = C_6$ generated by the polynomials

$$s_1 := 1 + x^3 \quad \text{and} \quad s_2 := x + x^2 + x^4 + x^5$$

over $F = \mathrm{GF}(4)$. Thus each f_i lifts to the 16 polynomials in the coset $f_i + C$ (where we use the same symbol for a polynomial in $R_{s,q}$ and its canonical pre-image in $R_{2s,q}$, by a convenient abuse of notation), and hence

$$OC(6,4) \leftrightsquigarrow \{(1 + \lambda s_1 + \mu s_2)f_i \colon i \in \{1,2,3\} \text{ and } \lambda, \mu \in F\}. \tag{7.86}$$

In order to construct $OC(12,4)$, we first determine the 4-dimensional vector space $C = C_{12}$ formed by the symmetric polynomials in $I = I_{12}$. Equation (7.76) now gives the conditions $a_0 = a_6$, $a_1 = a_5$ and $a_2 = a_4$, and thus C_{12} is the vector space generated by the four polynomials

$$w_1 := 1 + x^6, \ w_2 := x + x^5 + x^7 + x^{11}, \ w_3 := x^2 + x^4 + x^8 + x^{10} \text{ and } w_4 := x^3 + x^9.$$

We now need to decide which of the 48 polynomials $g_{i,\lambda,\mu} := (1+\lambda s_1+\mu s_2)f_i$ in (7.86) actually lift to orthogonal polynomials in $R_{12,4}$. Here $s = 6 \equiv 2 \bmod 4$, so this will be the case for exactly half of these polynomials. Equations (7.83) and (7.85) yield the criterion

$$b_3 = \alpha(\alpha+1) = 0 \quad \text{with } \alpha = u_0+u_2+u_4, \tag{7.87}$$

where the u_j denote the coefficients of $g_{i,\lambda,\mu}$. The reader should check that (7.87) holds if and only if $\lambda \in \{0,1\}$; this is a little tedious, but straightforward. Hence

$$OC(12,4) \leftrightsquigarrow \{(1+\lambda_1 w_1+\lambda_2 w_2+\lambda_3 w_3+\lambda_4 w_4)(1+\lambda s_1+\mu s_2)f_i\},$$

where $i \in \{1,2,3\}$, $\lambda \in \{0,1\}$ and $\mu, \lambda_1, \lambda_2, \lambda_3, \lambda_4 \in F$, and we have obtained a description for all orthogonal circulant $(12,12)$-matrices over F which is both concise and explicit.

We leave it to the reader to go on and determine the $4^6 \cdot 6144 = 25,165,824$ elements of $OC(24,4)$. In this case, $|C_{24}| = 4^7$ and exactly one quarter of the polynomials in $OC(12,4)$ lift to an orthogonal polynomial of degree 24, as now $s = 12 \equiv 0 \bmod 4$. □

We conclude this section with the promised result giving the number of self-dual normal bases.

Theorem 7.9.15. *Let q be a power of the prime p, and let n and b be positive integers. Assume that n is not divisible by p, let the irreducible factors of x^n-1 with roots of order ≥ 3 over $\mathrm{GF}(q)$ be labelled in such a way that*

$$f_i^* = f_i \text{ for } i=1,\dots,s \quad \text{and} \quad (f_{s+j})^\wedge = f_{s+t+j} \neq f_{s+j} \text{ for } j=1,\dots,t,$$

and denote their respective degrees as follows:

$$\deg f_i = 2d_i \text{ for } i=1,\dots,s \quad \text{and} \quad \deg f_{s+j} = \deg f_{s+t+j} = e_j \text{ for } j=1,\dots,t.$$

Finally, put

$$\varepsilon = \begin{cases} 1 & \text{if } q \text{ is even,} \\ 2 & \text{if } q \text{ and } n \text{ are odd,} \\ 0 & \text{if } q \text{ is odd and } n \text{ is even.} \end{cases}$$

Then the number $sdn(m,q)$ of self-dual normal bases for $\mathrm{GF}(q^m)$ over $\mathrm{GF}(q)$, where $m = p^b n$, is given by

$$sdn(m,q) = \begin{cases} \frac{\varepsilon}{n} \cdot \prod_{i=1}^{s}(q^{d_i}+1) \cdot \prod_{j=1}^{t}(q^{e_j}-1) & \text{if } b=0, \\ \frac{1}{p^b} q^{(p^b-1)n/2} \cdot sdn(n,q) & \text{if } p \neq 2 \text{ and } b \geq 1, \\ \frac{1}{2} q^{(n+1)/2} \cdot sdn(n,q) & \text{if } p=2 \text{ and } b=1, \\ 0 & \text{if } p=2 \text{ and } b \geq 2. \end{cases}$$

Proof. This follows from Theorem 7.8.1 and Corollary 7.9.2, together with Theorem 7.9.5 and iterative application of Theorems 7.9.11 and 7.9.13. The details are left to the reader. □

Exercises

Our first exercise establishes a connection between the material discussed in Example 7.9.6 (and its preceding comment) and Berlekamp's method for the factorization of polynomials over finite fields dealt with in Section 6.5:

Exercise 7.9.16. Let $F = \mathrm{GF}(q)$ be some finite field and n a positive integer which is relatively prime to q. Define an equivalence relation on the set $\{0, 1, \dots, n-1\}$ of residues modulo n as follows:

$$k \sim \ell \quad \Longleftrightarrow \quad \exists j\colon kq^j \equiv \ell \bmod n.$$

The equivalence classes are called the **cyclotomic cosets** (modulo n, w.r.t. q). Let $\mathscr{C}$ denote the set of all cyclotomic cosets, and for $C \in \mathscr{C}$ put

$$h_C(x) := \sum_{\ell \in C} x^\ell.$$

Prove that the polynomials $h_C(x)$ with $C \in \mathscr{C}$ form a basis of every Berlekamp algebra $\mathscr{B}_f^Q$, where $f(x) = x^n - 1$ and where Q is an arbitrary subfield of F.

Use this approach to determine a basis for the Berlekamp algebra $\mathscr{B}_{x^5-1}$ over $\mathrm{GF}(19)$. □

The subsequent three Exercises are based on [157].

Exercise 7.9.17. Consider the n-dimensional extension $E = \mathrm{GF}(q^n)$ of $F = \mathrm{GF}(q)$, where n is a multiple of the characteristic p of $\mathrm{GF}(q)$, say $n = sp$. In addition, assume the existence of a self-dual normal basis (SDNB) for K/F, where K is the intermediate field $\mathrm{GF}(q^s)$. Finally, let α and β generate SDNB's for E/F and K/F, respectively, and assume $\mathrm{Tr}_{E/K}(\alpha) = \beta$. Show that a circulant orthogonal (n,n)-matrix C over E transforms α into another SDNB generator for E/F with the same trace β over K if and only if the associated orthogonal polynomial $c(x) \in R_{sp,q}$ maps to the polynomial $1 \in R_{s,q}$ under the epimorphism π constructed in Proposition 7.9.7. □

Exercise 7.9.18. Let q be a power of an odd prime p, let s be an odd integer, and put $n := sp$. Use Exercise 7.9.17 together with a counting argument based on Theorem 7.9.11 and its proof to show that any SDNB generator β for $K = \mathrm{GF}(q^s)$ over $F = \mathrm{GF}(q)$ lifts to exactly $q^{(n-m)/2}$ SDNB generators α for $E = \mathrm{GF}(q^n)$ over F satisfying $\mathrm{Tr}_{E/K}(\alpha) = \beta$. Then use induction to establish the same result for values of n having the form $n = p^b s$. □

Exercise 7.9.19. Prove an analogue of Exercise 7.9.18 for $p = 2$. (Be careful about the admissible values for s and b here.) □

Chapter 8
Basis Representations and Arithmetics

Abstract The main topic of the present chapter is a detailed study of the interaction between theoretical results on various types of bases used to represent finite extension fields $E = \mathrm{GF}(q^n)$ over $\mathrm{GF}(q)$ and their application to efficiently performing the arithmetic operations in E, with particular emphasis on hardware implementations. Such questions are of fundamental importance for various practical applications, in particular, in Cryptography, Coding Theory, and Signal Processing. We also briefly discuss an alternative approach to the arithmetics of finite fields, namely discrete logarithms, which is likewise useful in some applications.

8.1 Basics on Computing in Finite Fields

If we want to construct a finite field $F = \mathrm{GF}(q)$ explicitly and perform actual arithmetical computations in F, we certainly have to master the arithmetics of the underlying prime field $P = \mathrm{GF}(p) = \mathbb{Z}_p$ first. This includes the following fundamental tasks: addition, multiplication and exponentiation modulo p, computation of the greatest common divisor of two integers, and computation of multiplicative inverses modulo p. This last operation is easily performed using the Extended Euclidean Algorithm: given an integer z, we not only obtain $d = \gcd(z, p)$ from this algorithm but also a representation $yz + ap = d$; see Algorithm 1.5.17. If z is invertible modulo p, that is, if $d = 1$, we therefore get $y = z^{-1} \bmod p$. From now on, we shall take the fundamental algorithms of modular arithmetics (which present no problems as long as p is not too large) for granted and refer the reader to Lüneburg [249] for a nice introductory treatment. We also recommend the more advanced monograph by Bach and Shallit [17] as a general reference for Algorithmic Number Theory.

It should be noted that there are two further important problems in modular arithmetics which are much harder: finding a primitive root w modulo p (see Section 5.6) and the computation of the so-called discrete logarithm of an invertible element z modulo p (given a fixed primitive root w), that is, the determination of the unique integer k with $0 \le k \le p-2$ satisfying $w^k \equiv z \bmod p$. This last problem seems to be

D. Hachenberger and D. Jungnickel, *Topics in Galois Fields*,
Algorithms and Computation in Mathematics 29,
https://doi.org/10.1007/978-3-030-60806-4_8

very hard for large primes p; in fact, it – and its analogue for $\mathrm{GF}(q)$ – form the basis for some of the standard algorithms proposed for Public Key Cryptography; see, for instance, Koblitz [220] and the surveys by Odlyzko [312] and McCurley [262]. We shall include a brief discussion of discrete logarithms at the end of this section.

We now turn our attention to the arithmetics of $F = \mathrm{GF}(p^n)$. Since F is an n-dimensional vector space over its prime subfield $P = \mathrm{GF}(p)$, we may choose a basis $\{\beta_1, \ldots, \beta_n\}$ of F over P; hence addition in F presents no problem since it is performed componentwise (and we assume that we know how to add modulo p). Up to this point, we can just consider F as the vector space of n-tuples over P. However, multiplication presents a problem in this representation. Of course, we have the formula

$$(a_1\beta_1 + \cdots + a_n\beta_n)(b_1\beta_1 + \cdots + b_n\beta_n) = \sum_{i,j=1}^{n} a_i b_j \beta_i \beta_j \tag{8.1}$$

and thus all we need to know is how to multiply two basis elements. Since the product of any two basis elements can again be expressed as a linear combination of the β's, there are n^2 equations of the following form:

$$\beta_i\beta_j = \sum_{k=1}^{n} c_{ijk}\beta_k \quad (i, j = 1, \ldots, n). \tag{8.2}$$

Of course, all this carries over to computing in extensions E of an arbitrary Galois field $F = \mathrm{GF}(q)$, provided we already know how to compute in that field. These considerations lead to the following terminology.

Definition 8.1.1. One calls a system $(c_{ijk})_{i,j,k=1,\ldots,n}$ of n^3 elements of $F = \mathrm{GF}(q)$ **explicit data** for $E = \mathrm{GF}(q^n)$, provided that the multiplication on the n-dimensional vector space F^n determined by Equations (8.1) and (8.2) turns E into a field. □

Remark 8.1.2. Of course, the determination of some set of explicit data for E is equivalent to the determination of some monic irreducible polynomial f of degree n over $\mathrm{GF}(q)$, since we may represent E as $F[x]/(f) \cong F(\alpha)$ (where α is a root of f) and choose any basis $\{\beta_1, \ldots, \beta_n\}$ of E. In Abstract Algebra, the standard choice is the **polynomial basis** $\{1, \alpha, \ldots, \alpha^{n-1}\}$, and then the coefficients c_{ijk} determining the products $\beta_i\beta_j$ $(i, j, k = 1, \ldots, n)$ are obtained by first multiplying the polynomials x^{i-1} and x^{j-1} associated with β_i and β_j and then computing the remainder of their product under division by f. This presents no fundamental difficulty, as we may take polynomial arithmetic over $F = \mathrm{GF}(q)$ for granted, since it reduces to the ordinary tasks of arithmetics over F.

The preceding considerations show that the explicit construction of irreducible polynomials studied in Chapter 5 is not only of theoretical interest, but also of real practical importance. The results obtained there indicate that it is possible to construct some irreducible polynomial of a prescribed degree n over $\mathrm{GF}(q)$ in a reasonably efficient way, even though we did not prove an explicit estimate on the number of operations over F required. The best known deterministic algorithms for this problem are due to Shoup [355] who augmented the basic ideas we have described by some other ingredients (like factoring polynomials of "small" degree) in a rather elaborate way. In particular, his results show that, for any fixed prime p, an irreducible polynomial of degree n over $\mathrm{GF}(p)$ can be constructed in a polynomial number of steps: more precisely, this task can be performed with $O(p^{1/2}(\log p)^3 n^{3+\varepsilon} + (\log p)^2 n^{4+\varepsilon})$ modular operations. With further effort, Shoup also obtained a comparable result for irreducible polynomials over $\mathrm{GF}(q)$.

For a more recent (short, but with a wealth of references) survey on the construction and arithmetics of finite fields, see Schost [342]. □

There are two further constructive problems related to explicit data which we shall not discuss. First, how can one decide whether or not any given system of n^3 elements of $\mathrm{GF}(q)$ forms a set of explicit data for $E = \mathrm{GF}(q^n)$? And, given two sets of explicit data for E, how does one find an isomorphism between the two representations of E determined by the corresponding multiplication rules on E? These two problems are less interesting for practical applications, and thus we just refer the reader to Lenstra [236] and Allombert [7] as well as the references cited there for this topic.

One might also ask how difficult it is to compute some set of explicit data for E over F (already given some appropriate irreducible polynomial f over F); of course, the answer to this question not only depends on f but also on the basis used. It is not surprising that polynomial bases allow to do so very efficiently; see Exercise 8.1.7. It should be noted that this task may be viewed as a pre-computation, so that one does not care all that much about the amount of time it takes: what one really is interested in for practical applications (for instance, in Coding Theory or Cryptography) is efficiently computing in E once this field has been constructed explicitly.

As Equations (8.1) and (8.2) show, the multiplication of two elements of E (using a given set of explicit data) may require up to $O(n^3)$ operations in F. It is therefore of considerable practical interest to construct explicit data which allow one to reduce the complexity of multiplication in E. The main topic studied in this chapter is the problem of finding efficient ways for performing multiplication in large fields – in particular, by using suitable hardware devices. As we will see, much depends indeed on the selection of the basis used, and the possibility of finding a basis with specified desirable properties usually boils down to the existence question for certain types of irreducible polynomials. For example, we shall see later in this chapter that the use of normal bases always allows to compute every coefficient in the product of two elements by using just one symmetric bilinear form, so that a set of explicit data of size only $O(n^2)$ instead of $O(n^3)$ will suffice. Quite often, by using an appropriate special type of normal basis (that is, of normal polynomial) it even is possible to achieve a complexity which is linear in n.

There is an alternative to using basis representations for finite fields which we will briefly discuss in the remainder of this section. If one represents the non-zero elements of E as the powers of a primitive element, multiplication is trivial. Of course, there is a prize to pay for this: addition will then become difficult. One such scheme exploits the so-called Zech logarithms; since this works as well for prime fields, we do not restrict our attention to extension fields here.

Definition 8.1.3. Let ω be a fixed primitive element for $F = \mathrm{GF}(q)$. For every element γ of F^*, we define the **discrete logarithm** $\log_\omega \gamma$ of γ (to the base ω) as the unique integer c with $0 \le c \le q-2$ satisfying $\omega^c = \gamma$. We also put $\log_\omega 0 := \infty$. □

If we now identify the elements of F with their discrete logarithms, multiplying two elements of F will indeed be trivial. As the formula $\omega^a \cdot \omega^b = \omega^{a+b}$ shows, this task reduces to just adding the corresponding discrete logarithms:

$$\log_\omega \gamma\delta = \log_\omega \gamma + \log_\omega \delta. \tag{8.3}$$

Of course, the addition in (8.3) has to be performed modulo $q-1$, with the convention $\infty + c = \infty$.

If we want to add in this representation, we will need to determine the discrete logarithm of $\gamma + \delta$ for $\gamma, \delta \in F$. Without loss of generality, we may assume $\gamma, \delta \neq 0$, that is, $\log_\omega \gamma$, $\log_\omega \delta \neq \infty$. Now note

$$\omega^c + \omega^d = \omega^c(1 + \omega^{d-c}).$$

Hence one may add two elements of F^* which are represented by their discrete logarithms c and d by adding c to the discrete logarithm of $1 + \omega^{d-c}$. Therefore it suffices to determine the discrete logarithms for all sums involving 1. This motivates the following definition.

Definition 8.1.4. The **Zech logarithm** $Z(e)$ of ω^e is defined as the discrete logarithm of $1 + \omega^e$. In other words, $Z(e)$ is determined from the equation

$$1 + \omega^e = \omega^{Z(e)}. \qquad \square$$

Using discrete logarithms in conjunction with Zech logarithms is a useful representation in practical applications where repeated computations over a relatively small finite field are required (with applications in Coding Theory being typical examples), since then the Zech logarithms can be pre-computed and, when needed for addition, retrieved by a simple table lookup. Let us give a small example for such a Zech log table:

Example 8.1.5. We construct $F = \mathrm{GF}(2^3)$ by adjoining to $\mathrm{GF}(2)$ a root ω of the primitive polynomial $x^3 + x + 1$. Thus $1 + \omega = \omega^3$, that is, $Z(1) = 3$ and $Z(3) = 1$. Now note

$$\omega^4 = \omega + \omega^2 \quad \text{and} \quad \omega^5 = \omega^3 + \omega^2 = 1 + \omega + \omega^2 = 1 + \omega^4,$$

and therefore $Z(4) = 5$ and $Z(5) = 4$. A similar computation gives $Z(2) = 6$ and $Z(6) = 2$, so that we obtain the Zech logarithms in Table 8.1. □

For a more detailed discussion of Zech logarithms and their applications (and more examples of logarithm tables), we refer to Vanstone and van Oorschot [383]

e	∞	0	1	2	3	4	5	6
$Z(e)$	0	∞	3	6	1	5	4	2

Table 8.1 Zech logarithms for GF(8)

and to Huber [192, 193]. We mention in passing that there are explicit formulae giving polynomial representations of discrete logarithms; see, for instance, Mullen and White [293], Niederreiter [301] and Meletiou and Mullen [264]. However, these formulae are of no practical interest.

It is clear that a table lookup of Zech logarithms becomes impractical for large finite fields. Thus the possibility of using this type of representation for large fields depends on the possibility of actually computing discrete logarithms, which is generally believed (but not proved) to be a very difficult problem. In fact, some cryptographic systems are based on the intractability of computing discrete logarithms in sufficiently large finite fields. We shall digress a little and give a brief description of such a cryptographic system.

Example 8.1.6. The following cryptographic system serves for the secure exchange (via an insecure channel) of a common key, which then determines an encryption function for the actual messages to be sent between two parties, traditionally called "Alice" and "Bob". For this scheme, which is due to Diffie and Hellman [107], one selects a primitive element ω for a large finite field $F = \mathrm{GF}(q)$; the possible keys are the elements of F^*.

In order to create a common key, Alice and Bob each select a random integer a and b, respectively, in the range $0 \leq a, b \leq q-2$. Then Alice computes $\alpha := \omega^a$ and sends it to Bob; similarly, Bob computes $\beta := \omega^b$ and sends it to Alice. The common key will be $\kappa := \omega^{ab}$.

Note that κ can be easily computed by both Alice and Bob, as $\kappa = \alpha^b = \beta^a$. However, there is no known way of computing κ from the information an eavesdropper might have, that is, from the knowledge of the field elements α, β and ω (but not the discrete logarithms a or b, which are kept secret), without first computing these discrete logarithms.[1] □

Clearly, the security of a cryptographic system as in Example 8.1.6 depends on the intractability of actually computing discrete logarithms in F. Since the field F and the primitive element ω are fixed, a huge amount of pre-computation (setting up a database) would be acceptable for breaking such a scheme (which might be used for commercial purposes, like electronic transfer of funds), provided only that individual discrete logarithms can afterwards be computed easily. An algorithm using

[1] Using similar approaches, one can also devise schemes for the actual encryption of messages and for digital signatures (that is, for a way of confirming without doubt who has sent a given message) based on the intractability of discrete logarithms, see ElGamal [111] or Koblitz [220].

this strategy is the so-called index calculus algorithm which in particular works for extension fields of GF(2); this algorithm is, for instance, presented in Koblitz [220] and Stinson [367].

That this type of attack indeed necessitates the use of rather large fields for cryptographic purposes was first demonstrated by Blake, Fuji-Hara, Mullin and Vanstone [40] who managed to solve the problem of computing discrete logarithms efficiently for a field as large as $\mathrm{GF}(2^{127})$. If one uses for F a prime field $\mathrm{GF}(p)$, where p is a large prime, other algorithms are known; already in 1990, LaMacchia and Odlyzko [227] provided an algorithm capable of solving the discrete logarithm problem for primes p of around 200 bits.

Various algorithms for the computation of discrete logarithms are discussed in Chapter 6 of Menezes, van Oorschot and Vanstone [266] and in Stinson [367]. For more information on the computation of discrete logarithms and their cryptographic significance, we refer to Odlyzko [312], McCurley [262] and van Oorschot [382]. In the special case of finite fields with small characteristic, considerable progress has been made resulting in quasi-polynomial algorithms to compute discrete logarithms in such fields; we refer the interested reader to the recent survey by Joux and Pierrot [209]. As these authors state, "the problem can no longer be considered for cryptographic use."

The advances in computing discrete logarithms in finite fields have made an alternative first suggested in 1987 by Koblitz [221] quite attractive: using discrete logarithms in groups which may be defined on elliptic curves over finite fields instead of discrete logarithms in the multiplicative groups of finite fields themselves. For an introduction to the increasingly important area of Elliptic Curve Cryptography (which is beyond the scope of our text), we refer to Koblitz [221] or Menezes [267]. In this situation, far smaller field sizes may be sufficient. A summary of early results can be found in Menezes et al. [266]. An early actual implementation for an elliptic curve cryptosystem over $\mathrm{GF}(2^{155})$ was given by Agnew, Mullin, Onyszchuk and Vanstone [1]. Nevertheless, some progress has also been made for computing discrete logarithms in this situation, though it does not yet preclude cryptographic uses; see Galbraith and Gaudry [122] for a recent survey.

As the preceding discussion indicates, cryptographic applications require the ability to compute efficiently in very large finite fields, in particular, in large extension fields of GF(2). In this situation, the use of Zech logarithms either makes no sense or is too time consuming. As an alternative, cleverly chosen basis representations for extension fields have become an essential tool in practical applications of Cryptography. The remainder of this chapter will be devoted to studying this task.

Exercises

Exercise 8.1.7. Assume that we construct $E = \mathrm{GF}(q^n)$ by using an irreducible polynomial f of degree n over $F = \mathrm{GF}(q)$. Show that the polynomial basis for E/F associated with f can be used to determine a set of explicit data for E/F with essentially $O(n^2)$ operations in F. □

Exercise 8.1.8. Zech logarithms may be used to simplify the solution of quadratic and cubic equations. Show that any such equation can (by a suitable substitution) be put into the respective standard form $x^2+x+c=0$ or $x^3+x+c=0$ and explain how one may solve such an equation with the aid of a Zech logarithm table. □

Exercise 8.1.9. Construct Zech logarithm tables for GF(9) and GF(16). □

8.2 Dual Basis Multipliers

This section presents an application of dual bases which was first suggested by Elwyn Berlekamp in 1982. In his seminal paper [29], Berlekamp showed that one may use a polynomial basis together with its dual basis to design a hardware implementation for multiplying an arbitrary element $\xi \in \mathrm{GF}(2^n)$ with a fixed element α by using a so-called linear feedback shift register (such objects will be studied in detail in Chapter 9) and gave a specific application in Coding Theory. Berlekamp's idea was later generalized to the multiplication of two arbitrary elements by Fumy [121] and by McEliece [263]. A detailed discussion of several types of dual basis multipliers may be found in the monograph by Gollmann [141].

Berlekamp's method makes use of the representation of elements of $E = \mathrm{GF}(q^n)$ in primal and dual coordinates already introduced in Definition 7.7.1. Indeed, the observations in Lemma 7.7.2 naturally lead to a hardware device for multiplying a variable element ξ by the fixed element β, where β generates a polynomial basis for E over $F = \mathrm{GF}(q)$, at least in the binary case $q = 2$. To this purpose, we first note that Lemma 7.7.2 gives the following formula for the dual coordinates of ξ with respect to the polynomial basis $B = (1, \beta, \dots, \beta^{n-1})$ considered here:

$$r_C(\xi) = ((\xi)_0, \dots, (\xi)_{n-1}) = \left(\mathrm{Tr}(\xi), \mathrm{Tr}(\xi\beta), \dots, \mathrm{Tr}(\xi\beta^{n-1})\right),$$

where C denotes the dual basis[2] of B and where Tr again is the trace mapping for E/F. Applying this observation also to $\beta\xi$ instead of ξ, it is easy to compute $\beta\xi$ from ξ in dual coordinates as follows:

$$(\beta\xi)_i = \mathrm{Tr}(\beta\xi\beta^i) = \mathrm{Tr}(\xi\beta^{i+1}) = (\xi)_{i+1} \quad \text{for } i = 0, \dots, n-2, \tag{8.4}$$

and

$$\begin{aligned} (\beta\xi)_{n-1} &= \mathrm{Tr}(\xi\beta^n) = \mathrm{Tr}\left(a_0\xi + \dots + a_{n-1}\xi\beta^{n-1}\right) \\ &= a_0(\xi)_0 + \dots + a_{n-1}(\xi)_{n-1}, \end{aligned} \tag{8.5}$$

where $m_\beta = x^n + a_{n-1}x^{n-1} + \dots + a_1x + a_0$ is the minimal polynomial of β – that is, the irreducible polynomial used to construct E.

In the binary case $q = 2$, the preceding observations suggest the hardware device drawn in Figure 8.1. The main part of this circuit is a linear feedback shift register

[2] In the context of polynomial bases, it will be more convenient to label the basis elements with the indices (or exponents) $0, \dots, n-1$ instead of $1, \dots, n$.

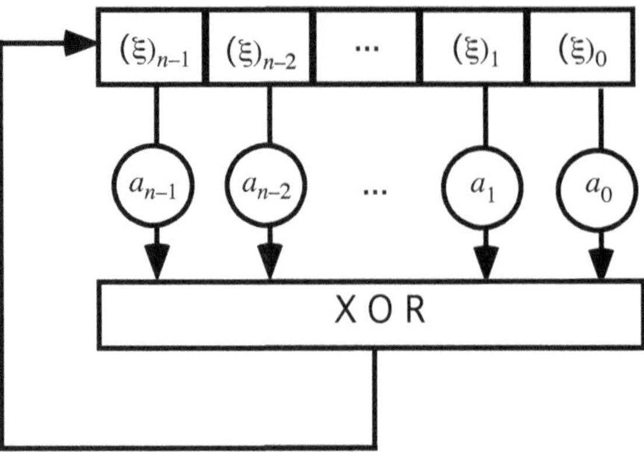

Fig. 8.1 Multiplication by β for $q = 2$

of length n originally holding the dual coordinate vector $r_C(\xi)$ of ξ. Here a **linear feedback shift register** (for short, an **LFSR**) of **length** n is a time-dependent device (running on a clock) of n cells each capable of holding a value from F, so that with each clock cycle the contents of the cells are shifted cyclically by one position (to the right, say). Thus the LFSR outputs the rightmost entry $(\xi)_0$ and replaces it with $(\xi)_1$, it replaces $(\xi)_1$ with $(\xi)_2$ etc., and it finally computes the new leftmost entry as the linear function $a_0(\xi)_0 + \cdots + a_{n-1}(\xi)_{n-1}$ of the initial **state vector** $((\xi)_0, (\xi)_1, \ldots, (\xi)_{n-1})$ and the **feedback coefficients** $(a_0, a_1, \ldots, a_{n-1})$.

In Figure 8.1, the box with the entry XOR stands for an adder modulo 2 (which can be realized by using an appropriate number of XOR-gates), and the circle with entry a_i indicates multiplication with a_i. Note that multiplying two elements of GF(2) could be realized by an AND-gate. Of course, there is no need to actually use such gates in the binary case, since one may also achieve the desired effect just by the presence or absence of a wire, depending on whether $a_i = 1$ or $a_i = 0$.[3]

In view of Equations (8.4) and (8.5), it is clear that the shift register will hold the dual coordinate vector $r_C(\beta\xi)$ after the first clock cycle. More generally, after j clock cycles, the shift register will hold the dual coordinate vector $r_C(\beta^j\xi)$. Also, the device first outputs the given dual coordinate $(\xi)_0$ and then, successively, the dual coordinates of $\beta\xi$ in the next n steps.

Next, let us see how one may use the basic device of Figure 8.1 to implement the multiplication of an arbitrary element ξ given in dual coordinates with an arbitrary element η given in primal coordinates $(y_0, \ldots, y_{n-1})$; the product $\pi = \xi\eta$ will be obtained in dual coordinates. The basic observation required is the following symmetry property for dual coordinates, which is an immediate consequence of Lemma 7.7.2:

$$(\beta^j\xi)_i = \mathrm{Tr}(\xi\beta^{i+j}) = (\beta^i\xi)_j \quad \text{for } i, j = 0, \ldots, n-1.$$

[3] Linear feedback shift registers over an arbitrary finite field will be studied in detail in Chapter 9.

Using this symmetry property, we compute

$$(\pi)_j = (\xi\eta)_j = \left(\xi \cdot \sum_{i=0}^{n-1} y_i\beta^i\right)_j = \sum_{i=0}^{n-1} y_i(\beta^i\xi)_j = \sum_{i=0}^{n-1} y_i(\beta^j\xi)_i. \tag{8.6}$$

As we already know how to compute the dual coordinate vectors

$$r_C(\beta\xi),\ r_C(\beta^2\xi),\ \ldots,\ r_C(\beta^j\xi),\ \ldots$$

successively with the circuit in Figure 8.1, Equation (8.6) suggests a way to also obtain the dual coordinates of π one by one; a corresponding circuit is given in Figure 8.2. Here the LFSR is again loaded with the dual coordinates of ξ, while the second register is loaded with the primal coordinates of η which are to be kept fixed throughout the computation.

According to Equation (8.6) with $j = 0$, this circuit originally produces the dual coordinate $(\pi)_0$. After one clock cycle, the LFSR holds the dual coordinates of $(\beta\xi)$, and hence Equation (8.6) with $j = 1$ shows that the circuit now yields the dual coordinate $(\pi)_1$, etc. The use of a second register (the contents of which are *not* shifted) allows to use the device in Figure 8.2 for variable input vectors η; if only multiplication by a fixed constant η is required, we can, of course, simplify the circuit by just using an XOR-block corresponding to the subset of the indices i with $y_i \neq 0$.

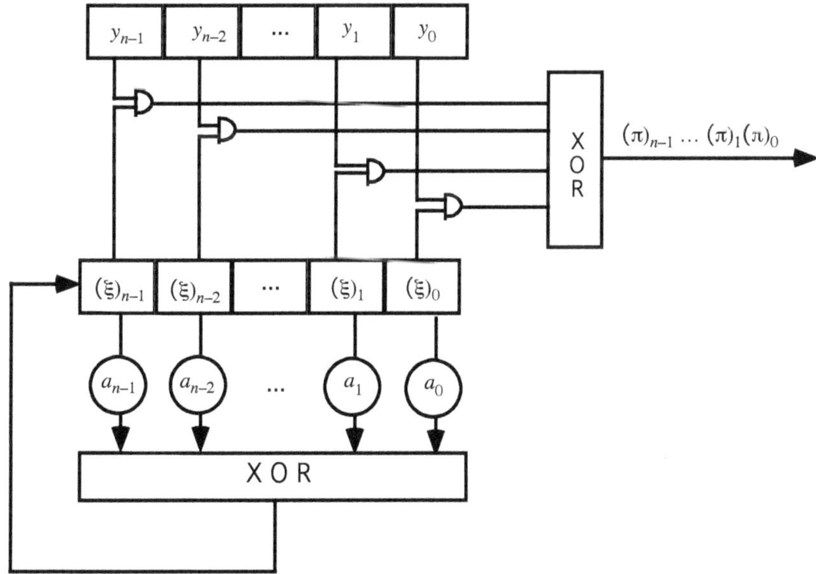

Fig. 8.2 Bit-serial dual basis multiplier

As the input of the coordinates of ξ and η is done in parallel, while the coordinates of π are computed bitwise (in successive clock cycles), an architecture as in Figure 8.2 is called a **bit-serial multiplier**. There are also **bit-parallel multipliers** where the input is bitwise, while the output is in parallel. For such an architecture using dual bases, one loads one shift register with the dual coordinates of ξ, whereas the input of the primal coordinates of η is done bitwise (with successive clock cycles); at the end of the computation (after $n-1$ clock cycles), a further register holds the dual coordinates of π. We refer the reader to Fumy [121] and Gollmann [141] who also discusses both bit-serial and bit-parallel multipliers based on representing both inputs in coordinates with respect to a polynomial basis.

Example 8.2.1. Let us give a specific example for the case $n = 6$. By Proposition 3.6.16, the cyclotomic polynomial $\Phi_9 = x^6 + x^3 + 1$ is irreducible over GF(2). A bit-serial dual basis multiplier corresponding to the polynomial basis generated by a root β of Φ_9 is given by the circuit in Figure 8.3. □

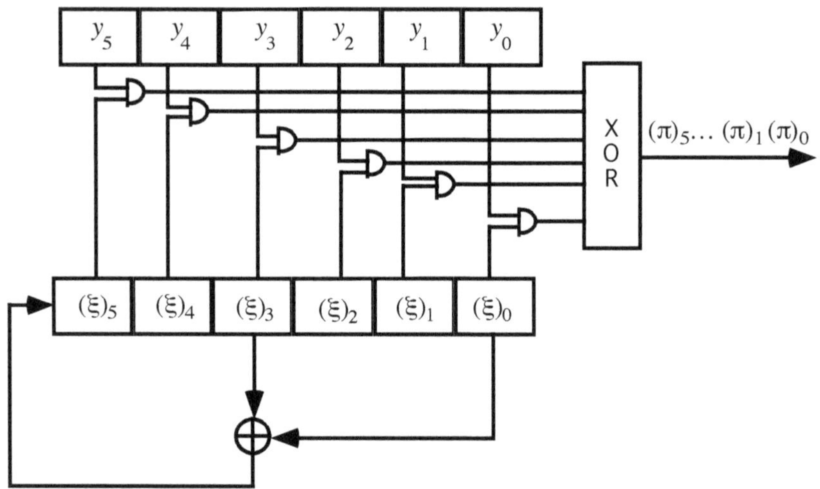

Fig. 8.3 Bit-serial dual basis multiplier for GF(64) corresponding to Φ_9

There remains one difficulty in using such **dual basis multipliers**: one needs to work with coordinate representations with respect to two different bases, which makes it necessary to be able to transform between primal and dual coordinates. The transformation from primal to dual coordinates can in fact be achieved in a comparatively simple way if one uses Lemma 7.7.2 to pre-compute the representation

$$r_C(1) = \left(\mathrm{Tr}(1), \mathrm{Tr}(\beta), \ldots, \mathrm{Tr}(\beta^{n-1})\right)$$

of the unit element 1 in dual coordinates, since the product $\xi = \xi \cdot 1$ – which may be computed using the given dual basis multiplier – then gives the dual coordinates $r_C(\xi)$ of ξ.

Unfortunately, the reverse transformation (from dual to primal coordinates) cannot be performed in such a simple manner and requires special circuitry which tends to make this approach complex and inefficient. Nevertheless, one may obtain simple circuits for specific examples; see, for instance, McEliece [263].

The obvious idea for overcoming these difficulties is to search for polynomial bases which coincide with their duals – then there would be no need at all for any coordinate transformations. If this approach turns out to be impossible (as it will), one may still search for polynomial bases leading to particularly simple types of coordinate transformations. These ideas motivate the investigation of self-dual polynomial bases and of somewhat weakened concepts in the next section.

8.3 Duality Theory of Polynomial Bases

We begin with the following explicit description of the dual basis of a given polynomial basis in terms of the corresponding minimal polynomial. Though the result in question may at first glance seem to have the character of a curiosity, we shall see that it has very interesting consequences. It follows from a more general theorem in Galois Theory; see, for instance, Artin [12]. However, in the special case of finite fields there is a particularly simple proof due to Imamura [201] which we will present now.

Theorem 8.3.1. *Let $B = (1, \beta, \dots, \beta^{n-1})$ be a polynomial basis for $E = \mathrm{GF}(q^n)$ over $F = \mathrm{GF}(q)$, and assume that the minimal polynomial $f = m_\beta$ of β splits over E as*

$$f = (x - \beta)(\alpha_0 + \alpha_1 x + \cdots + \alpha_{n-1} x^{n-1}). \tag{8.7}$$

Then the dual basis $C = (\gamma_0, \dots, \gamma_{n-1})$ of B satisfies

$$\gamma_i = \frac{\alpha_i}{f'(\beta)} \quad \text{for } i = 0, \dots, n-1, \tag{8.8}$$

where f' is the formal derivative of f.

Proof. The matrix M_B defined in Lemma 3.13.16 here takes the form

$$M_B = \begin{pmatrix} 1 & \beta & \dots & \beta^{n-1} \\ 1 & \beta^q & \dots & \beta^{(n-1)q} \\ \vdots & \vdots & \ddots & \vdots \\ 1 & \beta^{q^{n-1}} & \dots & \beta^{(n-1)q^{n-1}} \end{pmatrix}.$$

Since C is the dual basis of B, we have $M_B^T M_C = I$ by Lemma 3.13.16, and hence also $M_C M_B^T = I$. We now compute the entries in the first row of $M_C M_B^T$ and obtain

$$g(\beta^{q^j}) = \delta_{0j} \quad \text{for } j = 0, \dots, n-1,$$

where the polynomial g is defined by

$$g(x) = \gamma_0 + \gamma_1 x + \cdots + \gamma_{n-1} x^{n-1}. \tag{8.9}$$

By using Lagrange interpolation (see Observation 2.5.4), one derives the following formula for g:

$$g(x) = \frac{(x-\beta^q)(x-\beta^{q^2})\cdots(x-\beta^{q^{n-1}})}{(\beta-\beta^q)(\beta-\beta^{q^2})\cdots(\beta-\beta^{q^{n-1}})} = \frac{h(x)}{h(\beta)}, \tag{8.10}$$

where $h(x) = (x-\beta^q)(x-\beta^{q^2})\cdots(x-\beta^{q^{n-1}})$. On the other hand, $f = (x-\beta)h$, since the roots of f are the conjugates of β. This implies $f' = h + (x-\beta)h'$, and hence $f'(\beta) = h(\beta)$. Therefore (8.10) may be written as

$$g(x) = \frac{f(x)}{(x-\beta)f'(\beta)}$$

which – in view of (8.7) and (8.9) – proves the validity of the assertion (8.8). □

As we have seen in Section 8.2, it is desirable to find polynomial bases leading to particularly simple coordinate transformations. Unfortunately, the ideal case – namely, that of self-dual polynomial bases – turns out to be impossible. This will be an immediate consequence of the following result, which solves the more general question when the dual of a polynomial basis is again polynomial; cf. Gollmann [141] and Geiselmann and Gollmann [138].

Theorem 8.3.2. *Let $B = (1, \beta, \dots, \beta^{n-1})$ be a polynomial basis of $E = \mathrm{GF}(q^n)$ over $F = \mathrm{GF}(q)$, where $n \geq 2$. Then the dual basis C of B is likewise a polynomial basis if and only if the minimal polynomial m_β of β is a binomial and $n \equiv 1 \bmod p$, where p denotes the characteristic of E.*

Proof. First assume that C is a polynomial basis for E/F, say $C = (1, \gamma, \dots, \gamma^{n-1})$. Then

$$\mathrm{Tr}(\beta^i \gamma^j) = \delta_{ij} \quad \text{for } i, j = 0, \dots, n-1.$$

In particular,

$$\mathrm{Tr}(\beta^i) = \mathrm{Tr}(\gamma^i) = \begin{cases} 1 \text{ for } i = 0 \\ 0 \text{ for } i = 1, \dots, n-1. \end{cases} \tag{8.11}$$

Next we compute $\mathrm{Tr}(\beta^n)$ and $\mathrm{Tr}(\gamma^n)$. Let us denote the constant coefficients of the minimal polynomials m_β and m_γ by b_0 and c_0, respectively. Then (8.11) immediately gives

$$\mathrm{Tr}(\beta^n) = -b_0 \quad \text{and} \quad \mathrm{Tr}(\gamma^n) = -c_0. \tag{8.12}$$

By (8.11), one has $\mathrm{Tr}(1) = 1$. On the other hand, $1 \in F$ shows $\mathrm{Tr}(1) = n$, and hence indeed $n \equiv 1 \bmod p$. In particular, $n \geq 3$.

We now use Lemma 7.7.2 to represent γ and γ^2 with respect to B. In view of (8.11) and (8.12), we obtain

$$\gamma = \sum_{i=0}^{n-1} \mathrm{Tr}(\gamma\gamma^i)\beta^i = -c_0\beta^{n-1} \tag{8.13}$$

and

$$\gamma^2 = \sum_{i=0}^{n-1} \mathrm{Tr}(\gamma^2\gamma^i)\beta^i = -c_0\beta^{n-2} + \mathrm{Tr}(\gamma^{n+1})\beta^{n-1}. \tag{8.14}$$

Because of $n \geq 3$, we can determine $\mathrm{Tr}(\gamma^{n+1})$ from (8.11) and (8.14) as follows:

$$0 = \mathrm{Tr}(\beta\gamma^2) = -c_0\mathrm{Tr}(\beta^{n-1}) + \mathrm{Tr}(\gamma^{n+1})\mathrm{Tr}(\beta^n) = -b_0\mathrm{Tr}(\gamma^{n+1}),$$

that is, $\mathrm{Tr}(\gamma^{n+1}) = 0$. Substituting this in (8.14), we obtain from (8.13)

$$c_0^2\beta^{2n-2} = \gamma^2 = -c_0\beta^{n-2},$$

hence $\beta^n = -c_0^{-1}$. Thus the minimal polynomial m_β of β has to divide $x^n - c$, where $c := -c_0^{-1}$. As m_β is an irreducible polynomial of degree n, we conclude $m_\beta = x^n - c$, so that m_β indeed is a binomial.

Conversely, let $n \equiv 1 \bmod p$ and assume that m_β is a binomial, say $m_\beta = x^n - c$. We apply Theorem 8.3.1 to compute the dual basis $C = (\gamma_0, \ldots, \gamma_{n-1})$ of B. Dividing m_β by $x - \beta$ gives

$$m_\beta = (x-\beta)(x^{n-1} + \beta x^{n-2} + \cdots + \beta^{n-2}x + \beta^{n-1}).$$

Thus, in the notation of Theorem 8.3.1,

$$\alpha_j = \beta^{n-1-j} \quad \text{for } j = 0, \ldots, n-1.$$

Using this and $n \equiv 1 \bmod p$, we obtain

$$\gamma_j = \frac{\beta^{n-1-j}}{f'(\beta)} = \frac{\beta^{n-1-j}}{n\beta^{n-1}} = \beta^{-j} \quad \text{for } j = 0, \ldots, n-1, \tag{8.15}$$

so that C is in fact the polynomial basis generated by $\gamma := \beta^{-1}$. □

Before we characterize those extensions E/F which actually admit a dual pair of polynomial bases, we note two simple but important consequences of Theorem 8.3.2 [201, 141]. The first of these is an immediate consequence of (8.15), and the second one follows from the trivial fact that there are no non-linear irreducible binomials over GF(2).

Corollary 8.3.3. *Let $n \geq 2$. Then* $\mathrm{GF}(q^n)$ *does not admit a self-dual polynomial basis over* $\mathrm{GF}(q)$. □

Corollary 8.3.4. *Let $B = (1, \beta, \dots, \beta^{n-1})$ be a polynomial basis for* $\mathrm{GF}(2^n)$ *over* $\mathrm{GF}(2)$, *where $n \geq 2$. Then the dual basis B^* of B cannot be polynomial.* □

Theorem 8.3.5. *There exists a dual pair of polynomial bases for $E = \mathrm{GF}(q^n)$ over $F = \mathrm{GF}(q)$, where $n \geq 2$ and where the characteristic of E equals p, if and only if the following three conditions are satisfied:*

(1) $n \equiv 1 \bmod p$;

(2) *every prime r dividing n also divides $q-1$;*

(3) $n \equiv 0 \bmod 4$ *implies* $q \equiv 1 \bmod 4$.

Proof. By Theorem 8.3.2, the existence of a dual pair of polynomial bases for E/F is equivalent to the validity of condition (1) and the existence of an irreducible binomial of degree n over F. In Theorem 5.2.6, we have obtained necessary and sufficient conditions for a binomial $x^n - c$ (with $c \in F$) to be irreducible. In particular, (3) must be satisfied, and every prime divisor of n has to divide the order e of c and hence also $q-1$.

Conversely, assume that conditions (2) and (3) hold. Then Theorem 5.2.6 shows that the binomial $x^n - \omega$ is irreducible for every primitive element ω of F^*. □

Example 8.3.6. There is no dual pair of polynomial bases for $\mathrm{GF}(3^n)$ over $\mathrm{GF}(3)$. On the other hand, there exists a dual pair of polynomial bases for $\mathrm{GF}(4^n)$ over $\mathrm{GF}(4)$ if and only if n is a power of 3, and there exists a dual pair of polynomial bases for $\mathrm{GF}(5^n)$ over $\mathrm{GF}(5)$ if and only if n is a power of 16. □

Of course, the preceding results are quite disappointing if one wants to construct bit-serial dual basis multipliers, since Corollary 8.3.3 shows that some coordinate transformation is unavoidable; moreover, in view of Corollary 8.3.4, the transformation from dual to primal coordinates cannot even be performed by using another bit-serial dual basis multiplier – whereas the transformation from primal to dual coordinates can be performed with the original multiplier, as noted at the end of Section 8.2.

Nevertheless, there is no reason to give up yet, because it turns out that quite often very simple coordinate transformations are possible. The first result in this direction was given in 1986 by Fumy [121] who proved that an irreducible trinomial over $\mathrm{GF}(2)$ of the special form $x^n + x + 1$ leads to a polynomial basis B for which the dual basis C is simply given by a permutation of B. We will present more general results, which are due to Morii, Kasahara and Whiting [287], Gollmann [141], and Wang and Blake [396]. To do so, we require the following generalization of the concept of self-duality. We then show that this approach can indeed be used to achieve a transformation between dual and primal coordinates which reduces to a permutation.

Definition 8.3.7. A polynomial basis $B = (1, \beta, \ldots, \beta^{n-1})$ for $E = \mathrm{GF}(q^n)$ over $\mathrm{GF}(q)$ with associated dual basis $C = (\gamma_0, \ldots, \gamma_{n-1})$ is called **weakly self-dual** if there exist a permutation σ of $\{0, \ldots, n-1\}$ and an element $\delta \in E$ such that the following condition is satisfied:

$$\gamma_j = \delta \beta^{\sigma(j)} \quad \text{for } j = 0, \ldots, n-1. \tag{8.16}$$

Proposition 8.3.8. *Let $B = (1, \beta, \ldots, \beta^{n-1})$ be a weakly self-dual polynomial basis for $E = \mathrm{GF}(q^n)$ over $\mathrm{GF}(q)$ with respect to the parameters δ and σ, and let $\xi \in E$. Then the dual coordinates of the product $\xi\delta$ are obtained by permuting the primal coordinates of ξ according to σ.*

Proof. Let $\xi \in E$, with primal coordinates $r_B(\xi) = (x_0, \ldots, x_{n-1})$ and dual coordinates $r_C(\xi) = ((\xi)_0, \ldots, (\xi)_{n-1})$. Using (8.16), we obtain

$$\xi = \sum_{i=0}^{n-1} x_i \beta^i = \sum_{j=0}^{n-1} x_{\sigma(j)} \beta^{\sigma(j)} = \sum_{j=0}^{n-1} x_{\sigma(j)} \delta^{-1} \gamma_j = \delta^{-1} \cdot \sum_{j=0}^{n-1} x_{\sigma(j)} \gamma_j,$$

and therefore

$$\sum_{j=0}^{n-1} x_{\sigma(j)} \gamma_j = \xi\delta, \tag{8.17}$$

as desired. □

It is not difficult to check that the converse of Proposition 8.3.8 also holds, see Exercise 8.3.26. We now explain how weakly self-dual polynomial bases may be used to design efficient bit-serial multipliers.

Observation 8.3.9. In the situation of Proposition 8.3.8, let η be a further element of E given in primal coordinates, say $r_B(\eta) = (y_0, \ldots, y_{n-1})$. We want to compute the product $\pi := \xi\eta$.

Note that the dual basis multiplier associated with B as in Section 8.2 can be used to compute the product $\pi\delta = (\xi\eta)\delta = (\xi\delta)\eta$ (in dual coordinates) from $\xi\delta$ (in dual coordinates) and η (in primal coordinates). By Equation (8.17), the dual coordinates of $\xi\delta$ can be obtained by just permuting the given primal coordinates of ξ according to σ. Applying the same formula to $\pi\delta$ instead of ξ, we may then obtain the primal coordinates of the product $\pi = \xi\eta$ by permuting the dual coordinates of $\pi\delta$ according to σ^{-1}.

This gives the desired method for building a dual basis multiplier which takes two elements given in primal coordinates as input and outputs their product again in primal coordinates, and for which the necessary coordinate transformations reduce to permutations. Moreover, as we will see soon, the permutations arising are in fact very simple ones. With this approach, the dual basis multiplier of Figure 8.2 takes the form displayed in Figure 8.4. □

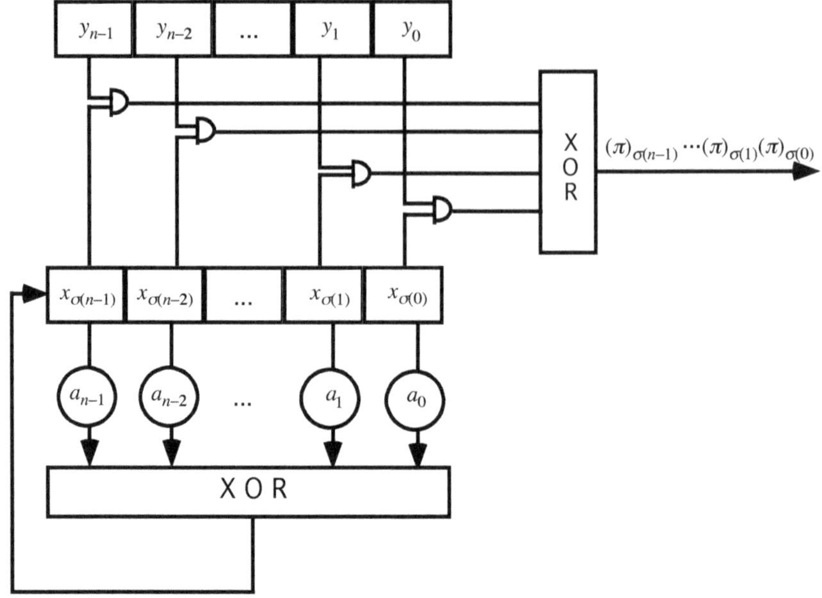

Fig. 8.4 Generalized bit-serial multiplier with respect to a weakly self-dual basis

Example 8.3.10. As in Example 8.2.1, we consider the cyclotomic polynomial $\Phi_9 = x^6 + x^3 + 1$ over $\mathrm{GF}(2)$. A bit-serial dual basis multiplier corresponding to the polynomial basis $B = \{1, \beta, \ldots, \beta^5\}$ generated by a root β of Φ_9 was given by the circuit in Figure 8.3. We now show that B is actually a weakly self-dual basis by verifying condition (8.16) with $\delta = \beta^4$ and $\sigma = (0,2)(3,5)$. To this end, we define elements γ_j of E as follows:

$$(\gamma_0, \ldots, \gamma_5) = (\beta^6, \beta^5, \beta^4, 1, \beta^8, \beta^7).$$

One easily sees $\mathrm{Tr}(\beta^3) = \mathrm{Tr}(\beta^6) = 1$ and $\mathrm{Tr}(1) = 0$; moreover,

$$\mathrm{Tr}(\beta) = \mathrm{Tr}(\beta^2) = \mathrm{Tr}(\beta^4) = \mathrm{Tr}(\beta^8) = \mathrm{Tr}(\beta^7) = \mathrm{Tr}(\beta^5) = 0,$$

by Propositions 3.12.2 and 3.12.3. Now it is not difficult to check that the elements γ_j defined above indeed satisfy the condition $\mathrm{Tr}(\beta^i \gamma_j) = \delta_{ij}$ for $i, j = 0, \ldots, 5$ and therefore form the dual basis C of B. The corresponding dual basis multiplier is given in Figure 8.5, where the p_i denote the primal coordinates of the product $\pi = \xi\eta$. □

Observation 8.3.9 was the motivation for investigating the existence of weakly self-dual polynomial bases, which was done in the papers already cited above. We now present the relevant results, concentrating on criteria for weak self-duality. Let us begin with the particularly simple case of binomials.

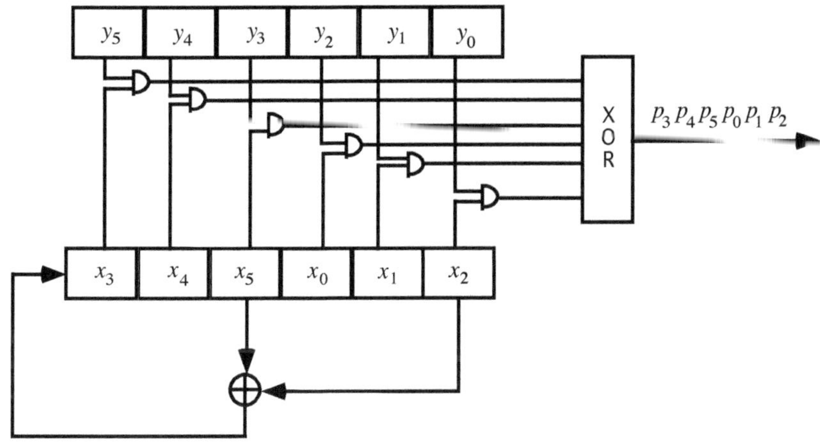

Fig. 8.5 Bit-serial multiplier for the weakly self-dual basis associated with $\Phi_9 = x^6 + x^3 + 1$

Observation 8.3.11. Let $B = (1, \beta, \ldots, \beta^{n-1})$ be a polynomial basis for $\mathrm{GF}(q^n)$ over $\mathrm{GF}(q)$, and assume that the minimal polynomial $f = m_\beta$ of β is a binomial, say $f = x^n - a$. We divide m_β by $x - \beta$ and apply Theorem 8.3.1 to compute the dual basis $(\gamma_0, \ldots, \gamma_{n-1})$ of B. This results in

$$\gamma_j = \delta\beta^{n-1-j}, \quad \text{where } \delta := \frac{1}{f'(\beta)} = \frac{1}{n\beta^{n-1}}, \tag{8.18}$$

(for $j = 0, \ldots, n-1$) and shows that B is always weakly self-dual in this case. The details are left to the reader as Exercise 8.3.27. □

We now prove the following major result: except for the binomials, the only other irreducible polynomials leading to weakly self-dual polynomial bases are trinomials of a special type. This will require a little more effort.

Theorem 8.3.12. *A polynomial basis $B = (1, \beta, \ldots, \beta^{n-1})$ for $\mathrm{GF}(q^n)$ over $\mathrm{GF}(q)$ is weakly self-dual (for suitable parameters δ and σ) if and only if the minimal polynomial m_β of β is either a binomial or a trinomial with constant term -1.*

Moreover, if $m_\beta(x)$ is the trinomial $x^n + ax^k - 1$, then

$$\delta = \frac{1}{\beta^k \cdot m'_\beta(\beta)} \quad \text{and} \quad \gamma_j = \delta\beta^{\sigma(j)} \text{ with } \sigma(j) = k - 1 - j \bmod n \tag{8.19}$$

for all j; in particular, $\gamma_{k-1} = \delta$.

Proof. In view of Observation 8.3.11, we may assume that $f = m_\beta$ is not a binomial. Then we may write f in the form

$$f = x^n + ax^k - r(x) \quad \text{with } a, k \neq 0 \text{ and } \deg r < k < n.$$

First let B be weakly self-dual. Then the assertion amounts to proving $r(x)=1$. As before, we divide m_β by $x-\beta$ and apply Theorem 8.3.1. This gives

$$f=(x-\beta)\Big(x^{n-1}+\beta x^{n-2}+\cdots+\beta^{n-k-1}x^k+(a+\beta^{n-k})x^{k-1}+\text{ lower terms}\Big).$$

Using the notation of Theorem 8.3.1, this means

$$\alpha_{n-1-j}=\beta^j \quad \text{for } j=0,\dots,n-k-1 \quad \text{and} \quad \alpha_{k-1}=\beta^{n-k}+a.$$

Put $i:=\sigma(n-1)$. Then Theorem 8.3.1 and condition (8.16) imply successively

$$\begin{aligned}
\gamma_{n-1}&=\frac{\alpha_{n-1}}{f'(\beta)}=\frac{1}{f'(\beta)}=\delta\beta^i,\\
\gamma_{n-2}&=\frac{\alpha_{n-2}}{f'(\beta)}=\frac{\beta}{f'(\beta)}=\delta\beta^{i+1},\\
&\vdots\\
\gamma_k&=\frac{\alpha_k}{f'(\beta)}=\frac{\beta^{n-k-1}}{f'(\beta)}=\delta\beta^{i+n-k-1},\\
\gamma_{k-1}&=\frac{\alpha_{k-1}}{f'(\beta)}=\frac{\beta^{n-k}+a}{f'(\beta)}=\delta(\beta^{i+n-k}+a\beta^i).
\end{aligned}$$

Since the elements $\delta^{-1}\gamma_j$ all belong to the polynomial basis B, a minute's thought shows that these equations force $i=k$ and $1=\delta^{-1}\gamma_{k-1}=\beta^n+a\beta^k=r(\beta)$, which implies $r(x)=1$ (since the minimal polynomial f of β has degree $n>\deg r$). Therefore $f=x^n+ax^k-1$, as claimed.

Conversely, assume $f=m_\beta=x^n+ax^k-1$, where $k\neq 0$. Again, we divide m_β by $x-\beta$. A short computation yields $f=(x-\beta)(\alpha_0+\alpha_1x+\cdots+\alpha_{n-1}x^{n-1})$ with

$$\alpha_{n-1-j}=\beta^j \quad \text{for } j=0,\dots,n-k-1$$

and

$$\alpha_{k-1-j}=\beta^{n-k+j}+a\beta^j \quad \text{for } j=0,\dots,k-1.$$

Hence Theorem 8.3.1 gives (with the index substitution $i:=k-1-j$)

$$\begin{aligned}
\gamma_i&=\frac{\alpha_i}{f'(\beta)}=\frac{\beta^{n-1-i}+a\beta^{k-1-i}}{f'(\beta)}=\frac{\beta^n+a\beta^k}{\beta^{i+1}f'(\beta)}\\
&=\frac{1}{\beta^{i+1}f'(\beta)}=\frac{\beta^{k-1-i}}{\beta^kf'(\beta)} \quad \text{for } i=0,\dots,k-1
\end{aligned}$$

and similarly (with the index substitution $i:=n-1-j$),

$$\gamma_i = \frac{\alpha_i}{f'(\beta)} = \frac{\beta^{n-1-i}}{f'(\beta)} = \frac{\beta^{n-1+k-i}}{\beta^k f'(\beta)} \quad \text{for } i = k, \ldots, n-1.$$

With $\delta := 1/(\beta^k f'(\beta))$, these observations show

$$(\gamma_0, \ldots, \gamma_{n-1}) = (\delta\beta^{k-1}, \delta\beta^{k-2}, \ldots, \delta\beta, \delta, \delta\beta^{n-1}, \delta\beta^{n-2}, \ldots, \delta\beta^k). \tag{8.20}$$

This proves that B is indeed weakly self-dual and also establishes the validity of Equation (8.19). □

For the actual construction of dual basis multipliers, it is worthwhile to state the following consequence of Theorem 8.3.12 for the binary case explicitly:

Theorem 8.3.13. *Let $f = x^n + x^k + 1$ be an irreducible trinomial over* GF(2)*, let β be a root of f, and let $C = (\gamma_0, \ldots, \gamma_{n-1})$ be the dual basis of the polynomial basis B generated by β. Then B is weakly self-dual with parameters as described in Equation* (8.19)*, and γ_k is given explicitly as follows:*

$$\gamma_k = \begin{cases} \beta^{n-k} + 1 & \text{if } n \text{ is odd and } k \text{ is even}, \\ \beta^{n-k}(\beta^{n-k} + 1) & \text{if } n \text{ is even and } k \text{ is odd}, \\ \beta^{n-k} & \text{if both } n \text{ and } k \text{ are odd}. \end{cases} \tag{8.21}$$

Proof. In view of Theorem 8.3.12, it only remains to establish the value of γ_k. Equation (8.19) gives

$$\gamma_k = \delta\beta^{\sigma(k)} = \delta\beta^{n-1} = \frac{\beta^{n-k-1}}{f'(\beta)},$$

so that the representation of γ_k will depend on the parity of n and k. For instance, when n is odd and k is even, we obtain

$$\gamma_k = \frac{\beta^{n-k-1}}{\beta^{n-1}} = \frac{1}{\beta^k} = \frac{\beta^n + \beta^k}{\beta^k} = \beta^{n-k} + 1.$$

We leave it to the reader to check the remaining two cases of (8.21). □

As a further consequence of practical interest, we characterize those polynomial bases B for GF(2^n) over GF(2) for which the dual basis is actually just a permutation of B.

Corollary 8.3.14. *Let $B = (1, \beta, \ldots, \beta^{n-1})$ be a polynomial basis for* GF(2^n) *over* GF(2)*. Then the dual basis C of B is a permutation of B if and only if n is odd and the minimal polynomial $f = m_\beta$ of β is the trinomial $f = x^n + x + 1$.*

Proof. Note that we are asking for a weakly self-dual basis B with $\delta = 1$. By Theorem 8.3.12, f necessarily is an irreducible trinomial, say $f = x^n + x^k + 1$, and we have to check when the additional condition $\delta = 1$ is satisfied. In view of Equation (8.19), this is equivalent to requiring $\beta^k f'(\beta) = 1$. This gives the condition

$$\beta^n + \beta^k = 1 = n\beta^{n+k-1} + k\beta^{2k-1},$$

that is,

$$1 + \beta^{n-k} = n\beta^{n-1} + k\beta^{k-1}.$$

As B is a basis, this holds if and only if n is odd and $k = 1$. □

The preceding results show that the weakly self-dual polynomial bases are those polynomial bases for which the transformation between primal and dual coordinates can be realized with a minimum of complexity. If one wants to build dual basis multipliers, the irreducible polynomials one should use are therefore (if available) the binary irreducible trinomials. Fortunately, these exist in many (though not all) cases:

Remark 8.3.15. There are extensive lists of irreducible trinomials over GF(2), indicating that weakly self-dual polynomial bases of GF(2^n) over GF(2) are quite common. A table of irreducible binary polynomials of lowest weight up to degree $n = 1025$ can be found in Section 2.2.1 of the Handbook of Finite Fields [292], and – as mentioned there – more extensive tabulations (up to $n = 120{,}000$) have been made. These tables support the interesting conjecture of Seroussi [347] that there exists an irreducible binary trinomial or an irreducible binary pentanomial for all extension degrees n.

As for non-existence results, there are no irreducible binary trinomials with degree $n \equiv 0 \bmod 8$, since $x^n + x^k + 1$ then splits into an even number of irreducible factors; this follows from a result of Swan [369], see also Corollary 6.696 in [30]. Moreover, there are quite a few further values of n for which there is no irreducible binary trinomial: up to 50, these are $n = 13, 19, 26, 27, 34, 37, 38, 42, 43, 44, 45, 50$. Unfortunately, there is no theoretical result which would explain these cases. □

In view of Remark 8.3.15, it is an interesting problem to generalize the previous approach (at least for $q = 2$) in order to deal with cases where no irreducible trinomial is available. As we shall see, special pentanomials can also be very useful.

Let us again consider a polynomial basis $B = (1, \beta, \ldots, \beta^{n-1})$ and a scalar multiple $\frac{1}{\delta} \cdot C$ of its dual basis C. According to Definition 8.3.7, B is a weakly self-dual basis if and only if the coordinate transformation from $\frac{1}{\delta} \cdot C$ to B is just a permutation; equivalently, we require the matrix S associated with this change of basis to have weight n (where, as usual, the **weight** $w(S)$ of a matrix S is the number of non-zero entries). If no weakly self-dual basis exists, we still would like to find a polynomial basis B and a suitable element δ for which the associated transformation matrix S is as simple as possible, meaning it should have the lowest possible weight. Trivially, the minimum weight is always at least n, which motivates the following definition.

Definition 8.3.16. Let S be an invertible (n,n)-matrix. The **excess** of S is defined to be $e(S) := w(S) - n$. □

Now let S be a binary matrix, and assume that S describes the coordinate transformation from a given basis B for GF(q^n) over GF(q) to the basis $\frac{1}{\delta} \cdot C$, where C is the dual basis of B. Then $e(S)$ is the number of XOR-gates required for computing the primal coordinates from the **generalized dual coordinates**, that is, the coordinates with respect to $\frac{1}{\delta} \cdot C$, with a hardware device.

In analogy to the case of weakly self-dual bases, one may then perform **generalized bit-serial multiplication** by computing the product $\pi\delta = (\xi\eta)\delta = (\xi\delta)\eta$ (in dual coordinates) from $\xi\delta$ (in dual coordinates) and η (in primal coordinates). Equivalently, one may think of working with ξ and π in generalized dual coordinates.

At this point, an example should be helpful; the following one is taken from [287].

Example 8.3.17. Consider the irreducible polynomial $f = x^8 + x^4 + x^3 + x^2 + 1$ over $F = \mathrm{GF}(2)$, and let β be a root of f. Then $f = (x - \beta)g$, where

$$g = \beta^{-1} + \beta^{-2}x + (1 + \beta + \beta^5)x^2 + (1 + \beta^4)x^3 + \beta^3 x^4 + \beta^2 x^5 + \beta x^6 + x^7.$$

By Theorem 8.3.1, the dual basis $C = (\gamma_0, \ldots, \gamma_{n-1})$ of the polynomial basis B defined by β is obtained by dividing the coefficients of g by $f'(\beta) = \beta^2$. We now choose $\delta := \left(\beta^3 f'(\beta)\right)^{-1} = \beta^{-5}$ and obtain $\frac{1}{\delta} \cdot C = (\delta_0, \ldots, \delta_{n-1})$, where

$$\begin{aligned} &\delta_0 = \beta^2,\ \delta_1 = \beta,\ \delta_2 = 1 + \beta^2,\ \delta_3 = \beta^3 + \beta^7, \\ &\delta_4 = \beta^6,\ \delta_5 = \beta^5,\ \delta_6 = \beta^4, \text{ and } \delta_7 = \beta^3. \end{aligned} \tag{8.22}$$

As noted at the end of Section 8.2, the transformation from primal to dual coordinates presents no real problem, since it can always be done by using the corresponding dual basis multiplier after precomputing the dual coordinates of 1. Similarly, precomputing the dual coordinates of δ – that is, the generalized dual coordinates of 1 – makes it possible to use the associated dual basis multiplier for computing the dual coordinates of $\xi\delta$ – that is, the generalized dual coordinates of ξ – from the primal coordinates of ξ. In our example, we get $1 = \delta_0 + \delta_2$. Here the coordinate change is so simple that it might as well be performed directly; indeed, (8.22) immediately yields the inverse coordinate transformation

$$\begin{aligned} &1 = \delta_0 + \delta_2,\ \beta = \delta_1,\ \beta^2 = \delta_0,\ \beta^3 = \delta_7, \\ &\beta^4 = \delta_6,\ \beta^5 = \delta_5,\ \beta^6 = \delta_4 \text{ and } \beta^7 = \delta_3 + \delta_7. \end{aligned} \tag{8.23}$$

If we neglect the hardware required to perform the simple coordinate changes described by Equations (8.22) and (8.23), we obtain a generalized bit-serial multiplier based on f as in Figure 8.6. □

In Example 8.3.17, the transformation matrix S from generalized dual to primal coordinates had excess only 2 for our choice of δ, which results in a rather simple transformation. Thus the generalization of Berlekamp's original method for bit-serial multiplication can indeed be quite useful.

However, as Exercise 8.3.29 indicates, the coordinate transformation from dual to primal coordinates may be much more involved for less fortunate choices of δ.

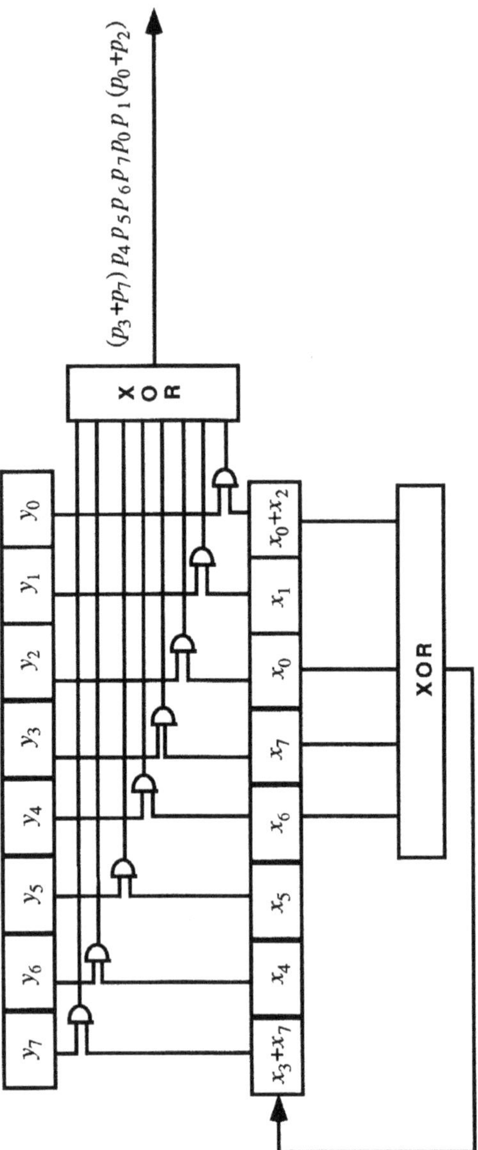

Fig. 8.6 Bit-serial multiplier for the weakly self-dual basis associated with $f = x^8 + x^4 + x^3 + x^2 + 1$

The problem of a suitable choice for δ was studied by Stinson [366] who managed to obtain the following general result for the excess of S in terms of the minimal polynomial m_β of β:

Theorem 8.3.18. *Let $B=(1,\beta,\dots,\beta^{n-1})$ be a polynomial basis for $E=\mathrm{GF}(2^n)$ over $\mathrm{GF}(2)$ with dual basis $C=(\gamma_0,\dots,\gamma_{n-1})$, and let*

$$f=x^n+x^{n_t}+\cdots+x^{n_1}+1 \quad \textit{with } 0<n_1<n_2<\ldots<n_t<n$$

be the minimal polynomial of β (so that t is odd). Put

$$\delta:=\frac{1}{f'(\beta)\beta^{n_s}}, \quad \textit{where } s:=\frac{t+1}{2}.$$

Then the transformation matrix S from $\frac{1}{\delta}\cdot C$ to B has excess

$$e(S)=\sum_{j=s+1}^{t} n_j-\sum_{j=1}^{s-1} n_j.$$

Proof. Using that β is a root of f and writing $n_0:=1$, the reader may check that $f=(x-\beta)(\alpha_0+\alpha_1x+\cdots+\alpha_{n-1}x^{n-1})$, where

$$\alpha_i=\sum_{\{j:n_j\le i\}}\beta^{n_j-i-1}=\beta^{n-i-1}+\sum_{\{j:n_j>i\}}\beta^{n_j-i-1}.$$

By Theorem 8.3.1, the dual basis $C=(\gamma_0,\dots,\gamma_{n-1})$ of the polynomial basis B is obtained by dividing the coefficients of g by $f'(\beta)$. With our choice of δ, we therefore obtain $\frac{1}{\delta}\cdot C=(\delta_0,\dots,\delta_{n-1})$, where

$$\delta_i=\beta^{n_s}\alpha_i=\begin{cases}\sum_{\{j:n_j\le i\}}\beta^{n_s+n_j-i-1} & \text{if } 0\le i\le n_s-1,\\ \beta^{n_s+n-i-1}+\sum_{\{j:n_j>i\}}\beta^{n_s+n_j-i-1} & \text{if } n_s\le i\le n-1.\end{cases} \tag{8.24}$$

One easily checks that all powers of β appearing in (8.24) have exponents j in the range $0\le j\le n-1$. Thus, in order to determine the weight of the associated transformation matrix S, we need to count the total number of powers of β in all n equations in (8.24). Let us write $n_{t+1}:=n$ and $k_i:=\left|\{j:\, n_j\le i\}\right|$ for $i=0,\dots,n-1$. Then (8.24) gives

$$w(S)=\sum_{\{i:i\le n_s-1\}}k_i+\sum_{\{i:i\ge n_s\}}(t+2-k_i). \tag{8.25}$$

Clearly, $k_i=h$ if $n_{h-1}\le i<n_h$, so that

$$\sum_{\{i:i\le n_s-1\}}k_i=n_1-n_0+2(n_2-n_1)+\cdots+s(n_s-n_{s-1})=sn_s-\sum_{j=0}^{s-1}n_j.$$

As $t+1-s=s$, also

$$\sum_{\{i:i\geq n_s\}} (t+2-k_i) = n_{t+1}-n_t+2(n_t-n_{t-1})+\cdots+s(n_{s+1}-n_s) = n-sn_s+\sum_{j=s+1}^{t} n_j.$$

In view of the preceding two equations, (8.25) gives the desired formula for the weight (respectively the excess) of S. □

Two special cases of Theorem 8.3.18 – namely, when f is either a trinomial or a pentanomial of the special form $f=x^n+x^{k+1}+x^k+x^{k-1}+1$ – were already obtained earlier by Morii, Kasahara and Whiting [287]. In view of Seroussi's conjecture that there always exists a binary trinomial or a pentanomial which is irreducible, it is worthwhile to state the following consequence of Theorem 8.3.18 for pentanomials explicitly.

Corollary 8.3.19. *In the case of an irreducible binary pentanomial*

$$f=x^n+x^{n_3}+x^{n_2}+x^{n_1}+1,$$

the associated transformation matrix has excess n_3-n_1. Moreover, the irreducible polynomials leading to a matrix of excess 2 are precisely the pentanomials of the special form $x^n+x^{k+1}+x^k+x^{k-1}+1$. □

Stinson also exhibited irreducible pentanomials of small excess for the seven values $n\leq 30$ for which no irreducible binary trinomial exists. For all but two of these cases, an excess of 2 is possible; for $n=24$ and $n=27$, one can only achieve an excess of 3 and 4, respectively. Moreover, he also gave the following interesting analogue of Corollary 8.3.14 for even values of n. We shall leave the proof as a more demanding exercise to the reader, since it uses similar methods as the proofs of Theorem 8.3.12 and Corollary 8.3.14.

Theorem 8.3.20. *Let $B=(1,\beta,\ldots,\beta^{n-1})$ be a polynomial basis for* GF(2^n) *over* GF(2). *Then the dual basis C of B leads to a transformation matrix of excess 1 (with $\delta=1$) if and only if n is even and the minimal polynomial $f=m_\beta$ of β is the trinomial $f=x^n+x+1$.* □

We conclude this section with some results on polynomial bases corresponding to matrices with small excess over a general field GF(q), which are due to Morgan, Mullen and Živković [284]. These generalize the results for the binary case which we have considered in detail, but seem to have no practical applications. Therefore, we shall omit all proofs.

For $q\neq 2$, a matrix of excess 0 is not necessarily a permutation matrix. This leads to the following generalization of weakly self-dual bases.

Definition 8.3.21. A basis $B=(1,\beta,\ldots,\beta^{n-1})$ for $E=$ GF(q^n) over $F=$ GF(q) with associated dual basis $C=(\gamma_0,\ldots,\gamma_{n-1})$ is called **almost weakly self-dual** if there exist an element $\delta\in E^*$, elements $c_0,\ldots,c_{n-1}\in F^*$, and a permutation σ of $\{0,\ldots,n-1\}$ such that the following condition holds:

$$\gamma_j = c_j \delta \beta^{\sigma(j)} \quad \text{for } j = 0, \ldots, n-1. \tag{8.26}$$

In particular, the almost weakly self-dual bases are precisely those bases which correspond to a transformation matrix S with $e(S) = 0$; note that such a basis is actually weakly self-dual if and only if $c_0 = \cdots = c_{n-1}$. □

Result 8.3.22 *A polynomial basis $B = (1, \beta, \ldots, \beta^{n-1})$ for* GF(q^n) *over* GF(q) *is almost weakly self-dual if and only if the minimal polynomial m_β of β is either a trinomial or a binomial.* □

There are no proper almost weakly self-dual bases which belong to irreducible binomials: in this case, the basis is even weakly self-dual by Theorem 8.3.12. Hence it suffices to consider the case of irreducible trinomials in the following generalization of Theorem 8.3.13.

Result 8.3.23 *Let $f(x) = x^m - ax^k - d$ be an irreducible trinomial over* GF(q)*, let β be a root of f, and $C = (\gamma_0, \ldots, \gamma_{n-1})$ the dual basis of the polynomial basis $B = (1, \beta, \ldots, \beta^{n-1})$ generated by β. Then B is almost weakly self-dual, and Equation* (8.26) *is satisfied with*

$$\delta = \gamma_{k-1} = \frac{d}{\beta^k f'(\beta)}, \qquad c_i = \begin{cases} 1 & \text{if } i \le k, \\ d^{-1} & \text{if } i > k \end{cases}$$

and $\sigma(i) = k - 1 - i \bmod n$ for $i = 1, \ldots, n-1$. □

There also is a generalization of Theorem 8.3.20 which determines the excess $e(S)$ in the general case:

Result 8.3.24 *Let β be a root of a monic irreducible polynomial f of degree n over $F =$* GF(q)*, let $B = (1, \beta, \ldots, \beta^{n-1})$ be the corresponding polynomial basis of* GF(q^n) *over F, and C the dual basis of B. Write*

$$f(x) = x^n + a_{n_t} x^{n_t} + \cdots + a_{n_1} x^{n_1} + a_{n_0} x^{n_0},$$

where $0 = n_0 < n_1 < \ldots < n_t < n$ and $a_i \neq 0$ for $i = 0, \ldots, t$, and put

$$\delta := \frac{1}{f'(\beta)\beta^{n_s}}, \qquad \text{where } s := \lceil t/2 \rceil.$$

Then the transformation matrix S from $\frac{1}{\delta} \cdot C$ to B has excess

$$e(S) = \begin{cases} \sum_{j=s+1}^{t} n_j - \sum_{j=1}^{s-1} n_j & \text{if } t \text{ is odd}, \\ \sum_{j=s+1}^{t} n_j - \sum_{j=1}^{s} n_j & \text{if } t \text{ is even}. \end{cases}$$ □

Finally, we note some particularly interesting consequences of Result 8.3.24:

Corollary 8.3.25. *With the assumptions of Result 8.3.24, the transformation matrix S from $\frac{1}{\delta} \cdot C$ to B has excess*

$$e(S) = \begin{cases} 0 & \text{if } f \text{ is either a trinomial or a binomial,} \\ 1 & \text{if } f \text{ is of the form } f(x) = x^m + a_k x^k + a_{k-1}x^{k-1} + a_0, \\ 2 & \text{if } f \text{ is of the form } f(x) = x^m + a_{k+1}x^{k+1} + a_{k-1}x^{k-1} + a_0 \\ & \qquad \text{or } f(x) = x^m + a_{k+1}x^{k+1} + a_k x^k + a_{k-1}x^{k-1} + a_0. \end{cases} \quad \square$$

Exercises

Exercise 8.3.26. Let $B = (1, \beta, \ldots, \beta^{n-1})$ be a polynomial basis for $E = \mathrm{GF}(q^n)$ over $\mathrm{GF}(q)$ and assume the existence of some $\delta \in E$ such that the transformation from the primal coordinates of an arbitrary element $\xi \in E$ to the dual coordinates of the element $\xi\delta$ is given by a permutation σ. Then B is weakly self-dual. (This was observed by Gollmann [141] in the special case $q = 2$.) □

Exercise 8.3.27. Provide the details for Observation 8.3.11, including the explicit form of the permutation σ involved. □

Exercise 8.3.28. Generalize Corollary 8.3.14 to arbitrary q by proving the following result: *Let $B = (1, \beta, \ldots, \beta^{n-1})$ be a polynomial basis for* $\mathrm{GF}(q^n)$ *over* $\mathrm{GF}(q)$*, where q is a power of the prime p. Then the dual basis C of B is a permutation of B if and only if n is not divisible by p and the minimal polynomial $f = m_\beta$ of β is a trinomial of the form $f = x^n + ax - 1$.* □

Exercise 8.3.29. Show that choosing $\delta = f'(\beta)^{-1} = \beta^{-2} = 1 + \beta + \beta^2 + \beta^6$ in Example 8.3.17 leads to a transformation matrix with excess 16. □

Exercise 8.3.30. Prove Theorem 8.3.20. □

8.4 Massey-Omura Multipliers

In the next few sections, we are going to discuss the problem of computing in the extension field $E = \mathrm{GF}(q^n)$ over $F = \mathrm{GF}(q)$ when E is represented with respect to a normal basis over F.[4]

Observation 8.4.1. In what follows, we let B be the normal basis for $E = \mathrm{GF}(q^n)$ over $F = \mathrm{GF}(q)$ generated by a specified normal element α. Then all elements of E have the form

$$\xi = x_0\alpha + x_1\alpha^q + \cdots + x_{n-1}\alpha^{q^{n-1}} \tag{8.27}$$

[4] The customary restriction to prime ground fields is not necessary – but, of course, the more general case is not really of practical interest.

with $x_i \in F$ for all i. Of course, as for any basis representation of E/F, addition is trivial also in the representation given by (8.27). However, now one further operation becomes trivial: computing the image of an element ξ under the Frobenius automorphism σ – that is, raising ξ to the q-th power – turns out to be just a cyclic shift of the coordinates to the right, as is easily seen from (8.27):

$$\xi^q = x_{n-1}\alpha + x_0\alpha^q + x_1\alpha^{q^2} + \cdots + x_{n-2}\alpha^{q^{n-1}}.$$

Equivalently, the image of ξ under σ^{-1} is obtained by a cyclic shift of the coordinates to the left, that is

$$\xi^{q^{n-1}} = x_1\alpha + x_2\alpha^q + \cdots + x_{n-1}\alpha^{q^{n-2}} + x_0\alpha^{q^{n-1}}. \tag{8.28}$$

It is customary to neglect the cost of additions and exponentiations with powers of q when estimating the complexity of arithmetic operations in a normal basis representation, since these two operations only require $O(n)$ steps each, where one usually considers additions and multiplications in the given ground field, comparisons, value assignments, etc. as single steps causing unit costs – an assumption which makes sense at least if q is a reasonably small prime. □

In contrast, multiplications are substantially more difficult; as formula (8.2) shows, a multiplication might require $O(n^3)$ steps in general, depending on the choice of basis. Note that using a polynomial basis – which amounts to using modular arithmetic modulo a fixed irreducible polynomial f – allows to perform a multiplication in just $O(n^2)$ steps. Thus the use of normal bases will only be of any real interest if one can find a way of multiplying elements which is at least as efficient.

Before we tackle this problem, we first consider the problem of exponentiating to an arbitrary exponent and hence, in particular, also of finding the inverse of an element.[5] We have already discussed the standard approach to exponentiation in an arbitrary monoid, namely the square and multiply algorithm; see Algorithm 1.6.25. In our situation, Proposition 1.6.26 gives the following result:

Proposition 8.4.2. *Algorithm 1.6.25 allows to perform an arbitrary exponentiation in $E = \mathrm{GF}(q^n)$ over $F = \mathrm{GF}(q)$ using $O(n\log_2 q)$ multiplications, with respect to any given basis representation for E/F. In particular, one may invert an arbitrary element in E^* using $O(n\log_2 q)$ multiplications.*

Proof. The first assertion is immediate from Proposition 1.6.26, as $e \le q^n - 2$. The problem of computing inverses reduces to exponentiation by the trivial observation $\alpha^{-1} = \alpha^{q^n-2}$. □

Observation 8.4.3. Proposition 8.4.2 shows that inversion and, more generally, arbitrary exponentiations may be performed in $O(n^3)$ steps when one uses a polynomial

[5] We note that exponentiation is one of the most important operations needed for cryptographic applications based on the difficulty of computing discrete logarithms; see the remarks at the end of Section 8.1.

basis representation for E/F; here q is considered as fixed, so that it no longer appears in the complexity estimate.

Now let $q = 2$. In contrast to the polynomial basis approach, a normal basis representation reduces the average number of multiplications required for an arbitrary exponentiation from about $3n/2$ to about $n/2$, since squaring then becomes a trivial shift operation. For practical applications, this is a substantial saving.[6] In the particularly important case of inversions, we note the following explicit formula:

$$\alpha^{-1} = \alpha^{2^n-2} = \left(\alpha^2\right)^{2^{n-1}-1} = \left(\alpha^2\right)^{2^{n-2}+2^{n-3}+\cdots+2+1}. \tag{8.29}$$

As Equation (8.29) shows, inverting α by exponentiation unfortunately yields a particularly bad case in the binary situation, since it requires $n-2$ multiplications. □

We shall see that normal bases can be used to reduce the number of multiplications required for computing inverses and general powers substantially, at least in the case $q = 2$ – which clearly is the most important case for practical applications. Such a speedup may be achieved by combining Algorithm 1.6.25 with a divide-and-conquer approach. We will only give an example for this approach; for general results in this direction, we refer the reader to Agnew, Mullin and Vanstone [2] and to Stinson [365]. For instance, Stinson shows that one can perform an arbitrary exponentiation in $GF(2^n)$ with about $n/(\log_2 n - \log_2 \log_2 n)$ multiplications. As a specific application, he mentions that at most 129 multiplications suffice to perform an arbitrary exponentiation in $GF(2^{593})$, a field which had been proposed for cryptographic use and for which hardware to perform arithmetic had been developed in 1988, see Calmos [58].

Example 8.4.4. Let $q = 2$ and $n = 57$. Then

$$\alpha^{-1} = \alpha^{2^{57}-2} = \left(\left(((\alpha^2)^{2^2-1})^{2^2+1}\right)^{2^4+1}\right)^{\sum_{i=0}^{6} 2^{8i}}.$$

In this way, the number of multiplications needed for inversion may be reduced from 55 to 10. □

We now turn our attention to the study of multiplication using a normal basis representation.

Observation 8.4.5. Let ξ be as in Equation (8.27), and let

$$\eta = y_0\alpha + y_1\alpha^q + \cdots + y_{n-1}\alpha^{q^{n-1}}$$

be a further element of E. Then

[6] For odd values of q, one might modify Algorithm 1.6.25 by using the q-adic representation of the exponents and exponentiations by q in step (5). However, this would only result in a minor improvement.

$$\pi := \xi\eta = \sum_{i,j=0}^{n-1} x_i y_j \alpha^{q^i+q^j} = \sum_{i,j=0}^{n-1} x_i y_j (\alpha^{q^{i-j}+1})^{q^j}. \tag{8.30}$$

It is clear from (8.30) that one can evaluate an arbitrary product as soon as one knows how to express the products of an arbitrary basis element α^{q^m} with α. We therefore define an (n,n)-matrix $T = (t_{mk})_{m,k=0,\dots,n-1}$ by

$$\alpha\alpha^{q^m} = \alpha^{q^m+1} = \sum_{k=0}^{n-1} t_{mk}\alpha^{q^k}. \tag{8.31}$$

Substituting (8.31) in (8.30) gives the following identity:

$$\pi = \xi\eta = \sum_{i,j,k=0}^{n-1} t_{i-j,k} x_i y_j \alpha^{q^{k+j}} = \sum_{i,j,m=0}^{n-1} t_{i-j,m-j} x_i y_j \alpha^{q^m}, \tag{8.32}$$

where the indices are taken modulo n. In contrast to the general setup in Section 8.1, where we required a 3-dimensional array of order n to describe the multiplication in an arbitrary basis representation (namely, the explicit data), we now only need an ordinary matrix, namely T. □

Next, we derive an interesting consequence of formula (8.32). The coefficient p_m of α^{q^m} in the product $\pi = \xi\eta$ is

$$p_m = \sum_{i,j=0}^{n-1} t_{i-j,m-j} x_i y_j,$$

and that of $\alpha^{q^{m+1}}$ is

$$p_{m+1} = \sum_{i,j=0}^{n-1} t_{i-j,m+1-j} x_i y_j = \sum_{i,j=0}^{n-1} t_{i-j,m-j} x_{i+1} y_{j+1}.$$

Thus p_{m+1} also equals the coefficient of α^{q^m} in the product of the field elements obtained from ξ and η by cyclically shifting the coordinates one position to the left, that is, of the elements $\xi^{q^{n-1}}$ and $\eta^{q^{n-1}}$, see (8.28). This observation shows that we may compute all coefficients $p_0, p_1, \dots, p_{n-1}$ recursively, by applying the formula for p_0 to successive cyclic shifts of the elements ξ and η. This leads to the following result:

Proposition 8.4.6. *Let α be a normal element for $\mathrm{GF}(q^n)$ over $\mathrm{GF}(q)$, let the matrix $T = (t_{mk})_{m,k=0,\dots,n-1}$ be determined by formula (8.31), and define a further matrix*

$$A = (a_{ij})_{i,j=0,\dots,n-1} \quad \text{with} \quad a_{ij} := t_{i-j,-j}.$$

Then A is symmetric and the product $\xi\eta$ of two arbitrary elements $\xi, \eta \in E$ may be computed using the symmetric bilinear form f associated with A as follows:

$$\xi\eta = f(\xi,\eta)\alpha + f(\xi^{q^{n-1}},\eta^{q^{n-1}})\alpha^q + f(\xi^{q^{n-2}},\eta^{q^{n-2}})\alpha^{q^2} + \cdots + f(\xi^q,\eta^q)\alpha^{q^{n-1}}.$$

Proof. The above formula for the product $\xi\eta$ in terms of the bilinear form f associated with A is clear from the preceding discussion. Now $\xi\eta = \eta\xi$ shows that f has to be symmetric, and hence A is likewise symmetric. □

Proposition 8.4.6 gives a simple and elegant description of the multiplication in $\mathrm{GF}(q^n)$ using a normal basis representation and reduces the size of the explicit data needed to $O(n^2)$. Nevertheless, one might still need $O(n^3)$ steps to perform the corresponding computation. Obviously, the number of steps will depend on the number of non-zero entries in A respectively T. This poses the question whether normal bases exist for which A is a sparse matrix. We shall consider this problem in the next section.

Next, we present an application of Proposition 8.4.6 to the construction of hardware devices for performing multiplications in $\mathrm{GF}(2^n)$ which was first proposed by Massey and Omura in 1981 [259]. More precisely, these authors suggested to construct a parallel-input-serial-output multiplier in the following way:

- The coordinates (with respect to the normal basis generated by α) of the elements ξ and η to be multiplied are loaded into two feedback shift registers.
- For every entry $a_{ij} = 1$, the cells i and j of the first and second shift register, respectively, are connected to an AND-gate.
- The outputs of all the AND-gates are summed up (modulo 2) via XOR-gates.
- Then, with each clock cycle, the output of the device gives one coordinate of the product $\xi\eta$; thus one successively obtains all coordinates in n clock cycles.

Such devices are now usually called **Massey-Omura multipliers**.

Example 8.4.7. Let α be a root of the irreducible (and hence primitive, as $2^5 - 1 = 31$ is a prime) polynomial $f = x^5 + x^4 + x^2 + x + 1$ over $F = \mathrm{GF}(2)$. We claim that α is also a normal element. To see this, we note that $x^5 - 1$ splits as $(x-1)\Phi_5$ and that Φ_5 is irreducible over F. Thus the additive 2-order of α has to be either Φ_5 or $x^5 - 1$, and the former possibility cannot occur, as the trace of α (that is, the coefficient of x^4 in f) is not 0. Thus α is indeed normal.

One can compute the following representations for the five products of α with its conjugates:[7]

$$\alpha^3 = \alpha + \alpha^8,\ \alpha^5 = \alpha^8 + \alpha^{16},\ \alpha^9 = \alpha^2 + \alpha^4,\ \alpha^{17} = \alpha^4 + \alpha^{16}$$

(and, trivially, $\alpha^2 = \alpha^2$). This results in the matrix

[7] Given our present knowledge, it is somewhat inconvenient to perform the necessary computations by hand, so the reader might want to use a computer algebra system for this task. We shall see in the next section that the irreducible polynomial in Example 8.4.7 actually is part of a series which can be obtained by theoretical reasoning. This will allow the determination of both the polynomial f and the matrix T in a much more efficient way: no explicit computations in the extension field GF(32), that is, no arithmetics modulo f, will be needed, and hand computation will suffice.

$$T = \begin{pmatrix} 0 & 1 & 0 & 0 & 0 \\ 1 & 0 & 0 & 1 & 0 \\ 0 & 0 & 0 & 1 & 1 \\ 0 & 1 & 1 & 0 & 0 \\ 0 & 0 & 1 & 0 & 1 \end{pmatrix}.$$

As we shall see in Section 8.5, it is by no means a coincidence that we have $A = T$ in this example. A corresponding Massey-Omura multiplier for GF(32) is given in Figure 8.7. □

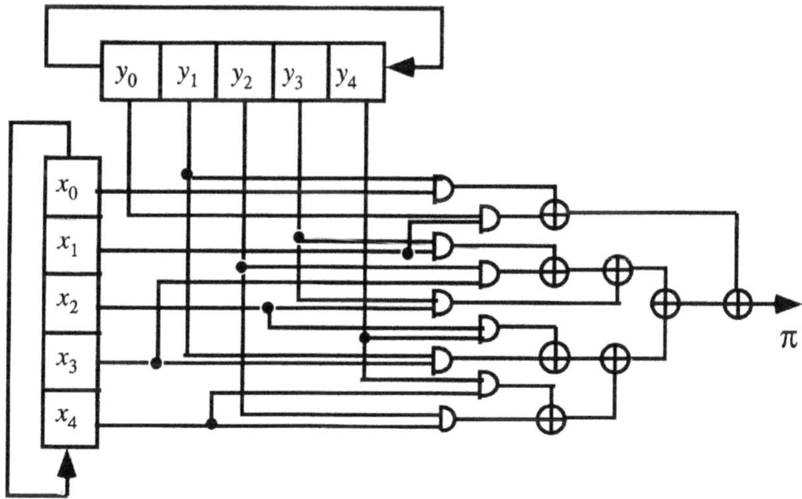

Fig. 8.7 Massey-Omura multiplier for GF(32)

The first VLSI implementation of a Massey-Omura multiplier was described in 1985 by Wang et al. [395]. Neither these authors nor Massey and Omura addressed the question of how the normal basis should be selected. Unfortunately, with a "random" normal basis, their approach becomes infeasible for the large values of n (in the order of magnitude of 1000 or more) which are often needed in cryptographic applications.

It is obvious that the number of gates required for a Massey-Omura multiplier based on the normal basis B only depends on the number d of non-zero entries in the corresponding matrix A. In fact, one needs precisely d AND-gates and $d-1$ XOR-gates. Thus the hardware realization of the multiplication using the normal basis B will become much more efficient if the corresponding matrix A is sparse. We will study this problem in the next section, where we shall see that sparse matrices can be selected in many cases. For instance, this made it possible to actually develop a Massey-Omura multiplier for $\mathrm{GF}(2^{593})$, see Calmos [58]. A description of the resulting encryption processor was given by Rosati [328].

In this connection, we remark that some even more efficient architectures (which have the same number of gates but are "modular" or simpler with regard to wiring, or require less area or have very small "fanout" – that is, the number of gates to which each cell of the shift registers has to be connected is small) than the original one by Massey and Omura have been considered. In Section 8.7, we will discuss a modification of the Massey-Omura architecture which can result in reducing the total number of gates needed even further.

We conclude this section with two simple but useful results on normal bases due to Gao [125]. It is interesting to note that both these results involve the elements $\alpha\alpha^{q^i}$ and therefore the rows of the multiplication matrix T introduced in Observation 8.4.5. More precisely, one uses the traces of these elements and thus, in essence, the row sums s_i of T:

$$\mathrm{Tr}(\alpha\alpha^{q^i}) = \mathrm{Tr}\Big(\sum_{j=0}^{n-1} t_{ij}\alpha^{q^j}\Big) = \sum_{j=0}^{n-1} t_{ij}\mathrm{Tr}(\alpha^{q^j}) = s_i\mathrm{Tr}(\alpha),$$

where we simply write Tr instead of $\mathrm{Tr}_{E/F}$. Gao's first result is the following characterization of normal elements, which complements the criterion given in Corollary 3.9.3.

Theorem 8.4.8. *Let $F = \mathrm{GF}(q)$ and let α be an element of $E = \mathrm{GF}(q^n)$. Put*

$$\alpha_i := \alpha^{q^i} \quad \textit{and} \quad t_i := \mathrm{Tr}(\alpha_0\alpha_i) \quad (i = 0,\ldots,n-1).$$

Then α generates a normal basis for E/F if and only if the polynomials $f = x^n - 1$ and

$$h(x) := t_{n-1}x^{n-1} + \cdots + t_1x + t_0$$

are relatively prime in $E[x]$.

Proof. By Theorem 3.13.9, the elements $\alpha_0,\ldots,\alpha_{n-1}$ form a (normal) basis if and only if the matrix

$$D := \begin{pmatrix} \mathrm{Tr}(\alpha_0\alpha_0) & \mathrm{Tr}(\alpha_0\alpha_1) & \ldots & \mathrm{Tr}(\alpha_0\alpha_{n-1}) \\ \mathrm{Tr}(\alpha_1\alpha_0) & \mathrm{Tr}(\alpha_1\alpha_1) & \ldots & \mathrm{Tr}(\alpha_1\alpha_{n-1}) \\ \vdots & \vdots & \ddots & \vdots \\ \mathrm{Tr}(\alpha_{n-1}\alpha_0) & \mathrm{Tr}(\alpha_{n-1}\alpha_1) & \ldots & \mathrm{Tr}(\alpha_{n-1}\alpha_{n-1}) \end{pmatrix}$$

is non-singular. Note that D is a circulant matrix with first row $(t_0,\ldots,t_{n-1})$, as $\mathrm{Tr}(\alpha_{i+1}\alpha_{j+1}) = \mathrm{Tr}(\alpha_i\alpha_j)$ for $j = 0,\ldots,n-1$. Hence D corresponds – under the isomorphism τ described in Theorem 7.3.2 – to the polynomial h defined above, and thus the assertion is an immediate consequence of Proposition 7.3.3. □

Gao then used Theorem 8.4.8 for the following explicit description of the dual basis of a normal basis:

Theorem 8.4.9. *Assume that α generates a normal basis B for $E = \mathrm{GF}(q^n)$ over $F = \mathrm{GF}(q)$, put*

$$\alpha_i := \alpha^{q^i} \text{ and } t_i := \mathrm{Tr}(\alpha_0\alpha_i) \quad (i = 0, \dots, n-1),$$

and consider the polynomial $h(x) := t_{n-1}x^{n-1} + \dots + t_1x + t_0$, which is invertible modulo $x^n - 1$ by Theorem 8.4.8.

Now let $g(x) = d_{n-1}x^{n-1} + \dots + d_1x + d_0$ be the unique polynomial of degree $< n$ satisfying $g(x)h(x) \equiv 1 \bmod x^n - 1$, and put

$$\beta := d_0\alpha_0 + \dots + d_{n-1}\alpha_{n-1}.$$

Then β generates the normal basis dual to B.

Proof. The condition $g(x)h(x) \equiv 1 \bmod x^n - 1$ means $\sum_{k=0}^{n-1} d_k t_{i-k} = \delta_{i0}$. Hence

$$\begin{aligned}\mathrm{Tr}(\alpha_i\beta^{q^j}) &= \mathrm{Tr}\Big(\alpha_i \cdot \sum_{k=0}^{n-1} d_k\alpha_{k+j}\Big) = \sum_{k=0}^{n-1} d_k \mathrm{Tr}(\alpha_i\alpha_{k+j}) \\ &= \sum_{k=0}^{n-1} d_k \mathrm{Tr}(\alpha_{i-j-k}\alpha_0) = \sum_{k=0}^{n-1} d_k t_{i-j-k} = \delta_{ij},\end{aligned}$$

and therefore the conjugates of β form the normal basis dual to B. □

8.5 Low Complexity Normal Bases

In this section, we study the notion of the complexity of a normal basis, which is motivated by Proposition 8.4.6 and the description of the Massey-Omura multipliers based on that result.

Definition 8.5.1. Let α be a normal element for $\mathrm{GF}(q^n)$ over $\mathrm{GF}(q)$, and let B be the normal basis B generated by α. The **complexity** C_B of B is the weight of the associated matrix T defined by Equation (8.31), that is, the number of non-zero entries of T. □

As we have seen in Section 8.4, the design of efficient Massey-Omura multipliers requires normal bases of small complexity. This problem was first studied around 1989 in the seminal paper by Mullin, Onyszchuk, Vanstone and Wilson [295]. The next result and the subsequent three constructions are all taken from that paper. We begin with the following simple restrictions on the complexity.

Lemma 8.5.2. *Let α be a normal element for $\mathrm{GF}(q^n)$ over $\mathrm{GF}(q)$, and let B be the normal basis B generated by α. Then $C_B \geq 2n-1$. Moreover, in the special case $q=2$, C_B is odd and $C_B \leq n^2-n+1$.*

Proof. Recall that the i-th row of the matrix T defined by Equation (8.31) is the coordinate vector of the product $\alpha\alpha^{q^i}$ with respect to the normal basis B. Hence the sum of all rows of T equals the coordinate vector of

$$\alpha\left(\alpha+\alpha^q+\cdots+\alpha^{q^{n-1}}\right) = \alpha\mathrm{Tr}(\alpha)$$

and is therefore the row vector $(\mathrm{Tr}(\alpha),0,\ldots,0)$.

Note that $\mathrm{Tr}(\alpha) \neq 0$, as the elements of the normal basis B are linearly independent. On the other hand, the rows of T are likewise linearly independent, since the vectors $\alpha,\alpha\alpha^q,\ldots,\alpha\alpha^{q^{n-1}}$ again form a basis for $\mathrm{GF}(q^n)$ over $\mathrm{GF}(q)$. Hence the columns of T are also linearly independent; in particular, T cannot contain any zero column. As the entries in each column of T – with the exception of the first column – sum to 0, each such column contains at least two non-zero entries. This establishes the lower bound in the assertion.

Now assume $q=2$, so that the matrix T has entries 0 and 1 only. Then the preceding arguments also show that the first column of T has to contain an odd number of entries 1 (since its sum is $\mathrm{Tr}(\alpha) \neq 0$), whereas all other columns have to contain an even number of entries 1. Therefore C_B is odd. Also, the first row of T now is the coordinate vector of α^2, that is, the row vector $(0,1,0,\ldots,0)$, which gives the desired upper bound. □

The following three constructions show that the lower bound in Lemma 8.5.2 can often be achieved. Normal bases for which equality is realized are said to be **optimal**.

Theorem 8.5.3. *Assume that q is a primitive root modulo $n+1$, where q is a prime power and $n+1$ a prime. Then the cyclotomic polynomial Φ_{n+1} is irreducible over $\mathrm{GF}(q)$, and its roots form an optimal normal basis for $\mathrm{GF}(q^n)$ over $\mathrm{GF}(q)$.*

Proof. We first note that Φ_{n+1} is irreducible over $F=\mathrm{GF}(q)$ by Proposition 3.6.16, since $\mathrm{ord}_{n+1}(q)=\phi(n+1)=n$ by hypothesis. Now let ζ be a root of Φ_{n+1}. Then the conjugates of ζ are all the roots of Φ_{n+1}, that is, the n primitive $(n+1)$-th roots of unity $\zeta,\zeta^2,\ldots,\zeta^n$ in $E=\mathrm{GF}(q^n)$. We now check that these elements are linearly independent, and hence form a normal basis $B=(\zeta,\zeta^q,\ldots,\zeta^{q^{n-1}})$ for E/F. Thus assume $a_1\zeta+a_2\zeta^2+\cdots+a_n\zeta^n=0$ with $a_1,\ldots,a_n \in F$. Then ζ is a root of the polynomial $f := a_1+a_2x+\cdots+a_nx^{n-1}$, which implies $f=0$ since the minimal polynomial Φ_{n+1} of ζ has degree n.

It remains to compute the complexity of the normal basis B. For this, we need to determine the representation of the n elements $\zeta\zeta^i$ $(i=1,\ldots,n)$ with respect to B. Clearly, $\zeta\zeta^n=1$, whereas $\zeta\zeta^i=\zeta^{i+1}$ is again an element of B for $i \neq n$. Hence each product $\zeta\zeta^i$ with $i \neq n$ contributes exactly one non-zero entry to T. Finally, the product $\zeta\zeta^n=1$ contributes n such entries, since the sum of all the conjugates of ζ gives the non-zero element $\mathrm{Tr}(\zeta) \in F$, as noted in the proof of Lemma 8.5.2. □

In cryptographic applications, one usually requires an extension of GF(2) of prime degree n. Unfortunately, Theorem 8.5.3 cannot yield an optimal normal basis for $\mathrm{GF}(q^n)$ over $\mathrm{GF}(q)$ when n is an odd prime. However, the following two (somewhat more involved) constructions for $q = 2$ can often be used to find an optimal normal basis in this situation.

Theorem 8.5.4. *Let* $2n+1$ *be a prime, and assume that* 2 *is a primitive root modulo* $2n+1$. *Then there exists an optimal normal basis for* $\mathrm{GF}(2^n)$ *over* GF(2).

Proof. By Theorem 8.5.3, we may use a root ζ of the cyclotomic polynomial Φ_{2n+1} to generate an optimal normal basis B for the degree 2 extension $K = \mathrm{GF}(2^{2n})$ of $E = \mathrm{GF}(2^n)$ over $F = \mathrm{GF}(2)$. By hypothesis, 2 is a primitive root modulo $2n+1$, so that $2^n \equiv -1 \bmod 2n+1$. Therefore $\zeta^{-1} = \zeta^{2^n}$ and hence

$$\alpha := \zeta + \zeta^{-1} = \zeta + \zeta^{2^n} = \mathrm{Tr}_{K/E}(\zeta)$$

is an element of E. We now show that α generates the desired optimal normal basis for E/F.

By Proposition 3.12.9, α is a normal element for E/F, and it only remains to determine the complexity of the associated normal basis N. Thus we need to compute the coordinate vectors of the n elements $\alpha\alpha^{2^k}$ with respect to N. For $k = 0$, we obtain α^2, which is itself an element of N. For $k \neq 0$, one gets

$$\alpha\alpha^{2^k} = (\zeta+\zeta^{-1})(\zeta^{2^k}+\zeta^{-2^k}) = (\zeta^{1+2^k}+\zeta^{-(1+2^k)}) + (\zeta^{1-2^k}+\zeta^{-(1-2^k)}).$$

Now $\zeta^{2^k} \neq \zeta^{\pm 1}$, as $1 \le k \le n-1$ and as 2 is a primitive root modulo $2n+1$. Hence

$$\zeta^{1+2^k}+\zeta^{-(1+2^k)} = \alpha^{2^s} \quad \text{and} \quad \zeta^{1-2^k}+\zeta^{-(1-2^k)} = \alpha^{2^t}$$

for suitable values of s and t. Thus each product $\alpha\alpha^{2^k}$ with $1 \le k \le n-1$ is the sum of two elements of N, and hence $C_N = 2n-1$. □

Theorem 8.5.5. *Let* $2n+1$ *be a prime* $\equiv 3 \bmod 4$, *and assume that* 2 *generates the quadratic residues modulo* $2n+1$. *Then there exists an optimal normal basis for* $\mathrm{GF}(2^n)$ *over* GF(2).

Proof. We first note that the hypothesis $2n+1 \equiv 3 \bmod 4$ forces n to be odd. The modified hypothesis that 2 generates the quadratic residues modulo $2n+1$ means $\mathrm{ord}_{2n+1}(2) = n$. Thus Proposition 3.6.16 shows that the cyclotomic polynomial Φ_{2n+1} now splits into two irreducible polynomials of degree n over $F = \mathrm{GF}(2)$, and hence $E = \mathrm{GF}(2^n)$ is the splitting field of Φ_{2n+1}.

The remainder of the proof will be similar to that of Theorem 8.5.4, but more involved, as we can no longer make use of any trace function. As before, we choose a primitive $(2n+1)$-th root of unity ζ and claim that $\alpha := \zeta + \zeta^{-1}$ generates the desired optimal normal basis for E/F. This time, $\alpha \in E$ holds trivially, but it is not at all obvious whether the conjugates of α are indeed linearly independent.

Thus assume $a_0\alpha + a_1\alpha^2 + \cdots + a_{n-1}\alpha^{2^{n-1}} = 0$ with $a_1, \ldots, a_n \in F$. Substituting $\alpha = \zeta + \zeta^{-1}$ gives

$$a_0(\zeta + \zeta^{-1}) + a_1(\zeta^2 + \zeta^{-2}) + \cdots + a_{n-1}(\zeta^{2^{n-1}} + \zeta^{-2^{n-1}}) = 0. \tag{8.33}$$

We now consider the polynomial $f = b_1 + b_2 x + \cdots + b_{2n} x^{2n-1}$, where

$$b_j := a_i \quad \text{if } j \equiv \pm 2^i \bmod 2n+1.$$

Note that this makes sense: all b_j are uniquely determined by the preceding requirement, since 2 generates the quadratic residues modulo $2n+1$ and since -1 is a quadratic non-residue (because of the hypothesis $2n+1 \equiv 3 \bmod 4$). Moreover, in view of (8.33), both ζ and ζ^{-1} are roots of f. Therefore f is divisible by the minimal polynomials m_ζ and $m_{\zeta^{-1}}$ of these two primitive $(2n+1)$-th roots of unity. As ζ^{-1} is not conjugate to ζ in E/F, we conclude $\Phi_{2n+1} = m_\zeta m_{\zeta^{-1}}$, and thus f is divisible by Φ_{2n+1}. In view of $\deg f = 2n-1 < \deg \Phi_{2n+1}$, this shows $f = 0$ and hence $a_i = 0$ for $i = 1, \ldots, n$, as desired.

It remains to determine the complexity of N, which can be done exactly as in the proof of Theorem 8.5.4. For this, one just observes that again $\zeta^{2^k} \neq \zeta^{\pm 1}$ for $1 \leq k \leq n-1$, since 2 generates the quadratic residues and since -1 is a quadratic non-residue modulo $2n+1$. □

Sometimes, one would like to know the minimal polynomials of the normal elements constructed in the proofs of Theorems 8.5.4 and 8.5.5. These minimal polynomials may be easily computed recursively as follows. Put

$$f_0 := 1, \quad f_1 := x+1 \quad \text{and} \quad f_k := x f_{k-1} + f_{k-2} \text{ for } k \geq 2. \tag{8.34}$$

Now let n be a positive integer satisfying the hypothesis of Theorem 8.5.4 or 8.5.5, and let α be the optimal normal basis generator for $GF(2^n)$ over $GF(2)$ constructed there. We claim $m_\alpha = f_n$. To see this, note

$$f_k(y + y^{-1}) = 1 + \sum_{i=1}^{k} (y^i + y^{-i}) \quad \text{for } k \geq 1,$$

which follows from (8.34) by induction. For $k = n$, substituting $y = \zeta$ gives

$$f_n(\alpha) = f_n(\zeta + \zeta^{-1}) = 1 + \sum_{i=1}^{n} (\zeta^i + \zeta^{-i}) = 1 + \sum_{j=1}^{2n} \zeta^j = 0,$$

since ζ is a primitive $(2n+1)$-th root of unity. This proves the following result:

Theorem 8.5.6. *Let $2n+1$ be a prime, and assume either that 2 is a primitive root modulo $2n+1$ or that $2n+1 \equiv 3 \bmod 4$ and 2 generates the quadratic residues modulo $2n+1$. Then the binary polynomial f_n defined by (8.34) is irreducible, and* *its roots form the optimal normal basis for* $\mathrm{GF}(2^n)$ *over* $\mathrm{GF}(2)$ *constructed in the proof of Theorem 8.5.4 or Theorem 8.5.5, respectively.* □

We shall now show how one may obtain Example 8.4.7 with very little computation from the preceding results.

Example 8.5.7. Note that 2 is a primitive root modulo 11. By Theorem 8.5.6, the roots of the polynomial f_5 form an optimal normal basis N for $E = \mathrm{GF}(32)$ over $F = \mathrm{GF}(2)$. Using (8.34), one checks that f_5 coincides with the polynomial f defined in Example 8.4.7. According to the proof of Theorem 8.5.4, the roots of f_5 are the conjugates of $\alpha = \zeta + \zeta^{-1}$, where ζ is a primitive 11-th root of unity over F. Thus the roots of f_5 are as follows:

$$\alpha = \zeta + \zeta^{-1},\ \alpha^2 = \zeta^2 + \zeta^{-2},\ \alpha^4 = \zeta^4 + \zeta^{-4},\ \alpha^8 = \zeta^8 + \zeta^{-8},\ \alpha^{16} = \zeta^5 + \zeta^{-5}.$$

Now it is easy to compute the values $\alpha\alpha^{2^i}$ given in Example 8.4.7 by only calculating with powers of ζ. For instance,

$$\alpha\alpha^4 = (\zeta + \zeta^{-1})(\zeta^4 + \zeta^{-4}) = \zeta^5 + \zeta^3 + \zeta^{-3} + \zeta^{-5} = \alpha^8 + \alpha^{16}.$$

In this way, one may determine the matrix T – and then also the matrix A defined in Proposition 8.4.6 – without performing any explicit computations in the extension field E. Of course, the same method can be applied for all optimal normal bases constructed via Theorems 8.5.4 and 8.5.5. □

Remark 8.5.8. In order to apply the preceding Theorems 8.5.3 and 8.5.4, one needs to know whether or not 2 is a primitive root modulo some given odd prime. This seems to happen quite often: for instance, according to the Wikipedia, this holds for the following 38 of the 95 primes below 500:

$$\begin{gathered} 3, 5, 11, 13, 19, 29, 37, 53, 59, 61, 67, 83, 101, 107, \\ 131, 139, 149, 163, 173, 179, 181, 197, 211, 227, 269, 293, \\ 317, 347, 349, 373, 379, 389, 419, 421, 443, 461, 467, 491. \end{gathered}$$

Of course, it would be very desirable to have a characterization of the primes for which 2 (respectively, for Theorem 8.5.5, -2) is a primitive root. Unfortunately, this remains an unsolved problem: one does not even know whether or not 2 is a primitive root for infinitely many primes. In this context, we mention one of the major open problems in Number Theory, namely **Artin's conjecture**:

An integer $b \neq 1, -1$ *which is not a perfect square*
is a primitive root modulo infinitely many primes.

It is known that this would be a consequence of the (still unproven) generalized Riemann hypothesis. Without this hypothesis, there is a striking non-constructive result proved by Heath-Brown [181] in 1986: Artin's conjecture holds for all primes b with at most two exceptions. However, one does not know whether it holds for any given *specific* prime, and no progress seems to have been made since then. For more details, the reader is referred either to the nice article by Murty [297] in the *Mathematical Intelligencer* or to the extensive survey by Moree [279]. □

Nevertheless, there are some useful sufficient conditions for 2 (or -2) to be a primitive root modulo some prime, which facilitate the application of Theorems 8.5.3 to 8.5.5. Since these criteria are not too difficult, they will be included. We begin with some well-known auxiliary facts from Number Theory.

Definition 8.5.9. Let p be a prime and a any integer not divisible by p. Then the **Legendre symbol** (a/p) is defined as follows:

$$(a/p) := \begin{cases} 1 & \text{if } a \text{ is a quadratic residue modulo } p, \\ -1 & \text{otherwise.} \end{cases} \quad \square$$

We first note a simple observation usually attributed to Euler and then also a less obvious result on the quadratic behavior of 2. For the sake of completeness (and its beauty), we will provide a short and elegant proof for the latter, which is taken from Serre [350] and makes essential use of finite fields.

Proposition 8.5.10 (Euler's criterion). *Let p be a prime and a any integer not divisible by p. Then*

$$(a/p) \equiv a^{(p-1)/2} \bmod p. \tag{8.35}$$

Proof. This is an immediate consequence of the fact that the multiplicative group of the field of residues modulo p is cyclic. $\square$

For the second result, we note that $p^2 \equiv 1 \bmod 8$ for every odd prime p.

Proposition 8.5.11. *Let p be an odd prime. Then $(2/p) = (-1)^{(p^2-1)/8}$, that is,*

$$(2/p) = \begin{cases} 1 & \textit{if } p \equiv \pm 1 \bmod 8, \\ -1 & \textit{if } p \equiv \pm 3 \bmod 8. \end{cases} \tag{8.36}$$

Proof. Let ζ be a primitive 8-th root of unity over $\mathrm{GF}(p)$. We consider the element $\alpha := \zeta + \zeta^{-1}$ in the extension field $\mathrm{GF}(p)(\zeta)$ and observe $\alpha^2 = \zeta^2 + \zeta^{-2} + 2 = 2$, which follows from $\zeta^4 = -1$. Now let $p \equiv \pm 1 \bmod 8$. Then

$$\alpha^p = \zeta^p + \zeta^{-p} = \alpha \text{ and thus } (2/p) = 2^{(p-1)/2} = \alpha^{p-1} = 1.$$

Similarly, for $p \equiv \pm 3 \bmod 8$,

$$\alpha^p = \zeta^5 + \zeta^{-5} = \zeta^4 \alpha = -\alpha \text{ and thus } (2/p) = 2^{(p-1)/2} = \alpha^{p-1} = -1. \quad \square$$

Using the preceding facts, we can prove the following criteria:

Proposition 8.5.12. *Let p and q be odd primes. Then:*

- *If $p = 4q + 1$, then 2 is a primitive root modulo p.*
- *If $q \equiv 1 \bmod 4$ and $p = 2q + 1$, then 2 is a primitive root modulo p.*

- *If* $q \equiv 3 \bmod 4$ *and* $p = 2q+1$, *then* -2 *is a primitive root modulo* p, *and hence* 2 *generates the quadratic residues modulo* p.

Proof. We will only deal with the first case, where $p = 4q+1$. (The proofs for the other two cases are similar and will be left to the reader as Exercise 8.5.27.) Here 2 is a primitive root modulo p if and only if $2^{4q/r} \not\equiv 1 \bmod p$ for every prime divisor r of $4q$, that is, for $r \in \{2, q\}$. This condition holds for $r = q$, as $2^4 = 16 \equiv 1 \bmod p$ would give the contradiction $p = 5$ and $q = 1$; and for $q = 2$, it is a direct consequence of Equations (8.35) and (8.36):

$$2^{2q} = 2^{(p-1)/2} \equiv (2/p) = (-1)^{(p^2-1)/8} = (-1)^{2q^2+q} = -1 \bmod p.$$

This establishes the assertion. □

Remark 8.5.13. It was conjectured by Mullin, Onyszchuk, Vanstone and Wilson [295] that the constructions in Theorems 8.5.3 to 8.5.5 already yield *all* optimal normal bases over GF(2). In other words, any optimal normal basis is one of the bases constructed in these results. Later this conjecture was extended to arbitrary finite fields by Mullin [294]. Here one needs to be a little more careful, as there are two simple modifications which will generate "new" optimal normal basis from the ones already constructed:

- If α generates an optimal normal basis for E/F, then so does $c\alpha$ for every $c \in F^*$.[8]
- Assume that α generates an optimal normal basis for E/F, where $F = \mathrm{GF}(q)$ and $E = \mathrm{GF}(q^n)$, and let $k \in \mathbb{N}$ with $\gcd(k,n) = 1$. Then α also generates an optimal normal basis for $L = \mathrm{GF}(q^{kn})$ over $K = \mathrm{GF}(q^k)$.

The first of these cases is trivial, and the second one follows from Theorem 3.5.9, as the restriction of the Galois group of L/K to E is the Galois group of E/F (so that α is also a normal element for L/K). □

Mullin also made some progress towards a proof of these conjectures. Finally, Lenstra and Gao [127] showed that the extended conjecture is indeed true; we shall give a proof for their "optimal normal basis theorem" in Section 8.8. For the time being, we just state the special case for optimal normal bases over GF(2):

Result 8.5.14 (Optimal normal basis theorem, binary case) *All optimal normal bases for some extension* $\mathrm{GF}(2^n)$ *of* GF(2) *arise from the constructions in Theorems 8.5.3 to 8.5.5.* □

Definition 8.5.15. The optimal normal bases constructed in Theorem 8.5.3 are called **type I** optimal normal bases, while those constructed in Theorems 8.5.4 and 8.5.5 are said to be **type II** optimal normal bases. □

---

[8] Two bases for E/F which arise from each other by multiplication with a constant element in F^* are often called **equivalent**.

As an application of Result 8.5.14 together with the criteria given in Proposition 8.5.12, we exhibit all values $n \leq 1,300$ for which there exists an optimal normal basis for $\mathrm{GF}(2^n)$ over $\mathrm{GF}(2)$ in Table 8.2. Here an entry marked with one or two stars $*$ indicates the existence of a type I or of both type I and type II optimal normal bases, respectively. In all other cases, only a type II basis exists.

In those cases where, according to Result 8.5.14, no optimal normal basis can exist, it is still of interest to construct a **low complexity** normal basis. This term is not defined strictly, but one usually calls a series of normal bases for $\mathrm{GF}(2^n)$ over $\mathrm{GF}(2)$ for a sequence of values of n a series of low complexity bases if their complexities are bounded by cn for some (preferably small) constant c. Ash, Blake and Vanstone [13] obtained series of such bases by generalizing the constructions for optimal normal bases given above; see Remark 8.5.19. Later, Wassermann [398] and Beth, Geiselmann and Meyer [34] independently observed that the same approach also works for arbitrary ground fields $\mathrm{GF}(q)$ and obtained the following general result, which we state without proof.

Result 8.5.16 *Consider a prime power q and an extension degree n. Assume that k and c are positive integers with the following three properties:*

- *$r := kn+1$ is a prime which divides neither q nor c;*
- $\mathrm{ord}_r(c) = k$;
- *q and c generate the multiplicative group of all non-zero residues modulo r.*

Now let ζ be any primitive r-th root of unity in $\mathrm{GF}(q^{kn})$. Then the element

$$\alpha := \zeta + \zeta^c + \cdots + \zeta^{c^{k-1}}$$

generates a normal basis N for $\mathrm{GF}(q^n)$ over $\mathrm{GF}(q)$. Moreover, $C_N \leq (k+1)n - k$, and if n is a multiple of $k \neq 1$, this bound may be improved to $C_N \leq kn - 1$. □

In the binary case, one can say more. The lower bounds in the following result are due to [13], whereas the upper bounds come from [34].

Result 8.5.17 *Let $q = 2$ in Result 8.5.16. Then*

$$kn - (k^2 - 3k + 3) \leq C_N \leq kn - k + 1 \quad \text{if } k \text{ is even},$$

and

$$(k+1)n - (k^2 + k + 1) \leq C_N \leq (k+1)n - 2k + 1 \quad \text{if } k \text{ is odd}.$$ □

The proofs for these two results are similar to that of Theorem 8.5.5, though more involved; the special case $k = 2$ reduces to the constructions of Theorems 8.5.4 and 8.5.5. We refer the reader to the original papers and to Meyer [273] for details. Using rather lengthy and technical arguments, one can also show that the lower bounds in Result 8.5.17 are quite often achieved; for instance, this holds for all sufficiently large n, provided that k is a power of 2, an odd prime, twice an odd prime, or four times an odd prime. Also, for $k \leq 7$ the lower bound is met, with just a handful of exceptions:

2**	3	4*	5	6	9	10*	11	12*	14
18**	23	26	28*	29	30	33	35	36*	39
41	50	51	52*	53	58*	60*	65	66*	69
74	81	82*	83	86	89	90	95	98	99
100*	105	106*	113	119	130*	131	134	135	138*
146	148*	155	158	162*	172*	173	174	178	179
180*	183	186	189	191	194	196*	209	210**	221
226*	230	231	233	239	243	245	251	254	261
268*	270	271	273	278	281	292*	293	299	303
306	309	316*	323	326	329	330	338	346*	348*
350	354	359	371	372*	375	378**	386	388*	393
398	410	411	413	414	418*	419	420*	426	429
431	438	441	442*	443	453	460*	466*	470	473
483	490*	491	495	508*	509	515	519	522*	530
531	540*	543	545	546*	554	556*	558	561	562*
575	585	586*	593	606	611	612*	614	615	618**
629	638	639	641	645	650	651	652*	653	658
659	660*	676*	683	686	690	700*	708*	713	719
723	725	726	741	743	746	749	755	756*	761
765	771	772*	774	779	783	785	788*	791	796*
803	809	810	818	820*	826*	828*	831	833	834
846	852*	858*	866	870	873	876*	879	882*	891
893	906*	911	923	930	933	935	938	939	940*
946*	950	953	965	974	975	986	989	993	998
1013	1014	1018*	1019	1026	1031	1034	1041	1043	1049
1055	1060*	1065	1070	1090*	1103	1106	1108*	1110	1116*
1118	1119	1121	1122*	1133	1134	1146	1154	1155	1166
1169	1170*	1178	1185	1186*	1194	1199	1211	1212*	1218
1223	1228*	1229	1233	1236*	1238	1251	1258*	1265	1269
1271	1274	1275	1276*	1278	1282*	1289	1290*	1295	1300*

Table 8.2 Values of $n \leq 1,300$ for which $\mathrm{GF}(2^n)$ over $\mathrm{GF}(2)$ has an optimal normal basis

Result 8.5.18 *Let $q = 2$ in Result 8.5.16. Then*

$$C_N = \begin{cases} 4n-7 & \textit{for } k=3 \textit{ and } k=4; \\ 6n-21 & \textit{for } k=5,\ n>2 \textit{ and for } k=6,\ n>12; \\ 8n-43 & \textit{for } k=7 \textit{ and } n>6. \end{cases} \qquad \square$$

Remark 8.5.19. In applications of Result 8.5.16 and its refinements, one considers a given pair (q,n) and has to search for pairs (k,c) with the three required properties. There may be many such pairs, but one is, of course, really only interested in the solution giving the smallest possible value of k. We note that the third requirement can be rephrased as follows:

- n is the smallest positive integer ℓ satisfying $q^\ell \equiv c^j \bmod r$ for some integer j.

We now explain our previous comment that Result 8.5.16 contains the three constructions for optimal normal bases:

- If $n+1$ is a prime with $\mathrm{ord}_{n+1}(q) = n$, we may choose $k = c = 1$ to obtain Theorem 8.5.3.
- Both Theorems 8.5.4 and 8.5.5 correspond to $q = 2$, $k = 2$ and $c = -1$, with a difference caused by the respective assumption on the order of q:
  - In Theorem 8.5.4, $\mathrm{ord}_r(q) = r-1 = 2n$, so that q already generates all nonzero residues modulo r; thus c is not required for this. However, ζ lies in a larger field than desired in this case, and the parameter c is used to ensure that the sum α is in the correct field $E = \mathrm{GF}(q^n)$.
  - In Theorem 8.5.5, n is odd and $\mathrm{ord}_r(q) = (r-1)/2 = n$. Note that the two subgroups of $\mathbb{Z}_r^*$ generated by q and c, respectively, are complementary. Now ζ is in the correct field E, but possibly not normal over F. Here the choice of c adjusts ζ to an element α with the desired properties.

All these constructions rest on the use of so-called *Gauss periods* and can lead to (at least theoretically) very efficient implementations of finite field arithmetics. For instance, it was shown by Gao, von zur Gathen, Panario and Shoup [129] that, for a small prime power q and infinitely many integers n, the following asymptotic complexities can be reached using a suitable normal basis of $\mathrm{GF}(q^n)$ over $\mathrm{GF}(q)$:

- multiplication with $O(n \log n \log\log n)$ operations in $\mathrm{GF}(q)$;
- division with $O(n \log^2 n \log\log n)$ operations in $\mathrm{GF}(q)$;
- exponentiation of an arbitrary element in $\mathrm{GF}(q^n)$ with $O(n^2 \log\log n)$ operations in $\mathrm{GF}(q)$.

These authors also gave an interesting explicit form for the generator of the dual normal basis of a normal bases constructed via Gauss periods. $\square$

Next, we mention a further interesting construction method, which applies to composite extension degrees and uses traces of optimal normal elements or the duals of such elements. We only state one particularly interesting special case, where the exact complexity has been determined; in general, one obtains only upper bounds.

We refer the reader to the original paper by Christopoulou, Garefalakis, Panario and Thomson [75] for more details and proofs.

Result 8.5.20 *Assume that α generates a type II optimal normal basis for* $\mathrm{GF}(2^{kn})$ *over* $\mathrm{GF}(2)$*, and let* $\beta := \mathrm{Tr}_{\mathrm{GF}(2^{kn})/\mathrm{GF}(2^n)}(\alpha)$*, where* $k \leq n$*. Then the complexity of the normal basis for* $\mathrm{GF}(2^n)$ *over* $\mathrm{GF}(2)$ *generated by* β *is* $2kn - 2k + 1$*.* □

The existing tables indicate that Result 8.5.20 and its generalizations often yield the lowest known complexities – but only for somewhat larger values of n, the only cases with $n \leq 100$ being $n = 54$, 59 and 71.

A direct construction of low complexity normal bases of $\mathrm{GF}(q^n)$ over $\mathrm{GF}(q)$ in the case where n either equals the characteristic of $\mathrm{GF}(q)$ or divides $q-1$ or $q+1$ is due to Blake, Gao and Mullin [41]. Since their results are mainly of interest for large values of q, we omit stating them here.

Next, we present a simple product construction, which was observed independently by several authors [212, 319, 345, 346].

Proposition 8.5.21. *Let α and β generate normal bases A and B for* $K = \mathrm{GF}(q^m)$ *and* $L = \mathrm{GF}(q^n)$ *over* $F = \mathrm{GF}(q)$*, respectively, where m and n are relatively prime. Then $\gamma := \alpha\beta$ generates a normal basis N for* $E = \mathrm{GF}(q^{mn})$ *over F, and the complexity of N satisfies $C_N = C_A C_B$.*

Proof. By Proposition 3.13.21, γ generates a normal basis N for E over F. Thus N is the set of conjugates of $\alpha\beta$ under the Galois group of E/F, and the reader may easily check that

$$N = \left\{\alpha^{q^c}\beta^{q^d} : c = 0,\ldots,m-1 \text{ and } d = 0,\ldots,n-1\right\}.$$

Now let $\xi = x_0\alpha + x_1\alpha^q + \cdots + x_{m-1}\alpha^{q^{m-1}}$ and $\eta = y_0\beta + y_1\beta^q + \cdots + y_{n-1}\beta^{q^{n-1}}$ be any two elements of K and L, respectively, so that

$$\xi\eta = \sum_{c=0}^{m-1}\sum_{d=0}^{n-1} x_c y_d \alpha^{q^c}\beta^{q^d}$$

is the representation of $\xi\eta$ with respect to the basis N. By definition, the complexity C_N of N is the sum of the weights of all the products $\gamma\gamma^{q^h}$ $(h = 0,\ldots,mn-1)$, that is, of all the products

$$(\alpha\beta)\left(\alpha^{q^c}\beta^{q^d}\right) \quad \text{with } c = 0,\ldots,m-1 \text{ and } d = 0,\ldots,n-1.$$

Applying these observations to $\xi := \alpha\alpha^{q^c}$ and $\eta := \beta\beta^{q^d}$, it is clear that the weight of $(\alpha\beta)\left(\alpha^{q^c}\beta^{q^d}\right)$ is the product of the weights of $\alpha\alpha^{q^c}$ and $\beta\beta^{q^d}$. In view of the ranges of c and d, this proves the product formula for the complexity C_N of N. □

Corollary 8.5.22. *Let $q = 2$ in Proposition 8.5.21, and assume that α and β both generate optimal normal bases. Then the normal basis N generated by $\gamma = \alpha\beta$ has complexity $C_N = 4mn - 2m - 2n + 1$.* □

Definition 8.5.23. One defines the **complexity** $C_q(n)$ of the field extension $\mathrm{GF}(q^n)$ over $\mathrm{GF}(q)$ in the obvious manner:

$$C_q(n) := \min\{C_N\colon N \text{ is a normal basis for } \mathrm{GF}(q^n) \text{ over } \mathrm{GF}(q)\}. \qquad \square$$

With this notation, we note the following immediate consequence of Proposition 8.5.21:

Corollary 8.5.24. *Assume that m and n be relatively prime. Then one has*

$$C_q(mn) \leq C_q(m)C_q(n). \qquad \square$$

n	number of normal bases	$\min C_N$	$\max C_N$	$\mathrm{avg}\, C_N$	Construction
2*	1	3	3	3.00	Theorems 8.5.3, 8.5.4
3*	1	5	5	5.00	Theorem 8.5.5
4*	2	7	9	8.00	Theorem 8.5.3
5*	3	9	15	11.67	Theorem 8.5.4
6*	4	11	17	15.00	Theorem 8.5.4
7	7	19	27	23.00	
8	16	21	35	29.00	
9*	21	17	45	35.57	Theorem 8.5.4
10*	48	19	61	44.83	Theorem 8.5.3
11*	93	21	71	55.82	Theorem 8.5.5
12*	128	23	83	64.13	Theorem 8.5.3
13	315	45	101	78.38	Result 8.5.18 ($k=4$)
14*	448	27	135	91.07	Theorem 8.5.4
15	675	45	137	105.89	Proposition 8.5.21
16	2048	85	157	115.82	
17	3825	81	177	132.77	Result 8.5.18 ($k=6$)
18*	5376	35	243	153.51	Theorems 8.5.3, 8.5.4
19	13797	117	229	172.00	
20	24576	63	257	190.80	Theorem 8.5.4

Table 8.3 Complexities of normal bases for $\mathrm{GF}(2^n)$ over $\mathrm{GF}(2)$ with $2 \leq n \leq 20$

In Tables 8.3 and 8.4, we present more information on the complexities of normal bases for $E = \mathrm{GF}(2^n)$ over $F = \mathrm{GF}(2)$ with $n \leq 39$. These tables contain results of computer searches reported in [260, 292, 295]. They list the number of normal bases for E/F, which can be computed from Theorem 7.3.10, as well as the minimum,

n	number of normal bases	$\min C_N$	$\max C_N$	$\text{avg}\, C_N$	Construction
21	27783	95	277	210.97	Proposition 8.5.21
22	95232	63	363	231.93	Proposition 8.5.21
23*	182183	45	325	254.02	Theorem 8.5.5
24	262144	105	375	276.89	Proposition 8.5.21
25	629145	93	383	301.01	Result 8.5.18 ($k=4$)
26*	1290240	51	555	325.96	Theorem 8.5.4
27	1835001	141	443	351.99	Result 8.5.18 ($k=4$)
28*	3670016	55	517	378.98	Theorem 8.5.3
29*	9256395	57	521	407.00	Theorem 8.5.4
30*	11059200	59	759	435.95	Theorem 8.5.4
31	28629151	237	587	466.00	
32	67108864	361	621	497.00	
33*	97327197	65	693	529.00	Theorem 8.5.4
34	250675200	243	819	562.00	Proposition 8.5.21
35*	352149515	69	779	596.00	Theorem 8.5.5
36*	704643060	71	1017	630.99	Theorem 8.5.5
37	1857283155	141	823	667.00	Result 8.5.18 ($k=4$)
38	3616800703	207	1131	704.00	Result 8.5.18 ($k=6$)
39*	5282242828	77	933	742.00	Theorem 8.5.5

Table 8.4 Complexities of normal bases for $GF(2^n)$ over $GF(2)$ with $21 \leq n \leq 39$

the maximum, and the average complexities of a normal basis for E/F. A star $*$ indicates the existence of an optimal normal basis, and the comment given under the heading "Construction" gives a theoretical construction for a normal basis with the minimum complexity if such a construction is available. Recall that the complexity of a normal basis is always odd in the binary case, by Lemma 8.5.2.

It is clear from Result 8.5.14 that there are at most two optimal normal bases for $GF(2^n)$ over $GF(2)$. But even when no such basis exists, the computer searches suggest that normal bases with minimum complexity are very rare: for $n \leq 27$, there are two such bases for $n = 19$, and only one example exists in all other cases. This was pointed out by Menezes [265], and according to the results of Masuda, Moura, Panario, and Thomson [260] his observation remains valid for $n \leq 39$.

The smallest known complexities of normal bases for $E = GF(2^n)$ over $F = GF(2)$ in the range $40 \leq n \leq 577$ can be found in [292, Table 2.2.10]. The known computational results have led to some interesting conjectures. We shall state two of these here, which were proposed in [260] and [407], respectively.

Conjecture 8.5.25. For $n \geq 8$, the average complexity of a normal basis for $\mathrm{GF}(2^n)$ over $\mathrm{GF}(2)$ is at most $(n^2-n+3)/2$. □

Conjecture 8.5.26. Assume that $\mathrm{GF}(2^n)$ does not admit an optimal normal basis over $\mathrm{GF}(2)$, that is, $C_2(n) \neq 2n-1$. Then actually $C_2(n) \geq 3n-3$. □

Clearly, a lot remains to be done regarding the theoretical understanding of the known computational results.

As already mentioned, optimal normal bases have real world applications in Cryptography. An early discussion of questions arising in the implementation of such systems was given by Agnew, Mullin, Onyszchuk and Vanstone [1]. We once more refer to [58, 328] for details of the actual implementation for a system relying on discrete logarithms in the multiplicative group of $\mathrm{GF}(2^{593})$.

Nowadays, one no longer uses systems based on discrete logarithms in $\mathrm{GF}(2^n)^*$, but rather elliptic curve cryptosystems. Here an early implementation – again using arithmetics with respect to an optimal normal basis – over $\mathrm{GF}(2^{155})$ was described in [3]. For detailed treatments of this topic, we refer the reader to the books [77, 179, 267].

Another application of optimal normal bases concerns finding the roots of polynomials which split over $\mathrm{GF}(q^n)$; as shown by Menezes, van Oorschot and Vanstone [268], this approach can lead to faster algorithms for the root finding problem.

Exercises

Exercise 8.5.27. In the proof of Theorem 8.5.5, we constructed a normal element α by using the trace function from K to E and by appealing to Proposition 3.12.9. Give a more direct proof by showing that α belongs to E and is a normal element for E/F without any use of the trace function. □

Exercise 8.5.28. Prove the second and third assertions in Proposition 8.5.12. □

8.6 The Complexity of Self-dual Normal Bases

As we have seen in the preceding sections, both normal bases and dual bases are important for implementing finite field arithmetics. Thus it is not surprising that the combination of both these topics has generated considerable interest, too. In this section, we will mainly be concerned with the complexity of self-dual and, more generally, trace-orthogonal normal bases. In particular, we provide a characterization of trace-orthogonal normal bases in terms of the matrices associated with their multiplicative properties and show that any such basis is actually equivalent to a self-dual normal basis. Following this, we will present some results on the complexity of self-dual normal bases over $\mathrm{GF}(2)$.

Throughout this section, we will use the following notation: given any element ξ of $E = \mathrm{GF}(q^n)$ and any basis $B = (\alpha_0, \ldots, \alpha_{n-1})$ for E over $F = \mathrm{GF}(q)$, we put

$$\sigma_B(\xi) := x_0 + \cdots + x_{n-1},$$

where, as usual, $r_B(\xi) = (x_0, \ldots, x_{n-1})$ is the coordinate vector of ξ with respect to B. We also write $w_B(\xi)$ for the **weight** of the vector $r_B(\xi)$, that is, the number of entries $x_i \neq 0$ in $r_B(\xi)$.

Moreover, whenever α generates a normal basis B for E/F, we will denote the elements of B as $\alpha_i = \alpha^{q^i}$ with $i = 0, \ldots, n-1$. As before, we will also abbreviate the notation for the trace function of E/F to Tr.

We begin with two almost trivial but useful facts:

Lemma 8.6.1. *Let α generate a normal basis B for $E = \mathrm{GF}(q^n)$ over $F = \mathrm{GF}(q)$. Then*

$$\mathrm{Tr}(\xi) = \sigma_B(\xi)\mathrm{Tr}(\alpha) \quad \textit{for all } \xi \in E,$$

and

$$\mathrm{Tr}(\alpha_i^2) = c \quad \textit{for } i = 0, \ldots, n-1$$

for some $c \in F$, where $c \neq 0$ provided that B is trace-orthogonal.

Proof. As B consists of the conjugates of α, all elements of B have the same trace, namely $\mathrm{Tr}(\alpha)$. Thus

$$\mathrm{Tr}(\xi) = \sum_{i=0}^{n-1} x_i \mathrm{Tr}(\alpha_i) = \sigma_B(\xi)\mathrm{Tr}(\alpha),$$

where $\xi \in E$ with $r_B(\xi) = (x_0, \ldots, x_{n-1})$.

Similarly, the elements α_i^2 are the conjugates of α^2, so that $\mathrm{Tr}(\alpha_i^2)$ also is a constant c. Now assume that B is trace-orthogonal. Then $c \neq 0$, since otherwise

$$\mathrm{Tr}(\alpha\xi) = \sum_{i=0}^{n-1} x_i \mathrm{Tr}(\alpha\alpha_i) = x_0 \mathrm{Tr}(\alpha^2) = 0$$

for all $\xi \in E$, contradicting the non-degeneracy of the trace bilinear form (see Proposition 3.13.7). □

The following result of [210] gives a somewhat surprising connection between the arithmetical properties of normal bases studied in Section 8.4 and the notion of trace-orthogonality.

Theorem 8.6.2. *Let α generate a normal basis B for $E = \mathrm{GF}(q^n)$ over $F = \mathrm{GF}(q)$, and let A and T be the corresponding matrices considered in Proposition 8.4.6. Then the following four conditions are all equivalent:*

(a) *B is trace-orthogonal;*

(b) $T = A$*;*

(c) *T is symmetric;*

(d) $\sigma_B(\alpha^2) = \mathrm{Tr}(\alpha)$ *and* $\sigma_B(\alpha\alpha_i) = 0$ *for* $i = 1, \ldots, n-1$.

Proof. First assume the validity of (a) and put $\gamma_i := \alpha_i/\mathrm{Tr}(\alpha^2)$ for $i = 0,\dots,n-1$. Then Lemma 8.6.1 shows $\mathrm{Tr}(\alpha_i\gamma_j) = \delta_{ij}$ for all $i, j = 0,\dots,n-1$, and thus $\gamma := \gamma_0$ generates the dual normal basis $C = (\gamma_0,\dots,\gamma_{n-1})$ of B. By Lemma 7.7.2, we may use the elements of C to compute the coordinate vector $r_B(\xi)$ of any $\xi \in E$ with respect to B. According to Equation (8.31), the (i, j)-entry t_{ij} of T is the coefficient of α_j in $r_B(\alpha\alpha_i)$, and we obtain

$$t_{ij} = \mathrm{Tr}(\alpha_0\alpha_i\alpha_j)/\mathrm{Tr}(\alpha^2) \quad \text{for } i, j = 0,\dots,n-1. \tag{8.37}$$

Hence, according to the definition of A in Proposition 8.4.6,

$$a_{ij} = t_{i-j,-j} = \mathrm{Tr}(\alpha_0\alpha_{i-j}\alpha_{-j})/\mathrm{Tr}(\alpha^2) \quad \text{for } i, j = 0,\dots,n-1.$$

We now apply the unique Galois automorphism mapping $\alpha = \alpha_0$ to α_j to this equation. This gives

$$a_{ij} = \mathrm{Tr}(\alpha_j\alpha_i\alpha_0)/\mathrm{Tr}(\alpha^2) = \mathrm{Tr}(\alpha_0\alpha_i\alpha_j)/\mathrm{Tr}(\alpha^2) = t_{ij},$$

and thus the validity of (a) indeed implies that of (b).

Trivially, (b) implies (c), since A is a symmetric matrix. Now assume that (c) holds. As we have seen in the proof of Lemma 8.5.2, the rows of T sum to $\mathbf{s} := (\mathrm{Tr}(\alpha), 0,\dots,0)$. In other words, $\mathbf{s}$ is the vector of column sums of T. Since T is assumed to be symmetric, $\mathbf{s}$ must also be the vector of row sums of T, which gives the validity of (d).

Finally, assume that (d) holds. Then, by Lemma 8.6.1,

$$\mathrm{Tr}(\alpha\alpha_i) = \sigma_B(\alpha\alpha_i)\mathrm{Tr}(\alpha) = \delta_{0i}\mathrm{Tr}(\alpha)^2 \quad \text{for } i = 0,\dots,n-1. \tag{8.38}$$

As B is a normal basis, this immediately implies that B is indeed trace-orthogonal, which establishes (a). □

We note that the case $i = 0$ of the final Equation (8.38) in the preceding proof yields a rather curious multiplicative property of the trace function, which is then applied to show that the seemingly more general concept of trace-orthogonal normal bases reduces to that of self-dual normal bases, a result observed independently in both [212] and [273].

Lemma 8.6.3. *Let α generate a trace-orthogonal normal basis B for* $\mathrm{GF}(q^n)$ *over* $\mathrm{GF}(q)$. *Then* $\mathrm{Tr}(\alpha^2) = \mathrm{Tr}(\alpha)^2$. □

Theorem 8.6.4. *Every trace-orthogonal normal basis for* $\mathrm{GF}(q^n)$ *over* $\mathrm{GF}(q)$ *is equivalent to a self-dual normal basis.*

Proof. In view of Lemma 8.6.3, the element $\beta := \frac{1}{\mathrm{Tr}(\alpha)} \cdot \alpha$ generates a self-dual normal basis. □

It is also worthwhile to specialize Theorem 8.6.2 to the binary case:

Corollary 8.6.5. *Let α generate a normal basis B for $E = \mathrm{GF}(2^n)$ over $F = \mathrm{GF}(2)$, and let A and T be the corresponding matrices considered in Proposition 8.4.6. Then the following conditions are equivalent:*

(a) *B is self-dual;*

(b) *$T = A$;*

(c) *T is symmetric;*

(d) *$w_B(\alpha\alpha_0) \equiv 1 \bmod 2$ and $w_B(\alpha\alpha_i) \equiv 0 \bmod 2$ for $i = 1,\ldots,n-1$.* □

The proofs of Theorems 8.5.4 and 8.5.5 show that every type II optimal normal basis for $\mathrm{GF}(2^n)$ over $\mathrm{GF}(2)$ satisfies condition (4) in Corollary 8.6.5. This gives the following interesting consequence, which – as we shall see in the next section – is also useful for practical applications.

Theorem 8.6.6. *Every type II optimal normal basis is self-dual.* □

Theorem 8.6.6 shows that self-dual normal bases which have the smallest possible complexity among *all* normal bases for $\mathrm{GF}(2^n)$ over $\mathrm{GF}(2)$ are quite frequent. In view of Corollary 8.6.5 and Theorem 8.6.6, it is now also clear why we had $A = T$ for the normal basis considered in Examples 8.4.7 and 8.5.7.

In the context of Theorem 8.6.6, we also state the following characterization of all trace-orthogonal optimal normal bases, which was noted in [212]. This will be an immediate consequence of Theorem 8.6.2 and the optimal normal basis theorem to be proved later (see Theorem 8.8.1 below), since a type I optimal normal basis (as constructed in Theorem 8.5.3) does not satisfy condition (c) in Theorem 8.6.2.

Theorem 8.6.7. *Let α generate a trace-orthogonal optimal normal basis B for $\mathrm{GF}(q^n)$ over $\mathrm{GF}(q)$. Then q is even, and B is equivalent to a type II optimal normal basis for $\mathrm{GF}(2^n)$ over $\mathrm{GF}(2)$ as constructed in Theorems 8.5.4 and 8.5.5. In particular, $2n+1$ must be a prime, and n either is a primitive root modulo $2n+1$ or $2n+1 \equiv 3 \bmod 4$ and n generates the quadratic residues modulo $2n+1$.* □

Next, we mention an application of Theorem 8.6.6 which gives us a simple product construction for low complexity self-dual normal bases.

Corollary 8.6.8. *Let α and β generate type II optimal normal bases A and B for $K = \mathrm{GF}(2^m)$ and $L = \mathrm{GF}(2^n)$ over $F = \mathrm{GF}(2)$, respectively, and assume that m and n are relatively prime. Then $\gamma := \alpha\beta$ generates a self-dual normal basis N for $E = \mathrm{GF}(q^{mn})$ over F with complexity $C_N = 4mn - 2m - 2n + 1$.*

Proof. This follows by combining Theorem 8.6.6 with Corollaries 3.13.22 and 8.5.22. □

Example 8.6.9. Corollary 8.6.8 shows that self-dual normal bases with the smallest possible complexity among all normal bases for $\mathrm{GF}(2^n)$ over $\mathrm{GF}(2)$ (according to Tables 8.3 and 8.4) exist for $n = 15$, 21 and 22. □

We now study a simple general method for constructing new (self-dual) normal bases from known examples. In some cases, we will even be able to relate their complexities. These results are taken from Jungnickel [212].

Thus, let α generate a normal basis B for $E = \mathrm{GF}(q^n)$ over $F = \mathrm{GF}(q)$. Trivially, $a\alpha$ is likewise normal for each choice of $a \in F^*$. Of course, the normal bases generated by these two elements are equivalent; in particular, they have the same complexity. We now discuss a more general way of using α to construct further normal elements, which will usually lead to bases of different complexity. For this, we put $\gamma := a + b\alpha$, where a and b are non-zero elements of F, and investigate when such an element γ is again a normal basis generator; a similar result will then be obtained under the additional assumption of trace-orthogonality.

Proposition 8.6.10. *Let α generate a normal basis B for $E = \mathrm{GF}(q^n)$ over $F = \mathrm{GF}(q)$, and let $a, b \in F^*$. Then $\gamma := a + b\alpha$ is likewise a normal element for E/F if and only if*

$$na + b\mathrm{Tr}(\alpha) \neq 0. \tag{8.39}$$

Proof. Assume first that (8.39) is violated. Then

$$\mathrm{Tr}(\gamma) = \mathrm{Tr}(a + b\alpha) = na + b\mathrm{Tr}(\alpha) = 0,$$

and hence the conjugates of γ are linearly dependent.

Conversely, assume the validity of (8.39). We need to show that the conjugates $\gamma_i = \gamma^{q^i}$ of γ (with $i = 0, \ldots, n-1$) are linearly independent. Suppose otherwise, say

$$x_0\gamma_0 + \cdots + x_{n-1}\gamma_{n-1} = 0, \quad \text{where not all } x_i = 0.$$

Substituting for the γ_i gives

$$x_0\alpha_0 + \cdots + x_{n-1}\alpha_{n-1} = -\frac{a}{b}(x_0 + \cdots + x_{n-1}) =: y.$$

Because of $\alpha_0 + \cdots + \alpha_{n-1} = \mathrm{Tr}(\alpha) \neq 0$, we can also write

$$y = \frac{y\mathrm{Tr}(\alpha)}{\mathrm{Tr}(\alpha)} = \frac{y(\alpha_0 + \cdots + \alpha_{n-1})}{\mathrm{Tr}(\alpha)},$$

and hence the uniqueness of the representation of y with respect to the normal basis B generated by α gives $x_0 = \cdots = x_{n-1} = y/\mathrm{Tr}(\alpha)$. This shows

$$y = -\frac{a}{b}(x_0 + \cdots + x_{n-1}) = -\frac{nay}{b\mathrm{Tr}(\alpha)}.$$

But $y \neq 0$, as the conjugates of α are linearly independent and as at least one $x_i \neq 0$, so that the preceding identity contradicts the validity of (8.39). □

We note that there always exist elements $a, b \in F^*$ for which (8.39) is satisfied, except in the case where $q = 2$ and n is odd. Next, we prove a similar result under the additional assumption that the normal basis generated by α is actually trace-orthogonal.

Proposition 8.6.11. *Assume that α generates a trace-orthogonal normal basis B for $E = \mathrm{GF}(q^n)$ over $F = \mathrm{GF}(q)$, and let $a, b \in F^*$. Then $\gamma := a + b\alpha$ likewise generates a trace-orthogonal normal basis if and only if*

$$na + 2b\mathrm{Tr}(\alpha) = 0. \tag{8.40}$$

Proof. Note first that the validity of (8.40) implies that of (8.39), since $b\mathrm{Tr}(\alpha) \neq 0$. Hence (8.40) can only hold if γ at least generates a normal basis. Now assume this to be the case and denote the conjugates of γ by $\gamma_i = \gamma^{q^i}$, as usual. By definition,

$$\gamma\gamma_i = (a + b\alpha)(a + b\alpha_i) = a^2 + ab(\alpha + \alpha_i) + b^2\alpha\alpha_i$$

and therefore, as B is trace-orthogonal, $\mathrm{Tr}(\gamma\gamma_i) = na^2 + 2ab\mathrm{Tr}(\alpha)$ for $i = 1, \ldots, n-1$. Because of $a \neq 0$, the normal basis generated by γ is again trace-orthogonal if and only if (8.40) holds. □

We leave the question when elements $a, b \in F^*$ exist for which condition (8.40) holds to the reader; see Exercise 8.6.27. Propositions 8.6.10 and 8.6.11 immediately yield the following result for the particularly interesting binary case:

Corollary 8.6.12. *Let $\alpha \in \mathrm{GF}(2^n)$, where n is even, and put $\gamma := 1 + \alpha$. Then γ generates a (self-dual) normal basis if and only if α does.* □

In Propositions 7.3.5 and 7.6.3, we have seen that one may generate all (self-dual) normal bases for $\mathrm{GF}(q^n)$ over $\mathrm{GF}(q)$ – provided that one such basis B is known – by transforming B with all (orthogonal) circulant matrices in $GL(n,q)$. Unfortunately, in general there seems to be no way of relating the complexities of the original basis and of the transformed basis obtained from a specified circulant matrix.

The interest of Corollary 8.6.12 is due to the fact that in this special case the complexities of the two normal bases can often be related, as we shall show next. This will then be used to obtain some results concerning the complexity of self-dual normal bases for $\mathrm{GF}(2^n)$ over $\mathrm{GF}(2)$, provided that n is even (and hence congruent to 2 modulo 4, by Proposition 3.13.20). Our main interest will be in an upper bound which quite surprisingly coincides with the maximum possible complexities given in Tables 8.3 and 8.4 for several values of n.

Theorem 8.6.13. *Let α generate a self-dual normal basis B for $E = \mathrm{GF}(2^n)$ over $F = \mathrm{GF}(2)$, and assume that n is even. Put $\gamma := 1 + \alpha$, and let B' denote the self-dual normal basis generated by γ, according to Corollary 8.6.12. Then the complexities of B and B' are related as follows:*

$$C_{B'} = n^2 - 3n + 8 - C_B.$$

Proof. We begin with two auxiliary observations concerning some entries of the multiplication matrix T for B introduced in Observation 8.4.5, both of which follow easily from Equation (8.38) in the proof of Theorem 8.6.2:

$$t_{i0} = \mathrm{Tr}(\alpha\alpha\alpha_i) = \mathrm{Tr}(\alpha_1\alpha_i) = \delta_{1i} \quad \text{for } i = 0, \dots, n-1 \tag{8.41}$$

and, similarly,

$$t_{ii} = \mathrm{Tr}(\alpha_i\alpha\alpha_i) = \mathrm{Tr}(\alpha\alpha_i^2) = \mathrm{Tr}(\alpha\alpha_{i+1}) = \delta_{0,i+1}, \tag{8.42}$$

where indices are taken modulo n.

In order to compute the complexity of B', we need to compute the weights $w_{B'}(\gamma\gamma_i)$ for $i = 0, \dots, n-1$. Trivially,

$$w_{B'}(\gamma\gamma_0) = w_{B'}(\gamma_1) = 1. \tag{8.43}$$

Now let $i \neq 0$. Then $\gamma\gamma_i = (1+\alpha)(1+\alpha_i) = 1 + \alpha + \alpha_i + \alpha\alpha_i$. Using $\mathrm{Tr}(\alpha) = 1$ and the fact that the coordinate vector $r_B(\alpha\alpha_i)$ is the i-th row of the matrix T, this may be written as

$$\gamma\gamma_i = (1+\alpha_0) + (1+\alpha_i) + (\alpha_0 + \cdots + \alpha_{n-1}) + (t_{i,0}\alpha_0 + \cdots + t_{i,n-1}\alpha_{n-1}).$$

By hypothesis and by Corollary 8.6.5, both n and $w_B(\alpha\alpha_i)$ are even, and we may re-write the preceding equation as follows:

$$\begin{aligned}\gamma\gamma_i &= (1+\alpha_0) + (1+\alpha_i) + \Big((1+t_{i,0})(1+\alpha_0) + \cdots + (1+t_{i,n-1})(1+\alpha_{n-1})\Big)\\ &= \gamma_0 + \gamma_i + \Big((1+t_{i,0})\gamma_0 + \cdots + (1+t_{i,n-1})\gamma_{n-1}\Big).\end{aligned}$$

By (8.41) and (8.42), $t_{i,0} = t_{ii} = 0$ if $i \neq 1, n-1$, so that the terms involving γ_0 and γ_i, respectively, cancel for $i \neq 1, n-1$; hence γ_j then has coefficient 1 if and only if $t_{i,j} = 0$ and $j \neq 0, i$. This shows

$$w_{B'}(\gamma\gamma_i) = n - 2 - w_B(\alpha\alpha_i) \quad \text{for } i \neq 0, 1, n-1. \tag{8.44}$$

A similar argument gives

$$w_{B'}(\gamma\gamma_i) = n - w_B(\alpha\alpha_i) \quad \text{for } i \in \{1, n-1\}. \tag{8.45}$$

Substituting (8.43), (8.44) and (8.45) for the weights $w_{B'}(\gamma\gamma_i)$, we obtain

$$\begin{aligned}C_{B'} &= w_{B'}(\gamma\gamma_0) + w_{B'}(\gamma\gamma_1) + \cdots + w_{B'}(\gamma\gamma_{n-1})\\ &= 1 + (n - w_B(\alpha\alpha_1)) + (n - w_B(\alpha\alpha_{n-1})) +\\ &\qquad + ((n-2-w_B(\alpha\alpha_2)) + \cdots + (n-2-w_B(\alpha\alpha_{n-2}))\\ &= 1 + 2n + (n-3)(n-2) - (C_B - 1),\end{aligned}$$

which gives the desired formula. □

Corollary 8.6.14. *Let* $n \equiv 2 \bmod 4$. *Then the average complexity of a self-dual normal basis for* $\mathrm{GF}(2^n)$ *over* $\mathrm{GF}(2)$ *is* $\frac{1}{2}(n^2 - 3n + 8)$. □

As Corollary 8.6.14 shows, Conjecture 8.5.25 is true for even values of n if we restrict attention to self-dual normal bases; indeed, then a somewhat stronger bound holds, and one actually has equality.

We now use Theorem 8.6.13 to obtain a further result on the complexity of self-dual normal bases over $\mathrm{GF}(2)$.

Theorem 8.6.15. *Let* B *be a self-dual normal basis for* $\mathrm{GF}(2^n)$ *over* $\mathrm{GF}(2)$, *where* $n \equiv 2 \bmod 4$. *Then*

$$2n - 1 \leq C_B \leq n^2 - 5n + 9. \tag{8.46}$$

Equality holds in one of these bounds if and only if either B *or the normal basis* B' *constructed in Corollary 8.6.12 is optimal; in this case,* $2n+1$ *is a prime and* 2 *is a primitive root modulo* $2n+1$.

Proof. By Lemma 8.5.2, the lower bound in (8.46) holds for any normal basis. Since B is self-dual, so is B'. Note that $(B')' = B$; hence we may apply Theorem 8.6.13 to B' and obtain

$$C_B = C_{(B')'} = n^2 - 3n + 8 - C_{B'} \leq n^2 - 3n + 8 - (2n - 1),$$

which gives the upper bound in (8.46).

Clearly, equality in one of the two bounds in (8.46) means that either B or B' is optimal. Because of the self-duality of these two bases, either B or B' is constructed as in Theorem 8.5.4 or Theorem 8.5.5, by Theorem 8.6.7. In particular, $2n+1$ is a prime satisfying the restrictions stated there. But $2n+1 \equiv 1 \bmod 4$ rules out the examples constructed in Theorem 8.5.5, which gives the assertion. □

Let us compare the preceding results with the relevant cases satisfying $n \equiv 2 \bmod 4$ in Tables 8.3 and 8.4. As we will see, we sometimes get theoretical constructions for maximum complexity normal bases.

Example 8.6.16. We note that the upper bound in (8.46) agrees with the maximum complexity of *any* normal basis in the cases $n = 2$, 14, 18, 26 and 30. Moreover, in each of theses five cases, there exists a type II optimal normal basis B, which is self-dual by Theorem 8.6.6; hence the corresponding self-dual basis B' is a maximum complexity normal basis, by Theorem 8.6.13. □

Example 8.6.17. By Theorem 8.6.6, there exist self-dual optimal normal bases for the degrees $m = 2$ and $n = 11$. Hence Corollary 8.6.8 gives the existence of a self-dual normal basis B with complexity 63 for $\mathrm{GF}(2^{22})$ over $\mathrm{GF}(2)$. The associated self-dual normal basis B' has complexity 363, by Theorem 8.6.6. According to Table 8.4, these are the minimum and maximum complexities of normal bases for the extension degree 22. □

Example 8.6.17 provides a further extension degree where we know theoretical constructions for normal bases with the extremal complexities. We can obtain two additional such cases by using the following strengthening of Result 8.5.17, which was proved in the Appendix of [13]:

Result 8.6.18 *Let B be one of the low complexity normal bases considered in Result 8.5.17. Then B is self-dual if and only if k is even.* □

Example 8.6.19. First let $n = 38$. By Result 8.6.18, the normal basis B with minimum complexity 207 listed in Table 8.4 is actually self-dual. By Theorem 8.6.13, the associated self-dual normal basis B' has complexity 1131 and thus is a maximum complexity normal basis.

Now let $n = 34$; this case is a little more involved. By Result 8.6.18, the normal basis with minimum complexity 81 listed in Table 8.4 for the degree $k = 17$ is actually self-dual. We use this basis together with the self-dual optimal normal basis of degree $m = 2$ in Corollary 8.6.8 to obtain a self-dual normal basis B of degree $n = mk = 34$ with complexity 243. Then the associated self-dual normal basis B' has complexity 819, again by Theorem 8.6.13. According to Table 8.4, these are indeed the extremal complexities for $n = 34$. □

Remark 8.6.20. As the preceding examples show, the bound in (8.46) yields a normal basis of maximum complexity for all $n \equiv 2 \bmod 4$ with $14 \le n \le 38$. In contrast, this does *not* hold for the two smallest values 6 and 10. Thus it seems rather difficult to guess what might happen for $n \ge 42$.

Indeed, for $n = 6$, the upper bound in (8.46) is only 15, which may be realized using an optimal normal basis in Theorem 8.6.13. However, according to Table 8.3, the maximum complexity among the four normal bases with degree 6 is 17.

For $n = 10$, there is no self-dual normal basis meeting the upper bound 59 in Theorem 8.6.15, since 21 is not prime. By Lemma 8.5.2, any normal basis has odd complexity, and thus the largest conceivable complexity for a self-dual normal basis with degree 10 is 57. According to Table 8.3, there is a normal basis with complexity 61 among the 48 normal bases for $n = 10$.

Remark 8.6.21. The preceding discussion shows that the assumption of self-duality in Theorems 8.6.13 and 8.6.15 is indeed important. For arbitrary normal bases B, it is not known how the complexities of B and B' are related. Reviewing the proof of Theorem 8.6.13, it seems difficult to obtain a general result, since the distribution of even and odd weights $w_B(\alpha\alpha_i)$ among the rows of the multiplication matrix T is, in general, unknown – which presumably prevents the required computations. It is also an open problem whether anything similar to Theorem 8.6.13 can be done if either $q = 2$ and n is odd, or if $q \neq 2$.

There is one exception to the preceding comment: Jungnickel [212] also determined the complexity of the normal basis B' generated by $\gamma = 1 + \alpha$ for the case where B is a binary type I optimal normal basis. In this case, B' turns out to be essentially – that is, up to a re-ordering – the dual basis of the original basis B generated by α, see Remark 8.6.23 below. Later, Wang and Zhou [394] managed to determine the complexity of the dual of an *arbitrary* type I optimal normal basis. We shall now present this general result, giving a considerably simplified version of their proof.

Theorem 8.6.22. *Let q be a prime power, and let $r = n+1$ be an odd prime for which q is a primitive root modulo r. Moreover, let B be the type I optimal normal basis for* $\mathrm{GF}(q^n)$ *over* $F = \mathrm{GF}(q)$ *generated by a primitive r-th root of unity α, as in the proof of Theorem 8.5.3.*

Then the dual normal basis B^* of B (and hence also the dual of every normal basis equivalent to B) has complexity either $3n-3$ or $3n-2$, depending on whether q is even or odd, respectively.

Proof. As usual, we write $B = (\alpha_0, \ldots, \alpha_{n-1})$ with $\alpha_i = \alpha^{q^i}$; similarly, we write the dual normal basis of B as $B^* = (\gamma_0, \ldots, \gamma_{n-1})$ with $\gamma_i := \gamma^{q^i}$. We want to compute $\gamma = \gamma_0$ via Theorem 8.4.9, and thus we need to determine the values $\mathrm{Tr}(\alpha\alpha_i)$. By hypothesis, $r = n+1$ is an odd prime, so that n is even, say $n = 2m$. Also $\gcd(q,r) = 1$, as q is a primitive root modulo r; hence r is invertible in F.

According to the construction in the proof of Theorem 8.5.3, the elements of B are the n non-trivial r-th roots of unity, that is, the roots of the self-reciprocal cyclotomic polynomial $\Phi_r = x^n + \cdots + x + 1$. By Theorem 5.7.2, $\alpha^{q^m} = \alpha^{-1}$, and hence $\mathrm{Tr}(\alpha\alpha_m) = \mathrm{Tr}(1) = n = r-1$. For all other indices, $\alpha\alpha_i$ is again a root of Φ_r, that is,

$$\alpha\alpha_i = \alpha_{\pi(i)} \quad \text{for } i = 0, \ldots, n-1,\ i \neq m, \tag{8.47}$$

where $\pi \colon \{0, \ldots, n-1\} \setminus \{m\} \to \{1, \ldots, n-1\}$ is a bijection with $\pi(i) \neq i$ for all i. Hence $\mathrm{Tr}(\alpha\alpha_i) = \mathrm{Tr}(\alpha_{\pi(i)}) = \mathrm{Tr}(\alpha) = \alpha + \alpha^2 + \cdots + \alpha^n = -1$ for $i \neq 0$, so that the polynomial h defined in Theorem 8.4.9 here is

$$h = -\left(1 + x + \cdots + x^{m-1}\right) + nx^m - \left(x^{m+1} + \cdots + x^{n-1}\right) = -f + rx^m,$$

where $f := x^{n-1} + \cdots + x + 1$. Using the trivial observation $xf \equiv f \pmod{x^n - 1}$, it is easily checked that $g := r^{-1}(f + x^m)$ is the unique polynomial of degree $< n$ satisfying $g(x)h(x) \equiv 1 \pmod{x^n - 1}$. Therefore,

$$\gamma_0 = \gamma = r^{-1}\left((\alpha_0 + \cdots + \alpha_{n-1}) + \alpha_m\right) = r^{-1}\left(\mathrm{Tr}(\alpha) + \alpha_m\right) = r^{-1}(\alpha_m - 1),$$

and hence

$$\gamma_i = r^{-1}(\alpha_{i+m} - 1) \quad \text{for } i = 0, \ldots, n-1.$$

This gives

$$\gamma\gamma_i = r^{-2}\left(1 - \alpha_m - \alpha_{i+m} + \alpha_m\alpha_{i+m}\right) = r^{-2}\left(1 - \alpha_m - \alpha_{i+m} + (\alpha\alpha_i)^{q^m}\right) \tag{8.48}$$

for $i = 0, \ldots, n-1$. We now use Equation (8.48) to determine the rows of the multiplication matrix T^* for B^*, that is, the coordinate vectors $r_{B^*}(\gamma\gamma_i)$.

First assume $i \neq m$. Then we can re-write Equation (8.48) as follows, using (8.47):

$$\begin{aligned}\gamma\gamma_i &= r^{-2}\left(1-\alpha_m-\alpha_{i+m}+\alpha_{\pi(i)+m}\right)\\ &= -r^{-1}\left(r^{-1}(\alpha_m-1)+r^{-1}(\alpha_{i+m}-1)-r^{-1}(\alpha_{\pi(i)+m}-1)\right)\\ &= -r^{-1}\left(\gamma+\gamma_i-\gamma_{\pi(i)+m}\right),\end{aligned}$$

so that $r_{B^*}(\gamma\gamma_i)$ has weight 3 for $i \neq 0, m$. For $i = 0$, the weight $r_{B^*}(\gamma\gamma_0)$ is either 1 or 2, depending on the characteristic of $\mathrm{GF}(q)$, since then the two terms involving $\gamma = \gamma_0$ cancel if q is even and simplify to $-2r^{-1}\gamma \neq 0$ if q is odd.

It remains to consider the case $i = m$. Then $\alpha_m = \alpha^{-1}$, and Equation (8.48) gives

$$\gamma\gamma_m = r^{-2}(1-\alpha_m-\alpha+1) = -r^{-1}(\gamma+\gamma_m),$$

so that $r_{B^*}(\gamma\gamma_m)$ has weight 2. Combining these observations gives the complexity of B^* as stated in the assertion. □

Remark 8.6.23. As mentioned in Remark 8.6.21, in the binary case the normal basis B' generated by $\gamma = 1+\alpha$ is essentially the dual of the type I optimal normal basis B. Using the proof of Theorem 8.6.22, this is easy to check:

$$\mathrm{Tr}(\gamma\alpha_i) = \mathrm{Tr}(\alpha_i)+\mathrm{Tr}(\alpha\alpha_i) = 1+\mathrm{Tr}(\alpha\alpha_i) = \delta_{im},$$

in view of Equation (8.47) and $\alpha_m = \alpha^{-1}$. □

Wang and Zhou also obtained the following weak converse of Theorem 8.6.22, which we state without proof.

Result 8.6.24 *Consider a dual pair B and B^* of normal bases for $\mathrm{GF}(q^n)$ over $F = \mathrm{GF}(q)$, and assume that B^* has complexity either $3n-3$ or $3n-2$, depending on whether q is even or odd, respectively. Assume in addition that the first column of the multiplication matrix T^* associated with B^* is $(2a, a, \ldots, a)^T$ for some $a \in F^*$. Then B is equivalent to a type I optimal normal basis.* □

The case $q = 2$ of Result 8.6.24 was obtained earlier by Panario and Young [407] who also posed the still unresolved problem of classifying all non-optimal normal bases for $\mathrm{GF}(2^n)$ over $\mathrm{GF}(2)$ with complexity at most $3n$. For $n > 15$, all known examples are either duals of type I optimal normal bases, or arise from type II optimal normal bases of degrees 2 and m via Proposition 8.5.21, where $m = n/2$ is odd. In particular, this includes the following interesting conjecture:

Conjecture 8.6.25. Assume that $E = \mathrm{GF}(2^n)$ does not admit an optimal normal basis over $F = \mathrm{GF}(2)$, that is, n satisfies neither of the two conditions in Theorem 8.8.1 below. Then every normal basis B for E/F has complexity $C_B \geq 3n-3$. □

We conclude this section with computational results of Geiselmann [136] who listed, for all $n \leq 60$, irreducible polynomials leading to a normal basis respectively a self-dual normal basis for $\mathrm{GF}(2^n)$ over $\mathrm{GF}(2)$ of the lowest *known* complexity among all such bases; recall that a self-dual normal basis exists if and only if n is not a multiple of 4.

n	C_B	irreducible polynomial	Remarks
2	3	111	sd, optimal
3	5	1101	sd, optimal
4	7	11111	optimal
5	9	110111	sd, optimal
6	11	1110011	sd, optimal
7	19 21	11100101 11010011	best best self-dual
8	21	110101001	best
9	17	1101110011	sd, optimal
10	19 27	11111111111 11100101011	optimal best self-dual
11	21	110100011101	sd, optimal
12	23	1111111111111	optimal
13	45	11010010011001	sd, best
14	27	111001100000011	sd, optimal
15	45	1101001010110101	sd, best
16	85	11011110110101111	best
17	81	110111111010101011	sd, best
18	35	1110011011100000111	sd, optimal
19	117	11010000100101010011	sd, best
20	63	111001101100001001011	best

Table 8.5 Irreducible polynomials giving low complexity normal bases for $\mathrm{GF}(2^n)$ over $\mathrm{GF}(2)$ with $2 \leq n \leq 20$

In Tables 8.5 to 8.7, such irreducible polynomials are given by exhibiting corresponding binary sequences; for instance, the sequence 1011 stands for x^3+x+1. An entry "optimal" or "best" in the column headed "Remarks" indicates that the irreducible polynomial displayed generates an optimal normal basis or a minimum complexity normal basis, respectively. An entry "sd" means that the resulting basis is self-dual, and an entry "best self-dual" indicates that the polynomial exhibited generates a self-dual normal basis with the minimum complexity among all self-dual normal bases (but not necessarily a minimum complexity normal basis).

Exercises

Exercise 8.6.26. Assume that α and β generate a pair of dual normal bases for $E = \mathrm{GF}(q^n)$ over $F = \mathrm{GF}(q)$. Prove that $\mathrm{Tr}(\alpha)\mathrm{Tr}(\beta) = 1 = \mathrm{Tr}(\alpha\beta)$. □

n	C_B	irreducible polynomial	Remarks
21	95	1110101110001000110001	best
	105	1101111000111000101101	best self-dual
22	63	11101011110010010100101	sd, best
23	45	110100010000000111010001	sd, optimal
24	105	1101111100000001110101001	best
25	93	11010010011001111000011111	sd, best
26	51	111001101110000000000110111	sd, optimal
27	141	1101001001000101111000000111	sd, best
28	55	11111111111111111111111111111	optimal
29	57	110111000001110000000000000111	sd, optimal
30	59	1110011000000110000000000000011	sd, optimal
31	237	11010000011000100001100010010001	sd, best
32	361	110111100110101011111010100100001	best
33	65	1101110011000000110000000000000011	sd, optimal
34	243	11100101100011110110010100110110111	sd
35	69	110100011101000011010000000000001101	sd, optimal
36	71	1111111111111111111111111111111111111	optimal
37	141	11010010011111001011000100011110001001	sd, best
38	207	111010111100011100100011110000010101101	sd
39	77	1101000100000000111010001000000011010001	sd, optimal
40	189	11010010010000101101011010000001000100001	

Table 8.6 Irreducible polynomials giving low complexity normal bases for GF(2^n) over GF(2) with $21 \leq n \leq 40$

Exercise 8.6.27. Determine for which pairs (q,n) elements $a,b \in \mathrm{GF}(q)^*$ exist which satisfy condition (8.40). □

8.7 Modified Massey-Omura Multipliers

In this section, which is based on Gollmann [141], we return to the topic of Section 8.4 and consider once more Massey-Omura multipliers. As we shall see, a modification of the obvious architecture discussed before can result in a further reduction of the number of gates required. We begin by giving a different interpretation of the matrix A occurring in Proposition 8.4.6, which also takes into account the concept of duality.

n	C_B	irreducible polynomial	Remarks
41	81	110111001100000000011100110000001101110011	sd,optimal
42	135	1110101100011000100110111001001000000001101	sd
43	165	11010010101011111011010101000101101110101001	best self-dual
44	147	111010000101001011111101011100001111000101101	
45	153	1101111110001001000000100101010111101001000111	best self-dual
46	135	11101011001000100110100100100001010010010010001	sd
47	261	110100101001011110100111111010011011011111101101	best self-dual
48	425	1101001100000110101011001001100010011000111111101	
49	189	11010010011111011101001000011110010011111110010011	sd
50	99	111001101110000011100000000000000000110111000000111	sd, optimal
51	101	1101000111010000110100000000000000000001110100001101	sd, optimal
52	103	11111111111111111111111111111111111111111111111111111	optimal
53	105	110111000001110011011100000000000000000000011100110111	sd, optimal
54	209 423	1100111100111100000001110101000001101100101101010010 1001 1110101100101100110110010011110111101001010000000011011	 sd
55	189	11010010101011101100001010100000000100111000110001110 0101	sd
56	399	111010110101111000100010001001001011011100111010000100001	
57	497	1101110101110101110000100010001110011111010100101011001111	sd
58	115 171	11111111111111111111111111111111111111111111111111111111111 11100101011110110110001101100110000000110110000000000011011	optimal sd
59	597	110100011110011010101110101000011100111111010011111100110001	sd
60	119	1111111111111111111111111111111111111111111111111111111111111	optimal

Table 8.7 Irreducible polynomials giving low complexity normal bases for $\mathrm{GF}(2^n)$ over GF(2) with $41 \leq n \leq 60$

In what follows, let B be a normal basis for $E = \mathrm{GF}(q^n)$ over $F = \mathrm{GF}(q)$ generated by α, and let B^* be its dual basis. As before, we write $B = (\alpha_0, \ldots, \alpha_{n-1})$ with $\alpha_i = \alpha^{q^i}$ for $i = 0, \ldots, n-1$ and let β denote the dual normal element generating B^*. Recall that β may be computed explicitly from the multiplication matrix T via Theorem 8.4.9, if desired. Moreover, let A be the symmetric matrix and f the associated symmetric bilinear form determined by the basis B as in Proposition 8.4.6, so that $f(\xi, \eta)$ is the coefficient of α in the product $\pi := \xi\eta$ (for all $\xi, \eta \in E$). The following observation is simple but crucial.

Lemma 8.7.1. *Assume that α and β generate a pair B and B^* of dual normal bases for $E = \mathrm{GF}(q^n)$ over $F = \mathrm{GF}(q)$. Let ξ and η be elements of E, given in primal* *coordinates* $r_B(\xi) = (x_0, \ldots, x_{n-1})$ *and* $r_B(\eta) = (y_0, \ldots, y_{n-1})$, *respectively. Finally, let A be the symmetric matrix and f its associated symmetric bilinear form defined in Proposition 8.4.6. Then*

$$f(\xi,\eta) = \mathrm{Tr}(\xi\eta\beta) \quad \textit{and} \quad r_B(\xi)A = r_{B^*}(\beta\xi).$$

Proof. This is an easy consequence of Lemma 7.7.2. Indeed, the first identity holds as $f(\xi,\eta)$ is the coefficient of α in the product $\pi = \xi\eta$. For the second identity, one notes

$$(\beta\xi)_j = \mathrm{Tr}(\beta\xi\alpha_j) \quad \text{for } j = 0,\ldots,n-1.$$

Together with the first identity and the trivial fact $a_{ij} = f(\alpha_i, \alpha_j)$, this gives

$$\begin{aligned}(\beta\xi)_j &= \mathrm{Tr}\Big(\beta\alpha_j\sum_{i=0}^{n-1} x_i\alpha_i\Big) = \sum_{i=0}^{n-1} x_i \mathrm{Tr}(\alpha_i\alpha_j\beta)\\ &= \sum_{i=0}^{n-1} x_i f(\alpha_i,\alpha_j) = \sum_{i=0}^{n-1} x_i a_{ij} \quad \text{for } j = 0,\ldots,n-1,\end{aligned}$$

that is, the desired second identity. □

By Lemma 8.7.1, the matrix A describes the linear transformation $\xi \mapsto \beta\xi$, with a simultaneous change from primal to dual coordinates. Moreover, the computation of the primal coordinate

$$p_0 = f(\xi,\eta) = r_B(\xi)Ar_B(\eta)^T = r_{B^*}(\beta\xi)r_B(\eta)^T = \mathrm{Tr}(\xi\eta\beta) \tag{8.49}$$

of the product $\pi = \xi\eta$ may be interpreted as either forming the scalar product of the coordinate vectors of $\beta\xi$ (in dual coordinates) and η (in primal coordinates), or as computing the trace of the product $\xi\eta\beta$.

Of course, there is no need to perform any coordinate changes in Lemma 8.7.1 and Equation (8.49), provided that α actually generates a self-dual normal basis. In particular, this holds for type II optimal normal bases, by Theorem 8.6.6.

This re-interpretation of the computation of the primal coordinate p_0 of $\pi = \xi\eta$ – and then, by cyclic shifts of the input vectors, of *all* primal coordinates of π – leads to a modification of the Massey-Omura architecture considered before when one uses a self-dual normal basis. Then Equation (8.49) and Proposition 8.4.6 give the following set of equations:

$$p_k = f(\xi^{q^{n-k}}, \eta^{q^{n-k}}) = r_B(\alpha\xi^{q^{n-k}})r_B(\eta^{q^{n-k}})^T \quad \text{for } k = 0,\ldots,n-1. \tag{8.50}$$

Based on these equations, one may now design a Massey-Omura multiplier by combining an architecture which multiplies a variable input vector ξ by the normal basis generator α with the trivial circuitry needed to compute a scalar product. An example should make this idea clear.

Example 8.7.2. As in Examples 8.4.7 and 8.5.7, we again consider a root α of the irreducible polynomial $f = x^5 + x^4 + x^2 + x + 1$ over GF(2). Recall that the associated multiplication matrix for the self-dual normal basis B generated by α is

$$A = T = \begin{pmatrix} 0 & 1 & 0 & 0 & 0 \\ 1 & 0 & 0 & 1 & 0 \\ 0 & 0 & 0 & 1 & 1 \\ 0 & 1 & 1 & 0 & 0 \\ 0 & 0 & 1 & 0 & 1 \end{pmatrix}.$$

Now the second identity in Lemma 8.7.1 gives

$$\begin{aligned} r_B(\alpha\xi) = r_B(\xi)A &= (x_0, x_1, x_2, x_3, x_4) \begin{pmatrix} 0 & 1 & 0 & 0 & 0 \\ 1 & 0 & 0 & 1 & 0 \\ 0 & 0 & 0 & 1 & 1 \\ 0 & 1 & 1 & 0 & 0 \\ 0 & 0 & 1 & 0 & 1 \end{pmatrix} \\ &= (x_1, x_0 + x_3, x_3 + x_4, x_1 + x_2, x_2 + x_4). \end{aligned}$$

Equation (8.50) with $k = 0$ shows that the multiplier in Figure 8.8 initially computes the value

$$p_0 = f(\xi, \eta) = r_B(\alpha\xi) r_B(\eta)^T.$$

Using Equation (8.50) with $k = 1$, one checks that the shift register will hold the coordinate $p_1 = f(\xi^{q^{n-1}}, \eta^{q^{n-1}})$ of the product $\pi = \xi\eta$ after one clock cycle. Thus, the device will yield the coordinate p_k of the product after k clock cycles. □

It is easily checked that a **modified Massey-Omura multiplier** based on a self-dual normal basis B for GF(2^n) over GF(2) needs exactly $C_B - 1$ XOR-gates and n AND-gates, which is, in comparison to the original architecture discussed in Section 8.4, a saving of exactly n AND-gates. (Of course, the XOR-block in Figure 8.8 is an abbreviation for an ensemble of five respectively, in general, $n - 1$ XOR-gates.) In particular, this approach reduces the total number of gates required by about 25 % if B is a type II optimal normal basis.

Lemma 8.7.1 was also used for a different architecture for normal basis multipliers by Agnew, Mullin, Onyszchuk and Vanstone [1]. Their device computes with each clock cycle exactly one (still missing) term of each of the scalar products $p_k = r_B(\alpha\xi^{q^{n-k}}) r_B(\eta^{q^{n-k}})^T$ and adds it in a further accumulation register to the previously computed partial sum. After $n - 1$ clock cycles, the k-th cell of the accumulation register will hold the value p_k for $k = 0, \ldots, n-1$. Thus the architectures which we have discussed are bit-serial multipliers, whereas the device of [1] is a bit-parallel normal basis multiplier. These authors also show how one may achieve a fanout of at most 4 if one uses a (type I or type II) optimal normal basis. Bit-parallel multipliers were also discussed by Gollmann [141]. A further reduction

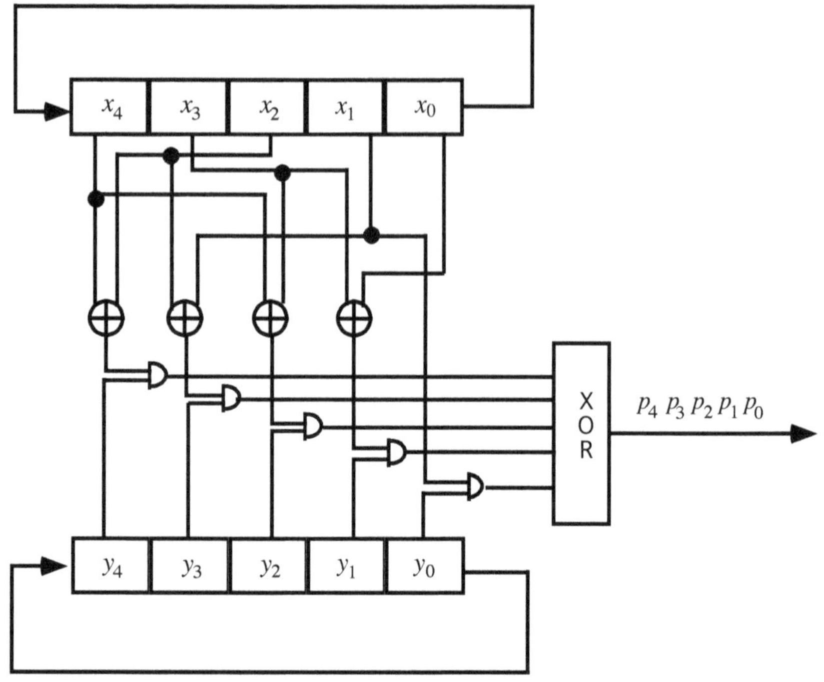

Fig. 8.8 Modified Massey-Omura multiplier for GF(32)

in the number of gates required can be achieved by exploiting the symmetry of the matrix A, see [137, 141].

The use of self-dual normal bases in the construction of Massey-Omura multipliers was also recommended by Wang [396] who obtained some results similar to those discussed here. In his paper, one also finds the multiplication matrices corresponding to self-dual normal bases – which were constructed by his method discussed in Theorem 7.8.9 – for the extension degrees $n = 9$ and $n = 17$; however, the resulting complexities are not particularly good, namely 29 and 117, respectively.

Wang also claimed that the complexity of a self-dual normal basis is less than that of an arbitrary normal basis; unfortunately, this was based solely on some computer searches. As the results of Section 8.6 show, his claim is not correct as stated for even values of n – even though there certainly exist many self-dual normal bases with a very favorable complexity. Nevertheless, the discussion in Section 8.6 suggests that one might (for even n) indeed get a better result than by just randomly selecting *any* normal basis B if one always chooses the lower complexity basis from the pair $\{B, B'\}$ for a randomly selected self-dual normal basis B.

8.8 The Optimal Normal Basis Theorem

In this final section, we provide a proof of the characterization theorem for optimal normal bases due to Gao and Lenstra [127], which was already mentioned before: the type I and type II optimal normal bases constructed in Section 8.5 essentially give all optimal normal bases. More precisely:

Theorem 8.8.1 (Optimal normal basis theorem). *Let α generate a normal basis B for $E = \mathrm{GF}(q^n)$ over $F = \mathrm{GF}(q)$. Then B is an optimal normal basis for E/F if and only if there exist a prime r, a primitive r-th root of unity ζ (in a suitable extension field of E), and an element $c \in F^*$ such that one of the following two conditions is satisfied:*

(a) $r = n+1$, q is a primitive root modulo r, and $\alpha = c\zeta$;

(b) $q = 2^k$ with $\gcd(k,n) = 1$, $r = 2n+1$, either 2 or -2 is a primitive root modulo r, and $\alpha = c(\zeta + \zeta^{-1})$.

Proof. In view of Theorems 8.5.3 to 8.5.5 (and their proofs) and Remark 8.5.13, either of the two conditions above is sufficient for α to generate an optimal normal basis.

Conversely, assume that α generates an optimal normal basis $B = (\alpha_0, \dots, \alpha_{n-1})$ for E/F, where $\alpha_i = \alpha^{q^i}$ for $i = 0,, \dots, n-1$. For $n = 2$, the assertion holds trivially. Thus let $n \geq 3$, and let the multiplication matrix $T = (t_{ij})$ associated with B be defined as before:

$$\alpha\alpha_i = \sum_{j=0}^{n-1} t_{ij}\alpha_j \quad \text{for } i = 0, \dots, n-1. \tag{8.51}$$

The remainder of the proof makes essential use of the dual normal basis B^* of B; as usual, we write $B^* = (\beta_0, \dots, \beta_{n-1})$ with $\beta_j = \beta^{q^j}$. By Lemma 7.7.2, we may compute the dual coordinate vectors $r_{B^*}(\alpha\beta_j)$ of the products $\alpha\beta_j$ as follows:

$$(\alpha\beta_j)_i = \mathrm{Tr}\big((\alpha\beta_j)\alpha_i\big) = \mathrm{Tr}\big((\alpha\alpha_i)\beta_j\big) = \mathrm{Tr}\Big(\sum_{k=0}^{n-1} t_{ik}\alpha_k\beta_j\Big) = t_{ij},$$

and hence

$$\alpha\beta_j = \sum_{i=0}^{n-1} t_{ij}\beta_i \quad \text{for } j = 0, \dots, n-1. \tag{8.52}$$

Throughout the remainder of the proof, all indices are to be considered modulo n. We will often apply a suitable power σ^k of the Frobenius automorphism $\sigma : x \mapsto x^q$ to an identity involving the α's and the β's, which, of course, just means adding k modulo n to all the indices in the given identity. For example, applying σ^{n-i} to Equation (8.51) establishes the following useful property of the multiplication matrix T:

$$t_{ij} = t_{-i,j-i} \quad \text{for } i, j = 0, \dots, n-1. \tag{8.53}$$

We next note that one may assume

$$\operatorname{Tr}(\alpha) = \operatorname{Tr}(\beta) = -1. \tag{8.54}$$

Indeed, we can certainly assume the validity of (8.54) for α, as we may replace the given α by $c\alpha$, where $c = -1/\operatorname{Tr}(\alpha)$. Then the validity of (8.54) for β is an immediate consequence of Exercise 8.6.26.

Since B is optimal, we see from the proof of Lemma 8.5.2 that all columns of T except for the first one contain exactly two non-zero entries, which have to add up to 0. Using Equation (8.52), we conclude that the elements $\alpha\beta_j$ with $j = 1,\dots,n-1$ have a very special form:

$$\alpha\beta_j = d_j\big(\beta_{k(j)} - \beta_{k'(j)}\big) \text{ with } d_j \in F^* \text{ and distinct indices } k(j), k'(j). \tag{8.55}$$

The proof of Lemma 8.5.2 also shows that the first column of T contains exactly one non-zero entry, which necessarily equals $\operatorname{Tr}(\alpha)$. Thus there is some index m with

$$t_{m0} = \operatorname{Tr}(\alpha) = -1 \quad \text{and} \quad t_{i0} = 0 \text{ for all } i \neq m. \tag{8.56}$$

Equations (8.56) and (8.52) yield

$$\alpha\beta = -\beta_m, \quad \text{where } m \neq 0. \tag{8.57}$$

(Note that $m = 0$ in (8.57) would give $\alpha = -1$ and hence $E = F$, contradicting the hypothesis $n \geq 2$.) We now put $\mu := \sigma^m$, that is, μ is the power of the Frobenius automorphism σ mapping β to β_m. Then $\mu \neq \mathrm{id}$, as $m \neq 0$.

We first consider the case where μ is an involution, that is, $\mu^2 = \mathrm{id}$. By (8.57), $\alpha = -\mu(\beta)/\beta$, so that $\mu(\alpha) = -\beta/\mu(\beta) = 1/\alpha$. Thus, in view of (8.54),

$$\alpha\alpha_m = \alpha\mu(\alpha) = 1 = -\operatorname{Tr}(\alpha) = -\sum_{j=0}^{n-1} \alpha_j,$$

and therefore $t_{mj} = -1$ for all j. Now (8.52) and (8.55) imply that, for each $j \neq 0$, there exists a uniquely determined index $j^* \neq m$ such that $\alpha\beta_j = \beta_{j^*} - \beta_m$, that is,

$$t_{j^*j} = -t_{mj} = 1 \quad \text{and} \quad t_{kj} = 0 \text{ for } k \neq j^*, m \quad (j = 1,\dots,n-1); \tag{8.58}$$

together with Equation (8.56), this determines all entries of T.

Moreover, $\alpha\beta_j \neq \alpha\beta_k$ shows $j^* \neq k^*$ for $j \neq k$, so that the mapping $j \mapsto j^*$ is a bijection from $\{1,\dots,n-1\}$ to $\{0,\dots,n-1\} \setminus \{m\}$. Thus each row of T – except for that indexed by m – contains a unique entry 1 and entries 0 otherwise. Now an application of (8.51) shows $\alpha\alpha_{j^*} = \alpha_j$ for $j^* \neq m$; also, as noted above, $\alpha\alpha_m = 1$. Hence the set

$$H := \{1\} \cup \{\alpha_j \colon j = 0,\dots,n-1\}$$

is closed under multiplication by α and, trivially, under the action of the Frobenius automorphism σ. Therefore H has to be a group, as it contains $\alpha_j^{-1} = \mu(\alpha_j)$ for every j. (Note that $\alpha^{-1} = \mu(\alpha)$ implies $\alpha_j^{-1} = \sigma^j(\mu(\alpha)) = \mu(\alpha_j)$.) Now it is clear that H is a cyclic subgroup of order $n+1$ of E^* which is generated by α, so that α is a primitive r-th root of unity, where $r := n+1$.

In particular, α is a root of the polynomial $f := x^n + \cdots + x + 1$. Note that $E = F(\alpha)$ has dimension n over F, and therefore f is the minimal polynomial m_α of α and hence irreducible. Thus $r = n+1$ is a prime, and $m_\alpha = \Phi_r$. Hence q has to be a primitive root modulo r, by Proposition 3.6.16. This verifies condition (a) in the statement of the theorem, provided that μ is an involution.

It remains to prove that condition (b) holds if μ is not an involution. Applying (8.53) to (8.56) shows

$$t_{-m,-m} = -1 \quad \text{and} \quad t_{ii} = 0 \quad \text{for } i \neq -m. \tag{8.59}$$

Therefore $\alpha\beta_{-m}$ involves the term β_{-m}, by (8.52), and hence (8.55) yields

$$\alpha\beta_{-m} = \beta_\ell - \beta_{-m} \quad \text{for some } \ell \neq -m. \tag{8.60}$$

We now claim

$$\alpha\beta_{\ell+m} = \beta_{2m} - \beta_m. \tag{8.61}$$

To see this, we compute the term $\alpha\alpha_m\beta$ in two ways. Equation (8.57) gives

$$\alpha_m(\alpha\beta) = -\alpha_m\beta_m = -\mu(\alpha\beta) = \mu(\beta_m) = \beta_{2m};$$

but also, using (8.60) and once more (8.57),

$$\alpha(\alpha_m\beta) = \alpha\mu(\alpha\beta_{-m}) = \alpha\mu(\beta_\ell - \beta_{-m}) = \alpha(\beta_{\ell+m} - \beta) = \alpha\beta_{\ell+m} + \beta_m.$$

Comparing the preceding two equations immediately yields (8.61). As μ is not an involution, $m \neq -m$, and thus $t_{mm} = 0$ by (8.59). In view of (8.52) and (8.61), this implies $\ell \neq 0$.

Our next intermediate goal is to establish

$$\text{char } F = 2 \quad \text{and} \quad \alpha\beta_m = \beta_{\ell+m} + \beta. \tag{8.62}$$

To this purpose, we first show

$$\alpha\beta_{-\ell-m} = \beta_{-\ell} - \beta_k \quad \text{for some index } k \neq -\ell, -\ell-m. \tag{8.63}$$

Because of (8.60) and (8.52), $t_{\ell,-m} = 1$ and hence also $t_{-\ell,-\ell-m} = 1$, by (8.53). Now (8.52) and (8.55) show that $\alpha\beta_{-\ell-m}$ has the desired form. Note $k \neq -\ell-m$, which follows from $\ell \neq 0$ because of (8.59).

Next, we compute the term $\alpha\alpha_\ell\beta_{-m}$ in two ways. Let us write $\lambda := \sigma^\ell$. Then, using (8.60) and (8.63),

$$\begin{aligned} \alpha_\ell(\alpha\beta_{-m}) &= \alpha_\ell(\beta_\ell - \beta_{-m}) = \lambda(\alpha\beta - \alpha\beta_{-\ell-m}) \\ &= \lambda(-\beta_m - \beta_{-\ell} + \beta_k) = -\beta_{m+\ell} - \beta + \beta_{k+\ell}. \end{aligned}$$

On the other hand, (8.57) and (8.63) give

$$\alpha(\alpha_\ell\beta_{-m}) = \alpha\lambda(\alpha\beta_{-\ell-m}) = \alpha\lambda(\beta_{-\ell}-\beta_k) = \alpha\beta-\alpha\beta_{k+\ell} = -\beta_m-\alpha\beta_{k+\ell},$$

and comparing the preceding two equations yields

$$\alpha\beta_{k+\ell} = -\beta_m+\beta_{\ell+m}+\beta-\beta_{k+\ell}. \tag{8.64}$$

In view of (8.59), the term $-\beta_{k+\ell}$ cannot appear in $\alpha\beta_{k+\ell}$, as $k+\ell\neq -m$. Hence one of the remaining three terms on the right hand side of (8.64) has to cancel $-\beta_{k+\ell}$. As $k+\ell\neq 0$, the term in question cannot be β; we claim that it cannot be $\beta_{\ell+m}$ either. Otherwise $k=m$, and then (8.64) reads $\alpha\beta_{\ell+m} = -\beta_m+\beta$. Together with (8.61), this would give $\beta=\beta_{2m}=\mu^2(\beta)$, contradicting the assumption $\mu^2\neq \mathrm{id}$.

Therefore $-\beta_{k+\ell}$ has to be cancelled by $-\beta_m$, implying $k+\ell=m$ and $2\beta_m=0$. Hence E has characteristic 2, and (8.64) reduces to the desired equation $\alpha\beta_m = \beta_{\ell+m}+\beta$. This finishes the proof of (8.62).

Our next aim is to show that B is self-dual. For this, we first apply μ to equation (8.60) which gives $\alpha_m\beta=\beta_{\ell+m}+\beta$, as char $F=2$. Together with (8.62), we obtain the identity

$$\alpha\beta_m = \alpha_m\beta. \tag{8.65}$$

Multiplying equations (8.57) and (8.65) gives the condition $\alpha^2=\alpha_m=\mu(\alpha)$. Now induction yields

$$\mu^d(\alpha) = \alpha^{2^d} \quad \text{for every integer } d\geq 0. \tag{8.66}$$

In particular, choosing d as the order b of μ in (8.66) shows $\alpha=\alpha^{2^b}$, and hence α belongs to $\mathrm{GF}(2^b)$. Since μ is a power of the Frobenius automorphism σ (which has order n), we see

$$b=\mathrm{ord}(\mu)\mid n=[E:F]=[F(\alpha):F]\mid[\mathrm{GF}(2)(\alpha):\mathrm{GF}(2)]\mid b.$$

Thus equality holds throughout, which shows that μ is a generator for the Galois group G of E/F, too:

$$\langle\mu\rangle = \langle\sigma\rangle. \tag{8.67}$$

Since we may write Equation (8.65) as $\mu(\alpha/\beta)=\alpha/\beta$, we conclude $\sigma(\alpha/\beta)=\alpha/\beta$, so that α/β belongs to the ground field F. But α and β have the same trace by (8.54), which implies $\alpha=\beta$. Hence B is indeed a self-dual normal basis, and therefore the matrix T is symmetric by Theorem 8.6.2.

Now let ζ be a root of the polynomial $x^2+\alpha x+1$ in a suitable extension field K of E, so that $\alpha=\zeta+\zeta^{-1}$. Since K is a finite field of characteristic 2, the order of ζ has to be odd, say $\mathrm{ord}(\zeta)=2s+1$. For each integer i, we write

$$\gamma_i := \zeta^i+\zeta^{-i}, \tag{8.68}$$

so that $\gamma_0=0$, $\gamma_1=\alpha$, and $\gamma_2=\alpha^2=\mu(\alpha)$. In general, (8.66) shows

$$\mu^d(\alpha) = \alpha^{2^d} = (\zeta + \zeta^{-1})^{2^d} = \gamma_{2^d} \quad \text{for every integer } d \geq 0. \tag{8.69}$$

Note that there are exactly s different non-zero elements in (8.68), namely $\gamma_1, \ldots, \gamma_s$, since $\gamma_i = \gamma_j$ if and only if the roots ζ^i and ζ^{-i} of the polynomial $x^2 + \gamma_i x + 1$ coincide with the roots ζ^j and ζ^{-j} of the polynomial $x^2 + \gamma_j x + 1$, and this in turn holds if and only if $i \equiv \pm j \pmod{2s+1}$.

We next claim that the n conjugates of α coincide with $\gamma_1, \ldots, \gamma_s$. In view of (8.69), it suffices to show that each γ_i is a conjugate of α, which will be established by induction on i. We have already seen that this assertion holds for $i = 1$ and $i = 2$, so that we may assume $3 \leq i \leq s$. Note that

$$\alpha\gamma_{i-2} = (\zeta + \zeta^{-1})(\zeta^{i-2} + \zeta^{2-i}) = \gamma_{i-1} + \gamma_{i-3}, \tag{8.70}$$

where, by the induction hypothesis, both γ_{i-2} and γ_{i-1} are conjugates of α. Therefore, the conjugate γ_{i-1} of α occurs with coefficient 1 in the coordinate vector $r_B(\alpha\gamma_{i-2})$, which is some row of T. As noted above, the multiplication matrix T associated with B is symmetric, and we conclude that γ_{i-2} has to occur with coefficient 1 in the coordinate vector $r_B(\alpha\gamma_{i-1})$. Now condition (8.55) and $\beta = \alpha$ imply that $\alpha\gamma_{i-1}$ equals the sum of γ_{i-2} and a suitable further conjugate of α. On the other hand, Equation (8.70) with i replaced by $i+1$ also shows $\alpha\gamma_{i-1} = \gamma_i + \gamma_{i-2}$, and thus that further conjugate has to be γ_i. This completes the inductive proof that each γ_i is conjugate to α.

Since the n conjugates of α coincide with the s different elements $\gamma_1, \ldots, \gamma_s$, necessarily $s = n$, and ζ is a primitive r-th root of unity with $r := 2n+1$. In view of (8.69), for each integer $d \in \{0, \ldots, n-1\}$ there has to be some integer $i \in \{1, \ldots, n\}$ with $\alpha^{2^d} = \gamma_{2^d} = \gamma_i = \gamma_{-i}$, that is, $2^d \equiv \pm i \pmod{2n+1}$. Hence all non-zero residues are powers or negative powers of 2 modulo $r = 2n+1$ and thus invertible modulo r, which shows that r is a prime.

Moreover, multiplication by 2 induces either an n-cycle or a $2n$-cycle on the residues modulo r, that is, either 2 or -2 is a primitive root modulo r. By (8.67) and (8.69), the mapping $x \mapsto x^2$ induces a permutation of the n conjugates of α, so that α actually belongs to $GF(2^n)$. By hypothesis, the minimum polynomial of α over $GF(2)$ stays irreducible over $F(\alpha) = E = GF(q^n)$, and hence $F = GF(2^k)$ for some k with $\gcd(k, n) = 1$, by Theorem 3.5.9. Altogether, this shows the validity of condition (b) when μ is not an involution, completing the proof. □

We remark that Theorem 8.8.1 generalizes to arbitrary finite dimensional Galois extensions E/F. The proofs carry over with suitable (not too difficult) modifications; of course, one can no longer use the fact that the Galois group of E/F is cyclic (this becomes part of the assertion), but has to work with an arbitrary (not necessarily abelian) finite Galois group G. For details, the reader is referred to Gao and Lenstra [127].

Let us also note the following immediate consequence of Theorem 8.8.1 and Lemma 8.5.2 for the particularly interesting binary case, which should be compared with Conjecture 8.6.25.

Corollary 8.8.2. *Let B be a normal basis for $E = \mathrm{GF}(2^n)$ over $F = \mathrm{GF}(2)$, and assume the validity of the following two conditions:*

- *If $n+1$ is a prime, then 2 is not a primitive root modulo $n+1$.*
- *If $2n+1$ is a prime, then neither 2 nor -2 is a primitive root modulo $2n+1$.*

Then B cannot be optimal, and hence $C_B \geq 2n+1$. □

In this context, we mention a result on the parity of C_B which also applies in the non-binary case, generalizing the corresponding remark in Lemma 8.5.2. We note that it is easy to find examples for both possible parities.

Proposition 8.8.3. *Let α generate a normal basis B for $\mathrm{GF}(q^n)$ over $\mathrm{GF}(q)$. Then the parity of C_B equals that of the weight $w_B(\alpha^2)$ of the coordinate vector $r_B(\alpha^2)$.*

Proof. Equation (8.53) gives $w_B(\alpha\alpha_i) = w_B(\alpha\alpha_{-i})$ for $i = 1,\ldots,n-1$. This shows that the weights of the distinct rows i and $-i$ of T agree. Moreover, if $n = 2m$ is even, then the weight of row m of T is even. Hence the parity of C_B indeed agrees with that of row 0 of T. □

Remark 8.8.4. In this chapter, we have concentrated on the problem of efficiently performing multiplications (and exponentiations) in $\mathrm{GF}(q^n)$ over $\mathrm{GF}(q)$, mainly with the aid of normal bases and of polynomial bases satisfying some type of duality condition. There exists a general area of mathematics where one studies the complexity of all kinds of algebraic computations in a field extension E/F, namely Algebraic Complexity Theory; see, for instance, the surveys by Strassen [368] or von zur Gathen [386] and, for a detailed treatment, the standard text book by Bürgisser, Clausen and Shokrollahi [55].

One of the central topics in this area is the study of bilinear problems, the most prominent of which is no doubt matrix multiplication; see [55] and also de Groote [103] for thorough treatments of bilinear problems. Of course, multiplication in an extension field E over a ground field F is also a specific bilinear problem. There has been great interest in determining the **bilinear complexity** (or **rank**) $R(E/F)$ which is defined as the smallest positive integer r for which there exist linear forms $u_1,\ldots,u_r$ and $v_1,\ldots,v_r$ in the dual vector space of E over F and elements $\alpha_1,\ldots,\alpha_r$ in E such that

$$\xi\eta = \sum_{i=1}^{r} u_i(\xi)v_i(\eta)\alpha_i \quad \text{for all } \xi,\eta \in E.$$

It is known that $R(E/F) \leq 2n-1$, where $n = [E:F]$, with equality if and only if $|F| \geq 2n-2$; in particular, the bilinear rank is determined for all finite extensions of infinite fields. However, the corresponding problem for finite fields is still open; we refer the interested reader to to the books already cited and to the Ph.D. thesis of Shokrollahi [354]. □

8.9 Concluding Remarks

We conclude this chapter with an outlook on further aspects of the arithmetics of finite fields and their applications in practice. There is a vast literature on these topics, and we will have to restrict ourselves to just a few comments.

The reader interested in a very elementary introduction to finite fields and a selection of applications should find the slim monograph by Mullen and Mummert [291] helpful. On the other end of the spectrum, Part III of the *Handbook of finite fields* [292] provides extensive surveys on various applications of finite fields, including a wealth of further references. In particular, Section 16.17 gives an interesting overview of various approaches to the hardware realization of arithmetic operations over binary extension fields.

As should have become clear by now, there is a close interaction between computational requirements arising from diverse application areas and the investigation of structural properties of finite fields. As already emphasized in the preface, finite fields are an indispensable tool in all areas of information and communication theory, in particular, in Coding Theory, Cryptography[9] and Signal Processing. Many practical problems in these areas admit alternative statements in the language of combinatorial designs and/or Galois geometries (that is, affine and projective geometries with coordinates from a Galois field). Therefore, the applications of finite fields in the more theoretical areas of Design Theory and Finite Geometry also play a very important role for practical problems arising in the more directly applied areas listed before.

While the present book contains quite a few examples of interesting applications from the disciplines just mentioned, an in-depth treatment is out of the scope of our text; this would require several additional volumes! Fortunately, there are many books concerned with these areas, of which we now collect a (by no means exhaustive) selection.

- There is a rather large number of books dealing with Coding Theory. Among these, the classic monograph by van Lint [381] gives a particularly nice concise introduction; alternatively, good introductory treatments are also provided by Hill [185] and Pless [320]. Remarkably, the encyclopaedic monograph by MacWilliams and Sloane [255] still serves as the standard reference; a more recent comprehensive treatment is due to Huffman and Pless [197]. Three other notable textbooks are those of Berlekamp [30], Bierbrauer [37] and Blahut [39].

---

[9] Historically, the term *Cryptography* only refers to (mathematical) techniques for designing secure communication in the presence of adversaries and for constructing protocols achieving various other goals in information security, including such diverse tasks as data confidentiality, data integrity, electronic signatures, authentication, non-repudiation, time stamping, secret sharing and access control. In contrast, analyzing and possibly breaking cryptographic systems was usually referred to as *Cryptanalysis*, with *Cryptology* being used as an umbralla term for both these fields; see Chapter 1 of the *Handbook of Applied Cryptography* [269]. Nowadays, it has become quite standard to consider the terms *Cryptography* and *Cryptology* interchangeable, with *Cryptography* used perhaps even more commonly.

Finally, the two volume *Handbook of coding theory* [321] provides an outstanding collection of survey articles on virtually all aspects of the area, giving the state of the art up to 1998.

- Comprehensive treatments of Cryptography can be found in the excellent textbooks by Stinson and Paterson [367] and by von zur Gathen [387], whereas nice less extensive introductions are provided by Buchmann [52] and Koblitz [220]. Two books emphasizing algebraic aspects are due to Koblitz [222] and Buchmann and Vollmer [53]; the latter of these focuses on (the algorithmic treatment) of binary quadratic forms and their applications in Cryptography. Readers interested in just a concise introduction to Cryptography may wish to consult the survey of Massey [258]. Naturally, the *Handbook of Applied Cryptography* [269] provides a wealth of further information and references.
- For an introduction to Digital Signal Processing in general, the reader may consult the recent textbook by Rao and Swamy [326]. The traditional methods in this subject rely on the complex field, and the use of finite fields is a comparatively recent development. Perhaps the first textbook dealing with this approach is due to Morgera and Krishna [285]; in particular, these authors consider connections to Algebraic Coding Theory. From the point of view of Galois fields, the most pertinent application in Signal Processing is sequence design. For this topic, we strongly recommend the outstanding monograph on the design of sequences with good correlation properties (and related objects) by Golomb and Gong [147], which provides an extensive coverage.
- The two volume monograph by Beth, Jungnickel and Lenz [35] gives an encyclopaedic treatment of Design Theory, reflecting the state of the art before 1999 and containing an abundance of references. Nice introductory treatments are provided in the books of Hughes and Piper [199], of Lindner and Rodger [244] and of Wallis [392]. The *CRC handbook of combinatorial designs* [97] provides extensive further information and many references.
- Two interesting monographs studying the links between designs, graphs and codes are due to Cameron and van Lint [59] and to Tonchev [374]. The recent books by Ding [108, 109] study more specific connections between Coding Theory and Design Theory and emphasize constructions. The standard reference on the links between codes and designs still is the outstanding text by Assmus and Key [14]; in particular, this books provides an extensive treatment of the codes arising from the classical designs defined by subspaces in affine and projective geometries.
- The series of three monographs by Hirschfeld [186, 187] and by Hirschfeld and Thas [189] comprises the standard textbooks on Galois Geometry. Beutelspacher and Rosenbaum [36] and Casse [69] provide introductory treatments of general projective spaces, with considerable emphasis on the finite case. Various other (algebraic or combinatorial) structures contained in affine and projective geometries have important applications in Coding Theory and/or Cryptography; the most prominent examples are subgeometries, curves and surfaces.

- In particular, algebraic curves (and, to a lesser degree, also surfaces) over finite fields constitute an extremely useful tool for applications to codes, designs and cryptosystems. In other words, Algebraic Geometry and Algebraic Number Theory (in the finite case) are important advanced tools in these areas. The monograph by Hirschfeld, Korchmaros and Torres [188] is probably the best available treatment of algebraic curves over a finite field, while the book of Stichtenoth [363] is the standard reference for algebraic function fields and their application in Coding Theory. The recent textbook of Pellikaan, Wu, Bulygin and Jurrius [316] provides connections between codes, cryptology and curves, including computational aspects. Niederreiter and Xing [308] give a good introduction to the application of Algebraic Geometry in Coding Theory and Cryptography. The two volumes of Tsfasman, Vlăduţ and Nogin [375, 376] constitute an extensive in-depth treatment of algebraic geometry codes by some of the masters in the area.
- More specifically, elliptic (and also hyperelliptic) curves over finite fields are of the utmost importance to modern Cryptography; this topic was already briefly discussed in Section 8.1. The slim monographs by Enge [113] and by Shemanske [353] provide introductory (not very technical) treatments of elliptic curves and their application, whereas the quite technical monograph by Hankerson, Menezes and Vanstone [179] focuses on the implementation of elliptic curve cryptosystems. Both the book by Ling, Wang and Xing [245] and the collection of surveys edited by Murty [298] are concerned with the links between algebraic curves and cryptography in some generality. For this area, there are two handbooks providing extensive treatments and containing a wealth of references for further reading, namely the *Handbook of applied cryptography* [269] and the *Handbook of elliptic and hyperelliptic curve cryptography* [77].
- Several types of functions over finite fields, often collectively referred to as *cryptographic functions*, also have important applications in all areas discussed here. These include the so-called *planar* and *(almost) perfect nonlinear* as well as the *(almost) bent functions*. There are several recent monographs on such functions, for instance, Budaghyan [54], Mesnager [272] and Tokareva [373].

Chapter 9
Shift Register Sequences

Abstract In Chapter 8, we already encountered linear feedback shift registers in connection with the hardware implementation of arithmetics in $\mathrm{GF}(2^n)$. In the present chapter, we will study the sequences produced by such devices – that is, shift register sequences – in detail. As we shall see, these sequences are just the solutions of linear recurrence relations over $\mathrm{GF}(2)$. In view of this connection, we will consider shift register sequences over $F = \mathrm{GF}(q)$ for general q, and initially even over arbitrary fields F. Even though shift registers (as hardware devices) are of little practical interest for other fields but $\mathrm{GF}(2)$, they are a useful concept for visualizing linear recurrences over general fields.

In the first two sections, we present some fundamental results which are valid over arbitrary fields F, including characterizations of shift register sequences and results about (ultimately) periodic sequences. After that, we shall restrict ourselves to the case of finite fields; in particular, we will consider shift register sequences associated with irreducible polynomials in Section 9.3. In the subsequent two sections, we discuss two applications, namely the construction of pseudo-random sequences and of some cyclic difference sets.

We then return to the general theory by considering the notions of the linear complexity and, more strongly, the linear complexity profile of an arbitrary sequence with entries from $\mathrm{GF}(q)$. We will describe methods for determining these quantities, in particular the Berlekamp-Massey algorithm. Finally, in Section 9.9, we give a further application and study the so-called GMW-sequences, a class of periodic binary sequences which combine good randomness properties with a comparatively high linear complexity.

9.1 Basic Results and Characterizations

Let us begin by recalling that a **linear feedback shift register** (for short, an **LFSR**) of **length** n is a time-dependent device (running on a clock) of n cells, each capable of holding a value from a field F, so that with each clock cycle the contents of the

D. Hachenberger and D. Jungnickel, *Topics in Galois Fields*,
Algorithms and Computation in Mathematics 29,
https://doi.org/10.1007/978-3-030-60806-4_9

cells are shifted cyclically by one position (to the right, say). It will be convenient to use a slightly different notational setup than in Chapter 8; see Figure 9.1.

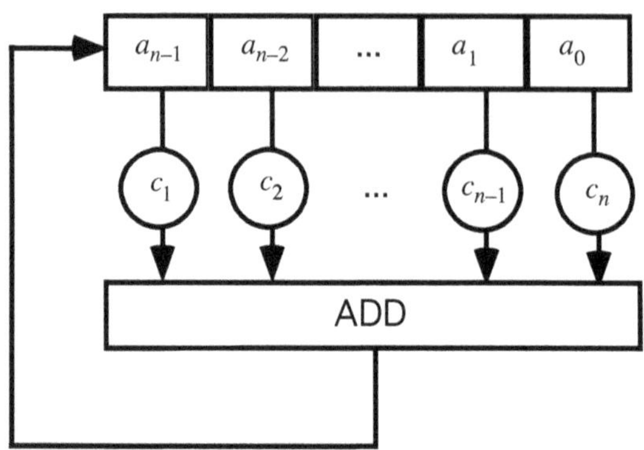

Fig. 9.1 Linear feedback shift register (LFSR)

Recall that the LFSR depicted in Figure 9.1 discards (or outputs) the rightmost entry a_0 and replaces it by a_1; it replaces a_1 by a_2; etc. Finally, it also computes a new leftmost entry, namely the linear function $c_1a_{n-1} + c_2a_{n-2} + \cdots + c_na_0$ of the given **state vector**[1] $(a_0, a_1, \ldots, a_{n-1})$ and the **feedback coefficients** $(c_1, c_2, \ldots, c_n)$. Thus the box with the entry ADD stands for an adder over F, and a circle with entry c indicates multiplication with $c \in F$.[2]

We refer to the specific values $(a_0, a_1, \ldots, a_{n-1})$ initially loaded into the LSFR as the **initial conditions**. Hence, the LFSR holds the vector $(a_1, a_2, \ldots, a_n)$, where

$$a_n := c_1a_{n-1} + c_2a_{n-2} + \cdots + c_na_0,$$

after the first clock cycle. More generally, after t clock cycles the LFSR will hold the vector $(a_t, a_{t+1}, \ldots, a_{t+n-1})$, where

$$a_{t+n-1} := c_1a_{t+n-2} + c_2a_{t+n-3} + \cdots + c_na_{t-1}. \tag{9.1}$$

Thus the **shift register sequence** $\mathbf{a} := (a_k)_{k\geq 0}$ produced by the LFSR satisfies the **linear recurrence relation**

$$a_k = \sum_{i=1}^{n} c_ia_{k-i} \quad \text{for } k \geq n \tag{9.2}$$

---

[1] Note that we write the components of the state vector according to the natural ordering of their indices, even though they fill the cells of the LFSR in Figure 9.1 from right to left.

[2] We do not care about the question how all this can be realized in hardware; readers who are interested in this topic might consult Tietze, Schenk and Gamm [372] and, for VLSI-design, Weste and Eshraghian [403].

of **order** n. With the convention $c_0 := -1$, we may rewrite (9.2) as follows:

$$\sum_{i=0}^{n} c_i a_{k-i} = 0 \quad \text{for } k \geq n. \tag{9.3}$$

There are several mathematical objects which can serve for the description of LFSR's or, equivalently, linear recurrence relations, which we will introduce next.

Definition 9.1.1. The **feedback polynomial** (sometimes also called the **reciprocal characteristic polynomial**) of the LFSR in Figure 9.1 is defined as

$$f(x) := -c_0 - c_1 x - \cdots - c_n x^n;$$

thus f is a polynomial of degree at most n with constant term $+1$. (Note that we do not require $c_n \neq 0$.) Let us call the vector $\mathbf{a}^{(t)} := (a_t, a_{t+1}, \ldots, a_{t+n-1})$ the t-th **state vector** of the LFSR ($t \geq 0$). Then we may rewrite (9.1) as

$$\mathbf{a}^{(t)} = \mathbf{a}^{(t-1)} A \quad \text{for } t \geq 1,$$

where the **feedback matrix** A is defined by

$$A := \begin{pmatrix} 0 & 0 & & & 0 & c_n \\ 1 & 0 & & & 0 & c_{n-1} \\ 0 & 1 & \ddots & & \vdots & \vdots \\ \vdots & 0 & \ddots & \ddots & & \vdots \\ \vdots & \vdots & & \ddots & 0 & c_2 \\ 0 & 0 & & & 1 & c_1 \end{pmatrix}.$$

Hence

$$\mathbf{a}^{(t)} = \mathbf{a}^{(0)} A^t \quad \text{for } t \geq 1. \tag{9.4}$$

Note that A is the companion matrix of the **reciprocal polynomial**[3] of f:

$$f^*(x) = x^n - c_1 x^{n-1} - \cdots - c_{n-1} x - c_n.$$

In view of the following simple lemma (which is immediate using a well-known fact from Linear Algebra, see Hoffman and Kunze [190, p.230]), f^* is usually called the **characteristic polynomial** of the LFSR in Figure 9.1. □

Lemma 9.1.2. *Let f be the feedback polynomial of an LFSR of length n over a field F. Then the feedback matrix A satisfies*

[3] In the context of shift register sequences and their feedback polynomials, it is convenient to use a somewhat different notion of reciprocal polynomials than the one introduced in Definition 5.1.9 (see also Remark 5.1.10), as we now consider polynomials of the form $f(x) = 1 - c_1 x - \cdots - c_n x^n$ where $c_n = 0$ is allowed. Throughout this chapter, we will use the subsequent definition for f^*, which agrees with the previous notion if and only if $c_n \neq 0$; in particular, then always $(f^*)^* = f$.

$$m_A = \chi_A = f^*,$$

where m_A and χ_A denote the minimal and the characteristic polynomial of A, respectively. □

Each of the three objects f, f^* and A can equally well serve for describing an LFSR with feedback coefficients $(c_1, c_2, \dots, c_n)$ or, equivalently, the recurrence relation (9.2); we shall generally use the feedback polynomial f for defining an LFSR. Note that the initial conditions $(a_0, a_1, \dots, a_{n-1})$ are not considered to be part of the definition of an LFSR, which certainly makes sense if we imagine LFSR's as hardware devices. Thus there always corresponds an entire family of shift register sequences to any given feedback polynomial f, obtained by using the corresponding LFSR with various initial conditions.

We now turn our attention to the problem of characterizing those sequences over F which may be obtained as shift register sequences. To this purpose, we identify an arbitrary sequence (a_k) over F with the formal power series[4]

$$a(x) := \sum_{k=0}^{\infty} a_k x^k \in F[[x]].$$

By Theorem 2.1.5, such a formal power series has a multiplicative inverse $a^{-1}(x)$ in $F[[x]]$ if and only if $a_0 \neq 0$, since F is a field. In particular, every polynomial $f(x) \in F[x]$ with $f_0 = 1$ is invertible when viewed as a formal power series. Then the rational function $g(x)/f(x)$ is again a formal power series for every polynomial $g(x) \in F[x]$.

Recall from Section 6.9 that the quotient field $F((x))$ of the ring $F[[x]]$ of formal power series is the field of formal Laurent series and that the elements of $F((x))$ have the form

$$c(x) = c_0(x) + c_1(x^{-1}) \quad \text{with } c_0(x) \in F[x] \text{ and } c_1(x) \in F[[x]].$$

Since $F[x] \subseteq F[[x]]$ and since the quotient field of $F[x]$ is the field $F(x)$ of rational functions in x, we have $F(x) \subseteq F((x))$.

We will soon establish the fundamental result that the subset $F(x) \cap F[[x]]$ of $F((x))$ is precisely the set of those formal power series which can be realized as shift register sequences over F. This will be a consequence of the following characterization of the shift register sequences belonging to an LFSR with a specified feedback polynomial:

Theorem 9.1.3. *Let $\mathbf{a} = (a_k)$ be a sequence over a field F with associated formal power series $a(x) \in F[[x]]$, let n be a positive integer, and let $f \in F[x]$ be a polynomial of degree at most n with constant coefficient 1. Then $\mathbf{a}$ can be realized as a shift register sequence resulting from an LFSR of length n associated with the feedback polynomial f if and only if $a(x)$ has the form*

[4] The use of formal power series for studying linear recurrence relations over finite fields goes back to Golomb [142].

$$a(x) = \frac{g(x)}{f(x)} \tag{9.5}$$

for a suitable polynomial $g \in F[x]$ with $\deg g < n$. Moreover, (9.5) gives a bijective correspondence between shift register sequences belonging to f and polynomials with degree less than n.

Proof. Let us view both f and g as formal power series over F:

$$f(x) = -\sum_{k=0}^{\infty} c_k x^k \quad \text{and} \quad g(x) = \sum_{k=0}^{\infty} b_k x^k,$$

where $c_0 = -1$ and $c_k = 0$ for $k > n$. In view of (9.5), we consider the equation

$$a(x)f(x) = g(x) \quad \text{in } F[[x]]. \tag{9.6}$$

Given $a(x)$, Equation (9.6) determines $g(x)$ uniquely:

$$b_k = -\sum_{i=0}^{k} c_i a_{k-i} \quad \text{for all } k \geq 0. \tag{9.7}$$

Assume first that $\mathbf{a}$ actually belongs to an LFSR with feedback polynomial f. Then $b_k = 0$ for $k \geq n$, by (9.3), and g is indeed a polynomial of degree $< n$.

Conversely, let g be any polynomial of degree $< n$. From the remarks preceding the theorem, it is clear that Equation (9.6) has a unique solution for $a(x)$ and that we will obtain a unique associated shift register sequence $\mathbf{a}$ from $g(x)$. Nevertheless, we will include an argument which shows how one may compute $\mathbf{a}$ explicitly. Thus we define $a(x)$ via (9.6) or, equivalently, (9.7). As $b_k = 0$ for $k \geq n$, condition (9.7) reduces to the linear recurrence relation (9.3). We now show how we may determine the initial values $a_0, a_1, \ldots, a_{n-1}$. Note that these values satisfy the matrix equation

$$(b_0, \ldots, b_{n-1}) = (a_0, \ldots, a_{n-1}) \begin{pmatrix} 1 & -c_1 & -c_2 & & -c_{n-2} & -c_{n-1} \\ 0 & 1 & -c_1 & & -c_{n-3} & -c_{n-2} \\ \vdots & 0 & 1 & \ddots & \vdots & \vdots \\ & \vdots & 0 & \ddots & -c_1 & \vdots \\ & & \vdots & \ddots & 1 & -c_1 \\ 0 & 0 & 0 & \ldots & 0 & 1 \end{pmatrix}.$$

Since the matrix on the right hand side has determinant 1, this indeed determines the a_k uniquely. □

Theorem 9.1.4. *Let $\mathbf{a} = (a_k)$ be a sequence over a field F, with associated formal power series $a(x) \in F[[x]]$. Then $\mathbf{a}$ is a shift register sequence if and only if $a(x)$ belongs to the field $F(x)$ of rational functions over F.*

Proof. The necessity of the criterion above is clear from Theorem 9.1.3. Conversely, assume that $a(x)$ belongs to $F(x)$, say

$$a(x) = \frac{g(x)}{f(x)}, \quad \text{where } \gcd(f,\, g) = 1.$$

As noted in Observation 2.3.4, the extended Euclidean algorithm gives polynomials p and q over F satisfying $pf + qg = 1$. Substituting $g = af$, we obtain $f(p + qa) = 1$; hence f is invertible in $F[[x]]$ and thus has constant term $\neq 0$. Without loss of generality,

$$f(x) = -\sum_{k=0}^{\infty} c_k x^k \quad \text{with } c_0 = -1 \text{ and } c_k = 0 \text{ for almost all } k.$$

Now put $n := \max(\deg f, \deg g + 1)$. Then another application of Theorem 9.1.3 shows that $a(x)$ can be generated by an LFSR of length n having as its feedback polynomial the formal power series describing f, truncated after the x^n-term. □

Example 9.1.5. Let us consider what is probably the most famous shift register sequence, namely the **Fibonacci sequence**. Here $F = \mathbb{Q}$, and $\mathbf{a} = (a_k)$ satisfies the recursion

$$a_k = a_{k-1} + a_{k-2} \tag{9.8}$$

with initial conditions $(a_0, a_1) = (1, 1)$. Thus the associated feedback polynomial is $f(x) = 1 - x - x^2$, and the polynomial $g(x)$ defined by the initial conditions according to (9.7) is $g(x) = 1$. Hence, written as a rational function, the formal power series describing $\mathbf{a}$ is

$$a(x) = \frac{1}{1 - x - x^2}.$$

Expanding $a(x)$ gives the expected result:

$$a(x) = 1 + x + 2x^2 + 3x^3 + 5x^4 + 8x^5 + 13x^6 + 21x^7 + 34x^8 + 55x^9 + 89x^{10} + \cdots \quad \square$$

Next we present a further interesting application of Theorem 9.1.3. It should be clear that any given shift register sequence can be obtained from many different shift registers. As a trivial example, the zero sequence is obtained from an arbitrary shift register over F by using the initial conditions $(0, \ldots, 0)$. To give a less trivial example, the Fibonacci sequence of Example 9.1.5 also satisfies the recurrence relation

$$a_k = a_{k-1} + a_{k-3} + a_{k-4} \quad \text{with initial conditions } (a_0, a_1, a_2, a_3) = (1, 1, 2, 3).$$

Note that the associated feedback polynomial $f_1(x) = 1 - x - x^3 - x^4$ is a multiple of the feedback polynomial f used in Example 9.1.5: $f_1(x) = f(x)(x^2 + 1)$. The following result shows that this is not a coincidence.

Theorem 9.1.6. *Let $\mathbf{a} = (a_k)$ be a shift register sequence over a field F, with associated formal power series $a(x) \in F[[x]]$. Then there exists a uniquely determined* *polynomial* f_0 *with constant term* $+1$ *such that* **a** *can be obtained from an LFSR with feedback polynomial* f *if and only if* f *is a multiple of* f_0 *with constant term* $+1$.

Proof. Choose an arbitrary feedback polynomial associated with an LFSR producing **a**, say f. By Theorem 9.1.3, there exists a polynomial g such that $a(x) = g(x)/f(x)$. Let $d = \gcd(f,g)$, where we normalize d to have constant term $+1$, and write $f = df_0$ and $g = dg_0$. We will show that f_0 is the desired polynomial with constant term $+1$.

By Theorem 9.1.3, any multiple hf_0 of f_0 with constant term $+1$ corresponds to an LFSR producing **a**, as

$$\frac{hg_0}{hf_0} = \frac{g_0}{f_0} = \frac{dg_0}{df_0} = \frac{g}{f} = a(x).$$

Conversely, let f_1 be a any feedback polynomial associated with some LFSR producing **a**. Again using Theorem 9.1.3, we obtain a polynomial g_1 such that $a(x) = g_1/f_1$. Hence $g_1 f_0 = g_0 f_1$ which shows that f_0 indeed divides f_1, since f_0 and g_0 are relatively prime. □

Corollary 9.1.7. *Let* $\mathbf{a} = (a_k)$ *be a shift register sequence over a field* F, *with associated formal power series* $a(x) \in F[[x]]$. *Then there exists a uniquely determined monic polynomial* m *such that* **a** *can be obtained from an LFSR with characteristic polynomial* f^* *if and only if* f^* *is a multiple of* m.

Proof. This follows from Theorem 9.1.6 by passing to reciprocal polynomials, with $m := f_0^*$. □

The polynomial m in Corollary 9.1.7 is called the **minimal polynomial** of the shift register sequence **a**. In other words, m is the characteristic polynomial of the linear recurrence relation of least order satisfied by **a**. Note that the degree of the polynomial f_0 described in Theorem 9.1.6 may be smaller than the length of the associated shift register of least length producing **a**, whereas the degree of the minimal polynomial always equals this length, since f_0 has constant coefficient $+1$ and $m = f_0^*$. Let us illustrate this with an explicit (though rather trivial) example:

Example 9.1.8. The sequence $\mathbf{a} = (0,1,1,1,\ldots)$ satisfies the second order linear recurrence relation $a_k = a_{k-1}$ with initial conditions $(0,1)$. Obviously the least length of an LFSR producing **a** is 2; the corresponding feedback polynomial is $f_0 = 1 - x$, whereas the minimal polynomial is the polynomial $m = x^2 - x$. □

The following result gives a simple criterion which allows us to decide whether some specified characteristic polynomial for a given shift register sequence already is the minimal polynomial of the sequence.

Proposition 9.1.9. *Let* $\mathbf{a} = (a_k)$ *be a shift register sequence over a field* F *belonging to an LFSR of length* n *with characteristic polynomial* f^*. *Then* f^* *actually is the minimal polynomial of* **a** *if and only if the first* n *state vectors* $\mathbf{a}^{(0)}, \mathbf{a}^{(1)}, \ldots, \mathbf{a}^{(n-1)}$ *are linearly independent.*

Proof. First let f^* be the minimal polynomial of $\mathbf{a}$ and assume that the first n state vectors are linearly dependent, say

$$\lambda_0 \mathbf{a}^{(0)} + \lambda_1 \mathbf{a}^{(1)} + \cdots + \lambda_{n-1} \mathbf{a}^{(n-1)} = \mathbf{0}.$$

Multiplying this equation with the power A^t of the feedback matrix A then shows

$$\lambda_0 \mathbf{a}^{(t)} + \lambda_1 \mathbf{a}^{(t+1)} + \cdots + \lambda_{n-1} \mathbf{a}^{(t+n-1)} = \mathbf{0}$$

for all $t \geq 0$. Thus $\mathbf{a}$ would also satisfy the linear recurrence relation

$$\lambda_0 a_t + \lambda_1 a_{t+1} + \cdots + \lambda_{n-1} a_{t+n-1} = 0 \quad (t \geq 0)$$

which clearly has order at most $n-1$, contradicting the defining property of the minimal polynomial.

Conversely, a similar argument shows that the state vectors $\mathbf{a}^{(0)}, \mathbf{a}^{(1)}, \ldots, \mathbf{a}^{(n-1)}$ must be linearly dependent if the minimal polynomial has degree $< n$. □

Example 9.1.8 shows that one cannot replace the first n state vectors in Proposition 9.1.9 by an arbitrary set of n consecutive state vectors $\mathbf{a}^{(t)}, \mathbf{a}^{(t+1)}, \ldots, \mathbf{a}^{(t+n-1)}$.

Next, we explain how one may obtain some solutions of a linear recurrence relation in terms of the associated characteristic polynomial.

Theorem 9.1.10. *Consider the linear recurrence relation*

$$a_k = \sum_{i=1}^{n} c_i a_{k-i} \quad \text{for } k \geq n \tag{9.9}$$

of order n with characteristic polynomial

$$f^*(x) = x^n - c_1 x^{n-1} - \cdots - c_{n-1} x - c_n$$

over a field F. If $\alpha_1, \ldots, \alpha_t$ are distinct roots of f^* in some extension field E of F, then

$$s_k := \lambda_1 \alpha_1^k + \cdots + \lambda_t \alpha_t^k \quad (k \in \mathbb{N}_0) \tag{9.10}$$

defines a solution $\mathbf{s} = (s_k)$ of (9.9) over E for every choice of $\lambda_1, \ldots, \lambda_t \in E$. Moreover, all solutions of this type form a vector space of dimension t over E.

Proof. Trivially, all sequences of the form (9.10) form a vector space. Thus it suffices to prove the first assertion in the special case $t = 1$ and $\lambda_1 = 1$. Write $\alpha := \alpha_1$. Then

$$0 = f^*(\alpha) = \alpha^n - \sum_{i=1}^{n} c_i \alpha^{n-i},$$

and therefore

$$\alpha^{t+n} = \sum_{i=1}^{n} c_i \alpha^{t+n-i} \quad \text{for all } t \geq 0.$$

Thus the sequence (α^k) is indeed a solution of (9.9).

Clearly, the vector space over E formed by all sequences of type (9.10) has dimension at most t. Hence the second assertion will follow if we can show that the solutions $(\alpha_1^k),\ldots,(\alpha_t^k)$ are linearly independent. Assume

$$\lambda_1\alpha_1^k+\cdots+\lambda_t\alpha_t^k=0 \quad \text{for all } k\geq 0.$$

Then $\lambda_1,\ldots,\lambda_t$ satisfy the system of linear equations

$$x_1\alpha_1^k+\cdots+x_t\alpha_t^k=0 \quad \text{for } k=0,\ldots,t-1.$$

The coefficient matrix

$$M=\begin{pmatrix} 1 & 1 & \ldots & 1\\ \alpha_1 & \alpha_2 & \ldots & \alpha_t\\ \vdots & \vdots & \ddots & \vdots\\ \alpha_1^{t-1} & \alpha_2^{t-1} & \ldots & \alpha_t^{t-1}\end{pmatrix}$$

of this system has determinant $\neq 0$, as M is a Vandermonde matrix and as $\alpha_1,\ldots,\alpha_t$ are pairwise distinct. Hence indeed $\lambda_1=\cdots=\lambda_t=0$. □

Corollary 9.1.11. *If the characteristic polynomial f^* of the linear recurrence relation (9.9) has distinct roots $\alpha_1,\ldots,\alpha_n$ over its splitting field E, then all solutions of (9.9) over E are of the form (9.10) with $t=n$.*

Proof. It suffices to note that all solutions of (9.9) over E form a vector space of dimension n. □

It is clear that all solutions of (9.9) over the given field F are among the solutions over E obtained in (9.10), provided that f^* has distinct roots in its splitting field E. Of course, unless f^* should already split over F, this leaves the problem of singling out the solutions over F among the solutions over E and of putting them into a suitable form. In the case of finite fields, there is a simple way for doing so, as we shall see in Section 9.3.

Example 9.1.12. Consider the Fibonacci recursion (9.8) with characteristic polynomial $f^*=x^2-x-1$ over $F=\mathbb{Q}$. The roots of f^* are $\alpha_{1,2}=(1\pm\sqrt{5})/2$. Hence all real solutions of (9.8) have the form

$$s_k=\lambda\left(\frac{1+\sqrt{5}}{2}\right)^k+\mu\left(\frac{1-\sqrt{5}}{2}\right)^k. \tag{9.11}$$

In particular, one may obtain the classical Fibonacci numbers in this form; see Exercise 9.1.15. □

We conclude this section with a characterization of the shift register sequences in $F[[x]]$ in terms of certain determinants, which is essentially due to Kronecker [226].[5]

[5] Kronecker only considered the case $F=\mathbb{R}$, but his proof carries over to arbitrary fields.

We will prove the easier part of his result later, see Corollary 9.6.4; a complete proof may be found in Lidl and Niederreiter [243, Chapter 6.6]. To state Kronecker's result, we first need to introduce the determinants in question:

Definition 9.1.13. Given any sequence $\mathbf{a} = (a_k)$ over a field F, the **Hankel determinants** $D^{(r)}(\mathbf{a})$ are defined as follows:

$$D^{(r)}(\mathbf{a}) := \det \begin{pmatrix} a_0 & a_1 & \dots & a_{r-1} \\ a_1 & a_2 & \dots & a_r \\ \vdots & \vdots & \ddots & \vdots \\ a_{r-1} & a_r & \dots & a_{2r-2} \end{pmatrix}. \qquad \square$$

Result 9.1.14 *Let* $\mathbf{a} = (a_k)$ *be a sequence over a field* F*. Then* $\mathbf{a}$ *is a shift register sequence if and only if* $D^{(r)}(\mathbf{a}) = 0$ *for all but finitely many values of* r*. Moreover, the degree of the minimal polynomial of* $\mathbf{a}$ *is the smallest positive integer* n *such that* $D^{(r)}(\mathbf{a}) = 0$ *for all* $r \geq n+1$. □

Exercises

Exercise 9.1.15. Determine the values λ and μ for which Equation (9.11) results in the classical Fibonacci numbers. □

9.2 Ultimately Periodic Sequences

In this section, we study sequences which are essentially periodic. More precisely, we introduce the following

Definition 9.2.1. A sequence $\mathbf{a} = (a_k)$ over some set S is called **ultimately periodic** with **period** r if it satisfies the condition

$$a_{k+r} = a_k \quad \text{for all sufficiently large values of } k;$$

if this even holds for all $k \geq 0$, one says that $\mathbf{a}$ is **periodic** with period r. The smallest number r_1 among all periods of an ultimately periodic sequence is called its **least period**, and the smallest integer N for which

$$a_{k+r_1} = a_k \quad \text{for all } k \geq N$$

is the **preperiod** of $\mathbf{a}$. In particular, an ultimately periodic sequence $\mathbf{a}$ is actually periodic if and only if it has preperiod 0. □

Proposition 9.2.2. *Let* $\mathbf{a} = (a_k)$ *be an ultimately periodic sequence over a field* F*, and let* r_1 *be its least period. Then the periods of* $\mathbf{a}$ *are precisely the multiples of* r_1.

Moreover, if **a** *is actually periodic for some period r, it is also periodic for its least period r_1.*

Proof. Trivially, every multiple of r_1 is a period for **a**. Conversely, let r be any period of **a**, and write $r = mr_1 + u$ with $0 \le u < r$; we have to show $u = 0$. Assume otherwise. Since both r and mr_1 are periods for **a**, there exists a positive integer N such that

$$a_{k+r} = a_k = a_{k+mr_1} \quad \text{for all } k \geq N,$$

and therefore

$$a_{k+u} = a_{k+r-mr_1} = a_{k+r} = a_k \quad \text{for all } k \geq N + mr_1.$$

Hence u is also a period for **a**, contradicting the choice of r_1 as the least period of **a**. This proves the first assertion.

Now let **a** be periodic for some period r. Then, by the first assertion, $r = mr_1$ for a suitable value of m, and

$$a_k = a_\ell \quad \text{for all } k, \ell \geq 0 \text{ with } \ell \equiv k \bmod mr_1.$$

Also, $a_\ell = a_{\ell+r_1}$ for all sufficiently large values of ℓ, say for $\ell \geq M$. Given an arbitrary index $k \geq 0$, choose any integer $\ell \geq M$ with $\ell \equiv k \bmod mr_1$. Then

$$a_k = a_\ell = a_{\ell+r_1} = a_{k+r_1},$$

and thus **a** is indeed periodic with period r_1. □

As the Fibonacci sequence discussed in Example 9.1.5 shows, shift register sequences need not be ultimately periodic. However, every ultimately periodic sequence with entries from a field F necessarily is a shift register sequence. This will be a consequence of the following characterization of periods in terms of formal power series.

Theorem 9.2.3. *Let $\mathbf{a} = (a_k)$ be a sequence over a field F, with associated formal power series $a(x) \in F[[x]]$. Then* **a** *is ultimately periodic with period r if and only if $(1 - x^r)a(x)$ is a polynomial over F.*

Proof. First assume that r is a period for **a**, say $a_{k+r} = a_k$ for all $k \geq N$. In terms of formal power series, this may be written as

$$\begin{aligned} a(x) &= (a_0 + a_1 x + \cdots + a_{N-1} x^{N-1}) \\ &\quad + (a_N + a_{N+1} x + \cdots + a_{N+r-1} x^{r-1})(x^N + x^{N+r} + x^{N+2r} + \cdots) \end{aligned}$$

Using the well-known identity

$$1 + x^r + x^{2r} + \cdots = (1 - x^r)^{-1}, \tag{9.12}$$

one computes

$$(1-x^r)a(x) = (a_0+a_1x+\cdots+a_{N-1}x^{N-1})(1-x^r) \\ + (a_N+a_{N+1}x+\cdots+a_{N+r-1}x^{r-1})x^N,$$

which proves $(1-x^r)a(x) \in F[x]$.

Conversely, let $(1-x^r)a(x)$ be a polynomial g and write

$$(1-x^r)a(x) = g(x) = (1-x^r)u(x)+v(x) \quad \text{with } 0 \le \deg v < r.$$

Then another application of (9.12) gives

$$a(x) = u(x)+v(x)(1+x^r+x^{2r}+\cdots),$$

which shows $a_{k+r} = a_k$ for all $k \ge \deg u + 1$. Hence **a** is indeed ultimately periodic with period r. □

Corollary 9.2.4. *Any ultimately periodic sequence over a field is a shift register sequence.*

Proof. This is an immediate consequence of Theorems 9.1.4 and 9.2.3. □

If the sequence **a** is actually periodic, we can use the preceding approach and obtain an interesting polynomial identity essentially due to Ward [397].

Theorem 9.2.5. *Let* $\mathbf{a} = (a_k)$ *be a sequence over a field* F*, with associated formal power series* $a(x) \in F[[x]]$*, and assume that* **a** *is periodic with period* r*. Then the following polynomial identity holds:*

$$f(x)s(x) = (1-x^r)g(x), \tag{9.13}$$

where $f(x) = -c_0 - c_1x - \cdots - c_nx^n$ *is the feedback polynomial of an associated LFSR and where the polynomials* s *and* g *are defined as follows:*

$$s(x) = a_0+a_1x+\cdots+a_{r-1}x^{r-1};$$

$$g(x) = \sum_{k=0}^{n-1}\Big(-\sum_{i=0}^{k} c_i a_{k-i}\Big)x^k.$$

Proof. Since **a** is periodic with period r, an application of (9.12) gives

$$a(x) = s(x)(1+x^r+x^{2r}+\cdots) = s(x)(1-x^r)^{-1}.$$

On the other hand, Theorem 9.1.3 and its proof show $a(x) = g(x)f(x)^{-1}$. Combining these two equations yields the desired identity (9.13). □

Theorem 9.2.5 is equivalent to Theorem 8.25 of Lidl and Niederreiter [243] who give a corresponding identity in terms of the characteristic polynomial instead of the feedback polynomial.

We conclude this section with the following strengthening of Proposition 9.1.9 for periodic sequences; as Example 9.1.8 shows, this result does not carry over to arbitrary shift register sequences.

Proposition 9.2.6. *Let* $\mathbf{a} = (a_k)$ *be a periodic shift register sequence over a field* F*, and assume that* $\mathbf{a}$ *belongs to an LFSR of length* n *with characteristic polynomial* f^**. Then* f^* *is the minimal polynomial of* $\mathbf{a}$ *if and only if any* n *consecutive state vectors are linearly independent.*

Proof. As in the proof of Proposition 9.1.9, the state vectors $\mathbf{a}^{(t)}, \mathbf{a}^{(t+1)}, \dots, \mathbf{a}^{(t+n-1)}$ have to be linearly dependent if the minimal polynomial of $\mathbf{a}$ has degree $< n$.

Conversely, let $m = f^*$ be the minimal polynomial of $\mathbf{a}$. Assume that the state vectors $\mathbf{a}^{(t)}, \mathbf{a}^{(t+1)}, \dots, \mathbf{a}^{(t+n-1)}$ are linearly dependent. As in the proof of Proposition 9.1.9, any n consecutive state vectors $\mathbf{a}^{(u)}, \mathbf{a}^{(u+1)}, \dots, \mathbf{a}^{(u+n-1)}$ with $u \geq t$ are then likewise linearly dependent. Let r be the least period of $\mathbf{a}$, and select any integer $u \geq t$ which a multiple of r. Then $\mathbf{a}^{(u+k)} = \mathbf{a}^{(k)}$ for all $k > 0$, and therefore $\mathbf{a}^{(0)}, \mathbf{a}^{(1)}, \dots, \mathbf{a}^{(n-1)}$ would also be linearly dependent, contradicting Proposition 9.1.9. □

9.3 Shift Register Sequences over Finite Fields

In the remainder of this chapter, we will restrict ourselves to shift register sequences over finite fields. In this case, the converse of Corollary 9.2.4 also holds. In fact, one can be more specific:

Proposition 9.3.1. *Let* $\mathbf{a} = (a_k)$ *be a shift register sequence over* $F = \mathrm{GF}(q)$ *with minimal polynomial* m *of degree* n*. Then* $\mathbf{a}$ *is ultimately periodic with least period* $r_1 \leq q^n - 1$.

Proof. By assumption, the sequence $\mathbf{a}$ can be obtained from an LFSR of length n with feedback polynomial $f = m^*$. Let A be the corresponding feedback matrix. Then the state vectors $\mathbf{a}^{(t)}$ are obtained from the initial conditions $\mathbf{a}^{(0)}$ according to Equation (9.4). Assume first that the zero vector is a state vector, say $\mathbf{a}^{(t)} = \mathbf{0}$. Then all subsequent state vectors are also $\mathbf{0}$, and $\mathbf{a}$ is ultimately periodic with least period 1.

Now assume that no state vector equals $\mathbf{0}$, and note that there are only $q^n - 1$ possible state vectors. Hence we must have an identity of the form

$$\mathbf{a}^{(s)} = \mathbf{a}^{(t)} \quad \text{with } 0 \leq s < t \leq q^n - 1.$$

Using (9.4) again, this implies $\mathbf{a}^{(s+k)} = \mathbf{a}^{(t+k)}$ for all $k \geq 0$, and hence $\mathbf{a}$ is ultimately periodic with period $r = t - s \leq q^n - 1$. □

Combining this result with Corollary 9.2.4 immediately gives the following nice characterization of the ultimately periodic sequences over finite fields.

Theorem 9.3.2. *The ultimately periodic sequences over a finite field F coincide with the shift register sequences over F.* □

Let us stress that shift register sequences over finite fields do not have to be periodic. As a trivial example, any second order recurrence relation

$$a_{k+2} = a_{k+1} \quad \text{with initial conditions } (a_0, a_1),\ a_0 \neq a_1$$

leads to a non-periodic shift register sequence **a**, even though **a** is ultimately periodic with least period 1. The following result shows that in fact every non-periodic shift register sequence over a finite fields is necessarily somewhat degenerate: the length of the corresponding LFSR has to be larger than the degree of the associated feedback polynomial, which means that the values a_k with $k \geq 1$ do not depend on the choice of the value a_0.

Proposition 9.3.3. *Let* $\mathbf{a} = (a_k)$ *be a shift register sequence over* $F = \mathrm{GF}(q)$ *belonging to an LFSR with feedback polynomial* $f(x) = -c_0 - c_1x - \cdots - c_nx^n$*, and assume* $c_n \neq 0$*. Then* **a** *is periodic with least period* $r_1 \leq q^n - 1$*. Moreover, the associated feedback matrix A is invertible, and* r_1 *divides the order of A.*

Proof. Note that the feedback matrix A introduced in Definition 9.1.1 has determinant $\pm c_n \neq 0$ and is therefore invertible. Using (9.4), we get

$$\mathbf{a}^{(t)} = \mathbf{a}^{(0)}A^t = \mathbf{a}^{(0)}A^{t+m} = \mathbf{a}^{(t+m)} \quad \text{for all } t \geq 0,$$

where m denotes the order of A. Thus **a** is a periodic sequence with period m. The remaining assertions are now immediate from Propositions 9.2.2 and 9.3.1. □

As already mentioned in Section 7.1, the possible orders of invertible (n,n)-matrices over $\mathrm{GF}(q)$ were determined by Niven [309].

Note that there are exactly q^n distinct shift register sequences which belong to a given LFSR of length n over $\mathrm{GF}(q)$, corresponding to the q^n possible initial conditions $\mathbf{a}^{(0)} = (a_0, a_1, \ldots, a_{n-1})$. We now describe a way of obtaining such a sequence for which the least period achieves the maximal value over all these q^n sequences. To this purpose, we give the following

Definition 9.3.4. Consider an LFSR of length n with feedback polynomial f. The shift register sequence **d** determined by f and the initial conditions $(0,\ldots,0,1)$ is called the **impulse response sequence** for the given LFSR.[6] □

Theorem 9.3.5. *Consider an LFSR with feedback polynomial f and feedback matrix A over* $\mathrm{GF}(q)$*, and let* $\mathbf{d} = (d_k)$ *be the impulse response sequence belonging to this LFSR. Then, for any two state vectors* $\mathbf{d}^{(s)}$ *and* $\mathbf{d}^{(t)}$*,*

$$\mathbf{d}^{(s)} = \mathbf{d}^{(t)} \iff A^s = A^t.$$

[6] This name is motivated by thinking of the LFSR of Figure 9.1 as being started by sending the "impulse" 1 through the left-most cell, where initially each cell is "empty".

Moreover, the least period of any shift register sequence **a** *which can be obtained from the given LFSR divides the least period of* **d**.

Proof. According to (9.4), $\mathbf{d}^{(t)} = \mathbf{d}^{(0)}A^t$ for all $t \geq 0$ and thus $A^s = A^t$ implies $\mathbf{d}^{(s)} = \mathbf{d}^{(t)}$. Conversely, assume $\mathbf{d}^{(s)} = \mathbf{d}^{(t)}$ and therefore $\mathbf{d}^{(s+k)} = \mathbf{d}^{(t+k)}$ for all $k \geq 0$. Again using (9.4), we conclude

$$\mathbf{d}^{(k)}A^s = \mathbf{d}^{(k)}A^t \quad \text{for all } k \geq 0. \tag{9.14}$$

Since $\mathbf{d}^{(0)}, \ldots, \mathbf{d}^{(n-1)}$ are obviously linearly independent, (9.14) implies $A^s = A^t$.

Now let r_1 be the least period and N the preperiod of **d**, respectively. Then $d_{k+r_1} = d_k$ and hence $\mathbf{d}^{(k+r_1)} = \mathbf{d}^{(k)}$ for all $k \geq N$, so that $A^{k+r_1} = A^k$ for all $k \geq N$. Applying (9.4) to an arbitrary shift register sequence **a** belonging to the given LFSR shows $\mathbf{a}^{(k+r_1)} = \mathbf{a}^{(k)}$ for all $k \geq N$, and hence r_1 likewise is a period for **a**. In view of Proposition 9.2.2, the least period of **a** has to divide r_1. □

In the situation considered in Proposition 9.3.3, we get an even stronger result:

Theorem 9.3.6. *Consider an LFSR of length n with feedback polynomial* $f(x) = -c_0 - c_1x - \cdots - c_nx^n$ *and feedback matrix A over* $F = \mathrm{GF}(q)$, *and assume* $c_n \neq 0$. *Then the associated impulse response sequence* $\mathbf{d} = (d_k)$ *is periodic, and the least period* r_1 *of* **d** *equals the order of A. Moreover,* r_1 *achieves the upper bound* $q^n - 1$ *in Proposition 9.3.1 if and only if f is primitive.*

Proof. By Proposition 9.3.3, **d** is periodic with period r_1 dividing the order of A. Thus $\mathbf{d}^{(r_1)} = \mathbf{d}^{(0)}$, and hence $A^{r_1} = I$ by (9.14). Thus r_1 indeed equals the order of A, establishing the first assertion.

By Theorems 7.1.6 and 7.2.5, the order of A equals $q^n - 1$ if and only if A is similar to the companion matrix of a primitive polynomial of degree n over F. As noted in Definition 9.1.1, A is itself the companion matrix of the characteristic polynomial f^* of the given LFSR, and by Theorem 5.7.1, f^* is primitive if and only if f is primitive. This proves the second assertion. □

Using Example 7.1.11, we see from Theorem 9.3.6 that it is always possible to realize the upper bound given in Proposition 9.3.1:

Corollary 9.3.7. *Let q be a prime power, and let n be any positive integer. Then there exists a periodic shift register sequence which belongs to an LFSR of length n over* $\mathrm{GF}(q)$ *and has least period* $q^n - 1$. □

Definition 9.3.8. A periodic shift register sequence belonging to an LFSR of length n over $\mathrm{GF}(q)$ with least period $q^n - 1$ is called a **maximum period sequence** or, for short, an m-**sequence**. Alternatively, the terms **pseudo-noise sequence** or, for short, **PN-sequence** are also common.[7] □

---

[7] This terminology will become clear in view of the results of the next section.

Our next result shows that all m-sequences are essentially impulse response sequences belonging to LFSR's with primitive feedback polynomials. In order to make this precise, we need to introduce the following terminology:

Definition 9.3.9. Two periodic sequences $\mathbf{a} = (a_k)$ and $\mathbf{b} = (b_k)$ are said to be **cyclically equivalent** if there exists an integer r such that $b_{k+r} = a_k$ for all $k \geq 0$. □

Proposition 9.3.10. *Let* $\mathbf{a} = (a_k)$ *be an m-sequence over* $F = \mathrm{GF}(q)$ *belonging to an LFSR with feedback polynomial* f *and feedback matrix* A*. Then* f *is a primitive polynomial, and* $\mathbf{a}$ *is cyclically equivalent to the impulse response sequence* $\mathbf{d}$ *for the given LFSR. Moreover, with the exception of the zero sequence* $\mathbf{0}$*, all shift register sequences which can be generated from this LFSR are cyclically equivalent.*

Proof. Since $\mathbf{a}$ is periodic with least period $q^n - 1$, its state vectors must contain every non-zero vector of length n over F; this follows from the proof of Proposition 9.3.1. Thus any vector $\mathbf{a}^{(0)} \neq \mathbf{0}$ of initial conditions generates a sequence cyclically equivalent to $\mathbf{a}$. In particular, the vector $(0,\dots,0,1)$ is a state vector for $\mathbf{a}$, and therefore $\mathbf{a}$ is cyclically equivalent to $\mathbf{d}$. □

Remark 9.3.11. Let $\mathbf{s} = (s_0,\dots,s_{N-1})$ be a finite sequence of length $N = m^k$ with entries from an m-set S. Then $\mathbf{s}$ is called an (m,k)-**de Bruijn sequence** provided that the k-tuples

$$(s_t, s_{t+1},\dots,s_{t+k-1}) \quad \text{with } t = 0,\dots,N-1,$$

where all indices are considered modulo N, contain each k-tuple with entries from S exactly once. Clearly, the initial entry $s_0 := 0$ followed by the first $q^n - 1$ entries of the impulse response sequence determined by a primitive polynomial of degree n over $\mathrm{GF}(q)$ yields a (q,n)-de Bruijn sequence.

There are also quite different constructions for de Bruijn sequences (for arbitrary values of m and k) using graph theoretic methods, which even allows one to count the number of such sequences; see, for instance, [214]. The interested reader is also referred to the 1982 survey paper by Fredericksen [120]. □

We now turn our attention to the problem of actually determining all shift register sequences belonging to an LFSR with given feedback polynomial f. In view of the preceding results for the case of primitive polynomials f, it is natural to consider the case where f is irreducible first.

Theorem 9.3.12. *Consider an LFSR of length* n *over* $F = \mathrm{GF}(q)$ *with feedback polynomial* f*, where* f *is an irreducible polynomial of degree* n*, and let* α *be a root of* f^* *in the extension field* $E = \mathrm{GF}(q^n)$*. Then the* q^n *shift register sequences belonging to the given LFSR are precisely the sequences* $\mathbf{s} = (s_k)$ *of the form*

$$s_k = \mathrm{Tr}_{E/F}(\theta\alpha^k) \quad (k \geq 0), \tag{9.15}$$

where θ *runs over the elements of* E*; here the element* θ *is uniquely determined by the sequence* $\mathbf{s}$*. Except for the trivial sequence* $\mathbf{0}$ *belonging to* $\theta = 0$*, all these* *sequences are periodic with least period $r_1 = \operatorname{ord} f$ and split into $(q^n - 1)/r_1$ classes of r_1 sequences each under cyclic equivalence.*

Proof. First note that the sequences of the form (9.15) indeed have entries from F. By Proposition 3.5.1, E is the splitting field of f^*, and the roots of f are the n conjugates $\alpha, \alpha^q, \ldots, \alpha^{q^{n-1}}$. Thus we can apply Theorem 9.1.10 and Corollary 9.1.11 to conclude that all solutions of the associated linear recurrence relation (9.9) over the extension field E form a vector space of dimension n over E and that each such solution $\mathbf{s}$ can be written uniquely in the form

$$s_k = \lambda_0 \alpha^k + \lambda_1 \alpha^{kq} + \cdots + \lambda_{n-1} \alpha^{kq^{n-1}} \quad (k \geq 0)$$

for suitable elements $\lambda_0, \ldots, \lambda_{n-1} \in E$. Note that the special choice

$$\lambda_0 = \theta,\ \lambda_1 = \theta^q, \ldots, \lambda_{n-1} = \theta^{q^{n-1}}$$

results in the sequence $\mathbf{s} = \mathbf{s}(\theta)$ defined in (9.15). This proves the first assertion, since we obtain q^n distinct solutions $\mathbf{s}(\theta)$ in this way.

By Proposition 9.3.3, all the sequences $\mathbf{s}(\theta)$ are periodic. Using the representation (9.15), we have

$$\begin{aligned} s_{k+r} = s_k \quad &\Longleftrightarrow \quad \operatorname{Tr}_{E/F}(\theta \alpha^{k+r}) = \operatorname{Tr}_{E/F}(\theta \alpha^k) \\ &\Longleftrightarrow \quad \operatorname{Tr}_{E/F}(\theta \alpha^k (1 - \alpha^r)) = 0. \end{aligned}$$

Since the trace bilinear form is non-degenerate, this holds for all $k \geq 0$ if and only if $\alpha^r = 1$ or $\theta = 0$. Therefore the least period of $s(\theta)$ with $\theta \neq 0$ is the order of α, that is, the order of f. This also gives the final assertion. □

Theorem 9.3.12 generalizes without any problems to the case where f is square-free. We just state the result one obtains and leave the details of the proof to the reader as Exercise 9.3.17.

Theorem 9.3.13. *Consider an LFSR of length n over $\mathrm{GF}(q)$ with a square-free feedback polynomial f of degree n, say $f = f_1 \cdots f_m$, where the f_i are distinct irreducible polynomials of degrees $n_1, \ldots, n_m$, respectively. For $i = 1, \ldots, m$, let α_i be a root of f_i^* in the extension field $E_i := \mathrm{GF}(q^{n_i})$. Then the shift register sequences $\mathbf{s} = (s_k)$ belonging to the given LFSR can be uniquely written in the form*

$$s_k = \operatorname{Tr}_{E_1/F}(\theta_1 \alpha_1^k) + \cdots + \operatorname{Tr}_{E_m/F}(\theta_m \alpha_m^k) \quad (k \geq 0),$$

where $\theta_1, \ldots, \theta_m$ run over the elements of $E_1, \ldots, E_m$, respectively. Except for the trivial sequence $\mathbf{0}$ (which belongs to $\theta_1 = \cdots = \theta_m = 0$), the sequence $\mathbf{s} = \mathbf{s}(\theta_1, \ldots, \theta_m)$ is periodic, and its least period r_1 is the least common multiple of all orders $\operatorname{ord} f_i$ with $\theta_i \neq 0$. □

In view of Theorem 9.3.5, we obtain the following immediate consequence of Theorem 9.3.13, which generalizes the case of primitive polynomials considered in Theorem 9.3.6.

Corollary 9.3.14. *Let* **d** *be the impulse response sequence belonging to an LFSR of length n over* $\mathrm{GF}(q)$ *with a square-free feedback polynomial f of degree n, say $f = f_1 \cdots f_m$, where the f_i are distinct irreducible polynomials. Then the least period of* **d** *equals the least common multiple of the orders* $\operatorname{ord} f_i$ $(i = 1, \ldots, m)$. □

It is also possible to give an explicit description for all solutions of a linear recurrence relation over $\mathrm{GF}(q)$ for which the feedback polynomial has repeated roots; for this, we refer the reader to Fillmore and Marx [117].

The preceding results illustrate the importance of the roots of the characteristic polynomial for describing the shift register sequences associated with a given LFSR. The following observation implies that they are also quite helpful in determining the possible periods, at least in the case where the characteristic polynomial is irreducible.

Theorem 9.3.15. *Consider an LFSR of length n over $F = \mathrm{GF}(q)$ with feedback polynomial $f(x) = -c_0 - c_1x - \cdots - c_nx^n$ of degree n. Assume that f is irreducible, and let α be a root of f^* in the extension field $E = \mathrm{GF}(q^n)$. Then the least period of every shift register sequence belonging to the given LFSR divides the order of α in E^*, and this value occurs.*

Proof. In view of the hypothesis $c_n \neq 0$, Proposition 9.3.3 and Theorem 9.3.6 apply. Hence the period of any shift register sequence belonging to f divides the order of the (invertible) feedback matrix A, and the impulse response sequence associated with f achieves this bound.

Thus it suffices to show that the orders of A and α agree. As noted in Definition 9.1.1, A is the companion matrix of the characteristic polynomial f^*. Let M_α be the linear mapping induced by α on E via multiplication. It is easily seen that A represents M_α with respect to the polynomial basis generated by α; see Examples 7.1.11 and 7.2.4. Hence the orders of A and α indeed agree. □

Example 9.3.16. Let us illustrate the results of this section by investigating the behavior of the Fibonacci recursion defined in Example 9.1.5 over $F = \mathrm{GF}(p)$.[8] Contrary to the case $F = \mathbb{Q}$ studied in Examples 9.1.5 and 9.1.12, now all solutions of Equation (9.8) (for arbitrary initial conditions) are periodic, by Proposition 9.3.3. We are interested in obtaining information about the possible periods of the resulting shift register sequences.

To this purpose, we first factor the characteristic polynomial $f^* = x^2 - x - 1$ over its splitting field. Note that the discriminant of f^* is $5 \bmod p$. Now assume that

[8] It seems that this question was first studied in 1960 by Wall [391], who actually dealt with the more general case of the Fibonacci recursion over $\mathbb{Z}_m$. As the coefficients of the characteristic polynomial are already in the prime field $\mathrm{GF}(p)$, it is not particularly interesting to consider this question over $\mathrm{GF}(q)$ in general; we will leave this as Exercise 9.3.18.

$p \neq 5$ is odd, let $\pm u$ be the two roots of $x^2 - 5$ in its splitting field E over GF(p), and let ζ be the inverse of 2 modulo p. Then

$$f^* = x^2 - x - 1 = (x - \alpha)(x - \beta) \quad \text{with } \alpha, \beta = \zeta \cdot (1 \pm u), \tag{9.16}$$

so that E is also the splitting field of f^*. Hence the solutions of the Fibonacci recurrence (9.8) over E have the form

$$s_k = \lambda \alpha^k + \mu \beta^k \quad (k \geq 0) \tag{9.17}$$

with $\lambda, \mu \in E$, by Theorem 9.1.10.

Let us assume in addition that f^* is irreducible over F, so that $E = \mathrm{GF}(p^2)$. Thus we require that 5 is a non-square modulo p. By the law of quadratic reciprocity (see Theorem 10.4.3), this holds if and only if $p \equiv \pm 2 \bmod 5$. Then Theorems 9.3.12 and 9.3.15 apply, and we conclude that every solution of (9.8) over F can be written in the form

$$s_k = \mathrm{Tr}_{E/F}(\theta \alpha^k) = \theta \alpha^k + \theta^p \alpha^{pk} \quad (k \geq 0)$$

for some $\theta \in E$; moreover, the least period of $\mathbf{s}$ divides the order of α. Since $\beta = \alpha^p$ in this case, we obtain $-1 = \alpha\beta = \alpha^{p+1}$ from (9.16), and hence the least period of $\mathbf{s}$ divides $2(p+1)$. In particular, the Fibonacci recursion can yield an m-sequence only if $p = 3$ (for odd p).

Now assume $p \equiv \pm 1 \bmod 5$, so that f^* is reducible and hence $E = F$. Then the sequence (α^k) is a solution of the trivial recurrence $a_k = \alpha a_{k-1}$, and another application of Theorem 9.3.15 shows that (α^k) has least period dividing $p-1$; of course, an analogous statement holds for the sequence (β^k). Because of (9.17), we conclude that every solution $\mathbf{s}$ then has period dividing $p-1$. For example, when $p = 11$ one indeed obtains a period of 10.

Finally, we briefly discuss the exceptional case $p = 5$, so that $\alpha = 3$. In Exercise 9.3.18, the reader is asked to check that all solutions of the Fibonacci recurrence then have the form

$$s_k = (a + 3b) \cdot 3^k \quad (k \geq 0)$$

and that the least period has to divide 20, which is realized by the Fibonacci sequence itself.

For many primes $p \neq 5$, the respective upper bounds $2(p+1)$ and $p-1$ on the period length are achieved by the Fibonacci sequence itself. Wall [391] gave a list of all primes $p < 2000$ for which this does *not* hold, including the corresponding least periods. There are 99 such primes, the ones below 200 being 29, 47, 89, 101, 107, 113, 139, 151, 181, and 199. □

Exercises

Exercise 9.3.17. Provide a proof for Theorem 9.3.13. □

Exercise 9.3.18. Discuss the Fibonacci recursion over GF(2) and also over GF(q) in general. Check the details for GF(5) and show that one can obtain an m-sequence for $q \in \{2,3\}$. □

9.4 Binary Pseudorandom Sequences

In many applications – for instance, in Electrical Engineering and in Cryptography – one requires binary random sequences. Theoretically, such a sequence may be defined as a sequence of independent random variables which attain the values 0 and 1 with probability $\frac{1}{2}$ each. For practical purposes, one might generate a (nearly) random sequence by repeatedly flipping a fair coin.

Let us digress for a moment and explain why binary random sequences are of importance in Cryptography. One way of secretly communicating a message is as follows. One first transforms the message into a string of 0's and 1's, for instance, by encoding each letter of the alphabet plus the "blank" indicating the end of a word as binary quintuples. There are several well-known schemes serving this purpose, even if one additionally requires certain special symbols. We may therefore consider a message to be a finite binary sequence $\mathbf{m}$ of length r, say. Now one adds (bitwise) a random binary sequence $\mathbf{r}$ of length r to $\mathbf{m}$, where addition is performed modulo 2, to obtain the enciphered message $\mathbf{c} := \mathbf{m} + \mathbf{r}$.

Then, trivially, $\mathbf{m} = \mathbf{c} + \mathbf{r}$, and thus the recipient can recover the original message by again adding $\mathbf{r}$. While this scheme (called the **one time pad**, since each such random sequence $\mathbf{r}$ is to be used only once) is unconditionally secure by a famous result of Shannon [352], it is certainly not very practical, as it requires the generation and transportation (by courier) of very long binary random sequences. Therefore, its application is in practice restricted to messages of the utmost (often military) importance. For this reason, one would like to replace the random sequence in the one time pad scheme by a deterministically generated sequence that still guarantees a reasonable amount of security. The resulting cryptographic schemes are called **stream ciphers**; see, for instance, Rueppel [329, 330].

The preceding remarks provide some motivation why one would like to generate long binary sequences deterministically – preferably, from a small amount of input data – which look like true random sequences, that is, which exhibit some of the statistical behavior that one would expect from such a sequence. The statistical properties usually required were first formulated by Golomb [142] in 1955 and are therefore often called the **Golomb axioms**. These axioms deal with the particular case of periodic binary sequences. Of course, periodicity is definitely not a random property, and thus the consideration of periodic sequences under the aspect of randomness makes only sense if one uses merely (a small part of) one period.

Therefore, only sequences for which the least period is large can be of any practical interest in this context. We shall now give a somewhat modified account of Golomb's approach, which is based on the presentation by Lidl and Niederreiter

[243], as this seems to us slightly preferable to the original version of the Golomb axioms.

Thus let $\mathbf{a} = (a_k)$ be a periodic sequence over $F = \mathrm{GF}(2)$ with least period r. If $\mathbf{a}$ would be a true random sequence, one would expect to have about as many entries 0 as entries 1 among the values $a_0, \dots, a_{r-1}$. More generally, for a given m-tuple $\mathbf{b} = (b_1, \dots, b_m) \in F^m$, we would expect the number

$$Z_{\mathbf{a}}(\mathbf{b}) := \left|\{t : 0 \le t \le r-1 \text{ and } (a_t, a_{t+1}, \dots, a_{t+m-1}) = \mathbf{b}\}\right|$$

to be independent of $\mathbf{b}$ as long as this makes sense; clearly, r has to be large enough in comparison to m, say $2^m \le r$.

Finally, if we compare a truly random binary sequence $\mathbf{a} = (a_k)$ with a shifted version $\mathbf{a}' = (a_{k+h})$ of itself, we would expect about as many agreements as disagreements: each of the pairs $(0,0)$, $(0,1)$, $(1,0)$ and $(1,1)$ should appear with probability about $1/4$. In order to measure the discrepancy between the number of agreements and disagreements of $\mathbf{a}$, we require the following

Definition 9.4.1. Let $\mathbf{a} = (a_k)$ be a periodic sequence over $F = \mathrm{GF}(2)$ with least period r. The **(periodic) autocorrelation function** $C_{\mathbf{a}}$ of $\mathbf{a}$ is defined as follows:

$$C_{\mathbf{a}}(h) := \sum_{k=0}^{r-1} (-1)^{a_k - a_{k+h}} \quad \text{for } h = 0, \dots, r-1. \qquad \square$$

Definition 9.4.2. Let $\mathbf{a} = (a_k)$ be a periodic sequence over $F = \mathrm{GF}(2)$ with least period r satisfying the following three axioms:

Distribution test: $\left|Z_{\mathbf{a}}(0) - Z_{\mathbf{a}}(1)\right| \le 1;$ (9.18)

Serial test: $\left|Z_{\mathbf{a}}(\mathbf{b}) - Z_{\mathbf{a}}(\mathbf{b}')\right| \le 1$ for any two distinct binary m-tuples $\mathbf{b}$ and $\mathbf{b}'$, provided that $2 \le m \le \log_2 r$; (9.19)

Autocorrelation test: $C_{\mathbf{a}}(h) = c$ if $h \not\equiv 0 \bmod r$, where c is some constant. (9.20)

Then one calls $\mathbf{a}$ a (binary) **pseudorandom sequence**. □

Remark 9.4.3. Note that the distribution test is just the serial test for the case $m = 1$; we have stated this requirement separately for historical reasons.

The subsequent theorem will show that m-sequences over $\mathrm{GF}(2)$ pass all three tests in Definition 9.4.2, a result essentially due to Golomb [142]. Our treatment differs from Golomb's in regard to the serial test. Golomb restricted attention to **runs**, that is, strings of consecutive 0's included between two 1's, or consecutive 1's included between two 0's, and required that half of all runs in a period of the sequence have length 1, a quarter of all runs have length 2, one eighth of all runs have length 3 etc. (as long as this makes sense). Also, in each of these cases there should be equally many runs of 0's and of 1's.

Then **u** *is likewise an m-sequence if and only if* $\gcd(d,q^n-1)=1$*, and* **u** *is actually cyclically equivalent to* **a** *if and only if* $d \equiv q^j \bmod q^n-1$ *for some j with* $0 \le j \le n-1$.

Moreover, every shift register sequence over F for which the minimal polynomial has non-zero constant term and degree some divisor m of n can be obtained from **a** *via a suitable decimation.*

Proof. By Proposition 9.3.10 and Theorem 9.3.12, the feedback polynomial f is primitive and, with $E = \mathrm{GF}(q^n)$,

$$a_k = \mathrm{Tr}_{E/F}(\theta\alpha^k) \quad (k \ge 0), \tag{9.24}$$

where θ is a suitable element of E^* and α a root of the characteristic polynomial f^* in E (and therefore a primitive element for E^*). Because of Equations (9.23) and (9.24), we obtain

$$u_k = a_{h+kd} = \mathrm{Tr}_{E/F}(\psi\beta^k) \quad (k \ge 0),$$

where $\beta := \alpha^d$ and $\psi := \theta\alpha^h$. Denote the minimal polynomial of β by g^*. Then another application of Theorem 9.3.12 shows that **u** belongs to an LFSR with feedback polynomial g and characteristic polynomial g^*. By Theorem 9.3.6, **u** is an m-sequence if and only if g is a primitive polynomial, that is, if and only if β is a primitive element for E^*, which gives the desired criterion $\gcd(d,q^n-1)=1$. Moreover, **u** is actually cyclically equivalent to **a** if and only if the characteristic polynomials f^* and g^* coincide, which means that $\beta = \alpha^d$ has to be a root of f^*. According to Proposition 3.5.1, this holds precisely under the condition stated in the assertion.

Finally, consider any shift register sequence $\mathbf{s} = (s_k)$ which has a minimal polynomial g of degree m dividing n and satisfying $g(0) \ne 0$. Then g splits in E^*, and thus there exists an integer $d \ge 1$ such that $\gamma := \alpha^d$ is a root of g. (The hypothesis $g(0) \ne 0$ guarantees that $\gamma \ne 0$ which means $g \ne x$, as g is irreducible.) Put $K := F(\gamma)$. Again using Theorem 9.3.12, we get

$$s_k = \mathrm{Tr}_{K/F}(\eta\gamma^k) \quad (k \ge 0),$$

where η is a suitable element of K^*. Now choose any element ψ in E^* with $\mathrm{Tr}_{E/K}(\psi) = \eta$ and write $\psi\theta^{-1} = \alpha^h$. Then the transitivity formula for the trace gives

$$a_{h+kd} = \mathrm{Tr}_{E/F}(\theta\alpha^{h+kd}) = \mathrm{Tr}_{K/F}(\mathrm{Tr}_{E/K}(\psi\gamma^k)) = \mathrm{Tr}_{K/F}(\eta\gamma^k) = s_k,$$

so that s is indeed a decimation of **a**. □

The special case $m = n$ of Theorem 9.4.7 establishes a particularly interesting fact, namely, that there is essentially just one m-sequence of any given order over $\mathrm{GF}(q)$:

Corollary 9.4.8. *Any two m-sequences belonging to LFSR's of degree n over* $\mathrm{GF}(q)$ *can be obtained from each other via decimation.* □

This section has shown that shift register sequences of maximal period lead to binary sequences with good randomness properties (according to the criteria introduced by Golomb), by restricting attention to one period of the sequence. It is also interesting to study the randomness properties of other shift register sequences or of sequences comprising only part of a period of an m-sequence. However, the results in the general situation are not as satisfactory, and the techniques required are considerably more involved. We refer the reader to Lidl and Niederreiter [243, §§ 6.7 and 7.4] for some results in this direction.

Finally, it is of considerable interest to investigate the "crosscorrelation" properties between two distinct binary sequences with the same period, in particular between two m-sequences; some elementary results in this direction are in McEliece [263, Chapter 11]. We also refer the reader to the survey by Sarwate and Pursley [332] and the literature cited there; a wealth of references to the distribution and correlation properties of shift register sequences can also be found in the notes to Chapter 8.7 of Lidl and Niederreiter [242].

9.5 Periodic Binary Sequences and Difference Sets

Motivated by the results of Section 9.4, we will now discuss the autocorrelation properties of binary sequences in more detail. In particular, we study an interesting connection to an important topic in Discrete Mathematics, namely the theory of cyclic difference sets. As we shall see, periodic binary sequences satisfying the autocorrelation test and cyclic difference sets are essentially the same objects.

Throughout this section, we will denote the least period of a periodic sequence $\mathbf{a}$ by v instead of r, in order to match the standard notation for difference sets.

Remark 9.5.1. We begin with an alternative definition of the autocorrelation function. Quite often, binary sequences are defined over $\{+1,-1\}$ instead of $F = \mathrm{GF}(2)$. Both approaches are, of course, equivalent. While one may think of sequences over F as being represented by an electrical current which can be either "on" or "off", sequences of ± 1's may be viewed as positive or negative pulses of unit amplitude. If we use a sequence $\mathbf{a}$ with entries ± 1, the formula for the autocorrelation function of a binary sequence with least period v in Definition 9.4.1 should be adjusted as follows:

$$D_{\mathbf{a}}(h) := \sum_{k=0}^{v-1} a_k a_{k+h}.$$

However, sometimes the preceding formula is also used to define the autocorrelation function of a sequence over F. This is unfortunate, as $C_{\mathbf{a}}$ and $D_{\mathbf{a}}$ result in different values over F; nevertheless, the following lemma shows that these values are closely related, and therefore it does not matter too much which definition is used. □

Lemma 9.5.2. *Let* $\mathbf{a} = (a_j)$ *be a periodic sequence with period* v *over* $F = \mathrm{GF}(2)$, *and let* h *be any integer which is not divisible by* v. *Then*

$$C_{\mathbf{a}}(h) = v - 4\big(k - D_{\mathbf{a}}(h)\big),$$

where k denotes the number of entries 1 *in one period of* **a**.

Proof. By definition, $D_{\mathbf{a}}(h)$ counts the number of pairs $(a_t, a_{t+h}) = (1,1)$, where $0 \le t \le v-1$, whereas $C_{\mathbf{a}}(h)$ counts the number of pairs $(a_t, a_{t+h}) \in \{(0,0),(1,1)\}$ minus the number of pairs $(a_t, a_{t+h}) \in \{(0,1),(1,0)\}$.

We now fix a value of h and simplify notation by putting $\mu := D_{\mathbf{a}}(h)$. Then there are $k-\mu$ pairs $(a_t, a_{t+h}) = (1,0)$ and also $k-\mu$ pairs $(a_t, a_{t+h}) = (0,1)$. Consequently, there are precisely $v - 2(k-\mu) - \mu = v - 2k + \mu$ pairs $(a_t, a_{t+h}) = (0,0)$, which gives

$$C_{\mathbf{a}}(h) = \big(\mu + (v - 2k + \mu)\big) - 2(k-\mu) = v - 4(k - D_{\mathbf{a}}(h)),$$

proving the assertion. □

In particular, we obtain the following important fact.

Corollary 9.5.3. *Let* $\mathbf{a} = (a_k)$ *be a periodic sequence with period v over* GF(2), *let k denote the number of entries* 1 *in one period of* **a**, *and let λ be a positive integer. Then the following two conditions are equivalent:*

$$D_{\mathbf{a}}(h) = \begin{cases} k & \text{for } h \equiv 0 \bmod v, \\ \lambda & \text{otherwise} \end{cases} \tag{9.25}$$

and

$$C_{\mathbf{a}}(h) = \begin{cases} v & \text{for } h \equiv 0 \bmod v, \\ v - 4(k-\lambda) & \text{otherwise.} \end{cases} \tag{9.26}$$
□

Definition 9.5.4. Binary sequences satisfying the equivalent conditions in Corollary 9.5.3 are called sequences with a **two-level autocorrelation function**. □

Such sequences have been studied extensively, since they are of considerable importance in Signal Processing. Moreover, they turn out to be equivalent to cyclic difference sets, a class of combinatorial objects of central interest in Finite Geometry and Design Theory, which also has interesting applications in Coding Theory. In order to explain this connection, we need the following

Definition 9.5.5. Let D be a k-subset of some (additively written) group G of order v. Then D is called a (v,k,λ)-**difference set** in G if, for every element $h \neq 0$ of G, the equation

$$h = d - d' \quad \text{with } d, d' \in D \tag{9.27}$$

has exactly λ solutions.

Moreover, D is said to be **abelian** if G is abelian, and D is called a **cyclic** difference set if G is (isomorphic to) the cyclic group $(\mathbb{Z}_v, +, 0)$ of residues modulo v. □

The following useful observation shows that one may always assume $k < v/2$ in the study of difference sets:

Lemma 9.5.6. *Let D be any (v,k,λ)-difference set D (in some group G). Then the parameters of D satisfy the equation*

$$\lambda(v-1) = k(k-1). \tag{9.28}$$

In particular, $v \neq 2k$. Moreover, the complement $\overline{D} := G \backslash D$ of D in G is again a difference set, with parameters

$$(\overline{v}, \overline{k}, \overline{\lambda}) = (v, v-k, v-2k+\lambda).$$

Proof. Note that D gives rise to $k(k-1)$ non-zero differences of the form (9.27). As each of the $v-1$ elements $h \neq 0$ of G has exactly λ such difference representations, one obtains (9.28). In particular, $v \neq 2k$, since λ is an integer.

Clearly, each $h \neq 0$ has exactly v representations as a difference of two elements of G; of these, λ are of the form (9.27), whereas $2(k-\lambda)$ have exactly one of the two group elements involved in D. The remaining $v-2k+\lambda$ representations belong to $G \backslash D$, proving the second assertion. □

Note that

$$k-\lambda = (v-k)-(v-2k+\lambda) = \overline{k}-\overline{\lambda},$$

which explains why this number – which is an invariant under complementation – is a further, quite important parameter for a difference set. One calls $k-\lambda$ the **order** of a (v,k,λ)-difference set.[9]

We next prove a simple but fundamental result which establishes the equivalence of cyclic difference sets with periodic sequences with a two-level autocorrelation function:

Proposition 9.5.7. *The existence of a periodic sequence with period v over $F = \mathrm{GF}(2)$ with k entries 1 per period and with a two-level autocorrelation function $C_{\mathbf{a}}(h)$ as in (9.26) is equivalent to that of a cyclic (v,k,λ)-difference set.*

Proof. In what follows, we let $[t]$ denote the residue class of an integer t modulo v. Given a (v,k,λ)-difference set D in the cyclic group $G = \mathbb{Z}_v$ of residues modulo v, we define a periodic sequence $\mathbf{a} = (a_t)$ with period v over F by putting

$$a_t := \begin{cases} 1 & \text{if } [t] \in D, \\ 0 & \text{otherwise.} \end{cases}$$

As D is a k-set, $\mathbf{a}$ has k entries 1 per period. Also, the number $D_{\mathbf{a}}(h)$ of pairs $(t, t+h)$ satisfying $a_t = a_{t+h} = 1$ equals the number of representations of an element $h \neq 0$

[9] In Design Theory, it is customary to denote the order $k-\lambda$ of a difference set by n, which collides with our usage of n for the degree of the primitive polynomial associated with an m-sequence. Therefore, we will avoid this notation as much as possible.

of G as in (9.27). In view of Corollary 9.5.3, the difference set property of D now translates into the validity of (9.26).

The converse is similar: given a binary periodic sequence $\mathbf{a}$ with period v and with k entries 1 per period which satisfies (9.26), one defines a k-subset D of G as follows:

$$D := \{[t] \in G\colon a_t = 1,\, t = 0,\dots,v-1\}.$$

Appealing again to Corollary 9.5.3, it is easy to check that D is the desired difference set. □

Remark 9.5.8. It should be emphasized that the binary sequences associated with cyclic difference sets are not, in general, pseudorandom sequences, since there is no reason why they should satisfy either the distribution or the serial test. We note that the impression given by Golomb [142] that every pseudorandom sequence (as defined by his axioms) is actually an m-sequence is not correct; see also [144]. The following counter-example with period $v = 127$ (which corresponds to the transformation of the $(127, 63, 31)$-difference set of type E given by Baumert [20] with the automorphism $x \mapsto 19x$) was noted by Cheng and Golomb [74]:

1111101111001111111001001011101010111100011000001001101110011000
110110111010010001101000010101001101001010001110110000101000000

Note that a periodic binary sequence with two-level autocorrelation function satisfies the distribution test (9.18) if and only if

$$2k-1 \le v \le 2k+1. \quad \square \tag{9.29}$$

Example 9.5.9. Together with the proof of Theorem 9.4.4, the preceding results show that the binary m-sequences belonging to an LFSR of length d with primitive characteristic polynomial correspond to cyclic difference sets with parameters

$$v = 2^d - 1,\ k = 2^{d-1} \text{ and } \lambda = 2^{d-2}$$

or, after complementation,

$$v = 2^d - 1,\ k = 2^{d-1} - 1 \text{ and } \lambda = 2^{d-2} - 1. \tag{9.30}$$

These difference sets turn out to be part of a much larger family of examples which we present next. □

Theorem 9.5.10. *For every prime power q and every positive integer d, there exists a cyclic difference set with parameters*

$$v = \frac{q^{d+1}-1}{q-1},\quad k = \frac{q^d - 1}{q-1} \text{ and } \lambda = \frac{q^{d-1}-1}{q-1}. \tag{9.31}$$

Proof. Let $\mathbf{a} = (a_k)$ be the m-sequence over $F = \mathrm{GF}(q)$ defined by

$$a_k := \mathrm{Tr}_{E/F}(\alpha^k) \quad (k \geq 0), \tag{9.32}$$

where α is a root of a primitive polynomial f of degree $d+1$ over F and therefore a primitive element for the multiplicative group of $E := \mathrm{GF}(q^{d+1})$; see Theorem 9.3.12. Note that $\mathbf{a}$ is essentially determined already by its first v entries, since

$$a_{iv+g} = \mathrm{Tr}_{E/F}(\alpha^{iv+g}) = \gamma^i \mathrm{Tr}_{E/F}(\alpha^g) \quad \text{for } i \geq 0 \text{ and } g = 0,\dots,v-1,$$

where $\gamma := \alpha^v$ is a primitive element for F^*. In particular,

$$a_{iv+g} = 0 \iff a_g = 0 \quad \text{for } i \geq 0 \text{ and } g = 0,\dots,v-1. \tag{9.33}$$

We now identify the cyclic group G of residues modulo v with the indices $0,\dots,v-1$ and claim that the set

$$D := \{g \in G\colon a_g = 0\} \tag{9.34}$$

is the desired difference set in G. In view of (9.33), it suffices to show that the set

$$C := \{g\colon g = 0,\dots,q^{d+1}-1,\ a_g = 0\}$$

contains exactly $k(q-1) = q^d - 1$ elements and meets each of its translates

$$C + x = \{g + x\colon g = 0,\dots,q^{d+1}-1,\ a_g = 0\} \quad (x = 1,\dots,v-1)$$

in exactly $\lambda(q-1) = q^{d-1} - 1$ elements. By (9.32),

$$C = \{g : g = 0,\dots,q^{d+1}-1,\ \mathrm{Tr}_{E/F}(\alpha^g) = 0\}.$$

Hence C corresponds to the non-zero points of the hyperplane

$$H := \{z \in E\colon \mathrm{Tr}_{E/F}(z) = 0\}$$

of the F-vector space E; in particular, C has the correct number of elements. Similarly, the set $C+x$ corresponds to the non-zero points of the hyperplane

$$H_x := \{z \in E\colon \mathrm{Tr}_{E/F}(\alpha^{-x} z) = 0\},$$

and thus H and H_x meet in exactly q^{d-1} elements, which establishes the desired result on the intersection of C and $C+x$. (Note that H and H_x are indeed distinct hyperplanes, since the linear forms $\mathrm{Tr}_{E/F}(z)$ and $\mathrm{Tr}_{E/F}(\alpha^{-x} z)$ have the same kernel if and only if x is a multiple of v.) □

Remark 9.5.11. In view of Corollary 9.4.5, the choice of the particular m-sequence in the proof of Theorem 9.5.10 is not important, since any two m-sequences can be obtained from each other by decimation. This implies that the difference sets D and D' associated with two different m-sequences as in (9.34) are **equivalent**: one has $D' = aD + g$ for some integer a co-prime with v and some element $g \in G$. □

The difference sets associated with m-sequences are usually called **Singer difference sets**, since they were first discovered – in a rather different setting – by Singer [357] in 1938 when studying certain collineation groups of finite projective spaces. In fact, they are equivalent to the classical symmetric designs formed by the points and hyperplanes of some finite projective space over $\mathrm{GF}(q)$.[10] This geometrical interpretation is very useful for studying the properties of m-sequences and also for constructing ternary sequences with interesting correlation properties; see, for instance Games [124]. The standard proof of Theorem 9.5.10 likewise uses the connection to projective spaces; see, for instance, Beth, Jungnickel and Lenz [35] or Lidl and Niederreiter [243]. While the geometric proof and the proof given here via m-sequences are essentially equivalent, the latter approach makes it possible to avoid the explicit use of finite projective spaces and symmetric designs.

A wealth of further difference sets with parameters (9.31) was constructed by Gordon, Mills and Welch [152]. The particularly important binary case will be discussed in Section 9.9.

As noted in (9.29), cyclic difference sets corresponding to binary sequences which satisfy the distribution test have their parameters related by $v = 2k + 1$ and hence $\lambda = (k-1)/2$, up to complementation. Such difference sets are of particular interest, since they also have a very small (in absolute value) autocorrelation coefficient, namely $c = v - 4(k - \lambda) = -1$, which is a highly desirable property for applications in Signal Processing.

Definition 9.5.12. Difference sets satisfying the condition $v - 4(k - \lambda) = -1$ are called **Hadamard difference sets**; in terms of their order $n := k - \lambda$, the parameters may be written as

$$v = 4n - 1, \quad k = 2n - 1 \quad \text{and} \quad \lambda = n - 1;$$

this follows using Equation (9.28). □

In view of (9.30) and (9.31), we obtain Hadamard difference sets with order a power of 2 from binary m-sequences. The next two results show how one may use finite fields to construct further examples. The first of these constructions was given by Paley [315] in 1933 (in different form).

Theorem 9.5.13. *Let $q = 4n - 1$ be a prime power. Then the set D of non-zero squares in $F = \mathrm{GF}(q)$ is a $(4n-1, 2n-1, n-1)$-difference set in the additive group G of F. In particular, a cyclic Hadamard difference set of order n exists whenever $4n - 1$ is a prime.*

Proof. Trivially, D is invariant under multiplication with non-zero squares in F, and hence the number of difference representations of the form (9.27) of an element $h \neq 0$ of G only depends on whether or not h is a square. Since -1 is a non-square in F by Proposition 3.2.13, the trivial equivalence

[10] We refer the interested reader to Baumert [20], Lander [229] or Beth, Jungnickel and Lenz [35] for background from Finite Geometries. Even though it is partially outdated, the first of these books is still an important reference for the theory of cyclic difference sets.

$$h = d - d' \iff -h = d' - d$$

shows that squares and non-squares have the same number λ of difference representations. Thus D is indeed a $(4n-1, 2n-1, \lambda)$-difference set, and the value for λ follows from (9.28). □

Definition 9.5.14. The difference sets constructed in Theorem 9.5.13 are usually called **Paley difference sets**. In the case where q is a prime, the associated periodic binary sequences are generally called **Legendre sequences**, since they may be described in terms of the Legendre symbol (see Definition 8.5.9):

$$a_k = 1 \iff (k/p) = 1 \quad \text{for all } k \geq 0. \qquad \square$$

It should be mentioned that Legendre sequences generally neither satisfy the serial test (9.19) nor Golomb's second axiom on runs, except for the smallest case $n=2$, where the Paley difference set and the Singer difference set in the cyclic group of order 7 coincide; we refer the reader to Perron [318] who studied the distribution of runs of length two in Legendre sequences, which behave as expected according to Golomb's axiom.

Our second construction for Hadamard difference sets gives the **twin prime power difference sets**; it is usually attributed to Stanton and Sprott [360], even though the cyclic case was known earlier.

Theorem 9.5.15. *Let q and $q+2$ be odd prime powers, and put $n := \frac{1}{4}(q+1)^2$. Then there exists a $(4n-1, 2n-1, n-1)$-difference set. In particular, a cyclic Hadamard difference set of order n exists whenever q and $q+2$ are twin primes.*

Proof. Let G be the direct sum $(F,+) \oplus (K,+)$ of the additive groups of the fields $F := \mathrm{GF}(q)$ and $K := \mathrm{GF}(q+2)$. Then

$$\big\{(x,y) \in F \oplus K\colon\ y = 0 \text{ or } x \text{ and } y \text{ are either both squares or both non-squares}\big\}$$

is the desired difference set D. The verification that D is indeed a difference set in G proceeds along similar lines as in the proof of Theorem 9.5.13, though it is somewhat more involved. We leave the details to the reader as the more demanding Exercise 9.5.21 (or refer to [35, Theorem VI.8.2]). □

Another family of Hadamard difference sets is due to Hall [177] and uses certain cosets of the sextic residues in $\mathrm{GF}(q)$, where q is a prime power of the form $q = 4x^2 + 27$; thus is does not give further orders. There are also a few other families of examples which duplicate the parameters in (9.30), among them the Gordon-Mills-Welch difference sets, which we will treat in Section 9.9. Apart from the families just discussed, only sporadic examples are known.

As mentioned before, sequences with a small (in absolute value) autocorrelation coefficient $c = v - 4(k-\lambda)$ are of particular interest. Thus it is very natural to wonder what can be said about the case $c = 0$, that is, about cyclic difference sets with $v = 4(k-\lambda)$. The sequence $(1\,1\,1\,0\,1\,1\,1\,0\ldots)$ provides an example with period 4.

Before discussing the existence question, let us note that the parameters of such a difference set must have a rather special form, namely

$$v = 4u^2, \;\; k = 2u^2 \pm u \;\; \text{and} \;\; \lambda = u^2 \pm u \tag{9.35}$$

for some integer u; see Exercise 9.5.22.

Remark 9.5.16. Sometimes, difference sets with parameters (9.35) are called **Menon difference sets**; in the cyclic case, the corresponding periodic binary sequences are called **perfect binary sequences**. The term "Menon difference set" was championed by Beth, Jungnickel and Lenz [35], since Menon [271] was the first author who studied such difference sets. However, many authors used the term "Hadamard difference set" instead, which is unfortunate, since difference sets with parameters (9.30) were usually also referred to by this name. To avoid confusion, difference sets with parameters (9.30) are now often called **Paley-Hadamard difference sets**.

In this context, we also warn the reader that Golomb [143] uses the term "perfect sequence" to denote any binary sequence associated with a cyclic difference set. □

Unfortunately, the sequence with period 4 shown above is the only known example of a perfect binary sequence. In fact, it is generally believed that there are no further examples.

Conjecture 9.5.17. There are no perfect binary sequences – equivalently, no cyclic difference sets with parameters (9.35) – for any $v > 4$. □

While this conjecture is still open, there is by now considerable evidence for its validity. In 1968, Turyn proved in his seminal paper [379] that any perfect binary sequence with $v > 4$ has period $v = 4u^2$ for some odd integer $u \geq 55$. We digress and mention a (possibly) stronger notion than perfect sequences, which is defined using an aperiodic variant of the autocorrelation concept:

Definition 9.5.18. A finite binary sequence $(b_1, \ldots, b_v)$, traditionally written with entries ± 1, is called a **Barker sequence** if it satisfies the condition

$$c_j := \sum_{i=1}^{v-j} b_i b_{i+j} \in \{0, -1, 1\} \quad \text{for } j = 1, \ldots, v-1. \quad \square$$

For instance, $(1\;1\;1\;-1\;-1\;1\;-1)$ is a Barker sequence of length 7. The special case where the c_j are restricted to 0 and -1 was introduced in 1953 by Barker [18] in connection with a problem in digital communication. Barker sequences are only known if the length v is one of 2, 3, 4, 5, 7, 11 or 13; moreover, they all give rise to cyclic difference sets.

Indeed, Turyn and Storer [380] proved that any example with $v > 13$ "is" actually a perfect sequence; in our notation, this interpretation requires both replacing all entries -1 by 0 and extending the resulting binary sequence over GF(2) to a periodic sequence with period v. It is still open whether or not the converse also holds.

In particular, there is no Barker sequence of odd length $v > 13$; a simpler proof of this result was recently given by Schmidt and Willms [339]. It is conjectured that the same result also holds for even length sequences:

Conjecture 9.5.19. There exists no Barker sequence of length $v > 13$ (and hence $v = 4u^2$ for some odd integer u). □

The evidence for Conjecture 9.5.19 is even larger. Eliahou, Kervaire and Saffari [112] proved its validity whenever u has a prime divisor $p \equiv 3 \bmod 4$. Jedwab and Lloyd [208] used this result and the fundamental work of Turyn [378] to check the validity of Conjecture 9.5.19 for all $v = 4u^2$, where $u < 5000$, with five possible exceptions. A breakthrough occurred around 1995 when Bernhard Schmidt invented his "field descent method" [337], see also [338]. After further refinements in joint work of Leung and Schmidt [239, 240, 241], the validity of Conjecture 9.5.19 now seems virtually certain:

Result 9.5.20 *There is no Barker sequence of length* v *for* $13 < v \leq 4 \cdot 10^{33}$. □

Strong results hold also with regard to Conjecture 9.5.17, though they are more difficult to state. We just mention that there are now only about 1500 open cases for the existence of a perfect binary sequence of length $4m^2$ with $m \leq 10^{13}$, the smallest of which is $m = 11715$. We refer the reader to [240] and the references given there for details.

Let us conclude this section with some remarks on difference sets in general groups. Difference sets over non-cyclic abelian groups are also of considerable interest in Signal Processing; in particular, there is a correspondence between difference sets with parameters (9.35) in non-cyclic abelian groups and a certain type of binary arrays, traditionally written with entries $+1$ and -1, the so-called *perfect binary arrays*. The possible dimensions d of the associated arrays A equal the number of direct cyclic summands in some decomposition of the abelian group G in question. (Thus the smallest possible dimension d is the number of invariant factors of G.) Then the array A is indexed by the group elements, written as d-tuples with respect to an appropriate basis of G; one puts $a_g = 1$ if and only if $g \in D$. It can be shown that A has a perfect autocorrelation function:

$$\sum_{g \in G} a_g a_{g+h} = 0 \quad \text{for all } h \neq 0.$$

In analogy to the case of perfect binary sequences, every perfect binary array arises from a difference set in this way; see Chan, Siu and Tong [72] for the 2-dimensional and Kopilovich [224] for the general case. The study of perfect binary arrays was started by Calabro and Wolf [57]. In Communications Engineering, the 2-dimensional case of these arrays is of particular interest; see, for instance, Lüke [246]. A table for this case was given by Chan and Siu [71] and later updated by Jedwab [206] who used results of Turyn [378] to obtain some new non-existence criteria.

We also mention an interesting survey by Siu [358], where connections between binary sequences on the one hand and combinatorial designs on the other hand are discussed, as well as a fundamental paper by Jedwab [207]. Finally, we note that the theory of difference sets is discussed in considerable detail in the book of Beth, Jungnickel and Lenz [35].

Exercises

Exercise 9.5.21. Give a detailed proof for Theorem 9.5.15. □

Exercise 9.5.22. Use Equation (9.28) to prove that a difference set with $v = 4(k-\lambda)$ indeed has parameters of the form given in Equation (9.35). □

9.6 The Linear Complexity of a Shift Register Sequence

While we have seen in Section 9.4 that binary m-sequences have some very good randomness properties, we now show that they also exhibit a behavior which is far from being random – a phenomenon typical for shift register sequences.

Trivially, knowledge of some shift register generating a given sequence $\mathbf{a}$ (over any field F) completely determines the sequence. As the following simple result shows, a rather small number of consecutive elements of $\mathbf{a}$ suffices to produce an associated LFSR.

Theorem 9.6.1. *Let $\mathbf{a} = (a_k)$ be a shift register sequence over a field F, let m be the minimal polynomial of $\mathbf{a}$, and put $n := \deg m$. Then m may be computed from the first $2n$ elements of $\mathbf{a}$. Moreover, if $\mathbf{a}$ is periodic, m may in fact be computed from any $2n$ consecutive elements of $\mathbf{a}$.*

Proof. Write $m(x) = x^n - c_1x^{n-1} - \cdots - c_{n-1}x - c_n$, so that $\mathbf{a}$ satisfies the linear recurrence relation (9.1):

$$a_k = \sum_{i=1}^{n} c_i a_{k-i} \quad \text{for } k \geq n.$$

Assuming that the elements $a_t, a_{t+1}, \ldots, a_{t+2n-1}$ are known, we obtain the following matrix equation:

$$\begin{pmatrix} a_{t+n} \\ a_{t+n+1} \\ \vdots \\ a_{t+2n-1} \end{pmatrix} = \begin{pmatrix} a_{t+n-1} & \cdots & a_{t+1} & a_t \\ a_{t+n} & \cdots & a_{t+2} & a_{t+1} \\ \vdots & \ddots & \vdots & \vdots \\ a_{t+2n-2} & \cdots & a_{t+n} & a_{t+n-1} \end{pmatrix} \begin{pmatrix} c_1 \\ \vdots \\ c_{n-1} \\ c_n \end{pmatrix}. \tag{9.36}$$

Note that the rows of this matrix are just the n state vectors $\mathbf{a}^{(t)}, \mathbf{a}^{(t+1)}, \ldots, \mathbf{a}^{(t+n-1)}$, up to reversing the order of their entries. If $\mathbf{a}$ is periodic, these vectors must be linearly independent by Proposition 9.2.6, as m is the minimal polynomial of $\mathbf{a}$. Hence

Equation (9.36) can be solved uniquely for $c_1,\dots,c_n$. If we drop the assumption that $\mathbf{a}$ is periodic, the same argument still applies provided that $t=0$, since we may then use Proposition 9.1.9 instead of Proposition 9.2.6. □

Definition 9.6.2. In view of Theorem 9.6.1, the degree of the minimal polynomial of a shift register sequence $\mathbf{a}$ is also called the **linear complexity** $L(\mathbf{a})$ of $\mathbf{a}$. Thus $L(\mathbf{a})$ is the smallest length of any LFSR producing $\mathbf{a}$, see Corollary 9.1.7. Alternative terms for $L(\mathbf{a})$ are **linear equivalent** and **linear span**. □

As the binary m-sequences show, a periodic shift register sequence may combine a very large least period with a very small linear complexity. It is questionable whether such a sequence should really be considered quasi-random, since the knowledge of a very short part of the sequence should not enable us to determine all of a "random" sequence. For some applications this defect of binary m-sequences is irrelevant; for instance, the most important feature for applying a binary sequence in Signal Processing is a good correlation behavior. However, for others applications, this may render the sequence totally unsuitable; for instance, such a sequence should definitely not be used as a stream cipher in the context of Cryptography. We shall return to this problem later.

If we want to apply Equation (9.36) to compute the minimal polynomial of a given shift register sequence $\mathbf{a}$, we need to know the linear complexity $L(\mathbf{a})$, since otherwise our argument for the invertibility of the matrix formed by the given consecutive state vectors breaks down. It is intuitively clear that one should be able to find $L(\mathbf{a})$ by successively trying larger and larger values of n in Equation (9.36). Indeed, the following result shows how $L(\mathbf{a})$ may be computed if just an upper bound N on $L(\mathbf{a})$ is given – that is, if it is known that $\mathbf{a}$ can be produced by some LFSR of length N.

Theorem 9.6.3. *Let $\mathbf{a}=(a_k)$ be a shift register sequence over a field F belonging to some LFSR of length N. Then the linear complexity $L(\mathbf{a})$ equals the maximum number of linearly independent vectors among the state vectors*

$$\mathbf{b}^{(t)} = (a_t, a_{t+1}, \dots, a_{t+N-1}) \qquad (t \geq 1).$$

Alternatively, $L(\mathbf{a})$ can also be obtained as the largest value of n for which the first n state vectors $\mathbf{b}^{(0)},\mathbf{b}^{(1)},\dots,\mathbf{b}^{(n-1)}$ are linearly independent.

Proof. Let $m(x)=x^n-c_1x^{n-1}-\dots-c_{n-1}x-c_n$ be the minimal polynomial of $\mathbf{a}$. We use the standard notation for the state vectors of $\mathbf{a}$ when viewed as a shift register sequence produced by the LFSR with characteristic polynomial m:

$$\mathbf{a}^{(t)} := (a_t, a_{t+1}, \dots, a_{t+n-1}) \qquad (t \geq 0).$$

Note that $\mathbf{b}^{(0)},\mathbf{b}^{(1)},\dots,\mathbf{b}^{(n-1)}$ are linearly independent, since the n-tuples of their first n coordinates form the vectors $\mathbf{a}^{(0)},\mathbf{a}^{(1)},\dots,\mathbf{a}^{(n-1)}$, and these are linearly independent by Proposition 9.1.9. Hence it suffices to check that every state vector

$\mathbf{b}^{(t)}$ with $t \geq n$ is a linear combination of $\mathbf{b}^{(0)}, \mathbf{b}^{(1)}, \ldots, \mathbf{b}^{(n-1)}$. We use the fact that $\mathbf{a}$ satisfies the linear recurrence relation associated with m:

$$a_{t+n-1} = c_1 a_{t+n-2} + c_2 a_{t+n-3} + \cdots + c_n a_{t-1} \quad \text{for all } t \geq 1,$$

which translates into the corresponding identity

$$\mathbf{b}^{(t+n-1)} = c_1 \mathbf{b}^{(t+n-2)} + c_2 \mathbf{b}^{(t+n-3)} + \cdots + c_n \mathbf{b}^{(t-1)} \quad \text{for all } t \geq 1$$

for the state vectors. Now an easy induction argument establishes the assertion. □

Using Theorems 9.6.1 and 9.6.3, we have an obvious (though tedious) way of computing both the linear complexity and the minimal polynomial of any given shift register sequence **a**. Making use of the special structure of the matrix appearing in Equation (9.36) leads to a better algorithm for a more general problem, which will be discussed in Section 9.7.

As a consequence of Theorem 9.6.3, we also obtain one half of Kronecker's Result 9.1.14.

Corollary 9.6.4. *Let* $\mathbf{a} = (a_k)$ *be a shift register sequence over a field* F. *Then* $D^{(r)}(\mathbf{a}) = 0$ *for all but finitely many values of* r. *Moreover, the linear complexity of* $\mathbf{a}$ *is the smallest positive integer* n *such that* $D^{(r)}(\mathbf{a}) = 0$ *for all* $r \geq n+1$.

Proof. Note that the rows of the Hankel determinant $D^{(r)}(\mathbf{a})$ introduced in Definition 9.1.13 are just the first r "state vectors" of length r of **a**. By Theorem 9.6.3, $L(\mathbf{a})$ equals the largest integer for which these vectors are linearly independent, that is, the largest integer for which $D^{(r)}(\mathbf{a}) \neq 0$. □

In the remainder of this section, we concentrate on periodic sequences. In particular, we wish to discuss the linear complexity of the periodic binary sequences associated with cyclic difference sets. Let us first introduce some notation.

Definition 9.6.5. Given a periodic sequence **a** with least period v over a field F, let us denote the (v, v)-matrix whose determinant is the v-th Hankel determinant $D^{(v)}(\mathbf{a})$ by M. Thus M is the following matrix:

$$M = \begin{pmatrix} a_0 & a_1 & \ldots & a_{v-1} \\ a_1 & a_2 & \ldots & a_v \\ \vdots & \vdots & \ddots & \vdots \\ a_{v-1} & a_v & \ldots & a_{2v-2} \end{pmatrix}. \tag{9.37}$$

We shall call M the **incidence matrix** of **a**. □

The following result is basic and explains why $L(\mathbf{a})$ is sometimes referred to as the **linear span** of **a**.

Remark 9.6.6. We stress that the terminology introduced in Definition 9.6.5 is *not* standard. It is motivated by the special case where **a** is the periodic binary sequence associated with a cyclic difference set D, since then M may be viewed (in design theoretic language) as the incidence matrix of the symmetric design defined by D. We refer the reader to Beth, Jungnickel and Lenz [35] or to Lander [229] for the connection between (cyclic) difference sets and symmetric designs.

In Design Theory, incidence matrices are, by definition, originally given over the field $\mathbb{Q}$: they have integer entries m_{ij} which are either 1 or 0, indicating whether or not the i-th point is on the j-th block. As any field F contains distinguished elements also denoted by 0 and 1, one may then consider any such incidence matrix also as a matrix over F. In particular, this is of considerable interest for the prime fields $\mathbb{Z}_p$; see Remark 9.6.8. □

Proposition 9.6.7. *Let* $\mathbf{a} = (a_k)$ *be a periodic sequence with period* v *over a field* F. *Then the linear complexity* $L(\mathbf{a})$ *of* $\mathbf{a}$ *equals the rank of the incidence matrix* M *of* $\mathbf{a}$ *over* F.

Proof. Since $\mathbf{a}$ is periodic with period v, it can be generated using the trivial LFSR of length v with characteristic polynomial $f^* = x^v - 1$ by choosing the initial conditions as $(a_0, a_1, \dots, a_{v-1})$. Note that the rows of M are just the state vectors corresponding to this LFSR. Therefore, the assertion is an immediate consequence of Theorem 9.6.3, with $N := v$. □

Remark 9.6.8. If we want to apply Proposition 9.6.7 to the periodic binary sequence associated with a cyclic difference set D, we need to compute the rank of the associated incidence matrix M over $F = \mathrm{GF}(2)$. This is a standard problem in Design Theory, where one actually studies a more general concept and considers also the rank of M over $\mathrm{GF}(p)$. This invariant is usually called the p-**rank** of D and denoted by $\mathrm{rank}_p D$.

We will now present a few simple results concerning the p-rank of a cyclic difference set which are due to MacWilliams and Mann [254]. In fact, the results in 9.6.10, 9.6.12 and 9.6.13 below hold, more generally, for symmetric designs, since they only depend on the validity of Equation (9.38), which is satisfied by the incidence matrix of any symmetric design; see Beth, Jungnickel and Lenz [35]. □

Lemma 9.6.9. *Let* $\mathbf{a}$ *be a periodic binary sequence associated with a cyclic* (v,k,λ)-*difference set. Then the incidence matrix* M *of* $\mathbf{a}$ *satisfies*

$$S = MM^T = \lambda J + (k - \lambda) I, \tag{9.38}$$

where J *denotes the* (v,v)-*matrix with all entries* 1.

Proof. Note that each entry of S is the inner product of two shifts of one period of $\mathbf{a}$. Hence the assertion is an immediate consequence of the autocorrelation property (9.25). □

Proposition 9.6.10. *Let* D *be a cyclic* (v,k,λ)-*difference set, and let* p *be a prime not dividing the order* $n := k - \lambda$ *of* D. *Then:*

$$\mathrm{rank}_p D = \begin{cases} v & \text{if } p \text{ does not divide } k, \\ v-1 & \text{if } p \text{ divides } k. \end{cases}$$

Proof. It suffices to show that the matrix S defined in Lemma 9.6.9 is similar to the diagonal matrix $\mathrm{diag}(k^2, n, \ldots, n)$. To see this, note that $(1, \ldots, 1)^T$ is an eigenvector for S corresponding to the eigenvalue $\lambda v + n = k^2$, where the equality follows from Equation (9.28). Moreover, the $v-1$ vectors of the form

$$(1,0,\ldots,0,-1,0,\ldots,0)^T \quad \text{(with exactly one entry } -1)$$

are linearly independent eigenvectors for S with eigenvalue n. □

Let us give a simple application of the preceding results. In particular, it is often easy to determine the linear complexity of a Legendre sequence.

Corollary 9.6.11. *Let* **a** *be the periodic binary sequence with period v associated with the complement of a Singer difference set as in Theorem 9.5.10, where q is odd, or with a cyclic Hadamard difference set of odd order (in particular, a Paley difference set of odd order). Then* **a** *has linear complexity v.*

Proof. In order to apply Proposition 9.6.7, we need to determine the 2-rank of the incidence matrix M. By Proposition 9.6.10, this rank is v, since both k and the order $n = k - \lambda$ of the given difference set are odd: the complement of a Singer difference set has order q^{d-1} and $k = q^d$, and a Hadamard difference set of order n has $k = 2n - 1$. □

If the order n is even, the exact determination of the linear complexity of **a** is usually a difficult problem. Of course, this also applies for the more general problem of computing the p-rank of a difference set D in the case where the order is a multiple of p. Nevertheless, there is a simple upper bound:

Proposition 9.6.12. *Let D be a cyclic (v,k,λ)-difference set, and let p be a prime dividing its order $n = k - \lambda$. Then one has the following:*

$$\mathrm{rank}_p D \le \begin{cases} \frac{v}{2} & \text{if } p \text{ divides } k, \\ \frac{v+1}{2} & \text{if } p \text{ does not divide } k. \end{cases}$$

Proof. Let C be the $\mathrm{GF}(p)$-vector space generated by the rows of the incidence matrix M, let $C^\perp$ be the associated orthogonal space (with respect to the usual inner product), and note

$$v = \dim C + \dim C^\perp. \tag{9.39}$$

If p divides k and n, it also divides λ; then Lemma 9.6.9 shows that C is contained in $C^\perp$, and hence (9.39) implies the assertion. If p divides n but not k, Lemma 9.6.9 shows that the difference of any two rows of M belongs to $C^\perp$, whereas the rows of M are not in $C^\perp$. This implies that $C \cap C^\perp$ has codimension 1 in C, and again (9.39) gives the assertion. □

A slight improvement of Proposition 9.6.12 – which actually makes use of the assumption that we deal with the incidence matrix of a cyclic difference set – in the case where $p = 2$ and k is even was given by Bromfield and Piper [50]. These authors did not discuss the connection to designs and previous work on the p-rank; it seems to be an open problem whether or not their improvement of Proposition 9.6.12 also holds for symmetric designs in general.

Under a more restrictive assumption on p, one may even compute the p-rank exactly. The following result (which we state without proof) is also due to MacWilliams and Mann [254].

Result 9.6.13 *Let D be a cyclic (v,k,λ)-difference set, and let p be a prime strictly dividing its order $n = k-\lambda$; that is, p divides n, but p^2 does not. Then one has the following:*

$$\operatorname{rank}_p D = \begin{cases} \frac{v-1}{2} & \text{if } p \text{ divides } k, \\ \frac{v+1}{2} & \text{if } p \text{ does not divide } k. \end{cases} \quad \square$$

In particular, Result 9.6.13 yields the linear complexity of the binary sequence associated with any cyclic difference set of order $n \equiv 2 \bmod 4$.

Next, we state a result which gives the p-rank of the Singer difference sets and was independently obtained by Goethals and Delsarte [140], MacWilliams and Mann [254] and Smith [359] around 1968; its proof will be postponed to Section 9.8.

Theorem 9.6.14. *Let $q = p^r$ be a power of a prime p and let D be the Singer difference set with parameters* (9.31) *constructed in the proof of Theorem 9.5.10. Then*

$$\operatorname{rank}_p D = \binom{p+d-1}{d}^r + 1. \quad \square$$

Corollary 9.6.15. *The linear complexity of the periodic binary sequence* **a** *associated with a Singer difference set D with parameters* (9.31)*, where $q = 2^r$, is as follows:*

$$\operatorname{rank}_2 D = 1 + (d+1)^r. \quad \square$$

Example 9.6.16. Corollary 9.6.15 has an interesting application in the special case $q = 2$: it shows that the linear complexity of the complementary sequence of a binary m-sequence **a** of length $2^d - 1$ is equal to $d+1$. Here the **complementary sequence** of **a** is defined in the obvious way, namely as the sequence $\overline{\mathbf{a}}$ obtained from **a** by interchanging entries 0 and 1 – that is, by adding 1 to each entry.

Note that the m-sequence **a** itself has linear complexity d, as its minimal polynomial is a primitive polynomial of degree d over GF(2). For the special case $d = 3$, this agrees with Result 9.6.13. □

The following result explains the observations in Example 9.6.16:

Proposition 9.6.17. *The linear complexities of a periodic binary sequence* **a** *and of its complementary sequence* $\overline{\mathbf{a}}$ *differ by at most* 1*. Moreover, both sequences actually have the same linear complexity if and only if the vector* **j** *with all entries* 1 *belongs to either both or none of* $\langle \mathbf{a} \rangle$ *and* $\langle \overline{\mathbf{a}} \rangle$*, where* $\langle \mathbf{a} \rangle$ *denotes the linear span of the rows of the incidence matrix of* **a** *given in (9.37).*

Proof. It suffices to note that the state vectors of $\overline{\mathbf{a}}$ are contained in $\langle \mathbf{a} \rangle$ if and only if $\mathbf{j} \in \langle \mathbf{a} \rangle$. □

We now use Proposition 9.6.17 to give an alternative proof for the result on the linear complexity of the complementary sequence of a binary m-sequence obtained in Example 9.6.16, which does not rely on the rather involved Theorem 9.6.14 and its Corollary 9.6.15.

Corollary 9.6.18. *The complementary sequence $\overline{\mathbf{a}}$ of a binary m-sequence $\mathbf{a}$ belonging to a primitive feedback polynomial f of degree d over* GF(2) *– that is, the sequence $\overline{\mathbf{a}}$ associated with a Singer difference set with parameters* (9.30) *– has linear complexity $d+1$.*

Proof. As mentioned before, $L(\mathbf{a}) = d$. Note that $\mathbf{j}$ cannot belong to $\langle \mathbf{a} \rangle$: otherwise, $\mathbf{j}$ would have to satisfy the linear recurrence relation associated with f, which contradicts Proposition 9.3.10, as the impulse response sequence clearly does not contain a run of v entries 1.

On the other hand, $\mathbf{j}$ lies in $\langle \overline{\mathbf{a}} \rangle$, since $\overline{\mathbf{a}}$ contains an odd number of entries 1 per period, namely $k = 2^d - 1$, so that the sum of all rows of the incidence matrix of $\overline{\mathbf{a}}$ is $\mathbf{j}$. In view of Proposition 9.6.17, we conclude that the linear complexities of $\mathbf{a}$ and $\overline{\mathbf{a}}$ have to differ by 1. Therefore, $\langle \mathbf{a} \rangle$ is contained in (but is different from) $\langle \overline{\mathbf{a}} \rangle$, and hence $\overline{\mathbf{a}}$ has linear complexity $d+1$. □

Remark 9.6.19. In Section 10.6, we will show that the linear complexity of the Legendre sequences (see Definition 9.5.14) and the sequences belonging to the twin prime difference sets (see Theorem 9.5.15) always equals $(v+1)/2 = 2n$ if the order $n = k - \lambda$ is even. Note that the case of odd orders is covered by Corollary 9.6.11. Again, this agrees with Result 9.6.13 when n is not a multiple of 4.

In particular, if $v = 2^d - 1$ is a Mersenne prime, the Legendre sequence of period v and the complementary sequence of the m-sequence of this period – which both belong to difference sets with parameters (9.30) – may be distinguished by their linear complexities. In view of Theorems 9.6.1 and 9.3.12, this is not too surprising. A stronger result was obtained by Hamada and Ohmori [178] who proved that the symmetric design associated with a Singer difference set with parameters (9.30) is characterized among *all* symmetric designs with these parameters by its 2-rank. □

We conclude this section with some additional comments. First of all, we stress that shift register sequences are – in spite of the remarks following Definition 9.6.2 – nevertheless useful in constructing quasi-random sequences.

Example 9.6.20. Let us return to Theorem 9.5.10, where we now assume q to be odd. In the proof of this result, we used the m-sequence defined in Equation (9.32), namely

$$a_k := \mathrm{Tr}_{E/F}(\alpha^k) \quad (k \geq 0),$$

where α is a root of a primitive polynomial f of degree $d+1$ over $F = \mathrm{GF}(q)$. We then proved that the first $v = (q^{d+1} - 1)/(q-1)$ entries of $\mathbf{a}$ essentially determine all of **a** and that the positions where one has an entry 0 yield a cyclic Singer difference set D with parameters (9.31).

According to Proposition 9.5.7, D gives rise to a periodic binary sequence **b** with period v and two-level autocorrelation. In view of Corollary 9.6.11 and Proposition 9.6.17, **b** also has a very large linear complexity (compared to $d+1$), and thus exhibits at least several desirable randomness properties. Moreover, the m-sequence **a** is easily generated (as described above), and then the corresponding binary sequence **b** is also easy to obtain – for instance, by raising each element of **a** to the $(q-1)$-th power. □

Note that one may view the process of generating **b** from **a** described in Example 9.6.20 as a sort of projection: every entry a_k of **a** is replaced in **b** by $\pi(a_k)$, where π denotes the map sending every non-zero element of F to 1, and 0 to 0.

This suggests the question what happens if one uses more general mappings for projecting **a**. This question was answered by Chan and Games [70] who showed that $L(\mathbf{a})$ is always quite large and can be calculated comparatively easily; we merely state their result without proof.

Result 9.6.21 *Let* **a** *be an m-sequence defined by a primitive polynomial of degree n over* $F=\mathrm{GF}(q)$ *as in Equation* (9.32)*, where q is odd, and let* **b** *be the periodic binary sequence obtained from* **a** *by replacing each entry* a_k *with* $\rho(a_k)$*, where* ρ *is some mapping from F to* $\mathrm{GF}(2)$ *with* $\rho(0)=0$.

Then **b** *has linear complexity* $L(\mathbf{b})=vL(\mathbf{c})$*, where* $v=(q^n-1)/(q-1)$ *and where* **c** *is the periodic binary sequence*

$$(\rho(1),\rho(\beta),\rho(\beta^2),\ldots,\rho(\beta^{q-2}),\ \rho(1),\rho(\beta),\ldots)$$

with $\beta=\alpha^v$. □

For even values of q, the corresponding problem is more involved; this was studied by Brynielsson [51].

Let us finally remark that the periodic binary sequences obtained from m-sequences over $\mathrm{GF}(q)$ do not satisfy the distribution test. If one desires a binary sequence with two-level autocorrelation (preferably with a small correlation coefficient c) and a large linear span which also satisfies the distribution test, our results suggest considering a Legendre sequence. Such a sequence is, of course, not as easily generated as a shift register sequence, since it involves computing all the Legendre symbols (k/p) for a large prime p. Note, however, that the calculation of Legendre symbols is much simplified by using the law of quadratic reciprocity; see Theorem 10.4.3 and Remark 10.4.4.

Another class of sequences with good correlation properties and comparatively large linear span which moreover have better balance properties (since they also satisfy a weak version of the serial test) is given by the GMW-sequences which will be considered in Section 9.9.

Exercises

Exercise 9.6.22. Check that the hypothesis that $\mathbf{a}$ is periodic is really needed to prove the second assertion in Theorem 9.6.1.

Hint: consider sequences which differ only in their preperiods. □

Exercise 9.6.23. Confirm Corollary 9.6.18 for the special case $d = 3$: use the primitive polynomial $f = 1 + x + x^3$ with initial conditions $(1,0,0)$ and compute the first period of the corresponding m-sequence $\mathbf{a}$ as well as its linear complexity. Then proceed similarly for $\bar{\mathbf{a}}$; also determine the minimal polynomial of $\bar{\mathbf{a}}$. □

9.7 The Linear Complexity Profile of a Sequence

In this section we consider the problem of computing the linear complexity of an arbitrary finite sequence over a finite field. To do so, we require some definitions.

Definition 9.7.1. Let $\mathbf{a}$ denote either a finite sequence $(a_k)_{k=0,\dots,N-1}$ of length N or an infinite sequence $(a_k)_{k\geq 0}$ over $F = \mathrm{GF}(q)$; in the latter case, we write $N := \infty$.

For every positive integer $k \leq N$, denote by $\Lambda_k(\mathbf{a})$ an[11] LFSR of least length over F capable of producing a shift register sequence $\mathbf{s}^{(k)}$ which agrees with $\mathbf{a}$ for the first k entries $a_0, \dots, a_{k-1}$. The characteristic polynomial of $\Lambda_k(\mathbf{a})$ will be denoted by $m_k(\mathbf{a})$, and its degree by $L_k(\mathbf{a})$.

In this way, one obtains a sequence $\mathbf{L} = (L_k(\mathbf{a}))$ over the positive integers which has the same length as $\mathbf{a}$. One calls $\mathbf{L}$ the **linear complexity profile** of $\mathbf{a}$. □

Let us give a rather trivial example and then mention two obvious properties of the linear complexity profile.

Example 9.7.2. Let $\mathbf{a}$ be the sequence $(0,\dots,0,\lambda)$ of finite length N over $F = \mathrm{GF}(q)$, where $\lambda \neq 0$. Then

$$L_k(\mathbf{a}) = \begin{cases} 1 & \text{for } k = 1,\dots,N-1, \\ N & \text{for } k = N. \quad \square \end{cases}$$

Lemma 9.7.3. *Let* $\mathbf{a}$ *be a sequence of length* N *over* $F = \mathrm{GF}(q)$. *Then*

$$L_{k-1}(\mathbf{a}) \leq L_k(\mathbf{a}) \leq k \quad \textit{for } k = 2,\dots,N,$$

and $L(\mathbf{a}) = L_{r+s}(\mathbf{a})$ *if* $\mathbf{a}$ *is ultimately periodic with period* r *and preperiod* s. □

[11] As we shall see in the proof of Theorem 9.7.7 and in Example 9.7.10 below, $\Lambda_k(\mathbf{a})$ and hence $m_k(\mathbf{a})$ are, in general, not uniquely determined by k and $\mathbf{a}$. However, all that is really needed for the subsequent development is the minimality of the *length* $L_k(\mathbf{a})$ of $\Lambda_k(\mathbf{a})$ – which is, of course, unique by definition.

Definition 9.7.4. In view of Lemma 9.7.3, one defines the **linear complexity** $L(\mathbf{a})$ of an arbitrary sequence $\mathbf{a}$ as the maximum value of all $L_k(\mathbf{a})$, provided that these values are bounded, and as ∞ otherwise. □

By Theorem 9.3.2 and Lemma 9.7.3, $L(\mathbf{a}) = \infty$ if and only if $\mathbf{a}$ is an infinite sequence which is not ultimately periodic. Moreover, the preceding definition agrees with the one given in 9.6.2 in the case where $\mathbf{a}$ is ultimately periodic and hence a shift register sequence. Finally, $L(\mathbf{a}) = L_N(\mathbf{a})$ if $\mathbf{a}$ is a finite sequence of length N. Thus the linear complexity profile constitutes a refinement of the linear complexity of a sequence.

Remark 9.7.5. As the trivial examples in 9.7.2 show, a high linear complexity by itself does not guarantee any randomness properties: in this example, the linear complexity profile is constant for $k \leq N-1$ and then jumps to the value N.

The linear complexity profiles of binary random sequences are analyzed in Rueppel [329, Chapter 4] who shows that a binary random sequence $\mathbf{a}$ of length N usually has linear complexity very close to $N/2$, with the complexity profile growing in a roughly (but not exactly!) continuous manner: $L_k(\mathbf{a})$ tends to be close to $k/2$. Moreover, if one extends $\mathbf{a}$ to a periodic sequence with period N, the resulting linear complexity will be close to N, provided that N is a power of 2 or a Mersenne prime. These results suggest requiring that a periodic binary sequence with good randomness properties should have complexity close to the period length and a profile growing more or less smoothly. □

In view of Corollary 9.6.4, it is in general a more difficult task to compute the linear complexity profile of a binary sequence than just its linear complexity; this is illustrated by the sequences considered in Section 9.6. The proof of Theorem 9.7.7 below will suggest an efficient algorithm capable of solving this problem, at least if $L(\mathbf{a})$ is not too large. We first need a lemma which is of independent interest.

Lemma 9.7.6. *Let* $\mathbf{a}$ *and* $\mathbf{b}$ *be two sequences of length* N *over* $F = \mathrm{GF}(q)$. *Then*

$$L_k(\mathbf{a}+\mathbf{b}) \leq L_k(\mathbf{a}) + L_k(\mathbf{b}) \quad \text{for } k = 1, \ldots, N.$$

Proof. Let f_a and f_b denote the feedback polynomials of suitable shift registers $\Lambda_k(\mathbf{a})$ and $\Lambda_k(\mathbf{b})$ of least lengths corresponding to $\mathbf{a}$ and $\mathbf{b}$, respectively, as in Definition 9.7.1. By Theorem 9.1.3, there exist polynomials g_a and g_b such that

$$s_a(x) = \frac{g_a(x)}{f_a(x)} \quad \text{and} \quad s_b(x) = \frac{g_b(x)}{f_b(x)},$$

where $s_a(x)$ and $s_b(x)$ denote the formal power series corresponding to the shift register sequences $\mathbf{s}_a^{(k)}$ and $\mathbf{s}_b^{(k)}$ produced by $\Lambda_k(\mathbf{a})$ and $\Lambda_k(\mathbf{b})$, respectively, as well as

$$\deg g_a < L_k(\mathbf{a}) \quad \text{and} \quad \deg g_b < L_k(\mathbf{b}).$$

Then

$$s_a(x)+s_b(x) = \frac{g_a(x)f_b(x)+g_b(x)f_a(x)}{f_a(x)f_b(x)}$$

and

$$\deg\left(g_a(x)f_b(x)+g_b(x)f_a(x)\right) < L_k(\mathbf{a})+L_k(\mathbf{b}),$$

as $\deg f_a \le L_k(\mathbf{a})$ and $\deg f_b \le L_k(\mathbf{b})$. Again using Theorem 9.1.3, we conclude that the sequence $\mathbf{s}_a^{(k)}+\mathbf{s}_b^{(k)}$ – which agrees with $\mathbf{a}+\mathbf{b}$ for its first k entries – can be obtained from an LFSR of length $L_k(\mathbf{a})+L_k(\mathbf{b})$ with feedback polynomial f_af_b, which gives the desired upper bound on the least length of an LFSR capable of producing the first k entries of the sequence $\mathbf{a}+\mathbf{b}$. □

We can now prove the following fundamental result which gives a recursive way of computing the linear complexity profile:

Theorem 9.7.7. *Let* $\mathbf{a}$ *be a sequence of length* N *over* $F=\mathrm{GF}(q)$*, let* k *be a positive integer with* $k+1\le N$*, and let* $\mathbf{s}=\mathbf{s}^{(k)}$ *be a shift register sequence which agrees with* $\mathbf{a}$ *for its first* k *entries* $a_0,\dots,a_{k-1}$ *and belongs to some LFSR* $\Lambda_k(\mathbf{a})$ *of length* $L_k(\mathbf{a})$*. Then*

$$L_{k+1}(\mathbf{a}) = \begin{cases} L_k(\mathbf{a}) & \text{if } a_k=s_k, \\ \max\left(L_k(\mathbf{a}),(k+1)-L_k(\mathbf{a})\right) & \text{if } a_k\ne s_k. \end{cases} \tag{9.40}$$

Proof. As noted in Lemma 9.7.3, $L_{k+1}(\mathbf{a})\ge L_k(\mathbf{a})$. In particular, this shows that the case $a_k=s_k$ in Equation (9.40) holds trivially.

Hence we may assume $a_k\ne s_k$. Put $\lambda := s_k-a_k$, and let $\mathbf{b}$ denote the sequence $(0,\dots,0,\lambda)$ of length $k+1$. Then $\mathbf{b}$ agrees with the first $k+1$ terms of the sequence $\mathbf{s}-\mathbf{a}$. Using Example 9.7.2 and Lemma 9.7.6 gives

$$k+1 = L_{k+1}(\mathbf{b}) = L_{k+1}(\mathbf{s}-\mathbf{a}) \le L_{k+1}(\mathbf{s})+L_{k+1}(-\mathbf{a}) = L_k(\mathbf{a})+L_{k+1}(\mathbf{a}),$$

since the sequences $\mathbf{a}$ and $-\mathbf{a}$ have the same linear complexity profile and $L_{k+1}(\mathbf{s}) = L_k(\mathbf{s}) = L_k(\mathbf{a})$. Therefore,

$$L_{k+1}(\mathbf{a}) \ge \max\left(L_k(\mathbf{a}),(k+1)-L_k(\mathbf{a})\right).$$

In order to establish the validity of (9.40), it now suffices to construct an LFSR $\Lambda_{k+1}(\mathbf{a})$ of length $\max\left(L_k(\mathbf{a}),(k+1)-L_k(\mathbf{a})\right)$ which produces the first $k+1$ entries of $\mathbf{a}$. This will be achieved using induction on k. More precisely, we will establish the existence of LFSR's $\Lambda_i(\mathbf{a})$ (for $i=1,\dots,k+1$) of length $L_i(\mathbf{a})$ which generate the first i entries $a_0,\dots,a_{i-1}$ of $\mathbf{a}$, where $L_1(\mathbf{a})=1$ and

$$L_{i+1}(\mathbf{a}) = \begin{cases} L_i(\mathbf{a}) & \text{if } a_i=s_i, \\ \max\left(L_i(\mathbf{a}),(i+1)-L_i(\mathbf{a})\right) & \text{if } a_i\ne s_i \end{cases}$$

(for $i=1,\dots,k$). In what follows, we denote the feedback polynomial of $\Lambda_i(\mathbf{a})$ by

$$f_i(x) := 1 - c_1^{(i)} x - c_2^{(i)} x^2 - \cdots - c_{L_i(\mathbf{a})}^{(i)} x^{L_i(\mathbf{a})}.$$

The induction base $k = 1$ is obvious, since the trivial LFSR $\Lambda_1(\mathbf{a})$ of length 1 with feedback polynomial $f_1(x) := 1 - x$ and initial conditions (a_0) generates the constant sequence $(a_0, a_0, \ldots)$.

For the induction step, we also put $L_0(\mathbf{a}) := 0$ and $f_0(x) := 1$. (If we would not do so, the first change of the LFSR would have to be described separately.) Thus assume $k > 1$, let m be the largest index $i \leq k - 1$ with $L_i(\mathbf{a}) < L_{i+1}(\mathbf{a})$, and put $n := L_{m+1}(\mathbf{a})$ and $r := L_m(\mathbf{a})$. Thus

$$n = L_k(\mathbf{a}) = \cdots = L_{m+1}(\mathbf{a}) > L_m(\mathbf{a}) = r.$$

In view of the induction hypothesis, this implies

$$n = \max(r, m+1-r) = m+1-r. \tag{9.41}$$

By definition,

$$\sum_{i=1}^{n} c_i^{(k)} a_{j-i} = \begin{cases} a_j & \text{for } j = n, \ldots, k-1, \\ s_k & \text{for } j = k, \end{cases} \tag{9.42}$$

as $a_k \neq s_k$. Similarly,

$$\sum_{i=0}^{r} c_i^{(m)} a_{j-i} = \begin{cases} a_j & \text{for } j = r, \ldots, m-1, \\ t_m & \text{for } j = m \end{cases} \tag{9.43}$$

for some $t_m \neq a_m$, since $\Lambda_{m+1}(\mathbf{a}) \neq \Lambda_m(\mathbf{a})$. Now put $\mu := t_m - a_m$ and define the polynomial f_{k+1} as follows:

$$f_{k+1}(x) := f_k(x) - \lambda \mu^{-1} x^{k-m} f_m(x),$$

where (as before) $\lambda = s_k - a_k$. Note that f_{k+1} can serve as the feedback polynomial of an LFSR $\Lambda_{k+1}(\mathbf{a})$ of length

$$\max(n, k-m+r) = \max\big(n, (k+1) - (m+1-r)\big) = \max(n, k+1-n) =: M,$$

by (9.41). Thus $\Lambda_{k+1}(\mathbf{a})$ has the desired length and it remains to verify that this LFSR indeed generates the first $k+1$ elements of $\mathbf{a}$, which can be done via a direct computation. One first checks that the feedback polynomial $x^{k-m} f_m(x)$ of length M and the initial conditions $(a_0, \ldots, a_{m-1})$ result in the sequence

$$b_j := -a_{j+m-k} + \sum_{i=0}^{r} c_i^{(m)} a_{j+m-k-i} \quad \text{for } j \geq M. \tag{9.44}$$

Also note that $j \geq M$ implies $j+m-k \geq r$, and that $j = k$ is equivalent to $j+m-k = m$. Hence (9.43) and (9.44) give

$$b_j = \begin{cases} 0 & \text{for } j = M, \dots, k-1 \\ \mu & \text{for } j = k. \end{cases}$$

Combining this with (9.42) gives the first $k+1$ elements of the shift register sequence $\mathbf{u}$ produced by $\Lambda_{k+1}(\mathbf{a})$ under the initial conditions $(a_0, a_1, \dots, a_{M-1})$:

$$u_j = -\lambda\mu^{-1}b_j + \sum_{i=1}^{n} c_i^{(k)} a_{j-i} = a_j \quad \text{for } j = M, \dots, k,$$

which proves the assertion. □

Theorem 9.7.7 is the basis for the celebrated Berlekamp-Massey algorithm for the determination of an LFSR $\Lambda_N(\mathbf{a})$ of least degree which is capable of producing a given finite sequence $\mathbf{a}$ of length N over $\mathrm{GF}(q)$. The proof of Theorem 9.7.7 leads to the following recursive algorithm for computing the linear complexity profile of $\mathbf{a}$ and corresponding feedback polynomials for LFSR's $\Lambda_k(\mathbf{a})$ of least length for all $k \le N$. We leave it to the reader to formally verify that the algorithm is indeed correct and just state this fact as Theorem 9.7.9.

Algorithm 9.7.8 (Berlekamp-Massey algorithm). Let $\mathbf{a}$ be a sequence of finite length N over $F = \mathrm{GF}(q)$. The following algorithm computes integers L_k and polynomials

$$f_k(x) = 1 - c_1^{(k)}x - c_2^{(k)}x^2 - \dots - c_{L_k(\mathbf{a})}^{(k)} x^{L_k(\mathbf{a})}$$

for all $k \le N$.

(1) $L_0 \leftarrow 0,\ L_1 \leftarrow 1,\ f_0 \leftarrow 1,\ f_1 \leftarrow 1 - x$
(2) **for** $k = 1$ **to** $N-1$ **do**
(3) $\quad \delta_k \leftarrow -a_k + \sum_{i=1}^{L_k} c_i^{(k)} a_{k-i}$
(4) $\quad$ **if** $\delta_k = 0$
(5) $\quad$ **then** $f_{k+1} \leftarrow f_k,\ L_{k+1} \leftarrow L_k$
(6) $\quad$ **else** $m \leftarrow \max\{i : i \le k-1, L_i < L_{i+1}\},\ L_{k+1} \leftarrow \max(L_k, k+1-L_k)$,
$\qquad\qquad \delta_m \leftarrow -a_m + \sum_{i=0}^{L_m} c_i^{(m)} a_{m-i},\ f_{k+1} \leftarrow f_k - \delta_k\delta_m^{-1}x^{k-m}f_m(x)$
(7) $\quad$ **fi**
(8) **od**

Theorem 9.7.9. *Let $\mathbf{a}$ be a sequence of finite length N over $F = \mathrm{GF}(q)$. Then the Berlekamp-Massey Algorithm 9.7.8 computes the values $L_k(\mathbf{a}) = L_k$ (and thus the linear complexity profile $\mathbf{L}$ of $\mathbf{a}$) together with appropriate feedback polynomials f_k for LFSR's $\Lambda_k(\mathbf{a})$ of length $L_k(\mathbf{a})$ generating the first k elements of $\mathbf{a}$ (for all $k = 1, \dots, N$).* □

It should be helpful to see an example for the application of the Berlekamp-Massey algorithm. To simplify the computations, we will consider a binary sequence. Then there is no need to distinguish between + and − signs; moreover, the k-th **discrepancy** δ_k always equals 1 or 0, so that the definition of f_{k+1} in Step (6) of Algorithm 9.7.8 simplifies as follows:

$$f_{k+1} := f_k + x^{k-m} f_m.$$

Example 9.7.10. Let us apply the Berlekamp-Massey algorithm to the binary sequence

$$\mathbf{a} = (1\,1\,0\,1\,0\,1\,1\,1\,0\,1) \text{ of length } N = 10.$$

In what follows, only those stages k are listed explicitly for which the shift register needs to be changed. We will give the current values of m, L_k and f_k, draw the associated LFSR $\Lambda_k(\mathbf{a})$ (basically as in Figure 9.1, but with the obvious simplifications for the binary case), and write down the shift register sequence $\mathbf{s}_k$ generated by this LFSR under the appropriate initial conditions. If a preperiod occurs, it is enclosed in square brackets, whereas periods of $\mathbf{s}_k$ are given in round brackets. Also, if the LFSR remains unchanged for the next value(s) of k, we indicate that L_k and f_k remain as before. With these conventions, the algorithm proceeds as follows.

Initialization and Stage $k = 1$

$f_1 = 1 + x$
$L_1 = 1$
$f_2 = f_1$, $L_2 = L_1$

1 1 1 1 ...

Stage $k = 2$

$m = 0$
$f_3 = f_2 + x^2 f_0 = 1 + x + x^2$
$L_3 = \max(1, 3-1) = 2$
$f_4 = f_3$, $L_4 = L_3$

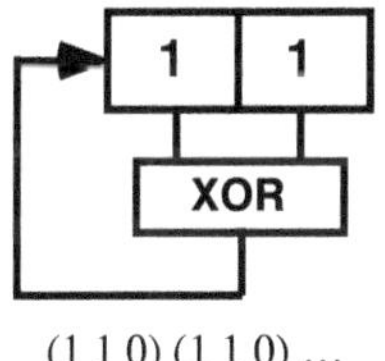

(1 1 0) (1 1 0) ...

Stage $k = 4$

$m = 2$
$f_5 = f_4 + x^2 f_2 = 1 + x + x^3$
$L_5 = \max(2, 5-2) = 3$

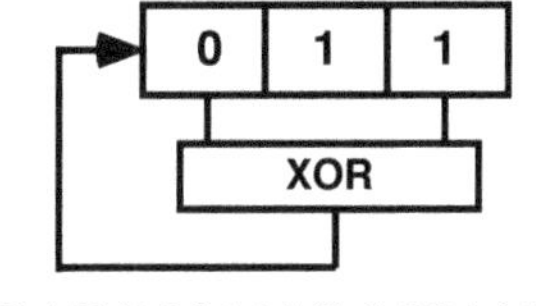

[1 1 0] (1 0 0 1 1 1 0) (1 0 0 1 1 1 0)...

Stage $k=5$

$m=4$
$f_6 = f_5 + xf_4 = 1 + x^2$
$L_6 = \max\,(3, 6-3) = 3$

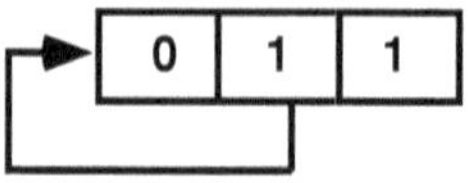

[1 1] (0 1) (0 1) (0 1)...

Stage $k=6$

$m=4$
$f_7 = f_6 + x^2 f_4 = 1 + x^3 + x^4$
$L_7 = \max\,(3, 7-3) = 4$
$f_8 = f_7, L_8 = L_7$

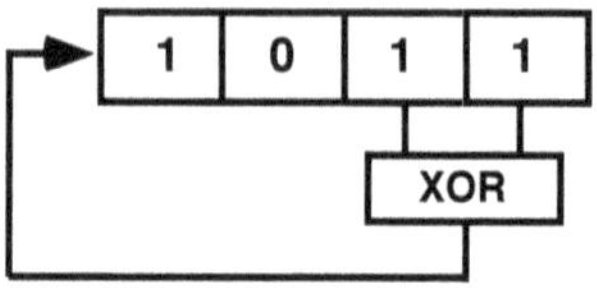

(1 1 0 1 0 1 1 1 1 0 0 0 1 0 0) ...

Stage $k=8$

$m=6$
$f_9 = f_8 + x^2 f_6 = 1 + x^2 + x^3$
$L_9 = \max\,(4, 9-4) = 5$

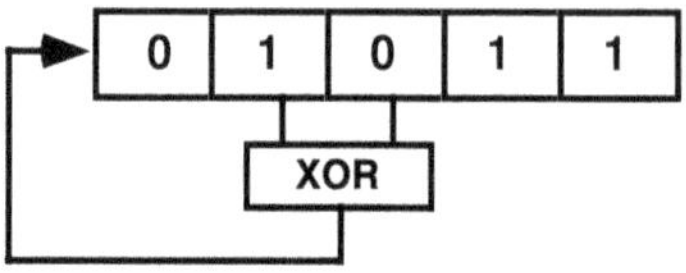

[1 1] (0 1 0 1 1 1 0) (0 1 0 1 1 1 0) ...

Stage $k=9$

$m=8$
$f_{10} = f_9 + xf_8 = 1 + x + x^2 + x^3 + x^4 + x^5$
$L_{10} = \max\,(5, 10-5) = 5$

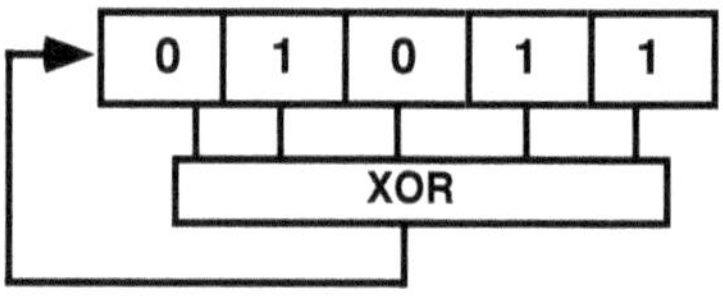

(1 1 0 1 0 1) (1 1 0 1 0 1) ...

Note that the linear complexity may remain unchanged, even though the LFSR changes: $\Lambda_{k+1}(\mathbf{a}) \neq \Lambda_k(\mathbf{a})$, but $L_{k+1} = L_k$. In our example, this phenomenon occurs for $k=4$ and $k=9$. □

Remark 9.7.11. The Berlekamp-Massey algorithm was first proposed (in a different form) by Berlekamp in his 1968 book *Algebraic Coding Theory* in order to solve a problem in this area; see [30] for the latest edition of this classic. A little later, Massey [257] recognized that the best way to derive this algorithm is in terms of the design of linear shift registers and gave an appropriate variation of the original algorithm.

In fact, Berlekamp devised his algorithm to perform the central stage in decoding BCH-codes, which involves solving the so-called "key equation". Here the sequence **a** consists of the known "syndromes" of a possibly corrupted codeword, which may be computed from the (corrupted) word actually received, and the problem consists of determining the unknown errors that (most likely) occurred. As Massey recognized, solving the key equation means determining an LFSR of least length generating the sequence of syndromes.

We refer to Blahut [39, Chapter 7] for a very nice presentation of the coding theoretic context of Algorithm 9.7.8. Blahut also discusses methods of further improving the performance of the Berlekamp-Massey algorithm; see his Section 11.6. □

In the important special case of binary codes, the syndrome sequences **a** are defined over some extension field F of GF(2), have even length, say $N = 2t$, and satisfy the following strong restriction:

$$a_{2k} = a_k^2 \quad \text{for } k = 1, \ldots, t. \tag{9.45}$$

(It is customary – and more natural – to label the sequence elements starting with a_1 in Coding Theory. Thus we will adopt this slight change of notation at this point.) In this case, the Berlekamp-Massey algorithm may be simplified considerably. As the following result shows, certain even values of k do not need to be considered in Algorithm 9.7.8 if (9.45) holds.

Proposition 9.7.12. *Let* $\mathbf{a} = (a_1, a_2, \ldots, a_N)$ *be a sequence of finite length* $N = 2t$ *over a field* F *of characteristic 2 satisfying condition (9.45). Assume* $L_{2k-1}(\mathbf{a}) \leq k$ *for some* $k \in \{1, \ldots, t\}$,[12] *and let* $\Lambda_{2k-1}(\mathbf{a})$ *be any LFSR generating the first* $2k-1$ *elements of* **a**. *Then* $\Lambda_{2k-1}(\mathbf{a})$ *actually generates the first* $2k$ *elements of* **a**.

Proof. Let $f(x) = 1 - c_1 x - \cdots - c_n x^n$ be the feedback polynomial of $\Lambda_{2k-1}(\mathbf{a})$, where $n := L_{2k-1}(\mathbf{a})$. By hypothesis, $n \leq k$ and therefore

$$\sum_{i=1}^{n} c_i a_{m-i} = a_m \quad \text{for } m = k, k+1, \ldots, 2k-1. \tag{9.46}$$

In view of (9.45), the special case $m = k$ of Equation (9.46) shows

$$a_{2k} = a_k^2 = \Big(\sum_{i=1}^{n} c_i a_{k-i}\Big)^2 = \sum_{i=1}^{n} c_i^2 a_{2k-2i}.$$

On the other hand, Equation (9.46) for $m = 2k-1, 2k-2, \ldots, 2k-n$ gives

$$\sum_{i=1}^{n} c_i a_{2k-i} = \sum_{i=1}^{n} c_i \Big(\sum_{j=1}^{n} c_j a_{2k-i-j}\Big) = \sum_{i,j=1}^{n} c_i c_j a_{2k-i-j} = \sum_{i=1}^{n} c_i^2 a_{2k-2i},$$

as the double sum is symmetric in i and j and as F has characteristic 2. This establishes the validity of Equation (9.46) for $m = 2k$. □

[12] In the coding theoretic context, this assumption is always satisfied: **a** consists of the $2t$ syndromes determined by a word of a binary t-error correcting BCH-code which has, by hypothesis, $v \leq t$ errors. But then the linear complexity of **a** is known to be exactly v; see, for instance, Blahut [39, Theorem 7.2.2].

Feng and Tzeng [116] gave an interesting generalization of the Berlekamp-Massey algorithm to the problem of synthesizing t minimum length LFSR's which are capable of generating t given sequences of length at most N. Again, this has applications to the decoding of cyclic codes, this time beyond the BCH-bound. Feng and Tzeng derived their algorithm as a special case of an algorithm for solving the more general problem of finding the smallest initial set of linearly dependent columns in a matrix over an arbitrary field.

9.8 An Application of the Discrete Fourier Transform

In this section, we show how one may use the Discrete Fourier Transform to determine the linear complexity of a periodic sequence; actually, there are two different – though basically equivalent – ways of doing so.

Our first result is essentially – that is, after translation into the language of periodic sequences – a special case of a theorem due to MacWilliams and Mann [254]. A more general version of this result will be discussed later in Section 10.6.

Theorem 9.8.1. *Let* $\mathbf{a} = (a_k)$ *be a periodic sequence with period* v *over* $F = \mathrm{GF}(q)$*. Assume that the characteristic* p *of* F *does not divide* v*, and let* ζ *be a primitive* v*-th root of unity in some extension field* E *of* F*. Then the linear complexity of* $\mathbf{a}$ *equals the number of non-zero coordinates*

$$A_j = \sum_{i=0}^{v-1} a_i \zeta^{ij} = a(\zeta^j) \quad (j = 0, \ldots, v-1)$$

of the discrete Fourier transform $\mathbf{A} = (A_0, A_1, \ldots, A_{v-1})$ *of the polynomial*

$$a(x) := a_{v-1}x^{v-1} + \cdots + a_1 x + a_0 \in E[x]/(x^v - 1).$$

Proof. By Proposition 9.6.7, the linear complexity $L(\mathbf{a})$ of $\mathbf{a}$ equals the p-rank of the incidence matrix

$$M = \begin{pmatrix} a_0 & a_1 & \ldots & a_{v-1} \\ a_1 & a_2 & \ldots & a_v \\ \vdots & \vdots & \ddots & \vdots \\ a_{v-1} & a_v & \ldots & a_{2v-2} \end{pmatrix} = \begin{pmatrix} a_0 & a_1 & \ldots & a_{v-1} \\ a_1 & a_2 & \ldots & a_0 \\ \vdots & \vdots & \ddots & \vdots \\ a_{v-1} & a_0 & \ldots & a_{v-2} \end{pmatrix}$$

of $\mathbf{a}$. Up to a row permutation, M coincides with the circulant matrix $A = (a_{ih})$ with first row $(a_0, \ldots, a_{v-1})$ associated with the polynomial $a(x) \in E[x]/(x^v - 1)$; see Theorem 7.3.2. Hence it suffices to determine the rank of A over E. Since A is circulant, it satisfies the condition

$$a_{ij} = a_{i+k,j+k} \quad \text{for all } k = 0, \ldots, v-1, \tag{9.47}$$

where all indices are taken modulo v. Now consider the v vectors

$$\mathbf{e}_j := \begin{pmatrix} 1 \\ \zeta^j \\ \zeta^{2j} \\ \vdots \\ \zeta^{(v-1)j} \end{pmatrix} \qquad (j = 0, \ldots, v-1),$$

that is, the columns of the Fourier matrix Z introduced in Definition 7.4.1. Using (9.47), the i-th entry of $A\mathbf{e}_j$ turns out to be

$$\sum_{h=0}^{v-1} a_{ih}\zeta^{hj} = \sum_{h=0}^{v-1} a_{0,h-i}\zeta^{(h-i)j}\zeta^{ij} = a(\zeta^j)\zeta^{ij},$$

and thus the vector $\mathbf{e}_j$ is an eigenvector of A with eigenvalue $a(\zeta^j) = A_j$. By Corollary 7.4.3, the Fourier matrix Z is invertible, and therefore the v eigenvectors $\mathbf{e}_0, \ldots, \mathbf{e}_{v-1}$ of A are linearly independent. Hence A is similar to the diagonal matrix $\operatorname{diag}(A_0, \ldots, A_{v-1})$ defined by the v Fourier coefficients of $a(x)$, which proves that the rank of A is indeed the number of non-zero Fourier coefficients. □

Theorem 9.8.1 also leads to the following alternative – though essentially equivalent – description of the linear complexity of a periodic sequence in terms of a polynomial representation:

Theorem 9.8.2. *Let $\mathbf{a} = (a_k)$ be a periodic sequence with period v over $F = \mathrm{GF}(q)$, where the characteristic p of F does not divide v, and let ζ be a primitive v-th root of unity in some extension field E of F. Then there exists a unique polynomial $c(x) := c_{v-1}x^{v-1} + \cdots + c_1x + c_0$ in $E[x]/(x^v - 1)$ such that*

$$a_i = \sum_{j=0}^{v-1} c_j\zeta^{ij} = c(\zeta^i) \qquad (i = 0, \ldots, v-1).$$

Moreover, the linear complexity of **a** *equals the number of non-zero coefficients c_j.*

Proof. This follows easily from Theorem 9.8.1 by applying the inverse DFT, see Corollary 7.4.3. □

The binary case of Theorem 9.8.2 was stated explicitly by Scholtz and Welch [341], where it is attributed to Key who employed this method already in 1976 in her paper [219] for determining the linear complexity of certain periodic binary sequences. Rueppel [330] attributes both Theorem 9.8.1 and Theorem 9.8.2 to work of Blahut [38] (who used them implicitly) and, for an explicit statement, to Massey [256].

We now use the method of Theorem 9.8.2 to give the promised proof of Theorem 9.6.14. In fact, we will derive this theorem as a rather simple consequence of the following considerably more general result due to Antweiler and Bömer [9], which establishes the linear complexity of a large class of periodic sequences over $\mathrm{GF}(q)$.

Theorem 9.8.3. *Consider the periodic sequence* $\mathbf{a} = (a_k)$ *over* $F = \mathrm{GF}(q)$ *with period* $v = q^{d+1} - 1$ *which is defined as follows:*

$$a_i := \left(\mathrm{Tr}_{E/F}(\zeta^i)\right)^m, \tag{9.48}$$

where $m \leq q-1$ *is a positive integer, where* $q = p^r$ *for some prime* p *not dividing* v*, and where* ζ *is a primitive element for the extension field* $E = \mathrm{GF}(q^{d+1})$ *of* F*. Furthermore, let*

$$m = \sum_{i=0}^{r-1} m_i p^i \quad \text{with } m_i \in \{0,\ldots,p-1\} \text{ for } i = 0,\ldots,r-1 \tag{9.49}$$

be the p*-ary representation of* m*. Then the linear complexity of* $\mathbf{a}$ *is given by*

$$L(\mathbf{a}) = \prod_{i=0}^{r-1} \binom{d+m_i}{d}.$$

Proof. We will apply Theorem 9.8.2 with $q = p^r$ and $v = q^{d+1} - 1$, so that ζ is a primitive v-th root of unity. Note first that the periodic sequence $\mathbf{a}$ defined in (9.48) is indeed a sequence over F.[13] Now consider the polynomial

$$c^{(m)}(x) := \left(x + x^q + x^{q^2} + \cdots + x^{q^d}\right)^m$$

of degree $q^d m \leq q^d(q-1) < v$, so that we may view $c^{(m)}(x)$ as a polynomial in $E[x]/(x^v - 1)$, and note

$$c^{(m)}(\zeta^i) = (\mathrm{Tr}_{E/F}(\zeta^i))^m = a_i.$$

Thus Theorem 9.8.2 indeed applies, and we need to determine the number of nonzero coefficients of $c^{(m)}(x)$. Using the p-ary expansion of m given in (9.49), we re-write $c^{(m)}(x)$ as follows:

$$c^{(m)}(x) = \prod_{i=0}^{r-1} \left(x^{p^i} + x^{p^{r+i}} + \cdots + x^{p^{dr+i}}\right)^{m_i} =: \sum_{j=0}^{mq^d} c_j x^j. \tag{9.50}$$

Let us first discuss the expansion of the individual terms appearing in the product in (9.50). An application of the multinomial theorem (see, for instance, Tucker [377]) gives

---

[13] Of course, this holds for all positive integers m. However, in view of $z^{q-1} = 1$ for all $z \in F^*$, it makes sense to restrict to $m \leq q-1$. The particular choice $m = q-1$ will lead to the promised proof of Theorem 9.6.14.

$$\left(x^{p^i}+x^{p^{r+i}}+\cdots+x^{p^{dr+i}}\right)^{m_i}=\sum\frac{m_i!}{b_{i_0}!b_{i_1}!\dots b_{i_d}!}\,x^{b_{i_0}p^i+b_{i_1}p^{r+i}+\dots+b_{i_d}p^{dr+i}},$$

where the summation runs over all choices of integers $b_{i_0},\dots,b_{i_d}$ satisfying

$$b_{i_0}+b_{i_1}+\cdots+b_{i_d}=m_i \quad\text{and}\quad 0\le b_{i_h}\le m_i \text{ for } h=0,\dots,d.$$

Since the p-ary representation of an arbitrary positive integer j is unique and since none of the multinomial coefficients involved can be 0 modulo p, the preceding equation shows that the number of non-zero coefficients in the expansion of the term

$$\left(x^{p^i}+x^{p^{r+i}}+\cdots+x^{p^{rd+i}}\right)^{m_i},$$

that is, the number of terms appearing in the sum

$$S_i:=\sum\frac{m_i!}{b_{i_0}!b_{i_1}!\dots b_{i_d}!}\,x^{b_{i_0}p^i+b_{i_1}p^{r+i}+\dots+b_{i_d}p^{dr+i}},$$

equals the number $n(d+1,m_i)$ of representations of m_i as an ordered sum of $d+1$ non-negative integers. It is well-known that this number is given by

$$n(d+1,m_i)=\binom{d+m_i}{d}; \tag{9.51}$$

see Exercise 9.8.4. Now the polynomial $c^{(m)}(x)$ in (9.50) is the product of the sums S_i, where i runs from 0 to $r-1$. Therefore, the powers x^j which have a non-zero coefficient in $c^{(m)}(x)$ belong precisely to those j which can be written as sums $j=j_0+\cdots+j_{r-1}$ of exponents j_i occurring in S_i. However, an exponent j_i for which x^{j_i} appears in any specific sum S_i has to have a p-ary representation which involves only powers of p with exponents in

$$T_i:=\{i,r+i,2r+i,\dots,dr+i\}\qquad (i=0,\dots,r-1).$$

Since these r sets T_i are pairwise disjoint and since the p-ary representation of an arbitrary positive integer j is unique, the number of non-zero coefficients of $c^{(m)}(x)$ is just the product of the numbers of terms appearing in the sums S_i. In view of (9.51), this gives the assertion. □

Proof of Theorem 9.6.14. We continue with the setup in Theorem 9.8.3 and denote by **a** and **b** the sequences obtained from (9.48) for the special choices $m=1$ and $m=q-1$, respectively. By Theorem 9.3.12, **a** is an m-sequence over $F=\mathrm{GF}(q)$, whereas the sequence **b** consists of entries 0 and 1 only, as $z^{q-1}=1$ for all $z\in F^*$. More precisely, **b** arises from **a** by replacing every non-zero entry of **a** with 1.

Now Equation (9.33) in the proof of Theorem 9.5.10 establishes the following fact:

$$b_{iv_0+g}=1 \iff b_g=1 \quad\text{for } i\ge 0 \text{ and } g=0,\dots,v_0-1,$$

where $v_0 := (q^{d+1}-1)/(q-1)$. (Note that v_0 is the value which was denoted as v in Theorem 9.5.10, and that the primitive element of $\mathrm{GF}(q^{d+1})$ was called α there.)

Thus $\mathbf{b}$ is also periodic with respect to the smaller period v_0, and the first period of length v_0 of $\mathbf{b}$ (in fact, any period of that length) corresponds to the complement $\overline{D}$ of the classical Singer difference set D constructed in Theorem 9.5.10. By Proposition 9.6.7, the p-rank of $\overline{D}$ equals the linear complexity of $\mathbf{b}$. In view of

$$q-1 = (p-1)+(p-1)p+\cdots+(p-1)p^{r-1},$$

an application of Theorem 9.8.3 gives

$$\operatorname{rank}_p \overline{D} = L(\mathbf{b}) = \binom{d+p-1}{d}^r. \tag{9.52}$$

In order to deal with the Singer difference set D itself in the same manner, we simply replace the polynomial $c^{(q-1)}(x)$ by $c(x) := 1 - c^{(q-1)}(x)$ and note that substituting the powers of ζ in $c(x)$ indeed results in the complementary sequence $\overline{\mathbf{b}}$ of $\mathbf{b}$. Compared to $c^{(q-1)}(x)$, the polynomial $c(x)$ has exactly one additional non-zero coefficient, namely the coefficient of x^0. Using this observation together with (9.52) in Theorem 9.8.2 then yields the desired formula for the p-rank of D:

$$\operatorname{rank}_p D = \binom{p+d-1}{d}^r + 1. \quad \square$$

Let us conclude this section with a remark. In geometric language, Theorem 9.6.14 gives the p-rank of the classical symmetric design formed by the points and hyperplanes of the d-dimensional projective geometry over $\mathrm{GF}(q)$; see, for instance, Beth, Jungnickel and Lenz [35]. The method employed in our proof may also be used to obtain an analogous result for the p-rank of the classical affine design formed by the points and hyperplanes of the d-dimensional affine geometry over $\mathrm{GF}(q)$; see Pott [325].

Exercises

Exercise 9.8.4. Let m and d be positive integers. Prove that the number $n(d,m)$ of representations of m as an ordered sum of d non-negative integers is given by

$$\binom{m+d-1}{d-1}.$$

Hint: view such representations as sequences of length $m+d-1$ over an alphabet with two elements. $\square$

9.9 GMW-sequences

We conclude this chapter with discussing a class of binary sequences which share most of the good randomness properties of the binary m-sequences (though they only satisfy a weakened version of the serial test), but tend to have considerably larger linear complexity. The sequences in question correspond to cyclic difference sets with parameters (9.30) which were discovered by Gordon, Mills and Welch [152] in 1962. The presentation in terms of binary sequences and the analysis of their balance and complexity properties is due to Scholtz and Welch [341]. Following Antweiler and Bömer [9] and Pott [325], we first consider a more general class of sequences which correspond to the construction given in [152] for a more general class of cyclic difference sets, namely with Singer parameters (9.31).

Definition 9.9.1. Consider three fields $K = \mathrm{GF}(p^s)$, $F = \mathrm{GF}(q)$ and $E = \mathrm{GF}(q^d)$ with $q = p^{rs}$, where r and d are positive integers $\neq 1$. Put $v := q^d - 1$, and let ζ be a primitive element of E^*. Moreover, let $m < q-1$ be a positive integer which is relatively prime to $q-1$ and define a periodic sequence $\mathbf{b} = (b_k)$ over K with period v as follows:

$$b_k := \mathrm{Tr}_{F/K}\Big(\big(\mathrm{Tr}_{E/F}(\zeta^k)\big)^m\Big). \tag{9.53}$$

Any sequence of the form (9.53) is called a **GMW-sequence**. □

Because of Theorem 9.3.12 and the transitivity formula for the trace function (see Theorem 3.12.8), $\mathbf{b}$ reduces to an ordinary m-sequence over K in the special case $m = 1$. In general, we may therefore think of a GMW-sequence as a twisted m-sequence, where the twisting is due to the inner trace being raised to the m-th power.

We now apply the method introduced in the preceding section to determine the linear complexities of the GMW-sequences:

Theorem 9.9.2. *Let $K = \mathrm{GF}(p^s)$, $F = \mathrm{GF}(q)$ and $E = \mathrm{GF}(q^d)$ with $q = p^{rs}$, where r and d are positive integers $\neq 1$. Put $v := q^d - 1$, let ζ be a primitive element of E^*, and let $m < q-1$ be a positive integer satisfying $\gcd(m, q-1) = 1$. Then the GMW-sequence $\mathbf{b}$ over $K = \mathrm{GF}(p^s)$ defined in (9.53) has linear complexity*

$$L(\mathbf{b}) = r \cdot \prod_{i=0}^{rs-1} \binom{d+m_i-1}{d-1},$$

where

$$m = \sum_{i=0}^{rs-1} m_i p^i \quad \text{with } m_i \in \{0,\ldots,p-1\} \text{ for } i = 0,\ldots,rs-1 \tag{9.54}$$

is the p-ary representation of m.

Proof. Let us write b_k as

$$b_k = \mathrm{Tr}_{F/K}(a_k) \quad \text{with} \quad a_k := \left(\mathrm{Tr}_{E/F}(\zeta^k)\right)^m, \tag{9.55}$$

and note that the *inner sequence* $\mathbf{a} := (a_k)$ is one of the sequences studied in Theorem 9.8.3 (where we used $q = p^r$ instead of $q = p^{rs}$, and where the extension degree was denoted by $d+1$ instead of d). As in the proof of that result, $\mathbf{a}$ belongs to the polynomial

$$c^{(m)}(x) := \left(x + x^q + x^{q^2} + \cdots + x^{q^{d-1}}\right)^m,$$

and hence $\mathbf{b}$ belongs to the polynomial

$$f^{(m)}(x) := c^{(m)}(x) + \left(c^{(m)}(x)\right)^{p^s} + \cdots + \left(c^{(m)}(x)\right)^{p^{(r-1)s}}, \tag{9.56}$$

considered as a polynomial in $E[x]/(x^v - 1)$. From Theorem 9.8.3 and its proof, we know the number $L(\mathbf{a})$ of non-zero coefficients of $c^{(m)}(x)$ and hence also of each of the powers of $c^{(m)}(x)$ appearing in (9.56):

$$L(\mathbf{a}) = \prod_{i=0}^{rs-1} \binom{d + m_i - 1}{d-1}. \tag{9.57}$$

Therefore the assertion will follow from Theorem 9.8.2, provided we can show that the **supports** of the r polynomials $\left(c^{(m)}(x)\right)^{p^{ts}}$ appearing in (9.56) – that is, the respective sets of powers of x with coefficients $\neq 0$ – are pairwise disjoint.

By way of contradiction, let us assume that a power x^j appears in the support of both $(c^{(m)}(x))^{p^{ts}}$ and $(c^{(m)}(x))^{p^{us}}$ for some integers t, u with $0 \le t < u \le r-1$. As the proof of Theorem 9.8.3 shows, j then has two representations

$$j \equiv \sum_{k=0}^{drs-1} y_k p^{k+ts} \equiv \sum_{k=0}^{drs-1} z_k p^{k+us} \mod p^{drs} - 1, \tag{9.58}$$

where the (non-negative) coefficients y_k and z_k have to satisfy the restrictions

$$y_i + y_{rs+i} + \cdots + y_{(d-1)rs+i} = z_i + z_{rs+i} + \cdots + z_{(d-1)rs+i} = m_i \tag{9.59}$$

for $i = 0, \ldots, rs-1$. Note that the congruence in (9.58) also holds modulo $p^{rs} - 1$. Using this and (9.59) gives

$$j \equiv \sum_{i=0}^{rs-1} m_i p^{i+ts} \equiv \sum_{i=0}^{rs-1} m_i p^{i+us} \mod p^{rs} - 1$$

and hence, from (9.54),

$$mp^{ts} \equiv mp^{us} \qquad \mod p^{rs} - 1.$$

As m was assumed to be relatively prime to $p^{rs} - 1$, we conclude $ts = us$, which is the desired contradiction. □

We now restrict attention to binary GMW-sequences, that is, we specify $p^s = 2$. The following result of Scholtz and Welch [341] shows that the binary GMW-sequences share most of the good randomness properties of the binary m-sequences. In this result, the number of coefficients 1 in the binary representation

$$m = \sum_{i=0}^{r-1} m_i 2^i \quad \text{with } m_i \in \{0,1\} \text{ for } i = 0,\dots,r-1$$

of a positive integer $m < 2^r$ is called the **weight** of m and denoted by $w(m)$.

Theorem 9.9.3. *Consider the fields $K = \mathrm{GF}(2)$, $F = \mathrm{GF}(q)$ and $E = \mathrm{GF}(q^d)$ with $q = 2^r$, where r and d are positive integers $\neq 1$. Put $v := q^d - 1$, let ζ be a primitive element of E^*, and let $m < q-1$ be a positive integer which is relatively prime to $q-1$. Then the binary GMW-sequence $\mathbf{b} = (b_k)$ defined as in (9.53) has linear complexity*

$$L(\mathbf{b}) = r \cdot d^{w(m)}.$$

Moreover, $\mathbf{b}$ has a two-level autocorrelation function and corresponds to a cyclic difference set with parameters

$$v = 2^{dr} - 1, \; k = 2^{dr-1} \text{ and } \lambda = 2^{dr-2}, \tag{9.60}$$

and hence the complementary sequence $\overline{\mathbf{b}}$ corresponds to a cyclic difference set with parameters (9.30) *(with d replaced by dr).*

Finally, $\mathbf{b}$ satisfies the following restricted version of the serial test (9.19), *where $\mathbf{c} = (c_1,\dots,c_\ell)$ is an arbitrary ℓ-tuple in K^ℓ for some $\ell \in \{1,\dots,d\}$:*

$$Z_{\mathbf{a}}(\mathbf{c}) = \begin{cases} 2^{dr-\ell} & \text{for } \mathbf{c} \neq \mathbf{0}, \\ 2^{dr-\ell} - 1 & \text{for } \mathbf{c} = \mathbf{0}. \end{cases} \tag{9.61}$$

Proof. The formula for the linear complexity of $\mathbf{b}$ holds by Theorem 9.9.2.

Regarding the autocorrelation function of $\mathbf{b}$, the parameters stated in (9.61) require to verify the validity of

$$C_{\mathbf{b}}(h) = \begin{cases} 2^{dr} - 1 & \text{for } h \equiv 0 \bmod 2^{dr} - 1, \\ -1 & \text{otherwise.} \end{cases} \tag{9.62}$$

We first note an auxiliary result concerning the inner sequence $\mathbf{a}$ defined in (9.55):

> Each segment of $t := (q^d - 1)/(q-1)$ consecutive symbols of $\mathbf{a}$ contains exactly $(q^{d-1} - 1)/(q-1)$ entries 0. (9.63)

This is an easy consequence of the proof of Theorem 9.5.10 (where we used $d+1$ instead of d), in particular of Equation (9.33).

We now express the indices $k = 0,\dots,2^{dr} - 2$ in the form $k = i + jt$, where $i = 0,\dots,t-1$ and $j = 0,\dots,2^r - 2$, and obtain

$$\begin{aligned}
C_{\mathbf{b}}(h) &= \sum_{k=0}^{2^{dr}-2} (-1)^{b_k - b_{k+h}} \\
&= \sum_{i=0}^{t-1} \sum_{j=0}^{2^r-2} (-1)^{\mathrm{Tr}_{F/K}\left((\mathrm{Tr}_{E/F}(\zeta^{i+jt}))^m - (\mathrm{Tr}_{E/F}(\zeta^{i+jt+h}))^m\right)}.
\end{aligned}$$

Because of $\zeta^t \in F$, this may be written in the form

$$C_{\mathbf{b}}(h) = \sum_{i=0}^{t-1} \sum_{j=0}^{2^r-2} (-1)^{\mathrm{Tr}_{F/K}(\zeta^{jtm}\delta(h,i))}, \tag{9.64}$$

where

$$\delta(h,i) := (\mathrm{Tr}_{E/F}(\zeta^i))^m - (\mathrm{Tr}_{E/F}(\zeta^{i+h}))^m. \tag{9.65}$$

As m is relatively prime to $q-1$, ζ^{tm} is a primitive element for F^*, and therefore ζ^{jtm} runs over all elements of F^*. Hence, by including the element $0 \in F$ in the inner summation, we may write (9.64) as

$$C_{\mathbf{b}}(h) = -t + \sum_{i=0}^{t-1} \sum_{\beta \in F} (-1)^{\mathrm{Tr}_{F/K}(\beta\delta(h,i))}. \tag{9.66}$$

The inner sum in (9.66) vanishes whenever $\delta(h,i) \neq 0$, since then $\beta\delta(h,i)$ runs over all elements of F. Hence (9.66) reduces to

$$C_{\mathbf{b}}(h) = -t + 2^r N(h), \tag{9.67}$$

where $N(h)$ denotes the number of indices $i = 0,\ldots,t-1$ with $\delta(h,i) = 0$. Since m is relatively prime to $q-1$, it has an inverse modulo $q-1$, and therefore (9.65) shows

$$\delta(h,i) = 0 \quad \Longleftrightarrow \quad \mathrm{Tr}_{E/F}\left((\zeta^h - 1)\zeta^i\right) = 0.$$

Using (9.63), we conclude

$$N(h) = \begin{cases} t & \text{for } h \equiv 0 \mod 2^{dr}-1, \\ (q^{d-1}-1)/(q-1) & \text{otherwise.} \end{cases}$$

Substituting this in (9.67) gives the desired formula (9.62).

It remains to prove the validity of (9.61). Note that the sequence $\mathbf{s} := (\mathrm{Tr}_{E/F}(\zeta^k))$ is an m-sequence over F with linear complexity d. The proof of the serial test (9.19) for binary m-sequences in Theorem 9.4.4 yields the following result on the distribution of ℓ-tuples $\mathbf{z} = (z_1,\ldots,z_\ell)$ in one period of $\mathbf{s}$ (for $\ell = 1,\ldots,d$):

$$Z_{\mathbf{s}}(\mathbf{z}) = \begin{cases} q^{d-\ell} & \text{for } \mathbf{z} \neq \mathbf{0}, \\ q^{d-\ell}-1 & \text{for } \mathbf{z} = \mathbf{0}. \end{cases} \tag{9.68}$$

As m is relatively prime to $q-1$, the mapping $\mu: x \mapsto x^m$ is a bijection on F. Therefore (9.68) carries over to the inner sequence $\mathbf{a}$, since $\mathbf{a}$ results from $\mathbf{s}$ by applying μ to all entries of that sequence:

$$Z_{\mathbf{a}}(\mathbf{z}) = \begin{cases} q^{d-\ell} & \text{for } \mathbf{z} \neq \mathbf{0}, \\ q^{d-\ell} - 1 & \text{for } \mathbf{z} = \mathbf{0}. \end{cases} \tag{9.69}$$

Now let $\mathbf{c}$ be any ℓ-tuple in K^ℓ, and denote by $M(\mathbf{c})$ the set of all ℓ-tuples in F^ℓ with image $\mathbf{c}$ under the trace mapping $\mathrm{Tr}_{F/K}$ (applied coordinate-wise). Then the number of occurrences of $\mathbf{c}$ in one period of $\mathbf{b}$ is

$$Z_{\mathbf{b}}(\mathbf{c}) = \sum_{\mathbf{z} \in M(\mathbf{c})} Z_{\mathbf{a}}(\mathbf{z}). \tag{9.70}$$

As $\mathrm{Tr}_{F/K}$ is a surjective mapping, it is easily checked that each set $M(\mathbf{c})$ has cardinality $2^{\ell(r-1)}$. Thus (9.69) and (9.70) imply the desired formula (9.61). □

It is possible to give a criterion for the cyclic equivalence of two binary GMW-sequences (for the same choice of d and r) and hence to determine the number of such sequences which are cyclically inequivalent. For a proof of the following result, see Scholtz and Welch [341].

Result 9.9.4 *Under the assumptions of Theorem 9.9.3, consider a second GMW-sequence* $\mathbf{c} = (c_k)$ *defined by*

$$c_k := \mathrm{Tr}_{F/K}\big((\mathrm{Tr}_{E/F}(\zeta^{ek}))^n\big),$$

where $n < q-1$ *is a positive integer which is relatively prime to* $q-1$*, and where* $e < q^d - 1$ *is a positive integer which is relatively prime to* $q^d - 1$*. Then* $\mathbf{b}$ *and* $\mathbf{c}$ *are cyclically equivalent if and only if*

$$n \equiv 2^k m \mod q-1 \quad \text{for some integer } k \text{ with } 0 \le k < r, \text{ and}$$
$$e \equiv 2^f \mod q^d - 1 \quad \text{for some integer } f \text{ with } 0 \le f < dr.$$

Hence the number of cyclically inequivalent GMW-sequences (for the given values d *and* r*) is*

$$N_{GMW}(d,r) = p(dr)p(r),$$

where $p(a)$ *denotes the number of primitive polynomials with degree* a *over* $K =$ GF(2). □

As the formula for the linear complexity of binary GMW-sequences in Theorem 9.9.3 shows, one may obtain different linear complexities depending on the weight of the twisting parameter m. Thus there will be a considerable number of essentially distinct such sequences – and hence of non-isomorphic difference sets with parameters (9.60). Let us illustrate this with an example:

Example 9.9.5. Take $r = 7$ in Theorem 9.9.3. Then there are six possibilities for the linear complexity of $\mathbf{b}$, corresponding to the 6 possible weights of m. If we choose $d = 4$, we obtain the possibilities exhibited in Table 9.1, where we also list the number $N_{GMW}(4,7)$ of cyclically inequivalent sequences with the corresponding linear complexity; these values are taken from [341]. We remark that all resulting sequences have an **autocorrelation ratio** – that is, the ratio between the peak values for $h \equiv 0 \mod 2^{28} - 1$ and the off-phase values – of magnitude $2^{28} - 1 : 1$, and satisfy the serial test for $\ell = 1,\dots,4$. (Of course, the case $m = 1$ gives an ordinary m-sequence for which we may even take $\ell = 1,\dots,28$.) □

$w(m)$	possible values of m	$L(\mathbf{b})$	$N_{GMW}(4,7)$
1	1	28	4741632
2	3, 5, 9	112	14224896
3	7, 11, 13, 19, 21	448	23708160
4	15, 23, 27, 29, 43	1792	23708160
5	31, 47, 55	7168	14224896
6	63	28672	4741632

Table 9.1 Design parameters for binary GMW-sequences with $r = 7$ and $d = 4$

One may use the general GMW-sequences over $K = \mathrm{GF}(p^s)$ studied at the beginning of this section (with d replaced by $d+1$) to obtain the non-classical cyclic difference sets with parameters (9.31) due to Gordon, Mills and Welch [152]. This can be done via a projection process similar to the one used at the end of the preceding section to prove Theorem 9.5.10; see Pott [325]. Unfortunately, there seem to be no nice formulas for the p-rank of these difference sets. Of course, at least in principle, the p-rank could be obtained using Theorem 9.9.2, by raising the polynomial $f^{(m)}(x)$ defined in (9.56) to its $(q-1)$-th power. Some computational results along these lines are given in [325].

From the point of view of cryptographic application as stream ciphers, the periodic sequences we have discussed do not offer sufficient security. Many authors have studied more involved ways of obtaining suitable sequences, for instance by combining several shift registers and submitting the output to some "filtering" function or by using irregularly clocked shift registers. Discussing these techniques is well beyond the scope of the present text. For an overview, we refer the interested reader to Rueppel [330] (in particular Section 3) and to the *Handbook of Applied Cryptography* [269, Chapter 6]; for more details, one may consult the many original sources cited in these two references.

There are also cryptographic applications of certain difference sets in elementary abelian groups corresponding to so-called "bent functions"; here we refer to the recent survey by Carlet and Mesnager [62].

We also recommend the excellent monograph on the design of sequences with good correlation properties (and related objects) for application in Wireless Communication, Cryptography and Radar by Golomb and Gong [147].

Finally, it should be emphasized that the systematic study of shift register sequences and their engineering applications is to a large extent the creation of Solomon Golomb. His vastly influential classical textbook *Shift Register Sequences* first appeared in 1967; now in its third edition [146], it is still very much worth studying.

Chapter 10
Characters, Gauss Sums, and the DFT

Abstract One of the major problems often encountered in working with finite fields is the transition between their additive and multiplicative structures. We have seen a concrete example of this type of problem when we studied the arithmetics in finite fields in Chapter 8. Depending on the choice of representation, either addition (in some basis representation) or multiplication (using Zech logarithms) is trivial, while the other operation is not at all easy to perform. The same type of problem arises also in theoretical investigations concerning both the additive and multiplicative structure simultaneously, for instance, when proving the existence of primitive normal bases or that of primitive elements with a prescribed value of the trace (into a specified subfield); these two problems will the topic of the final two chapters.

Such questions often require the use of tools from Representation Theory which we did not introduce up to now, that is, characters and character sums and, in particular, Gauss sums. These tools will be presented in the current chapter, where we first consider characters of finite abelian groups in general and then specialize to the case of finite fields and introduce the basic properties of Gauss sums.

We then give a few interesting applications of the quadratic character; in particular, we shall prove the law of quadratic reciprocity and consider solutions of quadratic equations in several variables over finite fields with odd characteristic. Following this, we shall prove three more advanced identities for Gauss sums; we will also obtain an interesting connection to the eigenvalues of the matrix of the Discrete Fourier Transform defined in Section 7.4. Finally, we extend the DFT to abelian groups in general and present some applications to abelian difference sets and periodic sequences.

10.1 Characters of Abelian Groups

Throughout this section, we will write groups multiplicatively and denote the unit element by 1 (unless stated otherwise). We start with the following generalization of Definition 3.8.2:

D. Hachenberger and D. Jungnickel, *Topics in Galois Fields*,
Algorithms and Computation in Mathematics 29,
https://doi.org/10.1007/978-3-030-60806-4_10

Definition 10.1.1. Let G be a finite abelian group and L a field. Then any homomorphism χ from G into the multiplicative group L^* of L is called a **character** of G over L. □

If $\chi : G \to L^*$ is a character, then the image of χ is a finite subgroup C of L^* and therefore cyclic; see Theorem 3.2.3. Moreover, $|C|$ divides the exponent of G. Throughout, let m denote the exponent and n the order of G; thus all values $\chi(g)$ are m-th roots of unity in L^*. In general, we will assume that L is a **splitting field** of G, which means that L contains primitive (and therefore all) m-th roots of unity. In particular, the characteristic of L does not divide m. The case where $L = \mathbb{C}$ is the field of complex numbers is referred to as the **classical case**.

Every character of G over L is by definition a mapping from G to L, that is, an element of the L-vector space L^G. Note that the proof of Theorem 3.8.3 remains valid in the present situation, and hence distinct characters are linearly independent over L; see also Corollary 10.1.7 for an alternative proof under our assumption that L is a splitting field of G. Therefore, the set $\widehat{G}$ of all characters of G over L has at most n elements. As we will see soon, equality holds, and hence $\widehat{G}$ is a basis of the vector space L^G. In fact, we will establish a considerably stronger result: $\widehat{G}$ is a group isomorphic to G. In view of this, $\widehat{G}$ is called the **character group** of G.

We first observe that $\widehat{G}$ is indeed a group under elementwise multiplication: clearly, the mapping $\chi\lambda \in L^G$ defined by

$$(\chi\lambda)(g) := \chi(g)\lambda(g) \quad \text{for } g \in G$$

is again a character for all $\chi, \lambda \in \widehat{G}$. The identity element of $\widehat{G}$ is the **trivial character**[1] χ_0 given by $\chi_0(g) = 1$ for all $g \in G$, and the inverse of a character χ is the mapping

$$\chi^{-1} : G \to L^*, \quad g \mapsto \chi(g)^{-1}.$$

Note that $\chi(g)^{-1} = \chi(g^{-1})$, as χ is a homomorphism.[2] Moreover,

$$\chi(g)^{-1} = \overline{\chi(g)}$$

in the classical case, where $\overline{a+bi} = a - bi$ is the complex conjugate of $a+bi \in \mathbb{C}$ (with $a,b \in \mathbb{R}$).

Now let $\operatorname{ord}(g) = \ell$. Then $\operatorname{ord}(\chi(g))$ divides ℓ, that is, $\chi(g)$ is an ℓ-th root of unity. For the classical case we recall that the ℓ-th roots of unity in $\mathbb{C}$ are all of the form

$$e^{2\pi ik/\ell} = \cos(\tfrac{2\pi k}{\ell}) + \sin(\tfrac{2\pi k}{\ell}) \cdot i,$$

where e denotes the base of the natural logarithm and where $0 \le k \le \ell - 1$; moreover, $e^{2\pi ik/\ell}$ is a primitive ℓ-th root of unity if and only if ℓ and k are relatively prime.

[1] In most text books, the term **principal character** is used instead.

[2] As mentioned at the start, we use multiplicative notation for G; in additive notation, this property would read $\chi(g)^{-1} = \chi(-g)$ instead.

Theorem 10.1.2. *Let G be a finite abelian group and L a splitting field of G. Then the group $\widehat{G}$ of characters of G over L is isomorphic to G.*

Proof. It is well-known that G can be written as a direct product of cyclic subgroups (see, for instance, Jacobson [204, Theorem 3.11]), say

$$G = \langle g_1 \rangle \times \cdots \times \langle g_r \rangle. \tag{10.1}$$

Let $m_i := \mathrm{ord}(g_i)$ for $i = 1, \ldots, r$. As noted above, $\chi(g_i)$ then is an m_i-th root of unity for each character χ and all $i = 1, \ldots, r$. The group C_i of m_i-th roots of unity in L^* is a cyclic group of order m_i, as L is a splitting field for G, and hence G is isomorphic to $C := C_1 \times \cdots \times C_r$. Thus it suffices to show that the mapping $\alpha\colon \widehat{G} \to C$ defined by

$$\alpha(\chi) := \big(\chi(g_1), \ldots, \chi(g_r)\big) \tag{10.2}$$

is an isomorphism. Clearly, α is a homomorphism with kernel $\{\chi_0\}$, so that α is injective. It remains to show that α is also surjective. For this, let ζ_i be an arbitrary element in C_i, for $i = 1, \ldots, r$. Then there exists a (unique) homomorphism $\chi_i\colon \langle g_i \rangle \to C_i$ mapping g_i to ζ_i. Using (10.1), one can combine these r homomorphisms to define a character χ of G by putting

$$\chi\Big(\prod_{i=1}^{r} g_i^{t_i}\Big) := \prod_{i=1}^{r} \chi_i(g_i^{t_i}) = \prod_{i=1}^{r} \chi_i(g_i)^{t_i}.$$

By our choice of the χ_i, we obtain

$$\alpha(\chi) = \big(\chi_1(g_1), \ldots, \chi_r(g_r)\big) = (\zeta_1, \ldots, \zeta_r),$$

as desired. □

We remark that the isomorphism α defined in (10.2) depends on the representation of G given in (10.1) and is therefore not canonical. There is, however, a canonical isomorphism between G and the character group $\widehat{\widehat{G}}$ of $\widehat{G}$: given any $g \in G$, we can define a character $\hat{g}$ of the group $\widehat{G}$ via

$$\hat{g}\colon \widehat{G} \to L^*, \quad \chi \mapsto \chi(g). \tag{10.3}$$

It will become clear later that the mapping $g \mapsto \hat{g}$ indeed gives an isomorphism between G and $\widehat{\widehat{G}}$.

We also note that the proof of Theorem 10.1.2 does not really use the additive structure of L and would also work if we replace L^* with an arbitrary group which contains a cyclic group of order $\exp G$. In contrast, the field property of L is needed to view the character group $\widehat{G}$ as a basis of the L-algebra L^G, as we have done at the beginning of this section. Of course, $\widehat{G}$ is *not* the canonical basis of L^G, which is – as for mappings from some finite set G into a field L in general – given by the characteristic functions

$$c_g\colon G\to L,\quad h\mapsto\begin{cases}1 & \text{if } h=g,\\ 0 & \text{otherwise}\end{cases}\tag{10.4}$$

(for all $g\in G$). In what follows, we will also need the **characteristic function** c_U of an arbitrary subset U of G:

$$c_U:=\sum_{g\in U}c_g,\tag{10.5}$$

that is, $c_U(h)=1$ if and only if $h\in U$, and $c_U(h)=0$ otherwise.

We now turn to some fundamental properties of characters, which may be interpreted as the transformation of the basis $\widehat{G}$ of L^G into the canonical basis. To do so, we require some notation. For a subgroup H of G and for $\chi\in\widehat{G}$, let $\chi_{|H}$ denote the restriction of χ to H. One now defines a fundamental pairing $\perp$ between subgroups of G and subgroups of $\widehat{G}$ as follows:

Definition 10.1.3. For a subgroup H of G and a subgroup S of $\widehat{G}$, let

$$H^\perp:=\{\chi\in\widehat{G}\colon \chi_{|H}=\chi_{0|H}\},\tag{10.6}$$

and

$$S^\perp:=\{g\in G\colon \lambda(g)=1 \text{ for all } \lambda\in S\}.\tag{10.7}$$

Obviously, $H^\perp$ is a subgroup of $\widehat{G}$ and $S^\perp$ is a subgroup of G; these subgroups are called the **duals** of H and S, respectively. □

Lemma 10.1.4. *Let L be a splitting field for the finite abelian group G of order n. Then the following identities hold for all subgroups H of G and S of $\widehat{G}$:*

$$\sum_{\chi\in S}\chi=|S|\cdot c_{S^\perp},\ \textit{that is}\ \sum_{\chi\in S}\chi(g)=\begin{cases}|S| & \textit{if } g\in S^\perp,\\ 0 & \textit{otherwise;}\end{cases}\tag{10.8}$$

$$\sum_{h\in H}\chi(h)=\begin{cases}|H| & \textit{if } \chi\in H^\perp,\\ 0 & \textit{otherwise.}\end{cases}\tag{10.9}$$

Proof. The first case of (10.8) is trivial. Thus let $g\notin S^\perp$. Then there is a character $\lambda\in S$ such that $\lambda(g)\neq 1$. In view of $\{\lambda\chi\colon\chi\in S\}=S$, we obtain

$$\sum_{\chi\in S}\chi(g)=\sum_{\chi\in S}(\lambda\chi)(g)=\lambda(g)\cdot\Big(\sum_{\chi\in S}\chi(g)\Big),$$

and therefore

$$(1-\lambda(g))\cdot\Big(\sum_{\chi\in S}\chi(g)\Big)=0.$$

As $\lambda(g)\neq 1$, this yields the desired result. The proof of Equation (10.9) is similar and may be left to the reader. □

The identities in Lemma 10.1.4 have the following very useful consequence:

Proposition 10.1.5. *Let L be a splitting field for the finite abelian group G of order n, and let H and S be subgroups of G and $\widehat{G}$, respectively. Then one has*

$$|S \cap H^{\perp}| \cdot |H| = |H \cap S^{\perp}| \cdot |S|. \tag{10.10}$$

In particular,

$$|S| \cdot |S^{\perp}| = n = |H| \cdot |H^{\perp}|. \tag{10.11}$$

Proof. In order to prove Equation (10.10), it suffices to evaluate $\sum_{g \in H} \sum_{\chi \in S} \chi(g)$ in two ways. Using Equation (10.8), we obtain

$$\sum_{g \in H} \sum_{\chi \in S} \chi(g) = \sum_{g \in H \cap S^{\perp}} |S| = |H \cap S^{\perp}| \cdot |S|;$$

on the other hand, Equation (10.9) gives

$$\sum_{\chi \in S} \sum_{g \in H} \chi(g) = \sum_{\chi \in S \cap H^{\perp}} |H| = |S \cap H^{\perp}| \cdot |H|.$$

We now put $H = G$ in Equation (10.10) and use that the dual group of G is just the trivial subgroup $\{\chi_0\}$ of $\widehat{G}$, by definition. This yields one half of Equation (10.11), namely

$$n = |G| = |S^{\perp}| \cdot |S|.$$

Next, we choose $S = \widehat{G}$ in the preceding equation to obtain

$$n = |\widehat{G}| \cdot |\widehat{G}^{\perp}| = n \cdot |\widehat{G}^{\perp}|,$$

which shows that $\widehat{G}^{\perp}$ is the trivial subgroup $\{1\}$ of G. Using this in Equation (10.10) (with the same choice of S) gives

$$|H^{\perp}| \cdot |H| = |\widehat{G}| = |G| = n,$$

which establishes also the second half of Equation (10.11). □

Lemma 10.1.4 and Proposition 10.1.5 lead to the following extremely important result:

Theorem 10.1.6 (Orthogonality relations). *Let L be a splitting field for the finite abelian group G. Then the following identities hold for all $g, h \in G$ and all $\chi, \lambda \in \widehat{G}$:*

$$\sum_{\chi \in \widehat{G}} \chi(g)\chi^{-1}(h) = \begin{cases} |G| & \text{if } g = h, \\ 0 & \text{otherwise,} \end{cases}$$

and

$$\sum_{g \in G} \chi(g)\lambda^{-1}(g) = \begin{cases} |G| & \text{if } \chi = \lambda, \\ 0 & \text{otherwise.} \end{cases}$$

Proof. As observed in the proof of Proposition 10.1.5, $\widehat{G}^{\perp} = \{1\}$. Applying Equation (10.8) with $S = \widehat{G}$ yields the first assertion:

$$\sum_{\chi \in \widehat{G}} \chi(g)\chi^{-1}(h) = \sum_{\chi \in \widehat{G}} \chi(gh^{-1}) = \begin{cases} |G| & \text{if } g = h \\ 0 & \text{otherwise.} \end{cases}$$

The second assertion follows in a similar manner. □

Using the second identity in Theorem 10.1.6, we can give the promised alternative proof for Theorem 3.8.3:

Corollary 10.1.7 (Dedekind independence theorem). *Let L be a splitting field for the finite abelian group G. Then distinct characters of G into L^* are linearly independent over L.*

Proof. We need to show that all characters of G form a set of linearly independent mappings in L^G. Thus assume $\sum_{\chi \in \widehat{G}} a_\chi \chi(g) = 0$ for all $g \in G$, and let $\lambda \in \widehat{G}$ be any character. Multiplying the preceding equation by $\lambda^{-1}(g)$ and summing over all $g \in G$, we obtain

$$0 = \sum_{g \in G} \sum_{\chi \in \widehat{G}} a_\chi \chi(g)\lambda^{-1}(g) = \sum_{\chi \in \widehat{G}} a_\chi \Big(\sum_{g \in G} \chi(g)\lambda^{-1}(g) \Big) = a_\lambda \cdot |G|,$$

and hence $a_\lambda = 0$, since the characteristic of L does not divide $|G|$. □

The final result of this section concerns the structure of the character groups of subgroups and factor groups of G and $\widehat{G}$, respectively, and establishes the canonical isomorphism between G and $\widehat{\widehat{G}}$ mentioned earlier.

Theorem 10.1.8 (Duality theorem). *Let L be a splitting field for the finite abelian group G, and let H, H_1, H_2 and S, S_1, S_2 be subgroups of G and $\widehat{G}$, respectively. Then the following assertions hold:*

(1) $(H^{\perp})^{\perp} = H$ *and* $(S^{\perp})^{\perp} = S$.
(2) $(H_1H_2)^{\perp} = H_1^{\perp} \cap H_2^{\perp}$ *and* $(H_1 \cap H_2)^{\perp} = H_1^{\perp}H_2^{\perp}$.
(3) $(S_1S_2)^{\perp} = S_1^{\perp} \cap S_2^{\perp}$ *and* $(S_1 \cap S_2)^{\perp} = S_1^{\perp}S_2^{\perp}$.
(4) $\widehat{H}$ *is isomorphic to* $\widehat{G}/H^{\perp}$ *and* $\widehat{S}$ *is isomorphic to* $G/S^{\perp}$.
(5) *The character group of* G/H *is isomorphic to* $H^{\perp}$ *and the character group of* $\widehat{G}/S$ *is isomorphic to* $S^{\perp}$.
(6) *The mapping* α *with* $\alpha(g) := \hat{g}$, *where* $\hat{g}$ *is defined according to Equation* (10.3), *is an isomorphism between* G *and* $\widehat{\widehat{G}}$.

Proof. By definition, $H \subseteq (H^{\perp})^{\perp}$. Proposition 10.1.5 gives

$$|(H^{\perp})^{\perp}| = \frac{|G|}{|H^{\perp}|} = \frac{|G|}{|G|/|H|} = |H|,$$

and therefore $H = (H^\perp)^\perp$. The second assertion in (1) is proved in the same way.

The first assertion in (2) and the inequality $H_1^\perp H_2^\perp \subseteq (H_1 \cap H_2)^\perp$ are immediate from the definitions. Using the first of these facts together with Exercise 1.3.15 and Proposition 10.1.5, we obtain

$$\begin{aligned} |H_1^\perp H_2^\perp| &= \frac{|H_1^\perp| \cdot |H_2^\perp|}{|H_1^\perp \cap H_2^\perp|} = \frac{|H_1^\perp| \cdot |H_2^\perp|}{|(H_1 H_2)^\perp|} \\ &= \frac{(|G|/|H_1|) \cdot (|G|/|H_2|)}{|G|/|H_1 H_2|} = \frac{|G| \cdot |H_1 H_2|}{|H_1| \cdot |H_2|} \\ &= \frac{|G|}{|H_1 \cap H_2|} = |(H_1 \cap H_2)^\perp|, \end{aligned}$$

which implies the desired equality $H_1^\perp H_2^\perp = (H_1 \cap H_2)^\perp$. The assertions in (3) are proved in a similar way.

In order to prove the first assertion in (4), we consider the **restriction homomorphism**

$$\mathrm{rest}_H \colon \widehat{G} \to \widehat{H}, \quad \chi \mapsto \chi_{|H} \,.$$

By definition, the kernel of rest_H is just $H^\perp$, the dual group of H. The homomorphism theorem shows that $\widehat{G}/H^\perp$ is isomorphic to a subgroup of $\widehat{H}$. As

$$|\widehat{H}| = |H| = |G|/|H^\perp| = |\widehat{G}|/|H^\perp| = |\widehat{G}/H^\perp|,$$

the mapping rest_H is an epimorphism, so that $\widehat{G}/H^\perp$ is indeed isomorphic to $\widehat{H}$. (Note that this proves that any character of H can be extended to a character of G.) The second assertion in (4) is established in a similar manner.

We now turn to the proof of the first assertion in (5). Given any character η of G/H, one defines a character $\mathrm{inf}_H(\eta)$ of G by

$$\mathrm{inf}_H(\eta)(g) := \eta(gH).$$

Then

$$\mathrm{inf}_H \colon \widehat{G/H} \to \widehat{G}, \quad \eta \mapsto \mathrm{inf}_H(\eta)$$

is a homomorphism, which is called the **inflation homomorphism**. By definition, the image of inf_H is contained in $H^\perp$. Since inf_H is injective and $\widehat{G/H}$ and $H^\perp$ have the same cardinality, the first assertion follows. Again, the second assertion in (5) is proved in a similar way.

It remains to establish (6). Given any $g \in G$, we first note that $\hat{g} = \alpha(g)$ is indeed a character of $\widehat{G}$, as

$$\hat{g}(\lambda\chi) = (\lambda\chi)(g) = \lambda(g) \cdot \chi(g) = \hat{g}(\lambda) \cdot \hat{g}(\chi)$$

for any two characters λ and χ of G. Moreover, α is a homomorphism, since

$$\widehat{gh}(\chi) = \chi(gh) = \chi(g) \cdot \chi(h) = \hat{g}(\chi) \cdot \hat{h}(\chi)$$

for any two elements g and h and any character χ of G. As noted in the proof of Proposition 10.1.5, $\widehat{G}^{\perp} = \{1\}$. Thus the kernel of α is trivial, that is, α is injective. As G and $\widehat{\widehat{G}}$ have the same cardinality, α is indeed an isomorphism. □

Remark 10.1.9. Of course, one may also consider homomorphisms $G \to L^*$ if G is a non-abelian group. As in the abelian case, such homomorphims are called *characters*. Even though we will not need this generalization, we mention two specific examples, since the reader will be familiar with these:

- the mapping associating with each permutation in the symmetric group S_n its sign;
- the mapping sending a matrix in $GL(n,q)$ to its determinant in $\mathrm{GF}(q)^*$.

Actually, in the non-abelian case one requires a much more general – and considerably more involved – notion of characters, of which homomorphisms are merely very special examples. The interested reader may consult any good introduction to Representation Theory, say Jacobson [205, Chapter 5] or the book by Curtis and Reiner [100]. □

Exercises

Exercise 10.1.10. Let G be a finite abelian group and consider $\widehat{G}$ in the classical case $L = \mathbb{C}$. Check that the inverse of a character χ is the character $\overline{\chi}$ defined by

$$\overline{\chi}(g) := \overline{\chi(g)}.$$

Using the orthogonality relations 10.1.6, show that

$$\langle \chi, \lambda \rangle := \frac{1}{|G|} \cdot \sum_{g \in G} \chi(g) \overline{\lambda(g)}$$

defines a non-degenerate Hermitian form on $\mathbb{C}^G$ such that $\widehat{G}$ is an orthonormal basis for $\mathbb{C}^G$ with respect to this form. □

10.2 Characters of Finite Fields

After having introduced the character group of a finite abelian group in general, we now turn to the characters of the multiplicative and the additive group of a finite field $F = \mathrm{GF}(q)$. For most later purposes it would be sufficient to restrict attention to the classical case, that is, where the splitting field (of both groups) is $L = \mathbb{C}$.

Let us start with the group of **multiplicative characters**. As F^* is a cyclic group of order $q-1$, we have to assume that the characteristic of L does not divide $q-1$

and that L contains a primitive $(q-1)$-th root of unity, say ζ. Let ω be a fixed primitive element for F^*. Then every multiplicative character ψ of F is uniquely determined by the image of ω under ψ. We shall denote the character sending ω to ζ^j by ψ_j, for $j = 0, \dots, q-2$, so that

$$\psi_j(\omega^a) = \zeta^{aj} \quad \text{for } a = 0, \dots, q-2. \tag{10.12}$$

With this notation, the character group of F^* is generated by ψ_1, one has $\psi_j^{-1} = \psi_{-j}$ (taking indices modulo $q-1$), and ψ_0 is the trivial multiplicative character.

Of course, ψ_j depends on the choice of both the primitive root ζ and the primitive element ω. While there is no canonical way to select ω, we can remove the ambiguity about ζ in the classical case $L = \mathbb{C}$ by choosing $\zeta = e^{2\pi i/(q-1)}$.

It is often convenient to extend the multiplicative characters ψ_j to multiplicative mappings from F into L by putting

$$\psi_j(0) := 0 \text{ for } j \neq 0 \quad \text{and} \quad \psi_0(0) := 1; \tag{10.13}$$

then $\widehat{F^*}$ becomes (formally) a subset of L^F, and ψ_0 agrees with the trivial additive character χ_0 via this extension.

Occasionally, it can be useful to consider also multiplicative characters of F into a field E which satisfies the assumption on the characteristic but is not yet a splitting field, since it does not contain the required roots of unity. This is only a minor generalization and poses no real problems, since we may always extend E to a splitting field L of F^*. We will see an example for this approach in Section 10.4.

We now turn to the group of **additive characters**. As $(F,+)$ is in general not cyclic, the description of the additive characters is a little more involved. Nevertheless, there is a natural way of labelling the additive characters with elements of F by using the absolute trace mapping $\mathrm{Tr}_{F/P}$, where P is the prime field of F; for simplicity, we will just write Tr for this mapping. We now have to assume that the characteristic of L is different from the characteristic p of F and that L contains a primitive p-th root of unity, say ξ; in the classical case $L = \mathbb{C}$ we will always take $\xi = e^{2\pi i/p}$. (Recall that $(F,+)$ is a vector space over P and thus has exponent p.) Because of the additivity of the trace function, the mapping

$$\chi_1 : F \to L^*, \ u \mapsto \xi^{\mathrm{Tr}(u)}$$

is an additive character of F; note that this definition makes sense, since the elements of the prime field $P = \mathrm{GF}(p)$ can be identified with the integers $0, 1, \dots, p-1$ modulo p. In view of the following result, χ_1 is called the **canonical additive character**.

Proposition 10.2.1. *Let L be a splitting field for the additive group $(F,+)$ of the finite field $F = \mathrm{GF}(q)$, and let ξ be a fixed primitive p-th root of unity in L. Then the additive characters of F are precisely the mappings χ_b with $b \in F$ defined by*

$$\chi_b(u) := \xi^{\mathrm{Tr}(bu)} = \chi_1(bu) \quad \textit{for } u \in F. \tag{10.14}$$

Proof. It is trivial to check that the mappings χ_b are additive characters. As the trace function is surjective, every χ_b with $b \neq 0$ is a non-trivial character. Obviously, $\chi_a(u)\chi_b(u) = \chi_{a+b}(u)$, and therefore the mapping $a \mapsto \chi_a$ is an injective group homomorphism of $(F,+)$ into its character group $\widehat{F}$. The finiteness of F now implies that $\{\chi_b : b \in F\}$ coincides with the group of additive characters of F. □

In analogy to the multiplicative case, one has $\chi_b^{-1} = \chi_{-b}$, and χ_0 is the trivial additive character (agreeing with our general notation).

In what follows, we will often require the orthogonality relations 10.1.6 specialized to the case of a finite field. In order to facilitate their application, we state these relations explicitly using the notations just introduced:

Corollary 10.2.2. *Let L be a splitting field for the multiplicative group of the finite field* $F = \mathrm{GF}(q)$*, and let the multiplicative characters of F be labelled as in* (10.12)*. Then the following identities hold for all* $c,d \in F^*$ *and all* $j,k = 0,\ldots,q-2$*:*

$$\sum_{j=0}^{q-2} \psi_j(c)\psi_{-j}(d) = \begin{cases} q-1 & \text{if } c=d, \\ 0 & \text{otherwise;} \end{cases}$$

$$\sum_{c\in F^*} \psi_j(c)\psi_{-k}(c) = \begin{cases} q-1 & \text{if } j=k, \\ 0 & \text{otherwise;} \end{cases}$$

$$\sum_{c\in F^*} \psi_j(c) = 0 \quad \text{for } j \neq 0. \qquad \square$$

Corollary 10.2.3. *Let E be a splitting field for the additive group of the finite field* $F = \mathrm{GF}(q)$*, and let the additive characters of F be labelled as in* (10.14)*. Then the following identities hold for all* $a,b,c,d \in F$*:*

$$\sum_{b\in F} \chi_b(c)\chi_{-b}(d) = \begin{cases} q & \text{if } c=d, \\ 0 & \text{otherwise;} \end{cases}$$

$$\sum_{c\in F} \chi_b(c)\chi_{-a}(c) = \begin{cases} q & \text{if } a=b, \\ 0 & \text{otherwise;} \end{cases}$$

$$\sum_{c\in F} \chi_b(c) = 0 \quad \text{for } b \neq 0. \qquad \square$$

We conclude this section with a first application showing how characters (in this case, multiplicative characters) can be useful in the theoretical study of the structural properties of finite fields. In Chapters 13 and 14, we will consider the existence problem for primitive normal elements and for primitive elements with a prescribed value of the trace, respectively. For both problems, the solution makes essential use of a connection between character sums and primitivity which seems to go back to Ivan Matveevich Vinogradov. This should provide some motivation for studying character sums – in particular, Gauss sums – in considerable detail in later sections.

We begin by stating the following fundamental criterion for the primitivity of an element of a finite field; this result may be found in the book by Landau [228, pp. 178–180], where it is attributed to Vinogradov.

Proposition 10.2.4 (Vinogradov criterion). *An element x of $F = \mathrm{GF}(q)$ is primitive if and only if the following condition holds:*

$$\sum_{d|q-1}\left(\frac{\mu(d)}{\phi(d)}\sum_{\psi:d}\psi(x)\right)\neq 0,$$

where μ and ϕ denote the Möbius and Euler function, respectively, and where the notation $\psi\!:\!d$ indicates that the inner summation runs over all complex multiplicative characters ψ of order d of F. □

We shall actually prove the following quantitative version of the preceding result, which immediately implies the validity of Proposition 10.2.4. This variant appears as Exercise 5.14 in Lidl and Niederreiter [242].

Proposition 10.2.5 (Vinogradov formula). *Let U be a subset of F^*, where $F = \mathrm{GF}(q)$, and denote by g_U the number of primitive elements of F contained in U. Then*

$$\sum_{d|q-1}\left(\frac{\mu(d)}{\phi(d)}\cdot\sum_{\psi:d}\sum_{x\in U}\psi(x)\right)=\frac{q-1}{\phi(q-1)}\cdot g_U,\tag{10.15}$$

where μ and ϕ denote the Möbius and Euler function, respectively, and where the notation $\psi\!:\!d$ indicates that the inner summation runs over all complex multiplicative characters ψ of order d of F. □

Proof. Note first that we may restrict the outer summation in Equation (10.15) to square-free divisors d of $q-1$, as otherwise $\mu(d)=0$. Now let ω be a fixed primitive element for F, and consider any element $x=\omega^k$ of F. Using that both the Möbius and the Euler function are multiplicative, we can rewrite the sum in the Vinogradov criterion as follows:

$$\begin{aligned}\sum_{d|q-1}\left(\frac{\mu(d)}{\phi(d)}\cdot\sum_{\psi:d}\psi(x)\right)&=\prod_{r|q-1}\left(1+\frac{\mu(r)}{\phi(r)}\cdot\sum_{\psi:r}\psi(x)\right)\\&=\prod_{r|q-1}\left(1-\frac{1}{r-1}\cdot\sum_{\psi:r}\psi(x)\right),\end{aligned}$$

where the product runs over all distinct prime divisors r of $q-1$, and where (for a fixed r) the sum runs over all $r-1$ multiplicative characters ψ with order r. Note that $\psi(\omega)$ is of the form ξ^j with $j\in\{1,2,\ldots,r-1\}$ for some primitive r-th root of unity $\xi\in\mathbb{C}^*$ when $\mathrm{ord}(\psi)=r$. Therefore,

$$\sum_{\psi:r} \psi(x) = \sum_{\psi:r} \psi(\omega^k) = \sum_{j=1}^{r-1} \xi^{jk},$$

since $\psi(\omega) \neq \lambda(\omega)$ for distinct multiplicative characters ψ and λ. If $\gcd(q-1,k) = 1$, then

$$\sum_{j=1}^{r-1} \xi^{jk} = \sum_{j=1}^{r-1} \xi^{j} = -1 \quad \text{for every prime divisor } r \text{ of } q-1,$$

while $\sum_{j=1}^{r-1} \xi^{jk} = r-1$ when $r \mid k$ (which holds for some r when $\gcd(q-1,k) \neq 1$). Altogether, this gives

$$\sum_{d|q-1} \left(\frac{\mu(d)}{\phi(d)} \cdot \sum_{\psi:d} \psi(x) \right) = \begin{cases} \prod_{r|q-1} \left(1 + \frac{1}{r-1}\right) & \text{if } \gcd(q-1,k) = 1, \\ 0 & \text{otherwise.} \end{cases} \tag{10.16}$$

Observe that the condition $\gcd(q-1,k) = 1$ just means that $x = \omega^k$ is a primitive element for F. Finally,

$$\prod_{r|q-1} \left(1 + \frac{1}{r-1}\right) = \prod_{r|q-1} \frac{r}{\phi(r)} = \frac{q-1}{\phi(q-1)}.$$

Hence the assertion follows by summing Equation (10.16) over all $x \in U$. □

We now put $F = \mathrm{GF}(q)$, $E = \mathrm{GF}(q^n)$ and write $U_a := \{x \in E \colon \mathrm{Tr}_{E/F}(x) = a\}$ for $a \in F$. Since the trace is a linear operator, $U_a = aU_1$ for all $a \in F^*$. Finally, given any multiplicative character ψ of E, we put

$$S_a(\psi) := \sum_{x \in U_a} \psi(x). \tag{10.17}$$

Applying Lemma 10.2.5 to the field E and the subset U_a yields the following result:

Corollary 10.2.6. *The number g_a of primitive elements of $E = \mathrm{GF}(q^n)$ with (E,F)-trace equal to a is given by*

$$g_a = \frac{\phi(q^n-1)}{q^n-1} \cdot \sum_{d|q^n-1} \left(\frac{\mu(d)}{\phi(d)} \cdot \sum_{\psi:d} S_a(\psi) \right),$$

where the character sums $S_a(\psi)$ are defined as in (10.17) and where the inner summation runs over all complex multiplicative characters of E which have order d. □

One cannot expect Corollary 10.2.6 to lead to a nice closed formula for the number of primitive elements ω of E for which $\mathrm{Tr}_{E/F}(\omega)$ takes the prescribed value a. Nevertheless, this result forms the basis for settling the existence problem for such elements. To do so, one clearly requires information about the character sums $S_a(\psi)$. Again, one cannot expect to determine these sums explicitly, but one may at least compute their absolute values by using Gauss sums, which we will introduce in the next section and study in more detail in Section 10.5. This already suffices to establish an asymptotic existence result for primitive elements with prescribed trace, as done by Jungnickel and Vanstone [217]. A little later, a complete solution was given by Cohen [82], but this requires more detailed work and additional ideas. We shall investigate the existence of primitive elements with prescribed (generalized) trace in detail in Chapter 14.

10.3 Basics on Gauss Sums

We now introduce an important concept combining multiplicative and additive characters of a finite field, namely Gauss sums. These are an essential tool for studying the interaction between the multiplicative and the additive structure.

Throughout this section, we let $F = \mathrm{GF}(q)$ be a finite field with characteristic p and L a field which is a splitting field for both the additive and the multiplicative group of F; in particular, the characteristic of L does not divide $q(q-1)$. Henceforth, we will simply say that L is a **splitting field** for $\mathrm{GF}(q)$.

Definition 10.3.1. Let χ be an additive character and ψ a multiplicative character of F into L^*. Then

$$G(\psi,\chi) := \sum_{u\in F^*} \psi(u)\chi(u) \tag{10.18}$$

is called the **Gauss sum** associated with ψ and χ. □

Note that the sum in Equation (10.18) may as well be taken to run over all of F provided that ψ is non-trivial, as then $\psi(0) = 0$ by convention. The classical case is, of course, once again $L = \mathbb{C}$. We will see an application where L is likewise a finite field in Section 10.4. The following simple but important result gives some basic information on the possible values of Gauss sums:

Proposition 10.3.2. *Let L be a splitting field for $F = \mathrm{GF}(q)$, and let χ be an additive character and ψ a multiplicative character of F into L^*. Then the Gauss sum $G(\psi,\chi)$ satisfies*

$$G(\psi,\chi) = \begin{cases} q-1 & \text{for } \psi = \psi_0 \text{ and } \chi = \chi_0, \\ -1 & \text{for } \psi = \psi_0 \text{ and } \chi \neq \chi_0, \\ 0 & \text{for } \psi \neq \psi_0 \text{ and } \chi = \chi_0. \end{cases}$$

Moreover,

$$G(\psi,\chi)G(\psi^{-1},\chi^{-1}) = q \quad \text{for } \psi \neq \psi_0 \text{ and } \chi \neq \chi_0. \tag{10.19}$$

Proof. The first assertion is an immediate consequence of Corollaries 10.2.2 and 10.2.3. Now let both ψ and χ be non-trivial characters. Then one computes

$$\begin{aligned}
G(\psi,\chi)G(\psi^{-1},\chi^{-1}) &= \sum_{u,v\in F^*} \psi(u)\chi(u)\psi^{-1}(v)\chi^{-1}(v) \\
&= \sum_{u,v\in F^*} \psi(uv^{-1})\chi(u-v) \\
&= \sum_{v,w\in F^*} \psi(w)\chi(v(w-1)).
\end{aligned}$$

Using $\sum_{w\in F^*} \psi(w) = 0$ from Corollary 10.2.2 shows

$$\begin{aligned}
G(\psi,\chi)G(\psi^{-1},\chi^{-1}) &= \sum_{w\in F^*} \Big(\psi(w)\cdot \sum_{v\in F^*} \chi(v(w-1))\Big) + \sum_{w\in F^*} \psi(w)\cdot\chi(0) \\
&= \sum_{w\in F^*} \Big(\psi(w)\cdot \sum_{v\in F} \chi(v(w-1))\Big).
\end{aligned}$$

We now observe that $w=1$ gives the summand $\psi(1)\cdot\sum_{v\in F}\chi(0) = \psi(1)q = q$, whereas any other element $w\in F^*$ (for $q>2$) gives

$$\psi(w)\cdot\sum_{v\in F}\chi(v(w-1)) = \psi(w)\cdot\sum_{c\in F}\chi(c) = 0,$$

by Corollary 10.2.3. This establishes also the second assertion. □

In the classical case, one can be more specific, as then the inverse of a character is given by its complex conjugate; see Exercise 10.1.10. We leave the simple proof of the following result as a further Exercise 10.3.7 to the reader.

Corollary 10.3.3. *If we have $L=\mathbb{C}$ in Proposition 10.3.2, then*

$$G(\psi^{-1},\chi^{-1}) = G(\overline{\psi},\overline{\chi}) = \overline{G(\psi,\chi)},$$

and therefore

$$|G(\psi,\chi)| = \sqrt{q} \quad \textit{for } \psi\neq\psi_0 \textit{ and } \chi\neq\chi_0. \qquad \square$$

The next result collects a number of useful identities for Gauss sums which are obtained under various transformations of the characters involved.

Proposition 10.3.4. *Let L be a splitting field for $F=\mathrm{GF}(q)$, and let ψ be a multiplicative character and χ an additive character of F into L^*. Moreover, assume that the additive characters of F into L^* are labelled as in* (10.14). *Then the following identities hold for all $a,b\in F$ with $a\neq 0$:*

$$
\begin{aligned}
G(\psi,\chi_{ab}) &= \psi^{-1}(a)G(\psi,\chi_b); && (1)\\
G(\psi,\chi^{-1}) &= \psi(-1)G(\psi,\chi); && (2)\\
G(\psi^{-1},\chi) &= \psi(-1)G(\psi^{-1},\chi^{-1}); && (3)\\
G(\psi,\chi)G(\psi^{-1},\chi) &= \psi(-1)q \quad \textit{for } \psi \neq \psi_0 \textit{ and } \chi \neq \chi_0; && (4)\\
G(\psi^p,\chi_b) &= G(\psi,\chi_{b^p}), \textit{ where } p = \operatorname{char} F. && (5)
\end{aligned}
$$

Proof. By (10.14), $\chi_{ab}(u) = \chi_1(abu) = \chi_b(au)$. Thus

$$
\begin{aligned}
G(\psi,\chi_{ab}) &= \sum_{u\in F^*} \psi(u)\chi_{ab}(u) = \sum_{u\in F^*} \psi(u)\chi_b(au)\\
&= \sum_{v\in F^*} \psi(a^{-1}v)\chi_b(v) = \psi^{-1}(a) \sum_{v\in F^*} \psi(v)\chi_b(v)\\
&= \psi^{-1}(a)G(\psi,\chi_b),
\end{aligned}
$$

proving (1). In order to check (2), we write $\chi = \chi_b$, so that $\chi^{-1} = \chi_{-b}$. Applying (1) with $a = -1$ gives

$$G(\psi,\chi^{-1}) = G(\psi,\chi_{-b}) = \psi^{-1}(-1)G(\psi,\chi_b) = \psi(-1)G(\psi,\chi),$$

since $\psi(-1) = \pm 1$ implies $\psi^{-1}(-1) = \psi(-1)^{-1} = \psi(-1)$. This establishes (2), which in turn yields the validity of (3) as follows:

$$G(\psi^{-1},\chi) = \psi^{-1}(-1)G(\psi^{-1},\chi^{-1}) = \psi(-1)G(\psi^{-1},\chi^{-1}).$$

Then (4) is an immediate consequence of (3) and Equation (10.19). Finally, note that the invariance of the absolute trace under the Frobenius automorphism $x \mapsto x^p$ (over the prime field of F) gives $\chi_1(u) = \chi_1(u^p)$ and thus

$$\chi_b(u) = \chi_1(bu) = \chi_1(b^p u^p) = \chi_{b^p}(u^p)$$

for all $u \in F$. Therefore,

$$G(\psi^p,\chi_b) = \sum_{u\in F^*} \psi^p(u)\chi_b(u) = \sum_{u\in F^*} \psi(u^p)\chi_{b^p}(u^p) = G(\psi,\chi_{b^p}),$$

which establishes also (5). □

In view of the preceding identities, it is of interest to determine the values $\psi(-1)$ explicitly:

Lemma 10.3.5. *Let L be a splitting field for the multiplicative group of the finite field $F = \mathrm{GF}(q)$, where q is odd, and let ψ be a multiplicative character of order e of F into L^*. Then*

$$\psi(-1) = \begin{cases} -1 & \text{if } e \text{ is even and } (q-1)/e \text{ is odd,} \\ 1 & \text{otherwise.} \end{cases}$$

Proof. As noted before, $\psi(-1) = \pm 1$. By hypothesis, the values of ψ are e-th roots of unity, so that $\psi(-1) = -1$ can only occur if e is even. Thus assume this to be the case. Note that e divides $q-1$, since $\widehat{F^*}$ is a cyclic group of order $q-1$, by Theorem 10.1.2. Let ω be a primitive element for F^*; then $\zeta := \psi(\omega)$ is a primitive e-th root of unity in L^*. Hence $\psi(-1) = \psi(\omega^{(q-1)/2}) = \zeta^{(q-1)/2}$, and this equals -1 if and only if $(q-1)/2 \equiv e/2 \bmod e$, which gives the desired criterion. □

We conclude this section by making our introductory remark that Gauss sums connect the additive and multiplicative structures of finite fields more precise: we will show that any multiplicative character of F can be expanded in terms of the additive characters, and vice-versa.

Proposition 10.3.6. *Let L be a splitting field for $F = \mathrm{GF}(q)$, and let ψ be a multiplicative character and χ an additive character of F into L^*. Moreover, let the characters of F into L^* be labelled as in Equations* (10.12) *and* (10.14)*, and assume that the multiplicative characters are extended to all of F. Then the following identities hold:*

$$\psi = \frac{1}{q} \cdot \sum_{b \in F} G(\psi, \chi_{-b}) \chi_b,$$

and

$$\chi = \frac{1}{q-1} \cdot \sum_{j=0}^{q-2} G(\psi_{-j}, \chi) \psi_j.$$

Proof. Using Corollary 10.2.3, we obtain

$$\begin{aligned} \psi(u) &= \frac{1}{q} \cdot \sum_{v \in F^*} \psi(v) \Big(\sum_{b \in F} \chi_b(u) \chi_{-b}(v) \Big) \\ &= \frac{1}{q} \cdot \sum_{b \in F} \chi_b(u) \Big(\sum_{v \in F^*} \psi(v) \chi_{-b}(v) \Big) \\ &= \frac{1}{q} \cdot \sum_{b \in F} G(\psi, \chi_{-b}) \chi_b(u) \end{aligned}$$

for all $u \in F$, which proves the first assertion. The second assertion follows from a similar computation, now using Corollary 10.2.2. □

Exercises

Exercise 10.3.7. Prove Corollary 10.3.3. □

Exercise 10.3.8. Confirm the first formula in Proposition 10.3.6 for the classical case by using Exercise 10.1.10, where we now take G as the additive group of the finite field $F = \mathrm{GF}(q)$. Thus let ψ be a non-trivial multiplicative character (extended to all of F) and expand ψ with respect to the orthonormal basis $\widehat{G}$ of $\mathbb{C}^G$, say $\psi = \sum_{\gamma \in \widehat{G}} a_\chi \chi$. Prove that $a_\chi = \frac{1}{q} \cdot G(\psi, \overline{\chi})$ for all χ. □

10.4 The Quadratic Character

In this section, we study a specific multiplicative character of a finite field $F = \mathrm{GF}(q)$ of odd characteristic and give three interesting applications of the theory developed up to now. Using the notation introduced in Equation (10.12), we define the **quadratic character** of F into L^* as the character $\eta := \psi_{(q-1)/2}$. Thus η is the unique involution in the cyclic group of multiplicative characters of F into L^*. Since the non-zero squares in F form the unique subgroup of F^* of order $(q-1)/2$, we get the following explicit description of η:

$$\eta(u) := \begin{cases} 1 & \text{if } u \text{ is a square in } F^*, \\ -1 & \text{otherwise;} \end{cases} \tag{10.20}$$

in view of this description, it makes sense to consider the quadratic character of F even if L is not a splitting field for F, provided that L has characteristic $\neq 2$. However, if we want to use Gauss sums to study η, we will need to retain the assumption that L is a splitting field for F (or extend L accordingly).

Lemma 10.4.1. *Let L be a splitting field for $F = \mathrm{GF}(q)$, where q is odd, and let η be the quadratic character and χ_1 the canonical additive character of F into L^*. Then the Gauss sum $\theta := G(\eta, \chi_1)$ satisfies*

$$\theta^2 = (-1)^{(q-1)/2} q.$$

Proof. Since η has order 2, we get $\theta^2 = \eta(-1)q$ from part (4) of Proposition 10.3.4, and then Lemma 10.3.5 yields the assertion. □

Note that Lemma 10.4.1 determines θ up to a sign in the classical case $L = \mathbb{C}$. It is in fact possible to compute the precise value, provided that one follows our convention to use the particular primitive p-th root of unity $\xi = e^{2\pi i/p}$ to define the canonical additive character; the resulting Gauss sum is then called a **normed Gauss sum**. For the time being, we just state the formula in question; its proof requires a more advanced result on Gauss sums and will be postponed to Theorem 10.5.4.

Result 10.4.2 *Let η be the quadratic character and χ_1 the canonical additive character of $F = \mathrm{GF}(q)$ into $\mathbb{C}$. Then the normed Gauss sum θ satisfies*

$$\theta = \begin{cases} (-1)^{n-1}\sqrt{q} & \textit{if } p \equiv 1 \bmod 4, \\ (-1)^{n-1} i^n \sqrt{q} & \textit{if } p \equiv 3 \bmod 4. \end{cases}$$

Here p denotes the characteristic of F, and $q = p^n$. □

We now turn to the promised applications of the quadratic character: we will prove the law of quadratic reciprocity, exhibit Paley's construction of Hadamard matrices, and finally consider quadratic equations over finite fields of odd characteristic.

Recall from Definition 8.5.9 that the Legendre symbol is given by

$$(a/p) := \begin{cases} 1 & \text{if } a \text{ is a quadratic residue modulo } p, \\ -1 & \text{otherwise,} \end{cases}$$

where p is a prime and a any integer not divisible by p. In view of (10.20), the Legendre symbol is nothing but the quadratic character η (into a suitable field L of characteristic $\neq 2$) in the special case where q is a prime p. We shall use this observation to prove the celebrated law of quadratic reciprocity, which was first established by Gauss in 1801 in his famous *Disquisitiones Arithmeticae* [132]; see also [134] for a translation into English. One of the proofs Gauss gave for this result actually uses the classical case of what is now called *Gauss sums*; however, the proof becomes simpler if one applies Gauss sums over finite fields instead. We refer the reader to the notes in Lidl and Niederreiter [242, Chapter 5.2] for more detailed references on the law of quadratic reciprocity and its various proofs.

Theorem 10.4.3 (Law of quadratic reciprocity). *Let p and q be any two distinct odd primes. Then one has*

$$(p/q) = (-1)^{(p-1)(q-1)/4}(q/p).$$

Proof. Without loss of generality, assume $q < p$. We take $F = \mathrm{GF}(q)$ and consider the quadratic character η from F^* into $K = \mathrm{GF}(p)$. As p does not divide $q(q-1)$, we may extend K to a (finite) splitting field L for F by adjoining a primitive $q(q-1)$-th root of unity, which allows us to apply Lemma 10.4.1. Since the absolute trace in (10.12) becomes trivial in the case of a prime field, the Gauss sum θ may now be written explicitly as

$$\theta = \sum_{u \in F^*} (u/q)\xi^u,$$

where ξ is a primitive q-th root of unity in L. From this, we compute

$$\theta^p = \sum_{u \in F^*} (u/q)\xi^{up} = \sum_{u \in F^*} (up^2/q)\xi^{up} = (p/q) \sum_{u \in F^*} (up/q)\xi^{up} = (p/q)\theta,$$

so that $\theta^{p-1} = (p/q)$. We now combine this identity with Lemma 10.4.1 to obtain the assertion as follows:

$$(p/q) = (\theta^2)^{(p-1)/2} = (-1)^{(p-1)(q-1)/4} q^{(p-1)/2} = (-1)^{(p-1)(q-1)/4}(q/p),$$

as $(q/p) \equiv q^{(p-1)/2} \bmod p$ by Proposition 8.5.10. □

We mention in passing that the Gauss sum θ (considered as an element of a finite field $\mathrm{GF}(p)$) also has an interesting application in Coding Theory, where it is used to describe the idempotent generators of the so-called quadratic residue codes; see, for instance, MacWilliams and Sloane [255].

Example 10.4.4. Repeated application of Theorem 10.4.3 together with Proposition 8.5.11 and the special case $(-1/p) \equiv (-1)^{(p-1)/2} \bmod p$ of Proposition 8.5.10 allows an easy evaluation of the Legendre symbol (a/p) for all values of a and p. As an example, we compute $(29/43)$ as follows:

$$(29/43) = (43/29) = (2/29)(7/29) = -(29/7) = -(1/7) = -1,$$

where $(2/29) = -1$ holds by Proposition 8.5.11. □

We remark that Gauss sums also play an important role in proving higher reciprocity laws, in particular the considerably more involved laws of cubic and biquadratic reciprocity. The interested reader may consult the book by Ireland and Rosen [203].

Our second application originated in the study of complex matrices. In 1893, Hadamard [176] proved the following famous inequality; see Craigen [99] for a particularly short and elegant proof.

Result 10.4.5 (Hadamard inequality) *Any complex matrix $A = (a_{ij})_{i,j=1,\ldots,n}$ satisfies*

$$|\det A|^2 \leq \prod_{i=1}^{n} \left(|a_{i1}|^2 + \cdots + |a_{in}|^2\right),$$

with equality if and only if either A contains a zero row or AA^* is a diagonal matrix (where, as usual, A^* denotes the transpose of the complex conjugate of A). □

Result 10.4.5 has the following consequence for real-valued matrices; we leave the simple derivation as Exercise 10.4.18 to the reader.

Corollary 10.4.6. *Let $A = (a_{ij})_{i,j=1,\ldots,n}$ be an invertible matrix with real entries of absolute value ≤ 1. Then A satisfies*

$$|\det A| \leq n^{n/2} \quad \text{with equality if and only if } AA^T = nI.$$

In particular, A has entries ± 1 in the case of equality. □

Definition 10.4.7. An (n,n)-matrix A with entries ± 1 is said to be a **Hadamard matrix** of order n if it satisfies $AA^T = nI$. □

Example 10.4.8. Here are trivial examples of Hadamard matrices of orders $n = 1$, 2 and 4:

$$(1) \qquad \begin{pmatrix} 1 & 1 \\ 1 & -1 \end{pmatrix} \qquad \begin{pmatrix} -1 & 1 & 1 & 1 \\ 1 & -1 & 1 & 1 \\ 1 & 1 & -1 & 1 \\ 1 & 1 & 1 & -1 \end{pmatrix}$$ □

Using the orthogonality property $AA^T = nI$, one may show that a Hadamard matrix of order $n \geq 3$ can only exist if n is a multiple of 4; for the sake of completeness, we include the simple proof below. It is a long-standing (still unresolved) conjecture that such a matrix indeed exists whenever n is a multiple of 4; the smallest undecided case is at present $n = 668$. Since Hadamard matrices are equivalent to certain "symmetric designs", they are usually studied in the context of Design Theory; we refer the interested reader to the book of Beth, Jungnickel and Lenz [35] for these connections.

Proposition 10.4.9. *A Hadamard matrix of order $n \geq 3$ can only exist if n is a multiple of* 4.

Proof. Let A be any Hadamard matrix of order n. Multiplying rows and columns of A with -1 does not affect the defining equation $AA^T = nI$, and thus we may assume that all entries in the first row and in the first column of A are $+1$. Then any other row contains as many entries -1 as entries $+1$, since it is orthogonal to the first row. Therefore, n is even, say $n = 2m$. Now consider the second and third rows of A, and denote the number of columns in which

- row 2 and row 3 both have entry $+1$ by x;
- row 2 has entry $+1$ and row 3 has entry -1 by y;
- row 2 has entry -1 and row 3 has entry $+1$ by z;
- row 2 and row 3 both have entry -1 by w.

Using the preceding observation and the orthogonality of rows 2 and 3 of A, we obtain the following system of equations:

$$\begin{aligned} x+y+z+w &= 2m \\ x+y &= m \\ x+z &= m \\ x-y-z+w &= 0, \end{aligned}$$

with the unique solution $x = y = z = w = m/2$. □

Next, we use the quadratic character to construct certain Hadamard matrices discovered by Paley [315] in 1933. We begin with a simple lemma:

Lemma 10.4.10. *Let L be a splitting field for the multiplicative group of the finite field $F = \mathrm{GF}(q)$, where q is odd, and let ψ be any non-trivial multiplicative character of F into L^*. Then, with the usual convention $\psi(0) = 0$,*

$$\sum_{u \in F} \psi(u+a)\psi^{-1}(u+b) = -1$$

for any two distinct elements $a, b \in F$.

Proof. Clearly, it suffices to consider the special case $b = 0$. Let us denote the resulting sum by $S(a)$. Then, for all $x \neq 0$,

$$\begin{aligned} S(a) &= \sum_{u \in F} \psi(u+a)\psi^{-1}(u) = \sum_{u \in F} \psi(x)\psi^{-1}(x)\psi(u+a)\psi^{-1}(u) \\ &= \sum_{u \in F} \psi(ux+ax)\psi^{-1}(ux) = \sum_{v \in F} \psi(v+w)\psi^{-1}(v) = S(w), \end{aligned}$$

where $v = ux$ and $w = ax \neq 0$. Hence the sum $S := S(a)$ does not depend on the choice of $a \neq 0$, and an application of Corollary 10.2.2 yields

$$\begin{aligned} 0 &= \Big(\sum_{v \in F} \psi(v)\Big)\Big(\sum_{u \in F} \psi^{-1}(u)\Big) \\ &= \sum_{u \in F} \psi(u)\psi^{-1}(u) + \sum_{a \in F^*} \sum_{u \in F} \psi(u+a)\psi^{-1}(u) \\ &= (q-1) + (q-1)S = (q-1)(1+S), \end{aligned}$$

which establishes the assertion. □

Theorem 10.4.11 (Paley's theorem). *Let q be an odd prime power. Then there exists a Hadamard matrix of order $q+1$ or $2(q+1)$, respectively, depending on whether $q \equiv 3 \bmod 4$ or $q \equiv 1 \bmod 4$.*

Proof. First let $q \equiv 3 \bmod 4$ and define a (q,q)-matrix $M = (m_{cd})$ with rows and columns indexed by the elements of $F = \mathrm{GF}(q)$ as follows:

$$m_{cd} := \begin{cases} -1 & \text{if } c = d, \\ \eta(d-c) & \text{otherwise.} \end{cases}$$

Note that every row of M has sum -1, and that the inner product of a row with itself equals q. By Proposition 3.2.13, $\eta(-1) = -1$. We now use this fact together with Lemma 10.4.10 for the quadratic character η and the convention $\eta(0) = 0$. Then $\psi = \psi^{-1} = \eta$, and we may compute the inner product of any two distinct rows of M as follows:

$$\sum_{d \in F} m_{cd} m_{bd} = -\eta(c-b) \quad \eta(b-c) + \sum_{d \in F} \eta(d-c)\eta(d-b) = -1.$$

Adjoining a new row and a new column with all entries $+1$ to M then yields the desired Hadamard matrix of order $q+1$.

Now assume $q \equiv 1 \bmod 4$. This time, we define a $(q+1,q+1)$-matrix M with rows and columns indexed by the elements of $F \cup \{\infty\}$ as follows:

$$m_{cd} := \begin{cases} 1 & \text{if exactly one of } c,d \text{ equals } \infty, \\ 0 & \text{if } c = d, \\ \eta(d-c) & \text{otherwise.} \end{cases}$$

By Proposition 3.2.13, $\eta(-1) = 1$, so that M is a symmetric matrix. Again using Lemma 10.4.10, one checks that any two distinct rows of M have inner product 0.

We now define two symmetric $(2,2)$-matrices A and B by

$$A := \begin{pmatrix} 1 & 1 \\ 1 & -1 \end{pmatrix} \quad \text{and} \quad B := \begin{pmatrix} 1 & -1 \\ -1 & -1 \end{pmatrix}$$

and replace each entry 0 of M by B and each entry ± 1 of M by $\pm A$. This yields a $(2q+2,2q+2)$-matrix H with entries ± 1. Using the preceding observations about M as well as

$$AA^T = BB^T = \begin{pmatrix} 2 & 0 \\ 0 & 2 \end{pmatrix} \quad \text{and} \quad AB^T = -BA^T = \begin{pmatrix} 0 & -2 \\ 2 & 0 \end{pmatrix},$$

the reader may check that H is the desired Hadamard matrix. □

Remark 10.4.12. Note that the first case of Theorem 10.4.11 is essentially equivalent to Theorem 9.5.13. We shall explain this in the special case where $q \equiv 3 \bmod 4$ is a prime. Let D be the Paley-Hadamard difference set in the cyclic group G of residues modulo q as defined in Theorem 9.5.13, and consider the associated periodic binary sequence $\mathbf{a} = (a_g)$. If we index the rows and columns of the matrix M in the proof of Theorem 10.4.11 in the natural order $0,\dots,q-1$, the first row of M is obtained from the first q entries of $\mathbf{a}$ by replacing each entry 0 by -1. More generally, the row of M indexed by $-h$ arises from the cyclically shifted sequence (a_{g+h}) in the same manner. Therefore, the assertion about the inner product of any two distinct rows of M made in the proof of Theorem 10.4.11 corresponds to the fact that $\mathbf{a}$ has autocorrelation coefficient $c = -1$. □

Our final application is a formula for the number of solutions of a non-degenerate quadratic equation over a finite field of odd characteristic. It will be convenient to use the notation $N(f(x_1,\dots,x_n) = c)$ for the number of solutions of the equation $f(x_1,\dots,x_n) = c$.

We will need a further auxiliary result on the quadratic character. To put this result into context, it is helpful to recall some well-known facts regarding the solutions of a quadratic equation

$$ax^2 + bx + c = 0 \ \text{(with } a \neq 0\text{)} \tag{10.21}$$

in just one variable over a field F of odd characteristic. One calls $d := b^2 - 4ac$ the **discriminant** of Equation (10.21), and the standard approach to solving a quadratic equation in one variable shows that (10.21) has a solution if and only if d is a square in F. Moreover, there is a unique solution in the case $d = 0$, and there are exactly two solutions if d is a non-zero square.

Lemma 10.4.13. *Consider the quadratic polynomial $f = ax^2 + bx + c$ over $F = \mathrm{GF}(q)$, where q is odd and $a \neq 0$. Let d be the discriminant of the associated Equation (10.21), and let η be the quadratic character of F into some field L of characteristic $\neq 2$. Then*

$$\sum_{u \in F} \eta(au^2 + bu + c) = \begin{cases} -\eta(a) & \text{if } d \neq 0, \\ (q-1)\eta(a) & \text{if } d = 0, \end{cases}$$

where we again use the convention $\eta(0) = 0$.

Proof. We multiply the sum in question by $\eta(4a^2) = 1$ and compute

$$(*) \qquad \begin{aligned} \sum_{u \in F} \eta(au^2 + bu + c) &= \eta(a) \sum_{u \in F} \eta(4a^2u^2 + 4abu + 4ac) \\ &= \eta(a) \sum_{u \in F} \eta((2au + b)^2 - d) \\ &= \eta(a) \sum_{u \in F} \eta(u^2 - d), \end{aligned}$$

which immediately gives the desired result for the case $d = 0$.

Now let $d \neq 0$ and note that $1 + \eta(u^2 - d)$ is the number of solutions of the equation $y^2 = u^2 - d$. Hence

$$\sum_{u \in F} \eta(u^2 - d) = -q + \sum_{u \in F} \left(1 + \eta(u^2 - d)\right) = -q + s(d),$$

where $s(d)$ denotes the number of ordered pairs (u, y) with $d = u^2 - y^2$. Putting $v := u + y$ and $w := u - y$, one sees that the number of ordered pairs (v, w) with $d = vw$ is also given by $s(d)$, since q is odd. This shows $s(d) = q - 1$ and then $(*)$ yields the assertion. □

We are now almost ready to deal with the next special case, namely quadratic equations in two variables. To state the result in question, we introduce an integer valued function ν on $F = \mathrm{GF}(q)$ via

$$\nu(c) := \begin{cases} q - 1 & \text{if } c = 0, \\ -1 & \text{if } c \neq 0 \end{cases} \tag{10.22}$$

and note the trivial identity

$$\sum_{u \in F} \nu(u) = 0. \tag{10.23}$$

Lemma 10.4.14. *Let* $F = \mathrm{GF}(q)$, *where* q *is odd, and let* η *be the quadratic character of* F *into some field* L *of characteristic* $\neq 2$.

Then the number $N = N\left(ax^2 + by^2 = c\right)$ *is given by* $N = q + \eta(-ab)\nu(c)$ *for all* $c \in F$ *and all* $a, b \in F^*$.

Proof. Note that $1 + \eta(e)$ is the number of solutions of the equation $x^2 = e$. Using this together with Corollary 10.2.2 and Lemma 10.4.13, we compute

$$
\begin{aligned}
N &= \sum_{u_1+u_2=c} N(ax^2 = u_1)N(by^2 = u_2) \\
&= \sum_{u_1+u_2=c} \left(1+\eta(a^{-1}u_1)\right)\left(1+\eta(b^{-1}u_2)\right) \\
&= q+\eta(a)\sum_{u\in F}\eta(u)+\eta(b)\sum_{u\in F}\eta(c-u)+\eta(ab)\sum_{u\in F}\eta(-u^2+cu) \\
&= q+\eta(ab)\nu(c)\eta(-1),
\end{aligned}
$$

which establishes the assertion. □

Before we can prove the desired general result, we need one further auxiliary fact:

Lemma 10.4.15. *Let* $F = \mathrm{GF}(q)$, *where* q *is odd, and let* ν *be the function defined in Equation* (10.22). *Then one has*

$$
\sum_{u_1+\cdots+u_m=c} \nu(u_1)\cdots\nu(u_k) = \begin{cases} 0 & \text{if } 1 \le k < m, \\ \nu(c)q^{m-1} & \text{if } k = m, \end{cases}
$$

for all $c \in F$.

Proof. Note that the sum in question runs over all $u_1,\ldots,u_m \in F$ summing to c, but that the products involve only the first k values $\nu(u_i)$.

Using the identity (10.23), we obtain the assertion in the case $k < m$ as follows:

$$
\begin{aligned}
\sum_{u_1+\cdots+u_m=c} \nu(u_1)\cdots\nu(u_k) &= \sum_{u_1,\ldots,u_k\in F} \Bigg(\nu(u_1)\cdots\nu(u_k)\cdot \sum_{\substack{u_{k+1}+\cdots+u_m=\\ c-(u_1+\cdots+u_k)}} 1\Bigg) \\
&= q^{m-k-1}\cdot\prod_{i=1}^{k}\sum_{u_i\in F}\nu(u_i) = 0.
\end{aligned}
$$

For the case $k = m$, we use induction on m. The induction base $m = 1$ is trivial. Now assume that the assertion holds for some $m \ge 1$. Using the first case with $m+1$ instead of m and $k = m$, we compute

$$
\begin{aligned}
\sum_{u_1+\cdots+u_{m+1}=c} \nu(u_1)\cdots\nu(u_{m+1}) &= \sum_{u_1+\cdots+u_{m+1}=c} \nu(u_1)\cdots\nu(u_m)(1+\nu(u_{m+1})) \\
&= \sum_{u_1,\ldots,u_m} \nu(u_1)\cdots\nu(u_m)\left(1+\nu(c-u_1-\cdots-u_m)\right) \\
&= q\cdot \sum_{u_1+\cdots+u_m=c} \nu(u_1)\cdots\nu(u_m) = \nu(c)q^m.
\end{aligned}
$$

(Note that $1+\nu(c-u_1-\cdots-u_m) \ne 0$ only if $c = u_1+\cdots+u_m$, in which case one obtains the value q.) □

We are now in a position to settle the general case of quadratic equations in an arbitrary number of variables. Recall that any non-degenerate symmetric bilinear form over a field of characteristic $\neq 2$ can be represented by a diagonal matrix; see Proposition 3.13.4. Hence it suffices to consider quadratic equations of the special type

$$a_1x_1^2 + \cdots + a_nx_n^2 = 0 \quad \text{with } a_1, \ldots, a_n \in F^*.$$

Theorem 10.4.16. *Let $f(x_1, \ldots, x_n)$ be a non-degenerate quadratic form over $F = \mathrm{GF}(q)$, where q is odd. Moreover, let η be the quadratic character of F into some field L of characteristic $\neq 2$ and put $\Delta := \det f$. Then the number $N = N\big(f(x_1, \ldots, x_n) = c\big)$ satisfies*

$$N = \begin{cases} q^{n-1} + \nu(c)q^{(n-2)/2}\eta\big((-1)^{n/2}\Delta\big) & \text{if } n \text{ is even,} \\ q^{n-1} + q^{(n-1)/2}\eta\big((-1)^{(n-1)/2}c\Delta\big) & \text{if } n \text{ is odd,} \end{cases}$$

where ν is defined as in Equation (10.22).

Proof. As noted above, we may assume

$$f(x_1, \ldots, x_n) = a_1x_1^2 + \cdots + a_nx_n^2 \quad \text{with } a_1, \ldots, a_n \in F^*,$$

and hence $\Delta = a_1 \cdots a_n$. First assume that n is even, say $n = 2m$. Using Lemmas 10.4.13 and 10.4.14, we obtain the assertion as follows:

$$\begin{aligned} N &= \sum_{u_1 + \cdots + u_m = c} N\big(a_1x_1^2 + a_2x_2^2 = u_1\big) \cdots N\big(a_{2m-1}x_{2m-1}^2 + a_{2m}x_{2m}^2 = u_m\big) \\ &= \sum_{u_1 + \cdots + u_m = c} \big(q + \nu(u_1)\eta(-a_1a_2)\big) \cdots \big(q + \nu(u_m)\eta(-a_{2m-1}a_{2m})\big) \\ &= q^{m-1}q^m \; + \; \eta\big((-1)^m a_1 \cdots a_{2m}\big) \cdot \sum_{u_1 + \cdots + u_m = c} \nu(u_1) \cdots \nu(u_m) \\ &= q^{2m-1} + \nu(c)q^{m-1}\eta\big((-1)^m\Delta\big). \end{aligned}$$

This establishes the first case of the assertion.

Now let n be odd, and note that the assertion holds trivially for $n = 1$. Thus let $n = 2m + 1$ with $m \neq 0$. We now apply the first case and obtain, using Corollary 10.2.2 and the identity (10.23),

$$\begin{aligned}
N &= \sum_{u_1+u_2=c} N\big(a_1x_1^2+\cdots+a_{2m}x_{2m}^2=u_1\big)N\big(a_{2m+1}x_{2m+1}^2=u_2\big)\\
&= \sum_{u_1+u_2=c} \big(q^{2m-1}+\nu(u_1)q^{m-1}\eta\big((-1)^m a_1\cdots a_{2m}\big)\big)\big(1+\eta(u_2a_{2m+1})\big)\\
&= q^{2m} + q^{2m-1}\eta(a_{2m+1})\sum_{u_2\in F}\eta(u_2)\\
&\qquad + q^{m-1}\eta\big((-1)^m a_1\cdots a_{2m}\big)\sum_{u_1\in F}\nu(u_1)\\
&\qquad + q^{m-1}\eta\big((-1)^m a_1\cdots a_{2m+1}\big)\sum_{u_1+u_2=c}\nu(u_1)\eta(u_2)\\
&= q^{2m} + q^{m-1}\eta\big((-1)^m\Delta\big)\sum_{u\in F}\nu(c-u)\eta(u).
\end{aligned}$$

Another application of Corollary 10.2.2 gives

$$\sum_{u\in F}\nu(c-u)\eta(u) = \sum_{u\in F}(1+\nu(c-u))\eta(u) = (1+\nu(0))\eta(c) = q\eta(c),$$

which establishes also the second case of the assertion. □

An alternative proof of Theorem 10.4.15 using Gauss and Jacobi sums can be found in the book by Lidl and Niederreiter [242]. We also refer the reader to Chapter 6 of that book for a corresponding result in the case of even characteristic and for further results on equations over finite fields which can be proved by elementary (though intricate) means – that is, for instance, by using character sums, but without the methods of Algebraic Geometry or Algebraic Number Theory. For a much more complete (but still elementary in the sense just explained) treatment, the interested reader may consult the monograph by Wolfgang Schmidt [340].

Exercises

Exercise 10.4.17. Show that $(3/p)=1$ if and only if $p\equiv\pm1 \bmod 12$. A similar application of quadratic reciprocity was already used in Example 9.3.16. □

Exercise 10.4.18. Prove Corollary 10.4.6. □

Exercise 10.4.19. Use Lemma 10.4.10 with $\psi=\eta$, the quadratic character, to give an alternative proof for Lemma 10.4.1. □

Exercise 10.4.20. Write down two Hadamard matrices of order 12 using the two constructions given in the proof of Theorem 10.4.11. □

10.5 More on Gauss Sums

In this section, we shall prove three important advanced results on Gauss sums. The first of these concerns Gauss sums of characters "lifted" to some extension field. Formally, we introduce the following notations:

Definition 10.5.1. Let $K = \mathrm{GF}(q^n)$ be any extension field of the field $F = \mathrm{GF}(q)$. Then every multiplicative character ψ of F (into a suitable field L) can be lifted to a multiplicative character ψ' of K by putting

$$\psi'(u) := \psi(N_{K/F}(u)) \quad \text{for all } u \in K^*,$$

where $N_{K/F}$ denotes the norm function from K^* onto F^*. Similarly, every additive character χ of F can be lifted to an additive character χ' of K by putting

$$\chi'(u) := \chi(\mathrm{Tr}_{K/F}(u)) \quad \text{for all } u \in K,$$

where $\mathrm{Tr}_{K/F}$ is the trace function. It is trivial to check that ψ' and χ' are indeed characters of K; they are called the **lifted characters** of ψ and χ, respectively. □

Note that our description of the additive character group $\widehat{F}$ in Section 10.2 may be viewed as lifting the additive character group $\widehat{P}$ of the underlying prime field P, which shows that the preceding definition is – at least in the additive case – a quite natural generalization of this approach. In analogy, the multiplicative case then leads to using the norm function.

Example 10.5.2. Let us denote the canonical additive character of K by κ_1. Then the canonical additive character χ_1 of F defined as in (10.14) lifts to κ_1, as

$$\mathrm{Tr}_{K/P}(u) = \mathrm{Tr}_{F/P}(\mathrm{Tr}_{K/F}(u)) \quad \text{for all } u \in K,$$

where P is the prime field of F, by the transitivity formula for the trace. More generally, the character χ_b of F defined in (10.14) lifts to the corresponding character κ_b of K associated with b, for every $b \in F$.

Clearly, a multiplicative character ψ of F with order d has to lift to a multiplicative character ψ' of K with the same order d. In particular, let q be odd. Then the quadratic character η of F lifts to the quadratic character η' of K, since both character groups are cyclic and hence contain a unique involution. □

The following result was first proved in 1935 by Davenport and Hasse [102], using rather advanced methods; the completely elementary proof we shall present here follows Schmid [336].

Theorem 10.5.3 (Davenport-Hasse theorem). *Let ψ and χ be a multiplicative and an additive character of $F = \mathrm{GF}(q)$, respectively (into some splitting field E for F), and assume that at least one of ψ and χ is non-trivial. Moreover, let $K = \mathrm{GF}(q^n)$ and consider the corresponding lifted characters ψ' and χ', as in Definition 10.5.1. Then*

$$G(\psi',\chi') = (-1)^{n-1} G(\psi,\chi)^n.$$

Proof. If $\chi = \chi_0$ is the trivial additive character of F, then χ' is the trivial additive character of K; in this case, the assertion is immediate, since then both Gauss sums equal 0, by Proposition 10.3.2.

Now assume $\chi \neq \chi_0$. By Proposition 10.2.1, $\chi = \chi_b$ for some $b \in F^*$, so that $\chi' = \kappa_b$, as noted in Example 10.5.2. Applying part (1) of Proposition 10.3.4 to both χ and χ' gives

$$G(\psi, \chi_b)^n = \left(\psi^{-1}(b)\right)^n G(\psi, \chi_1)^n \text{ and } G(\psi', \kappa_b) = \left(\psi^{-1}(b)\right)^n G(\psi', \kappa_1),$$

as $\psi'(b^{-1}) = \psi(N_{K/F}(b^{-1})) = \psi\left((b^{-1})^n\right) = \left(\psi(b^{-1})\right)^n$. Therefore, it suffices to prove the assertion for the special case $\chi = \chi_1$. We shall do so by using induction on n, the case $n = 1$ being trivial.

Thus consider $L := \mathrm{GF}(q^{n+1})$, and let ψ^* and χ^* be the corresponding lifted characters, that is, $\psi^*(z) = \psi(N_{L/F}(z))$ and $\chi^*(z) = \chi(\mathrm{Tr}_{L/F}(z))$ for all $z \in L$. We need to show

$$(*) \quad G(\psi^*, \chi^*) = (-1)^n G(\psi, \chi)^{n+1}.$$

By definition,

$$G(\psi^*, \chi^*) = \sum_{z \in L^*} \psi(N_{L/F}(z))\chi(\mathrm{Tr}_{L/F}(z)).$$

Let us denote the number of solutions $z \in L^*$ of

$$N_{L/F}(z) = u \quad \text{and} \quad \mathrm{Tr}_{L/P}(z) = v, \tag{10.24}$$

where P is the prime field of F, by $r(u,v)$. Using the definition of $\chi = \chi_1$ in Proposition 10.2.1, we get

$$G(\psi^*, \chi^*) = \sum_{u \in F^*} \psi(u)\Big(\sum_{v=0}^{p-1} r(u,v)\xi^v\Big), \tag{10.25}$$

where p is the characteristic of F and where ξ is a primitive p-th root of unity in E. On the other hand, the induction hypothesis yields

$$\begin{aligned}
(-1)^n G(\psi, \chi)^{n+1} &= -G(\psi, \chi)G(\psi', \chi') \\
&= -\Big(\sum_{x \in F^*} \psi(x)\chi(x)\Big)\Big(\sum_{y \in K^*} \psi(N_{K/F}(y))\chi(\mathrm{Tr}_{K/F}(y))\Big) \\
&= -\sum_{x \in F^*}\sum_{y \in K^*} \psi\big(xN_{K/F}(y)\big)\chi\big(x + \mathrm{Tr}_{K/F}(y)\big).
\end{aligned}$$

Now denote the number of solutions $(x,y) \in F^* \times K^*$ of

$$xN_{K/F}(y) = u \quad \text{and} \quad \mathrm{Tr}_{F/P}\big(x + \mathrm{Tr}_{K/F}(y)\big) = v \tag{10.26}$$

by $s(u,v)$. Then the preceding equation can be written as

$$(-1)^n G(\psi,\chi)^{n+1} = -\sum_{u\in F^*} \psi(u)\Big(\sum_{v=0}^{p-1} s(u,v)\xi^v\Big). \tag{10.27}$$

Subtracting (10.27) from (10.25) gives

$$G(\psi^*,\chi^*) - (-1)^n G(\psi,\chi)^{n+1} = \sum_{u\in F^*} \psi(u)\Big(\sum_{v=0}^{p-1} \big(r(u,v)+s(u,v)\big)\xi^v\Big).$$

In order to prove (*), it now suffices to show that the sum $r(u,v)+s(u,v)$ does not depend on v, as the powers of ξ sum to 0. In fact, we shall prove the following stronger result, where we write $q = p^a$:

$$r(u,v)+s(u,v) = p^{a-1}\big(1+q+\cdots+q^{n-2}+2q^{n-1}\big) =: e \tag{10.28}$$

for all (u,v) with $u\in F^*$ and $v=0,\ldots,p-1$. For this, we define a rational function $f_u \in F(t)$ for every $u\in F^*$ as follows:

$$f_u(t) := \sum_{i=0}^{a-1}\Big(t+t^q+\cdots+t^{q^{n-1}}+\frac{u}{t^{1+q+\cdots+q^{n-1}}}\Big)^{p^i}.$$

Then (10.24) can be written as

$$N_{L/F}(z) = u \quad \text{and} \quad f_u(z) = v \quad (\text{for } z\in L^*), \tag{10.29}$$

since $u = N_{L/F}(z) = z^{1+q+\cdots+q^n}$ shows

$$z^{q^n} = \frac{u}{z^{1+q+\cdots+q^{n-1}}}.$$

Similarly, by solving $xN_{K/F}(y) = u$ for x and substituting, we rewrite (10.26) as

$$f_u(y) = v \quad (\text{for } y\in K^*). \tag{10.30}$$

Thus we are interested in the number of solutions of the equation $f_u(t) = v$ in a suitable extension field of F. By multiplying this equation with $\big(t^{1+q+\cdots+q^{n-1}}\big)^{p^{a-1}}$, we obtain the equivalent polynomial equation

$$g_u(t) = vt^{dp^{a-1}}, \tag{10.31}$$

where $d := 1+q+\cdots+q^{n-1}$ and where g_u is the polynomial

$$g_u(t) := \sum_{i=0}^{a-1}\Big(t^{p^i+p^{a-1}d}+t^{qp^i+p^{a-1}d}+\cdots+t^{q^{n-1}p^i+p^{a-1}d}+u^{p^i}t^{d(p^{a-1}-p^i)}\Big). \tag{10.32}$$

As g_u has degree $p^{a-1}(d+q^{n-1}) = e$, Equation (10.31) can have at most e solutions in the composite field $\mathrm{GF}(q^{n(n+1)})$ of K and L, for every fixed value of v. In view of the preceding observations, substituting any $y \in K^*$ or any $z \in L^*$ satisfying $u = N_{L/F}(z)$ into g_u will give a solution of (10.31) for some value $v \in P$, and therefore

$$\sum_{v=0}^{p-1} \big(r(u,v)+s(u,v)\big) = q^n - 1 + \frac{q^{n+1}-1}{q-1}.$$

Hence the average value of $r(u,v)+s(u,v)$ over all $v \in P$ equals

$$\frac{2q^n + q^{n-1} + \cdots + q}{p} = e,$$

for every fixed $u \in F^*$. This implies the validity of (10.28) (and hence establishes the induction step (*)), provided that we can show

$$r(u,v)+s(u,v) \leq e \quad \text{for all } v. \tag{10.33}$$

Note that this inequality is trivial if we assume that Equations (10.29) and (10.30) have no common solution for the element $u \in F^*$ under consideration. Therefore, it only remains to check that (10.33) also holds for all values of u for which these two equations do have a common solution, say $y = z =: w$. But then w belongs to $K^* \cap L^* = F^*$, and substituting $w = z \in F$ in (10.29) shows

$$u = N_{L/F}(w) = w^{n+1} \quad \text{and} \quad v = \mathrm{Tr}_{F/P}(\mathrm{Tr}_{L/F}(w)) = (n+1)\mathrm{Tr}_{F/P}(w).$$

In this situation, we claim that w is a repeated root of Equation (10.31), and thus (10.33) indeed still holds. To see this, it suffices to check that w is a root of the formal derivative of the polynomial

$$h(t) = h_{u,v}(t) := g_u(t) - vt^{dp^{a-1}}.$$

First assume $a > 1$. Then (10.32) yields

$$h'(t) = t^{dp^{a-1}} - ut^{d(p^{a-1}-1)-1},$$

and $w^d = ww^q \cdots w^{q^{n-1}} = w^n$ and $u = w^{n+1}$ give $h'(w) = w^{np^{a-1}} - uw^{n(p^{a-1}-1)-1} = 0$, as claimed. Finally, let $a = 1$. Then (10.32) yields, using $d \equiv 1 \bmod p$,

$$h'(t) = 2t^d + t^{d+q-1} + \cdots + t^{d+q^{n-1}-1} - vt^{d-1},$$

and $w^q = w$ and $v = (n+1)w$ again give $h'(w) = w^{n-1}\big((n+1)w - v\big) = 0$. This finishes the proof. □

A non-trivial application of the Davenport-Hasse theorem in Algebraic Number Theory (concerning hypersurfaces over GF(q)) may be found in the book of Ireland and Rosen [203, Chapter 11].

Next, we give the promised proof for Result 10.4.2 and determine the sign of the normed Gauss sum θ. The evaluation of θ (in the essential case $n = 1$) is due to Gauss [133], see also [134]. We will use the proof of Schur [344], which rests on an interesting connection with the Discrete Fourier Transform studied in Section 7.4.

Theorem 10.5.4. *Let η be the quadratic character and χ_1 the canonical additive character of $F = \mathrm{GF}(q)$ into $\mathbb{C}$. Then the normed Gauss sum θ satisfies*

$$\theta = \begin{cases} (-1)^{n-1}\sqrt{q} & \text{if } p \equiv 1 \bmod 4, \\ (-1)^{n-1} i^n \sqrt{q} & \text{if } p \equiv 3 \bmod 4. \end{cases}$$

Here the odd prime p is the characteristic of F, and $q = p^n$.

Proof. In view of the Davenport-Hasse theorem, it suffices to consider the special case $n = 1$, since the canonical additive character χ_1 of $\mathrm{GF}(p)$ lifts to the canonical additive character κ_1 of $\mathrm{GF}(p^n)$ and the quadratic character η of $\mathrm{GF}(p)$ lifts to the quadratic character η' of $\mathrm{GF}(p^n)$, as noted in Example 10.5.2. Thus we have to show

$$\theta = \sum_{a=1}^{p-1} (a/p)\xi^a = \begin{cases} \sqrt{p} & \text{if } p \equiv 1 \bmod 4, \\ i\sqrt{p} & \text{if } p \equiv 3 \bmod 4, \end{cases}$$

where $\xi := e^{2\pi i/p}$. Using the fact $1 + \xi + \cdots + \xi^{p-1} = 0$, we may write θ in the following form:

$$\theta = \sum_{\substack{a=1 \\ (a/p)=1}}^{p-1} \xi^a - \sum_{\substack{a=1 \\ (a/p)=-1}}^{p-1} \xi^a = 1 + 2 \cdot \sum_{\substack{a=1 \\ (a/p)=1}}^{p-1} \xi^a = \sum_{k=0}^{p-1} \xi^{k^2}.$$

Hence θ equals the trace of the matrix

$$Z = \begin{pmatrix} 1 & 1 & 1 & \ldots & 1 \\ 1 & \xi & \xi^2 & \ldots & \xi^{p-1} \\ 1 & \xi^2 & \xi^4 & \ldots & \xi^{2(p-1)} \\ \vdots & \vdots & \vdots & \ddots & \vdots \\ 1 & \xi^{p-1} & \xi^{2(p-1)} & \ldots & \xi^{(p-1)^2} \end{pmatrix}, \tag{10.34}$$

which is the matrix of the Discrete Fourier Transform as introduced in Definition 7.4.1, in this case with $E = \mathbb{C}$, $n = p$, and with ζ replaced by ξ. Since the trace of Z equals the sum of its eigenvalues, the assertion will be an immediate consequence of Theorem 10.5.5 below, which determines the characteristic polynomial and hence the eigenvalues of Z. □

Theorem 10.5.5. *Let p be an odd prime and put $\xi := e^{2\pi i/p}$. Then the matrix Z of the Discrete Fourier Transform as in* (10.34) *has characteristic polynomial*

$$(x-\sqrt{p})^{\frac{p+3}{4}}(x-i\sqrt{p})^{\frac{p-1}{4}}(x+\sqrt{p})^{\frac{p-1}{4}}(x+i\sqrt{p})^{\frac{p-1}{4}}$$

for $p \equiv 1 \bmod 4$ *and*

$$(x-\sqrt{p})^{\frac{p+1}{4}}(x-i\sqrt{p})^{\frac{p+1}{4}}(x+\sqrt{p})^{\frac{p+1}{4}}(x+i\sqrt{p})^{\frac{p-3}{4}}$$

for $p \equiv 3 \bmod 4$.

Proof. One first checks the identity

$$Z^2 = \left(\begin{array}{c|cccc} p & 0 & \dots & 0 & 0 \\ \hline 0 & 0 & \dots & 0 & p \\ 0 & 0 & \dots & p & 0 \\ \vdots & & \iddots & & \\ 0 & p & \dots & 0 & 0 \end{array}\right),$$

which is sometimes called the "quasi-inversion property" of the DFT. In particular, all diagonal entries of Z^2 – with the exception of the top one – are 0. Then the characteristic polynomial g of Z^2 is given by

$$g(x) = (x-p)^{\frac{p+1}{2}}(x+p)^{\frac{p-1}{2}},$$

and hence Z^2 has eigenvalues p and $-p$ with respective multiplicities $\frac{p+1}{2}$ and $\frac{p-1}{2}$. As the eigenvalues of Z^2 are the squares of the eigenvalues of Z, the possible eigenvalues of Z are $\sqrt{p}$, $-\sqrt{p}$, $i\sqrt{p}$ and $-i\sqrt{p}$; let us denote their respective multiplicities by a, b, c, and d. Then

$$a+b+c+d=p, \;\; a+b = \frac{p+1}{2} \;\text{ and }\; c+d = \frac{p-1}{2} \tag{10.35}$$

and hence, as already shown in the proof of Theorem 10.5.4,

$$\theta = (a-b)\sqrt{p}+(c-d)i\sqrt{p}. \tag{10.36}$$

By Lemma 10.4.1, we already know θ up to a sign:

$$\theta = \begin{cases} \pm\sqrt{p} & \text{if } p \equiv 1 \bmod 4, \\ \pm i\sqrt{p} & \text{if } p \equiv 3 \bmod 4, \end{cases}$$

and a comparison with (10.36) shows

$$\begin{cases} a-b = \pm 1 \text{ and } c=d & \text{if } p \equiv 1 \bmod 4, \\ a=b \text{ and } c-d = \pm 1 & \text{if } p \equiv 3 \bmod 4. \end{cases} \tag{10.37}$$

In view of Equation (10.37), we need just one further relation between a, b, c, and d to determine the multiplicities exactly, and this will be obtained by evaluating the determinant of Z in two ways. A direct computation shows

$$\det Z^2 = p^p(-1)^{p(p-1)/2},$$

while taking the product of all eigenvalues of Z gives

$$\det Z = (-1)^b i^c (-i)^d p^{p/2} = i^{2b+c-d} p^{p/2}. \tag{10.38}$$

Hence

$$\det Z = \pm i^{p(p-1)/2} p^{p/2}. \tag{10.39}$$

In order to determine the correct sign in (10.39), we put $\zeta := e^{\pi i/p}$ and use that Z is a Vandermonde matrix. We get

$$\begin{aligned}\det Z &= \prod_{r<s} (\xi^s - \xi^r) = \prod_{r<s} (\zeta^{2s} - \zeta^{2r}) = \prod_{r<s} \zeta^{r+s}(\zeta^{s-r} - \zeta^{-(s-r)}) \\ &= \prod_{r<s} \left(\zeta^{r+s} \cdot 2i \sin \tfrac{(s-r)\pi}{p}\right) = (2i)^{p(p-1)/2} \prod_{r<s} \sin \tfrac{(s-r)\pi}{p},\end{aligned}$$

where r and s are always in the range $0, \ldots, p-1$. Note that the final equality follows from $\prod\limits_{r<s} \zeta^{r+s} = 1$, which in turn holds as

$$\begin{aligned}\sum_{r<s} (r+s) &= \sum_{s=1}^{p-1} \sum_{r=0}^{s-1} (r+s) = \sum_{s=1}^{p-1} \left(s^2 + \frac{s(s-1)}{2}\right) \\ &= \frac{3}{2} \cdot \frac{(p-1)p(2p-1)}{6} - \frac{(p-1)p}{4} = 2p\left(\frac{p-1}{2}\right)^2\end{aligned}$$

is a multiple of $2p$ and $\zeta^{2p} = 1$. Since $\sin\big((s-r)\pi/p\big)$ is positive for $0 \le r < s \le p-1$, we must have the plus sign in (10.39). Then a comparison with (10.38) shows

$$2b + c - d \equiv \frac{p(p-1)}{2} \mod 4.$$

Together with (10.35) and (10.37), we obtain

$$a - b = \frac{p+1}{2} - 2b \equiv \frac{p+1}{2} - \frac{p-1}{2} = 1 \bmod 4 \quad \text{if } p \equiv 1 \bmod 4,$$

and

$$c - d = \frac{p-1}{2} + 2b \equiv -\frac{p-1}{2} + \frac{p+1}{2} = 1 \bmod 4 \quad \text{if } p \equiv 3 \bmod 4.$$

These two congruences show that we indeed have the plus sign in (10.37) in both cases. Together with (10.35), this gives

$$a = \frac{p+3}{4} \quad \text{and} \quad b = c = d = \frac{p-1}{4} \qquad \text{for} \quad p \equiv 1 \bmod 4,$$

and

$$a = b = c = \frac{p+1}{4} \quad \text{and} \quad d = \frac{p-3}{4} \qquad \text{for} \quad p \equiv 3 \bmod 4,$$

as claimed. This finishes the proof of the assertion and also establishes Theorem 10.5.4. □

Remark 10.5.6. A somewhat simplified proof based on Schur's method was given by Waterhouse [399] and is also presented in Theorem 5.15 of Lidl and Niederreiter [243]. We have preferred the slightly more involved version because of the interesting connection to the eigenvalues of the DFT-matrix Z. An alternative proof due to Kronecker may be found in the book of Ireland and Rosen [203, §6.3].

We also remark that Theorem 10.5.4 – and its analogue for residue rings which we will discuss next – have interesting applications in Algebraic Number Theory, more precisely, to class numbers of quadratic number fields; see, for instance, Borevich and Shafarevich [46]. □

In Algebraic Number Theory, one also studies the characters of both the additive group and the multiplicative group of units for the rings $\mathbb{Z}_n$ of residues modulo n. Not surprisingly, one may also investigate Gauss sums in this more general setting; see, for instance, Hasse's book [180, Chapter 15]. In particular, there is the following analogue of Equation (10.34) which is already due to Gauss [133] for odd n:

Result 10.5.7 *Let n be a positive integer and put $\xi := e^{2\pi i/n}$. Then*

$$\sum_{k=0}^{n-1} \xi^{k^2} = \begin{cases} (1+i)\sqrt{n} & \text{if } n \equiv 0 \bmod 4, \\ \sqrt{n} & \text{if } n \equiv 1 \bmod 4, \\ 0 & \text{if } n \equiv 2 \bmod 4, \\ i\sqrt{n} & \text{if } n \equiv 3 \bmod 4. \end{cases} \qquad \square$$

A proof is given in Landau's book [228], and Gauss' original proof for odd n can be found in Nagell [299, pp.174–180]. Again, Result 10.5.7 is essentially equivalent to the following analogue of Theorem 10.5.5 giving the eigenvalues of the matrix of the DFT for arbitrary n; this is due to Carlitz [67].

Result 10.5.8 *Let n be a positive integer and put $\xi := e^{2\pi i/n}$. Then the characteristic polynomial of the DFT-matrix Z as in (7.20) is given by*

$$\begin{cases} (x-\sqrt{n})^{\frac{n+4}{4}}(x-i\sqrt{n})^{\frac{n}{4}}(x+\sqrt{n})^{\frac{n}{4}}(x+i\sqrt{n})^{\frac{n-4}{4}} & \text{if } n \equiv 0 \bmod 4, \\ (x-\sqrt{n})^{\frac{n+3}{4}}(x-i\sqrt{n})^{\frac{n-1}{4}}(x+\sqrt{n})^{\frac{n-1}{4}}(x+i\sqrt{n})^{\frac{n-1}{4}} & \text{if } n \equiv 1 \bmod 4, \\ (x-\sqrt{n})^{\frac{n+2}{4}}(x-i\sqrt{n})^{\frac{n-2}{4}}(x+\sqrt{n})^{\frac{n+2}{4}}(x+i\sqrt{n})^{\frac{n-2}{4}} & \text{if } n \equiv 2 \bmod 4, \\ (x-\sqrt{n})^{\frac{n+1}{4}}(x-i\sqrt{n})^{\frac{n+1}{4}}(x+\sqrt{n})^{\frac{n+1}{4}}(x+i\sqrt{n})^{\frac{n-3}{4}} & \text{if } n \equiv 3 \bmod 4. \end{cases} \qquad \square$$

Remark 10.5.9. The eigenvectors of the DFT-matrix are also known; they were determined independently by McClellan and Parks [261] and by Morton [289]. An interesting survey discussing various aspects of the Discrete Fourier Transform – for instance, the relation to Gauss sums, several ways of proving Result 10.5.7, as well as connections to pure mathematics on one hand and computational questions on the other hand – was given by Auslander and Tolimieri [15]. □

The final result of this section was first proved in 1890 by Stickelberger [364]. His result is similar in spirit to Theorem 10.5.4, as it also evaluates certain Gauss sums explicitly. While it leaves more choice regarding the order of the multiplicative character used, it restricts the order of the underlying field to squares. Our presentation follows the proof given in the book of Lidl and Niederreiter [243].

Theorem 10.5.10 (Stickelberger's theorem). *Let L be a splitting field for $K = \mathrm{GF}(q^2)$, let ψ be any non-trivial multiplicative character of an order m dividing $q+1$, and let χ_1 be the canonical additive character of K into L^*. Then*

$$G(\psi,\chi_1) = \begin{cases} -q & \text{if } m \text{ is even and } (q+1)/m \text{ is odd,} \\ q & \text{otherwise.} \end{cases}$$

Proof. Let ω be a primitive element of K^* and put $w := \omega^{q+1}$. Then w is a primitive element for $F = \mathrm{GF}(q)$, and every element $\alpha \in K^*$ has a unique representation

$$\alpha = w^j \omega^k \quad \text{with } 0 \le j \le q-2 \text{ and } 0 \le k \le q.$$

By hypothesis, $\psi(w) = \psi(\omega^{q+1}) = 1$, and therefore

$$G(\psi,\chi_1) = \sum_{j=0}^{q-2}\sum_{k=0}^{q} \psi(w^j\omega^k)\chi_1(w^j\omega^k) = \sum_{k=0}^{q} \psi(\omega)^k \Big(\sum_{b\in F^*} \chi_1(b\omega^k)\Big). \tag{10.40}$$

Let us denote the canonical additive character of F by τ_1. As noted in Example 10.5.2, τ_1 lifts to χ_1, so that $\chi_1(b\omega^k) = \tau_1\big(\mathrm{Tr}_{K/F}(b\omega^k)\big)$. Then Corollary 10.2.3 gives

$$\sum_{b\in F^*} \chi_1(b\omega^k) = \sum_{b\in F^*} \tau_1\big(b\mathrm{Tr}_{K/F}(\omega^k)\big) = \begin{cases} -1 & \text{if } \mathrm{Tr}_{K/F}(\omega^k) \ne 0, \\ q-1 & \text{otherwise.} \end{cases} \tag{10.41}$$

Now $\mathrm{Tr}_{K/F}(\omega^k) = \omega^k + \omega^{kq}$, and hence $\mathrm{Tr}_{K/F}(\omega^k) = 0$ if and only if $\omega^{k(q-1)} = -1$.

Assume first that q is odd. Then the condition $\omega^{k(q-1)} = -1$ gives $k = (q+1)/2$. Therefore, by (10.40) and (10.41),

$$\begin{aligned} G(\psi,\chi_1) &= (q-1)\psi(\omega)^{(q+1)/2} - \sum_{\substack{k=0 \\ k\ne (q+1)/2}}^{q} \psi(\omega)^k \\ &= q\psi(\omega)^{(q+1)/2} - \sum_{k=0}^{q} \psi(\omega)^k = q\psi(\omega)^{(q+1)/2}, \end{aligned}$$

as $\psi(\omega) \neq 1$ and $\psi(\omega^{q+1}) = 1$. Note that $\psi(\omega)^{(q+1)/2} = -1$ if and only if m is even and $(q+1)/m$ is odd. This proves the assertion when q is odd.

Finally, let q be even. Then $\omega^{k(q-1)} = 1$ implies $k = 0$, and we obtain

$$G(\psi, \chi_1) = (q-1) + \sum_{k=1}^{q} \psi(\omega)^k = q + \sum_{k=0}^{q} \psi(\omega)^k = q.$$

Thus the assertion holds also in this case. □

As an application of Gauss sums, we have already seen the proof of the quadratic residue law in Section 10.4. The setup for an application to the existence problem for primitive elements of $\mathrm{GF}(q^n)$ over $\mathrm{GF}(q)$ with prescribed trace was outlined at the end of Section 10.2 and will be presented in detail in Chapter 14.

There are many further interesting applications of Gauss sums, not only in the study of finite fields, but also to Number Theory, Coding Theory and Design Theory. The study of such sums and more general character and exponential sums is a vast topic; a detailed treatment is outside the scope of this book. The interested reader may consult Chapter 6 of the *Handbook of finite fields* [292] and the references given there. Here we just mention a few advanced monographs dealing with this topic: Berndt, Evans and Williams [31], Katz [218], Konyagin and Shparlinski [223] and Korobov [225].

Exercises

Exercise 10.5.11. Use Stickelberger's theorem to give an alternative proof for Theorem 10.5.4 in the special case where n is even. □

10.6 The Discrete Fourier Transform, II

This section deals with a generalization of the Discrete Fourier Transform studied in Section 7.4 to finite abelian groups in general. For this, we require the concept of a group algebra, which we now briefly describe. In this context (and throughout the entire section), groups will be written multiplicatively.

Definition 10.6.1. Let G be a (multiplicatively written) group of order n, and let E be a field. We consider the free E-vector space EG with basis G, that is, the set of all formal sums $\sum_{g\in G} a_g g$, where all a_g come from E, together with the obvious coordinatewise addition and scalar multiplication:

$$\sum_{g\in G} a_g g + \sum_{g\in G} b_g g := \sum_{g\in G} (a_g + b_g) g \quad \text{and} \quad c \cdot \Big(\sum_{g\in G} a_g g\Big) := \sum_{g\in G} c a_g g.$$

We also define a multiplication on EG as follows:

$$\Big(\sum_{g\in G} a_g g\Big)\cdot\Big(\sum_{h\in G} b_h h\Big) := \sum_{k\in G}\Big(\sum_{\substack{g,h\in G\\ gh=k}} a_g b_h\Big)k.$$

Simple computations show that this turns the vector space EG into an E-algebra, the **group algebra** of G over E. □

By abuse of notation, we use the following conventions resulting in more legible formulas. First of all, we identify the units of E, G and EG and denote them by 1. Consequently, the element of EG obtained by multiplying the unit element of G with the element $c \in E$ is denoted by c, which just amounts to embedding E into EG in the natural way. Similarly, a positive integer z has to be interpreted as the sum of z copies of the field unit 1 when considered as an element of EG, and $-z$ stands for the sum of z copies of -1.

From now on, we assume that G is abelian and that E is a splitting field for G. By Theorem 10.1.2, the set of all characters χ of G into E^* forms a group $\widehat{G}$ isomorphic to G. Note that each χ extends to a homomorphism from the group algebra EG into E by linearity, where we put $\chi(0) := 0$ (generalizing what we did for multiplicative characters of a finite field before):

$$\chi\Big(\sum_{g\in G} a_g g\Big) := \sum_{g\in G} a_g\chi(g). \tag{10.42}$$

As an example, let us reconsider the Discrete Fourier Transform studied in Section 7.4 in the light of the concepts just introduced:

Example 10.6.2. Let G be a cyclic group of order n, and let g be a generating element for G. Then the elements of the group algebra EG may be written in the form

$$A := \sum_{i=0}^{n-1} a_i g^i, \tag{10.43}$$

and identifying powers of g and powers of x shows that the group algebra EG is isomorphic to the polynomial algebra $R := E[x]/(x^n - 1)$. Under this isomorphism, the element A corresponds to the polynomial[3]

$$a(x) := a_{n-1}x^{n-1} + \cdots + a_1x + a_0$$

in R. We now choose a primitive n-th root of unity ζ in E^*. Then every character $\chi : G \to E^*$ maps g to a unique power of ζ; let us label the characters in such a way that χ_j maps g to ζ^j. Using (10.42), one sees that the group algebra element A defined in (10.43) is mapped to

$$\chi_j(A) = \sum_{i=0}^{n-1} a_i\chi_j(g^i) = \sum_{i=0}^{n-1} a_i\zeta^{ij} = a(\zeta^j).$$

[3] As before, we shall denote both the polynomial f in $E[x]$ and its image in R by the same symbol and refer to the elements of R simply as polynomials.

Thus the character values $\chi(A)$ are just the Fourier coefficients A_j introduced in Definition 7.4.1. □

In view of Example 10.6.2, there is a natural generalization of the Discrete Fourier Transform to finite abelian groups:

Definition 10.6.3. Let G be an abelian group of order n, and let E be a splitting field for G. Label the n characters of G as $\chi_0, \dots, \chi_{n-1}$; for the sake of consistency, χ_0 is taken to be the trivial character. Given any group algebra element $A \in EG$, we define the **Fourier vector** of A by

$$\mathbf{A} := \big(\chi_0(A), \chi_1(A), \dots, \chi_{n-1}(A)\big).$$

As in the cyclic case, this transformation from EG into E^n is called the **Discrete Fourier Transform** or, for short, the **DFT** of G. □

We then have the following analogue of Theorem 7.4.2:

Theorem 10.6.4. *Let G be an abelian group of order n, let E be a splitting field for G, and let $\chi_0, \dots, \chi_{n-1}$ be all characters from G to E^*. Then the DFT introduced in Definition 10.6.3 is an isomorphism from the group algebra EG to the algebra E^n (with componentwise operations).*

Proof. As in the proof of Theorem 7.4.2, the only part of the assertion which is not entirely obvious is the fact that the DFT is a bijection. However, this is immediate from the following analogue of Corollary 7.4.3, which allows one to recover any group algebra element from its Fourier vector. □

Lemma 10.6.5 (Inversion formula). *Let E be a splitting field for the finite abelian group G, and let $A = \sum_{g \in G} a_g g \in EG$. Then one may recover the coefficients of A from the character values $\chi(A)$ as follows:*

$$a_g = \frac{1}{|G|} \sum_{\chi \in \widehat{G}} \chi(A)\chi(g^{-1}).$$

In particular, if $A, B \in EG$ satisfy the condition $\chi(A) = \chi(B)$ for all characters χ of G, then necessarily $A = B$.

Proof. This is a simple consequence of the orthogonality relations for characters in Theorem 10.1.6:

$$\begin{aligned} \frac{1}{|G|} \sum_{\chi \in \widehat{G}} \chi(A)\chi(g^{-1}) &= \frac{1}{|G|} \sum_{\chi \in \widehat{G}} \Big(\sum_{h \in G} a_h \chi(h) \Big) \chi(g^{-1}) \\ &= \frac{1}{|G|} \sum_{h \in G} \Big(a_h \sum_{\chi \in \widehat{G}} \chi(h)\chi(g^{-1}) \Big) = a_g. \quad \square \end{aligned}$$

Remark 10.6.6. In Section 9.8, we obtained two results concerning the linear complexity of a periodic sequence $\mathbf{a} = (a_k)$ with period n over $F = \mathrm{GF}(q)$ by using the cyclic version of the DFT over a splitting field E for the cyclic group G of order n. In order to generalize these results to the abelian case, we need to decide what notion should replace that of a periodic sequence.

In the proof of Theorem 9.8.1, we have seen that the linear complexity of $\mathbf{a}$ equals the rank of the circulant matrix with first row $(a_0, \dots, a_{n-1})$ over E which is associated to the polynomial $a(x) \in R$ as in Example 10.6.2; let us denote this matrix by C here (since we now use the symbol A to denote the element of the group algebra EG corresponding to $a(x)$ as in Example 10.6.2). Then the rows of C correspond to the elements Ah $(h \in G)$ of EG, and hence the rank of C is just the dimension of the principal ideal (A) generated by A in EG, considered as a vector space over E.

Thus the appropriate generalization of periodic sequences $\mathbf{a}$ are just the elements A of the group algebra EG (where G now is an arbitrary abelian group of order n), and the notion of the linear complexity of $\mathbf{a}$ should be generalized to the dimension of the ideal generated by A in EG. □

These observations motivate the following definition:

Definition 10.6.7. Let E be a splitting field for the finite abelian group G, and let $A \in EG$. Then the dimension of the ideal (A) is called the p-**rank** of A, where p is the characteristic of E. For $p = 0$, we simply speak of the **rank** of A. □

We remark that the preceding definition agrees with the terminology for difference sets and symmetric designs used in Design Theory; see, for instance, Jungnickel [210, Section 15].

In view of Example 10.6.2, the following generalization of Theorem 9.8.1 is not surprising; indeed, the proof we gave for the special case can be adapted to the more general situation.

Theorem 10.6.8. *Let G be a finite abelian group of order n, and let F be any field whose characteristic p does not divide n (note that $p = 0$ is allowed). Moreover, let E be some extension field of F which is a splitting field for G, and let $A = \sum_{g \in G} a_g g$ be an element of the group algebra FG. Then the p-rank of A equals the number of characters $\chi : G \to E$ with $\chi(A) \neq 0$.*

Proof. Define a square matrix $C = (c_{g,h})$ indexed by the elements of G by putting

$$c_{g,h} := a_{hg^{-1}} \quad \text{for all } g, h \in G.$$

Thus the rows of C correspond to the elements $Ah \in FG$ with $h \in G$, and hence the p-rank of A equals the rank of C over F. Note that this agrees with the rank of C over E. For each character χ of G, let $\mathbf{e}_\chi = (\chi(g))_{g \in G}$ be the column vector using the same indexing as for the matrix C. Then the g-entry of $C\mathbf{e}_\chi$ turns out to be

$$\sum_{h \in G} c_{g,h}\chi(h) = \sum_{h \in G} a_{hg^{-1}}\chi(hg^{-1})\chi(g) = \chi(A)\chi(g),$$

which means that the vector $\mathbf{e}_\chi$ is an eigenvector with eigenvalue $\chi(A)$ for C. It follows from Theorem 3.8.3 (but is also easily checked directly) that the n eigenvectors $\mathbf{e}_\chi$ with $\chi \in \widehat{G}$ are linearly independent. Therefore C is similar to the diagonal matrix $\mathrm{diag}\big(\chi_0(A), \chi_1(A), \dots, \chi_{n-1}(A)\big)$ defined by the n Fourier coefficients of A, which establishes the assertion. □

Remark 10.6.9. Note that the matrix C introduced in the proof of Theorem 10.6.8 satisfies the condition $c_{g,h} = c_{gk,hk}$ for all $g, h, k \in G$. Such matrices are called **group invariant** and generalize the circulant matrices studied in Section 7.3. □

Exercises

Exercise 10.6.10. Verify that the group algebra EG introduced in Definition 10.6.1 is indeed an E-algebra. □

Exercise 10.6.11. Give a generalization of Theorem 9.8.2 to the DFT for arbitrary abelian groups. □

10.7 Some Applications of Characters to Difference Sets

We conclude this chapter with some applications of characters and the DFT to the theory of difference sets – and thus, in the special case of cyclic groups, also to periodic binary sequences. Of course, we have to restrict ourselves to a few illustrative examples, since the theory of difference sets is a vast area.

We first show that difference sets are equivalent to solutions of certain equations in suitable group algebras. For this, we need some additional notation. Let G be a (multiplicatively written) group of order v, and let E be a field. Given any subset S of G, we will – by a convenient abuse of notation – use the same symbol S to denote also the corresponding element of the group algebra EG:

$$S := \sum_{g \in S} g;$$

in particular, the formal sum of all group elements is simply denoted by G again. Now let t be any integer; then one puts

$$A^{(t)} := \sum_{g \in G} a_g g^t \quad \text{for } A = \sum_{g \in G} a_g g.$$

With these conventions, we have the following simple but fundamental result:

Lemma 10.7.1. *Let D be a k-subset of a multiplicatively written group G of order v, and let E be a field. If D is a (v,k,λ)-difference set in G, the following identity holds in EG:*

$$DD^{(-1)} = k - \lambda + \lambda G, \tag{10.44}$$

The converse also holds, provided that E has characteristic 0.

Proof. Since G is written multiplicatively, the defining condition (9.27) for a difference set D now takes the form of an assertion about the list of quotients of elements in D. Note that the left hand side of Equation (10.44) is just the formal sum of all quotients de^{-1} which may be formed from D. Clearly, we get the unit element 1 exactly k times, whereas each element $g \neq 1$ appears precisely λ times. This gives the validity of (10.44). The converse is similar and left to the reader. □

In most applications, one takes for E the field $\mathbb{C}$ of complex numbers, an algebraic number field (usually a cyclotomic field), or a finite field of characteristic p dividing the order $n = k - \lambda$ of the difference set. The following consequence of Lemma 10.7.1 indicates why Algebraic Number Theory is an important tool in the study of difference sets.

Proposition 10.7.2. *Let D be a k-subset of a multiplicatively written abelian group G of order v satisfying $k(k-1) = \lambda(v-1)$. Then D is a (v,k,λ)-difference set if and only if the following condition holds for all complex characters χ of G:*

$$\left|\chi(D)\right|^2 = \begin{cases} k - \lambda & \text{if } \chi \neq \chi_0, \\ k^2 & \text{if } \chi = \chi_0. \end{cases} \tag{10.45}$$

Proof. Recall that the hypothesis $k(k-1) = \lambda(v-1)$ is a necessary condition for the existence of a (v,k,λ)-difference set; see Lemma 9.5.6. By Theorem 10.1.6, $\chi(G) = 0$. Also,

$$\chi\left(D^{(-1)}\right) = \sum_{d \in D} \chi(d^{-1}) = \sum_{d \in D} \overline{\chi(d)} = \overline{\chi(D)},$$

and hence

$$\chi(DD^{(-1)}) = \chi(D)\chi(D^{(-1)}) = \left|\chi(D)\right|^2.$$

Therefore, applying χ to Equation (10.44) yields (10.45). Conversely, by Lemma 10.6.5, (10.45) in turn implies (10.44). Thus the assertion is an immediate consequence of Lemma 10.7.1. □

Clearly, the number $\chi(D)$ appearing in (10.45) is an integral combination of complex v-th roots of unity and thus belongs to a **cyclotomic field**, that is, a field $\mathbb{Q}(\zeta)$, where ζ is a primitive v-th root of unity. Results relying on this observation are outside the scope of this book; nevertheless, this approach is indispensable in developing an advanced non-existence theory for difference sets. The interested reader should consult the monographs by Pott [325] and Schmidt [338] and – for a more recent method of this type – the paper by Leung and Schmidt [241] as well as the references given in these sources.

But even without the use of Algebraic Number Theory, a wealth of interesting results can be obtained by just using group algebras, characters, and the DFT; see

Jungnickel [210] for a survey of the theory of difference sets emphasizing these methods. As an illustration, we will apply Theorem 10.6.8 to determine the p-ranks of the Hadamard difference sets constructed in Theorems 9.5.13 and 9.5.15 also for those primes p which divide the order $n = k - \lambda$ of the difference set; recall that the cases where p does not divide n are covered by Proposition 9.6.10. We begin with the rather simple case of Paley-Hadamard difference sets, which is due to MacWilliams and Mann [254]:

Theorem 10.7.3. *Let $q = 4n - 1$ be a prime power, and let p be a prime dividing n. Then the Paley-Hadamard difference set given in Theorem 9.5.13 has p-rank $2n$.*

Proof. By construction, the difference set D in question consists of the non-zero squares in the field $F = \mathrm{GF}(q)$. By Proposition 3.2.13, -1 is a non-square and thus F is the disjoint union of the set D of squares, the set $-D$ of non-squares, and the zero element 0. Now let E be a (finite) splitting field of characteristic p for G, where G denotes the multiplicatively written additive group of F. Then the preceding observation translates into the equation

$$G = 1 + D + D^{(-1)}$$

in the group algebra EG, which yields

$$\chi(D) + \chi\left(D^{(-1)}\right) = -1 \tag{10.46}$$

for every non-trivial character χ of G. This equation shows that at least one of $\chi(D)$ and $\chi\left(D^{(-1)}\right)$ has to be distinct from 0 for every character $\chi \neq \chi_0$; for the trivial character, $\chi_0(D) = \chi_0(D^{(-1)}) = 2n - 1 \neq 0$ in E.

According to Proposition 9.6.12, both D and $D^{(-1)}$ have p-rank at most $2n$. By Theorem 10.6.8, the p-rank of D is just the number of characters χ of G with $\chi(D) \neq 0$, and hence there can be at most $2n$ characters χ with $\chi(D) \neq 0$; of course, the analogous observation also holds for $D^{(-1)}$. Combining all these facts shows $\chi(D) \neq 0$ for exactly $2n$ characters, which implies the assertion. □

In the case where q is a prime, Theorem 10.7.3 has the following immediate consequence for the associated periodic binary sequences, which complements the result obtained in Corollary 9.6.11 for odd values of n:

Corollary 10.7.4. *Let* **a** *be a Legendre sequence with period q, where q is a prime of the form $q = 4n - 1$ for an even integer n. Then* **a** *has linear complexity $2n$.* □

Example 10.7.5. Let $v = 2^n - 1$ be a Mersenne prime. As already mentioned in Remark 9.6.19, the Legendre sequence of period v and the complementary sequence of the m-sequence of the same period – which both belong to difference sets with parameters (9.31) – can be distinguished by their linear complexities. Indeed, by Corollary 9.6.15, the complementary sequence of the m-sequence has linear complexity $n+1$, and by Corollary 10.7.4, the Legendre sequence has linear complexity 2^{n-1}. □

The analogue of Theorem 10.7.3 for twin prime power difference sets is due to Pott [324] and rather more involved:

Theorem 10.7.6. *Let q and $q+2$ be odd prime powers, put $n := (q+1)^2/4$, and let p be a prime dividing n. Then the Hadamard difference set of twin prime power type constructed in Theorem 9.5.15 has p-rank $2n$.*

Proof. Recall that the difference set D in question was defined in the direct sum G of the additive groups of $F = \mathrm{GF}(q)$ and $K = \mathrm{GF}(q+2)$ as follows:

$$D = \{(x,y) \in F \times K \colon y = 0 \text{ or both } x \text{ and } y \text{ are either squares or non-squares}\}.$$

Of course, in order to use group rings and characters in our proof, we will have to switch to multiplicative notation for G.

As in the proof of Theorem 10.7.3, we choose a splitting field E of characteristic p for G and apply Theorem 10.6.8. Let us write C_0 and C_1 for the elements of the group algebra EG which correspond to the set of squares and the set of non-squares in F, respectively. The analogous sets for K will be denoted by C_0' and C_1'. With this notation, we may write D as the element

$$D = C_0C_0' + C_1C_1' + H \in EG, \tag{10.47}$$

where H denotes the multiplicatively written additive group of the field F. In order to apply Theorem 10.6.8, we need to determine the number of characters χ of G with $\chi(D) \neq 0$. As in the proof of Theorem 10.7.3, it will actually not be necessary to determine the values $\chi(D)$ exactly.

Since G is – in multiplicative language – the direct product of H and the multiplicatively written additive group L of the field K, the corresponding character groups satisfy the analogous relation

$$\widehat{G} = \widehat{H} \times \widehat{L}, \tag{10.48}$$

which is clear from the proof of Theorem 10.1.2. Because of (10.48), the characters χ of G have the form $\chi = \beta\gamma$, where β and γ run over the characters of H and L, respectively. Fortunately, it turns out that these characters cannot take too many different values on D. To see this, let h be an element of F^*. For each non-trivial character β of H, define a mapping β_h by

$$\beta_h(x) := \beta(h \circ x) \quad \text{for every } x \in H, \tag{10.49}$$

where we use the symbol $\circ$ to denote the multiplication in the field F, as the multiplication in H corresponds to the addition in F. It is easy to check that β_h is again a non-trivial character of H, since $\circ$ is distributive over the multiplication in H; for reasons of cardinality, we obtain all non-trivial characters of H in this way. (Essentially, this is just a different way of phrasing Proposition 10.2.1.) Note that

$$\begin{array}{llll} \text{either} & h\circ C_0 = C_0 & \text{and} & h\circ C_1 = C_1, \\ \text{or} & h\circ C_0 = C_1 & \text{and} & h\circ C_1 = C_0, \end{array}$$

depending on whether h is a square or a non-square in F. Using (10.49), we conclude that the non-trivial characters of H can only take two distinct values on the elements C_0 and C_1 of EG. Moreover, (10.46) shows that these two values have to add to -1. Hence there exists an element $c \in E$ such that

$$\begin{cases} \beta_h(C_0) = c \text{ and } \beta_h(C_1) = -1-c & \text{if } h \text{ is a square in } F^*, \\ \beta_h(C_0) = -1-c \text{ and } \beta_h(C_1) = c & \text{if } h \text{ is a non-square in } F^*. \end{cases} \tag{10.50}$$

Similarly, every non-trivial character of L can be written in the form γ_ℓ with $\ell \in K^*$, using a fixed non-trivial character γ of L, and there exists an element $d \in E$ such that

$$\begin{cases} \gamma_\ell(C_0') = d \text{ and } \gamma_\ell(C_1') = -1-d & \text{if } \ell \text{ is a square in } K^*, \\ \gamma_\ell(C_0') = -1-d \text{ and } \gamma_\ell(C_1') = d & \text{if } \ell \text{ is a non-square in } K^*. \end{cases} \tag{10.51}$$

In view of (10.48), we therefore need to consider exactly nine different types of characters of H, if we also take into account the trivial characters.

First let $\beta = \beta_0$ be trivial on H, but γ non-trivial on L. Then we obtain from (10.47), (10.51) and $\gamma(L) = 0$

$$(\beta_0\gamma)(D) = \frac{q-1}{2}\gamma(C_0') + \frac{q-1}{2}\gamma(C_1') + q = \frac{q-1}{2}\cdot(-1) + q = \frac{q+1}{2} = 0,$$

since p divides n. Similarly, if β is non-trivial on H but $\gamma = \gamma_0$ is trivial on L, we have

$$(\beta\gamma_0)(D) = \frac{q+1}{2}\beta(C_0) + \frac{q+1}{2}\beta(C_1) = \frac{q+1}{2}\cdot(-1) = 0.$$

Also, if both $\beta = \beta_0$ and $\gamma = \gamma_0$ are trivial, we just get the cardinality of D:

$$(\beta_0\gamma_0)(D) = 2n - 1 \neq 0.$$

It remains to consider the cases where both β and γ are non-trivial. Then there are just two possible values for $(\beta\gamma)(D)$, as the reader may easily check:

$$(\beta\gamma)(D) \in \{cd + (1+c)(1+d),\ -c(1+d) - (1+c)d\}, \tag{10.52}$$

and both possible values occur with multiplicity $(q^2-1)/2$. We can now apply Theorem 10.6.8. If both values in (10.52) were equal to 0 in E, the rank of D would be 1, which is clearly absurd. On the other hand, if both these values were distinct from 0, the rank of D would be q^2, contradicting Proposition 9.6.12. Hence exactly one of the two values in (10.52) has to be distinct from 0, which shows that the rank of D is indeed $1 + (q^2-1)/2 = 2n$. □

Finally, we note the following immediate consequence of the cyclic case of Theorem 10.7.6 for the associated periodic binary sequences; again, this complements the much simpler result for odd values of n obtained in Corollary 9.6.11.

Corollary 10.7.7. *Let both q and $q+2$ be odd primes, assume that $n := (q+1)^2/4$ is even, and let **a** be the periodic binary sequence with period $v = q(q+2)$ associated with the corresponding cyclic Hadamard difference set of twin prime power type. Then **a** has linear complexity $2n$.* □

Chapter 11
Normal Bases and Cyclotomic Modules

Abstract The aim of this chapter is to determine a normal element for every extension E/F of finite fields. For this, it will be convenient to work (theoretically) in an algebraic closure $\widehat{F}$ of a fixed ground field $F = \mathrm{GF}(q)$ with q elements. While this will simplify the presentation, it is not actually required for constructing normal elements in specific cases: then all that is really needed are finite extensions of F.

Recall from Chapter 4 that for every $n \in \mathbb{N}^*$ there is exactly one intermediate field E_n of $\widehat{F}/F$ with degree n over F. We will explicitly describe a normal element w_n for the extension E_n/F in terms of certain roots of unity whose multiplicative orders depend on q and n.

Moreover, we give an inductive construction for a trace-compatible sequence $(w_n)_{n\in\mathbb{N}^*}$ over $\widehat{F}$ for which every w_n is a normal element for E_n/F. Finally, we also present two reasonably efficient algorithms for computing a normal element for an arbitrary extension $E = \mathrm{GF}(q^n)$ of $F = \mathrm{GF}(q)$.

11.1 Cyclotomic Modules

In this section, we first introduce the most fundamental concepts required for the desired explicit construction of normal elements over $F = \mathrm{GF}(q)$. This will be followed by an outline of the entire chapter and by two introductory examples.

Observation 11.1.1. We start our investigation by recalling some notations and results from earlier chapters which are fundamental for the theory of normal bases for Galois field extensions:

(1) As in Chapter 4, we denote the Frobenius automorphism of $\widehat{F}/F$, where $\widehat{F}$ is the algebraic closure of F, by σ. Then the q-order $\mathrm{Ord}_q(v)$ of an arbitrary element $v \in \widehat{F}$ is the monic polynomial $g(x) \in F[x]$ of least degree such that $g(\sigma)(v) = 0$; see Notation 3.10.1.

(2) The possible q-orders are precisely the monic polynomials of $F[x]$ which are not divisible by x. Let $h(x)$ be such a polynomial. Since x is a unit modulo $h(x)$,

D. Hachenberger and D. Jungnickel, *Topics in Galois Fields*,
Algorithms and Computation in Mathematics 29,
https://doi.org/10.1007/978-3-030-60806-4_11

there is a minimal positive integer m, namely the order $\mathrm{ord}_{h(x)}(x)$ of x modulo $h(x)$, such that $x^m \equiv 1 \bmod h(x)$. If $v \in \widehat{F}$ has q-order $h(x)$, then v has degree m over F, that is, $F(v)$ is the m-dimensional extension E_m of F in $\widehat{F}$.

(3) Observation 3.10.3 clarifies the structure of the (F,σ)-submodules of $\widehat{F}$ (that is, the finite σ-invariant subspaces): every monic $h(x) \in F[x]$ with $\gcd(h(x),x) = 1$ corresponds to the (F,σ)-submodule

$$V_h^F := \{v \in \widehat{F} \colon h(\sigma)(v) = 0\}.$$

Obviously, $v \in V_h^F$ if and only if the q-order of v divides $h(x)$. Moreover, $\mathrm{Ord}_q(v) = h(x)$ if and only if v generates V_h^F as an (F,σ)-module, that is, when

$$V_h^F = \{f(\sigma)(v) \colon f(x) \in F[x]\}.$$

Throughout this chapter, we will denote the set of all generators of V_h^F by $G_q(h)$; this replaces the notation Ω_h used in Definition 3.11.7, which had not indicated the value of q. The F-dimension of V_h^F equals $\deg h$, and one has $|G_q(h)| = \phi_q(h)$, where $\phi_q(h)$ is the number of units in the residue ring $F[x]/(h)$. In particular, any such submodule is cyclic.

(4) The extension fields E_m of F constitute a special class of (F,σ)-submodules of $\widehat{F}$, namely the submodules of the form

$$E_m = V_{x^m-1}^F,$$

and the $\phi_q(x^m-1)$ generators of E_m as such a submodule coincide with the normal elements for E_m/F; see Theorem 3.10.2. □

We now introduce the notion of a cyclotomic module over $F = \mathrm{GF}(q)$, which turns out to be fundamental for our description of normal elements. As before, we let p denote the characteristic of F and write m' for the p-free part of any positive integer m, that is, for the largest divisor of m which is not divisible by p.

Definition 11.1.2. Let $m = m'p^a \in \mathbb{N}^*$. Then the finite (F,σ)-submodule of $\widehat{F}$ corresponding to the p^a-th power of the cyclotomic polynomial $\Phi_{m'}(x)$, namely

$$\mathscr{C}_m^F := \{v \in \widehat{F} \colon \Phi_{m'}(\sigma)^{p^a}(v) = 0\},$$

will be called the **cyclotomic module** corresponding to m over F. □

As we shall see, $\mathscr{C}_m^F$ is closely related to the cyclotomic module $\mathscr{C}_{m'}^F$, since we have the identity

$$\Phi_k(x)^{p^a} = \Phi_k(x^{p^a}), \tag{11.1}$$

which holds as every cyclotomic polynomial over F has coefficients in the prime field of F, by Theorem 3.6.10.

Remark 11.1.3. As pointed out in Observation 3.10.4, the factorization

$$x^m - 1 = (x^{m'} - 1)^{p^a} = \prod_{k|m'} \Phi_k(x)^{p^a}$$

gives rise to a decomposition of E_m into the direct sum of the corresponding cyclotomic modules:

$$E_m = \bigoplus_{k|m'} \mathscr{C}^F_{kp^a}.$$

Now let $w \in E_m$, and decompose w accordingly:

$$w = \sum_{k|m'} w_k \quad \text{with } w_k \in \mathscr{C}^F_{kp^a} \text{ for all } k \mid m'.$$

Then w is normal for E_m/F if and only if each w_k generates $\mathscr{C}^F_{kp^a}$ as an (F,σ)-module, which holds if and only if $\mathrm{Ord}_q(w_k) = \Phi_k(x)^{p^a}$ for every divisor k of m'. This induces a corresponding additive decomposition

$$G_q(x^m - 1) = \sum_{k|m'} G_q(\Phi_k^{p^a}) \tag{11.2}$$

of the set of all normal basis generators for the m-dimensional extension E_m/F. □

Thus we wish to determine, for every $m \in \mathbb{N}^*$, an element $u_m \in \widehat{F}$ which has q-order $\Phi_{m'}(x)^{m/m'}$. When m is not divisible by p, this will essentially be achieved by explicitly describing such an element u_m in terms of certain roots of unity whose multiplicative orders depend on the specified values of q and m. Then an appropriate additive composition of such elements u_m together with a reduction argument (see Section 11.3) will result in a normal element for a specific finite extension over F as mentioned in Remark 11.1.3. Some fundamental ideas concerning this topic can be traced back to Semaev [346], but our exposition mainly follows Hachenberger [161]. We now outline the major steps:

- We will begin our investigation by providing a product construction for generators of cyclotomic modules in Section 11.2. Theorem 11.2.1 implies

$$G_q(x^{p^a} - 1) \cdot G_q(\Phi_k) \subseteq G_q(\Phi_k(x)^{p^a}) \tag{11.3}$$

for all $k \in \mathbb{N}^*$ which are not divisible by p and for all $a \in \mathbb{N}$. We also derive the following identity:

$$G_q(\Phi_k(x)^{p^a}) = u \cdot G_q(\Phi_k) + V^F_{\Phi_k(x)^{p^a-1}}, \tag{11.4}$$

which holds for all $k \in \mathbb{N}^*$ which are not divisible by p, for all $a \in \mathbb{N}$ and for all $u \in G_q(x^{p^a} - 1)$, that is, for all u which are normal elements for E_{p^a}/F. Moreover,

$$G_q(\Phi_k) \cdot G_q(\Phi_\ell) \subseteq G_q(\Phi_{k\ell}), \tag{11.5}$$

for all $k, \ell \in \mathbb{N}^*$ which are relatively prime and not divisible by p; see Theorem 11.2.3.

- In Section 11.3, we are going to study normal elements for extensions E_n over $F = \mathrm{GF}(q)$, where n is a power of the characteristic p of F. In view of the foregoing remarks, this (in principle) reduces the problem to the construction of generators for cyclotomic modules of the form $\mathscr{C}_{r^\ell}^F$, where r is a prime which is distinct from the characteristic p of F.
- In Section 11.4, we introduce the class of *strongly regular extensions* of F. This will allow us an explicit description of generators for all cyclotomic modules $\mathscr{C}_n^F$ – and, consequently, of normal elements for all n-dimensional extensions over F – whenever the radical $\mathrm{rad}(n)$ divides $q-1$, where $n \not\equiv 0 \bmod 4$ when $q \equiv 3 \bmod 4$. In Section 11.5, we construct corresponding normal polynomials (see Definition 3.9.1) for all strongly regular extensions satisfying $q \equiv 1 \bmod 4$ when n is even.
- In Section 11.6, we first prove a useful fundamental result on additive orders. This will allow us to extend some of the results of Section 11.4 to the considerably larger class of *regular extensions*; that is, we study those extensions E_n of F where $\mathrm{ord}_{\mathrm{rad}(n')}(q)$ and n are relatively prime, and where $q \equiv 1 \bmod 4$ when n is even. In particular, this covers the extensions with prime power degree.
- Our results are then completed in Section 11.7 by considering extensions of degree r^ℓ of F, where r is a prime distinct from p, with emphasis on 2-power extensions when $q \equiv 3 \bmod 4$.
- All these results eventually even lead to the explicit description of trace-compatible sequences of normal elements for the algebraic closure $\widehat{F}$ of F (as introduced in Definition 4.5.1) in Section 11.8. See also Remark 11.1.4 below, where we provide a trace-compatible sequence of normal elements for the r-primary closure E_{r^∞} of F in $\widehat{F}$ for every prime $r \neq p$.
- After presenting two specific algorithms for determining normal elements for arbitrary Galois field extensions in Section 11.9, we will conclude this chapter with a few additional remarks in Section 11.10.

The following rather simple example should suffice to motivate a closer study of cyclotomic modules:

Remark 11.1.4. Let r be a prime which is distinct from the characteristic p of the underlying field $F = \mathrm{GF}(q)$, and let ℓ be any positive integer. As noted in Remark 11.1.3, the decomposition $(x-1)\Phi_r(x)\Phi_{r^2}(x)\cdots\Phi_{r^\ell}(x)$ of $x^{r^\ell}-1$ gives rise to the decomposition

$$E_{r^\ell} = F \oplus \mathscr{C}_r^F \oplus \mathscr{C}_{r^2}^F \oplus \cdots \oplus \mathscr{C}_{r^\ell}^F$$

of the r^ℓ-dimensional extension E_{r^ℓ} of F into a direct sum of cyclotomic modules. An application of Proposition 3.6.8 gives

$$\Phi_{r^j}(x) = \Phi_r\left(x^{r^{j-1}}\right) = \sum_{i=0}^{r-1} x^{ir^j}$$

for every positive integer j, which shows that the cyclotomic module $\mathscr{C}_{r^j}^F$ is the kernel of the $(E_{r^j}, E_{r^{j-1}})$-trace mapping. Since

$$E_{r^\ell} = E_{r^{\ell-1}} \oplus \mathscr{C}^F_{r^\ell} \quad \text{for all } \ell \geq 1,$$

every normal element w for E_{r^ℓ}/F has the form $w = u + v$, where u is a normal element for $E_{r^{\ell-1}}/F$ and where v is a generator of the cyclotomic module $\mathscr{C}^F_{r^\ell}$. These observations can be used to construct trace compatible sequences of normal elements for the r-**primary closure**

$$E_{r^\infty} := \bigcup_{\ell \in \mathbb{N}} E_{r^\ell}$$

of F in $\widehat{F}$ as follows. Choose an element $v_0 \in F^*$ (for instance, simply take $v_0 = 1$), let $(v_\ell)_{\ell \in \mathbb{N}^*}$ be a sequence in $\widehat{F}$ such that $\mathrm{Ord}_q(v_\ell) = \Phi_{r^\ell}(x)$ for every $\ell \in \mathbb{N}^*$, and put

$$w_\ell := \sum_{j=0}^{\ell} v_j \quad \text{for all } \ell \in \mathbb{N}^*.$$

Then every w_ℓ is normal in E_{r^ℓ} over F, and the $(E_{r^\ell}, E_{r^{\ell-1}})$-trace of w_ℓ equals

$$\mathrm{Tr}_{E_{r^\ell}/E_{r^{\ell-1}}}(w_{\ell-1} + v_\ell) = r \cdot w_{\ell-1},$$

since $\mathscr{C}^F_{r^\ell}$ is the kernel of the $(E_{r^\ell}, E_{r^{\ell-1}})$-trace mapping. Hence the sequence $(\overline{w}_\ell)_{\ell \in \mathbb{N}}$, where $\overline{w}_\ell$ is recursively defined by $\overline{w}_0 := v_0$ and

$$\overline{w}_\ell := \frac{1}{r} \cdot \overline{w}_{\ell-1} + v_\ell \quad \text{for all } \ell \in \mathbb{N}^*,$$

is a trace-compatible sequence of normal elements for the r-primary closure E_{r^∞} of F. □

The construction in Remark 11.1.4 will be an important ingredient when we determine trace-compatible sequences of normal elements for the entire algebraic closure $\widehat{F}$ of F in Section 11.9. A generalization of this construction is given in Exercise 11.1.6.

We conclude this section with a further introductory example:

Example 11.1.5. Let us consider the simplest non-trivial case, namely the quadratic extension $E = \mathrm{GF}(q^2)$ of $F = \mathrm{GF}(q)$. According to whether the characteristic p of F is 2 or an odd prime, the polynomial $x^2 - 1$ splits as $(x-1)^2$ or into the two distinct linear factors $x - 1$ and $x + 1$, respectively.

Let v be any element of E and note that $\mathrm{Ord}_q(v)$ divides $x - 1$ if and only if $v \in F$. This shows that *every* element of $E \setminus F$ is a normal element for E/F when $p = 2$. Now let p be an odd prime. Here an element $v \in E \setminus F$ has q-order $x + 1$ if and only if $v \neq 0$ and $\sigma(v) + v = 0$. This implies $v^{q-1} = -1$ and therefore $v^{2(q-1)} = 1$, which means that $\mathrm{ord}(v)$ divides $2(q-1)$ (but not $q - 1$). It is clear that $\lambda + v$ is a normal element for E/F for every element v with q-order $x + 1$ and for every choice of $\lambda \in F^*$ (for instance, for $\lambda = 1$). Let us also mention how v may be chosen explicitly:

- When $q \equiv 3 \bmod 4$, we can take v as a primitive fourth root of unity;
- and when $q \equiv 1 \bmod 4$, one may take for v a primitive 2^{c+1}-th root of unity, where $2^c = \mathrm{pt}_2(q-1)$ is the largest power of 2 dividing $q-1$ (in particular, $c \geq 2$).

For odd p, our arguments show that the normal elements for E/F are exactly the elements of E^* which are not contained in the subgroup $U_{2(q-1)}$ of order $2(q-1)$. In particular, every primitive element of E^* is normal for E/F; cf. Proposition 13.1.1. Also, any primitive 8-th root of unity is normal for E/F when $q \equiv 3 \bmod 4$. □

Exercises

Exercise 11.1.6. Consider a field extension $E = \mathrm{GF}(q^n)$ over $F = \mathrm{GF}(q)$, where n is not divisible by the characteristic p of F, let k be a divisor of n, and denote the intermediate field of E/F with degree k over F by K. Express the kernel of the (E,K)-trace mapping as a direct sum of cyclotomic modules. □

11.2 A Product Construction for Generators of Cyclotomic Modules

The topic of the present section is an extension of Theorem 3.9.7 to general cyclotomic modules. We begin by stating the product construction for normal elements given in that result in our present setting: if u and v are normal elements for E_k/F and E_ℓ/F, respectively, where k and ℓ are relatively prime, then uv is normal for $E_{k\ell}/F$. Using the notation introduced in part (3) of Observation 11.1.1, this may be written as follows:

$$G_q(x^k - 1) \cdot G_q(x^\ell - 1) \subseteq G_q(x^{k\ell} - 1). \tag{11.6}$$

Let us determine how many normal elements for $E_{k\ell}/F$ result from this simple construction. For this, let u_1, u_2 be normal for E_k/F, let v_1, v_2 be normal for E_ℓ/F, and suppose that $u_1 v_1 = u_2 v_2$. Then $u_1 u_2^{-1} = v_1^{-1} v_2$ is an element of $E_k^* \cap E_\ell^* = F^*$, and we obtain

$$\left| G_q(x^k - 1) \cdot G_q(x^\ell - 1) \right| = \frac{\phi_q(x^k - 1) \cdot \phi_q(x^\ell - 1)}{q-1} \tag{11.7}$$

for the cardinality of the product set $G_q(x^k - 1) \cdot G_q(x^\ell - 1)$. We remark that this is considerably smaller than the total number $\phi_q(x^{k\ell} - 1)$ of normal elements for $E_{k\ell}/F$ when $k, \ell > 1$; see Exercise 11.2.8.

The subsequent two theorems provide the desired extension of these results to general cyclotomic modules:

Theorem 11.2.1. *Consider a finite field $F = \mathrm{GF}(q)$ with characteristic p, let k be a positive integer which is not divisible by p, and let $a \in \mathbb{N}$.*

If u is a normal element for E_{p^a}/F and if v is an (F,σ)-generator of the cyclotomic module $\mathscr{C}_k^F$, then uv is an (F,σ)-generator of the cyclotomic module $\mathscr{C}_{kp^a}^F$.

Proof. For every proper divisor d of k, choose some element $v_d \in \widehat{F}$ with $\mathrm{Ord}_q(v_d) = \Phi_d(x)$, and put $v_k := v$. Then $w := \sum_{d|k} v_d$ is a normal element for the k-dimensional extension E_k over F, by Remark 11.1.3. Hence uw is a normal element for $E_{p^a k}$ over F, by Theorem 3.9.7.

Now note that the q-order of uv_d has to divide $\Phi_d(x)^{p^a} = \Phi_d(x^{p^a})$ for every divisor d of k, since

$$\Phi_d(\sigma)^{p^a}(uv_d) = \Phi_d(\sigma^{p^a})(uv_d) = u \cdot \Phi_d(\sigma^{p^a})(v_d) = u \cdot \Phi_d(\sigma)^{p^a}(v_d) = 0,$$

and therefore the q-order of $uw = \sum_{d|k} uv_d$ divides

$$\prod_{d|k} \Phi_d(x)^{p^a} = (x^k - 1)^{p^a} = x^{kp^a} - 1.$$

Since uw is a normal element for $E_{p^a k}$, we actually have $\mathrm{Ord}_q(uw) = x^{kp^a} - 1$, which forces $\mathrm{Ord}_q(uv_d) = \Phi_d(x)^{p^a}$ for all d, where we again use Remark 11.1.3. In particular, we have $\mathrm{Ord}_q(uv) = \Phi_k(x)^{p^a}$, as claimed. □

Remark 11.2.2. We remark that (11.3) is just another way of stating Theorem 11.2.1. In analogy to Equation (11.7), one checks that this product construction yields exactly

$$\left|G_q(x^{p^a} - 1) \cdot G_q(\Phi_k)\right| = \frac{\phi_q(x^{p^a} - 1) \cdot \phi_q(\Phi_k)}{q - 1} \tag{11.8}$$

(F,σ)-generators of the cyclotomic module $\mathscr{C}_{kp^a}^F$.

The identity for the set $G_q(\Phi_k(x)^{p^a})$ of all generators for this module stated in Equation (11.4) still requires a proof. An application of Proposition 3.11.12 with $g(x) = \Phi_k(x)^{p^a}$ and $f(x) = \Phi_k(x)$ shows that $G_q(\Phi_k(x)^{p^a})$ is the preimage of $G_q(\Phi_k)$ under the mapping

$$\Pi_f^g : V_{\Phi_k(x)^{p^a}}^F \to V_{\Phi_k(x)}^F, \quad v \mapsto \Phi_k(\sigma)^{p^a - 1}(v).$$

Also note that $G_q(\Phi_k(x)^{p^a})$ and the set $u \cdot G_q(\Phi_k) + V_{\Phi_k(x)^{p^a - 1}}^F$, where u is any normal element for E_{p^a}/F, have the same cardinality, namely

$$\phi_q(\Phi_k(x)^{p^a}) = \phi_q(\Phi_k) \cdot q^{(p^a - 1)\cdot \deg \Phi_k} = \phi_q(\Phi_k) \cdot \left|V_{\Phi_k(x)^{p^a - 1}}^F\right|.$$

Now let $v \in G_q(\Phi_k)$ and $y \in V_{\Phi_k(x)^{p^a - 1}}^F$. In view of $g(x)/f(x) = \Phi_k(x)^{p^a - 1}$, we see that

$$\Pi_f^g(uv + y) = \Pi_f^g(uv) + \Pi_f^g(y) = \Pi_f^g(uv)$$

has q-order $\Phi_k(x)$, as uv has q-order $\Phi_k(x)^{p^a}$. Hence both sets in question indeed coincide. □

Theorem 11.2.3. *Consider a finite field $F = \mathrm{GF}(q)$ with characteristic p, and let k and ℓ be two positive integers which are relatively prime and not divisible by p.*

If u and v are (F,σ)-generators of the cyclotomic modules $\mathscr{C}_k^F$ and $\mathscr{C}_\ell^F$, respectively, then uv is an (F,σ)-generator of $\mathscr{C}_{k\ell}^F$.

Proof. For every proper divisor d of k, choose some element $u_d \in \widehat{F}$ with $\mathrm{Ord}_q(u_d) = \Phi_d(x)$, and put $u_k := u$. Similarly, choose some $v_e \in \widehat{F}$ with $\mathrm{Ord}_q(v_e) = \Phi_e(x)$ for every proper divisor e of ℓ, and put $v_\ell := v$. Note that $u_d \in E_d \subseteq E_k$ and $v_e \in E_e \subseteq E_\ell$ for every $d \mid k$ and every $e \mid \ell$.

By Remark 11.1.3, the elements $y := \sum_{d|k} u_d$ and $z := \sum_{e|\ell} v_e$ are normal for E_k/F and E_ℓ/F, respectively. Thus yz is a normal element for $E_{k\ell}/F$, by Theorem 3.9.7.

Now consider a pair (d,e) with $d \mid k$ and $e \mid \ell$. Since d and e are relatively prime by our hypothesis on k and ℓ, Proposition 3.6.12 shows that $\Phi_d(x) = \mathrm{Ord}_q(u_d)$ divides $\Phi_d(x^e) = \prod_{t|e} \Phi_{dt}(x)$ and that $\Phi_e(x) = \mathrm{Ord}_q(v_e)$ divides $\Phi_e(x^d) = \prod_{s|d} \Phi_{es}(x)$. This yields

$$\Phi_d(\sigma^e)(u_d v_e) = v_e \cdot \Phi_d(\sigma^e)(u_d) = 0 \ \text{ and } \ \Phi_e(\sigma^d)(u_d v_e) = u_d \cdot \Phi_e(\sigma^d)(v_e) = 0,$$

and hence the q-order of $u_d v_e$ divides

$$\gcd\left(\Phi_d(x^e), \Phi_e(x^d)\right) = \Phi_{de}(x).$$

Therefore, the q-order of $yz = \sum_{d|k} \sum_{e|\ell} u_d v_e$ is a divisor of

$$\prod_{d|k} \prod_{e|\ell} \Phi_{de}(x) = \prod_{a|k\ell} \Phi_a(x) = x^{k\ell} - 1.$$

On the other hand, the q-order of yz has to be exactly $x^{k\ell} - 1$, since yz is a normal element for $E_{k\ell}/F$. In view of Remark 11.1.3, this implies $\mathrm{Ord}_q(u_d v_e) = \Phi_{de}(x)$ for all (d,e). In particular, we have $\mathrm{Ord}_q(uv) = \Phi_{k\ell}(x)$, as claimed. □

Remark 11.2.4. Note that (11.5) is just another way of stating Theorem 11.2.3. In analogy to Equations (11.7) and (11.8), this product construction yields exactly

$$\left| G_q(\Phi_k) \cdot G_q(\Phi_\ell) \right| = \frac{\phi_q(\Phi_k) \cdot \phi_q(\Phi_\ell)}{q-1}. \tag{11.9}$$

(F,σ)-generators of the cyclotomic module $\mathscr{C}_{k\ell}^F$. □

Let us conclude this section with a comparison of the product constructions we have encountered in a specific situation:

Example 11.2.5. Let $F = \mathrm{GF}(q)$, where $q = 2^s$ with s odd, and consider the 15-dimensional extension E_{15} of F. We first apply the original reduction theorem for normal elements. By Equations (11.6) and (11.7), $G_q(x^3-1) \cdot G_q(x^5-1)$ is a subset of cardinality

$$A := \frac{\phi_q(x^3-1)\cdot\phi_q(x^5-1)}{q-1} = \frac{(q-1)(q^2-1)\cdot(q-1)(q^4-1)}{q-1}$$

of $G_q(x^{15}-1)$, which yields exactly $(q-1)(q^2-1)(q^4-1)$ normal elements for the extension E_{15}/F.

On the other hand, we may also apply Remark 11.1.3 in conjunction with Theorem 11.2.3. This gives the inclusion

$$G_q(x-1)+G_q(\Phi_3)+G_q(\Phi_5)+G_q(\Phi_3)\cdot G_q(\Phi_5) \subseteq G_q(x^{15}-1)$$

and (using Equation (11.9)) results in

$$B := \phi_q(x-1)\cdot\phi_q(\Phi_3)\cdot\phi_q(\Phi_5)\cdot\frac{\phi_q(\Phi_3)\cdot\phi_q(\Phi_5)}{q-1} = (q^2-1)^2(q^4-1)^2$$

normal elements for E_{15}/F. Note that the quotient B/A equals

$$\frac{(q^2-1)^2(q^4-1)^2}{(q-1)(q^2-1)(q^4-1)} = (q+1)(q^4-1),$$

which indicates that the more elaborate construction based on Theorem 11.2.3 is considerably more powerful than the original reduction theorem for normal elements. For a general result in this direction, see Exercise 11.2.8.

We remark that even this more powerful construction still falls short of the total number

$$\phi_q(x^{15}-1) = (q-1)(q^2-1)(q^4-1)^3$$

of normal elements for E_{15}/F. □

Exercises

Exercise 11.2.6. Consider a finite field $F = \mathrm{GF}(q)$ with characteristic p, let $f(x) \in F[x]$ be a monic square-free polynomial which is not divisible by x, and let $a \in \mathbb{N}$. Show that the identity

$$G_q(f^{p^a}) = u\cdot G_q(f) + V_{f(x)^{p^a-1}}$$

holds for every $u \in G_q(x^{p^a}-1)$. □

Exercise 11.2.7. Consider a finite field $F = \mathrm{GF}(q)$ with characteristic p, let m and n be two positive integers which are relatively prime, and let $w \in \widehat{F}$ be an (F,σ)-generator of the cyclotomic module $\mathscr{C}_m^F$.

Prove that w is also an (E_n,σ^n)-generator of the cyclotomic module $\mathscr{C}_m^{E_n}$, that is, show $\mathrm{Ord}_{q^n}(w) = \mathrm{Ord}_q(w)$. □

Exercise 11.2.8. Let k and ℓ be two positive integers which are relatively prime and not divisible by the characteristic p of the field $F = \mathrm{GF}(q)$, and let A and B denote the cardinality of the two sets

$$G_q(x^k-1)\cdot G_q(x^\ell-1) \quad \text{and} \quad \sum_{d|k}\sum_{e|\ell} G_q(\Phi_d)\cdot G_q(\Phi_e),$$

respectively. Prove that the quotient B/A is at least $(q-1)^t$, where

$$t = 1+(k-1)(\delta(\ell)-1)+(\ell-1)(\delta(k)-1)-(\delta(k)-1)(\delta(\ell)-1)$$

and where $\delta(k)$ and $\delta(\ell)$ denote the number of positive divisors of k and ℓ, respectively. □

11.3 Extensions with Degree a Power of the Characteristic

In this section, we study the special case where the extension degree n is a power of the characteristic, say $n = p^a$. By definition, the corresponding cyclotomic module $\mathscr{C}_{p^a}^F$ then coincides with the p^a-dimensional extension E_{p^a}/F.

Since $x^{p^a}-1 = (x-1)^{p^a}$, the (F,σ)-submodules of E_{p^a} form a chain corresponding to the divisor chain

$$1 \mid (x-1) \mid (x-1)^2 \mid \cdots \mid (x-1)^j \mid \cdots \mid (x-1)^{p^a};$$

this is in remarkable contrast to the case of extensions with a prime power degree r^a, where $r \neq p$, see Remark 11.1.4. Note that the subchain

$$(x-1) \mid (x-1)^p \mid \cdots \mid (x-1)^{p^j} \mid \cdots \mid (x-1)^{p^a}$$

yields the intermediate fields of E/F:

$$F \subset E_p \subset E_{p^2} \subset \cdots \subset E_{p^a}.$$

In particular, the unique maximal (F,σ)-submodule of E_{p^a} corresponds to the polynomial

$$(x-1)^{p^a-1} = \frac{x^{p^a}-1}{x-1}$$

and is therefore just the kernel of the (E_{p^a},F)-trace mapping. This observation already proves the following classical result, which is usually attributed to Perlis [317].

Theorem 11.3.1. *Let E be an extension with degree p^a of the Galois field $F = \mathrm{GF}(q)$, where p is the characteristic of F, and let $w \in E$.*

Then w is a normal element for E/F if and only if $\mathrm{Tr}_{E/F}(w) \neq 0$. Moreover, the number of normal elements of E over F equals $q^{p^a-1}(q-1)$. □

In view of $\Phi_1(x) = x-1$ and $V_{x-1}^F = F$, the special case $k = 1$ of the identity in Equation (11.4) now yields

$$G_q((x-1)^{p^a}) = u \cdot G_q(x-1) + V^F_{(x-1)^{p^a-1}} = u \cdot F^* + \ker(\mathrm{Tr}_{E_{p^a}/F}), \qquad (11.10)$$

where u is any normal element for E/F.

Remark 11.3.2. The preceding observations show that the task of determining a normal element for E/F, where F is finite and the degree of E/F is a power of the characteristic p, is essentially covered by the results in Section 5.1.

For the convenience of the reader, we summarize Theorems 5.1.3 and 5.1.5 as follows. Let $\beta_0 \in F$ be an element with absolute trace $\mathrm{Tr}_{F/P}(\beta_0) \neq 0$ (where P denotes the prime field of $F = \mathrm{GF}(q)$, as usual). Then $g_0(x) = x^p - x - \beta_0$ is irreducible in $F[x]$. The proof of Theorem 5.1.3 shows that β_0 is the (E_p, F)-trace of $\beta_1 := -\beta_0 v_0^{p-1}$, where $v_0 \in E_p$ is any root of $g_0(x)$. Hence β_1 is normal in E_p/F; moreover, $\mathrm{Tr}_{E_p/P}(\beta_1) \neq 0$ by the transitivity of the trace mappings.

We now assume that $\beta_0, \ldots, \beta_k$ is a trace-compatible sequence of normal elements, where β_0 has non-zero absolute trace. Thus every β_j is normal for the extension E_{p^j}/F and the $(E_{p^j}, E_{p^{j-1}})$-trace of β_j equals β_{j-1}, for $j = 1, \ldots, k$. This sequence can be extended by $\beta_{k+1} := -\beta_k v_k^{p-1}$, where $v_k \in E_{p^{k+1}}$ is a root of the irreducible polynomial $g_k(x) = x^p - x - \beta_k \in E_{p^k}[x]$, since v is normal in $E_{p^{k+1}}$ over F and has $(E_{p^{k+1}}, E_{p^k})$-trace equal to β_k; see Theorem 5.1.5.

This provides a recursive construction for a trace-compatible sequence of normal elements for the p-primary closure E_{p^∞} over F. □

Although the preceding results already settle our principal task, we will investigate this class of extensions in more detail and derive a generalization of Theorem 11.3.1 which provides further insight. For this, we require two preliminary facts, which actually hold for arbitrary cyclic Galois extensions.

Thus let E/F be a cyclic Galois extension with degree n, and let σ be a generator of the Galois group of E/F. Given any element v of some intermediate field K of E/F, we define a polynomial $g_{v,K}(x) \in K[x]$ as follows:

$$g_{v,K}(x) := \sum_{i=0}^{[K:F]-1} \sigma^i(v) x^i. \qquad (11.11)$$

Our first auxiliary result provides a characterization of normal elements for E/F in terms of such polynomials:

Proposition 11.3.3. *Consider a cyclic Galois extension E/F with degree n, and let σ be a generator for* $\mathrm{Gal}(E/F)$. *Then an element $w \in E$ is normal over F if and only if $g_{w,E}(x)$ and $x^n - 1$ are relatively prime.*

Proof. We apply Corollary 3.9.3 to the present situation, where $G = \mathrm{Gal}(E/F) = \{\sigma^i : i = 0, \ldots, n-1\}$. Thus w is a normal element for E/F if and only if

$$M(w) := \left(\sigma^{i+j}(w)\right)_{i,j=0,\ldots,n-1}$$

is an invertible matrix in $E^{(n,n)}$. Of course, this is equivalent to requiring that the matrix

$$M'(w) := \left(\sigma^{n-i+j}(w)\right)_{i,j=0,\ldots,n-1}$$

is invertible. This reordering of the rows of $M(w)$ has the advantage that $M'(w)$ is a circulant matrix, since $\sigma^{n-(i+1)+(j+1)}(w) = \sigma^{n-i+j}(w)$ for all i, j. In particular, $M'(w)$ is already determined by its first row $(w, \sigma(w), \ldots, \sigma^{n-1}(w))$, which consists of the conjugates of w under G.

We now apply the basic theory of circulant matrices developed in Section 7.3. According to Theorem 7.3.2, the circulant matrix $M'(w)$ corresponds to the polynomial $w + \sigma(w)x + \cdots + \sigma^{n-1}(w)x^{n-1} = g_{w,E}(x)$ in the residue ring $E[x]/(x^n - 1)$. Moreover, $M'(w)$ is invertible if and only if this polynomial and $x^n - 1$ are relatively prime, by Proposition 7.3.3. □

Our second auxiliary result provides a connection between polynomials $g_{w,E}(x)$ and corresponding polynomials with respect to an intermediate field K:

Lemma 11.3.4. *Consider a cyclic Galois extension E/F with degree n, let σ be a generator for* $\mathrm{Gal}(E/F)$*, and let w be an element of E. Suppose that $n = k\ell$, and let K be the intermediate field of E/F with degree k over F and $u := \mathrm{Tr}_{E/K}(w)$.*

Then $g_{w,E}(x)$ is congruent to $g_{u,K}(x)$ modulo $x^k - 1$ in $E[x]$.

Proof. According to Equation (11.11), we have

$$g_{w,E}(x) = \sum_{h=0}^{n-1} \sigma^h(w)x^h = \sum_{i=0}^{\ell-1}\sum_{j=0}^{k-1} \sigma^{ik+j}(w)x^{ik+j}.$$

Modulo $x^k - 1$, this yields the desired congruence:

$$\begin{aligned} g_{w,E}(x) &\equiv \sum_{j=0}^{k-1} \sigma^j\Big(\sum_{i=0}^{\ell-1} (\sigma^k)^i(w)\Big) \cdot x^j \\ &= \sum_{j=0}^{k-1} \sigma^j(\mathrm{Tr}_{E/K}(w))x^j \\ &= \sum_{j=0}^{k-1} \sigma^j(u)x^j = g_{u,K}(x). \end{aligned}$$ □

We now consider the special case where F has positive characteristic p and obtain the following strong generalization of Theorem 11.3.1:

Theorem 11.3.5. *Let E be a cyclic Galois extension with degree $n = kp^a$ of the Galois field $F =$* $\mathrm{GF}(q)$*, where p is the characteristic of F, let $w \in E$, and let K be the intermediate field of E/F with degree k over F.*

Then w is a normal element for E/F if and only if $\mathrm{Tr}_{E/K}(w)$ *is a normal element for K/F.*

Proof. By Proposition 3.12.9, the normality of w for E/F implies that of $u := \mathrm{Tr}_{E/K}(w)$ for K/F, even without any assumption on the characteristic, the Galois group or the degree of E/K.

Assume conversely that u is a normal element for K/F. Then $g_{u,K}$ and $x^k - 1$ have to be relatively prime, by Proposition 11.3.3. According to Lemma 11.3.4, the polynomial $g_{w,E}(x)$ is congruent to $g_{u,K}(x)$ modulo $x^k - 1$. This gives

$$\begin{aligned}\gcd\left(g_{w,E}(x), x^k - 1\right) &= \gcd\left(x^k - 1, g_{w,E}(x) \bmod x^k - 1\right)\\ &= \gcd\left(x^k - 1, g_{u,K}(x)\right) = 1,\end{aligned}$$

which shows that $g_{w,E}(x)$ and $x^k - 1$ are also relatively prime. In view of

$$x^n - 1 = x^{kp^a} - 1 = \left(x^k - 1\right)^{p^a},$$

we conclude that $g_{w,E}(x)$ and $x^n - 1$ are likewise relatively prime. Now another application of Proposition 11.3.3 gives the normality of w for E/F. □

Remark 11.3.6. Let us apply Theorem 11.3.5 in the special case where $F = \mathrm{GF}(q)$ and where k is not divisible by p. Here Proposition 3.11.12 implies

$$\mathrm{Tr}_{E/K}^{-1}(G_q(x^k - 1)) = G_q(x^{p^a k} - 1). \tag{11.12}$$

Now let $u \in E_{p^a}$ be any normal element over F, that is, let $u \in G_q(x^{p^a} - 1)$; then (11.6) gives

$$u \cdot G_q(x^k - 1) \subseteq G_q(x^{p^a k} - 1).$$

Since $V^F_{(x^k-1)p^a-1}$ is the kernel of $\mathrm{Tr}_{E/K}$, we infer from Theorem 11.3.5 that

$$S := u \cdot G_q(x^k - 1) + V^F_{(x^k-1)p^a-1} \subseteq G_q(x^{p^a k} - 1).$$

Obviously, $\mathrm{Tr}_{E/K}(uy) = y\mathrm{Tr}_{E/K}(u)$ and $\mathrm{Tr}_{E/K}(uz) = z\mathrm{Tr}_{E/K}(u)$ are distinct whenever y and z are distinct elements in $G_q(x^k - 1)$. Therefore S has cardinality

$$\phi_q(x^k - 1) \cdot q^{k(p^a - 1)} = \phi_q(x^{p^a k} - 1),$$

by Equation (11.12), which is the total number of normal elements for the extension E/F. Altogether, this establishes the identity

$$G_q(x^{p^a k} - 1) = u \cdot G_q(x^k - 1) + V^F_{(x^k-1)p^a-1} \tag{11.13}$$

for every element $u \in E_{p^a}$ which is normal over F. □

Exercises

Exercise 11.3.7. Consider the Galois field $F = \mathrm{GF}(q)$ with prime field $P = \mathrm{GF}(p)$, and let $\beta \in F$. By Theorem 5.1.1, the trinomial $f(x) = x^p - x - \beta \in F[x]$ is irreducible if and only if $a := \mathrm{Tr}_{F/P}(\beta) \neq 0$. Assume that this holds, and let $v \in E_p$ be any root of $f(x)$.

Prove the following two assertions for $j=1,\dots,p-1$, where σ is the Frobenius automorphism of E_p/F and where $h_j(x) := (x-1)^j$:

(1) $h_j(\sigma)(v^j) = j!\cdot a^j$;

(2) $\mathrm{Ord}_q(v^j) = (x-1)^{j+1}$.

In particular, this shows that v^{p-1} is a normal element for E_p/F. □

11.4 Strongly Regular Extensions

In this section, we will determine normal elements for a particular class of extensions which are called *strongly regular*. As before, we consider a finite ground field $F=\mathrm{GF}(q)$, where q is a power of the prime p, together with an algebraic closure $\widehat{F}$, and denote the Frobenius automorphism of $\widehat{F}/F$ by σ.

Definition 11.4.1. The n-dimensional extension E_n of $F=\mathrm{GF}(q)$ in $\widehat{F}$, as well as the pair (q,n), are said to be **strongly regular** provided that the following two conditions are satisfied:

- The radical $\mathrm{rad}(n)$ of n divides $q-1$; in particular, n is not divisible by p.
- If n is even, then $q\equiv 1 \bmod 4$.

In view of Remark 11.1.3, it will be useful to extend the notion of strong regularity to cyclotomic modules: for every strongly regular pair (q,n), we also call $\mathscr{C}_n^F$ a **strongly regular** cyclotomic module over $F=\mathrm{GF}(q)$. □

In the remainder of this chapter, $\mathscr{S}_q$ will denote the set of all $n\in\mathbb{N}$ such that (q,n) is strongly regular. It is easy to see that any such set $\mathscr{S}_q$ is a Steinitz number; cf. Section 4.2. By definition, we have $\mathscr{S}_q\neq\{1\}$ if and only if $q\neq 2,3$, and then $\mathscr{S}_q$ is clearly an infinite set. For example, one obtains $\mathscr{S}_{11}=\{5^a : a\in\mathbb{N}\}$ and $\mathscr{S}_{13}=\{2^a\cdot 3^b : a,b\in\mathbb{N}\}$.

Remark 11.4.2. The numerical conditions in Definition 11.1.2 are closely related to those which we have encountered in Section 5.2 in our study of irreducible binomials. As in that section, let B_q denote the set of all $n\in\mathbb{N}^*$ such that there exists an irreducible binomial with degree n in $F[x]$. Then actually

$$B_q = \begin{cases} \mathscr{S}_q & \text{when } q \text{ is even or } q\equiv 1 \bmod 4, \text{ by Corollary 5.2.7;} \\ \mathscr{S}_q\cup 2\mathscr{S}_q & \text{when } q\equiv 3 \bmod 4, \text{ by Corollary 5.2.8.} \end{cases}$$

As a consequence, irreducible binomials will be an essential tool for the constructions of normal bases in this section. □

Notation 11.4.3. According to Definition 1.7.2, the n-part $\mathrm{pt}_n(N)$ of N is the largest divisor k of N such that $\mathrm{rad}(k)$ divides $\mathrm{rad}(n)$. If (q,m) is a strongly regular pair, it will be convenient to introduce the following notations:

$$\ell = \ell(q,m) := \mathrm{pt}_m(q-1) \quad \text{and} \quad a = a(q,m) := \gcd(\ell,m);$$

note that $\mathrm{rad}(m) \mid a \mid \ell \mid q-1$. In addition, we will represent the set of units in the residue ring $\mathbb{Z}_a$ by I_a, that is, we put

$$I_a := \{j \in \mathbb{N} : 1 \le j \le a,\ \gcd(j,a) = 1\}. \qquad \square$$

Our main tool for investigating strongly regular extensions is the following explicit factorization of the relevant cyclotomic polynomials:

Proposition 11.4.4. *Assume that the pair (q,m) is strongly regular, and let $\lambda \in F^*$ be a primitive a-th root of unity, where $F = \mathrm{GF}(q)$ and where $a = a(q,m)$ is as in Notation 11.4.3. Then the canonical factorization of $\Phi_m(x)$ over F is given by*

$$\Phi_m(x) = \prod_{j \in I_a} \left(x^{m/a} - \lambda^j\right).$$

Proof. By Proposition 3.6.16, the cyclotomic polynomial $\Phi_m(x)$ splits in $F[x]$ into $\phi(m)/\mathrm{ord}_m(q)$ monic irreducible polynomials of degree $\mathrm{ord}_m(q)$. Since $m \in \mathscr{S}_q$ by hypothesis, part (1) of Proposition 1.7.10 gives

$$\mathrm{ord}_m(q) = \frac{m}{\gcd(\mathrm{pt}_m(q-1),m)} = \frac{m}{\gcd(\ell,m)} = \frac{m}{a}.$$

Now let $\eta \in \widehat{F}$ be any primitive m-th root of unity in $\widehat{F}^*$. Then $\eta^{m/a}$ is a primitive a-th root of unity, say $\eta^{m/a} = \lambda^j$ for $j \in I_a$, and thus η is a root of the monic polynomial

$$f_j(x) := x^{m/a} - \lambda^j \in F[x].$$

Note that f_j actually is the minimal polynomial of η over F, as η has degree $\mathrm{ord}_m(q) = m/a$ over F, and hence an irreducible divisor of Φ_m. Clearly, the polynomials f_j and f_k are distinct – and therefore relatively prime – whenever $j,k \in I_a$ are distinct. Consequently, $\prod_{j \in I_a} f_j(x)$ is a monic divisor of $\Phi_m(x)$. Now it suffices to check that these two polynomials have the same degree:

$$\deg\Big(\prod_{j \in I_a} f_j\Big) = |I_a| \cdot \mathrm{ord}_m(q) = \phi(a) \cdot \frac{m}{a} = \phi(m) = \deg \Phi_m,$$

where the penultimate equality holds since $\mathrm{rad}(m)$ divides a. $\square$

Thus the m-th cyclotomic polynomial splits over $F = \mathrm{GF}(q)$ into irreducible binomials for every strongly regular pair (q,m), which will allow us to determine normal bases for strongly regular extensions rather easily in Theorem 11.4.7.[1] For this, we first note an immediate consequence of Proposition 11.4.4, namely the fol-

[1] This essentially generalizes part of the work of Semaev [346], where the construction of normal elements is based on extensions of prime power degree for some prime divisor r of $q-1$, and where $q \equiv 1 \bmod 4$ when $r = 2$.

lowing explicit factorization of $x^n - 1$ over $F = \mathrm{GF}(q)$, whenever (q,n) is strongly regular:

Corollary 11.4.5. *Assume that (q,n) is strongly regular. For every divisor m of n, write (using the notation introduced in 11.4.3) $\ell(m) = \ell(q,m)$, $a(m) = a(q,m)$, and $I_{a(m)} = \{j \in \mathbb{N} : 1 \leq j \leq a(m), \gcd(j,a(m)) = 1\}$. Then the canonical factorization of $x^n - 1$ over $F = \mathrm{GF}(q)$ is given by*

$$x^n - 1 = \prod_{m|n} \prod_{j \in I_{a(m)}} \left(x^{m/a(m)} - \lambda^{j \cdot a(n)/a(m)}\right),$$

where λ is an arbitrary primitive $a(n)$-th root of unity in F^*.

Proof. Note that (q,m) is strongly regular for every divisor m of n, as $\mathrm{pt}_m(q-1)$ divides $\mathrm{pt}_n(q-1)$. Thus $\ell(m) \mid \ell(n)$, and therefore also $a(m) \mid a(n)$. Moreover, $\lambda^{a(n)/a(m)}$ is a primitive $a(m)$-th root of unity in F^*. Now the assertion follows from Proposition 11.4.4, since $x^n - 1 = \prod_{m|n} \Phi_m(x)$. □

We are now ready to determine (F,σ)-generators for all strongly regular cyclotomic modules:

Theorem 11.4.6. *Let (q,m) be a strongly regular pair, let u be a primitive $(m\ell)$-th root of unity in $\widehat{F}$, and put $v := \sum_{t \in I_a} u^t$, where we use the same notation as in Proposition 11.4.4 (and in 11.4.3). Then one has the following:*

(1) *$F(u)$ is the m-dimensional extension E_m of F in $\widehat{F}$.*

(2) *$\mathrm{Ord}_q(u)$ is an irreducible divisor of $\Phi_m(x)$ over F.*

(3) *$\mathrm{Ord}_q(v) = \Phi_m(x)$, that is, v is a generator of the strongly regular cyclotomic module $\mathscr{C}_m^F$ over F.*

Proof. For (1), it suffices to check that $F(u)$ has the correct degree over F, which follows from part (1) of Proposition 1.7.10:

$$[F(u) : F] = \mathrm{ord}_{m\ell}(q) = \frac{m\ell}{\gcd(\ell, m\ell)} = m.$$

In order to deal with (2), we put $A := \mathrm{pt}_m(q^{m/a} - 1)$ and write $q^{m/a} - 1 = k \cdot A$, where k and m are relatively prime. By Proposition 1.7.9,

$$A = \mathrm{pt}_m(q^{m/a} - 1) = \frac{m}{a} \cdot \mathrm{pt}_m(q-1) = \frac{m}{a} \cdot \ell = \frac{m}{\gcd(m,\ell)} \cdot \ell = \mathrm{lcm}(m,\ell).$$

Hence $\lambda := u^A = u^{m\ell/a}$ has multiplicative order a, so that $\lambda \in F^*$. By Proposition 11.4.4, $x^{m/a} - \lambda^s$ is an irreducible divisor of $\Phi_m(x)$ for $s := k \bmod a$, because of $s \in I_a$. Then

$$\begin{aligned}
\left(x^{m/a} - \lambda^s\right)(\sigma)(u) &= u^{q^{m/a}} - \lambda^s u = u \cdot \left(u^{q^{m/a}-1} - \lambda^s\right) \\
&= u \cdot \left(u^{Ak} - \lambda^s\right) = u \cdot \left(\lambda^k - \lambda^s\right) \\
&= u \cdot \left(\lambda^{k \bmod a} - \lambda^s\right) = 0,
\end{aligned}$$

since $\operatorname{ord}(\lambda) = a$ and $k \bmod a = s$. Thus $\operatorname{Ord}_q(u)$ divides $x^{m/a} - \lambda^s$, which is irreducible over F. This establishes the validity of (2).

Finally, let t be any positive integer satisfying $\gcd(t,a) = 1$ and therefore also $\gcd(t,m) = 1$. Then $\operatorname{ord}(u^t) = \operatorname{ord}(u)$, and thus u^t is again a primitive $(m\ell)$-th root of unity. A similar calculation as before gives $\operatorname{Ord}_q(v) = x^{m/a} - \lambda^{st}$. Since the mapping $t \mapsto st \bmod a$ is a bijection on I_a, we see that the q-order of $v = \sum_{t\in I_a} u^t$ is given by

$$\prod_{t\in I_a} \operatorname{Ord}_q(u^t) = \prod_{t\in I_a} \left(x^{m/a} - \lambda^{st}\right) = \prod_{j\in I_a} \left(x^{m/a} - \lambda^{j}\right) = \Phi_m(x),$$

which also establishes (3). □

As an immediate consequence, we obtain the following explicit description of certain normal elements for arbitrary strongly regular extensions:

Theorem 11.4.7. *Consider the finite field $F = \mathrm{GF}(q)$, and let (q,n) be a strongly regular pair. For every divisor m of n, let $\ell(m) := \mathrm{pt}_m(q-1)$, $a(m) := \gcd(\ell(m),m)$ and $I_{a(m)} := \{j \in \mathbb{N}\colon 1 \le j \le a(m),\ \gcd(j,a(m)) = 1\}$. Finally, let u be any primitive $(m\ell)$-th root of unity in $\widehat{F}$. Then*

$$w := \sum_{m|n} \sum_{t\in I_{a(m)}} \left(u^{n\ell(n)/m\ell(m)}\right)^t$$

is a normal element for the n-dimensional extension E_n of F in $\widehat{F}$.

Proof. For every divisor m of n, put $u_m := u^{n\ell(n)/m\ell(m)}$. Then u_m is a primitive $m\ell(m)$-th root of unity, and thus $v_m := \sum_{t\in I_{a(m)}} u_m^t$ has q-order $\Phi_m(x)$, by Theorem 11.4.6. Hence $w := \sum_{m|n} v_m$ has q-order $\prod_{m|n} \Phi_m(x) = x^n - 1$, as claimed. □

Example 11.4.8. Let us determine a normal element for the 18-dimensional extension of GF(13). Note that the pair $(q,n) = (13,18)$ is indeed strongly regular, as $13 \equiv 1 \bmod 4$ and $\operatorname{rad}(18) = 6$; thus $n\ell(n) = 18 \cdot 12 = 216$ in this case. The following table summarizes the data required for the application of Theorem 11.4.7, where u is a primitive 216-th root of unity and where we write $u_m := u^{216/m\ell(m)}$, as before:

$m \mid 18$	$\ell(m) = \mathrm{pt}_m(q-1)$	$m\ell(m)$	$\frac{216}{m\ell(m)}$	$a(m)$	u_m
18	12	216	1	6	u
9	3	27	8	3	u^8
6	12	72	3	6	u^3
3	3	9	24	3	u^{24}
2	4	8	27	2	u^{27}
1	1	1	216	1	1

By Theorem 11.4.7 and its proof, the sum $w := v_1 + v_2 + v_3 + v_6 + v_9 + v_{18}$ is a normal element for E_{18}/F, where the elements v_m with q-order $\Phi_m(x)$ are as follows:

$$
\begin{aligned}
v_{18} &:= \textstyle\sum_{j\in I_6} u_{18}^j &&= u+u^5 \\
v_9 &:= \textstyle\sum_{j\in I_3} u_9^j &&= u_9+u_9^2 = u^8+u^{16} \\
v_6 &:= \textstyle\sum_{j\in I_6} u_6^j &&= u_6+u_6^5 = u^3+u^{15} \\
v_3 &:= \textstyle\sum_{j\in I_3} u_3^j &&= u_3+u_3^2 = u^{24}+u^{48} \\
v_2 &:= \textstyle\sum_{j\in I_2} u_2^j &&= u_2 = u^{27} \\
v_1 &:= u_1 &&= 1.
\end{aligned}
$$

As a sum of powers of u, we have

$$w = 1+u^{27}+(u^{24}+u^{48})+(u^3+u^{15})+(u^8+u^{16})+(u+u^5). \qquad \square$$

The preceding results also allow us to deal with the second situation occurring in Remark 11.4.2, that is, with extensions with degree $n=2m$ of $F=\mathrm{GF}(q)$, where $q\equiv 3 \bmod 4$ and where the pair (q,m) is strongly regular. The simplest case, namely $m=1$, was already discussed in Example 11.1.5, and the general case follows easily:

Theorem 11.4.9. *Consider a finite field $F=\mathrm{GF}(q)$, where $q\equiv 3 \bmod 4$, and an n-dimensional extension E_n over F, where $n=2m$ and where the pair (q,m) is strongly regular (so that m is odd). Assume that w is any normal element for the strongly regular extension E_m/F, and let θ be a primitive 8-th root of unity in $\widehat{F}^*$. Then both θw and $w+\theta^2 w$ are normal elements for E_n/F.*

Proof. Note that $\theta\in E_2$, as q^2-1 is a multiple of 8. By Example 11.1.5, θ even is a normal element for E_2/F. Moreover, the primitive fourth root of unity $\varepsilon:=\theta^2$ has q-order $x+1$, again by Example 11.1.5. In view of $\mathrm{Ord}_q(1)=x-1$, we see that $1+\varepsilon$ is likewise a normal element for E_2/F. Now the reduction theorem (11.6) for normal elements shows that both θw and $(1+\varepsilon)w$ are normal for E_{2m}/F. $\square$

We conclude this section with a closer look at the situation in Theorems 11.4.6 and 11.4.7:

Remark 11.4.10. Consider a strongly regular pair (q,m). As we have seen in Theorem 11.4.6, certain primitive roots of unity then have irreducible q-orders. This suggests to investigate which positive integers arise as multiplicative orders of elements in $\widehat{F}$ having an irreducible divisor of $\Phi_m(x)$ as their q-order.

By Proposition 11.4.4, all irreducible monic divisors of $\Phi_m(x)$ over F have the form $f(x)=x^{m/a}-\lambda$, where $\ell=\mathrm{pt}_m(q-1)$ and $a=\gcd(m,\ell)$, and where λ is some primitive a-th root of unity in F^*. Consider such a binomial f, and let u be any element in $\widehat{F}$ with q-order f. Then u is an eigenvector of the F-endomorphism $\sigma^{m/a}$ with eigenvalue λ, and the corresponding eigenspace $V_f^F=\{v\in\widehat{F}: f(\sigma)(v)=0\}$ has dimension m/a. Since

$$u^{a(q^{m/a}-1)} = \lambda^a = 1,$$

every non-zero $u\in V_f^F$ has multiplicative order dividing $a(q^{m/a}-1)$. Now write $q^{m/a}-1=kA$, where $A=\mathrm{pt}_m(q^{m/a}-1)=\mathrm{lcm}(m,\ell)$, as in the proof of Theorem

11.4.6. Observe that the multiplicative order of u has to be a multiple of $m\ell$: otherwise, its kA-th power would not have multiplicative order $a = \operatorname{ord}(\lambda)$. This suggests that the hypothesis that u is a primitive $(m\ell)$-th root of unity in Theorem 11.4.6 may be replaced by the less restrictive condition

$$m\ell \mid \operatorname{ord}(u) \mid a(q^{m/a} - 1) = km\ell.$$

In order to prove that this is indeed the case, we consider the subgroup $U_{km\ell}$ of order $km\ell$ of E_m^*. Since this group decomposes as $U_{km\ell} = U_k \cdot U_{m\ell}$ (as k and $m\ell$ are relatively prime), the elements having order divisible by $m\ell$ form the subset $S := U_k \cdot C_{m\ell}$ of $U_{km\ell}$ with cardinality $k\phi(m\ell)$, where $C_{m\ell}$ denotes the set of generators of the group $U_{m\ell}$ (as in Section 3.6). We now check that S consists precisely of those elements in $\widehat{F}$ which have an irreducible divisor of $\Phi_m(x)$ as their q-order.

Choose a fixed primitive a-th root of unity λ. Then the primitive a-th roots of unity in $\widehat{F}$ are the powers λ^j, where $j \in I_a = \{j \in \mathbb{N} : 1 \le j \le a, \gcd(j,a) = 1\}$. Let W_j denote the kernel of $\sigma^{m/a} - \lambda^j \cdot \mathrm{id}$, that is, the eigenspace corresponding to the eigenvalue λ^j of the endomorphism $\sigma^{m/a}$. As we have seen, the non-zero elements of W_j form a subset of S for all $j \in I_a$. Thus the union $\bigcup_{j \in I_a} W_j$ contains exactly

$$\phi(a)(q^{m/a} - 1) = kA\phi(a) = k \cdot \frac{m\ell}{a} \cdot \phi(a) = k\phi(m\ell)$$

non-zero elements, where we have used $\operatorname{rad}(a) = \operatorname{rad}(m\ell)$. This yields the identity

$$\bigcup_{j \in I_a} W_j \setminus \{0\} = U_k \cdot C_{m\ell}$$

and establishes our claim.

Finally, we provide an alternative description of the elements in E_m whose q-order is an irreducible F-divisor of $\Phi_m(x)$. For this, we consider the intermediate field $K = E_{m/a}$ of the extension E_m/F and the eigenspace W_1 of λ with respect to the F-endomorphism $\sigma^{m/a}$. Obviously, both K and W_1 are vector spaces over F with dimension m/a. However, one can say a lot more: W_1 is in fact a 1-dimensional vector space over K, as

$$\sigma^{m/a}(\alpha u) = \sigma^{m/a}(\alpha) \cdot \sigma^{m/a}(u) = \alpha \cdot \lambda u = \lambda \cdot (\alpha u)$$

whenever $u \in W_1$ and $\alpha \in K$. In general, we obtain $W_j = Ku^j$ for every $j \in I_a$, and therefore

$$\bigcup_{j \in I_a} W_j \setminus \{0\} = \bigcup_{j \in I_a} K^* u^j = K^* \cdot \{u^j : j \in I_a\}.$$

We can summarize the preceding results in the following identity:

$$G_q(\Phi_m) = \sum_{j \in I_a} E_{m/a}^* u^j \tag{11.14}$$

for every $u \in E_m^*$ such that $m\ell \mid \mathrm{ord}(u) \mid a(q^{m/a}-1)$. □

The observations in Remark 11.4.10 provide a nice example for the interplay between the multiplicative and additive structures of finite fields. Theorem 11.4.7 provides another instance of this phenomenon:

Remark 11.4.11. In Theorem 11.4.7, we have used some primitive $n\ell(n)$-th root of unity (that is, some generator u of the subgroup $U_{n\ell(n)}$ of E_n^*) for the explicit construction of a normal element w for E_n/F as a suitable sum of powers of u. This shows that certain normal elements for E_n/F can be composed additively using elements of the (small) multiplicative group $U_{n\ell(n)}$, provided that (q,n) is a strongly regular pair. In this context, we remark that the set

$$\widehat{U} = \widehat{U}(q) := \bigcup_{m \in \mathscr{S}_q} U_m \subset \widehat{F}^*$$

is actually a subgroup of the multiplicative group of the algebraic closure $\widehat{F}$ of F; cf. Theorem 4.4.1. It is worth emphasizing that *all* strongly regular extensions of $\mathrm{GF}(q)$ admit suitable sums of elements of $\widehat{U}$ as normal basis generators. □

Exercises

Exercise 11.4.12. (See Brochero Martínez, Giraldo Verarga, de Oliveira [49].) Let $n \in \mathbb{N}^*$ be relatively prime to the characteristic of the field $F = \mathrm{GF}(q)$. Prove that the following two assertions are equivalent.

(i) Every monic irreducible factor $f(x) \in F[x]$ of $x^n - 1$ is a binomial.
(ii) $\mathrm{rad}(n)$ divides $q-1$, and $8 \nmid n$ when $q \equiv 3 \bmod 4$. □

Exercise 11.4.13. Consider a prime power $q \equiv 3 \bmod 4$, and let $m \in \mathscr{S}_q$. Determine the canonical factorization of the cyclotomic polynomials $\Phi_{2m}(x)$ and $\Phi_{4m}(x)$ over $\mathrm{GF}(q)$. □

Exercise 11.4.14. Consider an extension $E = \mathrm{GF}(q^r)$ of $F = \mathrm{GF}(q)$, where r is an odd prime dividing $q-1$. Let r^a be the maximal power of r dividing $q-1$, and let $u \in E$ be a primitive r^{a+1}-th root of unity. Show that

$$\left\{\alpha_0 + \sum_{j=1}^{r-1} \alpha_j u^j : \alpha_j \in F \text{ and } \alpha_j \neq 0 \text{ for all } j = 0, 1, \ldots, r-1\right\}$$

is the set of all normal elements for E/F. □

Exercise 11.4.15. Consider the strongly regular pair $(q,n) = (29,196)$. Determine a normal element for E_n/F according to Theorem 11.4.7; cf. Example 11.4.8. □

Exercise 11.4.16. Consider the pair $(q,n) = (7,54)$. Determine a normal element for E_n/F according to Theorem 11.4.9. □

11.5 Explicit Normal Polynomials for Strongly Regular Extensions

In the preceding section, we have constructed normal elements for strongly regular extensions in terms of suitable roots of unity. Often it is also of interest to have corresponding irreducible polynomials available, and therefore we continue our investigation of strongly regular extensions with some explicit constructions for (sequences of) normal polynomials. We use the notation introduced in Section 11.4 (in particular, that of 11.4.3) and begin with an observation summarizing the situation investigated in Remark 11.4.10:

Observation 11.5.1. Consider a strongly regular pair (q,n), and let e be a divisor of $(q-1)/\ell$, where $\ell = \mathrm{pt}_n(q-1)$. Moreover, let θ be an arbitrary primitive $n\ell e$-th root of unity. Then $F(\theta) = E_n$, and the q-order of θ is an irreducible divisor of $\Phi_n(x)$. Since $\zeta := \theta^n$ is a primitive ℓe-th root of unity, we have $\zeta \in F^*$, and hence the minimal polynomial of θ over F is the binomial $x^n - \zeta$. □

The following result of Blake, Gao, and Mullin [41] gives a fairly simple description of normal elements and normal polynomials for strongly regular extensions, using the setup in Observation 11.5.1:

Theorem 11.5.2. *Consider a strongly regular extension $E = \mathrm{GF}(q^n)$ of $F = \mathrm{GF}(q)$, where $n \geq 2$. Let $\theta \in E^*$ be a primitive $n\ell e$-th root of unity, where $\ell = \mathrm{pt}_n(q-1)$ and e is some divisor of $(q-1)/\ell$, and put $\zeta := \theta^n$.*

Then both $v := \sum_{i=0}^{n-1} \theta^i$ and $w := (1-\theta)^{-1}$ are normal elements for E/F. Moreover, the minimal polynomial of w over F equals

$$g(x) := \frac{1}{\zeta - 1} \cdot \left(\zeta x^n - (x-1)^n\right).$$

Proof. Note that $q \geq 4$, since $n \geq 2$ and (q,n) is strongly regular, so that $\zeta \neq 1$. According to Observation 11.5.1, $E = F(\theta)$, and hence also $E = F(w)$. We first prove the assertion on the minimal polynomial of w. It is easy to check that w is indeed a root of g:

$$g(w) = \frac{\zeta w^n}{\theta^n - 1} - \frac{(w-1)^n}{\theta^n - 1} = \frac{\theta^n}{(\theta^n - 1) \cdot (1-\theta)^n} - \frac{(w-1)^n}{\theta^n - 1} = 0,$$

since

$$(w-1)^n = \left(\frac{1}{1-\theta} - 1\right)^n = \frac{\theta^n}{(1-\theta)^n}.$$

As g is a monic polynomial of degree $n = [F(w) : F]$, it has to be the minimal polynomial of w over F; in particular, g is irreducible.

We still have to prove that the two elements v and w are normal for E_n/F. Note that

$$(1-\zeta)w = \frac{\theta^n - 1}{\theta - 1} = \sum_{i=0}^{n-1} \theta^i = v,$$

so that v and w are scalar multiples over F. Thus also $E = F(v)$, and the normality of w is equivalent to that of v; we shall check the normality of v. For this, we consider a monic divisor $f(x) \neq x-1$ of x^n-1 which is irreducible over F, say $f(x) \mid \Phi_m(x)$ for some $m \mid n$ with $m \neq 1$. Put

$$\theta_m := \theta^{n/m}, \ \ell(m) := \mathrm{pt}_m(q-1) \ \text{and} \ a(m) := \gcd(\ell(m), m).$$

Then

$$\mathrm{ord}(\theta_m) = \frac{\ell ne}{\gcd(\ell ne, n/m)} = \frac{\ell ne}{n/m} = \ell me = \ell(m)m \cdot es,$$

where $s := \ell/\ell(m)$. Observe that es divides $(q-1)/\ell(m)$, since $e\ell$ divides $q-1$. According to Remark 11.4.10, the q-order of θ_m is an irreducible F-divisor of $\Phi_m(x)$, and therefore there exists a j such that $1 \leq j \leq a(m)$ and $\gcd(j, a(m)) = 1$ and such that θ_m^j has q-order $f(x)$. Moreover, $\theta_m^j = \theta^{j \cdot n/m}$ and

$$j \cdot \frac{n}{m} < a(m) \cdot \frac{n}{m} = \gcd(\ell(m), m) \cdot \frac{n}{m} \leq m \cdot \frac{n}{m} = n.$$

Altogether, this shows that, for every divisor $m \neq 1$ of n and for every monic irreducible divisor $f(x)$ of $\Phi_m(x)$, there exists an integer $t \in \{1, \ldots, n-1\}$ such that $\mathrm{Ord}_q(\theta^t) = f(x)$.

In view of $\mathrm{Ord}_q(\theta^0) = x-1$, we see that every irreducible factor of x^n-1 occurs as the q-order of some term in $v = \sum_{i=0}^{n-1} \theta^i$. Note that there cannot be any cancellations of summands with the same q-order in $\sum_{i=0}^{n-1} \theta^i$, since $1, \theta, \theta^2, \ldots, \theta^{n-1}$ is a polynomial basis for E_n/F. Hence v has q-order x^n-1, as claimed. □

Corollary 11.5.3. *Consider a strongly regular extension $E = \mathrm{GF}(q^n)$ of $F = \mathrm{GF}(q)$, where $n \geq 2$, and let ζ be a primitive element of F. Then*

$$\frac{1}{\zeta - 1} \cdot \left(\zeta x^n - (x-1)^n\right)$$

is a normal polynomial over F. □

Remark 11.5.4. We return to Theorem 11.5.2 and have a closer look at the normal elements v and w constructed there. Let d be a proper divisor of n, and let L be the intermediate field of E/F with degree d over F. Since E/L is likewise strongly regular and since $L(\theta) = E$, we can apply Theorem 11.5.2 to that extension, which shows that the element $\sum_{i=0}^{n/d-1} \theta^i$ is normal for E/L. It is somewhat surprising that the same assertion also holds for v (and w, because of $v = (1-\zeta)w$), as

$$v = \sum_{i=0}^{n-1} \theta^i = \sum_{j=0}^{n/d-1} \sum_{k=0}^{d-1} \theta^{j+k \cdot \frac{n}{d}} = \sum_{k=0}^{d-1} \left(\theta^{n/d}\right)^k \cdot \sum_{j=0}^{n/d-1} \theta^j$$

and as $\sum_{k=0}^{d-1} (\theta^{n/d})^k \in L$, which follows from $\mathrm{ord}(\theta^{n/d}) = d\ell$ and $\mathrm{ord}_{d\ell}(q) = d$.

Since this holds for all d, both v and w are normal for every intermediate field extension E/L of E/F. Elements with this remarkable property are said to be **completely normal** for E/F and will be studied in detail in Chapter 12. □

The paper [41] contains further families of polynomials whose roots are normal (or even completely normal) elements. The first of these concerns polynomials over a ground field $F = \mathrm{GF}(q)$, where $q \equiv 3 \bmod 4$, which have degree a power of 2. Since this case is more involved than the one considered in Theorem 11.5.2, we refer the interested reader to the original paper. We shall consider 2-power extensions of $\mathrm{GF}(q)$ with $q \equiv 3 \bmod 4$ in Section 11.7. The second construction in [41] deals with polynomials of degree a power of p, where p is the characteristic of F; this situation was already discussed in Sections 5.1 and 11.2.

Our next aim is the generalization of some results in two papers by Meyn [276] and Chapman [73], who provided iterative constructions of irreducible polynomials with (completely) normal roots. We start with a variation of Theorem 11.5.2, which is inspired by [73]; our exposition follows Section 27 of Hachenberger [161].

Theorem 11.5.5. *Consider a strongly regular extension $E = \mathrm{GF}(q^n)$ of $F = \mathrm{GF}(q)$, where $n \geq 2$ and where F has odd characteristic. Let $\zeta \in F^*$ be an element with multiplicative order ℓe, where $\ell = \mathrm{pt}_n(q-1)$ and where e divides $(q-1)/\ell$. Then*

$$f(x) := -\frac{1}{\zeta - 1} \cdot (x+1)^n + \frac{\zeta}{\zeta - 1} \cdot (x-1)^n$$

is a monic irreducible polynomial over F, and its roots form a normal basis for E/F.

Proof. By Proposition 3.6.18, there exists an element $u \in \widehat{F}$ such that $\mathrm{ord}(u) = n\ell e$ and $u^n = \zeta$; it is clear from Observation 11.5.1 that $F(u) = E_n$. Now put

$$w := \frac{u+1}{u-1};$$

then $w \neq 1$ and $u = \frac{w+1}{w-1}$. Hence, $F(w) = F(u) = E_n$. Using $w+1 = \frac{2u}{u-1}$ and $w-1 = \frac{2}{u-1}$ as well as $\zeta = u^n$, we compute

$$f(w) = -\frac{1}{\zeta - 1} \cdot \frac{2^n u^n}{(u-1)^n} + \frac{\zeta}{\zeta - 1} \cdot \frac{2^n}{(u-1)^n} = 0.$$

Obviously, f is a monic polynomial with degree $n = [F(w) : F]$, and hence f is the minimal polynomial of w over F; in particular, f is irreducible. Now note

$$w = 1 + (w-1) = 1 + \frac{2}{u-1} = 1 + \frac{2}{u^n - 1} \cdot \left(\sum_{i=0}^{n-1} u^i\right) = 1 + \frac{2}{\zeta - 1} \cdot \left(\sum_{i=0}^{n-1} u^i\right).$$

Since $\sum_{i=0}^{n-1} u^i$ is normal for E_n/F (by Theorem 11.5.2), we conclude that $w-1$ is also normal. Then $w = (w-1)+1$ is likewise normal, since its F-component in the decomposition $E_n = \oplus_{m|n} \mathscr{C}_m^F$ is

$$1 + \frac{2}{\zeta - 1} = \frac{\zeta + 1}{\zeta - 1} \neq 0. \qquad \square$$

Our next aim is a generalization of Theorem 11.5.5 which yields sequences of normal polynomials with growing degrees. For this, we start with a field $F = \mathrm{GF}(q)$ with odd characteristic, where $q \neq 3$. For simplicity, we will denote the corresponding set $\mathscr{S}_q$ of strongly regular pairs by $\mathscr{S}$. Note that

$$E_{\mathscr{S}} := \bigcup_{n \in \mathscr{S}} E_n \tag{11.15}$$

is an (infinite) intermediate field of $\widehat{F}/F$, which will be called the **strongly regular closure** of F. Given any $n \in \mathscr{S} \setminus \{1\}$, we choose a primitive $n\ell(n)$-th root of unity $u_n \in E_n \subset E_{\mathscr{S}}$, where $\ell(n) = \mathrm{pt}_n(q-1)$, and put $\zeta_n := u_n^n$; in view of Theorem 4.4.9, we may assume

$$u_d = u_n^{n\ell(n)/d\ell(d)} \quad \text{for all } d, n \in \mathscr{S} \setminus \{1\} \text{ with } d \mid n.$$

Now consider the rational function

$$\rho(x) := \frac{x+1}{x-1} \in F(x), \tag{11.16}$$

acting as a linear fractional mapping on $E_{\mathscr{S}} \cup \{\infty\}$; for later use, we note $\rho(\rho(x)) = x$, that is, $\rho^{-1} = \rho$. The proof of Theorem 11.5.5 shows that $w_n := \rho(u_n)$ is normal in E_n/F for every $n \in \mathscr{S} \setminus \{1\}$, with minimal polynomial

$$f_n(x) = -\frac{1}{\zeta_n - 1} \cdot (x+1)^n + \frac{\zeta_n}{\zeta_n - 1} \cdot (x-1)^n$$

over F. We now consider subsets $\mathscr{S}(t)$ of $\mathscr{S}$ of the form

$$\mathscr{S}(t) := \{n \in \mathscr{S} : \mathrm{rad}(n) = t\},$$

where $t \neq 1$ is some square-free divisor of $q-1$. Then $\zeta := u_t^t$ satisfies

$$\zeta = \left(u_n^{n\ell(n)/t\ell(t)}\right)^t = u_n^n \quad \text{for all } n \in \mathscr{S}(t).$$

Finally, given any $n \in \mathscr{S}(t)$, we use the rational function

$$\alpha_n(x) := \rho(\rho(x)^n) = \rho(\rho^{-1}(x)^n),$$

which satisfies

$$\alpha_n(x) = \frac{(x+1)^n + (x-1)^n}{(x+1)^n - (x-1)^n}, \tag{11.17}$$

to define an operator $\mathscr{A}_n$ as follows:

$$\mathscr{A}_n(h(x)) := \frac{((x+1)^n - (x-1)^n)^{\deg h}}{2^{\deg h}} \cdot h(\alpha_n(x)). \tag{11.18}$$

Then a straightforward (but somewhat lengthy) calculation leads to the following result:

Theorem 11.5.6. *With the notation just introduced, one has*

$$\mathscr{A}_n(f_m(x)) = f_{nm}(x)$$

for all $m, n \in \mathscr{S}(t)$. □

Remark 11.5.7. Let us first discuss the special case $t = 2$ of the preceding results, where $q-1$ has to be a multiple of 4, due to the assumption of strong regularity. In fact the operator $\mathscr{A}_2$ reduces to the ***R*-transform**, which was introduced by Cohen [83] and used by Meyn [276] and Chapman [73] for the explicit construction of sequences of polynomials over $F = \mathrm{GF}(q)$, where each member is irreducible of degree a power of 2 and has (completely) normal roots over F. More precisely: after choosing a suitable start polynomial $f(x)$ of degree 2 over the ground field F, the sequence is obtained by the iterated application of the operator $\mathscr{A}_2$ which may be written as

$$\mathscr{A}_2(h(x)) = (2x)^{\deg(h)} \cdot h\left(\tfrac{x^2+1}{2x}\right) \tag{11.19}$$

(after some simplification). A suitable start polynomial is

$$f(x) := \frac{-1}{\zeta - 1} \cdot (x+1)^2 + \frac{\zeta}{\zeta - 1} \cdot (x-1)^2 = x^2 - \frac{2(\zeta+1)}{\zeta - 1} \cdot x + 1,$$

where $\zeta \in F^*$ is a primitive 2^s-th root of unity and $2^s = \mathrm{pt}_2(q-1)$. We also remark that the R-transform was already mentioned at the end of Section 5.7, where we studied a similar transform – namely the Q-transform – in detail in the context of constructing self-reciprocal irreducible polynomials.

Finally, if we choose t as an odd prime divisor of $q-1$, we recover the results in section 4 of Chapman [73]. In addition, [73] also provides sequences of polynomials with degree a power of 2 for the case where $q \equiv 3 \bmod 4$; again, this case is more involved. □

Reviewing Theorems 11.5.2 and 11.5.5 suggests that iterative constructions similar to the one in Theorem 11.5.6 might arise from using suitable variations of the rational function ρ defined in (11.16). In order to obtain normal elements based on the results in Section 11.4, one only has to assure that the image of a suitable root of unity u under ρ is *dense* in the following sense (compare the proof of Theorem 11.5.2): every irreducible (F,σ)-module has to be covered by some power of u, and no cancellations are allowed. We close this section with an example taken from [161, Section 27], which considers the normal polynomials $g(x)$ in Theorem 11.5.2 from this point of view.

Example 11.5.8. Let $F = \mathrm{GF}(q)$, where $q \geq 4$. We retain the notation in the proof of Theorem 11.5.5, and let n be an arbitrary integer in $\mathscr{S} \setminus \{1\}$ throughout. In particular, u_n denotes a primitive $n\ell(n)$-th root of unity, where $\ell(n) = \mathrm{pt}_n(q-1)$. We now consider the rational function

$$\tau(x) := \frac{1}{1-x}. \tag{11.20}$$

Then Theorem 11.5.2 shows that $w_n := \tau(u_n)$ is normal for E_n/F for all n. Next, define the rational function

$$\beta_n(x) := \tau(\tau^{-1}(x)^n) \in F(x);$$

since $\tau^{-1}(x) = (x-1)/x$, we obtain

$$\beta_n(x) = \frac{x^n}{x^n - (x-1)^n}. \tag{11.21}$$

This time, we introduce the following operators on $F[x]$:

$$\mathscr{B}_n(h(x)) := \left(x^n - (x-1)^n\right)^{\deg h} \cdot h(\beta_n(x)). \tag{11.22}$$

Finally, we again consider any subset $\mathscr{S}(t)$ of $\mathscr{S}$ for some square-free divisor $t \neq 1$ of $q-1$. Then the polynomial

$$g_n(x) = \frac{\zeta}{\zeta - 1} \cdot x^n - \frac{1}{\zeta - 1} \cdot (x-1)^n$$

is the minimal polynomial of w_n for all $n \in \mathscr{S}(t)$, where $\zeta = u_t^t = u_n^n$. These minimal polynomials are connected as follows:

$$\mathscr{B}_n(g_m(x)) = g_{nm}(x) \quad \text{for all } m,n \in \mathscr{S}(t). \tag{11.23}$$

We leave the details to the reader. □

We conclude this section with a final remark: Scheerhorn [335] has used Dickson polynomials (of the second kind) for the explicit construction of normal polynomials with degree n over $F = \mathrm{GF}(q)$, where n is odd and $\mathrm{rad}(n)$ divides $q+1$. Again, the normal elements determined in this manner are actually completely normal for the corresponding field extension.

Exercises

Exercise 11.5.9. Consider a strongly regular extension $E = \mathrm{GF}(q^n)$ of $F = \mathrm{GF}(q)$, where $n \geq 2$ and where F has odd characteristic, and let ζ be a primitive element of F. Show that the roots of the polynomial

$$-\frac{1}{\zeta - 1} \cdot (x+1)^n + \frac{\zeta}{\zeta - 1} \cdot (x-1)^n$$

are completely normal elements for the extension E/F. □

Exercise 11.5.10. Consider a field $F = \mathrm{GF}(q)$, where $q \equiv 1 \bmod 4$, and let $\zeta \in F^*$ be a primitive 2^s-th root of unity, where $2^s = \mathrm{pt}_2(q-1)$. Put

$$h_1(x) := x^2 + \frac{2}{\zeta - 1} \cdot x - \frac{1}{\zeta - 1}$$

and, recursively,

$$h_{k+1}(x) := (2x-1)^{2^k} \cdot h_k\left(\tfrac{x^2}{2x-1}\right) \quad \text{for } k \in \mathbb{N}.$$

Show that this yields normal polynomials h_k with degree 2^k over F. □

11.6 Regular Extensions

In this section, we study the *regular extensions* introduced by Hachenberger in [161], which form a considerably larger class than the strongly regular extensions considered in Sections 11.4 and 11.5. We begin with a fundamental, but quite technical, result on additive orders, which is a slight generalization of Theorem 14.5 in [161] and will be used several times throughout the present chapter as well as in Chapter 12.

The general setup for this result and its applications is as follows. As before, we consider a finite field $F = \mathrm{GF}(q)$ and let $m \geq 2$ be an integer which is not divisible by the characteristic p of F. Moreover, we let

- $f(x) \in F[x]$ be a monic irreducible divisor of $\Phi_m(x)$;
- s a divisor of $\mathrm{ord}_m(q)$ and $L = E_s$ the s-dimensional extension of F in $\widehat{F}$;
- and $h(x) \in L[x]$ a monic divisor of $f(x)$ which is irreducible over L.

Proposition 11.6.1. *In the situation just described, let $v \in \widehat{F}$ be an element with q^s-order $h(x)^c$ for some positive integer c, and put $w := \sum_{i=0}^{s-1} v^{q^i}$. Then the following hold:*

(1) $\mathrm{Ord}_q(v) = f(x^s)^c$;
(2) *the q^s-order of w is $f(x)^c$, whereas* $\mathrm{Ord}_q(w) = \mathrm{Ord}_q(v) = f(x^s)^c$.

Proof. We will make repeated use of the results on vector spaces with endomorphisms developed in Section 2.6, applied to powers of the Frobenius automorphism σ of $\widehat{F}/F$; see Observation 11.1.1. We first consider the $L[x]$-module homomorphism

$$\Omega_L\colon\ L[x] \to \widehat{F},\ \ \delta(x) \mapsto \delta(\sigma^s)(v)$$

with respect to σ^s. The hypothesis $\mathrm{Ord}_{q^s}(v) = h(x)^c$ implies that the kernel of Ω_L is the ideal generated by $h(x)^c$, while the image of Ω_L is the (L, σ^s)-submodule $V_{h^c}^L$ of $\widehat{F}$ generated by v; moreover,

$$V^L_{h^c} = \{y \in \widehat{F} : h(\sigma^s)^c(y) = 0\}.$$

Note that $V^L_{h^c}$ has L-dimension $c \cdot \deg h$ and F-dimension $sc \cdot \deg h$. As s divides $\mathrm{ord}_m(q)$, we have $\deg h = \mathrm{ord}_m(q^s) = \mathrm{ord}_m(q)/s = (\deg f)/s$, so that

$$\dim_F V^L_{h^c} = c \cdot \deg f.$$

Next, we also consider the restriction of Ω_L to the polynomial ring $F[x]$, that is,

$$\Omega_F : F[x] \to \widehat{F},\ \beta(x) \mapsto \beta(\sigma^s)(v).$$

Trivially, the image W of Ω_F is an F-subspace of $V^L_{h^c}$. Note that $f(x)^c$ belongs to the kernel of Ω_F, since $h(x)^c$ divides $f(x)^c$; thus this kernel is generated by a polynomial of the form $f(x)^b$ with $b \le c$, as $f(x)$ is irreducible over F. Now observe that the kernel of Ω_F is contained in the kernel of Ω_L, which is generated by $h(x)^c$. Therefore, $h(x)^c$ has to divide $f(x)^b$, and we conclude $b = c$. Hence W has F-dimension $c \cdot \deg f$, which coincides with the F-dimension of $V^L_{h^c}$, so that $W = V^L_{h^c}$.

Finally, we consider a third module homomorphism, namely

$$\Gamma : F[x] \to \widehat{F},\ \gamma(x) \mapsto \gamma(\sigma)(v).$$

The kernel of Γ is generated by the q-order of v, say $g(x) \in F[x]$, and the image of Γ is the (F,σ)-submodule

$$V^F_g = \{y \in \widehat{F} : g(\sigma)(y) = 0\}$$

of $\widehat{F}$. Note that the image W of Ω_F has to be a subspace of the image V^F_g of Γ, as $a(x^s) \in F[x]$ whenever $a(x) \in F[x]$; using $W = V^L_{h^c}$, this gives $V^L_{h^c} \subseteq V^F_g$. Since V^F_g is invariant under σ, we even have $\sigma^j(V^L_{h^c}) \subseteq V^F_g$ for all j.

As v has q^s-order $h(x)^c$, the q^s-order of $\sigma^j(v)$ is $\sigma^j(h(x))^c$, where σ is extended to $\widehat{F}[x]$ by acting on coefficients (as usual). By hypothesis, s divides $\mathrm{ord}_m(q)$, which gives

$$f(x) = \prod_{i=0}^{s-1} \sigma^i(h(x)),$$

so that $\sum_{i=0}^{s-1} V^L_{\sigma^i(h)^c}$ is an F-subspace of V^F_g. On the other hand,

$$\sum_{i=0}^{s-1} V^L_{\sigma^i(h)^c} = \bigoplus_{i=0}^{s-1} V^L_{\sigma^i(h)^c} = V^L_{f^c}.$$

Since $L = E_s$ and $f(x) \in F[x]$, we conclude that

$$V^L_{f^c} = \{y \in \widehat{F} : f(\sigma^s)^c(y) = 0\} = V^F_{f(x^s)^c}.$$

Altogether, this shows that $V^F_{f(x^s)^c}$ is an F-subspace of V^F_g, and therefore $f(x^s)^c$ has to divide $g(x) = \mathrm{Ord}_q(v)$. Conversely, also $g(x) \mid f(x^s)^c$, because of $f(\sigma^s)^c(v) = 0$

(which follows from $h(\sigma^s)^c(v) = 0$ and $h(x) \mid f(x)$). Since both $g(x)$ and $f(x^s)^c$ are monic, they have to coincide, proving (1).

Let us now turn to (2). The first assertion is easy: as we have seen above, $\sigma^i(v)$ has q^s-order $\sigma^i(h(x))^c$, and then $f(x) = \prod_{i=0}^{s-1} \sigma^i(h(x))$ gives

$$\mathrm{Ord}_{q^s}(w) = \mathrm{Ord}_{q^s}\Big(\sum_{i=0}^{s-1} \sigma^i(v)\Big) = \prod_{i=0}^{s-1} \mathrm{Ord}_{q^s}(\sigma^i(v)) = \prod_{i=0}^{s-1} \sigma^i(h(x))^c = f(x)^c.$$

For the proof of the second assertion in (2), we note that w may be written as $w = S(\sigma)(v)$, where

$$S(x) = \frac{x^s - 1}{x - 1} = \sum_{i=0}^{s-1} x^i \in F[x].$$

Thus the assertion $\mathrm{Ord}_q(w) = \mathrm{Ord}_q(v)$ is equivalent to proving that $S(x)$ and $\mathrm{Ord}_q(v) = f(x^s)^c$ are relatively prime. Since $f(x)$ is an irreducible divisor of $\Phi_m(x)$, we see that $f(x^s)^c$ divides $\Phi_m(x^s)^c$. On the other hand, $S(x)$ divides $x^s - 1$, and thus it suffices to show that $x^{s'} - 1 = \Phi_1(x^{s'})$ and $\Phi_m(x^{s'})$ are relatively prime, where s' denotes the p-free part of s. As $m \geq 2$, this holds in view of the basic properties of cyclotomic polynomials established in Section 3.6; see, in particular, Theorem 3.6.13. □

We now apply Proposition 11.6.1 to generalize Theorem 11.4.6. In the general setup for Proposition 11.6.1, we make the following additional specifications:

- We choose $s := \mathrm{ord}_{\mathrm{rad}(m)}(q)$, and put $\ell := \mathrm{pt}_m(q^s - 1)$ and $b := \gcd(\ell, m)$.
- We let $Q_b := \{q^j \bmod b \colon j \in \mathbb{Z}\}$ denote the subgroup of the group I_b of units modulo b generated by $q \bmod b$, and J_b a system of representatives for the cosets of Q_b in I_b.

We wish to determine an element in $\widehat{F}$ with q-order $\Phi_m(x^s)$, which will require an additional assumption when m is even.

Theorem 11.6.2. *Let $m \geq 2$ be an integer which is not divisible by the characteristic p of the underlying field $F = \mathrm{GF}(q)$, and assume that $q \equiv 1 \bmod 4$ when m is even. Using the notation introduced above, let $u \in \widehat{F}$ be a primitive $m\ell$-th root of unity and put $v := \sum_{j \in J_b} u^j$. Then the following hold:*

(1) *$F(u)$ is the sm-dimensional extension E_{sm} of F in $\widehat{F}$.*
(2) *The q-order of u has the form $\mathrm{Ord}_q(u) = f(x^s)$, where $f(x)$ is a monic irreducible divisor of $\Phi_m(x)$ over F.*
(3) *$\mathrm{Ord}_q(v) = \Phi_m(x^s)$, that is, v is an (F, σ)-generator of the module $V^F_{\Phi_m(x^s)}$.*

Proof. We first observe that the special case $s = 1$ reduces to the result on strongly regular extensions proved in Theorem 11.4.6: then $\ell = \mathrm{pt}_m(q - 1)$ and $b = a$, so that $J_b = I_a$ is the group of units modulo a, since Q_b is trivial in this case. Thus we may assume $s \neq 1$ in what follows. In fact, we will need to appeal to Theorem 11.4.6

for the pair (q^s, m) which is strongly regular, since $\mathrm{rad}(m)$ divides $q^s - 1$ and as $q^s \equiv 1 \bmod 4$ when m is even (by hypothesis).

According to Propositions 1.7.9 and 1.7.10, we have

$$\mathrm{pt}_m(q^{sm} - 1) = m \cdot \mathrm{pt}_m(q^s - 1) = m\ell \quad \text{and} \quad \mathrm{ord}_{m\ell}(q^s) = m,$$

which gives $\mathrm{ord}_{m\ell}(q) = sm$. Hence $F(u) = E_{sm}$, proving (1).

For part (2), we consider the s-dimensional extension field $K := E_s$ of F in $\widehat{F}$. Obviously, $K(u) = E_{sm}$ and therefore $[E_{sm} : K] = m$. Since (q^s, m) is strongly regular, we may apply Theorem 11.4.6 to conclude that the q^s-order of u is some monic divisor $h(x)$ of $\Phi_m(x)$ which is irreducible over K. As the parameter b defined above coincides with the corresponding parameter a for the pair (q^s, m) (see Notation 11.4.3), Proposition 11.4.4 and its proof show that

$$h(x) = \mathrm{Ord}_{q^s}(u) = x^{m/b} - \zeta$$

for some primitive b-th root of unity $\zeta \in K^*$, where $m/b = \mathrm{ord}_m(q^s)$. According to 11.4.3, $\mathrm{rad}(m)$ divides b, which implies $\phi(m) = \frac{m}{b}\phi(b)$. By Proposition 3.6.16, the cyclotomic polynomial $\Phi_m(x)$ splits over K into the product of

$$\frac{\phi(m)}{\mathrm{ord}_m(q^s)} = \frac{\phi(m)}{m/b} = \phi(b)$$

irreducible factors of degree $\mathrm{ord}_m(q^s) = m/b$, one of which is $h(x)$. Similarly, $\Phi_m(x)$ splits over F into

$$\frac{\phi(m)}{\mathrm{ord}_m(q)} = \frac{\phi(m)}{s \cdot \mathrm{ord}_m(q^s)} = \frac{\phi(b)}{s}$$

irreducible factors of degree $\mathrm{ord}_m(q) = s \cdot \mathrm{ord}_m(q^s)$ (which holds as $s = \mathrm{ord}_{\mathrm{rad}(m)}(q)$ divides $\mathrm{ord}_m(q)$). Consequently, every monic factor $f(x)$ of $\Phi_m(x)$ which is irreducible over F has to split over K into s irreducible factors. In particular, the polynomial

$$f(x) := \prod_{i=0}^{s-1} \sigma^i(h(x)) = \prod_{i=0}^{s-1} (x^{m/b} - \zeta^{q^i})$$

is a monic irreducible F-divisor of $\Phi_m(x)$. Now Proposition 11.6.1 yields $\mathrm{Ord}_q(u) = f(x^s)$, since u has q^s-order $h(x)$. This concludes the proof of (2).

For part (3), we note that $\mathrm{ord}_b(q) = s$, which follows from $\mathrm{rad}(m) \mid b \mid q^s - 1$ (see Notation 11.4.3). Thus Q_b has s elements, and J_b consists of $\phi(b)/|Q_b| = \phi(b)/s = \phi(m)/\mathrm{ord}_m(q)$ elements. Therefore, J_b corresponds bijectively to the monic divisors of $\Phi_b(x)$ which are irreducible over F, as well as to the monic divisors of $\Phi_m(x)$ which are irreducible over F.

By the definition of J_b, the powers u^i and u^j of u are not conjugate over F whenever $i, j \in J_b$ are distinct. In view of part (2) and its proof, this implies that the q-orders of u^i and u^j have the form $\mathrm{Ord}_q(u^i) = f_i(x^s)$ and $\mathrm{Ord}_q(u^j) = f_j(x^s)$, where $f_i(x)$ and $f_j(x)$ are distinct irreducible F-divisors of $\Phi_m(x)$. Hence

$$\mathrm{Ord}_q(v) = \mathrm{Ord}_q\Big(\sum_{j\in J_b} u^j\Big) = \prod_{j\in J_b} f_j(x^s) = \Phi_m(x^s),$$

which proves (3). □

Next, we will show that it is sometimes possible to construct an (F,σ)-generator of the cyclotomic module $\mathscr{C}_m^F$ from a generator of the module $V^F_{\Phi_m(x^s)}$ (under suitable assumptions on m and s). In particular, this applies to the element v constructed in Theorem 11.6.2, under an additional hypothesis on q and m. To do so, we require the following simple observation on cyclotomic polynomials:

Lemma 11.6.3. *Let s and m be positive integers, where m is not divisible by the characteristic p of the field $F = \mathrm{GF}(q)$. Then $\Phi_m(x)$ divides $\Phi_m(x^s)$ in $F[x]$ if and only if s and m are relatively prime.*

Proof. Write $s = s'p^\beta$, where s' is the p-free part of s, and decompose s' as $s_m \cdot \bar{s}$, where $s_m = \mathrm{pt}_m(s)$ (so that $\gcd(\bar{s},m) = 1$). Then

$$\Phi_m(x^s) = \Phi_{ms_m}(x^{\bar{s}})^{p^\beta} = \prod_{d\mid\bar{s}} \Phi_{ms_md}(x)^{p^\beta},$$

by Theorem 3.6.13. Thus $\Phi_m(x)$ divides $\Phi_m(x^s)$ if and only if $s_m = 1$, which means that s and m are relatively prime. □

Theorem 11.6.4. *Let $m \geq 2$ be an integer which is not divisible by the characteristic p of $F = \mathrm{GF}(q)$. Assume that $s := \mathrm{ord}_{\mathrm{rad}(m)}(q)$ and m are relatively prime, and let $H(x)$ be the polynomial $\Phi_m(x^s)/\Phi_m(x)$ (cf. Lemma 11.6.3). Finally, let v be any element in $E_{ms} \subset \widehat{F}$ satisfying $\mathrm{Ord}_q(v) = \Phi_m(x^s)$. Then the elements*

$$H(\sigma)(v) \quad \textit{and} \quad w := \mathrm{Tr}_{E_{ms}/E_m}(v)$$

both have q-order $\Phi_m(x)$.

Proof. The q-order of $H(\sigma)(v)$ is easily computed:

$$\frac{\mathrm{Ord}_q(v)}{\gcd(H(x),\mathrm{Ord}_q(v))} = \frac{\Phi_m(x^s)}{\gcd(H(x),\Phi_m(x^s))} = \frac{\Phi_m(x^s)}{H(x)} = \Phi_m(x).$$

In order to evaluate $\mathrm{Ord}_q(w)$, we write w in the following form:

$$w = \sum_{i=0}^{s-1} v^{q^{mi}} = \Big(\frac{x^{sm}-1}{x^m-1}\Big)(\sigma)(v).$$

Thus the q-order of w satisfies

$$\mathrm{Ord}_q(w) = \frac{\mathrm{Ord}_q(v)}{\gcd\big(\frac{x^{sm}-1}{x^m-1},\mathrm{Ord}_q(v)\big)} = \frac{\Phi_m(x^s)}{\gcd\big(\frac{x^{sm}-1}{x^m-1},\Phi_m(x^s)\big)}.$$

With the notation in the proof of Lemma 11.6.3, we now have $s = s' p^\beta = \bar{s} p^\beta$ (since $\gcd(m,s) = 1$). Applying Equation (3.9) for both $n = s'm$ and $n = m$ gives

$$\frac{x^{sm}-1}{x^m-1} = \frac{(x^{s'm}-1)^{p^\beta}}{x^m-1} = A(x)\cdot \Phi_m(x)^{p^\beta-1}\cdot B(x)\cdot C(x),$$

where

$$A(x) = \prod_{\substack{e\mid m\\ e\neq m}} \Phi_e(x)^{p^\beta-1},\quad B(x) = \prod_{\substack{d\mid s'\\ d\neq 1}} \Phi_{md}(x)^{p^\beta} \text{ and } C(x) = \prod_{\substack{f\mid s'm\\ f\nmid m,\, m\nmid f}} \Phi_f(x)^{p^\beta}.$$

Moreover, $\Phi_m(x^s) = \Phi_m(x)^{p^\beta}\cdot B(x)$, by Proposition 3.6.12. Substituting these identities, we obtain

$$\mathrm{Ord}_q(w) = \frac{\Phi_m(x)^{p^\beta}\cdot B(x)}{\gcd\left(A(x)\cdot \Phi_m(x)^{p^\beta-1}\cdot B(x)\cdot C(x),\ \Phi_m(x)^{p^\beta}\cdot B(x)\right)} = \Phi_m(x),$$

as claimed. □

The condition on s and m occurring in Theorem 11.6.4 motivates the following definition:

Definition 11.6.5. Let $F = \mathrm{GF}(q)$ and $E = \mathrm{GF}(q^n)$, and let n' denote the p-free part of n, where p is the characteristic of F. Then the extension E/F as well as the pair (q,n) are called **regular** provided that

- $\mathrm{ord}_{\mathrm{rad}(n')}(q)$ and n are relatively prime.

It is useful to extend the notion of regularity to cyclotomic modules: for every regular pair (q,n), the cyclotomic module $\mathscr{C}_n^F$ is also said to be **regular** over F. □

For a given prime power q, we will denote the set of all positive integers n such that (q,n) is regular by $\mathscr{R}_q$. We conclude this section with a few remarks and examples which should provide a better understanding of this important concept. We begin with a comparison of the sets $\mathscr{R}_q$ and $\mathscr{S}_q$:

Remark 11.6.6. Trivially, every strongly regular pair (q,n) is also regular: one then has $\mathrm{rad}(n) \mid q-1$ and therefore $\mathrm{ord}_{\mathrm{rad}(n)}(q) = 1$. Thus we have $\mathscr{S}_q \subseteq \mathscr{R}_q$. Moreover, if $q \equiv 3 \bmod 4$ and $n \in \mathscr{S}_q$, then the pair $(q,2n)$ is not strongly regular (by definition); however, it is still regular, because of $\mathrm{ord}_{\mathrm{rad}(2n)}(q) = \mathrm{ord}_{\mathrm{rad}(n)}(q) = 1$.

Thus $\mathscr{S}_q$ is, at least in these cases, a proper subset of $\mathscr{R}_q$. In fact, this always holds, as the set $\mathscr{R}_q$ actually contains all prime powers (for every q). This is an immediate consequence of the fact that $\mathrm{ord}_r(q)$ divides $\phi(r) = r-1$ for every prime r which is distinct from the characteristic p of $F = \mathrm{GF}(q)$, whereas $\mathrm{ord}_{p'}(q) = 1$. Thus $\mathscr{R}_q$ is a much larger set than $\mathscr{S}_q$, which contains the powers of a prime r if and only if r divides $\mathrm{rad}(q-1)$, where additionally $q \equiv 1 \bmod 4$ when $r = 2$. In particular, $\mathscr{R}_q$ is also an infinite set when $q \in \{2,3\}$, where $\mathscr{S}_q = \{1\}$.

In contrast to $\mathscr{S}_q$, however, the set $\mathscr{R}_q$ is never a Steinitz number. To see this, we choose some prime $r \neq p$ which does not divide $q-1$, and let s be any prime divisor of $(q^r-1)/(q-1)$. Note that then $\mathrm{ord}_s(q) = r$: clearly, $\mathrm{ord}_s(q)$ has to divide r, which is a prime; and we cannot have $q \equiv 1 \bmod s$, since this would imply

$$0 \equiv \frac{q^r-1}{q-1} = \sum_{i=0}^{r-1} q^i \equiv r \bmod s,$$

which would force $r = s \mid q-1$, contradicting our choice of r. We conclude that $n = rs = \mathrm{lcm}(r,s)$ does not belong to $\mathscr{R}_q$, since $\mathrm{ord}_{\mathrm{rad}(n')}(q) = \mathrm{ord}_{rs}(q)$ is divisible by $\mathrm{ord}_s(q) = r$, even though $r, s \in \mathscr{R}_q$.

On the other hand, one can show that $\mathscr{R}_q$ always contains an infinite set $\mathscr{P}$ of primes such that $\prod_{r\in\mathscr{F}} r$ belongs to $\mathscr{R}_q$ for every finite subset $\mathscr{F}$ of $\mathscr{P}$; see Exercise 11.6.11. □

Example 11.6.7. For a specific example, let us consider the prime $q = 2$, where $\mathscr{S}_q = \{1\}$ is trivial. We already know from Remark 11.6.6 that all prime powers belong to $\mathscr{R}_2$. We will now give some examples illustrating how such prime powers may be combined.

Let us first consider integers of the form $n = 3^a \cdot 5^b$. Then n and $\mathrm{ord}_{\mathrm{rad}(n')}(2) = \mathrm{ord}_{15}(2) = 4$ are relatively prime, so that $3^a \cdot 5^b \in \mathscr{R}_2$ for all $a, b \in \mathbb{N}$. Note that this may also be checked without evaluating $\mathrm{ord}_{15}(2)$ explicitly, which generalizes to a useful method for dealing with values of n where $\mathrm{rad}(n')$ has many prime factors. Since $\mathrm{ord}_{15}(2)$ divides $\phi(15) = (3-1)\cdot(5-1) = 8$, it suffices to observe that 8 is not divisible by 3 or 5.

More generally, one may combine powers of primes whenever none of these primes divides $\phi(r) = r-1$ for any of the larger primes r selected. Note that his does not apply for the primes 3 and 7; in fact, $\mathrm{ord}_7(2) = 3$, so that the product $3 \cdot 7$ indeed does not belong to $\mathscr{R}_2$. However, 7 can be combined with 5 (as 5 does not divide $7-1=6$), and therefore $5^a \cdot 7^b \in \mathscr{R}_2$ for all $a, b \in \mathbb{N}$.

Using this approach, it is easy to check much larger examples; for instance, one may combine all primes in $\{5,7,13,17,19,23,37\}$, which shows that any number of the form

$$5^a \cdot 7^b \cdot 13^c \cdot 17^d \cdot 19^e \cdot 23^f \cdot 37^g$$

belongs to $\mathscr{R}_2$; of course, we may then also multiply such a number with an arbitrary power of the characteristic 2. □

Remark 11.6.8. Assume that $n \in \mathscr{R}_q$ is not a multiple of the characteristic p of $F = \mathrm{GF}(q)$, where $q \equiv 1 \bmod 4$ when n is even, and let m be any divisor of n. Then we can apply Theorems 11.6.2 and 11.6.4 to construct an element $w_m \in \widehat{F}$ satisfying $\mathrm{Ord}_q(w_m) = \Phi_m(x)$, that is, a generator of the cyclotomic module $\mathscr{C}_m^F$ over F. As observed in Remark 11.1.3, we can then also construct an element w which is normal for the n-dimensional extension E_n/F, namely $w = \sum_{m|n} w_m$. □

Remark 11.6.9. Let $q \equiv 3 \bmod 4$, and assume that $m \in \mathscr{R}_q$ is odd and not a multiple of the characteristic p of $F = \mathrm{GF}(q)$. As in Remark 11.6.8, we can apply Theorems 11.6.2 and 11.6.4 to construct an element $w_m \in \widehat{F}$ satisfying $\mathrm{Ord}_q(w_m) = \Phi_m(x)$.

Now let $\varepsilon \in E_2$ be a primitive fourth root of unity, so that $\mathrm{Ord}_q(\varepsilon) = x+1 = \Phi_2(x)$, by Example 11.1.5. Then εw_m has q-order $\Phi_{2m}(x)$, by Theorem 11.2.3.

This shows that we may still construct an element w_m satisfying $\mathrm{Ord}_q(w_m) = \Phi_m(x)$ for every divisor m of n (and hence a normal element for E_n/F) when $n \in \mathscr{R}_q$ is not a multiple of the characteristic p of $F = \mathrm{GF}(q)$, where $q \equiv 3 \bmod 4$ and $n \equiv 2 \bmod 4$. □

Exercises

Exercise 11.6.10. Let r be an odd prime and consider a field $F = \mathrm{GF}(q)$, where $\mathrm{ord}_{r^2}(q) = (r-1)r$. Moreover, let k be a positive integer and $u \in \widehat{F}$ a primitive r^{k+1}-th root of unity, and put $w := \sum_{i=0}^{k-1} u^{r^i}$.

(1) Show that w is a normal element for the $(r-1)r^k$-dimensional extension $E_{(r-1)r^k}$ over F.
(2) Show that w is a normal element for the r^k-dimensional extension $E_{(r-1)r^k}$ over K, where $K = E_{r-1}$ is the $(r-1)$-dimensional extension over F. □

Exercise 11.6.11. Let q be any prime power. Show that $\mathscr{R}_q$ contains an infinite set $\mathscr{P}$ of primes such that $\prod_{r \in \mathscr{F}} r$ belongs to $\mathscr{R}_q$ for every finite subset $\mathscr{F}$ of $\mathscr{P}$. □

11.7 Extensions with Prime Power Degree

In this section, we consider the explicit construction of normal elements for extensions of $F = \mathrm{GF}(q)$ with prime power degree $m = r^k$, as these are – in view of the reduction result 3.9.7 for normal elements – of particular interest.

The case where r is the characteristic p of F was already dealt with in Section 11.2, and thus we assume $r \neq p$. The results on regular extensions in Section 11.6 settle all cases where r is odd; nevertheless, we will work out more details in Example 11.7.1 below. Finally, for $r = 2$, all cases where $q \equiv 1 \bmod 4$ are covered by the results on strongly regular extensions in Section 11.4. We will resolve the last remaining case of 2-power extensions for $q \equiv 3 \bmod 4$ in Example 11.7.2 and Theorem 11.7.3.

Example 11.7.1. Let r be an odd prime which is distinct from the characteristic p of $F = \mathrm{GF}(q)$, and let $s := \mathrm{ord}_r(q)$ and $\mathrm{pt}_r(q^s - 1) = r^c$. Given any positive integer k, choose some primitive r^{c+k}-th root of unity u in $\widehat{F}$. Then $u^{r^{k-i}}$ is a primitive r^{c+i}-th root of unity for $i = 1, \ldots, k$, and therefore $F(u^{r^{k-i}})$ is the sr^i-dimensional extension E_{sr^i} of F in $\widehat{F}$.

Now put $b_i := \gcd(r^c, r^i) = r^{\min(c,i)}$ for $i = 1, \ldots, k$, and note that $b_i = r^c$ whenever $i \geq c$. Furthermore, let Q_{b_i} denote the subgroup of the group I_{b_i} of units modulo b_i generated by $q \bmod b_i$, let J_{b_i} be a system of representatives for the cosets of Q_{b_i} in I_{b_i}, and put

$$v_i := \sum_{j \in J_{b_i}} u^{jr^{k-i}}$$

for $i = 1, \ldots, k$ and $v := \sum_{i=1}^{k} v_i$. According to Theorem 11.6.2, $\mathrm{Ord}_q(v_i) = \Phi_{r^i}(x^s)$ for all i, which gives

$$\mathrm{Ord}_q(v) = \prod_{i=1}^{k} \Phi_{r^i}(x^s) = \frac{x^{sr^k} - 1}{x^s - 1}.$$

Consequently, if $\alpha \in E_s$ is any normal element for E_s/F, then

$$w := \alpha + v = \alpha + \sum_{i=1}^{k} \sum_{j \in J_{b_i}} u^{jr^{k-i}}$$

is normal for the extension E_{sr^k}/F. Hence $\mathrm{Tr}_{E_{sr^k}/E_{r^k}}(w)$ is a normal element for E_{r^k} over F, by Proposition 3.12.9.

The drawback of this argument is that it requires knowing some normal element for the extension E_s/F. However, this can be circumvented by considering $\mathrm{Tr}_{E_{sr^k}/E_{r^k}}(v)$ instead of $\mathrm{Tr}_{E_{sr^k}/E_{r^k}}(w)$. By Theorem 11.6.4, $\mathrm{Tr}_{E_{sr^i}/E_{r^i}}(v_i)$ has q-order $\Phi_{r^i}(x)$ for all i. Now note that the power σ^{r^k} of the Frobenius automorphism σ of E_{sr^k}/F generates the Galois group of E_{sr^k}/E_{r^k}. Moreover, the restriction of σ^{r^k} to E_{sr^i} generates the Galois group of E_{sr^i}/E_{r^i} for all i, since s and r are relatively prime. This shows that $\mathrm{Tr}_{E_{sr^k}/E_{r^k}}(v_i)$ likewise has q-order $\Phi_{r^i}(x)$ for all i, and hence $\mathrm{Tr}_{E_{sr^k}/E_{r^k}}(v)$ has q-order

$$\prod_{i=1}^{k} \Phi_{r^i}(x) = \frac{x^{r^k} - 1}{x - 1},$$

and then the element

$$\overline{w} := 1 + \mathrm{Tr}_{E_{sr^k}/E_{r^k}}(v) = 1 + \sum_{i=1}^{k} \mathrm{Tr}_{E_{sr^k}/E_{r^k}}(v_i)$$

is normal for E_{r^k}/F. □

The remainder of this section is devoted to 2-power extensions E_{2^k} of $F = \mathrm{GF}(q)$ for $q \equiv 3 \bmod 4$. Note that the case $k = 1$ has already been dealt with in Example 11.1.5. We now settle the case $k = 2$ in Example 11.7.2 and deal with the cases $k \geq 3$ in Theorem 11.7.3.

Example 11.7.2. Let us consider the 4-dimensional extension E_4 over a field $F = \mathrm{GF}(q)$, where $q \equiv 3 \bmod 4$. In this case, $\Phi_4(x) = x^2 + 1$ is irreducible over F, and

x^4-1 splits as

$$x^4-1 = (x-1)(x+1)(x^2+1).$$

Let $K=E_2$ denote the unique proper intermediate field of E_4/F, and let $\zeta \in K$ be a primitive fourth root of unity. Then $F(\zeta)=K$, and

$$\Phi_4(x) = (x-\zeta)(x-\zeta^q) = (x-\zeta)(x-\zeta^{-1}).$$

Note that $\mathrm{pt}_2(q^2-1)=2^c$ with $c \geq 3$, so that $\mathrm{pt}_2(q^4-1)=2\cdot 2^c=2^{c+1}$. Now let u be any primitive 2^{c+1}-th root of unity. Then $F(u)=E_4$ and $\mathrm{Ord}_q(u)=\Phi_4(x)$, which follows by checking that

$$(x^2+1)(\sigma)(u) = u^{q^2}+u = u\cdot\left(u^{q^2-1}+1\right) = 0.$$

For this, we write $q^2-1=2^c\cdot t$, where t is odd, and note that $u^{2^c}=-1$, which shows $u^{q^2-1}=-1$, as desired. Thus u is a generator for the cyclotomic module $\mathscr{C}_4^F$.

Now let v be any normal element for E_2/F, for instance, a primitive 8-th root of unity; see Example 11.1.5. Then $\mathrm{Ord}_q(v)=x^2-1$, and therefore

$$\mathrm{Ord}_q(u+v) = (x^2-1)(x^2+1) = x^4-1,$$

providing the required normal element for E_4/F. □

We finally consider the case $k \geq 3$, where we will determine a particular generator for the cyclotomic module $\mathscr{C}_{2^k}^F$ which will be useful in Chapter 12; this was our motivation for separating the cases $k=2$ and $k\geq 3$. As explained in Remark 11.1.4, this also leads to a recursive construction of normal elements for all extensions E_{2^k}/F with $k \geq 3$.

Theorem 11.7.3. *Let $E=E_{2^k}$ be the 2^k-dimensional extension of $F=\mathrm{GF}(q)$ in $\widehat{F}$, where $q \equiv 3 \bmod 4$ and $k \geq 3$, and write $\mathrm{pt}_2(q^2-1)=2^c$ and $b=\min(k-1,c)$. Also, let Q_{2^b} be the subgroup of the group I_{2^b} of units modulo 2^b generated by $q \bmod 2^b$, and J_{2^b} a system of representatives for the cosets of Q_{2^b} in I_{2^b}.*

Furthermore, let u be any primitive 2^{k-1+c}-th root of unity in $\widehat{F}$, and put $v:=\sum_{j\in J_{2^b}} u^j$ and $w:=v+v^q$. Then one has

$$\mathrm{Ord}_q(v)=\mathrm{Ord}_q(w)=\Phi_{2^k}(x) \quad \text{and} \quad \mathrm{Ord}_{q^2}(w)=\Phi_{2^{k-1}}(x).$$

Proof. We first observe that $c\geq 3$ and thus $b \geq 2$. Now let $K=E_2$ be the intermediate field of E/F which is quadratic over the ground field F, and note that the extension E/K is strongly regular with degree 2^{k-1}. Proposition 3.6.8 gives $\Phi_{2^k}(x)=\Phi_{2^{k-1}}(x^2)$, and hence the two cyclotomic modules $\mathscr{C}_{2^k}^F$ and $\mathscr{C}_{2^{k-1}}^K$ are equal as sets. This suggests to investigate the 2^{k-1}-th cyclotomic polynomial over K.

We can apply Proposition 11.4.4 to determine how $\Phi_{2^{k-1}}$ splits over K. In the present situation, $m=2^{k-1}$ gives

$$\ell=\ell(q^2,2^{k-1}) = \mathrm{pt}_{2^{k-1}}(q^2-1) = 2^c \;\text{ and }\; a=\gcd\left(2^c,2^{k-1}\right)=2^b.$$

Hence we obtain

$$\Phi_{2^{k-1}}(x) = \prod_{\substack{j=1 \\ j \, odd}}^{2^b} (x^{2^{k-1-b}} - \lambda^j),$$

where λ is any primitive 2^b-th root of unity in K.

Because of $\ell m = 2^{k-1+c}$ and our choice of u, Theorem 11.4.6 shows that $K(u) = E$ and that the q^2-order of u is one of the irreducible divisors of $\Phi_{2^{k-1}}$ in the complete factorization over K given above, say $\mathrm{Ord}_{q^2}(u) = x^{2^{k-1-b}} - \lambda$. Now put

$$f(x) := (x^{2^{k-1-b}} - \lambda) \cdot (x^{2^{k-1-b}} - \lambda^q) = x^{2^{k-b}} - (\lambda + \lambda^q) \cdot x^{2^{k-1-b}} + \lambda\lambda^q.$$

Since $F(\lambda) = K$ and $\lambda + \lambda^q = \mathrm{Tr}_{K/F}(\lambda)$ and $\lambda \cdot \lambda^q = \mathrm{Norm}_{K/F}(\lambda)$, the polynomial f has coefficients in F; in fact, it is an irreducible divisor of $\Phi_{2^{k-1}}(x)$ over F, since $\mathrm{ord}_{2^{k-1}}(q) = 2 \cdot \mathrm{ord}_{2^{k-1}}(q^2)$. Now Proposition 11.6.1 yields $\mathrm{Ord}_q(u) = f(x^2)$.

As $q^2 \equiv 1 \bmod 2^b$, we see that Q_{2^b} is a group of order 2. Also, $\mathrm{Ord}_q(u^j) = f_j(x^2)$ for all $j \in J_{2^b}$, where

$$f_j(x) := (x^{2^{k-1-b}} - \lambda^j) \cdot (x^{2^{k-1-b}} - \lambda^{jq}).$$

Therefore $v = \sum_{j \in J_{2^b}} u^j$ satisfies

$$\mathrm{Ord}_q(v) = \prod_{j \in J_{2^b}} f_j(x^2) = \Phi_{2^{k-1}}(x^2) = \Phi_{2^k}(x),$$

as claimed. Now observe that

$$w = v + v^q - \sum_{\substack{i=1 \\ i \, odd}}^{2^b} u^i$$

indeed has q^2-order $\Phi_{2^{k-1}}(x)$, by Theorem 11.4.6. Finally, we also obtain

$$\mathrm{Ord}_q(w) = \mathrm{Ord}_q(v) - \Phi_{2^k}(x),$$

as $w = (x+1)(\sigma)(v)$ and as the cyclotomic polynomials $\Phi_2(x) = x+1$ and $\Phi_{2^k}(x)$ are relatively prime. □

Exercises

Exercise 11.7.4. Determine normal elements for the 8- and for the 16-dimensional extensions over the ternary field GF(3). □

11.8 Trace-Compatible Sequences of Normal Elements

The constructions in the preceding sections lead to the explicit determination of trace-compatible sequences for the r-primary closures E_{r^∞} of $F = \mathrm{GF}(q)$, where r can be any prime (including the characteristic p of F); see Remarks 11.1.4 and 11.3.2. In the present section, we show that these results even allow us to determine a trace-compatible sequence of normal elements for the entire algebraic closure $\widehat{F}$ of F, that is, a normal element for $\widehat{F}/F$; see Definition 4.5.1. This will provide a constructive version of Theorem 4.5.2.

Construction 11.8.1. Let $\widehat{F}$ be the algebraic closure of the Galois field $F = \mathrm{GF}(q)$. We use induction to construct a trace-compatible sequence $(w_n)_{n\in\mathbb{N}^*}$ over $\widehat{F}$, where every element w_n is normal for E_n/F.

For the induction basis $n = 1$, we let p denote the characteristic and P the prime subfield of F, as usual. In view of Remark 11.3.2, we start with an arbitrary element w_1 of F with absolute trace $\mathrm{Tr}_{F/P}(w_1) \neq 0$; in particular, when $F = P$, any element $w_1 \neq 0$ is suitable.

For the induction step, we let $n \geq 2$ and assume that we have already constructed a partial sequence $(w_1,\ldots,w_{n-1})$ with the desired properties:

- w_i is normal for E_i/F for $i = 1,\ldots,n-1$;
- $\mathrm{Tr}_{E_j/E_i}(w_j) = w_i$, whenever $i, j \in \{1,\ldots,n-1\}$ and $i \mid j$.

We now distinguish three cases:

Case 1. $n = r^k$ is a power of some prime $r \neq p$. In this case, we first determine a generator v_{r^k} of the cyclotomic module $\mathscr{C}^F_{r^k}$ as an (F,σ)-module, that is, any element v_{r^k} with q-order $\Phi_{r^k}(x)$. As we have seen, this can always be done:

- If r divides $q-1$, where r is odd or $q \equiv 1 \bmod 4$, then the pair (q, r^k) is strongly regular, and hence we may appeal to Theorem 11.4.6.
- If r is odd and does not divide $q-1$, we may apply Theorem 11.6.2 in combination with Theorem 11.6.4; see also Example 11.7.1.
- If $n = 2$, we apply Example 11.1.5, and if $n = 4$ and $q \equiv 3 \bmod 4$, we apply Example 11.7.2.
- If $n = 2^k$ with $k \geq 3$ and $q \equiv 3 \bmod 4$, we apply Theorem 11.7.3.

We then put $w_{r^k} := \frac{1}{r} \cdot w_{r^{k-1}} + v_{r^k}$. According to Remark 11.1.4, w_{r^k} is normal for E_{r^k}/F and satisfies

$$\mathrm{Tr}_{E_{r^k}/E_{r^{k-1}}}(w_{r^k}) = w_{r^{k-1}}.$$

In view of the transitivity of the trace mappings (see Theorem 3.12.8), we see that $w_n = w_{r^k}$ can be used to extend the partial sequence $(w_1,\ldots,w_{n-1})$.

Case 2. $n = p^k$. With induction, we may assume that $w_{p^{k-1}}$ has absolute trace $\mathrm{Tr}_{E_{p^{k-1}}/P}(w_{p^{k-1}}) \neq 0$. We now proceed as in Remark 11.3.2 and let y be a root of the polynomial

$$x^p - x - w_{p^{k-1}} \in E_{p^{k-1}}[x].$$

Then $F(y) = E_{p^{k-1}}(y) = E_{p^k}$, and the element $w_{p^k} := -w_{p^{k-1}}y^{p-1}$ has absolute trace $\mathrm{Tr}_{E_{p^k}/P}(w_{p^k}) \neq 0$. Thus also $\mathrm{Tr}_{E_{p^k}/F}(w_{p^k}) \neq 0$, and Theorem 11.3.1 shows that w_{p^k} is normal for E_{p^k}/F. Moreover, by the proof of Theorem 5.1.3,

$$\mathrm{Tr}_{E_{p^k}/E_{p^{k-1}}}(w_{p^k}) = w_{p^{k-1}}.$$

As in Case 1, we see that $w_n = w_{p^k}$ can be used to extend the partial sequence $(w_1, \dots, w_{n-1})$.

Case 3. n is not a prime power. Here we put

$$w_n := \prod_{j=1}^{t} w_{r_j^{a_j}},$$

where $n = \prod_{j=1}^{t} r_j^{a_j}$ is the prime power factorization of n. In view of Theorem 3.9.7, w_n is a normal element for E_n/F, as each $w_{r_j^{a_j}}$ is normal for $E_{r_j^{a_j}}/F$, by the induction hypothesis. We claim that this normal element can be used to extend the partial sequence $(w_1, \dots, w_{n-1})$. Due to the transitivity of the trace mappings (see Theorem 3.12.8), it suffices to check that w_n has the correct trace into every maximal intermediate field of E_n/F, namely

$$\mathrm{Tr}_{E_n/E_{n/r_j}}(w_n) = w_{n/r_j} \quad \text{for } j = 1, \dots, t.$$

Thus let r be one of the primes dividing n, say $r = r_1$ (without loss of generality). For better readability, we write

$$a := a_1, \quad m := n/r, \quad z := w_{r^a} \quad \text{and} \quad y := \prod_{j=2}^{t} w_{r_j^{a_j}}.$$

Thus $w_n = yz$, and $y \in E_m$ gives

$$\mathrm{Tr}_{E_n/E_m}(w_n) = y \cdot \mathrm{Tr}_{E_n/E_m}(z).$$

Note that the Galois group of E_n/E_m has order r, and that its restriction to the field E_{r^a} gives rise to the Galois group of $E_{r^a}/E_{r^{a-1}}$; see the proof of Proposition 3.9.5. Therefore, by induction,

$$\mathrm{Tr}_{E_n/E_m}(z) = \mathrm{Tr}_{E_{r^a}/E_{r^{a-1}}}(z) = w_{r^{a-1}}.$$

Altogether, this shows

$$\mathrm{Tr}_{E_n/E_m}(w_n) = y \cdot w_{r^{a-1}} = w_{r_1^{a_1-1}} \cdot \prod_{j=2}^{t} w_{r_j^{a_j}} = w_{n/r_1},$$

in view of the underlying inductive construction principle. □

Finally, we remark that trace-compatible sequences of normal elements have also been studied by Scheerhorn [333, 334]. As an example, we cite one of the main results of [333], which should be compared with Construction 11.8.1.

Result 11.8.2 *Let $n \geq 2$, and let $(w_1, \ldots, w_{n-1})$ be a partial sequence in the algebraic closure $\widehat{F}$ of a Galois field $F = \mathrm{GF}(q)$ with characteristic p. Assume that every w_i is normal for E_i/F and that $\mathrm{Tr}_{E_j/E_i}(w_j) = w_i$, whenever $i, j \in \{1, \ldots, n-1\}$ and $i \mid j$. Then there exists a normal element w_n for E_n/F such that the extended sequence $(w_1, \ldots, w_{n-1}, w_n)$ is still trace-compatible.*

Moreover, the number $N_q^{tr}(n)$ of normal elements w_n for E_n/F with this property satisfies

$$N_q^{tr}(n) = \begin{cases} q^{\phi(n)} \cdot \left(1 - q^{-d}\right)^{\phi(n)/d} & \text{if } n \not\equiv 0 \bmod p, \text{ where } d = \mathrm{ord}_n(q), \\ q^{\phi(n)} & \text{if } p \mid n. \end{cases} \qquad \square$$

An analogous result for norm-compatible sequences of primitive elements can also be found in [333].

11.9 Two Algorithms for Determining Normal Elements

In this section, we present two reasonably efficient algorithms for computing a normal element for an arbitrary extension $E = \mathrm{GF}(q^n)$ of $F = \mathrm{GF}(q)$. For this purpose, we obviously need some form of explicit data for E when viewed as an F-vector space, for instance, a polynomial basis $B = \{1, \beta, \ldots, \beta^{n-1}\}$, where $\beta \in E$ has degree n over F; see Definition 8.1.1 and Remark 8.1.2. We begin with a simple observation:

Observation 11.9.1. Let $B = \{\beta_0, \ldots, \beta_{n-1}\}$ be any F-basis for the extension $E = \mathrm{GF}(q^n)$ of $F = \mathrm{GF}(q)$. As $x^n - 1$ is the minimal polynomial of the Frobenius automorphism σ of E/F, the least common multiple of the q-orders $\mathrm{Ord}_q(\beta_i)$ for $i = 0, \ldots, n-1$ has to be equal to that polynomial. □

We first present an algorithm which is due to Heinz Lüneburg [248] and relies on the polynomial ring version of his Algorithm r, which was discussed in Section 1.7 for the ring of integers; see Algorithm 1.7.12 and Remark 1.7.14. Given two (monic) polynomials $f(x), g(x) \in F[x]$, Algorithm r determines the largest (monic) divisor $r(x) \in F[x]$ of $f(x)$ such that $r(x)$ is relatively prime to $g(x)$. Thus $r(x)$ satisfies the following three properties:

(i) $r(x)$ divides $f(x)$;
(ii) $r(x)$ and $g(x)$ are relatively prime;
(iii) every irreducible divisor of the quotient $f(x)/r(x)$ divides $g(x)$.

In analogy to the notion of the m-part of N, where m, N are positive integers (see Definition 1.7.2), the quotient $f(x)/r(x)$ will be called the g-**part** of $f(x)$, whereas $r(x)$ is said to be the **complementary** g-**part** of $f(x)$.

Using this approach, the proof of part (2) of Theorem 1.9.10 can be made constructive in the special case of the $F[x]$-module E with respect to σ: we will show in Theorem 11.9.3 that it is not necessary to use the canonical factorizations of the polynomials $f(x)$ and $g(x)$. For this, we need the following preliminary result, which is the polynomial analogue of a corresponding result for the case of integers (see Exercise 1.7.19).

Proposition 11.9.2. *Let $f(x)$ and $g(x)$ be monic polynomials in $F[x]$. Determine two monic polynomials $f_1(x)$ and $g_2(x)$ as follows:*

- *let $f_0(x) := f(x)/\gcd(f,g)$ and $g_0(x) := g(x)/\gcd(f,g)$;*
- *let $f_1(x)$ be the complementary g_0-part of $f(x)$, and let $g_1(x)$ be the complementary f_0-part of $g(x)$;*
- *finally, let $g_2(x) := g_1(x)/\gcd(f_1,g_1)$.*

Then f_1 and g_2 are relatively prime, and their product is the least common multiple of f and g.

Proof. We use the canonical factorizations of $f(x)$ and $g(x)$ into irreducible polynomials, say

$$f(x) = \prod_{i=1}^{s} h_i(x)^{a_i} \quad \text{and} \quad g(x) = \prod_{i=1}^{s} h_i(x)^{b_i},$$

where $a_i, b_i \geq 0$ and $\max(a_i, b_i) \geq 1$ for all i. Then the polynomials in question are easily determined as follows:

$$f_0(x) = \prod_{i=1}^{s} h_i(x)^{a_i - \min(a_i,b_i)} \quad \text{and} \quad g_0(x) = \prod_{i=1}^{s} h_i(x)^{b_i - \min(a_i,b_i)},$$

which implies

$$f_1(x) = \prod_{\substack{i=1 \\ a_i \geq b_i}}^{s} h_i(x)^{a_i} \quad \text{and} \quad g_1(x) = \prod_{\substack{i=1 \\ a_i \leq b_i}}^{s} h_i(x)^{b_i},$$

and hence

$$g_2(x) = \prod_{i=1, a_i < b_i}^{s} h_i(x)^{b_i}.$$

Now the assertion is immediate. □

Theorem 11.9.3. *Consider an extension E/F of Galois fields, where $F = \mathrm{GF}(q)$ and $E = \mathrm{GF}(q^n)$. Assume that $u, v \in E$ have q-orders $f(x)$ and $g(x)$, respectively, and let $f_1(x)$ and $g_2(x)$ be as in Proposition 11.9.2. Then*

$$y := \tfrac{f}{f_1}(\sigma)(u) + \tfrac{g}{g_2}(\sigma)(v)$$

has q-order $\mathrm{Ord}_q(y) = \mathrm{lcm}(f(x), g(x))$.

Proof. According to Lemma 1.6.15, the q-order of $(f/f_1)(\sigma)(u)$ is $f_1(x)$, and the q-order of $(g/g_2)(\sigma)(v)$ is $g_2(x)$. By Proposition 11.9.2, f_1 and g_2 are relatively prime, so that y has q-order $f_1 g_2 = \operatorname{lcm}\big(f(x), g(x)\big)$; see Lemma 1.6.19. □

These considerations lead to the following high-level description of a method for constructing normal elements:

Algorithm 11.9.4 (Lüneburg's algorithm).

- *Input:* A field extension E/F where $E = \mathrm{GF}(q^n)$ and $F = \mathrm{GF}(q)$, given by explicit data in form of an F-basis $B = \{b_0, \ldots, b_{n-1}\}$ of E.
- *Output:* A normal element w for E/F.

(1) $i \leftarrow 0, \quad u \leftarrow b_i$;
(2) determine the q-order $f(x)$ of u;
(3) **for** $i := 1$ **to** $n-1$ **do**
(4) $\quad v \leftarrow b_i$;
(5) $\quad$ determine the q-order $g(x)$ of v;
(6) $\quad$ determine an element y with q-order $\operatorname{lcm}(f, g)$;
(7) $\quad u \leftarrow y$
(8) **od**
(9) $w \leftarrow u$.

Remark 11.9.5. We note that Steps (2) and (5) of Algorithm 11.9.4, where the q-order of an element $v \in E$ is required, can be performed efficiently by applying the Gaussian algorithm to determine the smallest integer $k \geq 1$ such that $v, \sigma(v), \ldots, \sigma^k(v)$ are linearly dependent over F. Then a further application of that algorithm produces a linear representation of $\sigma^k(v)$ in terms of $v, \sigma(v), \ldots, \sigma^{k-1}(v)$, say $\sigma^k(v) = \sum_{i=0}^{k-1} \lambda_i \sigma^i(v)$, and one obtains $\mathrm{Ord}_q(v) = x^k - \sum_{i=0}^{k-1} \lambda_i x^i$.

Furthermore, one can give an efficient implementation of Step (6) according to Proposition 11.9.2 and Theorem 11.9.3, where ones uses Algorithm r to compute the required complementary parts; see Exercise 11.9.9.

With these specifications, the correctness of Algorithm 11.9.4 follows from Observation 11.9.1, Proposition 11.9.2 and Theorem 11.9.3. More formally, one uses induction on i to check that

$$\mathrm{Ord}_q(u) = \operatorname{lcm}\big(\mathrm{Ord}_q(b_j) : j = 0, \ldots, i\big)$$

is a loop invariant. □

For more applications and a formulation of Algorithm r for Bezout domains (in particular, for principal ideal domains), we refer the reader to [248]. Further detailed information can also be obtained from Lüneburg's book [250], where all algorithms are given in `PASCAL`-code. There Algorithm r is used as a basic tool for computing the rational normal form of an endomorphism, which in the cyclic case reduces to the determination of a generator of the underlying vector space.

Our second algorithm is due to Hendrik Lenstra [237]. His method constructs elements with q-orders of successively increasing degree, starting from an arbitrary element, and rests on the following result:

Proposition 11.9.6. *Consider an extension E/F of Galois fields, where $F = \mathrm{GF}(q)$ and $E = \mathrm{GF}(q^n)$, and let $u \neq 0$ be an element of E with q-order $h(x)$. Then there exist elements $v \in E$ such that $g(\sigma)(v) = u$, where σ is the Frobenius automorphism of E/F and where $g(x) = (x^n - 1)/h(x)$.*

For any such v, either $f(x) := \mathrm{Ord}_q(v)$ satisfies $\deg f > \deg h$, or $f = h$; moreover, in the latter case h is a proper divisor of $\mathrm{Ord}_q(u+y)$ for every element $y \in E^*$ satisfying $g(\sigma)(y) = 0$.

Proof. We consider the (F,σ)-module V_h^F generated by u, that is,

$$V_h^F = \{y \in E : h(\sigma)(y) = 0\};$$

see Observation 11.1.1. Note that V_h^F can also be written as the image of E under the F-linear mapping $g(\sigma)$. To see this, we let z be any normal element for E/F; then $g(\sigma)(z)$ also has q-order h and thus generates V_h^F (by Remark 2.6.7, since z has q-order $x^n - 1$).

Now let $u = g(\sigma)(v)$ and assume that $f(x) = \mathrm{Ord}_q(v)$ satisfies $\deg f \leq \deg h$. Then necessarily $f = h$, since $\mathrm{Ord}_q(u)$ has to divide $\mathrm{Ord}_q(v)$, as u is contained in the (F,σ)-submodule of E generated by v. Hence $g(x)$ and $h(x)$ have to be relatively prime; see Observation 2.6.8. Let $y \neq 0$ be any element of E satisfying $g(\sigma)(y) = 0$, so that $\mathrm{Ord}_q(y)$ is a monic divisor of g with degree at least 1. Then another application of Observation 2.6.8 shows that

$$\mathrm{Ord}_q(u+y) = \mathrm{Ord}_q(u) \cdot \mathrm{Ord}_q(y)$$

is indeed a proper multiple of $\mathrm{Ord}_q(u)$. □

Algorithm 11.9.7 (Lenstra's algorithm).

- *Input:* A field extension E/F where $E = \mathrm{GF}(q^n)$ and $F = \mathrm{GF}(q)$, and some element $u \neq 0$ of E.
- *Output:* A normal element w for E/F.

(1) determine the q-order $h(x)$ of u;
(2) **while** $\deg h < n$ **do**
(3) $\quad g(x) \leftarrow (x^n - 1)/h(x)$;
(4) $\quad$ determine an element $v \in E$ such that $g(\sigma)(v) = u$;
(5) $\quad f(x) \leftarrow \mathrm{Ord}_q(v)$;
(6) $\quad$ **if** $\deg f > \deg h$
(7) $\quad\quad$ **then** $u \leftarrow v$; $h(x) \leftarrow f(x)$
(8) $\quad\quad$ **else** determine some element $y \neq 0$ in V with $g(\sigma)(y) = 0$;
(9) $\quad\quad\quad u \leftarrow u + y$; $h(x) \leftarrow \mathrm{Ord}_q(u)$
(10) $\quad$ **fi**

(11) **od**;
(12) $w \leftarrow u$.

Remark 11.9.8. Using Proposition 11.9.6, it is easy to see that Algorithm 11.9.7 is correct: an element v as required in Step (4) always exists, and the degree of the polynomial $h(x) = \mathrm{Ord}_q(u)$ increases during every execution of the **while** loop. This is trivial when $\deg f > \deg h$; and if $\deg f \leq \deg h$, Steps (8) and (9) update h so that the requirement holds, according to 11.9.6. Thus the algorithm terminates with an element w which has q-order of degree n, and hence $\mathrm{Ord}_q(w) = x^n - 1$, as claimed.

We remark that the necessary computational tasks – that is, the determination of the elements v and y required in Steps (4) and (8), respectively, and the evaluation of q-orders – can again be performed efficiently via the Gaussian algorithm, as in the case of Lüneburg's algorithm. □

A further deterministic algorithm for computing a normal element for an extension $\mathrm{GF}(q^n)/\mathrm{GF}(q)$ was given by Bach, Driscoll, and Shallit in [16], who also proved that this task can be performed with $\mathrm{O}\big((n^2 + \log q)(n \log q)^2\big)$ bit operations, provided that q is a prime and that E is given as $E = F[x]/(f)$ for some monic irreducible polynomial f with degree n. It is also mentioned in [16] that both Lüneburg's and Lenstra's algorithms have the same complexity as their algorithm.

Efficient probabilistic and deterministic algorithms for determining a normal basis over an extension of finite fields are due to von zur Gathen and Giesbrecht in [389]; see Remark 7.3.11. Further deterministic algorithms can be found in Semaev [346] and in Poli [323]; the complexity of Poli's algorithm is given by $O\big((n^3 + n \log n \log\log n) \log q\big)$. The algorithms in these three references all rely on decompositions of the additive group of the extension field under consideration; compare with Remark 11.1.3.

Exercises

Exercise 11.9.9. Write down the details for Step (6) of Algorithm 11.9.4 as a subroutine relying on (the polynomial version of) Algorithm 1.7.12, Proposition 11.9.2 and Theorem 11.9.3. □

11.10 Concluding Remarks

The search for normal bases (of special types) for extensions of Galois fields is a central topic in finite field theory, and also in this book. This includes

- self-dual normal bases (see Section 7.8),
- low complexity and optimal normal bases (see Sections 8.5 and 8.8),
- complete normal bases (see Chapter 12),
- primitive normal bases (see Chapter 13).

For a recent survey and further references, the interested reader is referred to Chapter 5 of the *Handbook of Finite Fields* [292]. In particular, we refer to [292, Section 5.3.2] for the connections between *Gauss periods* and normal bases, which generalize optimal and low complexity normal bases. In a sense, the construction of normal bases via Gauss periods is complementary to the point of view we have taken in the present chapter.

We conclude this chapter with stating just one result in this direction, which is essentially [292, Theorem 5.3.17]; it is due to Wassermann [398] and Gao, von zur Gathen, Panario and Shoup [129, 130].

Result 11.10.1 *Let n and k be positive integers such that $r = nk + 1$ is a prime, and consider a finite field $F = \mathrm{GF}(q)$ with characteristic $p \neq r$. Let $\beta \in L = \mathrm{GF}(q^{kn})$ be a primitive r-th root of unity, and let G be the subgroup of order k of the group of units modulo r. Then the element $v := \sum_{a \in G} \beta^a$ has the following properties:*

(1) *$v \in E = \mathrm{GF}(q^n)$;*
(2) *v is normal for E/F if and only if $\gcd(n, nk/e) = 1$, where $e = \mathrm{ord}_r(q)$.* □

Elements v of the type considered in Result 11.10.1 are called **Gauss periods** of type (n, k) over $\mathrm{GF}(q)$. For the description of normal bases via general(ized) Gauss periods, we refer to Feisel, von zur Gathen and Shokrollahi [115].

Chapter 12
Complete Normal Bases and Generalized Cyclotomic Modules

Abstract In the present chapter, we consider a particular class of normal elements for extensions of finite fields. We introduce the topic by presenting two examples of normal elements for Galois extensions E/F which are *not* normal for some intermediate extension E/K of E/F. This suggests to study *completely normal* elements for E/F, that is, normal elements which are also normal over every intermediate field K of E/F.

We first prove the existence of such elements in the more general setting of Galois extensions, before we focus on the construction of completely normal elements in the special case of finite fields. In fact, the main parts of this chapter may be viewed as variations on the themes encountered in Chapter 10; for instance, the class of regular extensions will again play a prominent role. As a result of these deeper investigations, we will obtain an interesting structure theory for extensions of Galois fields.

12.1 A Strengthening of the Normal Basis Theorem

This section contains an introductory discussion of completely normal elements for arbitrary Galois extensions E/F and motivates the subject via a few first examples. We begin by setting up the necessary terminology, reviewing and extending the concepts discussed in Section 3.8. Let $G = \mathrm{Gal}(E/F)$ be the Galois group of E/F. Then the additive group of E becomes a module over the group algebra FG by defining the scalar multiplication via

$$\Big(\sum_{\gamma\in G} a_\gamma \gamma\Big) * w := \sum_{\gamma\in G} a_\gamma \gamma(w), \tag{12.1}$$

where $w \in E$ and $a_\gamma \in F$ for all $\gamma \in G$; we leave the simple verification as Exercise 12.1.8. Using this, the normal basis theorem (Theorem 3.9.4) may be rephrased as follows: there exists an element $w \in E$ such that

D. Hachenberger and D. Jungnickel, *Topics in Galois Fields*,
Algorithms and Computation in Mathematics 29,
https://doi.org/10.1007/978-3-030-60806-4_12

$$E = \{f * w \colon f \in FG\}.$$

Thus the additive group of E is actually a cyclic FG-module (isomorphic to FG), and the generators of this module are exactly the normal elements for E/F.

By the fundamental theorem of Galois Theory (see Theorem 3.8.8 or, for instance, Jacobson [204, Section 4.5]), the subgroups of G correspond bijectively to the intermediate fields of E over F: if H is a subgroup of G, then the set of elements of E which are fixed by every element of H is an intermediate field K of E/F, the **fixed field** of H. Conversely, if K is some intermediate field of E/F, then E is a Galois extension over K, and the Galois group of E/K is the subgroup H of G which fixes K elementwise. Consequently, for every intermediate field K of E/F with Galois group $H = \mathrm{Gal}(E/K)$, the additive group of E also carries the structure of a KH-module. We now apply the normal basis theorem to E/K and conclude that $(E,+)$ is also a cyclic KH-module (isomorphic to KH); moreover, the generators of this module are precisely those elements of E whose H-conjugates form a basis for E over K, that is, the normal elements for E/K. Note that $[K:F]\cdot|H| = [E:F] = |G|$, since $|H| = [E:K]$.

Now let w be any generator for E/F, that is, $F(w) = E$. Trivially, w generates E also over every intermediate field K of E/F, and therefore the initial subset $\{1, w, \dots, w^{[E:K]-1}\}$ of the polynomial basis $\{1, w, \dots, w^{[E:F]-1}\}$ for E/F is a polynomial basis for E/K. It is natural to ask whether or not an analogous result holds for normal elements: is an arbitrary normal element w for E/F also a normal element for E/K? More precisely, is the set $\{\beta(w) : \beta \in H = \mathrm{Gal}(E,K)\}$ of H-conjugates of w a (necessarily normal) basis for E/K? The following two examples will show that this is not always the case.

Example 12.1.1. Consider the binary field $F = \mathrm{GF}(2)$ and its 21-dimensional extension $E = \mathrm{GF}(2^{21})$. Note that the polynomial $f_1(x) := x^3 + x^2 + 1$ is an irreducible divisor of the cyclotomic polynomial $\Phi_7(x)$ over F and that

$$g(x) := f_1(x^7) = x^{21} + x^{14} + 1$$

is an irreducible divisor of $\Phi_7(x^7) = \Phi_{49}(x)$, since $\mathrm{ord}_{49}(2) = 21$. Thus any root u of g is a generator for E/F. As u is a primitive 49-th root of unity, $\zeta := u^7$ is a primitive 7-th root of unity. The minimal polynomial of ζ over F is f_1, and $K = F(\zeta)$ is the intermediate field $\mathrm{GF}(2^3)$ of E/F. We will show that

$$w := \zeta + u + u^3$$

is a normal element for E/F, but not for E/K. This requires a few steps:

- Note first that $x^3 - 1$ splits over F into the two irreducible factors $x - 1$ and $x^2 + x + 1$. Because of $\zeta \in K \setminus F$ and $\mathrm{Tr}_{K/F}(\zeta) = 1$, the 2-order of ζ is neither $x - 1$ nor $x^2 + x + 1$. Hence $\mathrm{Ord}_2(\zeta) = x^3 - 1$, and thus ζ is a normal element for K/F.
- Now we consider the extension E/K with degree 7; for the sake of clarity, we will use a different indeterminate to denote polynomials over K, say y. Here the

polynomial $y^7 - 1$ splits completely into linear factors, one of which is $y - \zeta$. This polynomial turns out to be the 8-order of u, since

$$(y-\zeta)(\sigma^3)(u) = u^8 - \zeta u = u \cdot (u^7 - \zeta) = 0,$$

where σ is the Frobenius automorphism of E/F. (Note that this is a special case of part (2) of Theorem 11.4.6.) Therefore, an application of part (2) of Theorem 11.6.2 (and its proof) yields $\mathrm{Ord}_2(u) = f_1(x^3) = x^9 + x^6 + 1$, since $f_1(y) = (y-\zeta)(y-\zeta^2)(y-\zeta^4)$. Similarly, one obtains $\mathrm{Ord}_8(u^3) = y - \zeta^3$ and $\mathrm{Ord}_2(u^3) = f_2(x^3) = x^9 + x^3 + 1$, where $f_2(x) = x^3 + x + 1$.

- The preceding facts combine to give

$$\mathrm{Ord}_2(w) = (x^3-1) \cdot f_1(x^3) \cdot f_2(x^3) = \Phi_1(x^3) \cdot \Phi_7(x^3) = x^{21} - 1,$$

so that w is a normal element for E/F, as claimed.

- Finally, $\mathrm{Ord}_8(\zeta) = y - 1$ yields

$$\mathrm{Ord}_8(w) = \mathrm{Ord}_8(\zeta) \cdot \mathrm{Ord}_8(u) \cdot \mathrm{Ord}_8(u^3) = (y-1) \cdot (y-\zeta) \cdot (y-\zeta^3),$$

which is a proper divisor of $y^7 - 1$. Therefore w is not normal for E/K. □

Example 12.1.2. As a further example, we consider the 8-dimensional extension $E = \mathrm{GF}(3^8)$ of the ternary field $F = \mathrm{GF}(3)$ and its intermediate field $K = \mathrm{GF}(3^2)$. Again, we determine a normal element w for E/F which is not normal for E/K.

- First, we apply Theorem 11.7.3 with $c = k = 3$ and $b = 2$, hence $k - 1 + c = 5$. Thus we let u be a primitive 32-th root of unity. Then $F(u) = E$, and $\mathrm{Ord}_3(u) = \Phi_8(x)$, since $Q_{2^b} = I_{2^b}$ in this case.
- The element u^2 is a primitive 16-th root of unity, and therefore $\mathrm{Ord}_3(u^2) = x^2 + 1$; see Example 11.7.2.
- The element u^4 is a primitive 8-th root of unity, and therefore $\mathrm{Ord}_3(u^4) = x^2 - 1$; see Example 11.1.5.
- Combining these facts, we see that $w := u^4 + u^2 + u$ is a normal element for the extension E/F:

$$\mathrm{Ord}_3(w) = (x^2-1) \cdot (x^2+1) \cdot \Phi_8(x) = (x^4-1) \cdot (x^4+1) = x^8 - 1.$$

It remains to check that the 9-order of w is a proper divisor of $y^4 - 1$, so that w is not normal for E/K. For this, we need to determine the 9-orders of the three components of w. It is easy to see $\mathrm{Ord}_9(u^4) = y - 1$ and $\mathrm{Ord}_9(u^2) = y + 1$. Finally, we note that $\iota := u^8$ is a primitive fourth root of unity and that $\mathrm{Ord}_9(u) = y - \iota$, since

$$(y-\iota)(\sigma^2)(u) = u^9 - \iota u = u \cdot (u^8 - \iota) = 0;$$

see part (2) of Theorem 11.4.6. This gives $\mathrm{Ord}_9(w) = (y-1)(y+1)(y-\iota)$. □

These two examples motivate the following definition, cf. Remark 11.5.4:

Definition 12.1.3. Let E/F be a Galois extension. If $w \in E$ is simultaneously normal over every intermediate field of E/F, then w is said to be a **completely normal element** for E/F. □

The *complete normal basis theorem*, which was first proved in full generality by Blessenohl and Johnsen [44] in 1986, states that completely normal elements always exist:

Theorem 12.1.4 (Complete normal basis theorem). *For every Galois extension E/F, there exists a completely normal element.*

In the remainder of this chapter, we shall prove this major result and also develop a structure theory for completely normal elements for Galois field extensions. Here is a brief overview:

- In Section 12.2, we prove the existence of a completely normal element for Galois extensions E/F over an infinite ground field F; this special case of Theorem 12.1.4 was already established in 1957 by Faith [114].
- The remainder of the chapter deals with the finite case, where $F = \mathrm{GF}(q)$. In Section 12.3, we characterize all those extensions E/F of finite fields for which all normal elements are actually completely normal. In particular, this holds when the degree of E/F is a power of the characteristic of F.
- In Section 12.4, we strengthen the reduction theorem for normal elements to a corresponding reduction theorem for completely normal elements. This reduces the proof of the complete normal basis theorem to the case where the degree of E/F is a power of a prime r which is distinct from the characteristic of F. Such extensions are then considered in Section 12.5, where we complete the proof of Theorem 12.1.4.
- In Section 12.6, we will extend the construction given in Section 11.8 and establish the existence of trace-compatible sequences for the algebraic closure $\widehat{F}$ of $F = \mathrm{GF}(q)$ which consist entirely of completely normal elements over F.
- The remaining three sections, which are to a large extent based on Hachenberger [161], develop a structure theory for completely normal elements over finite fields. For this purpose, we introduce the notions of a generalized cyclotomic module as well as the module character and complete generators of such an object in Section 12.7 and prove a decomposition theorem in Section 12.8. Finally, we revisit the class of regular extensions in Section 12.9, where we will investigate it from the complete point of view.

The proof of the complete normal basis theorem which is presented in this book follows [160]. It is arguably simpler and certainly more elementary than the original proof, since it relies on cyclotomic polynomials rather than the representation theory of abelian groups. Moreover, this approach forms the basis for the general structure theory developed in the final three sections of this chapter.

We conclude this introductory section by returning to the two specific extensions considered in Examples 12.1.1 and 12.1.2.

Example 12.1.5. We continue Example 12.1.1 and exhibit a completely normal element for E/F, namely

$$w' := \zeta + v_1 + v_2, \text{ where } v_1 := u + u^2 + u^4 \text{ and } v_2 := u^3 + u^6 + u^{12}.$$

Thus we have to check the normality of w' over the three non-trivial subfields F, K and L of E, where $L = \mathrm{GF}(2^7)$. For this, we use the 2- and 8-orders which were already computed in Example 12.1.1 as follows:

- We begin with K. Here we have

$$\mathrm{Ord}_8(v_1) = (y-\zeta)(y-\zeta^2)(y-\zeta^4) = y^3 + y^2 + 1 = f_1(y),$$

  and similarly

$$\mathrm{Ord}_8(v_2) = (y-\zeta^3)(y-\zeta^6)(y-\zeta^5) = y^3 + y + 1 = f_2(y).$$

  Because of $\mathrm{Ord}_8(\zeta) = y - 1$, we obtain $\mathrm{Ord}_8(w') = (y-1)f_1(y)f_2(y) = y^7 - 1$, and hence w' is normal for E/K.
- Next, we consider w' over F. Note that $v_1 = (x^2 + x + 1)(\sigma)(u)$ and

$$\gcd\left(x^2 + x + 1, \mathrm{Ord}_2(u)\right) = \gcd\left(\Phi_3(x), f_1(x^3)\right) = 1,$$

  so that $\mathrm{Ord}_2(v_1) = \mathrm{Ord}_2(u) = f_1(x^3)$. Similarly, $\mathrm{Ord}_2(v_2) = \mathrm{Ord}_2(u^3) = f_2(x^3)$. In view of $\mathrm{Ord}_2(\zeta) = x^3 - 1$, we obtain the required result

$$\mathrm{Ord}_2(w') = (x^3 - 1) \cdot f_1(x^3) \cdot f_2(x^3) = x^{21} - 1.$$

- Finally, we turn our attention to L, where we use the indeterminate z to describe 128-orders. Note that $[E : L] = 3$ and that $z^3 - 1$ splits over L into the two irreducible factors $z - 1$ and $z^2 + z + 1$. It is clear that the 128-order of w' cannot be one of these two polynomials, as $w' \notin L$ and $\mathrm{Tr}_{E/L}(w') \neq 0$ (which is a trivial consequence of $\mathrm{Tr}_{E/F}(w') \neq 0$). This proves that w' is also normal over L. □

Example 12.1.6. We continue Example 12.1.2 and show that

$$w' := u^4 + u^2 + v, \text{ where } v := u + u^3,$$

is a completely normal element. Again, E has three non-trivial subfields, namely F, K and $L = \mathrm{GF}(3^4)$. Using the results already computed in Example 12.1.2, we proceed as follows:

- We begin with K. Here $\mathrm{Ord}_9(u^3) = y - \iota^3 = y + \iota$, and therefore

$$\mathrm{Ord}_9(v) = \mathrm{Ord}_9(u) \cdot \mathrm{Ord}_9(u^3) = (y - \iota)(y + \iota) = y^2 + 1.$$

  Since $\mathrm{Ord}_9(u^4) = y - 1$ and $\mathrm{Ord}_9(u^2) = y + 1$, we obtain $\mathrm{Ord}_9(w') = x^4 - 1$, so that w' is normal for E/K.

- For the ground field F, we observe that $v = (x+1)(\sigma)(u)$ and that $x+1$ and $\mathrm{Ord}_3(u) = \Phi_8(x)$ are relatively prime, which gives $\mathrm{Ord}_3(v) = x^4+1$. Because of $\mathrm{Ord}_3(u^2) = x^2+1$ and $\mathrm{Ord}_3(u^4) = x^2-1$, we see that w' has 3-order x^8-1, as required.
- Finally, we consider L and argue as in Example 12.1.5. Now $[E:L] = 2$, and z^2-1 splits over L into $z-1$ and $z+1$. Again, $w' \notin L$ and $\mathrm{Tr}_{E/L}(w') \neq 0$, and therefore the 81-order of w' is z^2-1, as desired. □

We conclude this section with the following interesting result concerning normal elements for the algebraic closure of Galois fields, which is likewise due to Blessenohl and Johnsen; see [44, Satz 2.9].

Result 12.1.7 *Consider the algebraic closure $\widehat{F}$ of a finite field $F = \mathrm{GF}(q)$. An element w of $\widehat{F}$ is said to be* ***totally normal*** *over F, provided that $K(w)$ is normal over K for every intermediate field K of $\widehat{F}/F$. Then the following two conditions are equivalent for $w \in \widehat{F}$:*

(i) *w is totally normal over F;*
(ii) *w is a completely normal element for the extension $F(w)/F$.* □

Exercises

Exercise 12.1.8. Let E/F be an arbitrary Galois extension with Galois group $G = \mathrm{Gal}(E/F)$. Verify that the scalar multiplication defined in Equation (12.1) turns $(E,+)$ into an FG-module. □

Exercise 12.1.9. Prove Result 12.1.7. □

Exercise 12.1.10. Consider the quartic extension $E = \mathrm{GF}(q^4)$ of the Galois field $F = \mathrm{GF}(q)$. Prove that every element of E which is normal over F is in fact completely normal over F. □

12.2 Extensions over an Infinite Field

In the present section, which is based on Faith [114] and Blessenohl and Johnsen [44], we prove the complete normal basis theorem for Galois extensions over an infinite ground field F. This proof is a generalization of Artin's proof of the normal basis theorem presented in Section 3.9. As in this classical case, the arguments also apply to extensions of finite fields whenever the cardinality of the ground field is sufficiently large. In particular, for every fixed value of n, this leaves only finitely many prime powers q where the existence of a completely normal element for the n-dimensional extension $\mathrm{GF}(q^n)/\mathrm{GF}(q)$ is in doubt. In fact, there are no exceptions, but this cannot be shown using Artin's method and requires a deeper investigation, which will be the topic of the subsequent three sections.

Theorem 12.2.1. *Let E be a Galois extension with degree $n>1$ over F, and assume that $|F|\geq n(n-1)\cdot\mathrm{rad}(n)$. Then there exists a completely normal element for E/F.*

Proof The proof will be a more elaborate version of the one given for Case 2 of Theorem 3.9.4, using the same terminology (with a few minor adjustments). In particular, we let v be some element of E satisfying $F(v)=E$ and denote the minimal polynomial of v over F by $g(x)$. Since we now have to deal with all subgroups of the Galois group $G=\mathrm{Gal}(E/F)$, it is preferable to change the notation e_k used in the old proof slightly and denote the idempotent of the factor ring $E[x]/(g)$ corresponding to the root $\gamma(v)$ of $g(x)$ by $e_\gamma(x)$ (where $\gamma\in G$), that is,

$$e_\gamma(x) = \frac{1}{g'(\gamma(v))}\cdot\frac{g(x)}{x-\gamma(v)}.$$

Applying the Galois group G (by letting it act on the coefficients of the polynomials involved, as usual) gives

$$\beta(e_\gamma(x)) = e_{\beta\gamma}(x) \quad \text{for all } \beta\in G.$$

Given any intermediate field K of E/F, we abbreviate $\mathrm{Gal}(E/K)$ as G_K, define a matrix M_K with entries from $E[x]$ via

$$M_K := \left(e_{\gamma\delta}(x)\right)_{\gamma,\delta\in G_K},$$

and let $D_K(x)$ denote the determinant of M_K; note that we had only needed the special case $K=F$ in the proof of Theorem 3.9.4. As in that proof, we now show that $D_K(x)\neq 0$, by considering the matrix $M_K M_K^T$ with determinant $D_K(x)^2$ and (σ,τ)-entry

$$\sum_{\gamma\in G_K} e_{\sigma\gamma}(x)e_{\tau\gamma}(x).$$

Since the polynomials $e_\gamma(x)$ (with $\gamma\in G$) form a system of pairwise orthogonal idempotents of $E[x]/(g)$ decomposing the unit element, the (σ,τ)-entry of $M_K M_K^T$ is congruent to 0 modulo $g(x)$ when $\sigma\neq\tau$, and to $e_K(x)$ modulo $g(x)$ otherwise, where $e_K(x)$ denotes the idempotent

$$e_K(x) := \sum_{\gamma\in G_K} e_\gamma(x)$$

of $E[x]/(g)$ associated with the intermediate field K. This implies

$$D_K(x)^2 = \det\left(M_K M_K^T\right) \equiv e_K(x)^{|G_K|} \equiv e_K(x) \mod g(x).$$

Now let $\mathscr{R}_K$ be some system of representatives for the left cosets of G_K in G. Then

$$\sum_{\beta\in\mathscr{R}_K}\beta(e_K(x)) = \sum_{\beta\in\mathscr{R}_K}\sum_{\gamma\in G_K}\beta(e_\gamma(x)) = \sum_{\delta\in G} e_\delta(x) \equiv 1 \mod g(x),$$

which proves our claim that $D_K(x)$ cannot be the zero polynomial.

As in the proof of Theorem 3.9.4, the well-known formula for the determinant shows that the degree of $D_K(x)$ is at most $k(n-1)$, where $k := |G_K| = [E:K]$ is the degree of E/K, since M_K is a (k,k)-matrix and since the polynomials e_γ all have degree $n-1$. By hypothesis, $|F| > \deg D_K$, and hence there exists an element λ in F such that $D_K(\lambda) \neq 0$. This implies that $v := e_{\mathrm{id}}(\lambda)$ is a normal element for the extension E/K, as in the proof of Theorem 3.9.4.

It remains to show that there even exists an element in E which is completely normal over F. For this, we consider the polynomial

$$D(x) := \prod_K D_K(x),$$

where the product runs over the (finitely many) intermediate fields K of E/F. The above arguments show that $D(x)$ is a non-zero polynomial with degree at most $(n-1) \cdot \sum_K [E:K]$. If $|F|$ exceeds this integer, the existence of an element $\alpha \in F$ satisfying $D(\alpha) \neq 0$ follows, and then $w := e_{\mathrm{id}}(\alpha)$ is the desired completely normal element for E/F.

Trivially, this argument establishes the existence of completely normal elements in the case where F is infinite. Now let F be finite. In this case, there exists exactly one intermediate field K_d of E/F with degree d over F for every divisor d of n. Therefore, we wish to estimate the sum

$$t(n) := \sum_{d|n} \frac{n}{d} = \sum_{d|n} d$$

of the degrees $[E:K_d]$ over all intermediate fields. For this, we let $\prod_{i=1}^{\ell} r_i^{a_i}$ be the prime power factorization of n and compute

$$t(n) = \prod_{i=1}^{\ell} t(r_i^{a_i}) = \prod_{i=1}^{\ell} \Big(\sum_{j=0}^{a_i} r_i^j \Big) = \prod_{i=1}^{\ell} \frac{r_i^{a_i+1}-1}{r_i-1} < \prod_{i=1}^{\ell} r_i^{a_i+1} = n \cdot \mathrm{rad}(n).$$

Hence the polynomial $D(x)$ satisfies $\deg D < (n-1)n \cdot \mathrm{rad}(n) \leq |F|$, as required. □

Hachenberger [170] improved the bound $(n-1)n \cdot \mathrm{rad}(n)$ in Theorem 12.2.1 to $n \cdot \mathrm{rad}(n)$. Subsequently, Garefalakis and Kapatenakis [131] obtained a further improvement which we will present now. Both results require only elementary arguments.

Theorem 12.2.2. *The extension* $E = \mathrm{GF}(q^n)$ *over* $F = \mathrm{GF}(q)$ *admits a completely normal element whenever* $q \geq n+1$.

Proof. Let $\mathscr{C}$ denote the set of all elements of E which are completely normal over F. We will show that

$$|\mathscr{C}| \geq q^n \cdot \Big(1 - \frac{n(q+1)}{q^2}\Big);$$

it is clear that this quantity is indeed positive for $q \geq n+1$.

For every divisor k of n, let $\mathcal{N}_k$ be the set of all elements of E which are normal over the k-dimensional extension E_k over F, so that $\mathcal{C} = \bigcap_{k|n} \mathcal{N}_k$. Hence

$$E \setminus \mathcal{C} = \bigcup_{k|n} (E \setminus \mathcal{N}_k)$$

and therefore

$$q^n - |\mathcal{C}| = |E \setminus \mathcal{C}| \leq \sum_{k|n} |E \setminus \mathcal{N}_k| = \sum_{k|n} \left(q^n - \phi_{q^k}(x^{n/k} - 1)\right),$$

which immediately implies

$$|\mathcal{C}| \geq q^n + \sum_{k|n} \left(\phi_{q^k}(x^{n/k} - 1) - q^n\right).$$

Now fix some divisor k of n, and write $Q = q^k$ and $\frac{n}{k} = \ell p^b$, where p is the characteristic of F and ℓ is not divisible by p. Using Proposition 3.6.19, we obtain

$$\begin{aligned}
\phi_{q^k}(x^{n/k} - 1) &= \phi_Q(x^{\ell p^b} - 1) = Q^{\ell(p^b-1)} \cdot \prod_{d|\ell} \left(Q^{\operatorname{ord}_d(Q)} - 1\right)^{\phi(d)/\operatorname{ord}_d(Q)} \\
&\geq Q^{\ell(p^b-1)} \cdot \prod_{d|\ell} (Q-1)^{\phi(d)} = Q^{\ell(p^b-1)} \cdot (Q-1)^{\sum_{d|\ell} \phi(d)} \\
&= Q^{\ell(p^b-1)} \cdot (Q-1)^{\ell} = Q^{\ell p^b} \cdot \left(1 - \tfrac{1}{Q}\right)^{\ell}.
\end{aligned}$$

Consequently,

$$\phi_{q^k}(x^{n/k} - 1) \geq q^n \cdot (1 - \tfrac{1}{q^k})^{\ell} \geq q^n \cdot (1 - \tfrac{1}{q^k})^{n/k} \geq q^n \cdot (1 - \tfrac{n}{kq^k}),$$

where the final estimate follows from Bernoulli's inequality. This implies

$$|\mathcal{C}| \geq q^n + \sum_{k|n} \left(q^n \cdot (1 - \tfrac{n}{kq^k}) - q^n\right) = q^n \cdot \left(1 - \sum_{k|n} \tfrac{n}{kq^k}\right).$$

Finally, we compute the following rough estimate:

$$\begin{aligned}
\sum_{k|n} \frac{1}{kq^k} &\leq \frac{1}{q} + \sum_{k=2}^{n} \frac{1}{kq^k} \leq \frac{1}{q} + \frac{1}{2} \cdot \sum_{k=2}^{n} \frac{1}{q^k} \\
&= \frac{1}{q} + \frac{1}{2q^2} \cdot \sum_{\ell=0}^{n-2} \frac{1}{q^\ell} \leq \frac{1}{q} + \frac{1}{q^2} = \frac{1}{q^2} \cdot (q+1),
\end{aligned}$$

since $\sum_{\ell=0}^{n-2} q^{-\ell} \leq 2$ in view of the trivial inequality $q^d + \cdots + q + 1 \leq 2q^d$ (for all d). This yields the desired lower bound for $|\mathcal{C}|$. □

12.3 Completely Basic Extensions

In this section, we characterize those extensions of Galois fields where all normal elements are in fact completely normal, a situation first considered by Faith [114].

Definition 12.3.1. A Galois extension E/F is said to be **completely basic** provided that every normal element for E/F is in fact completely normal. □

Blessenohl [42] considered completely basic extensions which are cyclic of prime power degree. Subsequently, Blessenohl and Johnsen [45] characterized all completely basic abelian Galois extensions, and Meyer [274] extended these results to a certain class of not necessarily abelian extensions.

We will only deal with the case of finite fields (see Theorem 12.3.5 below), using an elementary approach based on Hachenberger [161, Section 15]. We first prove a fundamental fact on additive orders:

Proposition 12.3.2. *Consider a finite field $F = \mathrm{GF}(q)$ with characteristic p, let k and m be two positive integers, and assume that m is not divisible by p and that $\mathrm{ord}_m(q)$ and k are relatively prime.*

Moreover, let $f(x)$ be some monic divisor of $x^m - 1$ over F, and let v be any element in the algebraic closure $\widehat{F}$ of F which satisfies $\mathrm{Ord}_q(v) = f(x^k)^c$ for some positive integer c. Then one has $\mathrm{Ord}_{q^k}(v) = f(x)^c$.

Proof. By hypothesis, $f(x)$ is a square-free polynomial, say $f(x) = \prod_{i=1}^t g_i(x)$ with distinct monic irreducible polynomials $g_i(x) \in F[x]$. Then v can be decomposed accordingly, that is, $v = \sum_{i=1}^t u_i$ where each u_i has q-order $g_i(x^k)^c$.

Now let $g(x)$ be one of these irreducible divisors of $f(x)$, and note that g has to divide $\Phi_d(x)$ for some divisor d of m. By hypothesis, k and $\mathrm{ord}_d(q)$ are relatively prime, as $\mathrm{ord}_d(q)$ divides $\mathrm{ord}_m(q)$. Let u be the component of v which has q-order $g(x^k)^c$. Then $g(\sigma^k)^c(u) = 0$, and therefore the q^k-order of u divides $g(x)^c$. Since $\mathrm{ord}_d(q)$ and k are relatively prime, we have $\mathrm{ord}_d(q) = \mathrm{ord}_d(q^k)$, and hence every monic factor of $\Phi_d(x)$ which is irreducible over F remains irreducible over the k-dimensional extension K of F in $\widehat{F}$. Consequently, the q^k-order of u has the form $g(x)^b$ for some $b \le c$, and then $g(\sigma^k)^b(u) = 0$ implies that $\mathrm{Ord}_q(u) = g(x^k)^c$ divides $g(x^k)^b$. This forces $b = c$, that is, $\mathrm{Ord}_{q^k}(u) = g(x)^c$. Since this holds for all choices of $g = g_i$ and hence for all components u_i of v, we conclude that

$$\mathrm{Ord}_{q^k}(v) = \prod_{i=1}^t g_i(x)^c = f(x)^c,$$

as claimed. □

Remark 12.3.3. Let us give a first application of Proposition 12.3.2. For this, we consider a field $F = \mathrm{GF}(q)$ with characteristic p and an integer $n = n'p^a$, where n' is not divisible by p. Let m be a divisor of n', put $k := \frac{n'}{m} \cdot p^b$ with $b \le a$, and let $K = \mathrm{GF}(q^k)$. Take $f(x) = x^m - 1$, let $v \in E = \mathrm{GF}(q^n)$ be a normal element for E/F, that is,

$$\mathrm{Ord}_q(v) = x^n - 1 = \left((x^k)^m - 1\right)^{p^{a-b}} = f(x^k)^{p^{a-b}},$$

and assume that $\mathrm{ord}_m(q)$ and k are relatively prime. Then

$$\mathrm{Ord}_{q^k}(v) = f(x)^{p^{a-b}} = x^{mp^{a-b}} - 1 = x^{n/k} - 1,$$

and hence v is also normal for E/K. □

The following application of Proposition 12.3.2 provides the basis for proving Theorem 12.3.5; as usual, we denote the p-free part of any positive integer N by N', where p is the characteristic of the fields involved.

Proposition 12.3.4. *Consider the extension E/F where $F = \mathrm{GF}(q)$ and $E = \mathrm{GF}(q^n)$, let r be a prime divisor of n and let $K = E_r$ be the intermediate field of E/F with degree r over F. Then the following are equivalent:*

(i) $\mathrm{ord}_{(n/r)'}(q)$ *is not divisible by r;*

(ii) *every normal element for E/F is also normal for E/K.*

Proof. For the implication (i) ⇒ (ii), we proceed as in Remark 12.3.3 and write n as $n'p^a$ and let $m := (n/r)'$ and $f(x) := x^m - 1$. Note that $m = n'/r$ when $r \neq p$ and $m = n'$ for $r = p$; also, by hypothesis, r and $\mathrm{ord}_m(q)$ are relatively prime. Finally, let v be an arbitrary normal element for E/F, that is, $\mathrm{Ord}_q(v) = x^n - 1$.

In both possible cases, a direct application of Proposition 12.3.2 shows that v is also normal for E/K:

- First assume $r \neq p$. Then

$$\mathrm{Ord}_q(v) = x^n - 1 = \left((x^r)^{n'/r} - 1\right)^{p^a} = f(x^r)^{p^a},$$

  and we obtain the desired result

$$\mathrm{Ord}_{q^r}(v) = f(x)^{p^a} = \left(x^{n'/r} - 1\right)^{p^a} = x^{n/r} - 1.$$

- Now let $r = p$. Then

$$\mathrm{Ord}_q(v) = x^n - 1 = \left((x^p)^{n'} - 1\right)^{p^{a-1}} = f(x^p)^{p^{a-1}},$$

  and hence

$$\mathrm{Ord}_{q^p}(v) = f(x)^{p^{a-1}} = \left(x^{n'} - 1\right)^{p^{a-1}} = x^{n/p} - 1.$$

Conversely, assume that r divides $\mathrm{ord}_m(q)$, where again $m = (n/r)'$. In order to establish the implication (ii) ⇒ (i), we need to show the existence of some normal element for E/F which is not normal for E/K. For this, let $g(x)$ be a monic irreducible divisor of $\Phi_m(x)$ over F, and note that g splits into r irreducible factors over K. Let $h(x)$ be one of these factors. By Proposition 11.6.1, every $u \in E$ with $\mathrm{ord}_{q^r}(u) = h(x)$ satisfies $\mathrm{Ord}_q(u) = g(x^r)$. Any such element can serve as a component of some element v which is normal over F, but not normal over K. This is similar to the constructions given in Examples 12.1.1, 12.1.2, 12.1.5 and 12.1.6; we leave the details to the reader. □

After these preparations, we can now give the following simple number theoretic characterization of all completely basic extensions; see [161, Theorem 15.7].

Theorem 12.3.5. *Consider the extension E/F where $F = \mathrm{GF}(q)$ and $E = \mathrm{GF}(q^n)$, and denote the intermediate field of E/F with degree k over F by E_k (for all divisors k of n). Then the following are equivalent:*

(i) *E is completely basic over F;*
(ii) *for every prime divisor r of n, every normal element for E/F is also normal for E/E_r;*
(iii) *for every prime divisor r of n, the order $\mathrm{ord}_{(n/r)'}(q)$ is not divisible by r.*

Proof. By the definition of a completely basic extension, the implication (i) $\Rightarrow$ (ii) is trivial. Moreover, (ii) and (iii) are equivalent by Proposition 12.3.4.

It remains to show that the two equivalent conditions (ii) and (iii) imply that E/F is completely basic. Because of (ii), this amounts to showing that all extensions E/E_r are completely basic. We do so using induction on the number of (not necessarily distinct) prime divisors of n; the induction base, where n is itself a prime, is trivial. For the induction step, it suffices to check the validity of condition (iii) for all pairs $(\frac{n}{r}, q^r)$, that is, we need to check that

$$s \nmid \mathrm{ord}_{(n/rs)'}(q^r) \quad \text{for all primes } r, s \text{ with } r \mid n \text{ and } s \mid \tfrac{n}{r}.$$

In view of the divisor chain

$$\mathrm{ord}_{(n/rs)'}(q^r) \mid \mathrm{ord}_{(n/s)'}(q^r) \mid \mathrm{ord}_{(n/s)'}(q),$$

this is an immediate consequence of the validity of (iii) for the pair (n,q). □

Example 12.3.6. Let $\mathrm{GF}(q)$ be any finite field. Then $\mathrm{GF}(q^n)$ is completely basic over $\mathrm{GF}(q)$ in each of the following cases:

(1) $n = r$ or $n = r^2$, where r is a prime;
(2) n divides $q-1$;
(3) $n = p^a$ is a power of the characteristic p of $\mathrm{GF}(q)$.

We leave the (easy) verification as Exercise 12.3.8. □

Regarding case (3) of Example 12.3.6, one has the following stronger result, which actually extends to arbitrary cyclic Galois extensions with degree a power of the characteristic; see [161, Section 5].

Theorem 12.3.7. *Consider a finite field $F = \mathrm{GF}(q)$ with characteristic p and an extension $E = \mathrm{GF}(q^n)$, where n is a power of p. Let w be any element of E. Then the following are equivalent:*

(i) *w is completely normal for E/F;*
(ii) *w is normal for E/F;*
(iii) $\mathrm{Tr}_{E/F}(w) \neq 0$.

Proof. The implications (i) $\Rightarrow$ (ii) $\Rightarrow$ (iii) are trivial and even hold for arbitrary Galois extensions and arbitrary extension degrees n.

It remains to show that (iii) implies (i). Thus let K be any intermediate field of E/F. Because of the transitivity formula $\mathrm{Tr}_{E/F}(w) = \mathrm{Tr}_{K/F}(\mathrm{Tr}_{E/K}(w))$, the hypothesis $\mathrm{Tr}_{E/F}(w) \neq 0$ implies $\mathrm{Tr}_{E/K}(w) \neq 0$. Now Theorem 11.3.1 shows that w is normal for E/K, as $[E:K]$ is a power of p. □

Exercises

Exercise 12.3.8. Verify that the extensions listed in Example 12.3.6 are in fact completely basic. □

Exercise 12.3.9. Consider a field extension $E = \mathrm{GF}(q^n)$ over $F = \mathrm{GF}(q)$. Let r be the largest prime divisor of n and assume $\mathrm{pt}_r(n) \leq r^2$. Show that every normal element for E/F is normal over $\mathrm{GF}(q^r)$ and, if r^2 divides n, also over $\mathrm{GF}(q^{r^2})$. □

12.4 The Complete Reduction Theorem

In this brief section, we reduce the proof of the complete normal basis theorem to the special case of extensions of prime power degree. More precisely, we establish the following analogue of Theorem 3.9.7 for completely normal elements:

Theorem 12.4.1. *Consider the n-dimensional extension $E = \mathrm{GF}(q^n)$ of $F = \mathrm{GF}(q)$, where $n = k\ell$ and where k and ℓ are relatively prime. Let K and L be the intermediate fields of E/F with degrees k and ℓ over F and assume that $u \in K$ and $v \in L$ are completely normal for K/F and L/F, respectively. Then $w := uv$ is a completely normal element for E/F.*

Proof. Let m be an arbitrary divisor of n and denote the intermediate field of E/F with degree m over F by E_m; we have to show that w is normal for E/E_m. Write m as

$$m = \gcd(m,k) \cdot \gcd(m,\ell) = m_k \cdot m_\ell,$$

where $m_k = \gcd(m,k)$ and $m_\ell = \gcd(m,\ell)$ are relatively prime. Since u is completely normal for K/F, it is normal for K/E_{m_k}. Now note that $\gcd(m,k) = m_k$ yields

$$K \cap E_m = E_{m_k} \quad \text{and} \quad KE_m = E_s, \text{ where } s = \mathrm{lcm}(m,k) = km_\ell;$$

see Remark 3.4.3. Hence u is also normal for E_s/E_m, by an application of Proposition 3.9.5. Similarly, v is normal for both L/E_{m_ℓ} and E_t/E_m, where $t = \ell m_k$.

Now Theorem 3.9.7 shows that $w = uv$ is normal for E/E_m, since $[E_s : E_m] = k/m_k$ and $[E_t : E_m] = \ell/m_\ell$ are relatively prime and since

$$(k/m_k) \cdot (\ell/m_\ell) = n/m = [E_n : E_m]. \qquad \square$$

We remark that Theorem 12.4.1 also holds for arbitrary Galois extensions; see [161, Section 4].

Exercises

Exercise 12.4.2. Consider a field extension $E = \mathrm{GF}(q^n)$ over $F = \mathrm{GF}(q)$ and assume that $n/\mathrm{rad}(n')$ divides $\mathrm{rad}(n')$. Prove the existence of a completely normal element for E/F. $\square$

12.5 Extensions with Prime Power Degree

In view of Theorem 12.4.1, the proof of the complete normal basis theorem is now reduced to the case where the degree n of E/F is a power of some prime r. Moreover, Theorem 12.3.7 settles the case where r is the characteristic p of F. Thus we are left with the cases where $n = r^\ell$ with $r \neq p$, which present the real difficulty. Therefore, the goal of this section is the following theorem:

Theorem 12.5.1. *Consider the Galois field $F = \mathrm{GF}(q)$, let r be a prime which is distinct from the characteristic p of F, and let $E = E_{r^\ell}$ be the r^ℓ-dimensional extension of F. Then there exists an element $w \in E$ which is completely normal for E/F.*

The proof of Theorem 12.5.1 requires some effort and distinguishes several cases. Because of Example 12.3.6, we may assume $\ell \geq 3$ whenever convenient. Using the setup in Remark 11.1.4, we will show that the additive decomposition

$$E_{r^\ell} = E_{r^{\ell-1}} \oplus \mathscr{C}^F_{r^\ell} \tag{12.2}$$

of the extension field $E = E_{r^\ell}$ into the maximal intermediate field $L := E_{r^{\ell-1}}$ of E/F and the cyclotomic module $\mathscr{C}^F_{r^\ell}$ leads to an efficient description of all completely normal elements for E/F in terms of their components. This will allow us to prove Theorem 12.5.1 using induction on ℓ.

As usual, we let σ denote the Frobenius automorphism of E/F. By induction, we may assume that there exists a completely normal element $w_{\ell-1}$ for L/F. We recall from Remark 11.1.4 that

$$x^{r^\ell} - 1 = \prod_{j=0}^{\ell} \Phi_{r^j}(x) = \left(x^{r^{\ell-1}} - 1\right) \cdot \Phi_{r^\ell}(x)$$

and that $C := \mathscr{C}^F_{r^\ell}$ is the kernel of the (E,L)-trace mapping; thus the direct decomposition in Equation (12.2) can also be written as

$$E = L \oplus C, \text{ where } E = E_{r^\ell},\ L = E_{r^{\ell-1}} \text{ and } C = \ker(\mathrm{Tr}_{E/L}). \tag{12.3}$$

The description of C as $\ker(\mathrm{Tr}_{E/L})$ shows that C is a $\sigma^{r^{\ell-1}}$-invariant L-vector space, that is, an $(L, \sigma^{r^{\ell-1}})$-submodule of E. More generally, C is also an (E_{r^j}, σ^{r^j})-submodule of E for every j with $0 \leq j \leq \ell - 1$, where E_{r^j} again denotes the extension of F with degree r^j in E/F. In particular,

$$C = \mathscr{C}_{r^\ell}^F = \mathscr{C}_{r^{\ell-j}}^{E_{r^j}}$$

for every such j. This observation leads to the following characterization of completely normal elements for E/F:

Proposition 12.5.2. *With the assumptions of Theorem 12.5.1, let w be an element of E and consider its decomposition $w = u + v$ with $u \in L$ and $v \in C$, according to Equation* (12.3). *Then w is completely normal for E/F if and only if the following two conditions are satisfied:*

(a) *u is completely normal for L/F;*

(b) *v has q^{r^j}-order $\Phi_{r^{\ell-j}}(x)$ for $j = 0, \ldots, \ell - 1$.*

Proof. Let $j \in \{0, \ldots, \ell - 1\}$ and consider the intermediate field $K = E_{r^j}$, so that E has degree $r^{\ell-j}$ over K. Also, $\rho := \sigma^{r^j}$ is the Frobenius automorphism for E/K, and E is annihilated by $x^{r^{\ell-j}} - 1$ over K (with respect to ρ). Therefore, considered as (K, ρ)-modules, the decomposition $E = L \oplus C$ in (12.3) corresponds to the decomposition

$$x^{r^{\ell-j}} - 1 = \left(x^{r^{\ell-j-1}} - 1\right) \cdot \Phi_{r^{\ell-j}}(x)$$

of $x^{r^{\ell-j}} - 1$ over K, where L and C are annihilated by $x^{r^{\ell-j-1}} - 1$ and $\Phi_{r^{\ell-j}}(x)$, respectively. In view of Remark 11.1.4 (applied to the extension E/K), we conclude that $w = u + v$ is normal for E/K if and only if the following two conditions are satisfied:

(a$_j$) u is normal for L/K;

(b$_j$) v has q^{r^j}-order $\Phi_{r^{\ell-j}}(x)$.

Now the assertion is immediate: w is completely normal for E/F if and only if these conditions hold for all $j \in \{0, \ldots, \ell - 1\}$. □

Using induction, we may assume that we know a completely normal element $w_{\ell-1}$ for L/F, and it "only" remains to prove the existence of an element $v \in C$ satisfying condition (b) in Proposition 12.5.2 in order to establish Theorem 12.5.1: then $w_\ell = w_{\ell-1} + v$ is a completely normal element for E/F. It is worthwhile introducing a name for the required elements:

Definition 12.5.3. Let $F = \mathrm{GF}(q)$, and let r be a prime which is distinct from the characteristic of F and ℓ a positive integer. Then an element v of the cyclotomic module $C := \mathscr{C}_{r^\ell}^F$ is said to be a **complete generator** for C over F if and only if it satisfies the following condition:

$$\mathrm{Ord}_{q^{r^j}}(v) = \Phi_{r^{\ell-j}}(x) \quad \text{for } j = 0, \ldots, \ell-1, \tag{12.4}$$

that is, if v generates C as an (E_{r^j}, σ^{r^j})-module for every $j = 0, \ldots, \ell-1$. □

Thus we have to prove the existence of complete generators for all cyclotomic modules of the form $\mathscr{C}_{r^\ell}^F$ over $F = \mathrm{GF}(q)$, where we may assume that $\ell \geq 3$. Of course, condition (12.4) consists of the ℓ individual conditions (b_j) used in the proof of Proposition 12.5.2. The main idea now is to reduce the number of these conditions which actually need to be checked in order to decide whether or not a given element $v \in \mathscr{C}_{r^\ell}^F$ is a complete generator. Somewhat surprisingly, it turns out that usually just one of these conditions suffices; in the few exceptional situations, one has to check two of the conditions. As mentioned before, we will require a case distinction to establish these results.

We first deal with the cases where r is odd or $q \equiv 1 \bmod 4$; here one suitable condition (b_τ) will suffice. We specify the appropriate value of τ as follows:

- put $s := \mathrm{ord}_r(q)$, and write $r^a := \mathrm{pt}_r(q^s - 1)$;
- let $t := \ell - \min(a, \ell)$, so that $\mathrm{ord}_{r^\ell}(q) = sr^t$ by Exercise 1.7.18;
- finally, put $\tau = \lfloor t/2 \rfloor$, and note $\tau \leq \ell - 1$.

Theorem 12.5.4. *Consider $F = \mathrm{GF}(q)$ and the extension E/F with degree r^ℓ, where r is a prime which is distinct from the characteristic p of F and where $\ell \geq 2$. Assume that r is odd or $q \equiv 1 \bmod 4$ and define τ as just described. Then an element v of $C := \mathscr{C}_{r^\ell}^F$ is a complete generator of C over F if and only if it satisfies the condition*

(b_τ) $\mathrm{Ord}_{q^{r^\tau}}(v) = \Phi_{r^{\ell-\tau}}(x)$,

that is, if and only if v generates C as an $\left(E_{r^\tau}, \sigma^{r^\tau}\right)$-module.

Thus there always exists a complete generator for $\mathscr{C}_{r^\ell}^F$ over F and therefore also a completely normal element for E/F.

Proof. We need to show that the validity of condition (b_τ) forces v to be a complete generator; in other words, the validity of condition (b_τ) implies that of condition (b_j) for $j = 0, \ldots, \ell-1$. This will be done in three steps.

Step 1: Let us first consider the case where $t \leq 1$, so that $a \geq \ell - 1$ and $\tau = 0$. Thus we have to check that any element v with q-order $\Phi_{r^\ell}(x)$ already is a complete generator for the cyclotomic module $C = \mathscr{C}_{r^\ell}^F$ over F. Note that here $\mathrm{ord}_{r^{\ell-1}}(q) = s$, again by Exercise 1.7.18:

$$\mathrm{ord}_{r^{\ell-1}}(q) = s \cdot r^{\ell-1-\min(\ell-1,a)} = s \cdot r^0 = s.$$

As s is not a multiple of r, this even gives

$$\mathrm{ord}_{r^{\ell-j}}(q) = s \quad \text{for } j = 1, \ldots, \ell-1.$$

In view of

$$\mathrm{Ord}_q(v) = \Phi_{r^\ell}(x) = \Phi_{r^{\ell-j}}(x^{r^j}),$$

Proposition 12.3.2 shows that $\mathrm{Ord}_{q^{r^j}}(v) = \Phi_{r^{\ell-j}}(x)$. Thus v generates C as (E_{r^j}, σ^{r^j})-module for all $j = 1, \ldots, \ell-1$, as claimed.

Step 2: Now let $t \geq 2$, so that $\ell \geq 3$ and $t = \ell - a$. Let $K = E_r$ be the intermediate field with degree r over F in E/F. Then the extension E/K has degree $r^{\ell-1}$ and Frobenius automorphism $\rho = \sigma^r$. We claim that every generator of C as a (K,ρ)-module also generates C as an (F,σ)-module; this will be an essential ingredient for proving the theorem with induction on ℓ in Step 3.

Thus let $v \in C$ have q^r-order $\Phi_{r^{\ell-1}}(x)$. We will make use of the complete factorization of this cyclotomic polynomial over F. Because of $t \geq 2$ and Exercise 1.7.18, one has

$$\mathrm{ord}_{r^{\ell-1}}(q) = sr^{t-1} \quad \text{and} \quad \mathrm{ord}_{r^{\ell-1}}(q^r) = sr^{t-2}.$$

Hence Proposition 3.6.16 shows that $\Phi_{r^{\ell-1}}$ splits over F into

$$m := \frac{\phi(r^{\ell-1})}{\mathrm{ord}_{r^{\ell-1}}(q)} = \frac{(r-1)r^{\ell-2}}{sr^{t-1}} = \frac{r-1}{s} \cdot r^{a-1}$$

irreducible factors, say $\Phi_{r^{\ell-1}}(x) = \prod_{i=1}^m g_i(x)$. This induces a corresponding decomposition

$$\Phi_{r^\ell}(x) = \Phi_{r^{\ell-1}}(x^r) = \prod_{i=1}^m g_i(x^r)$$

of the cyclotomic polynomial Φ_{r^ℓ} over F. By Proposition 3.6.16, this decomposition is already the complete factorization of Φ_{r^ℓ} over F, since

$$\frac{\phi(r^\ell)}{\mathrm{ord}_{r^\ell}(q)} = \frac{r \cdot \phi(r^{\ell-1})}{sr^t} = \frac{r \cdot \phi(r^{\ell-1})}{r \cdot \mathrm{ord}_{r^{\ell-1}}(q)} = m.$$

Now write the given element v with q^r-order $\Phi_{r^{\ell-1}}(x)$ in the form $v = \sum_{i=1}^m v_i$, where each v_i has q^r-order $g_i(x)$. Since $g_i(\sigma^r)(v_i) = g_i(\rho)(v_i) = 0$, the q-order of v_i has to divide $g_i(x^r)$, which is an irreducible polynomial over F. Thus actually $\mathrm{Ord}_q(v_i) = g_i(x^r)$ for all i, and we obtain

$$\mathrm{Ord}_q(v) = \prod_{i=1}^m \mathrm{Ord}_q(v_i) = \prod_{i=1}^m g_i(x^r) = \Phi_{r^{\ell-1}}(x^r) = \Phi_{r^\ell}(x),$$

as claimed.

Step 3: We now use induction on ℓ to prove the theorem. Note that the induction basis $\ell = 2$ is covered by Step 1, as $\mathrm{ord}_{r^2}(q)$ divides $(r-1)r$. Now let $\ell \geq 3$ and (in view of Step 1) $t \geq 2$. Thus we may continue with our arguments in Step 2, where we have already seen that every generator of C as a (K,ρ)-module also generates C as an (F,σ)-module. Hence it suffices to show that any element $v \in C$ which satisfies condition (b_τ) is a complete generator of C as a (K,ρ)-module. This amounts to showing that v satisfies the following $\ell-1$ conditions for the extension E/K:

(b_j^K) $\mathrm{Ord}_{(q^r)^{r^j}}(v) = \Phi_{r^{\ell-1-j}}(x)$ for $j=0,1,\ldots,\ell-2$.

It should be noted that the condition (b_j) for E/F is identical with condition (b_{j-1}^K) for $j=1,\ldots,\ell-1$. In particular, condition (b_τ) coincides with condition $(\mathrm{b}_{\tau-1}^K)$, and therefore we can apply the induction hypothesis to E/K to obtain the desired result, provided that we can show that the "τ-value" for E/K is $\tau_K=\tau-1$. We therefore compute the required parameters for E/K, where we always use a subscript K to distinguish them from the original parameters for E/F:

- $s_K=\mathrm{ord}_r(q^r)=\mathrm{ord}_r(q)=s$, as $\gcd(r,s)=1$;
- $a_K=a+1$, as $\mathrm{pt}_r(q^{rs}-1)=r\cdot\mathrm{pt}_r(q^s-1)$ by Lemma 1.7.3;
- $t_K=(\ell-1)-\min(a_K,\ell-1)=(\ell-1)-\min(a+1,\ell-1)=\ell-a-2=t-2$;
- $\tau_K=\lfloor t_K/2\rfloor=\lfloor (t-2)/2\rfloor=\tau-1$,

as claimed. □

It remains to show the existence of a complete generator of the cyclotomic module $C=\mathscr{C}_{2^\ell}^F$ over F for the cases where $q\equiv 3 \bmod 4$ and $\ell\geq 3$. Here the intermediate field $K=E_2$ of F with degree 2 in E/F will play a key role. Since $q^2\equiv 1 \bmod 4$, the extension E/K (or, more precisely, C considered as the (K,ρ)-module $\mathscr{C}_{2^{\ell-1}}^K$, where $\rho=\sigma^2$) is covered by Theorem 12.5.4. Essentially, we just have to combine our results for E/K with the (F,σ)-module structure of C. Unfortunately, we will need to distinguish two subcases, depending on whether or not 2^ℓ divides q^2-1.

Let us begin with the subcase where 2^ℓ does not divide q^2-1, which turns out to be quite similar to the case dealt with in Theorem 12.5.4: again, it will suffice to check just one suitable condition (b_τ). We now specify the appropriate value of τ as follows:

- Let $\mathrm{pt}_2(q+1)=2^b$ and thus $\mathrm{pt}_2(q^2-1)=2^{b+1}$, and note $b\geq 2$ and $b+1<\ell$.
- Put $t':=\ell-\min(b+1,\ell)=\ell-b-1$ and $t:=t'+1=\ell-b$, so that

$$\mathrm{ord}_{2^\ell}(q)=2^t \quad\text{and}\quad \mathrm{ord}_{2^\ell}(q^2)=2^{t'};$$

  see Corollary 1.7.6 and its proof.
- Finally, put $\tau=\lfloor t/2\rfloor$, and note $\tau\leq\ell-1$.

Theorem 12.5.5. *Consider $F=\mathrm{GF}(q)$ and the extension E/F with degree 2^ℓ, where $q\equiv 3 \bmod 4$ and $\ell\geq 3$. Assume that 2^ℓ does not divide q^2-1, and define τ as just described. Then an element v of $C:=\mathscr{C}_{2^\ell}^F$ is a complete generator of C over F if and only if it satisfies the condition*

(b_τ) $\mathrm{Ord}_{q^{2^\tau}}(v) = \Phi_{2^{\ell-\tau}}(x)$,

that is, if and only if v generates C as an $\left(E_{2^\tau},\sigma^{2^\tau}\right)$-module.

Thus there exists a complete generator for $\mathscr{C}_{2^\ell}^F$ over F and therefore also a completely normal element for E/F.

Proof. We need to show that the validity of condition (b_τ) forces v to be a complete generator of C. For this, one first proves that every generator of C as a (K,ρ)-module also generates C as an (F,σ)-module; consequently, every complete generator of C over K is already a complete generator of C over F. This can be done by reasoning as in Step 2 of the proof of Theorem 12.5.4; we will leave this to the reader as Exercise 12.5.11.

It now suffices to check that the "τ-value" for the extension E/K is $\tau_K = \tau - 1$, as in Step 3 in the proof of Theorem 12.5.4. Because of $\mathrm{ord}_{2^{\ell-1}}(q^2) = 2^{t'-1}$ (see Exercise 12.5.11), we have $t_K = t' - 1 = t - 2$ and hence

$$\tau_K = \lfloor t_K/2 \rfloor = \lfloor (t-2)/2 \rfloor = \tau - 1,$$

as required. □

We now turn our attention to the second subcase, which turns out to be the only case where it is necessary to check two of the conditions (b_j) to guarantee that an element v of some cyclotomic module $\mathscr{C}^F_{r^\ell}$ is a complete generator. Therefore, we say that $\mathscr{C}^F_{r^\ell}$ is an **exceptional cyclotomic module** when

$$r = 2, \;\; q \equiv 3 \bmod 4, \;\; \ell \geq 3 \;\text{ and }\; 2^\ell \mid q^2 - 1.$$

For this situation, we have the following result:

Theorem 12.5.6. *Consider $F = \mathrm{GF}(q)$ and the extension E/F with degree 2^ℓ, where $q \equiv 3 \bmod 4$ and $\ell \geq 3$. Assume that $C = \mathscr{C}^F_{2^\ell}$ is an exceptional cyclotomic module, that is, 2^ℓ divides $q^2 - 1$. Then an element v of C is a complete generator for C over F if and only if it satisfies the two conditions* (b_0) *and* (b_1)*:*

$$\mathrm{Ord}_q(v) = \Phi_{2^\ell}(x) \quad \textit{and} \quad \mathrm{Ord}_{q^2}(v) = \Phi_{2^{\ell-1}}(x); \tag{12.5}$$

in other words, v has to generate C both as an (F,σ)- and a (K,σ^2)-module, where K is the intermediate field with degree 2 in E/F.

Moreover, there always exists a complete generator for $\mathscr{C}^F_{2^\ell}$ over F and therefore also a completely normal element for E/F.

Proof. We need to show that the validity of (12.5) forces v to be a complete generator of C. As already noted, we can apply Theorem 12.5.4 to the extension E/K. Since 2^ℓ divides $q^2 - 1$, we have $\mathrm{ord}_{2^{\ell-1}}(q^2) = 1$, which gives $t = 1$ and $\tau = 0$ in this case. Hence any element with q^2-order $\Phi_{2^{\ell-1}}(x)$ already is a complete generator of $C = \mathscr{C}^K_{2^{\ell-1}}$ over K. Now the first assertion is obvious.

Finally, the existence of elements satisfying the criterion in (12.5) is guaranteed by Theorem 11.7.3. □

This finishes the proof of the complete normal basis theorem, as outlined at the beginning of this section. For later reference, we state the following result which follows by combining Theorem 12.5.6 with Theorem 11.7.3:

Corollary 12.5.7. *Consider $F = \mathrm{GF}(q)$ and the extension E/F with degree 2^ℓ, where $q \equiv 3 \bmod 4$ and $\ell \geq 3$. Let $\mathrm{pt}_2(q^2-1) = 2^c$ and assume that $c \geq \ell$, so that $C = \mathscr{C}_{2^\ell}^F$ is an exceptional cyclotomic module. Furthermore, let Q be the subgroup of the group U of units modulo 2^c generated by $q \bmod 2^c$, and J a system of representatives for the cosets of Q in U.*

Now let u be any primitive $2^{c+\ell-1}$-th root of unity in the algebraic closure $\widehat{F}$ of F, and put $v := \sum_{j\in J} u^j$. Then $v + v^q$ is a complete generator for C over F. □

We conclude this section with a closer look at the exceptional case. We start with a few simple observations:

Remark 12.5.8. We first note that one really needs both conditions (b_0) and (b_1) for exceptional cyclotomic modules: there are elements in $C = \mathscr{C}_{2^\ell}^F$ which generate C as (F,σ)-module but not as (K,ρ)-module, and vice versa. This will be a consequence of Theorem 12.5.9 below (and its proof).

As noted in the proof of Theorem 12.5.6, one has $\mathrm{ord}_{2^{\ell-1}}(q^2) = 1$ in the exceptional case, and therefore $\mathrm{ord}_{2^\ell}(q) = 2$. In other words, $q \bmod 2^\ell$ is an involution in the group of units modulo 2^ℓ. By Theorem 1.7.7 and its proof, this group contains exactly three involutions, namely

$$5^{2^{\ell-3}} \bmod 2^\ell, \quad -1 \bmod 2^\ell \text{ and } -5^{2^{\ell-3}} \bmod 2^\ell.$$

Since $q \equiv 3 \bmod 4$ and $5^{2^{\ell-3}} \equiv 1 \bmod 4$, we cannot have $q \equiv 5^{2^{\ell-3}} \bmod 2^\ell$. The other two cases may occur: for instance, $q \equiv -5 \bmod 8$ when $q = 3$ and $\ell = 3$; and $q \equiv -1 \bmod 8$ when $q = 7$ and $\ell = 3$. □

Finally, we determine the number of complete generators for all exceptional cyclotomic modules:

Theorem 12.5.9. *Consider $F = \mathrm{GF}(q)$ and the extension E/F with degree 2^ℓ, where $q \equiv 3 \bmod 4$ and $\ell \geq 3$. Assume that 2^ℓ divides q^2-1, and let $f(x)$ be any monic irreducible divisor of $\Phi_{2^{\ell-1}}(x)$ over F.*

Then there are exactly $q^4 - 4q^2 + 3 > 0$ elements u in the exceptional cyclotomic module $C = \mathscr{C}_{2^\ell}^F$ which satisfy $\mathrm{Ord}_q(u) = f(x^2)$ and $\mathrm{Ord}_{q^2}(u) = f(x)$. Moreover, the total number of complete generators for C is

$$(q^4 - 4q^2 + 3)^{2^{\ell-3}}.$$

Proof. As in Step 2 of the proof of Theorem 12.5.4, we consider the complete factorizations of the relevant cyclotomic polynomials, both over F and over its extension K with degree 2. Note that $\Phi_{2^{\ell-1}}$ splits over F into $\phi(2^{\ell-1})/2 = 2^{\ell-3} =: d$ irreducible factors with degree 2 (because of $\mathrm{ord}_{2^{\ell-1}}(q) = 2$) and that each of these irreducible factors $f \in F[x]$ splits over K into two linear factors, since 2^ℓ divides q^2-1. Let us write these two complete factorizations of $\Phi_{2^{\ell-1}}$ as

$$\Phi_{2^{\ell-1}}(x) = f_1(x)\cdots f_d(x) = h_1(x)\cdots h_{2d}(x), \text{ where } f_i(x) = h_i(x)\cdot h_{d+i}(x).$$

Clearly, each polynomial $f_i(x^2)$ has degree 4 and divides $\Phi_{2^{\ell-1}}(x^2) = \Phi_{2^\ell}(x)$. Because of $\mathrm{ord}_{2^\ell}(q) = 2$, this cyclotomic polynomial splits over F into $\phi(2^\ell)/2 = 2^{\ell-2} = 2d$ irreducible factors with degree 2, say

$$\Phi_{2^\ell}(x) = g_1(x)\cdots g_{2d}(x), \text{ where } f_i(x^2) = g_i(x)\cdot g_{d+i}(x).$$

Now consider the decomposition of $C = \mathscr{C}^F_{2^\ell}$ as

$$C = \bigoplus_{i=1}^{d} C_i \quad \text{with } C_i = \{v \in E\colon f_i(\sigma^2)(v) = 0\} \text{ for } i = 1,\ldots,d,$$

and note that each of the C_i is both an (F,σ)-module and a (K,σ^2)-module. Moreover, as an (F,σ)-module, $C_i = A_{i,0} \oplus A_{i,1}$, where $A_{i,0}$ is annihilated by $g_i(x)$ and $A_{i,1}$ by $g_{d+i}(x)$ (over F). Similarly, as a (K,σ^2)-module, $C_i = B_{i,0} \oplus B_{i,1}$, where $B_{i,0}$ is annihilated by $h_i(x)$ and $B_{i,1}$ by $h_{d+i}(x)$ (over K).

These observations yield the following characterization of complete generators v for C over F: an element $v = \sum_{i=1}^{d} v_i$ with $v_i \in C_i$ for $i = 1,\ldots,d$ is a complete generator for C over F if and only if

$$\mathrm{Ord}_q(v_i) = f_i(x^2) \quad \text{and} \quad \mathrm{Ord}_{q^2}(v_i) = f_i(x) \tag{12.6}$$

for $i = 1,\ldots,d$. This shows that the second assertion is an immediate consequence of the first one.

Now let f_i be one of the irreducible factors of $\Phi_{2^{\ell-1}}$ over F and consider the (F,σ)-module

$$C' := \{v \in C\colon \mathrm{ord}_q(v) \mid f_i(x^2)\} = \{v \in C\colon \mathrm{ord}_{q^2}(v) \mid f_i(x)\}.$$

We have to determine the number of non-zero elements in C' whose q-order is not an irreducible F-divisor of $f_i(x^2)$ and whose q^2-order is not an irreducible K-divisor of $f_i(x)$. Because of the respective factorizations

$$f_i(x^2) = g_i(x)g_{d+i}(x) \quad \text{and} \quad f_i(x) = h_i(x)h_{d+i}(x),$$

the number of non-zero elements in C' which do not generate C' as an (F,σ)-module is given by $2\cdot(|F|^2-1) = 2(q^2-1)$, while the number of non-zero elements in C' which do not generate C' as a (K,σ^2)-module is $2\cdot(|K|-1) = 2(q^2-1)$.

Observe that these two subsets of C' with equal cardinality $2(q^2-1)$ have to be disjoint: by Proposition 11.6.1, any element with q^2-order $h_i(x)$ or $h_{d+i}(x)$ has q-order $f_i(x^2)$, which also implies that any element with q-order $g_i(x)$ or $g_{d+i}(x)$ has q^2-order $f_i(x)$. Therefore the number of elements of C' satisfying condition (12.6) is given by

$$|C'| - 1 - 4\cdot(q^2-1) = q^4 - 4q^2 + 3 = (q^2-1)(q^2-3),$$

which is obviously positive. □

Exercises

Exercise 12.5.10. Consider the situation in Step 1 in the proof of Theorem 12.5.4 and assume that $t = 1$. Show that an element $v \in C$ generates C as an (F,σ)-module if and only if v generates C as a (K,ρ)-module (with K and ρ as in Step 2). □

Exercise 12.5.11. Complete the proof of Theorem 12.5.5 by showing that every generator of C as a (K,ρ)-module also generates C as an (F,σ)-module.

Hint: First show that $t' \geq 1$ and

$$\mathrm{ord}_{2^{\ell-1}}(q) = 2 \cdot \mathrm{ord}_{2^{\ell-1}}(q^2) = 2^{\ell-b-1} = 2^{t'}.$$

Then check

$$\frac{\phi(2^{\ell-1})}{\mathrm{ord}_{2^{\ell-1}}(q)} = 2^{\ell-2-t'} = \frac{\phi(2^{\ell})}{\mathrm{ord}_{2^{\ell}}(q)}$$

and proceed as in Step 2 in the proof of Theorem 12.5.4. □

Exercise 12.5.12. Let F be any finite field and r a prime number. Determine the number of completely normal elements in E/F, where E is the extension of degree r^3 over F. Of course, case distinctions are necessary. □

Exercise 12.5.13. Determine a completely normal element for E/F, where $F = \mathrm{GF}(5)$ and E has degree 27 over F. □

Exercise 12.5.14. Determine a completely normal element for E/F, where $F = \mathrm{GF}(3)$ and E has degree 16 over F. □

12.6 Trace-Compatible Sequences of Completely Normal Elements

In this section, we combine the results of Sections 11.8 and 12.5 to show that the algebraic closure $\widehat{F}$ of a finite field $F = \mathrm{GF}(q)$ always contains a trace-compatible sequence which entirely consists of completely normal elements. More precisely, we will use a complete version of Construction 11.8.1 to prove the following theorem:

Theorem 12.6.1. *Let $F = \mathrm{GF}(q)$ be a Galois field and $\widehat{F}$ its algebraic closure. Then there exists a sequence $(w_n)_{n\in\mathbb{N}^*}$ in $\widehat{F}$ with the following properties:*

- *w_n is completely normal for E_n/F for all n;*
- *$\mathrm{Tr}_{E_n/E_m}(w_n) = w_m$ whenever $m \mid n$.*

Proof. Let P denote the prime subfield of F. As in Construction 11.8.1, we start with an arbitrary element $w_1 \in F$ such that $\mathrm{Tr}_{F/P}(w_1) \neq 0$. For the induction step, we let $n \geq 2$ and assume that we have already constructed a partial sequence $(w_1,\ldots,w_{n-1})$ with the desired properties:

- w_i is completely normal for E_i/F for $i = 1,\ldots,n-1$;
- $\mathrm{Tr}_{E_j/E_i}(w_j) = w_i$, whenever $i, j \in \{1,\ldots,n-1\}$ and $i \mid j$.

We now distinguish two cases:

Case 1. $n = r^\lambda$ is a power of a prime r.

- If r is the characteristic p of F, we proceed exactly as in Case 2 of Construction 11.8.1 and determine a normal element w_{p^k} for E_{p^k}/F which can serve to extend the partial sequence $(w_1,\ldots,w_{n-1})$ in a trace-compatible manner. Appealing to Theorem 12.3.7 instead of Theorem 11.3.1 shows that w_{p^k} is even completely normal for E_{p^k}/F.
- Let $r \neq p$ and assume that r is odd or $q \equiv 1 \bmod 4$. We let $s := \mathrm{ord}_r(q)$ and $r^a = \mathrm{pt}_r(q^s - 1)$, put $t = k - \min(a,k)$ and $\tau := \lfloor t/2 \rfloor$, and consider $L = E_{r^\tau}$. Now we apply Theorem 11.6.4 with $m = r^{k-\tau}$ to the field extension E_{r^k}/E_{r^τ} and determine an element $v_{r^k} \in \widehat{F}$ with q^{r^τ}-order $\Phi_{r^{a-\tau}}(x)$. By Theorem 12.5.4, v_{r^k} is a complete generator of $\mathscr{C}^F_{r^k}$. Then

$$w_{r^k} := \tfrac{1}{r} w_{r^{k-1}} + v_{r^k}$$

  is a completely normal element for E_{r^k}/F, which follows from Proposition 12.5.2 and the induction hypothesis. As in Case 1 of Construction 11.8.1, one shows that $(w_1,\ldots,w_{r^k-1},w_{r^k})$ remains trace-compatible.
- Finally, let $r = 2$ and $q \equiv 3 \bmod 4$. Note that the cases where $k \leq 2$ are covered by Examples 11.1.5 and 11.7.2. Now assume $k \geq 3$ and let $\mathrm{pt}_2(q^2-1) = 2^c$. We first determine a complete generator v_{2^k} for $\mathscr{C}^F_{2^k}$ over F.
  - If $c \geq k$, we simply apply Corollary 12.5.7.
  - If $c < k$, let $\mathrm{ord}_{2^k}(q) = 2^t$ and $\tau := \lfloor t/2 \rfloor$; then $\tau \geq 1$, so that $q^{2^\tau} \equiv 1 \bmod 4$. Similar to the previous case, we apply Theorem 11.4.6 and determine an element v_{2^k} with q^{2^τ}-order $\Phi_{2^{k-\tau}}(x)$. By Theorem 12.5.5, v_{2^k} is the desired complete generator for $\mathscr{C}^F_{2^k}$ over F.

  Now we choose

$$w_{2^k} := \tfrac{1}{2} w_{2^{k-1}} + v_{2^k}$$

  in order to extend the sequence appropriately.

Case 2. n is not a prime power. Here we put

$$w_n := \prod_{j=1}^{t} w_{r_j^{a_j}},$$

where $n = \prod_{j=1}^t r_j^{a_j}$ is the prime power factorization of n. Theorem 12.4.1 shows that w_n is a completely normal element for E_n/F, since each $w_{r_j^{a_j}}$ is completely normal for $E_{r_j^{a_j}}/F$ by induction. As in Case 3 of Construction 11.8.1, the extended sequence $(w_1,\ldots,w_{n-1},w_n)$ remains trace-compatible. □

12.7 Generalized Cyclotomic Modules

In the final three sections of this chapter, we will develop a structure theory for completely normal elements over finite fields. As we have seen in Chapter 11, the concept of a cyclotomic module is extremely useful for the construction of normal elements for extensions of Galois fields, mainly because of the decomposition of the additive group of an extension field into a direct product of such modules. This approach was also instrumental in proving the complete normal basis theorem in the preceding sections.

In order to derive general results on the structure of arbitrary completely normal elements, we first need to generalize the notion of a cyclotomic module, which is the topic of the present section. We have already used elements with q-order of the form $\Phi_m(x^s)$ in earlier arguments; the important Theorems 11.6.2 and 11.6.4 provide good examples. The modules associated with such polynomials are just what we need for the desired general structure theory:

Definition 12.7.1. Consider a Galois field $F = \mathrm{GF}(q)$ and its algebraic closure $\widehat{F}$, and let m and k be two positive integers, where m is not divisible by the characteristic p of F. Then

$$\mathscr{G}_{m,k}^F := \{v \in \widehat{F} \colon \Phi_m(\sigma^k)(v) = 0\}$$

is called a **generalized cyclotomic module** over F, where σ denotes the Frobenius automorphism of $\widehat{F}/F$.

Thus the generalized cyclotomic module for the pair (m,k) is the (F,σ)-module of $\widehat{F}$ which is annihilated by the **generalized cyclotomic polynomial** $\Phi_m(x^k)$. □

Remark 12.7.2. Since $\Phi_m(x)$ has coefficients in the prime field of F, we have the following identity:

$$\Phi_m(x^k) = \Phi_m(x^{k'})^{k/k'},$$

where, as usual, k' denotes the p-free part of k.

As suggested by the terminology, every cyclotomic module is also a generalized cyclotomic module; indeed, by the respective definitions,

$$\mathscr{C}_m^F = \mathscr{G}_{m',p^a}^F, \quad \text{where } m = m'p^a.$$

In particular, all extension fields E_n of F in $\widehat{F}$ are also generalized cyclotomic modules: $E_n = \mathscr{G}_{1,n}^F$, since $\Phi_1(x^n) = x^n - 1$.

Finally, we emphasize that several different index pairs may give rise to the same generalized cyclotomic module:

$$\mathscr{G}_{m,k}^F = \mathscr{G}_{mk_0,k_1}^F \quad \text{when } k = k_0k_1 \text{ with } \mathrm{rad}(k_0) \mid m, \tag{12.7}$$

since then $\Phi_m(x^k) = \Phi_{mk_0}(x^{k_1})$ by Proposition 3.6.8. In particular, we obtain $\mathscr{G}_{m,k}^F = \mathscr{G}_{mk,1}^F = \mathscr{C}_{mk}^F$ if $\mathrm{rad}(k)$ divides m; thus the generalized cyclotomic module $\mathscr{G}_{m,k}^F$ is actually cyclotomic in this case. □

Remark 12.7.3. In view of Remark 12.7.2, it is natural to ask whether there is a canonical choice among all index pairs describing the same generalized cyclotomic module $\mathscr{G}^F_{m,k}$. An obvious possibility arises from requiring that m and k should be relatively prime, which may always be achieved by shifting $\mathrm{pt}_m(k)$ from the k-index to the m-index according to Equation (12.7). In this situation, one obtains a nice decomposition of $\mathscr{G}^F_{m,k}$ into cyclotomic modules: if $\gcd(m,k)=1$ and $k=k'p^a$, one has

$$\mathscr{G}^F_{m,k} = \bigoplus_{d|k'} \mathscr{C}^F_{mdp^a},$$

which follows from Proposition 3.6.12 and Definition 11.1.2. This choice is often advantageous in proofs.

Nevertheless, from a conceptual point of view, the opposite approach of shifting as much of the m-index as possible to the k-index is even more interesting. Here we use the identity

$$\mathscr{G}^F_{m,k} = \mathscr{G}^F_{\mathrm{rad}(m),\ell} \quad \text{with } \ell := mk/\mathrm{rad}(m), \tag{12.8}$$

which also follows from Equation (12.7) and corresponds to the identity

$$\Phi_m(x^k) = \Phi_{\mathrm{rad}(m)}(x^\ell) \quad \text{with } \ell := mk/\mathrm{rad}(m) \tag{12.9}$$

for generalized cyclotomic polynomials; see again Proposition 3.6.8. Indeed, the parameter $mk/\mathrm{rad}(m)$ plays an important role for the generalized cyclotomic module $\mathscr{G}^F_{m,k}$; see Proposition 12.7.6 and Definition 12.7.7 below. □

By definition, generalized cyclotomic modules over $F=\mathrm{GF}(q)$ are σ-invariant F-subspaces, that is, (F,σ)-submodules of $\widehat{F}$. Proposition 12.7.6 below describes the maximal intermediate field K of $\widehat{F}/F$ for which a generalized cyclotomic module is a K-vector space. This relies on the following two auxiliary results, the first of which is quite obvious and has already been used a few times (for instance, in the proofs of Propositions 11.6.1 and 12.5.2).

Lemma 12.7.4. *Consider a polynomial $f(x)$ over a finite field $F=\mathrm{GF}(q)$ which is not a multiple of x and an (F,σ)-submodule of $\widehat{F}$ of the form $V=V^F_{f(x^k)}$.*

Then V is also the (K,σ^k)-submodule V^K_f of $\widehat{F}$, where $K=E_k$ is the k-dimensional extension of F in $\widehat{F}$. □

Proof. By definition, $v\in V^F_{f(x^k)}$ if and only if $f(\sigma^k)(v)=0$, which also is the defining condition for $v\in V^K_f$; see part (3) of Observation 11.1.1. □

The converse of Lemma 12.7.4 is less obvious:

Lemma 12.7.5. *Consider a finite field $F=\mathrm{GF}(q)$ and a finite (F,σ)-submodule V of $\widehat{F}$. Assume that $\lambda v\in V$ holds for all $\lambda\in K$ and all $v\in V$, where $K=E_k$ is the k-dimensional extension of F in $\widehat{F}$.*

Then there exists a unique monic polynomial $f(x)\in F[x]$ (which is not a multiple of x) such that $V=V^F_{f(x^k)}=V^K_f$.

Proof. Let $g(x) \in F[x]$ be the unique monic polynomial with $g(0) \neq 0$ such that

$$V = V_g^F = \{v \in \widehat{F} \colon g(\sigma)(v) = 0\}.$$

By hypothesis, V is a K-vector space. As V is invariant under (powers of) σ, it is also a (K, σ^k)-submodule of $\widehat{F}$. Hence there is a unique monic polynomial $h(x) \in K[x]$ with $h(0) \neq 0$ such that

$$V = V_h^K = \{v \in \widehat{F} \colon h(\sigma^k)(v) = 0\}.$$

Now let w be any (K, σ^k)-generator of V, that is, an element with q^k-order $h(x)$. Then $w^q = \sigma(w)$ has q^k-order $\sigma(h(x))$, where σ acts on $K[x]$ by applying it to the coefficients of a polynomial (as usual). Combining these observations gives

$$V_h^K = \sigma(V_h^K) = V_{\sigma(h)}^K,$$

which implies $\sigma(h) = h$ and shows that h has coefficients in F. As $g(\sigma)(v) = 0$ if and only if $h(\sigma^k)(v) = 0$, we conclude that $g(x) = h(x^k)$. □

Proposition 12.7.6. *Consider a generalized cyclotomic module $G := \mathcal{G}_{m,k}^F$ over $F = \mathrm{GF}(q)$, and let F_G be the set of all $\lambda \in \widehat{F}$ such that $\lambda v \in G$ for all $v \in G$. Then F_G is the intermediate field of $\widehat{F}/F$ with degree $\ell := mk/\mathrm{rad}(m)$ over F.*

Proof. We first use Theorem 1.4.22 to check that F_G is a field. Thus let $\alpha, \beta, \gamma \in F_G$, where $\gamma \neq 0$. By hypothesis, the mapping $v \mapsto \gamma v$ permutes G, and hence $\gamma^{-1} v \in G$ for all v. This gives $\gamma^{-1} \in F_G$, and therefore $(\alpha - \beta)\gamma^{-1} v \in G$ for all $v \in G$, so that the criterion in Theorem 1.4.22 indeed holds.

Trivially, F_G is an extension field of F. We claim that we even have

$$L \subseteq F_G \subseteq K,$$

where $L = E_\ell$ and $K = E_{mk}$ are the ℓ- and mk-dimensional extensions of F in $\widehat{F}$, respectively. For the first inclusion, we apply Lemma 12.7.4 with $f = \Phi_{\mathrm{rad}(m)}$. In view of Equations (12.8) and (12.9), we conclude that G is an (L, σ^ℓ)-submodule of $\widehat{F}$ and that

$$G = \mathcal{G}_{m,k}^F = \mathcal{G}_{\mathrm{rad}(m),\ell}^F = V_{\Phi_{\mathrm{rad}(m)}}^L.$$

In particular, this shows $L \subseteq F_G$. For the second inclusion, we first note $G \subseteq K$, which holds since $\Phi_m(x^k)$ divides

$$x^{mk} - 1 = \prod_{n|m} \Phi_n(x^k).$$

Now let $\gamma \in F_G$ and choose any non-zero element v of G. Then $\gamma v \in G \subseteq K$ gives $\gamma = (\gamma v) \cdot v^{-1} \in K$, which proves $F_G \subseteq K$.

Finally, it remains to show that actually $F_G = L$. For simplicity, let us write $M = F_G$ and $a = [M : L]$. An application of Lemma 12.7.5 (with M/L instead of K/F) gives a monic polynomial $h(x) \in M[x]$ such that $\Phi_{\mathrm{rad}(m)}(x) = h(x^a)$ and

$G = V^L_{\Phi_{\mathrm{rad}(m)}} = V^M_h$. Note that $h(x)$ has to divide $x^b - 1$, where $b := \mathrm{rad}(m)/a$, since $\Phi_{\mathrm{rad}(m)}(x) = h(x^a)$ divides $x^{\mathrm{rad}(m)} - 1 = (x^a)^b - 1$.

Now let $\eta \in \widehat{F}$ be any primitive $\mathrm{rad}(m)$-th root of unity. Then $h(\eta^a) = 0$, since η^a is a primitive b-th root of unity (as a and $b = \mathrm{rad}(m)/a$ are relatively prime). An application of Proposition 3.6.18 with $n = b$ and $m = ab$ shows that all primitive b-th roots of unity arise in this manner, and hence $\Phi_b(x)$ has to divide $h(x)$. Using Proposition 3.6.12, we obtain

$$\prod_{d|a} \Phi_{bd}(x) = \Phi_b(x^a) \mid h(x^a) = \Phi_{\mathrm{rad}(m)}(x) = \Phi_{ab}(x).$$

Obviously, this cannot hold when $a \neq 1$, which proves our claim $L = F_G$. □

Definition 12.7.7. Consider a generalized cyclotomic module $G := \mathscr{G}^F_{m,k}$ over a Galois field $F = \mathrm{GF}(q)$. Then

$$\kappa(\mathscr{G}^F_{m,k}) := \frac{mk}{\mathrm{rad}(m)}$$

is called the **module character** of G (and also of any associated generalized cyclotomic polynomial) over F. □

Remark 12.7.8. We first observe that the notion of a module character is well-defined, as $\Phi_m(x^k) = \Phi_n(x^\ell)$ if and only if $\mathrm{rad}(m) = \mathrm{rad}(n)$ and $mk/\mathrm{rad}(m) = n\ell/\mathrm{rad}(n)$. We also note that $\kappa(\mathscr{G}^F_{m,k}) = mk$ holds if and only if $m = 1$, which means that $\mathscr{G}^F_{m,k}$ is the field extension E_k with degree k over F.

Finally, we point out the following rather obvious but important fact: if d is any divisor of the module character $\kappa(G)$ of a generalized cyclotomic module $G = \mathscr{G}^F_{m,k}$ over $F = \mathrm{GF}(q)$, then

$$G = V^{E_d}_h \quad \text{with } h(x) = \Phi_{\mathrm{rad}(m)}(x^{\kappa(G)/d})$$

as an (E_d, σ^d)-submodule, where E_d is the d-dimensional extension of F in $\widehat{F}$. □

Definition 12.7.7 and Remark 12.7.8 motivate introducing the following concept:

Definition 12.7.9. Let $G = \mathscr{G}^F_{m,k}$ be a generalized cyclotomic module over $F = \mathrm{GF}(q)$. An element $\theta \in G$ is called a **complete generator** for G over F if θ generates G as an (E_d, σ^d)-module for every divisor d of its module character $\kappa(G)$, that is, if

$$\mathrm{Ord}_{q^d}(\theta) = \Phi_{\mathrm{rad}(m)}(x^{\kappa(G)/d}) \quad \text{for all } d \mid \kappa(G). \qquad \square$$

The following result generalizes the complete normal basis theorem to a *complete generator theorem* for generalized cyclotomic modules. However, the proof below requires that we already know the existence of completely normal elements.

Theorem 12.7.10 (Complete generator theorem). *Every generalized cyclotomic module G over some finite field F has a complete generator over F. If G is actually* *a finite extension field E of F, then the complete generators of G are the completely normal elements for E/F.*

Proof. The second assertion is clear. For the first assertion, let $G = \mathcal{G}_{m,k}^F$ and consider the extension $E = E_{mk}$ of degree mk over F. We use the module character $\kappa = \kappa(G)$ to decompose the additive group of E according to the factorization

$$x^{mk} - 1 = (x^\kappa)^{\mathrm{rad}(m)} - 1 = \prod_{d \mid \mathrm{rad}(m)} \Phi_d(x^\kappa),$$

that is, we write

$$E = \bigoplus_{d \mid \mathrm{rad}(m)} W_d \quad \text{with } W_d = \mathcal{G}_{d,\kappa}^F.$$

Note that every component W_d is a generalized cyclotomic module, with module character $d\kappa/\mathrm{rad}(d) = \kappa = \kappa(G)$ (since $\mathrm{rad}(m)$ is square-free). Now let w be any completely normal element for E/F, write w as

$$w = \sum_{d \mid \mathrm{rad}(m)} w_d \quad \text{with } w_d \in W_d \text{ for all } d \mid \mathrm{rad}(m),$$

and let ℓ be an arbitrary divisor of κ. By our choice of w as a completely normal element, w has to be normal for E/E_ℓ. This requires that each component w_d of w generates W_d as an (E_ℓ, σ^ℓ)-module and thus has q^ℓ-order $\Phi_d(x^{\kappa/\ell})$. As this holds for all divisors ℓ of κ, each w_d is in fact a complete generator for W_d over F. In particular, $G = W_{\mathrm{rad}(m)} = \mathcal{G}_{m,k}^F$ admits a complete generator over F. □

Exercises

Exercise 12.7.11. Consider a finite field $F = \mathrm{GF}(q)$ and let k, ℓ, s, t be positive integers, where $k\ell$ is not divisible by the characteristic p of F. Assume that ks and ℓt are relatively prime, and let u and v be complete generators for the generalized cyclotomic modules $\mathcal{G}_{k,s}^F$ and $\mathcal{G}_{\ell,t}^F$, respectively. Prove that uv is a complete generator for the generalized cyclotomic module $\mathcal{G}_{k\ell,st}^F$. □

12.8 Cyclotomic Decompositions

We now apply the notion of complete generators for generalized cyclotomic modules introduced in the previous section to provide a general decomposition theory for these structures. Our main goal is to use this theory for a deeper study of extensions E/F of Galois fields: we wish to determine non-trivial decompositions of E into a direct sum of generalized cyclotomic modules with the property that *every* completely normal element for E/F can be written as the sum of complete generators for the components of the decomposition.

Definition 12.8.1. Consider a generalized cyclotomic polynomial $\Phi_k(x^t)$ over a field $F = \mathrm{GF}(q)$ with characteristic p (so that $k \not\equiv 0 \bmod p$). A set $\Delta \subset F[x]$ of generalized cyclotomic polynomials which are pairwise relatively prime is called a **cyclotomic decomposition** of $\Phi_k(x^t)$ provided that

$$\Phi_k(x^t) = \prod_{\Psi(x)\in\Delta} \Psi(x).$$

A corresponding set of index pairs (ℓ,s) for the polynomials $\Psi(x) = \Phi_\ell(x^s)$ in Δ will be denoted by $i(\Delta)$; recall from Remark 12.7.2 that these index pairs are in general not unique. We will also refer to the associated decomposition

$$\mathscr{G}_{k,t}^F = \bigoplus_{(\ell,s)\in i(\Delta)} \mathscr{G}_{\ell,s}^F \tag{12.10}$$

as a **cyclotomic decomposition** of the generalized cyclotomic module $\mathscr{G}_{k,t}^F$ associated with $\Phi_k(x^t)$. □

In view of the previous section, the following result (taken from [161, Section 19]) is what one would expect. Nevertheless, the proof requires some effort; it adapts the arguments used for Theorem 12.7.10.

Proposition 12.8.2. *Let Δ be a cyclotomic decomposition of $\Phi_k(x^t)$ over $F = \mathrm{GF}(q)$, and consider the associated cyclotomic decomposition of $G = \mathscr{G}_{k,t}^F$ in Equation* (12.10)*. Given any element $w \in G$, let*

$$w = \sum_{(\ell,s)\in i(\Delta)} w_{\ell,s} \quad \text{with } w_{\ell,s} \in \mathscr{G}_{\ell,s}^F \tag{12.11}$$

be the corresponding decomposition of w into Δ-components. If w is a complete generator for $\mathscr{G}_{k,t}^F$ over F, then every component $w_{\ell,s}$ is a complete generator for $\mathscr{G}_{\ell,s}^F$ over F.

Proof. Without loss of generality, we may assume that k and t are relatively prime; see Remark 12.7.3. As usual, we write $t = t'p^a$, where p is the characteristic of F. Now consider an index pair $(\ell,s) \in i(\Delta)$, where we also assume $\gcd(\ell,s) = 1$. For simplicity, we write G' and w' instead of $\mathscr{G}_{\ell,s}^F$ and $w_{\ell,s}$. Thus we have to prove that w' is a complete generator of G' over F. For this, let d be an arbitrary divisor of the module character $\kappa' := \ell s/\mathrm{rad}(\ell)$ of G'; we need to show that $\Phi_{\mathrm{rad}(\ell)}(x^{\kappa'/d})$ is the q^d-order of w'.

Since $\Phi_\ell(x^s)$ divides $\Phi_k(x^t)$, it is clear that $t/t' = s/s'$. We may therefore concentrate on the p-free parts of the parameters in question; for simplicity, we may as well assume $a = 0$, so that $s = s'$ and $t = t'$. Using Proposition 3.6.12, we note that

$$\prod_{d|s} \Phi_{\ell d}(x) = \Phi_\ell(x^s) \;\text{ divides }\; \Phi_k(x^t) = \prod_{e|t} \Phi_{ke}(x).$$

In particular, this implies

$$\Phi_\ell = \Phi_{ke}, \quad \Phi_{\ell s} = \Phi_{keb} \text{ and } t = ebc$$

for suitable divisors e, b and c of t. Hence κ' divides the module character κ of G:

$$\frac{\kappa}{\kappa'} = \frac{kt \cdot \mathrm{rad}(\ell)}{\ell s \cdot \mathrm{rad}(k)} = \frac{kebc \cdot \mathrm{rad}(ke)}{keb \cdot \mathrm{rad}(k)} = c \cdot \mathrm{rad}(e),$$

since k and $t = ebc$ are relatively prime. In particular, the divisor d of κ' under consideration also divides κ. Now Lemma 12.7.5 gives $\Phi_k(x^t) = g(x^d)$ and $\Phi_\ell(x^s) = f(x^d)$ for suitable polynomials $f(x)$ and $g(x)$ in $F[x]$.

Since $f(x^d)$ divides $g(x^d)$, we conclude that $g(x)$ is a multiple of $f(x)$, and therefore $g(x^d) = f(x^d)h(x^d)$ with $h(x) := g(x)/f(x)$. In fact, the cyclotomic decomposition Δ of $\Phi_k(x^t)$ gives

$$h(x^d) = \prod_{(m,r)\in i(\Delta)'} \Phi_m(x^r), \quad \text{where } i(\Delta)' := i(\Delta) \setminus \{(\ell,s)\}.$$

We now consider the associated decomposition $G = G' \oplus D$ of G, that is, we put $D := \{v \in \widehat{F} : h(\sigma^d)(v) = 0\}$, and decompose the complete generator w of G accordingly as $w = w' + v$ with $v \in D$. Since both G' and D are E_d-vector spaces and since w is an (E_d, σ^d)-generator of G, we see that w' and v are (E_d, σ^d)-generators for G' and D, respectively. This proves that w' is an (E_d, σ^d)-generator for G' for every divisor d of the module character κ' of G'; thus w' is indeed a complete generator of G'. □

We emphasize that the converse of Proposition 12.8.2 does not hold in general. This stems from the fact that the module characters of the components of a non-trivial decomposition are *proper* divisors of the original module character κ, and therefore relevant information about a complete generator for the entire module G might be lost. We will investigate this problem in detail throughout the remainder of this section. In order to do so, we require some additional terminology:

Definition 12.8.3. Let Δ be a non-trivial cyclotomic decomposition of $\Phi_k(x^t)$ over $F = \mathrm{GF}(q)$. Given any element w of the corresponding generalized cyclotomic module $\mathcal{G}_{k,t}^F$, we decompose w into Δ-components as in Equation (12.11).

Then w is said to be a **Δ-generator** of $\mathcal{G}_{k,t}^F$ provided that every $w_{\ell,s}$ is a complete generator for its component $\mathcal{G}_{\ell,s}^F$. Moreover, one calls Δ an **agreeable decomposition** of $\Phi_k(x^t)$ if every Δ-generator of $\mathcal{G}_{k,t}^F$ is actually a complete generator for $\mathcal{G}_{k,t}^F$ over F. □

Remark 12.8.4. Consider a generalized cyclotomic polynomial $\Phi_k(x^t)$ over a field $F = \mathrm{GF}(q)$, and let $\phi_q^c(\Phi_k(x^t))$ denote the number of complete generators of the corresponding generalized cyclotomic module $\mathcal{G}_{k,t}^F$. Then ϕ_q^c is multiplicative in the following sense: the product formula

$$\phi_q^c(\Phi_k(x^t)) = \prod_{(\ell,s)\in i(\Delta)} \phi_q^c(\Phi_\ell(x^s))$$

holds for every agreeable decomposition Δ of $\Phi_k(x^t)$. This observation provides one of the main reasons for studying agreeable decompositions. □

Example 12.8.5. Let r be a prime different from the characteristic p of $F = \mathrm{GF}(q)$. In the terminology just introduced, Proposition 12.5.2 states that

$$\{x^{r^{n-1}} - 1,\ \Phi_r(x^{r^{n-1}})\}$$

is an agreeable decomposition of $x^{r^n} - 1$ over F. □

Observation 12.8.6. Let $\Phi_k(x^t)$ be a generalized cyclotomic polynomial over $F = \mathrm{GF}(q)$, and let Δ be a cyclotomic decomposition of $\Phi_k(x^t)$. Assume that Σ is a cyclotomic decomposition of some polynomial $g(x) \in \Delta$. Trivially,

$$\Gamma := (\Delta \setminus \{g(x)\}) \cup \Sigma$$

is again a cyclotomic decomposition of $\Phi_k(x^t)$; we say that Γ is a **refinement** of Δ.

Moreover, Γ is agreeable over F if and only if both Δ and Σ are agreeable over F, which is an easy but extremely important consequence of Definition 12.8.3. It allows the recursive construction of agreeable decompositions with a large number of factors from "small" decompositions. □

Example 12.8.7. Let us illustrate Observation 12.8.6 for the case of extensions with prime power degree r^n, where r is distinct from the characteristic of $F = \mathrm{GF}(q)$. Using the agreeable decompositions for the polynomials $x^{r^{n-j}} - 1$ provided in Example 12.8.5 recursively results in the agreeable decomposition

$$\Delta(n) := \{\Phi_{r^j}(x) \colon j = 0, \ldots, n\} \tag{12.12}$$

of $x^{r^n} - 1$ over F. Obviously, $\Delta(n)$ is the finest cyclotomic decomposition of $x^{r^n} - 1$ and is therefore called the **canonical decomposition** of $x^{r^n} - 1$. □

In order to obtain more general recursive constructions, we need to extend Example 12.8.5 to the case where the extension degree is not a prime power:

Theorem 12.8.8. *Consider an extension $E = \mathrm{GF}(q^n)$ of $F = \mathrm{GF}(q)$, where n is not a power of the characteristic p of F. Let r be the largest prime divisor of the p-free part n' of n, and let $R = \mathrm{pt}_r(n')$ be the largest power of r dividing n'. Then*

$$\Delta_r := \{x^{n/r} - 1,\ \Phi_r(x^{n/r})\} = \{x^{n/r} - 1,\ \Phi_R(x^{n/R})\}$$

is an agreeable decomposition of $x^n - 1$ over F.

Proof. It is easily checked that Δ_r is a cyclotomic decomposition of $x^n - 1$, even for an arbitrary prime divisor r of n'. Let w be any Δ_r-generator of E. We have to show that w is completely normal over F, that is, we have to verify that w has q^d-order $x^{n/d} - 1$ for every divisor d of n. Since the module characters of the two generalized cyclotomic polynomials in Δ_r are both n/r, the definition of a Δ-generator guarantees that this condition holds whenever d divides n/r.

Now assume that n/r is not a multiple of d, which means that R divides d. Put $\ell := d/r$ and $Q := q^\ell$, and consider the extension E/L, where $L = E_\ell$ is the intermediate field of E/F with degree ℓ over F. Note that ℓ divides n/r, so that w is normal for E/L in view of our preceding observation, and that the degree n/ℓ of E/L has the form rm, where m is not divisible by r. We now use that r is the largest prime dividing n' to observe that r cannot divide $s-1$ for any prime divisor s of m'. Hence $\mathrm{ord}_{m'}(Q)$ is not divisible by r, which allows us to apply Proposition 12.3.4: since w is normal for E/L, it is also normal for E/K, where K is the intermediate field of E/L with degree r over L. As E/K has degree n/d, we obtain $\mathrm{Ord}_{q^d}(w) = x^{n/d} - 1$, as required. □

Example 12.8.9. We use Theorem 12.8.8 to investigate the extension $E = \mathrm{GF}(q^{252})$ of $F = \mathrm{GF}(q)$, where q is relatively prime to $252 = 2^2 \cdot 3^2 \cdot 7$.

- Applying Theorem 12.8.8 with $r = 7$ shows that $\{x^{36} - 1, \Phi_7(x^{36})\}$ is an agreeable decomposition of $x^{252} - 1$ over $\mathrm{GF}(q)$.
- Applying Theorem 12.8.8 with $r = 3$ to $x^{36} - 1$ yields the agreeable decomposition $\{x^{12} - 1, \Phi_3(x^{12})\}$ of $x^{36} - 1$.
- A further application of Theorem 12.8.8 with $r = 3$ gives the agreeable decomposition $\{x^4 - 1, \Phi_3(x^4)\}$ of $x^{12} - 1$ over $\mathrm{GF}(q)$.
- Finally, $\{x - 1, \Phi_2(x), \Phi_4(x)\}$ is an agreeable decomposition of $x^4 - 1$ over $\mathrm{GF}(q)$, by Example 12.8.7.

Combining these decompositions according to Observation 12.8.6 shows that

$$\Delta = \{x - 1, \Phi_2(x), \Phi_4(x), \Phi_3(x^4), \Phi_3(x^{12}), \Phi_7(x^{36})\}$$

is an agreeable decomposition of $x^{252} - 1$ over F. □

It is clear that Theorem 12.8.8 does not allow to refine the agreeable decomposition Δ of $x^{252} - 1$ obtained in Example 12.8.12 any further, since it only applies to generalized cyclotomic polynomials of the form $x^n - 1$. Nevertheless, it is possible to refine Δ for specific choices of q, for instance, for $q = 5$; see Example 12.8.12 below. Obtaining such refinements obviously requires agreeable decompositions for *arbitrary* generalized cyclotomic polynomials. We now prove a fundamental result taken from [161, Section 19]:

Theorem 12.8.10 (Decomposition theorem). *Consider a generalized cyclotomic polynomial* $\rho(x) := \Phi_k(x^t)$ *over* $F = \mathrm{GF}(q)$, *where* k *is not divisible by the characteristic* p *of* F *and where, without loss of generality,* $\gcd(k,t) = 1$. *Let* r *be a prime divisor of the* p*-free part* t' *of* t, *let* $R = \mathrm{pt}_r(t')$ *be the largest power of* r *dividing* t', *and consider*

$$\delta(x) := \Phi_k(x^{t/r}) \quad \text{and} \quad \varepsilon(x) := \Phi_{kr}(x^{t/r}) = \Phi_{kR}(x^{t/R}).$$

Then $\Delta := \{\delta(x), \varepsilon(x)\}$ *is a cyclotomic decomposition of* $\rho(x)$ *over* F, *and the module characters of the three polynomials involved satisfy*

$$\kappa(\delta) = \kappa(\varepsilon) = \frac{\kappa(\rho)}{r}.$$

Moreover, Δ is an agreeable decomposition provided that $\mathrm{ord}_{\mathrm{rad}(kt')}(q)$ *is not a multiple of R.*

Proof. By hypothesis, r does not divide k, and an application of Proposition 3.6.9 gives (with $y := x^{t/r}$)

$$\rho(x) = \Phi_k(y^r) = \Phi_k(y) \cdot \Phi_{kr}(y) = \delta(x) \cdot \varepsilon(x);$$

thus Δ is a cyclotomic decomposition of ρ. Moreover, the assertion on the module characters is a direct consequence of Definition 12.7.7.

Now suppose that $\mathrm{ord}_{\mathrm{rad}(kt')}(q)$ is not a multiple of R, and let w be a Δ-generator of the generalized cyclotomic module $\mathscr{G}_{k,t}^F$ and d any divisor of $\kappa(\rho(x)) = kt/\mathrm{rad}(k)$. We have to show that w has q^d-order $\Phi_{\mathrm{rad}(k)}(x^{\kappa(\rho)/d})$. We will use the same strategy as in the proof of Theorem 12.8.8.

Thus we first consider the case where d divides $\kappa(\rho)/r$. Since w is a Δ-generator, the δ-component of w has q^d-order

$$\Phi_{\mathrm{rad}(k)}(x^{\kappa(\delta)/d}) = \Phi_{\mathrm{rad}(k)}(x^{\kappa(\rho)/dr}),$$

while the ε-component of w has q^d-order

$$\Phi_{\mathrm{rad}(k)r}(x^{\kappa(\varepsilon)/d}) = \Phi_{\mathrm{rad}(k)r}(x^{\kappa(\rho)/dr}).$$

Applying Proposition 3.6.9 again, we conclude that w has q^d-order $\Phi_{\mathrm{rad}(k)}(x^{\kappa(\rho)/d})$, as desired.

Now assume that $\kappa(\rho)/r$ is not a multiple of d, that is, R divides d. Let $Q := q^{d/r}$. By the previous case, the Q-order of w is $f(x^r)$, where $f(x) := \Phi_{\mathrm{rad}(k)}(x^{\kappa(\rho)/d})$. The assumption on $\mathrm{ord}_{\mathrm{rad}(kt')}(q)$ shows that r cannot divide $\mathrm{ord}_{\mathrm{rad}(kt')}(Q)$; trivially, this implies that $\mathrm{ord}_{\mathrm{rad}(kt'/d')}(q)$ is also not a multiple of r. As kt'/d' is not divisible by r, Proposition 1.7.10 and Remark 1.7.11 imply that $\mathrm{ord}_{kt'/d'}(Q)$ is not divisible by r either. Now the assertion follows from Proposition 12.3.2: every element with Q-order equal to $f(x^r)$ has Q^r-order equal to $f(x)$. □

Remark 12.8.11. Note that the hypothesis on the multiplicative order of q modulo $\mathrm{rad}(kt)$ in Theorem 12.8.10 is equivalent to the condition that $\mathrm{ord}_{\mathrm{rad}(kt')/r}(q)$ is not divisible by R. Furthermore, it is also equivalent to the condition that $\mathrm{ord}_{\mathrm{rad}(kt')}(q^{R/r})$ is not divisible by r.

We remark that the converse of Theorem 12.8.10 also holds: if Δ is agreeable, the condition on $\mathrm{ord}_{\mathrm{rad}(kt')}(q)$ has to be satisfied. For the rather technical proof, we refer to Hachenberger [163], where it is also shown that the recursive application of Theorem 12.8.10 to a generalized cyclotomic polynomial $\Phi_k(x^t)$ (as in Example 12.8.12 below) always leads to a unique non-refinable agreeable decomposition; that is, the final decomposition is independent of the order of the selection of the refinement steps during this process. This results from the fact that the number theoretical condition involving a prime r is not lost by deriving a refinement using a prime s different from r. □

Example 12.8.12. We continue Example 12.8.9 and consider the special case $q = 5$. Note that $\mathrm{rad}(252) = 2 \cdot 3 \cdot 7 = 42$ and

$$\mathrm{ord}_{42}(5) = \mathrm{lcm}\big(\mathrm{ord}_2(5), \mathrm{ord}_3(5), \mathrm{ord}_7(5)\big) = \mathrm{lcm}(1,2,6) = 6.$$

Applying the formulas provided in Section 1.7, one computes $\mathrm{ord}_{252}(5) = 6$ since $\mathrm{pt}_{252}(5^6 - 1) = 2 \cdot 252$. In the same way, one can obtain the multiplicative order of 5 modulo any specified divisor of 252.

Now we apply Theorem 12.8.10 to the factors of the agreeable decomposition Δ of $x^{252} - 1$ given in Example 12.8.9. Since $\mathrm{ord}_6(5) = 2$ and $\mathrm{ord}_{42}(5)$ are not divisible by 4 and 9, respectively, we conclude that

- $\{\Phi_3(x^4)\}$ is refinable to $\{\Phi_3(x^2), \Phi_{12}(x)\}$;
- $\{\Phi_9(x^4)\}$ is refinable to $\{\Phi_9(x^2), \Phi_{36}(x)\}$;
- $\{\Phi_7(x^{36})\}$ is refinable to $\{\Phi_7(x^{12}), \Phi_{63}(x^4)\}$;
- $\{\Phi_7(x^{12})\}$ is refinable to $\{\Phi_7(x^6), \Phi_{28}(x^3)\}$;
- $\{\Phi_{63}(x^4)\}$ is refinable to $\{\Phi_{63}(x^2), \Phi_{252}(x)\}$.

This results in the following refinement of Δ over $F = \mathrm{GF}(5)$:

$$\Delta' = \{x-1, \Phi_2, \Phi_4, \Phi_3(x^2), \Phi_{12}, \Phi_9(x^2), \Phi_{36}, \Phi_7(x^6), \Phi_{28}(x^3), \Phi_{63}(x^2), \Phi_{252}\}.$$

The reader may check that Theorem 12.8.10 cannot be applied to refine any of the members of Δ' over GF(5). However, over GF(25) as ground field, the decomposition theorem applies for $\Phi_3(x^2)$ and $\Phi_9(x^2)$, as well as for $\Phi_7(x^6)$ and $\Phi_{63}(x^2)$. After that, it is no longer applicable to the components of the resulting refinement Δ'' of Δ'. Finally, with $\mathrm{GF}(5^6)$ as ground field, Δ'' can even be refined to the finest possible cyclotomic decomposition of $x^{252} - 1$, namely $\{\Phi_d(x) : d \mid 252\}$. □

As the results of this section show, one can work separately on the components corresponding to the factors of an agreeable decomposition of $x^n - 1$ in order to describe arbitrary completely normal elements for an extension $E = \mathrm{GF}(q^n)$ over $F = \mathrm{GF}(q)$. Suppose we are given a factor $\Phi_k(x^t)$ of such a decomposition, where k and t are relatively prime, and let $G := \mathcal{G}_{k,t}^F$ be the corresponding component in the associated decomposition of $(E, +)$. The module structures which have to be considered in order to check whether an element of G is a complete generator correspond to the divisors of the module character $kt/\mathrm{rad}(k)$.

Now assume that we can apply Theorem 12.8.10 to G, that is, we can decompose $\Phi_k(x^t)$ into $\delta(x) \cdot \varepsilon(x)$ (using a suitable prime dividing t). Then G splits into the sum of the two associated generalized cyclotomic modules, and the module characters of these parts are strictly smaller than that of G. This simplifies the situation, as one has to consider fewer module structures simultaneously. For instance, in Example 12.8.12, the module characters of the members of the agreeable decomposition Δ' of $x^{252} - 1$ over GF(5) given above equal

$$1, 1, 2, 2, 2, 3, 6, 6, 6, 6, 6,$$

respectively. Therefore, instead of the 18 module structures corresponding to the divisors of 252, at most 4 module structures have to be considered simultaneously when we work with Δ': the complete information on the module structure of $\mathrm{GF}(5^{252})$ is already encoded in Δ'.

Of course, it is possible that the decomposition theorem cannot be applied to split the generalized cyclotomic polynomial $\Phi_k(x^t)$, which means that the p-free part t' of t divides the multiplicative order of q modulo $\mathrm{rad}(kt')$. Then the enumeration of all complete generators of $\mathscr{G}^F_{k,t}$ over F seems an extremely difficult problem, which is not yet solved.

Clearly, the best possible situation for the task of describing all completely normal elements for E/F arises when the number theoretic condition in Theorem 12.8.10 is *always* satisfied for any possible cyclotomic decomposition of a factor arising during the recursive application of the decomposition process to $x^n - 1$, since the **canonical decomposition**

$$\{\Phi_k(x)^{n/n'}: k \mid n'\}$$

of $x^n - 1$ is then agreeable. We will characterize the pairs (q,n) for which this holds in the next section.

Exercises

Exercise 12.8.13. Provide the missing details for the successive refinements given in Example 12.8.12. □

Exercise 12.8.14. Consider the extension $E = \mathrm{GF}(2^n)$ over the binary field $F = \mathrm{GF}(2)$, where $n = 11025$. Determine the unique finest agreeable decomposition of $x^n - 1$ over F by a recursive application of Theorem 12.8.10. □

12.9 Regular Extensions Revisited

Consider a generalized cyclotomic polynomial $\Phi_k(x^t)$ over a field $F = \mathrm{GF}(q)$ with characteristic p, where k and t are relatively prime and where t' is the p-free part of t. The results of the previous section suggest to investigate when the **canonical decomposition**

$$\Delta_{k,t} := \{\Phi_{kd}(x)^{t/t'}: d \mid t'\} \tag{12.13}$$

of $\Phi_k(x^t)$ into cyclotomic polynomials according to Proposition 3.6.12 is agreeable. We first give a sufficient general criterion and then a precise characterization for the special case corresponding to completely normal elements, that is, for the generalized cyclotomic polynomials of the form $\Phi_1(x^n) = x^n - 1$. This will lead us to reconsider the class of regular extensions E/F already studied in Section 11.7 from the complete point of view.

Theorem 12.9.1. *Let k, t and q be as in Theorem 12.8.10, and consider the generalized cyclotomic polynomial $\Phi_k(x^t)$ over $F = \mathrm{GF}(q)$. If t' and $\mathrm{ord}_{\mathrm{rad}(kt')}(q)$ are relatively prime, then the canonical decomposition $\Delta_{k,t}$ is an agreeable decomposition of $\Phi_k(x^t)$ over F.*

Proof. Let $\lambda(x)$ be any generalized cyclotomic polynomial which divides $\rho(x) = \Phi_k(x^t)$. Note that $\lambda(x)$ has the form $\Phi_\ell(x^{sp^a})$, where $p^a := t/t'$ and where s and ℓs divide t' and kt', respectively. Thus the assumption that t' and $\mathrm{ord}_{\mathrm{rad}(kt')}(q)$ are relatively prime implies that $\mathrm{ord}_{\mathrm{rad}(\ell s)}(q)$ and s are likewise relatively prime; that is, also the "smaller" generalized cyclotomic polynomial $\lambda(x)$ satisfies the hypothesis of the theorem. This shows that every prime divisor r of s satisfies the number theoretic condition in the decomposition theorem 12.8.10, and therefore $\lambda(x)$ can be decomposed accordingly. Now the assertion follows with induction on the degrees of the generalized cyclotomic polynomials under consideration. □

The following result concerning the case of completely normal elements is due to Hachenberger [161, Theorem 19.11]:

Theorem 12.9.2. *Consider a finite field $F = \mathrm{GF}(q)$ with characteristic p and an extension $E = \mathrm{GF}(q^n)$, where $n = n'p^a$. Then the canonical decomposition*

$$\Delta_n := \left\{\Phi_d(x)^{p^a} \colon d \mid n'\right\}$$

of $x^n - 1$ is agreeable over F if and only if n' and $\mathrm{ord}_{\mathrm{rad}(n')}(q)$ are relatively prime.

Proof. If n' and $\mathrm{ord}_{\mathrm{rad}(n')}(q)$ are relatively prime, then Δ_n is an agreeable decomposition over F by Theorem 12.9.1.

For the converse, suppose that Δ_n is agreeable over F, even though the number theoretic condition above is violated; we will obtain a contradiction by constructing a Δ_n-generator of E over F which is not completely normal for E/F. As we assume $\gcd\left(n', \mathrm{ord}_{\mathrm{rad}(n')}(q)\right) \neq 1$, there are two (distinct) prime divisors r and s of n' such that r divides $\mathrm{ord}_s(q)$.

We first deal with the special case where $n' = rs$, so that $n = rsp^a$. Let $K = E_r$ be the r-dimensional extension of F in E. Applying Proposition 3.6.16 to both F and K shows that every factor f in the canonical factorization of $\Phi_s(x)$ over F splits over K into r irreducible factors, since $r \mid \mathrm{ord}_s(q)$; we choose some irreducible K-divisor h_f of f for each such f. We also choose an element $v_f \in V^K_{h_f(x)^{p^a}}$ with q^{rp^b}-order $h_f(x)^{p^{a-b}}$ for $b = 0, \ldots, a$, which is possible as every complete generator of $\mathscr{C}^K_{sp^a}$ over K is a sum of elements of this type; then the q^{p^b}-order of v_f is $f(x)^{p^{a-b}}(x^r)$ for all b, by Proposition 11.6.1. Hence $v := \sum_f v_f$ (where the sum runs over all monic irreducible F-divisors of $\Phi_s(x)$) has q^{p^b}-order $\Phi_s^{p^{a-b}}(x^r)$ for all b.

Now let $v = v_s + v_{sr}$ be the decomposition of v induced by the decomposition $\{\Phi_s(x)^{p^a}, \Phi_{sr}(x)^{p^a}\}$ of $\Phi_s(x^r)^{p^a}$ over F. By construction, the component v_s is a complete generator of the cyclotomic module $\mathscr{C}^F_{sp^a}$ over F, while v_{sr} is a complete generator of the cyclotomic module $\mathscr{C}^F_{srp^a}$ over F. Finally, let $v' \in E_{rp^a}$ be a completely normal element over F; then $w := v + v'$ is a Δ_n-generator of E over F. However, w is not normal for E/K, since v has q^r-order $\prod_f h_f(x)^{p^a} \neq \Phi_s(x)^{p^a}$, which is the desired contradiction.

It only remains to reduce the case of arbitrary n to the special case just considered. For each factor

$$\lambda(x) \in \Delta_n \setminus \{x^{p^a} - 1, \Phi_r(x)^{p^a}, \Phi_s(x)^{p^a}, \Phi_{sr}(x)^{p^a}\},$$

we choose a complete generator u_λ for the corresponding generalized cyclotomic module and let $u := \sum_\lambda u_\lambda$ be the sum of all these components; also, let K and $w \in E_{rsp^a}$ be as in the special case above. Then $u + w$ is a Δ_n-generator of E over F which is not normal for E/K, and hence not completely normal for E/F. □

In particular, Theorem 12.9.2 shows that the canonical decomposition of $x^n - 1$ is agreeable over $F = \mathrm{GF}(q)$ whenever $E = \mathrm{GF}(q^n)$ is a regular extension over F; see Definition 11.6.5. This motivates the following extension of the notion of regularity to generalized cyclotomic modules:

Definition 12.9.3. Consider a generalized cyclotomic module $G = \mathscr{G}^F_{k,t}$ over $F = \mathrm{GF}(q)$, where k and t are relatively prime. Then G and the corresponding generalized cyclotomic polynomial $\Phi_k(x^t)$ are called **regular** over F if $\mathrm{ord}_{\mathrm{rad}(kt')}(q)$ and kt are relatively prime. □

In the remainder of this section, we provide an efficient characterization of complete generators for regular generalized cyclotomic modules, which generalizes the results that were obtained for extensions with prime power degree in Section 12.5. Since Theorem 12.9.1 applies to $\Phi_k(x^t)$ in the regular case, we may restrict attention to regular cyclotomic modules over F; that is, we may assume $t = 1$. The characterization in Theorem 12.9.6 below requires some preparation:

Notation 12.9.4. Consider two integers $q, m \geq 2$ with $\gcd(m, q) = 1$, and assume that m and $\mathrm{ord}_{\mathrm{rad}(m)}(q)$ are relatively prime. Let $\pi(m)$ denote the set of all prime divisors of m, and write $m_r = \mathrm{pt}_r(m)$ for $r \in \pi(m)$. By the results in Section 1.7 (see, in particular, Proposition 1.7.10 and Remark 1.7.11), the order of q modulo m has the form

$$\mathrm{ord}_m(q) = \mathrm{ord}_{\mathrm{rad}(m)}(q) \cdot \prod_{r \in \pi(m)} r^{\alpha(r)},$$

where $\prod_{r \in \pi(m)} r^{\alpha(r)}$ divides $m/\mathrm{rad}(m)$ and $r^{\alpha(r)}$ is the maximal power of r dividing $\mathrm{ord}_m(q)$. We now define

$$\tau(q, m) := \prod_{r \in \pi(m)} r^{\lfloor \alpha(r)/2 \rfloor}. \tag{12.14}$$

We emphasize that this adjusts the notation used in Section 12.5, where we dealt with the special case where $m = r^\ell$ is a power of a prime r. For instance, the values defined directly before Theorem 12.5.4 are connected to the new notation via $\alpha(r) = t$ and $\tau(q,r^\ell) = r^{\lfloor t/2 \rfloor} = r^\tau$. □

We also extend the notion of exceptional cyclotomic modules $\mathscr{C}_{r^\ell}^F$ introduced in Section 12.5 as follows:

Definition 12.9.5. A regular cyclotomic module $\mathscr{C}_n^F$ over $F = \mathrm{GF}(q)$ is called **exceptional** if the following conditions are satisfied:

$$q \equiv 3 \bmod 4 \quad \text{and} \quad 8 \mid n_2 \mid q^2 - 1,$$

where $n_2 = \mathrm{pt}_2(n)$; we also call the regular pair (q,n) **exceptional** in this case. □

We can now state the following major result, which is taken from Hachenberger [161, Section 20]. We give a full outline of the proof, but leave it to the reader to fill in some details.

Theorem 12.9.6. *Let $C := \mathscr{C}_n^F$ be a regular cyclotomic module over $F = \mathrm{GF}(q)$. Write $n = n'p^a$, where p is the characteristic of F and n' the p-free part of n, and let $\tau := \tau(q,n')$ be as in Notation 12.9.4. Then the following hold:*

(1) *if C is not exceptional over F, then $v \in C$ is a complete generator of C over F if and only if v has q^τ-order $\Phi_{n'/\tau}(x)^{p^a}$;*
(2) *if C is exceptional over F, then $v \in C$ is a complete generator of C over F if and only if v has q^τ-order $\Phi_{n'/\tau}(x)^{p^a}$ and $q^{2\tau}$-order $\Phi_{n'/2\tau}(x)^{p^a}$.*

Proof. For simplicity, we write m instead of n'. We begin with some preparatory remarks, where we again use the results in Section 1.7:

(a) Let $m = \prod_{j=1}^d r_j^{c_j}$ be the prime power factorization of m, and put $s_j := \mathrm{ord}_{r_j}(q)$. Then $s := \mathrm{ord}_{\mathrm{rad}(m)}(q) = \mathrm{lcm}(s_j : 1 \le j \le d)$. As C is regular, s is relatively prime to m and hence

$$\mathrm{ord}_m(q) = s \cdot \mathrm{ord}_m(q^s) = s \cdot \prod_{j=1}^{d} \mathrm{ord}_{R_j}(q^s) = s \cdot \prod_{j=1}^{d} \mathrm{ord}_{R_j}(q^{s_j}),$$

where we write $R_j = r_j^{c_j}$ for better readability.

(b) Now let r be a prime dividing m, and let $\mathrm{pt}_r(m) = r^c$. Assume that r divides $\mathrm{ord}_{r^c}(q)$ (thus $c \ge 2$). Then

$$\mathrm{ord}_{r^{c-1}}(q) = \begin{cases} \frac{1}{r} \cdot \mathrm{ord}_{r^c}(q) & \text{if } (q,r^c) \text{ is not exceptional} \\ \mathrm{ord}_{r^c}(q) & \text{otherwise;} \end{cases}$$

by definition, $r = 2,\ c \ge 3,\ q \equiv 3 \bmod 4$ and $2^c \mid \mathrm{pt}_2(q^2 - 1)$ in the exceptional case.

(c) Finally, let r be a prime dividing $\tau = \tau(q,m)$, which requires that r^2 divides $\mathrm{ord}_m(q)$. Note that this can only happen for $r = 2$ when the pair (q,m) is not exceptional (as $\mathrm{ord}_{2^c}(q) = 2$ in the exceptional case). Therefore, $\mathrm{ord}_{m/r}(q) = \frac{1}{r} \cdot \mathrm{ord}_m(q)$ by remarks (a) and (b), which implies $\mathrm{ord}_{m/r}(q^r) = \frac{1}{r^2} \cdot \mathrm{ord}_m(q)$. Then induction on the number of prime divisors of τ yields

$$\mathrm{ord}_{m/t}(q^t) = \frac{\mathrm{ord}_m(q)}{t^2} \quad \text{for every divisor } t \text{ of } \tau.$$

Before we turn to the proof of the theorem, we also note that the assertion makes sense: by Equation (12.14), τ always divides $\kappa(C) = m/\mathrm{rad}(m)$, and 2τ divides $m/\mathrm{rad}(m)$ in the exceptional case.

Now let $v \in C = \mathscr{C}_n^F = \mathscr{C}_{mp^a}^F$ have q^τ-order $\Phi_{m/\tau}(x)^{p^a}$, and assume in addition that the $q^{2\tau}$-order of v is $\Phi_{m/2\tau}(x)^{p^a}$ when (q,m) is an exceptional pair. We have to show that v then already is a complete generator for C over F, that is,

$$\mathrm{ord}_{q^{kp^b}}(v) = \Phi_{m/k}(x)^{p^{a-b}} \quad \text{for all } k,m \text{ with } k \mid \frac{m}{\mathrm{rad}(m)} \text{ and } b \le a.$$

We will distinguish four cases: first, we let $b = 0$ and consider the three cases where either k divides τ, or τ divides k, or neither of these two conditions holds; finally, we deal with the situation where $b \ge 1$.

Case 1: $b = 0$ and $k \mid \tau$.

Here we apply Proposition 3.6.16 to determine how the polynomials $\Phi_{m/\tau}(x)$ and $\Phi_{m/k}(x)$ split over the k-dimensional extension $K = E_k$ of F in E. Using (c) above, we compute the degrees of the irreducible divisors of these two polynomials over K as follows:

$$\mathrm{ord}_{m/\tau}(q^k) = \mathrm{ord}_{m/\tau}(q^\tau) \cdot \frac{\tau}{k} = \frac{\mathrm{ord}_m(q)}{\tau^2} \cdot \frac{\tau}{k} = \frac{\mathrm{ord}_m(q)}{\tau k}$$

and

$$\mathrm{ord}_{m/k}(q^k) = \frac{\mathrm{ord}_m(q)}{k^2} = \frac{\tau}{k} \cdot \frac{\mathrm{ord}_m(q)}{\tau k}.$$

This shows that every irreducible monic divisor $f(x)$ of $\Phi_{m/\tau}(x)$ gives rise to a monic divisor $f(x^{\tau/k})$ of $\Phi_{m/\tau}(x^{\tau/k}) = \Phi_{m/k}(x)$ which in fact is likewise irreducible over K. From this we conclude $\mathrm{Ord}_{q^k}(v) = \Phi_{m/k}(x)^{p^a}$, as claimed.

Case 2: $b = 0$ and $\tau \mid k$.

Put $Q := q^\tau$ and $M := m/\tau$, and note that (Q,M) is regular; moreover, one may check that (Q,M) is exceptional if and only if (q,m) has this property. Since $\mathrm{ord}_M(Q) = \mathrm{ord}_m(q)/\tau^2$ by (c), we see that $r \in \pi(m)$ divides $\mathrm{ord}_M(Q)$ if and only if $\alpha(r) \ge 1$ is odd and that r^2 cannot divide $\mathrm{ord}_M(Q)$; see Notation 12.9.4. Using the regularity of (Q,M), one can then show that $\mathrm{ord}_{M/r}(Q)$ is not divisible by r if r is odd, and also not by $r = 2$ provided that (Q,M) is not exceptional. In these cases, the M-

dimensional extension of $\mathrm{GF}(Q)$ is completely basic by Theorem 12.3.5. From this, one can derive that v has q^k-order $\Phi_{m/k}(x)^{p^a}$ in this case.

In the remaining subcase where (Q,M) is exceptional, a similar argument can be applied to the regular pair $(Q^2,M/2)$.

Case 3: $b=0,\ k\nmid\tau$ and $\tau\nmid k$.

In order to reduce this situation to the first two cases, we first show the auxiliary result

$$\tau_k := \tau(q^k,m/k) = \frac{\tau}{\gcd(k,\tau)} \quad \text{for all } k \mid \frac{m}{\mathrm{rad}(m)}. \tag{12.15}$$

For this, let r be any prime dividing k. If r divides τ, Equation (12.14) gives $\tau(q^r,m/r)=\tau/r$. On the other hand, if $r\nmid\tau$, then r^2 cannot divide $\mathrm{ord}_m(q)$, since $\mathrm{ord}_{\mathrm{rad}(m)}(q)$ is not divisible by r; hence $\mathrm{ord}_{m/r}(q^r)$ is not divisible by r. Since $\mathrm{ord}_{m/r}(q^r)$ is either $\mathrm{ord}_m(q)$ or $\mathrm{ord}_m(q)/r$ in this case, we then have $\tau(q^r,m/r)=\tau$. Altogether, we obtain

$$\tau\left(q^r,\frac{m}{r}\right) = \frac{\tau}{\gcd(r,\tau)} \quad \text{for each prime } r\mid k.$$

Now Equation (12.15) follows with induction on the number of prime divisors of k, counted with multiplicity.

Thus $k\tau_k=\mathrm{lcm}(\tau,k)$ divides $m/\mathrm{rad}(m)$ and is a multiple of τ, and hence we can apply Case 2 to this divisor of $m/\mathrm{rad}(m)$. We conclude that v has $q^{k\tau_k}$-order $\Phi_{m/k\tau_k}(x)^{p^a}$. Finally, we apply Case 1 to the field extension E/K for the divisor τ_k of $M/\mathrm{rad}(M)$, where K again denotes the k-dimensional extension of F and $M:=m/k$ is the degree of E over K. This shows that v has q^k-order $\Phi_{m/k}(x)^{p^a}$, as claimed.

Case 4: $b\geq 1$.

By the previous cases, the q^k-order of v is

$$\Phi_{m/k}(x)^{p^a} = \Phi_{m/k}(x^{p^b})^{p^{a-b}}.$$

Moreover, since p does not divide $\mathrm{ord}_q(m)$, Proposition 12.3.2 shows that v indeed has q^{kp^b}-order $\Phi_{m/k}(x)^{p^{a-b}}$. □

In combination with a generalization of Theorem 12.5.9 to arbitrary exceptional cyclotomic modules, Remark 12.8.4 and Theorem 12.9.6 lead to the following lower bound for the number of completely normal elements for any regular extension. We leave the details as an exercise, cf. [161, Theorem 21.3].

Result 12.9.7 *Let (q,n) be a regular pair, and write $n=n'p^a$, where q is a power of the prime p and n' the p-free part of n. Then the number of completely normal elements for* $\mathrm{GF}(q^n)/\mathrm{GF}(q)$ *satisfies*

$$\phi_q^c(x^n-1) \geq (q-1)^{n'}\cdot q^{n'(p^a-1)}.$$

Moreover, equality holds if and only if n' divides $q-1$, in which case the given extension is completely basic.

It is conjectured in [161] that the assertion of Result 12.9.7 actually holds without the hypothesis that (q,n) is a regular pair. This is one of the major open problems concerning completely normal elements for finite fields.

Exercises

Exercise 12.9.8. Provide the missing details in the proof of Theorem 12.9.6. □

Exercise 12.9.9. Let (q,n) be a regular pair. Determine a formula for the number $\phi_q^c(x^n-1)$ of completely normal elements for $\mathrm{GF}(q^n)/\mathrm{GF}(q)$, and derive the lower bound stated in Result 12.9.7. □

Chapter 13
Primitive Normal Bases

Abstract The topic of this chapter is the celebrated **primitive normal basis theorem**: for every extension E/F of Galois fields, there exists a primitive element of E which is normal over F. We will provide a complete proof for this fundamental result, which does not rely on the use of computers at all; all necessary calculations can easily be checked using a standard pocket calculator. Nevertheless, such a proof requires a lot of detailed work; an outline of the argument will be provided at the end of the introductory first section.

13.1 Introduction and a Density Result

The study of primitive normal elements in Galois fields has attracted a lot of attention over the years. The history of this research topic started in 1952, when Carlitz [63, 64] proved that there exists a primitive element of $\mathrm{GF}(q^n)$ which also generates a normal basis over $\mathrm{GF}(q)$ for all but a finite number of Galois field extensions $\mathrm{GF}(q^n)/\mathrm{GF}(q)$. In 1968, Davenport [101] could prove the existence of such a primitive normal basis generator for all values of n when q is a prime, and in 1987 the primitive normal basis theorem was finally established in general by Lenstra and Schoof [238].

The fundamental structures involved in the proof of this result are finite cyclic modules over principal ideal domains, and the proof machinery for the most part relies on the theory of group characters and estimates of Gauss sums, using a clever mixture of combinatorial and number theoretic arguments. Over the years, the techniques have been refined enormously. In fact, stronger versions of the primitive normal basis theorem have been established; we shall discuss this in the final section of this chapter.

Throughout the entire chapter, we let $F = \mathrm{GF}(q)$ be a Galois field with characteristic p, and $E = \mathrm{GF}(q^n)$ the n-dimensional extension of F. We will denote the number of all primitive elements of E which are normal over F by $PN_n(q)$; thus it is our aim to show that $PN_n(q) > 0$ for all pairs (q,n). We also recall that $\phi(q^n - 1)$ is

D. Hachenberger and D. Jungnickel, *Topics in Galois Fields*,
Algorithms and Computation in Mathematics 29,
https://doi.org/10.1007/978-3-030-60806-4_13

the number of all primitive elements of $E = \mathrm{GF}(q^n)$, while $\phi_q(x^n-1)$ is the number of all elements of E which are normal over $F = \mathrm{GF}(q)$; see Sections 3.2 and 3.10 (or 7.3), respectively.

The main goal of this introductory section, which is based on Hachenberger [169, Section 8], is to provide a basic counting argument and an estimate for the Euler function which together yield the following somewhat surprising density result: for every fixed n, the proportion of normal elements of $\mathrm{GF}(q^n)$ over $\mathrm{GF}(q)$ in the set of primitive elements of $\mathrm{GF}(q^n)$ tends to 1 when q goes to infinity.

For the sake of completeness, let us first mention the trivial case $n = 1$, which means $E = F$: obviously, every non-zero element of F is normal over F, and therefore every primitive element of F is normal over F. This gives $PN_1(q) = \phi(q-1)$ for every prime power q. In fact, the same result still holds for all quadratic extensions of a finite field, though this is a little less obvious:

Proposition 13.1.1. *For every prime power q, all primitive elements of $E = \mathrm{GF}(q^2)$ are normal over $F = \mathrm{GF}(q)$, and hence*

$$PN_2(q) = \phi(q^2-1).$$

Proof. Let us first consider the case where the characteristic is $p = 2$. Then $x^2 - 1 = (x-1)^2$, and therefore $u \in E$ is normal over F if and only if its q-order does not divide $x-1$, which means that u does not belong to F. Trivially, this holds for all primitive elements of E.

Next, let p be odd. Then $x^2-1 = (x-1)(x+1)$ is square-free. If $u \in E$ is any primitive element, then $\mathrm{Ord}_q(u)$ does not divide $x-1$, since $u \notin F$. Assume that $\mathrm{Ord}_q(u) = x+1$, that is $u^q + u = 0$. Dividing by u gives $u^{q-1} = -1$, and therefore $u^{2(q-1)} = 1$, which means that the order of u divides $2(q-1)$. Since $\mathrm{ord}(u) = q^2-1$, by the assumption that u is primitive, $q+1$ has to divide 2, a contradiction. □

From now on, we may assume that $n \geq 3$. The study of primitive elements in (affine) hyperplanes of E over F in the subsequent Chapter 14 will show that the additional assumption of normality is a true restriction in this case: one has $PN_n(q) < \phi(q^n-1)$ for all pairs (q,n) with $n \geq 3$. See Theorem 14.1.7 and observe that $\phi(q^n-1) = 36 > 27 = \phi_q(x^n-1)$ when $(q,n) = (4,3)$.

Our next result gives a simple lower bound for $PN_n(q)$ involving the number of all generators of E over F; cf. Lenstra and Schoof [238]. We shall refer to this result as the **trivial lower bound** for $PN_n(q)$.

Proposition 13.1.2. *Let q be a prime power and $n \geq 2$ an integer. Then*

$$\begin{aligned} PN_n(q) &\geq \phi(q^n-1) + \phi_q(x^n-1) - \sum_{d \mid n} \mu\left(\frac{n}{d}\right) q^d \\ &\geq \phi(q^n-1) + (q-1)^n - (q^n - q). \end{aligned} \tag{13.1}$$

Proof. Let $\mathscr{N}$ denote the set of normal elements of $E = \mathrm{GF}(q^n)$ over $F = \mathrm{GF}(q)$, and let $\mathscr{P}$ be the set of all primitive elements of E. Furthermore, let $\mathscr{G}$ denote the set of all generators of E over F, that is, $v \in \mathscr{G}$ if and only if $F(v) = E$. Obviously, $\mathscr{P} \cup \mathscr{N} \subseteq \mathscr{G}$, and therefore

$$\begin{aligned} PN_n(q) &= |\mathscr{P} \cap \mathscr{N}| = |\mathscr{P}| + |\mathscr{N}| - |\mathscr{P} \cup \mathscr{N}| \\ &\geq |\mathscr{P}| + |\mathscr{N}| - |\mathscr{G}| = \phi(q^n - 1) + \phi_q(x^n - 1) - |\mathscr{G}|. \end{aligned}$$

This gives the assertion, as $\phi_q(x^n - 1) \geq (q-1)^n$ (with equality if and only if n divides $q-1$, by Theorem 3.10.5) and as

$$|E| - |F| = q^n - q \geq |\mathscr{G}| = \sum_{d|n} \mu\left(\frac{n}{d}\right) q^d,$$

see Remark 3.4.7. □

We next provide a basic lower bound for the Euler function ϕ, which is taken from Ribenboim [327] and turns out to be efficient enough to establish the density result mentioned at the beginning of this section.

Lemma 13.1.3. *Let N be any positive integer. Then*

$$\phi(N) \geq \frac{\ln(2) \cdot N}{\ln(2N)},$$

where ln *denotes the natural logarithm.*

Proof. Let $N = \prod_{i=1}^{\ell} r_i^{a_i}$ be the prime power factorization of N. By Equation (1.21), $\phi(N) = \prod_{i=1}^{\ell} r_i^{a_i - 1}(r_i - 1)$ and thus $\phi(N)/N = \prod_{i=1}^{\ell}(r_i - 1)/r_i$. Let $p_1, p_2, \ldots, p_\ell$ be the sequence of the first ℓ prime numbers. Then we obtain

$$\frac{\phi(N)}{N} \geq \prod_{i=1}^{\ell} \frac{p_i - 1}{p_i} \geq \prod_{i=1}^{\ell} \frac{i}{i+1} = \frac{1}{\ell+1}.$$

On the other hand,

$$\frac{\ln(2)}{\ln(2N)} \leq \frac{\ln(2)}{\ln(2 \cdot \mathrm{rad}(N))} \leq \frac{\ln(2)}{\ln(2 \cdot 2^\ell)} = \frac{\ln(2)}{(\ell+1) \cdot \ln(2)} = \frac{1}{\ell+1},$$

as the radical $\mathrm{rad}(N)$ of N satisfies $\mathrm{rad}(N) = \prod_{i=1}^{\ell} r_i \geq 2^\ell$. Combining these two inequalities gives the assertion. □

Combining Proposition 13.1.2 and Lemma 13.1.3, we obtain the following bound:

$$PN_n(q) > \frac{\ln(2) \cdot (q^n - 1)}{\ln(2) + n\ln(q)} + (q-1)^n - (q^n - q). \tag{13.2}$$

This simple estimate already suffices to prove the desired density result on primitive normal elements:

Theorem 13.1.4. *Let q be a prime power and $n \geq 2$ an integer, and let $\pi_n(q)$ denote the proportion of normal elements of* GF(q^n) *over* GF(q) *in the set of primitive elements of* GF(q^n)*:*

$$\pi_n(q) = \frac{PN_n(q)}{\phi(q^n-1)}. \tag{13.3}$$

Then $\lim_{q\to\infty} \pi_n(q) = 1$ *for all n.*

Proof. As noted before, we may assume that $n \geq 3$. By Proposition 13.1.2,

$$\begin{aligned} PN_n(q) &\geq \phi(q^n-1) - \left(q^n - q - (q-1)^n\right) \\ &= \phi(q^n-1) - \left(nq^{n-1} - A(q)\right), \end{aligned}$$

where A is a polynomial of degree $n-2$ and $A(q) < nq^{n-1}$ for all q, and hence

$$\pi_n(q) \geq 1 - \frac{nq^{n-1} - A(q)}{\phi(q^n-1)}.$$

Using Lemma 13.1.3, we conclude that

$$\frac{nq^{n-1} - A(q)}{\phi(q^n-1)} \leq \frac{(nq^{n-1} - A(q)) \cdot (\ln(2) + n\ln(q))}{\ln(2) \cdot (q^n-1)}.$$

As the fraction

$$\frac{q^{n-1}\ln(q)}{q^n-1}$$

tends to 0 for $q \to \infty$ (and fixed n), we obtain the assertion. □

As stated before, the main aim of this chapter is to show that $PN_n(q) > 0$ holds for all pairs (q,n). Since this requires a lot of detailed work, it may be useful to provide the following outline of the proof:

- The basic counting arguments given in the present section suffice to handle all cubic extensions in Section 13.2.
- In Section 13.3, we use an approach introduced by Hachenberger [169] and study quartic extensions within the framework of projective geometries, which actually yields much stronger results than just $PN_4(q) > 0$, namely strong lower bounds for $PN_4(q)$.
- In Section 13.4, we prove that the character group of the additive group of a cyclic torsion module M over a principal ideal domain is actually a cyclic R-module isomorphic to M. For the classical case of complex characters, we then expand the characteristic functions of certain subsets of M – in particular, of the set of generators of M – as character sums.
- In Section 13.5, we apply the general results of Section 13.4 to describe the characteristic function of the set of all primitive normal elements in terms of additive and multiplicative characters of finite fields. Using Gauss sums, this

character approach leads to a sufficient number theoretic criterion for the existence of primitive normal bases in terms of the number $\omega(q^n-1)$ of distinct prime divisors of q^n-1 and the number $\omega_q(x^n-1)$ of distinct monic irreducible F-factors of x^n-1.

- Section 13.6 contains various bounds for $\omega(q^n-1)$ and $\omega_q(x^n-1)$, which are used to derive Carlitz' asymptotic existence result for primitive normal bases and to settle the case of 6-dimensional extensions with just three exceptions.
- In Section 13.7, we will establish the primitive normal basis theorem when the ground field is the binary or the ternary field, with only one possible exception, namely $(q,n)=(3,8)$.
- In Section 13.8, we strengthen the existence criterion for primitive normal bases given in Section 13.5. This allows us to conclude the proof of the primitive normal basis theorem when $q \geq 4$ in Section 13.9, leaving only one possible exception, namely $(q,n)=(7,6)$.
- Finally, Section 13.10 is devoted to the two specific pairs (7,6) and (3,8) not covered by the previous results. This relies on different arguments, in particular, a simple special case of a sieve method going back to Cohen [85].

Overall, the detailed proof of the primitive normal basis theorem presented here relies to a large extent on ideas of Lenstra and Schoof [238], though our presentation is quite different. Although it still requires rather extensive computations, all technical arguments are reproducible with the help of a simple pocket calculator.

Exercises

Exercise 13.1.5. Show that $\phi_q(x^n-1) \geq q^n-1-n(q^{n-1}-1)$, and use this to prove

$$\pi_n(q) \geq 1 - \frac{n(q^{n-1}-1)\cdot(\ln(2)+n\ln(q))}{\ln(2)\cdot(q^n-1)}. \qquad \square$$

13.2 Primitive Normal Elements for Cubic Extensions

In this section, we consider cubic extensions of a Galois field $F=\mathrm{GF}(q)$ (that is, the case $n=3$), which can still be handled using elementary arguments based on the trivial lower bound given in Proposition 13.1.2.

Theorem 13.2.1. *Let q be a prime power. Then there exists a primitive normal element for the cubic extension $\mathrm{GF}(q^3)$ over $\mathrm{GF}(q)$.*

Proof. As $(q-1)^3-(q^3-q)=-3q^2+4q-1$, Inequality (13.2) gives

$$PN_3(q) > \frac{\ln(2)\cdot(q^3-1)}{\ln(2)+3\ln(q)} - 3q^2+4q-1.$$

Hence $PN_3(q)$ is certainly positive whenever

$$\ln(2)q^3 \geq 3q^2\big(\ln(2)+3\ln(q)\big)+\ln(2)-(4q-1)\big(\ln(2)+3\ln(q)\big).$$

Since $\ln(2)-(4q-1)(\ln(2)+3\ln(q))$ is negative, we see that $PN_3(q)>0$ provided that $\ln(2)q \geq 3\ln(2)+9\ln(q)$. This condition can be examined using standard analytic arguments for the function (on the set of positive real numbers)

$$g(x) = x - \frac{9}{\ln(2)}\ln(x) - 3$$

with first derivative

$$g'(x) = 1 - \frac{9}{\ln(2)x}.$$

Note that $g'(x)>0$ if and only if $x>9/\ln(2)$, which holds for all $x \geq 13$, so that g is strictly increasing on the interval $[13,\infty)$. As $g(57)$ is positive, we conclude that $PN_3(q)>0$ for all $q \geq 57$.

It remains to examine cubic extensions over $F=\mathrm{GF}(q)$ with $q \leq 53$ more carefully, by using the precise values of $\phi_q(x^3-1)$ and, if necessary, also of $\phi(q^3-1)$. This requires a case distinction depending on $q \bmod 3$; cf. Exercise 3.10.6.

Case 1. Assume first that $q \equiv 0 \bmod 3$. Then $\phi_q(x^3-1)=q^3-q^2$, and therefore Proposition 13.1.2 in combination with Lemma 13.1.3 gives the stronger bound

$$PN_3(q) \geq \frac{\ln(2)\cdot(q^3-1)}{\ln(2(q^3-1))} - q^2 + q.$$

The right hand side of this inequality is non-negative for all $q \geq 27$, but not for $q \in \{3,9\}$. For these two values of q, we apply Proposition 13.1.2 with the exact value of $\phi(q^3-1)$, which is easily determined by using the multiplicativity of the Euler function and the prime power factorization of q^3-1. The required data are summarized in Table 13.1; in both cases, $PN_3(q)>0$.

Table 13.1 Evaluation of Proposition 13.1.2 for $q \leq 11$ and $q \not\equiv 1 \bmod 3$

q	q^3-1	prime power factorization of q^3-1	$\phi(q^3-1)$	$\phi(q^3-1)+\phi_q(x^3-1)-q^3+q$
2	7	7	6	3
3	26	$2\cdot 13$	12	6
5	124	$2^2\cdot 31$	60	36
8	511	$7\cdot 73$	432	369
9	728	$2^3\cdot 7\cdot 13$	288	216
11	1330	$2\cdot 5\cdot 7\cdot 19$	432	312

Case 2. When $q \equiv 2 \bmod 3$, we proceed in the same manner. Here $\phi_q(x^3-1) = (q-1)(q^2-1)$, and Proposition 13.1.2 and Lemma 13.1.3 now yield

$$PN_3(q) \geq \frac{\ln(2)\cdot(q^3-1)}{\ln(2(q^3-1))} - q^2 + 1.$$

The right hand side of this inequality is non-negative for all $q \geq 17$, but not for $q \in \{2,5,8,11\}$. For these remaining four values, we check the condition in Proposition 13.1.2 as in Step 1. Again, the required data are given in Table 13.1, and we have $PN_3(q) > 0$ in all four cases.

Case 3. Now let $q \equiv 1 \bmod 3$. Then $\phi_q(x^3-1)$ attains its minimal value, namely $(q-1)^3$, and thus we do not get an improvement of the lower bound (13.2) in this case. We therefore apply Proposition 13.1.2 for the ten prime powers $q \leq 53$ with $q \equiv 1 \bmod 3$ directly, that is, for

$$q \in \{4,7,13,16,19,25,31,37,43,49\}.$$

In analogy to Table 13.1, the relevant data are summarized in Table 13.2; all required factorizations can easily be obtained using a pocket calculator (or even by hand computation). We see that only one case, namely $q = 7$, fails the test based on Proposition 13.1.2; for the remaining nine cases, $PN_3(q)$ is indeed positive.

Table 13.2 Evaluation of Proposition 13.1.2 for $q \leq 49$ and $q \equiv 1 \bmod 3$

q	q^3-1	prime power factorization of q^3-1	$\phi(q^3-1)$	$\phi(q^3-1)+(q-1)^3-q^3+q$
4	63	$3^2\cdot 7$	36	3
7	342	$2\cdot 3^2\cdot 19$	108	-12
13	2196	$2^2\cdot 3^2\cdot 61$	720	264
16	4095	$3^2\cdot 5\cdot 7\cdot 13$	1728	1023
19	6858	$2\cdot 3^3\cdot 127$	2268	1260
25	15624	$2^3\cdot 3^2\cdot 7\cdot 13$	4320	2544
31	29790	$2\cdot 3^2\cdot 5\cdot 331$	7920	5160
37	50652	$2^2\cdot 3^3\cdot 7\cdot 67$	14256	10296
43	79506	$2\cdot 3^2\cdot 7\cdot 631$	22680	17304
49	117648	$2^4\cdot 3^2\cdot 19\cdot 43$	36288	29280

Case 4. It remains to examine the exceptional case $q = 7$ individually. Note first that any field $F = \mathrm{GF}(q)$ with $q \equiv 1 \bmod 3$ contains a primitive third root of unity, say λ. Hence $x^3 - 1$ splits over F as $(x-1)(x-\lambda)(x-\lambda^2)$. By way of contradiction, suppose that $E = \mathrm{GF}(q^3)$ does not contain a primitive normal element. Then the q-order of any primitive element w of E has to be a divisor of x^3-1 with degree at most 2. We first check that the case of degree 1 cannot occur. Indeed, $\mathrm{Ord}_q(w) = x-1$ would mean $w \in F$, whereas $\mathrm{Ord}_q(w) = x-\lambda$ means $w^{q-1} = \lambda$, so that $\mathrm{ord}(w)$ divides $3(q-1) < q^3-1$. Similarly, $\mathrm{Ord}_q(w) = x-\lambda^2$ is also impossible.

Thus $\mathrm{Ord}_q(w)$ has to have degree 2, and the number of such elements $w \in E$ is $3(q-1)^2$, which follows easily from Observation 3.10.3 and Corollary 7.3.8. For $q = 7$, this gives $3\cdot 6^2 = 108$, the total number of primitive elements of $\mathrm{GF}(7^3)$ (see Table 13.2). Thus every element $w \in \mathrm{GF}(7^3)$ with $\deg(\mathrm{Ord}_7(w)) = 2$ would have to be primitive. We now show that this condition leads to a contradiction.

For this, let w be any primitive element for $E = \mathrm{GF}(7^3)$. In view of the factorization $7^3 - 1 = 2 \cdot 171$, we see that w^{171} has order 2, that is, $w^{171} = -1$. Now consider the element $u := -w$. Obviously, $\mathrm{Ord}_7(u) = \mathrm{Ord}_7(w)$. However, u is not a primitive element:

$$u^{171} = (-w)^{171} = (-1)^{171} \cdot w^{171} = (-1)^{171+1} = 1,$$

which is the desired contradiction. □

We conclude this section with a stronger bound on $PN_3(q)$ taken from Hachenberger [169]. His result is based on a geometric approach which will be outlined in the next section, where it plays an essential role in establishing the primitive normal basis theorem for the case of quartic extensions.

Result 13.2.2 *Let q be a prime power. Then the number $PN_3(q)$ of primitive normal elements of $\mathrm{GF}(q^3)$ over $\mathrm{GF}(q)$ is at least*

$$G_3(q) := \phi(q^3 - 1) - f(q) \cdot \phi(q - 1),$$

where

$$f(q) = \begin{cases} \frac{9}{2} \cdot (q-1) & \text{if } q \equiv 1 \bmod 3, \\ q+1 & \text{if } q \equiv 2 \bmod 3, \\ q & \text{if } q \equiv 0 \bmod 3. \end{cases}$$ □

We examine the quality of the bound $G_3(q)$ for the prime powers $q \leq 32$ in Table 13.3, comparing it with the trivial lower bound $TLB_3(q)$ for $PN_3(q)$ taken from Proposition 13.1.2:

$$TLB_3(q) := \phi(q^3 - 1) + \phi_q(x^3 - 1) - (q^3 - q). \tag{13.4}$$

We also list the exact values for $PN_3(q)$ in Table 13.3; these values were calculated by Thomas Gruber [156], using the computer algebra system Sage [331]. It is remarkable that the equality $G_3(q) = PN_3(q)$ holds in 8 of these 18 cases, namely for $q \in \{2, 3, 5, 8, 11, 17, 23, 27\}$.

Exercises

Exercise 13.2.3. Let $q = 4$ or $q \geq 13$ be a prime power satisfying $q \equiv 1 \bmod 3$. Give an alternative proof for $PN_3(q) > 0$ based on the approach in Case 4 of the proof of Theorem 13.2.1, by showing that $\phi(q^3 - 1) > 3(q-1)^2$. □

Exercise 13.2.4. Use Result 13.2.2 to deduce the following bound on the density $\pi_3(q)$ introduced in Theorem 13.1.4:

$$\pi_3(q) \geq 1 - \frac{g(q)}{\phi(q^2 + q + 1)}, \text{ where } g(q) := \begin{cases} 3 \cdot (q-1) & \text{if } q \equiv 1 \bmod 3, \\ q+1 & \text{if } q \equiv 2 \bmod 3, \\ q & \text{if } q \equiv 0 \bmod 3. \end{cases}$$

Table 13.3 Primitive normal elements for cubic extensions

q	$\phi(q^3-1)$	$\phi_q(x^3-1)$	$PN_3(q)$	$G_3(q)$	$TLB_3(q)$
2	6	3	3	3	3
3	12	18	9	9	6
4	36	27	18	9	3
5	60	96	48	48	36
7	108	216	72	54	−12
8	432	441	378	378	369
9	288	648	264	252	216
11	432	1200	384	384	312
13	720	1728	576	504	264
16	1728	3375	1440	1188	1023
17	2448	4608	2304	2304	2160
19	2268	5832	1944	1782	1260
23	4680	11616	4440	4440	4152
25	4320	13824	3888	3456	2544
27	9072	18954	8748	8748	8370
29	9504	23520	9180	9144	8664
31	7920	27000	7200	6840	5160
32	27000	31713	26100	26010	25977

In particular, show that $G_3(q) \geq TLB_3(q)$ holds for all prime powers q, with equality only for $q=2$, and that $G_3(q)$ is always positive. □

13.3 Primitive Normal Elements for Quartic Extensions

The main aim of the present section is a proof of the primitive normal basis theorem for quartic extensions (that is, for $n=4$). Following Hachenberger [169], we will actually derive a fairly strong lower bound for $PN_4(q)$ which relies on facts from Projective Geometry and is a considerable refinement of the basic counting arguments used so far. More precisely, we shall establish the following major result:

Theorem 13.3.1. *Let q be a prime power. Then the number $PN_4(q)$ of primitive normal elements in $\mathrm{GF}(q^4)$ over $\mathrm{GF}(q)$ is at least*

$$G_4(q) := \phi(q^4-1) - 2^{\delta} \cdot \phi(q-1) \cdot \phi(q^2+1),$$

where

$$\delta = \begin{cases} 3 & \text{if } q \equiv 1 \bmod 4, \\ 2 & \text{if } q \equiv 3 \bmod 4, \\ 0 & \text{if } q \equiv 0 \bmod 2. \end{cases} \qquad \square$$

Let us first demonstrate the strength of the bound $G_4(q)$ for the 18 instances where $q \leq 32$; see Table 13.4. Note that the equality $G_4(q) = PN_4(q)$ holds for six of these cases, namely for $q \in \{2,3,4,7,16,31\}$, and that several other values come quite close to $PN_4(q)$. Again, the exact values for $PN_4(q)$ have been calculated by Gruber [156] using Sage [331]. We also compare $G_4(q)$ with the trivial lower bound $TLB_4(q)$ for $PN_n(q)$ taken from Proposition 13.1.2:

$$TLB_4(q) := \phi(q^4-1)+\phi_q(x^4-1)-(q^4-q^2). \tag{13.5}$$

Table 13.4 Primitive normal elements for quartic extensions (1)

q	$\phi(q^4-1)$	$\phi_q(x^4-1)$	$PN_4(q)$	$G_4(q)$	$TLB_4(q)$
2	8	8	4	4	4
3	32	32	16	16	−8
4	128	192	96	96	80
5	192	256	64	0	−152
7	640	1,728	480	480	16
8	1,728	3,584	1,512	1,440	1,280
9	2,560	4,096	1,536	1,280	176
11	3,840	12,000	3,200	2,880	1,320
13	6,144	20,736	4,352	4,096	−1,512
16	32,768	61,440	30,720	30,720	28,928
19	34,560	116,640	31,104	30,240	21,240
23	66,560	255,552	60,640	58,240	42,800
25	119,808	331,776	101,376	99,840	61,584
27	165,888	492,128	154,368	152,064	127,304
29	161,280	614,656	139,776	120,960	69,496
31	221,184	864,000	207,360	207,360	162,624
32	480,000	1,015,808	465,000	456,000	448,256

In fact, $G_4(q) \geq TLB_4(q)$ for all q, with equality only for $q=2$; see Exercise 13.3.8. In contrast to $TLB_4(q)$, the number $G_4(q)$ is always non-negative, and one has $G_4(q)=0$ only for $q=5$; this will be an immediate consequence of Corollary 13.3.3 below. For the exceptional case $q=5$, we will use a variation of our arguments to show $PN_4(5) \geq 32$; see Proposition 13.3.4. Altogether, these results will establish the validity of the primitive normal basis theorem for quartic extensions:

Theorem 13.3.2. *Let q be a prime power. Then there exists a primitive normal element for the quartic extension* $\mathrm{GF}(q^4)$ *over* $\mathrm{GF}(q)$. □

In the proof of Theorem 13.3.1, we make use of the result on generators for finite cyclic groups in Lemma 1.10.10. We will apply this result to the multiplicative group $G=E^*$ of the extension field $E=\mathrm{GF}(q^4)$, and the subgroup considered will be either $U=F^*$ or $U=K^*$, where $F=\mathrm{GF}(q)$ denotes the ground field (as usual) and where $K=\mathrm{GF}(q^2)$ is the intermediate field of degree 2 of E/F.

For the convenience of the reader, we also summarize the required notions from Projective Geometry. The d-dimensional **projective geometry** $\mathrm{PG}_d(\mathbb{F})$ over a field $\mathbb{F}$ is based on a vector space of dimension $d+1$ over $\mathbb{F}$, say $V=\mathbb{F}^{d+1}$. The **points** and **lines** of $\mathrm{PG}_d(\mathbb{F})$ are the 1-dimensional and the 2-dimensional subspaces of V, respectively, while the 3-dimensional and the d-dimensional subspaces of V (for

$d \geq 3$) are called **planes** and **hyperplanes**, respectively. In the special case $\mathbb{F} = \mathrm{GF}(q)$, one usually writes $\mathrm{PG}(d,q)$ instead of $\mathrm{PG}_d(\mathbb{F})$. For a detailed treatment of projective geometries over Galois fields, we refer the interested reader to Hirschfeld [187].

In our situation, we have two possible ways of considering $E = \mathrm{GF}(q^4)$ as a projective geometry:

- On the one hand, we may view E as (the underlying vector space of) the projective geometry (or **projective space**) $\mathrm{PG}(3,q)$ over F, which we will denote by Γ.
- Alternatively, we may also view E as (the underlying vector space of) the projective geometry $\mathrm{PG}(1,q^2)$ over K, which we will denote by Λ; it is common to refer to Λ as the **projective line** over K.

It will be crucial in our arguments to use both of these interpretations of E as a geometry simultaneously. For the sake of clarity, a point of Λ will also be called a **K-point** (or a **point of type** K), while a point of Γ will be referred to as an **F-point** (or **point of type** F). Note that any K-point ℓ can also be viewed as a projective line over F; we speak of the **induced F-line** of the K-point ℓ. (Of course, most lines of Γ are *not* induced.) In view of Proposition 7.1.4, the total number of points (of type F) of Γ is

$$\frac{q^4-1}{q-1} = q^3+q^2+q+1 = (q^2+1)\cdot(q+1);$$

similarly, Λ contains exactly q^2+1 points (of type K), and any line of Γ (whether induced or not) contains precisely $q+1$ points (of type F).

We also need the following classical result, which will be essential for our arguments: the induced F-lines of Γ form a **spread** of Γ, that is, a system of q^2+1 two dimensional F-subspaces of E any two of which have trivial intersection, and such that every point of Γ is contained in (exactly one) of these subspaces. In geometric terminology, a spread is a system of pairwise skew lines of Γ which covers all points of Γ. The projective space Γ contains, in general, many (and different types of) spreads; the particular spread constructed from the intermediate field K is called **regular** or **Desarguesian** and will be denoted by S in what follows.

After these preparations, we are ready to prove Theorem 13.3.1 with a series of three steps.

Step 1. In order to deal with normality, we describe certain configurations of points, lines and planes of Γ related to this condition. As usual, we will denote the Frobenius automorphism of E/F by σ. Recall that the minimal polynomial of σ (as an endomorphism of the F-vector space E) is equal to x^4-1. If $g(x) \in F[x]$ is any monic divisor of x^4-1, we let V_g denote the corresponding σ-invariant F-subspace of E, that is, V_g is the kernel of $g(\sigma)$. Now let v be any non-zero element of v. If v is a normal element for E/F, we will call the point $N = Fv$ of Γ likewise **normal**. Of course, if v is normal, then so is λv for all $\lambda \in F^*$, and therefore any normal point $N = Fv$ contains $q-1$ normal elements (as a set).

As the normal elements for E/F are precisely the elements of E with q-order x^4-1, a point N of Γ is normal if and only if it is not contained in any non-trivial σ-invariant subspace of E. This simple observation turns out to be of fundamental importance. Obviously, the configuration formed by the σ-invariant subspaces depends on the decomposition of x^4-1 into irreducible factors, which in turn depends on the characteristic of F and the residue of q modulo 4. We need to distinguish three cases:

1. First let $q \equiv 3 \bmod 4$. Then

$$x^4 - 1 = (x-1)(x+1)(x^2+1),$$

and x^2+1 is irreducible. Here the non-trivial σ-invariant subspaces of E are as follows. We write A for the F-point V_{x-1} in Γ (which in fact is equal to F), B for the F-point V_{x+1} of Γ, and T for the line V_{x^2+1} of Γ. Since T is invariant under multiplication with elements in K^* (it is the kernel of the (E,K)-trace mapping), the line T is induced from a K-point. The line of Γ through A and B is equal to $K = V_{x^2-1}$, and is therefore likewise induced. In particular, both K and T belong to the spread S. Finally, we have the subspaces corresponding to the divisors $(x-1)(x^2+1)$ and $(x+1)(x^2+1)$ of x^4-1,

$$H_1 = V_{(x-1)(x^2+1)} \text{ and } H_{-1} = V_{(x+1)(x^2+1)},$$

which are planes of Γ.[1] Obviously,

$$H_1 \cap H_{-1} = T.$$

In the case under consideration, a point N of Γ is therefore normal if and only if it is not contained in the union $H_1 \cup H_{-1} \cup K$.[2]

2. Next let $q \equiv 1 \bmod 4$. In this case, we obtain further σ-invariant subspaces (compared to the first case), as the factor x^2+1 of x^4-1 now splits over F; using a primitive 4-th root of unity $\iota \in F^*$, we get

$$x^4 - 1 = (x-1)(x+1)(x-\iota)(x+\iota).$$

With the same notation as in the previous case, the additional invariant subspaces are as follows. We obtain two further points, namely $C = V_{x-\iota}$ and $D = V_{x+\iota}$ (so that T is the line through these two F-points). Similarly, we have two additional planes, namely $H_\iota = V_{(x^2-1)(x-\iota)}$ and $H_{-\iota} = V_{(x^2-1)(x+\iota)}$, which intersect as

$$H_\iota \cap H_{-\iota} = K.$$

---

[1] Note that these subspaces are **projective planes** over F, that is, they are isomorphic to $\mathrm{PG}(2,q)$.

[2] This observation can also be used to determine the number N_F of normal points in Γ by computing how many points of Γ are not contained in the union $H_1 \cup H_{-1} \cup K$; alternatively, one can apply Corollary 7.3.8 and the factorization of x^4-1 above. As we will not need to know N_F to complete the proof, we will leave its determination to the reader; see Exercise 13.3.13.

In this case, a point N of Γ is normal if and only if it is not contained in the union $H_1 \cup H_{-1} \cup H_\iota \cup H_{-\iota}$ of the four invariant planes.

3. Finally, let q be even, so that $x^4 - 1 = (x-1)^4$. Now the non-trivial σ-invariant subspaces of E form a chain: they are (in descending order) $H = V_{(x-1)^3}$, the kernel of the (E,F)-trace mapping; $K = V_{(x-1)^2}$, and $F = V_{x-1}$. Thus a point is normal if and only if it is not contained in H.

Step 2. Next, we deal with primitivity. Given any element $v \in E^*$, we call the F-point $P = Fv$ of Γ **primitive** provided that P contains at least one primitive element of E (as a set). Similarly, we call the K-point $Q = Kv$ in Λ **primitive** if Q contains at least one primitive element of E (as a set); in this case, the line of Γ induced by Q is said to be a **primitive F-line**.

Note that it is not at all obvious how many primitive elements are contained in a given primitive point (in contrast to the case of normality); settling this problem will require some effort. Observe first that

$$q^4 - 1 = (q-1)\cdot(q+1)\cdot(q^2+1),$$

and write

$$q-1 = 2^a \cdot u_0, \quad q+1 = 2^b \cdot u_1 \quad \text{and} \quad q^2+1 = 2^c \cdot u_2,$$

where u_1, u_2 and u_3 are odd and (pairwise) relatively prime. Using the multiplicativity of the Euler function, the number of primitive elements of E is then given by

$$\phi(q^4-1) = 2^{a+b+c-\omega} \cdot \phi(u_0)\phi(u_1)\phi(u_2),$$

where $\omega = 0$ if F has characteristic 2 and $\omega = 1$ otherwise. We shall simplify this formula later, depending on the characteristic of F and the residue of q modulo 4.

Now let v be any primitive element of E, and denote the number of primitive elements in Fv and Kv by M_F and M_K, respectively. Note that this makes sense, as these numbers do not depend on the choice of v, by Lemma 1.10.10.

1. Assume first that q is odd. Then
   - $a = 1 = c$ and $b \geq 2$ if $q \equiv 3 \bmod 4$,
   - $b = 1 = c$ and $a \geq 2$ if $q \equiv 1 \bmod 4$.

   In both cases, we have

$$M_F = 2^a \cdot \phi(u_0) \quad \text{and} \quad M_K = 2^{a+b} \cdot \phi(u_0)\phi(u_1),$$

   by Lemma 1.10.10. Consequently, the number of primitive K-points in Λ is given by

$$P_K := \frac{\phi(q^4-1)}{M_K} = \frac{2^{a+b} \cdot \phi(u_0)\phi(u_1)\phi(u_2)}{2^{a+b} \cdot \phi(u_0)\phi(u_1)} = \phi(u_2),$$

and the number of primitive F-points on any primitive F-line (induced by a primitive K-point) is

$$\frac{M_K}{M_F} = \frac{2^{a+b} \cdot \phi(u_0)\phi(u_1)}{2^a \cdot \phi(u_0)} = 2^b \cdot \phi(u_1).$$

2. Now assume that q is even, so that $a = b = c = 0$. Then Lemma 1.10.10 leads (formally) to the same results as in the case of odd q, if we include redundant factors 2^0:

$$M_F = \phi(u_0) = 2^a \cdot \phi(u_0) \quad \text{and} \quad M_K = \phi(u_0)\phi(u_1) = 2^{a+b} \cdot \phi(u_0)\phi(u_1).$$

Again, the number of primitive K-points is

$$P_K := \frac{\phi(q^4-1)}{M_K} = \frac{\phi(u_0)\phi(u_1)\phi(u_2)}{\phi(u_0)\phi(u_1)} = \phi(u_2),$$

and the number of primitive F-points on each primitive F-line is

$$\frac{M_K}{M_F} = \frac{\phi(u_0)\phi(u_1)}{\phi(u_0)} = \phi(u_1) = 2^b \cdot \phi(u_1).$$

Step 3. We now come to the crucial step, where we combine the conditions for normality and primitivity. To this end, let ℓ be any primitive F-line, that is, $\ell = Kv$ for some primitive element v of E. Then ℓ belongs to the spread S, which (as we have seen before) also contains the lines K and T, and hence ℓ has no point in common with either of these lines. Note that K and T actually coincide when q is even; otherwise, they are disjoint. Therefore, ℓ cannot be contained in any of the σ-invariant planes H (in the even case) or H_i (in the two odd cases, with $i \in \{1, -1, \iota, -\iota\}$), as any two lines in a (projective) plane have an intersection point.

Consequently, ℓ intersects each of the relevant planes in exactly one F-point. None of these intersection points is normal; in the worst case, all of them might be distinct and primitive. This observation yields a lower bound for the number of primitive F-points on ℓ which are simultaneously normal, namely the number of primitive F-points on ℓ minus the number h of σ-invariant planes. Multiplying by the number M_F of primitive elements in any primitive F-point then gives a lower bound on the number of primitive normal elements contained in any specified primitive line. Finally, multiplying this bound by the total number P_K of primitive K-points (that is, the number of primitive F-lines), we obtain the following lower bound for the total number of primitive normal elements in E/F, since no two distinct primitive F-lines can intersect (as they belong to the spread S):

$$PN_4(q) \geq G_4(q) := P_K \cdot \left(\frac{M_K}{M_F} - h\right) \cdot M_F,$$

where h is the number of σ-invariant planes for E/F (which depends on the characteristic of F and the residue of q modulo 4). Substituting the formulas derived in

Step 2, we obtain

$$G_4(q) = \phi(u_2) \cdot \left(2^b \cdot \phi(u_1) - h\right) \cdot 2^a \cdot \phi(u_0).$$

In order to complete the proof of Theorem 13.3.1, we still have to evaluate this formula explicitly for the three possible cases:

1. Assume first that $q \equiv 3 \bmod 4$, hence $h = 2$ and $a = c = 1$. This gives the desired result:

$$\begin{aligned} G_4(q) &= 2\phi(u_0)\phi(u_2) \cdot \left(2^b \phi(u_1) - 2\right) \\ &= 2^{b+1}\phi(u_0)\phi(u_1)\phi(u_2) - 4\phi(u_0)\phi(u_2) \\ &= \phi(q^4-1) - 4\phi(q-1)\phi(q^2+1). \end{aligned}$$

2. Next, let $q \equiv 1 \bmod 4$. Then $h = 4$ and $b = c = 1$ and $a \geq 2$ imply

$$\begin{aligned} G_4(q) &= 2^a \phi(u_0)\phi(u_2) \cdot \left(2\phi(u_1) - 4\right) \\ &= 2^{a+1}\phi(u_0)\phi(u_1)\phi(u_2) - 8 \cdot 2^{a-1}\phi(u_0)\phi(u_2) \\ &= \phi(q^4-1) - 8\phi(q-1)\phi(q^2+1). \end{aligned}$$

3. Finally, let q be even. Then $h = 1$ and $a = b = c = 0$ give

$$\begin{aligned} G_4(q) &= \phi(u_0)\phi(u_2) \cdot (\phi(u_1) - 1) \\ &= \phi(u_0)\phi(u_1)\phi(u_2) - \phi(u_0)\phi(u_2) \\ &= \phi(q^4-1) - \phi(q-1)\phi(q^2+1). \end{aligned}$$

This completes the proof of Theorem 13.3.1. □

Now let $\pi_4(q) := PN_4(q)/\phi(q^4-1)$ denote the proportion of normal elements for $\mathrm{GF}(q^4)$ over $\mathrm{GF}(q)$ among the primitive elements of $\mathrm{GF}(q^4)$, as in Theorem 13.1.4.

Corollary 13.3.3. *For every prime power q, one has*

$$\pi_4(q) \geq 1 - \frac{2^\varepsilon}{\phi(q+1)} \quad \textit{with } \varepsilon = \begin{cases} 1 & \textit{if } q \equiv 1 \bmod 4, \\ 0 & \textit{otherwise.} \end{cases}$$

Proof. From Theorem 13.3.1, we obtain

$$\pi_4(q) = \frac{PN_4(q)}{\phi(q^4-1)} \geq \frac{G_4(q)}{\phi(q^4-1)} = 1 - \frac{2^\delta \cdot \phi(q-1)\phi(q^2+1)}{\phi(q^4-1)}.$$

It remains to evaluate the right hand side, by substituting the values for δ obtained in Theorem 13.3.1 (where we also use the same representation of $q-1$, $q+1$ and q^2+1 as given in the proof of that result). First let $q \equiv 1 \bmod 4$. Then $\delta = 3$ yields

$$1-\frac{2^3\cdot 2^{a-1}\cdot\phi(u_0)\phi(u_2)}{2^{a+1}\cdot\phi(u_0)\phi(u_1)\phi(u_2)}=1-\frac{2}{\phi(u_1)}=1-\frac{2}{\phi(q+1)}.$$

For $q\equiv 3 \bmod 4$, we have $\delta=2$ and obtain

$$1-\frac{2^2\cdot\phi(u_0)\phi(u_2)}{2^{1+b}\cdot\phi(u_0)\phi(u_1)\phi(u_2)}=1-\frac{1}{2^{b-1}\cdot\phi(u_1)}=1-\frac{1}{\phi(q+1)}.$$

Finally, $\delta=0$ when $q\equiv 0 \bmod 2$, which again gives $1-\frac{1}{\phi(q+1)}$. □

In particular, Corollary 13.3.3 shows that $G_4(q)$ is always non-negative. Moreover, $G_4(q)=0$ holds if and only if $2^\varepsilon=\phi(q+1)$, which means $\varepsilon=1$ (so that $q\equiv 1 \bmod 4$) and $q=5$. This simple argument establishes Theorem 13.3.2, except for $q=5$. We settle this final case in the following result, which then completes the proof of Theorem 13.3.2.

Proposition 13.3.4. *There are at least* 32 *primitive normal elements for* $E=\mathrm{GF}(5^4)$ *over* $F=\mathrm{GF}(5)$.

Proof. Again, we apply Lemma 1.10.10 to the multiplicative group $G=E^*$ of E. This time, we choose U as the subgroup of order 16 of G. Note that there are

$$\kappa:=\frac{|G|}{|U|}=\frac{5^4-1}{16}=39$$

cosets of U. In analogy to our earlier terminology, such a coset will be called **normal** if it contains a normal element for E/F, and **primitive** if it contains a primitive element of E^*. By Lemma 1.10.10, any primitive coset contains exactly 8 primitive elements.

In order to deal with normality, we require an elementary result due to Lenstra and Schoof [238], see Exercise 13.3.12: multiplication by any element $\lambda\in U$ induces a permutation on the set of normal elements for E/F.[3] Thus any normal coset consists entirely of normal elements.

Therefore, the numbers of primitive and of normal cosets are given by

$$P:=\frac{\phi(5^4-1)}{8}=24 \quad\text{and}\quad N:=\frac{(5-1)^4}{16}=16,$$

respectively. As $P+N=40>39=\kappa$, we see that there is at least one coset which is both primitive and normal, so that $PN_4(5)\geq 8$.

In order to improve this bound to the one claimed in the assertion, we can argue as follows. If $\eta\in E$ is a primitive third root of unity, then the three cosets U, $U\eta$ and $U\eta^2$ are neither normal nor primitive, as the union of these cosets is the subgroup of E^* of order 48 and as any element of this subgroup has 5-order dividing x^2-1 or x^2+1. Therefore, the number of cosets that are both primitive and normal is at least

[3] A more general result will be established later; see Step 2 in the proof of Proposition 13.8.2.

$$P + N - (\kappa - 3) = 24 + 16 - 36 = 4,$$

and thus there are at least $4 \cdot 8 = 32$ primitive elements in E^* which are also normal over F. □

We conclude this section with two further results concerning quartic extensions of Galois fields. Hachenberger [172] uses a different geometric approach, by partitioning the point set of the projective space Γ into **ovoids** instead of lines.[4] This leads to the following alternative lower bound for the number of primitive normal elements in quartic extensions:

Result 13.3.5 *The number of primitive normal elements of* GF(q^4) *over* GF(q) *is at least*

$$O_4(q) := \phi(q^4 - 1) - \omega(q)\phi(q^2 - 1),$$

where

$$\omega(q) = \begin{cases} 4(q-1) & \text{if } q \equiv 1 \bmod 4, \\ 2(q-1) & \text{if } q \equiv 3 \bmod 4, \\ q & \text{if } q \equiv 0 \bmod 2. \end{cases}$$ □

In Table 13.5, we list both the bounds $G_4(q)$ and $O_4(q)$ and the exact value $PN_4(q)$ of primitive normal elements in quartic extensions for the 40 prime powers q for which the exact value is known from Table 13.4 and the computations of Hackenberg [175], that is, for all $q \leq 100$ and for $q \in \{11^2, 5^3, 2^7, 13^2, 3^5\}$.[5] Examining the data in this table reveals a couple of noteworthy facts:

- The two lower bounds coincide in 5 of the 40 cases: $O_4(q) = G_4(q)$ holds for $q \in \{2, 3, 4, 8, 16\}$. Except for $q = 8$, the bounds then even coincide with the exact value $PN_4(q)$.
- In the majority of the 40 cases, the O-bound is stronger: $O_4(q) > G_4(q)$ holds for the 24 prime powers

$$\begin{aligned} q \in \{5,9,17,25,29,41,49,61,89,121,125,169\} \quad (q \equiv 1 \bmod 4), \\ q \in \{11,19,23,47,53,59,71,79,83\} \quad (q \equiv 3 \bmod 4), \\ q \in \{32,64,128\} \quad (q \equiv 0 \bmod 2). \end{aligned}$$

  Moreover, the O-bound even yields the precise result for 13 of these 24 instances: we have $PN_4(q) = O_4(q) > G_4(q)$ for

$$\begin{aligned} q \in \{11,19,59,71,79\} \quad (q \equiv 3 \bmod 4), \\ q \in \{5,9,25,29,49,61,121,169\} \quad (q \equiv 1 \bmod 4). \end{aligned}$$

[4] An **ovoid** in PG$(3, q)$ is a set of $q^2 + 1$ points, such that no three distinct of these points are on a common line.

[5] We have also listed the numbers of primitive and of normal elements for these extensions, that is, the values $\phi(q^4 - 1)$ and $\phi_q(x^4 - 1)$.

Table 13.5 Primitive normal elements for quartic extensions (2)

q	$\phi(q^4-1)$	$\phi_q(x^4-1)$	$PN_4(q)$	$G_4(q)$	$O_4(q)$
2	8	8	4	4	4
3	32	32	16	16	16
4	128	192	96	96	96
5	192	256	64	0	64
7	640	1,728	480	480	448
8	1,728	3,584	1,512	1,440	1,440
9	2,560	4,096	1,536	1,280	1,536
11	3,840	12,000	3,200	2,880	3,200
13	6,144	20,736	4,352	4,096	3,840
16	32,768	61,440	30,720	30,720	30,720
17	21,504	65,536	16,896	14,336	15,360
19	34,560	116,640	31,104	30,240	31,104
23	66,560	255,552	60,640	58,240	59,520
25	119,808	331,776	101,376	99,840	101,376
27	165,888	492,128	154,368	152,064	150,912
29	161,280	614,656	139,776	120,960	139,776
31	221,184	864,000	207,360	207,360	205,824
32	480,000	1,015,808	465,000	456,000	460,800
37	470,016	1,679,616	420,864	417,792	407,808
41	623,616	2,560,000	564,224	519,680	562,176
43	691,200	3,259,872	659,712	656,640	650,880
47	1,081,344	4,672,128	1,036,288	1,013,760	1,016,576
49	1,536,000	5,308,416	1,413,120	1,382,400	1,413,120
53	1,935,360	7,311,616	1,794,816	1,720,320	1,755,648
59	3,118,080	11,706,720	3,014,144	2,923,200	3,014,144
61	3,571,200	12,960,000	3,340,800	3,333,120	3,340,800
64	6,635,520	16,515,072	6,531,840	6,497,280	6,524,928
67	4,587,520	19,549,728	4,453,760	4,444,160	4,418,560
71	5,806,080	24,696,000	5,644,800	5,564,160	5,644,800
73	6,635,520	26,873,856	6,279,168	6,266,880	6,137,856
79	9,584,640	37,964,160	9,345,024	9,285,120	9,345,024
81	15,728,640	40,960,000	14,962,688	14,942,208	14,909,440
83	9,584,640	46,314,912	9,351,040	9,185,280	9,269,760
89	14,254,080	59,969,536	13,620,480	13,066,240	13,578,240
97	20,213,760	84,934,656	19,390,976	19,251,200	19,181,568
121	56,217,600	207,360,000	54,374,400	54,343,680	54,374,400
125	62,208,000	236,421,376	60,235,200	58,752,000	60,065,280
128	132,765,696	266,338,304	131,721,408	131,185,152	131,410,944
169	175,472,640	796,594,176	171,343,872	169,989,120	171,343,872
243	1,246,080,000	3,458,087,072	1,235,872,000	1,235,696,000	1,233,302,400

In particular, the O-bound yields equality even for the exceptional case $q = 5$, where the G-bound was 0 (which would have made Proposition 13.3.4 redundant).

- On the other hand, the G-bound is stronger in the remaining 11 cases: we have $O_4(q) < G_4(q)$ for

$$q \in \{13, 37, 73, 81, 97\} \quad (q \equiv 1 \bmod 4),$$
$$q \in \{7, 27, 31, 43, 67, 243\} \quad (q \equiv 3 \bmod 4).$$

Here the G-bound gives the precise result for just two instances: $PN_4(q) = G_4(q) > O_4(q)$ holds for $q = 7$ and $q = 31$.

It is especially remarkable that the O-bound provides the exact value of $PN_4(q)$ considerably more often than the G-bound does (for 17 versus 6 out of the 40 instances). This is by no means accidental: the superb behavior of the O-bound for certain instances of q is explained by the following theoretical result also contained in [172].

Result 13.3.6 *Let q be a prime power, and assume that either $\frac{1}{2}(q^2+1)$ or q^2+1 is a prime, depending on whether q is odd or even, respectively. Then*

$$PN_4(q) = O_4(q) = \begin{cases} (q-1)(q-3)\cdot\phi(q^2-1) & \text{if } q \equiv 1 \bmod 4, \\ (q-1)^2\cdot\phi(q^2-1) & \text{if } q \equiv 3 \bmod 4, \\ q(q-1)\cdot\phi(q^2-1) & \text{if } q \equiv 0 \bmod 2. \end{cases} \quad \square$$

Note that the hypothesis on q in Result 13.3.6 is satisfied for *all* the 17 instances in Table 13.5 where $O_4(q) = PN_4(q)$. In order to discuss its strength, we write $q = p^m$, where p is the characteristic of the underlying fields. From basic Number Theory, the primality of q^2+1 or $\frac{1}{2}(q^2+1)$, respectively, requires that $m = 2^k$ for some integer $k \geq 0$ (though this is in general not sufficient).

- If $p = 2$, then q^2+1 is the k-th Fermat number $F_k = 2^{2^k}+1$. There are only five Fermat primes known, namely F_k for $k \in \{0,1,2,3,4\}$. The values $k \in \{1,2,3\}$ give $q \in \{2,4,16\}$ and are covered by Table 13.5. For $k = 4$, we have $q = 2^8 = 256$ and Result 13.3.6 gives

$$PN_4(256) = 2{,}139{,}095{,}040.$$

- The situation is even more interesting when p is odd. For instance, every value $k \in \{0,1,2,4,5,6\}$ results in a prime number $\frac{1}{2}(q^2+1)$ for $p = 3$, and similarly for $k \in \{0,1,2\}$ and $p = 5$.
- Finally, we discuss the special case where $q = p$ is a prime. Here we find the following 23 primes with $101 \leq p \leq 1000$ satisfying the hypothesis of Result 13.3.6 (out of a total of 143 primes in that range); note that $p \bmod 5$ has to be 1 or 4.

  101, 131, 139, 181, 199, 271, 349, 379,
  409, 449, 461, 521, 569, 571, 631, 641,
  661, 739, 751, 821, 881, 929, 991.

  In the much larger range $3 \leq p \leq 1{,}000{,}000$, one finds 7019 primes which satisfy the hypothesis of Result 13.3.6. It is an open problem whether or not there are infinitely many primes p for which $\frac{1}{2}(p^2+1)$ is likewise prime. $\square$

Remark 13.3.7. The equality $PN_4(q) = G_4(q)$ holds in Theorem 13.3.1 whenever q is a Mersenne prime or $q+1$ is a Fermat prime.

- The first of these two cases, which requires $q \equiv 3 \bmod 4$, is immediate from the proof of Theorem 13.3.1 and explains the entries for $q \in \{3,7,31\}$ in Table 13.4. Here

$$PN_4(q) = (2q-2) \cdot \varphi(q-1) \cdot \varphi(q^2+1).$$

For example, this gives

$$PN_4(2^{13}-1) = PN_4(8191) = 917{,}070{,}336{,}000{,}000.$$

- The second case, which requires that q is even, is not obvious, but can be derived using arguments which are similar to those in the proof of Lemma 3.1 in Hachenberger [172]. In this case, one obtains

$$PN_4(q) = (q-1) \cdot \varphi(q-1) \cdot \varphi(q^2+1).$$

For example, this gives $G_4(2^8) = O_4(2^8)$ and

$$PN_4(2^{16}) = PN_4(65536) = 9{,}208{,}841{,}110{,}762{,}291{,}200,$$

which in fact is greater than $9{,}208{,}840{,}891{,}182{,}088{,}192 = O_4(2^{16})$. □

Exercises

Exercise 13.3.8. Show that $G_4(q) \geq TLB_4(q)$ (see the assertion of Theorem 13.3.1 and Equation (13.5)) holds for all prime powers $q \geq 2$, with equality if and only if $q = 2$. □

Exercise 13.3.9. Show that $TLB_4(q)$ is positive for all prime powers q distinct from 5 and 13. □

Exercise 13.3.10. Show that the following holds:

$$G_4(q) = \phi(q^4-1) - 2^{\gamma} \cdot \phi\left(\tfrac{q^4-1}{q+1}\right), \text{ where } \gamma = \begin{cases} 2 & \text{if } q \equiv 1 \bmod 4, \\ 1 & \text{if } q \equiv 3 \bmod 4, \\ 0 & \text{if } q \equiv 0 \bmod 2. \end{cases}$$ □

Exercise 13.3.11. Show that $\pi_4(q) > \frac{1}{2}$ when $q \not\equiv 1 \bmod 4$ and $q \geq 4$ and also when $q \equiv 1 \bmod 4$ and $q \geq 13$. Use this to prove the existence of a primitive element $v \in \mathrm{GF}(q^4)$ such that both v and v^{-1} are normal over $\mathrm{GF}(q)$ for these cases. □

Exercise 13.3.12. Let $\alpha \in E = \mathrm{GF}(5^4)$ be a normal element over $F = \mathrm{GF}(5)$, and let $\zeta \in E$ be any 16-th root of unity in E^*. Prove that $\zeta\alpha$ is again a normal element for E/F, that is, show that the conjugates $\zeta\alpha, (\zeta\alpha)^5, (\zeta\alpha)^{25}, (\zeta\alpha)^{125}$ of $\zeta\alpha$ are linearly independent over F. □

Exercise 13.3.13. Prove that the total number N_F of normal points in the projective space Γ investigated in our proof of Theorem 13.3.1 is as follows:

$$N_\Gamma = \begin{cases} (q-1)(q^2-1) & \text{for } q \equiv 3 \bmod 4, \\ (q-1)^3 & \text{for } q \equiv 1 \bmod 4, \\ q^3 & \text{if } q \text{ is even.} \end{cases}$$

Hint: Either use geometric arguments to determine the cardinality of the union of all (maximal) non-trivial σ-invariant subspaces of Γ, or apply Corollary 7.3.8 and the factorization of x^4-1 to evaluate $N_F = \phi_q(x^4-1)/(q-1)$. □

Exercise 13.3.14. Apply the geometric method used in this section to prove Result 13.2.2. □

13.4 Characters of Cyclic Modules

In order to attack the cases $n \geq 5$ of the primitive normal basis theorem, we will require a formula for the number $PN_n(q)$ of all primitive normal elements for the extension $E = \mathrm{GF}(q^n)$ over $F = \mathrm{GF}(q)$ in terms of character sums. This goes back to the pioneering work of Carlitz [63, 64] and will be presented in the next section. In the present section, we will provide some useful general background on characters of the additive group of a cyclic module over a principal ideal domain.

We have already seen in Theorem 10.1.2 that every finite abelian group G is isomorphic to its character group $\widehat{G}$ (over any splitting field L), so that G and $\widehat{G}$ are isomorphic as $\mathbb{Z}$-modules. We shall extend this as follows. Assume that M is any (additively written) finite abelian group which is also an R-module over some principal ideal domain R. Then the character group $\widehat{M}$ of M admits an induced R-module structure. In the important special case where M is a cyclic torsion module, both groups are actually also isomorphic as R-modules; see Theorem 13.4.1 below.

The motivation for considering this general result stems from its applicability to finite field extensions. Recall that the additive group of an extension field $E = \mathrm{GF}(q^n)$ of $F = \mathrm{GF}(q)$ becomes a cyclic module isomorphic to $F[x]/(x^n-1)$ over $F[x]$ with respect to the scalar multiplication defined by

$$f(x) \cdot w := f(\sigma)(w) \quad \text{for } w \in E \text{ and } f(x) \in F[x],$$

where σ is the Frobenius automorphism for E/F; see, for instance, the proof of Case 1 of the normal basis theorem (Theorem 3.9.4). Now let L be any splitting field for the additive group of E. Then the group $\widehat{(E,+)}$ of all additive characters of E will become an $F[x]$-module if we define the scalar multiplication by

$$f(x) \cdot \chi : E \to L^*, \quad u \mapsto \chi(f(x) \cdot u) \tag{13.6}$$

for all $\chi \in \widehat{(E,+)}$, all $f(x) \in F[x]$, and all $u \in E$.

For the following proofs, we need some technical setup. Throughout, we shall always work with a fixed system of representatives $\mathscr{R}$ modulo the unit group of R whenever we consider the divisibility and the factorization of elements in a given principal ideal domain R. Thus $\mathscr{R}$ is a system of pairwise non-associate divisors: $a \mid b$ and $b \mid a$ both hold for $a, b \in \mathscr{R}$ if and only if $a = b$. In the concrete situations where $R = \mathbb{Z}$ or $R = F[x]$, we always use the canonical choices for $\mathscr{R}$, namely the positive integers and the monic polynomials, respectively.

In what follows, we only require a (finite) set of representatives $\mathscr{R}$ for the set of all divisors of a specified non-zero element A of R, where we assume without loss of generality that $\mathscr{R}$ is closed under taking codivisors: if a divisor a of A belongs to $\mathscr{R}$, then so does A/a.

Theorem 13.4.1. *Let M be an (additively written) finite abelian group which also is an R-module over some principal ideal domain R, and let L be a splitting field for M. Then the character group $\widehat{M}$ of M becomes an R-module by defining*

$$(r\chi)(u) := \chi(ru) \quad \text{for } \chi \in \widehat{M},\ u \in M \text{ and } r \in R. \tag{13.7}$$

Moreover, M and $\widehat{M}$ are isomorphic as R-modules provided that M is a cyclic R-module with annihilator ideal $\mathscr{A}(M) = (A) \neq (0)$.

Proof. It is easy to check that $\widehat{M}$ indeed is an R-module with respect to the scalar multiplication defined in (13.7). Now assume that M is a cyclic R-module with non-trivial annihilator ideal $\mathscr{A}(M)$ generated by A (that is, M is a cyclic torsion module). The assertion will follow by showing that (A) is also the annihilator ideal of $\widehat{M}$. Observe first that

$$(A\chi)(u) = \chi(Au) = \chi(0) = 1$$

for all $u \in M$ and all $\chi \in \widehat{M}$, by the definition of A. Thus $A\chi$ is the trivial character χ_0 for all $\chi \in \widehat{M}$, and hence A is contained in the annihilator ideal of $\widehat{M}$.

Recall from Section 1.9 that the submodules of M (all of which are cyclic) correspond bijectively to a system $\mathscr{R}$ of pairwise non-associate divisors of A: for $a \in \mathscr{R}$, the submodule corresponding to a is given by $M_a := \{u \in M : au = 0\}$. We now define submodules of $\widehat{M}$ in an analogous manner:

$$M'_a := \{\chi \in \widehat{M} : a\chi = \chi_0\}$$

for every $a \in \mathscr{R}$. In a first step, we establish the identity (cf. Definition 10.1.3)

$$M'_a = M_{a^*}^{\perp} = \{\chi \in \widehat{M} : \chi(x) = 1 \text{ for all } x \in M_{a^*}\},$$

where $a^* := A/a$ is the codivisor of a in $\mathscr{R}$, according to our convention above. In order to do so, let $w \in M$ be a generator of M as an R-module. Then aw is a generator of M_{a^*}. Given any $\chi \in M'_a$, we have

$$\chi(aw) = (a\chi)(w) = \chi_0(w) = 1,$$

and hence $M'_a \subseteq M_{a^*}^{\perp}$. Conversely, if $\lambda \in M_{a^*}^{\perp}$, then $1 = \lambda(raw) = (a\lambda)(rw)$ holds for all $r \in R$, hence $(a\lambda)(u) = 1$ for all $u \in M$. This shows $a\lambda = \chi_0$, that is $\lambda \in M'_a$, also establishing $M_{a^*}^{\perp} \subseteq M'_a$.

We now apply part (5) of Theorem 10.1.8 with $G = M$ and $H = M_{a^*}$, which shows that $H^{\perp} = M'_a$ and the character group of M/M_{a^*} are isomorphic as groups. Moreover, the mapping $\psi\colon M \to M$, $u \mapsto a^*u$ is an R-module endomorphism on M, with kernel M_{a^*} and image M_a. Altogether, this yields

$$|M'_a| = \left|\widehat{M/M_{a^*}}\right| = |M/M_{a^*}| = |M_a|.$$

Since the R-modules M_a and $R/(a)$ are isomorphic, we obtain $|M'_a| = |R/(a)|$.

For each divisor a of A, let $\phi_R(a)$ be the (finite) number of units of the factor ring $R/(a)$. According to Remarks 1.9.12 and 1.9.17, $\phi_R(a)$ is the number of elements u of M whose order ideal satisfies $\mathrm{Ord}(u) = (a)$ and equals the number of generators of the cyclic R-module M_a. We therefore define $\phi'_R(a)$ to be the number of $\chi \in \widehat{M}$ such that the order ideal of χ is generated by a; in other words, ϕ'_R is the Euler function for the R-module $\widehat{M}$. We claim that ϕ_R and ϕ'_R coincide. Once we know this, we obtain in particular the existence of an element $\psi \in \widehat{M}$ with $\mathrm{Ord}_R(\Psi) = (A)$, which implies that the submodule of $\widehat{M}$ generated by ψ is isomorphic to $R/(A)$. Then the finiteness of M (and therefore that of $R/(A)$) shows that $\widehat{M}$ is cyclic (with ψ as a generator), see Proposition 1.9.18, which will finish the proof.

It remains to show that the two Euler functions ϕ_R and ϕ'_R indeed coincide. Because of the multiplicativity of both functions, it suffices to show that these mappings agree for prime power divisors of A. Thus let $s \in \mathscr{R}$ be a prime element of R and assume that s^k divides A for some $k \in \mathbb{N}^*$. By Remark 1.9.12, we conclude (using $|M_a| = |M'_a|$ for all $a \in \mathscr{R}$)

$$\phi_R(s^k) = |M_{s^k}| - |M_{s^{k-1}}| = |M'_{s^k}| - |M'_{s^{k-1}}| = \phi'_R(s^k),$$

as claimed. □

In the next section, we will use Theorem 13.4.1 to prove a twofold generalization of Vinogradov's formula from Proposition 10.2.5. For this, we have to introduce some additional terminology, in particular the abstract notion of *Möbius functions* for principal ideal domains:

Definition 13.4.2. Let R be any principal ideal domain. We denote the number of distinct pairwise non-associate prime divisors of any non-zero element z of R by $\omega_R(z)$. With this notation, the **Möbius function** μ_R of R is defined as follows:

- $\mu_R(z) := 1$ if z is a unit in R;
- $\mu_R(z) := 0$, if z is divisible by the square of some prime;
- $\mu_R(z) := (-1)^{\omega_R(z)}$, if z is not divisible by the square of a prime, that is, if z is square-free. □

In accordance with the notation used for the rings $\mathbb{Z}$ and $F[x]$ earlier, we define the radical $\mathrm{rad}_R(z)$ of $z \in R^*$ as the product of all distinct prime divisors of z (up to association). Throughout, we shall simply write $\mathrm{rad}(z)$ instead of $\mathrm{rad}_R(z)$, since it will be clear from the context which ring R is considered.

Similarly, we also generalize the notion of the a-**part** of b (which was introduced for integers in Definition 1.7.2) to elements $a,b \in R^*$ as follows: $\mathrm{pt}_a(b)$ is the largest divisor d of b such that $\mathrm{rad}(d) \mid \mathrm{rad}(a)$; in other words, $\mathrm{pt}_a(b)$ is the smallest divisor d of b for which b/d and a are relatively prime. Regarding the order ideals of elements of the R-module M, we will write $\mathrm{Ord}_R(u) = a$ instead of $\mathrm{Ord}_R(u) = (a)$ from now on.

Remark 13.4.3. In the situation of Theorem 13.4.1, if $u \in M$ and $t \mid A$ such that $\mathrm{pt}_t(A)$ divides $\mathrm{Ord}_R(u)$, then u is said to be an element which is **not any kind of** t**-th multiple** in M.[6] This terminology can be explained as follows: let r be some prime divisor of t, then there is no $v \in M$ satisfying $u = rv$. For otherwise, we would have $u = r(aw)$ for some $a \in R$ and some generator w of M, and then $\mathrm{Ord}_R(u)$ would be a divisor of

$$\frac{\mathrm{Ord}_R(w)}{\gcd(\mathrm{Ord}_R(w), ra)} = \frac{A}{\gcd(A, ra)},$$

hence also a divisor of A/r, contradicting the assumption that $r \mid t$ and $\mathrm{pt}_t(A)$ divides $\mathrm{Ord}_R(u)$.

Finally, given any divisor t of A, we let

$$\Omega_t^+ := \{u \in M\colon \ \mathrm{pt}_t(A) \text{ divides } \mathrm{Ord}_R(u)\} \tag{13.8}$$

be the set of all $u \in M$ which are not any kind of t-th multiple. Note that $\Omega_t^+ = \Omega_{\mathrm{rad}(t)}^+$, by our definitions; in particular, Ω_A^+ is the set of all generators of M considered as a cyclic R-module. □

From now on, all characters are assumed to be complex, that is, $\widehat{M}$ is viewed as a subset of the $\mathbb{C}$-algebra $\mathbb{C}^M$, equipped with pointwise operations. We can then prove a general representation for the characteristic functions of sets of the form Ω_t^+ in terms of characters:

Theorem 13.4.4. *Let M be a finite cyclic module over a principal ideal domain R, and assume that $\mathscr{A}(M) = (A) \neq (0)$. For every divisor t of A, put*

$$\Gamma_t := \frac{\phi_R(t)}{|M_t|} \cdot \sum_{\chi \in M_t'} \frac{\mu_R(\mathrm{Ord}_R(\chi))}{\phi_R(\mathrm{Ord}_R(\chi))} \chi,$$

where we use the same notation as in the proof of Theorem 13.4.1. Then Γ_t is the characteristic function of the set Ω_t^+, that is,

$$\Gamma_t(u) = \begin{cases} 1 & \textit{if } u \in \Omega_t^+, \\ 0 & \textit{otherwise.} \end{cases}$$

[6] This follows the terminology of Cohen [85].

Proof. We first show that the corresponding function Γ on the set of divisors of A is multiplicative, which will allow us to reduce the assertion to the case of prime divisors of A. Thus let r and s be any two divisors of A which are relatively prime. By definition,

$$\Gamma_r \cdot \Gamma_s = \frac{\phi_R(r)\phi_R(s)}{|M_r||M_s|} \cdot \sum_{\chi \in M'_r} \sum_{\lambda \in M'_s} \frac{\mu_R(\mathrm{Ord}_R(\chi))\mu_R(\mathrm{Ord}_R(\lambda))}{\phi_R(\mathrm{Ord}_R(\chi))\phi_R(\mathrm{Ord}_R(\lambda))} \chi\lambda.$$

As r and s are relatively prime, we have $M'_r M'_s = M'_{rs}$, in particular $|M_r||M_s| = |M_{rs}|$. Since both the Möbius function and the Euler function for R are multiplicative, and as $\mathrm{Ord}_R(\lambda)\mathrm{Ord}_R(\chi) = \mathrm{Ord}_R(\lambda\chi)$, we obtain

$$\Gamma_r \cdot \Gamma_s = \frac{\phi_R(rs)}{|M_{rs}|} \cdot \sum_{\psi \in M'_{rs}} \frac{\mu_R(\mathrm{Ord}_R(\psi))}{\phi_R(\mathrm{Ord}_R(\psi))} \psi = \Gamma_{rs},$$

as claimed. Since we also have $\Gamma_t = \Gamma_{\mathrm{rad}(t)}$ for every $t \mid A$ (by definition of the Möbius function on R), it will suffice to consider the case where t is a prime divisor r of A.

In this case, $\phi_R(r) = |M_r| - 1 = \phi_R(\mathrm{Ord}_r(\chi))$ and $-1 = \mu_R(\mathrm{Ord}_r(\chi))$ for every $\chi \in M_r$ distinct from χ_0 (since $\mathrm{Ord}_R(\chi) = r$ in this case), while $\phi_R(\mathrm{Ord}_r(\chi_0)) = 1 = \mu_R(\mathrm{Ord}_r(\chi_0))$, as $\mathrm{Ord}_R(\chi_0) = 1$. This gives

$$\begin{aligned} \Gamma_r &= \frac{|M_r| - 1}{|M_r|} \cdot \left(\chi_0 - \frac{1}{|M_r| - 1} \sum_{\substack{\chi \in M'_r \\ \chi \neq \chi_0}} \chi \right) \\ &= \frac{|M_r| - 1}{|M_r|} \cdot \left(\frac{|M_r|}{|M_r| - 1} \chi_0 - \frac{1}{|M_r| - 1} \sum_{\chi \in M'_r} \chi \right) \\ &= \chi_0 - \frac{1}{|M_r|} \sum_{\chi \in M'_r} \chi. \end{aligned}$$

By Lemma 10.1.4,

$$\sum_{\chi \in M'_r} \chi(u) = \begin{cases} |M'_r| - |M_r| & \text{if } u \subset (M'_r)^\perp, \\ 0 & \text{otherwise,} \end{cases}$$

and hence $\Gamma_r(u) = 0$ if $u \in (M'_r)^\perp$ and $\Gamma_r(u) = 1$ otherwise. Thus Γ_r is the characteristic function of the subset of all elements $u \in M$ which are not contained in $(M'_r)^\perp = M_{A/r}$, where the equality follows from the proof of Theorem 10.1.2. Note that $u \in M \setminus M_{A/r}$ if and only if u is not a multiple of r in M.

Combining these observations, we conclude that $\Gamma_t = \Gamma_{r_1} \cdots \Gamma_{r_k}$ (where $r_1, \ldots, r_k$ are the distinct prime divisors of t) is the characteristic function of the subset $\bigcap_{i=1}^k (M \setminus M_{A/r_i})$ of M, which by definition is the set Ω_t^+ of all elements of M which are not any kind of t-th multiple. $\square$

In the next section, we will apply the preceding general results to the Galois extension E/F with $F = \mathrm{GF}(q)$ and $E = \mathrm{GF}(q^n)$.

13.5 A Character Theoretic Existence Criterion

In this section, we first apply the abstract results obtained in Section 13.4 in two ways to the Galois extension E/F with $F = \mathrm{GF}(q)$ and $E = \mathrm{GF}(q^n)$, noting that both the multiplicative and the additive group of E can be viewed as cyclic modules in a natural way. This gives us a Vinogradov type representation of the characteristic function of all primitive normal elements for E/F, which we then use to derive a sufficient number theoretic condition for the existence of such elements.

For the multiplicative group E^* of E, we take $R = \mathbb{Z}$ and $M = E^*$ in the general setup; in this case, we will use the notation $\mathscr{P}_t$ instead of Γ_t, where t is any positive divisor of $q^n - 1$. Here M_t is the unique subgroup U_t of order t of E^*, and we obtain the following explicit form for $\mathscr{P}_t$:

$$\mathscr{P}_t = \frac{\phi(t)}{t} \cdot \sum_{\psi \in U_t'} \frac{\mu(\mathrm{ord}(\psi))}{\phi(\mathrm{ord}(\psi))} \psi = \frac{\phi(t)}{t} \cdot \sum_{d|t} \frac{\mu(d)}{\phi(d)} \cdot \sum_{\psi:d} \psi, \tag{13.9}$$

where the index $\psi : d$ indicates that the corresponding sum runs over all multiplicative characters ψ of $\widehat{E^*}$ with order d, as in Proposition 10.2.5.

Similarly, for the additive group of E, we take $R = F[x]$ and consider $(E,+)$ as a cyclic module M with respect to the Frobenius automorphism σ, as explained at the beginning of Section 13.4; here we will write $\mathscr{N}_g$ instead of Γ_g, where $g(x) \in F[x]$ is any monic divisor of $x^n - 1$. In this situation, M_g is the σ-invariant F-subspace of E which is annihilated by g (this was previously denoted by V_g; see, for instance, Observation 3.11.8 (4)), and we obtain

$$\mathscr{N}_g = \frac{\phi_q(g)}{q^{\deg g}} \cdot \sum_{\chi \in V_g'} \frac{\mu_q(\mathrm{Ord}_q(\chi))}{\phi_q(\mathrm{Ord}_q(\chi))} \chi = \frac{\phi_q(g)}{q^{\deg g}} \cdot \sum_{h|g} \frac{\mu_q(h)}{\phi_q(h)} \cdot \sum_{\chi:h} \chi, \tag{13.10}$$

where $\chi : h$ indicates that the corresponding sum runs over all additive characters χ of $\widehat{(E,+)}$ with q-order h.

For the particular choices $t = q^n - 1$ and $g(x) = x^n - 1$, Theorem 13.4.4 now gives the desired generalization of Vinogradov's formula:

Corollary 13.5.1. *Consider the n-dimensional extension $E = \mathrm{GF}(q^n)$ of the Galois field $F = \mathrm{GF}(q)$. Then:*

(1) *$\mathscr{P}_{q^n-1}$ is the characteristic function of the set of all primitive elements of E^*, while $\mathscr{N}_{x^n-1}$ is the characteristic function of the set of all elements of E which are normal over F.*

(2) *In particular, $u \in E^*$ is a primitive element for E which is normal over F if and only if $\mathscr{P}_{q^n-1}(u)\mathscr{N}_{x^n-1}(u) = 1$.*

(3) *The number of all primitive elements for E which are normal over F is given by*

$$PN_n(q) = \sum_{u \in E^*} \mathscr{P}_{q^n-1}(u)\mathscr{N}_{x^n-1}(u). \qquad \square$$

Next, we use the characteristic function $\mathscr{P}_{q^n-1}\mathscr{N}_{x^n-1}$ of all primitive normal elements for the extension E/F to derive a sufficient number theoretic criterion for the existence of such elements. In fact, we will prove a more general result for arbitrary divisors of $q^n - 1$ and $x^n - 1$. For this, we require some further notation:

Notation 13.5.2. Let ℓ be a positive divisor of $q^n - 1$, and let $g \in F[x]$ be a monic divisor of $x^n - 1$.

- We denote the number of distinct positive prime divisors of ℓ by $\omega(\ell)$; similarly, $\omega_q(g)$ denotes the number of distinct monic divisors of g which are irreducible over F.
- For such a pair (ℓ, g), we are interested in the set

$$\Omega_{\ell,g}^+ := \Omega_\ell^+ \cap \Omega_g^+.$$

In view of Remark 13.4.3, we have $u \in \Omega_{\ell,g}^+$ if and only if u is not any kind of ℓ-th power in E^* and not any kind of g-th multiple in E. More formally, $u \in \Omega_{\ell,g}^+$ if and only if $\mathrm{pt}_\ell(q^n - 1)$ divides $\mathrm{ord}(u)$ and $\mathrm{pt}_g(x^n - 1)$ divides $\mathrm{Ord}_q(u)$. $\square$

After these preparations, we can now prove the following powerful lower bound for the cardinality of such a set $\Omega_{\ell,g}^+$:

Proposition 13.5.3. *Let ℓ and g be as above. Then*

$$|\Omega_{\ell,g}^+| \geq \frac{\phi(\ell)}{\ell} \cdot \frac{\phi_q(g)}{q^{\deg g}} \cdot \left(q^n - (2^{\omega(\ell)} - 1)(2^{\omega_q(g)} - 1)q^{n/2}\right).$$

In particular, $\Omega_{\ell,g}^+ \neq \emptyset$ provided that

$$\sqrt{q^n} > (2^{\omega(\ell)} - 1)(2^{\omega_q(g)} - 1).$$

Proof. We first note the following more general version of part (3) of Corollary 13.5.1:

$$|\Omega_{\ell,g}^+| = \sum_{u \in E^*} \mathscr{P}_\ell(u)\mathscr{N}_g(u),$$

which is also immediate from applying Theorem 13.4.4 to both kinds of modules introduced above. The remainder of the proof essentially relies on the theory of Gauss sums introduced in Chapter 10. Compared to Definition 10.3.1, it will be convenient to adjust the definition of $G(\psi, \chi)$ slightly.

Let ψ be some multiplicative character and χ some additive character of E. As in Section 10.2, we extend ψ to all of E by putting $\psi(0)=0$ if $\psi \neq \psi_0$ and $\psi_0(0)=1$. We now redefine $G(\psi,\chi)$ as follows:

$$G(\psi,\chi) := \sum_{u\in E} \psi(u)\chi(u). \tag{13.11}$$

Then $G(\psi_0,\chi_0) = q^n$ and $G(\psi_0,\chi) = 0$ when $\chi \neq \chi_0$, whereas all other results in Proposition 10.3.2 and Corollary 10.3.3 remain unchanged: $G(\psi,\chi_0)=0$ when $\psi \neq \psi_0$ and $|G(\psi,\chi)| = \sqrt{q^n}$ when $\psi \neq \psi_0$ and $\chi \neq \chi_0$. (Note that we sum over all elements of the extension field E, and that we only work with complex characters.)

Let us introduce the abbreviation

$$\theta = \theta(\ell,g) := \frac{\phi(\ell)}{\ell} \cdot \frac{\phi_q(g)}{q^{\deg g}}.$$

Using Equations (13.9) and (13.10) now gives

$$\begin{aligned}
\frac{1}{\theta}\cdot|\Omega^+_{\ell,g}| &= \frac{1}{\theta}\cdot\Big(\sum_{u\in E}\mathscr{P}_\ell(u)\mathscr{N}_g(u)\Big)\\
&= \frac{1}{\theta}\cdot\sum_{u\in E}\Big(\frac{\phi(\ell)}{\ell}\cdot\sum_{d|\ell}\frac{\mu(d)}{\phi(d)}\cdot\sum_{\psi:d}\psi(u)\Big)\cdot\Big(\frac{\phi(g)}{q^{\deg g}}\cdot\sum_{f|g}\frac{\mu_q(f)}{\phi_q(f)}\cdot\sum_{\chi:f}\chi(u)\Big)\\
&= \sum_{d|\ell}\sum_{f|g}\frac{\mu(d)\mu_q(f)}{\phi(d)\phi_q(f)}\cdot\sum_{\psi:d}\sum_{\chi:f}\sum_{u\in E}\psi(u)\chi(u)\\
&= \sum_{d|\ell}\sum_{f|g}\frac{\mu(d)\mu_q(f)}{\phi(d)\phi_q(f)}\cdot\sum_{\psi:d}\sum_{\chi:f}G(\psi,\chi)\\
&= \sum_{d|\mathrm{rad}(\ell)}\sum_{f|\mathrm{rad}(g)}\frac{\mu(d)\mu_q(f)}{\phi(d)\phi_q(f)}\cdot\sum_{\psi:d}\sum_{\chi:f}G(\psi,\chi),
\end{aligned}$$

where the final equality holds as the Möbius functions involved take non-zero values only for square-free arguments. We will analyze the final term distinguishing the following four cases:

- $d=1$ and $f=1$,
- $d=1$ and $f\neq 1$,
- $d\neq 1$ and $f=1$,
- $d\neq 1$ and $f\neq 1$.

For this, let us write

$$\tau(d,f) := \frac{\mu(d)\mu_q(f)}{\phi(d)\phi_q(f)}\sum_{\psi:d}\sum_{\chi:f}G(\psi,\chi). \tag{13.12}$$

The only characters with order $d=1$ and q-order $f=1$ are the trivial characters ψ_0 and χ_0, respectively. Hence,

$$\tau(1,1) = G(\psi_0, \chi_0) = q^n.$$

Furthermore, we have $\tau(1,f) = 0$ if $f \neq 1$ (by the above modification for the Gauss sum); similarly, $\tau(d,1) = 0$ whenever $d \neq 1$. These observations give

$$\frac{1}{\theta} \cdot |\Omega^+_{\ell,g}| = q^n + \sum_{\substack{d|\text{rad}(\ell)\\ d\neq 1}} \sum_{\substack{f|\text{rad}(g)\\ f\neq 1}} \tau(d,f). \tag{13.13}$$

Now let $d \neq 1$ be any divisor of $\text{rad}(\ell)$, and $f \neq 1$ any divisor of $\text{rad}(g)$. Then $\mu(d)$ and $\mu_q(f)$ only take values in $\{-1,1\}$, and $|G(\psi,\chi)| = \sqrt{q^n}$ whenever $\text{ord}(\psi) = d$ and $\text{Ord}_q(\chi) = g$. This yields the inequality

$$|\tau(d,f)| \leq \frac{1}{\phi(d)\phi_q(f)} \sum_{\psi:d} \sum_{\chi:f} \sqrt{q^n},$$

by taking absolute values on both sides of Equation (13.12) and using the triangle inequality. Since there are exactly $\phi(d)$ multiplicative characters ψ with $\text{ord}(\psi) = d$ and exactly $\phi_q(f)$ additive characters χ with $\text{Ord}_q(\chi) = f$, the preceding inequality simplifies to

$$|\tau(d,f)| \leq \sqrt{q^n}. \tag{13.14}$$

Combining Equations (13.13) and (13.14) gives

$$\begin{aligned}\frac{1}{\theta} \cdot |\Omega^+_{\ell,g}| &\geq q^n - \sum_{\substack{d|\text{rad}(\ell)\\ d\neq 1}} \sum_{\substack{f|\text{rad}(g)\\ f\neq 1}} \sqrt{q^n} \\ &= q^n - (2^{\omega(\ell)} - 1)(2^{\omega_q(g)} - 1)\sqrt{q^n},\end{aligned}$$

where the equality holds as $\text{rad}(\ell)$ has exactly $2^{\omega(\ell)}$ divisors and as $\text{rad}(g)$ has exactly $2^{\omega_q(g)}$ monic divisors in $F[x]$ (by the definitions of $\omega(\ell)$ and $\omega_q(g)$, respectively). This establishes the first assertion; then the second assertion is an immediate consequence. □

In particular, Proposition 13.5.3 specializes to the following result on the existence of primitive normal elements:

Corollary 13.5.4. *Let q be a prime power and $n \geq 2$ an integer, and assume that*

$$(2^{\omega(q^n-1)} - 1) \cdot (2^{\omega_q(x^n-1)} - 1) < \sqrt{q^n}.$$

Then there exists a primitive element for $\text{GF}(q^n)$ *which is normal over* $\text{GF}(q)$. □

Exercises

Exercise 13.5.5. Adapt the proof of Proposition 13.5.3 to show the following upper bound on the cardinalities of the sets $\Omega^+_{\ell,g}$:

$$|\Omega^+_{\ell,g}| \le \frac{\phi(\ell)}{\ell} \cdot \frac{\phi_q(g)}{q^{\deg g}} \cdot \left(q^n + (2^{\omega(\ell)}-1)(2^{\omega_q(g)}-1)q^{n/2}\right). \qquad \square$$

13.6 Asymptotic Results and Extensions of Degree 6

The main goal of the present section is an asymptotic version of the primitive normal basis theorem, which goes back to work of Carlitz [63, 64]; see also Davenport [101]. Moreover, we will also establish several more specific results, for instance, for the sextic case and for extensions where the degree is a power of the characteristic. All these results are based on Corollary 13.5.4 and suitable estimates for the functions $\omega(\ell)$ and $\omega_q(g)$ introduced in Notation 13.5.2.

We begin with a formal statement of the two main results in this section:

Theorem 13.6.1. *There are at most finitely many pairs* (q,n) *such that there is no primitive normal element for the field extension* $\mathrm{GF}(q^n)/\mathrm{GF}(q)$. $\square$

Theorem 13.6.2. *Consider a field extension* $E = \mathrm{GF}(q^6)$ *over* $F = \mathrm{GF}(q)$. *Then there exists a primitive element for* E *which is normal over* F, *with the possible exceptions of* $q \in \{4,5,7\}$. $\square$

The proofs of these two results will rest on the sufficient condition in Corollary 13.5.4. For simplicity, we shall henceforth use the abbreviations

$$\omega := \omega(q^n-1) \text{ and } \Omega := \omega_q(x^n-1).$$

Obviously, if U is a strict upper bound for $(2^\omega-1)(2^\Omega-1)$, then $U \le \sqrt{q^n}$ is a sufficient condition for $PN_n(q) > 0$. The derivation of such upper bounds U requires upper bounds for ω and Ω. All upper bounds on Ω will have the form $\alpha n + \beta$; trivially, we can always take $\alpha = 1$ and $\beta = 0$. Before deriving various more interesting possibilities for (α,β), we consider upper bounds on ω.

Proposition 13.6.3. *For every real number* $\varepsilon > 0$, *there exists a positive constant* $c(\varepsilon)$ *such that* $2^{\omega(N)} \le c(\varepsilon)\cdot N^\varepsilon$ *for all* $N \in \mathbb{N}^*$. *In fact, the assertion holds for* $c(\varepsilon) = (2^{1-\varepsilon})^{2^{1/\varepsilon}}$.

Proof. Consider the function

$$f_\varepsilon : \mathbb{N}^* \to \mathbb{R}^*, \quad N \mapsto \frac{2^{\omega(N)}}{N^\varepsilon},$$

where ε is some positive real number. Obviously, this function is multiplicative; moreover, $f_\varepsilon(N) \le f_\varepsilon(\mathrm{rad}(N))$, where, as before, $\mathrm{rad}(N)$ denotes the radical of N. If $\pi(N)$ is the set of all prime divisors of N, then

$$f_\varepsilon(N) \le \prod_{r\in\pi(N)} \frac{2}{r^\varepsilon}.$$

Of course, $2 \leq r^{\varepsilon}$ if and only if $2^{1/\varepsilon} \leq r$. We now consider the sets P_{ε} of all primes s such that $s < 2^{1/\varepsilon}$ and $I_{\varepsilon} := \pi(N) \cap P_{\varepsilon}$ and obtain the following estimate:

$$f_{\varepsilon}(N) \leq \prod_{r \in I_{\varepsilon}} \frac{2}{r^{\varepsilon}} \leq \prod_{r \in I_{\varepsilon}} \frac{2}{2^{\varepsilon}} = (2^{1-\varepsilon})^{|I_{\varepsilon}|} \leq (2^{1-\varepsilon})^{|P_{\varepsilon}|}.$$

Trivially, $|P_{\varepsilon}| \leq 2^{1/\varepsilon}$, and hence

$$f_{\varepsilon}(N) \leq c(\varepsilon), \text{ where } c(\varepsilon) = (2^{1-\varepsilon})^{2^{1/\varepsilon}}. \tag{13.15}$$

This yields the assertion. □

Example 13.6.4. For instance, the choice $\varepsilon = \frac{1}{4}$ leads to $c(\frac{1}{4}) = 2^{12}$, which however is too weak to give results for moderate values of q and n. Nevertheless, we can use the same approach to establish an alternative upper bound for $\omega(N)$ due to Cohen and Hachenberger [88, Lemma 3.3], which turns out to be quite effective:

$$2^{\omega(N)} \leq 5 \cdot \sqrt[4]{N} \quad \text{for all } N \in \mathbb{N}^*. \tag{13.16}$$

To see this, it suffices to analyze the situation in the proof of Proposition 13.6.3 for $\varepsilon = \frac{1}{4}$ in more detail. Then $2^{1/\varepsilon} = 16$, so that $P_{\varepsilon} = \{2,3,5,7,11,13\}$. This implies

$$f_{1/4}(N) \leq \prod_{r \in I_{1/4}} \frac{2}{\sqrt[4]{r}} \leq \prod_{r \in P_{1/4}} \frac{2}{\sqrt[4]{r}} = \frac{2^6}{\sqrt[4]{2 \cdot 3 \cdot 5 \cdot 7 \cdot 11 \cdot 13}} = \frac{64}{\sqrt[4]{30030}},$$

which yields the desired inequality. (In fact, one even has $64/\sqrt[4]{30030} < 4.86$.) □

Remark 13.6.5. Note that the set I_{ε} of prime divisors s of N which are smaller than $2^{1/\varepsilon}$ may be empty. In this case, the proof of Proposition 13.6.3 shows $f_{\varepsilon}(N) < 1$, so that $2^{\omega(N)} \leq N^{\varepsilon}$. □

In view of the results from Sections 13.1, 13.2 and 13.3, we may from now on assume that $n \geq 5$. Before turning to the proofs of Theorems 13.6.1 and 13.6.2, we give a particularly simple application of Inequality (13.16) for extensions where the degree is a power of the characteristic:

Theorem 13.6.6. *Let q as well as $n \in \mathbb{N}^*$ be powers of a prime p. Then there exists a primitive element for* $\mathrm{GF}(q^n)$ *which is normal over* $\mathrm{GF}(q)$.

Proof. In the case under consideration, the polynomial $x^n - 1$ splits over $\mathrm{GF}(q)$ as $(x-1)^n$ and therefore $\Omega = \omega_q(x^n - 1) = 1$. Hence, using Inequality (13.16),

$$(2^{\omega} - 1) \cdot (2^{\Omega} - 1) = 2^{\omega} - 1 < 2^{\omega} \leq 5 \cdot \sqrt[4]{q^n - 1} < 5 \cdot \sqrt[4]{q^n}.$$

In view of Corollary 13.5.4, $PN_n(q)$ will certainly be positive if we can show

$$5 \cdot \sqrt[4]{q^n} \leq \sqrt{q^n}, \text{ that is } q^n \geq 625.$$

Because of $n \geq 5$, this condition holds for all $q \geq 4$; and when $q = 3$ or $q = 2$, it holds for all $n \geq 6$ and for all $n \geq 10$, respectively. This leaves only the pair $(q,n) = (2,8)$, where we can apply Corollary 13.5.4 directly: we have $\omega = 3$ (since $2^8 - 1 = 255 = 3 \cdot 5 \cdot 17$), so that $2^\omega - 1 = 7 < 16 = \sqrt{2^8}$. □

Let us now turn to the proofs of Theorems 13.6.1 and 13.6.2. We begin by showing an easy preliminary result: the primitive normal basis theorem holds whenever q is at least moderately large, namely for $q \geq 59$. For this, it still suffices to apply Inequality (13.16) together with the trivial upper bound on Ω.

Proposition 13.6.7. *Let $q > 53$ be a prime power and $n \geq 2$ an integer. Then there exists a primitive normal element for the field extension* $\mathrm{GF}(q^n)/\mathrm{GF}(q)$.

Proof. In view of Inequality (13.16) and the trivial bound $\Omega \leq n$, we have

$$(2^\omega - 1)(2^\Omega - 1) < 5 \cdot \sqrt[4]{q^n} \cdot 2^n =: U.$$

Hence the assertion follows from Corollary 13.5.4 if we can show $U \leq \sqrt{q^n}$, which holds if and only if

$$5 \cdot \left(\frac{2}{\sqrt[4]{q}} \right)^n \leq 1. \tag{13.17}$$

Because of $q > 16$, we have $2/\sqrt[4]{q} < 1$; in conjunction with $n \geq 5$, this yields the estimate

$$5 \cdot \left(\frac{2}{\sqrt[4]{q}} \right)^n \leq 5 \cdot \left(\frac{2}{\sqrt[4]{q}} \right)^5 = 160 \cdot q^{-5/4},$$

which is indeed smaller than 1, as $q \geq 58$ by hypothesis. □

For prime powers q in the range $[17,53]$, one may of course derive concrete bounds for n from Inequality (13.17); see Exercise 13.6.15.

Next, we prove Theorem 13.6.2. Here one can apply Corollary 13.5.4 directly, without using any estimates on ω or Ω, as Proposition 13.6.7 allows us to assume $q \leq 53$. This makes it possible to determine the prime power factorization of $q^6 - 1$, using

$$q^6 - 1 = \prod_{d|6} \Phi_d(q) = (q-1) \cdot (q+1) \cdot (q^2+q+1) \cdot (q^2-q+1),$$

so that we can also compute the value of ω explicitly. Moreover, Ω can likewise be calculated easily. Nevertheless, this direct approach obviously requires considerable computational effort. Fortunately, Inequality (13.16) can be used to reduce the number of cases which have to be investigated individually; we will summarize the necessary computational data for these cases in Table 13.6.

In the present situation, (13.16) gives

$$(2^\omega - 1)(2^\Omega - 1) < (2^\Omega - 1) \cdot 2^\omega \leq (2^\Omega - 1) \cdot 5 \cdot \sqrt[4]{q^6 - 1} < (2^\Omega - 1) \cdot 5 \cdot \sqrt{q^3},$$

and this is bounded by $\sqrt{q^6} = q^3$ (as required for the application of Corollary 13.5.4) if and only if

$$\sqrt[3]{25 \cdot (2^{\Omega} - 1)^2} \le q. \qquad (13.18)$$

We now use a case distinction according to the residue of q modulo 6:

1. If $q \equiv 2 \bmod 6$, then q is even and $x^6 - 1$ splits as $(x-1)^2(x^2+x+1)^2$. If $q \equiv 3 \bmod 6$, then q is a power of 3 and $x^6 - 1$ splits as $(x-1)^3(x+1)^3$. In both cases, $\Omega = 2$, so that (13.18) holds for all $q \ge 7$. This leaves the two cases $q = 2$ and $q = 3$, which are included in Table 13.6 below.
2. If $q \equiv 4 \bmod 6$, then q is even and $x^6 - 1$ splits as $(x-1)^2(x-\lambda)^2(x-\lambda^2)^2$, where λ is a primitive third root of unity. Thus $\Omega = 3$, and (13.18) holds for all $q \ge 11$. The remaining case $q = 4$ is again included in Table 13.6.
3. Now let $q \equiv 5 \bmod 6$. Then $x^6 - 1$ splits as $(x-1)(x+1)(x^2+x+1)(x^2-x+1)$, hence $\Omega = 4$ and (13.18) is satisfied for all $q \ge 18$. This leaves the cases $q \in \{5, 11, 17\}$, which are covered by Table 13.6.
4. Finally, let $q \equiv 1 \bmod 6$, where $x^6 - 1$ splits into linear factors. Then $\Omega = 6$ and (13.18) holds for all $q \ge 47$. This leaves the values $q \in \{7, 13, 19, 25, 31, 37, 43\}$, which are also all contained in Table 13.6.

Table 13.6 Evaluation of Corollary 13.5.4 for $q \in \{2,3,4,5,7,11,13,17,19,25,31,37,43\}$, when $n = 6$.

q	$q^6 - 1$	Ω	$(2^{\omega}-1)\cdot(2^{\Omega}-1)$	$\sqrt{q^6}$
2	$3^2 \cdot 7$	2	9	8 *
3	$2^3 \cdot 7 \cdot 13$	2	21	27
4	$3^2 \cdot 5 \cdot 7 \cdot 13$	3	105	64 *
5	$2^3 \cdot 3^2 \cdot 7 \cdot 31$	4	225	125 *
7	$2^4 \cdot 3^2 \cdot 19 \cdot 43$	6	945	343 *
11	$2^3 \cdot 3^2 \cdot 5 \cdot 7 \cdot 19 \cdot 37$	4	945	1331
13	$2^3 \cdot 3^2 \cdot 7 \cdot 61 \cdot 157$	6	1953	2197
17	$2^5 \cdot 3^3 \cdot 7 \cdot 13 \cdot 307$	4	465	4913
19	$2^3 \cdot 3^3 \cdot 5 \cdot 7^3 \cdot 127$	6	1953	6859
25	$2^4 \cdot 3^2 \cdot 7 \cdot 13 \cdot 31 \cdot 601$	6	3969	15625
31	$2^6 \cdot 3^2 \cdot 5 \cdot 7^2 \cdot 19 \cdot 331$	6	3969	29791
37	$2^3 \cdot 3^3 \cdot 7 \cdot 19 \cdot 31 \cdot 43 \cdot 67$	6	8001	50653
43	$2^3 \cdot 3^2 \cdot 7 \cdot 11 \cdot 13 \cdot 139 \cdot 631$	6	8001	79507

We see from Table 13.6 that the condition in Corollary 13.5.4 is not satisfied when $q \in \{2,4,5,7\}$. In order to complete the proof of Theorem 13.6.2, we now apply Proposition 13.1.2 for $q = 2$:

- $\phi(2^6 - 1) = \phi(3^2 \cdot 7) = 6 \cdot 6 = 36$,
- $\phi_2(x^6 - 1) = (2^2 - 2^1)(2^4 - 2^2) = 2 \cdot 12 = 24$,
- $\sum_{d|6} \mu(\frac{6}{d}) \cdot 2^d = 2^6 - 2^3 - 2^2 + 2^1 = 54$,

and hence $PN_6(2) \ge 36 + 24 - 54 = 6 > 0$. □

We remark that Proposition 13.1.2 fails for $q \in \{4,5,7\}$. We will deal with these cases in later sections, after deriving improved existence criteria for primitive normal elements.

We now turn to the proof of Theorem 13.6.1. By the previous results, we may assume $q \leq 53$. Settling the remaining cases requires non-trivial estimates for Ω. Anticipating these estimates for the moment, we can complete the proof rather quickly.

Thus assume $\Omega \leq \alpha n + \beta$. Substituting this together with the explicit bound for $2^{\omega(N)}$ given in Proposition 13.6.3 in Corollary 13.5.4 proves that $PN_n(q)$ is positive whenever

$$(2^{1-\varepsilon})^{2^{1/\varepsilon}} \cdot q^{n\varepsilon} \cdot 2^{\alpha n+\beta} \leq \sqrt{q^n},$$

which is equivalent to

$$\left(2^{1/\varepsilon}(1-\varepsilon)+\beta\right) \cdot \ln(2) \leq \left((\tfrac{1}{2}-\varepsilon)\ln(q) - \alpha\ln(2)\right) \cdot n.$$

We now check that the coefficient $(\frac{1}{2}-\varepsilon)\ln(q) - \alpha\ln(2)$ of n is positive for some suitable choice of the values α, β and ε (depending on the specific $q \leq 53$). This will then establish $PN_n(q) > 0$ for any fixed q and all sufficiently large n, as claimed in Theorem 13.6.1. We distinguish three cases:

- *Case 1:* $q \geq 5$. Here we may even use the trivial bound for Ω given by $\alpha = 1$ and $\beta = 0$, together with $\varepsilon = \frac{1}{20}$.
- *Case 2:* $q \in \{3,4\}$. Here we use Example 13.6.9 below and choose $\alpha = 1/2$ and $\beta = (q-1)/2$. Thus $\beta = 3/2$ for $q = 4$, and $\varepsilon = 1/5$ is a suitable choice in this case. Similarly, $\beta = 1$ for $q = 3$, and now $\varepsilon = 1/6$ works.
- *Case 3:* $q = 2$. Then we may take $\alpha = 1/3$ and $\beta = 1$, by Example 13.6.10 below; here any ε in the interval $(0, \frac{1}{6})$ is suitable. □

It remains to prove the validity of the upper bounds for Ω which we have just used. For this, we establish the following more general result of Lenstra and Schoof [238]. In fact, this result allows several useful variations, as we shall see later.

Lemma 13.6.8. *Let q be a power of some prime p, and $n \in \mathbb{N}^*$. For every $\ell \in \mathbb{N}^*$, let $D_\ell \subseteq \mathbb{N}^*$ denote the set of all divisors of* $\gcd(n, q^\ell - 1)$. *Moreover, let e be a positive integer and T_e some set of divisors of n' such that $D_\ell \subseteq T_e$ for all $\ell < e$, where n' denotes the p-free part of n. Then*

$$\omega_q(x^n - 1) \leq \frac{n'}{e} + \sum_{d \in T_e} \phi(d) \cdot \left(\frac{1}{\mathrm{ord}_d(q)} - \frac{1}{e}\right).$$

Proof. In view of $\omega_q(x^n-1) = \omega_q(x^{n'}-1)$, Equation (3.9) and Proposition 3.6.16 give the following formula for $\omega_q(x^n-1)$:

$$\omega_q(x^n - 1) = \sum_{d|n'} \frac{\phi(d)}{\mathrm{ord}_d(q)} = \sum_{\substack{d|n' \\ d \notin T_e}} \frac{\phi(d)}{\mathrm{ord}_d(q)} + \sum_{\substack{d|n' \\ d \in T_e}} \frac{\phi(d)}{\mathrm{ord}_d(q)}.$$

By hypothesis, an integer $d \notin T_e$ cannot belong to any of the sets D_ℓ with $\ell < e$, so that $\mathrm{ord}_d(q) \geq e$ in this case, whereas every $d \in T_e$ divides n'. This yields

$$\begin{aligned}\omega_q(x^n-1) &\leq \sum_{\substack{d|n'\\ d\notin T_e}} \frac{\phi(d)}{e} + \sum_{d\in T_e} \frac{\phi(d)}{\mathrm{ord}_d(q)}\\ &= \sum_{d|n'} \frac{\phi(d)}{e} + \sum_{d\in T_e} \left(\frac{\phi(d)}{\mathrm{ord}_d(q)} - \frac{\phi(d)}{e}\right).\end{aligned}$$

In view of $\sum_{d|n'} \phi(d) = n'$, this proves the assertion. □

Example 13.6.9. We apply Lemma 13.6.8 with $e = 2$ and $T_e = D_1$, so that T_e is just the set of all divisors of $\gcd(n', q-1)$. This gives the upper bound

$$\omega_q(x^n-1) \leq \frac{n'}{2} + \frac{\gcd(n',q-1)}{2}, \tag{13.19}$$

since here $\mathrm{ord}_d(q) = 1$ for all $d \in T_e$ and $\sum_{d\in T_e} \phi(d) = \gcd(n', q-1)$. In particular, $\omega_q(x^n-1) \leq \frac{n'}{2} + \frac{q-1}{2}$, which we have used in Case 2 above. We also note that $\omega_q(x^n-1) \leq \frac{3}{4}n$ whenever $q-1$ is not divisible by n. □

Example 13.6.10. We apply Lemma 13.6.8 with $e = 3$ and $T_e = D_2$. Then $D_1 \subseteq D_2$ and T_e is the set of all divisors of $\gcd(n', q^2-1)$. This gives

$$\begin{aligned}\omega_q(x^n-1) &\leq \frac{n'}{3} + \sum_{d\in T_e} \phi(d)\left(\frac{1}{\mathrm{ord}_d(q)} - \frac{1}{3}\right)\\ &= \frac{n'}{3} + \sum_{d\in D_1} \phi(d)\left(1-\frac{1}{3}\right) + \sum_{d\in D_2\setminus D_1} \phi(d)\left(\frac{1}{2}-\frac{1}{3}\right)\\ &= \frac{n'}{3} + \frac{2}{3}\cdot \gcd(n',q-1) + \frac{1}{6}\cdot\left((\gcd(n',q^2-1) - \gcd(n',q-1)\right),\end{aligned}$$

and therefore

$$\omega_q(x^n-1) \leq \frac{n'}{3} + \frac{\gcd(n',q-1)}{2} + \frac{\gcd(n',q^2-1)}{6}. \tag{13.20}$$

For $q = 2$, we obtain $\omega_2(x^n-1) \leq \frac{n}{3} + \frac{1}{2} + \frac{3}{6} = \frac{n}{3} + 1$, which was used for Case 3 above. □

In the next section, we will settle the existence of primitive normal elements over the binary field completely. This will require a slightly more involved application of Lemma 13.6.8 which gives the following improved bound on $\omega_2(x^n-1)$:

Corollary 13.6.11. *Let $q = 2$ and $n \in \mathbb{N}^*$. Then*

$$\omega_2(x^n-1) \leq \frac{n'}{4} + \frac{5}{4},$$

where n' denotes the odd part of n.

Proof. Applying Lemma 13.6.8 with $e = 4$ and any permissible choice of T_4 gives $\omega_2(x^n - 1) \leq \frac{n'}{4} + \beta$, where

$$\beta = \sum_{d \in T_4} \phi(d) \cdot \left(\frac{1}{\mathrm{ord}_d(2)} - \frac{1}{4} \right).$$

We now select T_4 depending on the value of n, where we take into account that $D_1 = \{1\}$, $D_2 \subseteq \{1,3\}$ and $D_3 \subseteq \{1,7\}$:

- Assume first that n is a multiple of 21. Then we choose $T_4 := \{1,3,7,21\}$ and obtain $\beta = \frac{3}{4} + 2 \cdot \frac{1}{4} + 6 \cdot \frac{1}{12} - 12 \cdot \frac{1}{12} = \frac{3}{4}$.
- When $3 \mid n$ and $n \not\equiv 0 \bmod 7$, let $T_4 := \{1,3\}$. Then $\beta = \frac{3}{4} + 2 \cdot \frac{1}{4} = \frac{5}{4}$.
- When $7 \mid n$ and $n \not\equiv 0 \bmod 3$, let $T_4 := \{1,7\}$. Then $\beta = \frac{3}{4} + 6 \cdot \frac{1}{12} = \frac{5}{4}$.
- Finally, when $n \not\equiv 0 \bmod 3$ and $n \not\equiv 0 \bmod 7$, then $T_4 := \{1\}$ gives $\beta = \frac{3}{4}$.

Thus $\beta \leq 5/4$ for all values of n, as desired. □

We close this section with an alternative bound for ω which is also due to Lenstra and Schoof [238] and allows a lot of flexibility. Let us first fix some notation. As in the proof of Proposition 13.6.3, we denote the set of all prime divisors of a positive integer N by $\pi(N)$. Given any further integer $\ell \geq 3$, we let π_ℓ denote the set of all primes $< \ell$ and put $\pi_\ell(N) := \pi(N) \cap \pi_\ell$. Thus $\pi_\ell(N)$ consists of all prime divisors of N which are strictly smaller than ℓ.

Lemma 13.6.12. *Let $N \in \mathbb{N}^*$, let $\ell \geq 3$ be an integer, and let Λ be any set satisfying $\pi_\ell(N) \subseteq \Lambda \subseteq \pi_\ell$. Put $L = L(\Lambda) := \prod_{r \in \Lambda} r$, with the convention $L = 1$ if Λ is empty. Then*

$$\omega(N) \leq |\Lambda| + \frac{\ln(N) - \ln(L)}{\ln(\ell)}.$$

Proof. We reconsider the proof of Proposition 13.6.3 with $\varepsilon := \ln(2)/\ln(\ell)$, that is, $2^{1/\varepsilon} = \ell$. Then $I_\varepsilon = \pi_\ell(N)$, and we obtain the estimate

$$\frac{2^{\omega(N)}}{N^\varepsilon} \leq \prod_{r \in \pi_\ell(N)} \frac{2}{r^\varepsilon} \leq \prod_{r \in \Lambda} \frac{2}{r^\varepsilon} = \frac{2^{|\Lambda|}}{L^\varepsilon},$$

which gives

$$\omega(N) \cdot \ln(2) \leq |\Lambda| \cdot \ln(2) + \varepsilon \cdot (\ln(N) - \ln(L)).$$

In view of $\varepsilon = \ln(2)/\ln(\ell)$, this is the assertion. □

Let us give a first application of Lemma 13.6.12, which corresponds to the choice $\varepsilon = \frac{1}{6}$ in Proposition 13.6.3:

Proposition 13.6.13. *Suppose that* $\min(q,n) \geq 16$. *Then there exists a primitive normal element for the field extension* $E = \mathrm{GF}(q^n)$ *over* $F = \mathrm{GF}(q)$.

Proof. By Proposition 13.6.7, we may assume $q \leq 53$. We apply Lemma 13.6.12 with $\ell = 64$ and

$$\Lambda := \pi_\ell = \{2,3,5,7,11,13,17,19,23,29,31,37,41,43,47,53,59,61\}.^7$$

For notational simplicity, we now switch to the logarithm with base 2 instead of ln. We have $|\Lambda| = 18$ and $\frac{\log_2(L)}{\log_2(\ell)} = \frac{\log_2(L)}{6} > \frac{51}{4}$ (which can be checked with a pocket calculator), which gives

$$\begin{aligned}\omega = \omega(q^n - 1) &\leq |\Lambda| + \frac{\log_2(q^n - 1) - \log_2(L)}{\log_2(\ell)} \\ &< \frac{\log_2(q^n)}{\log_2(\ell)} + |\Lambda| - \frac{\log_2(L)}{\log_2(\ell)} \\ &< \frac{n}{6}\log_2(q) + \frac{21}{4}.\end{aligned}$$

Using the trivial bound n for $\Omega = \omega_q(x^n - 1)$, we obtain $\omega + \Omega < \frac{n}{6}\log_2(q) + \frac{21}{4} + n$, which is bounded by $\frac{n}{2}\log_2(q)$ (as required for the application of Corollary 13.5.4) if and only if

$$\log_2(q) \geq 3 + \frac{63}{4n}.$$

Note that this indeed holds in view of the hypothesis $\min(q,n) \geq 16$:

$$3 + \frac{63}{4n} \leq 3 + \frac{63}{64} < 4 \leq \log_2(q). \qquad \square$$

Exercises

Exercise 13.6.14. Prove that

$$2^{\omega(N)} \leq \frac{45}{4} \cdot \sqrt[5]{N}$$

holds for every $N \in \mathbb{N}^*$. $\square$

Exercise 13.6.15. Let q be a prime power with $17 \leq q \leq 53$. Verify that Inequality (13.17) is satisfied in all the following cases:

q	17	19	23	25	27	29	31	32	37	41	43	47	49	53
$n \geq$	107	38	18	15	13	11	10	10	8	7	7	6	6	6

. $\square$

[7] Note that this is not an optimal choice of Λ for any fixed value of $q \leq 53$, as $\pi_{67} = \pi_{64}$ and the characteristic p belongs to π_{64}. Trivially, p is not a divisor of $q^n - 1$, and therefore could be excluded from Λ. Nevertheless, the choice above allows us to treat all relevant q simultaneously and results in an easier argument.

Exercise 13.6.16. Check that Proposition 13.1.2 fails for the three cases where $n=6$ and $q \in \{4,5,7\}$. □

Exercise 13.6.17. Try to show by hand computation that $\log_2(L) > 153/2$, where

$$L = 2\cdot 3\cdot 5\cdot 7\cdot 11\cdot 13\cdot 17\cdot 19\cdot 23\cdot 29\cdot 31\cdot 37\cdot 41\cdot 43\cdot 47\cdot 53\cdot 59\cdot 61. \quad \square$$

13.7 The Primitive Normal Basis Theorem for $q=2$ and $q=3$

In the present section, we will prove the primitive normal basis theorem for the binary and the ternary ground fields (with one possible exception), by following the basic strategy established in the previous two sections: in general, we apply Corollary 13.5.4 in conjunction with bounds on the numbers $\omega = \omega(q^n-1)$ and $\Omega = \omega_q(x^n-1)$ obtained in Section 13.6. For just a few specific cases, we either need to apply Corollary 13.5.4 directly or have to appeal to Proposition 13.1.2 instead (as for the case $(q,n)=(2,6)$ already settled in Theorem 13.6.2).

It is easily checked that the condition $\omega+\Omega \le n\ln(q)/\ln(4)$ is sufficient for the application of Corollary 13.5.4, as then $(2^\omega-1)(2^\Omega-1) < \sqrt{q^n}$. Based on Lemmas 13.6.8 and 13.6.12, we have various upper bounds of the form

$$\omega < \frac{n\ln(q)}{\ln(\ell)} + |\Lambda| - \frac{\ln(L)}{\ln(\ell)} \quad \text{and} \quad \Omega \le \alpha n + \beta$$

at our disposal (as $\ln(q^n-1) < n\ln q$). Given such a bound, it would therefore suffice to show that

$$|\Lambda| - \frac{\ln(L)}{\ln(\ell)} + \beta \le \left(\left(\tfrac{1}{\ln(4)} - \tfrac{1}{\ln(\ell)}\right)\ln(q) - \alpha\right)\cdot n. \tag{13.21}$$

For a concrete application of this idea, let us suppose that some value for q has been specified, for instance, $q=2$ or $q=3$. Depending on properties of the extension degree n, some specific parameters α and β will then be available for the Ω-bound, by the results in the last section. This leaves the task of choosing a value of ℓ for which the coefficient

$$\kappa(q,\alpha,\ell) := \left(\tfrac{1}{\ln(4)} - \tfrac{1}{\ln(\ell)}\right)\cdot\ln(q) - \alpha \tag{13.22}$$

of n in Inequality (13.21) is positive, where one should also try to choose ℓ in a manner which keeps the set Λ as small as possible. In order to exclude certain primes r from Λ, it will be important to have the order of q modulo r available. After a suitable specification of all parameters, $PN_n(q)$ will then be positive whenever

$$n \ge \frac{|\Lambda| - \frac{\ln(L)}{\ln(\ell)} + \beta}{\kappa(q,\alpha,\ell)}. \tag{13.23}$$

We shall now discuss this approach in detail for the values $q = 2$ and $q = 3$. As we will see, this is a rather effective method: we will be left with only a small number of cases which need to be considered individually, and only one of these will require further work later in this chapter.

Theorem 13.7.1. *For every field extension $E = \mathrm{GF}(2^n)$ of the binary field $F = \mathrm{GF}(2)$, there exists a primitive element for E which is normal over F.*

Proof. We begin by discussing a few specific extension degrees n. The cases where n is a power of 2 or $n \in \{3,6\}$ have already been settled, see Theorems 13.6.6, 13.2.1 and 13.6.2.

Next, suppose that $2^n - 1$ is a Mersenne prime, which requires that n is itself a prime (though this is not sufficient). Then every normal element of $\mathrm{GF}(2^n)$ over $\mathrm{GF}(2)$ is primitive. For instance, this holds when $n \in \{5,7,13\}$, as $2^5 - 1 = 31$ and $2^7 - 1 = 127$ and $2^{13} - 1 = 8191$ are prime numbers. For $n = 11$, however, $2^{11} - 1$ splits as $23 \cdot 89$. Here we can argue as follows: since $\mathrm{ord}_{11}(2) = 10$, the polynomial $x^{11} - 1$ splits over $\mathrm{GF}(2)$ into the two irreducible factors $x - 1$ and $\Phi_{11}(x)$. Therefore, $\omega = 2 = \Omega$ for $n = 11$, and since $(2^2 - 1)^2 = 9 < \sqrt{2^{11}}$, Corollary 13.5.4 shows that $PN_{11}(2)$ is also positive.

We now turn to the systematic examination of general extension degrees n, using the approach outlined above. Note that

$$\kappa(2,\alpha,\ell) = \frac{1-2\alpha}{2} - \frac{\ln(2)}{\ln(\ell)},$$

which is certainly positive for $\alpha \le 1/4$ and $\ell \ge 17$ (for instance). Note also that $\mathrm{ord}_r(2) \mid n \mid r-1$ holds for every prime divisor r of $2^n - 1$. It will be useful to collect the precise values $\mathrm{ord}_r(2)$ for small r, see Table 13.7.

Table 13.7 The order of $q = 2$ modulo odd primes r with $r \le 31$.

r	3	5	7	11	13	17	19	23	29	31
$\mathrm{ord}_r(2)$	2	4	3	10	12	8	18	11	28	5

For the remainder of the argument, we distinguish four cases, according to divisibility properties of n.

Case 1: n is even, but not a multiple of 4.

Then $n' = n/2$, and hence $\Omega \le \alpha n + \beta$ for $\alpha = \frac{1}{8}$ and $\beta = \frac{5}{4}$; see Corollary 13.6.11. As $\mathrm{ord}_r(2)$ is divisible by 4 for the primes $r \in \{5,13,17\}$ (see Table 13.7), none of these primes can divide $2^n - 1$, since n is not a multiple of 4. In view of $\kappa(2,\frac{1}{8},19) > 0$, we may then take $\ell = 19$ and $\Lambda = \{3,7,11\}$. With this specification of the necessary parameters, Inequality (13.23) holds for all $n \ge 18$. This leaves only the cases $n = 10$ and $n = 14$, for which we refer to Table 13.8.

Case 2: n is a multiple of 4.

Here Corollary 13.6.11 gives $\Omega \le \alpha n + \beta$ with $\alpha = \frac{1}{16}$ and $\beta = \frac{5}{4}$. Because of $\kappa(2,\frac{1}{16},11) > 0$, we may take $\ell = 11$ and $\Lambda = \{3,5,7\}$. With these choices,

Inequality (13.23) is satisfied for $n \geq 16$. This leaves only the case $n = 12$; again, we refer to Table 13.8 for this case.

Case 3: n is an odd prime.

We may assume $n \geq 17$, as the cases $n \in \{5, 7, 11, 13\}$ have already been dealt with at the beginning of the proof. Note that Corollary 13.6.11 now only gives us $\alpha = \frac{1}{4}$ in the Ω-bound, as n is odd. Nevertheless, we can obtain a better value for α by the following simple direct argument. As n is a prime, we have $x^n - 1 = (x-1) \cdot \Phi_n$, where Φ_n splits over F into irreducible factors of degree $\mathrm{ord}_n(2)$ each, by Proposition 3.6.16. Trivially, $\mathrm{ord}_n(2) \geq 5$ (as $n \geq 17$), and therefore $x^n - 1$ has at most $1 + (n-1)/5$ irreducible factors. Thus we may take $\alpha = \frac{1}{5}$ and $\beta = \frac{4}{5}$ in this case, as well as $\ell = 23$ and $\Lambda = \emptyset$ (according to Table 13.7). With these choices, Inequality (13.23) holds for all $n \geq 11$, which settles this case completely.

Case 4: n is odd, but not a prime.

Here we use the values $\alpha = \frac{1}{4}$ and $\beta = \frac{5}{4}$ provided by Corollary 13.6.11. According to Table 13.7, we may choose $\ell = 31$ and $\Lambda = \{7, 23\}$. Then Inequality (13.23) holds for all $n \geq 37$, and we are left with the cases $n \in \{9, 15, 21, 25, 27, 33, 35\}$, three of which are settled in Table 13.8.

It remains to deal with the ten exceptional values of n which we have encountered. Applying Corollary 13.5.4 directly, the data in Table 13.8 establish $PN_n(2) > 0$ for six of these ten cases, namely for $n \in \{9, 10, 12, 14, 21, 25\}$; we have also included $n = 15$ in Table 13.8, to demonstrate that Corollary 13.5.4 fails for this case. The remaining four extension degrees $n \in \{15, 27, 33, 35\}$ will be discussed individually in more detail.

Table 13.8 Evaluation of Corollary 13.5.4 for $q = 2$ and $n \in \{9, 10, 12, 14, 15, 21, 25\}$.

n	$2^n - 1$	$\omega = \omega(2^n - 1)$	$\Omega = \omega_2(x^n - 1)$	$(2^\omega - 1) \cdot (2^\Omega - 1)$	$\sqrt{2^n}$
9	$7 \cdot 73$	2	3	21	> 22
10	$3 \cdot 11 \cdot 31$	3	2	21	32
12	$3^2 \cdot 5 \cdot 7 \cdot 13$	4	2	45	64
14	$3 \cdot 43 \cdot 127$	3	3	49	128
15	$7 \cdot 31 \cdot 151$	3	5	217	< 182 *
21	$7^2 \cdot 127 \cdot 337$	3	6	441	> 1448
25	$31 \cdot 601 \cdot 1801$	3	3	49	> 5792

Checking the data in Table 13.8 is rather easy: one can derive the required factorizations with only moderate effort (even by hand calculation, if so desired). For instance, for $n = 21$

$$2^{21} - 1 = (2^7 - 1) \cdot (2^{14} + 2^7 + 1) = 127 \cdot 16513;$$

observing that $2^3 - 1 = 7$ divides $(2^{21} - 1)/(2^7 - 1)$ already leads to $2^{21} - 1 = 127 \cdot 7 \cdot 2359 = 127 \cdot 7^2 \cdot 337$. Similarly, for $n = 25$,

$$\begin{aligned}2^{25}-1 &= \Phi_5(2)\cdot\Phi_{25}(2) = \Phi_5(2)\cdot\Phi_5(2^5)\\ &= (2^4+2^3+2^2+2+1)\cdot(2^{20}+2^{15}+2^{10}+2^5+1)\end{aligned}$$

gives the partial factorization $31 \cdot 1082401$, where 1082401 is not divisible by 31. If r is any prime divisor of 1082401, then $\mathrm{ord}_r(2) = 25$ divides $r-1$. Hence $r \equiv 1 \bmod 25$, and since r is odd, even $r \equiv 1 \bmod 50$. Using this, one quickly reaches $r = 601$ as a prime divisor of $2^{25}-1$, resulting in the factorization given in Table 13.8.

As noted above, Corollary 13.5.4 fails for $n = 15$. For this case, we can apply Proposition 13.1.2 instead. With

- $\phi(2^{15}-1) = 6\cdot 30\cdot 150 = 27000$,
- $\phi_2(x^{15}-1) = (2^1-1)\cdot(2^2-1)\cdot(2^4-1)\cdot(2^4-1)^2 = 10125$,
- $\sum_{d|15}\mu\left(\frac{n}{d}\right)\cdot 2^d = 2^{15}-2^3-2^5+1 = 32729$,

we obtain that $PN_{15}(2)$ is indeed positive:

$$PN_{15}(2) \geq \phi(2^{15}-1)+\phi_2(x^{15}-1)-\sum_{d|15}\mu\left(\tfrac{n}{d}\right)\cdot 2^d = 4396.$$

For the final three cases $n \in \{27,33,35\}$, we can again apply Corollary 13.5.4 directly to show $PN_n(2) > 0$. For $n = 27$, we use the factorization

$$x^{27}-1 = (x-1)\cdot\Phi_3(x)\cdot\Phi_3(x^3)\cdot\Phi_3(x^9),$$

which gives

$$2^{27}-1 = (2^2+2+1)\cdot(2^6+2^3+1)\cdot(2^{18}+2^9+1) = 7\cdot 73\cdot 262657.$$

Every prime divisor r of $2^{18}+2^9+1 = 262657$ satisfies $r \equiv 1 \bmod 54$ (since $27 = \mathrm{ord}_r(2) \mid r-1$, as $\gcd(262657, 2^9-1) = 1$, and since r is odd). Hence $r \geq 109$. Note that 262657 has at most two distinct prime divisors, as $109^3 > 262657$, which shows $2^\omega - 1 \leq 15$.[8] Furthermore, $\Omega = \omega_2(x^{27}-1) = 4$, since $\mathrm{ord}_{3^k}(2) = 2\cdot 3^{k-1} = \phi(3^k)$ for all $k \in \mathbb{N}^*$. Hence $(2^\omega-1)(2^\Omega-1) \leq 15^2 = 225 < \sqrt{2^{27}}$, as required.

Similarly, for the case $n = 33$, we calculate

$$2^{33}-1 = (2^{11}-1)\cdot(2^{22}+2^{11}+1) = 2047\cdot 4196353,$$

where $2047 = 23\cdot 89$. Also, $2^3-1 = 7$ divides 4196353, whereas the resulting quotient 599479 is relatively prime to $7\cdot 23\cdot 89$. Therefore, any prime r dividing 599479 has to satisfy $r \equiv 1 \bmod 66$, so that $r \geq 67$. Hence 599479 has at most three distinct prime divisors (as $67^4 > 599479$),[9] which yields $\omega = \omega(2^{33}-1) \leq 6$. Moreover, $\Omega = \omega_2(x^{33}-1) = 1+1+1+2 = 5$. This shows $(2^\omega-1)(2^\Omega-1) \leq 63\cdot 31 = 1953 < \sqrt{2^{33}}$, as desired.

[8] In fact, 262657 is a prime, and hence $2^\omega - 1 = 7$, but the simple estimate suffices for our purposes.

[9] In fact, 599479 is also a prime.

Finally, let $n = 35$. Then

$$2^{35} - 1 = (2^7 - 1) \cdot (2^{28} + 2^{21} + 2^{14} + 2^7 + 1) = 127 \cdot 270549121.$$

Since $31 = 2^5 - 1$ is a divisor of $2^{35} - 1$, we get $270549121 = 31 \cdot 8727391$. In fact, $8727391 = \Phi_{35}(2)$, hence every prime divisor r of 8727391 satisfies $r \equiv 1 \bmod 70$, so that $r \geq 71$. We obtain $8727391 = 71 \cdot 122921$, and 122921 is neither divisible by 71 nor by 211 (observe that 141 is not a prime). Since 211^3 is greater than 122921, the latter can have at most two distinct prime divisors,[10] and hence $\omega \leq 5$. Also,

$$\Omega = \omega_2(x^{35} - 1) = 1 + \frac{\phi(5)}{\mathrm{ord}_5(2)} + \frac{\phi(7)}{\mathrm{ord}_5(2)} + \frac{\phi(35)}{\mathrm{ord}_{35}(2)} = 1 + \frac{4}{4} + \frac{6}{3} + \frac{24}{12} = 6.$$

This yields the required estimate $(2^\omega - 1)(2^\Omega - 1) \leq 31 \cdot 63 = 1953 < \sqrt{2^{35}}$ for this last special case, concluding the proof. □

Theorem 13.7.2. *With the possible exception of $n = 8$, every extension $E = \mathrm{GF}(3^n)$ of the ternary field $F = \mathrm{GF}(3)$ contains a primitive element for E which is normal over F.*

Proof. The proof is very similar to that for the binary case given in Theorem 13.7.1. Again, we use the approach outlined at the beginning of this section and distinguish four cases, according to divisibility properties of n. We will leave some details to the reader, for instance, checking that $\kappa(3, \alpha, \ell)$ is positive, and the corresponding evaluation of Equation (13.23). This time, we will require the values $\mathrm{ord}_r(3)$ for small primes r, see Table 13.9.

Table 13.9 The order of $q = 3$ modulo primes r with $5 \leq r \leq 43$.

r	5	7	11	13	17	19	23	29	31	37	41	43
$\mathrm{ord}_r(3)$	4	6	5	3	16	18	11	28	30	36	8	42

Case 1: n is a multiple of 3.

Then the 3-free part n' of n divides $n/3$, and hence Inequality (13.19) implies that $\Omega \leq \frac{1}{6}n + 1$, regardless of the exact value and the parity of n'. Thus we may choose $\alpha = \frac{1}{6}$ and $\beta = 1$, and then $\ell = 11$ and $\Lambda = \{2, 5, 7\}$. With these choices, Inequality (13.23) holds for all $n \geq 14$. In view of Theorems 13.6.2 and 13.6.6, this leaves only the case $n = 12$, which will be dealt with in Table 13.10.

Case 2: n is odd and not a multiple of 3.

Here Inequality (13.20) gives $\Omega \leq \alpha n + \beta$ with $\alpha = \frac{1}{3}$ and $\beta = \frac{1}{2} + \frac{1}{6} = \frac{2}{3}$. Letting $\ell = 47$, the data in Table 13.9 show that $\Lambda = \{2, 11, 23\}$ is a suitable choice. With these values of the parameters, Inequality (13.23) holds for all $n \geq 12$, leaving the cases $n \in \{5, 7, 11\}$.

The extension degrees $n = 5$ and $n = 7$ are easily settled by applying Corollary 13.5.4 directly. In both cases, we have $\omega = 2 = \Omega$, as

[10] In fact, 122921 is also a prime.

- $3^5 - 1 = 2 \cdot 11^2$ and $\mathrm{ord}_5(3) = 4$; and
- $3^7 - 1 = 2 \cdot 1093$ and $\mathrm{ord}_7(3) = 6$.

This gives the validity of the required estimate (13.14), as $(2^2 - 1)^2 = 9 < \sqrt{3^n}$ for $n \geq 5$. Corollary 13.5.4 may also be applied for $n = 11$: every prime divisor r of $(3^{11} - 1)/2 = 88573$ satisfies $r \equiv 1 \bmod 22$. This shows $88573 = 23 \cdot 3851$, and $23^3 > 3851$ implies $\omega \leq 4$.[11] Since $\Omega = 3$, we have $(2^\omega - 1) \cdot (2^\Omega - 1) \leq 15 \cdot 7 = 105$, which is less than $\sqrt{3^{11}}$.

Case 3: n is not a multiple of 3, and $n \equiv 2 \bmod 4$.

Now (13.20) gives $\Omega \leq \frac{1}{3}n + \beta$ with $\beta = \frac{4}{3}$, as $\gcd(n, q-1) = \gcd(n, 2) = 2$ and $\gcd(n, q^2 - 1) = \gcd(n, 8) = 2$ in this case. With $\ell = 47$ and $\Lambda = \{2, 11, 23\}$, Inequality (13.23) is satisfied for all $n \geq 16$. The remaining cases $n = 10$ and $n = 14$ are covered by Table 13.10.

Case 4: n is not a multiple of 3, and $n \equiv 0 \bmod 4$.

Here $\gcd(n, q-1) = \gcd(n, 2) = 2$ and $\gcd(n, q^2 - 1) = \gcd(n, 8) \leq 8$ lead to $\Omega \leq \frac{1}{3}n + \beta$, where $\beta = \frac{7}{3}$. With $\ell = 41$ and $\Lambda = \{2, 5, 11, 17, 23, 29\}$, Inequality (13.23) holds for all $n \geq 28$. This leaves the three cases $n \in \{8, 16, 20\}$, which are included in Table 13.10.

Table 13.10 Evaluation of Corollary 13.5.4 for $q = 3$ and $n \in \{8, 10, 12, 14, 16, 20\}$.

n	$3^n - 1$	$\Omega = \omega_3(x^n - 1)$	$(2^\omega - 1) \cdot (2^\Omega - 1)$	$\sqrt{3^n}$
8	$2^5 \cdot 5 \cdot 41$	5	217	27 *
10	$2^3 \cdot 11^2 \cdot 61$	4	105	243
12	$2^4 \cdot 5 \cdot 7 \cdot 13 \cdot 73$	3	217	729
14	$2^3 \cdot 547 \cdot 1093$	4	105	2187
16	$2^6 \cdot 5 \cdot 17 \cdot 41 \cdot 193$	7	3937	6561
20	$2^4 \cdot 5^2 \cdot 11^2 \cdot 61 \cdot 1181$	7	3937	59049

For the remaining six cases $n \in \{8, 10, 12, 14, 16, 20\}$, Table 13.10 shows that Corollary 13.5.4 only fails when $n = 8$. As in the binary situation, we briefly comment on the required factorizations:

- the factorization of $3^{10} - 1$ is obtained using $3^5 - 1 = 246 = 2 \cdot 11^2$ and $3^5 + 1 = 244 = 2^2 \cdot 61$;
- the factorization of $3^{20} - 1 = (3^{10} - 1) \cdot (3^{10} + 1)$ is obtained from the previous one, using $3^{10} + 1 = 59050 = 2 \cdot 5^2 \cdot 1181$;
- for $n = 8$, one uses $3^8 - 1 = (3 - 1) \cdot (3 + 1) \cdot (3^2 + 1) \cdot (3^4 + 1)$;
- then the factorization of $3^{16} - 1 = (3^8 - 1) \cdot (3^8 + 1)$ follows from $3^8 + 1 = 6562 = 2 \cdot 3281$, observing that $3281 = 17 \cdot 193$;
- for $n = 12$, one uses $3^{12} - 1 = (3^3 - 1) \cdot (3^3 + 1) \cdot (3^6 + 1) = 26 \cdot 28 \cdot 730$;
- the factorization of $3^7 - 1$ given in Case 2 leads to that of $3^{14} - 1$, as $3^7 + 1 = 4 \cdot 547$.

[11] In fact, 3851 is a prime.

This completes the proof of Theorem 13.7.2. □

We finally remark that Proposition 13.1.2 fails for the one remaining pair $(q,n) = (3,8)$; this instance will be treated in Section 13.10, using a different type of argument.

Exercises

Exercise 13.7.3. Verify that Proposition 13.1.2 fails for the pair $(q,n) = (3,8)$. □

13.8 Improved Existence Criteria for Primitive Normal Bases

In the present section, we consider ways of improving the fundamental sufficient condition used up to now, namely Corollary 13.5.4. Although our preceding results indicate that this criterion is quite efficient, it tends to fail for some smaller values of q and n; in particular, this is the case when $(q,n) \in \{(4,6),(5,6),(7,6),(3,8)\}$, the four instances which we had to leave unresolved in Theorems 13.6.2 and 13.7.1.

The basic idea is simple: try and find normal elements which are not primitive themselves, but can be used to construct a primitive normal element. More technically, determine proper divisors Δ of $q^n - 1$ for which $\mathscr{P}_\Delta(u) \cdot \mathscr{N}_{x^n-1}(u) = 1$ for some $u \in \mathrm{GF}(q^n)$ already suffices to conclude $PN_n(q) > 0$. Of course, this will only help if it leads to a sufficient criterion for the existence of such an element u which is less restrictive than the one provided by Corollary 13.5.4.[12]

The first (rather obvious) attempt to find a suitable divisor Δ for this approach goes back to the work of Davenport [101]. Let

$$D = D(q,n) := \frac{q^n - 1}{q-1}, \tag{13.24}$$

and write $q^n - 1 = k \cdot \overline{D}$, where k and D are relatively prime, that is, $\overline{D} = \mathrm{pt}_D(q^n - 1)$; thus $k = (q^n - 1)/\mathrm{pt}_D(q^n - 1)$ divides $q-1$.

Now suppose that $u \in \mathrm{GF}(q^n)$ satisfies $\mathscr{P}_D(u) \cdot \mathscr{N}_{x^n-1}(u) = 1$, so that u is a normal element for $E = \mathrm{GF}(q^n)$ over $F = \mathrm{GF}(q)$ and $\overline{D}$ divides $\mathrm{ord}(u)$. Trivially, λu is then likewise normal for every $\lambda \in F^*$, and it is easy to see that a suitable choice of λ with $\mathrm{ord}(\lambda) \mid k$ will result in a primitive element λu of E. Thus the condition

$$(2^{\omega(D)} - 1)(2^{\omega_q(x^n-1)}) < \sqrt{q^n}$$

already implies that $PN_n(q)$ is positive.

Obviously, this approach can only lead to an improvement of Corollary 13.5.4 if $k \neq 1$, that is, if $\mathrm{rad}(q-1) \nmid \mathrm{rad}(q^n - 1)/(q-1)$. Note that this condition cannot be

[12] It should be noted that this requires the existence of some prime divisor r of $(q^n - 1)/\Delta$ which does not divide Δ, as otherwise $\mathrm{pt}_\Delta(q^n - 1) = q^n - 1$ and hence $\mathscr{P}_\Delta = \mathscr{P}_{q^n-1}$.

satisfied whenever $\mathrm{rad}(q-1)$ divides n, as any prime dividing both $q-1$ and n also divides $q^{n-1}+\cdots+q+1$. Unfortunately, this holds in each of the four open cases above, so that Davenport's idea does not settle any of these cases.

Nevertheless, even this simple approach can produce interesting results, as the following example shows.

Example 13.8.1. Consider the pair $(q,n)=(4,5)$. Then

$$4^5-1 = 3\cdot\frac{4^5-1}{4-1} = 3\cdot 341 = 3\cdot(11\cdot 31),$$

so that $\omega=3$ and $\omega(D)=2$. Moreover, $\Omega=3$. Thus Corollary 13.5.4 fails, since $(2^3-1)\cdot(2^3-1)=49>32=\sqrt{4^5}$. On the other hand, $(2^2-1)\cdot(2^3-1)=21<32$, and hence $PN_5(4)$ is indeed positive.[13] □

A considerable improvement compared to Davenport's choice of Δ as the quotient D above is due to Lenstra and Schoof [238]. This rests on observing that $\gcd(q-1,D)=\gcd(q-1,n)$, which follows from $q^{n-1}+\cdots+q+1\equiv n \bmod d$ for every divisor d of $q-1$. Therefore, we put

$$C=C(q,n) := \frac{D}{\gcd(q-1,D)} = \frac{q^n-1}{(q-1)\cdot\gcd(q-1,n)}. \tag{13.25}$$

For simplicity, we also introduce the abbreviation

$$\delta=\delta(q,n) := (q-1)\cdot\gcd(q-1,n). \tag{13.26}$$

Finally, in analogy to the definition of $\overline{D}$ above, we write $q^n-1=\ell\cdot\overline{C}$, where ℓ and C are relatively prime, that is, $\overline{C}=\mathrm{pt}_C(q^n-1)$.

Proposition 13.8.2. *Consider the field extension $E=\mathrm{GF}(q^n)$ over $F=\mathrm{GF}(q)$ corresponding to the pair (q,n). Assume that there exists an element $w\in E$ which is normal over F and such that $\overline{C}=\mathrm{pt}_C(q^n-1)$ divides $\mathrm{ord}(w)$, where $C=C(q,n)$ is as in Equation (13.25). Then there exists a primitive normal element for E/F.*

The proof of this result relies on two auxiliary assertions, both of which are interesting in their own right. We begin with the following algebraic characterization of the number $\delta=\delta(q,n)$ defined in (13.26):

Step 1. The subgroup U_δ of order δ of E^* consists of the elements $\lambda\in E^*$ satisfying $\lambda^{q-1}\in F^*$.

Proof. Let $\gamma\in E^*$ such that $\gamma^{q-1}\in F^*$. Then $\gamma^{(q-1)^2}=1$, and hence $\mathrm{ord}(\gamma)$ divides

$$\gcd\left((q-1)^2,q^n-1\right) = (q-1)\cdot\gcd(q-1,D) = (q-1)\cdot\gcd(q-1,n) = \delta,$$

by the definition of D and δ. Thus $\gamma\in U_\delta$.

[13] Alternatively, this can also be derived from Proposition 13.1.2, as $\phi(4^5-1)+\phi_4(x^5-1)=2\cdot 10\cdot 30+(4-1)\cdot(4^2-1)^2=1275>1020=4^5-4$.

Conversely, let $\alpha \in U_\delta$. Then the order of α^{q-1} divides $\delta/(q-1) = \gcd(q-1,n)$, so that $\alpha^{q-1} \in F^*$.

Step 2. The subgroup U_δ of E^* acts on the set of all normal elements of E/F by multiplication.

Proof. Let $\lambda \in U_\delta$ and assume that $u \in E$ is a normal element over F. In order to show that λu is likewise normal over F, we consider a polynomial $f(x) \in F[x]$ with $\deg f < n$ which annihilates λu, that is, $f(\sigma)(\lambda u) = 0$. Let $f(x) = \sum_{i=0}^{m-1} f_i x^i \in F[x]$. Then

$$0 = \sum_{i=0}^{m-1} f_i(\lambda u)^{q^i} = \lambda \cdot \Big(\sum_{i=0}^{m-1} (f_i \lambda^{q^i-1}) u^{q^i} \Big),$$

and therefore $f_0 u + \sum_{i=1}^{m-1} (f_i \lambda^{q^i-1}) u^{q^i} = 0$. As $q-1$ divides $q^i - 1$ for every $i \geq 1$, Step 1 shows $f_i \lambda^{q^i-1} \in F$ for all i. Since u is normal over F and $\deg f < n$, we obtain $f_i \lambda^{q^i-1} = 0$ and therefore $f_i = 0$ for all i. Thus λu is indeed normal over F.

Step 3. Now it is easy to conclude the proof of Proposition 13.8.2. Assume that $w \in E$ is normal over F and that $\overline{C}$ divides $\operatorname{ord}(w)$. Then λw is a primitive element of E for some suitable $\lambda \in U_\ell$, where ℓ is the cofactor of $\overline{C}$ in $q^n - 1$. Since U_ℓ is a subset of U_δ, we also have that w is a normal element for E over F. □

Using the notation introduced in Equations (13.9) and (13.10), we conclude that $\sum_{u \in E} \mathscr{P}_C(u) \mathscr{N}_{x^n-1}(u) \neq 0$ already implies $\sum_{u \in E} \mathscr{P}_{q^n-1}(u) \mathscr{N}_{x^n-1}(u) \neq 0$. This gives the following improvement of Corollary 13.5.4:

Corollary 13.8.3. *Let q be a prime power and $n \in \mathbb{N}^*$. Define $C = C(q,n)$ as in Equation* (13.25)*, and assume that*

$$(2^{\omega(C)} - 1)(2^{\omega_q(x^n-1)} - 1) < \sqrt{q^n}.$$

Then there exists a primitive element for $\mathrm{GF}(q^n)$ *which is normal over* $\mathrm{GF}(q)$. □

Unfortunately, Corollary 13.8.3 gives no improvement over Corollary 13.5.4 when $q = 2$ or when $q = 3$ and n is even. Thus the extensive computations carried out in the last section have not been redundant.

In particular, the case $(q,n) = (3,8)$ still remains open. Nevertheless, we can try and apply Corollary 13.8.3 to the remaining three pairs $(q,n) \in \{(4,6),(5,6),(7,6)\}$; see Table 13.11 and compare with Table 13.6. This approach succeeds for the two pairs $(4,6)$ and $(5,6)$, but not for the pair $(7,6)$.

Table 13.11 Evaluation of Corollary 13.8.3 for $n = 6$ and $q \in \{4,5,7\}$.

q	C	$\omega(C)$	Ω	$(2^{\omega(C)}-1)\cdot(2^{\Omega}-1)$	$\sqrt{q^6}$
4	$455 = 5\cdot 7\cdot 13$	3	3	49	64
5	$1953 = 3^2\cdot 7\cdot 31$	3	4	105	125
7	$3268 = 2^2\cdot 19\cdot 43$	3	6	441	343 *

We now use Corollary 13.8.3 to settle the case of extensions of prime degree:

Theorem 13.8.4. *Let q be a prime power and n a prime. Then there exists a primitive element for* $\mathrm{GF}(q^n)$ *which is normal over* $\mathrm{GF}(q)$.

Proof. Because of Proposition 13.1.1 and Theorems 13.2.1 and 13.6.6, we may assume that the prime n is at least 5 and not the characteristic p of $\mathrm{GF}(q)$. By Theorems 13.7.1 and 13.7.2, we may also assume $q \geq 4$.

In order to apply Corollary 13.8.3, we need to investigate the prime divisors r of $C = C(q,n)$. We begin by deriving severe restrictions on the possible values of r, depending on the specified value of n. Obviously, we have $\mathrm{ord}_r(q) \in \{1,n\}$, since r divides $q^n - 1$ and n is a prime. We claim that the case $\mathrm{ord}_r(q) = 1$ cannot occur. Assume otherwise. Then r also divides $q-1$ and

$$r \mid \gcd(q-1,C) \mid \gcd(q-1,D) = \gcd(q-1,n),$$

where $D = D(q,n)$ is as above. As n is a prime, this gives $\gcd(q-1,n) = n = r$, which leads to a contradiction as follows: by Equation (13.25), r^2 would divide $D = (q^r - 1)/(q-1)$; however, this is impossible in view of part (1) of Lemma 1.7.3. Thus $\mathrm{ord}_r(q) = n$, as claimed. Since $\mathrm{ord}_r(q) \mid r-1$, we conclude $r \equiv 1 \bmod n$ and then even $r \equiv 1 \bmod 2n$ (as n and r are odd). In particular, $r \geq 11$ when $n = 5$, and $r \geq 23$ for $n \geq 7$.

We can use these restrictions on r to give a strong improvement of the bound

$$2^{\omega(C)} \leq 5 \cdot \sqrt[4]{C}$$

obtained in Example 13.6.4 (with $N := C$). For this, we simply go through the argument given there and note that the set $I_\varepsilon = P_\varepsilon \cap \pi(C)$ (with $\varepsilon = \frac{1}{4}$, as before) is now empty for $n \geq 7$, and the singleton $\{11\}$ for $n = 5$. Using Remark 13.6.5 (for $n \geq 7$) and the observation $2/\sqrt[4]{11} < \frac{11}{10}$ (for $n = 5$), we obtain

$$2^{\omega(C)} < \gamma \cdot \sqrt[4]{C} \quad \text{with } \gamma = \begin{cases} \frac{11}{10} & \text{if } n = 5, \\ 1 & \text{if } n \geq 7. \end{cases}$$

We will now use this improved bound in Corollary 13.8.3, together with the trivial estimate

$$C = \frac{D(q,n)}{s} = \frac{1}{s} \cdot \frac{q^n - 1}{q-1} < \frac{2}{s} \cdot q^{n-1},$$

where we write $s := \gcd(q-1,n)$.

Let us first consider the case $n = 5$. Then $\sqrt[4]{q^{n-1}} = q$, and we obtain

$$\begin{aligned}(2^{\omega(C)} - 1) \cdot (2^{\Omega} - 1) &< (2^{\Omega} - 1) \cdot \frac{11}{10} \cdot \sqrt[4]{C} \\ &< (2^{\Omega} - 1) \cdot \frac{11}{10} \cdot \sqrt[4]{\frac{2}{s}} \cdot q,\end{aligned}$$

which should be bounded by $\sqrt{q^5}$ if we want to apply Corollary 13.8.3. We now distinguish two cases:

- Assume first that $q \equiv 1 \bmod 5$. Then $\Omega = 5$ and $s = 5$, and the sufficient condition becomes
$$31 \cdot \frac{11}{10} \cdot \sqrt[4]{\frac{2}{5}} \le \sqrt{q^3},$$
which always holds (as $q \ge 11$ in this case).
- Now let $q \not\equiv 1 \bmod 5$. Then $\Omega \le 3$ and $s = 1$, and we obtain the sufficient condition
$$7 \cdot \frac{11}{10} \cdot \sqrt[4]{2} \le \sqrt{q^3},$$
which holds for all $q \ge 5$.

As we already know $PN_5(4) > 0$ from Example 13.8.1, this establishes the primitive normal basis theorem for $n = 5$.

Now let $n \ge 7$. A quite rough preliminary estimate (where we use the trivial bound $\Omega \le n$ and replace $2/s$ with 2 in the bound for C given above) yields

$$(2^{\omega(C)} - 1) \cdot (2^{\Omega} - 1) < 2^n \cdot \sqrt[4]{C} < 2^n \cdot \sqrt[4]{2} \cdot \sqrt[4]{q^{n-1}},$$

which is bounded by $\sqrt{q^n}$ if and only if $2^{4n+1} \le q^{n+1}$. Trivially, this condition is satisfied for all $q \ge 16$ (independent of n). Therefore, it only remains to investigate the seven prime powers $q \in \{4, 5, 7, 8, 9, 11, 13\}$.

As $q = 8$ is the only instance where $q - 1$ has a prime divisor $r \ge 7$, we begin by examining the particular pair $(q, n) = (8, 7)$. Using the factorization of $8^7 - 1 = 2^{21} - 1$ already given in Table 13.8, we have $C = (8^7 - 1)/7^2 = 127 \cdot 337$ and therefore $\omega(C) = 2$. Because of $\Omega = 7$, this gives

$$(2^{\omega(C)} - 1) \cdot (2^{\Omega} - 1) = 381 < 512 = \sqrt{8^6},$$

and hence Corollary 13.8.3 applies for this pair.

In all other cases, $\mathrm{ord}_n(q) \ge 2$ and therefore $\omega_q(x^n - 1) \le 1 + \frac{n-1}{2}$. Using a similar (rough) estimate as before, we now obtain

$$(2^{\omega(C)} - 1) \cdot (2^{\Omega} - 1) < \sqrt{2^{n+1}} \cdot \sqrt[4]{2} \cdot \sqrt[4]{q^{n-1}},$$

which is bounded by $\sqrt{q^n}$ if and only if $2^{2n+3} \le q^{n+1}$. This holds for all $q \ge 5$, as $n \ge 7$.

Thus we are left with the case $q = 4$, where we need a better estimate. Note that the powers $4^c - 1$ with $c \in \{1, 2, 3\}$ only admit the prime divisors 3, 5 and 7. Therefore, $\mathrm{ord}_n(4) \ge 4$ provided that $n \ge 11$. Then $\Omega \le 1 + \frac{n-1}{4}$, and we get the improved estimate

$$(2^{\omega(C)} - 1) \cdot (2^{\Omega} - 1) < \sqrt[4]{2^{n+3}} \cdot \sqrt[4]{2} \cdot \sqrt[4]{4^{n-1}} = \sqrt[4]{2^{3n+2}},$$

which is bounded by $\sqrt{4^n} = 2^n$ for all $n \neq 1$.

Finally, it remains to consider the case $(q,n) = (4,7)$. The factorization of $4^7 - 1 = 2^{14} - 1$ is contained Table 13.8 and gives $C = 43 \cdot 127$, hence $\omega(C) = 2$. Since $\mathrm{ord}_7(4) = 3$, we have $\Omega = 3$, so that

$$(2^{\omega(C)} - 1)(2^{\Omega} - 1) = 3 \cdot 7 = 21 < 64 = \sqrt{4^6}.$$

Thus Corollary 13.8.3 applies, which finishes the proof. □

Next, we apply Corollary 13.8.3 to settle the unpleasant case $q \equiv 1 \bmod n$, where $x^n - 1$ splits into distinct linear factors over $\mathrm{GF}(q)$, so that the trivial bound $\Omega = \omega_q(x^n - 1) = n$ is reached.

Theorem 13.8.5. *Let q be a prime power and n a divisor of $q-1$. Then there exists a primitive element for $\mathrm{GF}(q^n)$ which is normal over $\mathrm{GF}(q)$, with the possible exception of the case $(q,n) = (7,6)$.*

Proof. By Proposition 13.6.13, the assertion holds whenever $n \geq 16$, since n divides $q-1$ and therefore $\min(q,n) = n$. In view of Proposition 13.1.1 and Theorems 13.2.1, 13.3.1, 13.6.2 and 13.8.4, it remains to consider the six degrees $n \in \{8,9,10,12,14,15\}$.

By the proof of Proposition 13.6.13, the inequality

$$\log_2(q) \geq 3 + \frac{63}{4n} \tag{13.27}$$

is a sufficient condition for the existence of a primitive normal element in the corresponding field extension. The resulting lower bounds for q are listed in Table 13.12.

Table 13.12 Evaluation of Equation (13.27) for $n \subset \{8,9,10,12,14,15\}$.

n	8	9	10	12	14	15
$q \geq$	32	27	25	23	19	17

This leaves just seven remaining pairs with $q \equiv 1 \bmod n$, namely

$$(q,n) \in \{(9,8),\ (17,8),\ (25,8),\ (19,9),\ (11,10),\ (13,12),\ (16,15)\}.$$

We will investigate these cases by applying Lemma 13.6.12 to $\omega(C)$ instead of $\omega(q^n - 1)$. This gives (for appropriate choices of ℓ and Λ)

$$\begin{aligned} \omega(C) &\leq \frac{\ln(C)}{\ln(\ell)} + |\Lambda| - \frac{\ln(L)}{\ln(\ell)} \\ &< \frac{n\ln(q)}{\ln(\ell)} + |\Lambda| - \frac{\ln(L \cdot (q-1) \cdot n)}{\ln(\ell)}, \end{aligned}$$

as $\gcd(q-1,n) = n$ by hypothesis. Using the trivial bound $\Omega = \alpha n + \beta$ with $(\alpha,\beta) = (1,0)$, the approach introduced at the beginning of Section 13.7 then shows that

$$|\Lambda| - \frac{\ln(L \cdot (q-1) \cdot n)}{\ln(\ell)} \leq \left(\left(\tfrac{1}{\ln(4)} - \tfrac{1}{\ln(\ell)}\right)\ln(q) - 1\right) \cdot n \tag{13.28}$$

is a sufficient condition for the existence of a primitive normal element. This allows us to settle five of the seven pairs listed above; the computational details are left to the reader.

- For $q = 25$ and $n = 8$, condition (13.28) is satisfied with $\ell = 17$ and $\Lambda = \{2,3,7,11,13\}$.
- For $q = 17$ and $n = 8$, it is satisfied with $\ell = 19$ and $\Lambda = \{3,5,7,11,13\}$ (observe that C is odd in this case).
- For $q = 19$ and $n = 9$, it holds with $\ell = 23$ and $\Lambda = \{2,5,7,11,13,17\}$ (observe that C is not divisible by 3 in this case).
- For $q = 13$ and $n = 12$, it is satisfied with $\ell = 23$ and $\Lambda = \{5,7,11,17,19\}$ (observe that C is odd and not divisible by 3 in this case).
- For $q = 16$ and $n = 15$, it holds with $\ell = 19$ and $\Lambda = \{7,11,13\}$ (here C is odd, not divisible by 3 or 5, and also not divisible by 17, since $\mathrm{ord}_{13}(2) = 12$).

For the remaining two cases $(q,n) = (9,8)$ and $(q,n) = (11,10)$, we verify the condition in Corollary 13.8.3 directly:

- The factorization of $9^8 - 1 = 3^{16} - 1$ given in Table 13.10 shows that $C(9,8) = (9^8-1)/8^2 = 5 \cdot 17 \cdot 41 \cdot 193$, which gives $\omega(C) = 4$. This yields the required inequality

$$(2^{\omega(C)} - 1) \cdot (2^{\Omega} - 1) = 15 \cdot 255 = 3825 < 6561 = \sqrt{9^8}.$$

- Finally,

$$C(11,10) = \frac{\Phi_2(11) \cdot \Phi_5(11) \cdot \Phi_{10}(11)}{10} = \frac{\Phi_2(11) \cdot \Phi_5(11) \cdot \Phi_5(-11)}{10}$$

gives $C(11,10) = \frac{12 \cdot 16105 \cdot 13421}{10} = 2 \cdot 3 \cdot 3221 \cdot 13421$. Each prime divisor of 3221 and of 13421 has to be congruent to 1 modulo 10. Since 3221 is not divisible by 31 and 41, it is a prime number. The number 13421 is also a prime, since it is not divisible by 31, 41, 61, 71 and 101. Hence $\omega(C) = 4$, and therefore

$$(2^{\omega(C)} - 1) \cdot (2^{\Omega} - 1) = 15 \cdot 1023 = 15345 < 161051 = \sqrt{11^{10}}.$$

This completes the proof of Theorem 13.8.5. □

We conclude this section by showing that Corollary 13.8.3 may be improved even further in special situations. The following somewhat technical result is taken from Cohen and Huczynska [91].

Proposition 13.8.6. *Consider the field extension $E = \mathrm{GF}(q^n)$ over $F = \mathrm{GF}(q)$, and let r be any prime divisor of n. Then $R := C(q^{n/r}, r)$ divides $C = C(q,n)$.*

Now assume that R is likewise a prime and that there exists a normal element w for E/F for which $\mathrm{pt}_{C/R}(q^n-1)$ divides $\mathrm{ord}(w)$. Then there also exists a primitive normal element for E/F.

Proof Let $r_0 := \gcd(q^{n/r}-1, r)$. Since r is a prime, both r_0 and $\gcd(q-1,r)$ can only take the values 1 and r. Moreover, $\gcd(q-1,r) = r$ forces $r_0 = r$, too. As $(q-1)\cdot\gcd(q-1,\frac{n}{r})$ divides $q^{n/r}-1$, it follows easily that

$$R = C(q^{n/r},r) = \frac{q^n-1}{(q^{n/r}-1)\cdot r_0} \quad \text{divides} \quad \frac{q^n-1}{(q-1)\cdot\gcd(q-1,n)} = C(q,n).$$

Now assume that R is a prime, and let w be any normal element for E/F for which $\mathrm{pt}_{C/R}(q^n-1)$ divides $\mathrm{ord}(w)$. In view of Proposition 13.8.2, it suffices to show that $\mathrm{ord}(w)$ is then actually a multiple of $\mathrm{pt}_C(q^n-1)$.

We first show an auxiliary result: w cannot be an R-th power in E^*. Suppose otherwise, say $w = u^R$ for some $u \in E^*$. Then

$$w^{r_0} = u^{r_0 R} = \mathrm{Norm}_{E/K}(u) \in K,$$

where $K = \mathrm{GF}(q^{n/r})$. Because of the normality of w for E/F, this excludes the possibility $r_0 = 1$. Hence $r_0 = r$, so that r divides $q^{n/r}-1$, say $q^{n/r}-1 = r\cdot t$. This shows

$$w^{q^{n/r}} = ww^{q^{n/r}-1} = w(w^r)^t = \gamma w,$$

where $\gamma = w^{q^{n/r}-1} \in K$. As $R\cdot(q^{n/r}-1)\cdot r = q^n-1$, we have $\gamma^r = u^{R\cdot(q^{n/r}-1)\cdot r} = 1$. Thus γ is a primitive r-th root of unity, which implies

$$\mathrm{Tr}_{E/K}(w) = \Big(\sum_{j=0}^{r-1}\gamma^j\Big)w = 0.$$

Because of the transitivity of the trace mappings, we obtain $\mathrm{Tr}_{E/F}(w) = 0$, which contradicts the normality of w.

Now it is easy to finish the proof, noting that the R-th powers in E^* form the subgroup U of E^* of order $(q^n-1)/R$. Thus the normal element w does not belong to U, and it follows easily that $\mathrm{pt}_R(q^n-1)$ has to divide $\mathrm{ord}(w)$, since R was assumed to be a prime. (This is the only part of the proof where this hypothesis is needed.) By hypothesis, $\mathrm{pt}_{C/R}(q^n-1)$ also divides $\mathrm{ord}(w)$, and we conclude that $\mathrm{pt}_C(q^n-1)$ indeed divides $\mathrm{ord}(w)$. □

Proposition 13.8.6 immediately yields the following variation of Corollary 13.8.3:

Corollary 13.8.7. *Consider the field extension $E = \mathrm{GF}(q^n)$ over $F = \mathrm{GF}(q)$, and let r be a prime divisor of n for which $R := C(q^{n/r},r)$ is likewise a prime. Let $C = C(q,n)$ and assume that*

$$(2^{\omega(C/R)}-1)(2^{\omega_q(x^n-1)}-1) < \sqrt{q^n}.$$

Then there exists a primitive element for E which is normal over F. □

Unfortunately, it is not easy to apply Corollary 13.8.7, as the following example illustrates:

Example 13.8.8. We try to apply Corollary 13.8.7 to the two pairs still unresolved, starting with $(q,n) = (7,6)$. Here neither of the two primes r dividing n leads to a prime R:

- for $r = 2$, we obtain $R = C(7^3,2) = \frac{7^3+1}{2} = 2^2 \cdot 43$;
- for $r = 3$, we have $R = C(7^2,3) = \frac{7^4+7^2+1}{3} = 817 = 19 \cdot 43$.

For the second open pair $(q,n) = (3,8)$, the situation is similarly disappointing, even though the hypothesis of Corollary 13.8.7 is fulfilled in this case, as the only possible choice $r = 2$ results in the prime $R = C(3^4,2) = (3^4+1)/2 = 41$. However, now the sufficient criterion in Corollary 13.8.7 does not hold: here $\Omega = 5$ and $\omega(C/R) = 2$, as

$$\frac{C}{R} = \frac{C(3,8)}{C(3^4,2)} = \frac{3^8-1}{2 \cdot 2 \cdot 41} = 2^3 \cdot 5,$$

and $(2^2-1) \cdot (2^5-1) = 93 > \sqrt{3^8}$. □

Example 13.8.9. Let us also give an example where Corollary 13.8.7 works nicely. We consider the pair $(q,n) = (5,8)$, which will be important in the next section. Here the prime power factorization of $C = C(5,8)$ is

$$\frac{(5+1) \cdot (5^2+1) \cdot (5^4+1)}{4} = 2 \cdot 3 \cdot 13 \cdot 313,$$

and $\Omega = 1+1+2+2 = 6$. This gives

$$(2^4-1) \cdot (2^6-1) = 15 \cdot 63 = 945 > \sqrt{5^8},$$

so that the sufficient criterion in Corollary 13.8.3 is not satisfied. But Corollary 13.8.7 does apply, with $r = 2$, as

$$R = C(5^4,2) = \frac{5^8-1}{(5^4-1) \cdot 2} = \frac{5^4+1}{2} = 313$$

is a prime. Therefore, $\omega(C/R) = 3$ and now

$$(2^{\omega(C/R)}-1) \cdot (2^{\Omega}-1) = 7 \cdot 63 = 441 < \sqrt{5^8}$$

shows that $PN_8(5)$ is indeed positive. □

Exercises

Exercise 13.8.10. In Example 13.8.1, we have seen that $PN_5(4) > 0$. Give an alternative indirect proof for this result, by first showing that $PN_5(4) = 0$ would imply $\mathrm{ord}(w) \in \{11,31,33,93\}$ for every normal element w of $\mathrm{GF}(4^5)$ over $\mathrm{GF}(4)$. □

Exercise 13.8.11. Determine all pairs (q,n) with $n \in \{2,3,4\}$ which satisfy the condition of Corollary 13.8.3. □

Exercise 13.8.12. Try to determine all pairs (q,n) which satisfy the condition of Corollary 13.8.3, where n is divisible by the characteristic p of $\mathrm{GF}(q)$. □

Exercise 13.8.13. Let $C = C(q,n)$ be as in Equation (13.25). Derive conditions characterizing those prime divisors r of $q-1$ which do not divide C. □

Exercise 13.8.14. Reconsider the class of 6-dimensional extensions of finite fields and the proof of Theorem 13.6.2 in the light of Corollary 13.8.3, trying to avoid computing the factorization of $q^6 - 1$ as far as possible. □

13.9 The Primitive Normal Basis Theorem for $q \geq 4$

We can now complete the proof of the primitive normal basis theorem for $q \geq 4$ (except for the case $(q,n) = (7,6)$, which will still be left open). Because of earlier results in this chapter (namely, Proposition 13.1.1 and Theorems 13.2.1, 13.3.1, 13.6.2, 13.6.6, 13.8.4 and 13.8.5), we may assume that the degree n of the extension considered is at least 8, that $q \not\equiv 1 \bmod n$, and that n is neither a prime nor a power of the characteristic p of $\mathrm{GF}(q)$. Finally, by Proposition 13.6.7 and Theorems 13.7.1 and 13.7.2, we may also assume that q is in the range $4 \leq q \leq 53$.

We will use the method introduced at the beginning of Section 13.7 which allowed us to attack the binary and ternary case successfully. The only difference is that we now apply this approach with the improved existence criterion given in Corollary 13.8.3, that is, we take $\omega := \omega(C)$ (instead of $\omega = \omega(q^n - 1)$ in Section 13.7), where C is as in Equation (13.25):

$$C = C(q,n) = \frac{q^n - 1}{(q-1) \cdot s} \quad \text{with } s = \gcd(q-1,n);$$

note that $\ln(C) < n\ln(q) - \ln(q-1) - \ln(s)$. Using this observation, together with a suitable estimate of the form $\alpha n + \beta$ for $\Omega = \omega_q(x^n - 1)$, then gives the following improved variant of the sufficient condition (13.23) for $PN_n(q) > 0$:

$$n \geq \frac{|\Lambda| - \frac{\ln(L \cdot (q-1) \cdot s)}{\ln(\ell)} + \beta}{\kappa(q,\alpha,\ell)}, \tag{13.29}$$

where we have to assume, as in Section 13.7, that

$$\kappa(q,\alpha,\ell) = \left(\tfrac{1}{\ln(4)} - \tfrac{1}{\ln(\ell)}\right) \cdot \ln(q) - \alpha$$

is positive.

Let us start with a rather crude application of this criterion, which will already allow us to reduce our upper bound for q from 53 to 13. For this, we take $\ell = 67$

and $\Lambda = \pi_\ell \setminus \{p\}$, where $p \in [2,53]$ is the characteristic of $F = \mathrm{GF}(q)$ (as $q \leq 53$), so that $|\Lambda| = 17$. Since $p \leq 53$, we have

$$L = \prod_{r \in \Lambda} r \geq 2 \cdot 3 \cdot 5 \cdot 7 \cdot 11 \cdot 13 \cdot 17 \cdot 19 \cdot 23 \cdot 29 \cdot 31 \cdot 37 \cdot 41 \cdot 43 \cdot 47 \cdot 59 \cdot 61 =: L'.$$

For Ω, we choose the rather weak upper bound given by $\alpha = 3/4$ and $\beta = 0$; see Example 13.6.9. Moreover, we even use the weakened version of the criterion (13.29) which results by replacing $\ln(L \cdot (q-1) \cdot s)$ with the smaller term $\ln(L' \cdot (q-1))$. (Note that we have no control over $s = \gcd(q-1, n)$ in this general situation.) After that, we replace q by 16 and finally obtain the following sufficient condition for $q \geq 16$:

$$n \geq \frac{17 - \frac{\ln(L' \cdot (15))}{\ln(67)}}{\left(\frac{1}{\ln(4)} - \frac{1}{\ln(67)}\right) \cdot \ln(16) - \frac{3}{4}} \geq \frac{17 - \frac{\ln(L' \cdot (q-1))}{\ln(67)}}{\left(\frac{1}{\ln(4)} - \frac{1}{\ln(67)}\right) \cdot \ln(q) - \frac{3}{4}};$$

the reader should check that this indeed holds for $n \geq 8$. (Note that the right-hand side is a decreasing function of q, by elementary calculus.)

We use the same approach for the remaining seven prime powers $q \in [4,13]$, but with suitable specific choices of ℓ, Λ, α and β.

The case $q = 13$. As before, we take $\alpha = \frac{3}{4}$ and $\beta = 0$ and weaken Criterion (13.29) by replacing s with 1, but we now choose $\ell = 17$ and $\Lambda = \{2,3,5,7,11\}$. This gives the sufficient condition

$$n \geq \frac{5 - \frac{\ln(27720)}{\ln(17)}}{\left(\frac{1}{\ln(4)} - \frac{1}{\ln(17)}\right) \cdot \ln(13) - \frac{3}{4}},$$

which indeed holds for $n \geq 8$.

For each of the four cases $q \in \{7,8,9,11\}$, we apply Example 13.6.9 and generally choose $\alpha = \frac{1}{2}$ and $\beta = \frac{s}{2}$. In contrast, we will make specific choices for ℓ and Λ. Of course, taking the value of $s = \gcd(q-1, n)$ into account necessitates case distinctions depending on n.

The case $q = 11$. Here we take $\ell = 13$ and $\Lambda = \{2,3,5,7\}$.

- Let n be odd and not a multiple of 5. Then $s = 1$ and Inequality (13.29) holds for $n \geq 7$.
- Let n be even and not a multiple of 5. Then $s = 2$ and Inequality (13.29) holds for $n \geq 6$.
- Let n be an odd multiple of 5. Then $s = 5$ and Inequality (13.29) holds for $n \geq 10$. Since $5 \mid n$ and $n \geq 8$, this covers all cases.
- Let n be an even multiple of 5. Then Inequality (13.29) holds for $n \geq 18$. The remaining case $n = 10 = q - 1$ is covered by Theorem 13.8.5.

The case $q = 9$. Now we take $\ell = 17$ and $\Lambda = \{2,5,7,11,13\}$.

- For odd values of n, one has $s = 1$ and Inequality (13.29) holds for $n \geq 5$.
- Let n be even, but not a multiple of 4. Then $s = 2$ and Inequality (13.29) holds for $n \geq 6$.
- Let n be a multiple of 4, but not of 8. Then $s = 4$ and Inequality (13.29) holds for $n \geq 9$. Again, this covers all relevant cases
- Finally, let n be a multiple of 8. Then $s = 8$ and Inequality (13.29) holds for $n \geq 14$. The remaining case where $n = 8 = q - 1$ is covered by Theorem 13.8.5.

The case $q = 8$. Here we take $\ell = 19$ and $\Lambda = \{3, 5, 7, 11, 13, 17\}$.

- If n is not divisible by 7, then $s = 1$ and Inequality (13.29) holds for $n \geq 6$.
- If n is divisible by 7, then $s = 7$ and Inequality (13.29) holds for $n \geq 14$, which covers all relevant cases.

The case $q = 7$.

- Assume first that n is not a multiple of 3. Then $s \leq 2$, and 3 cannot divide C. Since $\text{ord}_{13}(7) = 12$, we see that C is also not divisible by 13. Therefore, we may choose $\ell = 17$ and $\Lambda = \{2, 5, 11\}$. As $s \leq 2$, we may certainly use $\beta = 1$ (and still $\alpha = \frac{1}{2}$). With these choices, Inequality (13.29) holds for $n \geq 8$.
- Next, let n be an odd multiple of 3. Then $s = 3$, and C is odd. Since $\text{ord}_5(7) = 4$ and $\text{ord}_{11}(7) = 10$ and $\text{ord}_{13}(7) = 12$ are all even, the primes 5, 11 and 13 cannot divide C. Therefore, we may now choose $\ell = 17$ and $\Lambda = \{3\}$. Then Inequality (13.29) holds for $n \geq 6$.
- Finally, let n be a multiple of 6, so that $s = 6$, and take $\ell = 17$ and $\Lambda = \{2, 3, 5, 11, 13\}$. Then Inequality (13.29) holds for $n \geq 18$. This leaves only the case $n = 12$, which will be settled by an explicit evaluation of the condition in Corollary 13.8.3; see Table 13.13 below.

For the final two cases $q \in \{4, 5\}$, we use Example 13.6.9 to select suitable values of α and β. According to Inequality (13.20),

$$\Omega \leq \frac{1}{3}n + \frac{\gcd(n, q-1)}{2} + \frac{\gcd(n, q^2-1)}{6}.$$

With the additional notation $t := \gcd(n, q^2 - 1)$, we may write this as $\Omega \leq \alpha n + \beta$ with $\alpha = \frac{1}{3}$ and $\beta = \frac{s}{2} + \frac{t}{6}$.

The case $q = 5$. Here $s \mid 4$ and $t \mid 24$.

- Assume first that n is not a multiple of 4. Then $s \leq 2$ and $t \leq 6$ allow us to choose $\alpha = \frac{1}{3}$ and $\beta = 2$. Note that C has to be odd in this case, regardless of the parity of n, since $5 \equiv 1 \bmod 4$. As both $\text{ord}_{13}(5) = 4$ and $\text{ord}_{17}(5) = 16$ are multiples of 4, C is not divisible by either of these two primes. Thus we may take $\ell = 19$ and $\Lambda = \{3, 7, 11\}$. With these choices and substituting the lower bound 1 for s, Inequality (13.29) holds for $n \geq 10$. This leaves just the case $n = 9$, for which we again refer to Table 13.13.

- Next, let n be a multiple of 4, but not of 3. Now $s = 4$ and $t \leq 8$, which leads to the choices $\alpha = \frac{1}{3}$ and $\beta = \frac{10}{3}$. Also, C is not a multiple of 7, as $\mathrm{ord}_7(5) = 6$. Thus we may take $\ell = 17$ and $\Lambda = \{2,3,11,13\}$. With these choices, Inequality (13.29) holds for $n \geq 16$. This leaves the case $n = 8$, which was already settled in Example 13.8.9.
- Finally, let n be a multiple of 12. Here $s = 4$ and $t \leq 24$. Therefore, we have to choose $\alpha = \frac{1}{3}$ and $\beta = 6$. With $\ell = 23$ and $\Lambda = \{2,3,7,11,13,17,19\}$, Inequality (13.29) holds whenever $n \geq 24$. This only leaves the case $n = 12$, which is dealt with in Table 13.13.

The case $q = 4$. Here $s \mid 3$ and $t \mid 15$, and we choose $\ell = 11$.

- Assume first that n is not a multiple of 3, so that $s = 1$. Here neither 3 nor 7 can divide C, since $\mathrm{ord}_7(4) = 3$, and we may take $\Lambda = \{5\}$.
  - If $5 \mid n$, then $t = 5$, and we obtain $\alpha = \frac{1}{3}$ and $\beta = \frac{4}{3}$. Then Inequality (13.29) holds for $n \geq 14$. The remaining case $n = 10$ is dealt with in Table 13.13.
  - If $5 \nmid n$, then $t = 1$ allows the better choice $\beta = \frac{2}{3}$, and Inequality (13.29) holds already for $n \geq 7$.
- Suppose next that n is an odd multiple of 3, so that $s = 3$. Since 5 cannot divide C (as $\mathrm{ord}_5(4) = 2$), we may take $\Lambda = \{3,7\}$.
  - If $5 \mid n$, then $t = 15$, and we have to choose $\alpha = \frac{1}{3}$ and $\beta = 4$. Here Inequality (13.29) holds whenever $n \geq 44$, which leaves the case $n = 15$; again, see Table 13.13.
  - If $5 \nmid n$, then $t = 3$ allows the better choice $\beta = 2$, and Inequality (13.29) holds for $n \geq 21$. The remaining case $n = 9$ is also dealt with in Table 13.13.
- Finally, let n be a multiple of 6. Then the odd part of n is at most $\frac{n}{2}$, and we use Inequality (13.19) instead of (13.20) to obtain the bound $\Omega \leq \frac{1}{4}n + \frac{3}{2}$. Thus we may choose $\alpha = \frac{1}{4}$ and $\beta = \frac{3}{2}$ in this case. With $\Lambda = \{3,5,7\}$, Inequality (13.29) is satisfied for all $n \geq 10$, which covers all relevant values of n.

It remains to settle the six cases not covered by the preceding arguments. The necessary data for the application of Corollary 13.8.3 are summarized in Table 13.13; fortunately, the required inequality holds in all cases. □

Table 13.13 Evaluation of Corollary 13.8.3 for the remaining pairs (q,n) with $q \geq 4$ and $n \geq 8$.

q	n	C	$\omega(C)$	Ω	$(2^{\omega(C)}-1)\cdot(2^{\Omega}-1)$	$\sqrt{q^n}$
4	9	$3\cdot 7\cdot 19\cdot 73$	4	5	465	512
4	10	$5^2\cdot 11\cdot 31\cdot 41$	4	3	105	1024
4	15	$7\cdot 11\cdot 31\cdot 151\cdot 331$	5	9	15841	32768
5	9	$19\cdot 31\cdot 829$	3	3	49	> 1397
5	12	$3^2\cdot 7\cdot 13\cdot 31\cdot 601$	5	8	7905	15625
7	12	$2^3\cdot 5^2\cdot 13\cdot 19\cdot 43\cdot 181$	6	9	32193	117649

Exercises

Exercise 13.9.1. Check the entries in Table 13.13. In particular, determine the factorization of $4^{15}-1$ using cyclotomic polynomials. □

13.10 The Two Exceptional Cases

In order to finish the proof of the primitive normal basis theorem, we still need to settle the existence of such elements for the two extensions $E = \mathrm{GF}(q^n)$ over $F = \mathrm{GF}(q)$ with $(q,n) \in \{(7,6),(3,8)\}$, for which all criteria applied so far have failed. Of course, one might just use computer searches to deal with these two comparatively small instances. For example, Davenport [101] gave the following explicit solutions:

- Let α be a root of the irreducible polynomial $x^6+x^5+x^4+x^3+x^2+x+3$ over $\mathrm{GF}(7)$. Then α^5 is a primitive element for $\mathrm{GF}(7^6)$ which is normal over $\mathrm{GF}(7)$.
- Similarly, let α be a root of the irreducible polynomial x^8-x^2-1 over $\mathrm{GF}(3)$. Then $(\alpha+1)^{19}$ is a primitive element for $\mathrm{GF}(3^8)$ which is normal over $\mathrm{GF}(3)$.

Nevertheless, it is interesting to note that one can settle the existence of primitive normal elements for these two extensions purely theoretically, too. This was already done by Lenstra and Schoof [238]. In fact, their argument for the pair $(3,8)$ is quite simple, see Exercise 13.10.9, whereas that for the pair (7,6) is similar, but more elaborate.

We will now resolve these two exceptional cases theoretically by exploiting a rather different tool (which we have not used so far): sieving techniques, applied to the additive group of the extension field E. For the pair (7,6), a very simple version suffices, whereas the case (3,8) will require a somewhat more sophisticated argument.

We have decided to use this approach (and not the arguments presented in [238]), as it will provide two relatively simple examples for a technique admitting elaborate refinements (collectively known as "sieve methods"). These methods can be used both for an alternative proof of the primitive normal basis theorem itself and also for establishing stronger versions of this result; see Section 13.11 for a brief overview.

We first introduce the following terminology:

Definition 13.10.1. Let $g_1(x)$, $g_2(x)$ and $g(x)$ be monic polynomials over $\mathrm{GF}(q)$. Then g_1 and g_2 are said to be **complementary divisors** of g provided that they are relatively prime and satisfy $g_1(x)g_2(x) = g(x)$. □

Recall that the fundamental sufficient criterion for the existence of primitive normal elements proved in Corollary 13.5.4 was a direct consequence of the estimate for the cardinality of sets of the type $\Omega^+_{\ell,g}$ given in Proposition 13.5.3. We now prove a simple recursive bound for $|\Omega^+_{\ell,g}|$ in terms of sets of the same type involving complementary divisors of g:

Lemma 13.10.2. *Let q be a prime power, n a positive integer, and ℓ a divisor of $q^n - 1$. Moreover, let $g(x)$ be a polynomial in $\mathrm{GF}(q)[x]$ which is a monic divisor of $x^n - 1$, and suppose that $g_1(x)$ and $g_2(x)$ are complementary divisors of $g(x)$. Then one has*

$$\left|\Omega^+_{\ell,g}\right| \geq \left|\Omega^+_{\ell,g_1}\right| + \left|\Omega^+_{\ell,g_2}\right| - \left|\Omega^+_{\ell,1}\right|.$$

Proof. By definition, the complementarity of $g_1(x)$ and $g_2(x)$ implies $\Omega^+_{\ell,g_1} \cap \Omega^+_{\ell,g_2} = \Omega^+_{\ell,g}$. Together with the trivial observation $\Omega^+_{\ell,g_1} \cup \Omega^+_{\ell,g_2} \subseteq \Omega_{\ell,1}$, this gives

$$\begin{aligned} \left|\Omega^+_{\ell,g}\right| &= \left|\Omega^+_{\ell,g_1}\right| + \left|\Omega^+_{\ell,g_2}\right| - \left|\Omega^+_{\ell,g_1} \cup \Omega^+_{\ell,g_2}\right| \\ &\geq \left|\Omega^+_{\ell,g_1}\right| + \left|\Omega^+_{\ell,g_2}\right| - \left|\Omega^+_{\ell,1}\right|, \end{aligned}$$

as claimed. □

Example 13.10.3. We apply Lemma 13.10.2 to settle the case $(q,n) = (7,6)$. For this, we use the complementary divisors $g_1(x) = x^3 - 1$ and $g_2(x) = x^3 + 1$ of $g(x) = x^6 - 1$, with $\ell = C(7,6)$. According to Table 13.11,

$$C = C(7,6) = \frac{7^6 - 1}{6^2} = 4 \cdot 19 \cdot 43.$$

In view of Proposition 13.8.2, it suffices to show that Ω^+_{C,x^6-1} is not empty. By Lemma 13.10.2,

$$\left|\Omega^+_{C,x^6-1}\right| \geq \left|\Omega^+_{C,g_1}\right| + \left|\Omega^+_{C,g_2}\right| - \left|\Omega^+_{C,1}\right|.$$

Note that an element $w \in E^*$ belongs to $\Omega_{C,1}$ if and only if $\overline{C} := \mathrm{pt}_C(7^6 - 1) = 4C$ divides $\mathrm{ord}(w)$. As $(7^6 - 1)/\overline{C} = 9$, we obtain

$$\left|\Omega^+_{C,1}\right| = \sum_{d|9} \phi(4C \cdot d) = \phi(4C) \cdot \sum_{d|9} \phi(d) = 8 \cdot 18 \cdot 42 \cdot 9 = 54432.$$

On the other hand, by Proposition 13.5.3,

$$\left|\Omega^+_{C,g_i}\right| \geq \frac{\phi(C)}{C} \cdot \frac{\phi_7(g_i)}{7^{\deg g_i}} \cdot \left(7^6 - (2^{\omega(C)} - 1)(2^{\omega_7(g_i)} - 1) \cdot 7^3\right),$$

where $i = 1,2$. Since both $g_1(x)$ and $g_2(x)$ have degree 3 and split into linear factors, this gives

$$\left|\Omega^+_{C,g_1}\right| = \left|\Omega^+_{C,g_2}\right| \geq \frac{2\cdot 18\cdot 42\cdot 6^3}{4\cdot 19\cdot 43}\cdot\left(7^3-(2^3-1)^2\right) > 29381.$$

Therefore,

$$\left|\Omega^+_{C,x^6-1}\right| > 2\cdot 29381 - 54432 > 0,$$

and hence $PN_6(7)$ is indeed positive. □

Example 13.10.4. We now try to apply Lemma 13.10.2 also to the case $(q,n) = (3,8)$. As we have seen in Example 13.8.8, the hypothesis of Corollary 13.8.7 holds in this case, with $r = 2$ and $R = C(3^4,2) = 41$, but the sufficient bound in this result (which was derived from Proposition 13.8.6) does not. However, it might be possible to use Lemma 13.10.2 to show $|\Omega^+_{\ell,x^8-1}| > 0$, where $\ell = C/R = 40$. Then a direct application of Proposition 13.8.6 would show the existence of a primitive normal element for $\mathrm{GF}(3^8)$ over $\mathrm{GF}(3)$, as desired.

For this, we have to select two complementary divisors g_1 and g_2 of x^8-1 in Lemma 13.10.2; then

$$\left|\Omega^+_{40,x^8-1}\right| \geq \left|\Omega^+_{40,g_1}\right| + \left|\Omega^+_{40,g_2}\right| - \left|\Omega^+_{40,1}\right|.$$

Note that $w \in \Omega^+_{40,1}$ if and only if $\mathrm{ord}(w)$ is a multiple of 160, as $\mathrm{pt}_{40}(3^8-1) = 160 = (3^8-1)/41$. Hence we obtain

$$\left|\Omega^+_{40,1}\right| = \sum_{d|41}\phi(160\cdot d) = \phi(160)\cdot\sum_{d|41}\phi(d) = 64\cdot 41 = 2624.$$

Furthermore, by Proposition 13.5.3,

$$\begin{aligned}\left|\Omega^+_{40,g_i}\right| &\geq \frac{\phi(40)}{40}\cdot\frac{\phi_3(g_i)}{3^{\deg g_i}}\cdot\left(3^8-(2^{\omega(40)}-1)(2^{\omega_3(g_i)}-1)\cdot 3^4\right)\\ &= \frac{2}{5}\cdot\frac{\phi_3(g_i)}{3^{\deg g_i}}\cdot 3^4\cdot\left(3^4-3\cdot(2^{\omega_3(g_i)}-1)\right),\end{aligned}$$

where $i = 1,2$. Since x^8-1 splits over $\mathrm{GF}(3)$ as

$$x^8-1 = (x-1)(x+1)(x^2+1)(x^2+x-1)(x^2-x-1),$$

there are several possibilities for selecting g_1 and g_2. Unfortunately, none of these results in a positive lower bound for $|\Omega^+_{\ell,x^8-1}|$; see Exercise 13.10.8. The best result is obtained by taking $g_1(x) = x^4-1$ and $g_2(x) = x^4+1$. Substituting the respective values $\deg g_i = 4$, $\omega(g_1) = 3$, $\phi_3(g_1) = (3-1)^2\cdot(3^2-1) = 32$, $\omega(g_2) = 2$ and $\phi_3(g_2) = (3^2-1)^2 = 64$, one obtains

$$\left|\Omega^+_{40,x^4-1}\right| \geq 768 \quad\text{and}\quad \left|\Omega^+_{40,x^4+1}\right| \geq \frac{9216}{5} > 1843,$$

which comes quite close to what would be needed, but still fails: $768 + 1844 = 2612 < 2624$. □

As mentioned in our introductory remarks above, a more sophisticated version of the sieving approach will in fact work for the case (3,8). For this, we require the following generalization of Lemma 13.10.2 due to Cohen and Huczynska [91]:

Lemma 13.10.5 (Sieving inequality). *Let q be a prime power, n a positive integer, and ℓ a divisor of $q^n - 1$. Moreover, let g and $g_0, \ldots, g_r$ be monic polynomials in $\mathrm{GF}(q)[x]$ dividing $x^n - 1$, where $r \geq 2$, and suppose $\mathrm{lcm}(g_1, \ldots, g_r) = g$ and $\gcd(g_i, g_j) = g_0$ for all i, j with $i \neq j$ and $i, j = 1, \ldots, r$. Then one has*

$$|\Omega^+_{\ell,g}| \geq \sum_{i=1}^{r} |\Omega^+_{\ell,g_i}| - (r-1)|\Omega^+_{\ell,g_0}|.$$

Proof. The case $r = 2$ is very similar to the proof of Lemma 13.10.2. The general case then follows using induction on r. We leave the details to the reader as Exercise 13.10.7. □

Example 13.10.6. We now use Lemma 13.10.5 with $\ell = 40$, $g = x^8 - 1$, $g_0 := x^2 - 1$, and

$$g_1 := (x^2-1)(x^2+1), \;\; g_2 := (x^2-1)(x^2+x-1), \;\; g_3 := (x^2-1)(x^2-x-1)$$

to settle the final remaining case $(q,n) = (3,8)$. To simplify the notation, we will use the abbreviation Ω_i for Ω^+_{40,g_i}, where $i = 0, \ldots, 3$.

Unfortunately, a naive application of Lemma 13.10.5 – that is, simply substituting the bounds following from Proposition 13.5.3 and Exercise 13.5.5 – again fails; see Exercise 13.10.8. However, a more promising approach suggests itself if one notices that some of the character sums $\tau(d,f)$ used in the proof of Proposition 13.5.3 will appear in both $|\Omega_0|$ and $|\Omega_i|$ (albeit with different coefficients), as trivially $\Omega_i \subseteq \Omega_0$ (for $i \neq 0$). Therefore, we rearrange the sieving inequality as

$$|\Omega^+_{40,x^8-1}| \geq |\Omega_1| - (|\Omega_0| - |\Omega_2|) - (|\Omega_0| - |\Omega_3|),$$

evaluate the representation of $\Delta_i := |\Omega_0| - |\Omega_i|$ in terms of character sums $\tau(d,f)$, and only then use the estimate based on Gauss sums. Fortunately, this idea really works. Here are the details, where we use the abbreviation

$$M := \frac{\phi(40)}{40} \cdot \frac{\phi_3(g_0)}{3^{\deg g_0}} = \frac{2}{5} \cdot \frac{(3-1)^2}{3^2} = \frac{8}{45}.$$

Then, by Equation (13.13) in the proof of Proposition 13.5.3,

$$|\Omega_0| = M \cdot \Big(3^8 + \sum_{d} \sum_{\substack{f \mid x^2-1 \\ f \neq 1}} \tau(d,f)\Big),$$

where all summations over d will run over the three divisors $d \neq 1$ of $\operatorname{rad}(40) = 10$. Similarly, for $i = 1, 2, 3$,

$$|\Omega_i| = M \cdot \frac{3^2 - 1}{3^2} \cdot \Big(3^8 + \sum_d \sum_{\substack{f \mid g_i \\ f \neq 1}} \tau(d,f)\Big),$$

and hence

$$\begin{aligned}\frac{1}{M} \cdot \Delta_i &= \Big(1 - \frac{8}{9}\Big) \cdot \Big(3^8 + \sum_d \sum_{\substack{f \mid x^2-1 \\ f \neq 1}} \tau(d,f)\Big) - \frac{8}{9} \cdot \sum_d \sum_{\substack{f \mid g_i \\ f \nmid x^2-1}} \tau(d,f) \\ &= 3^6 + \frac{1}{9} \sum_d \sum_{\substack{f \mid x^2-1 \\ f \neq 1}} \tau(d,f) - \frac{8}{9} \sum_d \sum_{\substack{f \mid g_i \\ f \nmid x^2-1}} \tau(d,f).\end{aligned}$$

By Inequality (13.14), we have $|\tau(d,f)| \leq 3^4$ for all $9 + 12$ pairs (d,f) occurring in the two sums, which implies the estimate

$$\frac{1}{M} \cdot \Delta_i \leq 3^6 + \frac{1}{9} \cdot 9 \cdot 3^4 + \frac{8}{9} \cdot 12 \cdot 3^4 = 1674,$$

so that $\Delta_i \leq (8 \cdot 1674)/45 = 297.6$. As we already know $|\Omega_1| \geq 768$ from Example 13.10.4, we conclude that

$$|\Omega^+_{40,x^8-1}| \geq 768 - 2 \cdot 297 > 0.$$

In view of Proposition 13.8.6, this finally establishes the existence of a primitive normal element for $\mathrm{GF}(3^8)$ over $\mathrm{GF}(3)$. □

Exercises

Exercise 13.10.7. Provide a detailed proof for Lemma 13.10.5. □

Exercise 13.10.8. Check in detail (for all possible choices of g_1 and g_2) that the approach in Example 13.10.4 cannot settle the case $(q,n) = (3,8)$. Also show that a direct application of Lemma 13.10.5 does not work either. □

Exercise 13.10.9. Consider the field extension $E = \mathrm{GF}(3^8)$ over $F = \mathrm{GF}(3)$ and its intermediate field $K = \mathrm{GF}(3^2)$. Let $\mathscr{N}$ and $\mathscr{P}$ denote the sets of all normal elements of E over F, respectively all primitive elements of E.

(a) Use the (E,K)-trace mapping to show that $\mathscr{N}$ and $\zeta \cdot \mathscr{N}$ have empty intersection, where ζ is a primitive element for K.
(b) Show that the sets $\mathscr{N} \cup \zeta \cdot \mathscr{N}$ and $\mathscr{P}$ have at least one element in common.
(c) Deduce that $PN_8(3)$ is positive. □

13.11 Concluding Remarks

As mentioned at the beginning of Section 13.1, the study of primitive normal basis generators for Galois fields has generated a lot of activity in the research on finite fields, especially after the publication of Lenstra and Schoof's work [238]. There have been many variations and strengthenings of the primitive normal basis theorem, which generally rely on sieve methods. We have seen two specific examples for this approach in the previous section.

Sieve methods are, of course, a classical tool in (Analytic) Number Theory; see, for instance, the nice introductory treatment by Cojocaru and Murty [96]. The application of such methods to finite fields – in particular, to their additive groups – is, however, comparatively recent and largely the creation of Stephen Cohen, who has used this approach in many papers; for instance, see [85], to mention just one specific example.

As hinted at before, it is in fact possible to give a proof of the primitive normal basis theorem itself which relies entirely on the sieve method; see Cohen and Huczynska [91]. In [92], the same authors applied refined sieving arguments for a proof of the following impressive strengthening of this result:

Result 13.11.1 (Strong primitive normal basis theorem) *Let q be a prime power and $n \geq 2$ an integer. Then there exists a primitive element α for* $\mathrm{GF}(q^n)$ *such that both α and its inverse α^{-1} are normal over* $\mathrm{GF}(q)$, *with the exception of the five pairs* $(q,n) \in \{(2,3),(2,4),(3,4),(4,3),(5,4)\}$. □

An asymptotic version of Result 13.11.1 (yielding the validity for $n \geq 32$) had been established a few years earlier by Tian and Qi [371].

A quite natural generalization of the primitive normal basis theorem is studied in Hachenberger [171]: the Frobenius automorphism σ is replaced by an arbitrary cyclic F-endomorhism τ of E; an element of $E = \mathrm{GF}(q^n)$ is then called a **primitive τ-generator** over $F = \mathrm{GF}(q)$, provided that it generates the multiplicative group of E as well as the additive group of E, regarded as an $F[x]$-module with respect to τ. A pair (q,n) and the corresponding field extension E/F are said to be **extensive** if there exists a primitive τ-generator for E over F for *every* cyclic F-vector space endomorphism τ of E. In order to formulate the main result of [171], let

$$\mathscr{C} = \{(2,2),(3,2),(5,2),(2,4),(2,6)\},$$

and let $\mathscr{U}$ be the set consisting of the following 19 pairs:

$$(2,8),\ (2,10),\ (2,12),\ (2,14),\ (2,15),\ (2,16),\ (2,18),\ (2,20),\ (2,24),$$
$$(3,8),\ (3,10),\ (3,12),\ (4,6),\ (4,9),\ (4,10),\ (4,12),\ (5,4),\ (7,6),\ (8,8).$$

Result 13.11.2 *Let q be a prime power and $n \geq 2$ an integer. Then the following hold:*

- *If $(q,n) \notin \mathscr{C} \cup \mathscr{U}$, then (q,n) is extensive.*

- *If $(q,n) \in \mathscr{C}$, then (q,n) is not extensive.*

For the 19 pairs in $\mathscr{U}$, the question whether or not the corresponding extension is extensive remains open. Result 13.11.2 covers the work of Hsu and Nan [191], at least when $(q,n) \neq (3,12)$, who have studied the existence of primitive elements which additionally generate a finite *Carlitz module*; we refer to the book of Goss [153] for this concept.

Another variation of the primitive normal basis theorem was introduced by Huczynska, Mullen, Panario and Thomson [196]: given an extension E/F of Galois fields, an element of E is called ***k*-normal**, if its conjugates over F generate an F-subspace of co-dimension k. One of their main results is as follows:

Result 13.11.3 *Let q be a power of the prime p and n a positive integer which is not a multiple of p, and assume that*

$$n \geq \begin{cases} 6 & \text{if } q \geq 11, \\ 3 & \text{if } q \in \{3,4,5,7,8,9\}. \end{cases}$$

Then there exists a primitive element for $\mathrm{GF}(q^n)$ *which is* 1*-normal over* $\mathrm{GF}(q)$. □

In 1996, Morgan and Mullen [283] conjectured that there always exists a primitive element for $E = \mathrm{GF}(q^n)$ which is completely normal over the field $F = \mathrm{GF}(q)$. Although there are serious obstacles before this common generalization of the primitive normal basis theorem and the complete normal basis theorem might be proved in full generality, there have been several partial results which provide strong evidence for this conjecture. Morgan and Mullen had based their conjecture on a computational search covering all pairs (q,n) where $q \leq 100$ is a prime and $q^n \leq 10^{50}$; they also computed the precise numbers of all completely normal and of all primitive completely normal elements for 66 small pairs (q,n), where the largest extension field considered was $\mathrm{GF}(3^{12})$.

The first theoretical results about primitive completely normal bases were provided by Hachenberger in [164, 173]. As before, we use the notation n' for the largest divisor of n which is not divisible by the characteristic of the underlying ground field F to state his main result:

Result 13.11.4 *Let q be a prime power and n a positive integer, and assume that the pair (q,n) is regular, that is, n and* $\mathrm{ord}_{\mathrm{rad}(n')}(q)$ *are relatively prime. Then there exists a primitive element for* $\mathrm{GF}(q^n)$ *which is completely normal over* $\mathrm{GF}(q)$. □

In particular, Result 13.11.4 settles the conjecture of Morgan and Mullen for all extensions with prime power degree. Its proof combines the structure theory for completely normal elements as presented in Chapter 12 with the use of finite field characters.

We also note that the density result of Hachenberger [169] presented in Theorem 13.1.4 can be strengthened to give a corresponding statement for completely normal elements. The following result was proved by Hachenberger [170], again using only elementary methods:

Result 13.11.5 *Let $PCN_n(q)$ denote the number of all primitive completely normal elements for $E = \mathrm{GF}(q^n)$ over $F = \mathrm{GF}(q)$, and let*

$$\pi_n^c(q) := \frac{PCN_n(q)}{\phi(q^n-1)}$$

be the proportion of completely normal elements for E/F in the set of primitive elements of $\mathrm{GF}(q^n)$. Then $\lim_{q\to\infty}\pi_n^c(q) = 1$ for all n. □

It was also shown in [170] that $PCN_n(q)$ is positive when $n \geq 7$ and $q \geq n^{7/2}$, and when $n \geq 37$ and $q \geq n^3$. Using stronger estimates for the Euler phi function and for the function τ for which $\tau(n)$ is the sum of all divisors of n together with the machinery of characters and Gauss sums, this was recently improved considerably by Garefalakis and Kapatenakis [131]:

Result 13.11.6 *With the notation of Result 13.11.5, $PCN_n(q) > 0$ holds whenever $q \geq n'$.* □

Using Result 13.11.6, the computational results of Morgan and Mullen could be improved dramatically by Hachenberger and Hackenberg [174]:

Result 13.11.7 *Let $\mathscr{G}$ denote the set of all positive integers n such that for every prime power q there exists a primitive element for $\mathrm{GF}(q^n)$ which is completely normal over $\mathrm{GF}(q)$. Then $n \in \mathscr{G}$ whenever $1 \leq n \leq 202$.*

Furthermore, the existence of a primitive element of $\mathrm{GF}(p^n)$ which is completely normal over $\mathrm{GF}(p)$ is settled for all pairs (p,n), where $p \leq 10000$ is a prime and $p^n \leq 10^{80}$, by providing corresponding irreducible polynomials. □

Moreover, the exact number $CN_n(q)$ of all completely normal elements for $E = \mathrm{GF}(q^n)$ over $F = \mathrm{GF}(q)$ as well as the number $PCN_n(q)$ of all primitive completely normal elements for E/F are determined in [174] for the pairs (q,n) listed in Tables 13.14 and 13.15 below. It should be noted that the identity $PCN_n(q) = PN_n(q)$ holds whenever $n \leq 5$, since E/F then is a completely basic extension, by Example 12.3.6. Thus the values for $PN_3(q)$ and $PN_4(q)$ listed in Tables 13.3 and 13.5 actually give the number of all primitive completely normal elements for all cases covered in these tables. In particular, Table 13.5 provides the value $PN_4(q)$ for all prime powers q listed in the corresponding row of Table 13.15.

There are many further results on primitive normal elements satisfying additional restrictions, such as a prescribed trace or norm into the ground field. We shall return to this topic at the end of the next chapter, where the existence of primitive elements in affine hyperplanes is studied.

Table 13.14 Complete enumeration of $CN_n(q)$ and $PCN_n(q)$, (a).

q		q		q	
2	$2 \le n \le 31$	3	$2 \le n \le 20$	4	$2 \le n \le 14$
5	$2 \le n \le 12$	7	$2 \le n \le 11$	8	$2 \le n \le 9$
9	$2 \le n \le 9$	11	$2 \le n \le 7$	13	$2 \le n \le 7$
16	$2 \le n \le 7$	17	$2 \le n \le 7$	19	$2 \le n \le 7$
23	$2 \le n \le 7$	25	$2 \le n \le 6$	29	$2 \le n \le 7$
31	$2 \le n \le 6$	43	$2 \le n \le 6$		

Table 13.15 Complete enumeration of $CN_n(q)$ and $PCN_n(q)$, (b).

n	q prime power
3	$2 \le q \le 97$ and $q \in \{121, 125, 128, 169, 243, 256, 289, 343, 361, 512, 529, 625, 729, 841, 961\}$
4	$2 \le q \le 97$ and $q \in \{121, 125, 128, 169, 243\}$
6	$2 \le q \le 43$

Chapter 14
Primitive Elements in Affine Hyperplanes

Abstract This final chapter deals with another important result on primitive elements: given an extension E/F of Galois fields with degree $n \geq 2$, usually every affine F-hyperplane of E contains a primitive element. The proof will take up almost all of this chapter; some motivation and a detailed outline is provided in the introductory first section.

In the final two sections, we consider an interesting application of finite fields for which the strongest known results rely on theorems concerning primitive elements that are quite similar in spirit to those considered in the main part of this chapter and in Chapter 13, and we also include a brief discussion of some existence results for primitive elements satisfying various additional requirements.

14.1 The Basic Problem

We start with some general facts on finite dimensional field extensions E/F. Recall that the set $\mathrm{Hom}_F(E,F)$ of all F-linear mappings from E to F is an F-vector space with dimension $n = [E:F]$. If E/F is a Galois extension, the trace mapping $\mathrm{Tr}_{E/F}$ from E to F induces a natural isomorphism from E (considered as an F-vector space) to $\mathrm{Hom}_F(E,F)$, given by

$$\beta \in E \mapsto L_\beta \quad \text{with } L_\beta(x) := \mathrm{Tr}_{E/F}(\beta x) \text{ for all } x \in E;$$

we will refer to L_β as a **generalized trace mapping**. Given any elements $\beta \in E^*$ and $a \in F$, the set

$$H_{\beta,a} := \{v \in E \colon L_\beta(v) = a\} = \{v \in E \colon \mathrm{Tr}_{E/F}(\beta v) = a\}$$

is an **affine hyperplane** of the F-vector space E, and one has $H_{\beta,a} = H_{\gamma,b}$ if and only if $\gamma = \lambda\beta$ and $b = \lambda a$ for some $\lambda \in F^*$; see Exercise 14.1.9.

D. Hachenberger and D. Jungnickel, *Topics in Galois Fields*,
Algorithms and Computation in Mathematics 29,
https://doi.org/10.1007/978-3-030-60806-4_14

From now on, we assume that $F = \mathrm{GF}(q)$ and $E = \mathrm{GF}(q^n)$ are finite fields, where $n \geq 2$. It is quite natural to ask whether a given affine hyperplane $H_{\beta,a}$ (with $(\beta, a) \in E^* \times F$) contains a primitive element of E; as we shall see, this holds in almost all cases. Similar to the proof of the primitive normal basis theorem in Chapter 13, this needs a lot of detailed work; moreover, some parts of the cases $n = 2$ and $n = 3$ still require the use of a computer algebra system.

The special case $\beta = 1$ is of particular interest, as any primitive element $u \in H_{1,a}$ (that is, with $\mathrm{Tr}_{E/F}(u) = a$) has a minimal polynomial of the form

$$x^n - ax^{n-1} + h(x) \quad \text{with } \deg h \leq n-2.$$

Thus the assertion that every hyperplane $H_{1,a}$ (with $a \in F$) contains a primitive element means that one can prescribe the second highest coefficient when searching for a primitive polynomial of degree n over F. This was the main motivation for investigating this existence problem, which was settled by Stephen Cohen in a series of papers [79, 81, 82]:

Result 14.1.1 (Existence theorem for primitive elements with prescribed trace) *Consider the n-dimensional extension $E = \mathrm{GF}(q^n)$ of the Galois field $F = \mathrm{GF}(q)$, where $n \geq 2$. Let $a \in F$, and assume that $a \neq 0$ for $n = 2$ and for $(q,n) = (4,3)$. Then there exists a primitive element $u \in E$ such that $\mathrm{Tr}_{E/F}(u) = a$.* □

Note that it is necessary to assume $a \neq 0$ in Result 14.1.1 for $n = 2$: then we cannot have $\mathrm{Tr}_{E/F}(u) = 0$ for any primitive element u, as u has to be normal over F in this case, by Proposition 13.1.1. We will show in Example 14.1.5 below that the hypothesis $a \neq 0$ is also needed in the case $(q,n) = (4,3)$.

The main part of Cohen's paper [82] deals with extensions of degree $n \geq 3$ in the case of traces $a \neq 0$. This part of Result 14.1.1 has also been proved (independently) by Jungnickel and Vanstone [217]; these authors also settled the case $n = 2$ up to 147 possible exceptions, the largest of which was $q = 3{,}847{,}271$. In fact, the case of quadratic extensions was already solved a few years earlier in Cohen's paper [79], where he actually established a more general result:

Result 14.1.2 *Consider the quadratic field extension $E = \mathrm{GF}(q^2)$ over the Galois field $F = \mathrm{GF}(q)$. Let $\alpha \in E^*$ and $\gamma \in E \setminus F$. Then there exists an element $b \in F$ such that $\alpha(\gamma + b)$ is a primitive element of E.* □

The reader should check that Result 14.1.2 is equivalent to a positive answer to the general problem about primitive elements in affine hyperplanes stated above (for $n = 2$); see Exercise 14.1.10.

Finally, the remaining case of extensions with degree $n \geq 3$ and trace $a = 0$ is covered by the results in [81], where Cohen established the following general theorem on cyclic difference sets (as studied in Section 9.5):

Result 14.1.3 *With the exception of the two $(21,5,1)$-difference sets $\{3,6,7,12,14\}$ and $\{7,9,14,15,18\}$ in $(\mathbb{Z}_{21},+)$, every (v,k,λ)-difference set in $(\mathbb{Z}_v,+)$ contains a residue which is relatively prime to v.* □

Remark 14.1.4. Since it is far from obvious that the preceding theorem on cyclic difference sets contains the case $n \geq 3$ and $a = 0$ of Result 14.1.1, we will now explain this connection in some detail. As in Section 13.3, we consider a projective space $PG(n-1,q)$, say Γ; recall that Γ has $Q = \frac{q^n-1}{q-1}$ points. Theorem 9.5.10 is equivalent to the statement that Γ admits a cyclic collineation group G of order Q which acts regularly on both the set of points and the set of hyperplanes; see also Remark 9.5.11. Therefore, each point p of Γ may, after choosing a "base point" p_0, be identified with the unique element $g \in G$ mapping p_0 to p; combining this with an isomorphism between G and $(\mathbb{Z}_Q, +)$ then shows that the points of Γ correspond to the residues in the ring $\mathbb{Z}_Q$. Under this correspondence, the point set of any hyperplane yields a difference set with "classical parameters"

$$v = \frac{q^n-1}{q-1}, \quad k = \frac{q^{n-1}-1}{q-1} \quad \text{and} \quad \lambda = \frac{q^{n-2}-1}{q-1}$$

(that is, with parameters (9.31) for $d = n-1$) in the additive group $(\mathbb{Z}_Q, +)$. Finally, primitive points for Γ (as defined in Section 13.3) correspond to residues in $\mathbb{Z}_Q$ which are relatively prime to Q, that is, to the units of the ring $(\mathbb{Z}_Q, +, \cdot)$.

As any hyperplane of $\Gamma = PG(n-1,q)$ is, by definition, a (linear) hyperplane H of the n-dimensional vector space $\mathrm{GF}(q)^n \cong E$ and as the Singer difference set in Theorem 9.5.10 was constructed from the elements of E with trace 0, we conclude that any unit in the difference set associated with H gives us a primitive element of E with trace 0 in H. Thus Result 14.1.3 is indeed a strong generalization of the case $n \geq 3$ and $a = 0$ of Result 14.1.1. □

When $q = 4$ and $n = 3$, the parameters (v, k, λ) are $(21, 5, 1)$, which explains why the 3-dimensional extension of the field with four elements plays a special role. We now check that there really is no primitive element with trace 0 in this case:

Example 14.1.5. Consider the extension $E = \mathrm{GF}(4^3)$ of $F = \mathrm{GF}(4)$, and observe that the elements $u \in E^*$ with $\mathrm{Tr}_{E/F}(u) = 0$ are precisely the roots of the polynomial

$$x^{4^2-1} + x^{4-1} + 1 = x^{15} + x^3 + 1 = y^5 + y + 1,$$

where we write $y = x^3$. Note that $y^5 + y + 1$ has to be a multiple of $\Phi_3(y) = y^2 + y + 1$, since every primitive third root of unity λ satisfies $\lambda^5 = \lambda^2$. A simple polynomial division gives $y^5 + y + 1 = (y^2 + y + 1) \cdot (y^3 + y^2 + 1)$, and therefore

$$x^{15} + x^3 + 1 = (x^6 + x^3 + 1) \cdot (x^9 + x^6 + 1).$$

Since $y^3 + y^2 + 1$ is a divisor of the cyclotomic polynomial $\Phi_7(y)$, we conclude (using the formulas in Section 3.6) that $x^{15} + x^3 + 1$ divides

$$\Phi_3(x^3) \cdot \Phi_7(x^3) = \Phi_9(x) \cdot \Phi_7(x) \cdot \Phi_{21}(x);$$

in fact, x^9+x^6+1 splits as $(x^3+x+1)\cdot(x^6+x^4+x^2+x+1)$. This proves that $\mathrm{ord}(u)\in\{7,9,21\}$ for every $u\in E^*$ with $\mathrm{Tr}_{E/F}(u)=0$, and hence no such element can be primitive. □

The existence problem for primitive elements of $\mathrm{GF}(4^3)$ which are mapped to $0\in\mathrm{GF}(4)$ under some *generalized* trace mapping is considered in Exercise 14.1.11. Similarly, it also makes sense to ask for a primitive element u such that $\mathrm{Tr}_{E/F}(\beta u)=0$ (for some $\beta\neq 0$) in the case of an arbitrary quadratic extension E/F; see Theorem 14.1.8 below.

The aim of the present chapter is to generalize Cohen's results on primitive elements with a prescribed trace by investigating when the image of a primitive element under a *generalized* trace mapping can also be prescribed. In other words: for which extensions E of a given Galois field F does every affine F-hyperplane of E contain a primitive element? We will solve this problem by establishing the following results.

Theorem 14.1.6 (Inhomogeneous case). *Consider the n-dimensional extension* $E=\mathrm{GF}(q^n)$ *of the Galois field* $F=\mathrm{GF}(q)$, *where* $n\geq 2$, *and let* $\beta\in E^*$ *and* $a\in F^*$. *Then there exists a primitive element* $u\in E$ *with* $\mathrm{Tr}_{E/F}(\beta u)=a$.

Theorem 14.1.7 (Homogenous case for $n\geq 3$). *Consider the n-dimensional extension* $E=\mathrm{GF}(q^n)$ *of the Galois field* $F=\mathrm{GF}(q)$, *where* $n\geq 3$. *Let* $\beta\in E^*$, *and assume that the order of* β *is divisible by* 9 *when* $n=3$ *and* $q=4$. *Then there exists a primitive element* $u\in E$ *with* $\mathrm{Tr}_{E/F}(\beta u)=0$.

As for the existence problem for primitive normal elements considered in Chapter 13, the proofs of Theorems 14.1.6 and 14.1.7 rely on an appropriate character sum formulation of the problem at hand. This characterization is given in Section 14.3, after pointing out a basic reduction result in the next section. We then investigate a special class of extensions in Section 14.4, before we prove Theorem 14.1.6 for the case $n\geq 3$ in Section 14.5. After that, Theorem 14.1.7 is proved in Sections 14.6 (for $n\geq 4$) and 14.7 (for $n=3$).[1] Finally, the case of quadratic extensions will be dealt with in Sections 14.8 and 14.9.

As in Chapter 13, it is again our aim to present the computational details to the greatest possible extent. Unfortunately, the investigation of the cases

- $n=3$, trace 0 and $q\equiv 1 \bmod 3$;
- $n=2$ and non-zero trace

requires the use of some computer algebra system. To the best of our knowledge, no computer-free proof is available for these cases.

We conclude this introductory section with the following result for the case of quadratic extensions and trace 0:

[1] Let us stress that our proof of Theorem 14.1.7 will not use difference sets at all, as opposed to Cohen's approach to the homogeneous case via Result 14.1.3, which was already discussed in Remark 14.1.4 above.

Theorem 14.1.8 (Homogenous case for $n = 2$). *Consider the quadratic extension $E = \mathrm{GF}(q^2)$ of the Galois field $F = \mathrm{GF}(q)$. Given any element $\beta \in E^*$, let $P_{\beta,0}(q)$ denote the number of primitive elements $u \in E$ such that $\mathrm{Tr}_{E/F}(\beta u) = 0$. In the case where q is odd, let s denote the odd part of $q+1$ and 2^m the 2-part of q^2-1.*

Then one has $P_{\beta,0}(q) \neq 0$ if and only if one of the following cases occurs:

- *q is even and $\mathrm{ord}(\beta)$ is a multiple of $q+1$;*
- *$q \equiv 1 \bmod 4$, s divides the order of β, but 2^m does not divide $\mathrm{ord}(\beta)$;*
- *$q \equiv 3 \bmod 4$ and $2^m s$ divides the order of β.*

In these cases, $P_{\beta,0}(q)$ satisfies

$$P_{\beta,0}(q) = \begin{cases} \phi(q-1) & \text{if } q \text{ is even,} \\ 2\phi(q-1) & \text{if } q \text{ is odd.} \end{cases}$$

Proof (for q even). Let $u \in E^*$ such that $\mathrm{Tr}_{E/F}(\beta u) = 0$. Then $\beta u + (\beta u)^q = 0$ shows that $(\beta u)^{q-1} = -1$. If q is even, this means $\beta u \in F^*$. Because of $\gcd(q-1, q+1) = 1$ for even q, we may decompose u (uniquely) as $u = u_0 u_1$, where u_1 belongs to the subgroup U_{q+1} of order $q+1$ of E^* and where u_0 is in the subgroup of order $q-1$, that is, $u_0 \in F^*$. Note that u is a primitive element for E if and only if $\mathrm{ord}(u_0) = q-1$ and $\mathrm{ord}(u_1) = q+1$.

In the same way, decompose β as $\beta = \beta_0 \beta_1$. Then $\beta u = (\beta_0 u_0) \cdot (\beta_1 u_1)$ is in F^* if and only if $u_1 = \beta_1^{-1}$, so that u_1 is determined by β_1 and the assumption $\mathrm{Tr}_{E/F}(\beta u) = 0$. As observed above, u can only be a primitive element for E if $\mathrm{ord}(u_1) = q+1$, which shows that $q+1$ has to divide the order of $\beta = \beta_0 u_1^{-1}$. If this holds, we have exactly $\phi(q-1)$ choices for u_0 which result in a primitive element u of E, since u_0 has to satisfy $\mathrm{ord}(u_0) = q-1$. □

The proof for the case where q is odd is similar (though more involved) and will be postponed to Section 14.9, where we consider quadratic extensions in more detail, using a group theoretic point of view; see Remark 14.9.7.

Exercises

Exercise 14.1.9. Consider a finite-dimensional Galois field extension E/F, and let $\beta, \gamma \in E^*$ and $a, b \in F$. Show that $H_{\beta,a} = H_{\gamma,b}$ holds if and only if there is some $\lambda \in F^*$ such that $\lambda\beta = \lambda\gamma$ and $\lambda a = \lambda b$. □

Exercise 14.1.10. Let $n = 2$. Show that the formulation of the problem in Result 14.1.2 then is equivalent to the formulation via generalized trace mappings at the beginning of this section. □

Exercise 14.1.11. Consider the 3-dimensional extension $E = \mathrm{GF}(4^3)$ of the Galois field $F = \mathrm{GF}(4)$, and let $\beta \in E^*$. Show that there exists a primitive element u of E with $\mathrm{Tr}_{E/F}(\beta u) = 0$ if and only if $\mathrm{ord}(\beta)$ is a multiple of 9. □

14.2 A Basic Reduction

The topic of this brief section is a basic observation which allows us to reduce our problem to extensions of prime degree. In order to state this result, we introduce the following convenient terminology:

Definition 14.2.1. An integer $n \geq 2$ is called **extensive** if the following condition holds:

- For every prime power q, for every non-zero $a \in F = \mathrm{GF}(q)$, and for every non-zero $\beta \in E = \mathrm{GF}(q^n)$ there exists a primitive element $u \in E$ such that $\mathrm{Tr}_{E/F}(\beta u) = a$.

Similarly, n is said to be 0-**extensive** if the following condition holds:

- For every prime power q and for every non-zero $\beta \in E = \mathrm{GF}(q^n)$ there exists a primitive element $u \in E$ such that $\mathrm{Tr}_{E/\mathrm{GF}(q)}(\beta u) = 0$. □

The following important result is a simple consequence of the transitivity of the trace mappings (see Theorem 3.12.8). It implies that (in principle) we may restrict our attention to extensions with prime degree.

Proposition 14.2.2. *Assume that an integer $r \geq 2$ is extensive or 0-extensive. Then every multiple of r is likewise extensive or 0-extensive, respectively.*

Proof. First assume that r is extensive. Let q be an arbitrary prime power and consider a field extension $E = \mathrm{GF}(q^n)$ of $F = \mathrm{GF}(q)$, where $n = rk$. Given any $a \in F^*$ and any $\beta \in E^*$, we need to show the existence of a primitive element $u \in E$ such that $\mathrm{Tr}_{E/F}(\beta u) = a$. For this, we will make use of the intermediate field K of degree k over F, so that E has degree r over K. Choose any element $b \in K$ satisfying $\mathrm{Tr}_{K/F}(b) = a$. Since r is extensive, there exists a primitive element $u \in E$ such that $\mathrm{Tr}_{E/K}(\beta u) = b$. Then the transitivity of the trace mappings immediately gives the desired result, as

$$\mathrm{Tr}_{E/F}(\beta u) = \mathrm{Tr}_{K/F}(\mathrm{Tr}_{E/K}(\beta u)) = \mathrm{Tr}_{K/F}(b) = a.$$

The case where r is 0-extensive follows in the same way, with $a = 0$. □

Remark 14.2.3. Let $n = rk$ be a proper multiple of r. Then the extensiveness of r also implies the 0-extensiveness of n. This follows as in the proof of Proposition 14.2.2, by choosing b as an element in $K \setminus F$ satisfying $\mathrm{Tr}_{K/F}(b) = 0$. □

We close this section with an interesting application of Result 14.1.1 to trace-compatible sequences of primitive normal elements, taken from Hachenberger [167]. As in the proof of Proposition 14.2.2, the transitivity of the trace mapping plays an essential role.

Theorem 14.2.4. *Let $F = \mathrm{GF}(q)$ be a finite field with characteristic p, and let r be a prime. Assume that either $r = p$, or that r is odd and q has order $(r-1)r$ modulo r^2. Then there exists a sequence $(y_m)_{m \in \mathbb{N}}$ in the r-primary closure $E_{r^\infty} = \bigcup_{m \in \mathbb{N}} E_{r^m}$ satisfying the following properties, where $E_{r^m} = \mathrm{GF}(q^{r^m})$:*

- *for every m, the element y_m is primitive in E_{r^m} and normal over F;*
- *if $k, \ell \in \mathbb{N}$ such that $k \leq \ell$, then the (E_{r^ℓ}, E_{r^k})-trace of y_ℓ equals y_k.*

Proof. We use induction on m to construct the desired sequence. The induction basis $m = 0$ is trivial: just take any primitive element y_0 of the ground field F.

Now assume that $y_0, y_1, \ldots, y_m$ is a partial sequence of primitive elements satisfying the two required properties of normality and trace-compatibility (for some $m \geq 0$). By Result 14.1.1 or Theorem 14.1.6 applied to the extension E/L, where $E = \mathrm{GF}(q^{r^{m+1}})$ and $L = \mathrm{GF}(q^{r^m})$, there exists a primitive element $u \in E$ satisfying $\mathrm{Tr}_{E/L}(u) = y_m$. We now show that u is normal over F, so that we may extend the sequence $y_0, y_1, \ldots, y_m$ by taking $y_{m+1} := u$ (using the transitivity of the trace mappings). In the case $r = p$, this is immediate from Theorem 11.3.1, since $\mathrm{Tr}_{E/F}(y_m) = y_0 \neq 0$.

Now let $r \neq p$ be an odd prime. Since y_m is a normal element for E/F, it has q-order $x^{r^m} - 1$. On the other hand, we also have

$$\mathrm{Ord}_q(y_m) = \frac{\mathrm{Ord}_q(u)}{\gcd\left(t(x), \mathrm{Ord}_q(u)\right)}, \quad \text{where } t(x) = (x^{r^{m+1}} - 1)/(x^{r^m} - 1)$$

is the trace polynomial for E/L, and hence $\mathrm{Ord}_q(u)$ is a multiple of $x^{r^m} - 1$. Note that $\mathrm{Ord}_q(u) \neq x^{r^m} - 1$, as $u \notin L$. Therefore, some irreducible factor of $\Phi_{r^{m+1}}(x)$ divides $\mathrm{Ord}_q(u)$. By Proposition 1.7.8, the hypothesis $\mathrm{ord}_{r^2}(q) = (r-1)r$ implies $\mathrm{ord}_{r^n}(q) = (r-1)r^{n-1}$ for all $n \geq 1$, so that all cyclotomic polynomials $\Phi_{r^n}(x)$ are irreducible over F. This shows

$$\mathrm{Ord}_q(u) = (x^{r^m} - 1) \cdot \Phi_{r^{m+1}}(x) = x^{r^{m+1}} - 1,$$

proving that u is normal over F also in this case. □

14.3 A Character Sum Approach to the Basic Problem

In this section, we derive a sufficient criterion for the existence of a primitive element β for $E = \mathrm{GF}(q^n)$ satisfying $\mathrm{Tr}_{E/F}(\beta u) = a$ (where $n \geq 2$ and a is a prescribed element of $F = \mathrm{GF}(q)$) which is based on the theory of finite field characters. For this, we fix $\beta \in E^*$ and $a \in F$, and consider the affine hyperplane $H_{\beta,a}$ of E.

We start by recalling the definition of the function $\mathscr{P}_t$ from Equation (13.9) in Section 13.5, where t is any divisor of $q^n - 1$:

$$\mathscr{P}_t = \frac{\phi(t)}{t} \cdot \sum_{\psi \in U_t'} \frac{\mu(\mathrm{ord}(\psi))}{\phi(\mathrm{ord}(\psi))} \psi = \frac{\phi(t)}{t} \cdot \sum_{d|t} \frac{\mu(d)}{\phi(d)} \cdot \sum_{\psi:d} \psi. \tag{14.1}$$

As before, U_t will denote the subgroup of order t of E^* and U_t' the corresponding subgroup of the (multiplicative) character group of E^*. Again, the notation $\psi : d$

indicates that the final sum runs over all multiplicative characters ψ with order d. As noted in Corollary 13.5.1, $\mathscr{P}_{q^n-1}$ is just the characteristic function of the set of all primitive elements of E. In general, $\mathscr{P}_t$ is the characteristic function of the set of all $u \in E$ such that the order of u is divisible by the t-part $\mathrm{pt}_t(q^n-1)$ of t in q^n-1, that is, by the largest multiple of $\mathrm{rad}(t)$ dividing q^n-1.

Consequently, the number of all primitive elements of E which are contained in the affine hyperplane $H_{\beta,a}$ is given by

$$\sum_{\substack{v\in H_{\beta,a}\\ v\neq 0}} \mathscr{P}_{q^n-1}(v) = \frac{\phi(q^n-1)}{q^n-1}\cdot \sum_{\psi\in\widehat{E^*}} \frac{\mu(\mathrm{ord}(\psi))}{\phi(\mathrm{ord}(\psi))} S_{\beta,a}(\psi), \tag{14.2}$$

where

$$S_{\beta,a}(\psi) := \sum_{\substack{v\in H_{\beta,a}\\ v\neq 0}} \psi(v). \tag{14.3}$$

Observation 14.3.1. Let H denote the kernel of the (E,F)-trace mapping, that is, $H = H_{1,0}$. For $a \in F$, let $\overline{a}$ be any element of E satisfying $\mathrm{Tr}_{E/F}(\overline{a}) = a$. Then we have $\mathrm{Tr}_{E/F}(\beta u) = a$ if and only if $u \in \beta^{-1}(H+\overline{a})$, and hence

$$S_{\beta,a}(\psi) = \sum_{\substack{v\in H\\ v+\overline{a}\neq 0}} \psi(\beta^{-1}(v+\overline{a})) = \psi(\beta^{-1})\cdot \sum_{\substack{v\in H\\ v+\overline{a}\neq 0}} \psi(v+\overline{a}) = \psi(\beta^{-1})S_{1,a}(\psi).$$

Now assume $a \neq 0$. Then the mapping $w \mapsto aw$ is a bijection between $H_{\beta,1}$ and $H_{\beta,a}$, and therefore $S_{\beta,a}(\psi) = \sum_{v\in H_{\beta,1}} \psi(av) = \psi(a)S_{\beta,1}(\psi)$. Altogether, one has

$$S_{\beta,0}(\psi) = \psi(\beta^{-1})S_{1,0}(\psi) \text{ and } S_{\beta,a}(\psi) = \psi(\beta^{-1}a)S_{1,1}(\psi) \text{ for } a \neq 0. \tag{14.4}$$

In particular, we see that the character sums $S_{\beta,a}(\psi)$ take at most two distinct absolute values (for a fixed character ψ):

$$|S_{\beta,0}(\psi)| = |S_{1,0}(\psi)| \text{ and } |S_{\beta,a}(\psi)| = |S_{1,1}(\psi)| \text{ for all } \beta \in E^* \text{ and all } a \in F^*.$$

It is trivial to determine $S_{\beta,a}(\psi_0)$, where ψ_0 is the trivial multiplicative character:

$$S_{\beta,a}(\psi_0) = q^{n-1} \quad \text{for all } \beta \in E^* \text{ and all } a \in F^*, \tag{14.5}$$

and

$$S_{\beta,0}(\psi_0) = q^{n-1}-1 \quad \text{for all } \beta \in E^*. \tag{14.6}$$

For the case $\psi \neq \psi_0$, we will soon see that all character sums $S_{\beta,a}(\psi)$ are restricted to just four absolute values, depending on whether $a=0$ and whether ψ belongs to the dual subgroup of F^* in $\widehat{E^*}$. □

Remark 14.3.2. Recall that the dual subgroup $(F^*)^{\perp}$ of F^* in $\widehat{E^*}$ consists of all multiplicative characters ψ satisfying $\psi(c) = 1$ for all $c \in F^*$, that is, of all ψ which restrict on F^* to the trivial character of F^*. By Proposition 10.1.5, $(F^*)^{\perp}$ has order

$$Q = Q(q,n) := \frac{|E^*|}{|F^*|} = \frac{q^n - 1}{q-1}, \tag{14.7}$$

and therefore $(F^*)^\perp$ is the subgroup U'_Q of order Q of $\widehat{E^*}$. □

As mentioned above, the character sums $S_{\beta,a}(\psi)$ are restricted to just four possible absolute values (for $\psi \neq \psi_0$). These values are determined in the following two results, which cover the cases $a = 0$ and $a \neq 0$, respectively. However, it will be convenient to give a common proof for these two results.

Lemma 14.3.3. *Consider the extension field* $E = \mathrm{GF}(q^n)$ *over* $F = \mathrm{GF}(q)$, *where* $n \geq 2$, *and let* $\beta \in E^*$ *and* $\psi \in \widehat{E^*}$. *Then:*

(1) $|S_{\beta,0}(\psi)| = (q-1)\sqrt{q^{n-2}}$ *for* $\psi \in (F^*)^\perp \setminus \{\psi_0\}$;

(2) $S_{\beta,0}(\psi) = 0$ *for* $\psi \notin (F^*)^\perp$.

Lemma 14.3.4. *Consider the extension field* $E = \mathrm{GF}(q^n)$ *over* $F = \mathrm{GF}(q)$, *where* $n \geq 2$, *and let* $a \in F^*$, $\beta \in E^*$ *and* $\psi \in \widehat{E^*}$. *Then:*

(1) $|S_{\beta,a}(\psi)| = \sqrt{q^{n-2}}$ *for* $\psi \in (F^*)^\perp \setminus \{\psi_0\}$;

(2) $|S_{\beta,a}(\psi)| = \sqrt{q^{n-1}}$ *for* $\psi \notin (F^*)^\perp$.

Proof of Lemmas 14.3.3 and 14.3.4. As $\psi \neq \psi_0$, we have $\sum_{u \in E^*} \psi(u) = 0$, by Corollary 10.2.2 (applied to the non-trivial character ψ of E^*). Since the collection of all affine hyperplanes $H_{\beta,a}$ with a fixed β forms a partition of E, this gives, together with Equation (14.4),

$$\begin{aligned} 0 = \sum_{u \in E^*} \psi(u) &= S_{\beta,0}(\psi) + \sum_{a \in F^*} S_{\beta,a}(\psi) \\ &= S_{\beta,0}(\psi) + S_{\beta,1}(\psi) \cdot \Big(\sum_{a \in F^*} \psi(a) \Big). \end{aligned}$$

Using Corollary 10.2.2 again (now applied to the restriction η of ψ to F^*), we note that

$$\sum_{a \in F^*} \psi(a) = \sum_{a \in F^*} \eta(a) = \begin{cases} 0 & \text{for } \psi \notin (F^*)^\perp, \\ q-1 & \text{for } \psi \in (F^*)^\perp. \end{cases}$$

In view of Remark 14.3.2, this shows $S_{\beta,0}(\psi) = 0$ if $\mathrm{ord}(\psi)$ does not divide Q (which is assertion (2) in Lemma 14.3.3), whereas

$$S_{\beta,0}(\psi) = -(q-1)S_{\beta,1}(\psi) \quad \text{if } \mathrm{ord}(\psi) \mid Q. \tag{14.8}$$

For the remaining assertions, we also need to consider additive characters and Gauss sums. Note first that every additive character λ of F can be lifted to an additive character χ of the extension field E (depending on β) by putting

$$\chi(u) := \lambda\big(\mathrm{Tr}_{E/F}(\beta u)\big) \quad \text{for } u \in E;$$

cf. Definition 10.5.1. In what follows, λ will always be a non-trivial character. We now consider the two related Gauss sums

$$G_E^*(\psi,\chi) := \sum_{u\in E^*} \psi(u)\chi(u) \quad \text{and} \quad G_F^*(\eta,\lambda) := \sum_{v\in F^*} \eta(v)\lambda(v),$$

where ψ, η and χ, λ are as above. Using (14.4) and $\chi(u) = \lambda(a)$ for $\mathrm{Tr}_{E/F}(\beta u) = a$, we compute

$$\begin{aligned} G_E^*(\psi,\chi) &= \sum_{\substack{u\in H_{\beta,0}\\ u\neq 0}} \psi(u)\chi(u) + \sum_{a\in F^*}\sum_{u\in H_{\beta,a}} \psi(u)\chi(u) \\ &= \lambda(0)\cdot \sum_{\substack{u\in H_{\beta,0}\\ u\neq 0}} \psi(u) + \sum_{a\in F^*} \lambda(a)\cdot \sum_{u\in H_{\beta,a}} \psi(u) \\ &= S_{\beta,0}(\psi) + \sum_{a\in F^*} \lambda(a)\psi(a) S_{\beta,1}(\psi) \\ &= S_{\beta,0}(\psi) + S_{\beta,1}(\psi) G_F^*(\eta,\lambda). \end{aligned}$$

Assume first that $\mathrm{ord}(\psi)$ does not divide Q. Then $\eta \neq \eta_0$, where η_0 denotes the trivial multiplicative character of F^*. As we have already seen, this implies $S_{\beta,0}(\psi) = 0$, so that the preceding identity simplifies to

$$G_E^*(\psi,\chi) = S_{\beta,1}(\psi) G_F^*(\eta,\lambda).$$

Taking absolute values, Corollary 10.3.3 and Equation (14.4) yield

$$|S_{\beta,a}(\psi)| = |S_{\beta,1}(\psi)| = \frac{|G_E^*(\psi,\chi)|}{|G_F^*(\eta,\lambda)|} = \frac{\sqrt{q^n}}{\sqrt{q}} = \sqrt{q^{n-1}},$$

which is assertion (2) in Lemma 14.3.4.

Now assume that $\mathrm{ord}(\psi)$ divides Q, so that $\eta = \eta_0$ and $G_F^*(\eta,\lambda) = -1$, by Proposition 10.3.2. In view of Equation (14.8), the identity above becomes

$$G_E^*(\psi,\chi) = -(q-1)S_{\beta,1}(\psi) + S_{\beta,1}(\psi)G_F^*(\eta,\lambda) = -qS_{\beta,1}(\psi)$$

for this case. Now Corollary 10.3.3 and Equation (14.4) yield

$$|S_{\beta,a}(\psi)| = |S_{\beta,1}(\psi)| = \frac{|G_E^*(\psi,\chi)|}{q} = \frac{\sqrt{q^n}}{q},$$

proving assertion (1) in Lemma 14.3.4. Finally, using this together with Equation (14.8) also gives

$$|S_{\beta,0}(\psi)| = (q-1)|S_{\beta,1}(\psi)| = (q-1)\sqrt{q^{n-2}},$$

which is assertion (1) in Lemma 14.3.3. □

Throughout this chapter, we will denote the number of all primitive elements for $E = \mathrm{GF}(q^n)$ which are contained in an affine hyperplane $H_{\beta,a}$ by $P_{\beta,a}$, and $Q = Q(n,q)$ will be as in Equation (14.7). As in Chapter 13, $\omega(N)$ will denote the number of distinct prime divisors of a positive integer N.

Lemmas 14.3.3 and 14.3.4 lead to the following estimates for the $P_{\beta,a}$; again, we will give a common proof for these two results.

Proposition 14.3.5. *Consider the extension field $E = \mathrm{GF}(q^n)$ over $F = \mathrm{GF}(q)$, where $n \geq 2$, and let $\beta \in E^*$. Then*

$$P_{\beta,0} \geq \frac{\phi(q^n-1)}{q^n-1} \cdot \left(q^{n-1} - 1 - (2^{\omega(Q)} - 1)(q-1)\sqrt{q^{n-2}}\right).$$

Proposition 14.3.6. *Consider the extension field $E = \mathrm{GF}(q^n)$ over $F = \mathrm{GF}(q)$, where $n \geq 2$, and let $a \in F^*$ and $\beta \in E^*$. Then*

$$P_{\beta,a} \geq \frac{\phi(q^n-1)}{q^n-1} \cdot \left(q^{n-1} - (2^{\omega(Q)} - 1)\sqrt{q^{n-2}} - (2^{\omega(q^n-1)} - 2^{\omega(Q)})\sqrt{q^{n-1}}\right).$$

Proof of Propositions 14.3.5 and 14.3.6. Let a be an arbitrary element of F. By Equation (14.2),

$$\begin{aligned}
\frac{q^n-1}{\phi(q^n-1)} \cdot P_{\beta,a} &= \sum_{\psi \in \widehat{E^*}} \frac{\mu(\mathrm{ord}(\psi))}{\phi(\mathrm{ord}(\psi))} S_{\beta,a}(\psi) \\
&= \sum_{d \mid q^n-1} \frac{\mu(d)}{\phi(d)} \cdot \sum_{\psi:d} S_{\beta,a}(\psi) \\
&= S_{\beta,a}(\psi_0) + \sum_{\substack{d \mid q^n-1 \\ d \neq 1}} \frac{\mu(d)}{\phi(d)} \cdot \sum_{\psi:d} S_{\beta,a}(\psi).
\end{aligned}$$

Taking absolute values on both sides, and using the inequality $|c_1 + \cdots + c_m| \geq |c_1| - |c_2| - \cdots - |c_m|$ for a list $c_1, \ldots, c_m$ of complex numbers, gives

$$\frac{q^n-1}{\phi(q^n-1)} \cdot P_{\beta,a} \geq |S_{\beta,a}(\psi_0)| - \sum_{\substack{d \mid q^n-1 \\ d \neq 1}} \frac{|\mu(d)|}{\phi(d)} \cdot \sum_{\psi:d} |S_{\beta,a}(\psi)|. \tag{14.9}$$

We will now apply Lemmas 14.3.3 and 14.3.4 to evaluate the right hand side of this inequality. Let us begin with two simple general observations, before considering the various cases which arise in detail. As $|S_{\beta,a}(\psi)|$ only depends on the order d of ψ and as there are exactly $\phi(d)$ characters ψ with order d, we can simplify Inequality (14.9) as follows: for every divisor d of $q^n - 1$, we have

$$\frac{|\mu(d)|}{\phi(d)} \cdot \sum_{\psi:d} |S_{\beta,a}(\psi)| = |\mu(d)| \cdot |S_{\beta,a}(\psi)|, \tag{14.10}$$

where ψ is an arbitrary character of order d. We will also use that

$$\sum_{d|N} |\mu(d)| = \sum_{d|\mathrm{rad}(N)} 1 = 2^{\omega(N)}$$

holds for every $N \in \mathbb{N}^*$, which follows from the basic properties of the Möbius function given in Proposition 2.1.12.

We first deal with the case $a \neq 0$, where Lemma 14.3.4 applies. If d divides Q, the term in (14.10) equals $|\mu(d)| \cdot \sqrt{q^{n-2}}$, which gives

$$\sum_{\substack{d|Q \\ d\neq 1}} |\mu(d)| \cdot \left|S_{\beta,a}(\psi)\right| = \sqrt{q^{n-2}} \cdot \sum_{\substack{d|Q \\ d\neq 1}} |\mu(d)| = \sqrt{q^{n-2}} \cdot \left(2^{\omega(Q)} - 1\right).$$

Similarly,

$$\begin{aligned}
\sum_{\substack{d|q^n-1 \\ Q\not\equiv 0 \bmod d}} |\mu(d)| \cdot \left|S_{\beta,a}(\psi)\right| &= \sqrt{q^{n-1}} \cdot \sum_{\substack{d|q^n-1 \\ Q\not\equiv 0 \bmod d}} |\mu(d)| \\
&= \sqrt{q^{n-1}} \cdot \Big(\sum_{d|q^n-1} |\mu(d)| - \sum_{d|Q} |\mu(d)| \Big) \\
&= \sqrt{q^{n-1}} \cdot \left(2^{\omega(q^n-1)} - 2^{\omega(Q)}\right).
\end{aligned}$$

In conjunction with the trivial Equation (14.5), this proves Proposition 14.3.6.

It remains to consider the case $a = 0$, where Lemma 14.3.3 applies. If d divides Q, the term in (14.10) equals $|\mu(d)| \cdot (q-1)\sqrt{q^{n-2}}$, and we obtain

$$\sum_{\substack{d|Q \\ d\neq 1}} |\mu(d)| \cdot \left|S_{\beta,0}(\psi)\right| = (q-1)\sqrt{q^{n-2}} \cdot \left(2^{\omega(Q)} - 1\right).$$

On the other hand, $\left|S_{\beta,0}(\psi)\right| = 0$ when Q is not a multiple of d. Together with Equation (14.6), this establishes Proposition 14.3.5. □

Propositions 14.3.5 and 14.3.6 immediately give the following sufficient conditions for the desired existence of primitive elements with prescribed (generalized) trace:

Corollary 14.3.7. *Consider the extension field $E = \mathrm{GF}(q^n)$ over $F = \mathrm{GF}(q)$, where $n \geq 2$, and let β be an arbitrary element in E^*. If*

$$1 + (2^{\omega(Q)} - 1)\sqrt{q^{n-2}}(q-1) < q^{n-1}, \tag{14.11}$$

then there exists a primitive element u for E such that $\mathrm{Tr}_{E/F}(\beta u) = 0$. □

Corollary 14.3.8. *Consider the extension field $E = \mathrm{GF}(q^n)$ over $F = \mathrm{GF}(q)$, where $n \geq 2$, and let β and a be arbitrary elements in E^* and F^*, respectively. If*

$$2^{\omega(Q)} - 1 + \left(2^{\omega(q^n-1)} - 2^{\omega(Q)}\right)\sqrt{q} < \sqrt{q^n}, \tag{14.12}$$

then there exists a primitive element u for E such that $\mathrm{Tr}_{E/F}(\beta u) = a$. □

We will apply these sufficient conditions in the subsequent sections, distinguishing the cases $a = 0$ and $a \neq 0$. Note that Corollary 14.3.7 is of no use for the case $n = 2$: then the left hand side of Inequality (14.11) equals

$$1 + (2^{\omega(Q)} - 1)(q-1) \geq 1 + (q-1) = q,$$

which is the right hand side of (14.11).

Remark 14.3.9. We close this section with a generalization of Propositions 14.3.5 and 14.3.6 which will be needed to handle a few exceptional pairs (q,n). As the proof proceeds along the same lines as before, we will leave it as an exercise.

Let t be a divisor of $q^n - 1$, and denote the number of all non-zero elements u in $H_{\beta,a}$ such that $\mathrm{pt}_t(q^n-1)$ divides the order of u by $P_{\beta,a}(t)$; thus $P_{\beta,a} = P_{\beta,a}(q^n-1)$. Then

$$P_{\beta,a}(t) = \sum_{\substack{v \in H_{\beta,a} \\ v \neq 0}} \mathscr{P}_t(v),$$

where $\mathscr{P}_t(v)$ is given by Equation (14.1). Moreover, one has the following estimates:

$$P_{\beta,0}(t) \geq \frac{\phi(t)}{t} \cdot \left(q^{n-1} - 1 - (2^{\omega(t \wedge Q)} - 1)(q-1)\sqrt{q^{n-2}}\right) \tag{14.13}$$

and, for $a \neq 0$,

$$P_{\beta,a}(t) \geq \frac{\phi(t)}{t} \cdot \left(q^{n-1} - (2^{\omega(t \wedge Q)} - 1)\sqrt{q^{n-2}} - (2^{\omega(t)} - 2^{\omega(t \wedge Q)})\sqrt{q^{n-1}}\right), \tag{14.14}$$

where $t \wedge Q$ denotes the greatest common divisor of t and Q. □

Exercises

Exercise 14.3.10. Prove Inequalities (14.13) and (14.14). □

14.4 A Preliminary Result

We now apply the sufficient conditions obtained in Section 14.3 to prove a preliminary result which will be useful throughout this chapter: Theorems 14.1.6 and 14.1.7 hold whenever the radical of $q-1$ divides the extension degree n. Trivially, this covers the binary case $q = 2$, and we will start with this rather simple special case. In fact, here we can prove a slightly stronger result:

Proposition 14.4.1. *The assertions of Theorems 14.1.6, 14.1.7 and 14.1.8 hold for* $q=2$.

Proof. With $Q=Q(2,n)=2^n-1$ and $\omega := \omega(Q)$, the conditions in Corollaries 14.3.7 and 14.3.8 become

$$1+(2^\omega-1)\sqrt{2^{n-2}} < 2^{n-1} \quad \text{and} \quad 2^\omega-1 < \sqrt{2^n},$$

respectively; note that the validity of the first condition implies that of the second one. By Example 13.6.4,

$$2^\omega-1 \le 5\cdot\sqrt[4]{2^n-1}-1 < 5\cdot\sqrt[4]{2^n}-1,$$

and hence

$$1+(2^\omega-1)\sqrt{2^{n-2}} < 5\cdot\sqrt[4]{2^n}\cdot\sqrt{2^{n-2}} = 5\cdot\sqrt[4]{2^{3n-4}} \le 2^{n-1}$$

for all n satisfying $5\le\sqrt[4]{2^n}$. Thus the sufficient conditions in question certainly hold for all $n\ge 10$.

Note that 2^n-1 is a Mersenne prime for $n\in\{3,5,7\}$; then every non-zero element is actually primitive, and there is nothing to prove. For the remaining four cases with $4\le n\le 9$, one can check Condition (14.11) directly; see Table 14.1.

Finally, the smallest instance $n=2$ is also easy to check, by using a primitive third root of unity, say λ. Then all assertions follows from the fact that λ and λ^2 have trace 1, whereas 0 and 1 have trace 0. □

Table 14.1 Evaluation of Corollary 14.3.7 for $q=2$ and $n\in\{4,6,8,9\}$.

n	2^n-1	ω	$1+(2^\omega-1)\cdot\sqrt{2^{n-2}}$	2^{n-1}
4	$3\cdot 5$	2	7	8
6	$3^2\cdot 7$	2	13	32
8	$3\cdot 5\cdot 17$	3	57	128
9	$7\cdot 73$	2	<35	256

We note in passing that the existence of a primitive element u in $E=\mathrm{GF}(2^n)$ with non-zero trace in the ground field $F=\mathrm{GF}(2)$ also follows from the primitive normal basis theorem, since any element in E which is normal over F has (E,F)-trace 1. However, one needs the results of Section 14.3 to deal with the homogeneous case and with arbitrary values $\beta\in E^*$.

Proposition 14.4.2. *Let q be a prime power and $n\ge 3$, and assume that n is a multiple of the radical of $q-1$. Then the assertions of Theorems 14.1.6 and 14.1.7 hold.*

Proof. In view of Proposition 14.4.1, we may assume $q\ge 3$. Note that every prime divisor of $q-1$ divides $Q=\frac{q^n-1}{q-1}$, by the hypothesis on (q,n), so that $\omega(Q)=\omega(q^n-1)=:\omega$. Then the conditions in Corollaries 14.3.7 and 14.3.8 become

$$1+(2^{\omega}-1)\sqrt{q^{n-2}}(q-1) < q^{n-1} \quad \text{and} \quad 2^{\omega}-1 < \sqrt{q^{n}},$$

respectively; as in the binary case, the validity of the first condition implies that of the second one.

Again, we use the estimate on ω in Example 13.6.4 to obtain

$$2^{\omega}-1 \leq 5\cdot\sqrt[4]{Q}-1 < 5\cdot\sqrt[4]{2q^{n-1}}-1 < 5\cdot\sqrt[4]{2}\cdot\sqrt[4]{q^{n-1}},$$

and hence

$$1+(2^{\omega}-1)\sqrt{q^{n-2}}(q-1) < 1+5\cdot\sqrt[4]{2}\cdot\sqrt[4]{q^{3n-1}}.$$

Therefore, the sufficient conditions in Corollaries 14.3.7 and 14.3.8 are satisfied whenever $5\cdot\sqrt[4]{2}\cdot\sqrt[4]{q^{3n-1}} \leq q^{n-1}$, which means

$$q^{n-3} \geq 1250; \tag{14.15}$$

note that this holds whenever $n \geq 10$ and $q \geq 3$.

For $n = 9$, Inequality (14.15) requires $q \geq 4$; however, the case $(q,n) = (3,9)$ does not arise in this context, as then the hypothesis that $\mathrm{rad}(q-1)$ divides n is violated. Similarly, (14.15) gives $q \geq 5$ for $n = 8$. Now the hypothesis is violated for $q = 4$, but holds for $q = 3$. Therefore, we check Condition (14.11) directly for the pair $(q,n) = (3,8)$; see Table 14.2.

When $n = 7$, we have to consider the cases $q \leq 5$; here the radical hypothesis is never satisfied. Similarly, the case $n = 5$ requires to check the prime powers $q < 37$; again, the hypothesis is always violated.

Now assume $n = 6$, so that (14.15) holds for all $q \geq 11$. Here the radical hypothesis only excludes one case, namely $q = 8$, and we are left with the five cases where $q \in \{3,4,5,7,9\}$. Similarly, for $n = 4$ it remains to investigate those prime powers $q < 1250$ for which $q-1$ is a power of 2, that is, $q \in \{3,5,9,17\}$. These nine exceptional pairs are likewise considered in Table 14.2. With the exception of the pair $(q,n) = (5,4)$, Condition (14.11) is always satisfied; this case will be settled in Example 14.4.3.

Table 14.2 Evaluation of Corollary 14.3.7 for pairs (q,n) with $\mathrm{rad}(q-1) \mid n$.

q	n	q^n-1	ω	$1+(2^{\omega}-1)\cdot\sqrt{q^{n-2}}\cdot(q-1)$	q^{n-1}
3	4	$2^4\cdot 5$	2	19	27
5	4	$2^4\cdot 3\cdot 13$	3	141	25 *
9	4	$2^5\cdot 5\cdot 41$	3	505	729
17	4	$2^6\cdot 3^2\cdot 5\cdot 29$	4	4081	4913
3	6	$2^3\cdot 7\cdot 13$	3	127	243
4	6	$3^2\cdot 5\cdot 7\cdot 13$	4	721	1024
5	6	$2^3\cdot 3^2\cdot 7\cdot 31$	4	1501	3125
7	6	$2^4\cdot 3^2\cdot 19\cdot 43$	4	4411	16807
9	6	$2^4\cdot 5\cdot 7\cdot 13\cdot 73$	5	20089	59049
3	8	$2^5\cdot 5\cdot 41$	3	379	2187

For $n = 3$, Inequality (14.15) gives no information, but the radical hypothesis requires that $q-1$ is a power of 3, so that q has to be even. Thus we need to consider the condition $2^b - 1 = 3^c$ for some $b, c \in \mathbb{N}^*$. It is well-known that the only solution of this Diophantine equation is $(b,c) = (2,1)$; see Exercise 14.4.4. Hence the only case arising for $n = 3$ is $q = 4$, which has already been dealt with in Example 14.1.5 and Exercise 14.1.11. □

We remark that we had to exclude the case $n = 2$ from the assertion of Proposition 14.4.2, since, for instance, $\mathrm{rad}(q-1) = 2$ for every Fermat prime q.

Example 14.4.3. We now deal with the one remaining pair $(q,n) = (5,4)$. Here at least Corollary 14.3.8 applies, as $2^\omega - 1 = 7 < 25 = \sqrt{5^4}$. This gives the validity of the inhomogeneous case for $(q,n) = (5,4)$, and then the homogeneous case can be settled via the approach used in Remark 14.2.3. For this, let $K = \mathrm{GF}(5^2)$ be the unique intermediate field of E/F and consider the field extension E/K with degree 2. Let $\beta \in E^*$ and $\overline{a} \in K^*$ be arbitrary. With $\overline{q} = 25$, $\overline{Q} = 25 + 1 = 26$ and $\overline{n} = 2$, the condition in Corollary 14.3.8 holds for E/K: as $\overline{Q} = 2 \cdot 13$ and $\overline{q}^2 - 1 = 624 = 2^3 \cdot 3 \cdot 13$, one has

$$(2^{\omega(\overline{Q})} - 1) + (2^{\omega(\overline{q}^2-1)} - 2^{\omega(\overline{Q})})\sqrt{\overline{q}} = 3 + 4 \cdot 5 < 25 = \sqrt{\overline{q}^{\overline{n}}}.$$

Hence there exists a primitive element u of E with $\mathrm{Tr}_{E/K}(\beta u) = \overline{a}$. We apply this for some $\overline{a} \in K^*$ satisfying $\mathrm{Tr}_{K/F}(\overline{a}) = 0$; of course, there are exactly four elements with this property (in fact, the four primitive 8-th roots of unity). Using the transitivity of the trace mappings, we obtain $\mathrm{Tr}_{E/F}(\beta u) = \mathrm{Tr}_{K/F}(\mathrm{Tr}_{E/K}(\beta u)) = \mathrm{Tr}_{K/F}(\overline{a}) = 0$, as desired. □

Exercises

Exercise 14.4.4. Show that the Diophantine equation $2^b - 1 = 3^c$ (with $b, c \in \mathbb{N}^*$) has the unique solution $a = 2$ and $b = 1$. □

14.5 The Inhomogeneous Case for $n \geq 3$

In this section, we establish Theorem 14.1.6 for all extension degrees $n \geq 3$: in the terminology of Definition 14.2.1, every integer $n \geq 3$ is extensive. This relies on showing that the sufficient condition (14.12) in Corollary 14.3.8 is satisfied in all these cases, with the single exception of the pair $(q,n) = (11,3)$, which is dealt with individually at the end of this section.

Thus let $E = \mathrm{GF}(q^n)$ and $F = \mathrm{GF}(q)$, where $n \geq 3$, and let $\beta \in E^*$ and $a \in F^*$. By Proposition 14.4.2, we may assume that $\mathrm{rad}(q-1)$ does not divide n; in particular, $q \neq 2$. Throughout, we will use the following abbreviations:

$$\omega_0 := \omega(q-1), \quad \omega_1 := \omega(Q), \quad \text{and} \quad \omega := \omega(q^n - 1).$$

Then Inequality (14.12) becomes

$$\left(2^{\omega_1} - 1\right) + \left(2^{\omega} - 2^{\omega_1}\right)\sqrt{q} < \sqrt{q^n}. \tag{14.16}$$

Since

$$(2^{\omega_1} - 1) + (2^{\omega} - 2^{\omega_1})\sqrt{q} = -1 + 2^{\omega_1}(1 - \sqrt{q}) + 2^{\omega}\sqrt{q} < 2^{\omega}\sqrt{q},$$

we may also use the weaker (but more convenient) condition

$$2^{\omega} \leq \sqrt{q^{n-1}} \tag{14.17}$$

to establish the existence of a primitive element with the desired property. By Example 13.6.4, $2^{\omega} < 5 \cdot \sqrt[4]{q^n - 1} < 5 \cdot \sqrt[4]{q^n}$. Therefore, Inequality (14.17) certainly holds whenever

$$\sqrt[4]{q^{n-2}} \geq 5. \tag{14.18}$$

Step 1. If $n \geq 8$, then Inequality (14.18) holds for all $q \geq 3$, so that every integer $n \geq 8$ is extensive. Moreover, for $n \in \{3,4,5,6,7\}$, Inequality (14.18) is satisfied for almost all q; thus there are only finitely many exceptional pairs (q,n) for which we have to investigate Inequality (14.16) more carefully.

Step 2. Assume that $n = 7$. Then Inequality (14.18) holds for all $q \geq 4$. For $q = 3$, we check Inequality (14.16) directly. The prime power factorization of $3^7 - 1$ is $2 \cdot 1093$, so that $\omega_1 = 1$ and $\omega = 2$, and the required condition holds, as $1 + 2 \cdot \sqrt{3} < \sqrt{3^7}$. Thus $n = 7$ is likewise extensive.

Step 3. Let $n = 6$. Here it suffices to recall that the extensiveness of $n = 6$ will follow from that of $n = 3$, by Proposition 14.2.2.

Step 4. For $n = 5$, Inequality (14.18) holds for all $q \geq 9$. Moreover, Inequality (14.16) – and, in fact, even the weaker condition in Inequality (14.17) – is satisfied for the remaining values of q; see Table 14.3. This shows that $n = 5$ is extensive.

Table 14.3 Evaluation of Inequality (14.16) for $n = 5$ and $q \in \{3,4,5,7,8\}$.

q	$q^5 - 1$	ω	ω_1	$(2^{\omega_1} - 1) + (2^{\omega} - 2^{\omega_1})\sqrt{q}$	$\sqrt{q^{5/2}}$
3	$2 \cdot 11^2$	2	1	< 5	> 15
4	$3 \cdot 11 \cdot 31$	3	2	11	32
5	$2^2 \cdot 11 \cdot 71$	3	2	< 12	> 55
7	$2 \cdot 3 \cdot 2801$	3	1	< 17	> 129
8	$7 \cdot 31 \cdot 151$	3	2	< 15	> 181

The cases $n = 4$ and $n = 3$ may, in principle, be solved as the case $n = 5$:

- when $n = 4$, Inequality (14.18) is satisfied for all $q \geq 25$,
- and when $n = 3$, it holds for all $q \geq 625$.

Instead of producing further extensive tables, we shall examine the instances in question using improved estimates. Note that the extensiveness of $n = 4$ also follows from that of $n = 2$, by Proposition 14.2.2. Nevertheless, we have decided to include a direct discussion of the quartic case, as the solution of the basic problem for quadratic extensions requires rather extensive computer searches; see Sections 14.8 and 14.9.

Step 5. Let $n = 4$. Assume first that q is odd, so that 16 divides $q^4 - 1$. For $\omega \geq 4$, this yields

$$q^4 > q^4 - 1 \geq 16 \cdot 3 \cdot 5 \cdot 7 \cdot 11^{\omega-4} = 1680 \cdot 11^{\omega-4},$$

which implies

$$\sqrt{q^3} = (q^4)^{3/8} > 1680^{3/8} \cdot (11^{3/8})^{\omega-4} > 16 \cdot 2^{\omega-4} = 2^{\omega},$$

since $11^{3/8} > 2$ and $1680^{3/8} > 16$. Thus the sufficient condition of Inequality (14.17) holds for $\omega \geq 4$. Now let $\omega \leq 3$. Then $2^{\omega} \leq 8$, and Inequality (14.17) is satisfied for all $q \geq 4$. This leaves only a case already excluded, namely $q = 3$, where $n = 4$ is divisible by $\mathrm{rad}(q-1)$.

Now assume that q is even. If $\omega \geq 5$, then

$$q^4 > q^4 - 1 \geq 3 \cdot 5 \cdot 7 \cdot 11 \cdot 13 \cdot 17^{\omega-5} = 15015 \cdot 17^{\omega-5}.$$

This shows

$$\sqrt{q^3} = (q^4)^{3/8} > 15015^{3/8} \cdot (17^{3/8})^{\omega-5} > 32 \cdot 2^{\omega-5} = 2^{\omega},$$

and again Inequality (14.17) is satisfied. Finally, let $\omega \leq 4$. Then $2^{\omega} \leq 16$, and Inequality (14.17) holds for all $q \geq 7$. Since q is even, it only remains to consider the case $q = 4$. Here $4^4 - 1 = 3 \cdot 5 \cdot 17$ gives $\omega = 3$, hence $2^{\omega} = 8 = \sqrt{4^3}$ shows that (14.17) holds with equality.

Altogether, we see that $n = 4$ is indeed extensive.

Step 6. Finally, it remains to show that $n = 3$ is likewise extensive; not surprisingly, this is considerably more involved. We follow the exposition of Cohen [82].

We first deal with the cases where $\omega = \omega(q^3 - 1) \leq 4$; let us denote the set of all prime powers q satisfying this condition by $\mathscr{F}$. As the sufficient criterion of Inequality (14.17) becomes $2^{\omega} \leq q$ in this case, we only need to consider the $q \in \mathscr{F}$ with $q < 16$ in more detail; for these, we determine the exact value of ω. As the data in Table 14.4 show, all $q < 16$ actually belong to $\mathscr{F}$. Moreover, (14.17) is satisfied for $q \in \{5, 8, 9, 13\}$, but not for $q \in \{3, 7, 11\}$. (The cases $q = 4$ and $q = 2$ were already excluded by Proposition 14.4.2.)

For the three remaining values $q \in \{3, 7, 11\}$, we use the data from Table 14.4 to test whether Inequality (14.16) holds. This is the case for $q = 3$ and $q = 7$, but not for $q = 11$:

- if $q = 3$, then $(2^{\omega_1} - 1) + (2^{\omega} - 2^{\omega_1})\sqrt{q} = 1 + 2\sqrt{3} < \sqrt{3^3}$;

- if $q = 7$, then $(2^{\omega_1} - 1) + (2^{\omega} - 2^{\omega_1})\sqrt{q} = 3 + 4\sqrt{7} < \sqrt{7^3}$;
- for $q = 11$, we have $(2^{\omega_1} - 1) + (2^{\omega} - 2^{\omega_1})\sqrt{q} = 3 + 12\sqrt{11} > \sqrt{11^3}$.

We will deal with the pair $(11,3)$ at the end of this section; see Example 14.5.1.

Table 14.4 Evaluation of Inequality (14.16) for $q < 16$ and $n = 3$.

q	$q^3 - 1$	2^{ω}
3	$2 \cdot 13$	4 *
5	$2^2 \cdot 31$	4
7	$2 \cdot 3^2 \cdot 19$	8 *
8	$7 \cdot 73$	4
9	$2^3 \cdot 7 \cdot 13$	8
11	$2 \cdot 5 \cdot 7 \cdot 19$	16 *
13	$2^2 \cdot 3^2 \cdot 61$	8

From now on, we may assume that $q \geq 16$ and $\omega \geq 5$. For these cases, we will work with Inequality (14.16), which we here rewrite as

$$2^{\omega} - 2^{\omega_1} + \frac{2^{\omega_1} - 1}{\sqrt{q}} < q.$$

Since $q \geq 16$, we have

$$2^{\omega} - 2^{\omega_1} + \frac{2^{\omega_1} - 1}{\sqrt{q}} \leq 2^{\omega} - 2^{\omega_1} + \frac{2^{\omega_1} - 1}{4} < 2^{\omega} - 3 \cdot 2^{\omega_1 - 2},$$

which gives the following sufficient condition for the existence of a primitive element with the desired property:

$$2^{\omega} - 3 \cdot 2^{\omega_1 - 2} \leq q. \tag{14.19}$$

This certainly holds whenever

$$q \geq 2^{\omega} - 2, \tag{14.20}$$

as $\omega_1 \geq 1$ and as both q and 2^{ω} are integers. At this point, we need some additional notation:

- for every $j \in \mathbb{N}^*$, let p_j denote the j-th prime, and let r_j be the j-th prime which is congruent to 1 modulo 6;
- for $m \in \mathbb{N}^*$, put $\pi_0(m) := p_1 \cdots p_m$ and $\pi_1(m) := r_1 \cdots r_m$.

We will also require an identity connecting the three values ω, ω_0 and ω_1 introduced at the beginning of this section, namely

$$\omega_1 = \omega - \omega_0 + \varepsilon, \quad \text{where } \varepsilon = \begin{cases} 1 & \text{if } q \equiv 1 \bmod 3, \\ 0 & \text{otherwise,} \end{cases}$$

which is immediate from

$$\gcd(q-1,Q) = \gcd(q-1,q^2+q+1) = \begin{cases} 3 & \text{if } q \equiv 1 \bmod 3, \\ 1 & \text{otherwise.} \end{cases}$$

Note also that every prime divisor $r \neq 3$ of Q is congruent to 1 modulo 6, since q^2+q+1 is odd and $\mathrm{ord}_r(q) = 3 \mid r-1$. Combining the preceding observations gives the following useful lower bounds for q and Q in terms of the numbers of distinct prime divisors of $q-1$ and Q, respectively:

$$q > q-1 \geq \pi_0(\omega_0) \quad \text{and} \quad Q \geq 3^{\varepsilon} \cdot \pi_1(\omega_1 - \varepsilon).$$

Let us note a simple but important consequence of this bound on q: Inequality (14.20) holds whenever $\omega_0 \geq m$ and $\pi_0(m) \geq 2^{\omega} - 2$ for some $m \in \mathbb{N}^*$, as then

$$2^{\omega} - 2 \leq \pi_0(m) \leq \pi_1(\omega_0) \leq q-1 < q.$$

We will now use this observation to settle the remaining case $\omega \geq 5$, by splitting it into five subcases.

(a) For $\omega = 5$, we choose $m = 3$. Then $2^{\omega} - 2 = 30 = 2 \cdot 3 \cdot 5 = \pi_0(m)$, and hence (14.20) holds for $\omega_0 \geq 3$. On the other hand,

$$\omega_1 = \omega - \omega_0 + \varepsilon = 5 - \omega_0 + \varepsilon \geq 3 + \varepsilon \quad \text{if } \omega_0 \leq 2,$$

which implies $Q \geq \pi_1(3) = 7 \cdot 13 \cdot 19 = 1729$. Hence $q \geq 43 > 2^{\omega} - 2$, so that (14.20) actually holds for all values of ω_0.

(b) Similarly, we choose $m = 4$ when $\omega = 6$. Then $2^{\omega} - 2 = 62 < 210 = 2 \cdot 3 \cdot 5 \cdot 7 = \pi_0(m)$, and (14.20) holds for $\omega_0 \geq 4$. As in case (a), we have

$$\omega_1 = \omega - \omega_0 + \varepsilon = 6 - \omega_0 + \varepsilon \geq 3 + \varepsilon \quad \text{if } \omega_0 \leq 3;$$

again, we will show that (14.20) also holds for these values of ω_0, but this time we need slightly more effort.

- If $q \equiv 1 \bmod 3$, then $\omega_1 \geq 4$ and therefore $Q \geq 3 \cdot \pi_1(3) = 3 \cdot 1729 = 5187$, which implies $q \geq 73 > 2^{\omega} - 2$.
- Now let $q \not\equiv 1 \bmod 3$. If $\omega_0 = 3$, then $q > q-1 \geq 2 \cdot 5 \cdot 7 = 70 > 2^{\omega} - 2$, as desired. Finally, we have $\omega_1 = \omega - \omega_0 = 6 - \omega_0 \geq 4$ for $\omega_0 \leq 2$, so that $Q \geq \pi_1(4) = 7 \cdot 13 \cdot 19 \cdot 31 = 53599$, which even gives $q \geq 233$.

(c) Now let $\omega = 7$, so that $2^{\omega} - 2 = 126$. As in Case (b), this gives $2^{\omega} - 2 < \pi_0(m)$ for $m = 4$, so that (14.20) again holds for $\omega_0 \geq 4$. For $\omega_0 \leq 3$, we now obtain $\omega_1 = \omega - \omega_0 + \varepsilon = 7 - \omega_0 + \varepsilon \geq 4 + \varepsilon \geq 4$, and the validity of condition (14.20) follows from the estimate $q \geq 233$, as in case (b).

(d) For $\omega = 8$, we choose $m = 5$. Then $2^{\omega} - 2 = 254 < 2 \cdot 3 \cdot 5 \cdot 7 \cdot 11 = \pi_0(m)$, and hence (14.20) is satisfied whenever $\omega_0 \geq 5$. We now check that (14.20) also holds when $\omega_0 \leq 4$. This is easy for $\omega_0 \leq 3$: then $\omega_1 = \omega - \omega_0 + \varepsilon = 8 - \omega_0 + \varepsilon \geq 5 + \varepsilon$, so that $Q \geq \pi_1(5) = \pi_1(4) \cdot 37 = 1983163$, which obviously implies the required inequality for q, namely $q > 254 = 2^{\omega} - 2$. This actually also holds when $\omega_0 = 4$, as is seen via the following simple case distinction:

- If $q \not\equiv 1 \bmod 3$, then $q > q-1 \geq 2 \cdot 5 \cdot 7 \cdot 11 = 770$.
- If $q \equiv 1 \bmod 3$, then $\omega_1 = \omega - \omega_0 + \varepsilon = 5$ implies $Q \geq 3 \cdot \pi_1(4) = 3 \cdot 53599 = 160797$, so that $q \geq 401$.

(e) We finally consider the case $\omega \geq 9$, where we choose m as the integral part of $(\omega+1)/2$; thus $2m-1 \leq \omega \leq 2m$ and $m \geq 5$. We require the following inequalities, which hold for all $\ell \geq 5$ and are easily proved by induction, see Exercise 14.5.2:

$$\pi_0(\ell) \geq 4^{\ell} \quad \text{and} \quad \pi_1(\ell) > \frac{6^{\ell} \cdot (\ell+1)!}{3} > 16^{\ell}.$$

Let us first assume that $\omega_0 \geq m$. Then

$$q > q-1 \geq \pi_0(\omega_0) \geq \pi_0(m) \geq 4^m = 2^{2m} \geq 2^{\omega},$$

so that Inequality (14.17) holds for this case. It remains to consider the case where $\omega_0 \leq m-1$. Here we have

$$\omega_1 = \omega - \omega_0 - \varepsilon \geq (2m-1) - (m-1) + \varepsilon = m + \varepsilon,$$

and hence

$$16^m < \pi_1(m) \leq \pi_1(\omega_1 - \varepsilon) \leq 3^{\varepsilon} \cdot \pi_1(\omega_1 - \varepsilon) \leq Q < (q+1)^2.$$

This gives

$$2^{\omega} \leq 4^m = \sqrt{16^m} < q+1,$$

and shows that condition (14.20) is satisfied. □

Example 14.5.1. We now consider the pair $(q,n) = (11,3)$ to complete the proof of Theorem 14.1.6 for $n \geq 3$; as we have seen, the sufficient condition (14.16) holds for all other cases. As in our treatment of the pair $(q,n) = (7,6)$ for the primitive normal basis theorem in Example 13.10.3, we will use a simple estimate based on the inclusion-exclusion principle. For $(q,n) = (11,3)$, we have

$$q-1 = 2 \cdot 5 \quad \text{and} \quad Q = \frac{q^3-1}{q-1} = 11^2 + 11 + 1 = 7 \cdot 19.$$

Given any divisor t of $q^3 - 1 = 1330$, we let $U(t)$ denote the set of all elements $u \in E^*$ such that $\mathrm{Tr}_{E/F}(\beta u) = a$ and $\mathrm{pt}_t(1330)$ divides $\mathrm{ord}(u)$. Note that $|U(t)| = P_{\beta,a}(t)$, where $P_{\beta,a}(t)$ is as in Remark 14.3.9. Obviously,

$$U(1330) = U(2 \cdot 5) \cap U(7 \cdot 19) = U(10) \cap U(133),$$

and hence

$$\begin{aligned} P_{\beta,a} &= P_{\beta,a}(1330) = |U(10) \cap U(133)| \\ &= |U(10)| + |U(133)| - |U(10) \cup U(133)| \\ &\geq |U(10)| + |U(133)| - |\{u \in E^* : \mathrm{Tr}_{E/F}(\beta u) = a\}| \\ &= |U(10)| + |U(133)| - 121. \end{aligned}$$

Applying Inequality (14.14) gives

$$|U(10)| = P_{\beta,a}(10) \geq \frac{\phi(10)}{10} \cdot \left(11^2 - (2^{\omega(10)} - 1) \cdot \sqrt{11^{3-1}}\right)$$

and

$$|U(133)| = P_{\beta,a}(133) \geq \frac{\phi(133)}{133} \cdot \left(11^2 - (2^{\omega(133)} - 1) \cdot \sqrt{11^{3-2}}\right).$$

Hence

$$\begin{aligned} P_{\beta,a} &\geq \frac{4}{10} \cdot (121 - 33) + \frac{6 \cdot 18}{133} \cdot \left(121 - 3 \cdot \sqrt{11}\right) - 121 \\ &= \frac{8283}{665} - \frac{324}{133} \cdot \sqrt{11} > 4, \end{aligned}$$

so that $P_{\beta,a}$ is indeed positive. □

Exercises

Exercise 14.5.2. For every $j \in \mathbb{N}^*$, let p_j denote the j-th prime and r_j the j-th prime which is congruent to 1 modulo 6; furthermore, put $\pi_0(m) := p_1 \cdots p_m$ and $\pi_1(m) := r_1 \cdots r_m$ for $m \in \mathbb{N}^*$. Prove that the following estimates hold for all $m \geq 5$:

$$\pi_0(m) \geq 4^m \quad \text{and} \quad \pi_1(m) \geq \frac{6^m \cdot (m+1)!}{3} > 16^m. \qquad \square$$

14.6 The Homogeneous Case for $n \geq 4$

After settling the inhomogeneous case for all extension degrees $n \geq 3$ in Section 14.5, we turn our attention to the homogeneous case, that is, we investigate the existence of primitive elements in hyperplanes through the origin in some extension $E = \mathrm{GF}(q^n)$ of $F = \mathrm{GF}(q)$. Recall that any such hyperplane has the form $H_{\beta,0} = \{v \in E : \mathrm{Tr}_{E,F}(\beta v) = 0\}$ for some $\beta \in E^*$.

We will prove Theorem 14.1.7 for $n \geq 4$ in the present section, and for $n = 3$ in the next section. In other words, n is 0-extensive, whenever $n \geq 3$. The proof will be quite similar to our treatment of the inhomogeneous case in the last section, at least for $n \geq 4$. While we will proceed along the same lines also for the case

$n = 3$, the details turn out to be considerably more demanding: we will have to deal individually with seven instances $(q, 3)$, and computer support will be required to exclude many further potential exceptions in certain quite large ranges of q. This is the reason why we have decided to deal with this case in a separate section.

In view of Proposition 14.4.2, we may again assume that $\mathrm{rad}(q-1)$ does not divide n, in particular, $q \neq 2$. As before, $P_{\beta,0}$ denotes the number of primitive elements contained in $H_{\beta,0}$. By Proposition 14.3.5,

$$P_{\beta,0} \geq \frac{\phi(q^n-1)}{q^n-1} \cdot \left(q^{n-1} - 1 - (2^{\omega_1} - 1)(q-1)\sqrt{q^{n-2}}\right), \tag{14.21}$$

where again $\omega_1 = \omega(Q)$ and $Q = (q^n-1)/(q-1)$ (and always $n \geq 3$). In the interest of better readability, we sometimes write t instead of $\omega_1 = \omega(Q)$.

Before we present the proof proper, we begin with a rather simple observation which will be needed to deal with the cases where $n \leq 5$.

Lemma 14.6.1. *Let $p_1 < p_2 < \cdots < p_t$ be the distinct prime divisors of Q, and suppose that $P_{\beta,0} = 0$. Then:*

$$\frac{p_1 p_2 \cdots p_t}{2} < q^{n-1} < (2^t - 1)^{2(n-1)/(n-2)} < 2^{2(n-1)t/(n-2)}. \tag{14.22}$$

Proof. The lower bound does not even require the hypothesis $P_{\beta,0} = 0$. In fact, it is rather trivial:

$$p_1 \cdots p_t \leq Q < 2q^{n-1}.$$

In view of Inequality (14.21), the hypothesis $P_{\beta,0} = 0$ implies

$$q^{n-1} - 1 \leq (2^t - 1)(q-1)\sqrt{q^{n-2}}.$$

Hence

$$q^{n-1} < Q - 1 = q \cdot \frac{q^{n-1} - 1}{q-1} \leq q(2^t - 1)\sqrt{q^{n-2}},$$

that is, $q^{(n-2)/2} < 2^t - 1$, which yields the upper bound for q^{n-1}. □

Remark 14.6.2. Usually, it will suffice to use the weaker upper bound

$$q^{n-1} < 2^{2(n-1)t/(n-2)}$$

when applying Lemma 14.6.1, which is easier to handle. However, we will actually need the sharper condition $q^{n-1} < (2^t - 1)^{2(n-1)/(n-2)}$ in some parts of the proof for the case $n = 3$. □

We now turn to the proof of Theorem 14.1.7 for $n \geq 4$ and begin by showing that only three small values of n need to be investigated in detail. For this, we use the estimates given in the proof of Proposition 14.4.2; in particular, Inequality (14.15) guarantees that $P_{\beta,0}$ is positive whenever $q^{n-3} \geq 1250$. Table 14.6 shows for which values of q this condition holds (depending on the value of n).

Table 14.5 Evaluation of Inequality (14.15).

n	≥ 10	9	8	7	6	5	4
q	≥ 3	≥ 4	≥ 5	≥ 7	≥ 11	≥ 37	≥ 1250

In view of $q \neq 2$, we conclude that every integer $n \geq 10$ is 0-extensive. By Remark 14.2.3, the 0-extensiveness of $n = 9$ and $n = 6$ follows from the extensiveness of $n = 3$ established in Section 14.5. Similarly, the 0-extensiveness of $n = 8$ follows from the extensiveness of $n = 4$.

Of course, the 0-extensiveness of $n = 4$ can be obtained in the same manner: it is implied by the extensiveness of $n = 2$. Nevertheless, in spite of the effort needed, we have decided to include an explicit proof for the quartic case, since the solution of the basic problem for quadratic extensions in Sections 14.8 and 14.9 relies heavily on using some computer algebra system (as already mentioned in Section 14.5).

Thus we are left with the extension degrees $n \in \{4,5,7\}$.

Case 1: $n = 7$. This case is easily settled using the sufficient criterion (14.11) in Corollary 14.3.7; see Table 14.6.

Table 14.6 Evaluation of Inequality (14.11) for $n = 7$.

q	Q	ω_1	$1+(2^{\omega_1}-1)(q-1)\sqrt{q^{n-2}}$	q^{n-1}
3	1093	1	< 33	729
4	$43 \cdot 127$	2	289	4096
5	19531	1	< 225	15625

Case 2: $n = 5$. Then Q is odd, and 5 divides Q if and only if $q \equiv 1 \bmod 5$. It is easily checked that every other prime divisor of Q has to be congruent to 1 modulo 10.

Suppose that $P_{\beta,0} = 0$. Then Lemma 14.6.1 gives

$$\frac{p_1 p_2 \cdots p_t}{2} < q^4 < 2^{8t/3}.$$

We now use a simple case analysis to show that there is no prime power q for which q^4 is in the required range, which will establish that $n = 5$ is indeed 0-extensive.

Assume first that $q \not\equiv 1 \bmod 5$. If $t = 1$, the upper bound $q^4 < 2^{8/3}$ is violated. For $t \geq 2$, we also obtain a contradiction as follows:

$$\frac{p_1 p_2 \cdots p_t}{2} \geq \frac{11 \cdot 31}{2} \cdot 41^{t-2} > 128 \cdot 16^{t-2} = 2^{4t-1} > 2^{8t/3}.$$

Thus let $q \equiv 1 \bmod 5$. If $t \leq 2$, we would need $q^4 \leq 2^{16/3}$, which is obviously impossible. For $t \geq 3$, we obtain

$$\frac{p_1 p_2 \cdots p_t}{2} \geq \frac{5 \cdot 11 \cdot 31}{2} \cdot 41^{t-3} > 512 \cdot 16^{t-3} = 2^{4t-3},$$

which is again a contradiction, as $4t - 3 > \frac{8}{3}t$ for $t \geq 3$.

Case 3: $n = 4$. This is similar to the previous case $n = 5$, but more involved. Supposing that $P_{\beta,0} = 0$, we now obtain from Lemma 14.6.1

$$\frac{p_1 p_2 \cdots p_t}{2} < q^3 < 2^{3t}.$$

Again, we undertake a (now rather elaborate) case analysis of this condition; this time, there is a unique prime power q for which q^4 is in the required range, namely $q = 13$. Hence this case has to be settled individually, which requires some effort; see Example 14.6.3.

(a) Assume first that q is even, so that Q is odd. It is a simple matter to obtain the desired contradiction if $t \geq 7$:

$$\frac{p_1 p_2 \cdots p_t}{2} \geq \frac{3 \cdot 5 \cdot 7 \cdot 11 \cdot 13 \cdot 17 \cdot 19}{2} \cdot 23^{t-7} > 2^{21} \cdot 16^{t-7} = 2^{4t-7} \geq 2^{3t}.$$

Thus let $t \leq 6$, so that we require $q^3 < 2^{18}$, that is, $q < 64$. As q is assumed to be even (but $q \neq 2$), we are left with the values $q \in \{4, 8, 16, 32\}$. These are easily seen to be out of range by computing the precise t-values. In fact, the upper bound $q < 2^t$ is always violated:

q	32	16	8	4
Q	$3 \cdot 5^2 \cdot 11 \cdot 41$	$17 \cdot 257$	$3^2 \cdot 5 \cdot 13$	$5 \cdot 17$
2^t	16	4	8	4

(b) Assume next that $q \equiv 3 \bmod 4$. Then $Q = (q+1)(q^2+1)$ is a multiple of 8, and we may therefore replace the term p_1 in Inequality (14.22) by 8, which results in the following sharper restriction for q:

$$4 p_2 \cdots p_t < q^3 < 2^{3t}.$$

From this, we easily obtain the desired contradiction whenever $t \geq 8$:

$$4 p_2 \cdots p_t \geq 4 \cdot 3 \cdot 5 \cdot 7 \cdot 11 \cdot 13 \cdot 17 \cdot 19 \cdot 23^{t-8} > 2^{24} \cdot 16^{t-8} = 2^{4t-8} \geq 2^{3t}.$$

Note that $t \leq 2$ forces $q = 3$, which is covered by Proposition 14.4.2. For the remaining values $t \in \{3, 4, 5, 6, 7\}$, we compute a concrete range for q from the above restriction, which implies

$$\sqrt[3]{4 p_2 \cdots p_t} < q \leq 2^t - 1$$

(as q is an integer). This results in the following intervals:

t	7	6	5	4	3
q	$101 \leq q \leq 127$	$40 \leq q \leq 63$	$17 \leq q \leq 31$	$8 \leq q \leq 15$	$4 \leq q \leq 7$

Altogether, this leaves 14 possibilities for q, namely

$$q \in \{7, 11, 19, 23, 27, 31, 43, 47, 59, 67, 103, 107, 119, 127\}.$$

All these values are actually out of range; as in case (a), this can be seen by by computing the precise t-values. Again, it turns out that the upper bound $q < 2^t$ is always violated; in fact, one has

- $t = 5$ for $q \in \{47, 119\}$;
- $t = 4$ for $q \in \{23, 27, 43, 59, 67, 103, 107\}$;
- $t = 3$ for $q \in \{11, 19, 31, 127\}$;
- $t = 2$ for $q = 7$.

We leave the details to the reader and refrain from providing a list of the respective factorizations of Q.

(c) Finally, let $q \equiv 1 \bmod 4$. This time, we may replace the term p_1 in Inequality (14.22) only by 4, as Q is now a multiple of 4, but not of 8. This results in the condition

$$2p_2 \cdots p_t < q^3 < 2^{3t},$$

and we easily obtain the desired contradiction whenever $t \geq 9$:

$$2p_2 \cdots p_t \geq 2 \cdot 3 \cdot 5 \cdot 7 \cdot 11 \cdot 13 \cdot 17 \cdot 19 \cdot 23 \cdot 29^{t-9} > 2^{27} \cdot 16^{t-9} = 2^{4t-9} \geq 2^{3t}.$$

Also note that $t \leq 2$ is clearly impossible. For the remaining values of t, we proceed as in Case (b) and compute concrete ranges for q from the above restriction:

t	8	7	6	5	4	3
q	$214 \leq q \leq 255$	$80 \leq q \leq 127$	$32 \leq q \leq 63$	$14 \leq q \leq 31$	$6 \leq q \leq 15$	$4 \leq q \leq 7$

Since the values $q \in \{5, 9, 17\}$ are covered by Proposition 14.4.2, we are left with the following 19 possibilities:

$$q \in \{13, 25, 29, 37, 41, 49, 53, 61, 81, 89, 97, 101, 109, 113, 121, 125, 229, 233, 241\}.$$

As in the previous two cases, one factors the respective numbers Q to obtain the exact t-values; again, the details will be left to the reader. As a result, we see that all but one of these 19 values of q are out of range:

- $t = 6$ for $q = 233$;
- $t = 5$ for $q \in \{89, 109, 113, 125, 229\}$;
- $t = 4$ for $q \in \{13, 29, 37, 41, 53, 81, 97, 101, 241\}$;
- $t = 3$ for $q \in \{25, 49, 61, 121\}$.

This leaves the case $q = 13$, where $Q = \frac{13^4 - 1}{12} = 2^2 \cdot 5 \cdot 7 \cdot 17$. We therefore check the sufficient condition (14.11); unfortunately, it does not apply here, as

$$q^{n-1} - 1 - (2^{\omega(Q)} - 1)(q - 1)\sqrt{q^{n-2}} = 2197 - 1 - 15 \cdot 12 \cdot 13 = -144.$$

Moreover, the trick used to settle the case $(q,n) = (5,4)$ in Example 14.4.3 likewise fails for $q = 13$. In fact, the pair $(13,4)$ requires a more detailed individual investigation; this is done in the subsequent example. □

Example 14.6.3. We finally settle the homogenous case for the pair $(q,n) = (13,4)$. As already noted, we here have

$$Q = 2380 = 2^2 \cdot 5 \cdot 7 \cdot 17 \text{ and thus } q^4 - 1 = 2^4 \cdot 3 \cdot 5 \cdot 7 \cdot 17 = 28560.$$

For every divisor d of $\text{rad}(Q) = 2 \cdot 5 \cdot 7 \cdot 17 = 1190$, we put

$$U(d) := \{u \in E^* : \text{pt}_d(28560) \mid \text{ord}(u) \text{ and } \text{Tr}_{E/F}(\beta u) = 0\}.$$

By Inequality (14.13) in Remark 14.3.9,

$$\begin{aligned} |U(d)| &\geq \frac{\phi(d)}{d} \cdot (13^3 - 1 - (2^{\omega(d \wedge Q)} - 1) \cdot 12 \cdot 13) \\ &= \frac{\phi(d)}{d} \cdot (2196 - (2^{\omega(d)} - 1) \cdot 156), \end{aligned}$$

as here $d \wedge Q = \gcd(d,Q) = d$.

We now show that it suffices to prove that $U(1190)$ cannot be empty. By definition, the order of any element $u \in U(1190)$ is a multiple of $\text{pt}_{1920}(28560) = 9520 = (13^4 - 1)/3$. Hence u is either primitive (and we are done) or $\text{ord}(u) = 9520$. In the latter case, we multiply u by a primitive third root of unity, say λ; note that $\lambda \in F^*$. Then λu is a primitive element and satisfies

$$\text{Tr}_{E/F}(\beta(\lambda u)) = \lambda \cdot \text{Tr}_{E/F}(\beta u) = 0,$$

as desired.

Thus it only remains to check that $U(1190) \neq \emptyset$. Note that the above estimate for $|U(d)|$ agrees with the one resulting from (14.11) when $d = \text{rad}(Q) = 1190$; as we have already seen, this does not work, as it leads to the negative lower bound -144. Fortunately, the approach used in Example 14.5.1 to deal with the inhomogeneous case for the pair $(11,3)$ can also be applied in the present situation. For this, we consider the complementary divisors

$$k := 2 \cdot 5 \cdot 7 \quad \text{and} \quad \ell := 17$$

of $1920 = \text{rad}(Q)$. Using the general estimate for $|U(d)|$, we see that $|U(1190)|$ is indeed positive:

$$\begin{aligned} |U(1190)| &= |U(k) \cap U(\ell)| = |U(k)| + |U(\ell)| - |U(k) \cup U(\ell)| \\ &\geq |U(k)| + |U(\ell)| - |\{u \in E^* : \text{Tr}_{E/F}(\beta u) = 0\}| \\ &\geq \tfrac{2 \cdot 6}{5 \cdot 7} \cdot (2196 - 7 \cdot 156) + \tfrac{16}{17} \cdot (2196 - 156) - 2196 \\ &> 378 + 1920 - 2196 = 102. \end{aligned}$$

□

Exercises

Exercise 14.6.4. Let $n \geq 4$. Show that the upper bound for q^{n-1} in Inequality (14.22) can be improved to $(2^t - 2)^{2(n-1)/(n-2)}$. □

Exercise 14.6.5. Improve the lower bound for q^{n-1} in Inequality (14.22) when $n = 4$ and $q \equiv 3 \bmod 4$, where $q \neq 3$. Use this to narrow the ranges containing potential exceptions for q. Proceed in a similar manner for $q \equiv 1 \bmod 4$. □

14.7 The Homogeneous Case for $n = 3$

In this section, we continue our proof of Theorem 14.1.7 by settling the case $n = 3$; we will proceed as in the previous section and also use the same notations. We first show that the sufficient condition (14.11) in Corollary 14.3.7 holds except when

$$q \in \{4, 16, 25, 37, 121, 163, 211, 919\}.$$

For this, we again assume that $P_{\beta,0} = 0$ (as in the last section when dealing with the cases $n = 4$ and $n = 5$). Now Lemma 14.6.1 gives

$$\frac{p_1 p_2 \cdots p_t}{2} < q^2 < (2^t - 1)^4 < 2^{4t},$$

where $p_1 < p_2 < \cdots < p_t$ are the distinct prime divisors of $Q = q^2 + q + 1$; trivially, this implies $t \geq 2$. Note that Q is odd, and that 3 divides Q if and only if $q \equiv 1 \bmod 3$; moreover, every other prime divisor of Q has to be congruent to 1 modulo 6.

Case 1: Assume first that $q \not\equiv 1 \bmod 3$. When $t \geq 6$, we obtain

$$\frac{p_1 p_2 \cdots p_t}{2} \geq \frac{7 \cdot 13 \cdot 19 \cdot 31 \cdot 37 \cdot 43}{2} \cdot 61^{t-6} > 2^{25} \cdot 32^{t-6} = 2^{5t-5} > 2^{4t},$$

which is a contradiction. For $t = 5$, we get

$$995 < \sqrt{\tfrac{7 \cdot 13 \cdot 19 \cdot 31 \cdot 37}{2}} < q < (2^5 - 1)^2 = 961,$$

again a contradiction. As in the previous section, we compute a concrete range for q for the remaining values $t \in \{2, 3, 4\}$ from the above restriction:

t	4	3	2
q	$164 \leq q \leq 224$	$30 \leq q \leq 48$	$7 \leq q \leq 8$

Altogether, this leaves the nine possibilities

$$q \in \{8, 32, 41, 47, 167, 173, 179, 191, 197\},$$

all of which are out of range: they violate the upper bound $q < (2^t - 1)^2$, as

- $t = 3$ for $q = 191$;
- $t = 2$ for $q \in \{32, 47, 179, 197\}$;
- $t = 1$ for $q \in \{8, 41, 67, 173\}$.

Case 2: Now let $q \equiv 1 \bmod 3$. For $t \geq 8$, we have the contradiction

$$\frac{p_1 p_2 \cdots p_t}{2} \geq \frac{3 \cdot 7 \cdot 13 \cdot 19 \cdot 31 \cdot 37 \cdot 43 \cdot 61}{2} \cdot 67^{t-8} > 2^{32} \cdot 64^{t-8} = 2^{6t-16} \geq 2^{4t}.$$

For the remaining values of t, we obtain the following concrete ranges for q:

t	7	6	5
q	$11310 \leq q \leq 16128$	$1725 \leq q \leq 3968$	$284 \leq q \leq 960$
t	4	3	2
q	$51 \leq q \leq 224$	$12 \leq q \leq 48$	$4 \leq q \leq 8$

In view of the large ranges occurring, one now requires computer support to factor $Q = q^2 + q + 1$ in order to determine the precise t-values. It turns out that all but the nine prime powers

$$q \in \{4, 7, 16, 25, 37, 121, 163, 211, 919\}$$

are out of range. Of these, the instance $q = 4$ is covered by Example 14.1.5 and Exercise 14.1.11. Moreover, the sufficient condition (14.11) in Corollary 14.3.7 is satisfied for $q = 7$, but not for the seven larger exceptional values of q (see Exercise 14.7.3); these will be settled in the subsequent two examples. □

Example 14.7.1. We now settle six of the remaining seven cases, namely

$$q \in \{25, 37, 121, 163, 211, 919\},$$

by applying the method already used for the instance $(q, n) = (13, 4)$ in Example 14.6.3.[2] Thus we put

$$U(d) := \{u \in E^* : \mathrm{pt}_d(q^3 - 1) \mid \mathrm{ord}(u) \text{ and } \mathrm{Tr}_{E/F}(\beta u) = 0\}$$

for every divisor d of $\mathrm{rad}(Q) = \mathrm{rad}(q^2 + q + 1)$. Again, we use Inequality (14.13) to obtain the estimate

$$|U(d)| \geq \frac{\phi(d)}{d} \cdot \left(q^2 - 1 - (2^{\omega(d)} - 1) \cdot (q - 1) \cdot \sqrt{q}\right).$$

As in Example 14.6.3, it suffices to prove that $U(\mathrm{rad}(Q))$ cannot be empty, that is, there exists an element $u \in E^*$ satisfying $\mathrm{Tr}_{E/F}(\beta u) = 0$ and $3^s \cdot Q \mid \mathrm{ord}(u)$, where 3^s is the largest power of 3 dividing $q - 1$. If u should not be primitive, we multiply

[2] As the final case $q = 16$ is more involved and requires additional arguments, it will be dealt with separately in Example 14.7.2 below.

u by a suitable element λ of order dividing $(q-1)/3^s$ to obtain a primitive element λu satisfying

$$\mathrm{Tr}_{E/F}\big(\beta(\lambda u)\big) = \lambda \cdot \mathrm{Tr}_{E/F}(\beta u) = 0.$$

As before, we always select a suitable pair of complementary divisors k and ℓ of $\mathrm{rad}(Q)$ and use the general estimate

$$\begin{aligned} |U(\mathrm{rad}(Q))| &= |U(k\cdot\ell)| = |U(k)\cap U(\ell)| \\ &= |U(k)|+|U(\ell)|-|U(k)\cup U(\ell)| \\ &\geq |U(k)|+|U(\ell)|-(q^2-1) \end{aligned}$$

to show that $|U(\mathrm{rad}(Q))|$ is indeed positive. For the convenience of the reader, we summarize the necessary computations.[3]

Case $q=25$:

$q-1=24=2^3\cdot 3$ and $Q=651=3\cdot 7\cdot 31$;
$|U(3\cdot 7)| \geq \frac{4}{7}\cdot(624-3\cdot 24\cdot 5) > 150$;
$|U(31)| \geq \frac{30}{31}\cdot(624-24\cdot 5) > 487$;
$|U(3\cdot 7)|+|U(31)|-624 \geq 15$.

Case $q=37$:

$q-1=36=2^2\cdot 3^2$ and $Q=1407=3\cdot 7\cdot 67$;
$|U(3\cdot 7)| \geq \frac{4}{7}\cdot(1368-3\cdot 36\cdot\sqrt{37}) > 406$;
$|U(67)| \geq \frac{66}{67}\cdot(1368-36\cdot\sqrt{37}) > 1131$;
$|U(3\cdot 7)|+|U(67)|-1368 \geq 171$.

Case $q=121$:

$q-1=120=2^3\cdot 3\cdot 5$ and $Q=14763=3\cdot 7\cdot 19\cdot 37$;
$|U(3\cdot 7\cdot 19)| \geq \frac{72}{133}\cdot(14640-7\cdot 120\cdot 11) > 2923$;
$|U(37)| \geq \frac{36}{37}\cdot(14640-120\cdot 11) = 12960$;
$|U(3\cdot 7\cdot 19)|+|U(37)|-14640 \geq 1244$.

Case $q=163$:

$q-1=162=2\cdot 3^4$ and $Q=26733=3\cdot 7\cdot 19\cdot 67$;
$|U(3\cdot 7\cdot 19)| \geq \frac{72}{133}\cdot(26568-7\cdot 162\cdot\sqrt{163}) > 6544$;
$|U(67)| \geq \frac{66}{67}\cdot(26568-162\cdot\sqrt{163}) > 24134$;
$|U(3\cdot 7\cdot 19)|+|U(67)|-26568 \geq 4112$.

Case $q=211$:

$q-1=210=2\cdot 3\cdot 5\cdot 7$ and $Q=44733=3\cdot 13\cdot 31\cdot 37$;
$|U(3\cdot 13\cdot 31)| \geq \frac{240}{403}\cdot(44520-7\cdot 210\cdot\sqrt{211}) > 13796$;

[3] For the resulting lower bounds for $|U(\mathrm{rad}(Q))|$, we also use the trivial fact that $U(k)$ and $U(\ell)$ have to be integers. It is interesting to note that one always obtains a considerably better estimate than the required lower bound 1.

$|U(37)| \geq \frac{36}{37} \cdot (44520 - 210 \cdot \sqrt{211}) > 40348;$
$|U(3 \cdot 13 \cdot 31)| + |U(37)| - 44520 \geq 9626.$

Case $q = 919$:

$q - 1 = 918 = 2 \cdot 3^3 \cdot 17$ and $Q = 845481 = 3 \cdot 7 \cdot 13 \cdot 19 \cdot 163;$
$|U(3 \cdot 7 \cdot 13 \cdot 19)| \geq \frac{864}{1729} \cdot (844560 - 15 \cdot 918 \cdot \sqrt{919}) > 213437;$
$|U(163)| \geq \frac{162}{163} \cdot (844560 - 918 \cdot \sqrt{919}) > 811720;$
$|U(3 \cdot 7 \cdot 13 \cdot 19)| + |U(163)| - 844560 \geq 180599.$ □

Example 14.7.2. It finally remains to consider the case $q = 16$,[4] where

$$q - 1 = 15 = 3 \cdot 5 \quad \text{and} \quad Q = 273 = 3 \cdot 7 \cdot 13.$$

We proceed as for the six cases in Example 14.7.1 and try to verify the sufficient condition $U(\mathrm{rad}(Q)) \neq \emptyset$. With the standard choice of complementary divisors of $\mathrm{rad}(Q)$, namely $k = 3 \cdot 7$ and $\ell = 13$, we obtain

$$|U(k)| \geq \tfrac{4}{7} \cdot (255 - 3 \cdot 15 \cdot 4) = 42 + \tfrac{6}{7}$$

and

$$|U(\ell)| > \tfrac{12}{13} \cdot (255 - 15 \cdot 4) = 180.$$

Hence our usual estimate

$$|U(k \cdot \ell)| \geq |U(k)| + |U(\ell)| - |U(k) \cup U(\ell)| \geq |U(k)| + |U(\ell)| - (q^2 - 1)$$

unfortunately results in a negative lower bound in this case.

It is an obvious idea to try and improve the trivial estimate $|U(k) \cup U(\ell)| \leq q^2 - 1$ (which was sufficient for all previous cases) by proving that certain elements $u \in E^*$ satisfying $\mathrm{Tr}_{E/F}(\beta u) = 0$ (that is, belonging to the hyperplane $H_{\beta,0}$) can not occur in $U(k) \cup U(\ell)$. We now show how this may be done when $\beta = 1$; after that, we will use a suitable trick to derive the result for general $\beta \in E^*$ from this special case.

For this, we let η be a primitive 9-th root of unity in E^*, that is, a root of the cyclotomic polynomial $\Phi_9 = x^6 + x^3 + 1$. Similarly, we let ζ be a root of Φ_7, which splits into two irreducible factors over F (see Proposition 3.6.16); specifically, we choose ζ as a root of $x^3 + x + 1$. Then it is not difficult to check the following facts; the details are left to the reader.

- $\mathrm{Tr}_{E/F}(\eta) = \mathrm{Tr}_{E/F}(\eta^2) = \mathrm{Tr}_{E/F}(\zeta) = 0$.
- $F^*\eta, F^*\eta^2$ and $F^*\zeta$ are distinct cosets of F^*, and hence their union S is a subset of $H_{1,0}$ with cardinality 45.
- Every element $u \in S$ satisfies $\mathrm{ord}(u) \in \{3^2, 5 \cdot 3^2, 7, 3 \cdot 7, 5 \cdot 7, 3 \cdot 5 \cdot 7\}$.
- No element of S belongs to $U(k)$ or $U(\ell)$.

---

[4] It is remarkable that this case was not dealt with explicitly by Cohen in [81], who referred the reader to Baumert's [20] computer classification of the corresponding difference sets with parameters $(v, k, \lambda) = (273, 17, 1)$ instead.

Hence we may replace the trivial estimate for $|U(k) \cup U(\ell)|$ with

$$|U(k) \cup U(\ell)| \leq |\{u \in E^* : \mathrm{Tr}_{E/F}(u) = 0\} \setminus S| = 255 - 45 = 210$$

(when $\beta = 1$), which then results in the positive lower bound

$$|U(\mathrm{rad}(Q))| = |U(k \cdot \ell)| \geq 43 + 180 - 210 = 13.$$

Thus there exists an element $u \in H_{1,0}$ such that $\mathrm{pt}_Q(q^3 - 1) = 3^2 \cdot 7 \cdot 13 = 819$ divides $\mathrm{ord}(u)$. As usual, we may even assume that u is a primitive element of E: if necessary, multiply u by a primitive fifth root of unity.

It remains to deal with the case of an arbitrary $\beta \in E^*$. Note that the twelve conjugates u^{2^j} $(j = 0, \ldots, 11)$ of u under the Galois group of E over its prime field GF(2) are again primitive elements with trace 0. Now consider an arbitrary element $\beta \in E^*$, and write β as a power of the given primitive element u, say $\beta = u^b$. Obviously,

$$\mathrm{Tr}_{E/F}(\beta \cdot (\beta^{-1} u^{2^j})) = \mathrm{Tr}_{E/F}(u^{2^j}) = 0 \quad \text{for } j = 0, \ldots, 11.$$

Moreover, $\beta^{-1} u^{2^j} = u^{2^j - b}$ is likewise a primitive element for E if and only if the condition $\gcd(2^j - b, q^3 - 1) = 1$ holds; thus we wish to determine such an element j, for every choice of $b = 1, \ldots, q^3 - 2$. Note that it suffices to check that there is some $j \in \{0, 1, \ldots, 11\}$ such that

$$\gcd(2^j - b, Q) = \gcd(2^j - b, 3 \cdot 7 \cdot 13) = 1$$

holds, since we may multiply $u^{2^j - b}$ by a primitive fifth root of unity (if necessary). Thus we want to show that there always exists some choice for j such that none of the primes $p \in \{3, 7, 13\}$ divides $2^j - b$ (for a specified b). For this, we note that the 12 residues $2^j - b \bmod 13$ have to be distinct, as 2 is a primitive root modulo 13; see Proposition 8.5.12. Therefore, the condition $2^j - b \not\equiv 0 \bmod 13$ rules out at most one choice of j. Similarly, 2 generates the quadratic residues modulo 7, so that the values $2^j - b \bmod 7$ consist of three distinct residues, each repeated four times, and hence the requirement $2^j - b \not\equiv 0 \bmod 7$ rules out at most 4 choices of j. Finally, the condition $2^j - b \not\equiv 0 \bmod 3$ rules out at most 6 possibilities for j. As $1 + 4 + 6 < 12$, we see that there is indeed at least one feasible choice for j. □

Exercises

Exercise 14.7.3. Check the sufficient condition (14.11) in Corollary 14.3.7 for the eight values $q \in \{7, 16, 25, 37, 121, 163, 211, 919\}$. □

Exercise 14.7.4. Try to improve Inequality (14.22) for $n = 3$ and under the assumption that $q \geq 13$, such that the intervals for the exceptional values of q become $15367 < q < 16129$ when $t = \omega(Q) = 7$, and $2343 \leq q < 3969$ when $t = 6$, where $q \equiv 1 \bmod 3$. □

14.8 An Asymptotic Result for Quadratic Extensions

It remains to prove Theorem 14.1.6 for quadratic extensions: given arbitrary non-zero elements $\beta \in E = \mathrm{GF}(q^2)$ and $a \in F = \mathrm{GF}(q)$, there exists a primitive element u for E such that $\mathrm{Tr}_{E/F}(\beta u) = a$. In the present section, we will use the sufficient condition (14.12) in Corollary 14.3.8 to establish an asymptotic version of this result. This will allow us to complete the proof for the case of even q without much effort. In contrast, extensive computations are required to show that exactly 234 possible exceptions remain in odd characteristic. These are far more difficult to settle and will be dealt with via a group theoretic approach in the next section.

Our approach to the asymptotic result for $n = 2$ is similar to that for the case of extension degrees $n \geq 3$ in Section 14.5; again, we use the abbreviations

$$\omega := \omega(q^2 - 1), \quad \omega_1 := \omega(Q) = \omega(q+1) \text{ and } \omega_0 := \omega(q-1).$$

By Corollary 14.3.8, Theorem 14.1.6 holds for the pair $(q, 2)$ provided that

$$(2^{\omega_1} - 1) + (2^{\omega} - 2^{\omega_1})\sqrt{q} < q.$$

We shall say that q is an **exceptional prime power**[5] if this criterion is *not* satisfied, that is, if

$$q \leq (2^{\omega_1} - 1) + (2^{\omega} - 2^{\omega_1})\sqrt{q}. \tag{14.23}$$

We begin with a rather simple preliminary result, which already suffices to establish the asymptotic validity of Theorem 14.1.6 for quadratic extensions.

Proposition 14.8.1. *There are only finitely many exceptional prime powers q. More precisely, if q is exceptional, then*

- $q \in \{4, 16, 64\}$ *when q is even;*
- $\omega \leq 14$ *and* $q < 2^{28}$ *when q is odd.*

Proof. Let q be an exceptional prime power. Then (14.23) gives the rough estimate

$$q \leq (2^{\omega_1} - 1) + (2^{\omega} - 2^{\omega_1})\sqrt{q} < 2^{\omega}\sqrt{q},$$

and thus $\sqrt{q} < 2^{\omega}$. In order to obtain an upper bound for ω, we will apply Lemma 13.6.12 with $\ell = 53$ and

$$\Lambda = \{3, 5, 7, 11, 13, 17, 19, 23, 29, 31, 37, 41, 43, 47\},$$

so that $|\Lambda| = 14$ and $L = \prod_{r \in \Lambda} r > 3.07444 \cdot 10^{17}$.

Assume first that q is even, say $q = 2^s$. Then Lemma 13.6.12 yields

---

[5] This terminology is not standard and is only introduced to simplify the presentation of our proof for the quadratic case.

$$\begin{aligned}
\omega &\le \frac{\log_2(q^2-1)}{\log_2(\ell)} + |\Lambda| - \frac{\log_2(L)}{\log_2(\ell)} \\
&< \frac{2s}{\log_2(53)} + 14 - \frac{\log_2(L)}{\log_2(53)} \\
&< 0.35s + 14 - 10.14 = 0.35s + 3.86.
\end{aligned}$$

Since $\sqrt{q} < 2^{\omega}$ means $\frac{s}{2} < \omega$ in this case, we conclude that $s < 0.7s + 7.72$, which gives $s \le 25$ (as s is an integer). Substituting this bound for s in the above inequality for ω shows $\omega < 12.61$, and hence $\omega \le 12$.

But then $\frac{s}{2} < \omega \le 12$ gives $s < 24$, and therefore $s \le 23$. Substituting this improved upper bound for s leads to the improved estimate $\omega \le 11$, and hence $s \le 21$. Substituting this new bound does not lead to any further improvement for ω.

For the remaining values $s \in \{2,3,\dots,21\}$, one determines the factorization of $q^2-1 = (2^s-1)\cdot(2^s+1)$ and checks directly whether 2^s is exceptional. We leave it to the reader to verify that this happens if and only if $s \in \{2,4,6\}$, that is, for $q \in \{4,16,64\}$.

Now let q be odd, so that q^2-1 is divisible by 8. As we want to apply Lemma 13.6.12 with the above choices of ℓ and Λ, we have to do so for the odd part of q^2-1. Denoting this integer by v and using $v < q^2/8$, we obtain the estimate

$$\begin{aligned}
\omega - 1 &< \frac{\log_2(v)}{\log_2(\ell)} + |\Lambda| - \frac{\log_2(L)}{\log_2(\ell)} \\
&< \frac{2\log_2(q) - 3}{\log_2(53)} + 14 - \frac{\log_2(L)}{\log_2(53)} \\
&< 0.35\log_2(q) - 0.52 + 14 - 10.14 \\
&= 0.35\log_2(q) + 3.34.
\end{aligned}$$

Using $\sqrt{q} < 2^{\omega}$, we conclude that

$$\log_2(q) < 2\omega < 0.7\log_2(q) + 8.68,$$

which gives $\log_2(q) < 29$. Substituting this rough bound in the above inequality for $\omega - 1$ shows that $\omega < 15$, and hence $\omega \le 14$ (since ω is an integer). Therefore $q < 2^{28}$. □

Example 14.8.2. Let us consider the three exceptional powers of 2 determined in Proposition 14.8.1. For this, we use Remark 14.3.9 together with an inclusion-exclusion argument (as several times before). Since $q-1$ and $q+1$ are relatively prime, we obtain

$$\begin{aligned}
P_{\beta,a} &\ge P_{\beta,a}(q-1) + P_{\beta,a}(q+1) - q \\
&\ge \frac{\phi(q-1)}{q-1}\cdot\left(q - (2^{\omega_0}-1)\cdot\sqrt{q}\right) + \frac{\phi(q+1)}{q+1}\cdot\left(q - (2^{\omega_1}-1)\right) - q.
\end{aligned}$$

For $q \in \{16, 64\}$, this approach yields the desired positive lower bound for $P_{\beta,a}$, whereas it unfortunately fails for $q = 4$:

$$P_{\beta,a} > \begin{cases} \frac{2\cdot 2}{3} + \frac{4\cdot 3}{5} - 4 = -\frac{4}{15} & \text{if } q = 4, \\ \frac{8\cdot 4}{15} + \frac{16\cdot 15}{17} - 16 = \frac{64}{255} & \text{if } q = 16, \\ \frac{4\cdot 40}{7} + \frac{48\cdot 61}{65} - 64 = \frac{1776}{455} & \text{if } q = 64. \end{cases}$$

It is possible to handle the case $q = 4$ via an improved criterion. Nevertheless, we will deal with this case directly in Example 14.8.3, as it will be instructive to consider a concrete field extension in detail at least once. □

Example 14.8.3. Consider the field extension $E = \mathrm{GF}(16)$ over $F = \mathrm{GF}(4)$, and let λ be a primitive element of F, so that $F = \{0, 1, \lambda, \lambda^2 = 1+\lambda\}$. In view of Theorem 5.1.1, the polynomial $f(x) := x^2 + x + \lambda$ is irreducible over F; of course, this also follows by checking that f has no roots in F:

$$f(0) = \lambda = f(1), \;\; f(\lambda) = 1 + \lambda = f(1+\lambda).$$

Then $E = F(\gamma)$, where γ is a root of $f(x)$, and $\{1, \lambda, \gamma, \lambda\gamma\}$ is a basis of E over its prime field $P = \mathrm{GF}(2)$. Using the relation $\gamma^2 = \lambda + \gamma$, we obtain

$$\gamma^3 = \lambda + \gamma + \lambda\gamma, \quad \gamma^4 = 1 + \gamma \quad \text{and} \quad \gamma^5 = \lambda,$$

which shows that γ is a primitive element of E. Then the remaining $\phi(15) - 1 = 7$ primitive elements of E are precisely the powers γ^k, where $2 \leq k \leq 14$ and $\gcd(k, 15) = 1$. It is easy to compute the representations of these powers in terms of the basis given above:

$$\gamma^2 = \lambda + \gamma, \quad \gamma^4 = 1 + \gamma, \quad \gamma^7 = 1 + \lambda + \lambda\gamma, \quad \gamma^8 = 1 + \lambda + \gamma,$$
$$\gamma^{11} = \gamma + \lambda\gamma, \quad \gamma^{13} = 1 + \lambda\gamma \quad \text{and} \quad \gamma^{14} = 1 + \lambda + \gamma + \lambda\gamma.$$

Now note that the five F-lines of E through the origin (that is, the five 1-dimensional linear subspaces of E/F) are given by

$$\begin{aligned} \langle\lambda\rangle &= \{0, 1, \lambda, 1+\lambda\}; \\ \langle\gamma\rangle &= \{0, \gamma, \lambda\gamma, \gamma+\lambda\gamma\}; \\ \langle 1+\gamma\rangle &= \{0, 1+\gamma, \lambda+\lambda\gamma, 1+\lambda+\gamma+\lambda\gamma\}; \\ \langle\lambda+\gamma\rangle &= \{0, \lambda+\gamma, 1+\lambda+\lambda\gamma, 1+\gamma+\lambda\gamma\}; \\ \langle 1+\lambda+\gamma\rangle &= \{0, 1+\lambda+\gamma, 1+\lambda\gamma, \lambda+\gamma+\lambda\gamma\}. \end{aligned}$$

Each (proper) affine line of E is parallel to exactly one of these five lines, and each of these 15 lines contains at least one primitive element:

$$\begin{array}{lll}
\gamma \in \gamma + \langle \lambda \rangle, & 1+\lambda\gamma \in \lambda\gamma + \langle \lambda \rangle, & \gamma+\lambda\gamma \in \gamma+\lambda\gamma+\langle \lambda \rangle; \\
1+\lambda\gamma \in 1+\langle \gamma \rangle, & \lambda+\gamma \in \lambda+\langle \gamma \rangle, & 1+\lambda+\lambda\gamma \in 1+\lambda+\langle \gamma \rangle; \\
\gamma \in 1+\langle 1+\gamma \rangle, & 1+\lambda+\gamma \in \lambda+\langle 1+\gamma \rangle, & \lambda+\gamma \in 1+\lambda+\langle 1+\gamma \rangle; \\
1+\lambda+\gamma \in 1+\langle \lambda+\gamma \rangle, & \gamma \in \lambda+\langle \lambda+\gamma \rangle, & 1+\gamma \in 1+\lambda+\langle \lambda+\gamma \rangle; \\
\lambda+\gamma \in 1+\langle 1+\lambda+\gamma \rangle, & 1+\gamma \in \lambda+\langle 1+\lambda+\gamma \rangle, & \gamma \in 1+\lambda+\langle 1+\lambda+\gamma \rangle.
\end{array}$$

Thus Theorem 14.1.6 holds also for the pair $(q,n) = (4,2)$. □

In the remainder of this section, we discuss the problem of determining the exceptional odd prime powers q. Heavily relying on computer support, it can be shown that there are exactly 234 exceptional values. According to Proposition 14.8.1, any such q has to satisfy $q < 2^{2\omega}$, where $\omega = \omega(q^2-1) \leq 14$.

First assume that actually $\omega = 14$. Then

$$q^2 - 1 \geq 8 \cdot 3 \cdot 5 \cdot 7 \cdot 11 \cdot 13 \cdot 17 \cdot 19 \cdot 23 \cdot 29 \cdot 31 \cdot 37 \cdot 41 \cdot 43$$

shows that $q \geq 228,759,799$. We now improve the upper bound $q < 2^{28}$, by observing that

$$2 \cdot 3 \cdot 5 \cdot 7 \cdot 11 \cdot 13 \cdot 17 \cdot 19 \cdot 23 \cdot 29 = 6,469,693,230$$

exceeds the preceding lower bound (considerably). This shows that both $q-1$ and $q+1$ can have at most 9 distinct prime divisors, that is, $\omega_0, \omega_1 \leq 9$. Thus $14 = \omega = \omega_0 + \omega_1 - 1 \leq 8 + \omega_1$, and hence $\omega_1 \geq 6$. Since q is assumed to be exceptional, this implies

$$q \leq (2^{\omega_1} - 1) + (2^{\omega} - 2^{\omega_1})\sqrt{q} < (2^9 - 1) + (2^{14} - 2^6) \cdot 2^{14} = 267,387,391.$$

Hence we have

- $228,759,799 \leq q \leq 267,387,389 \quad$ for $\omega = 14$.

Now let $\omega = 13$, so that $q < 2^{26}$. Proceeding exactly as for $\omega = 14$, we then obtain $q \geq 34,885,543$ and $\omega_0, \omega_1 \leq 8$. Again, this gives $\omega_1 \geq 6$, which then leads to the improved upper bound $q < (2^8 - 1) + (2^{13} - 2^6) \cdot 2^{13} = 66,584,831$. Thus

- $34,885,543 \leq q \leq 66,584,829 \quad$ for $\omega = 13$.

Continuing in the same fashion, one computes the next three ranges as follows:

- $5,448,207 \leq q \leq 16,515,197 \quad$ for $\omega = 12$;
- $895,681 \leq q \leq 4,128,893 \quad$ for $\omega = 11$;
- $160,869 \leq q \leq 1,015,869 \quad$ for $\omega = 10$.

Observe that the ranges for $\omega = 11$ and $\omega = 10$ overlap. The same phenomenon actually occurs for all ranges with $\omega \leq 11$, so that one has to investigate all odd prime powers $q \leq 4,128,893$, in addition to those in the ranges for $\omega \in \{12,13,14\}$.

Obviously, it requires a computer search to determine all prime powers q in the corresponding intervals having the required ω-value. It turns out that no such prime powers exist for $\omega \in \{12, 13, 14\}$, whereas one obtains exactly one candidate in the next range, namely $q = 3,847,271$. Here

$$q-1 = 2\cdot 5\cdot 7\cdot 17\cdot 53\cdot 61 \quad \text{and} \quad q+1 = 2^3\cdot 3\cdot 11\cdot 13\cdot 19\cdot 59,$$

so that $\omega_0 = \omega_1 = 6$ and $\omega = 11$; thus $q = 3,847,271$ is indeed exceptional. For $\omega \leq 10$, one discovers exactly 233 exceptional odd prime powers, which are listed in Table 14.7; we will deal with these exceptional cases in the next section. We remark that this table contains all 147 cases left unresolved in [217] for the case $n = 2$, $\beta = 1$.

Table 14.7 All exceptional odd prime powers $q < 3,847,271$.

							11	13	19
29	31	41	43	59	61	67	71	79	103
121	131	139	169	181	211	229	239	281	307
311	331	349	373	379	409	419	421	439	443
461	463	491	521	529	547	571	631	659	661
691	859	911	1021	1091	1231	1259	1289	1291	1301
1331	1427	1429	1481	1483	1559	1597	1609	1709	1721
1741	1847	1849	1861	1871	1931	1973	1979	2003	2029
2069	2089	2129	2131	2141	2209	2221	2243	2281	2311
2381	2437	2521	2531	2551	2729	2731	2851	2861	2927
3011	3037	3067	3109	3121	3191	3541	3571	3739	4003
4421	4523	4691	4759	5279	5641	5741	5851	6709	6733
6889	6971	7069	7309	7411	7481	7589	7591	7789	7853
8009	8059	8161	8581	8741	8779	8969	9029	9043	9239
9283	9349	9461	9491	9547	9619	9859	10429	10711	10739
10789	10949	11059	11131	11311	11593	11621	11731	11779	11831
11971	12211	12409	13399	14281	14629	17291	20021	20747	21319
23561	23869	24179	27259	27691	30211	34781	35531	36191	38011
38039	40699	41411	41539	42181	42979	44021	44269	44771	45319
46229	46411	46619	47059	47741	48179	48299	48619	49477	49531
49741	49939	50051	53923	55021	55931	56167	56239	77141	81509
102829	106261	147629	149731	154699	167441	174019	181609	183611	188189
195131	197539	197779	203321	214369	216061	216371	223211	225149	227011
242971	609179	614041	620311	647219	650761	684419	749891	786829	865591

14.9 A Group Theoretic Approach to Quadratic Extensions

The previous section has left Theorem 14.1.6 unresolved for 234 odd prime powers q in the case of quadratic extensions (since the sufficient condition in Corollary 14.3.8 was not satisfied), and in many of these cases q is quite large. Therefore, we will provide three alternative sufficient criteria in this section to resolve the 234 remaining cases. This will rest on using the inclusion-exclusion principle – similar to our treatment of the cases $q = 16$ and $q = 64$ in Example 14.8.2. In contrast to the original work of Cohen [79], our presentation uses a more group theoretic approach and exploits the particular algebraic nature of the basic problem for $n=2$.

Throughout, we let q be an odd prime power and write

$$q^2-1 = 2^{\ell+1}\cdot r\cdot s,$$

where r and s are the odd parts of $q-1$ and of $q+1$, respectively. Note that q^2-1 is a multiple of 8, so that $\ell \geq 2$; more precisely,

- $q-1=2^\ell r$ and $q+1=2s$ if $q\equiv 1 \bmod 4$;
- $q-1=2r$ and $q+1=2^\ell s$ if $q\equiv 3 \bmod 4$.

We begin with some basic results concerning the structure of the multiplicative group E^*:

Observation 14.9.1. Given any divisor d of q^2-1, we let U_d be the subgroup of order d of E^* and denote the set of all primitive d-th roots of unity by Ω_d. Note that the multiplicative group of E decomposes as $E^* = U_{2^{\ell+1}}\cdot U_r\cdot U_s$; accordingly, every element w of E^* has a unique representation in the form

$$w = w_2 w_r w_s \quad \text{with } w_2\in U_{2^{\ell+1}},\, w_r\in U_r \text{ and } w_s\in U_s.$$

Now let ζ be a primitive $2^{\ell+1}$-th root of unity in $\in E^*$, and θ a primitive 4-th root of unity – for instance, $\theta=\zeta^{2^{\ell-1}}$. In view of Exercise 14.9.15, the kernel of the (E,F)-trace mapping is then given by

$$\ker\left(\mathrm{Tr}_{E/F}\right) = \begin{cases} F\zeta & \text{if } q\equiv 1 \bmod 4;\\ F\theta & \text{if } q\equiv 3 \bmod 4.\end{cases}$$

Hence one has $\mathrm{Tr}_{E/F}(\beta u)=a$ (where $\beta\in E^*$ and $a\in F^*$) if and only if

- $u\in\beta^{-1}\cdot(F\zeta+\frac{a}{2})$ for $q\equiv 1 \bmod 4$;
- $u\in\beta^{-1}\cdot(F\theta+\frac{a}{2})$ for $q\equiv 3 \bmod 4$. □

Now let t be an arbitrary divisor of q^2-1 and put

$$U_{\beta,a}(t) := \left\{u\in E^* : \mathrm{Tr}_{E/F}(\beta u)=a \text{ and } \mathrm{pt}_t(q^2-1)\mid \mathrm{ord}(u)\right\}.$$

With the notation introduced in Section 14.3, we have

$$|U_{\beta,a}(t)| = P_{\beta,a}(t) = \frac{\phi(q-1)}{q-1}\cdot\sum_{d|t}\left(\frac{\mu(d)}{\phi(d)}\cdot\sum_{\psi:d}S_{\beta,a}(\psi)\right),$$

where $S_{\beta,a}(\psi)=\sum_{u\in H_{\beta,a}}\psi(u)$. Then most of the remaining 234 exceptions can be settled via the following simple inclusion-exclusion argument:

Lemma 14.9.2. *One has*

$$P_{\beta,a} \geq P_{\beta,a}(q+1) + P_{\beta,a}(q-1) - P_{\beta,a}(2).$$

Proof. In view of $\gcd(q+1, q-1) = 2$, we obtain

$$\begin{aligned} P_{\beta,a} &= \left|U_{\beta,a}(q^2-1)\right| = \left|U_{\beta,a}(q+1) \cap U_{\beta,a}(q-1)\right| \\ &= \left|U_{\beta,a}(q+1)\right| + \left|U_{\beta,a}(q-1)\right| - \left|U_{\beta,a}(q+1) \cup U_{\beta,a}(q-1)\right| \\ &\geq \left|U_{\beta,a}(q+1)\right| + \left|U_{\beta,a}(q-1)\right| - \left|U_{\beta,a}(2)\right|, \end{aligned}$$

as claimed. □

Therefore, we need to evaluate (or at least estimate) the three numbers $P_{\beta,a}(t)$, where $t \in \{2, q-1, q+1\}$. For $t = q-1$, we use the following simple estimate:

Proposition 14.9.3. *One has*

$$P_{\beta,a}(q-1) \geq \phi(q-1) \cdot \left(1 - \frac{(2^{\omega(q-1)} - 2) \cdot \sqrt{q}}{q-1}\right).$$

Proof. By Remark 14.3.9 (with $n = 2$),

$$P_{\beta,a}(q-1) \geq \frac{\phi(q-1)}{q-1} \cdot \left(q - 1 - (2^{\omega(q-1)} - 2) \cdot \sqrt{q}\right),$$

which gives the assertion. □

For $t = 2$, an exact evaluation is possible, by using the quadratic character η of E^* studied in detail in Section 10.4:

Proposition 14.9.4. *For every $a \in F^*$, one has*

$$P_{\beta,a}(2) = \frac{1}{2} \cdot (q - \eta_0(-1)\eta(\beta)) = \begin{cases} \frac{1}{2} \cdot (q - \eta(\beta)) & \text{if } q \equiv 1 \bmod 4, \\ \frac{1}{2} \cdot (q + \eta(\beta)) & \text{if } q \equiv 3 \bmod 4, \end{cases}$$

where η_0 denotes the quadratic character of F^*.

Proof. As η is the only multiplicative character with order 2, we have

$$P_{\beta,a}(2) = \frac{1}{2} \cdot (q - S_{\beta,a}(\eta)).$$

Using the general formula $S_{\beta,a}(\psi) = \psi(a) S_{\beta,1}(\psi)$ from Observation 14.3.1 for $\psi = \eta$, we obtain

$$S_{\beta,a}(\eta) = \eta(a) S_{\beta,1}(\eta) = S_{\beta,1}(\eta),$$

since every element of F^* is a square in E^*. Moreover, Equation (14.8) gives

$$S_{\beta,0}(\eta) = -(q-1) S_{\beta,1}(\eta),$$

as $\mathrm{ord}(\eta) = 2$ divides $Q = q+1$. Now put $\rho := \zeta$ if $q \equiv 1 \bmod 4$ and $\rho := \theta$ otherwise. Together with Observation 14.9.1, the preceding identities yield

$$\begin{aligned}(q-1)S_{\beta,\alpha}(\eta) &= (q-1)S_{\beta,1}(\eta) = -S_{\beta,0}(\eta) = -\sum_{\lambda\in F^*}\eta(\beta^{-1}(\lambda\rho)) \\ &= -\eta(\beta^{-1})\eta(\rho)\cdot\sum_{\lambda\in F^*}\eta(\lambda) = -\eta(\beta)\eta(\rho)(q-1)\end{aligned}$$

and hence $-S_{\beta,\alpha}(\eta) = \eta(\beta)\eta(\rho)$. Now note that ζ is a non-square, whereas θ is a square. Therefore

$$\eta(\beta)\eta(\rho) = \begin{cases}\eta(\beta)\eta(\zeta) = -\eta(\beta) & \text{if } q \equiv 1 \bmod 4,\\ \eta(\beta)\eta(\theta) = \eta(\beta) & \text{if } q \equiv 3 \bmod 4.\end{cases}$$

Substituting into the above formula for $P_{\beta,a}(2)$ gives the assertion, since $\eta_0(-1) = 1$ if and only $q \equiv 1 \bmod 4$. □

It remains to examine the case $t = q+1$, where again a precise determination is possible, though this will be a little more involved. By definition, $P_{\beta,a}(q+1)$ is the number of elements $u \in E^*$ satisfying $\mathrm{Tr}_{E/F}(\beta u) = a$ and such that $2^{\ell+1}s$ divides the order of u. The latter condition means that u belongs to the set $S := U_r \cdot \Omega_{2^{\ell+1}} \cdot \Omega_s$.

It is easily checked that F^* acts on S by multiplication: if $\lambda \in F^*$ and $u \in S$, then also $\lambda u \in S$. Thus the elements with non-zero trace are equally distributed over S, as $\mathrm{Tr}_{E/F}(\beta\cdot(\lambda u)) = \lambda\mathrm{Tr}_{E/F}(\beta u)$. In view of Exercise 14.9.16, this yields

$$P_{\beta,a}(q+1) = \frac{|S| - P_{\beta,0}(q+1)}{q-1} = \phi(q+1) - \frac{P_{\beta,0}(q+1)}{q-1} \quad \text{for } a \in F^*. \tag{14.24}$$

Thus it remains to evaluate the term $P_{\beta,0}(q+1)$, that is, the number of elements $u \in F^*$ satisfying $\mathrm{Tr}_{E/F}(\beta u) = 0$ and such that $2^{\ell+1}s$ divides the order of u. For this, we need a case distinction.

Proposition 14.9.5. *Assume that* $q \equiv 1 \bmod 4$. *Then*

$$P_{\beta,0}(q+1) = (q-1)\cdot\delta_\beta,$$

where

$$\delta_\beta = \begin{cases}1 & \text{if } \mathrm{ord}(\beta_2) \neq 2^{\ell+1} \text{ and } \mathrm{ord}(\beta_s) = s,\\ 0 & \text{otherwise.}\end{cases}$$

Proof. Let $u \in S = U_r \cdot \Omega_{2^{\ell+1}} \cdot \Omega_s$. By Observation 14.9.1, we have $\mathrm{Tr}_{E/F}(\beta u) = 0$ if and only if $(\beta_2 u_2)\cdot(\beta_s u_s) \in F\zeta = U_r \cdot \Omega_{2^{\ell+1}}$. This means $\beta_s u_s = 1$ and $\beta_2 u_2 \in \Omega_{2^{\ell+1}}$, which requires $\mathrm{ord}(\beta_s) = s$ and $\mathrm{ord}(\beta_2) \neq 2^{\ell+1}$. If this holds, $u_s = \beta_s^{-1}$ is uniquely determined, whereas there are precisely $\phi(2^{\ell+1})$ permissible choices for u_2, namely the elements in $\Omega_{2^{\ell+1}}$; note that there is no restriction on u_r. Altogether, this gives $\phi(2^{\ell+1})r = 2^\ell r = q-1$ possibilities for u. □

Proposition 14.9.6. *Assume that* $q \equiv 3 \bmod 4$. *Then*

$$P_{\beta,0}(q+1) = (q-1) \cdot \varepsilon_\beta,$$

where

$$\varepsilon_\beta = \begin{cases} 1 & \text{if } \operatorname{ord}(\beta_2) = 2^{\ell+1} \text{ and } \operatorname{ord}(\beta_s) = s, \\ 0 & \text{otherwise.} \end{cases}$$

Proof. The proof proceeds exactly as that of Proposition 14.9.5. Now $\mathrm{Tr}_{E/F}(\beta u) = 0$ holds if and only if $(\beta_2 u_2) \cdot (\beta_s u_s) \in F\theta$, which means $\beta_s u_s = 1$ and $\beta_2 u_2 \in \Omega_4$ and requires $\operatorname{ord}(\beta_s) = s$ and $\operatorname{ord}(\beta_2) = 2^{\ell+1}$. As before, u_s is then uniquely determined and there is no restriction on u_r, but now there are just two permissible choices for u_2, namely $\pm\beta_2^{-1}\theta$. Thus we obtain $2r = q-1$ possibilities for u in this case. □

Remark 14.9.7. The preceding two results also allow us to finish the proof of Theorem 14.1.8 by dealing with the case where q is odd. Suppose that there exists a primitive element $u \in E$ such that $\mathrm{Tr}_{E/F}(\beta u) = 0$. Then $P_{\beta,0}(q+1) \neq 0$, since $2^{\ell+1}s$ divides the order of u. Using Propositions 14.9.5 and 14.9.6, we conclude that s has to divide the order of β and that $2^{\ell+1}$ divides $\operatorname{ord}(\beta)$ when $q \equiv 3 \bmod 4$, whereas $\operatorname{ord}(\beta_2) \neq 2^{\ell+1}$ when $q \equiv 1 \bmod 4$.

Now assume that this condition is satisfied. We then count the possibilities for u by proceeding exactly as in the proofs of Propositions 14.9.5 and 14.9.6. The only difference is that we obtain an additional restriction: u_r has to have order r as u is required to be primitive, which leaves precisely $\phi(r)$ permissible choices for u_r. Altogether, this results in $2\phi(q-1)$ possibilities for u in both cases. □

Substituting the formulas in Propositions 14.9.5 and 14.9.6 into Equation (14.24) gives the following result:

Proposition 14.9.8. *One has*

$$P_{\beta,a}(q+1) = \begin{cases} \phi(q+1) - \delta_\beta & \text{if } q \equiv 1 \bmod 4, \\ \phi(q+1) - \varepsilon_\beta & \text{if } q \equiv 3 \bmod 4. \end{cases}$$

In particular, $P_{\beta,a}(q+1) \geq \phi(q+1) - 1 > 0$. □

Altogether, we obtain the following estimate for $P_{\beta,a}$ by combining Lemma 14.9.2 with Propositions 14.9.3, 14.9.4 and 14.9.8:

Theorem 14.9.9. *Let* $\beta \in \mathrm{GF}(q^2)^*$ *and* $a \in \mathrm{GF}(q)^*$, *where q is odd. Then*

$$\begin{aligned} P_{\beta,a} \geq\ & \phi(q+1) + \phi(q-1) \cdot \Big(1 - \frac{(2^{\omega(q-1)} - 2) \cdot \sqrt{q}}{q-1}\Big) \\ & - \frac{q+1}{2} + \frac{1 + \eta_0(-1)\eta(\beta) - 2\alpha}{2}, \end{aligned}$$

where $\alpha = \delta_\beta$ *if* $q \equiv 1 \bmod 4$ *and* $\alpha = \varepsilon_\beta$ *if* $q \equiv 3 \bmod 4$. *In particular,*

$$P_{\beta,a} \geq \phi(q+1) + \phi(q-1) \cdot \left(1 - \frac{(2^{\omega(q-1)} - 2) \cdot \sqrt{q}}{q-1}\right) - \frac{q+1}{2}.$$

Proof. It only remains to check the second assertion. Since every element of $U_r \cdot U_s$ is a square in E, we have

$$\eta(\beta) = \eta(\beta_2)\eta(\beta_r)\eta(\beta_s) = \eta(\beta_2),$$

which equals 1 if and only if $\operatorname{ord}(\beta_2) \neq 2^{\ell+1}$. Using this fact, it is easy to check that $1 + \eta_0(-1)\eta(\beta) - 2\alpha$ is always non-negative; see Exercise 14.9.17. □

Theorem 14.9.9 corresponds to Corollary 4.3 of Cohen [79]. Applying this lower bound to the remaining (exceptional) values for q, namely $q = 3,847,271$ and the 233 values listed in Table 14.7, proves $P_{\beta,a} > 0$ for 225 of these cases and leaves only the nine exceptional prime powers

$$q \in \{11, 31, 43, 71, 79, 131, 139, 211, 911\}$$

unresolved. In order to deal with these remaining instances, we will provide two further lower bounds for $P_{\beta,a}$. The main idea is to consider the distribution of the norms of the elements $u \in E^*$ satisfying $\operatorname{Tr}_{E/F}(\beta u) = a$. The simpler of the two bounds in question is as follows:

Proposition 14.9.10. *Let $\beta \in \mathrm{GF}(q^2)^*$ and $a \in \mathrm{GF}(q)^*$, where q is odd. Then*

$$P_{\beta,a} \geq \phi(q+1) + 2\phi(q-1) - (q-1) - \alpha,$$

where $\alpha = \delta_\beta$ if $q \equiv 1 \bmod 4$ and $\alpha = \varepsilon_\beta$ if $q \equiv 3 \bmod 4$. In particular,

$$P_{\beta,a} \geq \phi(q+1) + 2\phi(q-1) - q.$$

Proof. According to Observation 14.9.1, the set

$$A := \beta^{-1} \cdot \left(F\rho + \tfrac{a}{2}\right), \text{ where } \rho = \begin{cases} \zeta & \text{if } q \equiv 1 \bmod 4, \\ \theta & \text{if } q \equiv 3 \bmod 4, \end{cases}$$

consists of all elements $u \in E^*$ which satisfy $\operatorname{Tr}_{E/F}(\beta u) = a$. Since ρ has trace 0, we can compute the norm $\operatorname{Norm}_{E/F}(\gamma) = \gamma^{q+1}$ of an element $\gamma = \beta^{-1}(\lambda\rho + \frac{a}{2}) \in A$ as follows:

$$\begin{aligned} \operatorname{Norm}_{E/F}\left(\beta^{-1}(\lambda\rho + \tfrac{a}{2})\right) &= \beta^{-1-q} \cdot \left((\lambda\rho^q + \tfrac{a}{2}) \cdot (\lambda\rho + \tfrac{a}{2})\right) \\ &= \beta^{-1-q} \cdot \left(\rho^{q+1}\lambda^2 + \tfrac{a}{2}(\rho + \rho^q)\lambda + \tfrac{a^2}{4}\right) \\ &= \beta^{-1-q} \cdot \left(\rho^{q+1}\lambda^2 + \tfrac{a^2}{4}\right). \end{aligned}$$

Hence two elements $\beta^{-1}(\lambda\rho + \frac{a}{2})$ and $\beta^{-1}(\mu\rho + \frac{a}{2})$ of A have the same norm if and only if $\lambda^2 = \mu^2$, so that every element $b \in F^*$ has at most two preimages in

A under the norm mapping. This implies $|C| \leq 2 \cdot |N(C)|$ for every subset C of A, where $N(C)$ denotes the image of C under the norm mapping.

We now choose C as the set of all non-primitive elements $\gamma \in A$ such that $\mathrm{ord}(\gamma)$ is a multiple of $2^{\ell+1}s$. Then

$$|C| = P_{\beta,a}(q+1) - P_{\beta,a}.$$

Let $\gamma \in C$ and write $\gamma = \gamma_2 \gamma_r \gamma_s$, as in Observation 14.9.1; then $\gamma_r \in U_r \setminus \Omega_r$, $\gamma_s \in \Omega_s$ and $\gamma_2 = \zeta^i$ for some odd i. Note that

$$\mathrm{Norm}_{E/F}(\gamma) = \gamma^{q+1} = \zeta^{i(q+1)} \cdot \gamma_r^2$$

cannot be a primitive element for F, because of $\mathrm{ord}(\gamma_r^2) = \mathrm{ord}(\gamma_r) \neq r$. Moreover, $\mathrm{Norm}_{E/F}(\gamma)$ is not a square in F, since ζ^{q+1} is a non-square by Exercise 14.9.15 and i is odd. Thus $N(C)$ is contained in $N \setminus \Omega_{q-1}$, where N denotes the set of non-squares in F. In view of Proposition 3.2.13, we obtain the estimate

$$|C| \leq 2 \cdot |N(C)| \leq 2 \cdot \left(|N| - |\Omega_{q-1}|\right) = q - 1 - 2\phi(q-1),$$

and hence Proposition 14.9.8 gives

$$\begin{aligned} P_{\beta,a} &= P_{\beta,a}(q+1) - |C| \\ &= \phi(q+1) - \alpha - |C| \\ &\geq \phi(q+1) - \alpha - (q-1) + 2\phi(q-1), \end{aligned}$$

as claimed. □

Remark 14.9.11. Since $\delta_\beta = \varepsilon_\beta = 0$ for $\beta = 1$, Proposition 14.9.10 coincides with the results of Jungnickel and Vanstone [217, Proposition 4.3 and Remark 4.5] in the special case when $\beta = 1$ and q is odd. It also corresponds to the second lower bound given in Theorem 4.4 of Cohen [79]. □

Proposition 14.9.10 establishes $P_{\beta,a} > 0$ for the six values

$$q \in \{11, 31, 43, 71, 79, 131\};$$

see Exercise 14.9.18. Finally, the last three cases $q \in \{139, 211, 911\}$ are covered by the following variation of Proposition 14.9.10, which is essentially the first lower bound given in Theorem 4.4 of Cohen [79]; see Exercise 14.9.19.

Proposition 14.9.12. *Let $\beta \in \mathrm{GF}(q^2)^*$ and $a \in \mathrm{GF}(q)^*$, where q is odd, and let f be a proper divisor of the odd part r of $q-1$ such that $g = r/f$ $(\neq 1)$ and f are relatively prime. Then*

$$P_{\beta,a} \geq \tfrac{\phi(f)}{f} \cdot \tfrac{\phi(q+1)}{q+1} \cdot \left(q + 1 - 2^{\omega(q+1)} \cdot \left(1 + (2^{\omega(f)} - 1)\sqrt{q}\right)\right)$$
$$+ 2\phi(q-1) - \tfrac{\phi(f)}{f} \cdot (q-1).$$

In particular, one has

$$P_{\beta,a} \geq \tfrac{(f-1)\phi(q+1)}{f(q+1)} \cdot \left(q+1-2^{\omega(q+1)} \cdot (\sqrt{q}+1)\right) + 2\phi(q-1) - \tfrac{f-1}{f} \cdot (q-1),$$

provided that f is a prime.

Proof. We proceed as in the proof of Proposition 14.9.10, but now choose C as the set of all non-primitive elements $\gamma \in A$ whose order is a multiple of $2^{\ell+1}sf$. As before, we write $\gamma = \gamma_2\gamma_r\gamma_s$, where $\gamma_s \in \Omega_s$ and $\gamma_2 = \zeta^i$ for some odd i. Now $\gamma_r \in U_r$ has order divisible by f, but not equal to r, which means $\gamma_r \in U_g\Omega_f \setminus \Omega_r$; also,

$$|C| = P_{\beta,a}\big(f(q+1)\big) - P_{\beta,a}.$$

As in the proof of Proposition 14.9.10, $c := \mathrm{Norm}_{E/F}(\gamma) = \zeta^{i(q+1)} \cdot \gamma_r^2$ is a non-square in F but not a primitive element. Moreover, f now has to divide $\mathrm{ord}(c)$, as the order of $\zeta^{i(q+1)}$ is a power of 2, whereas the order of γ_r is an odd multiple of f. Therefore,

$$c \in U_g\Omega_f\Omega_{2^m} \setminus \Omega_{q-1}, \text{ where } m = \begin{cases} \ell & \text{if } q \equiv 1 \bmod 4, \\ 1 & \text{if } q \equiv 3 \bmod 4. \end{cases}$$

This yields

$$\begin{aligned} |C| \leq 2 \cdot |N(C)| &\leq 2 \cdot \big(|U_g| \cdot |\Omega_f| \cdot |\Omega_{2^m}| - |\Omega_{q-1}|\big) \\ &= 2g\phi(f)\phi(2^m) - 2\phi(q-1) \\ &= \tfrac{\phi(f)}{f} \cdot \big(fg \cdot 2\phi(2^m)\big) - 2\phi(q-1) \\ &= \tfrac{\phi(f)}{f} \cdot (q-1) - 2\phi(q-1). \end{aligned}$$

Substituting this lower bound for C and the upper bound for $P_{\beta,a}$ following from Remark 14.3.9 in

$$P_{\beta,a} = P_{\beta,a}\big(f(q+1)\big) - |C|$$

results in the assertion (after a little simplification). □

Even though the proof of Theorem 14.1.6 is now complete, we will conclude this section by showing that one may reduce the huge amount of computer work which we had needed considerably. This observation is due to Cohen [79] and relies on Theorem 14.9.9. We have already used that result to settle 225 of the 234 cases not covered by Corollary 14.3.8, leaving only nine comparatively small instances unresolved. Thus it is not all that surprising that Theorem 14.9.9 can also be used to kill the extremely large intervals determined at the end of the last section; in fact, it allows to handle all odd prime powers $q \geq 70{,}000$ theoretically, without much effort. The actual computations required for this can be verified easily with the aid of a standard pocket calculator, using the following (rather technical) result:

Proposition 14.9.13. *Let $I = [z, Z]$ be a specified real interval, where $z \geq 2$ and $Z \geq 5$, and introduce the following data:*

- *Let $t \in \mathbb{N}^*$ be maximal such that $\sqrt{8 \cdot p_2 \cdots p_t + 1} \in I$, where $p_2 < \cdots < p_t$ are the first $t-1$ odd primes.*
- *Put $\sigma(t) := \prod_{j=2}^{t} \frac{p_j - 1}{p_j}$.*
- *Let $\tau \in \mathbb{N}^*$ be maximal such that $2p_2 \cdots p_\tau + 1 \in I$.*
- *Define*

$$W = W(\sigma, z) := \frac{(2^\tau - 2) \cdot \sqrt{z} + \frac{1}{2\sigma}}{z - 1},$$

where σ may be chosen arbitrarily from the interval $(0, \sigma(t)]$.

Then one has $P_{\beta,a} > 0$ for all odd prime powers q contained in I, provided that W satisfies the condition $W < 1 - \frac{1}{4\sigma}$.

Proof. Consider an arbitrary odd prime power q contained in I, and let $r_2 < \cdots < r_\omega$ be the $\omega - 1$ distinct odd prime divisors of $q^2 - 1$, where $\omega = \omega(q^2 - 1)$ (as before). Note that this requires

$$\omega \leq t \text{ and } \omega(q-1) \leq \tau.$$

With the abbreviations $y := \frac{2\phi(q-1)}{q-1}$ and $x := \frac{2\phi(q+1)}{q+1}$, we obtain the estimate

$$yx = \frac{r_2 - 1}{r_2} \cdots \frac{r_\omega - 1}{r_\omega} \geq \frac{p_2 - 1}{p_2} \cdots \frac{p_t - 1}{p_t} = \sigma(t),$$

as $t \geq \omega$ and as $w \mapsto \frac{w}{w+1}$ is a strictly increasing function on $\mathbb{R}^+$. Therefore,

$$\frac{1}{xy} \leq \frac{1}{\sigma(t)} \leq \frac{1}{\sigma}.$$

The remainder of the proof rests on a technical trick, namely, on rewriting the term

$$\phi(q+1) + \phi(q-1) \cdot \left(1 - \frac{(2^{\omega(q-1)} - 2) \cdot \sqrt{q}}{q-1}\right) - \frac{q+1}{2}$$

appearing in Proposition 14.9.9 as

$$\frac{q-1}{2} \cdot (x + y(1-U) - 1),$$

where

$$U := \frac{(2^{\omega(q-1)} - 2) \cdot \sqrt{q} + \frac{2(1-x)}{y}}{q-1}.$$

The reader may verify that this is correct by an elementary – though quite tedious – calculation.

Our next goal is to show that $U \leq W$. For this, we first note that the parabola $w \mapsto w(1-w)$ takes the maximal value $\frac{1}{4}$, which implies

$$\frac{2(1-x)}{y} = \frac{2x(1-x)}{xy} \leq \frac{1}{2xy} \leq \frac{1}{2\sigma}.$$

Since $w \mapsto \frac{\sqrt{w}}{w-1}$ is a strictly decreasing function on $\mathbb{R}^+$, this leads to the desired inequality:

$$\begin{aligned} U &\leq \frac{(2^{\omega(q-1)}-2)\cdot\sqrt{q}+\frac{1}{2\sigma}}{q-1} \\ &\leq \frac{(2^{\tau}-2)\cdot\sqrt{q}+\frac{1}{2\sigma}}{q-1} \\ &\leq \frac{(2^{\tau}-2)\cdot\sqrt{z}+\frac{1}{2\sigma}}{z-1} = W. \end{aligned}$$

In view of the hypothesis on W, we conclude that U is in the interval $\left(0, 1-\frac{1}{4\sigma}\right)$. This yields

$$\begin{aligned} P_{\beta,a} &\geq \frac{q-1}{2}\cdot\big(x+y(1-U)-1\big) \\ &\geq \frac{z-1}{2}\cdot\big(x+y(1-U)-1\big) \\ &\geq \frac{z-1}{2}\cdot\big(x+y(1-W)-1\big) \\ &\geq \frac{z-1}{2}\cdot\big(\tfrac{\sigma}{y}+(1-W)y-1\big) \\ &\geq \frac{z-1}{2}\cdot\big(2\cdot\sqrt{\sigma(1-W)}-1\big), \end{aligned}$$

where we have used $xy \geq \sigma$ in the fourth step and

$$\frac{\sigma}{y}+(1-W)y \geq 2\cdot\sqrt{\sigma(1-W)}$$

(which follows from the inequality between the arithmetic and geometric means) in the final step. Now the hypothesis for W gives

$$\sigma\cdot(1-W) > \sigma\cdot\left(1-1+\frac{1}{4\sigma}\right) = \frac{1}{4},$$

and we conclude that

$$P_{\beta,a} > \frac{z-1}{2}\cdot\left(2\cdot\sqrt{\tfrac{1}{4}}-1\right) = 0. \quad \square$$

Example 14.9.14. Let us show that a single application of Proposition 14.9.13 suffices to reduce the size of the largest odd prime powers q which have to be examined from $q \leq 267{,}387{,}389$ to $q \leq 34{,}800{,}000$.

We have seen in the last section that $P_{\beta,a} > 0$ whenever $\omega(q^2-1) \geq 15$. Thus let $\omega(q^2-1) \leq 14$, and choose $Z := 270{,}000{,}000$, so that $t = 14$ and $\tau = 9$. The reader may check that $\sigma := 0.28$ satisfies $\sigma < \sigma(t) = \prod_{j=2}^{14} \frac{p_j - 1}{p_j}$.

With $z := 34{,}800{,}000$, the required condition $W(z,\sigma) < 1 - \frac{1}{4\sigma}$ holds; in fact, $1 - \frac{1}{4\sigma} - W > 0.02$. Thus $P_{\beta,a} > 0$ holds for all odd prime powers q contained in the interval $I = [z, Z]$. □

We leave it to the reader to check the details for the reduction to $q \leq 70{,}000$, using the data provided in Table 14.9 below, which is a minor variation on the table given by Cohen [79, p.227].

Table 14.8 Reduction to $q \leq 70{,}000$ via Proposition 14.9.13.

z	Z	t	τ	σ
5,440,000	34,800,000	12	8	0.29
510,000	5,440,000	11	7	0.305
160,000	510,000	10	6	0.315
70,000	160,000	9	6	0.327

Exercises

Exercise 14.9.15. Let ζ and θ be as in Observation 14.9.1. Verify that

- $\mathrm{Tr}_{E/F}(\zeta) = 0$ if $q \equiv 1 \bmod 4$;
- $\mathrm{Tr}_{E/F}(\theta) = 0$ if $q \equiv 3 \bmod 4$.

Also prove that $\mathrm{Norm}_{E/F}(\zeta) = \zeta^{q+1}$ cannot be a square in F. (Trivially, it is a square in E.) □

Exercise 14.9.16. Check that the set $S = U_r \cdot \Omega_{2^{\ell}+1} \cdot \Omega_s$ defined before Equation (14.24) has cardinality $(q-1) \cdot \phi(q+1)$, regardless whether $q \equiv 1 \bmod 4$ or $q \equiv 3 \bmod 4$. □

Exercise 14.9.17. Check that the term $\frac{1}{2} \cdot (1 + \eta_0(-1)\eta(\beta) - 2\alpha)$ appearing in Theorem 14.9.9 is always non-negative. □

Exercise 14.9.18. Apply Proposition 14.9.10 to settle the cases

$$q \in \{11, 31, 43, 71, 79, 131\}. \quad \square$$

Exercise 14.9.19. Apply Proposition 14.9.12 to settle the cases $q \in \{139, 211, 911\}$. Hint: choose f as the smallest odd prime dividing $q - 1$. □

Exercise 14.9.20. Consider the quadratic extension E over the field $F = \mathrm{GF}(11)$. Verify directly with calculations in that field extension that for every $a \in F^*$ and for every $\beta \in E^*$ such that the order of β is divisible by 24, there does exist a primitive element u of E such that $\mathrm{Tr}_{E/F}(\beta u) = a$. □

14.10 An Application: Costas Arrays

In this section, we discuss an interesting application of finite fields for which the strongest known results rely on theorems concerning primitive elements that are quite similar in spirit to those considered in the preceding sections and in Chapter 13; moreover, also the methods used to establish these theorems are largely those we have already encountered. As all known proofs again require a considerable amount of technical work, we will not present them here.

The application in question concerns problems arising from radar, sonar, physical alignment and time-position synchronization, as studied by Golomb and Taylor [149] in 1982. For instance, in radar one may want to generate a (periodically repeated) sequence of pulses of distinct frequencies so that a returning echo of this sequence which is shifted in both time and frequency (due to the distance of and the Doppler effect caused by a moving target) can be used to determine the correct range and velocity of the target. Clearly, this requires that the only translate of the original two-dimensional time-frequency pattern which has a significant correlation with the received configuration will be the one whose time shift corresponds to the correct range and whose frequency shift corresponds to the correct velocity of the target; similar problems are encountered in constructing sonar signal patterns. We refer the reader to the seminal papers by Golomb and Taylor [149, 150] or to the books of Golomb and Gong [147] and Schroeder [343] for more details on this topic.

A particular sonar problem which was first discussed by John Costas [98] in 1965 leads to the search for a specific type of two-dimensional agreement patterns, which have therefore been named after him.[6] Formally, one requires a permutation π of $\{1,\ldots,n\}$ satisfying the following condition:

$$\pi(i+k) - \pi(i) \neq \pi(j+k) - \pi(j) \quad \text{for all } i,j,k \text{ with } 1 \leq i < j \leq n-k. \tag{14.25}$$

One then associates with π the corresponding permutation matrix A, that is, the (n,n)-matrix $A = (a_{ij})$ with entries 0 and 1 satisfying $a_{ij} = 1$ if and only if $\pi(i) = j$. Such a permutation matrix is called a **Costas array** of size n. Frequently, a Costas array is graphically represented as an (n,n)-grid in the plane with dots in those cells for which one has $a_{ij} = 1$, whereas the remaining cells are left empty. Then condition (14.25) may be phrased geometrically as follows: any two of the $\binom{n}{2}$ vectors

[6] As pointed out in the *Nanoexplanations* blog of Aaron Sterling [362] in 2011, Costas arrays were independently discovered by Edgar Gilbert [139] (also in 1965, albeit in a different form). Moreover, Gilbert's paper even contains a result equivalent to the Welch construction in Theorem 14.10.2 below, which was only published much later, namely in [149] in 1982.

connecting a pair of distinct dots are distinct, that is, no two of them agree in both magnitude and slope. In Figure 14.1, we exhibit three examples of Costas arrays of size 6 from Golomb [145], and a fourth one will be given in Example 14.10.9.

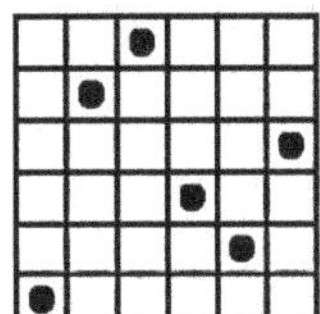
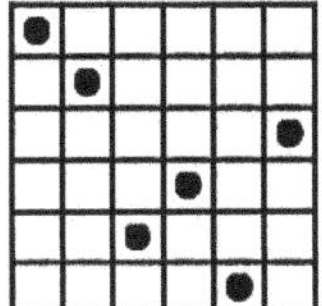
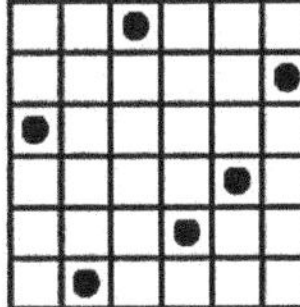

Fig. 14.1 Three Costas arrays of size 6

From the graphic interpretation of Costas arrays, one immediately sees the validity of the following simple but extremely useful result:

Lemma 14.10.1. *Assume the existence of a Costas array of size n which has a dot in one of its four corners. Then one obtains a Costas array of size $n-1$ by removing the corresponding row and column of the given array.* □

All known systematic constructions for Costas arrays involve the use of primitive elements in finite fields. We will present two general existence results in this section.

The simplest construction for Costas arrays is usually attributed to Lloyd R. Welch and was first reported in 1982 by Golomb and Taylor [149] (though it is equivalent to an earlier construction by Gilbert, as explained in Footnote 6). Two years later, the same authors [150] gave some further constructions and a table for the existence problem (up to $n = 360$).

Theorem 14.10.2. *For every prime p, there exist Costas arrays of sizes $p-1$ and $p-2$. If 2 is a primitive root modulo p, then there also exists a Costas array of size $p-3$.*

Proof. Let w be a primitive root modulo p. We claim that the permutation π of $\{1,\dots,p-1\}$ defined by

$$\pi(i) = h \quad \Longleftrightarrow \quad w^i \equiv h \bmod p$$

satisfies condition (14.25). Suppose that $\pi(i+k) - \pi(i) = \pi(j+k) - \pi(j)$ for some $k \neq 0$, that is, $w^{i+k} - w^i = w^{j+k} - w^j$ in $\mathbb{Z}_p$. Then $w^i(w^k - 1) = w^j(w^k - 1)$, and hence $i = j$, as $w^k - 1 \neq 0$. Thus π indeed corresponds to a Costas array A of size $p-1$. As $w^{p-1} = 1$, the array A has a dot in the lower left-hand corner. Therefore, removing the last row and the first column of A results in a Costas array B of size $p-2$, by Lemma 14.10.1.

Now assume that 2 is a primitive root modulo p, so that we can choose $w = 2$. Then the trivial equation $2^1 = 2$ shows that B has a dot in its upper left-hand corner, and another application of Lemma 14.10.1 produces a Costas array C of size $p-3$, as desired. □

The reader should check that the Welch construction with $p=7$ and $w=3$ yields the first Costas array of size 6 exhibited in Figure 14.1, and hence also an array of size 5. Note that 2 is not a primitive root modulo 7, so that we cannot obtain an array of size 4 in this way. The question under which conditions 2 is a primitive root modulo a prime p was already discussed in Section 8.5; see, in particular, Proposition 8.5.12.

Remark 14.10.3. Note that periodically repeating the columns of a Costas array constructed as in Theorem 14.10.2 results in an infinite pattern for which any n consecutive columns form a Costas array. Costas arrays with this property are called **singly periodic**. It was conjectured by Golomb and Taylor [150] in 1984 that the only examples are the arrays which arise from the Welch construction in Theorem 14.10.2. Stronger periodicity properties and a weakened version of the preceding conjecture were considered by Golomb and Moreno [148] in 1996; see also Moreno [281]. The validity of this weaker conjecture was established in 2015 by Muratović-Ribić, Pott, Thomson and Wang [296]. □

For our second construction for Costas arrays, we require the following result on primitive elements of finite fields which was conjectured by Golomb [145] and established independently by both Moreno and Sotero [282] and by Cohen and Mullen [94] (with a more satisfactory argument which requires considerably less machine computation).

Result 14.10.4 (Golomb's Conjecture A) *Let $q \neq 2$. Then* $\mathrm{GF}(q)$ *contains two primitive elements α and β satisfying $\alpha+\beta=1$.* □

As mentioned in our introductory remarks, the proof of Result 14.10.4 rests on methods similar to those used in the preceding sections to study the existence problem for primitive elements with a prescribed trace and will be omitted.

Remark 14.10.5. It should be noted that Result 14.1.1 implies the validity of Result 14.10.4 whenever q is a square, say $q=r^2$, since we may then select a primitive element α of $\mathrm{GF}(q)$ with trace 1 over $\mathrm{GF}(r)$ and take $\beta := \alpha^r$. This very special case of Result 14.1.1 was likewise conjectured by Golomb [145] (Golomb's Conjecture D); it was also established (independently of the work of Cohen on the general case) by Moreno [280] in 1989.

We also remark that Golomb's paper contains two further related conjectures on the existence of primitive elements (Golomb's Conjectures B and C) which have likewise generated much interest; see Cohen and Mullen [94] and Cohen [84] and the references cited there.

The study of interesting linear combinations of primitive elements is currently a quite active research topic. As an example, we mention the following two striking results obtained in a recent paper by Cohen, Oliveira e Silva, Sutherland and Trudgian [95]:

- Let $q \notin \{2,3,4,5,7,9,13,25,121\}$. Then there is a primitive element $\alpha \in \mathrm{GF}(q)$ such that $\alpha + \alpha^{-1}$ is also primitive.
- Let $q \notin \{2,3,4,5,7,9,13\}$. Then there are primitive elements $\alpha, \beta \in \mathrm{GF}(q)$ such that both $\alpha + \beta$ and $\alpha^{-1} + \beta^{-1}$ are also primitive.

Asymptotically, an even stronger result was obtained in Gupta, Sharma and Cohen [158]:

- Let $n \geq 5$, and let a be an arbitrary element of $\mathrm{GF}(q)$. Then there is a primitive element α for $\mathrm{GF}(q^n)$ such that $\alpha + \alpha^{-1}$ is also primitive and has trace a over $\mathrm{GF}(q)$.

For further results of a similar flavor, we refer the interested reader to [95] and the references cited there. □

Theorem 14.10.6. *For every prime power $q \neq 2$, there exist Costas arrays of sizes $q-2$ and $q-3$. If $q \geq 8$ is even, then there also exists a Costas array of size $q-4$.*

Proof. Let α and β be any two (not necessarily distinct) primitive elements for $F = \mathrm{GF}(q)$. We claim that the permutation π of $\{1, \ldots, q-2\}$ defined by

$$\pi(i) = h \quad \Longleftrightarrow \quad \alpha^i + \beta^h = 1 \tag{14.26}$$

satisfies condition (14.25). Suppose $\pi(i+k) - \pi(i) = \pi(j+k) - \pi(j)$ for some $k \neq 0$. Then

$$\frac{\beta^{\pi(i+k)}}{\beta^{\pi(i)}} = \frac{\beta^{\pi(j+k)}}{\beta^{\pi(j)}}$$

and hence, by (14.26),

$$(1-\alpha^{i+k})(1-\alpha^j) = (1-\alpha^{j+k})(1-\alpha^i),$$

or equivalently,

$$\alpha^i(1-\alpha^k) = \alpha^j(1-\alpha^k).$$

This shows $i = j$, as $\alpha^k \neq 1$, so that π indeed corresponds to a Costas array A (of size $q-2$).

Now assume that α and β satisfy the condition $\alpha + \beta = 1$, which is possible by Result 14.10.4. Then A has a dot in the upper left-hand corner, and thus Lemma 14.10.1 implies the existence of a Costas array B of size $q-3$ obtained by removing the first row and the first column of A. If q is a power of 2, the condition $\alpha + \beta = 1$ implies the equation $\alpha^2 + \beta^2 = 1$. Therefore, B also has a dot in the upper left-hand corner, and another application of Lemma 14.10.1 produces a Costas array of size $q-4$, as desired. □

The construction in the proof of Theorem 14.10.6 is due to Golomb [145] and is the reason why he was interested in his Conjecture A.

Remark 14.10.7. For $\alpha = \beta$, the Golomb construction reduces to a construction due to Abraham Lempel which was first reported by Golomb and Taylor [150]. This special case is closely related to the Zech logarithms discussed in Section 8.1, since then the definition of π in (14.26) can be written as follows:

$$1-\alpha^i = \alpha^{\pi(i)}. \tag{14.27}$$

If F has characteristic 2, we see that π is just the Zech logarithm as introduced in Definition 8.1.4 (where we used the notation ω instead of α). In the case of odd characteristic, π is not the Zech logarithm itself but a shifted version, since (14.27) then yields

$$1-\alpha^i = 1+\alpha^{i+(q-1)/2} = \alpha^{Z(i+(q-1)/2)}.$$

In view of Equation (14.27), the Lempel construction obviously yields **symmetric** Costas arrays. Unfortunately, these arrays cannot be reduced to give symmetric arrays of smaller size, as $2\alpha \neq 1$ (except for $q=3$). □

We conclude this section with another special case of the construction in Theorem 14.10.6 which is also due to Golomb [145]. It yields symmetric Costas arrays also for the smaller sizes occurring in Theorem 14.10.6, provided that q is a perfect square, and was the reason for Golomb's interest in his Conjecture D.

Corollary 14.10.8. *Let q be a prime power which is a perfect square, say $q=r^2$. Then there exist symmetric Costas arrays of sizes $q-2$ and $q-3$. If $q \neq 4$ is even, then there also exists a symmetric Costas array of size $q-4$.*

Proof. Choose α as a primitive element of $F=\mathrm{GF}(q)$ with trace 1 over $\mathrm{GF}(r)$ in the proof of Theorem 14.10.6, which is possible by Theorem 14.1.6, and take $\beta := \alpha^r$. Then the validity of the equation $\alpha^i+\beta^h = \alpha^i+\alpha^{hr}=1$ implies that of $\alpha^{ir}+\alpha^{hr^2} = \beta^i+a^h=1$, that is,

$$\pi(i)=h \iff \pi(h)=i.$$

Hence the Costas array A and therefore also the arrays derived from A according to Lemma 14.10.1 are indeed symmetric. □

Example 14.10.9. We apply Corollary 14.10.8 to construct a symmetric Costas array of size 7 using $F=\mathrm{GF}(9)$. Let α be a root of the irreducible polynomial $f=x^2+2x+2$ over $\mathrm{GF}(3)$ and note that α is a primitive element of F with trace 1. We exhibit the powers of α and their logarithms to the base α in Table 14.9 and the corresponding symmetric Costas array of size 7 in Figure 14.2. Omitting the first row and the first column of this array gives a (symmetric) Costas array of size 6, which is different from the three examples shown in Figure 14.1. □

This section provides a further example for the power of applying finite fields to the construction of interesting combinatorial objects. Nevertheless, the algebraic construction methods we have presented are very far away from settling the existence problem for Costas arrays: in the words of Solomon Golomb [145],

> The great majority of all integers n will not be included in these four classes of known constructions.

log	0	1	2	3	4	5	6	7
exp	1	α	$1+\alpha$	$1+2\alpha$	2	2α	$2+2\alpha$	$2+\alpha$

Table 14.9 Logarithms for GF(7)

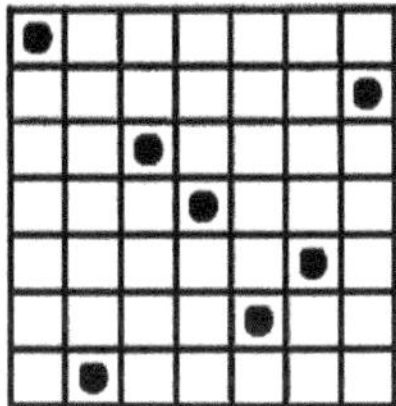

Fig. 14.2 A symmetric Costas array of size 7

For $n \leq 100$, there are still 17 sizes for which the existence of a Costas array is unknown; the five smallest open cases are at present 32, 33, 43, 48 and 49. On the positive side, all Costas arrays of size ≤ 29 have been enumerated. Relevant references can be found in the paper by Drakakis [110], which focuses on open problems but also serves as a good survey. A current database on Costas arrays up to order 1300 is maintained by James Beard [26].

Exercises

Exercise 14.10.10. Check that only the first of the three Costas arrays of size 6 exhibited in Figure 14.1 is singly periodic. □

Exercise 14.10.11. Use Theorem 14.10.2 to construct Costas arrays of all possible sizes between 8 and 12. □

Exercise 14.10.12. Check that the Golomb construction in Theorem 14.10.6 with $q=7$, α a root of x^3+x+1, and $\beta=\alpha^3$ produces the second Costas array of size 6 exhibited in Figure 14.1 (and hence also arrays of sizes 5 and 4).

Also check that the third (symmetric) Costas array of size 6 exhibited there is constructed from the Zech logarithm for GF(8) (as given in Table 8.1); see Remark 14.10.7. □

14.11 Concluding Remarks

There is a vast literature on the existence of primitive elements in finite fields admitting additional properties, too much to be reproducible in a short survey.

As mentioned in the introductory section of this chapter, the theorem on primitive elements with prescribed trace was motivated by the equivalent fact that the coefficient of the monomial x^{n-1} can be prescribed when searching for a primitive polynomial of degree n over a finite field $F = \mathrm{GF}(q)$. By now, there has been a lot of work on the existence of primitive polynomials with *some* or even *several* coefficient(s) prescribed. For concrete results as well as the description of advanced methods, we refer to the recent survey by Huczynska [194].

The formulation of Result 14.1.2 obviously leads to the question whether an n-dimensional extension $E = \mathrm{GF}(q^n)$ over $F = \mathrm{GF}(q)$ with $n \geq 3$ always contains an element $a \in F$ such that $\alpha(\theta + a)$ is a primitive element of E (for any given $\theta, \alpha \in E^*$ with $F(\theta) = \alpha$). Here we refer to Cohen [86], which includes a survey on results in this direction.

Now let $E = \mathrm{GF}(q^n)$ and $F = \mathrm{GF}(q)$, where $n \geq 2$. We mention a few variations and combinations of the topics studied in Chapters 13 and 14:

- The existence of a primitive element in E which is normal over F and which has a prescribed (E,F)-trace, necessarily non-zero, was proved by Cohen and Hachenberger [87].
- The existence of a primitive element in E which is normal over F and which has a prescribed (E,F)-norm (which is necessarily a primitive element of the ground field) is proved in Cohen and Hachenberger [89].
- Cohen and Hachenberger [89] also prove the existence of primitive normal elements with prescribed trace and norm into the ground field for $n \geq 7$ with at most eight possible exceptional pairs (q,n). The problem whether the trace and the norm of a primitive normal element can be prescribed has then been solved in a series of papers by Cohen [85] for $n \geq 5$, by Cohen and Huczynska [90] for $n = 4$, and by Huczynska and Cohen [195] for $n = 3$. The final result is as follows:

Result 14.11.1 *Consider a field extension E/F of Galois fields with degree at least 3, and let $a, b \in F^*$, where b is a primitive element. Then there exists a primitive element $v \in E$ which is normal over F, and such that $\mathrm{Tr}_{E/F}(v) = a$ and $\mathrm{Norm}_{E/F}(v) = b$.* □

The existence of primitive normal elements with prescribed trace and/or norm into some intermediate field of the given extension in the spirit of Theorem 14.2.4 is studied in Hachenberger [162, 165, 166, 168]. We just state one result from [166].

Result 14.11.2 *Let $F = \mathrm{GF}(q)$ be a finite field, and let r be a prime, where $r \geq 7$. Then there exists a universal generator for the r-primary closure $E_{r^\infty} = \bigcup_{m \in \mathbb{N}} E_{r^m}$ of F, where $E_{r^m} = \mathrm{GF}(q^{r^m})$ for every m; that is, there exists a sequence $(w_m)_{m \in \mathbb{N}}$ satisfying the following properties:*

- *for every m, the element w_m is a primitive element of E_{r^m} which is completely normal over F;*
- *if $k, \ell \in \mathbb{N}$ such that $k \leq \ell$, then the (E_{r^ℓ}, E_{r^k})-trace of w_ℓ and the (E_{r^ℓ}, E_{r^k})-norm of w_ℓ are both equal to w_k.* □

Finally, we mention one further recent variation on primitive elements with prescribed trace. In analogy to the notion of k-normal elements discussed in Section 13.11, one says that an element of $\mathrm{GF}(q)$ with order $(q-1)/k$ is k-**primitive**. Of course, this notion is of particular interest for small values of k, as discussed in Remark 5.6.13.

The possible traces of 2-primitive elements have been completely determined by Cohen and Kapetenakis [93]; note that such elements only exist when q is odd. Their main result is as follows:

Result 14.11.3 *Consider the n-dimensional extension* $E = \mathrm{GF}(q^n)$ *of the Galois field* $F = \mathrm{GF}(q)$, *where q is odd and* $n \geq 2$. *Let* $a \in F$, *and assume that* $a \neq 0$ *and* $q \notin \{3,5,7,9,11,13,31\}$ *when* $n = 2$. *Then there exists a 2-primitive element* $u \in E$ *such that* $\mathrm{Tr}_{E/F}(u) = a$. □

Cohen and Kapetenakis have also listed all possible traces for the exceptional values of q (when $n = 2$) occurring in the preceding result.

List of Symbols

We here list those symbols which recur throughout this book, together with short descriptions; more detailed explanations can be found in the corresponding definitions, which may be located using the index.

As much as possible, we have chosen to adopt standard notations generally appearing in the literature, with a few exceptions designed to make our presentation simpler or more precise. Also, for some more specialized concepts no standard notation is (as yet) established. In most of these cases, we have added a footnote (in the text proper) explaining our choice and discussing alternative notations used in the literature.

We begin by listing fundamental general symbols concerning sets, numbers, functions, basic algebraic structures and the like. Following this, we cover the more specialized symbols required for our study of Galois fields.

General notations

$x := y$ or $y =: x$	x is defined to be y
$x \leftarrow y$	x is assigned the value of y
$A \leftrightsquigarrow B$	A corresponds to B
$\square$	end of an item (like a proof, definition, example, etc.)

Sets

$\emptyset$	the empty set
$A \subset B$	A is a subset of B

D. Hachenberger and D. Jungnickel, *Topics in Galois Fields*,
Algorithms and Computation in Mathematics 29,
https://doi.org/10.1007/978-3-030-60806-4

$A \cup B$	the union of the sets A and B
$A \dot{\cup} B$	the disjoint union of the sets A and B
$\bigcup_{i \in I} A_i$	the union of a family of sets
$A \cap B$	the intersection of the sets A and B
$\bigcap_{i \in I} A_i$	the intersection of a family of sets
$\overline{A}$	the complement of A (w.r.t. a given superset)
$A \setminus B$	A without B: $A \cap \overline{B}$
$A \oplus B$	the symmetric difference of A and B: $(A \setminus B) \cup (B \setminus A)$
$A \times B$	the Cartesian product of the sets A and B
2^A or $\mathscr{P}(A)$	the power set of A
A^t	the set of all ordered t-tuples with elements from A
$\binom{A}{t}$	the set of all t-subsets of A
$\binom{n}{t}$	the number of t-subsets of an n-set: $\frac{n!}{t!(n-t)!}$
$\|A\|$	the cardinality of A

Mappings and relations

A^B	the set of all mappings from B to A
$\operatorname{map}(A)$	the set of all mappings from A to A
$f\colon A \to B$	f is a mapping from A to B
$f(a)$	the image of $a \in A$ under the mapping $f\colon A \to B$
$f\colon a \mapsto b$	f maps a to b, that is, $f(a) = b$
$f(X)$	$\{f(x)\colon x \in X\}$ for $f\colon A \to B$ and $X \subset A$
$f^{-1}(Y)$	$\{x\colon f(x) \in Y\}$ for $f\colon A \to B$ and $Y \subset B$
$\operatorname{im} f$	$\{f(a)\colon a \in A\}$ for $f\colon A \to B$
$f + g$	the pointwise addition of $f, g \in R^X$ (R a ring)
$f \cdot g$ or fg	the pointwise multiplication of $f, g \in R^X$ (R a ring)
rf	the multiplication of $f \in R^X$ by a scalar $r \in R$ (R a ring)
$(M, \preceq)$	generic notation for partially ordered sets
$\operatorname{prolim}_{M,\preceq}(X, \varphi)$	the projective limit of a projective system over $(M, \preceq)$

Sets of numbers

$\mathbb{N}$	the set of natural numbers (including 0)
$\mathbb{N}^*$	the set of positive integers
$\mathbb{N}_p^*$	the set of all positive integers not divisible by the prime p
D_n	the set of all divisors of $n \in \mathbb{N}^*$
$\pi(n)$	the set of all prime divisors of $n \in \mathbb{N}^*$
π_ℓ	the set of all primes $< \ell$
$\mathbb{Z}$	the set of integers
$\mathbb{Z}^*$	the set of integers $\neq 0$
$\mathbb{Z}_n$	the set of residue classes modulo n
$\mathbb{Z}_{[\mathcal{N}]}$	the projective limit of $(\mathcal{N}, (\mathbb{Z}_n)_{n \in \mathcal{N}}, \mathtt{mod})$ for the Steinitz number $\mathcal{N}$
$\mathbb{Q}$	the set of rational numbers
$\mathbb{Q}^+$	the set of positive rational numbers
$\mathbb{Q}_0^+$	the set of non-negative rational numbers
$\mathbb{R}$	the set of real numbers
$\mathbb{R}^+$	the set of positive real numbers
$\mathbb{R}_0^+$	the set of non-negative real numbers
$\mathbb{C}$	the set of complex numbers

Numbers

$\overline{c}$	the conjugate of a complex number c
$\lvert x \rvert$	the absolute value of a real number x
$\mathrm{sgn}(z)$	the sign of an integer z
$\lceil x \rceil$	the smallest integer $\geq x$ (for $x \in \mathbb{R}$)
$\lfloor x \rfloor$	the largest integer $\leq x$ (for $x \in \mathbb{R}$)
e	the base of the natural logarithm

i	the imaginary unit
$\gcd(x,y)$	the non-negative greatest common divisor of $x, y \in \mathbb{Z}$
$\operatorname{lcm}(x,y)$	the non-negative least common multiple of $x, y \in \mathbb{Z}$

Functions on $\mathbb{N}^*$

(a/p)	the Legendre symbol
n'	the p-free part of $n \in \mathbb{N}^*$ (w.r.t. a given prime p)
$n!$	$n(n-1)(n-2)\cdots 1$ (for $n \in \mathbb{N}$)
$a \texttt{ div } b$	the quotient of a under division by b
$a \texttt{ mod } b$	the remainder of a under division by b
$\operatorname{ord}_n(q)$	the order of q modulo n (for $n \geq 2$)
$\operatorname{pt}_n(N)$	the n-part of N (for $n, N \geq 2$)
$\operatorname{rad}(n)$	the radical of n
$\operatorname{sf}(n)$	the square-free part of n
$\mu(n)$	the Möbius function of n
$\phi(n)$	the Euler totient function of n
$\omega(n)$	the number of distinct prime divisors of n

Monoids

0 or 0_M	the identity element in an additively written monoid
1 or 1_M	the identity element in a multiplicatively written monoid
$C(M)$	the center of a monoid M
$U(M)$	the group of units of a monoid M
u^{-1} or $\frac{1}{u}$	the inverse of a unit u
$\prod_{i=1}^{n} a_i$	the product $a_1 a_2 \cdots a_n$ (in multiplicative notation)
a^n	the n-th power of an element a (in multiplicative notation)
$\sum_{i=1}^{n} a_i$	the sum $a_1 + \cdots + a_n$ (in additive notation)

$n \cdot a$ or na	the n-fold multiple of an element a (in additive notation)
$\langle a \rangle$	the submonoid (resp. subgroup) of M generated by a

Divisibility in a commutative monoid M with cancellation

$a \mid b$	a is a divisor of b
$a \approx b$	a and b are associates
$\mathrm{CD}(a,b)$	the set of all common divisors of a and b
$\mathrm{CM}(a,b)$	the set of all common multiples of a and b
$\mathrm{GCD}(a,b)$	the set of all greatest common divisors of a and b
$\mathrm{LCM}(a,b)$	the set of all least common multiples of a and b
$\overline{M}$	the factor monoid of M
$I(M)$	the set of irreducible elements of M
$P(M)$	the set of prime elements of M
$Q(M)$	the quotient group of M

Groups

$U \leq G$	U is a subgroup of the group G
$a \sim_U b$	$a^{-1}b \subset U$ (for $U \leq G$)
aU	the left coset of a w.r.t. a subgroup U
G/U	the set of all left cosets of U (for $U \leq G$)
$[G:U]$	the index of U in G (for $U \leq G$)
G/N	the factor group of a normal subgroup N of G
AB	the product of A and B (in multiplicative notation)
$A \times B$ or $A \otimes B$	the direct product of A and B (in multiplicative notation)
$A + B$	the sum of A and B (in additive notation)
$A \oplus B$	the direct sum of A and B (in additive notation)
$\langle S \rangle$	the subgroup of G generated by a subset S
$\langle g_1, \ldots, g_k \rangle$	the subgroup of G generated by the elements $g_1, \ldots, g_k$

$\exp G$	the exponent of a finite group G
$\ker \psi$	the kernel of a group homomorphism ψ
$\operatorname{ord}(g)$	the order of a group element g
U_d	the unique subgroup of order d in a given cyclic group G, where d divides $\lvert G \rvert$

Permutation groups

S_n	symmetric group acting on n elements
S_X or $\operatorname{Sym}(X)$	symmetric group acting on a set X
id_X	the identity permutation on a set X
gx	the image of $x \in X$ under $g \in G$ (for an action of G on X)
Gx	the orbit of $x \in X$ under G (for an action of G on X)
G_x	the stabilizer of $x \in X$ in G (for an action of G on X)
$\operatorname{sgn}(\pi)$	the sign of a permutation π

Rings

R^*	the set of non-zero elements of a ring R
$I(X)$	the ideal generated by a subset X of R
$Q(R)$	the quotient field of a commutative domain R
$U(R)$	the group of units of the multiplicative monoid $(R, \cdot, 1)$
$\prod_{j=1}^{\ell} I_j$	the product of ideals $I_1, \ldots, I_\ell$ in a commutative ring R
$Rx_1 + \cdots + Rx_k$	the ideal generated by the elements $x_1, \ldots, x_k \in R$
$(x_1, \ldots, x_k)$	the ideal generated by the elements $x_1, \ldots, x_k \in R$
(x) or Rx	the principal ideal generated by the element $x \in R$
$x \equiv y \bmod I$	x and y are congruent modulo the ideal I in R
$x \equiv y \bmod b$	x and y are congruent modulo the principal ideal bR in R
$\mathrm{pt}_a(b)$	the a-part of b for elements a, b in a principal ideal domain R
$\mathrm{rad}_R(z)$	the radical of an element z in a principal ideal domain R

μ_R	the Möbius function of a principal ideal domain R
$a \operatorname{\mathtt{div}} b$	a quotient of a under division by b (for Euclidean domains)
$a \operatorname{\mathtt{mod}} b$	a remainder of a under division by b (for Euclidean domains)

Fields and field extensions

F^*	the multiplicative group of the field F
$F^{\square}$	the set of squares in F^*
$F^{(n)}$	the n-th cyclotomic extension of F
U_n	the cyclic group of the n-th roots of unity in $F^{(n)}$
C_n	the set of all primitive n-th roots of unity in $F^{(n)}$
π_n^m	the canonical epimorphism $U_m \to U_n$, where $n \mid m$
$\Phi_n(x)$	the n-th cyclotomic polynomial over a specified field F
$\widehat{F}$	the algebraic closure of the field F
E/F	generic notation for field extensions
$[E:F]$	the degree of E/F
$F(S)$	the intermediate field of E/F generated by S
$F(v_1,\ldots,v_m)$	the intermediate field of E/F generated by $\{v_1,\ldots,v_m\}$
KL	the compositum of the intermediate fields K and L of E/F
$\operatorname{acl}(E/F)$	the algebraic closure of F in E
$\operatorname{char} F$	the characteristic of the field F
$\operatorname{Gal}(E/F)$	the Galois group of E/F
$\operatorname{Norm}_{E/F}$	the norm function of a Galois extension E/F
$\operatorname{Tr}_{E/F}$	the trace function of a Galois extension E/F
$\Delta_{E/F}(b_1,\ldots,b_n)$	the discriminant of $b_1,\ldots,b_n \in E$ (for a Galois extension E/F with degree n)

Modules

$\sum_{j\in J} U_j$	the sum of submodules U_j of a module V, where $j \in J$
$\oplus_{j\in J} U_j$	the direct sum of submodules U_j of a module V, where $j \in J$
$\mathscr{A}(X)$	the annihilator ideal of a subset X of an R-module V
$\mathscr{A}(V)$	the annihilator ideal of an R-module V
M_I	the submodule of an R-module V annihilated by the ideal I
M_a	the submodule of an R-module V annihilated by the principal ideal Ra
M_{p^∞}	the primary module of an R-module V w.r.t. the prime p (for principal ideal domains R)
$\mathrm{mdl}(X)$	the submodule of an R-module V generated by a subset X
$\mathrm{mdl}(x_1,\ldots,x_k)$	the submodule of an R-module V generated by $x_1,\ldots,x_k$
$\mathscr{O}(v)$	the order ideal of an element v in an R-module V (for principal ideal domains R)
$\mathrm{Ord}_R(v)$	the R-order of an element v in an R-module V (for principal ideal domains R)
$T(V)$	the torsion submodule of a module V over a domain R
ϕ_V and ϕ_R	analogues of the Euler phi function for modules (over principal ideal domains R)
Ω_t^+	see Equation (13.8) in Remark 13.4.3

Formal power series over a commutative ring R (w.r.t. a simple factorial monoid N)

$f \star g$	the convolution of two mappings $f, g \in R^N$
$R[[N]]$	the ring of formal power series over R and N
$R[N]$	the ring of polynomials over R and N
$R[[x]]$	the ring of formal power series over R (in the indeterminate x)
$F((x))$	the field of formal Laurent series over a field F (in the indeterminate x)
ε	the identity element of $R[[N]]$

μ	the Möbius function of $R[[N]]$
ζ	the zeta function of $R[[N]]$
S_f	the summation function of a formal power series f (see Theorem 2.1.10)
$\operatorname{supp} f$	the support of a formal power series f

Polynomials (over a commutative ring R resp. a field F)

$R[x]$	the ring of polynomials over R (in the indeterminate x)
$\deg(f)$ or $\deg f$	the degree of a polynomial $f \in R[x]$
$\operatorname{rad}(f)$	the radical of a polynomial $f \in F[x]$
$f \texttt{ div } g$	the quotient of f when dividing by g (for $f, g \in F[x]$)
$f \texttt{ mod } g$	the remainder of f when dividing by g (for $f, g \in F[x]$)
$f'(x)$	the formal derivative of a polynomial $f \in F[x]$
$F[x]_{\text{mon}}$	the set of all monic polynomials over F
$F[x]_{<n}$	the set of all polynomials with degree $< n$ over F
$\mathscr{B}_f$	the Berlekamp algebra of $f \in F[x]$
$\mathscr{B}_f^P$	the prime Berlekamp algebra of $f \in F[x]$
$\mathscr{B}_f^Q$	the Berlekamp algebra of $f \in F[x]$ over the subfield Q
C_h	the set of spectral coefficients of $h \in \mathscr{B}_f$
μ_F	the Möbius function of the ring $F[x]$
ϕ_F	the Euler function of the ring $F[x]$
ζ_F	the zeta function of the ring $F[x]$
$\omega_F(f)$	the number of distinct monic irreducible divisors of f (f a polynomial over F)

Rational functions (over a field F)

$F(x)$	the field of all rational functions $y = f(x)/g(x)$ over F (in the indeterminate x)
$y'(x)$	the formal derivative of $y = f(x)/g(x)$

$y^{(k)}(x)$	the k-th formal derivative of $y = f(x)/g(x)$
$\deg(y)$ or $\deg y$	the degree of $y = f(x)/g(x)$
$\mathcal{N}_f^Q$	the polynomial Niederreiter space of $f \in F[x]$ (over the subfield Q)
$\bar{\mathcal{N}}_f^Q$	the fractional Niederreiter space of $f \in F[x]$ (over the subfield Q)
W_f	the space of all rational functions $y = f(x)/g(x)$ satisfying $\deg g < \deg f$ (for $f \in F[x]$)
$\Psi\colon \widehat{F}(x) \to \widehat{F}(x)$	the Niederreiter mapping over F

Algebras A over a field F

$f(a)$	the evaluation of a polynomial $f \in F[x]$ at an element $a \in A$
Γ_a	the evaluation homomorphism at $a \in A$
$\mathrm{alg}_F(a)$	the F-subalgebra of A generated by a
$\deg_F(a)$	the degree of $a \in A$ over F (if $\ker \Gamma_a \neq \{0\}$)
$\mathrm{mpol}_a(x)$	the minimal polynomial of $a \in A$ (if $\ker \Gamma_a \neq \{0\}$)
$\mathrm{mpol}_{a,F}(x)$	the minimal polynomial of $a \in A$ over F (if $\ker \Gamma_a \neq \{0\}$)
FG	the group algebra of a group G over F

Vector spaces and endomorphisms

F^n	the n-dimensional vector space of n-tuples over the field F
$\mathrm{Hom}_F(V,W)$	the vector space of all F-linear mappings from V to W
$\mathrm{End}_F(V)$	the F-algebra of all homomorphisms from V into itself (for an F-vector space V)
$U^{\perp}$	the subspace orthogonal to U (w.r.t. a given bilinear form)
V^*	the dual space of V
$\mathbf{v}^f$	$f(\tau)(\mathbf{v})$ (for $\tau \in \mathrm{End}_F(V)$ and f in $F[x]$)
(V,τ)	V considered as an $F[x]$-module w.r.t. the endomorphism τ

$(V,\tau)_g$	the (F,τ)-submodule of (V,τ) annihilated by g (for a monic divisor g of $\mathrm{mpol}_\tau(x)$)
$\mathrm{mdl}_\tau(\mathbf{v})$	the $F[x]$-submodule of V generated by $\mathbf{v}$ (w.r.t. the endomorphism τ)
$\Gamma_{\tau,\mathbf{v}}$	the pairing homomorphism between the $F[x]$ modules $F[x]$ and (V,τ) (w.r.t. $\mathbf{v} \in V$)
$\mathrm{mpol}_{\tau,\mathbf{v}}(x)$	the τ-order of $\mathbf{v} \in V$ (for $\tau \in \mathrm{End}_F(V)$)
$\mathrm{Ord}_\tau(\mathbf{v})$	the τ-order of $\mathbf{v} \in V$ (for $\tau \in \mathrm{End}_F(V)$)

Matrices

$R^{(m,n)}$	the set of (m,n)-matrices over a ring R
A^T	the transpose of the matrix A
A^*	the conjugate transpose of a complex matrix A
I	an identity matrix
J	a matrix with all entries equal to 1
Z	a Fourier matrix over a field
(a_{ij})	the matrix with entries a_{ij}
$\mathrm{circ}(c_0,\dots,c_{n-1})$	the circulant matrix with first row $(c_0,\dots,c_{n-1})$
$\mathrm{diag}(a_1,\dots,a_n)$	the diagonal matrix with entries $a_{11}=a_1,\dots,a_{nn}=a_n$
$\det A$ or $\lvert A\rvert$	the determinant of the matrix A
$w(S)$	the weight of a matrix S

Galois fields F and extensions E/F

$\mathrm{GF}(p)$ or $\mathbb{Z}_p$	the field with p elements, p a prime
$\mathrm{GF}(q)$ or $\mathbb{F}_q$	the Galois field with q elements, q a prime power
$\mathscr{C}_m^F$	the cyclotomic module corresponding to m over F
E_n	the extension field of F with degree n in E/F (for $n \in N(E/F)$)
E_{r^∞}	the r-primary closure of F in $\widehat{F}$

$(E,\sigma)_g$ or V_g	the (F,σ)-submodule of E annihilated by a divisor g of x^n-1
$\mathcal{G}^F_{m,k}$	a generalized cyclotomic module over F (corresponding to (m,k))
$H_{\beta,a}$	an affine hyperplane in E/F
L_β	a generalized trace mapping of E/F
$\log_\omega \gamma$	the discrete logarithm of γ to the base ω
$\mathcal{N}_g$	characteristic functions related to normality, see Equation (13.10)
$N(E/F)$	the Steinitz number of E/F
$\mathrm{Ord}_q(v)$	the q-order of $v \in E$, where $F = \mathrm{GF}(q)$
$\mathcal{P}_t$	characteristic functions related to primitivity, see Equation (13.9)
$P_{\beta,a}$	the number of primitive elements in an affine hyperplane $H_{\beta,a}$
$PN_n(q)$	the number of primitive normal elements of $\mathrm{GF}(q^n)/\mathrm{GF}(q)$
$\mathcal{R}_q$	the set of degrees of regular extensions of $\mathrm{GF}(q)$
$\mathcal{S}_q$	the Steinitz number associated with the strongly regular extensions of $F = \mathrm{GF}(q)$
V^F_h	the (F,σ)-submodule of $\widehat{F}$ annihilated by h over $F = \mathrm{GF}(q)$
$\delta_q(d)$	the number of elements of $\mathrm{GF}(q^n)$ with degree d over $\mathrm{GF}(q)$
$\kappa(\mathcal{G}^F_{m,k})$	the module character of $\mathcal{G}^F_{m,k}$
μ_q	the Möbius function of the ring $\mathbb{Z}[[\mathrm{GF}(q)[z]_{\mathrm{mon}}]]$
ϕ_q	the ananlogue of the Euler phi function for $\mathrm{GF}(q)[x]$
Π^g_f	the canonical epimorphism $V_g \to V_f$, where $f \mid g$
σ	the Frobenius automorphism of E/F
Ω_h or $G_q(h)$	the set of all elements with q-order h over $F = \mathrm{GF}(q)$
$\Omega^+_{\ell,g}$	set of all $u \in \mathrm{GF}(q^n)$ satisfying $\mathrm{pt}_\ell(q^n-1) \mid \mathrm{ord}(u)$ and $\mathrm{pt}_g(x^n-1) \mid \mathrm{Ord}_q(u)$

Polynomials over Galois fields

$c^T(x)$	the transpose of a polynomial $c(x)$ in $\mathrm{GF}(q)[x]/(x^n-1)$
f^*	the reciprocal polynomial of a polynomial f

$f^{\wedge}$	the monic multiple of f^* for an irreducible polynomial f, see Theorem 5.7.1
f^Q	the Q-transform of a polynomial f
f^R	the R-transform of a polynomial f
$f \circledast g$	the root composition of the G-polynomials f and g
$A_g(x)$	the q-polynomial associated with $g \in \mathrm{GF}(q)[x]$
$I_{n,q}$	the product of all monic irreducible polynomials with degree n over $\mathrm{GF}(q)$
$i_q(n)$	the number of monic irreducible polynomials with degree n over $\mathrm{GF}(q)$
$\mathbb{M}_z^*$	the set of all monic polynomials in the indeterminate z over $\mathrm{GF}(q)$ which are not divisible by z
$n_q(n)$	the number of normal polynomials in $\mathrm{GF}(q)[x]$ with degree n
$\mathrm{ord}(f)$ or $\mathrm{ord}\, f$	the order of $f \in \mathrm{GF}(q)[x]$
$\mathrm{ord}_g(z)$	the order of z modulo $g(z)$, where $g \in \mathrm{GF}(q)[z]$, $z \nmid g$
$R_{n,q}$	the factor ring $\mathrm{GF}(q)[x]/(x^n - 1)$
$SI_{n,q}$	the product of all self-reciprocal monic irreducible polynomials with degree n over $\mathrm{GF}(q)$
$si_q(n)$	the number of self-reciprocal monic irreducible polynomials with degree n over $\mathrm{GF}(q)$
$Z(e)$	the Zech logarithm of ω^e
$\Phi_n(x^k)$	a generalized cyclotomic polynomial
$\Psi_g(x)$	the Psi polynomial of g over $\mathrm{GF}(q)$
$\omega_q(g)$	the number of distinct monic irreducible factors of g, where $g(x) \in \mathrm{GF}(q)[x]$

Vector spaces and matrices over Galois fields

$\begin{bmatrix} n \\ k \end{bmatrix}_q$	a Gaussian coefficient: the number of k-subspaces of an n-vector space over $\mathrm{GF}(q)$
B^*	the dual basis of a basis B for $\mathrm{GF}(q^n)/\mathrm{GF}(q)$
$A_B(\gamma)$	the matrix representing the linear mapping M_γ (see below) w.r.t. the basis B for $\mathrm{GF}(q^n)/\mathrm{GF}(q)$

$b(n,q)$	the number of ordered bases of F^n over $F = \mathrm{GF}(q)$
$C(n,q)$	the group of circulant invertible (n,n)-matrices over $\mathrm{GF}(q)$
C_B	the complexity of a normal basis B
$C_q(n)$	the minimal complexity of a normal basis for $\mathrm{GF}(q^n)/\mathrm{GF}(q)$
$\mathrm{GL}(n,q)$	the general linear group consisting of the invertible (n,n)-matrices over $\mathrm{GF}(q)$
M_B	the matrix associated with a basis B for $\mathrm{GF}(q^n)/\mathrm{GF}(q)$, see Lemma 3.13.16
M_γ	the linear mapping on the $\mathrm{GF}(q)$-vector space $\mathrm{GF}(q^n)$ induced by multiplication with $\gamma \in \mathrm{GF}(q^n)$
$N_q(n,r)$	the number of symmetric matrices of rank r over $\mathrm{GF}(q)$
$N_q^0(n,r)$	the number of skew-symmetric matrices of rank r over $\mathrm{GF}(q)$
$O(n,q)$	the group of orthogonal (n,n)-matrices over $\mathrm{GF}(q)$
$OC(n,q)$	the group of orthogonal circulant (n,n)-matrices over $\mathrm{GF}(q)$
$r_B(\xi)$	the (primal) coordinates of ξ w.r.t. a basis B of $\mathrm{GF}(q^n)$ over $\mathrm{GF}(q)$
$R(B)$	the matrix representation for $\mathrm{GF}(q^n)$ associated with the basis B of $\mathrm{GF}(q^n)$ over $\mathrm{GF}(q)$
$sd(n,q)$	the number of self-dual bases of $\mathrm{GF}(q^n)$ over $\mathrm{GF}(q)$
$sdn(n,q)$	the number of self-dual normal bases of $\mathrm{GF}(q^n)$ over $\mathrm{GF}(q)$
$\mathrm{SL}(n,q)$	the special linear group consisting of the invertible (n,n)-matrices with determinant 1 over $\mathrm{GF}(q)$

Sequences over Galois fields

$\mathbf{a}^{(t)}$	the t-th state vector of an LFSR
$C_{\mathbf{a}}(h)$	the autocorrelation function of a periodic sequence $\mathbf{a}$
$D^{(r)}(\mathbf{a})$	the r-th Hankel determinant of a sequence $\mathbf{a}$
$L(\mathbf{a})$	the linear complexity of a sequence $\mathbf{a}$
$\mathbf{L} = (L_k(\mathbf{a}))$	the linear complexity profile of a sequence $\mathbf{a}$
$\mathrm{rank}_p D$	the p-rank of a difference set D
$\mathbf{u}(\mathbf{a},d,h)$	a decimation of a periodic sequence $\mathbf{a}$

Characters of abelian groups G

$\widehat{G}$	the character group of G
χ_0	the trivial character of G
$H^{\perp}$	the dual of a subgroup H of G
$S^{\perp}$	the dual of a subgroup S of $\widehat{G}$

Characters of Galois fields F

$G(\psi,\chi)$	the Gauss sum associated with ψ and χ
χ_0	the trivial additive character of F
χ_1	the canonical additive character of F
χ_b	the additive character of F associated with b
η	the quadratic character of F (in odd characteristic)
ψ_0	the trivial multiplicative character of F
ψ_j	the multiplicative character of F sending a specified primitive element ω to ζ^j,where ζ is a fixed primitive $(q-1)$-th root of unity
θ	the normed Gauss sum of F (in odd characteristic)

References

Whenever possible, we have included the first names of all authors, even for those references where only initials had been used.

1. Gordon B. Agnew, Ronald C. Mullin, Ivan M. Onyszchuk, and Scott A. Vanstone, *An implementation for a fast public-key cryptosystem*, J. Cryptology **3** (1991), 63–79.
2. Gordon B. Agnew, Ronald C. Mullin, and Scott A. Vanstone, *Fast exponentiation in* GF(2^n), Advances in cryptology – EUROCRYPT 88, Lect. Notes Comput. Sci. **330**, Springer, New York, 1988, pp. 251–255.
3. ———, *An Implementation of Elliptic Curve Cryptosystems Over* $F_{2^{155}}$, IEEE Journal on Selected Areas in Communications **11** (1993), 804 – 813.
4. Alfred V. Aho, John E. Hopcroft, and Jeffrey D. Ullman, *The design and analysis of computer algorithms*, Addison-Wesley, Reading, Mass., 1974.
5. Martin Aigner, *Combinatorial theory*, Springer, New York, 1979.
6. A. Adrian Albert, *Symmetric and alternate matrices in an arbitrary field. I*, Trans. Amer. Math. Soc. **43** (1938), 386–436.
7. Bill Allombert, *Explicit computation of isomorphisms between finite fields*, Finite Fields Appl. **8** (2002), 332–342.
8. George E. Andrews, *Reciprocal polynomials and quadratic transformations*, Util. Math. **28** (1985), 255–64.
9. Markus Antweiler and Leopold Bömer, *Complex sequences over* GF(p^M) *with a two-level autocorrelation function and a large linear span*, IEEE Trans. Inform. Theory **38** (1992), 120–130.
10. Emil Artin, *Linear mappings and the existence of a normal basis*, Studies and essays presented to R. Courant on his 60th birthday, January 8, 1948, Interscience Publishers, New York, 1948, pp. 1–5.
11. ———, *Geometric algebra*, Interscience Publishers, New York, 1957.
12. ———, *Algebraic numbers and algebraic functions*, Gordon and Breach, New York, 1967.
13. David W. Ash, Ian F. Blake, and Scott A. Vanstone, *Low complexity normal bases*, Discrete Appl. Math. **25** (1989), 191–210.
14. Edward F. Assmus, Jr. and Jennifer D. Key, *Designs and their codes*, Cambridge University Press, Cambridge, 1992.
15. Louis Auslander and Richard Tolimieri, *Is computing with the finite fourier transform pure or applied mathematics?*, Bull. Amer. Math. Soc. (N.S.) **1** (1979), 847–897.
16. Eric Bach, James Driscoll, and Jeffrey Shallit, *Factor refinement*, J. Algorithms **15** (1993), 199–222.

D. Hachenberger and D. Jungnickel, *Topics in Galois Fields*,
Algorithms and Computation in Mathematics 29,
https://doi.org/10.1007/978-3-030-60806-4

17. Eric Bach and Jeffrey Shallit, *Algorithmic number theory, Vol. 1: Efficient algorithms*, The MIT Press, Cambridge, MA, 1996.
18. R. H. Barker, *Group synchronizing of binary digital systems*, Communication theory (Ed. Willis Jackson), Academic Press, New York, 1953, pp. 273–287.
19. Ulrich Baum and Michael Clausen, *Some lower and upper complexity bounds for generalized Fourier transforms and their inverses*, SIAM J. Comput. **20** (1991), 451–459.
20. Leonard D. Baumert, *Cyclic difference sets*, Springer, New York, 1971.
21. Eva Bayer-Fluckiger, *Self-dual normal bases*, Indag. Math. **51** (1989), 379–383.
22. Eva Bayer-Fluckiger and Hendrik W. Lenstra, Jr., *Forms in odd degree extensions and self-dual normal bases*, Am. J. Math. **112** (1990), 359–373.
23. Jacob T.B. Beard, Jr., *Matrix fields over finite extensions of prime fields*, Duke Math. J. **39** (1972), 475–484.
24. ———, *Matrix fields over prime fields*, Duke Math. J. **39** (1972), 313–321.
25. ———, *The number of matrix fields over* GF(q), Acta Arith. **25** (1974), 315–329.
26. James K. Beard, *Costas arrays and enumeration to order 1030*, IEEE Dataport, http://dx.doi.org/10.21227/H21P42, 2017.
27. Elwyn R. Berlekamp, *Factoring polynomials over finite fields*, Bell Systems Tech. J. **46** (1967), 1853–1859.
28. ———, *Factoring polynomials over large finite fields*, Math. Comp. **24** (1970), 713–735.
29. ———, *Bit-serial Reed-Solomon encoders*, IEEE Trans. Inform. Theory **28** (1982), 869–874.
30. ———, *Algebraic coding theory*, 3rd revised ed., World Scientific Publishing, Hackensack NJ, 2015.
31. Bruce C. Berndt, Ronald J. Evans, and Kenneth S. Williams, *Gauss and Jacobi sums*, Wiley, New York, 1998.
32. Thomas Beth, *Verfahren der schnellen Fourier-Transformation*, B. G. Teubner, Stuttgart, 1984.
33. Thomas Beth and Willi Geiselmann, *Selbstduale Normalbasen über* GF(q), Arch. Math. (Basel) **55** (1990), 44–48.
34. ———, *Finding (good) normal bases in finite fields*, Proceedings of the 1991 international symposium on Symbolic and algebraic computation ISSAC 91, ACM Press, New York, 1991, pp. 173–178.
35. Thomas Beth, Dieter Jungnickel, and Hanfried Lenz, *Design theory. Vol. I*, 2nd ed., Cambridge University Press, Cambridge, 1999.
36. Albrecht Beutelspacher and Ute Rosenbaum, *Projective geometry: From foundations to applications*, Cambridge University Press, Cambridge, 1998.
37. Jürgen Bierbrauer, *Introduction to coding theory*, Chapman & Hall/CRC, Boca Raton, FL, 2005.
38. Richard E. Blahut, *Transform techniques for error control codes*, IBM J. Res. Dev. **23** (1979), 299–315.
39. ———, *Theory and practice of error control codes*, Addison-Wesley, Reading, Mass., 1983.
40. Ian F. Blake, Ryoh Fuji-Hara, Ronald C. Mullin, and Scott A. Vanstone, *Computing logarithms in finite fields of characteristic two*, SIAM J. Algebraic Discrete Methods **5** (1984), 276–285.
41. Ian F. Blake, Shuhong Gao, and Ronald C. Mullin, *Explicit factorization of* $x^{2^k}+1$ *over* F_p *with prime* $p \equiv 3$ mod 4, Appl. Algebra Engrg. Comm. Comput. **4** (1993), 89–94.
42. Dieter Blessenohl, *Abelsche Erweiterungen, in denen jedes reguläre Element vollständig regulär ist*, Arch. Math. (Basel) **54** (1990), 146–156.
43. ———, *On the normal basis theorem*, Note Mat. **27** (2007), 5–10.
44. Dieter Blessenohl and Karsten Johnsen, *Eine Verschärfung des Satzes von der Normalbasis*, J. Algebra **103** (1986), 141–159.
45. ———, *Stabile Teilkörer galoisscher Erweiterungen und ein Problem von C. Faith*, Arch. Math. (Basel) **56** (1991), 245–253.
46. Zenon I. Borevich and Igor R. Shafarevich, *Number theory*, Academic Press, New York, 1966.

47. Joel V. Brawley and Leonard Carlitz, *Irreducibles and the composed product for polynomials over a finite field*, Discrete Math. **65** (1987), 115–139.
48. Joel V. Brawley and George E. Schnibben, *Infinite algebraic extensions of finite fields*, American Mathematical Society, Providence, RI, 1989.
49. Fabio Enrique Brochero Martínez, Carmen Rosa Giraldo Vergara, and Lilian Batista de Oliveira, *Explicit factorization of* $x^n - 1 \in \mathbb{F}_q[x]$, Des. Codes Cryptogr. **77** (2015), 277–286.
50. Anthony J. Bromfield and Fred C. Piper, *Linear recursion properties of uncorrelated binary sequences*, Discrete Appl. Math. **27** (1990), 187–193.
51. Lennart Brynielsson, *On the linear complexity of combined shift register sequences*, Advances in cryptology – EUROCRYPT 85, Lect. Notes Comput. Sci. **219**, Springer, New York, 1986, pp. 156–160.
52. Johannes Buchmann, *Introduction to cryptography*, 2nd ed., Springer, New York, 2004.
53. Johannes Buchmann and Ulrich Vollmer, *Binary quadratic forms. An algorithmic approach*, Springer, Berlin, 2007.
54. Lilya Budaghyan, *Construction and analysis of cryptographic functions*, Springer, Cham, 2014.
55. Peter Bürgisser, Michael Clausen, and Mohammad Amin Shokrollahi, *Algebraic complexity theory*, Springer, Berlin, 1997.
56. Kenneth A. Byrd and Theresa P. Vaughan, *Counting and constructing orthogonal circulants*, J. Combin. Theory Ser. A **24** (1978), 34–49.
57. Domenico Calabro and Jack K. Wolf, *On the synthesis of two-dimensional arrays with desirable correlation properties*, Inf. Control **11** (1967), 537–560.
58. Calmos, *Ca34c168 data encryption processor. Document 01-34168-500*, Calmos Semiconductor Inc., Kanata, Ont., Canada, 1988.
59. Peter J. Cameron and Jacobus H. van Lint, *Designs, graphs, codes and their links*, Cambridge University Press, Cambridge, 1991.
60. Oscar A. Cámpoli, *A principal ideal domain that is not a Euclidean domain*, Amer. Math. Monthly **95** (1988), 868–871.
61. David G. Cantor and Hans Zassenhaus, *A new algorithm for factoring polynomials over finite fields*, Math. Comp. **36** (1981), 587–592.
62. Claude Carlet and Sihem Mesnager, *Four decades of research on bent functions*, Des. Codes Cryptogr. **78** (2016), 5–50.
63. Leonard Carlitz, *Primitive roots in a finite field*, Trans. Amer. Math. Soc. **73** (1952), 373–382.
64. ———, *Some problems involving primitive roots in a finite field*, Proc. Nat. Acad. Sci. U.S.A. **38** (1952), 314–318; errata, 618.
65. ———, *Representation by quadratic forms in a finite field*, Duke Math. J. **21** (1954), 123–137.
66. ———, *Representations by skew forms in a finite field*, Arch. Math. (Basel) **5** (1954), 19–31.
67. ———, *Some cyclotomic matrices*, Acta Arith. **5** (1959), 293–308.
68. ———, *Some theorems on irreducible reciprocal polynomials over a finite field*, J. Reine Angew. Math. **227** (1967), 212–220.
69. Rey Casse, *Projective geometry: an introduction*, Oxford University Press, Oxford, 2006.
70. Agnes Hui Chan and Richard A. Games, *On the linear span of binary sequences obtained from q-ary m-sequences, q odd*, IEEE Trans. Inform. Theory **36** (1990), 548–552.
71. Wai-Kiu Chan and Man-Keung Siu, *Summary of perfect* $s \times t$ *arrays*, $1 \leq s \leq t \leq 100$, Electronics Letters **27** (1991), 709–710.
72. Yu-Ki Chan, Man-Keung Siu, and Po Tong, *Two-dimensional binary arrays with good autocorrelation*, Inf. Control **42** (1979), 125–130.
73. Robin Chapman, *Completely normal elements in iterated quadratic extensions of finite fields*, Finite Fields Appl. **3** (1997), 1–10.
74. Unjeng Cheng and Solomon W. Golomb, *On the characterization of PN sequences*, IEEE Trans. Inform. Theory **29** (1983), 600.

75. Maria Christopoulou, Theo Garefalakis, Daniel Panario, and David Thomson, *The trace of an optimal normal element and low complexity normal bases*, Des. Codes Cryptogr. **49** (2008), 199–215.
76. Krzysztof Ciesielski, *Set theory for the working mathematician*, Cambridge University Press, Cambridge, 1997.
77. Henri Cohen, Gerhard Frey, Roberto Avanzi, Christophe Doche, Tanja Lange, Kim Nguyen, and Frederik Vercauteren (eds.), *Handbook of elliptic and hyperelliptic curve cryptography*, Chapman & Hall/CRC, Boca Raton, FL, 2006.
78. Stephen D. Cohen, *On irreducible polynomials of certain types in finite fields*, Proc. Cambridge Philos. Soc. **66** (1969), 335–344.
79. ———, *Primitive roots in the quadratic extension of a finite field*, J. London Math. Soc., II. Ser **27** (1983), 221–228.
80. ———, *Polynomials over finite fields with large order and level*, Bull. Korean Math. Soc. **24** (1987), 83–96.
81. ———, *Generators in cyclic difference sets*, J. Combin. Theory Ser. A **51** (1989), 227–236.
82. ———, *Primitive elements and polynomials with arbitrary trace*, Discrete Math. **83** (1990), 1–7.
83. ———, *The explicit construction of irreducible polynomials over finite fields*, Des. Codes Cryptogr. **2** (1992), 169–174.
84. ———, *Primitive elements and polynomials: Existence results*, Finite fields, coding theory, and advances in communications and computing (Eds. Gary L. Mullen and Peter Jau-Shyong Shiue), Marcel Dekker, New York, 1993, pp. 43–55.
85. ———, *Gauss sums and a sieve for generators of Galois fields*, Publ. Math. Debrecen **56** (2000), 293–312.
86. ———, *Primitive elements on lines in extensions of finite fields*, Finite fields: Theory and applications, Contemp. Math., vol. 518, Amer. Math. Soc., Providence, RI, 2010, pp. 113–127.
87. Stephen D. Cohen and Dirk Hachenberger, *Primitive normal bases with prescribed trace*, Appl. Algebra Engrg. Comm. Comput. **9** (1999), 383–403.
88. ———, *The dynamics of linearized polynomials*, Proc. Edinburgh Math. Soc. (2) **43** (2000), 113–128.
89. ———, *Primitivity, freeness, norm and trace*, Discrete Math. **214** (2000), 135–144.
90. Stephen D. Cohen and Sophie Huczynska, *Primitive free quartics with specified norm and trace*, Acta Arith. **109** (2003), 359–385.
91. ———, *The primitive normal basis theorem—without a computer*, J. London Math. Soc. (2) **67** (2003), 41–56.
92. ———, *The strong primitive normal basis theorem*, Acta Arith. **143** (2010), 299–332.
93. Stephen D. Cohen and Giorgos Kapetanakis, *The trace of 2-primitive elements of finite fields*, Acta Arith. **192** (2020), 397–419.
94. Stephen D. Cohen and Gary L. Mullen, *Primitive elements in finite fields and Costas arrays*, Appl. Algebra Engrg. Comm. Comput. **2** (1991), 45–53.
95. Stephen D. Cohen, Tomás Oliveira e Silva, Nicole Sutherland, and Tim Trudgian, *Linear combinations of primitive elements of a finite field*, Finite Fields Appl. **51** (2018), 388–406.
96. Alina Carmen Cojocaru and M. Ram Murty, *An introduction to sieve methods and their applications*, Cambridge University Press, Cambridge, 2006.
97. Charles J. Colbourn and Jeffrey H. Dinitz (eds.), *The CRC handbook of combinatorial designs*, 2nd ed., Chapman & Hall/CRC, Boca Raton, FL, 2007.
98. John P. Costas, *Medium constraints on sonar design and performance*, Class 1 Report R65EMH33, G.E. Corporation, 1965.
99. Robert W. Craigen, *A direct approach to Hadamard's inequality*, Bull. Inst. Combin. Appl. **12** (1994), 28–32.
100. Charles W. Curtis and Irving Reiner, *Representation theory of finite groups and associative algebras*, Wiley, New York, 1962.
101. Harold Davenport, *Bases for finite fields*, J. London Math. Soc. **43** (1968), 21–39.

102. Harold Davenport and Helmut Hasse, *Die Nullstellen der Kongruenzzetafunktionen in gewissen zyklischen Fällen*, J. Reine Angew. Math. **172** (1934), 151–182.
103. Hans F. de Groote, *Lectures on the complexity of bilinear problems*, Springer, Berlin, 1987.
104. Richard Dedekind, *Abriß einer Theorie der höhern Congruenzen in Bezug auf einen reellen Primzahl-Modulus*, J. Reine Angew. Math. **54** (1857), 1–26.
105. Max Deuring, *Galoissche Theorie und Darstellungstheorie*, Math. Ann. **107** (1933), 140–144.
106. Jean Dieudonné, *La géométrie des groupes classiques*, Springer, Berlin, 1955.
107. Whitfield Diffie and Martin E. Hellman, *New directions in cryptography*, IEEE Trans. Inform. Theory **22** (1976), 644–654.
108. Cunsheng Ding, *Codes from difference sets*, World Scientific Publishing, Hackensack, NJ, 2015.
109. ______, *Designs from linear codes*, World Scientific Publishing, Hackensack, NJ, 2019.
110. Konstantinos Drakakis, *Open problems in Costas arrays*, arXiv:1102.5727, 2011.
111. Taher ElGamal, *A public key cryptosystem and a signature scheme based on discrete logarithms*, IEEE Trans. Inform. Theory **31** (1985), 469–472.
112. Shalom Eliahou, Michel Kervaire, and Bahman Saffari, *A new restriction on the lengths of Golay complementary sequences*, J. Combin. Theory Ser. A **55** (1990), 49–59.
113. Andreas Enge, *Elliptic curves and their applications to cryptography. An introduction.*, Kluwer Academic Publishers, Boston, MA, 1999.
114. Carl C. Faith, *Extensions of normal bases and completely basic fields*, Trans. Amer. Math. Soc. **85** (1957), 406–427.
115. Sandra Feisel, Joachim von zur Gathen, and Mohammad Amin Shokrollahi, *Normal bases via general Gauss periods*, Math. Comp. **68** (1999), 271–290.
116. Gui-Liang Feng and Kenneth K. Tzeng, *A generalization of the Berlekamp-Massey algorithm for multisequence shift-register synthesis with applications to decoding cyclic codes*, IEEE Trans. Inform. Theory **37** (1991), 1274–1287.
117. Jay P. Fillmore and Morris L. Marx, *Linear recursive sequences*, SIAM Rev. **10** (1968), 342–353.
118. Peter Fleischmann, *Connections between the algorithms of Berlekamp and Niederreiter for factoring polynomials over* $\mathbf{F}_q$, Linear Algebra Appl. **192** (1993), 101–108.
119. Gudmund Skovbjerg Frandsen, *On the density of normal bases in finite fields*, Finite Fields Appl. **6** (2000), 23–38.
120. Harold Fredricksen, *A survey of full length nonlinear shift register cycle algorithms*, SIAM Rev. **24** (1982), 195–221.
121. Walter Fumy, *Symmetry and duality in normal basis multiplication*, Applied algebra, algebraic algorithms and error-correcting codes AAECC-3, Lect. Notes Comput. Sci. **229**, Springer, New York, 1986, pp. 131–134.
122. Steven D. Galbraith and Pierrick Gaudry, *Recent progress on the elliptic curve discrete logarithm problem*, Des. Codes Cryptogr. **78** (2016), 51–72.
123. Évariste Galois, *Sur la théorie des nombres*, Bulletin des Sciences Mathématiques XIII (1830), 428–435, Reprinted in Écrits et Mémoires Matheématiques d'Évariste Galois, pp. 112-128.
124. Richard A. Games, *The geometry of quadrics and correlations of sequences*, IEEE Trans. Inform. Theory **32** (1986), 423–426.
125. Shuhong Gao, *Normal bases over finite fields*, Ph.D. thesis, University of Waterloo, Dept. of Combinatorics and Optimization, 1993.
126. ______, *Elements of provable high orders in finite fields*, Proc. Am. Math. Soc. **127** (1999), 1615–1623.
127. Shuhong Gao and Hendrik W. Lenstra, Jr., *Optimal normal bases*, Des. Codes Cryptogr. **2** (1992), 315–323.
128. Shuhong Gao and Joachim von zur Gathen, *Berlekamp's and Niederreiter's polynomial factorization algorithms*, Finite fields: theory, applications, and algorithms (Las Vegas, NV, 1993), Amer. Math. Soc., Providence, RI, 1994, pp. 101–116.

129. Shuhong Gao, Joachim von zur Gathen, Daniel Panario, and Victor Shoup, *Algorithms for exponentiation in finite fields*, J. Symbolic Comput. **29** (2000), 879–889.
130. ———, *Erratum: "Algorithms for exponentiation in finite fields"*, J. Symbolic Comput. **30** (2000), 491.
131. Theodoulos Garefalakis and Giorgos Kapatenakis, *On the existence of primitive completely normal bases of finite fields*, J. Pure Appl. Algebra **223** (2019), 909–921.
132. Carl Friedrich Gauss, *Disquisitiones arithmeticae*, Fleischer, Leipzig, 1801.
133. ———, *Summatio quarundam serierum singularium*, Comment. Soc. Reg. Sci. Gottigensis 1, 1811.
134. ———, *Disquisitiones arithmeticae. Transl. from the Latin by Arthur A. Clarke, Rev. by William C. Waterhouse*, Springer, New York, 1986.
135. Keith O. Geddes, Stephen R. Czapor, and George Labahn, *Algorithms for computer algebra*, Kluwer Academic Publishers, Dordrecht, 1992.
136. Willi Geiselmann, *Algebraische Algorithmenentwicklung am Beispiel der Arithmetik in endlichen Körpern*, Ph.D. thesis, Universität Karlsruhe, 1992.
137. Willi Geiselmann and Dieter Gollmann, *Symmetry and duality in normal basis multiplication*, Applied algebra, algebraic algorithms and error-correcting codes AAECC-6, Lect. Notes Comput. Sci. **357**, Springer, New York, 1989, pp. 230–238.
138. ———, *Self-dual bases in* $\mathrm{GF}(q^n)$, Des. Codes Cryptogr. **3** (1993), 333–345.
139. Edgar N. Gilbert, *Latin squares which contain no repeated digrams*, SIAM Rev. **7** (1965), 189–198.
140. Jean-Marie Goethals and Philippe Delsarte, *On a class of majority-logic decodable cyclic codes*, IEEE Trans. Inform. Theory **14** (1968), 182–188.
141. Dieter Gollmann, *Algorithmenentwurf in der Kryptographie*, B.I. Wissenschaftsverlag, Mannheim, 1994.
142. Solomon W. Golomb, *Sequences with randomness properties*, Reprinted in revised form as Chapter III of [146], Glenn L. Martin Company, Baltimore, MD, 1955.
143. ———, *Structural properties of PN sequences*, Reprinted as Chapter IV of [146], Jet Propulsion Laboratory, California Institute of Technology, Pasadena, CA, 1958.
144. ———, *On the classification of balanced binary sequences of period* $2^n - 1$, IEEE Trans. Inform. Theory **26** (1980), 730–732.
145. ———, *Algebraic constructions for Costas arrays*, J. Combin. Theory Ser. A **37** (1984), 13–21.
146. ———, *Shift register sequences. Secure and limited-access code generators, efficiency code generators, prescribed property generators, mathematical models*, 3rd revised ed., World Scientific Publishing, Hackensack, NJ, 2017.
147. Solomon W. Golomb and Guang Gong, *Signal design for good correlation. For wireless communication, cryptography, and radar*, Cambridge University Press, Cambridge, 2005.
148. Solomon W. Golomb and Oscar Moreno, *On periodicity properties of Costas arrays and a conjecture on permutation polynomials*, IEEE Trans. Inf. Theory **42** (1996), 2252–2253.
149. Solomon W. Golomb and Herbert Taylor, *Two-dimensional synchronization patterns for minimum ambiguity*, IEEE Trans. Inform. Theory **28** (1982), 600–604.
150. ———, *Construction and properties of Costas arrays*, Proc. IEEE **72** (1984), 1143–1163.
151. Frederick M. Goodman, *Algebra. Abstract and concrete,* Edition 2.6, SemiSimple Press, Iowa City, IA, 2014.
152. Basil Gordon, William H. Mills, and Lloyd R. Welch, *Some new difference sets*, Canad. J. Math. **14** (1962), 614–625.
153. David Goss, *Basic structures of function field arithmetic*, Springer, Berlin, 1996.
154. Rainer Göttfert, *An acceleration of the Niederreiter factorization algorithm in characteristic* 2, Math. Comp. **62** (1994), 831–839.
155. John Greene, *Principal ideal domains are almost Euclidean*, Amer. Math. Monthly **104** (1997), 154–156.
156. Thomas Gruber, *Eine Programmierumgebung zur Unterstützung der Grundlagenforschung im Bereich der endlichen Körper*, Masterarbeit, Institut für Mathematik, Universität Augsburg, 2013.

157. Anju Gupta and Rajendra Kumar Sharma, *A note on the distribution of self-dual normal bases generators of finite fields under trace map*, Beitr. Algebra Geom. **57** (2016), 573–578.
158. Anju Gupta, Rajendra Kumar Sharma, and Stephen D. Cohen, *Primitive element pairs with one prescribed trace over a finite field*, Finite Fields Appl. **54** (2018), 1–14.
159. Dirk Hachenberger, *On primitive and free roots in a finite field*, Appl. Algebra Engrg. Comm. Comput. **3** (1992), 139–150.
160. ———, *On completely free elements in finite fields*, Des. Codes Cryptogr. **4** (1994), 129–143.
161. ———, *Finite fields: Normal bases and completely free elements*, Kluwer Academic Publishers, Boston, MA, 1997.
162. ———, *Primitive normal bases for towers of field extensions*, Finite Fields Appl. **5** (1999), 378–385.
163. ———, *A decomposition theory for cyclotomic modules under the complete point of view*, J. Algebra **237** (2001), 470–486.
164. ———, *Primitive complete normal bases for regular extensions*, Glasg. Math. J. **43** (2001), 383–398.
165. ———, *Universal generators for primary closures of Galois fields*, Finite fields and applications (Eds. D. Jungnickel and H. Niederreiter), Springer, Berlin, 2001, pp. 208–223.
166. ———, *Generators for primary closures of Galois fields*, Finite Fields Appl. **9** (2003), 122–128.
167. ———, *Characterizing normal bases via the trace map*, Comm. Algebra **32** (2004), 269–277.
168. ———, *Primitive complete normal bases: existence in certain 2-power extensions and lower bounds*, Discrete Math. **310** (2010), 3246–3250.
169. ———, *Primitive normal bases for quartic and cubic extensions: a geometric approach*, Des. Codes Cryptogr. **77** (2015), 335–350.
170. ———, *Asymptotic existence results for primitive completely normal elements in extensions of Galois fields*, Des. Codes Cryptogr. **80** (2016), 577–586.
171. ———, *Primitive generators for cyclic vector spaces over a Galois field*, Australas. J. Combin. **64** (2016), 289–326.
172. ———, *Ovoids and primitive normal bases for quartic extensions of Galois fields*, J. Algebraic Combin. (2019), http://doi–org–443.webvpn.fjmu.edu.cn/10.1007/s10801–019–00920–8.
173. ———, *Primitive complete normal bases for regular extensions: exceptional cyclotomic modules*, arXiv:1912.04886v1 [math.NT], 2019.
174. Dirk Hachenberger and Stefan Hackenberg, *Computational results on the existence of primitive complete normal basis generators*, arXiv:1912.07541 [math.NT], 2019.
175. Stefan Hackenberg, *Theoretische und experimentelle Untersuchungen zu Normalbasen für Erweiterungen endlicher Körper*, Masterarbeit, Institut für Mathematik, Universität Augsburg, 2015.
176. Jacques Hadamard, *Résolution d'une question relative aux déterminants*, Bull. Sci. Math., II. Sér. **17** (1893), 240–246.
177. Marshall Hall, Jr., *A survey of difference sets*, Proc. Amer. Math. Soc. **7** (1957), 975–986.
178. Noboru Hamada and Hiroshi Ohmori, *On the BIB design having the minimum p-rank*, J. Combin. Theory Ser. A **18** (1975), 131–140.
179. Darrel Hankerson, Alfred Menezes, and Scott Vanstone, *Guide to elliptic curve cryptography*, Springer-Verlag, New York, 2004.
180. Helmut Hasse, *Vorlesungen über Zahlentheorie*, 2nd ed., Springer, New York, 1964.
181. Roger Heath-Brown, *Artin's conjecture for primitive roots*, Q. J. Math., II. Ser. **37** (1986), 27–38.
182. Kurt Hensel, *Über die Darstellung der Zahlen eines Gattungsbereiches für einen beliebigen Primdivisor*, J. Reine Angew. Math. **103** (1888), 230–237.
183. David Hilbert, *Über die Theorie der algebraischen Formen*, Math. Ann. **36** (1890), 473–534.
184. ———, *Die Theorie der algebraischen Zahlkörper*, Jahresber. Deutsch. Math.-Verein. **4** (1897), 175–546.

185. Raymond Hill, *A first course in coding theory*, Clarendon Press, Oxford, 1986.
186. James W. P. Hirschfeld, *Finite projective spaces of three dimensions*, The Clarendon Press, Oxford, 1985.
187. ———, *Projective geometries over finite fields*, 2nd ed., Clarendon Press, Oxford, 1998.
188. James W. P. Hirschfeld, Gábor Korchmáros, and Fernando Torres, *Algebraic curves over a finite field*, Princeton University Press, Princeton, 2008.
189. James W. P. Hirschfeld and Joseph A. Thas, *General Galois geometries*, 2nd ed., Springer, London, 2016.
190. Kenneth Hoffman and Ray Kunze, *Linear algebra*, Prentice-Hall, Englewood Cliffs, NJ, 1971.
191. Chih-Nung Hsu and Ting-Ting Nan, *A generalization of the primitive normal basis theorem*, J. Number Theory **131** (2011), 146–157.
192. Klaus Huber, *Some comments on Zech's logarithms*, IEEE Trans. Inform. Theory **36** (1990), 946–950.
193. ———, *Solving equations in finite fields and some results concerning the structure of* $\mathrm{GF}(p^m)$, IEEE Trans. Inform. Theory **38** (1992), 1154–1162.
194. Sophie Huczynska, *Existence results for finite field polynomials with specified properties*, Finite fields and their applications. Character sums and polynomials (Pascale Charpin, Alexander Pott, and Arne Winterhof, eds.), de Gruyter, Berlin, 2013, pp. 65–87.
195. Sophie Huczynska and Stephen D. Cohen, *Primitive free cubics with specified norm and trace*, Trans. Amer. Math. Soc. **355** (2003), 3099–3116.
196. Sophie Huczynska, Gary L. Mullen, Daniel Panario, and David Thomson, *Existence and properties of k-normal elements over finite fields*, Finite Fields Appl. **24** (2013), 170–183.
197. W. Cary Huffman and Vera S. Pless, *Fundamentals of error-correcting codes*, Cambridge University Press, Cambridge, 2003.
198. Daniel R. Hughes and Fred C. Piper, *Projective planes,* Corrected 2nd printing, Springer, New York, 1982.
199. ———, *Design theory*, Cambridge University Press, Cambridge, 1985.
200. Bertram Huppert, *Endliche Gruppen I*, Springer, New York, 1967.
201. Kyoki Imamura, *On self-complementary bases of* $\mathrm{GF}(q^n)$ *over* $\mathrm{GF}(q)$, Trans. IECE Japan, E **66** (1983), 717–721.
202. ———, *The number of self-complementary bases of a finite field of characteristic two*, IEEE Internat. Sympos. Inform. Theory, Kobe, Japan, IEEE Press, New York, 1988.
203. Kenneth Ireland and Michael Rosen, *A classical introduction to modern number theory*, 2nd ed., Springer, New York, 1990.
204. Nathan Jacobson, *Basic algebra I*, 2nd ed., W. H. Freeman and Company, New York, 1985.
205. ———, *Basic algebra II*, 2nd ed., W. H. Freeman and Company, New York, 1989.
206. Jonathan Jedwab, *Nonexistence of perfect binary arrays*, Electronics Letters **27** (1991), 1252–1254.
207. ———, *Generalized perfect arrays and Menon difference sets*, Des. Codes Cryptogr. **2** (1992), 19–68.
208. Jonathan Jedwab and Sheelagh Lloyd, *A note on the nonexistence of Barker sequences*, Des. Codes Cryptogr. **2** (1992), 93–97.
209. Antoine Joux and Cécile Pierrot, *Technical history of discrete logarithms in small characteristic finite fields. The road from subexponential to quasi-polynomial complexity*, Des. Codes Cryptogr. **78** (2016), 73–85.
210. Dieter Jungnickel, *Difference sets*, Contemporary design theory. A collection of surveys (Eds. Jeffrey H. Dinitz and Douglas R. Stinson), Wiley, New York, 1992, pp. 241–324.
211. ———, *Finite fields: Structure and arithmetics*, Bibliographisches Institut, Mannheim, 1993.
212. ———, *Trace-orthogonal normal bases*, Discrete Appl. Math. **47** (1993), 233–249.
213. ———, *On the order of a product in a finite abelian group*, Math. Mag **69** (1996), 53–57.
214. ———, *Graphs, networks and algorithms*, 4th ed., Springer, New York, 2013.
215. Dieter Jungnickel, Thomas Beth, and Willi Geiselmann, *A note on orthogonal circulant matrices over finite fields*, Arch. Math. (Basel) **62** (1994), 126–133.

216. Dieter Jungnickel, Alfred J. Menezes, and Scott A. Vanstone, *On the number of self-dual bases of* $\mathrm{GF}(q^m)$ *over* $\mathrm{GF}(q)$, Proc. Amer. Math. Soc. **109** (1990), 23–29.
217. Dieter Jungnickel and Scott A. Vanstone, *On primitive polynomials over finite fields*, J. Algebra **124** (1989), 337–353.
218. Nicholas M. Katz, *Gauss sums, Kloosterman sums, and monodromy groups*, Princeton University Press, Princeton, 1988.
219. Edwin L. Key, *An analysis of the structure and complexity of nonlinear binary sequence generators*, IEEE Trans. Inform. Theory **22** (1976), 732–736.
220. Neal Koblitz, *A course in number theory and cryptography*, Springer, New York, 1987.
221. ———, *Elliptic curve cryptosystems*, Math. Comp. **48** (1987), 203–209.
222. ———, *Algebraic aspects of cryptography. With an appendix on hyperelliptic curves by Alfred J. Menezes, Yi-Hong Wu, and Robert J. Zuccherato*, Springer, Berlin, 1998.
223. Sergei Vladimirovich Konyagin and Igor Shparlinski, *Character sums with exponential functions and their applications*, Cambridge University Press, Cambridge, 1999.
224. Lazarus E. Kopilovich, *On perfect binary arrays*, Electronics Letters **24** (1988), 566–567.
225. Nikolai Mikhailovich Korobov, *Exponential sums and their applications*, Kluwer Academic Publishers, Dordrecht, 1992.
226. Leopold Kronecker, *Zur Theorie der Elimination einer Variablen aus zwei algebraischen Gleichungen*, Monatsbericht der Königlich-Preussischen Akademie der Wissenschaften zu Berlin **1881** (1881), 535–600.
227. Brian A. LaMacchia and Andrew M. Odlyzko, *Computation of discrete logarithms in prime fields*, Des. Codes Cryptogr. **1** (1991), 47–62.
228. Edmund Landau, *Vorlesungen über Zahlentheorie*, S. Hirzel, Leipzig, 1927.
229. Eric S. Lander, *Symmetric designs: An algebraic approach*, Cambridge University Press, Cambridge, 1983.
230. Therese C.Y. Lee and Scott A. Vanstone, *Subspaces and polynomial factorizations over finite fields*, Appl. Algebra Engrg. Comm. Comput. **6** (1995), 147–157.
231. Abraham Lempel, *Matrix factorization over* $\mathrm{GF}(2)$ *and trace-orthogonal bases of* $\mathrm{GF}(2^n)$, SIAM J. Comput. **4** (1975), 175–186.
232. ———, *Characterization and synthesis of self-complementary normal bases in finite fields*, Linear Algebra Appl. **98** (1988), 331–346.
233. Abraham Lempel and Gadiel Seroussi, *Explicit formulas for self-complementary normal bases in certain finite fields*, IEEE Trans. Inform. Theory **37** (1991), 1220–1222.
234. Abraham Lempel and Marcelo J. Weinberger, *Self-complementary normal bases in finite fields*, SIAM J. Discrete Math. **1** (1988), 193–198.
235. Arjen K. Lenstra, Hendrik W. Lenstra, Jr., and László Lovász, *Factoring polynomials with rational coefficients*, Math. Ann. **261** (1982), 515–534.
236. Hendrik W. Lenstra, Jr., *Algorithms for finite fields*, Number theory and cryptography, London Math. Soc. Lecture Note Ser. **154**, Cambridge University Press, Cambridge, 1990, pp. 76–85.
237. ———, *Finding isomorphisms between finite fields*, Math. Comp. **56** (1991), 329–347.
238. Hendrik W. Lenstra, Jr. and René Schoof, *Primitive normal bases for finite fields*, Math. Comp. **48** (1987), 217–231.
239. Ka Hin Leung and Bernhard Schmidt, *The field descent method*, Des. Codes Cryptogr. **36** (2005), 171–188.
240. ———, *New restrictions on possible orders of circulant Hadamard matrices*, Des. Codes Cryptogr. **64** (2012), 143–151.
241. ———, *The anti-field-descent method*, J. Combin. Theory Ser. A **139** (2016), 87–131.
242. Rudolf Lidl and Harald Niederreiter, *Finite fields*, Cambridge University Press, Cambridge, 1983.
243. ———, *Introduction to finite fields and their applications (revised edition)*, Cambridge University Press, Cambridge, 1994.
244. Curt C. Lindner and Christopher A. Rodger, *Design theory*, CRC Press, Boca Raton, FL, 2009.

245. San Ling, Huaxiong Wang, and Chaoping Xing, *Algebraic curves in cryptography*, CRC Press, Boca Raton, FL, 2013.
246. Hans Dieter Lüke, *Sequences and arrays with perfect periodic correlation*, IEEE Trans. Aerosp. Electron. Syst. **24** (1988), 287–294.
247. Heinz Lüneburg, *Galoisfelder, Kreisteilungskörper und Schieberegisterfolgen*, Bibliographisches Institut, Mannheim, 1979.
248. ______, *On a little but useful algorithm*, Algebraic algorithms and error correcting codes (Grenoble, 1985), Lect. Notes Comput. Sci. **229**, Springer, Berlin, 1986, pp. 296–301.
249. ______, *Kleine Fibel der Arithmetik*, B.I.-Wissenschaftsverlag, Mannheim, 1987.
250. ______, *On the rational normal form of endomorphisms. a primer to constructive algebra*, B.I.-Wissenschaftsverlag, Mannheim, 1987.
251. Saunders MacLane and Garrett Birkhoff, *Algebra*, 3rd ed., Chelsea Publishing Company, New York, 1988.
252. Florence Jessie MacWilliams, *Orthogonal matrices over finite fields*, Amer. Math. Monthly **76** (1969), 152–164.
253. ______, *Orthogonal circulant matrices over finite fields, and how to find them*, J. Combin. Theory Ser. A **10** (1971), 1–17.
254. Florence Jessie MacWilliams and Henry B. Mann, *On the p-rank of the design matrix of a difference set*, Inf. Control **12** (1968), 474–488.
255. Florence Jessie MacWilliams and Neil J.A. Sloane, *The theory of error-correcting codes (3rd repr.)*, North-Holland (Elsevier). Amsterdam, 1985.
256. James L. Massey, *Cryptography and system theory*, Proc. 24th Annual Allerton Conf. on Communication, Control and Computing, 1986, pp. 1–8.
257. ______, *Shift-register synthesis and BCH decoding*, IEEE Trans. Inform. Theory **15** (1969), 122–127.
258. ______, *Contemporary cryptology. An introduction*, Contemporary cryptology (Ed. Gustavus J. Simmons), IEEE Press, New York, 1992, pp. 3–39.
259. James L. Massey and Jim K. Omura, *Computational method and apparatus for finite field arithmetic*, U. S. Patent application # US06418039, 1981.
260. Ariane M. Masuda, Lucia Moura, Daniel Panario, and David Thomson, *Low complexity normal elements over finite fields of characteristic two*, IEEE Trans. Comput. **57** (2008), 990–1001.
261. James H. McClellen and Thomas W. Parks, *Eigenvalue and eigenvector decomposition of the Discrete Fourier Transform*, IEEE Trans. Audio Electroacust. **20** (1972), 66–74.
262. Kevin S. McCurley, *The discrete logarithm problem*, Cryptology and computational number theory, Proc. Symp. Appl. Math. **42**, American Mathematical Society, Providence, RI, 1990, pp. 49–74.
263. Robert J. McEliece, *Finite fields for computer scientists and engineers*, Kluwer Academic Publishers, Boston, MA, 1987.
264. Gerasimos C. Meletiou and Gary L. Mullen, *A note on discrete logarithms in finite fields*, Appl. Algebra Engrg. Comm. Comput. **3** (1992), 75–78.
265. Alfred J. Menezes, *Representations in finite fields*, Master's thesis, University of Waterloo, 1989.
266. ______, *Elliptic curve cryptosystems*, Ph.D. thesis, University of Waterloo, 1992.
267. ______, *Elliptic curve public key cryptosystems*, Kluwer Academic Publishers, New York, 1993.
268. Alfred J. Menezes, Paul C. van Oorschot, and Scott A. Vanstone, *Some computational aspects of root finding in* $\mathrm{GF}(q^m)$, Symbolic and algebraic computation ISSAC 88, Lect. Notes Comput. Sci. **358**, Springer, New York, 1989, pp. 259–270.
269. ______, *Handbook of applied cryptography*, CRC Press, Boca Raton, FL, 1997.
270. Alfred J. Menezes (ed.), Ian F. Blake, XuHong Gao, Ronald C. Mullin, Scott A. Vanstone, and Tomik Yaghoobian, *Applications of finite fields*, Kluwer Academic Publishers, Boston, MA, 1993.
271. P. Kesava Menon, *On difference sets whose parameters satisfy a certain relation*, Proc. Amer. Math. Soc. **13** (1962), 739–745.

272. Sihem Mesnager, *Bent functions. Fundamentals and results*, Springer, Cham, 2016.
273. F. Meyer, *Normalbasismultiplikation in endlichen Körpern*, Diplomarbeit, Universität Karlsruhe, 1990.
274. Petra Meyer, *Eine Charakterisierung vollständig regulärer, abelscher Erweiterungen*, Abh. Math. Sem. Univ. Hamburg **68** (1998), 199–223.
275. Helmut Meyn, *On the construction of irreducible self-reciprocal polynomials over finite fields*, Appl. Algebra Engrg. Comm. Comput. **1** (1990), 43–53.
276. ———, *Explicit N-polynomials of 2-power degree over finite fields. I*, Des. Codes Cryptogr. **6** (1995), 107–116.
277. Robert L. Miller, *Necklaces, symmetries and self-reciprocal polynomials*, Discrete Math. **22** (1978), 25–33.
278. August Ferdinand Möbius, *Über eine besondere Art von Umkehrung der Reihen*, J. Reine u. Angew. Math. **9** (1832), 105–123.
279. Pieter Moree, *Artin's primitive root conjecture – a survey*, Integers **12** (2012), 1305–1416, A13.
280. Oscar Moreno, *On the existence of a primitive quadratic of trace* 1 *over* $\mathrm{GF}(p^m)$, J. Combin. Theory Ser. A **51** (1989), 104–110.
281. ———, *A shifting property of some Costas sequences*, Ars Comb. **33** (1992), 157–160.
282. Oscar Moreno and José Sotero, *Computational approach to Conjecture A of Golomb*, Proc. 20th Southeastern International Conference on Combinatorics, Graph Theory and Computing, Boca Raton/FL (USA) 1989, Utilitas Mathematica Publishing Inc., Winnipeg, 1990, pp. 7–16.
283. Ilene H. Morgan and Gary L. Mullen, *Completely normal primitive basis generators of finite fields*, Utilitas Math. **49** (1996), 21–43.
284. Ilene H. Morgan, Gary L. Mullen, and Miodrag Živković, *Almost weakly self-dual bases for finite fields*, Appl. Algebra Engrg. Comm. Comput. **8** (1997), 25–31.
285. Salvatore D. Morgera and Hari Krishna, *Digital signal processing. Applications to communications and algebraic coding theories*, Academic Press, Boston, 1989.
286. Masakatu Morii and Kyoki Imamura, *A theorem that* $\mathrm{GF}(2^{4m})$ *has no self-complementary normal bases over* $\mathrm{GF}(2)$ *for odd m*, Trans. IECE Japan, E **67** (1984), 655–656.
287. Masakatu Morii, Masao Kasahara, and Douglas L. Whiting, *Efficient bit-serial multiplication and the discrete-time Wiener-Hopf equation over finite fields*, IEEE Trans. Inform. Theory **35** (1989), 1177–1183.
288. Kent E. Morrison, *Integer sequences and matrices over finite fields*, J. Integer Seq. **9** (2006), Article 06.2.1, 28 pp.
289. Patrick Morton, *On the eigenvectors of Schur's matrix*, J. Number Theory **12** (1980), 122–127.
290. Theodore Motzkin, *The Euclidean algorithm*, Bull. Amer. Math. Soc. **55** (1949), 1142–1146.
291. Gary L. Mullen and Carl Mummert, *Finite fields and applications*, American Mathematical Society, Providence, RI, 2007.
292. Gary L. Mullen and Daniel Panario, *Handbook of finite fields*, CRC Press, Boca Raton, FL, 2013.
293. Gary L. Mullen and David White, *A polynomial representation for logarithms in* $\mathrm{GF}(q)$, Acta Arith. **47** (1986), 255–261.
294. Ronald C. Mullin, *A characterization of the extremal distributions of optimal normal bases*, Proceedings of the Marshall Hall conference on coding theory, design theory, group theory (Burlington, Vermont 1990), Wiley, New York, 1993, pp. 41–49.
295. Ronald C. Mullin, Ivan M. Onyszchuk, Scott A. Vanstone, and Richard M. Wilson, *Optimal normal bases in* $\mathrm{GF}(p^n)$, Discrete Appl. Math. **22** (1989), 149–161.
296. Amela Muratović-Ribić, Alexander Pott, David Thomson, and Qiang Wang, *On the characterization of a semi-multiplicative analogue of planar functions over finite fields*, Topics in finite fields. Proceedings 11th international conference on finite fields and their applications (Fq11), Magdeburg, Germany, 2013, Amer. Math. Soc., Providence, RI, 2015, pp. 317–325.
297. M. Ram Murty, *Artin's conjecture for primitive roots*, Math. Intelligencer **10** (1988), 59–67.

298. V. Kumar Murty (ed.), *Algebraic curves and cryptography*, American Mathematical Society, Providence, RI, 2010.
299. Trygve Nagell, *Introduction to number theory*, Wiley, New York, 1951.
300. Harald Niederreiter, *An enumeration formula for certain irreducible polynomials with an application to the construction of irreducible polynomials over the binary field*, Appl. Algebra Engrg. Comm. Comput. **1** (1990), 119–124.
301. ______, *A short proof for explicit formulas for discrete logarithms in finite fields*, Appl. Algebra Engrg. Comm. Comput. **1** (1990), 55–57.
302. ______, *Factorization of polynomials and some linear-algebra problems over finite fields*, Linear Algebra Appl. **192** (1993), 301–328.
303. ______, *A new efficient factorization algorithm for polynomials over small finite fields*, Appl. Algebra Engrg. Comm. Comput. **4** (1993), 81–87.
304. ______, *Factoring polynomials over finite fields using differential equations and normal bases*, Math. Comp. **62** (1994), 819–830.
305. ______, *New deterministic factorization algorithms for polynomials over finite fields*, Finite fields: theory, applications, and algorithms (Las Vegas, NV, 1993), Contemp. Math., vol. 168, Amer. Math. Soc., Providence, RI, 1994, pp. 251–268.
306. Harald Niederreiter and Rainer Göttfert, *Factorization of polynomials over finite fields and characteristic sequences*, J. Symbolic Comput. **16** (1993), 401–412.
307. ______, *On a new factorization algorithm for polynomials over finite fields*, Math. Comp. **64** (1995), 347–353.
308. Harald Niederreiter and Chaoping Xing, *Algebraic geometry in coding theory and cryptography*, Princeton University Press, Princeton, NJ, 2009.
309. Ivan Niven, *Fermat's theorem for matrices*, Duke Math. J. **15** (1948), 823–826.
310. Emmy Noether, *Normalbasis bei Körpern ohne höhere Verzweigung*, J. Reine Angew. Math. **167** (1932), 147–152.
311. Henri J. Nussbaumer, *Fast Fourier transform and convolution algorithms*, Springer, New York, 1981.
312. Andrew M. Odlyzko, *Discrete logarithms in finite fields and their cryptographic significance*, Advances in cryptology – EUROCRYPT 84, Lect. Notes Comput. Sci. **209**, Springer, New York, 1985, pp. 224–314.
313. Øystein Ore, *On a special class of polynomials*, Trans. Amer. Math. Soc. **35** (1933), 559–584.
314. ______, *Contributions to the theory of finite fields*, Trans. Amer. Math. Soc. **36** (1934), 243–274.
315. Raymond E.A.C. Paley, *On orthogonal matrices*, J. Math. Phys., Mass. Inst. Techn. **12** (1933), 311–320.
316. Ruud Pellikaan, Xin-Wen Wu, Stanislav Bulygin, and Relinde Jurrius, *Codes, cryptology and curves with computer algebra*, Cambridge University Press, Cambridge, 2018.
317. Sam Perlis, *Normal bases of cyclic fields of prime power degree*, Duke Math. J. **9** (1942), 507–517.
318. Oskar Perron, *Bemerkungen über die Verteilung der quadratischen Reste*, Math. Z **56** (1952), 122–130.
319. Antonio Pincin, *Bases for finite fields and a canonical decomposition for a normal basis generator*, Comm. Algebra **17** (1989), 1337–1352.
320. Vera S. Pless, *Introduction to the theory of error-correcting codes*, 3rd ed., Wiley, Chichester, 1998.
321. Vera S. Pless and W. Cary Huffman (eds.), *Handbook of coding theory*, Elsevier, Amsterdam, 1998.
322. Michael Pohst and Hans Zassenhaus, *Algorithmic algebraic number theory*, Cambridge University Press, Cambridge, 1989.
323. Alain Poli, *A deterministic construction of normal bases with complexity $O(n^3 + n \log n \log(\log n) \log q)$*, J. Symbolic Comput. **19** (1995), 305–319.
324. Alexander Pott, *On Abelian difference set codes*, Des. Codes Cryptogr. **2** (1992), 263–271.
325. ______, *Finite geometry and character theory*, Lecture Notes in Mathematics, vol. 1601, Springer, New York, 1995.

326. K. Deergha Rao and M. N. S. Swamy, *Digital signal processing. Theory and practice*, Springer, Singapore, 2018.
327. Paulo Ribenboim, *The new book of prime number records*, 3rd ed., Springer, New York, 1996.
328. Tony Rosati, *A high speed data encryption processor for public key cryptography*, Proceedings Custom Integrated Circuits Conference, IEEE Press, New York, 1989, pp. 12.3.1–12.3.5.
329. Rainer A. Rueppel, *Analysis and design of stream ciphers*, Springer, Berlin, 1986.
330. ———, *Stream ciphers*, Contemporary cryptology (Ed. Gustavus J. Simmons), IEEE Press, New York, 1992, pp. 65–134.
331. `Sage`, Free open-source mathematics software system, http://www.sagemath.org/.
332. Dilip Vishwanath Sarwate, *Crosscorrelation properties of pseudo-random and related sequences*, Proc. IEEE **68** (1980), 593–619.
333. Alfred Scheerhorn, *Trace- and norm-compatible extensions of finite fields*, Appl. Algebra Engrg. Comm. Comput. **3** (1992), 199–209.
334. ———, *Darstellungen des algebraischen Abschlusses endlicher Körper und spur-kompatible Polynomfolgen*, Arbeitsberichte des Instituts für Mathematische Maschinen und Datenverarbeitung (Informatik), vol. 26, Friedrich-Alexander-Universität zu Erlangen-Nürnberg, Erlangen, 1993.
335. ———, *Dickson polynomials, completely normal polynomials and the cyclic module structure of specific extensions of finite fields*, Des. Codes Cryptogr. **9** (1996), 193–202.
336. Hermann Ludwig Schmid, *Relationen zwischen verallgemeinerten Gaußschen Summen*, J. Reine Angew. Math. **176** (1937), 189–191.
337. Bernhard Schmidt, *Cyclotomic integers and finite geometry*, J. Amer. Math. Soc. **12** (1999), 929–952.
338. ———, *Characters and cyclotomic fields in finite geometry*, Springer, New York, 2002.
339. Kai-Uwe Schmidt and Jürgen Willms, *Barker sequences of odd length*, Des. Codes Cryptogr. **80** (2016), 409–414.
340. Wolfgang M. Schmidt, *Equations over finite fields. An elementary approach*, Springer, New York, 1976.
341. Robert A. Scholtz and Lloyd R. Welch, *GMW sequences*, IEEE Trans. Inform. Theory **30** (1984), 548–553.
342. Eric Schost, *Algorithms for finite field arithmetic*, Proceedings of the 40th international symposium on symbolic and algebraic computation (ISSAC 2015), Association for Computing Machinery (ACM), New York, NY, 2015, pp. 7–12.
343. Manfred R. Schroeder, *Number theory in science and communication. With applications in cryptography, physics, digital information, computing, and self-similarity. 5th ed.*, Springer, Berlin, 1986.
344. Issai Schur, *Über die Gaußschen Summen*, Nachr. Ges. Wiss. Göttingen, Math.-Phys. Kl. **1921** (1921), 147–153.
345. Gérald E. Séguin, *Low complexity normal bases for $F_{2^{mn}}$*, Discrete Appl. Math. **28** (1990), 309–312.
346. Igor A. Semaev, *Construction of polynomials irreducible over a finite field with linearly independent roots*, Math. USSR, Sb. **63** (1989), 507–519.
347. Gadiel Seroussi, *Table of low-weight binary irreducible polynomials*, Tech. Report HPL-98-135, Hewlett–Packard, 1998, Available at http://www.hpl.hp.com/techreports/98/HPL-98-135.pdf.
348. Gadiel Seroussi and Abraham Lempel, *Factorization of symmetric matrices and trace-orthogonal bases in finite fields*, SIAM J. Comput. **9** (1980), 758–767.
349. ———, *On symmetric representations of finite fields*, SIAM J. Algebraic Discrete Methods **4** (1983), 14–21.
350. Jean-Pierre Serre, *A course in arithmetic*, Springer, New York, 1973.
351. Joseph Alfred Serret, *Cours d'algèbre supérieure*, 3rd ed., Gauthier-Villars, Paris, 1866.
352. Claude E. Shannon, *Communication theory of secrecy systems*, Bell Systems Tech. J. **28** (1949), 656–715.

353. Thomas R. Shemanske, *Modern cryptography and elliptic curves. A beginner's guide*, American Mathematical Society, Providence, RI, 2017.
354. Mohammad Amin Shokrollahi, *Beiträge zur Codierungs- und Komplexitätstheorie mittels algebraischer Funktionenkörper*, Ph.D. thesis, Universität Bonn, 1991.
355. Victor Shoup, *New algorithms for finding irreducible polynomials over finite fields*, Math. Comp. **54** (1990), 435–447.
356. Igor Shparlinski, *On finding primitive roots in finite fields*, Theoret. Comput. Sci. **157** (1996), 273–275.
357. James Singer, *A theorem in finite projective geometry and some applications to number theory*, Trans. Amer. Math. Soc. **43** (1938), 377–385.
358. Man-Keung Siu, *From binary sequences to combinatorial designs*, J. Math. Res. Exposition **9** (1989), 605–621.
359. Kempton J.C. Smith, *On the p-rank of the incidence matrix of points and hyperplanes in a finite projective geometry*, J. Combin. Theory **7** (1969), 122–129.
360. Ralph G. Stanton and David A. Sprott, *A family of difference sets*, Canad. J. Math. **10** (1958), 73–77.
361. Ernst Steinitz, *Algebraische Theorie der Körper*, J. Reine Angew. Math. **137** (1910), 167–309.
362. Aaron Sterling, *An independent discovery of Costas arrays*, https://nanoexplanations.wordpress.com/2011/10/09/an-independent-discovery-of-costas-arrays/, 1965.
363. Henning Stichtenoth, *Algebraic function fields and codes*, 2nd ed., Springer, Berlin, 2009.
364. Ludwig Stickelberger, *Ueber eine Verallgemeinerung der Kreistheilung*, Math. Ann. **37** (1890), 321–367.
365. Douglas R. Stinson, *Some observations on parallel algorithms for fast exponentiation in* $\mathrm{GF}(2^n)$, SIAM J. Comput. **19** (1990), 711–717.
366. ______, *On bit-serial multiplication and dual bases in* $\mathrm{GF}(2^m)$, IEEE Trans. Inform. Theory **37** (1991), 1733–1736.
367. Douglas R. Stinson and Maura B. Paterson, *Cryptography: Theory and practice*, 4th ed., Chapman & Hall/CRC, Boca Raton, FL, 2019.
368. Volker Strassen, *Algebraische Berechnungskomplexität*, Perspectives in mathematics. Anniversary of Oberwolfach 1984, Birkhäuser, Basel, 1984, pp. 509–550.
369. Richard G. Swan, *Factorization of polynomials over finite fields*, Pacific J. Math. **12** (1962), 1099–1106.
370. `The Prime Page`, https://primes.utm.edu.
371. Tian Tian and Wen Feng Qi, *Primitive normal element and its inverse in finite fields*, Acta Math. Sinica (Chin. Ser.) **49** (2006), 657–668.
372. Ulrich Tietze, Christoph Schenk, and Eberhard Gamm, *Electronic circuits. Handbook for design and application*, 2nd ed., Springer, Berlin, 2008.
373. Natalia Tokareva, *Bent functions. Results and applications to cryptography*, Elsevier/Academic Press, Amsterdam, 2015.
374. Vladimir D. Tonchev, *Combinatorial configurations: designs, codes, graphs*, Wiley, New York, 1988.
375. Michael Tsfasman, Serge Vlăduţ, and Dmitry Nogin, *Algebraic geometric codes: basic notions*, American Mathematical Society, Providence, RI, 2007.
376. ______, *Algebraic geometry codes: advanced chapters*, American Mathematical Society, Providence, RI, 2019.
377. Alan Tucker, *Applied combinatorics*, Wiley, New York, 1980.
378. Richard J. Turyn, *Character sums and difference sets*, Pacific J. Math. **15** (1965), 319–346.
379. ______, *Sequences with small correlation*, Error correcting codes (Ed. Henry B. Mann), Wiley, New York, 1969, pp. 195–228.
380. Richard J. Turyn and James E. Storer, *On binary sequences*, Proc. Amer. Math. Soc. **12** (1961), 394–399.
381. Jacobus H. van Lint, *Introduction to coding theory*, 3rd ed., Springer, Berlin, 1999.

382. Paul C. van Oorschot, *A comparison of practical public-key cryptosystems based on integer factorization and discrete logarithms*, Advances in cryptology – CRYPTO '90. Lect. Notes Comput. Sci. **537**, Springer-Verlag, New York, 1991, pp. 576–581.
383. Scott A. Vanstone and Paul C. van Oorschot, *An introduction to error correcting codes with applications*, Kluwer Academic Publishers, Boston, MA, 1989.
384. Rom R. Varshamov, *Operator substitutions in a Galois field and their application*, Sov. Math. Dokl. **14** (1973), 1095–1099.
385. Rom R. Varshamov and G.A. Garakov, *On the theory of self-dual polynomials over a Galois field*, Bull. Math. Soc. Sci. Math. Républ. Soc. Roum., Nouv. Sér. **13** (1970), 403–415 (Russian).
386. Joachim von zur Gathen, *Algebraic complexity theory*, Ann. Rev. Comput. Sci. **3** (1988), 317–347.
387. ———, *CryptoSchool*, Springer, Berlin, 2015.
388. Joachim von zur Gathen and Jürgen Gerhard, *Modern computer algebra*, 3rd ed., Cambridge University Press, Cambridge, 2013.
389. Joachim von zur Gathen and Mark Giesbrecht, *Constructing normal bases in finite fields*, J. Symbolic Comput. **10** (1990), 547–570.
390. Joachim von zur Gathen and Victor Shoup, *Computing Frobenius maps and factoring polynomials*, Comput. Complexity **2** (1992), 187–224.
391. Donald D. Wall, *Fibonacci series modulo m*, Amer. Math. Monthly **67** (1960), 525–532.
392. Walter D. Wallis, *Introduction to combinatorial designs*, Chapman & Hall/CRC, Boca Raton, FL, 2007.
393. Daqing Wan, *Computing zeta functions over finite fields*, Finite fields: Theory, applications, and algorithms. Fourth international conference, Amer. Math. Soc., Providence, RI, 1999, pp. 131–141.
394. Zhe-Xian Wan and Kai Zhou, *On the complexity of the dual basis of a type I optimal normal basis*, Finite Fields Appl. **13** (2007), 411–417.
395. Charles C. Wang, Trieu-Kien Truong, Howard M. Shao, Leslie J. Deutsch, Jim K. Omura, and Irving S. Reed, *VLSI architectures for computing multiplications and inverses in* $\mathrm{GF}(2^m)$, IEEE Trans. Comput. **34** (1985), 709–717.
396. Muzhong Wang and Ian F. Blake, *Bit serial multiplication in finite fields*, SIAM J. Discrete Math. **3** (1990), 140–148.
397. Morgan Ward, *The arithmetical theory of linear recurring series*, Trans. Amer. Math. Soc. **35** (1933), 600–628.
398. Alfred Wassermann, *Konstruktion von Normalbasen*, Bayreuth. Math. Schr. **31** (1990), 155–164.
399. William C. Waterhouse, *The sign of the Gaussian sum*, J. Number Theory **2** (1970), 363.
400. Joseph Henry Maclagan Wedderburn, *A theorem on finite algebras*, Trans. Amer. Math. Soc. **6** (1905), 349–352.
401. Marcelo J. Weinberger, *Factorization of symmetric circulant matrices and self-complementary normal bases in finite fields*, Master's thesis, Technion, Haifa, 1987.
402. Marcelo J. Weinberger and Abraham Lempel, *Factorization of symmetric circulant matrices in finite fields*, Discrete Appl. Math. **28** (1990), 271–285.
403. Neil H.E. Weste and Kamran Eshraghian, *Principles of CMOS VLSI design*, 2nd ed., Addison-Wesley, Reading, Mass., 1994.
404. Doug Wiedemann, *An iterated quadratic extension of* $\mathrm{GF}(2)$, Fibonacci Quart. **26** (1988), 290–295.
405. Michael Willett, *Matrix fields over* $\mathrm{GF}(q)$, Duke Math. J. **40** (1973), 701–704.
406. Ernst Witt, *Über die Kommutativität endlicher Schiefkörper*, Abh. Math. Semin. Univ. Hamb. **8** (1931), 413.
407. Benjamin Young and Daniel Panario, *Low complexity normal bases in* $\mathbb{F}_{2^n}$, Finite Fields Appl. **10** (2004), 53–64.
408. Fuzhen Zhang, *Matrix theory. Basic results and techniques*, Springer, New York, 1999.

Index

D. Hachenberger and D. Jungnickel, *Topics in Galois Fields*,
Algorithms and Computation in Mathematics 29,
https://doi.org/10.1007/978-3-030-60806-4

GPSR Compliance
The European Union's (EU) General Product Safety Regulation (GPSR) is a set of rules that requires consumer products to be safe and our obligations to ensure this.

If you have any concerns about our products, you can contact us on

ProductSafety@springernature.com

In case Publisher is established outside the EU, the EU authorized representative is:

Springer Nature Customer Service Center GmbH
Europaplatz 3
69115 Heidelberg, Germany

www.ingramcontent.com/pod-product-compliance
Ingram Content Group UK Ltd.
Pitfield, Milton Keynes, MK11 3LW, UK
UKHW021007290726
14059UKWH00001BA/11